Endocrine pharmacology

Endocrine pharmacology

Physiological basis and
therapeutic applications

P. J. BENTLEY

MOUNT SINAI SCHOOL OF MEDICINE OF
THE CITY UNIVERSITY OF NEW YORK

CAMBRIDGE UNIVERSITY PRESS

CAMBRIDGE

LONDON   NEW YORK   NEW ROCHELLE

MELBOURNE   SYDNEY

CAMBRIDGE UNIVERSITY PRESS
Cambridge, New York, Melbourne, Madrid, Cape Town,
Singapore, São Paulo, Delhi, Tokyo, Mexico City

Cambridge University Press
The Edinburgh Building, Cambridge CB2 8RU, UK

Published in the United States of America by Cambridge University Press, New York

www.cambridge.org
Information on this title: www.cambridge.org/9780521279352

First published 1980
First paperback edition 2011

A catalogue record for this publication is available from the British Library

Library of Congress Cataloguing in Publication data
Bentley, P. J.
Endocrine pharmacology.
1. Hormone therapy.   2.   Endocrine glands – Drug
effects.   1.   Title.
RM288.B45   615'.36   79-19487

ISBN 978-0-521-22673-8 Hardback
ISBN 978-0-521-27935-2 Paperback

For my pharmacologist friends
in the New World, the Old World, and the Antipodes

Contents

Preface

This book presents a view of endocrinology as seen by a pharmacologist. It is concerned with the actions and uses of hormones as drugs, and with the drugs that influence endocrine functions in the body. As the practical application of such substances arises in health and disease, the physiological background for both these situations is provided. No prior knowledge of endocrinology or pharmacology is assumed, so the material should be understandable at the undergraduate level. No attempt is made to provide a detailed protocol (for instance, dosages) for the clinical use of such drugs, but an account is provided of the preparations that are available, their administration, expected therapeutic responses, side effects, and interaction with other drugs. Special emphasis has been placed on the mechanisms of action of such drugs and hormones, and the relationship of their chemical structure to their biological activities and structural analogues. The emphasis is on the human situation, but clearly our understanding of the working of the endocrine system depends largely on experiments carried out on animals. An extrapolation between human and nonhuman species is, however, made with care, especially when considering the use of endocrine-active drugs.

The clinical application of drugs can be a highly controversial practice, and changes arising from advances in knowledge are continual. I am not a physician and so am reticent to describe clinical procedures. I have therefore tried to remain objective and dispassionate in such descriptions and have attempted to quote the most recent and best authorities of whom I am aware. The editorial advice given by such eminent publications as the *British Medical Journal, The Lancet,* and *The New England Journal of Medicine* has therefore been proffered especially often.

This book is quite a long one, so it is unlikely that many people will read it in its natural sequence.

An attempt has therefore been made to make each section reasonably self-contained. A certain amount of repetition is thus inevitable but is limited by cross-referencing. When attempting to cover such a large scientific area, a problem inevitably arises regarding the quotation of references. For the earlier work in endocrinology, I have often sought refuge in reviews provided by others, but I have attempted to give original references to more recent papers. An especially useful source of basic material has been the endocrinology section of the *Handbook of Physiology,* prepared under the guidance of Drs. R. O. Greep and E. B. Astwood for the American Physiological Society. The background for many of the more clinically related aspects of the subject has been provided by numerous textbooks of medicine, but Dr. A. Labhart's excellent and comprehensive book, *Clinical Endocrinology: Theory and Practice* (Springer-Verlag, 1976), has been of special help. For more basic pharmacological information, I have often dipped into such volumes as *Drill's Pharmacology in Medicine* (edited by Dr. J. R. DiPalma; McGraw-Hill, 1971) and *The Pharmacological Basis of Therapeutics* (edited by Drs. L. S. Goodman and A. Gilman; Macmillan, 1975).

In these days, multiauthored scientific books are the rule rather than the exception, and they are considered by many to be a necessity. Some may therefore question my audacity and ability to overcome this prejudice; after writing this book I can certainly appreciate the problems that are involved. Nevertheless, I hope that the overall view of one person may provide some interest and cohesion, and a useful general background for what is a subject of quite general importance.

I am primarily indebted to the many basic and clinical scientists who have published their observations, and I apologize to the many authors who, because of my ignorance of their work or the necessity for literary continuity, remain anonymous. Some of the inspiration and information for this book resulted from my listening to lectures in courses on pharmacology, endocrinology, and reproduction, which my institution provides for its medical students. A lot of people, and their publishers, have allowed me to quote their work, and in many instances this has involved active help in providing material suitable for reproduction and suggestions for modifications and changes. I would also like to thank all those who have helped me, over a period of about three years, in the more onerous tasks of collecting references and preparing the manuscript. The Mount Sinai School of Medicine and The City University of New York succored me during the writing of this book. Special thanks are due my Chairmen, Professor J. P. Green and Professor S. A. Podos, for their moral and practical support during a period when some may have considered that my time could have been better spent.

P. J. B.

New York
January 1980

1
Introduction

1.1. Definition and scope

Endocrine pharmacology is a branch of both endocrinology and pharmacology. It is a happy marriage that has mutually benefited both of these important biological disciplines. *Endocrine pharmacology* may be defined as the study of hormones that are used as drugs, and drugs, including analogues of hormones, that are used as agonists and antagonists of endocrine functions. The scope of these studies includes an understanding of the various effects of these drugs and hormones as well as their metabolism, mechanism of action, and therapeutic use. (A *drug* is commonly defined as any substance used in the composition of medicine.)

The endocrine glands and the nervous system coordinate and control the multitude of bodily activities concerned with physiological homeostasis, growth, and reproduction. Their actions are mediated by chemical compounds which they can synthesize and release and which can influence the activity of other cells. Such substances are quite basic for the life of multicellular animals. Nerves and endocrine glands share the basic role of integrating the function and activity of the various types of cells and tissues in the body, although they act in somewhat different ways. Nerve transmission is relatively fast and generally can be directed more precisely for relatively short and specific periods of time. Endocrine glands, on the other hand, release their hormones into the circulation, so that the onset of their action is usually slower but their effects are often quite ubiquitous and of relatively prolonged duration. These properties are, however, only a broad generalization, and there are exceptions.

Hormones are especially well suited for the types of functions where chronic, relatively long-term stimulation of an organ or tissue may be necessary. As they are distributed widely by the circulation, they are also well designed for physiological situa-

tions in which stimulation at widespread and multiple types of sites is required.

The actions of hormones can be divided into four basic types, which usually reflect their special properties and propensities as described above (Figure 1.1):

a. Hormones regulate the various interconversions of nutrient fats, carbohydrates, and proteins, and the production of energy, which constitutes the body's *intermediary metabolism*.
b. Hormones can influence the *tone of smooth muscle* cells such as are present in the uterus, gut, and blood vessels.
c. Hormones also provide stimuli, which may arise as a result of changes in the external or internal environment, which *influence the activity of other endocrine glands*. Such tropic actions are especially important in regulating cyclical events such as reproduction, and in providing feedback control mechanisms for hormone secretion.
d. A quite general property of many hormones, which may also be related to the processes described above, is their ability to influence the *permeability* of biological membranes to such substances as water, ions, and metabolic solutes such as sugars and amino acids. Changes in the permeability of cells may be directly important for bodily homeostasis, as in promoting the conservation or excretion of salts, or they may contribute indirectly by providing signals, such as substrates and ions, that trigger metabolic mechanisms.

It should be emphasized that the nervous and endocrine systems cannot be considered as independent entities – they interact, support, and reinforce each other's actions in various ways. The present book, principally for practical reasons, is concerned with the functioning of the endocrine glands, but their many interactions with the nervous system will be emphasized.

Endocrinology has many practical applications in our everyday life. Apart from the treatment of human disease, it has provided methods for limit-

1

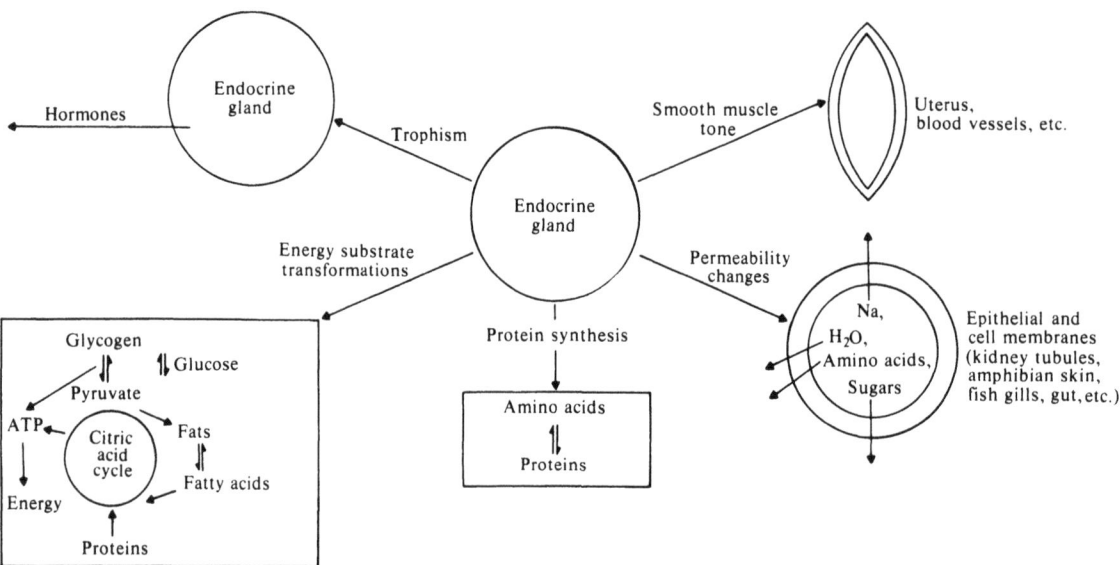

Fig. 1.1. Summary of the basic types of hormone actions. (From Bentley, 1976.)

ing the growth of the human population and promoting the supplies of food that are provided from animals. In 1967, the use of endocrine preparations in the United States amounted to $500 million, which was about 12 percent of the total value of all therapeutic drugs used.

Several pharmacological and therapeutic strategies are used to modify endocrine function in the body. In the instance of hyposecretion of hormones, replacement therapy may be instituted in several ways. Hormones from exogenous sources can be administered. These drugs may be the naturally occurring hormones that can be extracted from animal glands or may be made by chemical synthesis. In some instances, a surrogate compound can be provided which, although not identical to the endogenous hormones, may, nevertheless, exert a satisfactory effect. An early developed example of such a substance is diethylstilbestrol (DES), a synthetic compound with female sex hormone (estrogenic) activity but which, unlike the natural ovarian hormones, is not even a steroid. In some instances, a hypoactive endocrine gland can be stimulated to synthesize and/or release additional amounts of its secretion. Such an effect can be promoted by preparation of natural tropic hormones, such as corticotropin (ACTH), which stimulates the adrenal cortex, or even foreign chemical compounds such as tolbutamide, which stimulates the release of insulin. The action of some hormones can be potentiated by drugs and it seems likely that such an effect may account for the usefulness of chlorpropamide in the treatment of diabetes insipidus, which is due to insufficient an-

tidiuretic hormone. Hyperfunction of endocrine glands can also be treated pharmacologically, although usually surgical, or sometimes radiological procedures are ultimately used. An overactive endocrine gland can be suppressed by the administration of a drug that inhibits hormone synthesis, such as the action of propylthiouracil on the thyroid, or release, for instance the effect of diazoxide on the pancreatic B-cells, or the action of the hormone at its effector site, such as the antagonism of aldosterone's effect by spironolactone.

There are other occasions apart from hypo- and hyperendocrine conditions when pharmacological interference with a normal endocrine gland may be desirable. Oral contraceptive drugs can be used to prevent pregnancy, insulin to produce psychotherapeutic convulsions or to lower plasma K levels, androgens to promote muscle growth in athletes, and estrogens to reduce the proliferation of certain tumors.

Endocrine preparations may also be used for medical diagnostic purposes. These methods need not necessarily be of direct endocrine significance; for instance, the hypothalamic releasing factors and pituitary tropic hormones can be used to diagnose and locate brain tumors.

Hormones may also be utilized for the treatment of certain nonendocrine diseases. The best known example of such use is that of adrenocorticosteroids to treat inflammatory diseases such as rheumatoid arthritis.

Of importance to human welfare is the use of endocrine preparations to promote the health, reproduction, and growth of farm animals that provide us with food. The veterinarian, like the physician, uses endocrine preparations to control and cure animal diseases. Probably of even greater significance, however, is the use of hormones to speed growth and enhance the efficiency with which animals convert feed into bodily tissues. The use of such drugs has, however, recently been questioned, owing to uncertainties about the effects of such hormone residues when they are consumed by man.

No drug is completely specific in its action, but produces side effects whose number, incidence, and importance differ widely. Many nonendocrine drugs may have adverse effects in the body which can be related to their interference with normal endocrine function. The recognition and understanding of such effects may play an important role in assessing the possible significance of a drug's side effects and possibly suggest ways of antagonizing it. For instance, diazoxide is a potent hypotensive drug which also has a hyperglycemic action due to an inhibition of the release of insulin. It is interesting that this side effect of diazoxide has been utilized to treat hypoglycemia, so that what is a side effect for one application of a drug can in turn be utilized for its desirable endocrine effect.

1.2. History and synopsis

The roots of endocrine pharmacology lie in the application of knowledge of the workings of the endocrine glands. The word "hormone" was not coined until 1902, when W. M. Bayliss and E. H. Starling used it to describe the properties of a chemical excitant, secretin, which they had extracted from the intestine. During the 10 years prior to this semantic beginning of endocrinology, the biological activities of extracts from the thyroid, adrenal medulla, and the posterior pituitary gland were demonstrated. Renin, the midwife of angiotensin, was also discovered during that period. Endocrinology was thus born less than 100 years ago. The pharmacological application of the new information was almost immediate. In 1891, G. R. Murray treated hypothyroid patients with preparations of thyroid gland. In 1909, W. Blair Bell injected posterior pituitary gland extract into women to prevent postpartum uterine bleeding, and in 1911, J. Hofbauer used the same material (which contains the hormone oxytocin) to stimulate labor. The pharmacological applications of new knowledge about endocrine function has indeed often been quite prompt. The use of nonhormonal types of drugs that influence endocrine functions has, however, usually been much slower to occur and has largely depended on empirical types of studies with a good deal of help in the form of serendipity – happy accidental observations.

Knowledge of the endocrine glands as anatomical, but not hormone-secreting structures, origi-

nated, in many instances, several hundreds of years ago. In some cases, however, their discovery was quite recent. For instance: the islets of Langerhans were first described in 1864, the parathyroids in 1880, and the thyroid C-cells in 1876. The association of human diseases with pathological changes in such glandular tissues is also relatively recent. These include the description of adrenocortical insufficiency by T. Addison in 1855, exophthalmic goiter by R. J. Graves in 1835, hypothyroidism by H. Gull in 1874, and acromegaly by P. Marie in 1886. At these times, however, the essential endocrine nature of these tissues, and hence the disease, was unknown, so that specific treatment was not possible. It is, however, interesting that iodide was first used empirically, to treat hyperthyroidism, more than 100 years ago, and there was much speculation about the possible use of testicular extracts to promote masculine sexual vigor.

Organotherapy or opotherapy is an ancient branch of medicine based on the assumption that one can correct the disease of an organ by ingesting an equivalent organ from an animal or even another human being. The practice of eating the heart of one's enemy, to gain courage, as formerly practiced in some societies, is probably in the nature of such therapy. The self-administration of testicular extracts to improve sexual vigor was described by C. E. Brown-Sequard in 1889 and is a type of organotherapy. He reported favorable results, although subsequent tests of extracts of the type he used showed them to be devoid of a rational basis for the effects he described, as they contained no male androgenic hormones. This valiant attempt at hormone replacement therapy is in fact an early example of endocrine pharmacology. In 1891, G. R. Murray successfully administered thyroid gland extracts for the treatment of hypothyroidism, probably the first valid example of therapeutic endocrine pharmacology. This event was, however, soon followed by the use of posterior pituitary gland extracts in obstetrical practice. In 1913, R. van den Welden and F. Farini separately discovered the use of such glandular extracts for the treatment of diabetes insipidus. The most famous discovery of a hormone preparation that could be used to treat a specific human disease was that of insulin. This hormone was isolated from animal pancreases by F. G. Banting and C. H. Best in 1921 and was used to save the life of a 14-year-old boy suffering from diabetes mellitus. The event was followed by popular acclaim which was important at a time when endocrinology was still treated with some suspicion and was assumed to be mainly concerned with nefarious sexual activities.

In the succeeding decade, the importance of the anterior lobe of the pituitary became apparent. Largely as a result of the studies of P. E. Smith,

H. M. Evans, S. Ascheim and B. Zondek, O. Riddle, and others, the presence and actions of such hormones as the gonadotropins, thyrotropin, growth hormone, prolactin, and corticotropin were demonstrated in the pituitary gland. These discoveries were of immense importance for elucidating the mechanisms of the control of endocrine functions and gave a basis for understanding the nature of diseases that affect the pituitary. What was probably the first diagnostic endocrine test was concurrently developed: the Ascheim–Zondek pregnancy test, which utilized urinary gonadotropins to stimulate ovulation in rabbits or mice. The use of such hormones to treat human disease was, however, not of practical significance for more than 20 years. In 1950, corticotropin was used to stimulate the adrenal cortex in inflammatory diseases, especially rheumatoid arthritis, and in 1958 human growth hormone was introduced for the treatment of hypopituitary dwarfism in children. The gonadotropins have been used sporadically for many years in attempts to treat infertility in women, by inducing ovulation, but it is only recently that this has become an accepted practice.

Advances in endocrine pharmacology have largely depended on the application of new chemical techniques. Procedures developed by both organic and physical chemists have made possible the isolation of hormones in a pure form. The elucidation of their structure and, subsequently, their chemical synthesis is then made possible. The provision of suitable hormone preparations for therapeutic use is also facilitated. In addition, it then becomes feasible to modify the chemical structures of such hormones in a manner that may change their properties. These include a prolongation of their actions and alterations in the spectrum of their different effects in the body. The provision of pure hormones and knowledge of their chemical nature also allows the development of more convenient and accurate methods for their measurement and identification.

In 1914, E. C. Kendall prepared crystalline thyroxine. C. R. Harrington, in 1926, described its structure and in the next year chemically synthesized this hormone. The structures of many of the steroidal sex hormones, from the ovaries and testes, were also described during the next decade and their synthesis was then undertaken. These important contributions involved such pioneers in steroid hormone biochemistry as A. Butenandt, E. A. Doisy, G. F. Marrian, and L. Ruzicka. Of particular interest to endocrine pharmacology was the demonstration by Butenandt in 1937 that esters of estrone and testosterone had prolonged activity. The potent orally active female sex steroid ethinyl estradiol was synthesized by H. H. Inhoffen and W. Hohlweg in 1939. Just prior to this, in 1937, the

progestin ethisterone was synthesized by Ruzicka, but although, unlike progesterone, this was active orally, it had strong androgenic side effects that limited its use. Probably the most interesting contribution to endocrine pharmacology at the end of this decade was the synthesis of an artificial estrogen, diethylstilbestrol (DES), by E. C. Dodds in 1938. This chemical is not a steroid but has potent estrogenic properties, is active orally both in man, and of special interest, in farm animals. It is also cheap to produce. Dodds did not even take out a patent on this compound, an omission he bemoaned many years later, as it would have provided his laboratory with more liberal funds for research.

Such discoveries have provided hormone-like preparations which are useful for the treatment of endocrine disorders in man. They have also furnished drugs that are used to control fertility (oral contraceptives). The use of these compounds was developed by G. Pincus, starting in 1955, largely in response to prompting by Margaret Sanger. This application of estrogens and progestins has involved many millions of women throughout the world and has contributed in a major way to limiting human population. Estrogens are also of considerable direct economic importance as they have been used for more than 20 years to promote growth and improve the quality of meat in farm animals.

Research on the chemical nature of the steroid hormones produced by the adrenal cortex was also started in the 1930s. Deoxycorticosterone, an intermediate in the biosynthesis of these hormones, but which exhibits some activity, was synthesized by T. Reichstein in 1937. This compound had a limited use for replacement therapy in Addison's disease. However, the real impetus for the chemical study of the adrenocortical steroid hormones came during World War II, when it gained a priority associated with the Allies' war effort. It has been suggested that a rumor was spread that German pilots were taking these steroids in order to allow them to fly at high altitudes. If this is true, the efforts were misdirected but nevertheless had a productive outcome. In 1949, studies at the Mayo Clinic showed that cortisone had dramatic effects in the treatment of rheumatoid arthritis. This steroid was popularly called a "wonder drug" and ushered in the era of the antiinflammatory steroids. E. C. Kendall, T. Reichstein, and P. S. Hench jointly received a Nobel Prize in 1950 for their roles in this discovery. This was not the first such recognition for the work of endocrinologists, nor was it to be the last.

The determination of the structures and the chemical synthesis of the polypeptide hormones was ushered in 1951, when V. du Vigneaud described the structure of oxytocin. The structure of vasopressin was described soon thereafter, and both compounds were then synthesized. Several hundred analogues of these octapeptide hormones have since been prepared. The structure of the larger hormones was more of a problem, but F. Sanger described the amino acid sequence of insulin in 1955. This was the first such description of the structure of a protein. Human insulin was chemically synthesized in 1964 by P. Katsoyannis. This synthetic product, however, is too expensive to make for commercial therapeutic use. The large pituitary hormones require even more herculean efforts; C. H. Li has been preeminent for many years in such research, including the determination of the sequences of the nearly 200 amino acids in growth hormone. He has also made such measurements on prolactin, β-lipotropin, and corticotropin. Future provision of therapeutic preparations of such large hormones appears to rest in the possible use of smaller active fragments and the artificial genetic programming of bacteria to form them by biosynthesis. In December 1977, K. Itakura and his colleagues described the successful implantation of a chemically synthesized gene for the hormone somatostatin into *Escherichia coli* and the separation of the resulting tetradecapeptide product that was synthesized by the bacteria.

In the last 20 years there have been major advances in the elucidation of the nature of the relationship between the nervous and endocrine systems, which is often described under the discipline of neuroendocrinology. The brain is now known to have a direct influence on many endocrine glands. These effects are conveyed from the hypothalamus, at the base of the brain, to the pituitary gland and then, via the pituitary tropic hormones, to the thyroid, adrenal cortex, testes, and ovaries. G. W. Harris was an originator and the most persistent proponent of the concept that hormones are formed in the hypothalamus and are conveyed in the small portal blood vessels to the pituitary gland. The related supporting observation that nerve cells can secrete hormones (the process of neurosecretion) was proposed many years earlier by E. and B. Scharrer. Several hormones have been identified in the hypothalamus which mediate this function, and the structures of three of these – thyrotropin-releasing hormone (TRH), somatostatin, and luteinizing/follicle-stimulating hormone releasing hormone (LH/FSH-RH) – were described between 1969 and 1971 by R. Guillemin and A. V. Schally. These are peptides that have been chemically synthesized.

Such investigations have often involved laborious or, perhaps more aptly described, "heroic" methods. Butenandt had to extract 15,000 liters of urine to obtain 15 mg of the testosterone metabolite androsterone. Guillemin used the hypothalami of 5 million sheep to prepare 1 mg of pure

thyrotropin-releasing hormone. More recently, E. Rinderknecht and R. E. Humbel extracted 11,000 kg of the Cohn protein fraction of human plasma, representing nearly 1 million liters of blood, to obtain enough material for determination of the amino acid sequence of a somatomedin (insulin-like growth factor).

A considerable amount of information is now available regarding the basic mechanisms of hormone action. In 1958, E. W. Sutherland described the role of cyclic AMP as a "second messenger" in the functioning of glucagon and epinephrine. This mechanism has since been shown to apply to several other hormones, especially polypeptides. Steroid hormones, however, work differently and act on the cell nucleus to promote genetic transcription, the formation of mRNA, and the *de novo* synthesis of specific proteins which mediate their effects. The original hypothesis was proposed by U. Clever and P. Karlson in 1960. The development of this concept of the mechanism of action of a hormone is largely based on the theory of F. Jacob and J. Monod, which they conceived in 1959 to account for the genetic control of cell functions.

The hypothesis that drugs influence the functioning of cells by combining with specific components called "receptors" was first proposed by J. N. Langley in 1905 and was developed by A. J. Clark in the 1920s. This pharmacological concept has received considerable support with respect to the action of hormones. Macromolecules that specifically bind particular steroid and polypeptide hormones have been identified, described, and even isolated. This area of research is currently very active. Its recent advances depended largely on the technical ability to prepare radioactively labelled hormones that have a high ratio of label to biological activity. E. V. Jensen in 1961 provided suitable steroid hormone preparations while comparable labelling procedures for polypeptides were concurrently developed in several laboratories. The subject of hormone–receptor interactions is a special example of the happy marriage of endocrinology and pharmacology.

The binding of hormones to their receptors is a reversible process that involves a complementary association of fields of forces in their molecules. The arrangement and even existence of such properties depends on the three-dimensional structure or conformation of the molecules and the particular disposition of their reactive groups. Knowledge of the conformation of hormone molecules is thus of basic importance for understanding their function and may provide information that is of practical significance for the design of modified forms of hormones. In 1969, D. C. Hodgkin and her collaborators, using the technique of X-ray crystallographic analysis, provided a three-dimensional picture of insulin.

The initial discovery and identification of all hormones has depended on the technique of *bioassay*. This type of procedure utilizes various responses, *in vivo* and *in vitro*, of animals and tissues to provide what are usually sensitive and specific indicators of the presence, in extracts of bodily fluids and tissues, of unique biological substances, including hormones. Bioassay methods can be somewhat tedious and quantitatively not very accurate. However, in the hands of many pharmacologists, especially H. Dale, J. H. Gaddum, and H. O. Schild, they were provided with a technical and mathematical sophistication which improved their use. Bioassay procedures play an essential role in the initial study of hormones and, as a perusal of the U.S. and British Pharmacopoeias will show, still provide the standard "official" methods for quantitative measurements of some hormones. A major advance in methodology for identifying and measuring extant hormones was the introduction of the technique of *radioimmunoassay* by S. A. Berson and R. S. Yalow in 1957. This procedure, and its offshoots, competitive protein-binding assays and radioligand receptor-binding assays, have contributed in a major way to the recent rapid advances in endocrine research. Compared to bioassays, radioimmunoassays are usually more convenient, rapid, accurate, and sensitive. There are exceptions, however, and the use of radioimmunoassay procedures essentially depends on the availability of a known pure substance so that it provides a secondary, but not primary, role in the discovery of the actions of hormones.

Serendipity has played a notable role in the discovery of drugs that influence the activity of the endocrine glands. The two best examples of such substances are those that inhibit the activity of the thyroid gland, and the oral antidiabetic or hypoglycemic agents. The possibility that naturally occurring substances may inhibit the activity of the thyroid gland arose from observations of A. M. Chesney in 1928. He noted that rabbits which he was using to study experimental syphilis developed enlarged thyroid glands or goiters. It was subsequently shown that this condition resulted from the presence of a goiter-producing chemical in the cabbage on which they were fed. A number of other chemicals were subsequently shown to have such antithyroid properties, and this led to the introduction, by E. B. Astwood in 1942, of thiouracil for the treatment of hyperthyroidism. In that year, a group of French clinicians led by E. B. Janbon observed that some patients with typhoid fever who were being treated with a sulfonamide drug suffered from hypoglycemia, which sometimes even resulted in their death. In 1944, A. Loubatières showed that this effect was due to a stimulation of the release of insulin, an observation that led, in 1957, to the introduction of the orally active sulfonylurea drug tolbutamide for the treatment of diabetes mellitus.

Without the benefit of historical hindsight and the decisions of the Nobel Prize committees, it is difficult to select or predict the most important contemporary contributions to endocrine pharmacology. The discovery of the growth-regulating polypeptides or somatomedins that circulate in the plasma, the result of observations of V. Daughaday in 1957 and E. R. Froesch in 1963, appears to be of special interest. Quite recently, a number of intriguing observations have been made on the possible role of hormones or their metabolites on the activity of the brain, including the sensation of pain. Autoradiographic and immunohistochemical techniques have identified sites of accumulation of steroid and polypeptide hormones in the brain. In 1975, J. Hughes and H. W. Kosterlitz isolated two new peptides with morphine-like properties from brain tissue extracts. These were called enkephalins and were found to have similar amino acid sequences to fragments of the pituitary polypeptide β-lipotropin, which had been discovered by C. H. Li in 1964. A family of such related polypeptides, called endorphins, have been identified in the brain and the pituitary gland, but the possible relationships and physiological role of the substances at each site are not clear. At the present time, there is quite a lot of information available about the pharmacological effects of hormone-related peptides on behavior, the possible physiological significance of which is intriguing but unresolved.

2
Basic principles of pharmacology in relation to the endocrine function

The use and application of hormones and drugs that mimic and antagonize the endocrine system follow the same basic principles that apply to drugs generally. Such compounds must gain access to the systemic circulation and their effector tissues and have a finite time of existence in the body. Like all drugs, they have the propensity to act in an undesirable manner and at nonspecific sites, which may result in "side effects" that can even endanger life. An understanding of the general background that governs these processes in the body not only makes it easier to use such compounds but also to understand their actions, including their side effects and interactions. The processes involved include absorption into the circulation, their transport and distribution to the tissues, the nature and mechanisms of their effects, and the processes that control their metabolism and elimination from the body.

2.1. Absorption, distribution, and elimination of drugs: pharmacokinetics

When a drug, such as a hormone preparation, is administered, it will usually enter the circulation and be rapidly distributed to the remainder of the body fluids and tissues. To be therapeutically effective, it must reach and maintain a certain minimum concentration in the plasma. If, however, a very high level is attained, it may result in toxic effects.

The general principles that can be used to describe and predict the behavior of the levels of drugs in the body are called *pharmacokinetics*. This type of analysis facilitates the temporal description, in mathematical terms, of the effects of absorption, distribution, metabolism, and excretion on the levels of a drug in the body (see Greenblatt and Koch-Weser, 1975a,b). Such studies are useful, as they may provide data and methods by which the clinician can predict such important parameters as

the dosage of a drug that is necessary to attain a given therapeutic concentration and the frequency at which the drug must be administered to maintain such levels.

Certain assumptions are made and different models can be applied for such mathematical analyses of changing drug levels in the body. The particular ones used are undoubtedly simpler than the actual situation that exists in the body, but they can nevertheless provide useful information.

a. The simplest model (Figure 2.1a) assumes that the drug is present in a one-compartment system into which drugs can be absorbed, and from which they can be irreversibly eliminated as a result of metabolism and excretion (k_a and k_e are the absorption and elimination rate constants, respectively).

b. A slightly more complex model (Figure 2.1b), which approaches the *in vivo* situation more closely, is the two-compartment system. The central compartment (1) can be considered as being synonymous with the plasma and interstitial fluids of highly perfused organs such as the liver, kidney, and heart. The contents of this compartment are in equilibrium with a "peripheral compartment" (2), which corresponds to the tissues and the interstitial fluids in less well perfused organs. The rate constants for the movements of drug between these two compartments are called k_{12} and k_{21}. Absorption of the dose of the drug and its irreversible elimination are both considered to take place in the central compartment.

For mathematical treatment the rates of the exchanges are considered to be related to the concentration of the drug (first-order kinetics). However, some drugs do not behave in this way, such as when their elimination may involve facilitated diffusion or active transport, so that saturation occurs (these follow zero-order kinetics).

The basic information that is required to make mathematical predictions of pharmacokinetic pa-

7

(a)

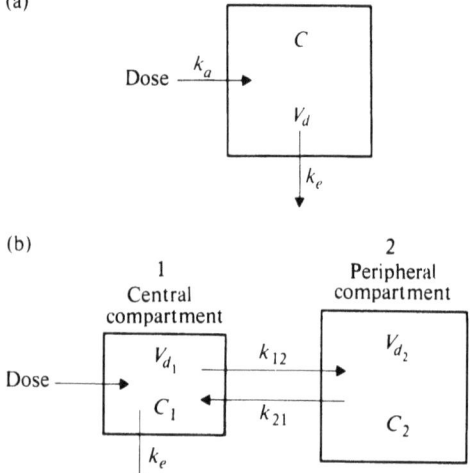

(b)

Fig. 2.1. Schematic models that can be used to describe the distribution and pharmacokinetic behavior of drugs. (a) One-compartment "open" model. (b) Two-compartment "open" model. k_a, absorption rate constant; C, drug concentration; V_d, apparent volume of drug distribution; k_e, elimination rate constant; k_{12} and k_{21}, rate constants for movements of drugs between compartments 1 and 2. First-order (i.e., concentration-dependent) kinetics are considered to apply to k_a, k_e, and k_{12} and k_{21}.

rameters is the administered dose (D or D_0) and subsequent serial measurements of changes in serum concentration of the drug. Such serum values, following oral administration of a drug, can be plotted graphically as shown in Figure 2.2a. The drug levels can be seen to show an initial, relatively fast rise; a peak concentration; and finally a slower decline. A slightly more detailed analysis following intravenous administration of a drug is shown in Figure 2.2b. In this instance, the drug concentration in the serum rises more rapidly and, after reaching a maximum, declines rapidly. This initial fast decrease reflects the redistribution of the drug between the central compartment and the peripheral compartment and has been called "phase alpha," with a slope (its units are min^{-1}) called α. Subsequently, there is an abrupt change in this slope and a slower decline ensues. This is called "phase beta" and reflects the irreversible elimination of the drug as a result of its excretion and metabolism (its slope is called β). This is the elimination rate constant k_e (which is expressed as units of min^{-1}).

Using such mathematical and graphical data, one can calculate (Table 2.1) such values as the total apparent volume of distribution (V_d), the elimination rate constant (k_e), the half-life ($t_{1/2}$), and the clearance rate. This information can be

useful clinically, such as in the calculation of dosage and the schedule for administration, so as to assure therapeutic but not toxic levels of a drug in the body. Multiple doses of a drug are often necessary to maintain effective therapeutic levels in the blood, so that correct spacing of the doses is important. If a drug is given too frequently, accumulation can occur and toxic levels thus attained; but, conversely, if the drug is given at long intervals, therapeutic levels may not be maintained. The half-life of a drug is an important consideration in relation to the dosage interval. Some of the patterns in serum concentration that may be expected from different regimens are shown in Figure 2.3. Generally speaking, if a drug is given at intervals that are substantially less than its half-life, it will accumulate. However, if the interval corresponds to the half-life, it will attain an equilibrium concentration after about four such doses. The "steady-state" concentration of a drug on different regimen can be calculated as shown in Table 2.1.

It should be emphasized that although most drugs behave in a manner that allows such calculations and predictions to be made, there are a number of notable exceptions. This problem may occur when the processes for elimination of a drug are saturable (they follow nonlinear or zero-order kinetics). The binding of drugs to plasma proteins, the formation of active metabolites, or persistent sequestration in other compartments, such as fat, can also complicate the use and interpretation of such pharmacokinetic methods.

2.2. Administration and bioavailability of drugs

The method of administering a drug is dictated by a number of considerations.

a. The chemical nature of drugs is of primary importance. Thus, proteins and polypeptides will be inactivated by the digestive juices and will be ineffective when given by mouth. Oral absorption across the gastrointestinal mucosa will be favored if the molecules are lipophilic, such as are the steroids. Despite such absorption, some drugs are still relatively ineffective, because when they pass into hepatic–portal circulation and go to the liver, they may be inactivated there. In some cases, such as that of the natural steroidal sex hormones, this process of metabolism can be so effective as to nearly destroy all of the activity before it can enter the general circulation. Such hormones can, however, often be chemically modified so as to block their metabolism by the liver.

A number of endocrine preparations can be absorbed across the mucosa of the mouth, nasal passages, rectum, vagina, and even the bronchial tree. Absorption will be favored in strongly lipophilic compounds, but weaker ones, such as certain peptides, can also be absorbed in this man-

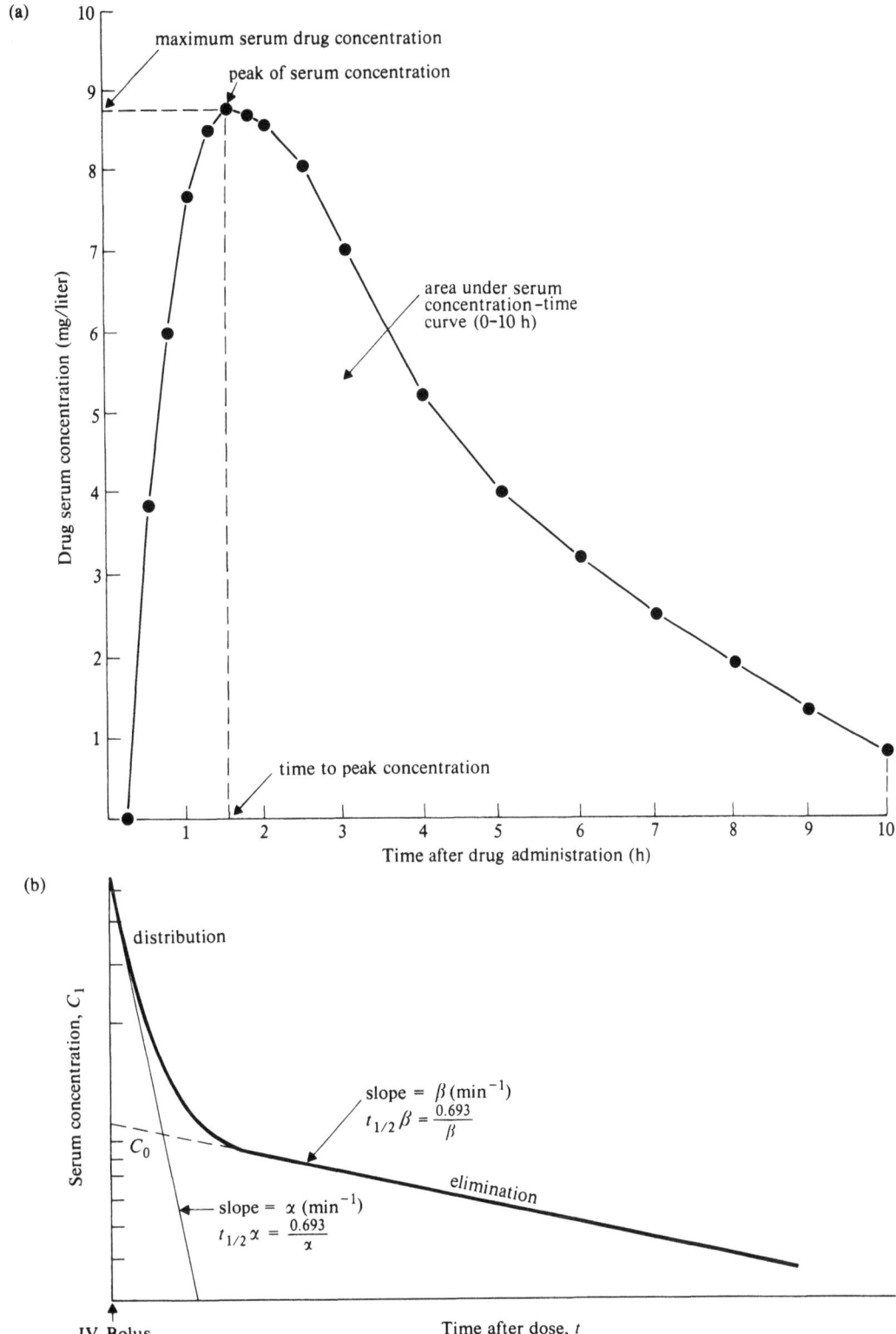

Fig. 2.2. (a) Serum concentration–time curve following oral administration of hypothetical drugs. (From Koch-Weser, 1974a. Reprinted by permission from *New England Journal of Medicine 291*, 234.) (b) Graphical representation of the pharmacokinetic behavior of the serum concentration (C_1) of a drug following the injection of a single intravenous bolus. The predicted behavior follows that of the two-compartment open-model system (Figure 2.1b). The initial rapid decline (slope α) reflects the redistribution of the drug into its compartments. The subsequent slower elimination phase (slope β, or k_e) reflects its metabolism and excretion. The $t_{1/2}$ is the half-life of the drug with respect to its elimination (β). (From Greenblatt and Koch-Weser, 1975a. Reprinted by permission from *The New England Journal of Medicine 293*, 703.)

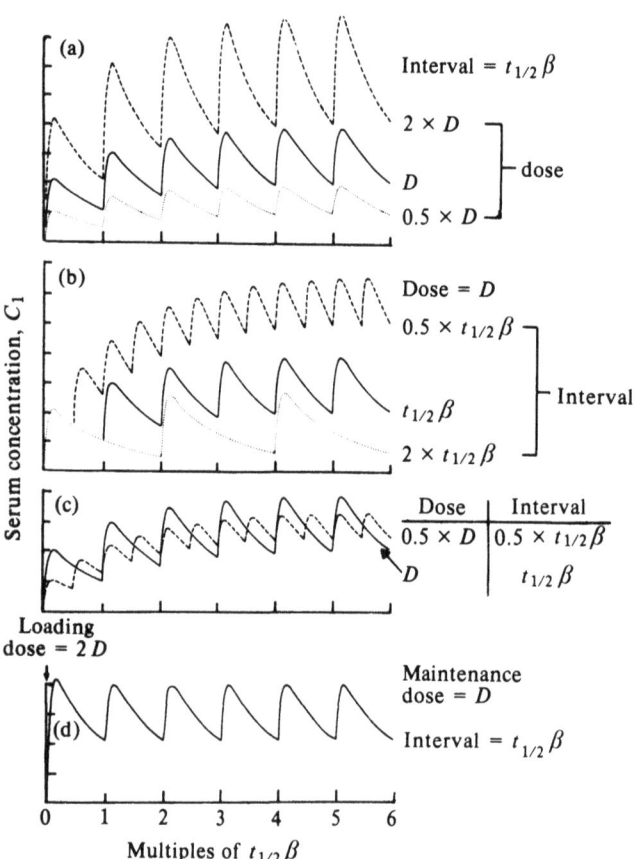

Fig. 2.3. Effects of varying doses and intervals between doses on drug accumulation during repeated administration. (a) Varying the dose but administering it at the same interval. (b) Varying the interval between the dosages but keeping the dose constant. (c) When the dose is halved but is given twice as often, the steady-state concentration is maintained but the fluctuation of the serum levels is reduced. (d) By administering an appropriate loading dose (1.5 to 2 times the maintenance dose), the steady-state serum concentration is achieved more rapidly. $t_{1/2}\beta$, half-life of the drug in the body; D, dose. (From Greenblatt and Koch-Weser, 1975b. Reprinted by permission from *The New England Journal of Medicine 293*, 967.)

ner. Such routes of administration can be utilized so as to avoid the action of the digestive and hepatic enzymes. For instance, the peptide hormones of the neurohypophysis and hypothalamus can be effectively administered as nasal snuff or nasal spray, the male steroidal sex hormone preparation methyltestosterone by sublingual buccal administration, the glucocorticoid beclomethasone by an inhalation spray, and prostaglandins by the rectal or vaginal route. Such routes for drug administration have the advantage that injection of the drug is unnecessary although total absorption is usually difficult to predict. Local inflammatory and allergic reactions at the site of the absorption may also occur, especially if the preparation is not pure. This problem occurred with early preparations of neurohypophysial hormones when given as nasal snuff.

Injection or parenteral administration of drugs may be necessary to avoid the digestive tract and to attain and maintain therapeutically effective levels in the body. For endocrine preparations this is especially useful for polypeptides and proteins, including all the pituitary hormones, insulin, and glucagon, as well as epinephrine. There are some disadvantages of parenteral administration, the most obvious being the difficulties associated with self-administration [subcutaneous, (s.c.) injection is the only really practical method] and the pain and inflammation that may occur at local sites. The latter often reflect the relative impurity of the preparation. Repeated injections at single sites, such as often occur with the use of insulin, can result in local changes, including degeneration of the tissues.

b. The desired speed of onset of the action of a

Table 2.1. *Some commonly used pharmacokinetic parameters of drugs*

Total absorbed dose or fractional absorption, f

$$f = \frac{\text{absorbed dose}}{\text{total administered dose } (D)}$$

$$f = \frac{\text{area under the curve (AUC) for time } (t) = \text{zero to } t = \infty \text{ after an oral or i.m. dose } (D)}{\text{AUC for } t = 0 \text{ to } t = \infty \text{ for i.v. dose } D}$$

Note that when absorption is complete, $f = 1$.

Total apparent volume of distribution, V_d

$$V_d = \frac{Df}{C_0} \quad \text{(liters)}$$

where C_0 is the serum concentration of the drug calculated from the intercept, at zero time, of the elimination curve (see Figure 2.2b).

Elimination rate constant, k_e

  This value can be calculated from observed changes in the serum concentration of the drug during the elimination phase (Figure 2.2b). Its units are usually expressed as min^{-1}.

Half-life for elimination, $t_{1/2}$ or $t_{1/2\beta}$

$$t_{1/2} = \frac{\ln 2}{k_e} = \frac{0.693}{k_e} \quad \text{(minutes)}$$

Clearance rate of a drug

$$\text{Clearance rate} = V_d k_e = \frac{V_d \cdot 0.693}{t_{1/2}} \quad \text{(liters/min)}$$

Mean steady-state serum concentrations, $\bar{C}$, with multiple doses of a drug

$$\bar{C} = \frac{1}{\text{clearance}} \frac{Df}{T} \quad \text{(mol/liter)}$$

or

$$\frac{Df}{V_d k_e T} = \frac{1.44 Df t_{1/2}}{V_d T}$$

where T is the interval between doses, in minutes.

Area under the curve of serum concentration versus time, AUC.

  The AUC can furnish information that can be used in the foregoing equations.

$$\text{AUC} = \frac{Df}{V_d k_e}$$

Notes: These formulas can be transposed, as shown for $\bar{C}$, in various ways in accordance with the different methods for calculating the parameters or their individual terms. Other types of units can, of course, also be used. For more detailed information, see Greenblatt and Koch-Weser (1975a,b).

drug will also influence the choice of a route of administration. Clearly, the intravenous method, such as a single bolus injection or a continual infusion, will result in the most rapid attainment of therapeutic levels. Such a procedure is usually necessary only in certain crisis situations, such as may involve the rapid and immediate restoration of normal levels of corticosteroids, insulin, or thyroid hormones, or in the use of oxytocin to promote labor. In some instances, such as if the patient is comatose or vomiting, oral administration will be impractical. Although not as rapid as the i.v. route, the subcutaneous or intramuscular injection of drugs also usually has quite a rapid onset of action. The speed will depend, however, on such factors as the vehicle in which it is contained and the pres-

ence of contaminants, such as proteins, to which it may be bound. The nature of the tissue into which it is injected will also influence its rate of absorption. For instance, it will usually be slowed if it is deposited in adipose tissue, whereas its absorption will be increased if there is a plentiful blood supply.

  c. A prolonged time of action may be desirable. This period may range from a day or two to several months and can be manipulated in several ways. The most common procedure is to modify the physical nature of the drugs, such as by compressing them into pellets or growing large crystals that have a relatively small surface area, or by combining them with vehicles to which they are physically adsorbed or in solution. Chemical combinations, such as the formation of esters, are also used for such purposes.

There are many endocrine examples of the use of such methods to prolong the period of a preparation's action. The steroidal sex hormones can be implanted as pellets, usually subcutaneously, from which they will continue to be absorbed and act for many weeks. Esters of such steroids are also prepared with periods of action that vary from a few days to 2 or 3 weeks. It is also possible to control the size of crystalline forms of hormone preparations, and this has been utilized in the preparation of the slowly absorbed, lente, insulins.

To prolong their action, drugs may also be included in special vehicles, such as oil, or mixed with proteins in which they become enmeshed and adsorbed. Steroids are sometimes prepared and injected in oil. This vehicle has also been used to create slowly absorbable depots for an ester of ADH (vasopressin tannate in oil) and for administering iodine in geographic areas (goiter belts) where a deficiency of this element results in thyroid problems. The combination of insulin with proteins, such as protamine and globin, is used to delay its absorption. Other hormones, such as ACTH, are administered in a protein gel. Such delayed absorption can create problems, as there is a loss in the amount of control and the vehicles themselves sometimes promote local adverse tissue reactions.

Another method of prolonging the effective absorption of a drug is to create a tissue reservoir by administering it in a form that is rapidly accumulated by a tissue, usually fat, from which it is slowly released. The drug itself may initially be an inactive form, such as the female sex hormone chorotrianisene, which is metabolized and activated after its release.

d. A localized administration of drug preparations is often used to treat a lesion directly. This may afford the most convenient method of attaining and maintaining high therapeutically effective concentrations. Such a procedure may indeed be essential to avoid general systemic concentrations of drugs that may have unacceptable side effects. Among endocrine-related drugs, the most important are lipophilic ones, especially the antiinflammatory corticosteroids, catecholamines, and sometimes estrogens. The corticosteroids are used in this way by direct topical application on the skin and cornea, injection intraarticularly into joints, insertion into the rectum, and inhalation into the bronchi. Catecholamines are administered by inhalation to dilate the bronchi. Estrogens have been used in cosmetic skin preparations, and suppositories are used for some vaginal disorders. Problems may sometimes occur due to the absorption of the drug from such sites into the general circulation. Such effects may, however, be rela-

tively localized, as may be seen in a degeneration of underlying cutaneous tissues or increases in intraocular pressure associated with prolonged topical use of corticosteroids. The manufacture of estrogens has thus been associated with disorders in reproductive function due to the absorption of these chemicals across the skin and, possibly, from the pulmonary tract. The use of corticosteroids to treat asthma has resulted in problems due to systemic absorption, which occurs even when these steroids are applied directly in the form of inhalation sprays. Chemical analogues have been developed which exhibit high local activity but are not appreciably absorbed.

Vasoconstrictor drugs are sometimes added to preparations, such as local anesthetics, which are injected into sites where localized action is required. The vasoconstrictor effects of catecholamines are generally used for such purposes, but analogues of vasopressin are also sometimes used.

Although successful intravenous administration of a drug assures absorption (fractional absorption $f = 1$), the process is not necessarily completed when other routes are used. The proportion, or fraction, of a drug that is absorbed into the circulation following its administration is called its *bioavailability* (see Koch-Weser, 1974a,b). The problem of partial absorption of a drug is perhaps most obvious when considering its oral administration, but it may also be incomplete, because of its local inactivation or binding, when it is given i.m. or s.c.

Some of the many factors that can influence a drug's bioavailability after oral administration are shown in Table 2.2. They include conditions in the gastrointestinal tract, such as the pH; motility; the actions of microorganisms that may metabolize the drug; the temporal relationship to meals; and the presence of other drugs.

Different species of drugs obviously may exhibit different characteristics with respect to the degree of completeness of their absorption from a particular site. In addition, there have been an increasing number of reports of variations in bioavailability among brands and formulations of the same generic drug. This may be due to the presence of different fillers, such as lactose or calcium carbonate; the size of the particles of the drug; or the ease with which the pill or capsule disintegrates. The latter can be influenced by its water content, age, or the degree of compression used to mold the tablet. In some instances, differences between preparations of a drug can be related to the time that is necessary for it to disintegrate and dissolve (*dissolution time*) *in vitro*.

Differences in bioavailability that are not readily predictable can lead to clinical problems, such as the administration of ineffective doses or unexpected toxicities. These can arise when simply

Table 2.2. *Factors influencing the bioavailability and absorption of drugs from the gastrointestinal tract*

Factors affecting the bioavailability of orally administered drugs
1. *Characteristics of drug*
   Inactivation before gastrointestinal absorption
   Incomplete absorption
   Biotransformation in intestinal wall or liver
2. *Formulation of drug product*
   State of the drug
   Excipients
3. *Interaction with other substances in gastrointestinal tract*
   Food
   Drugs
4. *Patient characteristics*
   Gastrointestinal pH
   Gastrointestinal motility
   Gastrointestinal perfusion
   Gastrointestinal flora
   Gastrointestinal structure
   Malabsorption states
   Hepatic function
   Genetic phenotype

Drug interactions that may occur in the gut which can influence absorption of drugs
1. Change in gastric or intestinal pH (antacids)
2. Change in gastrointestinal motility (cathartics, motility depressants)
3. Change in gastrointestinal perfusion (cardiovascular drugs)
4. Interference with mucosal function (neomycin, colchicine)
5. Chelation (tetracycline – calcium, magnesium, aluminum, iron)
6. Exchange resin binding (cholestyramine – acidic drugs)
7. Adsorption (charcoal, kaolin, antacids)
8. Solution in poorly absorbable liquid (mineral oil)
9. Unknown mechanisms (heptabarbital–bishydroxycoumarin; phenobarbital–griseofulvin; allopurinol–warfarin)

Source: Koch-Weser, 1974a. Adapted by permission from *The New England Journal of Medicine 291*, 235.

changing from one brand of a drug to another, even though the dose present is the same.

2.3. Distribution and binding of drugs

When a drug enters the circulation, either following absorption from the gut or by injection, it is rapidly diluted and dispersed, so that its local concentration declines rapidly. It does not, however, necessarily enter all the fluid compartments or tissues with equal facility, if at all. This process of distribution is an important practical consideration, because a drug may, for instance, have difficulty in maintaining an effective concentration or in reaching sites where its therapeutic effect may be de-

sired. Conversely, it may gain access to undesirable places, where it may exert toxic or side effects.

There are several barriers that can restrict the movement of drugs in the body, notably the capillaries, which limit movement into the interstitial fluids; the blood-brain barrier, which controls access to the brain; and the cell membrane. There are also barriers inside the cell, and the ability of drugs to enter the mitochondria or the lipid vesicles of the endoplasmic reticulum can influence their metabolism. Generally, access across such membranes is promoted if they are lipid-soluble; water-soluble molecules, especially large ones and those that are ionized, gain access with more difficulty. Binding to the tissues, especially to plasma proteins, will also limit dispersal, as will ready solubility in the fat stores of adipose tissue.

Many drugs and hormones bind to plasma proteins (Anton and Solomon, 1973; Dayton, Israili, and Perel, 1973; Gillette, 1973; Koch-Weser and Sellers, 1976a,b). In many instances where this occurs, less than 5 percent of the total drug in the plasma is in its "free," unbound state. The albumins are the most plentiful of the plasma proteins and provide the largest number of potential binding sites, but other plasma proteins, especially the globulins, are often more specific with respect to the substances with which they react. They therefore have only a low capacity to bind a drug or hormone, but they do so quite strongly (they have a high affinity). Binding to plasma proteins is usually a reversible process (covalent binding occurs, but rarely) and there is an equilibrium between the bound drug and that which is free in solution. The number of binding sites on a protein is limited, so that it is saturable. The binding of two or more substances may thus be interfered with because of competition among the substances for the same sites, or the occupation of adjacent positions, so that the substances interfere with each other.

Steroid and thyroid hormones (see Oppenheimer, 1973) are readily bound to plasma proteins. Information regarding possible binding of protein and polypeptide hormones is often controversial, although with some substances, such as somatomedins, it definitely seems to play an important role in their functioning. Drugs and hormones often share such binding sites, so that they can displace each other.

A. Consequences of displacement

The mutual displacement of drugs and hormones from binding sites on plasma proteins can have quite dramatic effects on their actions in the body. These effects are related not only to changes in levels of the free form of the drug or hormone, which is that which is biologically active, but they can also alter the rate of the elimination of the compound from the body. Initially, however, one

Table 2.3. *Clinically important interactions between drugs which result in their displacement from binding to plasma proteins*

Displaced drug	Displacing drug	Possible clinical consequences
Warfarin and other highly albumin bound coumarins (anticoagulants)	Clofibrate Ethacrynic acid Mefenamic acid Nalidixic acid Oxyphenbutazone Phenylbutazone Trichloroacetic acid (metabolite of chloral hydrate)	Excessive hypopro-thrombinemia Hemorrhage
Tolbutamide (oral hypoglycemic drug)	Phenylbutazone Sulfaphenazole Salicylates	Hypoglycemia

Source: Koch-Weser and Sellers, 1976b. Adapted by permission from *The New England Journal of Medicine* 294, 527.

may expect an exaggerated and even toxic effect when one drug displaces another from a binding site. As the active free fraction of the drug may only represent 1 or 2 percent of the total fraction in the blood, a quite modest displacement from the predominant remaining proportion attached to plasma proteins can have a large effect. Such a change can be of considerable importance clinically; for instance, it may result in hemorrhage in patients being treated with anticoagulants such as warfarin, or in hypoglycemia in diabetic patients who are taking tolbutamide. Both of these drugs are bound to plasma proteins, and they may mutually displace each other. Such an event may also occur in the presence of other drugs (Table 2.3).

Other changes occur subsequent to the displacement of a drug and may mask or modify the change. These responses are the result of attainment of a new steady state as a result of redistribution of the drug and changes in the rate of its elimination. The expected changes in serum concentration and the processes concerned with elimination of the drug are summarized in Table 2.4. When the amount of unbound drug in the serum is increased, the drug will be free to move into other fluid compartments; a decrease in concentration results from this redistribution. Filtration of the drug across the glomerulus will also be enhanced, so that drug excretion via the urine will be increased. The effects on renal tubular secretion of a drug are variable, but if active transport is involved, the secretion rates will be unchanged, or decreased. The rate of dissociation of a drug from plasma proteins is so fast that binding will not

Table 2.4. *Consequences of the displacement of a drug from binding to serum albumin*

	Immediately after displacement	New steady state
Free drug fraction in serum	Increased	Increased
Free drug concentration in serum	Increased	Unchanged
Total drug concentration in serum	Unchanged	Decreased
Pharmacologic activity	Increased	Unchanged
Glomerular filtration	Increased	Unchanged
Tubular secretion	Variable	Unchanged
Diffusion into liver cell	Increased	Unchanged
Active hepatic uptake	Variable	Unchanged

Source: Koch-Weser and Sellers, 1976b. Adapted by permission from *The New England Journal of Medicine* 294, 528.

normally limit the rate of active transport. Thus, no significant change will be expected. However, if as a result of redistribution, the total concentration of the drug in the serum declines, active transport-mediated excretion will decline. The same arguments apply to an active transport process that may be involved in the drug's access to other sites, especially the liver. Diffusion to metabolic sites in the liver will, however, be increased as a result of displacement from plasma proteins so that the drug's metabolism will be enhanced. Thus, a new steady-state equilibrium will be attained where the total drug in the serum is decreased, but the free, biologically active portion will be similar to its initial value. This does not mean that dosages of drugs need not be adjusted in the expectation of mutual displacement of drugs from plasma-protein binding; indeed, it is only by doing this that one can avoid possible *initial* toxic responses.

B. Changes in levels of binding proteins

The levels of *specific plasma-binding proteins* exhibit some lability, which is related to the physiological circumstances and may be altered in response to disease and therapy. Increases in plasma proteins, which can bind steroid hormones, occur in pregnancy. The therapeutic administration of estrogens may have similar effects. Growth hormone is thought to regulate the production of a plasma protein that binds somatomedin, the polypeptide that mediates many of its actions. *Plasma protein* levels, including the albumins, may decrease in certain conditions, especially in the presence of liver and renal disease. Changes in the particular protein composition of the plasma may also have qualitative effects which decrease the binding of drugs. As the unbound concentrations of many drugs are relatively small compared to the bound fraction, quite modest declines in binding can have large effects on the free therapeutically or physiologically active levels of a drug or hormone. It may thus be necessary to adjust the dosage of drugs in such diseases in order to maintain their therapeutic action or avoid toxic side effects. However, a number of occasions when the effectiveness to compensate for such effects, especially as changes in plasma protein levels are usually gradual and quite slow in their onset.

2.4. Metabolism, or biotransformation, of drugs

The chemical structures of drugs, including hormone preparations, can be altered as a result of enzymically controlled metabolic activities at various sites in the body. Such modifications in molecular structure change various properties of the molecule, including its biological activity and solubility. These changes usually result in a reduction or abolition of its actions (inactivation) and an increase in its water solubility, which facilitates its excretion in the urine and bile. There are, however, a number of occasions when the effectiveness of drugs (or hormones) is enhanced as a result of their metabolism, so that an activation is said to occur. In other instances, metabolites may retain an appreciable amount of the original activity, or substances that have other types of effects, including toxic actions, may be formed.

The metabolism of drugs and hormones may occur in any tissue in the body. Some of the enzymes involved, such as peptidases and esterases, often have quite nonspecific effects with respect to particular drugs and hormones and have a ubiquitous distribution. In other instances, however, special enzymes are found predominantly in certain tissues. The liver plays a central and predominant role in the metabolism of most drugs and hormones, but the kidney and gastrointestinal tract are also important and in certain instances, tissues as diverse as the brain, lungs, prostate, placenta, and the blood plasma may be involved. The target, or effector, tissues are often important sites for drug metabolism, especially when an activation is required, for instance in the conversion of testosterone to 5α-dihydrotestosterone. Metabolism of drugs and hormones may occur at either extracellular or intracellular sites. The latter are especially important and are associated with various cellular structures, including the smooth endoplasmic reticulum (SER, in the microsomal fraction), the mitochondria, or the "cell sap" (the supernatant fraction).

Drugs and hormones can be enzymically assaulted in a host of ways. Most compounds are usually, however, predominantly metabolized in a quite specific manner which is dictated by their chemical structure and the sites where they tend to accumulate in the body. In some instances, this may involve the action of a single enzyme, but usually there is a serial chain of reactions which result in the progressive availability of substrates that are formed stepwise from the drugs. Thus, a side chain may be progressively attacked and shortened, or a hydroxyl group may be formed which provides a site for the combination of the drug with such moieties as sugars, amino acids, and sulfuric acid. The types of metabolic changes in drugs can be divided into two main groups.

a. *Nonsynthetic* or *degradative reactions,* which involve such processes as oxidation, reduction, and hydrolysis. They include deamination, dealkylation, dehalogenation, and hydroxylation.
b. *Synthetic* or *conjugation reactions,* which involve the attachment of new components, such as glucuronide, sulfate, glycine, acetate, and methylation.

As synthetic reactions usually succeed nonsynthetic ones, they have also been referred to as phase I and II reactions.

A. Sites of drug metabolism

a. Extracellular metabolism of drugs occurs in the plasma and the interstitial fluids in the proximity or possibly at the surface of cells. Such processes would be expected to be relatively more important for substances that have difficulty in entering cells, especially large water-soluble molecules and those for which no specific mechanism for cellular uptake exists. Thus, many polypeptides are attacked by trypsin and chymotrypsin-like enzymes, carboxypeptidases, and aminopeptidases, which are extruded by cells and may be activated in the extracellular fluids. Such processes, however, also undoubtedly occur inside cells. Esterases are plentiful in the extracellular fluids of the gastrointestinal tract as a result of both the presence of digestive enzymes and the activities of bacteria. The latter are of special interest with respect to their actions in hydrolyzing conjugated compounds, notably glucuronides, which are secreted in the bile.

b. Intracellular metabolism of drugs and hormones is of predominant importance and is of course dependent on their ability to enter the cell. Lipid solubility or a specific uptake process, or "pump," will favor these processes. Such biotransformation may be due to soluble enzymes in the cytoplasm or it may take place in the mitochondria or as a result of reactions that occur with the components of the smooth endoplasmic reticulum. The activity of the latter structure is particularly important in the liver. Compounds must be lipid-soluble in order to gain access to the drug-metabolizing components of the endoplasmic reticulum.

c. The cytoplasm contains a number of soluble enzymes. These include methyltransferases, which add methyl groups to drugs and hormones. One of the best known is catechol-*O*-methyltransferase (COMT; Scheme 2.1), which plays an important role in the inactivation of catecholamines such as epinephrine. These methylation reactions involve utilization of the cofactor *S*-adenosyl methionine, which is formed from ATP and methionine. N- and S-methylation may also occur. (Methyl-transferases are also present in the microsomes.)

Sulfokinase (or sulfotransferase) enzymes have also been identified in the cytoplasm. These enzymes promote the synthesis of sulfate conjugates or esters of drugs and hormones, including the steroid and thyroid hormones. The reaction involves the formation of an intermediate, 3'-phosphoadenosine-5'-phosphosulfate, known as "active sulfate" or PAPS, formed from SO_4^{2-} and ATP (Scheme 2.2). Sulfation can occur at either the O or NH positions.

d. The mitochondria are the site of a number of oxidizing enzymes. These include monoamine oxidase (MAO), which acts on catecholamines, such as epinephrine and norepinephrine (Scheme 2.3).

e. The microsomal system that oxidizes drugs and the steroid hormones is unique, as it utilizes a reverse transport of electrons which reduces or "activates" molecular oxygen. This process has been best characterized in the liver, but all the details are not yet understood, and some are contentious and may differ from tissue to tissue. There are several components of this *mixed oxidase system* (as two substrates NADPH and the drug or hormone are oxidized), but the central one is a heme-iron protein called *cytochrome P-450*. It was given this name because, when it is combined with carbon monoxide, it exhibits a strong absorption band at 450 nm. Cytochrome P-450 can combine with many drugs, steroid hormones, and molecular oxygen. The latter is "reduced" and incorporated into the drug. The entire cycle of events is called the *cytochrome P-450 oxygenase system*.

The electrons, or reducing equivalents, are derived as a result of the oxidation of NADPH (or possibly also NADH). The sequence of events is shown in Figure 2.4. The drug (or steroid) combines with the oxidized form of the cytochrome P-450, which is reduced. This process involves the enzyme *NADPH-cytochrome c reductase*, which appears to be identical to what was formerly called NADPH-cytochrome P-450 reductase. It is thought that a microsomal *non-heme-iron protein*, which carries the electrons to the cytochrome P-450, is involved as an intermediary in this reaction. The reduced cytochrome P-450 drug complex then combines with molecular O_2, which is then said to be "activated." Additional electrons are passed from the NADPH or NADH system via cytochrome b_5, and upon the acquisition of $2H^+$ the complex dissociates into the oxidized drug, H_2O, and oxidized cytochrome P-450. This microsomal system usually mediates hydroxylation reactions, but oxidation at other sites, including C, N, and S, are possible, so that deamination, desulfuration, and dealkylation reactions occur. Cytochrome P-450-mediated reactions also occur, but to a lesser extent, in other tissues, including adrenalcortical mitochondria, the placenta, the kidney, and the gastrointestinal tract.

Several other types of cytochrome P-450 have been identified with different absorption bands. There thus appears to be a family of such proteins which may each be preferentially involved in the metabolism of certain types of drugs.

Glucuronide conjugation also occurs in the liver microsomes and involves the formation of activated glucuronic acid or uridine diphosphate glucuronic acid from glucose-1-phosphate and

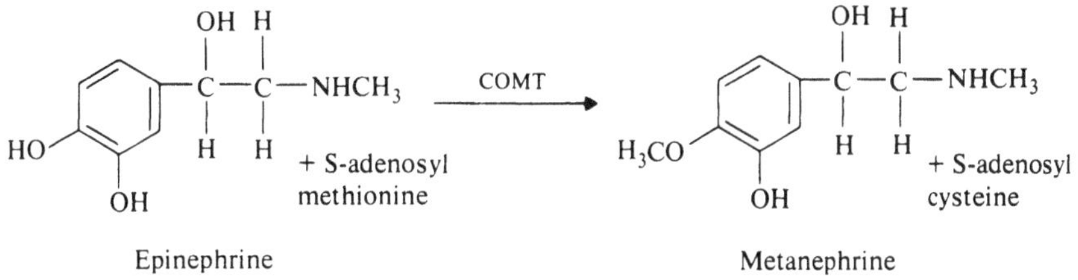

Scheme 2.1.

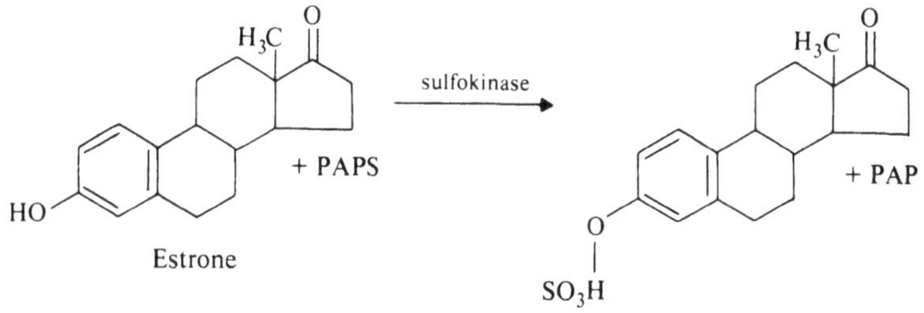

Scheme 2.2.

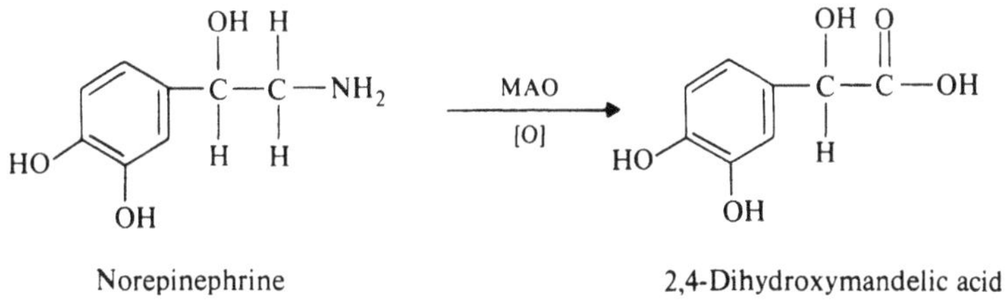

Scheme 2.3.

UTP in a series of enzymic reactions. The final reaction, which occurs in the microsomes, is

UDP—glucuronic acid + ROH $\xrightarrow{\text{glucuronyl transferase}}$

RO—glucuronide + UDP

This is an important reaction in the excretion of steroids and thyroid hormones which appear in the urine and bile. This type of conjugation can also occur at other sites on the molecules, such as NH, S, or ester groups.

1. Reduction reactions. The metabolism of drugs and hormones by reduction reactions is not common but can be important, especially for steroids. Saturation of double bonds and carbonyl groups in such compounds is an important process in their metabolism, including inactivation and activation. These types of reactions involve NADH or NADPH as an H$^+$ donor. The precise location of the enzymes in the cells is not always clear, but microsomal, nonmicrosomal, and nuclear sites have been identified. Two examples of such reactions are shown in Schemes 2.4 and 2.5. The 5β-

dihydro derivative can also be formed but is inactive. (Saturation of the double bond occurs as a result of the action of 5α-reductase; two carbonyl groups have also been reduced, probably by a nonmicrosomal reaction.)

Dehalogenation can involve reduction, if the halogen is replaced by a hydrogen, or oxidation, such as the removal of a hydrogen (dehydrohalogenation). A deiodinase enzyme has a ubiquitous distribution in the body, where it is concerned with the transformation of the tetraiodo compound thyroxine to the more active triiodothyronine. Removal of additional iodine moieties results in inactivation of this hormone.

2. Factors influencing drug and hormone metabolism. The rate of metabolism, and clearance, of drugs can vary considerably; in man, a six-fold range is thought to be normal. Such differences appear to reflect nutritional, environmental, and genetic factors. There may also be sexual differences. It should be emphasized that interspecific

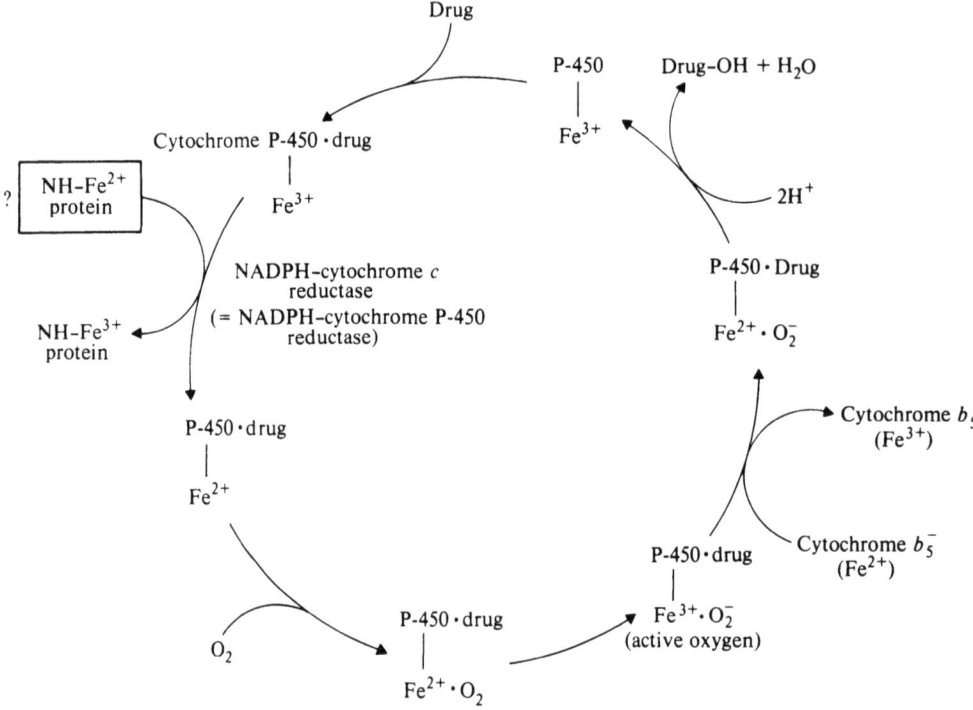

Fig. 2.4. Cytochrome P-450 oxygenase cycle. The electrons (or reducing equivalents) are derived from the oxidation of NADPH to NADP$^+$. The reduced non-heme-iron protein (NH–Fe^{2+} protein) is a microsomal constituent which *may be* an intermediary in this transfer from NADPH to cytochrome P-450. Further electrons are also derived from NADPH and are thought to be transferred from cytochrome b_5.

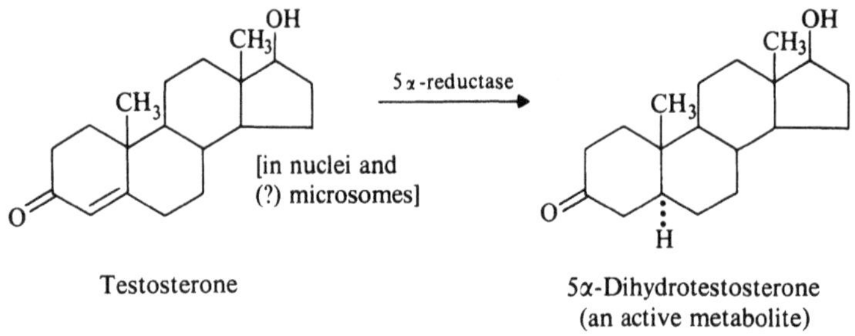

Testosterone → 5α-reductase [in nuclei and (?) microsomes] → 5α-Dihydrotestosterone (an active metabolite)

Scheme 2.4

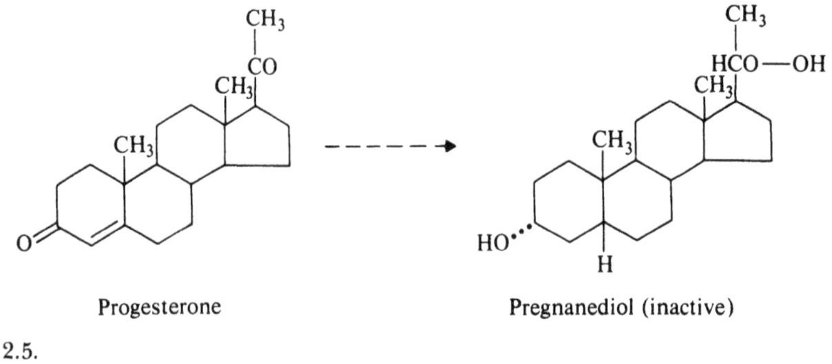

Progesterone - - - → Pregnanediol (inactive)

Scheme 2.5.

differences exist, so that one must be careful in extrapolating to man results gained from animal experiments. Considerable variation can occur in man, related to disease and prior exposure to drugs and environmental contaminants. Newborn infants have a very poor capacity to metabolize drugs, but this ability develops rapidly in the first year of life. Such differences in the ability to metabolize drugs and eliminate them from the body can have important consequences with respect to their therapeutic actions and possible side effects. One of the most obvious situations where possible changes in the elimination of a drug must be considered is in liver and kidney disease, when there often may be a dramatic decline in this process. The use of drugs in infants also needs considerable circumspection. In these days of polypharmacy, when several drugs are often administered concurrently, interactions that may influence their metabolism are an increasing problem.

Several factors can influence the rate of drug metabolism by a tissue, but two are predominant: its blood supply and its innate metabolic ability to carry out the biotransformational changes. The binding of drugs to plasma proteins and their tissue storage, such as in adipose tissue, also affect drug metabolism. In advanced liver disease, such as cirrhosis, all these factors may be involved: a decreased blood supply, an inhibition of tissue metabolic activity, and a decline in plasma proteins. Pregnancy can also alter drug metabolism due to changes in the production of plasma-binding proteins, and the additional activity of the placenta and the fetal liver.

The drug metabolizing activity of the smooth endoplasmic reticulum may be altered in a number of ways:

a. As these reactions are usually saturable, and exhibit an exponential course, competition between drugs for metabolism can occur. However, owing to the very large capacity of such enzymic processes for such substrates, this form of inhibition is rarely seen except under *in vitro* conditions. Attempts have been made to synthesize inhibitors of the enzymes concerned and such an experimental drug is SKF 525A (β-diethylaminoethyl-2,2-diphenylpentanoate). Such compounds may bind to essential sites, including cytochrome P-450, in a competitive or noncompetitive manner. As such enzyme reactions have a widespread physiological role, it is not surprising to find that such drugs are of doubtful therapeutic use. A sufficient degree of selectivity to influence a drug's metabolism without upsetting normal physiological processes may not be possible.

b. The enzymes and associated proteins and lipids of the smooth endoplasmic reticulum can increase in response to certain stimuli, including many drugs. This *induction* of the microsomal en-

Table 2.5. *Some drugs that can induce the formation of microsomal enzymes, especially in the liver, and so can enhance the rates of their own metabolism and that of other drugs and hormones*

Hypnotics, sedatives, and anticonvulsants
Phenobarbital and other barbiturates
Glutethimide
Ethanol[a]
Chloral hydrate[a]
Diphenylhydantoin
Primidone

Tranquilizers and antipsychotic drugs
Chlorpromazine
Meprobamate
Chlordiazepoxide
Imipramine

Antiinflammatory drugs
Phenylbutazone
Aminopyrine
Corticosteroids

Sulfonamides
Tolbutamide
Carbutamide
Sulfaphenazone

Steroids
Androgens
Estradiol[a]
Corticosteroids
Vitamins D_2 and D_3

Assorted
o,p'-DDD[a]
Metyrapone[a]
Bishydroxycoumarin
Insecticides (e.g., DDT, dieldrin)
Polycyclic hydrocarbons (e.g., cigarette smoke)

[a] First depress and then stimulate.

zyme system appears to be most important in the liver. More than 200 drugs and toxic compounds have been shown to elicit this response. Some, with this effect, are shown in Table 2.5. They include a range of types of drugs, including some hypnotics, sedatives, anesthetics, and steroids. A number of polycyclic hydrocarbon compounds, such as are present in tobacco smoke and insecticides (e.g., DDT and dieldrin), also can induce hepatic microsomal enzymes. The prototype of such compounds is the barbiturate phenobarbital. After several days' administration of this hypnotic drug, there is an increase in the size of the liver, and a large proliferation of the smooth endoplasmic reticulum can be observed. These changes are associated with an increase in many liver enzymes, but especially the microsomal ones associated with the cytochrome P-450 oxygenase system. The change reflects a *de novo* protein synthesis and can be inhibited by actinomycin D or puromycin. The response appears to be elicited by the binding of the drug to the cytochrome P-450. It does not, however, need to be metabolized in order to initiate the response. The effect is substantially nonspecific, as the metabolism of a variety of other drugs is also made possible by the additional microsomal enzymes. In man, this increase is usually about twofold. There are, however, instances when the effect may be relatively specific, such as seen with 3,4-benzpyrene, which induces enzymes that will metabolize the analgesic acetanilid but not the anesthetic hexobarbital. Drugs that induce microsomal enzymes must be lipid-soluble and preferably should have a relatively prolonged period of action.

The structure of drugs and hormones can be modified in order to change the rate of their metabolism and excretion and so alter the time of their action in the body. Thus, substitution of certain amino acids or the removal of terminal amino groups can limit the abilities of enzymes to attack polypeptide chains so that they are protected and their effects are prolonged. Substitution of a halogen, a Cl for a methyl group, results in a decreased metabolism of the oral hypoglycemic drug chlorpropamide as compared to tolbutamide. The ability of enzymes to change the structure of certain drugs and so activate them may also be utilized to prolong their action, such as is seen in the preparation of polypeptides with additional, protective, terminal chains of amino acids, and in the esterification of steroids. Such compounds can often persist in inactive form for relatively long periods of time in the body, but as a result of their metabolism, they are slowly changed to their active form.

2.5. Excretion of drugs

Drugs and their metabolites can be excreted through several pathways. The most important of these is the urine, followed by the bile, but the other gastrointestinal secretions, the sweat, and even the milk of lactating women can also contribute. The lungs may be important in certain instances, involving substances that are readily vaporized, such as many general anesthetics. Compounds that are completely dependent on such channels for their elimination from the body usually have a relatively long half-life compared to those drugs that undergo metabolism in the body.

The secretion of urine by the kidneys involves filtration of the plasma across the glomerulus, followed, in the renal tubule, by an assortment of processes of reabsorption from the filtrate back into the blood, and a secretion of solutes, by the renal tubular cells into the urine. Apart from the ultrafiltration at the glomerulus, these processes involve diffusion and active transport. The filtra-

tion of a substance will depend on its size, which will influence its ability to pass across the glomerulus. Drugs bound to plasma proteins will thus not be readily filtered. The lipid solubility of the compound will have an important effect on its ability to diffuse back into the blood, so that hydrophilic molecules and those salts that exist in a dissociated, electrically charged form will tend to be retained in the urine. The excretion of lipid-soluble drugs will, however, be very slow, and for excretion to occur, they must be metabolized to more-water-soluble forms. Some water-soluble solutes may, however, be reabsorbed from the glomerular filtrate via aqueous channels and by processes involving active transport. Secretion of a number of compounds can occur across the renal tubule, and this involves the water-soluble ionic forms of certain acids and bases. These secretion "pumps" are dependent on metabolic activity, they are saturable, and their actions can be inhibited competitively by related compounds. Renal diseases clearly may influence the urinary excretion of a drug from the body and necessitate reconsideration of its dose or even its use.

Some drugs, following their metabolic processing in the liver, are excreted in water-soluble form in the bile and thus pass into the intestine. Such compounds, which include a number of steroids, are usually in hydrophilic form, owing largely to their conjugation to glucuronide and sulfate. These conjugated compounds are not readily absorbed from the gut, and if unchanged they will be excreted in the feces. However, they may be altered by tissue and bacterial enzymes in the digestive tract and reassume their lipophilic characteristics so that they can be reabsorbed back into the circulation. Their elimination will thus be delayed as a result of their being trapped in this cycle of *enterohepatic circulation*. Small quantities will, however, inevitably "escape" and be excreted in either the feces or the urine.

A number of hormones and their metabolic products, some of which retain biological activity, are excreted in the urine. These include the steroid hormones, which are often in their conjugated water-soluble form, but other metabolites are also present. Some of the polypeptide hormones, the glycoprotein hormones, and catecholamines are also excreted in the urine. These processes are of practical importance, as they provide convenient methods for assessing the rates of production of certain hormones that may be important for the diagnosis of a disease. In some instances, such as the conjugated estrogen hormones, the glycoprotein hormones, and catechlolamines are also excreted in the urine. peutic use.

Drugs can sometimes be excreted in the milk, and the mammary tissue itself may contribute to their metabolism, so that lactation can influence their clearance from the circulation. It can, of course, also provide an undesirable source of drugs for the infant.

2.6. Types of responses to drugs

A drug ultimately acts to increase or decrease the activities of organs and cells. Such effects may be manifested in various ways, such as: (a) the contraction or relaxation of a muscle; (b) the increase or decrease in the synthesis or release of a secretion (either exocrine or endocrine); (c) the uptake or loss of an ion (e.g., Na, K, Ca) or metabolic substrate (such as a sugar, amino acid, or fatty acid); (d) the formation, activation, or degradation of an intracellular metabolite (e.g., cyclic AMP or GMP) or enzyme (e.g., RNA polymerase); and (e) changes in the rate of growth, maturation, and division of cells.

Such physiological processes are complex and invariably involve a number of sequential, but distinct, steps each of which may provide a possible locus for interference by a drug. They are also usually dependent on the general integrity of the cell, so that an action of drugs on one type of process may indirectly influence another.

2.7. General requirements of a drug

A drug has several required general properties which are dictated by its physicochemistry and the properties of its target tissues.

a. It should exhibit a propensity to react preferentially at certain tissue sites. This characteristic gives it a selectivity so that it can exert effects that are sufficiently specific to make it useful. Its side effects are thus limited.
b. An ability to act at relatively low concentrations is usually desirable.
c. Its effects on the cell should generally be reversible so that permanent damage, which is difficult to repair, does not occur. However, covalent interactions between certain drugs and tissues can be useful, such as in those compounds with chemotherapeutic cytotoxic actions.
d. Elimination from the body, either as a result of its metabolism and/or excretion, must be possible.

2.8. How drugs act

Drugs act to alter the excitability and metabolic activities of tissues. They may elicit their effects on cells in several general ways, which can be arranged in two main groups:

a. By reproducing, mimicking, or blocking the effects of natural excitants such as neurotransmitters and hormones. These effects may occur at any of a number of sites of reactions that are involved in a response, such

as an initial, terminal, or intermediate metabolic event.
b. By interfering with the normal formation, activation, degradation, or accumulation of natural metabolic substrates, enzymes, and ions.

2.9. Where drugs act in cells

Drugs usually act directly on their effector cells, but in some instances, the initial site of their action may be elsewhere in the body. Thus, the activity of a tissue can be influenced by changes in its blood supply, the availability of external metabolites and ions such as glucose and Ca^{2+}, and stimulation by tropic hormones that are released into the circulation at some distal site.

Drugs may act at several different types of sites in cells:

a. They may act at the specific sites that normally mediate responses to natural excitants. Such natural "receptors" (see below) have been identified at effector sites and appear to be present in the plasma membrane, the cytoplasm, and the nucleus. They may lie in close proximity or even be a component of enzymes. It seems likely that such receptors may also be associated with other organelles, such as cytoplasmic granules and mitochondria.

b. Changes may occur as a result of physicochemical interactions with structural components of the cell, especially the various membranes that sequester its contents. Such changes may alter the intracellular distribution of ions and metabolites and so evoke responses, including changes in muscular contractility and secretion. These interactions may occur at localized specific sites, such as the action of tetrodotoxin, which blocks sodium channels in nerve and muscle membranes. It may, however, be a quite general effect which need not have precise structural requirements but involve simply the saturation of a biophase of the cell such as the membrane lipids. This is called the "Ferguson principle." Many anesthetics appear to act in this manner.

c. A physical or chemical interaction with tissue enzymes and metabolic substrates may occur which alters their structural conformation and ability to take part in the normal processes in the cell. This may involve the substitution of a "fake" metabolite which can enter the system but will not undergo normal transformation. An interaction with an allosteric enzyme may change its conformation and initiate a response.

"Receptors" were originally hypothetical components of cells, the existence of which was proposed to account for the actions of drugs and natural physiological excitants that are present in the body. Over the last few years their presence has received dramatic confirmation as a result of the precise physicochemical identification of a number of such receptors, especially those for hormones, in tissues and extracts of cells. The term "receptors" has on occasion been used to describe any physical or chemical site with which a drug, neurotransmitter, or hormone can react in the cell. Such a definition is, however, now too broad to be useful and does not fulfill the requirements of the original "receptor hypothesis." The term "receptor" is at present generally reserved for specific natural sites that are present at low concentrations in cells. Such distinct components can be viewed as having been formed as a result of the normal processes of evolution and may even be considered as another type of cell "organelle." They are, however, only of molecular (or macromolecular) dimensions and usually exist as a part of, or in close association with, larger structures. Some receptors have, however, been "solubilized" and separated from other parts of the cell. The existence of such a receptor for a drug suggests the presence of natural homologues to the drug with which it combines, although these have not always been identified. A dramatic example of this type of assumption has, however, been provided by the identification of natural substances that can combine with "opiate receptors." The actions of the opiate drugs, morphine and heroin have been known for hundreds of years, and specific receptors for them were recently isolated from brain and intestine. Subsequently, starting in 1975 a group of naturally occurring peptides, called enkephalins and endorphins, which react with these opiate receptors were isolated from brain tissue. The nature and properties of receptors is described in more detail in succeeding sections. [An up-to-date summary of information about peptide and steroid hormone receptors has been provided by Catt and Dufau (1977), O'Malley and Schrader (1976), and Hollenberg and Cuatrecasas (1978).]

2.10. Role of receptors

The function of receptors in the body is to sort out information with which the cell may be provided by the blood and nerves that supply it and, when appropriate, to transmit a signal to the machinery of the cell. It thus acts as a transducer. As the energy provided by such an initial signal is usually very small compared to that involved in the final response, the receptor also plays a role in the process of the amplification which is necessary for a full response to occur. It should be emphasized that responses to drugs and hormones are usually quite complex and may involve a whole series of reactions of which the interaction of the drug or hormone with its receptor is only a distal, but still essential, primary event.

2.11. Receptor theory

Drugs and hormones usually act at relatively low concentrations, frequently at about 10^{-9} M, although levels as low as 10^{-12} M are not uncommon. Their effects are also usually remarkably specific in relation to their chemical structure. These properties of drugs have been recognized for about 100 years and led J. N. Langley in 1905 to suggest the presence of a specific "receptive substance" to account for the actions of curare and nicotine at the neuromuscular junction. P. Ehrlich made a similar suggestion in 1913 to explain the specific actions of certain dyes on bacteria. Although these receptors remained somewhat enigmatic hypothetical entities until quite recently, they provided a basis for the theoretical analysis of the quantitative actions of drugs, both in eliciting a response (agonists) and antagonizing it (antagonists). Several such theories to account for the actions of drugs have been proposed. The two major current ones are *occupation theory*, originally developed about 50 years ago by A. J. Clark at University College London, and *rate theory*, proposed by W. D. M. Paton at Oxford in 1961 (Paton, 1961). Occupation theory has been somewhat modified, especially by E. J. Ariëns (see Ariëns and Simonis, 1964a,b), and is currently the most favored. It should, however, be emphasized that no single theory can completely account for all the various phenomena associated with drug action, and it is possible that some groups of drugs act differently, so that a unitary theory may not be appropriate. In addition, conclusive experimental proof of such theories is probably not feasible, although they can provide a useful and productive framework for understanding the mechanisms by which drugs work.

A. Occupation theory

When the concentration of a drug is plotted in relation to its response, such as the contraction of a piece of smooth muscle in an organ bath, there is a rapid rise in the effects, once the threshold level has been achieved, in relation to the increasing level of the drug. Eventually, saturation occurs and a hyperbolic curve is seen (Figure 2.5a). Such a dose–response relationship is also frequently plotted using a logarithmic scale for the dose or concentration, and this gives a sigmoidal-shaped curve (Figure 2.5b). (The latter has certain practical advantages, as the central part of the curve is linear, which facilitates comparisons between the potency of drugs.) Clark proposed that the effects of a drug were proportional to the quantity of the receptor–drug complex, so that when the response is 50 percent of maximal, half of the receptors are bound to the drug. (As we shall see, this half-saturation is an oversimplification, as maximal re-

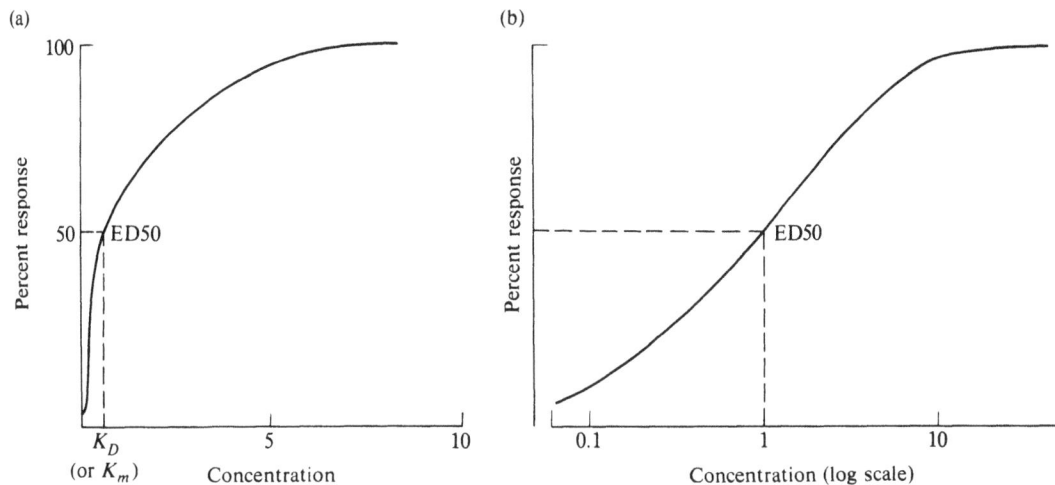

Fig. 2.5. Drug concentration or dose–response curves. (a) The response, plotted as a percent of the maximal one (n = 100 percent) versus the drug concentration. The curve has a hyperbolic shape. At a response equivalent to 50 percent of the maximal one (ED50; also called A_{50}), receptor "occupation theory" predicts that 50 percent of the receptors will be occupied. The concentration of the drug at which this occurs will then be expected to be equivalent to K_D or, by analogy with enzyme (for the drug–receptor interaction) kinetics, K_m. As we shall see, direct measurements of drug–receptor binding do not always confirm this relationship, so that this value is often called the $K_{D\text{ apparent}}$ or $K_{D\text{ activation}}$. (b) The dose of the drug is more often plotted in relation to the logarithm of its concentration, so that a sigmoidal-shaped curve is seen. As the central part of this line is linear, it facilitates comparison with other drugs and interactions with antagonists.

sponses can often be obtained when only a small proportion of the total receptors present are occupied.) Such a relationship can be described in terms of the law of mass action or the Langmuir adsorption isotherm. Thus, according to the mass action law:

[drug] + [receptor] $\underset{k_2}{\overset{k_1}{\rightleftharpoons}}$ [drug–receptor complex]

where k_1 and k_2 are the rate constants for the forward and backward reactions, respectively.

The ratio

$$\frac{k_2}{k_1} = \frac{[\text{drug}] \times [\text{receptor}]}{[\text{drug–receptor complex}]}$$

and is called the *equilibrium dissociation constant* – K_d, K_{diss}, or most often K_D (although the term K_A is also used for an agonist drug). K_D is expressed in moles per liter and is expected to be equivalent to the concentration of the drug that is present when the response is 50 percent of the maximum (ED50) attainable by that drug. Such an analysis of hormone–receptor interactions is analogous to that used to describe the combination of enzymes with their substrates, in which case the equilibrium constant k_2/k_1 is called K_m (the Michaelis–Menton constant). The equilibrium association constant k_1/k_2 can also be used to describe this reaction. It is expressed as K_a or K_s and has units of liters per mole.

Drugs interact with their receptors with differing degrees of readiness or strength, and this property

of a drug is called its *affinity* for its receptor. Affinity is inversely proportional to the dissociation constant ($1/K_D$ or K_a), so that affinity is said to be high the lower the concentration of a drug that results in the 50 percent saturation of the receptors. The affinity is also expressed as PD_2, which is the negative logarithm of the molar concentration producing a half-maximal response. Thus, if this concentration is 10^{-9} M, the PD_2 is 9. The higher the value, the greater the affinity the drug has for its receptor.

Different drugs, even when they are closely related chemically, can exhibit quite different abilities to elicit a response, and some may even act as antagonists. Such properties cannot be accounted for simply in terms of the law of mass action, and it is necessary to propose that drugs have other properties related to the strength of the effect they can elicit. Stephenson referred to the *efficacy* of a drug, and Ariëns termed this property *intrinsic activity* (α):

response = α(drug–receptor complex)

Alpha is a proportionality constant related to the drug's intrinsic ability to elicit a response once it has combined with the receptor. In terms of the law of mass action, it can be described as

(drug) + (receptor) $\underset{k_2}{\overset{k_1}{\rightleftharpoons}}$ drug–receptor complex $\xrightarrow{k_3}$

receptor + response

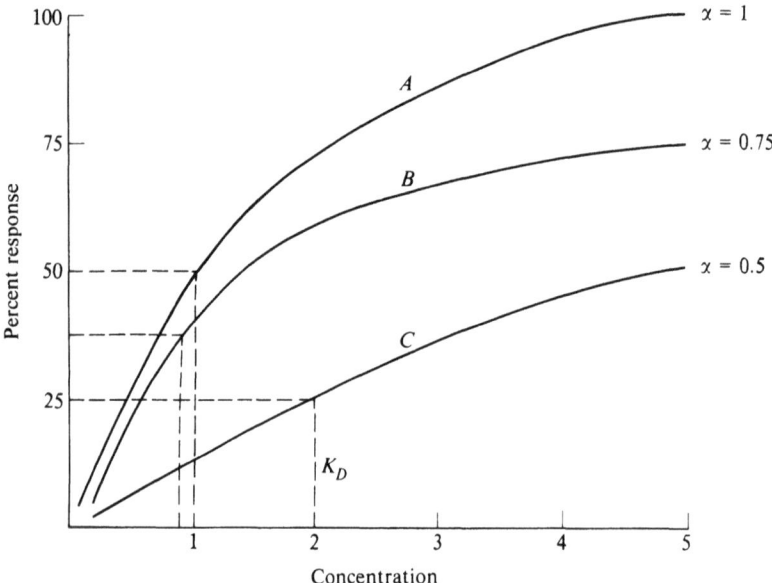

Fig. 2.6. Intrinsic activities (α) and affinities ($1/K_D$) of three drugs as illustrated from their concentration–percent response curves. The maximal responses differ in all three drugs, so that they have intrinsic activities of 1, 0.75, and 0.5. The affinities ($1/K_D$) for each drug do not, however, necessarily have the same relative relationship. In this instance drug *B* displays the greatest affinity for its receptor. This figure also illustrates the difficulties of predicting the potency of a drug from either its intrinsic activity or its affinity. Thus, if a response corresponding to 25 percent of the maximal one was chosen for comparison, drugs *A* and *B* would have the same potency and they would be about 4 times as potent as drug *C*. However, at 75 percent maximal response, drug *A* would clearly be the most potent.

In enzyme kinetics, k_3 is the rate constant of the final step of reaction and in the present analogy is equivalent to the intrinsic activity.

The efficacy or intrinsic activity of a drug is the maximal or ceiling effect that it can elicit. Different members of a group of drugs may differ in their ability to bring such a response about, but the one that exerts the highest effect ever observed is said to have an intrinsic activity of 1. Plotted graphically, this can be seen to correspond to the 100 percent response. A drug that can elicit only 50 percent of this maximal response has an α of 0.5. The affinity ($1/K_D$) may or may not be the same for drugs with differing intrinsic activities.

Intrinsic activity should not be equated with *potency*, which is a comparison of the abilities of drugs to elicit the same quantitative response. It can be seen in Figure 2.6 that when we compare the abilities of the three drugs to elicit the same response, for instance equivalent to 50 percent, *A* and *B* have similar potencies (despite their quite different intrinsic activities) while *C* is about 4 times less potent. It can be seen that the ED50, and thus the affinities ($1/K_D$) of each drug for its receptor, are in this instance different (ED50 is also not necessarily related to potency).

Some of these relationships, especially those of affinity, are seen more clearly when such results are plotted as the logarithm of the dose (Figure 2.7). In this instance, drugs *A* and *B* have the same intrinsic activity (α), but 10 times more *B* is needed to elicit the same response (it is 10 times less potent). *C* requires 10 to 1000 times the dose of *A*, depending on which point in the two nonparallel curves one chooses to make the comparison. It can be seen that while drugs *A* and *B* can elicit the same maximal response, the plot for *B* is displaced in parallel to the right. The ED50 is 10 times greater, so that its affinity ($1/K_D$) for its receptor is much less than that of *A*. This type of difference, where two drugs have the same intrinsic activity but differing affinities, is usually seen in drugs that are structurally related and suggests that they are probably interacting with the same receptor (there are, however, exceptions, and this likelihood is not a rule). Drug *C* is quite different from *A* and *B* as it not only has a lower intrinsic activity, but its ED50, and hence K_D, corresponds to the log dose of 1. Its dose–response plot is not only displaced far to the right, but in contrast to drugs *A* and *B* it is not parallel to the other drugs. These observations indicate that *C* is not acting in the same way as are *A* and *B*, and its structure will almost certainly be quite different from those of *A* and *B*.

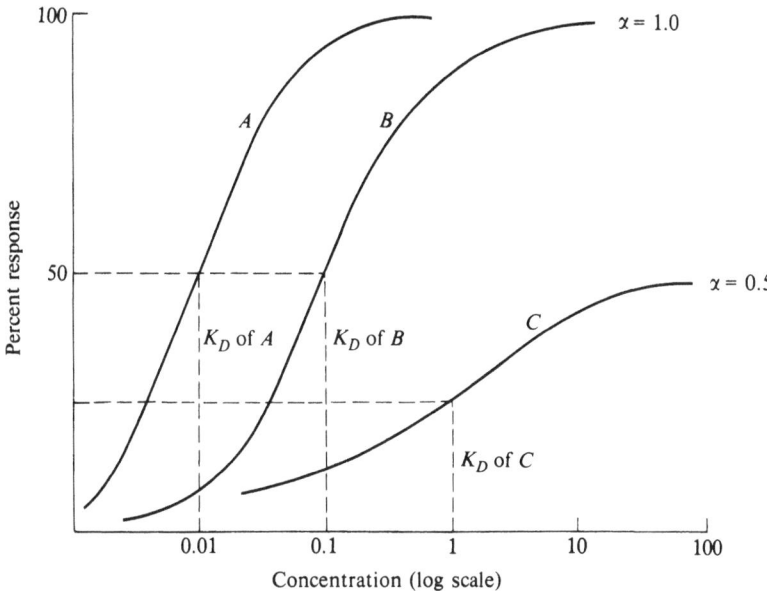

Fig. 2.7. Relationships of affinity, intrinsic activity, and potency of three drugs A, B, and C. The concentration is plotted on a log scale. It can be seen that A and B have the same intrinsic activity ($\alpha = 1$), but the dose–response curve for B is displaced in parallel to the right, so that the K_D for its drug–receptor complex appears to be 10 times greater. (It has a lower affinity for the receptor.) In this instance, drug B is also 10 times less potent than drug A. Drug C has a lower intrinsic activity ($= 0.5$) than drugs A and B, and the dose–response curve is displaced far to the right and is not parallel. Although it would appear that drugs A and B are chemically related to each other, drug C is probably quite different.

1. Antagonists. Drugs are widely used to antagonize the effects of other drugs and endogenous physiological responses such as those to neurotransmitters and hormones. They may exert such effects in various ways, including an opposing physiological response, such as activating the relaxation as opposed to the contraction of a muscle (physiological antagonism). Antagonists may also work more specifically by exerting an effect at the level of the response mechanism itself (pharmacological antagonism). As already described, the response to an excitant may be quite a complex process and involve numerous individual but related effects. Thus, it is feasible to block such an effect of a drug at various points. In the instance of smooth muscle contraction, this effect could involve the release of a neurotransmitter, a drug's access to or interaction with its receptor, the membrane depolarization process, coupling events that involve an increase in cytoplasmic Ca^{2+}, or the activity of the contractile proteins themselves. An endocrine example would be the events that result in the antidiuretic effect of vasopressin (ADH), which includes an interaction with its receptor, activation of adenyl cyclase, synthesis and destruction of cyclic AMP, activation of a protein kinase, and the integrity of the actual mechanism that in-

creases the permeability of the renal tubular cells to water.

Dose–response curves to a drug are plotted in Figure 2.8 in relation to the presence of different concentrations of two types of antagonists. It can be seen that when the excitant drug E, an agonist, is present with drug A (an antagonist), the response is depressed but, provided that the concentration of E is raised enough, the same effect can be elicited. In the log dose–response plot, the curve for E + drug A is displaced in parallel to the right. This form of antagonism is the *competitive* type. In its simplest interpretation, in terms of receptor theory, drugs E and A may be considered as competing for the same receptor, a situation that most likely reflects similarities in their structures. A change in the conformation of the receptor that reduces its affinity for the drug may also be occurring in the presence of the antagonist. It is difficult to exclude the possibility that the antagonist may be influencing a more distal process in the chain of the effector response, the rate of which is closely related to the primary drug–receptor interaction. In this instance, however, the two drugs would probably be structurally dissimilar (remarkable coincidences excepted). Competitive antagonists are considered to have an affinity for the receptor

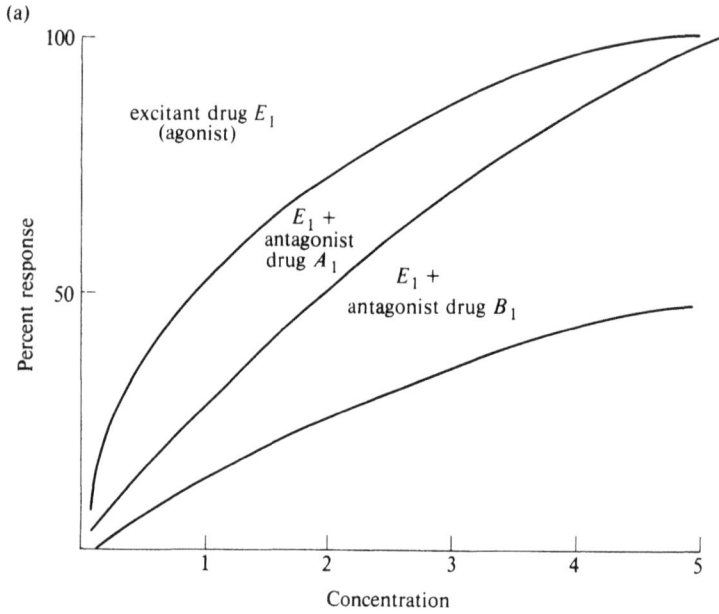

(a)

excitant drug E_1
(agonist)

E_1 +
antagonist
drug A_1

E_1 +
antagonist drug B_1

Percent response

Concentration

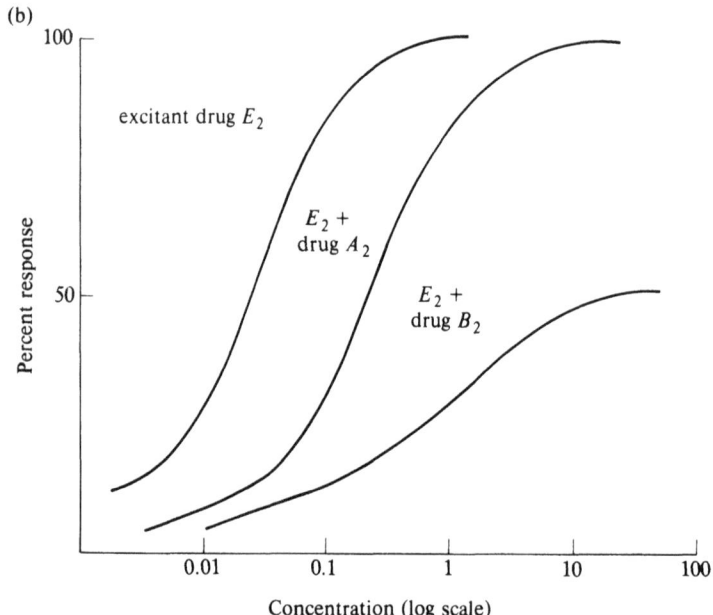

(b)

excitant drug E_2

E_2 +
drug A_2

E_2 +
drug B_2

Percent response

Concentration (log scale)

Fig. 2.8. Drug dose–response curves, illustrating competitive and noncompetitive antagonism. Two types of plots are made using an arithmetic (a) and a log concentration scale (b). It can be seen in both (a) and (b) that when a fixed concentration of drug A (an antagonist) is administered at the same time as drug E (the excitant or agonist), a higher concentration of E must be used to elicit the same effect. Provided that enough of E is administered, however, the same maximal response can be achieved. The antagonism can be overcome and is said to be competitive. In the instance of the second antagonist, drug B, however, the decrease in the effect cannot be overcome; its effect is insurmountable or noncompetitive.

but a low, or zero, intrinsic activity. They may still (although usually at relatively high concentrations) elicit responses themselves, and are then called *partial agonists,* but can usually produce only submaximal responses.

The antagonism of drug B for drug E (Figure 2.8) is quite different from that of A for E. In this instance, no matter how much the concentration of E is increased, its original response cannot be restored; the antagonism is insurmountable. This is called *noncompetitive antagonism.* A displacement of the response to the right can be seen in the log dose–response plot, but this is not parallel to the original effect of E. The effects of such an antagonist may occur at some rate-limiting step of the response, distal to the drug–receptor complex. However, it remains possible that the latter is affected in some way, such as if the antagonist combined with some adjacent moiety, and so excludes the drug from its receptor. The formation of an irreversible covalent linkage to the receptor or some other form of receptor inactivation is also expected to result in this type of effect (but not if there are "spare" receptors; see next paragraph).

As emphasized at the beginning of this section, occupation theory cannot account for all observations on the responses of tissues to drugs. Various modifications have therefore been proposed, and one of the most important of these is the concept of "receptor reserve" or *spare receptors.* This was made necessary as a result of several types of observations. On the basis of occupation theory, as originally proposed, the drug concentration–effect curve should ideally have a slope of unity, but in many instances this is not observed. The slope is often much steeper than this (see Rang, 1971). It has also been observed that whereas certain drugs (notably the β-haloalkylamines) that covalently bind to receptors, and so irreversibly inactivate them, result in a decline in the maximal response at high concentrations, at lower levels they simply produce a parallel displacement of the curve to the right. Thus, under the latter condition, a maximal response can still be attained, although with higher levels of the agonist. This effect would not be possible if 100 percent of the receptors are required to elicit the full effect. These observations led to the suggestion that, in fact, not *all* the receptors need be occupied and that in some instances only a few percent need to combine with the drug in order to elicit the normal maximal response. Thus, there may be an excess or reserve in the numbers of receptors. More direct counts of receptors have recently been obtained for several hormone-responsive tissues and the relationship of their occupancy to the response has confirmed this theory of spare receptors.

The potency of an antagonist can be expressed as the pA_2, as described by Schild (1957). This value is the negative logarithm of the concentration of the antagonist, which requires a doubling of the concentration of the agonist to overcome its inhibitory effects. The higher the value, the more effective the antagonist (pA_{10} values are also sometimes used).

B. Rate theory

Although most observations on drug–receptor interactions are consistent with the action of a drug being elicited while it actually occupies a site on the receptor, it cannot account for all such phenomena. In addition, as we have seen, the occupation theory has been modified in several ways, especially with the introduction of the concept of intrinsic activity to account for the fact that occupancy of a receptor can have either an agonistic or antagonistic effect. Other basic problems arise when one tries to explain how some drugs, for instance nicotine, first stimulate and then block a response. Many drugs also initially exert an effect which subsequently declines, or "fades," even though the drug remains in contact with the tissue. In 1961, Paton presented a theory to account for such anachronisms, which he called rate theory. It was proposed that excitation of the receptor only occurs in a quantal manner as a result of its initial combination, or collision, with the drug. Further effects are dependent on its ability to dissociate from the receptor so that it is free to be stimulated once again. It is therefore the rate of stimulation or number of successful collisions of the drug with its receptor that is important in eliciting a response. In terms of association (k_1) and dissociation (k_2) rate constants of the reaction

$$[\text{drug}] + [\text{receptor}] \underset{k_2}{\overset{k_1}{\rightleftharpoons}} [\text{drug–receptor complex}]$$

an effective agonist must also have a high k_2, whereas an antagonist will have a low k_2 and will so limit the rate of new collisions. The k_2 or rate of dissociation of the drug–receptor complex thus determines what in occupation theory is called the intrinsic activity while the affinity remains equivalent to the reciprocal of the dissociation constant k_2/k_1 ($1/K_D$). Stimulation followed by blockade or fading of a response can be accounted for by the difficulty a drug may have in dissociating from its drug–receptor complex following the initial successful collision that produces an effect. Clearly, rate theory has its attractions despite the current popularity of occupation theory.

C. Physicochemical nature of the interaction of a drug and its receptor

To trigger a biological response, a drug or hormone must not only interact with a receptor, but must do so in a manner that is sufficiently favorable to elicit or fulfill any change that may be required by the receptor. It is clear that most receptors have very specific requirements, as what appear to be quite small changes in the structure of a drug can result in dramatic changes in its activity. Thus, one form of a drug may be highly effective, whereas a stereoisomer of it may completely lack an ability to interact with a receptor. In other instances, some form of interaction between a drug and receptor may occur, but this may only elicit a partial response or even none at all. The precise reasons for such differences are not completely understood, but it is suspected that they ultimately could reflect the nature of the alignment and chemical interactions between the drug and its receptor. A favorable collision and combination most often appears to depend on a physicochemical interaction involving several sites on the molecules, those on the drug being complementary to those on the receptor. In some instances, such as those involving complex polypeptides, such interacting sites may be quite numerous.

Probably the primary requirement for a drug's success in combining with a receptor is its three-dimensional shape or conformation (tertiary structure). This may be important for several reasons.

a. The drug's accessibility to the receptor site is sometimes conceived as being by way of a specifically shaped cavity, such as a cleft or pore.

b. The receptor itself is also thought to have an intricate three-dimensional shape which may be complementary to the drug, so that the two fit into each other like a pair of oddly shaped wooden blocks. Probably of more basic importance, however, is that the shape of the molecule will determine the respective positions and alignments, and even the existence, of the various forces that are necessary for its successful interaction with its receptor. These areas of forces may not only be present individually on the receptor and drug, but may also be induced following their alignment, for example as a result of dipole–dipole interactions between them. The approach and proximity of two such molecules can result in changes in both of them, including alterations in their conformation.

The lipid and water solubility of a drug may also be important and may determine its ability to gain access to the region of its receptor. This will be influenced by such factors as the presence and distribution of polar and nonpolar chemical groups and the degree of ionization of the molecule at the ambient pH (its pK_a). (Ionized drugs are much less lipid-soluble.)

Several types of forces may be involved in the interactions of drugs and receptors, and several of these may take part in the formation of a single drug–receptor complex. The irreversible *covalent-type bond* is not typically involved, although it does occur and, like with the α-adrenergic blocker, phenoxybenzamine or the cholinesterase inhibitor DFP usually results in an inhibitory-

type response. Certain antineoplastic drugs which, for instance, bind to components in the nucleus also interact with tissues in this way. Toxic drugs and metabolites often bind covalently to tissues. Such responses and interactions are difficult to reverse and may depend on the normal processes of tissue replacement and growth. Weaker electrical forces are more usual and result in the reversible type of interactions that are essential for a receptor's normal activity. These may be simple *ionic bonds* which are provided by carboxyl $(-COO^-)$ or phosphoryl $\left(O = P \begin{smallmatrix} O^- \\ \diagdown \\ OH \end{smallmatrix} \right)$ moieties and by partially ionized sulfhydryl $(-SH)$ groups. Cationic sites include the guanidinium moiety, such as is present on arginine, ε-ammonium on lysine, as well as partially ionized α-ammonium $(-NH_2{}^+)$ and amide groups. Hydroxyl groups on phenolic or aromatic structures are not ionized at a physiological pH.

When the drug and receptor are in close proximity to each other, short-range, relatively weak electrical forces can operate. These result from *attraction between dipoles*. Dipoles arise due to disparities in the distribution of electrons between two adjacent sites, such as the carbon and the oxygen in the carbonyl $(C=O)$ group (the oxygen is more electronegative). The close proximity of two electron-dense regions in separate molecules may also be sufficient to produce a redistribution of electrons so that an *induced dipole* (van der Waals' forces) is formed. Such dipoles, utilizing either their cationic or anionic sites, can interact (dipole–dipole interactions) with a complementary adjacent charge and so contribute to a link between two molecules.

Hydrogen bonds are commonly utilized in drug–receptor interactions. Hydrogens can form cross-linkages between two partially negative electron-rich sites, both within molecules and between molecules. They can thus link peptide groups $\left(\diagdown C=O \cdots H-N \diagup \right)$, hydroxyls $\left(\begin{smallmatrix} OH \cdots O \\ \diagup \quad \mid \diagdown \diagup \\ C \quad H \quad C \end{smallmatrix} \right)$, or carboxy and hydroxy moieties.

Hydrophobic forces appear to be frequently involved in the interaction of drugs and receptors. In aqueous solutions, both of these components are usually surrounded by a halo of ordered water molecules. Interactions between hydrophobic groups in the drug and receptor results in the release of water molecules as the two molecules approach each other. The molecules of water will be pushed away from the region of the receptor and drug and interact with each other. The free energy of the receptor–drug system will thus be lowered as compared to that of its separated components in solution. Their aggregation will thus be favored, and this is referred to as a hydrophobic type of interaction.

D. Identification of receptors

The direct identification of specific receptors for drugs and hormones depends on the use of radioactively labelled compounds which have a high specific activity (i.e., radioactivity per mole of drug or hormone). Such substances, when labelled in the more usual way with ^{125}I or ^3H, should retain their biological activity. Tissues, usually studied *in vitro*, but also *in vivo*, can be shown to accumulate such compounds, but as this process may involve several types of sites, it is necessary to distinguish specific binding such as would be expected to occur to receptors, from nonspecific binding that occurs elsewhere in the tissue. To make this distinction, the tissue is exposed to a relatively low concentration of the labelled drug or hormone and, after allowing a period of time for equilibration, an excess (usually about 100 times greater concentration) of the unlabelled "cold" material is added. This procedure will be expected to displace the labelled drug from its specific binding sites and will give a measure of the receptors. This effect is shown in Figure 2.9, in which ^{125}I-insulin is displaced from binding to rat liver cell membranes. In this instance, only about 10 percent of the bound insulin was present at nonspecific sites. In many instances, however, this nonspecific fraction may be much greater and even the predominant one. It can be seen that other polypeptide hormones, such as ACTH, glucagon, and growth hormone, did not displace the ^{125}I-insulin from binding. Bovine and human insulin were as effective as "cold" porcine insulin in displacing the bound (labelled) porcine insulin. However, insulins from other species or modified preparations (analogues) or the separated A- or B-chains of the hormone were less effective, reflecting their relative abilities to combine with the receptors in the rat liver membranes (Figure 2.10a). These differences in affinity for the receptors can also be seen to be reflected in their biological responses (stimulation of glucose oxidation in fat cells Figure 2.10b). Such structural specificity for the binding sites and the quantitative relationship of this association to the biological response are important criteria in confirming that such sites are really the receptors.

As receptor sites are expected to be finite in number, they should be saturable with the drug or hormone. Examples of this property, using ^{125}I-human growth hormone, is shown in Figure 2.11. It can be seen that specific binding of this hormone

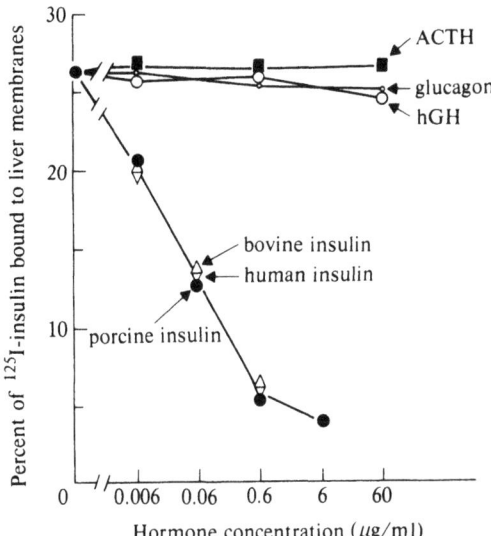

Fig. 2.9. Identification of "specific" receptor sites for insulin on rat liver membranes. These preparations were exposed to labelled porcine ^{125}I-insulin at a concentration of 0.7 nM (4 ng/ml). The radioactivity in the pellet of the membrane preparation is then plotted as a function of "cold" unlabelled hormone preparations. It can be seen that ACTH, glucagon, and human growth hormone (hGH) failed to alter the amount of radioactivity in the pellet, but three different preparations of insulin (human, bovine, and porcine) all displaced radioactivity from binding to the membrane preparations. This displaced radioactivity is considered to correspond to that which is specifically bound, most probably to the insulin receptors. With a 1000-fold excess of "cold" insulin, it can be seen that only about 10 percent of the originally bound material is still present, which corresponds to "nonspecific" binding. (From Freychet, Roth, and Neville, 1971.)

to rat liver membranes can be detected at about 10 pmol/liter and reaches saturation at 360 pmol/liter. Nonspecific binding of the hormones, however, did not display saturation at these concentrations.

A receptor is expected to have a high affinity constant $(1/K_D)$ for a drug, and the reaction occurs at low concentrations. This value can be estimated in several ways, including calculation from dose–response curves of the concentration at which 50 percent saturation of the receptors occurs. In 1949, G. Scatchard described a theoretical way of estimating the affinity of proteins for small molecules, as well as the number of sites on the protein that are occupied. This involves plotting the ratio bound/free molecule (e.g., a drug or hormone) on the ordinate, and plotting on the abcissa the total amount that is bound. The intercept with the abcissa gives an estimate of the number of binding sites, and the slope indicates the affinity.

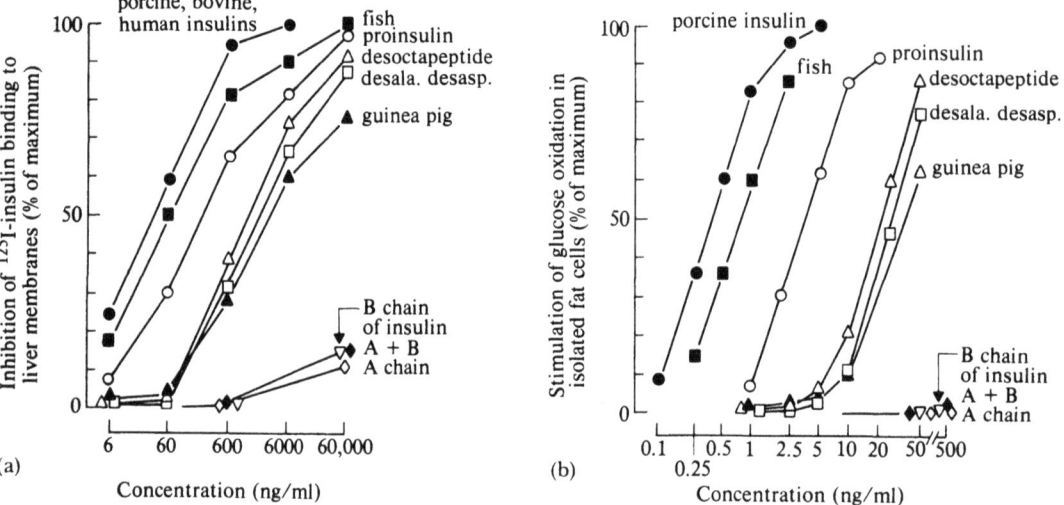

(a)   Concentration (ng/ml)

(b)   Concentration (ng/ml)

Fig. 2.10. Parallel abilities of different insulin preparations to bind to receptors and to elicit a biological response. As shown in Figure 2.9, specific binding of insulin to rat liver membrane preparations can be measured by recording their abilities to displace porcine ^{125}I-insulin from specific binding receptor sites. While the receptors in the rat liver membrane preparations (Figure 2.10a) apparently cannot distinguish between the human, porcine, and bovine insulin; fish and guinea pig insulin, proinsulin, and chemically modified fragments (desoctapeptide and desalanine, desaspartic acid–insulin, as well as the separated insulin A- and B-chains) are much less effective. In Figure 2.10b it can be seen that the biological response (stimulation of glucose oxidation) by fat cells displayed a parallel responsiveness to the receptor binding of the different preparations of insulin. (From Freychet, Roth, and Neville, 1971.)

An example that involves the binding of [^{3}H]aldosterone to toad bladder tissue (a model for the kidney) is shown in Figure 2.12. It can be seen that two groups of binding sites can be distinguished, one with a high affinity but with a relatively small number of sites and the other with a lower affinity but more numerous binding sites. Estimates of numbers of receptors per cell depend on the par-

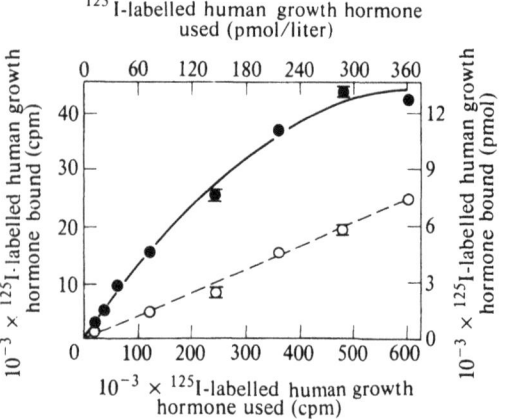

Fig. 2.11. Saturation of the specific binding sites for ^{125}I-human growth hormone to rat liver membranes (●—●). In contrast, it can be seen that at the concentrations used, nonspecific binding (○---○) did not display such saturation. (From Herington, Veith, and Burger, 1976.)

ticular system being studied but vary from about 1000 to more than 100,000.

Using such modern techniques for studying and isolating receptors, it is now possible to define such values as the equilibrium dissociation constant, K_D, of the drug–receptor complex by direct measurements. Thus, the concentration that corresponds to 50 percent saturation of the binding of a drug, or hormone, to its specific receptors is the K_D (sometimes called K_D *binding*). This determination involves measurements of the ability of unlabelled drug to displace labelled drug from its receptor sites. However, if a suitable labelled form of the drug is not available, this parameter can be obtained by measuring its ability to displace a labelled analogue from binding. This value is the same as the K_i, the inhibition equilibrium constant for an antagonist. It is the concentration necessary to displace 50 percent of the bound (labelled) drug. As described earlier (Section 2.11A) such values as the K_D, as well as the K_i, can be estimated less directly by recording, respectively, the concentrations necessary to elicit a half-maximal response or inhibit a maximal response by 50 percent. Such values are sometimes called the $K_{D\ apparent}$ ($K_{D\ app}$) or the $K_{D\ activation}$ ($K_{D\ act}$). The measured biological response may be that of a final effector, such as the contraction of a piece of smooth muscle, or it can involve an intermediate type of response, such as the activation of an enzyme such as adenyl cyclase. The two values for the K_D determined in these

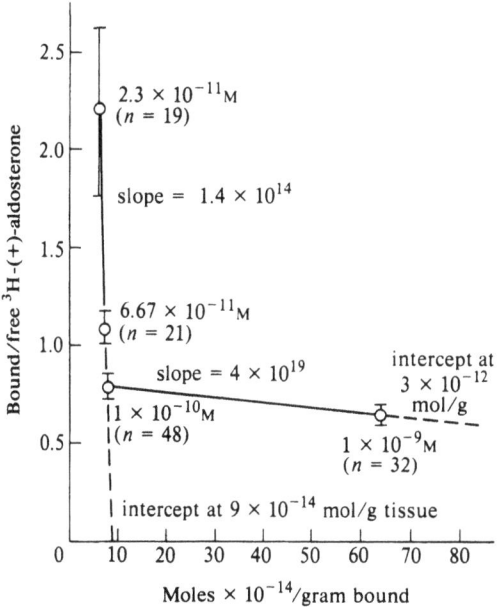

Fig. 2.12. Scatchard plot of the binding of [³H](+)aldo-sterone to toad urinary bladder *in vitro*. This preparation is commonly used as a model membrane to study the actions of hormones and drugs on the kidney. When plotted in this way, the binding of the aldosterone indicates that there are two sets of binding sites. The intercepts with the abscissa indicate the maximal numbers of each of these, while the slopes of the lines reflect the affinity of the aldosterone for each set of the receptors. It can be seen that the receptors indicated by the first part of the line are less numerous but have a higher affinity for the hormone than do the others. (From Sharp, Komack, and Leaf, 1966.)

ways ($K_{D\ binding}$ and $K_{D\ activation}$) do not always correspond. The possible reasons for this include the presence of spare receptors, in which case only a small proportion of the total may need to be activated to elicit a maximal response. It has also been viewed (Maguire, Ross, and Gilman, 1977) as reflecting differences in the "coupling efficiency" between the drug–receptor complex and the response, for instance the activation of adenyl cyclase. If the ratio $K_{D\ binding}/K_{D\ activation}$ is greater than 1, it suggests that relatively few receptors may need to be occupied to activate each unit of the effector substance (e.g., enzyme). The converse could occur if activation of multiple receptors is involved in triggering the activation of each such enzyme unit.

Such methods for identifying and counting receptors and the theoretical analysis of the results are not without criticisms and constraints. Some of these have been reviewed by Cuatrecasas et al. (1975). The original analysis of the molecular binding process which was performed by Scatchard in-

volved soluble proteins and relatively small molecules that were bound with a low affinity. The process of drug and hormone receptor interactions usually differs somewhat from the process described by Scatchard. The distinction between specific and nonspecific binding can also sometimes be difficult and depends on the conditions, including the concentrations used and even the amount of shaking and the nature of the containers in which the reaction is studied. What appears to be specific binding can occur on the walls of cellulose acetate tubes and Millipore filters. Insulin can even be shown to bind to talc in a manner suggesting the presence of a "specific receptor." Thus, the process of adsorption can imitate the binding of a drug or hormone to its receptor. Clearly, the identification and measurement of receptors requires, apart from theoretical knowledge, considerable experience, critical ability, and a sense of skepticism.

E. Nature of receptors

Until quite recently, receptors were hypothetical entities whose presence and properties were inferred on the basis of the special behaviors of various drugs, hormones, and neurotransmitters in eliciting or inhibiting biological responses. In 1960, Schueler described a receptor in physical terms as follows: "The drug-receptor is in general the pattern R of forces of diverse origin forming a part of some biological system and having roughly the same dimensions as a certain pattern M of forces presented by the drug molecule in such a way that between patterns M and R a relationship of complementarity for interaction exists." This concept is still valid, but today we have available more concrete information about the factors that determine these properties.

The availability of radioactively labelled drugs and hormones that retain their biological effects and have a high specific activity has facilitated the direct study of receptors. An important technical advance was the utilization of the principles of radioimmunoassays for competitive-binding studies. Receptors in various tissues and organs have been studied, using such methods, both *in vitro* and *in vivo*. Cell fractionation procedures have also been utilized to identify and study receptors in the microsomes, cytoplasm, and nuclei of cells. Soluble receptors that are normally present in the cytoplasm have been isolated, and in some instances those that are present in the plasma membrane have been solubilized with the aid of detergents such as Triton-X.

The receptors for hormones have provided the most dramatic advances in our knowledge of these constituents of cells. They are proteinaceous and usually have a molecular weight of about 60,000 to 300,000 and a hydrodynamic radius of 60 to 70 Å. Some of these receptors have been shown to be oligomeric and contain subunits, each of which may or may not bind an excitant molecule. Their protein nature is reflected in their sensitivity to proteolytic enzymes such as trypsin and to SH-active chemicals. Some hormone receptors have been shown to contain carbohydrate moieties and are thus probably glycoproteins. The properties of some receptors that are attached to membranes are altered following incubation with phospholipase enzymes. This observation may reflect the nature of their association with the plasma membrane, or they may contain a lipid component which is essential for their activity.

Steroid hormones have been shown to bind to receptors, or "acceptors," in the nucleus. These nuclear receptors are present in the chromatin, where they are associated with acidic, nonhistone proteins. To interact with these nuclear receptors, the steroid must, it appears, first be bound to the cytoplasmic receptor, which helps to effect its transfer into the nucleus and the binding process. Steroids can bind to naked nuclear DNA, but this process is not specific.

Receptors have a high degree of specificity with respect to the structure of the molecule they can combine with; stereoisomers of a drug usually cannot bind to the receptor. Usually, a receptor only binds one molecule of an agonist or antagonist, but in some instances, such as certain cytoplasmic steroid hormone receptor subunits, more than one molecule may be bound.

Receptors have a high affinity (a K_D of about 10^{-8} to 10^{-11} M) for such binding. The interaction of hormones with their receptors may depend on various external factors, including the pH and the presence of divalent ions, especially Ca^{2+} and Mg^{2+}. Such ions may alter the affinity of the receptor for a hormone, possibly by altering its conformation.

Membrane receptors are thought to consist of a hydrophobic base associated with the lipids in which it lies and a hydrophilic apex that projects into the external aqueous solution. Their position or function may be related to the cell microfilament system, as in some instances the cytochalasin drugs can block their functioning. It is considered that receptors may either have a relatively fixed position in the plasma membrane or they may be able to move about laterally in it ("mobile receptor theory"). Either type of receptor is expected to have a spatial relationship, either fixed or flexible, to another membrane component which helps relay its signal. In the instance of many polypeptide hormones and catecholamines, this constituent is adenyl cyclase. This enzyme system is thought to contain at least two other important components, apart from the receptor: a catalytic site, which is concerned with its interaction with ATP, and a regulatory site, which controls the conformation of the enzyme by its ability to combine with nucleotides, mainly GTP, and possibly also alter the drug–receptor interaction (see Section 6.14B).

By labelling hormones and their antibodies with radioactive atoms or fluorescent compounds it has been possible to trace their distribution following their interaction with cells. It has recently become apparent the "membrane" receptors are not necessarily confined to the plasma membrane. Various large polypeptides, including insulin, growth hormone, and prolactin, have been identified, following a period of incubation, within cells (see Kolata, 1978). Specific receptor sites for such hormones have also been identified associated with the Golgi apparatus. The mechanism and function of this entry of the hormone into the cell is not clear. It seems likely that it brings its membrane receptor with it, and this could involve a process of endocytosis following aggregation in areas called "coated pits." Functionally, this process could provide a method for inactivating the hormone or regulating the number of membrane receptors. It is also feasible that such "internalized" hormones may be involved in mediating some of the actions of the hormones, such as are known to involve more long-term intracellular processes.

Receptors do not appear to directly alter or metabolize the agonists or antagonists with which they bind. Dissociation can usually occur without permanent change in either component. Whether or not there is an intermediate structure linking the receptor and the enzyme is not clear. It is, however, possible to isolate membrane fragments containing the receptor with its attached adenyl cyclase, and this can be activated by hormones with the resulting formation of cyclic AMP. Solubilized membrane receptors, however, appear to be separated from the enzyme, as this response cannot then be elicited by the hormone. Such preparations, however, still retain their abilities to specifically interact with the hormones.

An excellent description of the methods used to identify and study hormone receptors is provided in a laboratory methods manual prepared by Schrader and O'Malley (1978). The relationships of the more classical pharmacological methods for analyzing the interactions of drugs and receptors from their dose–response relationships, and the analysis of the interactions of radioactively labelled drugs and their receptors, have been described by Furchgott (1978).

F. Nature of the receptor response

It appears that receptors usually undergo a conformational change as a result of their interaction with a drug that acts as an agonist. This effect can be likened to that of the interaction of ligand for the allosteric site on an allosteric protein or enzyme (see, e.g., Changeux et al., 1967; Karlin, 1967; Rang, 1971; Thron, 1973). However, it is doubtful, despite earlier speculation, that the receptor always constitutes a permanent component of an enzyme or is an enzyme per se. The receptor–drug complex sooner or later may interact with an enzymic system and alter its activity, but it has been shown in many instances to be a separate dissociable entity. A conformational change in the receptor may have several effects, including its subsequent dissociation into subunits, promotion of movement from the cytoplasm to the nucleus, a lateral migration within the plasma membrane, or even transfer from the latter into the cytoplasm. The change may also promote an association of the receptor with enzymes, such as in the chromatin or with membrane adenyl cyclase. In the instance of "fixed" receptors which may be a part (even a regulatory allosteric site) of an enzyme, the binding of the drug may have a more direct, allosteric-like effect on its conformation and hence change its activity.

G. Control of receptor function

There appears to be a continual process of degradation and renewal ("turnover") of many receptors. It is thus not surprising to observe that their numbers may vary in response to drugs, hormones, and in disease. Such changes may account for variations in the sensitivity of a target tissue to a drug or hormone. A deficiency of receptors for insulin, growth hormone, and androgenic steroids has been observed in certain diseases. This can be genetic in origin or reflect environmental changes. Thus, high circulating levels of growth hormone and insulin have been observed to bring about a decline in the number of their own, but not other types, of receptors. Hormones may control the synthesis of the heterologous receptors; for instance, estrogens can stimulate the formation of progesterone receptors and α-adrenergic receptors. Some drugs can inhibit the formation of receptors; the antiestrogen tamoxifen thus appears to block synthesis of estrogen receptors. The formation of new hormone receptors involves protein synthesis and can be inhibited by actinomycin D or puromycin.

The interaction between a hormone (or possibly also drug) and its receptor may be self-regulating as a result of changes in the affinity of the receptor. This phenomenon has been observed with such hormones as insulin, thyrotropin, and catecholamines. When the concentration of hormone is raised, their binding to the receptors declines, so that the K_D increases. This process is called "negative cooperativity" and it may result from a transformational change in the receptor following its interaction with the hormone. From the practical, functional standpoint it is best to consider the converse of this situation, in which the sensitivity of the receptors is greatest when the concentration of a hormone is low. This phenomenon also suggests that certain types of receptors may interact with each other so that they could be present in

close association in "clusters" on the cell membrane.

2.12. Development and testing of new drugs for human use

The provision of drugs for human use is a complex process both scientifically and administratively. The guidelines and requirements for such development have been provided by committees and organizations delegated to protect the interests of the public. In the United States this is the Food and Drug Administration (FDA). The review procedure will not be discussed in detail in this volume.

The *discovery* and *design* of new drugs to treat human disease can be the result of several circumstances, planned and otherwise. Accidental discovery and serendipity play a prominent role. Thus, observations that a certain drug exhibits a "side effect" that may in other circumstances be exploited to treat another disease are not uncommon. For instance, the drug diazoxide was first used to lower blood pressure in hypertensive emergencies, and one of its side effects is a hyperglycemia. This effect is apparently due to a blockade of the release of insulin, and the drug has subsequently been used for this purpose. Similar observations led to the development of the oral hypoglycemic drugs, which have been used to treat diabetes mellitus, and the modern diuretic drugs. Folk remedies, usually derived from plant sources, have also been exploited as sources of new drugs. More recently, these include the use of licorice to treat peptic ulcer, which led to the development of carbenoxolone. More rational approaches involve the testing of drugs with known types of biochemical actions which may be useful in certain diseases. This approach is often combined with judicious changes in their structure. In other instances, however, the approach may be more empirical and involves careful consideration of the structure–activity relationships of the drug, following which changes can be proposed that are calculated to alter its activity in some way. This method involves "brains" rather than "brawn" and has led to the development of such important new drugs as propranolol (a β-adrenergic blocking drug used to treat a variety of disorders including hypertension) and cimetidine [a H_2 (histamine)-receptor blocking drug for the treatment of peptic ulcer].

Novel ways of formulating drugs so as to facilitate their administration and action utilize well-prescribed methods. Thus, oral administration may be assisted by the addition of chemical groups that make the drug more lipid-soluble or protect it from enzymic degradation in the gut or liver. A drug's effect can also be extended by incorporating it into various types of vehicles such as oil (as for vasopressin) or gelatin (as for corticotropin) or, in the case of steroids, esterifying the molecule.

Drugs can also be prepared in particles or crystals of different size; the larger, which have the smallest relative surface area, have the more extended action. This procedure has been utilized in making long-acting insulin preparations.

The tests and screening of new drugs are initially performed on animals, especially rats, mice, rabbits, and dogs. On the basis of their nature, the tests can be divided into two major categories:

a. The therapeutic effectiveness of the drugs must be established. Preliminary experiments, such as measuring the ability of a drug to lower blood glucose levels, inhibit ovulation, produce an antidiuresis, or exert an anabolic action, can be carried out on animals. Diseases can also be induced experimentally in animals, so that one can test a drug's ability to promote the regression of a tumor, exert an antiinflammatory effect, or restore growth in a pituitary gland deficiency. Such experiments are basically important in the development of a drug. However, animals and human beings may differ in certain aspects of their physiology, so that ultimately, screening tests must be performed on man, and these procedures are in the realm of the clinical pharmacologist.

b. The drugs must be tested for toxic and adverse effects. Initially, such tests involve animal studies. The acute effects of a drug are studied following the i.v. or oral administration of increasing doses so that the toxic and lethal dose may be estimated. Some drugs are administered chronically over a period of many years or even a lifetime, so that colonies of animals are tested in a comparable way. Side effects are recorded simultaneously. Special tests are often required to determine if a drug has carcinogenic effects or produces deformities in the fetus (teratogenicity). Again, animals may behave differently from man, so that the estimated nontoxic dosages of the drugs must be carefully calculated in the clinical situation. Side effects observed in animals may also occur in man, but this is not always so; novel adverse effects may be observed in the clinical situation. Conversely, a side effect in an animal does not necessarily indicate that it will invariably occur in man, but it is taken as a warning. This problem is especially manifested with respect to possible carcinogenic effects of drugs, which may only appear a long time after exposure in man. Thus, in this instance the results of animal tests are given considerable weight. Often, it is only after several years of clinical experience with a drug that a reasonably complete dossier of its foibles becomes available.

2.13. Nature of the toxicities and side effects

Drug toxicities and side effects are a major clinical problem which have an alarmingly high incidence and, even in the hospital setting, result in many deaths. Such iatrogenic (drug-induced) diseases appear to have become more common, or at least have been recognized more frequently, since the dawn of the era of the "wonder drugs" about 30 to 40 years ago. The initial euphoria that followed the discovery of many such drugs led to the feeling that all drugs were good (or "all good"), and the possibility that they might have other, less desirable actions was somewhat neglected. This problem was compounded by the relative absence of governmental rules and guidelines regarding introduction of the use of drugs and the reporting of observed adverse reactions. An early dramatic example of such problems arose with the introduction of cortisone to treat rheumatoid diseases. This corticosteroid undoubtedly has dramatic antiinflammatory and palliative effects in such disorders, but it also has a wide range of other types of actions in the body which can result in toxicity and side effects, and even death. Such adverse responses are common, and indeed are usually unavoidable and occur quite early in such therapy. However, it took a surprisingly long time before they were recognized, reported, accepted, and acted upon. Currently, governmental groups, the purveyors of medicaments, and physicians are much more aware of the potential hazards of drugs, so that steps can be taken to avoid such problems or to rapidly correct the condition if it occurs.

Drugs can exert a variety of toxicities and side effects on different tissues and organs. Some of these effects are not unexpected and can be predicted on the basis of the known locus of the drug's action and the various types of effects it is known to have in the body. Thus, a β-adrenergic blocking drug such as propranolol is expected to act at many sites in the body, and as the heart and the bronchi have a β-adrenergic control mechanism, they may be expected to slow the heart or constrict the bronchi in certain situations. Corticosteroid hormones are known to increase Na retention by the kidneys, so it is not surprising that the use of hydrocortisone (cortisol) as an antiinflammatory drug can result in excessive accumulation of fluid in the body. The toxic effects of some drugs are merely an extension of their therapeutic effects; hence, the digitalis cardiac glycosides can be used to increase the strength of contraction and control the rhythms of the heart, but in excess can result in ventricular fibrillation and death. Androgenic steroids have anabolic effects that are utilized therapeutically, but high doses of such compounds can readily exert a virilizing action.

Many side effects of drugs, usually including the most frequent and troublesome ones, are not readily predictable. Some effects may be common, even invariable, occurrences associated with the use of drugs. Gastrointestinal disorders and skin rashes are common, but these may disappear with time or adjustment of the dosage, or they can often be tolerated or ameliorated in some way. Other types of conditions are less common and may include hypersensitivity reactions due to allergies. This type of problem may be localized, such as at the site of an injection, or is more general in effect. Some drugs may alter kidney function and, like many antihypertensive drugs, result in salt retention; or, like amphotericin B, promote salt loss; or increase or decrease the urine volume (as seen with some tetracyclines and chlorpropamide, respectively). Liver disorders, including necrosis, biliary stasis, and the occurrence of tumors, can be a serious problem, as they can be difficult to correct. Such disturbances in hepatic function may be the result of the formation of toxic metabolites of drugs that covalently bind to the tissue. Blood dyscrasias are usually rare, but feared, and include hemolytic anemia, aplastic anemia, and agranulocytosis. Their causes are uncertain but may involve allergic responses and the covalent binding of toxic metabolites to blood cell-forming tissues, especially bone marrow. In some cases of hemolytic anemia, a genetic deficiency of glucose-6-phosphate dehydrogenase in the erythrocytes is involved. Neurological and behavioral disorders may occur with the use of drugs that can cross the blood-brain barrier. Tumors, either benign or malignant, are also a relatively rare occurrence and, apart from the liver, have been described in the uterus and vagina following the use of estrogens. In animals, the use of drugs has been associated with a more ubiquitous distribution of tumors. Such effects may be related to the ability of drugs or their metabolites to covalently bind to DNA or in some instances, such as the estrogens, to promote cell activity in such a way as to increase the propensity of other agents to elicit such an effect. Some drugs, if given during certain stages of pregnancy, can influence the fetus and have teratogenic actions. These effects will be expected to depend on the drug or its metabolites being able to cross the placenta, but more indirect actions are also feasible. Examples of teratogenic effects include the virilizing effect of progesterone on female fetuses and malformations of the limbs associated with the use of estrogens.

2.14. Clinical parameters of a drug's effects

In its clinical use a drug is intended to have a therapeutic action but, as just described, it can also cause toxic reactions and side effects. Such responses can reflect deficiencies in a drug's selectivity and may even occur at the same or even smaller doses than those which are therapeutically effective. They can also result from excessive doses of drugs, which are then said to be at toxic concentrations and ultimately may even result in death. Such toxic effects are often merely extensions of therapeutic effects. Drug concentrations in the blood are a convenient way of predicting its effects. These can be divided into several general levels: the concentration of the drug which is the *threshold* of the therapeutic response, the *maximal* level of therapeutic effect, the concentration where *toxic* effects may be expected to occur, and the *lethal* concentration. The difference between the maximally effective therapeutic dose and that at which toxic effects may be expected to occur can be large or relatively small and depends mainly on the drug. The greater the difference between these two concentrations, the safer the drug is expected to be.

This "safety factor" may be expressed in terms of the *therapeutic index*, which is the ratio of the dose that has a lethal effect (LD) to the dose that is therapeutically effective (ED). Thus, the larger the value, the safer the drug. The therapeutic index has various meanings and, based on animal experiments, can be calculated as the ratio of the lethal and effective doses expressed as a percentage of the population tested: for instance, the median or 50 percent response, or LD_{50} and ED_{50}. The ratio LD_{50}/ED_{50} may be used, but a safer expression of this therapeutic index is LD_1/ED_{99} or $LD_{0.1}/ED_{99.9}$. In man, clinical experience may afford a measure of the toxic dose (TD), and one can use TD_1/ED_{99} as a measure of a drug's safety. It should be recalled, however, that some such calculations are based on animal experiments and one has to be careful about extrapolating such estimates to man. In addition, to complicate the problem, there can be considerable variation between the responses of individual people. These differences can be due to various factors: genetic, environmental, nutritional, disease, and prior treatment with other drugs. A note of caution is also necessary when comparing blood levels, as what may appear to be therapeutically effective in one individual may sometimes be toxic in another. There is no absolute blood level of a drug that can with certainty be stated to be therapeutic or toxic although such measurements can provide important clues as to the likely situation.

Graphical representation of the doses of drugs that have an effective therapeutic (or lethal) action versus the percentage of population responding are usually simple bell-shaped curves that would be expected in a population showing a normal distribution. In some instances, however, there may be more than one such component, each of which shows distinct separation. This type of distribution indicates that there is more than one such population and probably reflects genetic differences such as could be due to the presence or absence of enzymes that influence drug metabolism or differences in the sensitivity or the numbers of receptors.

3
The pituitary gland

The pituitary gland produces at least nine hormones, several of which exert a controlling influence on other endocrine glands. It has thus been dubbed the "master gland" or "conductor of the endocrine orchestra." It is now clear, however, that the pituitary itself is under the direct control of the hypothalamic region of the brain, and it has been suggested (McCann, Fawcett, and Krulich, 1974) that its title should be relegated to that of "concertmaster."

3.1. Functions and morphology

A. Functions

The integration of bodily functions depends partly on the rapid transmission of information along the morphological pathways of the nervous system, but the transport of hormones in the blood is of parallel importance. It is in the pituitary gland that the principal interaction of these two types of processes occurs. Pituitary function is thus often described under the discipline *neuroendocrinology*, which also includes the role of the hypothalamus and its functional relationships with the rest of the brain.

The pituitary gland is a discrete piece of tissue attached to the base of the brain in the region of the optic chiasma. In man, it weighs about 700 mg. Its presence has been known to anatomists for about 2000 years, but its endocrine function was only recognized at the turn of this century. Before that time the pituitary was thought to secrete fluids into the nasal passages (the term derives from *pituita*, phlegm or mucus). In 1886, Pierre Marie described how pathological changes in the pituitary were associated with the growth-deforming disease *acromegaly*. In 1900, C. Benda suggested that this condition was related to a tumor in the pituitary. Acromegaly is now known to be due to hypersecretion of growth hormone. The possible endocrine function of the pituitary had been rec-

ognized previously when G. Oliver and E. A. Schafer in 1895 showed that it contained a substance that increased the blood pressure of cats. This response is now known to be due to the hormone vasopressin, or antidiuretic hormone (ADH). It is secreted by an anatomically distinct region of the pituitary called the neurohypophysis, in contrast to the adenohypophysis, which is the site of origin of growth hormone.

Information about the endocrine role of the neurohypophysial region of the pituitary proceeded fairly rapidly. In 1905, Henry Dale identified a material that could contract the uterus (Dale, 1906); this was oxytocin, and it was also shown to initiate milk let-down in nursing goats (Ott and Scott, 1910). The disease diabetes insipidus (in which excessive amounts of urine are formed) was shown to be related to hypofunction of the pituitary, and it was found that this could be corrected by the injection of vasopressin or ADH (Von den Velden, 1913). Progress in understanding the more ubiquitous role of the adenohypophysial part of the pituitary gland was much slower. Pathologists recognized that a number of disorders involving dysfunction of the pituitary influenced, apart from growth and secretion of urine, the functioning of the thyroid, adrenal cortex, and gonads. The major breakthrough that contributed to our current knowledge of the many roles of the pituitary gland was of a technological nature. In 1926, P. E. Smith successfully removed the pituitary gland in rats, thus providing an experimental preparation in which the various physiological deficiencies associated with the absence of the pituitary could be identified and studied. The various pituitary hormones and their principal functions in the body are shown in Table 3.1.

The functional relationship of the pituitary gland to the brain was first suspected as a result of observations on its role in reproduction. Thus, the

Table 3.1. *Hormones secreted by the pituitary gland and their target organs and functions*

Hormone	Target organs	Physiological functions
Adenohypophysis		
Corticotropin, ACTH	Adrenal cortex	Secretion of cortisol
Melanocyte-stimulating hormone, α-MSH (melanotropin)	Melanocytes, sebaceous glands	? Pigmentation, sebum secretion
β-lipotropin, β-LPH	?	? Precursor for endorphins and β-MSH
β-endorphin	?	
β-MSH	? As for α-MSH	
Thyrotropin, TSH	Thyroid gland	Secretion of thyroid hormones
Follicle-stimulating hormone, FSH	Ovary and testis	Maturation Graafian follicle, spermatogenesis
Luteinizing hormone, LH	Ovary and testis	Ovulation, secretion of androgens and progesterone
Prolactin, PRL	Mammary glands	Secretion of milk (lactogenesis)
Growth hormone, GH (somatotropin)	Muscle, bone, liver, adipose tissue, etc.	Tissue growth and development
Neurohypophysis		
Vasopressin, antidiuretic hormone, ADH	Kidney	Reabsorption of water from glomerular filtrate (antidiuresis)
Oxytocin	Uterus, mammary gland	Contracts uterus, milk let-down (galactobolic)

effects of mood and emotion on the human menstrual cycle are well known. The seasonal patterns of reproduction that occur in many animals in response to external environmental changes were, however, of primary importance in recognizing that the brain may play a role in controlling endocrine function. Such a concept was made more tenable by the observations of Ernst and Berta Scharrer, who clearly demonstrated the phenomenon of *neurosecretion*. The showed that nerve cells not only released chemicals into the region of other nerve cells (neurotransmitters) but that some of these cells could synthesize and secrete substances into the circulation and thus function as hormones. The initial observations were made on invertebrates, but these subsequently extended to vertebrates. Such *neurosecretory cells* are now known to play an important role in maintaining the neuroendocrine relationships of the brain and the pituitary gland.

The nerve supply to the pituitary is confined to the posterior lobe and is absent in the anterior lobe, which is nevertheless the site of synthesis and release of six of its major hormones, including those that control reproduction, the adrenal cortex, the thyroid, and growth. G. E. Harris in England was primarily responsible for developing our concepts about how the brain can control the secretion of these hormones despite the absence of direct neural connections. When the pituitary is removed from its position at the base of the brain and transplanted elsewhere in the body, it functions in an autonomous fashion that is generally unrelated to

the normal physiological requirements of the body. If such a pituitary is retransplanted back to its original position, after a delay of several days its normal functioning may be resumed. Recovery does not occur, however, if a piece of waxed paper is inserted between the gland and the brain. Normally, vascular connections are restored and this process is prevented by the paper barrier. Harris proposed that the anterior lobe of the pituitary is controlled by secretions formed in the hypothalamic region of the brain and carried in the vascular portal system, which carries blood from the hypothalamus to the pituitary. This theory was initially quite a contentious one but has since been completely confirmed. A number of distinct chemical compounds are released into this portal blood system from neurosecretory cells in the hypothalamus, and these are carried to the pituitary, where they each initiate synthesis and release of specific pituitary hormones. In 1969, Roger Guillemin and Andrew Schally independently identified one of these "factors" or hormones as the thyrotropin-releasing hormone (TRH) and showed that it was a tripeptide. It was synthesized; the product had the same activity as the natural hormone. This achievement was the culmination of extensive efforts to understand the mechanism of control of the pituitary gland.

B. Embryonic origins and morphology

The pituitary gland has a dual embryonic origin (see Wingstrand, 1966a). It arises as a result of differentiation of what is usually considered to be

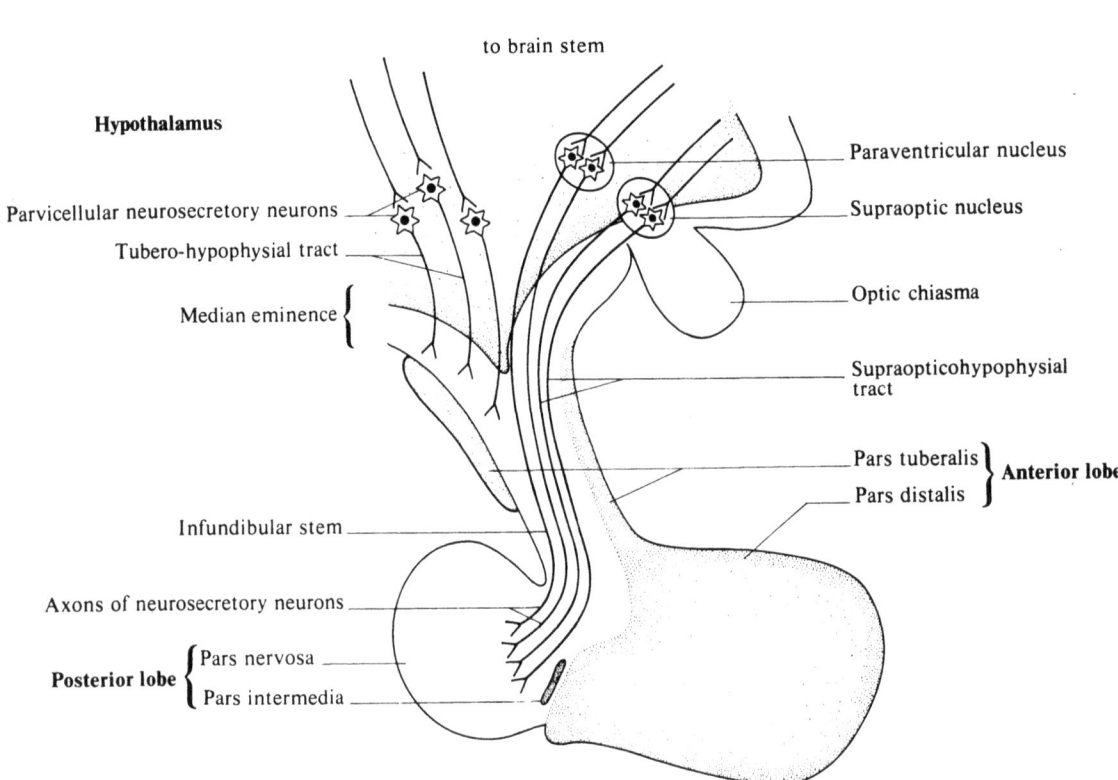

Third ventricle

to brain stem

Hypothalamus

Parvicellular neurosecretory neurons

Tubero-hypophysial tract

Median eminence {

Infundibular stem

Axons of neurosecretory neurons

Posterior lobe { Pars nervosa
Pars intermedia

Paraventricular nucleus

Supraoptic nucleus

Optic chiasma

Supraopticohypophysial tract

Pars tuberalis } **Anterior lobe**
Pars distalis

Fig. 3.1. Pituitary gland.

ectodermal epithelium in the dorsal region of the pharynx. The pharyngeal tissue forms a sac, called *Rathke's pouch*, which grows upward and meets a downgrowth of tissue from the *infundibulum*. The two tissues become associated, and together they make up the pituitary gland.

The embryonic origins are clearly reflected in the morphology of the pituitary. It is made up of two distinct regions (Figure 3.1 and Table 3.2): the *adenohypophysis*, which is the part derived from Rathke's pouch, and the *neurohypophysis*, which is neural tissue.

The cell bodies for the neurohypophysial tissue lie in the *paraventricular* and *supraoptic nuclei* of the hypothalamus. The hormones of the neurohypophysis are formed in these cell bodies and are carried down their axons, the *supraopticohypophysial tract*, which pass through a stalk of tissue, the *infundibular stem* (or pituitary stalk), to terminate in an expanded region (also containing glial cells and pituicytes) called the *infundibular process* or *pars nervosa* (neural lobe). The hormones are secreted from the axonal terminals into blood vessels and so pass into the general circulation. It has been suggested that the paraventricular and su-

praoptic nuclei may be specifically concerned with the formation of oxytocin and ADH, respectively, but this interesting suggestion has not been confirmed.

The *adenohypophysis* (Figure 3.1) consists of three main regions:

a. The *pars distalis*, which is the predominant part of the tissue and the site of synthesis of at least six hormones. These appear to be produced by a number of distinct types of cells which have been characterized by their histological staining properties (see Purves, 1966; Baker, 1974). Originally, two major types of cells were identified: *chromophobes*, cells that could not be stained with dyes, and *chromophils*, cells that could be stained. The latter were subdivided on the basis of staining with acid or basic dyes and hence were called *acidophils* (or α-cells) or *basophils* (β-cells). The protein hormones are often considered to be secreted by acidophils and glycoproteins by basophils. In man, however, the ACTH-secreting cells are basophils. More sophisticated methods using immunofluorescence and immunohistochemistry have been used to characterize the different types of cells in relation to the hormones they form. The

Table 3.2. *Terminology used to describe the morphological divisions of the hypophysis or pituitary gland*

Embryonic tissue	Morphological division[a]
Hypophysis (Rathke's pouch, ectodermal oral epithelium from the roof of the stomodeum)	*Adenohypophysis* (*lobus glandularis, glandular lobe*) *Pars tuberalis* (pars infundibularis) *Pars distalis* (pars glandularis of the adenohypophysis) *Pars intermedia* (intermediate lobe, zona intermedia)
Infundibulm (neural tissue from the floor of the diencephalon, saccus infundibuli)	*Neurohypophysis* *Median eminence* of the tuber cinereum (eminentia mediana) Infundibular stem (*neural stalk,* pars proximalis of the neurohypophysis) Infundibular process (*Neural lobe,* pars nervosa, lobus nervosus, pars distalis of the neurohypophysis)

[a] The intermediate lobe and the neural lobe are separated from the pars distalis by the hypophysial cleft and cannot readily be separated from each other. They are thus together often called the *posterior lobe* of the pituitary. The latter term, however, is sometimes used synonymously with the term "neural lobe." The pars tuberalis plus the pars distalis together constitute the *anterior lobe* of the pituitary. The pars distalis surrounds the infundibular stem and these together are sometimes referred to as the hypophysial or pituitary stalk. The most commonly used terms are italicized.

terminology is, unfortunately, not uniform, especially in man. The cell type associated with a particular hormone is given the suffix *troph,* so that those forming growth hormone, or somatotropin, are called somatotrophs; thyrotropin, thyrotrophs, and so on. The most widely used terminology is the Romeis one, in which Greek letters are used to indicate cells that stain in different ways: for example α-cells stain red and are somatotrophs (see Table 3.3).

b. In man, the *pars intermedia* is an indiscrete region lying between the pars distalis and the neurohypophysis (see Wingstrand, 1966b). In many animals, it forms a more distinct section of the pituitary. It is the site of formation of melanocyte-stimulating hormone (MSH) (also called melanotropin), but this hormone does not appear to occur naturally in man [Section 3.2 B(c)].

c. The *pars tuberalis* lies between the pars distalis and the hypothalamus and forms an envelope that folds around the pituitary stalk and overlies the neurohypophysial infundibular stem. It is associated with the portal vessels that run from the hypothalamus to the pars distalis but its precise physiological role has not been defined.

The pars distalis and pars tuberalis together are often referred to as the *anterior lobe* of the pituitary, and the neurohypophysis and pars intermedia make up the *posterior lobe.* The latter term is, however, also often used synonymously with "pars nervosa" or "neural lobe."

Whereas the hormones of the hypothalamus and neurohypophysis are undoubtedly formed in cells derived embryonically from nerve cells, it is widely considered that the adenohypophysis and most other endocrine glands are nonneural in their origins. This more classical view has, however, been recently challenged by A. G. E. Pearse, who has formulated his ideas in a theory called the *APUD concept.* This title is an acronym for "Amine content and/or Precursor Uptake and Decarboxylation" (see Pearse, 1968, 1975). A large number of cell types situated in different parts of the body have been shown to exhibit similar cytological characters often combined with certain common biochemical properties. The latter include the ability to take up amines, such as dopa, and 5-hydroxytryptophan, which may be stored, or converted, via their decarboxylation to such products as epinephrine and 5-hydroxytryptamine. These cells also invariably have the ability to synthesize polypeptides that may be secreted. In some instances, these APUD cells have lost their ability to metabolize amines, but they are, nevertheless, considered to be derived from cells that exhibited such properties ancestrally. The embryological ori-

Table 3.3. *Histological types of cells thought to be associated with secretion of hormones in the adenohypophysis*

Growth hormone[a]	Somatotrophs, α-cells (acidophils, eosinophils)[b]
Prolactin[a]	Mammotrophs, lactotrophs, epsilon-*eta* cells
Thyrotropin	Thyrotrophs, β-cells; also S_2, theta, and blue theta (basophils)
Gonadotropins	Gonadotroph, B-cell, cyanophil, delta, S_1; these cells have been differentiated into those that form FSH and LH (basophils, delta-basophils)
Corticotropin, β-lipotropin, and β-MSH	Corticotroph, zeta, purple-beta, beta, R, and β(R) (basophils)

[a] The cells that produce growth hormone and prolactin appear to arise from a single cell line and are interchangeable depending on the particular physiological conditions.
[b] Terms in parentheses indicate the type of cell based on *general* stainable properties.

gins of some of these APUD cells have been confirmed by exposing embryonic neuroectodermal cells to dopa or 5-hydroxytryptophan and using the fluorescent properties of the amines to follow the path of the cells migration during development. Another procedure for determining the origins of such cells involves the grafting of neuroectodermal tissues from a quail into a developing chick (allografting). The cells derived from the quail tissue are readily identifiable and can be traced cytologically. Using such methods (see Pearse and Takor, 1976), the neuroectodermal origins of the thyroid C-cells (which secrete calcitonin) and the adrenal medullary tissues have been confirmed. The cytological APUD cells in the adenohypophysis, pancreas, gut, and parathyroids are considered to be, as yet, of unproven origin, but there is substantial evidence available to suggest that they may also be derived from the neuroectoderm. Pearse considers that all cells that secrete polypeptide hormones, including those from the adenohypophysis, may actually have a neural origin. He has suggested that a large part of the endocrine system is in fact a part of the nervous system and he classifies it as "central neuroendocrine," which includes the pituitary, hypothalamus, and pineal gland, and "peripheral" components. The latter includes the pancreas, gut, thyroid C-cells, and the adrenal medulla.

The APUD concept may provide a uniform explanation of the nature of several endocrine diseases. It has been observed that a number of tumors, benign and malignant, can produce a variety of hormones or hormone-like substances. Such tumors may be situated in endocrine glands, such as pancreatic islet cells and the thyroid gland, or nonendocrine tissues, such as the lung and skin. These neoplasms may produce a single or often several types of hormones and so serve as ectopic sites for hormone production. (As these cells are

not usually subject to the normal processes regulating secretion, they can result in syndromes associated with excesses of their products.) Most such tumors are associated with the production of peptide and amine hormones. The facility by which various tumors in different tissues may each produce several types of hormones, such as ACTH, ADH, and calcitonin in an oat-cell carcinoma of the bronchus, has been a source of wonderment to endocrinologists. It has been suggested (Pearse, 1975; Tischler et al., 1977) that many of these tumors in fact arise from embryonic neuroectodermal cells, and they have been given the general title *APUDomas* or neuroendocrine neoplasms. Such cells are originally pluripotent, but they normally lose this facility following their migration and development at their ultimate resting places in the body. Inherited defects such as multiple endocrine adenomatoses or, in some instances, medullary thyroid carcinoma may result in failure of the cells to maintain their normal predestined function. Alternatively, local environmental conditions either during development, or possibly even subsequently, may influence their final differentiation and manner of functioning. Such cells may then produce the variety of hormones that was originally possible but not ultimately intended.

C. The nerve and blood supply

Apart from some nerves that supply its blood vessels, the pars distalis lacks a nerve supply (see Green, 1966; Porter, Ondo, and Cramer, 1974). The pars intermedia, on the other hand, is well supplied with nerves, which appear to be aminergic and, possibly, peptidergic. The neurohypophysis is essentially a neural structure containing the neurosecretory axons that arise from cells in the hypothalamic supraoptic and paraventricular nuclei (magnocellular neurons) (see Christ, 1966; Bargmann, 1968). The nerve

fibers from these cell bodies run along the supraopticohypophysial tract, which passes through
the median eminence region of the hypothalamus,
down the infundibular stem (pituitary stalk), terminating in the pars nervosa. Some of these
neurosecretory nerve fibers have also been observed to terminate in the wall of the third ventricle and in the median eminence. This relationship
is consistent with the identification of vasopressin
in the long hypophysial portal vessels and the cerebrospinal fluid.

As described earlier, the pars distalis is under the
humoral control of the hypothalamus. Neurosecretory hormones are carried to the gland by the
hypophysial portal system. The hypothalamus is
that region of the brain formed by the diencepha-
lon and lying below the third ventricle (see Knigge
and Silverman, 1974). In man it is about 2.5 cm in
length and weighs about 4 g. The infundibular
stem leaves the hypothalamus just posteriorly to
the optic chiasma, and this site provides a marker
to its division into anterior and posterior sections.
The hypothalamus is the site of many specialized
regions that control bodily functions, including
those of the pituitary gland. Several morphologically distinct conglomerations of nerve cells, or
nuclei, have been identified which are related to
certain hypothalamic functions. A map of some of
these is shown in Figure 3.2. Internal nerve tracts,
including the neurosecretory fibers and a
dopaminergic system, facilitate the transmission of
information within the hypothalamus. There are
also afferent and efferent nerve fibers which extend outside the hypothalamus, so that messages
can be passed to and from this, and other, regions
of the brain and the autonomic nervous system. At
the ventral edge of the hypothalamus lies a specialized zone, the median eminence, where neurosecretory fibers from a number of relatively small nuclei (parvicellular neurons) converge and may
terminate in a position adjacent to a mass of blood
vessels into which they can discharge their contents.
These vessels (see Green, 1966) supply the pars
distalis and are called the *primary plexus* of the
hypophysial portal system (or sometimes the *mantle
plexus*). The region of the hypothalamus that is involved in the control of adenohypophysial function
is called the *hypophysiotropic area.*

The pituitary receives a substantial blood supply
(Figure 3.3) from the internal carotid artery (see
Green, 1966; Christ, 1966; Porter et al., 1974). The
adenohypophysis and neurohypophysis receive
separate supplies of blood, but in man it is possible
that some blood, after passing through the pars
nervosa, may then enter the pars distalis. The
superior hypophysial artery supplies the vessels of
the primary plexus in the median eminence, and
these send "long portal" vessels down the pituitary
stalk to a second plexus in the pars distalis. The

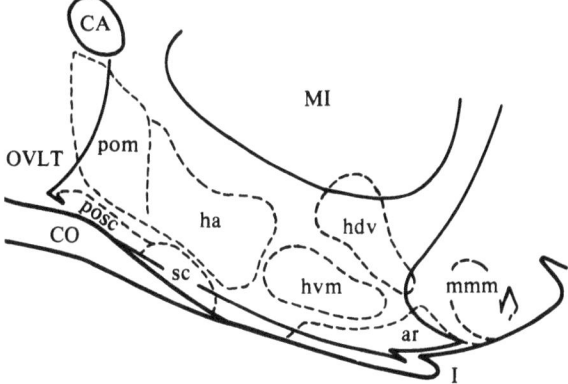

Fig. 3.2. Parasagittal section through the hypothalamic
region of the rat, showing the distribution of nuclei of
nerve cells. OVLT, organus vasculosum lamina terminalis; I, median eminence; ar, arcuate nucleus; ha, nucleus anterior hypothalamus; hdv, nucleus dorsomedialis
hypothalamus; hvm, nucleus ventromedialis hypothalamus; pom, nucleus preopticus medialis; posc, nucleus preopticus, pars suprachiasmatic; mmm, nucleus
mamillaris medialis, pars medialis; sc, nucleus suprachiasmaticus; CA, commissura anterior; CO, chiasma
opticum; MI, massa intermedia.

direction of the blood flow in the long portal vessels is usually considered to be only in one direction, from the hypothalamus to the pars distalis.
However, it has been suggested recently that a retrograde flow, in the opposite direction, may occur
in some of the long vessels (Oliver, Mical, and Porter, 1977; Bergland and Page, 1978). Such a pathway is of considerable theoretical interest, as pituitary secretions may then be able to enter the brain,
where they could not only help regulate their own
secretion (via a short-loop feedback) but could also,
possibly, influence brain function.

The trabecular artery supplies a second series of
short portal vessels in the lower regions of the
pituitary stalk. These vessels, however, only supply
about 20 to 30 percent of the blood that passes
through the pars distalis. In man, the trabecular
artery appears to anastomose with vessels from the
inferior hypophysial artery, which supplies the
pars nervosa. The pars intermedia appears to be
poorly vascularized and does not have a distinct
group of blood vessels.

3.2. The adenohypophysis

A. Functions
The endocrine role of the pituitary was first
identified as a result of its malfunction in the disease acromegaly. This condition is the result of
overproduction of growth hormone, which, when
it occurs in adults, where the epiphyses have

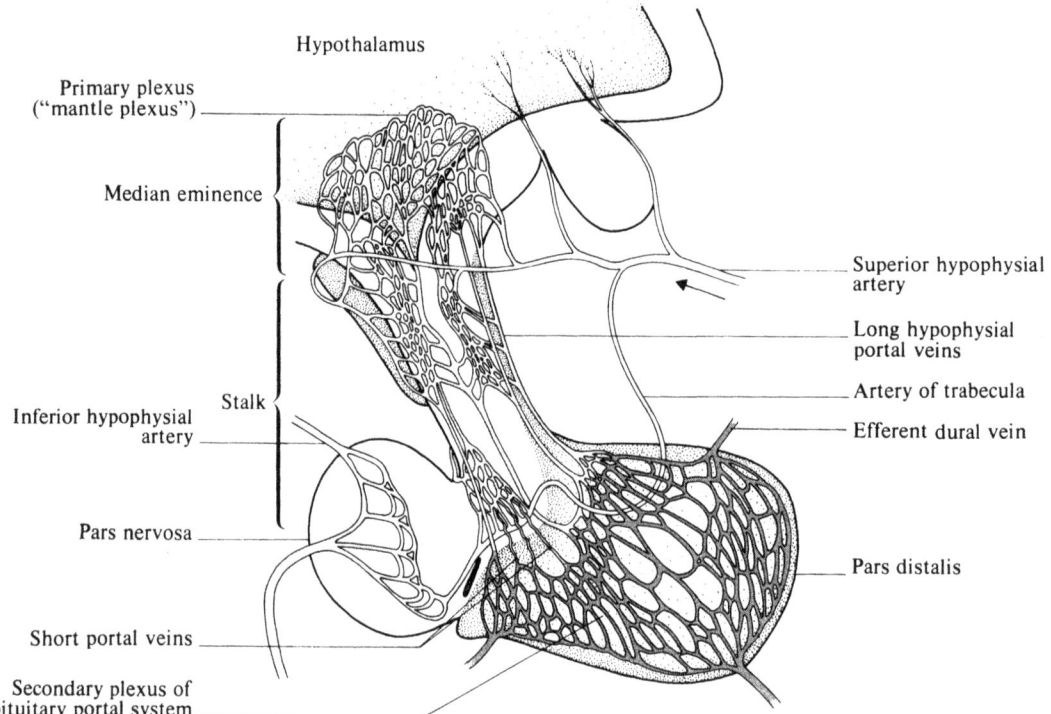

Fig. 3.3. Blood supply to the pituitary gland. The direction of the blood flow is indicated by the arrows. Arterial blood flows via the superior hypophysial artery to the hypothalamus, including the median eminence, before passing through the long hypophysial portal veins to the pars distalis. The latter also receives some blood from the trabecular artery, which is dispersed in the tissue through the short portal veins. In man, some of this blood may pass into the pars nervosa, which, however, receives its major blood supply from the inferior hypophysial artery. (Based on Krieger, 1971.)

closed, results in deformities in the growth of the bones, especially in the hands and feet and the chin. In children, this hormone controls growth and has numerous effects on bones, cartilage, and muscles, which are mediated by its effects on the intermediary metabolism of proteins, minerals, fats, and carbohydrates. These actions of growth hormone are described in more detail later [Section 3.2F(1)]. However, the adenohypophysis has several other important functions, which were discovered as a result of observations in human pituitary disease and following surgical interference or removal of the gland in man and animals. It was found to control the function of the gonads and reproduction in men and women (see Section 7.2C). The activity of the thyroid gland and the production of cortisol (but not aldosterone) by the adrenal cortex is also under the pituitary control of thyrotropin and corticotropin (ACTH), respectively (see Sections 4.2B and 6.6A). These actions on the gonads and adrenal and thyroid glands are trophic ones that control the synthesis and secretion of their hormones. An overproduction of pro-

lactin can result in inappropriate lactation, or galactorrhea [Section 7.6H(6)].

Corticotropin may exert some other actions apart from those on the adrenal cortex. It has under pharmacological conditions been shown to have a lipolytic effect *in vitro*. Fragments of ACTH have also, when injected intraventricularly into the brain of rats, been shown to elicit excessive grooming behavior (Gispen and Wiegant, 1976). They can influence memory and patterns of learning in these animals (de Wied, 1977a and b). The effects on grooming can be blocked by the morphine antagonist naloxone. There has thus been speculation that corticotropin-like peptides may influence the activity of the central nervous system, possibly by acting as neurotransmitters. This possibility is supported by the observation that the corticotropin fragment ACTH-(1–24) can bind to opiate receptors in the brain (Terenius, 1976). These effects may reflect structural similarities between ACTH and β-lipotropin. The recent identification of immunoreactive ACTH-like activity in various parts of the brain of rats is therefore of special interest

(Krieger, Liotta, and Brownstein, 1977). As this material was also seen in hypophysectomized animals, it is unlikely to be of pituitary origin. Its functions are unknown, but it is probably nonendocrine and may be concerned with the function of the central nervous system.

The *functions of MSH* in mammals are not clear (see Howe, 1973). This problem has been compounded by uncertainties as to whether it is even a natural hormone in man. A material that behaves immunologically like MSH has been identified in the circulation of man, especially in disease states (see Abe et al., 1969), but this is probably β-lipotropin or one of its fragments (Gilkes, Rees, and Besser, 1977). In many nonmammals, MSH mediates physiological color changes in response to the color or shade of the background on which it is lying. In these instances it promotes (via formation of cyclic AMP) the dispersion of melanin granules in cutaneous melanophores and so has a darkening effect on the skin. MSH can also influence the synthesis of melanin in the skin of vertebrates, including mammals, by an action upon the melanocyte cells. It appears to exert this effect by increasing tryosinase activity (possibly via activation of adenyl cyclase), which is necessary for the synthesis of melanin (Lee, Lee, and Lu, 1972; Wong and Pawelek, 1973). However, other hormones, such as the sex steroids, are also involved, and ACTH has a small (about 4 percent of MSH) but significant MSH-like activity. Some endocrine disturbances in man result in changes in cutaneous pigmentation which may be related to the presence of MSH or MSH-like activity. Certain hypofunctional pituitary disorders are thus associated with skin palor and an inability to suntan. A deficiency of adrenocortical steroids often results in excessive pigmentation of the skin, and it has been found that this effect is accompanied by high circulating MSH-like activity (Abe et al., 1969). Pregnancy in women may also be associated with increased local deposits of melanin in the skin. It is often difficult to be certain what the precise endocrine cause of such distrubances may be. The injection of α- or β-MSH into American blacks of African origin has been shown to increase skin pigmentation (Lerner and McGuire, 1961).

More recently, it has been proposed that MSH may increase sebum secretion from sebaceous glands (see Shuster and Thody, 1974; Ebling et al., 1975; Thody et al., 1976). In rats, hypophysectomy results in a decline of sebum secretion. This response is partly due to a deficiency of androgens, but β-MSH injection also contributes to a restoration of secretion. It is interesting that in patients suffering from parkinsonism, in which there is a lack of dopamine in the brain, the plasma MSH levels (MSH-like?) are elevated and seborrhea may occur (Shuster et al., 1973). It has thus been suggested that MSH should be renamed "sebotrophic hormone."

Intraventricular injections of MSH have also been shown to elicit certain behavioral responses (see Howe, 1973) such as "stretching reactions" in dogs and "attentive processes" and a reduction in drug-induced tremor in man. Its administration exacerbates parkinsonism (see Shuster et al., 1973). MSH can also induce certain experimental neural electrophysiological disturbances. It has thus been suggested that this peptide hormone may have a role to play in the central nervous system.

Radioactively labelled β-MSH has been identified in different parts of the brain of rats following its injection into the carotid artery (Kastin et al., 1976). Immunoreactive β-MSH, apparently of nonpituitary origin, has also been identified in CSF (Shuster et al., 1977). Biologically active and immunoreactive α-MSH has also been found in various parts of the brain of rats, including the hypothalamus, pineal gland, brain stem, cerebrum, and cerebellum (Oliver and Porter, 1978). It has been suggested that MSH may have a role in the maintenance of brain tyrosinase activity, which is necessary for the synthesis of L-dopa. An interesting suggestion regarding a hypothetical role for MSH in the pathogenesis of parkinsonism is shown in Figure 3.4. It must be emphasized that the existence of a separate MSH in man is doubtful, but the observed effects could reflect the activity of related hormones, especially ACTH and β-lipotropin.

A variety of other peptides, notably the *endorphins* (named from endogenous morphine), have been identified in the pituitary [(see Section 3.2 B(c)], but their function is not yet clear (see Goldstein, 1976; Guillemin, 1978a). In 1964, C. H. Li isolated a polypeptide from sheep pituitaries which, as it had a lipid-mobilizing effect in an *in vitro* system, was called β-lipotropin (β-LPH) (Li et al., 1965). This substance has also been identified in the pituitary of man (Figure 3.5) and is secreted into the circulation. It is thus a putative hormone but suffered relative neglect until recently when it was noted that it contained an amino acid sequence, β-LPH-(61–65), that is identical to an interesting brain peptide called *Met-enkephalin*.

Met-enkephalin is a pentapeptide which, together with a similar compound, *Leu-enkephalin*, was originally isolated from brain extracts of pigs (Hughes et al., 1975) and cattle (Simantov and Snyder, 1976). They were found to exert morphine-like effects on *in vitro* preparations of the guinea pig ileum and mouse vas deferens, in which they inhibit electrically induced contractions. These effects, like that of morphine, can be specifically inhibited by the morphine antagonist naloxone. A larger fragment of β-LPH, called β-endorphin [= LPH-(61–91)], which includes the

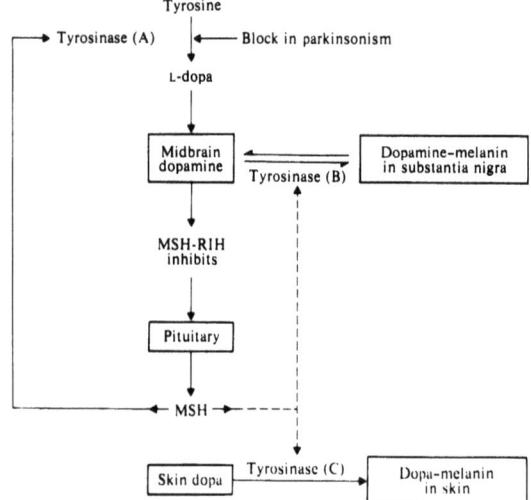

Fig. 3.4. Hypothesis proposed to account for a possible role of melanocyte-stimulating hormone (MSH) in the origin of Parkinson's disease. The proposed regulation of MSH secretion is shown by the continuous lines. Normally, MSH maintains midbrain dopamine by stimulating tyrosinase activity (A), excess dopamine being stored as dopamine melanin (B). MSH-RIH acts by inhibiting pituitary secretion of MSH. In parkinsonism, the conversion of tyrosine to L-dopa is blocked at A and midbrain dopamine tends to decrease. In the early stages, dopamine concentration is maintained by breakdown of dopamine-melanin in the substantia nigra, which depigments. When the store of dopamine-melanin is depleted, midbrain dopamine decreases, with a reduction in MSH-RIH release, and MSH secretion increases. The MSH converts residual midbrain dopamine into dopamine-melanin and so exacerbates the disease. (It should be noted that in man, MSH may not be secreted under normal conditions.) (From Shuster et al., 1973.)

sequence of Met-enkephalin, has similar properties (Ling and Guillemin, 1976). Intact β-LPH lacks such activity. It was also shown by Snyder (see Snyder, 1977) that the enkephalins can be specifically bound to opiate (morphine) receptors in the brain. When the natural enkephalins are injected directly into the brains of animals, however, they have only a weak morphine-like analgesic effect, whereas β-endorphin, the larger β-LPH fragment, was found to be more active than morphine (Jacquet and Marks, 1976; Li et al., 1977). Other fragments of lipotropin also exert these analgesic effects (Graf et al., 1976). A synthetic analogue of the enkephalins Tyr-D-Ala-Gly-mePhe-Met(o)-ol (33-824), which resists enzymic degradation, has analgesic properties in animals comparable to morphine and is even active after oral administration (Roemer et al., 1977). It is thus possible that the analgesic responses to β-endorphin reflect the activity of a smaller fragment, such as Met-

enkephalin, but whether these pharmacological tests reflect an endocrine role or that of locally formed neurotransmitters is unknown. β-Lipotropin has been identified immunohistologically in various parts of the brain (Watson, Barchas, and Li, 1977). Its presence is, however, not invariably associated with that of Met-enkephalin, and vice versa, so that its possible role as a precursor of such peptides is not clear.

The endorphins have been shown to exert some other effects. They can induce "excessive grooming behavior," "wet-dog shakes," and catatonic-like states when injected into the brains of rats (Gispen et al., 1976; Bloom et al., 1976; Jacquet and Marks, 1976). Analogues of Met-enkephalin and β-endorphin can also stimulate the release of prolactin and growth hormone in rats (Cusan et al., 1977; Dupont et al., 1977; Rivier et al., 1977). They thus could have a role in the regulation of the release of these pituitary hormones. β-Endorphin can also stimulate release of ADH in rats and rabbits, but this effect is not a direct one on the neurohypophysis (Weitzman et al., 1977).

It is considered likely that the endorphins, especially the enkephalins, have a role as neurotransmitters in the central nervous system, but their putative endocrine role is more obscure. Leucine-enkephalin has been identified immunohistochemically in neurons in the brain of man (Cuello, 1978). It has been suggested that endorphins or β-lipotropin which are found in the brain may be precursors of enkephalins found there. At present, however, it seems unlikely that they are derived from the pituitary (see Section 3.2C). Endogenous hypothalamic endorphins could nevertheless have an endocrine role and be acting as hypophysiotropic factors. A more peripheral action of released pituitary endorphins and β-lipotropin is also possible. β-Endorphin is released in parallel to ACTH, from the pituitary of rats, in response to stress (Guillemin et al., 1977). Its target sites, if any, are unknown. Normally, however, β-endorphin does not seem to appear in human plasma, nor has it been identified in the human pituitary or the rat's pars distalis (Suda, Liotta, and Krieger, 1978; Liotta, Suda, and Krieger, 1978). However, it is present in the plasma of patients suffering from diseases (Cushing's disease, Nelson's syndrome) associated with an increased release of β-lipotropin.

The administration of morphine antagonists to heat-stressed rats results in hyperthermia, and it has therefore been suggested (Holaday et al., 1978) that endorphins may contribute to temperature regulation during heat stress. Another suggestion (Lord et al., 1977) is that they may have a role analogous to the "flight-and-fright" response to epinephrine, and they could account for such phenomena as the lack of perception of pain that has been reported on the battlefield and during

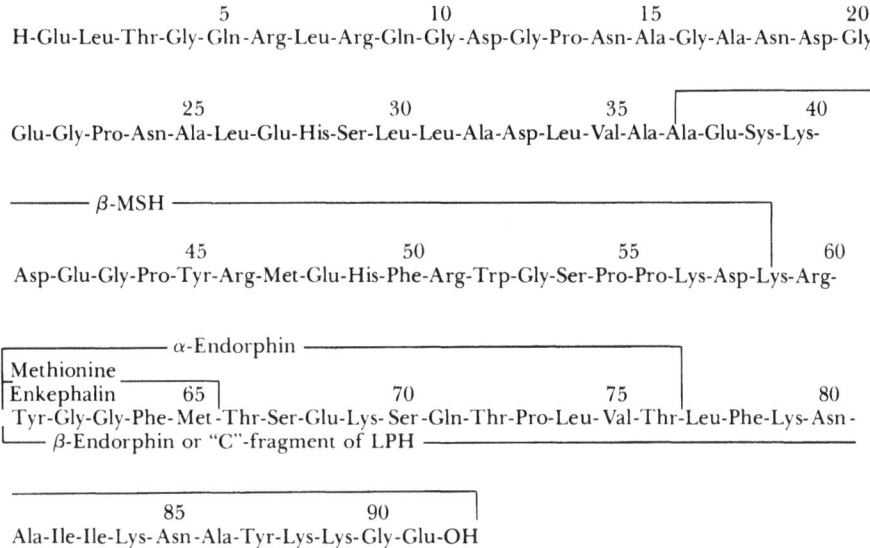

Fig. 3.5. Amino acid sequence of human β-lipotropin according to Li and Chung (1976). The sections of the molecule that correspond to some of the known active fragments have been superimposed.

childbirth. Placebo anesthesia, in response to post-operative dental pain, can be overcome by the morphine antagonist naloxone, which could be consistent with an endogenous action of endorphins under such conditions (Levine, Gordon, and Fields, 1978). Whether or not the endorphins that may act in such circumstances may be derived from the pituitary is, however, unknown.

B. Structure of the hormones

The adenohypophysial hormones, on the basis of their chemical structures, can be divided into three groups. Not surprisingly, there may be related similarities in their actions and it has been suggested that they may share pathways for their biosynthesis and have common evolutionary origins. These families of hormones are:

a. The glycoproteins: *thyrotropin* (TSH), *follicle-stimulating hormone* (FSH), and *luteinizing hormone* (LH). These hormones have a molecular weight of about 30,000 and contain 10 to 20 percent carbohydrates, including sialic acid. The hormones can be readily dissociated into two distinct units, called α- and β-chains (or sometimes CI and CII). A more detailed account of the structure and activities of these hormones is given later in the sections that deal with the thyroid hormones and gonads (Sections 4.2 and 7.3). It is, however, especially notable that it is the β-chain that appears to carry the main information that determines the action of each hormone, while the α-chains are interchangeable. The dissociation and recombinations of these chains from the three different hormones is readily carried out in a test tube. *Human chorionic*

gonadotropin (hCG), which is produced during pregnancy by the placenta (see Section 7.3) is also a glycoprotein with a similar general structure. It has been suggested that these hormones may share their ancestry and that the pituitary gonadotropins and thyrotropin may have evolved from a common parental hormone. The possibility that they share a common biosynthetic pathway has also been considered.

b. *Growth hormone* and *prolactin* are proteins containing about 190 amino acids and have a molecular weight of about 21,000. They share the majority of their amino acid residues, and these similarities are reflected in a crossover in their activities. Growth hormone thus has lactogenic activity on the mammary gland, but this is only about 10 percent that of prolactin. In many nonmammals, prolactin has been shown to exhibit some of the actions of growth hormone. Until less than 10 years ago there was considerable doubt as to whether, in fact, prolactin and growth hormone existed as distinct molecules in man or whether there was a single hormone exhibiting both types of biological activity. Careful bioassay procedures and radioimmunoassays, however, resolved the problem, and it is now clear that two distinct, but related, hormones are secreted by the human pituitary gland. *Human placental lactogen* (HPL) is a chemically related hormone secreted by the placenta which exhibits both growth hormone-like and prolactin-like activities [Section 7.6H(3)]. All three molecules show considerable similarity in their chemical structures, and there are also homologies within successive sections of each mol-

Hormone	Structure	Relative MSII potency *in vitro*
α-MSH =ACTH-(1-13):	Acetyl-Ser -Tyr-Ser- Met -Glu-His-Phe-Arg-Trp-Gly-Lys-Pro- Val -NH₂ 1 4 10 13	1.00
ACTH:	Ser -Tyr-Ser- Met -Glu-His-Phe-Arg-Trp-Gly -Lys-Pro-Val-Gly- 1 4 10 Lys-Lys-Arg-Arg · · · Phe 15 18 39	0.01
β-MSH =β-LPH-(37-58):	Ala-Glu-Lys-Lys-Asp-Glu-Gly-Pro-Tyr-Arg- 1 Met-Glu-His-Phe-Arg-Trp-Gly -Ser-Pro-Pro-Lys- Asp 11 17 22	0.23

Fig. 3.6. Amino acid sequences of human α-MSH and β-MSH and their relationship to ACTH and β-lipotropin. (*Note:* In man, MSH may be an artifact of the extraction procedure.)

ecule (see Figure 7.23). This has led to the suggestion that they evolved from a common smaller molecule as a result of a tandem-like self-replication.

c. *Corticotropin* (ACTH), *β-lipotropin* (β-LPH), α- and β-*melanocyte-stimulating hormones* (MSH), and *the endorphins* bear many analogies in their amino acid sequence. Corticotropin (ACTH) and β-lipotropin are both formed in the same type of cell and are stored in the same granules in the adenyhypophysis (Pelletier et al., 1977). It appears from studies on a line of pituitary tumor cells that they both have a common origin from a larger parent molecule (with a molecular weight of about 31,000) which is split to form each type of polypeptide (Mains, Eipper, and Ling, 1977). They are linear peptides; ACTH has 39 amino acids and β-LPH has 91 such residues. These products, in turn, may also act as precursors and give rise to α-MSH, β-MSH, and several of the endorphin peptides.

The amino composition of these hormones and their common sequences are shown in Figures 3.5 and 3.6. It can be seen that α-MSH is the same as ACTH-(1-13), while β-MSH is identical to β-LPH-(37-58). Similarly, β-endorphin = β-LPH-(61-91), α-endorphin = β-LPH-(61-76), and γ-endorphin = β-LPH-(61-77). The peptide methionine (or Met)-enkephalin is β-LPH-(61-65). Such fragments of the molecules may be formed as a result of the actions of tissue enzymes or even on occasion as a result of the not-too-tender attentions of the chemist.

It has been proposed (Lowry and Scott, 1975) that while ACTH and β-LPH are produced by the same type of cell, the secreted products may differ, depending on their site in the pituitary gland (see Figure 3.7). The cells in the pars distalis may secrete both hormones intact; however, in species with a distinct pars intermedia, selective cleavage of each polypeptide may occur (Figure 3.8) to pro-

duce α-MSH (ACTH fragment 1-13) and the corticotropin-like intermediate lobe peptide (CLIP) (see Scott et al., 1973), which has the amino acid sequence of the 18-39 portion of ACTH. In man, there is no distinct pars intermedia, and careful examination suggests that neither α-MSH, β-MSH, nor CLIP is secreted (Scott and Lowry, 1974). The isolation of such peptides from pituitary gland extracts appears to reflect the chemical breakdown of ACTH and β-LPH. In other vertebrates, the pars intermedia seems to be the site of

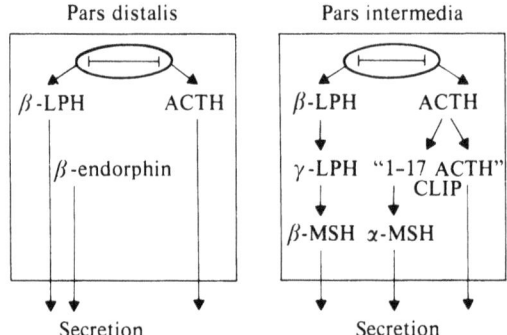

Fig. 3.7. Lowry–Scott hypothesis, which describes the formation of β-lipotropin, ACTH, and α- and β-MSH from a common parent precursor molecule. β-Endorphin has since been shown to be secreted from the same cells that form ACTH, so that it has been added. The corticotroph cells in the pars distalis synthesize and release intact ACTH, β-LPH, and, possibly in stress, β-endorphin. In animals that possess a distinct pars intermedia (not man), the same precursor molecules are present, but these can be split to α- and β-MSH and the corticotropin-like intermediate lobe peptide (CLIP) (see also Figure 3.8). The particular cells in the pars distalis and pars intermedia may be derived from a single cell type whose precise functioning depends on the position it takes during embryological development. (From Lowry and Scott, 1975.)

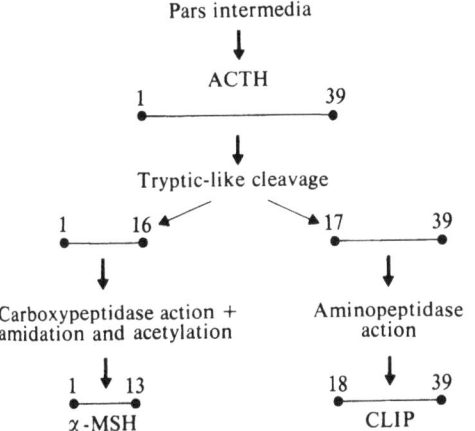

Fig. 3.8. Type of enzyme mechanism that may be involved in the formation of α-MSH and CLIP from ACTH in the pars intermedia. CLIP, corticotropin-like intermediate-lobe peptide. (From Scott et al., 1973.)

enzymes that cleave the ACTH and β-LPH to form α- and β-MSH.

Other active fragments of corticotropin and β-LPH have been prepared in the laboratory. ACTH-(1–24) and ACTH-(4–10) have been shown to influence behavior when injected into the brain of rats. Gamma-LPH [= β-LPH-(1–58)] has also been isolated, and this includes the sequence of β-MSH. The remaining part of the β-LPH molecule can give rise to the *endorphins*, which exhibit many of the properties of morphine (morphinomimetic peptides; see Lazarus, Ling, and Guillemin, 1976). Whether or not pituitary β-LPH actually undergoes such enzymic cleavage to produce all these peptides, which may then act at sites in the brain, is not clear (see Goldstein, 1976). Thus, hypophysectomy in rats does not influence the levels of such peptides in the brain, and it is difficult to envisage how such molecules could get from the pituitary to all these areas of the brain. However, β-endorphin can, at least under *in vitro* conditions, be converted by brain extracellular enzymes to smaller fragments, including γ-endorphin, α-endorphin, and Met-enkephalin (Austen, Smyth, and Snell, 1977), but the source of any such a precursor peptide probably does not involve the pituitary.

C. Synthesis of adenohypophysial hormones

The adenohypophysial hormones are synthesized in discrete types of cells in the pituitary. While the particular cells have usually each been identified on the basis of tinctorial and immunohistological differences, such pictures may only reflect the predominant type of hormone that is being stored, rather than indicating separate sites of synthesis. Thus, it is now apparent that a single type of pituitary cell may form more than one hormone and that the type of hormone which it synthesizes may change. It has, for instance, been shown in tissue culture that a single line of pituitary cells can produce both growth hormone and prolactin. The synthesis of growth hormone is favored by triiodothyronine. This thyroid hormone actually suppresses the formation of prolactin (see Samuels and Shapiro, 1976; Seo et al., 1977). Prolactin synthesis, however, increases in pregnancy and in response to estrogens (see Maurer, Stone, and Gorski, 1976). As described earlier, corticotropin (ACTH), β-lipotropin, and the melanocyte stimulating hormones are also formed by one type of cell. When they are present in the pars distalis, ACTH and β-lipotropin are the principal products, but in animals that possess a distinct pars intermedia (*not* man), the MSHs are also formed (Figure 3.7). The β-endorphins may also be secreted by such cells under some conditions. The biosynthetic relationships of the glycoprotein hormones TSH, LH, and FSH are not clear at this time.

The biosynthesis of the adenohypophysial hormones takes place in several stages, but the precise details for each hormone are not yet available. As proteins destined for "export" from the cell, it is considered that they follow the general type of pathway proposed by Palade (see Palade, 1975). This involves a synthesis by polysomes on the membranes of the rough endoplasmic reticulum, segregation into the cisternal space of these membranes, followed by transport to the Golgi apparatus, where they may be processed further and concentrated into storage granules. These events are shown diagrammatically in Figure 3.9, which refers principally to the "prolactin cell." The entire synthetic process for prolactin appears to take about 3 hours (Farquhar, Reid, and Daniell, 1978).

The primary initiation of synthesis involves the transcription of mRNA, which promotes the translation of a hormonal precursor by the ribosomes. Specific mRNAs for prolactin and growth hormone have been isolated (Maurer, Stone, and Gorski, 1976; Bancroft, Wu, and Zubay, 1973). The translated protein that is formed appears to be invariably larger than the hormones and is called a "pre-hormone" if its existence is quite transient, or a "pro-hormone" if it has a longer life such as a storage product. In the instances of prolactin and growth hormone, these parent molecules contain about 20 to 60 additional amino acids (Sussman, Tushinskj, and Bancroft, 1976; Maurer, Stone, and Gorski, 1976; Evans, Hucko, and Rosenfeld, 1977). A large glycoprotein molecule with a molecular weight of about 31,000 has been isolated from a culture of mouse pituitary tumor cells (Mains, Eipper, and Ling, 1977; Roberts and Her-

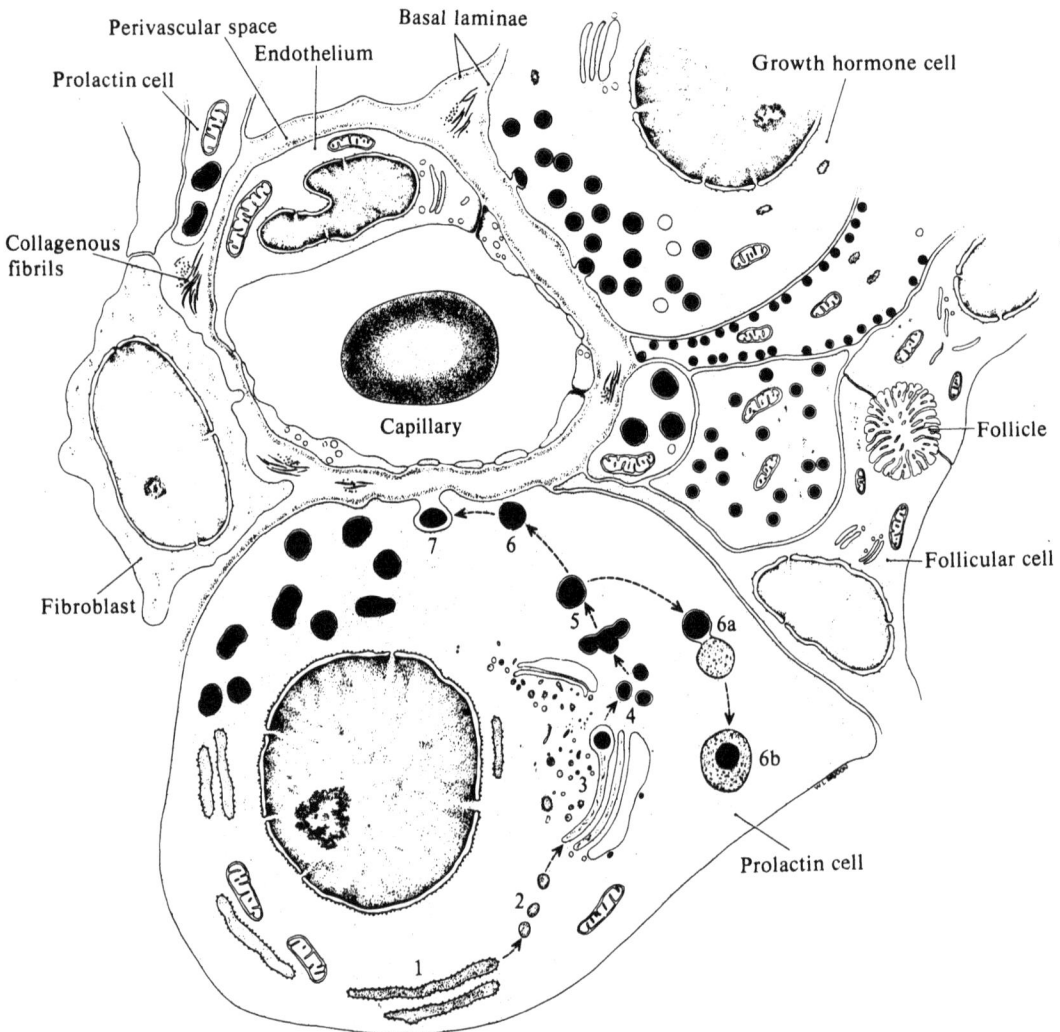

Fig. 3.9. Ultrastructural changes that take place in secretory cells in the pars distalis during hormone synthesis. This diagram refers mainly to the formation of prolactin in the rat pituitary. The lower prolactin cell illustrates the steps in the secretory process: (1) deposition of polypeptides in cisternae of endoplasmic reticulum; (2) transport of polypeptides to Golgi apparatus in vesicles; (3) condensation of polypeptides and granule formation in Golgi saccules; (4) liberation of membrane-enclosed granules from Golgi apparatus; (5) coalescence of granules; (6) movement of granules to peripheral plasmalemma; and (7) extrusion of the granule by exocytosis. Intracellular disposal of secretion is shown (6a) by fusion of a secretory granule with a dark body containing lysosomal hydrolytic enzymes and digestion of the contents (6b). (From Baker, 1974.)

bert, 1977; Mains and Eipper, 1978), and this appears to be the precursor for ACTH and β-lipotropin and possibly, subsequently the endorphins and the MSHs (see Scott et al., 1973; Crine et al., 1977). The nature of the precursors of the glycoprotein hormones FSH, LH, and TSH have not been described, but the considerable similarities in the structures of these hormones suggest that they may also share some common initial biosynthetic pathways.

The splitting and reassembly of such polypeptide precursor molecules to form the hormones could occur at various sites. Thus, the conversion of pregrowth hormone to growth hormone occurs on the membrane-bound polysomes (Spielman and Bancroft, 1977), while β-endorphin can be converted to Met-enkephalin at peripheral extracellular sites (Austen, Smyth, and Snell, 1977). The sites for the transfer of the carbohydrate moieties to the glycoprotein hormones is unknown, but it could occur in Golgi apparatus (Palade, 1975). ACTH and β-lipotropin are stored in the same granules and β-endorphin may, in some instances, also be secreted simultaneously with ACTH (Pelletier et al., 1977; Guillemin et al., 1977). The separation of the molecules from each other could thus be occurring at a number of sites, including the granules themselves. The selective cleavage of the polypeptide chains appears to depend largely on tryptic digestion at the sites of arginine residues, but other types of enzymes, including carboxypeptidases and aminopeptidases, may be involved (see, e.g., Figure 3.8).

There has been speculation regarding the possible functions of the extra peptide segment on the precursor molecules of protein hormones, and these include protection during the process of their segregation in the cell and the attachment of the polysomes to the membranes of the endoplasmic reticulum. They have been called *signal peptides* and usually contain a large proportion of hydrophilic amino acids, which appear to facilitate their binding to these cellular organelles.

D. Pituitary gland disorders

Dysfunction of the pituitary gland results in a variety of physiological problems, both endocrine and nonendocrine, depending on the primary nature of the disease. Increases or decreases in the endocrine activity of the gland may occur. The causes of such disorders may include the presence of tumors, inflammation as the result of disease, vascular accidents that produce a necrosis of the gland, endocrine disorders such as hypothyroidism, or an excess of adrenocorticosteroids that influence its functioning, radiological, and surgical treatment, such as for brain tumors and hypophysectomy. In many instances, specific lesions cannot be identified, even following postmortem histological examination.

Tumors may enhance or depress the endocrine activities of the pituitary. Thus, adenomas of the eosinophil type of acidophil cells of the adenohypophysis can result in an excessive secretion of growth hormone. Nonendocrine functioning tumors are, however, more common, and these usually depress the secretion of hormones as a result of increases in pressure, especially if the tumor is present in the sella or if it results in damage to the pituitary stalk, vascular supply, or the regions of the hypothalamus that control the activity of the gland.

Disorders of pituitary function may involve all or only some of its endocrine activities. Complete cessation of all function is called *panhypopituitarism* or Simmonds' disease. In other instances, the disorder may be an *isolated deficiency* and only influence a single hormone.

Panhypopituitarism (see Nabarro, 1972; Strong, 1976) is principally the result of:

a. The presence of tumors in either the sella or the hypothalamus. These are usually nonfunctional chromophobe adenomas, but craniopharyngiomas are also involved, especially before the age of 20 years. Secondary carcinomas occur but are relatively rare.
b. Necrosis of the pituitary following postpartum hemorrhage (Sheehan's syndrome), but this condition is now quite rare.
c. Hypophysectomy is sometimes performed for the treatment of breast carcinoma and has been utilized in the treatment of diabetes mellitus, especially when vascular changes in the retina threaten sight (retinopathy).

Pituitary disease may progress over a period of 10 to 15 years and result in a successive loss of pituitary hormones, usually starting with growth hormone and gonadotropins, followed by TSH and finally ACTH. The neurohypophysis may also be affected, but this does not necessarily result in the disorder of diabetes insipidus unless the hypophysial supraoptic and paraventricular nuclei are involved (Section 3.3C). Pituitary function can be maintained even when only about 25 percent of the glandular tissue is functioning, but the amount may decline to negligible levels in severe panhypopituitarism. Surgical removal of the pituitary results in symptoms of insufficiency in about 2 to 4 weeks.

Not surprisingly, pituitary insufficiency can have widespread symptoms and effects, especially as it can involve so many other endocrine glands. It has therefore also been called "pluri-glandular deficiency." In women the initial endocrine symptoms are usually a failure to menstruate (amenorrhea). Common initial nonendocrine disorders are headache and disturbances of vision, such as a loss of the visual field. These changes reflect increases in pressure due to expanding tumors; in the case of changes in vision they involve the optic chiasma. Endocrine-related disorders include psychic changes, apathy and depression, a loss of pubic and axillary hair, a decreased metabolic rate, muscular weakness, and a loss of libido. These changes reflect a loss of gonadotropins, TSH, and ACTH. Unless some concurrent disease occurs, there is no immediate threat to life and the patient may survive for 10 to 15 years or even longer. The quality of life is, however, severely impaired and the patient may become a complete invalid. The deficiency of TSH and ACTH places severe restriction on the ability to adapt to changing environmental conditions and to combat other diseases. The patient may lapse into coma as a result of the thyroid and adrenocortical deficiencies. In children there will also be a failure of growth, and sexual maturity will not be attained.

The availability of hormones for therapeutic use has dramatically improved the treatment and ability to diagnose panhypopituitarism. These preparations include the hypophysiotropic hormones, adenohypophysial and neurohypophysial hormones, and the hormones from the pituitary-responsive peripheral endocrine glands. These hormones and their use, actions, and side effects are dealt with in the appropriate sections of this book. Treatment of panhypopituitarism involves surgical and radiological as well as pharmacological procedures.

The diagnostic distinction between primary disorders that involve the hypothalamus or pituitary and those that involve the peripheral endocrine glands can be made with the aid of preparations of hormones from the adenohypophysis or hypothalamus. Those that are available and useful include TSH, ACTH, the gonadotropins, ADH, TRH, and LHRH. The effects of some drugs and hormones in releasing pituitary hormones are also used to test the integrity of the pituitary. These responses include the secretion of gonadotropins by the antiestrogen clomiphene, growth hormone by insulin hypoglycemia, ACTH by vasopressin, ADH by nicotine, and prolactin by sulpiride.

The replacement of hormones in hypopituitarism will depend on the precise occurrence of deficiencies ("partial hypopituitarism" may occur), the age of the patient, and whether it is desired to promote fertility. As pituitary hormones are proteinaceous and are thus readily destroyed in the digestive tract and the circulation, their chronic administration is not always very practical. The exceptions are growth hormone, which only needs to be administered once or twice a week [see Section 3.2 F(6)], and new synthetic preparations of vasopressin (ADH), which can be given as a nasal spray and act for about 20 hours [see Section 3.3 F(4)]. The trophic effects of TSH and ACTH are best substituted for by directly administering thyroid hormones and adrenocortical steriods. In the instance of the latter, synthetic analogues, usually glucocorticoids (the production of the mineralocorticoid aldosterone is little affected by lack of ACTH) such as prednisone, are used, but in crisis situations cortisol is administered. The sexual characters can be maintained by the administration of androgenic steroids in men and estrogens and progesterone in women. By giving the latter two hormones in a cyclical manner, a normal menstrual cycle can be mimicked but ovulation will not occur. If pregnancy is desired, more direct treatment that affects the hypothalamus or pituitary, using such preparations as clomiphene, LHRH, and gonadotropins [see Section 7.6H(7)], is used. It is necessary to replace growth hormone in children so as to resume normal growth. If diabetes insipidus occurs, it may be treated with various preparations of vasopressin or some drugs such as chlorpropamide [see Section 3.3 F(4)].

Hypersecretion of the pituitary hormones can result from the presence of hormone-secreting tumors (see Strang, 1976) or interference with the hypothalamic control mechanisms. The latter may be due to physical lesions, including the presence of tumors, such as a craniopharyngioma. Hypersecretion may also result from the action of drugs or a decline in the negative-feedback responses due to a reduced production of peripheral hormones such as the adrenocorticosteroids and sex steroids. An excess of growth hormone, usually due to a pituitary acidophil adenoma, can result in *gigantism* if it occurs in children or *acromegaly* in adults [see Section 3.2 F(6)]. Hyperprolactinemia is usually due to interference with the hypothalamic control mechanism, which inhibits release of this hormone [see Section 7.6 H(7)]. It can also result from the action of drugs, including reserpine and α-methyldopa, which are used to treat hypertension, and phenothiazines, which are administered for some psychic disorders. Prolactin-secreting chromophobe adenomas have been identified. Hyperprolactinemia results in galactorrhea and is usually accompanied by amenorrhea. It can be treated by bromocriptine, which inhibits the release of the hormone. Excessive secretion of vasopressin (*inappropriate secretion of ADH*) results from hypothalamic disorders that control the release of the hormone [see Section 3.3 F(1)]. Hypersecretion of the other pituitary hormones is less likely to result directly in endocrine problems. Thyrotropin-secreting tumors are only a rare cause of hyperthyroidism. Cushing's disease (hyperadrenocortisolism) is usually considered to result from a disorder in the hypothalamic mechanism controlling the release of ACTH. However, basophilic pituitary adenomas that secrete ACTH have been identified, although usually with great difficulty. Such tumors are more commonly seen following bilateral adrenalectomy for the treatment of Cushing's disease. This is thought to reflect an enhanced release of corticotropin-releasing factor (CRF) in the absence of adequate corticosteroids. An excessive secretion of ACTH can result in increased pigmentation. Hypersecretion of gonadotropins follows gonadectomy or the decline in ovarian activity that occurs at the menopause. It does not appear to have any adverse effects. However, if secretion of gonadotropins occurs too early in childhood, precocious sexual maturation will occur.

E. Control of hormone secretion by the adenohypophysis – role of the hypothalamus

As described in the previous section, the pars distalis lacks a nerve supply but is in close vascular contact with the median eminence region of the hypothalamus (see Knigge and Silverman, 1974). The control of its secretory activity is dependent on the integrity of the hypophysial portal network of blood vessels, which carry a number of secretory products from the median eminence to the pars distalis (see Guillemin, 1978 a,b; Schally, 1978). This vascular channel thus provides a pathway by which information both from the internal and external environments, which is processed by the brain, can be transmitted to various endocrine organs. It is a system in which signals can be amplified many thousands of times before they reach their ultimate effectors, such as the liver, muscles, and reproductive organs.

The secretion of the hormones of the pars distalis are closely regulated, presumably in response to physiological need. Release of the hormones may be relatively steady, occur in a sudden tonic surge, or take place in a cyclical manner, such as in a daily rhythm, or a pattern that extends over periods of time ranging from several days to weeks or even months. During such cycles the hormone concentrations in the plasma are relatively stable and the long-term oscillations are usually the result of quite gradual changes. There is thus a short-term type of control that is apparently predetermined by a "set point" in the control mechanism. This set point, however, is not inviolate and may be altered to give longer-term oscillations in hormone levels, which may be part of normal physiological cycles or the result of disease. The short-term control appears to be the result of the activity of a *negative-feedback mechanism* (see Figure 3.10). Thus, the release of the adenohypophysial hormones are each influenced by their own levels or those of other hormones or metabolites that are formed as a result of their action. For instance, a rise in plasma cortisol levels, in response to stimulation of the adrenal cortex by corticotropin, in turn inhibits further release of the latter hormone. A similar relationship exists between thyrotropin and thyroid hormones and the gonadotropins and steroid sex hormones. In the instance of growth hormone, metabolites such as glucose and amino acids can influence release. The site of the feedback response is usually the hypothalamus, but in some instances, especially thyroid hormones, the pars distalis itself is involved. "Higher" centers in the CNS may also be affected. There is evidence that another type of negative-feedback mechanism may exist which could involve a more direct effect of the adenohypophysial hormones on the hypothalamus. The latter is called a *short-loop* negative feedback in contrast to the *long-loop* feedback that arises from secretions and metabolites formed at more peripheral sites. *Positive-feedback mechanisms* are more rare in nature, but there is one well-established process in the pituitary-hypothalamic system. Luteinizing hormone stimulates the secretion of estrogens by the ovaries, and under certain conditions these steroid hormones can themselves trigger a sudden release of LH. This positive-feedback response helps to initiate ovulation.

Rhythmical and acute tonic changes in the release of adenohypophysial hormones that occur in response to external environmental stimuli or internal physiological changes are transmitted by afferent neural pathways to the appropriate regions of the hypothalamus, where they appear to modulate the activities of the parvicellular neurosecretory neurons. Such information may come from higher centers in the brain or via spinal pathways and the autonomic nervous system. It is thought that such nerve fibers form synapses with the cell bodies in the nuclei of the hypophysiotropic region, or they could even affect more peripheral regions of the hypothalamus, such as the median eminence. The nature of the neurotransmitters has been the subject of considerable study and speculation. The reason for the interest is not only academic but also practical, as certain drugs with specific agonist or antagonist properties may be useful therapeutic agents for controlling the secretion of the pituitary hormones.

The substances that are produced by the hypothalamus and which control the synthesis and release of the adenohypophysial hormones are sometimes called pituitropins or hypophysiotropic hormones. They may either increase or decrease the secretion of a hormone, and two may act jointly to effect control. Considering the size of the hypothalamus and the small quantities of such "factors" that are stored there, it is not surprising to observe that their identification and chemical and pharmacological characterization have been beset by considerable technical difficulties. As described earlier, the isolation of a few milligrams of a pure hypophysiotropic hormone may require the processing of literally millions of animal hypothalami. The hypothalamus, although not quantitatively rich in these "factors" (a few nanograms are usually present), contains a wide qualitative array of such pharmacologically active substances. It is therefore not unexpected to find that information about such putative releasing or regulating factors may often be confusing and controversial.

The bioassay systems used initially to identify these substances usually consisted of the isolated (*in vitro*) adenohypophysis, sometimes with the attached hypothalamus. These preparations can be exposed to various stimuli, including extracts of the hypothalamus, and the release of hormones into the incubation medium are measured. If larger quantities of active material are available, such as when they can be prepared by chemical synthesis, then they may also be tested *in vivo* by injecting them into the blood supplying the pituitary or into the brain fluids. Such hypothalamic materials that influence the release of the adenohypophysial hormones were initially called "factors," with the prefix "releasing" or "inhib-

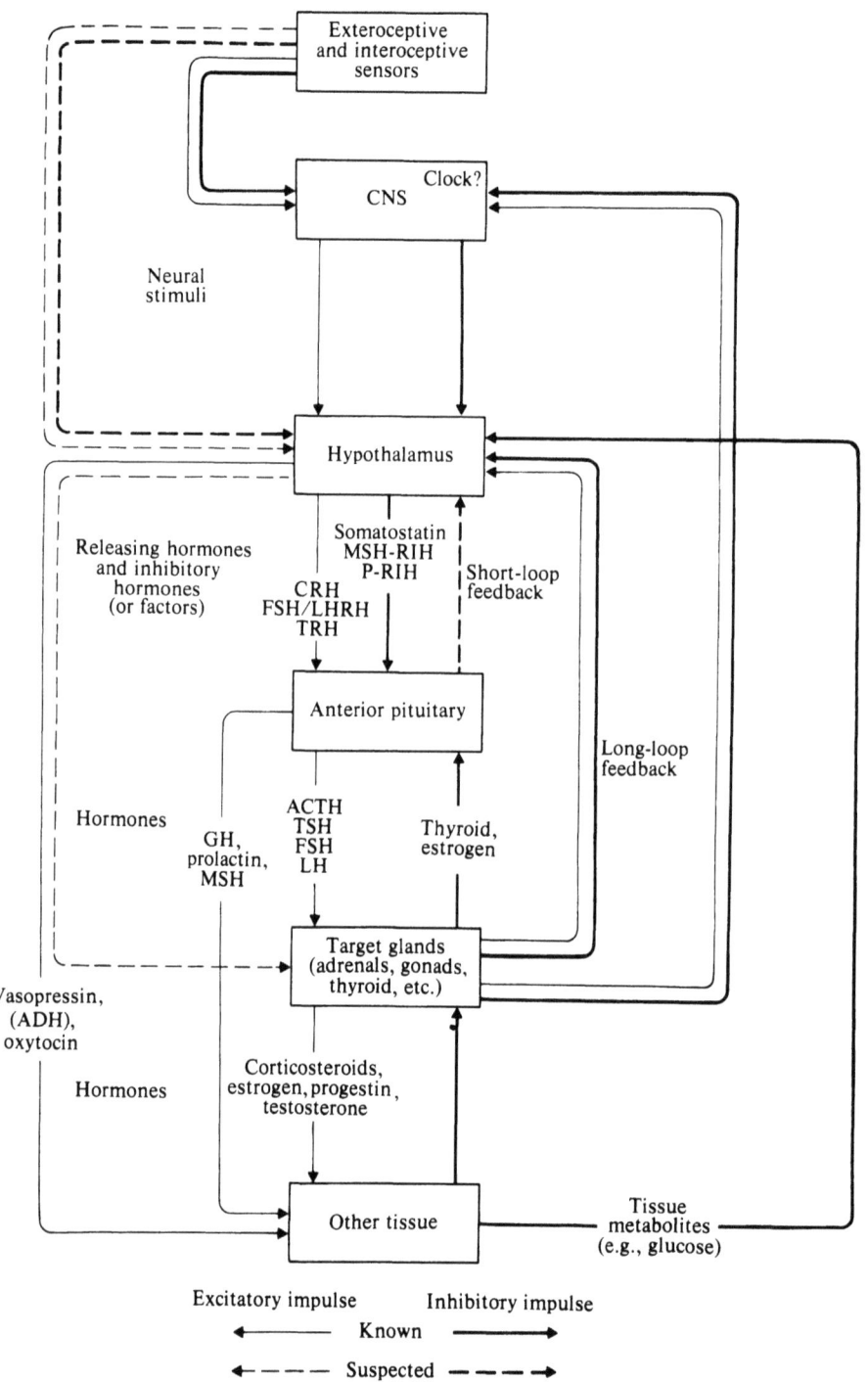

Fig. 3.10. Factors controlling the release of hormones from the anterior pituitary (adenohypophysis). Shown is the use of negative-feedback inhibition of secreted hormones and tissue metabolites in influencing further hormonal release. As can be seen, the hypothalamus (and its associated median eminence) plays a central role in this process. Stimulatory effects are shown by thin lines and inhibitory processes by thick lines. (From Krieger, 1971.)

Table 3.4. *Hypothalamic hormones believed to control the release of pituitary hormones*

Hypothalamic hormone (or factor)	Abbreviation
Corticotropin (ACTH)-releasing hormone	CRH or CRF
Thyrotropin (TSH)-releasing[a] hormone	TSH-RH or TRH or TRF
Luteinizing hormone (LH)-releasing[a] hormone	LHRH or LH-RF
Follicle-stimulating hormone (FSH)-releasing[a] hormone	FSH-RH or FSH-RF
Growth hormone (GH)-releasing[a] hormone	GH-RH or GH-RF
Growth hormone (GH) release-inhibiting hormone or somatostatin	GH-R-IH or GIF
Prolactin release-inhibiting hormone	PR-IH or PIF[b]
Prolactin-releasing hormone	PRH or PRF[c]
Melanocyte-stimulating hormone (MSH) release-inhibiting hormone	MSH-R-IH or MRIH or MIF
Melanocyte-stimulating hormone (MSH)-releasing hormone	MRH or MRF

Note: The evidence for the presence of some of these hormones is still equivocal.
[a] Or regulating hormone. [b] ? Dopamine. [c] TRH?
Source: Based on Schally, Arimura, and Kastin, 1973. Copyright 1973 by the American Association for the Advancement of Science.

iting," depending on their action. This terminology is still used, but in many cases, where there remains little doubt that a physiological role is being subserved, they have been elevated by some to the title "hormone." Reasonable evidence for the presence of about 10 such factors has been obtained (Table 3.4), but only three of these have been chemically characterized. The latter are peptides and are thyrotropin-releasing hormone (TRH); luteinizing hormone-releasing hormone (which also releases FSH), abbreviated LHRH or LH/FSH-RH; and growth hormone release-inhibiting hormone (GH-RIH) or somatostatin. Peptides that influence the release of MSH, called MSH-RII and MSH-R-III, are known but their definitive physiological role has not been established [Section 3.2E(7)]. The separate identities of some other pituitropins are not clear; thus, prolactin-releasing hormone, PRH, could be identical to TRH, while prolactin release-inhibiting hormone, PRIH, may be synonymous with dopamine.

The release of the hypophysiotropic hormones can be influenced by a number of neurotransmitters and their chemical analogues, which act as agonists or antagonists. Such responses may reflect the nature of the neurons that impinge from other areas of the brain onto the neurosecretory parvicellular nuclei or the median eminence. The hypothalamus is a rich source of such neurotransmitters, which makes precise identification of the relationships of each type of nerve cell difficult. Putative neurotransmitters that have been identified in the hypothalamus include acetylcholine, epinephrine, norepinephrine, dopamine, serotonin, histamine, substance P, and GABA. The quantities of some substances in the hypothalamus are shown in Table 3.5.

Direct localization of the hypophysiotropic hormones, especially in relation to the sites of neurotransmitters, has been attempted using histochemical and immunofluorescent techniques and immunoautoradiography. Microdissection and microassay of discrete regions of frozen sections of the hypothalamus have also been performed. Some of the results of such procedures are summarized in Table 3.5 (see also Krulich et al., 1977; Brownstein, 1977; Zimmerman, 1977; Goldsmith, 1977). While the hypophysiotropic hormones are usually relatively concentrated in the median eminence, they are also present in other regions of the hypothalamus. Thus, only 25 percent of the TRH in the brain is in the hypothalamus, while somatostatin is not only present in the brain but in many other tissues of the body, including the gut and the islets of Langerhans. TRH has even been identified in the skin of frogs, where it is present in much higher concentrations than in the hypothalamus (Jackson and Reichlin, 1977a). It has quite recently become apparent that such peptides may serve more general functions in the body, apart from controlling the release of the hormones of the anterior lobe of the pituitary.

While the various neurotransmitters and hypophysiotropic hormones are ubiquitous in the median eminence and other regions of the hypothalamus, their distribution is not necessarily uniform (Table 3.5); TRH is present in higher concentrations in the middle regions of the median eminence and the ventromedial nucleus, while somatostatin tends to be more localized in the arcuate and periventricular nuclei. The functional significance of such differences is not yet completely clear but probably represents a predominance in the number and activity of specialized neurosecretory neurons. The putative neuro-

Table 3.5. *Distribution of various active peptides and putative neurotransmitters in the median eminence and its associated hypothalamic nuclei (as pmol/mg protein)*

a. Concentrations in the median eminence of the rat

Substance	Content
LHRH	19.0
TRH	110.0
Somatostatin	189.0
Vasopressin	717.0
Oxytocin	416.0
Norepinephrine	118.0
Dopamine	523.0
Epinephrine	3.4
5-Hydorxytryptamine (serotonin)	87.0
Histamine	160.0

b. Concentrations in hypothalamic nuclei[a]

	Arcuate nucleus	Ventro-medial nucleus	Dorso-medial nucleus	Peri-ventricular nucleus	Supra-chiasmatic nucleus	Supra-optic nucleus	Para-ventricular nucleus
LHRH	2.5	0.3	<0.05	<0.05	~0.1	~0.1	<0.05
TRH	11.3	17.3	11.4	12.3	5.2	2.5	7.5
Somatostatin	27.3	8.9	3.3	14.5	4.9	2.0	2.7
Vasopressin	6.5	<0.2	<0.2	<0.2	37.0	65.0	39.0
Oxytocin	43.0	<0.2	<0.2	<0.2	<0.2	106.0	112.0
Norepinephrine	118.0	130.0	230.0	201.0	148.0	140.0	301.0
Dopamine	98.0	46.0	65.0	46.0	59.0	24.0	65.0
5-Hydroxytryptamine	114.0	48.0	77.0	62.0	144.0	54.0	77.0
Histamine	56.0	32.0	36.0	33.0	57.0	27.0	22.0

c. Distribution within the median eminence of the ox

	Rostral	Anterior external	Anterior internal	Middle external medial	Middle external lateral	Middle internal medial	Middle internal lateral	Caudal
LHRH	1.9	2.3	3.5	1.6	3.6	2.2	2.9	1.8
TRH	10.1	20.5	15.0	73.1	32.4	32.4	12.1	5.5
Norepinephrine	473.0	308.0	544.0	231.0	438.0	438.0	503.0	444.0
Dopamine	333.0	484.0	595.0	490.0	556.0	556.0	471.0	261.0
5-Hydroxytryptamine	33.5	29.0	23.0	19.3	19.3	19.3	26.1	21.6
Histamine	95.0	180.0	324.0	333.0	586.0	586.0	523.0	180.0

[a] For a map showing the positions of the various nuclei, see Figure 3.15.
Source: Based on Brownstein, 1977.

transmitters also lack a uniform distribution and the exclusive neuronal origins of some of these, such as histamine and GABA, are in doubt. The neuronal norepinephrine levels decline dramatically following deafferentation of the hypothalamus (from the rest of the brain), indicating that the cells that produce it lie outside this region. They are present in the brain stem. Dopamine levels are unaffected by such surgery; the dopaminergic neurons arise from the arcuate nucleus and reflect the presence of a nerve tract that terminates in the median eminence. Serotonin levels decline somewhat following hypothalamic deafferentation, but its synthesis continues in isolated hypothalami. Histamine levels also persist in the latter circumstances, and it seems likely that substantial amounts of this substance and also, possibly, serotonin are present in mast cells. The physiological role of these cells in the brain is not known. The precise relationships of the peptidergic and nonpeptidergic neurons in the hypothalamus have not yet been elucidated, but it seems likely that such information will be forthcoming before long. It is also possible that endogenous prostaglandins play a role in regulating hypothalamic–pituitary function.

The responsive *sites of the feedback mechanisms* that control the release of the adenohypophysial hormones have been the subject of intensive investigations. These studies involve such procedures as observing the effects of localized lesions and implants of hormones into the different regions of the hypothalamus. The histological localization of the hormones, especially in relation to possible neurosecretory neurons, has also been used in attempts to narrow down the sites of the feedback controls. Problems arise as it is uncertain how extensive the effects of such lesions are likely to be and diffusion of the hormones from the sites of the implants may occur. Several different sites may mediate such feedback mechanisms.

a. The *pituitary gland* itself responds to the presence of circulating hormones. In the instance of the thyroid hormones this appears to be the predominant site for negative-feedback control. Triiodothyronine has been identified in the nuclei of pituitary cells, and its presence appears to be related to the suppression of the release of TSH (Silva and Larsen, 1977). Estrogens and progesterone may also act on the pituitary to enhance or inhibit the release of gonadotropins. This effect appears to be due to a modulation of its responses to LHRH. Progesterone receptors have been identified in the pituitary (Kato and Onouchi, 1977). High levels of hydroxylated metabolites of estrogens, the catechol estrogens, have been identified in the pituitary (Paul and Axelrod, 1977).

b. The *hypothalamus* appears to be the more usual site of the feedback control mechanisms, but their exact location (see Knigge and Silverman, 1974) is not clear. They may be present in the median eminence or in more distal parts of the brain. They possibly influence release from the terminals of neurosecretory cells, either directly or via synaptic junctions with other neurons.

The responsiveness of the cell bodies of the neurosecretory neurons may be altered by a direct action or as the result of an effect on associated neurons. The distribution of estrogens and androgens in relation to that of LHRH is shown in Figure 3.11 and may provide clues to the site of their feedback mechanisms.

Progesterone receptors have been identified in the hypothalamus (Kato and Onouchi, 1977) and estrogens accumulate in the arcuate nucleus and around the preoptic nucleus in this tissue (Stumpf and Sar, 1977). High concentrations of catechol

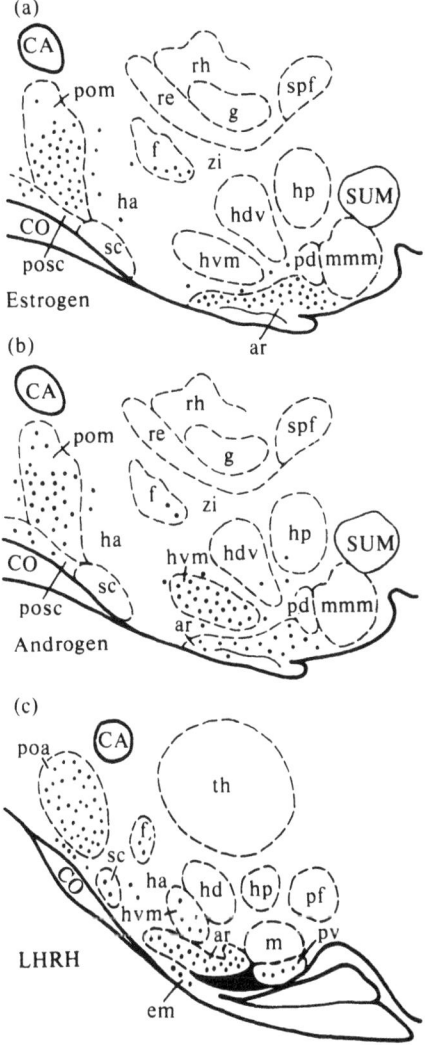

Fig. 3.11. Distribution in the rat hypothalamus of estrogen- and androgen-concentrating neurons and their relationship to the distribution of LHRH. ar, nucleus arcuatus; em, eminentia mediana; f, nucleus paraventricularis; ha, area hypothalamica anterior; hvm, nucleus ventromedialis; poa, area preoptica; pv, nucleus premamillaris ventralis; pom, nucleus preopticus medialis; posc, nucleus preopticus, pars suprachiasmatica; sc, nucleus suprachiasmaticus; CA, commissura anterior; CO, chiasma opticum. The mapping was made using autohistoradiographic methods. (From Stumpf and Sar, 1977.)

estrogens, especially 2-hydroxyestradiol and 2-hydroxyestrone, have also been identified there (Paul and Axelrod, 1977). It is considered likely that these estrogen metabolites are formed *in situ* by a catechol-estrogen-forming cytochrome P-450 enzyme, which has been identified in the brain (Paul, Axelrod, and Diliberto, 1977). Estradiol increases cyclic AMP levels in the hypothalamus, but the catechol estrogens oppose this effect and so act as antiestrogens (Paul and Skolnick, 1977).

c. More distal regions of the brain, such as the *central nervous system,* can also accumulate and respond to hormones that are involved in the feedback control mechanisms, so that these may also play a part, and many such hormones, for instance estrogens, progesterone, and corticosteroids, have been identified at various sites in the brain (see Stumpf and Sar, 1977).

Clearly, the nature of the feedback site may vary for different hormones, and more than one may be involved for a single such hormone. It has been suggested that the hormones and metabolites that initiate the feedback responses may do so by altering the thresholds for stimulation of the appropriate types of cells.

There is a reasonable amount of information regarding the sites of origin and the neuronal control of the release of LHRH (see McCann, 1977; Krulich et al., 1977; Stumpf and Sar, 1977; Goodman, 1978). This peptide is present in high concentrations in the median eminence (60 percent of the total in the hypothalamus). In other parts of the hypothalamus it is present in the arcuate nucleus (25 percent of the total), which lies dorsally to the median eminence and also anteriorly in the suprachiasmatic–preoptic region (Figure 3.12 and Table 3.5). Activity can also be detected between these two regions of the brain. Electrical stimulation of the preoptic region of the brain results in a release of luteinizing hormone, and this effect can be blocked by α-adrenergic inhibitors. It has thus been proposed that release of LHRH occurs as a result of stimulation of the parvicellular neurons via a noradrenergic pathway. Stimulation in the region of the arcuate nucleus also results in a release of LH, but this effect is not blocked by adrenergic-blocking agents, so that the response appears to be a direct one on the neurosecretory neurons. It has thus been proposed that the site of the positive-feedback control, by estrogen, of LH release lies in the preoptic region of the anterior hypothalamus, and its effects are facilitated or transmitted via a noradrenergic synapse. The site of the negative-feedback inhibition by estrogen may be more basal, in the region of the arcuate nucleus, but may also directly involve the anterior pituitary. This mechanism is summarized in Figure 3.13.

Similar processes appear to control the release of other pituitropins, but different hypothalamic nuclei and transmitters may be involved. The possible function of intrahypophysial dopaminergic neurons is especially interesting, as dopamine appears to play an important role in controlling the release of prolactin and growth hormone.

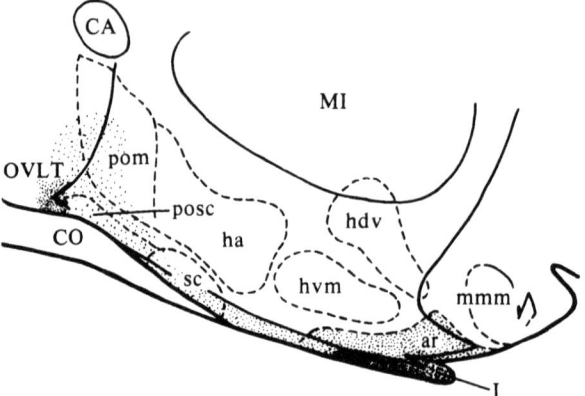

Fig. 3.12. Relative concentrations of LHRH in the hypothalamic region of the rat. The relative concentration of LHRH in a region is indicated by the intensity of the stippling. Note in particular the location of the organum vasculosum lamina terminalis (OVLT), the arcuate nucleus (ar) with overlying median eminence and pituitary stalk (I). Abbreviations for the remaining hypothalamic structures are: ha, nucleus anterior hypothalamus; hdv, nucleus dorsomedialis hypothalamus; hvm, nucleus ventromedialis hypothalamus; pom, nucleus preopticus medialis; posc, nucleus preopticus, pars suprachiasmatica; mmm, nucleus mamillaris medialis, pars medialis; sc, nucleus suprachiasmaticus; CA, commissura anterior; CO, chiasma opticum; MI, massa intermedia. (From McCann, 1977. Reprinted by permission from *The New England Journal of Medicine 296,* 798.)

It is possible that nonneuronal pathways are involved in the transfer of some hypophysiotropic hormones to the median eminence. The inner ependymal region of this tissue is in contact with the fluids of the third ventricle, into which transmitters can be discharged. The ependyma contains cells called *tanycytes,* which send processes deep into the median eminence. Such cells may be able to engulf molecules and transmit them to the more vascularized regions of the median eminence, from which they can be carried to the pituitary.

Prostaglandins may influence the release of pituitropins. Thus, it has been observed that indomethacin, which inhibits the prostaglandin synthetase, can prevent the release of gonadotropins. The intraventricular injection or implantation of prostaglandins, especially the E series, into the discrete regions of the hypothalamus that contain LHRH can elicit a release of LH and FSH (Ojeda, Jameson, and McCann, 1977a,b). This effect appears to reflect a stimulation of the neurosecretory neurons. Conversely, a prostaglandin antagonist (CN-0164), when injected intraventricularly, can block ovulation in rats (Botting, Linton, and Whitehead, 1977).

Arginine-vasotocin is a peptide that has an endocrine role in nonmammalian vertebrates which is

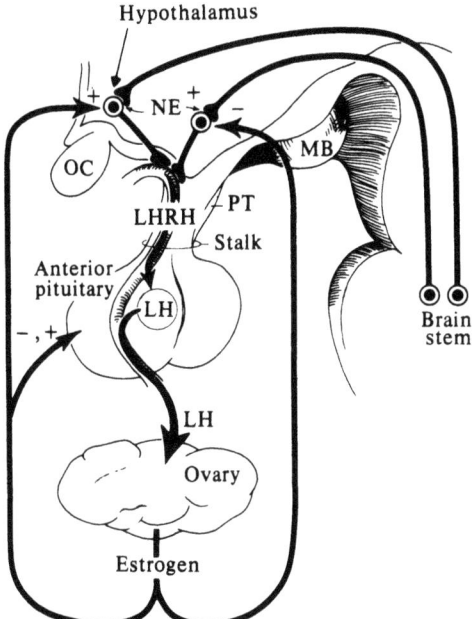

Fig. 3.13. Diagramatic representation of the stimulatory (+) and inhibitory (−) effects of estrogens in controlling the release of luteinizing hormone (LH). The principal sites of these effects are situated in the hypothalamus and probably involve noradrenergic (NE) neurons. Positive-feedback (+) control appears to be localized in the anterior parts of the hypothalamus, while negative-feedback (−) control occurs in the more basal regions of the tissue, in the area of the arcuate nucleus. The anterior pituitary itself is also directly involved in regulating the release of LH. OC, optic chiasma; MB, mamillary bodies; PT, pars tuberalis. Stalk refers to pituitary stalk. (From McCann, 1977. Reprinted by permission from *The New England Journal of Medicine 296*, 799.)

similar to that of ADH. It is found in the neurohypophysis of these species. In mammals, it has been identified in fetal neurohypophyses and in adults in certain parts of the brain, especially in the pineal gland. Its function in mammals in unknown. When this peptide is injected intraventricularly into rats, it inhibits the preovulatory "surge" in the release of LH (Cheesman, Osland, and Forsham, 1977; Osland, Cheesman and Forsham, 1977). Vasotocin is not effective when injected into the systemic circulation, and it does not antagonize the effects of prostaglandin E_2. It was thus suggested that it acts higher in the central nervous system. Whether this response reflects a physiological role or is merely a pharmacological effect is unknown.

The hypothalamus is involved in mediating the various *rhythms* that occur in the release of the pituitary hormones. Such changing patterns, or periodicity, of hormone secretion may occur during the normal day or extend over a period of sev-

eral weeks or even months. The corticosteroids show a daily pattern of release, and in man they are at their highest concentrations in the early morning waking hours. Growth hormone and prolactin levels rise during sleep, but this release appears to be an acute response rather than a rhythm. In women, the gonadotropins show a reproducible pattern of release extending over a period of about 28 days. In seasonally breeding animals, such changes may occur over even more prolonged periods of time. The "cues" for such rhythms may be endogenous ones or environmental stimuli, such as the length of the day, periods of activity, or mealtimes. Specifically placed lesions, the presence of drugs that influence neuronal function, and exposure to certain conditions during development can interrupt or abolish such rhythms. The initiating stimuli appear to undergo modulation in the brain before being passed to the hypothalamus and then the pituitary. The locus of these processes is sometimes referred to as the "biological clock." Whether there is only one such regulating center is unknown. The entrainment and ability to respond to suitable stimuli develops early in pre- and postnatal life. When newborn female rats are injected with male androgenic hormones, they fail to develop the normal female cyclical pattern of release of gonadotropins. Similarly, if they are given cortisol or dexamethasone for 2 to 4 days after birth, the normal daily patterns of corticosteroid release (reflecting that of pituitary ACTH) fails to develop (Figure 3.14). Antiserotoninergic drugs such as cyproheptadine, p-chloroamphetamine, and L-α-methyl-p-tyrosine, which deplete stores of norepinephrine, interfere with circadian changes in corticosteroid release (Krieger and Rizzo, 1969; Scapagnini et al., 1970). The cyclical release of gonadotropins can be interrupted in several ways. Clomiphene, which is an antiestrogen, appears to block the negative feedback of estrogen, which inhibits the release of LH and so promotes the release of this gonadotropin. Progestins and estrogens, on the other hand, by exerting a negative-feedback response in the hypothalamus, block the release of gonadotropins and so, by preventing ovulation, are effective oral contraceptive drugs. It is thus possible to interfere with the normal rhythmical release of pituitary hormones with drugs that affect different aspects of hypothalamic function.

1. Thyrotropin-releasing hormone, TRH. The TRH was the first hypophysial-regulating hormone to be isolated in a chemically pure form so that its structure could be determined (see Vale et al., 1976; Vale, Rivier, and Brown, 1977). It is a tripeptide with the structure (pyro) Glu-His-Pro-NH_2 and it has been prepared by chemical synthesis. Its distribution in the hypothalamus has been mapped and it is known to be present at high con-

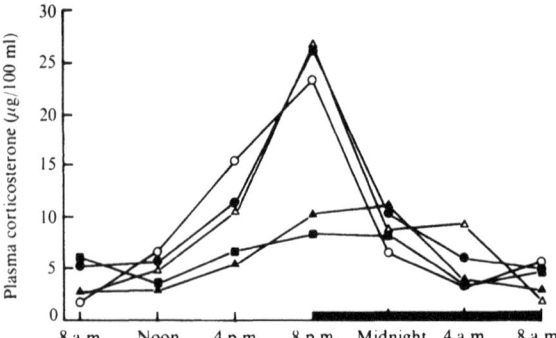

Fig. 3.14. Circadian pattern in the concentration of corticosteroids in the plasma of rats: the effects of the administration of exogenous corticosteroids to young, developing rats on the subsequent circadian periodicity in the endogenous corticosteroid levels. It can be seen that the administration of dexamethasone or hydrocortisone (cortisol) on days 2 to 4 after birth suppressed the rhythmical release. However, when dexamethasone was given on days 12 to 14 after birth, no effect was seen: ●, control; ○, saline, 0.1 ml, day 2–4; ■, hydrocortisone acetate, 500 μg, day 3; ▲, dexamethasone PO$_4$, 1 μg, day 2–4; △, dexamethasone PO$_4$, 1 μg, day 12–14. (From Krieger, 1972. Copyright 1972 by the American Association for the Advancement of Science.)

centrations in the median eminence, and it extends to the stria terminalis and dorsomedial nucleus (Figure 3.15). It has subsequently been shown to be ubiquitous in the brain and more than 80 percent of it is extrahypothalamic. TRH is also present in high concentrations in the pars nervosa (Jackson and Reichlin, 1977b) and appears in the systemic circulation. It not only triggers release of pituitary thyrotropin but also prolactin. Other pituitary hormones could also be involved. TRH can alter certain behavior patterns in experimental animals and it has been reported to alleviate the symptoms of depression in man (Prange et al., 1972), although such observations have not been confirmed (Coppen et al., 1974; Mountjoy et al., 1974). It can, however, influence the activity of certain neurons in the CNS (Yarbrough, 1976). Thus, the actions of TRH may not be confined to the endocrine system and it could have more general importance in the body, especially in the nervous system. The name TRH may thus be a historical misnomer.

TRH is released from the hypothalamus in response to exposure to cold, so that it mediates metabolic changes triggered by the thyroid hormones. Phentolamine, reserpine, and atropine interfere with this response, probably via effects in the central nervous system and/or the hypothalamus. The peptide is rapidly destroyed in the plasma, where it has a half-life of about 4 minutes. Cleavage of the (pyro) Glu-His linkage occurs. Specific recep-

tors for TRH have been identified in the adenohypophysis, where they are present in the cell membranes. They are also present in other parts of the brain. Pituitary hormones can be released by cyclic AMP or its analogues and drugs, such as theophylline, which inhibit the destruction of this nucleotide in the cell. Cyclic AMP levels in pituitary gland preparations have been found to increase in the presence of TRH, and it has been suggested that this nucleotide normally mediates the hormone's effects via a stimulation of membrane adenyl cyclase.

The clinical uses of TRH are described later (Section 4.10F). They principally involve tests for the integrity of hypophysial and pituitary function, (see, e.g., Hall et al., 1972). As this tripeptide is readily inactivated in the plasma, there has been a practical incentive to produce analogues that have a greater potency. The rationale for their design is twofold: (a) to increase the activity of the molecule at its effector site, and (b) to slow its rate of inactivation. There is not much room for maneuvers of this kind in such a small molecule, and quite small changes usually result in drastic declines in potency. These changes in activity are usually parallel to changes in binding to receptors rather than effecting alterations in the hormone's intrinsic activity (see Vale et al., 1976; Vale, Rivier, and Brown, 1977). The effects of a N-methyl imidazole group is interesting as [^{7}Me-His^2]TRH ([3-N-Me-His^2]TRH) is about 8 times as potent as TRH (Table 3.6). The precise reason for this effect is not clear, but it appears to be sterically specific, as its isomer [^{7}Me-His^2]TRH is almost devoid of activity. Modification of the first amino acid, such as [N-formyl-Pro^1]TRH, increases resistance to inactivation. It appears that the (pyro) glutamyl residue is involved in the interaction of the peptide with enzyme(s) that inactivate it. These two properties, which increase affinity for the receptor and decrease inactivation, have been combined in the molecule [N-formyl-Pro^1-^{7}Me-His^2]TRH, which has, however, only 40 percent of the activity of the parent molecule. TRH can be coupled to larger molecules, such as dextrans, and still retain activity. This observation is consistent with its action on the cell membrane. The properties of the receptors for the actions of TRH on the release of TSH and prolactin appear to be similar.

2. Gonadotropin-releasing hormones. It was originally proposed that there are two distinct hypophysial-releasing hormones: LH-RH and FSH-RH (see McCann, 1977; Vale et al., 1976; Vale, Rivier, and Brown, 1977). In 1971, Schally and his collaborators described the structure of what was originally thought to be LHRH and found that it also exhibited FSH-RH activity. This decapeptide was called LH/FSH-RH, although currently it is usually called LHRH. Its ability to release pituitary LH is more prominent, but it is still considered likely that it is a hormone with a dual function. It has thus been suggested that its effects may be modulated by different levels of the steroid sex hormones, which change the thresholds for stimulation and determine the appropriate response. Alternatively, two hormones could be present, the FSH-RH-like action of LHRH reflecting the similarities in their structures. This problem of whether there are one or two gonadotropin-releasing hormones has not yet been satisfactorily resolved.

As shown in Table 3.5, LHRH activity is present at high concentrations in the median eminence, in the arcuate nucleus, and anteriorly in the suprachiasmatic preoptic region. It is also found in other regions of the brain, including the organum vasculosum of the lamina terminalis, from which it appears to be secreted into the fluid of the third ventricle.

Release of LHRH can be influenced by electrical stimulation of the hypothalamus and by estrogens. There is both a positive- and a negative-feedback control by estrogens, depending on the circumstances and concentrations of the steroids. As described earlier [Section 3.2E(a)], catechol estrogens, which are formed locally as a result of metabolism of estrogenic hormones, may mediate these responses. Norepiniphrine appears to mediate the release of LHRH, and this effect can be blocked by α-adrenergic inhibitory drugs, reflecting adrenergic nerve mechanisms. Prostaglandins, especially E_2, and GABA can also stimulate secretion of hypothalamic LHRH, but the physiological significance of these effects is not clear (Ojeda, Jameson, McCann, 1977a,b; Ondo, 1974). Specific receptors for LHRH have been identified on cell membranes of the pituitary cells that secrete gonadotropins. The response is associated with increased levels of cyclic AMP, apparently resulting from activation of adenyl cyclase. Theophylline and exogenous cyclic AMP can mimic the effects of LHRH and bring about a release of pituitary LH. The nucleotide is thought to activate a protein kinase which changes membrane permeability resulting in an increased influx of Ca^{2+}. This divalent ion stimulates the extrusion of LH from its secretory granules by a process of exocytosis. These events are summarized in Figure 3.16. Inactivation of LHRH occurs principally in the tissues and only slowly in the plasma.

Hypophysial gonadotropin-releasing hormones have a number of established and potential practical uses. These include the clinical testing of hypophysial and pituitary function and the treatment of infertility. Analogues with antagonistic activity may also have potential use as contraceptives. It is therefore not surprising to observe that over 400 of these structural analogues of LHRH have

been prepared and tested. This hormone has 10 amino acid residues, so that the possibility for change is quite considerable. Increases in activity can depend on changes in the interaction of the hormone with its receptor tissue, but in the instance of LHRH it appears to be mainly related to an ability to resist inactivation. The hormone can be destroyed by cleavage of the bond between positions 6 and 7 and hydrolysis at position 10. Hence, the more potent analogues have substituted amino acids at these sites. The effects of modifications at position 6, where glycine is replaced by various D-amino acids, is shown in Table 3.7. Thus, the presence of D-alanine results in a fourfold increase in activity, but if L-alanine is present it has negligible activity. Aromatic amino acids, such as D-tyrosine and D-typtophan, are even more effective. Thus, [D-Trp^6]LHRH is 36 times more potent than the parent hormone. Changes at the C-terminal glycineamide, position 10, can also enhance activity. For instance, the nonapeptide des-Gly^{10}-NH$_2$-LHRH ethylamide is five times as active as LHRH (combinations of such changes, as in [D-Leu^6,des-Gly^{10}-NH$_2$]LHRH ethylamide, and [D-Ser(Bu^t)]^{6}des-Gly^{10}-NH$_2$-LHRH ethylamide results in compounds of even greater activity – an increase of as much as 50 times or even more, depending on the assay used.

Inhibitors can be produced by substitutions at position 2 (Table 3.8). This reduces the intrinsic activity, but the molecule can still interact with the receptor so that the antagonism is competitive. Aromatic L-amino acids favor agonist activity at this position, but substitution by the D-form produces an antagonist. Thus, [D-Phe^2]LHRH has little intrinsic LHRH activity but is an effective antagonist. The same applies to [D-Ala^2]LHRH. Removal of the histidine as in des-His^2-LHRH also results in an antagonist. As with agonists, substitution of D-amino acids at position 6 or 7 enhances the activity of the antagonists also apparently by decreasing the rate of their inactivation. Analogues of LHRH have also been prepared which are conjugated to larger molecules such as dextran and polyglutamate, for instance E-poly[Glu,D-Lys^6]LHRH (see Amoss, Monahon, and Verlander, 1974). This maneuver can enhance the molecule's activity, probably by decreasing its inactivation and excretion by the kidney. It is unknown if dissociation of the molecule must occur before the peptide can act.

It is interesting that despite the synthesis of a large number of analogues of LHRH, a separation of its actions on release of LH and FSH has not been observed.

3. Corticotropin-releasing factor (CRF) or hormone (CRH). The first hypothalamic-regulating hormone to be identified was CRF (or

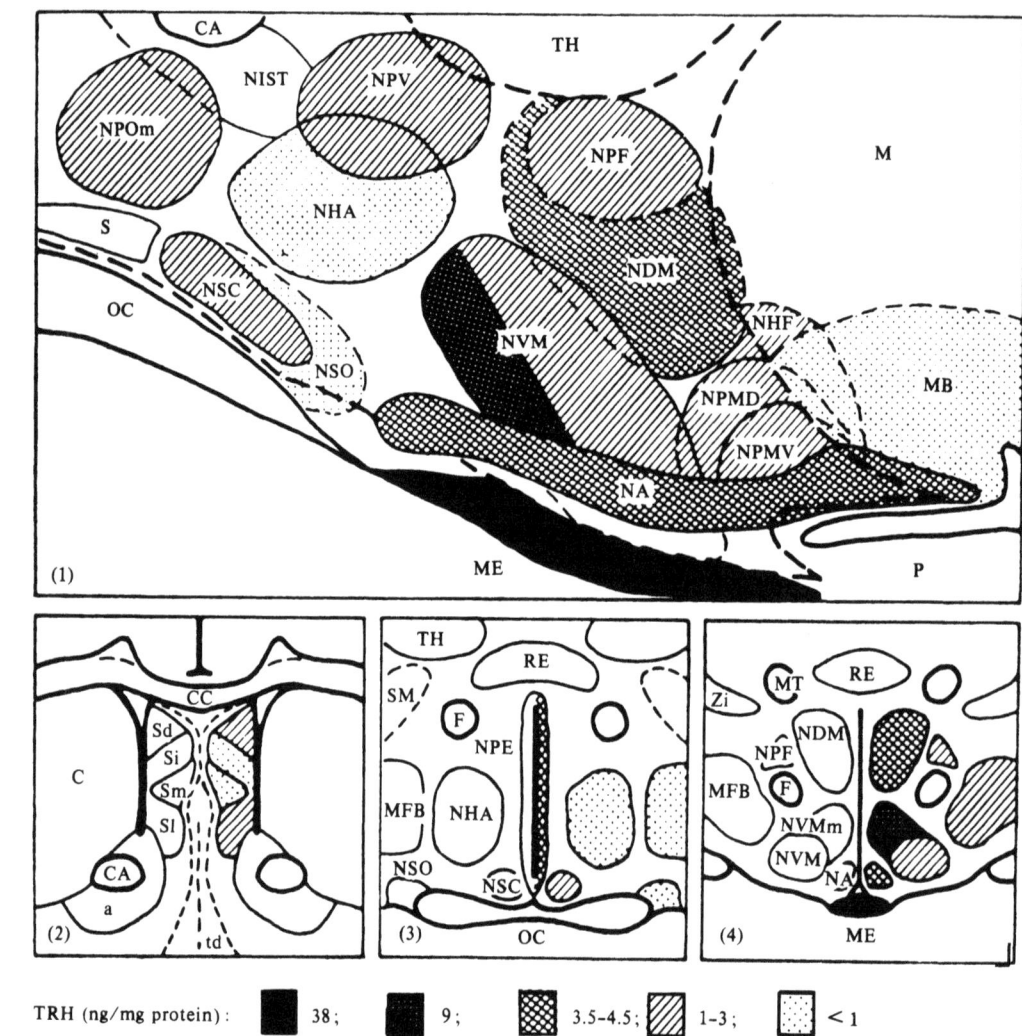

Fig. 3.15. (a) Quantitative distribution (ng/mg/protein) of thyrotropin-releasing hormone, TRH, in the hypothalamus of the rat. (From Brownstein et al., 1974. Copyright 1974 by the American Association of the Advancement of Science.) (1) is a parasagittal section; parts (2–4) are frontal sections through the anterior hypothalamus, the tuberal region, and the premamillary region, respectively. (b) Relative distribution of somatostatin in the region of the hypothalamus of the rat. The sections are the same as those described above. (From Brownstein et al., 1975.) Key: CA, anterior commissure; F, fornix; M, mesencephalon; MB, mamillary body; ME, median eminence; MFB, medial fore-brain bundle; MT, mamillothalamic tract; NA, arcuate nucleus; NDM, dorsomedial nucleus; NHA, anterior hypothalamic nucleus; NHP, posterior hypothalamic nucleus; NIST, nucleus interstialis striae terminalis; NPE, periventricular nucleus; NPF, perifornical nucleus; NPL, prelateral mamillary nucleus; NPMD, dorsal premamillary nucleus; NPMV, ventral premamillary nucleus; NPOm, medial preoptic nucleus; NPV, paraventricular nucleus; NSC, suprachiasmatic nucleus; NSO, supraoptic nucleus; NVM, ventromedial nucleus; OC, optic chiasm; P, pituitary; RE, nucleus reuniens thalami; S, preoptic suprachiasmatic nucleus; SM, stria medullaris; TH, thalamus; Zi, zona incerta.

CRH). It is probably a peptide, but its structure appears to be so fragile that for more than 20 years it has defied numerous attempts aimed at its purification and chemical characterization (see Seelig and Sayers, 1977; Jones, Gillham, and Hillhouse, 1977). It is present in the hypothalamus. It is in highest concentration in the median eminence and extends into the posterior lobe of the pituitary gland (Yasuda et al., 1977). The neurohypophysial hormone vasopressin (ADH) has CRF-like activity, and it was once suspected that the two substances may be synonymous, but this possibility has now

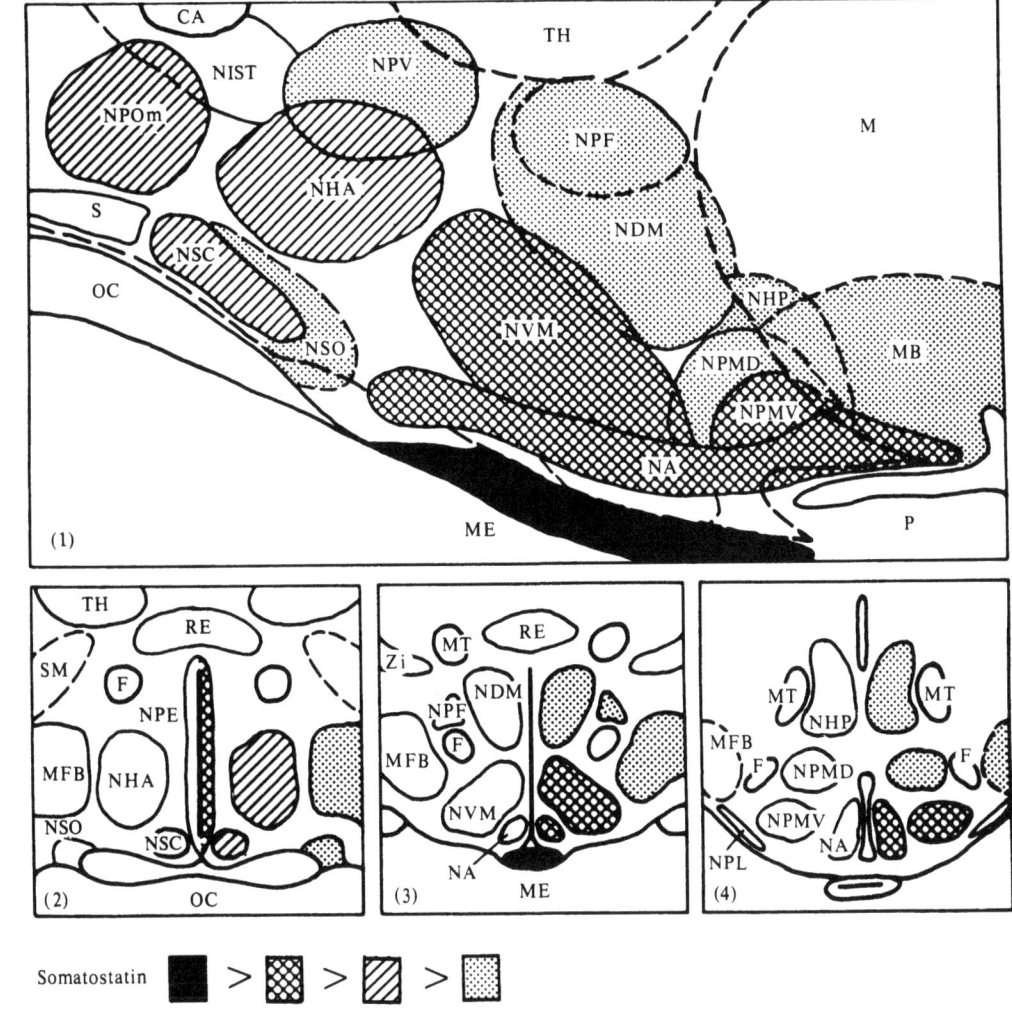

(b)

Somatostatin ■ > ▨ > ▧ > ▤

Table 3.6. *Biological activity (release of thyrotropin) of TRH analogues*

<Glu-His-Pro-NH$_2$	100%	TRH
N-Formyl-Pro-His-Pro-NH$_2$	8%	Resistant to plasma degradation
<Glu-$^\tau$Me-His-Pro-NH$_2$	800%	
N-Formyl-Pro-$^\tau$Me-His-Pro-NH$_2$	40%	
<Glu-His-Pro-NH-(CH$_2$)$_6$-NH$_2$	15%	Free amino group for coupling
<Glu-$^\tau$Me-His-Pro-NH-(CH$_2$)$_6$-NH$_2$	45%	Free amino group for coupling
<Glu-$^\tau$Me-His-Pro-NH-(CH$_2$)$_6$-NH$-\overset{\displaystyle O}{\underset{\displaystyle }{C}}-$(CH$_2$)$_2$-	20%	For nonhistidine iodination

Source: Based on Vale et al. 1976.

(a)

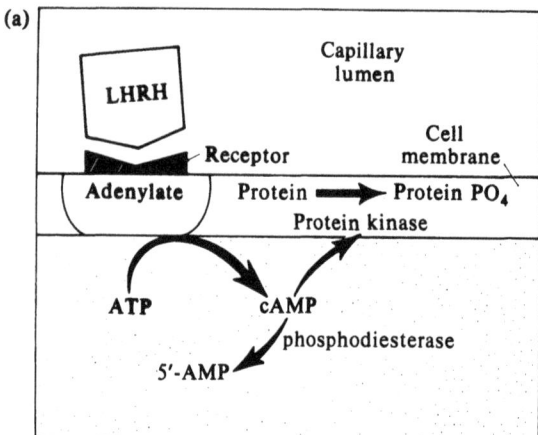

(b)

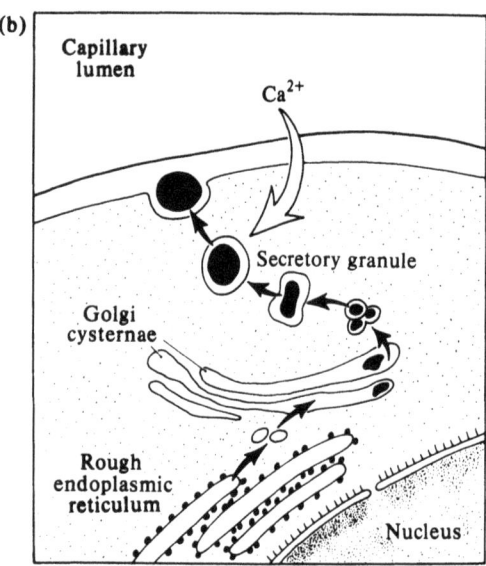

Fig. 3.16. Mechanism of action of LHRH in promoting the synthesis and release of luteinizing hormone (LH). (a) Upon interaction with the receptor in the plasma membrane of the gonadotroph cell in the pars distalis, the LHRH activates adenyl cyclase so that cyclic AMP (cAMP) is formed which converts a protein kinase to its functioning form, resulting in the phosphorylation of a membrane protein. (b) Ca^{2+} then enters the cell at an increased rate, and this is involved in the induction of the exocytotic extrusion of the LH from the secretory granules (in which it is stored). The LH is formed in the rough endoplasmic reticulum and is probably packaged into the secretory granules in the Golgi apparatus. Recent evidence (Nakano et al., 1978) suggests that cyclic GMP may also be involved in the response. (From McCann, 1977. Reprinted by permission from *The New England Journal of Medicine 296*, 801.)

been excluded although they may be closely related. In the absence of the preparation of authentic CRF, vasopressin is, however, used clinically to test hypophysial and pituitary corticotropic hormone function. It is interesting that vasopressin has been identified in the long hypophysial portal vessels of monkeys (Zimmermann, 1977). A nonhypothalamic corticotropin-releasing factor has been identified in peripheral blood (Brodish, 1977) which appears to differ from hypothalamic CRF and may originate in the liver and cerebral cortex. Jones, Gillham, and Hillhouse (1977) identified two hypophysial peptides with CRF activity, one with a molecular weight of 1300, the other of 2500.

The ability of vasopressin to release corticotropin has been compared, *in vitro*, to that of a crude extract of the median eminence tissue (Gillies, van Wimersma Greidanus and Lowry, 1978). The dose–response curves for the two were different, confirming the individual nature of each. Radioimmunoassay has indicated that about 30 percent of the corticotropin-releasing activity in the tissue extract was due to vasopressin. However, it was especially interesting to observe that vasopressin antiserum completely abolished the activity of such extracts. These observations suggested that while the vasopressin and CRF-like material are not identical, they apparently bear some remarkable similarities to each other.

The release of CRF from the hypothalamus *in vitro* is increased by acetylcholine, 5-hydroxytryptamine, and angiotensin II (Jones, Gillham, and Hillhouse, 1977). Norepinephrine, GABA, and corticosterone antagonize these effects. A model showing the possible relationships of these transmitters in the mechanism controlling the release of CRF is shown in Figure 3.17.

Preparations of hypothalamic CRF may not only increase the release of corticotropin but also that of β-lipotropin and β-endorphin. These effects were shown *in vitro* in a cultured preparation of rat adenohypophysial cells (Vale et al., 1978). It is thus possible that this hypothalamic substance may coordinate the release of several pituitary polypeptides.

4. Prolactin release-inhibiting factor (PRIF) and prolactin-releasing factor (PRF). The release of prolactin is under predominantly inhibitory control, so that if the pituitary stalk is cut or if the gland is transplanted elsewhere in the body, there is an increased release of the hormone. Extracts of the hypothalamus have been shown to inhibit the secretion of prolactin but their chemical nature is uncertain. It is considered by some (see Vale, Rivier, and Brown, 1977) to be a peptide like other hypophysiotropic hormones. Dopamine, however, inhibits the secretion of pro-

Table 3.7. *Relative potencies (determined in vitro) of analogues of LHRH in which the glycine at position 6 has been substituted*

1 2 3 4 5 6 7 8 9 10	
LHRH: <Glu - His - Trp - Ser - Tyr - Gly - Leu - Arg - Pro - Gly-NH$_2$	
Analogue	Relative potency
LHRH	1.0
[D-Ala^6]LHRH	3.7 (2.0–6.5)
[D-Leu^6]LHRH	3.2 (2.4–4.1)
[D-Lys^6]LHRH	3.8 (2.8–5.2)
[ε-Lauryl-D-Lys^6]LHRH	1.6 (1.3–2.1)
[ε-Dextran-D-Lys^6]LHRH	0.15 (0.10–0.21)
[D-Arg^6]LHRH	3.9 (2.9–5.1)
[D-Met^6]LHRH	1.9 (1.1–3.3)
[D-Tyr^6]LHRH	14.0 (8.6–22.0)
[D-Trp^6]LHRH	36 (26–50)

Source: Vale et al., 1976.

lactin and there are high concentrations of this amine in the hypothalamus and in the blood in the hypophysial pituitary stalk (Gibbs and Neill, 1978). Dopamine can act directly on the pituitary to inhibit release of prolactin, but it is also possible that it may act indirectly and mediate a release of a PRIF from the hypothalamus. At present this question has not been resolved.

A factor that can stimulate the release of prolactin has been identified in the hypothalamus, and it has been called PRF. However, TRH also exhibits this effect and is present in extracts of the hypothalamus. Histamine is another substance that can stimulate the release of prolactin and is present in the hypothalamus. In addition, endorphins can act, like morphine, to stimulate prolactin release *in vivo*. This effect has been demonstrated in response to β-endorphin (Rivier et al., 1977).

The precise nature of the "factors" controlling the release of prolactin remains to be unravelled.

Table 3.8. *Antagonist and agonist activities of a series of analogues of LHRH*

Compound	Antagonist activity, molar ratio analogue/LHRH: 50% inhibition of effect of LHRH (= MR$_{50}$)	Agonist activity as % LHRH
LHRH	—	100
[Des-His^2]LHRH	3,000	<0.001
[Des-His^2,Des-Gly-NH$_2$^{10}]LHRH ethylamide	>3,000	0.003
[Des-His^2,D-Leu^6]LHRH	500	<0.001
[Des-His^2,D-Phe^6]LHRH	60	<0.00001
[Des-His^2,D-Lys^6]LHRH	>30,000	0.003
[Des-His^2,Des-Gly-NH$_2$^{10}]LHRH pentafluoropropylamide	>30,000	0.1
[D-Phe^2]LHRH	1,000	<0.00001
[D-Phe^2,D-Leu^6]LHRH	150	<0.0001
[D-Phe^2,D-Phe^6]LHRH	25	0.003
[D-Phe^2,D-Glu^6]LHRH	2,000	<0.0001
[D-Phe^2,D-Leu^6,Des-Gly-NH$_2$^{10}]LHRH ethylamide	>30,000	0.001
[Des-His^2,Des-Gly-NH$_2$^{10}]LHRH propylamide	6,500	<0.0001
[Des-His^2,Des-Gly-NH$_2$^{10}]LHRH trifluoroethylamide	10,000	<0.0001
[D-Phe^2,D-Phe^6,D-Phe^7]LHRH	400	<0.0001
[D-Phe^2,Phe^5,D-Phe^6]LHRH	75	<0.0001
[D-Phe^2,Phe^3,D-Phe^6]LHRH	100	—

Source: Ferland et al., 1976.

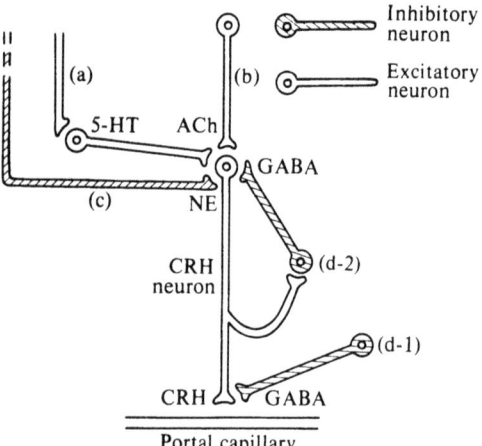

Fig. 3.17. Hypothalamic–neural model proposed by Jones, Hillhouse, and Burden, (1976) to account for observations on the effects of different neurotransmitter substances on the release of corticotropin-releasing hormone (CRH) from the hypothalamus. (a) 5-Hydroxytryptamine (5-HT) pathway stimulates CRH release via a cholinergic interneuron. (b) Cholinergic neurons stimulate CRH release synapsing at the dendritic or somal level (nicotinic receptors). (c) Noradrenergic neuroinhibitory pathway synapsing at the dendritic or somal level. (d) γ-Aminobutyric acid (GABA-containing neurons inhibit CRH release in one of two possible ways: (1) acting presynaptically at the terminal bouton of the CRH neuron; or (2) a recurrent collateral from the CRH neuron excites a GABA interneuron, which synapses with the CRH cell and forms a recurrent collateral feedback loop. ACh, acetylcholine; NE, norepinephrine. (From Jones, Hillhouse, and Burden, 1976.)

5. Growth hormone release-inhibiting hormone (GH-RIH) or somatostatin.

The hypothalamic factor that can inhibit the release of pituitary growth hormone was chemically identified by Guillemin and his collaborators in 1972 (see Guillemin, 1977, 1978a,b; Guillemin and Gerich, 1976; Vale et al., 1975). It is a tetradecapeptide that contains a cystine ring structure. This hormone is present in high concentrations in the median eminence (Figure 3.15). The neurosecretory neurons that form it are present in the periventricular nucleus (Pelletier, Dubé, and Puviani, 1977). It has also been found in the posterior lobe of the pituitary but is absent from the anterior lobe (Patel, Zingg, and Dreifuss, 1977). While this peptide was originally identified in the hypothalamus and characterized by its ability to inhibit the release of growth hormone, it has since been found to be ubiquitous in the body and has been identified in other parts of the brain, the islets of Langerhans, and the gut. Its actions are even more general and include an ability to inhibit the release of insulin, glucagon,

thyrotropic hormone (but not prolactin or gonadotropins), calcitonin, and parathyroid hormone. Somatostatin also reduces the release of gastrin and gastric acid secretion in the stomach, and secretin and exocrine secretions from the pancreas (Boden et al., 1975), and it can even prevent the release of acetylcholine from the myoenteric plexus (Guillemin, 1976). Antisera to somatostatin have been shown to antagonize such responses in the presence of endogenous somatostatin, confirming that the peptide probably has diverse physiological roles.

Release of growth hormone in response to such stimuli as sleep, exercise, insulin hypoglycemia, L-dopa, morphine, and electrical stimulation is inhibited by somatostatin. Growth hormone is also released in response to cyclic AMP and theophylline, which may reflect the mechanism that controls its normal release, and this response is also blocked by somatostatin. Its action is not inhibited by blockers of RNA or protein synthesis. Somatostatin specifically prevents release, not synthesis, of growth hormone. Its action appears to be at a level that is closely associated with the actual hormone release mechanism. In the instance of its blockade of secretion of insulin and glucagon, its effect can be prevented by the Ca ionophore A23187 (Fujimoto and Ensinck, 1976), which is consistent with the suggestion that it may reduce the entry of calcium into excitable cells. This divalent ion is necessary for many coupling reactions, including the release of many hormones.

Initially, there was considerable excitement about the possible therapeutic uses of somatostatin, such as in the treatment of acromegaly and diabetes mellitus. This euphoria has, however, been somewhat dissipated with the realization that its ubiquitous effects limit is selectivity and possible use for treatment of specific conditions. In addition, somatostatin is rapidly destroyed in the body, having a half-life of only about 4 minutes. The natural hormone, prepared by chemical synthesis, in order to be effective, must therefore be given by i.v. infusion. Attempts to prolong its action and allow s.c. or i.m. administration have involved its admixture with vehicles such as arachis oil, gelatin, and protamine zinc, but these preparations have not been very satisfactory. Such problems have prompted the search for structural analogues with prolonged and selective activity with a view to utilizing their action in clinical situations.

Numerous analogues of somatostatin have been prepared which provide some insight into the structure–activity relationship of the molecule (Table 3.9). The hormone can withstand a considerable number of deletions or additions and still retain biological activity. The size of the molecule is not critical. Thus, a lengthening of the N-terminal side chain, as in Gly-Gly-Tyr-Ala[1]-somatostatin, or

Table 3.9. *Relative agonist potencies of some analogues of somatostatin*

```
         S─────────────────────────────────────────S
         |                                          |
H-Ala-Gly-Cys-Lys-Asn-Phe-Phe-Trp-Lys-Thr-Phe-Thr-Ser-Cys-OH
   1   2   3   4   5   6   7   8   9  10  11  12  13  14
                 Somatostatin
```

a. Inhibition in vivo *of release of growth hormone in rats*

	Relative potency (% of somatostatin) 15 minutes after s.c. injection
Somatostatin	100
Des-Asn⁵-Somatostatin	1–20
Des-Ala¹,Gly²,Asn⁵-Somatostatin	5–15
Des-Ala¹,Gly²,Lys⁴-Somatostatin	1–15
Des-Ser¹³-Somatostatin	7–50
(Gly-Gly-Tyr-Ala¹)Somatostatin	5–40
[Nle⁹]Somatostatin	<0.5
[D-Lys⁹]Somatostatin	<0.5

b. Different spectra of the activities of analogues of somatostatin in three types of assay preparations

	% Potency based on inhibition of:		
	GH *in vitro*	Insulin *in vivo*	Glucagon *in vivo*
Somatostatin	100	100	100
[Ala²]Somatostatin	186	135	279
[Ala³]H₂Somatostatin	<0.5	<10	<10
[Ala⁵]Somatostatin	130	132	176
[Ala⁶]Somatostatin	1	<10	<10
[Ala⁸]Somatostatin	0.5	<10	<10
[D-Trp⁸]Somatostatin	848	821	639
[Ala¹⁰]Somatostatin	25	14	<10
[D-Cys¹⁴]Somatostatin	271	20	310
[D-Trp⁸-D-Cys¹⁴]Somatostatin	647	130	950

Source: (a) Sarantakis et al., 1976; (b) Vale et al., 1976.

deletion of three amino acids such as in des-Ala¹,Gly²,Asn⁵-somatostatin reduces, but does not abolish activity. The latter compound is of special interest because, although it inhibits the release of growth hormone and insulin, it has no effect on the secretion of glucagon. It thus has a selective action. Such studies indicate that the receptors at different sites do not necessarily have the same structural requirements. The oxidized cyclic form of somatostatin appears to be important for activity, as when the cystine ring is omitted, as in [Ala³]somatostatin or, if the cystine residue is blocked, so that it cannot form a ring, as in [S-MeCys³,¹⁴]somatostatin, then most activity is lost. The effects of substitution of various amino acids with alanine is shown in Table 3.9. Positions 5 and 10 appear to be somewhat less critical than positions 6 and 8. Position 9 is also very important (Ta-

ble 3.9), as a substitution of D- for L-lysine or the omission of an ε-amino group virtually abolishes activity.

It is interesting that substitution of D- for L-tryptophan at position 8 increases activity about eightfold. It is thought that somatostatin is destroyed by cleavage of the peptide bond at position 8–9, and it is possible that, as in LHRH, the substitution of the D-amino acid slows this reaction.

The analogue [D-Cys¹⁴]somatostatin has an enhanced ability to reduce the release of growth hormone and glucagon, but its effect on insulin is much less (Brown, Rivier, and Vale, 1977; Meyers et al., 1977). It thus has a selective effect which could be useful therapeutically. Additional substitution at position 8 to produce [D-Trp⁸,D-Cys¹⁴] somatostatin results in an analogue with a similar selective activity (glucagon to insulin ratio of 22 : 1

and GH to insulin of 100 : 1), but it is much more potent and so may be of greater potential therapeutic use. The relationship of structure to activity with respect to its inhibitory action on gastric acid secretion is described in Section 10.4.

A three-dimensional structure for somatostatin has been proposed (Holladay and Puett, 1976; Holladay, Rivier, and Puett, 1977). It appears to have an elliptical shape and contains a hydrophilic domain at one end and a hydrophobic one at the other. A study of the structure–activity relationships of a number of analogues of somatostatin containing additional covalent bridges suggests that changes in its conformation may occur in the region of its receptors (Veber et al., 1978). Substitution of the phenylalanine residues at positions 6 and 11 indicated that these amino acids are not directly important for the hormone's interaction with its receptor, but they may stabilize the molecule by providing hydrophobic bonding.

6. Growth hormone-releasing hormone (GHRH).
The presence of a hypothalamic growth hormone-releasing hormone has been inferred from physiological experiments and the effects of tissue extracts on the release of growth hormone. The activity appears to be associated with peptides, but it has not yet been chemically identified (see Vale, Rivier, and Brown, 1977). A decapeptide candidate with the structure H-Val-His-Leu-Ser-Ala-Glu-Glu-Lys-Glu-Ala-OH was proposed but, although it stimulated release of growth hormone as determined by a bioassay, it was found to be ineffective in changing secretion of immunoreactive GH. Growth hormone can be released by β-endorphin (Rivier et al., 1977; Dupont et al., 1977), Met-enkephalin (Cusan et al., 1977), as well as by catecholamines, including dopamine, so that there are a number of possible candidates for GHRH.

A search involving the biological scanning of 17 polypeptide "peaks" (separated on Sephadex G-25 columns) in extracts from 6000 bovine hypothalami has been described by Nair et al. (1978). Two fractions were found to contain activity that could release growth hormone, *in vivo*, in rats. The active molecule has not been fully described but appears to be a polypeptide which, like LHRH and TRH, has a pyroglutamyl residue at the N-terminus. It was closely associated in the purification procedure with somatostatin, and it was necessary to remove the latter peptide before the activity of the putative GHRH could be identified.

7. Melanocyte-stimulating hormone release-inhibiting hormone (MSH-RIH) and releasing hormone (MSH-RH).
The role of hypothalamic factors in controlling the release of MSH is very controversial (McCann et al., 1974; Hadley, Hruby, and Bower, 1975; Vale, Rivier, and Brown, 1977). The experimental observations are based on animal experiments, which is apt, as this hormone does not appear to be naturally present in man. The release of MSH from the pars intermedia of animals is predominantly under inhibitory control, so that cutting the pituitary stalk or transplanting the glandular tissue to other sites in the body results in increased secretion. The pars intermedia has a nerve supply, but its vascularization is poor and blood vessels that go directly to the tissue may even be absent.

Dopamine, epinephrine, and norepinephrine can inhibit release of MSH under either *in vivo* or *in vitro* conditions (see Bower, Hadley, and Hruby, 1974). These effects are α-adrenergic ones. In contrast, β-adrenergic stimuli can increase the release of MSH. The precise nature of the catecholamines that mediate these effects are not clear, but the pars intermedia has adrenergic nerves, while the hypothalamus is known to be a rich source of dopamine.

It has also been suggested that the hypothalamus may contain peptides that influence the release of MSH. A tripeptide, H-Pro-Leu-Gly-NH_2, has been identified which may inhibit release of MSH and be the MSH-RIH (Celis, Taleisnik, and Walter, 1971; Nair, Kastin, and Schally, 1971). Its structure is identical to the side chain of oxytocin, from which it may be formed under the influence of an enzyme that has been identified in the hypothalamus (Walter, Griffiths, and Hooper, 1973). The effects of this peptide have not been confirmed by all, and it has been suggested that "MSH-RIH" activity in hypothalamic extracts may be an artifact due to the enzymic degradation of the MSH, which is released in the assay system (see Hadley, Hruby, and Bower, 1975). Other fragments of oxytocin and vasopressin have been reported to exhibit MSH-RIH and MSH-RH activities, but these have not been identified in the hypothalamus, nor have their effects been consistently demonstrated in different laboratories.

The role of peptides in controlling the release of MSH is therefore equivocal. If such endogenous substances do act, they could enter the pars intermedia via a blood supply, which seems unlikely in view of its generally poor vascularization, or be secreted by peptidergic neurons, possibly in response to adrenergic stimuli.

8. Dysfunction of the hypothalamus.
The physiological importance of discrete regions of the hypothalamus are well known, mainly due to experimentally induced lesions (Table 3.10) but also as a result of human pathological experimentations. There are many effects of such damage, but

Table 3.10. *Disorders that can be induced experimentally by the production of lesions in parts of the hypothalamus and pituitary gland*

	Syndrome	Histological localization	Nature of manifestations[a]
Anterior hypothalamus	1. Diabetes insipidus	Supraoptic and paraventricular nuclei	1
	2. Failure of temperature regulation, hyperthermia, thyroid regulation	Region of the anterior hypothalamic nucleus	1 and 2
	3. Hemorrhagic pulmonary edema	Preoptic medioventricular area	2
	4. Insomnia, hyperactivity, excitability, manic states	? Suprachiasmatic area	2
Tuber cinereum	1. Obesity (hyperphagia)	Region of the ventromedial hypothalamic nucleus ("satiety center")	2
	2. Genital dystrophy (hypogonadism)	? Ventral nucleus of the tuber, infundibular nucleus	1
	3. Pathological inclination to outbursts of temper	Region of the ventromedial hypothalamic nucleus, and somewhat rostral to this	2
Posterior hypothalamus	1. Hypersomnia, coma	Laterodorsal zone	1
	2. Poikilothermia	Laterodorsal zone	1
	3. Anorexia	Lateral hypothalamus adjacent to the cerebral peduncle ("feeding center")	1
	4. Pubertas praecox (precocious puberty)	Infundibular–mammillary region	2?
Median eminence	1. Pituitary insufficiency		1
	2. Galactorrhea		2
Pituitary stalk	1. Galactorrhea		2
	2. Diabetes insipidus		1

[a] 1, failure; 2, stimulation.
Source: Based on Labhart, 1976.

the endocrine disturbances include diabetes insipidus, reproductive problems (such as infertility, hypogonadism, precocious puberty, and galactorrhea), inadequate or abnormal growth (dwarfism, giantism, and acromegaly), and disturbances in the regulation of thyroid and adrenocortical functions (see Jenkins, 1972; Besser, 1974a,b). Such effects are predictable on the basis of our knowledge of the hypothalamic mechanisms that control the function of the pituitary gland. However, their precise diagnosis was, until recently, very difficult, and their specific treatment is usually not feasible at this time.

Many endocrine diseases may be related to disorders of pituitary function which may be primary or secondary to a malfunction of the hypothalamus. The availability of various preparations of hypophysial-regulating hormones and drugs that influence their release has provided convenient ways to diagnose such conditions accurately. By appropriate comparison of responses (such as peripheral hormone or metabolite levels) to hypothalamic and pituitary hormone preparations, or drugs that influence the release of these hormones, it is often possible to decide at which site

the problem lies: the hypothalamus, pituitary, or end organ. These uses of such drugs and hormones will be described in subsequent sections that deal with specific endocrine dysfunctions.

Hypothalamic disorders can have a variety of causes and may be quite generalized or specifically related to certain functions. Basically, they appear to result from damage to the hypothalamus, either directly or to its afferent or efferent connections in other parts of the brain, such as the limbic system and brain stem, or afferent outputs in the hypophysial-portal system and pituitary stalk. Specific lesions frequently cannot be identified, even following postmortem examination. Certain disorders, such as some forms of diabetes insipidus, are familial, reflecting genetic problems. In other instances, physical damage may occur due to the presence of tumors, hemorrhage, surgery, head injuries, and birth trauma. Some diseases that result in inflammation (meningitis) can also damage the hypothalamus. A number of drugs, such as certain tranquilizers, antihypertensives, corticosteroids, and sex steroids (Table 3.11), can exhibit undesirable side effects that reflect action on the hypothalamus. Information on the effects of

Table 3.11. *Some drugs that exhibit side effects that are related to their actions on the hypothalamus*

Tranquilizers		
Phenothiazines	Inhibit release gonadotropins	Infertility
(e.g., chlorpromazine)	Increase release prolactin	Galactorrhea
Antihypertensive drugs		
Reserpine	Promote prolactin release	? Breast cancer
		Galactorrhea
α-Methyldopa	Promote prolactin release	Galactorrhea
Clonidine	Increase release growth hormone	
Ethanol	Inhibit release ADH	Diuresis
Steroids		
Corticosteroids	Inhibit ACTH	Hypoadrenocortisolism
Progestins	Inhibit gonadotropins	Infertility
Estrogens	Inhibit gonadotropins	Infertility

steroids on the embryonic development of hypothalamic functions, such as rhythms, suggests that congenital problems due to teratogenic effects of drugs may also occur. Many such disorders can be of a permanent nature, but some are reversible. Thus, the hypophysial–portal system has remarkable regenerative powers and the correction of inflammatory conditions and the removal of tumors is sometimes followed by a restoration of function. The effects of drugs are also usually reversible. Such regenerative changes may, however, only occur over a period of many months.

With the identification of the hypophysial regulatory hormones and drugs that can influence their release, the prospect of specific treatment of hypothalamic disorders has approached reality. These will be described later in the sections dealing with each endocrine gland. Many of these uses are, however, only at an experimental level and of an investigative nature. Some drugs, such as the antiestrogen clomiphene, do have a well-established use in the treatment of infertility in women. Bromocriptine is used to treat disorders associated with inappropriate release of prolactin and growth hormone. The best known examples of such drugs are, however, the oral contraceptives, estrogens and progestins, which can block the release of gonadotropins. Replacement therapy using hypothalamic peptides such as LHRH, TRH, and somatostatin does not have an established use. Such peptides need to be administered parenterally, usually i.v., and are readily inactivated in the body. These problems are, however, being solved by the provision of longer-acting and more potent analogues. In many instances, such treatment is not really essential, as secondary replacement of the pituitary or peripheral hormones is more convenient and effective.

F. Growth hormone (somatotropin)

Growth hormone is preeminent among the adenohypophysial hormones as the recognition of its effects on tissue growth at the turn of this century led to the discovery of the endocrine role of the pituitary gland. Despite its early impact on endocrine history and its relatively enormous concentrations in the pituitary, its chemical structure has only been described quite recently. The role of growth hormone in the body is also not yet fully understood, but there are some tantalizing and interesting observations which suggest that it may have a ubiquitous role in influencing the levels of a whole series of "growth factors" which are present in the plasma. Growth hormone is the only hormone whose actions in man cannot be imitated by hormone preparations obtained from animal pituitaries. It is thus only available for clinical study and use in relatively limited quantities. The unravelling of the niceties of the relationship of growth hormone to circulating "growth factors" in man holds the promise of providing a major advance in our understanding of the role of intermediary metabolism in the process of tissue differentiation, growth, and repair.

1. Functions and actions. The clearest effects of growth hormone are probably apparent in disease (see, e.g., Tanner, 1972; McGarry and Beck, 1972; Cheek and Hill, 1974). A deficiency of this hormone in children results in reduced growth and failure to obtain normal expected stature. This effect can be reproduced experimentally following hypophysectomy in young animals, such as rats, puppies, and kittens. The failure to grow is accompanied by several morphological disorders, including inadequate skeletal and muscle growth and an increased deposition of fat. These deficiencies can be corrected in children (or in experimental animals) by the injection of human growth hormone. Conversely, a hypersecretion of growth hormone can promote excessive growth. This effect may result in giantism (or gigantism) if it occurs during childhood. This is a rare disorder, however, and excessive height is not necessarily related to hypersecretion of this pituitary hormone. In adults, excessive secretion of growth hormone can result in acromegaly. As the epiphyses are fused, linear growth is not promoted but prominent skeletal changes occur due to stimulation of cartilaginous development and overgrowth, and thickening of bones in the face, the ribs, vertebrae, and tarsals and carpels. It thus can result in prominent morphological changes in the face, hands, and feet. This disorder is commonly accompanied by many other changes, including menstrual and psychic disorders and hyperglycemia (diabetes mellitus), which appears to reflect the ubiquitous effects of this hormone on intermediary metabolism.

The precise metabolic basis for these widespread general effects of growth hormone are not always easy to understand. Responses to injected growth hormone differ due to the general physiological and, especially, endocrinological conditions, whether or not the hormone preparation is a variety which is natural to that species, if the animal is mature or immature and whether the observations are performed *in vivo* or *in vitro*. The particular response may also differ with the dosage, the presence or absence of other hormones, and the time, following administration, when the particular response is measured.

The action of growth hormone is accompanied by a positive nitrogen, calcium, and phosphorus balance, which is consistent with the types of general responses to the hormone. It exerts a widespread anabolic action on cartilage, bone, muscle, and kidney. The effects of growth hormone on the nervous tissue are controversial, but it has been shown to promote brain growth in fetal rats (Sara et al., 1974). The effects of growth hormone are reflected by an increased rate of protein synthesis and tissue DNA content, due to cell multiplication.

Changes in the body fluids following administration of growth hormone may include hyperglycemia; an increase in free fatty acids and reduced amino acids in the plasma; reduced urinary P, K, and Na excretion; and calciuria. Such responses may, however, differ initially, within the first hour, such as when a hypoglycemia and decreased free fatty acid levels in the plasma may occur.

The nature of the effects of growth hormone on each tissue has been studied *in vivo* and more directly *in vitro*. Growth hormone *in vitro* at high concentrations promotes the uptake of amino acids and protein synthesis in skeletal and cardiac muscle and in liver (Kostyo and Nutting, 1974). This action is reminiscent of that of insulin.

In the skeleton growth hormone promotes the incorporation of sulfur to form chondroitin sulfate. It increases the activity of chondrocytes and osteocytes at the epiphyseal growth plate in bone, an effect that is seen especially in the proximal part of the tibia. The effect on incorporation of sulfate into cartilage cannot be elicited by growth hormone alone but is dependent on the formation of an intermediary peptide which was first called *sulfation factor* but is now called *somatomedin*. Its role in the mechanism of action of growth hormone will be discussed later [Section 3.2F(5)].

Growth hormone may reduce glucose tolerance and have a diabetogenic effect, resulting in hyperglycemia. This response can reflect several actions, including an inhibition of glucose uptake by muscle and fat cells and, with prolonged administration, a failure of insulin secretion. The β-cells of the islets of Langerhans may degenerate. The primary hyperglycemic effect is more apparent when the hormone is administered to hypopituitary patients and following hypophysectomy.

Growth hormone has a lipolytic effect which has been demonstrated in rat adipose tissue. This *in vitro* effect requires very high concentrations of growth hormone and the presence of corticosteroids. Its relationship to the *in vivo* increase in plasma free fatty acids is not clear. It has been suggested (Tanner, 1972) that the latter may be mediated by somatomedin. Growth hormone exhibits some prolactin-like activity which appears to reflect the similarities in the structure of these two hormones.

Many of the physiological effects of growth hormone have been utilized in order to pharmacologically identify and assay growth hormone. Apart from the rat tibia, uptake of amino acids by the rat diaphragm muscle, mobilization of fat from adipose tissue, and the diabetogenic action (as measured by the appearance of glucose in the urine) are utilized. Radioimmunoassays are now more routinely used for many types of investigations, including diagnostic assays on human plasma.

2. Structure. Growth hormone (see Wilhelmi, 1974) is a polypeptide which in man contains 191 amino acid residues. C. H. Li has been preeminent in elucidating the chemical structure of this hormone in animals and man (Li and Dixon, 1971) The peptide chain of GH is interconnected at two points by disulfide bridges formed by half-cystine residues at positions 53 and 165, and 183 and 189 (Figure 3.18). The molecular weight of human

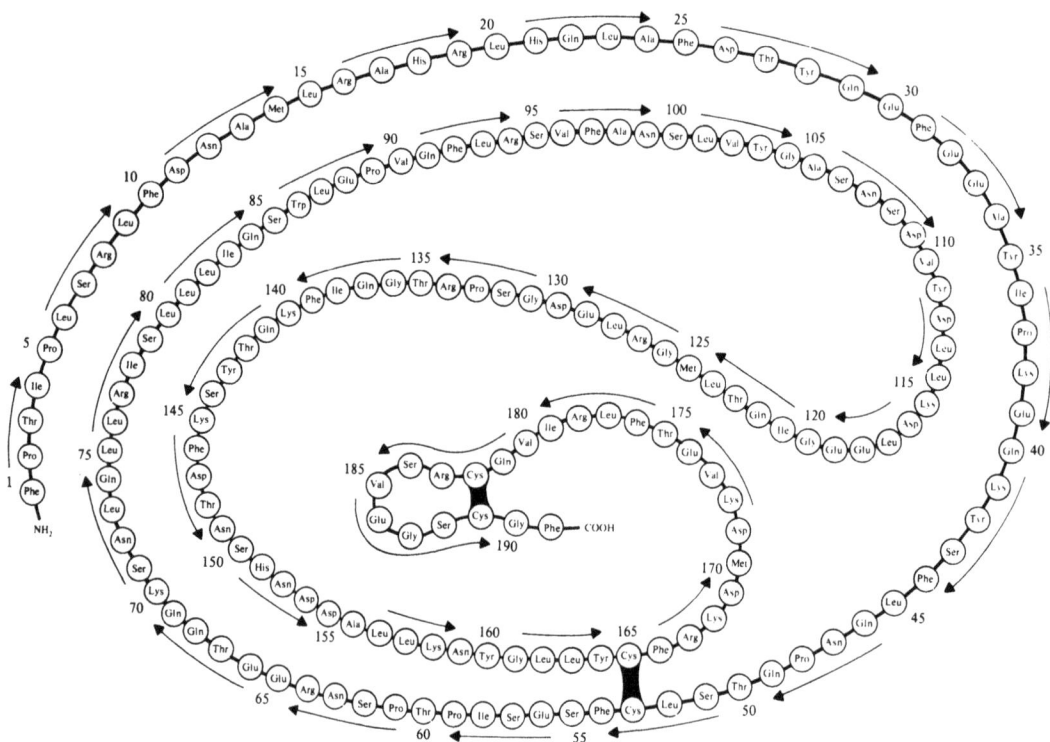

Fig. 3.18. Amino acid sequence of human growth hormone (hGH). Note the interchain disulfide bridges linking amino acids 53 and 165, and 182 and 189. (From Li and Gráf, 1974.)

growth hormone (hGH) is about 21,500. Growth hormone is present in animals, but this is not identical to the hormone in man, although it shows considerable immunochemical similarity (see, e.g., Hayashida, Farmer, and Papkoff, 1975). The differences, however, result in changes in biological activities, so that although growth hormone from one species is often effective in another, they are not equally active. It is perhaps unfortunate that although hGH is effective in many animals, the converse is not so. This situation appears to be unique and has the corollary that animal sources cannot supply effective hormones for clinical use in man. As hGH has not been synthesized in commercial amounts, its only source is the pituitaries from human cadavers. The GH content of the human pituitary is very high and makes up about 1 percent of its weight, or about 5 mg/gland.

Several forms of growth hormone have been isolated from pituitaries and identified in human plasma. These appear to be dimers or even trimers of the hormone which are linked by interpolypeptide chain disulfide bonds. Such forms are called "big" growth hormone, as opposed to the monomeric "little" growth hormone (see, e.g., Gorden et al., 1973; Benveniste et al., 1975). Exogenous monomeric hGH, when injected *in vivo*, is

converted into the larger variety, which suggests that the latter is not a prohormone, although both the big and little forms can apparently be released from the rat pituitary *in vitro* (Stachura and Frohman, 1975). Big growth hormone has a decreased potency and does not combine as readily with receptors for GH (Gorden et al., 1973). Partial dissociation into the more active monomeric form can be promoted chemically *in vitro* (Benveniste et al., 1975), but it is not clear to what extent this can occur *in vivo*.

The immunological and biological activities of growth hormone are not dependent on the molecule being intact. Fragments have been shown to exhibit activity, and in some instances they are even more effective than is the parent hormone (see Li and Gráf, 1974; Singh et al., 1974; Reagan et al., 1975a,b; Li, 1975; Reagan et al., 1978). Such fragments are produced as a result of incubation of the hormone with various proteinases, especially the human blood fibrinolytic enzyme, plasmin. This manipulation has been combined with chemical reduction and carbamidomethylation of the sulfurs in the half-cystine residues, thus breaking the disulfide links between parts of the polypeptide chain.

The types of fragments that are formed are

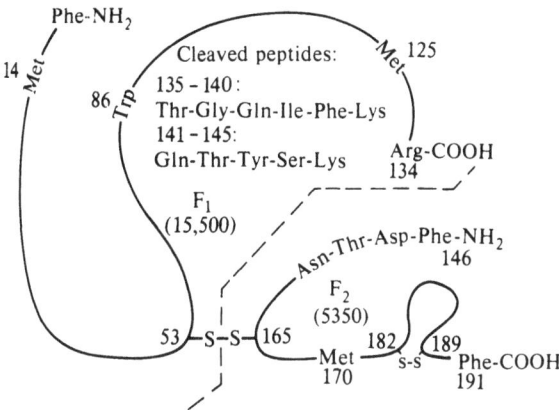

Fig. 3.19. Schematic representation of the molecule of human growth hormone, showing the types of fragments (F) which are formed upon treatment with proteolytic enzymes and the reduction of the disulfide bridges 53–165 and 182–189. (From Singh et al., 1974.)

shown in Figure 3.19. Removal of sections of the amino acids lying between positions 135 and 145 actually results in an increase in activity based on the rat tibia growth test. When the hexapeptide 135–140 was eliminated, activity was similar to the intact molecule. Such changes increase the negative net charge on the molecule. Smaller fragments, made by reducing the disulfide bridges, also retain some activity, including the amino acid sequence from the N-terminal of positions 1–134, 15–125, and 141–191. For optimal activity, how-

ever, the 1–134 fragment needs to be associated with the 141–191 fragment of hGH (Reagan et al., 1978). The smallest fragment to retain activity in man was isolated from tryptic digests of bovine growth hormone (Yamasaki et al., 1972; Yamasaki, Shimanaka, and Sonenberg, 1975). This contains 39 amino acids and has many similarities to positions 95–133 in human growth hormone and also has an analogous amino acid sequence in human placental lactogen (Figure 3.20). It is considered that this section of the molecule may contain a "common active site" and may form the basis for the preparation of therapeutical preparations that are useful in man. A fragment with comparable activity (2 to 10 percent that of the intact hormone) has been prepared from hGH (Reagan et al., 1978). Such fragments may also retain immunological activity. (Some differences in the results of the bioassays have been reported from various laboratories and the changes in potency are not identical in all of the apparently analogous types of preparations.) The 96–133 fragment of bovine growth hormone has also been shown to exhibit *direct effects, in vitro*, in stimulating DNA synthesis and the uptake of sulfate in rat cartilage preparations (Liberti and Miller, 1978). It could also bind to insulin receptors and so display a somatomedin-like activity. Such fragments of growth hormone could be formed during the metabolism of the hormone *in vivo* and may be the "active" forms of the hormone.

The 1-134 amino acid fragment of hGH retains the same intrinsic prolactin-like activity (assayed on mouse mammary gland) as the parent hormone

bGH: Val-Phe-Thr-Asn-Ser -Leu-Val-Phe-Gly-Thr-Ser-Asp-   -Arg-Val- Tyr -Glu-Lys-Leu-Lys-
    96   100   110

oGH: Val-Phe-Thr-Asp-Ser -Leu-Val-Phe-Gly-Thr-Ser-Asp-   -Arg-Val- Tyr -Glu-Lys-Leu-Lys-
    96   100   110

HPL: Met-Phe-Ala-Asn-Asn-Leu -Val-Tyr-Asp-Thr-Ser-Asp-Ser-Asp-Asp- Tyr -His-Leu-Leu-Lys-
    95   100   110

hGH: Val-Phe-Ala-Asn-Ser- Leu -Val-Tyr-Gly-Ala-Ser-Asn-Ser-Asp-Val- Tyr -Asp-Leu-Leu-Lys-
    95   100   110

bGH: Asp-Leu-Glu-Glu-Gly- Ile -Leu-Ala-Leu-Met-Arg-Glu-Leu-Glu-Asp-Gly-Thr-Pro-Arg
                120   133

oGH: Asp-Leu-Glu-Glu-Gly- Ile -Leu-Ala-Leu-Met-Arg-Glu-Leu-Glu-Asp-Val-Thr-Pro-Arg
                120   133

HPL: Asp-Leu-Glu-Glu-Gly- Ile -Gln-Thr-Leu-Met-Gly-Arg-Leu-Gly-Asp-Gly-Ser-Arg-Arg-
                120   133

hGH: Asp-Leu-Glu-Glu-Gly- Ile -Glu-Thr-Leu-Met-Gly-Arg-Leu-Glu-Asp-Gly-Ser-Pro-Arg
                120   133

Fig. 3.20. Comparison of the amino acid sequences of fragments of bovine growth hormone (bGH), ovine growth hormone (oGH), and human placental lactogen (HPL) with that of human growth hormone (hGH), 96–133. These fragments were obtained by enzymic tryptic digestion. The fragments of growth hormone retain biological activity and the common sequences in each molecule may reflect the presence of a "common active site." (From Yamasaki, Shimanaka, and Sonenberg, 1975.)

(Doneen, Bern, and Li, 1977). The 51 amino acid residue at the opposite, C-terminus, exhibited neither type of effect. Thus, it appears that the growth hormone and prolactin-like activities of the molecule are closely associated with each other.

3. Synthesis and release. Growth hormone is synthesized and stored in the adenohypophysial somatotrophs. These cells are acidophils of the eosinophil type. Thyroid hormone can stimulate the synthesis of growth hormone in cultured growth hormone cells, and this effect can be inhibited by actinomycin D and cycloheximide, indicating that genetic transcription and translation are involved (Samuels and Shapiro, 1976). Messenger RNA coded for growth hormone has been isolated from normal pituitaries and pituitary tumor cells and its synthesis can be stimulated by thyroxine, triiodothyronine, and the synthetic corticosteroid dexamethasone (Seo et al., 1977; Martial et al., 1977). The synthesis of growth hormone has also been demonstrated in cell-free extracts of ascites cells provided with mRNA from rat pituitaries (Bancroft, Wu, and Zubay, 1973). Growth hormone mRNA can also stimulate formation of growth hormone in cell-free wheat germ-extract preparations (Sussman, Tushinski, and Bancroft, 1976). This biochemical preparation has been especially useful, as it lacks the animal degradative enzymes that normally convert the precursor, pregrowth hormone, to the hormone. Pregrowth hormone has a molecular weight estimated to be about 4,500 greater than that of growth hormone. This difference reflects an additional chain of amino acids at the amino terminus of the hormone and corresponds to the "signal peptide."

The nucleotide sequence of the structural gene (the DNA) for rat growth hormone mRNA has been determined (Seeburg et al., 1977). This remarkable achievement not only allows a prediction of the amino acid sequence of the hormone itself but also that of the attached signal peptide. The coding sequence of the nucleotides indicates that the latter consists of 26 amino acids, 17 of which are of a hydrophilic nature.

The conversion of pregrowth hormone to growth hormone apparently occurs very rapidly during the membrane-associated synthesis of hormone (Spielman and Bancroft, 1977). The growth hormone appears to be stored in granules in the cell, and *in vitro* experiments suggest that it may be extruded in monomeric or dimeric forms (Stachura and Frohman, 1975). The latter appears to occur directly following synthesis, whereas release of the simple monomeric form may take place after intracellular processing, possibly during storage in the prominent granules of the somatotroph cells.

Growth hormone is released (see Reichlin, 1974) from the pituitary in response to a variety of stimuli (Table 3.12). This process is not a tonic one but occurs in bursts at certain periods of the day. At many times the hormone may be almost undetectable in the plasma. There is a nocturnal release of growth hormone which is not a circadian rhythm but a response to sleep and may also occur in response to naps during the day. Release occurs shortly after falling asleep; this response is greater in children than in adults but declines with age. It is not present in infants but appears at about 4 to 5 years of age. Other daily events that trigger the release of the hormone are exercise; acute hypoglycemia, such as that associated with the injection of insulin or possibly fasting; or a high-protein meal, which reflects elevated plasma amino acid levels. A high-carbohydrate meal, on the other hand, may inhibit release of growth hormone because of its hyperglycemic effects. These types of responses are blunted in children as compared with adults and are usually greater in women than in men. These age and sex differences may reflect the levels of the steroidal sex hormones, androgens, and estrogens. Thyroid hormones stimulate synthesis of growth hormone; it is depressed in thyroid deficiency. Growth hormone is secreted by the fetus and can be detected from about the age of 70 days (see Reichlin, 1974). It rises toward midterm after which it declines toward birth when it is, however, still relatively high. Premature infants have elevated plasma growth hormone levels at delivery.

Growth hormone is released from pituitaries transplanted away from the hypothalamus and so displays some degree of autonomy. The control mechanism concerned with normal release is situated in the hypothalamus (see Section 3.2E), in which the hormone somatostatin is formed and carried in the hypophysial portal vessels to the adenohypophysis. As described earlier, there also appears to be a GH-releasing hormone present in the hypothalamus. The dopaminergic drugs (Table 3.12) apomorphine and L-dopa can stimulate release of growth hormone. It is especially interesting that β-endorphin and its possible product Met-enkephalin can also initiate secretion. The neural input that influences release of growth hormone via the hypothalamus appears to be dopaminergic. However, as norepinephrine and serotonin are also effective, multiple neural pathways may be involved.

Only sporadic observations are available about the precise mechanism mediating the release of growth hormone from the somatotrophs. As in other such pituitary hormones, it seems likely that cyclic nucleotides and Ca^{2+} are involved. Thus (see Reichlin, 1974), theophylline and dibutyrylcyclic

Table 3.12. *Physiological and pharmacological factors influencing the release of growth hormone*

Increase release
Arginine infusion
Sleep
Exercise
Acute fear and stress (physical and psychic stimuli)
Hypoglycemia (e.g., insulin injection)
Glucagon[1]
Starvation and fasting
Protein meal
Apomorphine[2,3]
β-Endorphin[4]
α-Bromocriptine[5,a]
L-Dopa[6,a]
Clonidine[7]
L-5-Hydroxytryptophan, 5-HTP[8]
Dopamine (inhibited by TRH)[9]
TRH (in many acromegalic patients)[10]

Decrease release
Chronic emotional stress in children
Hyperglycemia
Thyroid deficiency
Somatostatin
α-Bromocriptine (in acromegaly)[5]
Ethanol (reduced to arginine infusion)[11]
L-Dopa (in acromegaly, brief effect)[12]
Cyproheptadine
Methysergide } inhibit response to insulin hypoglycemia[6,13] and 5-HTP
Fenfluramine[14]
Lergotrile mesylate (in acromegaly)[15]

Note: For a general review, see Reichlin, 1974.
References: [1] Editorial, 1973a; [2] Brown, Van Woert, and Ambani, 1973; [3] Maany, Frazer, and Mendels, 1975; [4] Dupont et al., 1977; [5] Wass et al., 1977; [6] Nakai et al., 1974; [7] Lal et al., 1975; [8] Imura, Nakai, and Yoshimi, 1973; [9] Steiner et al., 1977; [10] Liuzzi et al., 1974; [11] Tamburrano et al., 1976; [12] Chiodini ct al., 1974; [13] Bivens, Lebovitz, and Feldman, 1973; [14] Sulaiman and Johnson, 1973; [15] Kleinberg, Schaaf, and Frantz, 1978.
[a] Decreases in acromegaly, "paradoxical effect."

AMP (which penetrates cells more readily than cyclic AMP) can stimulate release, and Ca^{2+} is necessary for this response. The latter observation may reflect the well-known ability of this divalent ion to couple such excitatory–secretory events. It is thus interesting to observe that somatostatin can block the ability of the Ca ionophore A23187 to stimulate release of growth hormone *in vitro* (Bicknell and Schofield, 1977). This effect appears to reflect an interference with the movements of Ca^{2+}. It has also been suggested that cyclic GMP, rather than cyclic AMP, may be involved in the process of release of growth hormone.

A variety of drugs may act by interfering with normal hypothalamic processes that increase or decrease the release of growth hormone, both in normal and disease conditions (Table 3.12). These drugs are not only of theoretical interest, but they can be used to test pituitary function and suppress release of growth hormone in acromegaly. Stimuli used to test the ability of the pituitary to release growth hormone include acute insulin hypoglycemia and the infusion of the amino acid arginine. Glucagon injection has also been utilized, but its mechanism of action is uncertain. It does not appear to reflect the hypoglycemia that follows the hyperglycemic action of this hormone, and it has been suggested that it is due to the feeling of nausea it promotes. It may also have a more direct action. A number of dopaminergic drugs can also stimulate release, including apomorphine; L-dopa, which is a precursor of dopamine; and 5-hydroxytryptophan. Such drugs have not proved useful either in diagnostic tests or for treatment of growth hormone diseases. For instance, apomorphine has a potent emetic action and dopaminergic drugs, such as L-dopa, only have a very brief effect. The dopaminomimetic drugs bromocriptine and lergotrile mesylate often have paradoxical effects when the rate of release of growth hormone is high and reduce its secretion in patients suffering from acromegaly. The inhibitory effects of the 5-hydroxytryptamine antagonists cyproheptadine and methysergide are interesting, as they suggest that a serotoninergic mechanism may also be involved in release of growth hormone. Clonidine is an α-adrenergic drug which is used to treat hypertension and is thought to act in the central nervous system. This drug also stimulates release of growth hormone, an effect that is independent of its hypotensive action (Lal et al., 1975). The release of growth hormone can be inhibited by the α-adrenergic antagonist phentolamine. These observations may reflect the presence of an α-adrenergic mechanism controlling secretion of this hormone. Ethanol depresses the release of growth hormone in normal and acromegalic subjects, but its use in the treatment of the latter condition has not been suggested.

Intravenous infusion of somatostatin effectively inhibits release of growth hormone in patients with acromegaly (Hall et al., 1973; Yen, Siler, and De-Vane, 1974) but its effects are brief and this method of administration is not convenient for chronic use. It is possible that long-acting synthetic preparations may be more useful, but it will be recalled that somatostatin can have widespread actions in the body which may limit its usefulness for the treatment of specific diseases. Fenfluramine, a relative of amphetamine, which is used to promote weight loss, reduces secretion of growth hormone in normal and acromegalic subjects, but its suggested use (Sulaiman and Johnson, 1973) to treat the latter condition does not appear to have been heeded. Bromocriptine was developed to treat hyperprolactinemia [see Section 7.6G(3)] but has been effectively used to treat acromegaly [see Section 3.2F(6)] (Wass et al., 1977).

4. Metabolism. Growth hormone exists in the plasma in at least two forms, the simple monomeric (or "little") growth hormone and the "big" growth hormone. The latter undoubtedly reflects the presence of aggregates of the hormone, but it is also possible that binding to a plasma protein occurs (see Beitins, Rattazzi, and MacGillivray, 1977). In man, growth hormone has a relatively long half-life, about 25 minutes. Only small quantities are excreted in the urine. It is inactivated by enzymes that have a proteolytic action or can split the disulfide bridges. This process appears to occur in the liver and kidneys and is decreased in liver and renal failure and hypothyroidism (see Reichlin,

1974). Growth hormone does not cross the placenta.

5. Mechanism of action of growth hormone; the somatomedins [nonsuppressible insulin-like activity (NSILA) or insulin-like growth factor (IGF)]. Persistent difficulties have been experienced in persuading growth hormone to act *in vitro*. Very high concentrations are usually needed to elicit an effect, and different tissue preparations show considerable variability in their responsiveness. Growth hormone increases the growth of cartilage, and in rats this effect is reflected by an increased incorporation of $^{35}SO_4$ into the tissue. Although this latter response can be seen *in vivo*, it is not normally seen *in vitro*. In 1957, Salmon and Daughaday showed that plasma from normal rats could promote the incorporation of sulfate into cartilage. They called the active material *sulfation factor*. Plasma from hypophysectomized rats contained reduced activity, but it could be restored by the injection of growth hormone. The levels of sulfation factor were also observed to be decreased in patients suffering from hypopituitarism and were elevated in those with hyperpituitarism (acromegaly) (Daughaday, Salmon, and Alexander, 1959). It was proposed (see Daughaday, 1971; Daughaday et al., 1972) that growth hormone promoted the formation of an active intermediate possibly in the liver, which was renamed *somatomedin* (as it mediates the effects of somatotropin or GH). Several types of somatomedins have since been identified in plasma, and it seems likely that they mediate other responses to growth hormone (see Van Wyk et al., 1974). The effects of somatomedins include an insulin-like action on amino acid and glucose uptake in muscle and adipose tissue, as well as the promotion of growth.

In 1963, Froesch and his collaborators (Froesch et al., 1963; 1967) observed a discrepancy between the response of an *in vitro* preparation of rat adipose tissue to insulin-like material in plasma and to true insulin. Exogenous insulin promotes glucose uptake by the adipose tissue and its incorporation into fatty acids, and this response is accompanied by the evolution of CO_2. Using these parameters as a bioassay, the insulin-like activity (ILA) in serum was measured. However, when antiinsulin serum was allowed to react with the ILA, the response of the adipose tissue was barely affected; it decreased by only about 10 percent. The remaining insulin-like activity has been called *nonsuppressible insulin-like activity* or *NSILA*. This material displays many of the effects of "true" insulin but is not identical to it. Thus, it exhibits the metabolic properties of insulin on the tissue uptake of glucose and amino acids, but it is about 60 times less effective. Insulin is known to have growth-promoting activities which can be observed at high concentrations *in vitro,* but in this instance NSILA is 50 to 100 times more effective (see Rinderknecht and Humbel, 1976a). Insulin and NSILA have different receptors, but there is a crossover so that each substance can apparently interact with the ligand for the other, but they exhibit different abilities (affinities) to do this (Roth et al., 1975). The NSILA level is unaffected by the injection of insulin and is unchanged in diabetes mellitus, but it is increased by administered growth hormone.

It is apparent that there are similarities between NSILA and some somatomedins, and it has been observed that there are parallel changes in their activities when they are chemically extracted from plasma (see Hall, 1972; Shields, 1977). Somatomedins have also, like NSILA, been observed to bind to insulin receptors (Hintz et al., 1972). The activities of both NSILA and somatomedin increase in parallel in patients with hyperpituitarism (acromegaly) and decrease in hypopituitary dwarfism. The NSILA, like somatomedin, can also increase the incorporation of $^{35}SO_4$ into cartilage (Zingg and Froesch, 1973). It thus seems likely (see Schlumpf et al., 1976) that both types of substances, which are known to be polypeptides, are either identical or belong to a closely related family of hormones that regulate the growth of tissues. A number of such excitants have been identified (see Table 3.13 and Rechler and Nissley, 1977). The precise identities of each and the true extent of the "somatomedin family" awaits the chemical analysis of each.

Several such somatomedin-like molecules have been isolated; they are polypeptides with 50 to 70 amino acid residues and molecular weights of about 5000 to 8000. Three somatomedins, A, B, and C, with different biological profiles in their effects have been separated from human plasma (Van Wyk et al., 1974). The amino acid sequence of two such substances, which exhibit both NSILA-like and cell-growth-promoting activities, have been partially determined (Rinderknecht and Humbel, 1976a,b, 1978). These materials, which were separated from human plasma, were called NSILA I and II. They are slightly basic peptides, with molecular weights of about 7000. NSILA I consists of a single chain of 70 amino acids (molecular weight 7649) containing three intrachain disulfide bridges. The amino acid sequence of NSILA I and the first 31 residues of NSILA II are shown in Figure 3.21. They share at least 22 residues, 3–24 in NSILA I and 6–27 in NSILA II. The amino acid sequence of the A- and B-chains of human insulin share a number of characteristics with NSILA I and II. It seems likely that insulin and NSILA I and II originally evolved from a common ancestral molecule. It has been suggested (Rinderknecht and Humbel, 1976b) that NSILA be renamed *insulin-like growth factor* or IGF I and II.

Somatomedin and NSILA activity in the plasma is bound to a plasma protein (Moses et al., 1976; Cohen and Nissley, 1976; Zapf, Waldvogel, and Froesch, 1975). Its clearance from the plasma is considerably reduced as a result of this interaction, and it seems likely that while it is in this condition its growth-promoting effects are also limited. The

Table 3.13. *Some "broad-spectrum" growth factors that have been identified in blood plasma and extracts of various tissues*

	Molecular weight and chemistry	Responsive cell types	Control
Somatomedins[a]		Ectodermal origin	
Sm-A	6500–8500	GH$_3$ cells (hypophysial)	
Sm-C	Neutral or basic	Mesodermal origin	
NSILA I and II		Chondrocytes	Growth hormone
MSA	Homologous with insulin	Myoblasts	
		Ovarian tumor cells	
		Entodermal origin	
		Fetal liver	
Epidermal growth factor		Ectodermal origin	
Rodent salivary gland	6300	Epithelium of skin and cornea	
Human pregnancy urine	Acidic; tightly coiled	Mesodermal origin	Androgens (also growth hormone?)
Urogastrone		Fibroblasts	
Human pregnancy urine		Granulosa cells	

[a] Sm, somatomedin (human plasma); NSILA, nonsuppressible insulin-like activity (human plasma); MSA, multiplication stimulating activity (rat liver cell line).
Source: J. J. Van Wyk, personal communication.

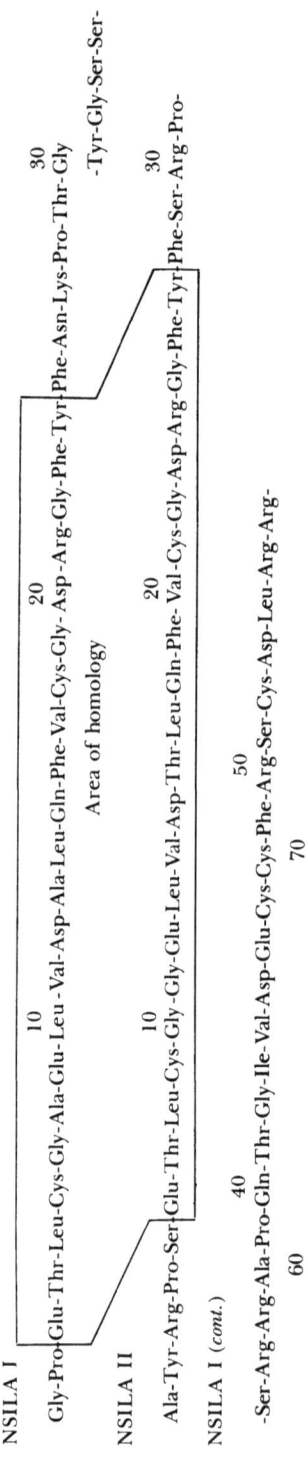

NSILA I

10   20   30

Gly-Pro-Glu-Thr-Leu-Cys-Gly-Ala-Glu-Leu - Val-Asp-Ala-Leu-Gln-Phe-Val-Cys-Gly- Asp-Arg-Gly-Phe-Tyr-Phe-Asn-Lys-Pro-Thr-Gly

Area of homology

-Tyr-Gly-Ser-Ser-

NSILA II

10   20   30

Ala-Tyr-Arg-Pro-Ser-Glu-Thr-Leu-Cys-Gly-Gly-Glu-Leu-Val-Asp-Thr-Leu-Gln-Phe- Val-Cys-Gly-Asp-Arg-Gly-Phe-Tyr-Phe-Ser- Arg-Pro-

NSILA I (*cont.*)

40   50

-Ser-Arg-Arg-Ala-Pro-Gln-Thr-Gly-Ile-Val-Asp-Glu-Cys-Cys-Phe-Arg-Ser-Cys-Asp-Leu-Arg-Arg-

60   70

-Leu-Glu-Met-Tyr-Cys-Ala-Pro-Leu-Lys-Pro-Ala-Lys-Ser-Ala

Fig. 3.2.1. Amino acid sequences of two somatomedins, NSILA I and NSILA II, extracted from human plasma. Apart from the similarities between each of these polypeptides, they also exhibit considerable homology to the amino acid sequences in the insulin B-chain (positions 1–29) and insulin A-chain (positions 42–62). The half-cystine and glycine residues of insulin are all conserved, which suggests that the three-dimensional structure of NSILA I is probably similar to that of insulin. It has been proposed that they be renamed insulin-like growth factor, or IGF I and II. (The complete amino acid sequence of NSILA II is not yet available.) (Based on Rinderknecht and Humbel, 1976a, 1976b, 1978.)

binding protein itself, like somatomedin, is also regulated by growth hormone, which thus has a dual control over the actions of its intermediary effector material.

The cellular mechanism of action of NSILA and the somatomedins is unknown. In cartilage, somatomedins (including IGF I and II and somatomedin A) increase the incorporation of thymidine into DNA and uridine into RNA, as well as increasing protein synthesis and cell replication (see Garland et al., 1972; Rechler and Nissley, 1977). Such changes could involve a stimulation of genetic transcription, but it appears unlikely that such an effect is a direct one, as the receptors for somatomedin appear to be present on the plasma membrane (Hintz et al., 1972). Like insulin, somatomedin has, under some conditions, been shown to inhibit the accumulation of cyclic AMP in tissues, especially following increases seen in response to epinephrine (see Van Wyk et al., 1974).

It is not clear at this time whether somatomedins act as intermediaries in all the responses to growth hormone *in vivo* or whether growth hormone has a dual action which also involves a direct effect (see Van Wyk et al., 1974; Talwar et al., 1975). The liver, either *in vivo* or *in vitro*, produces a somatomedin-like activity when stimulated by growth hormone (McConaghey, 1972; Uthne and Uthne, 1972; Ash and Francis, 1975). The kidney has also been reported to be a source of such a substance (McConaghey and Dehnel, 1972). Whether these are the sole sites of somatomedin synthesis is not clear, but growth hormone receptors have been identified in other tissues and fragments of growth hormone have been observed to exhibit somatomedin-like actions [Section 3.2F(2)].

Labelled ^{125}I-human growth hormone preparations have been observed to bind to several tissues. The specificity of such interactions have been established by measuring the ability of "cold" unlabelled growth hormone preparations, such as those obtained from man, cattle, and sheep, to displace the ^{125}I-hGH from its binding sites. Such sites are considered to be "receptors," although the confirmation of this possibility awaits the establishment of a clear relationship between the binding to the tissue and its biological response. The receptors appear to be present on the outer surface of the cells. However, isolated membranes prepared from the Golgi apparatus of liver cells have also been shown to specifically bind human growth hormone (Bergeron et al., 1978).

Growth hormone receptors have been identified in plasma membranes prepared from liver cells (Tsushima and Friesen, 1973; Posner et al., 1974; Herington, Veith, and Burger, 1976) and in a culture line of human lymphocytes (Lesniak et al., 1974). Specific binding has also been observed in a number of other tissues, including rabbit and sheep adrenal, mammary gland, and ovary and monkey heart, ovary, and diaphragm (Posner et al., 1974). The levels in these tissues are, however, usually lower than those observed in liver and lymphocytes. The speed of such binding of growth hormone is relatively slow and increases progressively over a period of several hours in liver, although in lymphocytes it reaches a steady state in 90 minutes. Binding is reversible, but it takes several hours to wash the hormone from its receptors. Growth hormone receptors solubilized with Triton X-100 have been prepared from rabbit liver (Herington and Veith, 1977), and this preparation should facilitate the study of their properties.

At this time most of our information about the precise nature of the growth hormone receptors is based on the studies of Roth and his collaborators using human lymphocytes (Lesniak et al., 1974; Lesniak and Roth, 1976; Van Obberghen, De Meyts, and Roth, 1976). There appear to be about 4000 such sites on one of these cells. Binding is little affected by changes in pH or Mg^{2+} and Ca^{2+} concentrations. Tryptic digestion destroys the ability of the cells to bind growth hormone. The disposition of the receptors appears to be related to the integrity of microfilaments in the cell, as cytochalasin A, B, and D, which disrupts these structures, reduces the number of such sites. Growth hormone itself can also influence the number of receptors, as incubation with this hormone reduced their number. When the hormone is removed from the external media, the number of receptors returns to normal. This restoration process, however, can be prevented by cycloheximide, suggesting that they are being resynthesized. Such autoregulation of its receptors by growth hormone may play an important role in regulating its action.

6. Disorders in growth hormone secretion. Increases or decreases in the secretion and action of growth hormone can result in human diseases. In children such changes can disturb growth: hyposecretion of growth hormone may produce pituitary dwarfism, and an excess can result in giantism. In adults a deficiency of growth hormone does not appear to be debilitating, although it can be associated with a lack of other pituitary hormones which may have more dramatic effects. An excess of growth hormone, however, results in the disease of acromegaly in adult life, the metabolic disturbances of which have been described earlier [see Section 3.2F(1)].

Short stature can have a variety of causes, most of which are not related to a deficiency of growth hormone (see Mason, 1972; Tanner et al., 1971; Tanner, 1972; Parkin, 1976). It can, for instance, be associated with malnutrition or with debilitating diseases such as cyanotic congenital heart disease, chronic renal or hepatic failure, coeliac disease, and cystic fibrosis. Short stature can be due to chromosomal disorders, as seen in Turner's syndrome in girls. It can, of course, also be genetic, related to the height of the parents.

The secretion of growth hormone may be reduced, such as in hypothyroidism or hypercortisolism (due to Cushing's disease or treatment with corticosteroids). In many instances, short stature may result from psychic disturbances arising from unhappy childhood home circumstances ("gross emotional or social deprivation"), which appear to reduce the normal secretion of growth hormone. A deficiency of growth hormone occurs most commonly in the absence of any other pituitary hormone abnormalities ("isolated deficiency"), but it may be accompanied by a more general hypopituitarism, which may involve some or all (panhypopituitarism) the other hormones. Initially, a single such hormonal deficiency may occur, but others may follow. A deficiency of growth hormone in children may directly involve either the adenohypophysis or its hypothalamic controlling mechanism. The latter is usually considered more commonly affected (see, e.g., Okada et al., 1978). The changes can result from physical damage, such as the presence of a craniopharyngioma, but most commonly the cause is unknown (idiopathic). In some instances it may be genetic and involve the inheritance of a recessive gene. Such dwarfs sometimes intermarry and have children who are also, invariably, dwarfs.

A growth hormone deficiency disease is usually due to low levels of the hormone, but other factors can be involved (see Cheek and Hill, 1974). The target end organs may be unresponsive, as seen in the African pygmies. In these people, levels of growth hormone and somatomedin are normal. In another form of genetic deficiency, the Laron type, somatomedin levels are depressed despite normal endogenous concentrations of growth hormone. Exogenous growth hormone also fails to stimulate somatomedin levels in these people. In these two types of genetic dwarfism, treatment with growth hormone is clearly of no avail, although it is possible that if human somatomedin becomes available, it may be useful in the Laron-type disorder.

The incidence of dwarfism due to growth hormone deficiency is uncertain, but it has been estimated as 1/10,000 (see Parkin, 1976). It is about 2 to 3 times more common in boys than in girls. Early diagnosis is important, as replacement therapy with growth hormone is ineffective once the epiphyses have closed and "catch-up" growth is incomplete. The usual aim is to commence treatment at about 5 years of age, but identification of affected children may be difficult. Charts are available which indicate the expected range of heights for children in a particular community, with adjustments for the height of the parents. If a child

falls below a certain level (usually 2 to 3 standard deviations below the mean), he or she is investigated further. Determinations of bone age, dental development, and skin thickness are used to more precisely assess the situation. When other possible explanations for the short stature are excluded, growth hormone levels can be measured in response to certain stimuli, such as exercise, the infusion of arginine, or insulin hypoglycemia. Correct diagnosis is important, as treatment with human growth hormone is a prolonged undertaking and supplies of the hormone are limited. Short stature due to other causes is not responsive to the administration of exogenous growth hormone.

The determination of human growth hormone involves radioimmunoassay, which once set up, has the merits of relative accuracy, simplicity, and convenience. The principal disadvantage is the determination of "normal" levels, as plasma concentrations may change dramatically during the course of the day. It has been proposed that the measurement of somatomedin activity in the plasma may provide a more useful index of growth hormone deficiency as well as a more rapid assessment of the response to the hormone (see Hall and Olin, 1972; Schwalbe et al., 1977). At present this procedure involves a bioassay, which measures the incorporation of ^{35}S-sulfate into embryonic chick cartilage. When pure preparations of human somatomedin become available, it would seem that radioimmunoassay may be possible. Somatomedin levels are depressed in hypopituitary dwarfs and children who suffer a retardation of growth associated with chronic renal insufficiency. In normal children the levels are related to age and are elevated in sexual precocity. Such an assay offers an attractive alternative or supplement to that for growth hormone. The parameters used for diagnosis appear to be reasonably predictable and afford an early insight into whether a response to administered growth hormone is occurring.

Growth hormone prepared from the pituitaries of animals is ineffective in man. It has therefore been necessary to extract growth hormone from the pituitaries of human cadavers for the treatment of children suffering from this hormonal deficiency. Fortunately, the human pituitary is a relatively rich source of growth hormone and with the widespread cooperation of pathologists it has been possible to collect enough material to treat many cases of pituitary dwarfism. Such a program for providing growth hormone was first started in the United Kingdom but has now been extended to several European countries, Australia, and the United States. These programs are organized by the National Institutes of Health in the United States and the Medical Research Council in England. Supplies are nevertheless limited and the process of collection and preparation is time consuming, so that the possibility of developing active synthetic preparations is very attractive. The possibility of "grafting" a gene or mRNA coded for human growth hormone into bacteria, which can then transcribe and/or translate it and produce the hormone, is rapidly becoming a practical possibility. As described elsewhere, this form of synthesis has already been achieved for somatostatin. Seeburg et al. (1978) have recently grafted rat growth hormone DNA onto genetic material in *Eschericia coli* and recovered growth hormone from the bacteria. Even more recently it has been reported (*New York Times*, 17 July, 1979) that this procedure has been accomplished for human growth hormone. The commercial utilization of this procedure may alleviate the current shortage of this hormone.

Active fragments of growth hormone prepared from more readily available animal growth hormones is another possible source of clinically useful material (Yamasaki et al., 1972). A patent for such a 38-amino acid segment of bovine growth hormone was recently granted in the United States (Jones, 1977), but no details appear to be available at this time.

The administration (Mason, 1972; Tanner, 1972; Tanner et al., 1971) of growth hormone for the treatment of hypopituitary dwarfism is best begun as early as possible and is continued until the epiphyses close, an event signalled by sexual maturity, usually at about 14 years of age. Even when therapy is started relatively late, there may be some catch-up growth, but this does not completely compensate for the preexisting deficiency. The hormone must be administered by i.m. injection, but its action is fortunately prolonged because of its relatively long half-life and the persistent formation and survival of its intermediary compound, somatomedin. Growth hormone is therefore usually administered only twice a week, and definite benefits have also been reported from a single weekly injection. An initial trial period of 1 year is observed to confirm that the treatment is having a beneficial effect on growth. The formation of antibodies, which oppose the action of the hormone, has been reported, but with better methods of purification these are now infrequent. The concurrent administration of estrogens and androgens is contraindicated, as they hasten closure of the epiphyses. The use of such steroids may however eventually be required if there is an associated gonadotropin deficiency. Corticosteroids have a catabolic action and oppose the effects of the growth hormone. No adverse side effects to growth hormone appear to have been reported, except possibly for occasional minor fluid retention (see Hutchings et al., 1959). Somatomedin levels in the plasma increase in children treated with human growth hormone, and the

changes observed parallel the rate of growth (Hall and Olin, 1972).

Hypersecretion of growth hormone usually results from the presence of pituitary tumors, most often involving the eosinophil-type acidophils (somatotrophs), which normally synthesize the hormone. It is, however, also considered likely that a disorder of the hypothalamic mechanism, which controls the activity of these pituitary somatotroph cells, may be involved. Occasionally, growth hormone may be secreted from tumors, such as lung carcinoma, at ectopic sites (Greenberg et al., 1972). When a pituitary tumor is involved, its progressive enlargement may compromise the activity of other hormone-secreting cells so that a deficiency of gonadotropins, TSH, and ACTH sometimes develops. The effects of an excess of growth hormone in pituitary giantism and acromegaly were described earlier [see Section 3.2F(1)]. Treatment is usually aimed at removing or controlling a tumor using surgical, radiological, or cryoscopic procedures. These are not always successful and can result in deficiencies of other pituitary hormones, so that appropriate replacement therapy may be necessary. A number of drugs are known to decrease the secretion of growth hormone in patients suffering from acromegaly. However, the use of most of these, which include L-dopa, chlorpromazine, somatostatin, and medroxyprogesterone, have not met with practical sustained success.

Bromocriptine has, however, been shown to be useful (Wass et al., 1977; Besser, Wass, and Thorner, 1978). In a clinical trial in the United Kingdom involving 73 patients with acromegaly, it resulted in objective and symptomatic improvement in 71 cases. A reversal of many of the effects, including altered facial appearance, occurred. This response was usually associated with a decline in plasma levels of growth hormone (Figure 3.22), though there were some interesting exceptions which suggested that the nature of the circulating hormone may be changed. For instance, the hormone could be present as a biologically inactive oligomer. The bromocriptine was given in divided doses every 6 hours, and these were gradually increased to attain satisfactory therapeutic levels. Side effects were not a serious problem. They included nausea, which could be minimized by taking the drug at mealtimes. Mild constipation and a digital spasm in response to cold sometimes occurred with high doses. It was suggested that this form of treatment is useful prior to surgical or radiological intervention, or as an additional treatment, or an alternative if such procedures are ineffective or not feasible. Comparable results have been reported in other clinical trials (Eskildsen et al., 1978; Lundin et al., 1978). It should be recalled that the action of bromocriptine in reducing growth hormone levels in acromegaly is paradoxical and that in normal

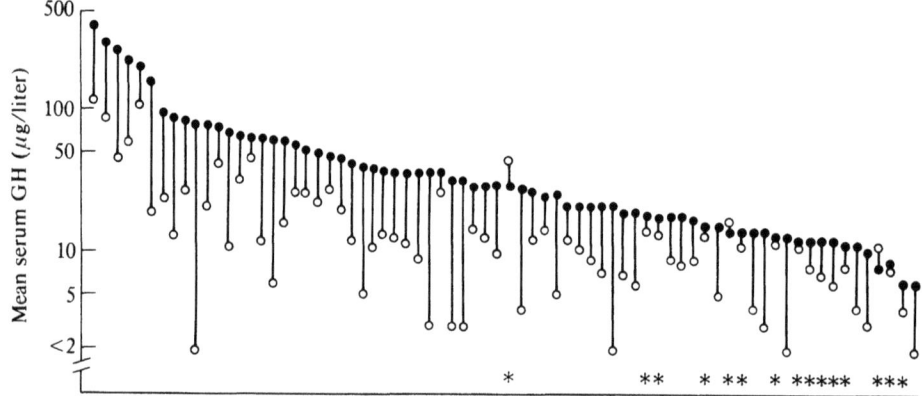

Fig. 3.22. Change in mean serum GH before (●) and during (○) bromocriptine treatment in 73 patients. Asterisks indicate nonresponders. For convenience, GH was plotted on a logarithmic scale. (From Wass et al., 1977.)

people or even those with elevated growth hormone levels due to other causes, such as chronic renal failure (Kanis et al., 1976), it usually increases the release of the hormone.

Lergotrile mesylate is a dopaminomimetic drug which was, like bromocriptine, chemically derived from ergot [it is an ergoline derivative; Section 7.6G(3)] for the possible treatment of hyperprolactinemia. Like bromocriptine, it has been shown to display a paradoxical response and lower growth hormone levels in patients suffering from acromegaly (Kleinberg, Schaaf, and Frantz, 1978; Thorner et al., 1978). The plasma levels of growth hormone returned to their previous high levels once use of the drug ceased, so that surgical and radiological treatment remains preeminent for the treatment of the disease. Its effect is shorter in onset and duration than that of bromocriptine. Side effects include drowsiness, hypotension, and nausea. Long-term use in rats resulted in tumors in the reproductive tract, so that the manufacturers (Eli Lilly) limited its use to acromegaly, breast cancer, and Parkinson's disease. Its use in man has, however, been associated with liver damage (which is reversible) in about one-third of those patients to whom it was clinically administered. It has been withdrawn from clinical use. This is unfortunate, as it is apparently cheaper than bromocriptine to produce.

3.3. The neurohypophysis

The hormones secreted by the neurohypophysis are the products of a process of neurosecretion by neurons with their cell bodies in the hypothalamic supraoptic and paraventricular nuclei. Our knowledge regarding the origins of the neurohypophysial hormones is principally due to the careful histological observations of W. Bargmann in Kiel (see

Bargmann, 1968; Lederis, 1974). The Gomori chrome-alum hematoxylin stain has a special affinity for sulfhydryl groups and stains the neurosecretory substances in the nerve tracts of the neurohypophysis. This material was observed to accumulate in the nerve endings, in proximity to blood vessels in the pars nervosa. It could also, however, be traced back along the connecting axons to the paraventricular and supraoptic nuclei. Parallel measurements of the neurohypophysial hormones vasopressin (ADH) and oxytocin showed that these peptides and the stainable neurosecretory material were closely associated with each other. It is now known that these hormones are attached to "carrier" molecules called neurophysins, and it is these proteins that are histologically stainable. More recently (see Zimmerman and Robinson, 1976), antibodies to the neurophysins, and the peptide hormones they carry, have been prepared so that these observations have been extended using immunohistochemical and immunoassay techniques. Vasopressin and oxytocin are formed by separate neurons and are stored in distinct neurosecretory granules (see also Swaab, Nijveldt, and Pool, 1975; Tasso, Picard, and Dreifuss, 1976). Contrary to previous suggestions, both hormones are formed in both the supraoptic and paraventricular nuclei, which apparently have distinct populations of "oxytocinergic" and "vasopressinergic" neurons (Zimmerman et al., 1974).

A. Functions and actions

The neurohypophysis is not essential for life but contributes to some physiologically useful responses (Table 3.14). In the absence of vasopressin (antidiuretic hormone, ADH), large volumes of urine are formed; this disease is called diabetes insipidus. It reflects an inability to reabsorb adequate

Table 3.14. *Summary of the physiological and pharmacological effects of vasopressin (antidiuretic hormone, ADH) and oxytocin*

Vasopressin
1. *Antidiuresis* (increased water absorption across renal tubule)
2. Vascular effects
   a. Constricts capillaries and arterioles
   b. Constricts coronary blood vessels
   c. Increase in blood pressure in anesthetized or pithed animals
3. Hyperglycemia
4. Releases some hypophysiotropic hormones, especially ACTH
5. Contracts nonvascular smooth muscle of the intestine (peristalsis) and uterus (at term)
6. Releases renin
7. Facilitates learning?

Oxytocin
1. *Increases contractility of uterus,* especially during parturition
2. *Milk let-down from mammary gland*
3. Vasodilatation, especially skin vessels
4. Transitory fall in blood pressure
5. Contracts intestine (weak)
6. Very weak antidiuretic action
7. Facilitates transport of sperm in female reproductive tract?

Notes: The known physiological effects are shown in italic type, but there may be others. The vasodilatatory responses to oxytocin may be partly due to the presence of chlorbutanol in the hormone preparations.

amounts of water from the glomerular filtrate across the distal segment and collecting ducts in the kidney. This condition is not fatal but necessitates an adequate intake of water, and as the volumes of urine may be quite large (20 to 30 liters/day have sometimes been reported), it imposes certain social inconveniences and disturbs sleep. Oxytocin, the other neurohypophysial hormone, is released during labor and may facilitate this event, although birth can still occur, even in the absence of this secretion. Oxytocin is necessary for adequate lactation and is released in response to suckling by the infant. By contracting the myoepithelial cells that surround the secretory alveoli in the mammary glands, it promotes the "draught" or "let-down" of milk. It is possible that these peptide hormones exhibit other physiological responses, such as contributing to the regulation of the secretion of adenohypophysial hormones, but there is no conclusive evidence that this occurs.

The neurohypophysial hormones when administered in relatively large doses exhibit a number of pharmacological actions (Table 3.14), some of which are of interest in relation of their therapeutic use. Vasopressin is so named because of its ability to contract blood vessels and increase the blood pressure. It was this response that led to its discovery by Oliver and Schäfer in 1895. The dosages required to elicit such effects are, however, many times greater than those which produce an antidiuresis. Vasopressin can contract many peripheral vessels (see Saameli, 1968; Nakano, 1974; Altura and Altura, 1977), especially arterioles, but even venules and capillaries respond. These effects have been demonstrated in several vascular beds, including the splanchnic circulation, the uterus, and the skin, where a prominent blanching response may be observed. It can also affect the renal vasculature, and both increases and decreases in glomerular filtration rate have been described with resulting parallel effects on urinary excretion of electrolytes, especially Na and Cl. Studies on synthetic analogues of the hormones indicate that they may have an additional natriuretic action which is independent of such vascular changes (see Cort et al., 1973).

A rather unique response for such vasoactive agents is the contraction of the coronary vessels. This is accompanied by decreased cardiac contractility and oxygen consumption and in certain instances, especially in people with heart problems, this response can have disastrous consequences. Vasopressin usually does not cause a rise in blood pressure in normal persons or animals. An increase is, however, usually seen under anesthesia as the result of a damping of compensatory cardiovascular reflexes. Thus, under normal conditions vasopressin can cause slowing of the heart rate, mainly due to the activation of baroreceptor-mediated reflexes. This effect can be blocked by atropine. The effects on the blood vessels are direct ones that cause a contraction of the vascular smooth muscle. They cannot, for instance, be prevented by specific adrenergic or angiotensinergic blocking drugs.

The mechanism of the vascular responses is unknown but appears to be unrelated to the hormones' effects on the permeability of the renal tubule. Thus, the two types of responses can be readily dissociated in different structural analogues of the hormones. The renal tubular response involves an activation of adenyl cyclase, but there is no real evidence that this is occurring in smooth muscle. Repeated injections of vasopressin (or in birds, oxytocin) often results in a progressive decline in the vascular response and the change in blood pressure. The cause of this tolerance, or tachyphylaxis, is unknown, but it is not as prominent if the doses are given at greater intervals of time.

The neurohypophysial hormones have also been observed to produce vasodilatation, but this response is not usually seen in man. A redistribution of blood, such as to the muscles, may, however, be seen following vasoconstriction in other vascular beds. In birds and some other lower vertebrates, vasopressin and oxytocin can produce a fall in blood pressure. This response led to the use of the chicken as a bioassay preparation for oxytocin, still a standard in the British and U.S. Pharmacopoeias. Observations have been made suggesting that oxytocin may exert a peripheral vasodilator effect in man. However, it appears that such results can often be accounted for by the presence of relatively high concentrations of the preservative chlorbutanol, which was present in the preparations of oxytocin used.

The contractility of other types of smooth muscle, apart from the uterus and vasculature, can be influenced by vasopressin and oxytocin. Vasopressin increases intestinal motility (*in vitro* and *in vivo*) and acts on both the ileum and colon. It has been used for this purpose therapeutically in man to treat paralytic ileus. Oxytocin is less effective.

The neurohypophysial hormones in pharmacologically active doses can exert metabolic effects (see Mirsky, 1968). Hyperglycemia occurs, and vasopressin has in the past been recommended for the treatment of insulin hypoglycemia in man. An alternative explanation for its antiinsulin action is an ability to delay the absorption of this hormone, which is due to its peripheral contrictor effect. In some species, notably the rabbit, the neurohypophysial hormones may lower plasma free fatty acid levels. The mechanisms of these effects are unknown, but the hyperglycemic action could be due to activation of liver adenyl cyclase and the promotion of glycogenolysis. The effects on plasma FFA have been likened to that of insulin.

There has been a considerable amount of speculation regarding the possible roles of the neurohypophysial hormones in controlling the release of adenohypophysial hormones (see Doepfner, 1968). Vasopressin can initiate the release of corticotropin under both *in vitro* and *in vivo* conditions. It is present in the median eminence and has been identified in the long portal vessels. In man there are also vascular connections, via the short-portal vessels, between the adeno- and neurohypophysis, so that such a role is possible. Apart from releasing corticotropin, there are reports that vasopressin can, under certain conditions, increase the release of gonadotropins, growth hormone, and thyrotropic hormone.

Recent interest in the possible roles of brain peptides in learning and memory (de Wied, 1977a,b) has included the neurohypophysial peptides. These observations were initiated by de Wied (1971, 1976) in The Netherlands. It was found that rats suffering from hereditary diabetes insipidus (the Brattleboro strain) failed to retain memory of an experimental avoidance-type response but could do so following the intraventricular administration of vasopressin or its analogues. These and other observations have led to speculation (Gold, Goodwin, and Reus, 1978) that a disordered metabolism of vasopressin may accompany some affective illnesses. There is also clinical evidence from pilot studies that the administration of neurohypophysial peptides to patients suffering from amnesia improves their memory (Oliveros et al., 1978). Vasopressin was also shown to improve memory and learning in a small group of older men aged 50 to 65 (Legros et al., 1978). It has been pointed out (Gold, Goodwin, and Reus, 1978) that many drugs that influence behavior, such as ethanol, carbamazepine, morphine, and lithium carbonate, also have effects on the release and action of vasopressin. The mechanism of these effects of vasopressin are unknown, but it may affect noradrenergic systems in the brain (Tanaka et al., 1977).

There is a significant degree of crossover in the effects of vasopressin and oxytocin, and these effects can be of therapeutic significance. The chemically separated fractions of each hormone prepared from animal pituitaries are not usually completely pure, so that a small amount of cross contamination – vasopressin in oxytocin preparations and vice versa – occurs. Even the pure hormones prepared by chemical synthesis exhibit some of the properties of their neurohypophysial partner, which reflects the similarities in their structure. Thus, in standard animal bioassays, oxytocin has about 1 percent of the antidiuretic and pressor activity of vasopressin, while the latter has significant oxytocic activity. Vasopressin can contract the human uterus at term. The infusion of large doses of oxytocin to assist labor may have a substantial antidiuretic effect.

Neurohypophysial hormones

```
          1   2    3    4    5    6    7    8    9
```

Oxytocin: Cys-Tyr-*Ile*-Gln-Asn-Cys-Pro-*Leu*-Gly-NH₂

Arginine vasopressin (AVP, ADH):   Cys-Tyr-*Phe*-Gln-Asn-Cys-Pro-*Arg*-Gly-NH₂

Lysine vasopressin (LVP):   Cys-Tyr-*Phe*-Gln-Asn-Cys-Pro-*Lys*-Gly-NH₂

Vasotocin (AVT):   Cys-Tyr-*Ile*-Gln-Asn-Cys-Pro-*Arg*-Gly-NH₂

Man and other mammals

Suinae (pig family)

Nonmammals and fetal mammals (? pineal)

Fig. 3.23. Amino acid sequences of naturally occurring neurohypophysial peptide hormones in mammals. Vasotocin is present in many nonmammals and has been identified in the fetal pituitary and the pineal glands of some mammals. Lysine vasopressin has an antidiuretic action when injected into man and is commonly used therapeutically for the treatment of diabetes insipidus.

B. Structure of the hormones

The neurohypophysial hormones oxytocin and vasopressin are octapeptides (Figure 3.23) consisting of a ring of five amino acids joined by a disulfide bridge, with a side chain of three amino acids. These two natural hormones differ from each other by the presence of two amino acids at positions 3 and 8. Seven other naturally occurring analogues of these hormones have been identified in other vertebrates. In some mammals belonging to the Suinae (pigs), the vasopressin molecule contains lysine at position 8, instead of the arginine that is present in man and other mammals. *Lysine-vasopressin (LVP)* is almost as active as *arginine-vasopressin (AVP)* in man, and both the natural hormone, obtained commercially from pig

pituitaries, and its synthetic form are used therapeutically. The commonest natural form of the neurohypophysial hormones in nonmammals consists of the ring structure of oxytocin and the side chain of AVP and is called *vasotocin*. It exhibits activities of both the natural hormones, although it is about half as potent with respect to each type of response (Table 3.15). This observation is an interesting demonstration of the importance of both segments of the peptide for each type of biological activity. According to the U.S. Pharmacopeia, pituitaries from slaughtered animals, including chickens, can be used to make commercial preparations of pituitary hormones. This possible use of bird pituitaries has, it seems, not been exploited, which is probably fortunate, as attempts to sepa-

Table 3.15. *Relationships of structure to biological activities in molecules of the natural hormones oxytocin and vasopressin: the importance of positions 3 and 8*

Hormone or analogue	Activity units/μmole			
	Rat anti-diuresis	Rat blood pressure	Rabbit mammary gland (galacto-bolic)	Rat oxytocin
Oxytocin (3-isoleucine, 8-leucine)	5	5	450	450
Substitution at position 3				
[Leu³]Oxytocin	~1	~0.03	~35	~4
[Val³]Oxytocin	~1	~0.03	206	59
[Phe³]Oxytocin (oxypressin)	~30	~3	~65	~20
Substitution at position 8				
[Arg⁸]Oxytocin (vasotocin)	260	255	220	120
[Lys⁸]Oxytocin	25	133	185	80
Substitution at positions 3 and 8				
[Phe³,Arg⁸]Oxytocin (arginine-vasopressin, ADH)	435	435	70	17
[Phe³,Lys⁸]Oxytocin (lysine-vasopressin)	260	285	63	5

Source: Based on Berde and Boissonnas, 1968.

rate the oxytocic from the antidiuretic or vasopressin activity would have encountered insurmountable problems.

The amino acid structures of oxytocin and vasopressin were proposed by V. Du Vigneaud in 1953 (see Du Vigneaud and Tripett, 1953; Du Vigneaud, Lawler, and Popenoe, 1953). This important work was followed by the chemical synthesis of these hormones. As they are relatively small molecules, it was feasible to produce a series of analogues that permit study of the activity of the molecule in relation to its structure. Even in an octapeptide, however, there are many hundreds of possible variations in structure, and it has been estimated that about 300 such analogues have been prepared. A substantial summary of these and a description of their relative potencies and effects has been prepared by Berde and Boissonnas (1968). Such observations on the neurohypophysial peptides have had far-reaching effects, as they have paved the way for a subsequent explosion of interest in the structure–activity relationship of many other types of biological peptides.

The synthesis of analogues of such peptide hormones has several aims, both theoretical and practical. By considering the relationships of the presence of certain amino acids to the potency and effects of the molecules, it may be possible to understand something about their mechanisms of action. This includes affinity for their receptors as well as their intrinsic activity. It has been recognized that a certain tertiary structure and amino acids are conducive to some of the effects of such peptides, but in many instances the information is empirical.

The practical aspects of changing the structure of the natural hormones are related to the production of more useful therapeutic compounds. This may involve a prolongation of the period of their activity in the body, which can be achieved by such changes as those that slow inactivation of the peptides by enzymes or the provision of inactive "hormonogens" which are slowly converted to an active form, so that circulating levels are maintained. The other major aim of varying the peptides' structure is to change the qualitative activity of the molecule, such as increasing the antidiuretic effects while decreasing the vasopressor action, or vice versa. There are many examples of such types of modification of the molecules of vasopressin and oxytocin, but their commercial application has been rather limited. Another aim in modifying the structure of these hormones has been to produce antagonists. Success in this area has also been rather limited, and no antagonists have been made available for routine clinical use.

There are several comprehensive descriptions of the activities of the many analogues of the neurohypophysial hormones, including those of Berde and Boissonnas (1968); Rudinger (1971);

Rudinger, Pliška, Krejči (1972); Sawyer and Manning (1973), and Manning et al. (1977).

Unlike larger hormones, the neurohypophysial hormones do not appear to have a precise "active center," although certain chemical groups may be more important than others. Structural changes at any position, including quite small modifications of the amino acid side chains, invariably result in some change in the hormones' potency or the spectrum of their actions.

The ring of oxytocin and vasopressin is a 20-membered structure (see Figure 3.24) which plays a primary role in determining the tertiary conformation of the molecule. The peptide side chain is not essential for activity. If the ring is broken to form a linear peptide, the molecule has little biological effect. No other amino acids can substitute for the half-cystines at positions 1 and 6, but the disulfide bridge that they contribute is not essential for activity. Thus, replacement of the sulfurs by methylene or selenium, so that the ring is left intact, reduces but does not abolish activity. A closed ring, rather than a reactive disulfide bridge, is necessary for activity. Increases or decreases in the size of the ring also decrease the molecule's effects. The isomers of the hormones, such as D-oxytocin, lack activity, although the substitution of D-amino acids is tolerated at certain loci and such changes can be used to advantage to alter the spectrum of its pharmacological actions. Such observations indicate that the particular shape of the molecule is important.

A model of the three-dimensional tertiary structure of oxytocin has been proposed by Urry and Walter (1971), and a recent version of this is shown in Figure 3.25. These studies have also been extended to arginine- and lysine-vasopressin (Figure 3.26; Walter et al., 1972, 1974, 1977). The initial construction of such conformational models is usually based on nuclear magnetic resonance (NMR) studies carried out in nonaqueous solvents, principally dimethylsulfoxide. Some differences clearly exist, however, compared to natural conditions, when the hormones are in aqueous solution. Nevertheless, many conclusions from such analyses are consistent with the effects of experimental changes in their structure on their biological activities. Such models are usually corrected on the basis of the latter type of observation. Urry and Walter found that the ring of oxytocin contained a right-handed helical twist, or β-turn, which involved the -Tyr-Ile-Gln-Asn sequence. This conformation (see Figure 3.25) is stabilized by an intramolecular hydrogen bond connecting the carbonyl, $C=O$, of the tyrosine (position 2) and the NH of the asparagine (position 5). The acyclic side chain of oxytocin has another β-turn, involving the -Cys-Pro-Leu-Gly-NH_2 sequence, and this is also stabilized by the asparagine, the carbonyl group of which

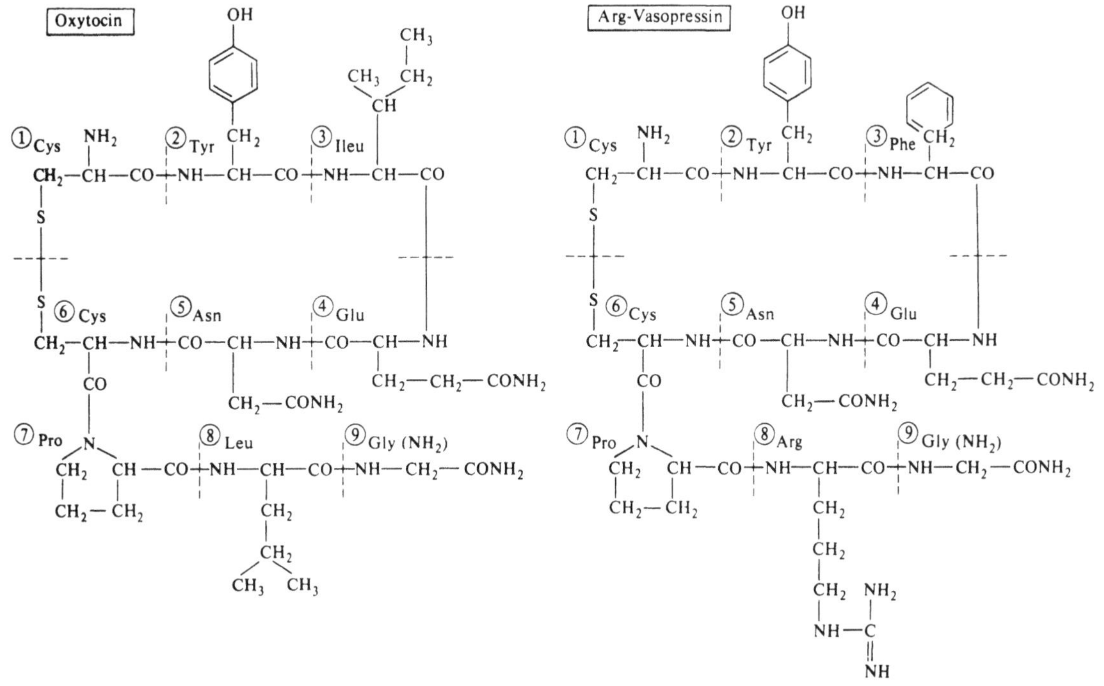

Fig. 3.24. Primary structures of oxytocin and vasopressin, showing the amino acid side chains. These hormones consist of a 20-membered ring which is closed by a disulfide bridge between the two half-cystines at positions 1 and 6, and a peptide side chain containing three amino acids. The two hormones differ by the presence of two amino acids at positions 3 and 8. (From Jard and Bockaert, 1975.)

hydrogen-bonds with the NH of the leucine (position 8). There is also on intramolecular bond between the NH of the glycine (position 9) and the carbonyl of the cystine (position 6). It was proposed that while the aromatic group of the tyrosine is free in the dimethylsulfoxide, in the "active" form, in aqueous solution, this folds over the ring.

The other natural hormones appear to have a similar general structure (Figure 3.26), but the peptide side chain of vasopressin exhibits a greater motility than does that of oxytocin (Walter et al., 1972, 1974, 1977). The presence of the second β-turn on the molecule's side chain is less certain. Also, in contrast to oxytocin, the side chain of the tyrosine in vasopressin appears to be oriented away from the ring, as its hydroxyl group does not appear to be involved in its reaction with the antidiuretic receptor. It is apparent that the general conformation of the molecules depends on the half-cystines at positions 1 and 6, the tyrosine at position 2, and, most important, the asparagine at position 5.

Information about the structure of neurohypophysial hormones can be useful in predicting the types of effects that certain amino acid substitutions may have on a hormone's biological activity (Walter et al., 1971).

a. Changes that alter the basic conformation of the molecule would be expected to have prominent, indeed drastic, effects on the molecule's affinity for its receptors and on its intrinsic activity. Thus, substitution for asparagine at position 5, even by closely related amino acids such as D-asparagine or glutamine, almost always abolishes activity. Breaking the —S—S— bridge has similar effects, but even subtle changes in the ring —S—S— junction, such as substitution of selenium or methylene for sulfur, reduce, although they do not abolish, activity.

b. It is possible to change the molecule in more subtle ways while leaving the peptide backbone conformation intact. Thus, substitution at less essential positions, such as 3, 4, and 8, can, by changing the local electronic, steric, lipophilic, and hydrophilic properties, alter the affinity and intrinsic activity of the molecule, as well as its spectrum of biological effects. The natural hormones exhibit differences of this nature, which confer on each of them their special properties.

c. The affinity for the receptors may be relatively unaffected by certain types of changes, but the molecule's intrinsic activity may be altered. This type of change, which may be due to alterations on the "active surface" of the

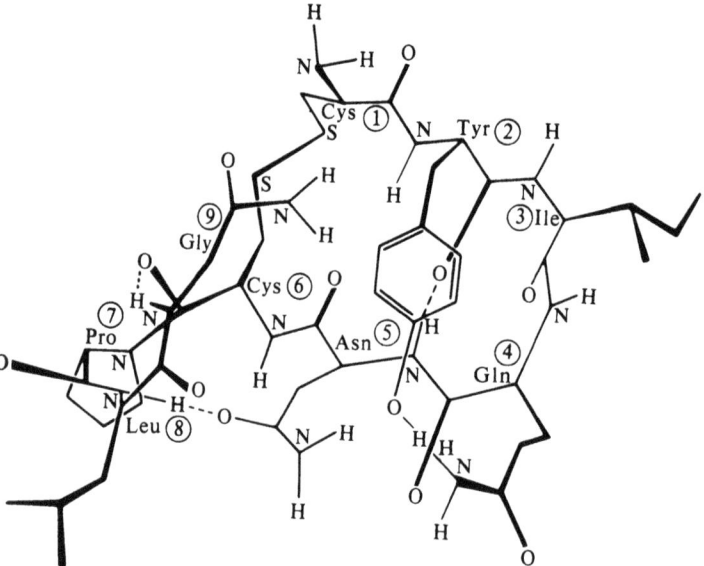

Fig. 3.25. Urry–Walter "cooperative model," showing the proposed teritary structure of oxytocin in its biologically active form in aqueous solution. One plane, the hydrophobic surface, is essentially featureless except for the protruding disulfide bond and the peptide N-H of glutamine. The other side, the hydrophilic surface, involves Tyr[2], Asn[5], Gln[4], and the linear tripeptide sequence of oxytocin. The model utilizes each of the constituent amino acids to its greatest effectiveness to achieve (i) formation and intramolecular stabilization of the backbone, to which the side chains of Cys[1], Cys[6], Tyr[2], and Asn[5] contribute. (ii) The side chains of Ile[3], Gln[4], Pro[7], and Leu[8] are considered to be free to engage in intermolecular interactions, while having a limited effect on the conformation of the peptide backbone, with the exception of the corner residue (Pro), in view of its relative rigidity. (From Walter, 1977.)

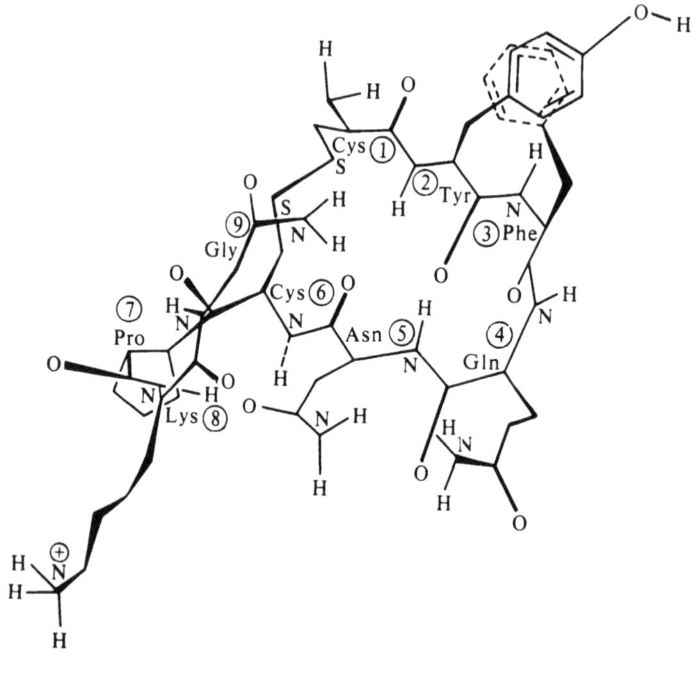

Fig. 3.26. Proposed tertiary structure of lysine-vasopressin in aqueous solution. The backbone structure also applies to arginine-vasopressin. (From Walter et al., 1977.)

Table 3.16. *Configuration of the side chains of amino acids at positions 3 and 8 in oxytocin and arginine- and lysine-vasopressin*

	Position 3	Position 8
Oxytocin	NH HC—CH (CH$_3$) CO (CH$_2$—CH$_3$)	NH HC—CH$_2$—CH (CH$_3$)(CH$_3$) CO
[Arg⁸]Vasopressin (ADH)	NH HC—CH$_2$—⟨phenyl⟩ CO	NH HC—CH$_2$—CH$_2$—CH$_2$—NH—C(NH$_2$)(=NH) CO
[Lys⁸]Vasopressin	NH HC—CH$_2$—⟨phenyl⟩ CO	NH HC—CH$_2$—CH$_2$—CH$_2$—CH$_2$—NH$_2$ CO

hormones, can result in the formation of antagonists. Such changes may not alter the configuration or be extensive enough to influence its ability to bind to the receptor, but once the hormone is on the receptor, it may not be able to trigger a response as readily. An example of such a change is seen in analogues of oxytocin, in which the phenolic hydroxyl group of the tyrosine is alkylated, such as in [2-O-methyl-tyrosine]oxytocin, which partially antagonizes the effects of oxytocin in contracting the uterus.

As referred to earlier, the potency of the neurohypophysial hormones can also be altered by changes that are unrelated to their interactions with their receptors. These include modifications that alter the rate of their clearance and enzymic inactivation and their absorption and distribution in the body. Such effects are usually seen more clearly *in vivo*.

The neurohypophysial hormones can exert a number of effects, such as their uterotonic, galactobolic, pressor, and antidiuretic actions, each of which have distinct requirements with respect to the structure of the molecules. Such preferences would appear primarily to reflect differences in their receptors, but the subsequent events that follow the hormone–receptor interaction may also differ. In the succeeding section, the main requirements for each type of response will be summarized together with the structural modifications that can be utilized to selectively change the peptide's activity.

1. Relation of structure to activity in oxytocin and vasopressin. Vasopressin and oxytocin differ by two amino acid substitutions, at positions 3 and 8. Oxytocin has isoleucine and leucine, whereas in vasopressin the respective amino acids are phenylalanine and arginine. The presence or absence of these amino acids results in the very different spectrum of activity of the two hormones (see Table 3.15). The more intimate nature of the changes in relation to the structure of the amino acid side chains are summarized in Table 3.16. The isoleucine at position 3 is quite critical for the oxytocic activity of the molecule, whereas arginine or a similar basic amino acid, such as lysine, at position 8 strongly favors pressor and antidiuretic effects. The substitution of phenylalanine for the isoleucine appears to depress oxytocic and galactobolic activities and only moderately increase the antidiuretic effects of the peptide (Table 3.15). The naturally occurring analogue [8-arginine]vasotocin, which has isoleucine at position 3 (as in oxytocin) but arginine at position 8, has a wide spectrum of activities; it is oxytocic, galactobolic, pressor, and antidiuretic. The combination of both substitutions, 3-phenylalanine and 8-arginine, results in the hormone vasopressin, which has only limited crossover with the actions of oxytocin. Other types of artificial changes can be utilized to change the ratios of the other activities of the molecules, such as their vasopressor and antidiuretic effects. These will be described in a later section.

Table 3.17. *Effects of removing the amino group from the 1-half cystine group of oxytocin and vasopressin*

| Analogue | Activity units/μmol | |
	Rat antidiuresis	Rat oxytocin
Oxytocin	5	450
1-Deaminooxytocin	~19	795
Vasopressin	435	17
1-Deaminovasopressin	1400	29

Source: Based on Berde and Boissonnas, 1968.

2. Enhancement of potency by influencing the metabolism of the peptides. The neurohypophysial hormones are rapidly destroyed by proteolytic enzymes and so have a half-life of only about 5 minutes in the circulation (see Section 3.3D). Their potency can be considerably enhanced *in vivo* by modifying the molecule so as to increase its resistance to enzymic attack. These changes are described more fully in Section 3.3D. Removal of the terminal amino group on the 1-half cystine reduces enzymic inactivation and considerably enhances the oxytocic and antidiuretic effects of the hormones (Table 3.17). Another way of prolonging the activity of the neurohypophysial peptides is to attach additional amino acids, either singly or in chains of two or three, at position 1 in the molecules (see Rudinger, Pliška, and Krejčí, 1972; Rudinger, 1971). These peptides are initially inactive but are "activated" as a result of enzymic cleavage of the additional amino acids (Pliška et al., 1976). Such precursors, or "hormonogens" have a prolonged effect; their actions develop and decline more slowly than do those of the natural hormones.

3. Enhancement of oxytocic activity. The presence of isoleucine at position 3 in oxytocin appears to be almost essential for oxytocic effects, and any change results in a drastic decline in this activity (Table 3.15). Changes at position 8 are also invariably detrimental to this response.

The effects of substitutions at *position 3* are summarized in Table 3.18, where the amino acid side chains are also depicted. It is apparent that quite small changes can have drastic effects on the molecule's activity. For instance, the allosteric form, [3-alloisoleucine]oxytocin, only has about $\frac{1}{20}$ the oxytocic activity of the natural hormone. Lengthening this amino acid side chain by one carbon, as seen when one compares [3-valine]- and [3-leucine]oxytocin, results in a 95 percent decline in activity in the latter. The presence of an aromatic ring, as in 3-phenylalanine-oxytocin, is also highly detrimental.

Position 8 is less sensitive to change (Table 3.18). Thus, [8-isoleucine]oxytocin is nearly as active as oxytocin and shortening the side chain, as in [8-valine]oxytocin, also leaves a molecule that has considerable oxytocic activity. Even the nature of the side chain does not appear to be especially important, as substitution of a basic amino acid, arginine, leaves a considerable oxytocic response. However, if the amino acid side chain is replaced by a hydrogen, as in [8-glycine]oxytocin, 97 percent of the activity is lost.

Many types of changes at *position 4* of oxytocin are tolerated with retention of its activity (Table 3.19). Indeed, substitution for the glutamine at this site occurs in some natural hormones which are present in fishes. One artificial analogue, [4-threonine]oxytocin, has almost *twice* the oxytocic activity of the natural hormone (Sawyer and Manning, 1971). Several properties of the amino side chain in position 4 have been related to its oxytocic activity. A two-carbon side chain, such as in [4α-aminobutyric acid]oxytocin, is optimal, and increasing or decreasing the length reduces activity. However, the addition of a branched methyl group as in [4-valine]oxytocin, doubles activity as compared to [Abu^4]oxytocin. The uterotonic effects are favored by hydrophilic groups in the side chain, such as the hydroxyl in [4-serine]- as compared to the methyl in [Abu^4]oxytocin.

The natural hormone also has such a moiety, the carboxamide terminal of this side chain. A lipophilic group, however, also appears to be advantageous in this position. Thus [4-threonine]oxytocin provides both of these, a methyl and a hydroxyl, on the β-carbon. A nonionizable hydrophilic group appears to be of primary importance for uterotonic activity in this position, but a lipophilic group also helps. [4-threonine]Oxytocin, apart from being a more potent oxytocic peptide, has other potential advantages over the natural hormone. It, for instance, has a relatively low endogenous antidiuretic and pressor activity. It would thus appear to offer practical advantages when used to assist labor [see Section 7.6 H(4)], although it has not yet been adopted for this purpose.

A more detailed model of the relationship of the structure to the activity of oxytocin has been proposed (see Hechter et al., 1975; Walter, 1977). Consensus is not complete, but it is agreed that the 20-membered ring and its conformation are of primary importance and that the peptide side chain also contributes receptor binding groups. The hormone's alignment with its receptors involves hydrogen and hydrophobic bonding. The β-turn in the ring structure results in exposure of two distinct surfaces. One of these is relatively featureless, or "planar" (except for the C=O of isoleucine at position 3). The other, "nonplanar," side has numerous projecting amino acid side chains, notably those of the tyrosine, glutamine, and asparagine at positions 2, 4, and 5, respectively. Walter has divided the contributions of the amino acids into two groups: (a) those necessary for binding (or the molecule's affinity), which are isoleucine, glutamine, proline, and leucine, at positions 3, 4, 7, and 8. (other groups may also contribute, but less directly); and (b) active elements that determine intrinsic activity, the hydroxyl group of the aromatic ring of the tyrosine (position 2) and the carboxamide group of the asparagine (position 5). Two possible models of the type of interaction that oxytocin may have with its receptors are shown in Figure 3.27.

4. Enhancement of vasopressor or antidiuretic activities. In contrast to oxytocic activity, the nature of the amino acid at position 8 of neurohypophysial peptides is critical to vasopressor and antidiuretic activity (Table 3.20). A basic amino acid, such as arginine at position 8, is essential for optimal activity. Substitution of citrulline, a neutral analogue of this amino acid, exhibits less than 10 percent of the activity of the natural hormone. Amino acids less basic than arginine, such as lysine and ornithine, are effective, but if the basic terminal amino group is masked by a formyl group, activity declines drastically. Neutral amino acids, such as leucine (which is present in oxytocin), reduce activity by more than 90 percent.

It is notable (Table 3.20) that the substitution of ornithine for arginine in position 8 of vasopressin leaves the pressor activity nearly intact but reduces the antidiuretic activity by about 80 percent. This indicates that the two responses can be dissociated and that their receptor requirements differ. Similarly, the replacement of L-arginine by its D-form also produces a relative change in the two ac-

tivities, but in this instance the antidiuretic effect persists and is nearly 30 times greater than the pressor one. Analogues with either selective pressor or antidiuretic activities are potentially useful, and several have been developed.

The analogue [8-ornithine]vasopressin has a pressor/antidiuretic activity ratio of 4:1, compared to 1:1 in [8-arginine]- and [8-lysine]vasopressin (Table 3.21). Changes in positions 2 and 3 also result in alterations in the ratio of activities of the 8-lysine and 8-ornithine analogues, but not in that of the 8-arginine compound. The substitution of isoleucine for phenylalanine in these analogues results in a modest increase in pressor relative to antidiuretic activity, but these molecules also have substantial oxytocic activity. Removal of the phenolic hydroxyl from the tyrosine in position 2 to produce phenylalanine also enhances the pressor/antidiuretic ratio, but the molecule has little oxytocic activity. The combination of these two changes produces the [Phe^2,Ile^3]analogues of [Lys^8]- or [Orn^8]vasopressin, which have selective pressor activity. In man, tests on [Phe^2,Lys^8]vasopressin (the generic name is *octapressin*) indicate that it has a high ratio of pressor/antidiuretic activity (see Berde and Boissonnas, 1968). Such compounds may be useful clinically as local vasoconstrictors to prevent local bleeding and delay the absorption of injected drugs, especially local anesthetics. At this time, however, they do not have an established clinical application.

It is also possible to change the type of vascular constrictor effect of analogues of vasopressin (Altura, 1976a,b). Thus, [1-deamino-Phe^2,Arg^8]vasopressin (DPAVP) and [Phe^3,Orn^8]vasopressin (POV) have more selective effects than does [Arg^8]vasopressin and partially constrict, but do not obstruct, the venules, so that they remain open. The peripheral circulation is thus improved and sustained. These effects have been demonstrated in rats, in which species they increase survival following experimental circulatory shock. There has been speculation that such analogues may be useful for the treatment of shock in man. Modification of DPAVP (Smith and Walter, 1978) by substituting 3,4-dehydroproline for proline at position 7 to produce [1-deamino-Phe^2, (3,4-dehydroproline^7), Arg^8] vasopressin (DPD-AVP), abolishes the vasopressor activity but results in an enhancement of the antidiuretic effect, so that it is 25 to 30 times more active than is the parent hormone, vasopressin.

Such analogues with selective antidiuretic activity have achieved therapeutic acceptance. Several types of analogues have been shown to have an enhanced ratio of antidiuretic/pressor activity (Table 3.22). These include [8-D-arginine]vasopressin,

Table 3.18. *Effects of substitution at positions 3 and 8 in oxytocin*

Analogue	Position 3	Position 8	Activity units/μmol	
			Rat oxytocic	Rat anti-diuretic
Oxytocin	NH–HC(CH₃–CH–CH₂–CH₃)–CO	NH–HC(CH₂–CH(CH₃)–CH₃)–CO	450 ([D-Leu⁸]Oxytocin = 20)	5
[Val³]Oxytocin	NH–HC(CH₃–CH–CH₃)–CO	As above	58	~0.8
[Leu³]Oxytocin	NH–HC(CH₂–CH(CH₃)–CH₃)–CO	As above	~4	~1
[aIle³]Oxytocin	NH–HC(CH₂–CH₃)(CH–CH₃)–CO	As above	~25	~0.05
[Phe³]Oxytocin (oxypressin)	NH–HC(CH₂–)–CO	As above	~20	~30

Compound	Structure	Structure		
[Ile⁸]Oxytocin	NH—CH(—CH₃)—... HC—CO, side chain CH₃, CH₂—CH₃	NH—CH(—CH₃)—... HC—CO, side chain CH₃, CH₂—CH₃	291	~1
[Val⁸]Oxytocin	As above	NH—CH—CO, HC, side chains CH₃, CH₃	199	0.8
[Gly⁸]Oxytocin	As above	NH—CH—CO, HC—H	15	~0.15
[Arg⁸]Oxytocin (vasotocin)	As above	NH—CH—CO, HC—CH₂—CH₂—CH₂—NH—C(=NH)—NH₂	120	260

Source: Based on Berde and Boissonnas, 1968.

Table 3.19. *Effects of substitution at position 4 on the uterotonic (oxytocic) effects of oxytocin*

Analogue	Structure of the amino acid at position 4	Oxytocin activity on rat uterus *in vitro* (units/mg)
Oxytocin (glutamine at position 4)	NH HĊ—CH$_2$—CH$_2$—CO—NH$_2$ ĊO	520
[Thr⁴]Oxytocin	NH CH$_3$ HĊ—CH ĊO OH	920
[Ser⁴]Oxytocin	NH HĊ—CH$_2$—OH ĊO	197
[Val⁴]Oxytocin	NH CH$_3$ HĊ—CH ĊO CH$_3$	140
[Abu⁴]Oxytocin	NH HĊ—CH$_2$—CH$_3$ ĊO	82
[Leu⁴]Oxytocin	NH CH$_3$ HĊ—CH$_2$—CH ĊO CH$_3$	13 (antagonist to ADH)

Source: Based on Sawyer and Manning, 1973.

analogues where the 1-amino group is removed, and those with substitutions at position 4. The effects usually result from a relatively strong suppression of the pressor responses. The antidiuretic effect can be moderately enhanced, or even, as in the [8-D-arginine] analogue, decreased.

It has been suggested that the deletion of the 1-amino group and the substitution of the 8-D-arginine acts by delaying the inactivation of the peptide in the vicinity of its receptors in the kidney (Rudinger, Pliška, and Krejčí, 1972). It seems likely that 8-D-arginine also has a differential effect in the interactions of the hormone with each type of receptor. Substitution at position 4 appears to influence interaction with the receptors. The amino acids α-aminobutyric acid, valine, and threonine at this site all increase the ratio of the antidiuretic/pressor activity. By the addition of methyl groups they make the molecule more lipophilic. There is, however, a steric restriction on

Table 3.20. *Effects of substitutions in position 8 in vasopressin*

Analogue	Structure of amino acid at position 8	Activity units/μmol	
		Rat pressor activity	Rat antidiuretic activity
[Arg[8]]Vasopressin	$HC-CH_2-CH_2-CH_2-NH-C$ (NH, NH$_2$; backbone NH, CO)	435	435
[Lys[8]]Vasopressin	$HC-CH_2-CH_2-CH_2-CH_2-NH_2$ (backbone NH, CO)	285	260
[Orn[8]]Vasopressin	$HC-CH_2-CH_2-CH_2-NH_2$ (backbone NH, CO)	375	92
[Cit[8]]Vasopressin	$HC-CH_2-CH_2-CH_2-NH-C(O)-NH_2$ (backbone NH, CO)	43	~16 (dog)
[Leu[8]]Vasopressin (oxypressin)	$HC-CH_2-CH(CH_3)_2$ (backbone NH, CO)	~3	~30
[D-Arg[8]]Vasopressin	$NH_2-C(NH)-NH-CH_2-CH_2-CH_2-CH$ (backbone NH, CO)	~4	~114

Source: Based on Berde and Boissonnas, 1968.

this effect, as leucine, which has two methyl groups but a side chain that contains one extra carbon, has very low activity. The increased lipophilic properties at position 4 appear to be mainly detrimental to the pressor response but to slightly enhance the antidiuretic action.

The effects of such changes at positions 1, 4, and 8 in the molecule of vasopressin are generally addi-

tive. These have thus been combined to produce analogues with high ratios of antidiuretic/pressor activity (Table 3.22). Thus [1-deamino, Val[4], D-Arg[8]]vasopressin (DVDAVP) has a ratio of antidiuretic/pressor activity which is greater than 125,000. This analogue has not been introduced into clinical practice, but [1-deamino,D-Arg[8]]vasopressin (DDAVP) is now being widely used to treat

(a)

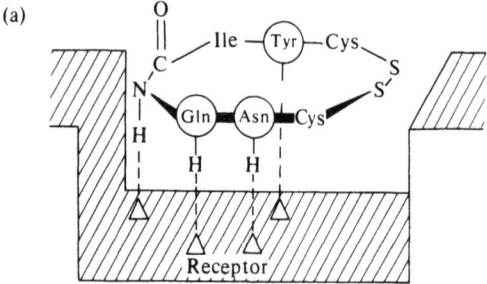

(b)

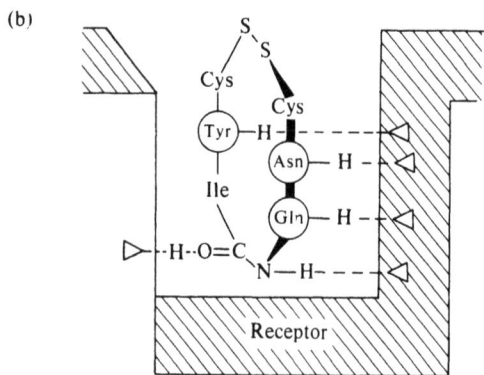

Fig. 3.27. Two models proposed by Hechter et al. (1975) to describe the interaction of the ring of oxytocin with its receptors. (The role of the peptide side chain is not shown.) (a) Interaction of the nonplanar surface of the tocin ring with a receptor pocket, where the planar surface of the peptide is exposed to the aqueous environment. Hydrogen bonds are illustrated; the other bonding interactions are not shown. (b) Hydrogen-bonding interactions of both surfaces of the tocin ring with a receptor pocket. Other types of bonding are not illustrated. (*Note:* The Urry–Walter model does not consider it likely that the asparagine contributes directly to the binding, and the tyrosine may only have a minor role. However, they are thought to make an important contribution to the hormone's intrinsic activity.) (From Hechter et al., 1975.)

diabetes insipidus. The estimates of its ratio of antidiuretic/pressor activity vary depending on its source (there is possible contamination with [8-D-arginine]vasopressin) but may be as high as 2000 : 1. Its oxytocic activity is also much less than that of vasopressin.

5. Molecular antagonists of the neurohypophysial hormones. Modification of the structure of the neurohypophysial peptides in order to produce specific molecular antagonists has met with only limited success (see Rudinger and Krejčí, 1968; Rudinger, 1971; Manning et al., 1977). Some analogues have been prepared which under certain conditions can antagonize the effects of

oxytocin and even those of vasopressin, but these responses have only been demonstrated in animal experiments and have not been extended to man.

The substitution of two methyl groups on the β-carbon in the 1-half-cystine residue of oxytocin produces an antioxytocic compound, 1-L-penicillamine-oxytocin (βMe_2-oxytocin). The 1-deamino-L-penicillamine analogue is even more effective. If only a single methyl group is inserted, antioxytocic activity is lost. Increasing the length of the side chains, however, can increase activity, as seen in βET_2-deaminooxytocin, [1-β-mercapto-(βET_2)] oxytocin, and [1-β-mercapto-(β-(CH_2)$_5$)]oxytocin (Chan et al., 1974; Nestor, Ferger, and Du Vigneaud, 1975). The substitution of threonine for glutamine at position 4 of 1-deamino-L-penicillamine-oxytocin (= [1-deaminopencillamine-4-threonine]oxytocin or dPTOT) results in the most potent inhibitor of rat oxytocic activity *in vitro* that is known at this time (Manning et al., 1977). Although these effects can be demonstrated on the rat uterus *in vitro*, their actions *in vivo* are relatively weak.

Substitutions at position 2 have resulted in a number of antagonists of oxytocin. These effects can be demonstrated *in vitro* and *in vivo* in animals, but the particular experimental conditions influence the response and they may also behave as partial agonists (see, e.g., Krejčí, Poláček, and Rudinger, 1967). Tyrosine, which has an aromatic ring with a phenolic hydroxyl group at the para position, is normally present at position 2. A series of substitutions at the para position has been made (Table 3.23). The bulk of the para group appears to be important and may contribute steric hindrance, which affects the intrinsic activity of the molecule rather than its ability to bind to its receptors. The 2-O-alkyltyrosine analogues of oxytocin- and lysine-vasopressin have also been shown to inhibit the pressor effects of vasopressin in animals (Krejčí, Kupková, and Vávra, 1967).

Other types of substitutions that may result in antagonistic effects include anti-ADH responses to [4-asparagine]-, [4-leucine]-, and [4-phenylalanine] oxytocin in rats (Datta and Chaudhury, 1970; Chan, 1976) and the inhibition of the antidiuretic effects of vasotocin in frogs by oxytocin. Antagonists of the neurohypophysial hormones are potentially useful, such as in the treatment of the syndrome of inappropriate secretion of ADH and even possibly for delaying parturition. It is therefore to be hoped that more practically useful analogues will be devised for such purposes.

C. Synthesis and release
The stimuli that effect the *release of vasopressin and oxytocin* are summarized in Table 3.24. The principal physiological stimulus for the release of ADH is an increase in the osmotic pressure

Table 3.21. *Analogues of vasopressin with selective vasopressor activity*

| Analogue | Activity units/μmol | | |
	Rat pressor	Rat antidiuretic	Ratio pressor/ antidiuretic
[Arg⁸]Vasopressin	435	435	1.0
[Lys⁸]Vasopressin	285	260	1.1
[Phe²,Lys⁸]Vasopressin (octapressin)	57	21	2.7
[Ile³,Lys⁸]Vasopressin (Lysine-vasotocin, [Lys⁸]oxytocin)	133	25	5.3
[Phe²,Ile³,Lys⁸]Vasopressin ([Phe²,Lys⁸]oxytocin)	32	1	32
[Orn⁸]Vasopressin	375	92	4.1
[Phe²,Orn⁸]Vasopressin	157	16	9.8
[Ile³,Orn⁸]Vasopressin ([Orn⁸]oxytocin)	104	2.5	4.6
[Phe²,Ile³,Orn⁸]Vasopressin ([Orn⁸]oxytocin)	123	0.55	224

Note: The naming of the molecules has for descriptive purposes been related to vasopressin, but the alternative names are given in parentheses.
Source: Based on Berde and Boissonnas, 1968.

of the plasma, such as that which occurs as a result of dehydration. The details of this process were described by E. B. Verney in his Croonian Lecture (Verney, 1947). Conversely, hypoosmotic conditions inhibit the release of ADH. The osmoreceptors are thought to be located in the vicinity of the internal carotid artery, possibly near the supraoptic nucleus. Their stimulation appears to evoke a neuronal discharge which is probably cholinergic. Thus, the release of ADH can be stimulated by cholinergic drugs and nicotine. The latter response is seen following the smoking of a couple of cigarettes. Adrenergic

mechanisms may also be involved and could be exerting an inhibitory action, but these effects are ill defined. A reduction in plasma volume, such as that which occurs in hemorrhage, is also a potent stimulus for the release of ADH. The receptors for this response are present in the left atrium and are associated with the baroreceptors in the aortic arch and carotid sinuses (see Share, 1974). Vasopressin is also readily released in response to stress and emotion, but the neural pathways involved have not been identified. A number of pharmacological stimuli also influence the release of ADH (Table 3.24). One of the more notable ones is an inhibition

Table 3.22. *Analogues of vasopressin with selective antidiuretic activity*

| Analogue | Activity units/mg | | |
	Rat anti- diuretic	Rat pressor	Ratio anti- diuretic/ pressor
[Arg⁸]Vasopressin	332	376	0.9
[D-Arg⁸]Vasopressin	114	4	28
[1-deamino,Arg⁸]Vasopressin	1390	370	3.8
[1-deamino-D-Arg⁸]Vasopressin (DDAVP)	870	11	79
	955	0.47	2000
[Abu⁴,Arg⁸]Vasopressin	760	38	20
[1-deamino-Abu⁴]Vasopressin	1020	11	95
[Val⁴]Vasopressin	738	32	23
[Val⁴-D-Arg⁸]Vasopressin	653	0.037	17,650
[1-deamino-Val⁴,-D-Arg⁸]Vasopressin (DVDAVP)	1230	<0.01	>125,000
[1-deamino,Phe²,Pro(3,4-dehydro)⁷,Arg⁸]Vasopressin (DPD-AVP)[a]	13,000	~0.0	

[a] Smith and Walter, 1978.
Source: Based on Sawyer et al., 1974; Manning et al., 1973; and László et al., 1975.

Table 3.23. *Antagonists of oxytocin: effects of substitution at position 2*

Analogue	Structure of amino acid at position 2	Rat uterus (activity units/μmol)
Oxytocin (tyrosine at position 2)	NH HC—CH$_2$—⟨ring⟩—OH CO	450
[Phe^2]Oxytocin	NH HC—CH$_2$—⟨ring⟩ CO	32
[p-MePhe^2]Oxytocin	NH HC—CH$_2$—⟨ring⟩—CH$_3$ CO	19 (Partial agonist and antagonist)
[O-MeTyr^2]Oxytocin (2-methyltyrosine-oxytocin, methyloxytocin)	NH HC—CH$_2$—⟨ring⟩—O—CH$_3$ CO	~2.0 (Partial agonist and antagonist)
[O-EtTyr^2]Oxytocin (2-ethyltyrosine-oxytocin)	NH HC—CH$_2$—⟨ring⟩—O—CH$_2$—CH$_3$ CO	Antagonist

Source: Based on Berde and Boissonnas, 1968; Rudinger and Krejčí, 1968.

of release resulting from rising levels of ethanol in the blood. Many anesthetics and morphine are potent stimulants of release. The effects of the latter can be blocked by the narcotic antagonist oxilorphan. It has been observed that certain drugs exhibit side effects that result in a low plasma sodium concentration, and this undesirable situation is associated with an increased release of ADH. Such drugs include the hypolipidemic agent clofibrate, and possibly carbamazepine, which is a tricyclic antidepressant sometimes used in the treatment of epilepsy. Chlorpromazine, a phenothiazine tranquilizer, and the antiepileptic drug diphenylhydantoin reduce the release of ADH. It is notable that most of these drugs act in the CNS, but the precise nature of their effect on the neuro-hypophysis is not clear. The possible uses of such drugs as therapeutic agents to control the release of ADH will be discussed later.

Oxytocin is released in response to a suckling response by the infant, and the afferent part of this reflex is a neural pathway to the brain. Stimulation of the female genitalia and parturition also result in a release of this hormone.

Usually, some oxytocin and vasopressin are released together in response to the same stimulus, but the relative amounts of each hormone differ depending on the precise nature of the stimulus. Thus, in rabbits it has been estimated that 100 times more oxytocin than vasopressin is released in response to suckling by the young. In other instances, such as dehydration, ADH predominates.

Table 3.24. *Stimuli that influence the release of antidiuretic hormone, ADH*

Physiological	
Increase in plasma osmotic pressure	↑
Reduced plasma volume (e.g., hemorrhage)	↑
Emotion and stress	↑
Dilution of plasma (e.g., hydration)	↓
Pharmacological	
Nicotine	↑
Morphine	↑
Angiotensin II	↑
Cholinergic drugs (not inhibited by atropine)	↑
Anesthetics (ether and barbiturates)	↑
Clofibrate	↑
Carbamazepine	↑
Chlorpropamide	↓
Ethanol	↓
Chloral hydrate	↓
Oxilorphan (narcotic antagonist)	↓
Diphenylhydantoin	↓
Chlorpromazine	↓

Note: ↑, increase; ↓, decrease.
Source: Based on Ginsburg, 1968. Other references are presented in the text.

Any differences that may exist in the release of the two hormones in response to drugs have not been precisely defined. However, there appear to be common relationships, and ethanol has also been used to decrease the release of oxytocin, in attempts to delay labor [Section 7.6H(4)].

The *synthesis of oxytocin and vasopressin* occurs in the cell bodies of the supraoptic and paraventricular nuclci. Our knowledge of the details of this process is due principally to the work of Sachs and his group (Sachs et al., 1969; Valtin, Stewart, and Sokal, 1974). Labelled ^{35}S-cysteine was infused into the cerebrospinal fluids and the labelled hormone and its precursors were recovered from the supraopticohypophysial tract and pars nervosa. Autoradiographic, as well as biochemical procedures have been used to follow the synthesis of the hormones. Labelled inactive precursor (but not hormone) appears in the hypothalamic nuclei after 1 to 2 hours. This response can be inhibited by puromycin, which inhibits protein synthesis by the ribosomes. The labelled hormone itself, however, does not appear for about another 4 hours, which suggests that the transition from precursor to hormone is a relatively slow process. If puromycin is given subsequent to the initial period when the precursor is being formed, it has no effect. It therefore appears that the final formation of the hormones themselves does not occur at the ribosomal level.

The *neurohypophysial hormones are stored in granules* which have been identified by electron microscopy and can be separated by differential centrifugation procedures (Barer, Heller, and Lederis, 1963). These granules are bounded by membranes and contain an ATPase. The hormones can be released from *in vitro* preparations by their disruption, using such methods as sonication or exposure to hypoosmotic solutions. Each hormone appears to be stored in a different type of granule. The granules have been identified at all levels of the supraopticohypophysial tract but accumulate in greatest numbers at nerve terminals abutting blood vessels in the pars nervosa.

In 1942, H. B. Van Dyke and his collaborators (Van Dyke et al., 1942) showed that vasopressin and oxytocin were bound in the pituitary to a protein with a molecular weight of about 30,000, now known as the "Van Dyke protein." Acher and his group (Acher, Chauvet, and Olivry, 1956) characterized this protein and showed that it could be broken down to release oxytocin, vasopressin, and a protein which they called *neurophysin*. This "carrier" protein is rich in cysteine and has been shown to exist in several forms (Hope and Pickup, 1974). It is the histochemically stainable neurosecretory material. The molecular weight of the nondimeric forms are about 10,000, reflecting the presence of about 100 amino acids, which are cross-linked. In a bovine neurophysin (II) (Figure 3.28), there are seven such disulfide bridges. Usually, as in man, two forms of neurophysin, I and II, appear to be present. Despite initial uncertainties it appears that, *in vivo*, each form binds a single molecule of

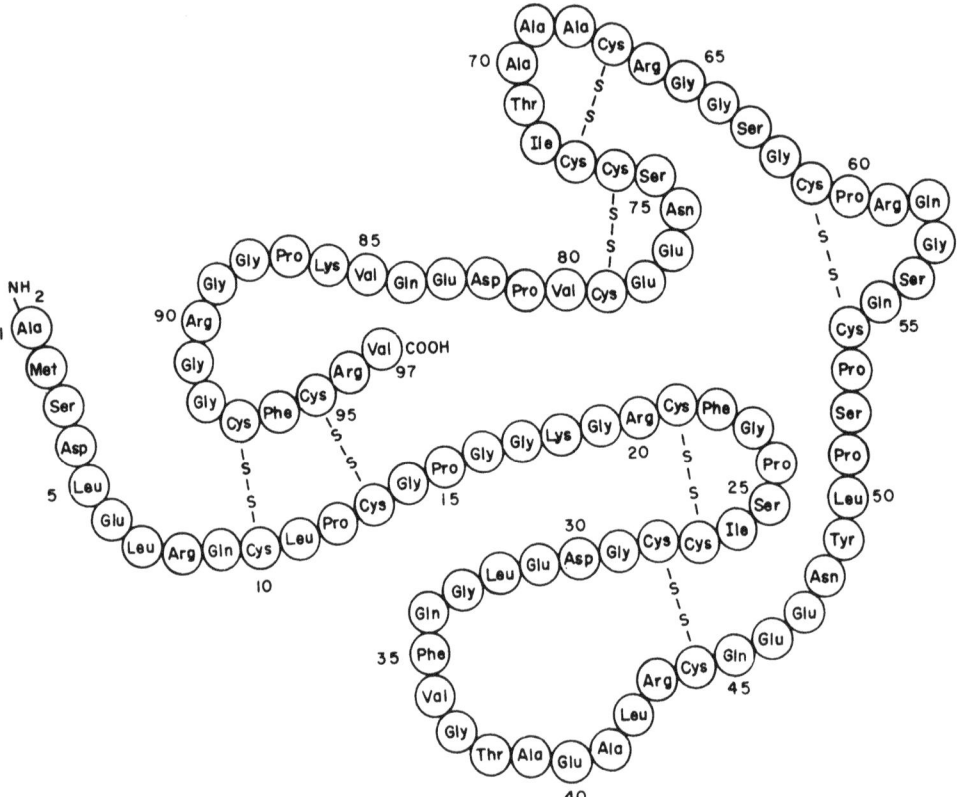

Fig. 3.28. Amino acid sequence of bovine neurophysin II, showing the disposition of the seven disulfide bridges. (From Schlesinger, Frangione, and Walter, 1972.)

the hormone. In cattle, it has been shown that neurophysin I is associated with oxytocin and neurophysin II with vasopressin. Under *in vitro* conditions, neurophysins can bind one molecule of either oxytocin or vasopressin. The linkage of the peptides and their carrier is noncovalent and can be readily disrupted. The binding is, nevertheless, quite specific for the neurohypophysial hormones, although it does not appreciably discriminate between them. Analogues lacking the α-amino group at position 1 do not bind to the neurophysins, but other changes in the molecule have little effect (see Breslow and Walter, 1972). The neurophysins are present in the storage granules, where they bind oxytocin or vasopressin. They are released into the circulation at the same time the hormones are released, but the two components are dissociated under these conditions.

The precise details of the process of packaging of the neurohypophysial hormones and the neurophysins into the neurosecretory granules are uncertain. The granules are formed in the perikaryon (Figure 3.29) in the Golgi apparatus. It appears likely that the hormones themselves are split

off from their precursor molecule within the granules during their movement down the axon. It has recently been shown (Gainer, Sarne, and Brownstein, 1977) that the neurophysin and the hormones are derived from a common precursor molecule with a molecular weight of about 20,000. Apart from the neurophysin, oxytocin, and vasopressin, six other cysteine-containing peptides are split off the precursor and are transported to the neurohypophysis.

The granules travel down the axons at a speed of about 4 mm/h, which is much faster than normal axonal flow. It appears that special processes, possibly involving the fiber structures in the axons, may be involved, but there is no specific information about this possibility.

Our knowledge regarding the nature of the process by which oxytocin or vasopressin and neurophysin are discharged from their granules at the nerve terminals is principally due to the observations of W. W. Douglas (see Douglas, 1974). It was found that release of the hormones could be promoted *in vitro* from pars nervosa tissue by depolarizing concentrations of external potassium or

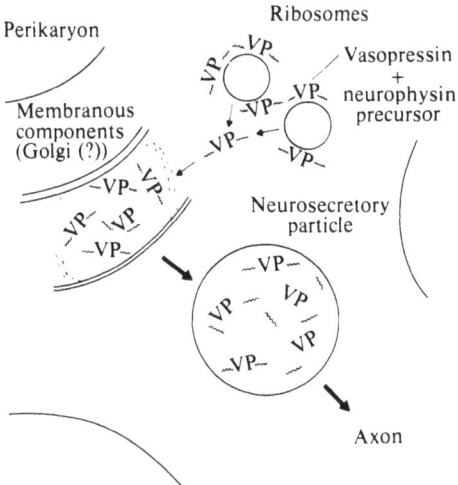

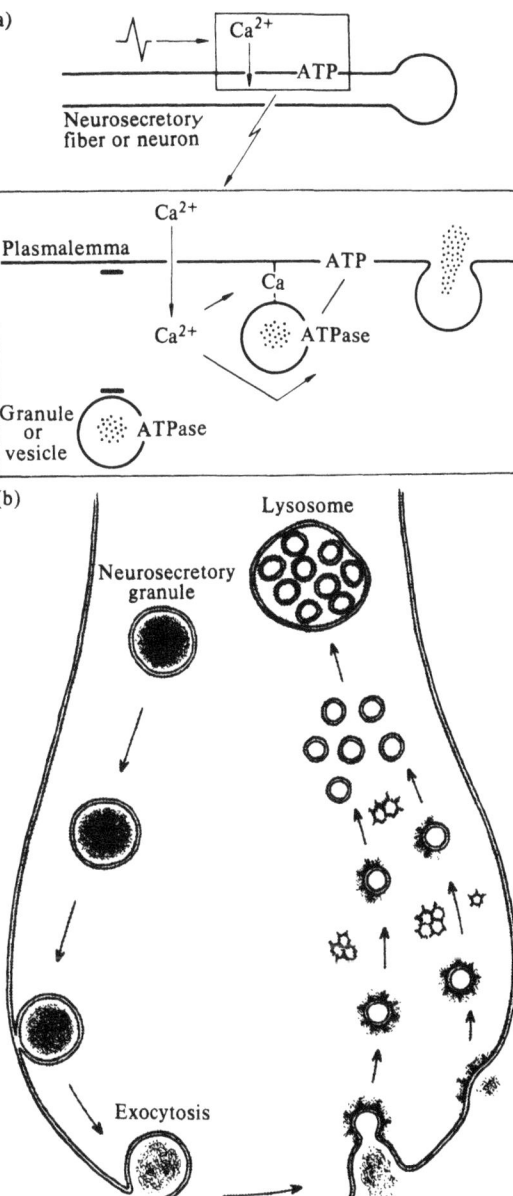

Fig. 3.29. Model describing the formation of vasopressin from a precursor molecule in the perikaryon of the neurosecretory cells of the supraoptic and paraventricular nuclei in the hypothalamus. The molecules are thought to be packaged into granules in the Golgi apparatus, and the active peptide hormone is probably split off the precursor molecule during the passage of the granule down the nerve axon toward the pars nervosa. (From Sachs and Takabatake, 1964.) This model has been confirmed by Gainer, Sarne, and Brownstein (1977), who also showed that neurophysin, as well as the hormones, are derived from the precursor, which has a molecular weight of 20,000.

by electrical stimulation. This secretion, however, did not occur if calcium was omitted from the bathing medium. This observation suggests that Ca^{2+} may be necessary for a coupling of the stimulus and response process, such as is well known in several other systems, including contraction of muscle and the processes of secretion of other exocrine and endocrine glands. An influx of Ca^{2+} into the neurohypophysial axons was directly demonstrated. The Ca ionophore A23187, which carries this divalent ion across cell membranes, also initiates such a release of ADH. It has been proposed that the contents of the secretory granules are discharged following their fusion to the cell membrane across the cell membrane by a process of exocytosis, which could involve the breakdown of ATP (Figure 3.30). Calcium could have several effects. These include the carrying of an electrical charge, which contributes to depolarization of the membrane, and it has a role in the linkage of the granular and plasma membranes. The latter may involve an activation of granular ATPase or the formation of cross-linkages between the two structures. It is possible that cell microtubules and microfilaments are involved in this process, because the drugs colchicine and cytochalasin B, which, re-

Fig. 3.30. Model to describe the release of neurophypophysial hormones from their neurosecretory storage granules at the neurosecretory nerve terminals in the pars nervosa. (a) Possible role of Ca^{2+} in initiating exocytosis. This may involve a neutralization of negative charges so that a cationic bridge is formed between the granular membrane and the plasma membrane. It is also possible that the Ca^{2+} plays a role in the activation of granular ATPase, which can then split ATP in the cell membrane. (From Douglas, 1973.) (b) Following exocytosis of the contents of the neurosecretory granules, the granular membranes may be retrieved by a process of vesiculation and are eventually incorporated into lysosomal bodies. (From Douglas, 1974.)

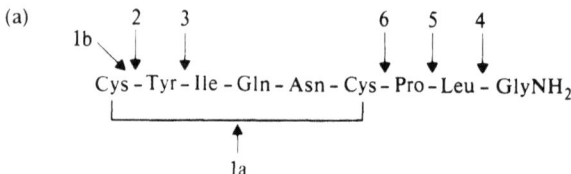

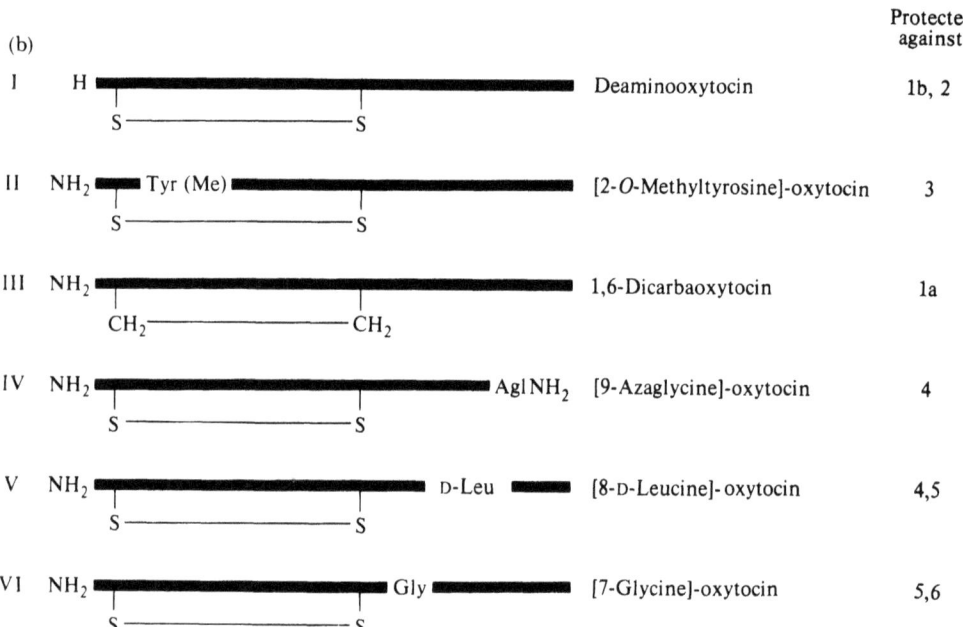

Fig. 3.31. (a) Sites in the molecule of oxytocin where enzymic inactivation mechanisms have been shown to act. (1a) SS-SH transhydrogenase or nonenzymic reduction; (1b) aminopeptidase splitting after 1a; (2) serum oxytocinase; (3) tyrosinase; (4) carboxamidopeptidase; (5 and 6) endopeptidases found in hypothalamic tissue. (b) Synthetic analogues of oxytocin that are resistant to attack at such sites. (From Pliška and Rudinger, 1976.)

spectively, inhibit the activity of these two types of structures, can block the release of neurohypophysial hormones. The specificity of their actions at the concentrations used to elicit such effects is, however, in doubt (see Douglas, 1974).

The *release of the neurohyopohysial hormones* thus appears to follow the following sequence. As a result of stimulation of specific receptors, the neurons arising in the supraoptic and paraventricular nuclei are depolarized. This response may be initiated by acetylcholine. When the electrical change reaches the nerve terminals, it produces an increased influx of Ca^{2+}. This ion couples the stimulus and response so that the storage granules fuse with the axonal membrane and there is an exocytotic extrusion of the hormones and their associated neurophysins into the adjacent blood vessels.

D. Metabolism

The neurohypophysial hormones in the blood are not bound to plasma proteins or the neuro-

physins that are simultaneously released with them. They are rapidly cleared from the circulation and have a half-life of about 5 minutes (see Ginsburg, 1968; Walter, 1973; Pliška and Rudinger, 1976). About 5 to 10 percent of the hormones appear unchanged in the urine. The remainder are inactivated, mainly by the kidney and to a slightly lesser extent by the liver. This process involves the actions of peptidases and the splitting of the disulfide bridge by an SS—SH transhydrogenase. Other tissues may contribute to the inactivation of the hormones, and these include effectors such as the uterus and the mammary glands. In pregnant women, an aminopeptidase is present in the plasma which inactivates oxytocin and is thus called oxytocinase. Its action is, however, very slow, so that the significance of its contribution *in vivo* is in doubt. Inactivation of oxytocin and vasopressin occurs as a result of breaking the molecule at specific points (Figure 3.31). The acyclic side chain is especially vulnerable and can be cleaved at three sites; between positions 8 and 9, such as occurs in the kidney, positions 7 and 8 in the uterus, and the positions 6 and 7 in the hypothalamus. The ring structure can be attacked via the amino group at position 1, or the disulfide may be directly broken. In the kidney it appears that the release of the C-terminal glycinamide moiety is of major importance (Walter and Bowman, 1973). The destruction of the neurohypophysial peptides can be slowed by changing certain groups in the molecule. Thus, the 1-deamino analogues have a prolonged activity (see Gazis and Sawyer, 1978). Similarly, substitution of glycine for proline at position 7 can enhance the half-life of these peptides. The substitution of ethylene for the sulfurs in the ring also stabilizes the molecules, as seen in [1,6-aminosuberic acid]oxytocin. This change, apart from preventing biological ring reduction, also enhances the peptides shelf life, as there appears to be a polymerization of the disulfide bond during prolonged storage. The effects of such substitutions on the resistance of oxytocin to inactivation are illustrated in Figure 3.31 (see also Walter, Yamanaka, and Sakakibara, 1974b).

E. Mechanism of action

The neurohypophysial hormones act at a number of sites in the body, but their effects can be divided into two general categories:

a. Actions that result in increased permeability of epithelia to water, as seen in the distal parts of the mammalian renal tubules. Epithelia from amphibians, such as the skin and urinary bladders of frogs and toads, respond in a similar manner to the mammalian renal tubules, and *in vitro* preparations of these membranes have been widely used as models to study the mechanism of action of these peptide hormones.
b. Contractile effects on smooth muscle, which include the uterus, the myoepithelial cells of the mammary glands, and the blood vessels.

A considerable amount of information is available about the mechanism of action on the permeability of epithelia, but their effects on smooth muscle are less well understood. It is, in addition, even uncertain whether all types of smooth muscle utilize the same mechanism in their response to the neurohypophysial peptides. Recent general reviews include those of Jard and Bochaert (1975), Dousa and Valtin (1976), and Hays (1976).

The receptors for the neurohypophysial peptides are thought to be present in the cell membrane, so that the hormones do not have to enter the cell in order to act. In the renal tubule and model amphibian epithelia, they have a unilateral distribution on the cells and are only present on the basal and possibly lateral surfaces, so that they are in contact with the extracellular fluids. ADH is thus ineffective when present on the luminal side of the renal tubule, such as when it appears in the urine.

The structure–activity relationships and the importance of the three-dimensional configuration of the neurohypophysial peptides have already been described. Clearly, the receptors have rather precise requirements with respect to their ability to interact with such hormones, as well as for the generation of a response (i.e., affinity and intrinsic activity). The receptors in various tissues and animal species show distinct differences, as reflected by the abilities of the organs to preferentially respond to different analogues of the hormones. There would, however, also appear to be some general similarities in the receptors, as each tissue can usually respond to some degree to a wide range of such analogues. Thus, a cross-reactivity between preferred hormones can occur, but because of a lack of optimal structure for a "best fit" in forming a particular hormone–receptor complex, the response may not be as great.

The identification, isolation, and study of the properties of the receptors for neurohypophysial hormones depends to a considerable extent on the availability of active radioactively labelled hormone preparations such as [³H]lysine-vasopressin and [³H]oxytocin, which have a high specific activity. The binding of the hormones to receptors can thus be measured, they can be "tagged" for isolation studies, and the specificity of the interaction can be assessed by the abilities of "cold" hormones and their analogues to displace their labelled, "hot" congeners.

1. Epithelial membranes. The *receptors* for the neurohypophysial peptides on epithelial membranes have been studied in intact tissues, such as the kidney, and *in vitro* in amphibian skin and urinary bladder. Membrane fragments of broken cells containing receptors that have been solubilized with the aid of the nonionic detergent Triton X-100 have also been prepared (see Jard and Bockaert, 1975). The properties of the receptors generally conform to those expected from observations on the tissues' response. Thus, binding is readily reversible and the K_D (concentration of the hormone when binding is half-maximal) is 10^{-8} to 10^{-9} M. The minimum effective concentrations in plasma are about 10^{-12} M. There is evidence that the responses are related to the degree of occupancy of these receptors, but the relationship is not linear and there appears to be a surfeit of receptors (spare receptors).

The vasopressin receptor has been separated in soluble form from membrane fractions prepared from the medulla of pig kidneys (Roy et al., 1975). It appears to be proteinaceous. In the plasma membranes this receptor is associated with adenyl cyclase. The precise nature of the link between the two is unknown, but it appears to involve hydrophobic forces, as the two components can be separated by detergents. When the linkage between the receptor and the enzyme is retained, exposure of the broken cell membrane preparations to neurohypophysial peptides results in an activation of the adenyl cyclase. The effectiveness of different structural analogues of vasopressin in eliciting an activation of adenyl cyclase *in vitro* parallels their potency on their effectors *in vivo* (see Roy, Barth, and Jard, 1975a,b). The linkage of the hormone and the receptor is not covalent and does not involve the disulfide bridge, as was formerly suggested. Some of these peptides may, however, be bound by the tissue via the disulfide group, especially at high concentrations, but this process is nonspecific and does not involve the receptors. Relatively high concentration of Ca^{2+}, up to 10 mM, can inhibit the binding of vasopressin to its epithelial receptors (Campbell, Woodward, and Borberg, 1972), which may account for the vasopressin insensitivity that is sometimes observed during hypercalcemia.

The receptor is thought to face outward on the basal plasma membrane while its associated adenyl cyclase lies on the inner surface so that it is in contact with the contents of the cell. The activation of this enzyme plays a vital role in the permeability response of epithelial membranes to neurohypophysial peptides. There is an increased conversion of ATP to cyclic AMP. The nature of the stimulus is unknown, but there may be an increased affinity of ATP for the enzyme or an increased maximal velocity of the enzyme reaction. Magnesium is a cofactor and increased concentrations of Ca^{2+} may inhibit the enzyme, but its precise relationship to the presence of the latter ion is obscure. Low concentrations of Ca^{2+} may be necessary for the reaction. Some prostaglandins, especially PGE_1, appear to inhibit the action of adenyl cyclase and may modulate the response to the hormone. This may explain the effects of indomethacin and aspirin, which inhibit prostaglandin synthetase, in enhancing the antidiuretic effects of vasopressin (Berl et al., 1977; Fejes-Tóth, Magyar, and Walter, 1977). Catecholamines may also inhibit antidiuretic responses, and this effect appears to be related to α-adrenergic effects. It is possible that this latter response is mediated via an action on adenyl cyclase, but this possibility does not yet appear to have been tested directly. Changes in the tissue osmolality such as those that accompany the increased admission of water into the cells may also help modulate the response.

Cyclic 3',5'-AMP acts as a "second messenger" in an incompletely defined series of events that results in an increase in permeability of epithelial cells to water. This nucleotide can be inactivated by conversion to 5'-AMP by a *phosphodiesterase*. This soluble enzyme is present in the epithelial cells and can be inhibited by xanthine drugs, such as theophylline and caffeine.

The exposure *in vitro* of certain preparations of epithelial membranes, such as the amphibian urinary bladder to exogenous cyclic AMP or theophylline, results in an increase in their permeability to water. The latter thus mimic the effects of vasopressin on the kidney. Direct measurements of the levels of cyclic AMP in such cells stimulated by vasopressin show that the levels rise, and this occurs prior to the response. There is thus little doubt that the activation of adenyl cyclase and the formation of cyclic AMP is related to the response of epithelia to ADH.

The precise nature of the effect, however, is unknown, but several processes that appear to be involved have been identified. Observations on isolated amphibian epithelia indicate that the response of the cell is mediated by a change in permeability which is confined to the plasma membrane at the luminal, or apical, border of the cell. The receptor and the adenyl cyclase are, however, present at the basal border of the cell (see Figure 3.32). The cyclic AMP that is generated as a result of the activation of adenyl cyclase has been observed, in turn, to activate a *protein kinase*. This enzyme mediates the transfer of phosphorus from ATP to threonine and serine chains on proteins and polypeptides. Such phosphorylation may change their structural configuration. This effect of cyclic AMP was first shown to occur in broken-cell preparations of epithelia but has recently also been demonstrated in the renal medulla *in vivo* (Dousa and Barnes, 1977). A protein has been identified in the plasma membranes of model amphibian epithelia which undergoes changes in phosphorylation in the presence of neurohypophysial peptides (DeLorenzo et al., 1973; Ferguson and Twite, 1974; Handler, Strewler, and Orloff, 1977). This may be closely involved with the final effector response. Thus, a phosphorylation or dephosphorylation of such a protein may change the configuration of local areas of the apical plasma membrane and result in an increase in diffusion of water into the cell. The phosphorylation may be reversed by a *phosphatase* which has also been demonstrated in these cells. Protein kinases are, however, relatively nonspecific in their choice of protein substrates, and a more direct identification of the effector has not yet been made. It is, however, interesting that local changes in the aggregation of particles on the luminal plasma membrane have been observed in amphibian urinary bladder epithelium exposed to ADH (Chevalier, Bourguet, and Hugon 1974; Kachadorian, Wade, and DiScala, 1975). These changes may reflect an alteration in the structure of the membrane.

It has been proposed that several types of processes may be involved in the influencing of the permeability of epithelia, which respond to

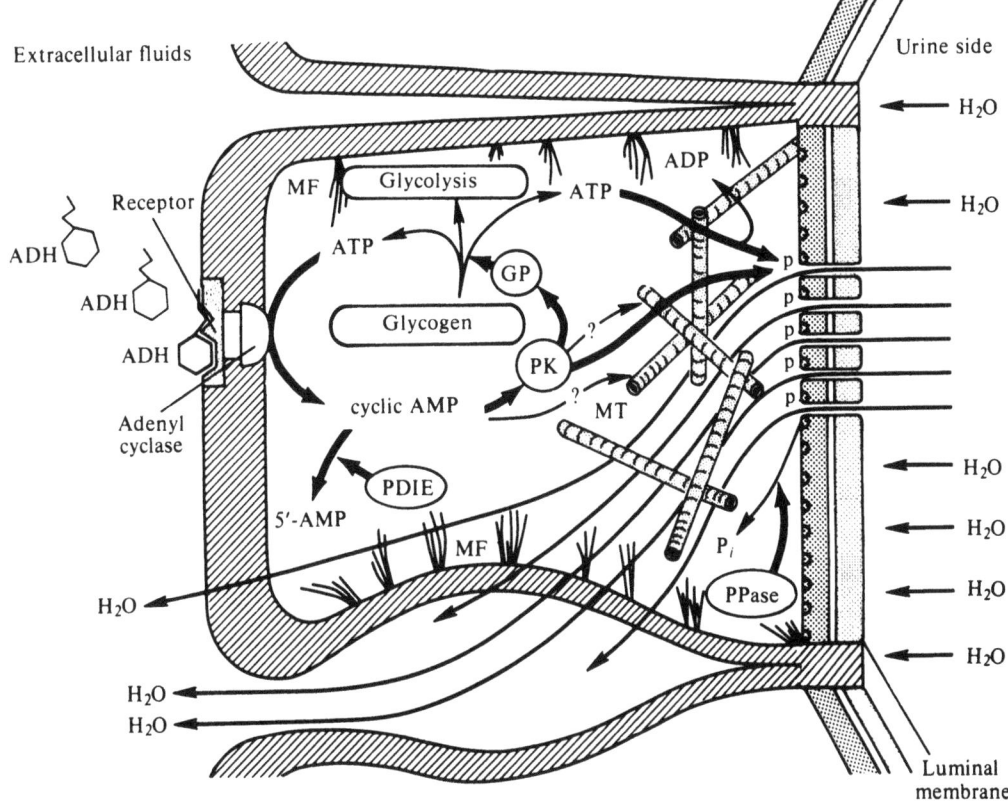

Epithelial cell of renal collecting duct

Fig. 3.32. Summary of some of the information regarding the mechanism of action of antidiuretic hormone (ADH) in increasing absorption of water from the urinary glomerular filtrate across the renal collecting ducts. The hormone acts, when present in the plasma, on receptors that are present in the basal plasma membranes of the renal epithelial cells. Cyclic AMP is formed as a result of the activation of adenyl cyclase. The cyclic AMP activates a protein kinase (PK). The active form of this enzyme, with the aid of ATP, phosphorylates a protein in the luminal plasma membrane, which results in a change in its configuration so that its permeability to water increases. This process is reversed (dephosphorylation) under the influence of a protein phosphatase (PPase) with the release of inorganic P (P_i). The protein kinase may also activate glycogen phosphorylase (GP), which can result in increased glycolysis in the cell. The cyclic AMP is converted to 5'-AMP under the influence of the soluble enzyme phosphodiesterse (PDIE). Microtubules (MT) and microfilaments (MF) also appear to play a role in the response. (From Dousa and Valtin, 1976; reprinted from *Kidney International*, with permission.)

neurohypophysial peptides. Thus, drugs that break microtubules, such as colchicine and vincristine, and those that break microfilaments, such as cytochalasin B, can inhibit this response (Taylor et al., 1975a,b; Taylor, 1977). These subcellular structures have been identified in the responsive cells. It is not known whether they are directly involved and are responsive to cyclic AMP or activated protein kinases, or whether they only provide the necessary basic cellular framework for the hormone effect. It has also been observed that ADH can stimulate the release of hydrolase enzymes from lysosomes across the luminal border of frog bladder epithelial cells (Pietras, Seeler, and

Szego, 1975). It is possible that such enzymes could be acting on the plasma membrane to increase its permeability to water and so mediate the response to ADH.

2. Smooth muscle cells. Receptors for neurohypophysial peptides have also been identified in the uterine myometrium and the mammary gland (see Soloff and Swartz, 1974; Soloff et al., 1977). The binding of [³H]oxytocin has been measured and has a K_D of about 10^{-9} M, which corresponds to the concentration necessary to bring about a half-maximal contraction. The effectiveness of different analogues in displacing bound

[³H]oxytocin is consistent with their relative abilities to produce a contraction.

The receptors for oxytocin are present on the plasma membrane of the smooth muscle cells, and as binding of the hormone can be inhibited by trypsin and substances that interfere with —S—S bridges, they appear to be proteinaceous. Binding is enhanced by certain divalent ions, especially Mn^{2+} and Mg^{2+} (but not Ca^{2+}), which may change the receptors' affinity for the hormone. Nucleotide triphosphates, such as ATP, inhibit binding. The sensitivity of the uterus to oxytocin is increased by estrogens, and this has been related to an increased binding of the peptide, suggesting that the numbers of receptors increase in response to the steroid.

Oxytocin produces a small reduction in the electrical polarization of uterine cells, but they are still able to respond to this peptide even when they are depolarized by high extracellular concentrations of potassium. External Ca^{2+} is, however, essential for the response, and it appears that oxytocin increases the influx of this ion into the cell, although an increased mobilization of this ion from membrane sites may also occur (see Soloff et al., 1977; Carsten and Miller, 1977).

The Ca^{2+} would appear to be necessary for the process of excitation–contraction coupling in the muscle. There has also been speculation that other processes may be involved in the response to oxytocin, including prostaglandins and nucleotides. However, there appears to be little evidence for this at present. Catecholamines relax the uterus, and this effect is related to an increased level of cyclic AMP. It was thus an attractive possibility that oxytocin may have the opposite effect and initiate a decline in the levels of this nucleotide, but such an effect has not been consistently demonstrated (see Dousa, 1977). Alternatively, it has been suggested that guanyl cyclase may be activated and that increased formation of cyclic GMP may initiate contraction, but evidence supporting this interesting possibility has not been forthcoming.

Oxytocin has been found to promote the release of prostaglandins from the uterus of animals (see Roberts et al., 1976). Contraction of the uterus results in a release of prostaglandins from the myometrium, but this effect is not directly related to the effects of oxytocin. The responses to each type of agonist can be separately blocked (Chan, 1977). The endometrium, however, does specifically respond to oxytocin by the release of prostaglandin $F_{2\alpha}$, and this effect may contribute to parturition and potentiate the effects of oxytocin in facilitating labor on this occasion.

The mechanism of action of vasopressin on vascular smooth muscle has received little attention (see Altura and Altura, 1977; Dousa, 1977). Receptors that show a distinct predilection for certain structural features of the neurohypophysial peptides are undoubtedly present, but they have not been precisely characterized. As in the uterus, Mg^{2+} appears to enhance the affinity of the receptors for the peptide, but it may also influence the responses at other sites in the muscle. External Ca^{2+} is also necessary for the response. The possible role of cyclic AMP in the action of vasopressin does not appear to have been investigated in vascular smooth muscle.

F. Diseases of the neurohypophysis

Dysfunction of the neurohypophysis can result in hyper- or hyposecretion of its hormones. Such diseases involving vasopressin have been well characterized because of their prominent effects on renal function. The physiological role of oxytocin is, however, more specialized, and although it is possible that such female reproductive functions as parturition and lactation may be affected by changes in the attainable levels of this hormone, there is no clear demonstration of such an effect in women.

1. Hyperactivity of ADH. When the effects of ADH are enhanced, a condition of dilutional hyponatremia may result. Plasma sodium concentration may fall to levels of less than 115 mequiv/liter. The osmotic concentration of the plasma is low and there is an expansion of the extracellular fluid volume, which appears to lead to an increased GFR and a resulting increased excretion of sodium in the urine. There is no edema, and a deficiency in adrenocortical function is not involved. Depending on the severity of the hyponatremia, the symptoms include anorexia, nausea, vomiting, and in more extreme situations, prominent CNS disorders, such as excitement, depression, lethargy, convulsions, and eventually coma (see Editorial, 1972a; Moses and Miller, 1974; Miller and Moses, 1976).

The basic cause of dilutional hyponatremia is an impaired ability to excrete water, which can result from the effects of various drugs (Table 3.25) as well as ADH, although there may be an interaction between these two factors. Thus, an overactivity of ADH can be seen in the presence of the oral hypoglycemic drug chlorpropamide (see Section 8.6A), which appears to enhance the effects of ADH on the renal tubule. Dilutional hyponatremia due, apparently, to a hypersecretion of ADH is associated with a number of conditions and is called the *syndrome of inappropriate secretion of ADH*, or the *Schwartz–Bartter syndrome*. The ADH level may rise as the result of an enhanced release by the pituitary, such as occurs in disorders related to brain function. These can result from the presence of tumors, meningitis, injury, and subarachnoid

Table 3.25. *Characteristics of drugs that promote water retention*

Type of agent	Effective for DI[a]	Induces hypo-natremia	Interferes with water-load excretion	Effective in nephro-genic DI	Effect inhibited by ethanol	Augments action of ADH	Releases ADH
Biguanides	+		−	?+		+	
Vincristine		+					+
Cyclophosphamide		+	+		+		
Carbamazepine	+	+		−	+	−	+
Clofibrate	+		+	−	+	−	+
Acetaminophen	+		−			+	
Diuretics	+	+	+	+			+
Diazoxide			+	+			
Chlorpropamide	+	+	+	−	+	+	+
Tolbutamide	−	+	−			+	

[a] Diabetes insipidus.
Source: Moses and Miller, 1974. Adapted by permission from *The New England Journal of Medicine 291,* 1236.

hemorrhage. This effect may also be seen in pulmonary diseases such as pneumonias and tuberculosis. It has been suggested that disturbance in the pulmonary circulation may lead to stimulation of the receptors in the left atrium, which is known to increase the release of ADH. Malignant carcinomas of the lung, duodenum, and pancreas are also known to be associated with the Schwartz–Bartter syndrome. Extracts of such tumors have been shown to contain an ADH-like material (Amatruda et al., 1963; de Sousa, Berde, and Mach, 1965; Bentley and Cobb, 1967), which if released into the plasma could account for the condition.

Drugs that can promote the release of ADH were described earlier (Section 3.3C). The therapeutic use of chlorpropamide; some tricyclic antidepressants, such as amitriptyline; and antiineoplastic agents, such as vincristine and cyclophosphamide, are sometimes associated with an impaired ability to excrete water and a dilutional hyponatremia (Table 3.25).

Treatment of dilutional hyponatremia principally consists of water restriction. Attempts to replace the sodium are invariably unsuccessful, as it is rapidly excreted. Specific antagonists of ADH are of potential use, but despite the development of some peptides with such an effect, they have not been utilized clinically. Drugs that can block the release of ADH, such as ethanol and diphenylhydantoin, are also generally unsuitable for the treatment of this condition. The restriction of water intake remains an effective and simple alternative.

Prolonged water restriction in cases of chronic disease can, however, be uncomfortable for the patient. As described earlier, the therapeutic use of lithium carbonate and the antibiotic demeclocycline have been associated with a polyuria due to an inhibition of the action of ADH. These observations have led to the use of these drugs to treat the syndrome of inappropriate secretion of ADH. The successful treatment of a patient with LiCO$_3$ has been described (White and Fetner, 1975). Some reservations have been expressed with regard to possible success and undesirable side effects of the routine use of LiCO$_3$ in such a disorder. A larger trial using seven patients treated with demeclocycline has been described (De Troyer, 1977). Urine flow increased, urinary Na excretion decreased, and the hyponatremia was corrected (Figure 3.33). Blood urea and creatinine levels rose, suggesting that a moderate impairment of renal function was occurring, and this indicates the need for surveillance of the drug's effects.

2. Diabetes insipidus. Inadequate secretion of vasopressin may result in diabetes insipidus. In this condition, urine volume is usually 4 to 8 liters/day, but amounts up to 30 liters have been reported. The urine is dilute and hypertonicity is not attained. The excretion of solutes, including Na, is normal. If the water is not replaced rapidly enough, severe dehydration may occur. There is a burning thirst, which contributes to the maintenance of hydration, although in babies the condition may be unrecognized, so that brain damage or death from dehydration may occur. Hydration can also be a problem if such people become comatose.

Diabetes insipidus can have several causes, apart from a deficiency in antidiuretic hormone.

a. Excessive drinking (potomania) may occur in some psychiatric disorders or as a result of a deficiency in the hypothalamic "thirst" center. This

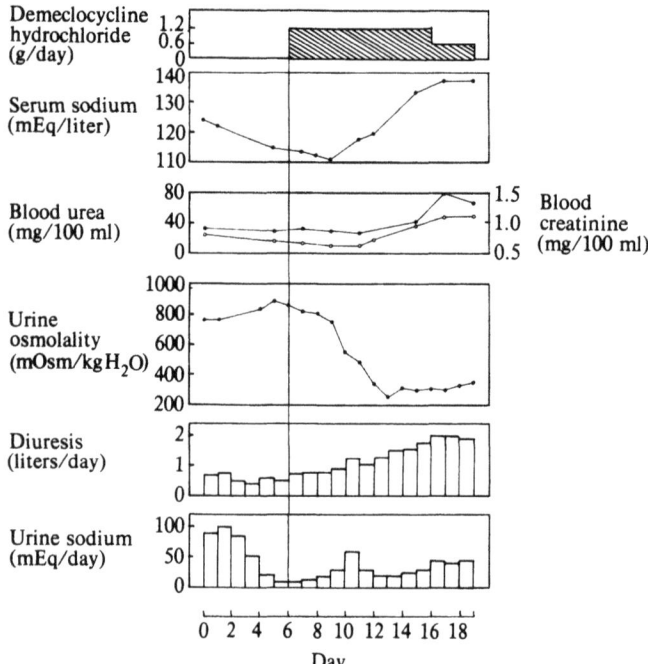

Fig. 3.33. Effects of treatment with demeclocycline hydrochloride on a patient with inappropriate secretion of ADH (Schwartz–Barrter syndrome). Serum sodium, blood urea (●), and creatinine values (○), urine osmolality, daily diuresis, and daily naturiuria before and during treatment. (From De Troyer, 1977.)

primary polydipsia can be distinguished from the renal-type deficiency that leads to secondary polydipsia, by the deprivation of water. In cases of compulsive drinking, the urine can be concentrated in response to fluid deprivation or injected ADH and the plasma concentration will not rise as readily to hyperosmotic levels. In instances of the renal-type disorder, the unabated urine flow will lead to dehydration. In practice, the distinction can, however, be difficult to demonstrate.

b. There may be adequate ADH secreted but the kidney may not respond to it, a condition called *nephrogenic diabetes insipidus.* This disease may be inherited or result from metabolic disorders that lead to hypokalemia or hypercalcemia. It can also be caused by drugs (see Singer and Forrest, 1976; Forrest and Singer, 1977). The use of lithium carbonate to treat manic depression has been related to a high incidence (about 12 percent of cases) of nephrogenic diabetes insipidus (see Editorial, 1972a; Forrest, 1975). The lithium appears to block the action of ADH at some step which occurs subsequent to the formation of cyclic AMP. A central interference with release of ADH is, however, also possible. Other drugs that have been reported to produce diabetes insipidus in man are certain tetracycline antibiotics, especially high doses of demeclocycline. The anesthetic methoxyflurane

and the oral hypoglycemic sulfonylurea drug glibenclamide can also produce polyuria, which is not responsive to the injection of ADH. It has been suggested that other drugs, such as the analgesic propoxyphene; colchicine, which is used to treat gout; and the antineoplastic agent isophosphamide, may also precipitate nephrogenic diabetes insipidus. Experimentally, it has been shown that high concentrations of magnesium, manganese, and cesium can inhibit the action of ADH.

Nephrogenic diabetes insipidus can be readily distinguished from diabetes insipidus that results from a deficiency in ADH as the urine flow fails to respond to the injection of the hormone. It can be treated with thiazide diuretic drugs such as chlorothiazide. The mechanism of action of thiazides in this condition has not been satisfactorily described but appears to be an indirect result of their natriuretic actions, which could be decreasing the GFR and the delivery of fluid to the distal parts of the renal tubule. They do not appear to mimic ADH by a direct action on the permeability of the renal tubules.

c. Disorders that are associated with a dysfunction of the anterior hypothalamus, the supraoptic nucleus, or the median eminence can result in an inadequate secretion of ADH and lead to "central" diabetes insipidus (see Hockaday, 1972; Maffly,

1977). It has been estimated that this disease is only manifested when 85 to 95 percent of the neurosecretory neurons are inactive. Removal of the neural lobe does not usually result in permanent diabetes insipidus, as sufficient ADH can be released from the cut ends of the remaining axons. Hypophysectomy also usually does not result in diabetes insipidus, although replacement therapy with thyroid hormones and corticosteroids may sufficiently restore and elevate renal hemodynamic function so that the condition can then be manifested. Most cases of neurohypophysial diabetes insipidus are idiopathic or the result of damage to the hypothalamus or pituitary stalk as a result of the presence of tumors, inflammatory diseases, or injuries such as those that can result form accidents. In some instances, the condition is not a permanent one. The disorder has also been identified in some families, where it is genetic. An interesting experimental model has been provided by the Brattleboro strain of rats, which, in the homozygous condition, completely lacks ADH, although oxytocin levels are normal (see Valtin, Stewart, and Sokol, 1974). In these rats, the neurons associated with the formation of ADH lack neurosecretory material, storage granules, and neurophysin.

3. Diagnostic tests for disorders of the neurohypophysis. Several types of tests are performed to help diagnose the primary nature of diabetes insipidus. Vasopressin (usually synthetic lysine-vasopressin) can be injected, and in normal individuals, those with hypothalamic diabetes insipidus or that which results from polydipsia, the urine volume will decrease and its osmotic concentration rises. Patients suffering from nephrogenic diabetes insipidus will be unresponsive. The vasopressin can be given s.c. or i.v., but special care must be taken, especially if the latter route is used, in patients with cardiac problems as the peptide can cause a constriction of the coronary vessels. Other tests include the i.v. infusion of hyperosmolar sodium chloride solution (Hickey–Hare test) or nicotine (a higher dose is needed in smokers than in nonsmokers), which will stimulate release of ADH and promote an antidiuresis if the neurohypophysis is normal and the kidneys are responsive. The simplest procedure is "water deprivation" which will fail to produce a reduced urine volume or increased urine concentration in patients with hypothalamic or nephrogenic diabetes insipidus. These two conditions can then be distinguished by applying the "vasopressin test."

4. Treatment of diabetes insipidus. The most important consideration in treating patients with diabetes insipidus is to ensure that they have adequate water. Normally, the thirst mechanism will take care of this problem, but in some situations, especially in infants, it may require a more positive approach. In some instances people suffering from this disease become so used to the necessary routine of maintaining their hydration by drinking frequently that they do not seek drug therapy or only utilize it in special circumstances.

There are several possible ways of treating hypothalamic or central diabetes insipidus. If the condition is only partial and some vasopressin is present, or if the deficiency involves a disorder in the ability to release the hormone, then several types of drugs can be utilized. The antiepileptic drug *carbamazepine,* which may stimulate release of ADH (Kimura et al., 1974) or enhance the response of the kidney to the hormone (Stephens, Coe, and Baylis, 1978), can be used to treat the disorder (Wales, 1975). Dizziness and lethargy may occur early in the treatment but usually subside. [Conversely, water intoxication has been reported as a side effect in the more conventional use of carbamazepine for the treatment of epilepsy (Ashton et al., 1977).] *Chlorpropamide* is also effective when some endogenous ADH is present and has been widely used (see Maffly, 1977). Not unexpectedly (as it is more traditionally an oral hypoglycemic drug used to treat diabetes mellitus), it can produce a hypoglycemia. As chlorpropamide appears to potentiate the effects of ADH on the kidney, it has on occasion been combined with carbamazepine or is given concomitantly with small doses of vasopressin. Diuretic drugs, especially *chlorothiazide* and *hydrochlorothiazide,* have been utilized to reduce urine volume but, as described earlier, their mechanism of action is uncertain. Such diuretics have a more established place in the treatment of nephrogenic diabetes insipidus. *Clofibrate* releases ADH, but as the more traditional use of this drug as an antihyperlidemic agent has been proscribed (an increased incidence of "cardiac accidents"), its utilization to treat diabetes insipidus would not currently appear to be favored. The cytotoxic drug *cyclophosphamide* has been shown to correct the diabetes insipidus associated with a case of Wegener's granulomatosis (Haynes and Fauci, 1978). This condition appears to be due to an inflammation in the region of the hypothalamus, but the precise locus of the action of the cyclophosphamide, with respect to the restoration of the antidiuretic response, is not clear.

Specific hormone replacement therapy with extracts of the posterior pituitary gland has been practiced for more than 60 years. For the treatment of diabetes insipidus, preparations of *vasopressin* can be administered parenterally, either s.c. or i.m., or for its absorption across the nasal mucosa, as a "snuff" or spray. The effects of vasopressin if given s.c. or nasally only persist for 3 to 4 hours and in an attempt to extend their life a "depot" preparation of *vasopressin tannate in oil* is available. When given i.m. its effects last for 1 to 3 days. Problems can, however, arise at the sites of the injection, and self-administration is difficult. Originally, whole extracts of the posterior lobe of the pituitary were used to make preparations that were usually given s.c. or as a nasal snuff. Subsequently, the oxytocic and vasopressor fractions were separated and purified so that more specific treatment was possible. However, despite considerable increases in purity, these preparations of vasopressin, which are made from animal glands, contained significant amounts of contaminants, including some oxytocin, neurophysin, and even a corticotropin-like peptide (Martin, 1971; Scott et al., 1972). These contaminants can result in the formation of antibodies and local reactions at the site of administration. The preparation of *synthetic lysine-vasopressin* was a considerable advance (synthetic arginine-vasopressin was not sufficiently stable for commercial use). Lysine-vasopressin given intranasally as a spray does not display the allergic hypersensitivity reactions commonly associated with the use of snuff preparations made from glandular extracts. Its action, however, remains relatively brief. The long-acting analogue [1-deamino-8-D-arginine]vasopressin (*DDAVP, desmopressin*) has been tested in both Europe and the United States (see Edwards et al., 1973; Robinson, 1976; Cobb, Spare, and Reichlin, 1978; Kosman, 1978) and when given as a nasal spray has been found to be ideal. Desmopressin acetate has recently been cleared for use in the United States (by the FDA). It acts for about 8 to 20 hours and in addition has a relatively low pressor and oxytocic activity. It is thus less likely to produce adverse vascular responses or contractions of the intestine or uterus. It has been used successfully in pregnant women and newborn infants. The related analogue [1-deamino-4-valine-8-arginine]vasopressin (DVDAVP), which has an even more favorable ratio of antidiuretic/pressor activity (this is virtually infinity), has undergone a more limited trial and was also found to be effective in patients suffering from hypothalamic diabetes insipidus (László et al., 1975).

5. Other uses of vasopressin. The treatment of hypothalamic diabetes insipidus is the major accepted use for vasopressins; however, it has also been utilized for some other purposes (Table 3.26). According to reports from Eastern Europe (Hospital Tribune, 1972), its ability to produce an antidiuresis and water retention have been utilized to treat "beer alcoholism." The subject is persuaded to take a dose of DDAVP prior to visiting the local beer house. The drug restricts the excretion of the imbibed fluid, resulting in an unpleasant feeling of malaise and nausea due to water intoxication. It is

Table 3.26. *Summary of various uses of vasopressin*

Replacement for ADH in hypothalamic diabetes insipidus
Treatment of enuresis in children
As a "beer disulfiram"
Bleeding esophageal varices (reduces hepatic portal pressure)
Paralytic ileus and intestinal distention
As a local hemostatic and to delay absorption of local anesthetics
Hemophilia and von Willebrand's disease
Affective disorders?
Test of renal concentrating ability, especially diagnosis of diabetes insipidus (hypothalamic or nephrogenic?)
Test pituitary ability to release ACTH

Note: It should be noted that some of these are controversial.

said that further drinking ceases and premature retirement to "sleep it off" occurs. DDAVP has thus been said to act as a *"beer disulfiram."* Clearly, care must be taken in patients with cardiac problems, because of the expansion of the volume of body fluids. The same report describes how it has also been used to treat *enuresis* ("bed wetting") in children. The administration of DDAVP limits urine formation and when treatment is ceased more normal habits follow in about half the children. Clinical trials do not appear to have been undertaken or reported upon in countries on the other side of the "iron curtain."

There has been considerable speculation about the possible utilization of the vasopressor effects of vasopressin. Its potential action in constricting the coronary vessels has resulted in considerable caution in such applications. In Europe it has been used in mixtures with local anesthetics so as to delay their absorption. The analogue [Phe^2,Lys^8]vasopressin (*octapressin*) has been used because of its selective vasopressor, as compared to antidiuretic, activity (see Berde and Boissonnas, 1968). Vasopressins have also been utilized as a local hemostatic and to stop bleeding of esophageal varices. The latter condition, which is life threatening, arises due to increase in hepatic portal pressure, such as those that accompany cirrhosis of the liver. Vasopressin decreases the portal pressure, possibly by a local constriction of the splanchnic vessels, although more general effects, including that on the heart, may be involved (see Saameli, 1968). In a European clinical trial (Aronsen et al., 1975) the long-acting hormonogen N-α-glycyl-glycyl-glycyl,Lys^8-vasopressin (triglycyl vasopressin) was found to be effective in treating gastrointestinal bleeding, including esophageal varices. No adverse cardiac effects were observed.

A novel use for DDAVP has recently been described by Mannucci et al. (1977). It was found that the administration of this peptide could effectively prevent bleeding in patients suffering from mild or moderate *hemophilia* and *von Willebrand's disease*. This effect, which was utilized in such patients who were undergoing surgery, was due to a release of factor VIII (anti-hemophilic-factor)-related proteins into the plasma. More effective vasoactive drugs, such as epinephrine and vasopressin, have a similar effect on the clotting factor(s), but their circulatory side effects are unacceptable. It was suggested that an analogue lacking the antidiuretic effects of DDAVP would be more acceptable for use in these bleeding disorders and that it may be possible to devise such a drug.

It has recently been suggested (Gold, Goodwin, and Reus, 1978) that DDAVP may be useful in treating certain affective illnesses in man. Some results (Legros et al., 1978; Oliveros et al., 1978) have indicated that vasopressin (lysine-vasopressin was used) may help to restore memory in patients suffering from amnesia and old age. This fascinating novel use for neurohypophysial peptides remains to be explored further. The demonstration (Walter, Van Ree, and de Wied, 1978) that the structure–activity relations for the effects of such peptides on water metabolism and behavior can be dissociated offers the prospect of preparing clinically useful analogues with selective behavioral effects.

4
The thyroid gland

The thyroid gland in man is the site of formation and release of two types of hormones: the "classical" thyroid hormones, thyroxine (T_4) and triiodothyronine (T_3), as well as calcitonin (or thyrocalcitonin), which arises from the parafollicular or C-cells. The two tissues are quite distinct; they have different embryological origins and secrete basically different hormones, and, indeed, in nonmammals they are morphologically separated. In such species the C-cells form the ultimobranchial bodies. Calcitonin plays a role in the regulation of calcium metabolism, which will be dealt with in Chapter 9.

The thyroid hormones have many unique characteristics. They are the only hormones to contain an inorganic moiety, iodine, which is essential for their functioning, they are partly synthesized and stored extracellularly, and their effects are probably the most ubiquitous of all the endocrine secretions, as they have important, but not necessarily identical, effects on most tissues in the body.

4.1. Morphology

The thyroid gland is situated in the neck and in man normally weighs about 30 g, but in disease it may increase enormously in size and appear in what has been described as a grapefruit-like proportion. It is present in all vertebrates and is derived embryologically from the floor of the pharynx. Histologically, it has a follicular arrangement (Figure 4.1), with single layers of cells surrounding a cavity that contains a colloidal material in which the thyroid hormones are partly synthesized and stored. The shape of the thyroid cells alters depending on the activity of the gland; they are high and have a columnar appearance when it is most active and are flattened when it is least active.

Diseases of the thyroid are numerous and among the endocrines are probably second in incidence to diabetes mellitus. It has, for instance,

been estimated that 200 million people suffer from endemic goiter (compensatory enlargement of the thyroid), which is most usually due to an iodine deficiency in certain geographic regions. Cretinism, a congenital abnormality in growth and development which especially affects the nervous system, is also prevalent in such regions. Varying degrees of hypothyroidism and hyperthyroidism also commonly occur and often necessitate medical intervention. Apart from disorders of growth and differentiation in children, the most distinct characteristic of hypothyroidism is a reduction of the basal metabolic rate and oxygen consumption, reflecting a general decline in metabolic processes. This condition is manifested by a slowness in bodily functions, including mental, cardiac, and intestinal ones. Hyperthyroidism has the converse effects. The diseases of the thyroid are discussed in more detail later (Section 4.10).

The two thyroid hormones that are synthesized and secreted by the thyroid gland are thyroxine (T_4) and triiodothyronine (T_3) (Figure 4.2). Triiodothyronine is 3 to 4 times as active as T_4 but has a shorter onset of action, about 4 to 6 hours compared to 4 days for T_4. Thyroxine, however, acts for much longer. Triiodothyronine usually is the active form of the hormone. Most of it is formed peripherally from T_4, which is deiodinated. In the human thyroid, T_4 and T_3 are normally present in a ratio of about 4 : 1 but this may change, such as in iodide deficiency, when the synthesis of T_3 increases.

4.2. Functions and actions

Thyroid hormones have ubiquitous effects in the body. In the later stages of prenatal life and in the newborn they play an important role in controlling *growth* and *development*. At this early stage in life their role is especially important in the development of the nervous system, and a deficiency of the hormones results in mental deficiency and deaf

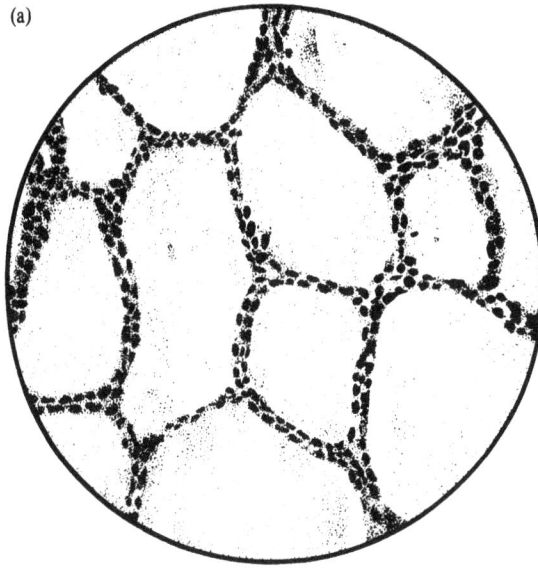

(a)

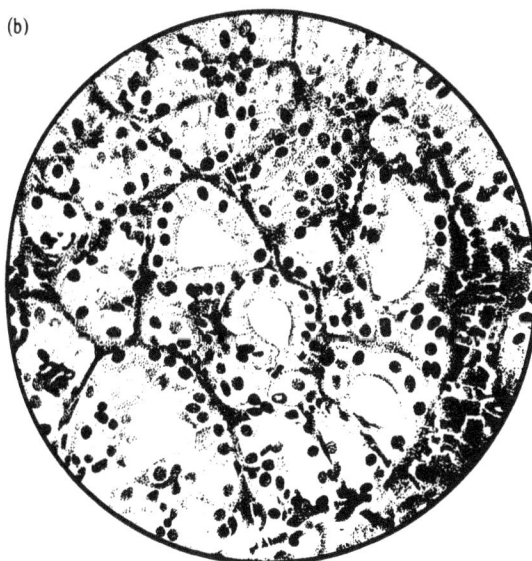

(b)

$$HO \underset{I}{\overset{R}{\langle\rangle}} \overset{I}{-O-} \underset{I}{\overset{I}{\langle\rangle}} CH_2CH(NH_2)COOH$$

Thyroxine (tetraiodothyronine, T_4) R = I
Triiodothyronine (T_3) R = H

Fig. 4.2. Chemical structures of the thyroid hormones, showing the conventional numbering of the carbon atoms.

Fig. 4.1. Thyroid gland of the laboratory rat, showing the follicles surrounded by epithelial cells. (a) The inactive condition, where the cells are flattened and the follicles are distended with "colloid," which contains the thyroglobulin. (b) The active condition, where the epithelial cells are columnar and little colloid is present. (From Bentley, 1976.)

mutism. Growth of muscles and bones is also inadequate, so that dwarfism can result. Complete maturation of the reproductive system may also be slower and puberty is delayed. These symptoms are typical of congenital cretinism, which is most commonly due to a lack of the iodine necessary for the hormone synthesis, but it may arise due to other causes. The defects in neural differentiation once developed are irreversible, even when exogenous thyroid hormones are subsequently supplied.

The second major role of thyroid hormones is in *metabolism*. The rate of oxygen consumption appears to be directly influenced. This function at least partly reflects the need for the control of heat production in response to homeothermic need as seen in mammals and birds. Increases in the release of thyroid hormones occur in response to cold, and long-term adaptation to hot climates may involve a reduced rate of secretion. Changes in oxygen consumption principally occur in the muscles and liver and are not seen in the brain. An altered basal metabolic rate is also involved in the effects of thyroid hormones on the functioning of several tissues, especially the heart, intestine, and skeletal muscles. Thyroid hormones increase the heart rate (chronotropic effect) and result in the tachycardia, arrythmias, and high-output cardiac failure that may occur in hyperthyroidism. Intestinal motility is also increased, so that diarrhea may result. There is increased activity of the sympathetic nervous system and responses to catecholamines, which contribute to the cardiac effects. Muscular weakness is associated with both hypo- and hyperthyroidism. Renal function declines in hypothyroidism; there is a lower glomerular filtration rate, sodium reabsorption from the nephron decreases, and the responses to aldosterone and antidiuretic hormone (ADH) are reduced. Thyroid hormones are necessary for optimal reproductive function. Either a lack or an excess may result in menstrual abnormalities, and at one time the administration of thyroid hormones was used in efforts to overcome infertility in women. Lactation is also dependent on adequate thyroid function, and administered thyroid hormones have been shown to increase milk production in cows. Indeed, thyroid hormones appear to be necessary for optimal functioning of almost all the tissues in the body, and this diversity in their effects has led to the proposal that they exert a general "permissive role" in maintaining the optimal metabolic functioning of cells. These effects include the functioning of other hormones, such as the catecholamines, corticosteroids and antidiuretic hormone. Thyroid hormones have, for instance, been shown to influence levels of hormone receptors, their rates of synthesis and metabolism, and the level of responsiveness of target organs.

The mechanism for such effects is not completely understood. The molecular mechanism of action of thyroid hormone will be described later, but it appears to involve the synthesis of new proteins as a result of nuclear transcription of DNA and ribosomal translation activities of mRNA in cells. Such proteins probably have an enzymic role and they are diverse in their number and actions. Enzyme and coenzyme functions that have been related to actions of thyroid hormones include adenyl cyclase, NaK-activated ATPase, pyridine nucleotide coenzymes, coenzyme Q, and coenzyme A.

4.3. Thyrotropic hormone (thyroid-stimulating hormone, TSH, thyrotropin)

The thyroid gland does not function adequately and degenerates in the absence of the pituitary gland. This deficiency reflects the absence of the thyrotropic hormone, which is synthesized and released by basophilic cells of the adenohypophysis.

A. Structure

The thyrotropic hormone is a glycoprotein with a molecular weight of about 28,000 (see Pierce, 1974). It has about 200 amino acids and contains 13 to 14 percent carbohydrate. The latter consists mainly of mannose, glucosamine, and galactosamine. In human TSH, sialic acid is also a prominent component. The carbohydrate moieties are attached to asparagine residues in the molecule. Thyrotropic hormone consists of two nonidentical subunits, TSH-α and TSH-β, which can be dissociated and recombined *in vitro*. The structure of bovine TSH is shown in Figure 4.3 (see also Liao and Pierce, 1971). The TSH-α has 96 amino acids and TSH-β 113 such residues. The β-subunit carries the information to which the thyroid gland can respond (hormone-specific subunit) but it has little or no activity in the absence of the α-subunit. The latter shows considerable interspecific variability in its structure. In man, TSH-α and LH-α are identical in their amino acid sequences (Cornell and Pierce, 1973; Sairam and Li, 1973). The TSH-α is also similar to the α-subunits in hCG and FSH. Thus, TSH-β can be combined with LH-α to produce a molecule which has identical activity to TSH. The reverse combination TSH-α and LH-β, however, lacks significant TSH activity. The α-subunits contain the highest proportion of carbohydrate and have five intrachain disulfide bonds compared with six in the β-unit.

Substances that can stimulate the activity of the thyroid gland can also be formed at a number of

(a) Bovine TSH-α

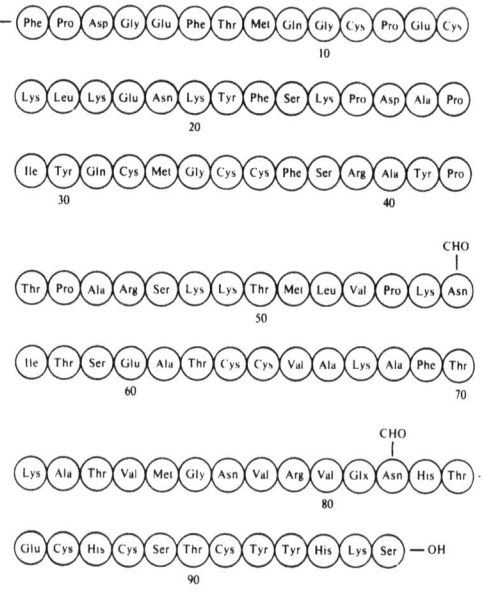

(b) Bovine TSH-β

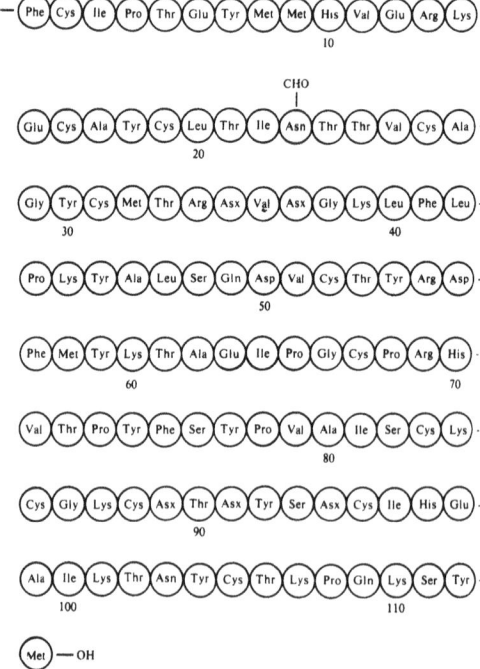

Fig. 4.3. Amino acid sequences of the α- and β-subunits of bovine thyrotropic hormone. The carbohydrate moieties (CHO) can be seen attached to the asparagine (Asn) residues. (Based on Papkoff, 1972.)

nonpituitary sites. In 1956 Adams and Purves showed that the serum from patients suffering from hyperthyroidism (Grave's disease) contained a substance that could stimulate the activity of the thyroid gland. As it has a delayed and prolonged effect, it was called *long-acting thyroid stimulator* or *LATS*. Subsequently another type of activity, which could prevent the binding of LATS to human thyroid tissue, was identified in human serum and called LATS-protector or LATS-P. These substances were shown to be immunoglobulins and appear to be antibodies that can stimulate the thyroid gland. They are thus called generically *thyroid-stimulating antibodies* or *TSAb,* and they have been identified in the serum of at least 90 percent of patients suffering from Grave's disease (Adams, Kennedy, and Stewart, 1974).

The human placenta also appears to contain two substances which have thyrotropic activity. One of them displays an immunological cross-reactivity to pituitary TSH and appears to be similar in size. It has been called human chorionic thyrotropin or *HCT.* A substance that appears to account for the hyperthyroidism associated with hydatidiform mole (a benign neoplasm) in pregnancy has been extracted from the tumor tissue and is also present in pregnancy urine. It has TSH-like activity which appears to be identical to human chorionic gonadotropin (hCG) (Kenimer, Hershman, and Higgins, 1975; Nisula and Ketelslegers, 1974). This glycoprotein is described in Section 7.3B and has structural similarities to TSH. On a molar basis it has about 1/4000 the thyrotropic activity of TSH, but as it can be present in very high concentrations in hydatidiform mole, this appears to be sufficient to result in hyperthyroidism. It is considered possible that all the thyrotrophic activity found in the placenta is in fact due to hCG.

B. Release

The release of TSH is influenced by a number of conditions (see Florsheim, 1974; Sterling and Lazarus, 1977), which may act directly on the pituitary gland or via the hypothalamus and the secretion of TSH-releasing hormone [TRH; see Section 4.3E(1)]. If the pituitary stalk is cut or if the gland is transplanted to a site that is distal to the hypothalamus, thyroid function is reduced but is not abolished. Pituitary TSH function can thus be partly autonomous, but it is modulated by TRH, which passes to it in the hypophysial portal vessels. The TRH influences the "set point," which determines when TSH is released.

Thyrotropic hormone can be secreted in response to cold adaptation. This effect is observed in many animals and newborn human infants and children but not adults. It is thought that this response is mediated by a release of TRH from the hypothalamus. Relatively acute stress, such as sur-

gery, can result in an increase, followed by a depression of release of TSH. Chronic stress, however, can result in increased secretion. The site of these effects is uncertain but probably involves the hypothalamus. There is a diurnal rhythm in the release of TSH; it is at its lowest in late afternoon and highest at about 4 A.M. This pattern may also be determined by TRH secretion.

The principal factor influencing TSH release is the circulating level of the thyroid hormones, and T_3 and T_4 receptors have been identified in the pituitary. These hormones exert a negative-feedback effect to block release of TSH. There has been considerable speculation as to whether thyroid hormones also exert a similar effect on TRH release in the hypothalamus, but the evidence for such a role is equivocal.

Electrical stimulation in the region of the hypothalamus elicits a release of TSH which can be mimicked by the injection of TRH. This tripeptide hormone is thought to act by inhibiting the negative feedback response to the thyroid hormones in the pituitary gland. The effect appears to be mediated by an activation of pituitary adenyl cyclase and the formation of cyclic AMP. Uridine uptake by pituitary cells is also rapidly increased by TRH (Martin, Cort, and Tashjian, 1978), but the possible relationship of this response to the release of TSH is not yet known.

The action of TRH on the release of TSH can be inhibited by somatostatin (Carr et al., 1975). High levels of corticosteroids can also inhibit TSH release (Wilber and Utiger, 1969), possibly also by reducing the sensitivity to TRH. Bromocriptine [an ergot alkaloid; see Section 7.6G(3)] reduces serum TSH in human patients suffering from hypothyroidism and may act on the hypothalamus or pituitary gland (Miyai et al., 1974).

C. Transport and metabolism

In man, the half-life of TSH in the plasma is 30 to 90 minutes. It seems likely that this somewhat extended period of activity is influenced, as in other glycoproteins, by protection of the molecule against inactivation by the carbohydrate moieties present in it. A thermolabile, nondialyzable component has been identified in normal human sera which can bind TSH (Lee et al., 1977). This substance prevents the interaction of TSH with plasma membranes from the thyroid gland and so could play a role in regulating TSH activity.

D. Mechanism of action

Thyrotropic hormone not only initiates the release of thyroid hormones but also stimulates all the activities of the thyroid gland, including uptake of I^-, the synthesis of the hormones, the deiodination of released iodotyrosine, and the morphological growth and development of the endocrine tissue. These effects principally involve an activation of adenyl cyclase in the plasma membranes of the glandular cells. It is also possible that some effects of TSH may not involve cyclic AMP.

The TSH has been labelled with ^3H and ^{125}I, and these molecules retain biological activity so that they can be used to study the receptor functions of the hormone (see Amire et al, 1973; Moore and Wolff, 1974; Winand and Kohn, 1975; Tate et al., 1975a,b). Binding of ^3H-TSH and ^{125}I-TSH has been demonstrated in plasma membrane preparations made from thyroid glands. This process occurs very rapidly and at low concentrations ($\approx 10^{-11}$ M). The amount of binding can be correlated with increases in membrane adenyl cyclase activity. Natural analogues of TSH show a parallel relationship to their activity: the ratio of binding of TSH, LH, hCG, TSH-β, and TSH-α is $100 : 10 : 5 : 2 : < 0.5$. The binding sites thus appear to be synonymous with specific hormone receptors. Two types of sites with different properties have been identified: one has a high affinity for the TSH but low capacity (apparent $K_D = 3 \times 10^{-9}$ M, binding 1.6 pmol TSH/mg membrane protein), and the other has a lower affinity and a higher capacity (apparent K_D about 10^{-7} M, capable of binding 340 pmol TSH/mg membrane protein) (Moore and Wolff, 1974). Both sets of receptors appear to exhibit similar properties with respect to the ability to stimulate the thyroid gland.

A number of properties of the receptors have been studied (Moore and Wolff, 1974; Winand and Kohn, 1975). They exhibit a negative cooperativity, so that binding decreases at high TSH concentrations. Binding is also decreased by divalent ions (e.g., Mg^{2+} and Ca^{2+} at 0.1 mM) and monovalent cations (Na^+, K^+, Li^+ at concentrations greater than 10 to 15 mM). Phospholipase A enhances binding. Exposure of the membranes to trypsin results in a loss of receptors, which then appear (minus their associated adenyl cyclase) in the external bathing media. The glandular tissue can, however, subsequently regenerate more receptors.

Solubilized receptors separated from thyroid glands of dogs, cattle, and man by treatment with trypsin or lithium diiodosalicylate have been studied. They have a molecular weight of 15,000 to 30,000 and contain about 30 percent carbohydrate and 10 percent sialic acid (see Tate et al., 1975a,b; Winand and Kohn, 1975). Similar preparations have been made from porcine and human thyroid glands by treating them with the detergent Triton X-100 (Dawes et al., 1978). Thyrotropin receptors isolated in this way and subsequently separated by SDS gel electrophoresis had a molecular weight of about 50,000. Thyrotropic hormone and its analogues bind to these soluble receptor units in a similar manner to the membrane preparations.

Human chorionic gonadotropin has also been shown to display a high affinity for human thyroid gland plasma membrane preparations (Azukizawa et al., 1977). This observation is consistent with the observed thyrotropic activity of this placental hormone. Thyroid-stimulating antibodies, TSAb (see Section 4.3A) have been shown to compete with TSH for binding to human, thyroid gland, plasma membrane preparations and even to solubilized receptors (Rees Smith and Hall, 1974; Mehdi, Badger and Kriss, 1977). Like TSH, in these membrane receptor preparations, they activate adenyl cyclase (Rees Smith et al., 1977).

E. Nonthyroidal effects

Hyperthyroidism is commonly accompanied by a protrusion of the eyeball which is called exophthalmos. This condition can have two causes. It may be due to an increased tone of the orbital muscles, which is called *spastic exophthalmos*. This, in turn, may be the result of an increased sensitivity to catecholamines due to high circulatory levels of thyroid hormones. *Infiltrative exophthalmos* is more serious and is due to a buildup of tissue, including enlargement of muscles in the orbit. This condition can also occur in the absence of hyperthyroidism. It is thought to be due to the presence of a TSH-like material called *exophthalmic-producing substance* or *EPS*. It has been isolated from the pituitary. Incubation of TSH with pepsin has been shown to decrease its TSH activity but leaves EPS activity relatively unchanged (Kohn and Winand, 1971). There has been speculation that EPS activity resides in a part of the molecule of TSH and could arise due to changes in its metabolism, especially in the pituitary. TSH receptors have been identified in retroorbital muscle, but they have different characteristics than those in the thyroid. The interaction of TSH and EPS with the eye muscle, but not the thyroid gland, can be enhanced by LATS from exophthalmic patients (Amir et al., 1973).

4.4. Synthesis and release of thyroid hormones

The thyroid hormones are synthesized from tyrosine and iodine. There are several distinct steps involved, which each limit the formation of the hormone. These processes may be compromised in disease and can be selectively blocked by a variety of chemical substances and drugs, which may constitute undesirable chemical toxicities and side effects, or they may be utilized for therapeutic purposes.

A. The iodide pump

An adequate supply of dietary iodine is essential for the formation of optimal amounts of the thyroid hormones. This anion is usually obtained in the form of I^-, or organic I, which is reduced to I^- in the gut. The I^- is absorbed and is carried in the blood to the thyroid gland. The basal border of the thyroid cells have a remarkable capacity to "trap" this I^- and transfer it into the cell. This ability has been called the iodide "trap" or "pump," and enables the thyroid cells to accumulate this anion against an electrochemical gradient. Iodide can, for instance, be accumulated in the thyroid cells even when the concentration gradient, cell/plasma, is 300 to 400:1, although normally it is much less than this. It thus appears to be a process of active transport and requires a supply of energy. Such an iodide pump also occurs in some other tissues, including the salivary glands, choroid plexus, stomach, and mammary glands. Its discrete biochemical existence is indicated by the observation that a genetic deficiency has been described in which the ability to "pump" I^- is deficient at all these sites in the body. Apart from a supply of metabolic substrates, the I^- pump requires the presence of Na,K-activated ATPase. This enzyme can be inhibited by cardiac glycosides, such as ouabain, but the use of therapeutic concentrations of such drugs, as in the treatment of congestive heart failure, does not appear to compromise thyroid hormone synthesis. The precise molecular nature of the I^- pump is unknown, but it has been suggested that it may involve a "carrier" containing lecithin. A number of other anions can substitute in varying degrees for I^- in this mechanism. These ions, in increasing order of effectiveness, include $Br^- < OCN^- < NO_2^- < NO_3 < I^- < SCN^- < ClO_4^- < TCO_4^-$ (Wolff, 1964). The structural requirements are thus not especially precise and include univalency and a general limitation involving "ionic size." These observations are of pharmacological significance, as they provide substances that can interfere with I^- uptake. They are discussed in detail in Section 4.9B. The I^- pump not only promotes the uptake of dietary I^- but also limits the losses of thyroidal iodine, particularly those which occur during the process of hormone release due to the deiodination of associated iodoamino acids.

B. Synthesis of thyroglobulin

Thyroglobulin is a glycoprotein (see Ui, 1974), with a molecular weight of about 660,000, which plays a central role in the gland's ability to synthesize thyroxine and triiodothyronine. Its important properties include the presence of tyrosine. It contains 120 tyrosyl groups, which are the sites where iodination of the thyroglobulin occurs. While this protein is usually considered to be unremarkable in its composition, as compared to other proteins, its amino acid sequence is a precise one and characteristic of each particular animal species. Thyroglobulin is assembled in two stages: its constituent polypeptide chains are assembled by the polysomes in the thyroid cell, while the carbohydrate moiety is added as a result of the activity of the Golgi apparatus and the endoplasmic reticulum. The completed thyroglobulin molecule then passes out across the apical border of the thyroid cell into the follicular lumen, where it becomes a part of the mass of intrafollicular colloid.

C. Synthesis of thyroxine and triiodothyronine

The process of synthesis of the thyroid hormone is summarized in Figure 4.4. The iodination of the tyrosyl groups on the thyroglobulin occurs outside the thyroid cells in the follicular lumen. However, some *in vitro* experiments using thyroid cells in the absence of follicular colloid have shown that iodoamino acids are still formed, so that the possibility of intracellular iodination cannot be completely excluded. The iodide must be oxidized to iodine before it can react with the tyrosyl groups. This process is accomplished with the aid of H_2O_2 generated as a result of the action of thyroid peroxidase. The action of this enzyme also appears to be necessary for subsequent steps in the hormonal synthesis, including the "coupling" of the iodoamino acids. It is possibly the site of action of certain antithyroid drugs which act to prevent the "organic binding" of I (see Section 4.9). Autoradiographic studies indicate that the iodination of the tyrosyl groups of the thyroglobulin occurs at the surface of the microvilli at the apical border of the thyroid cells, or possibly even more deeply in the follicular lumen.

The iodoamino acids present in the thyroid include diiodothyronine (DIT) and monoiodothyronine (MIT), which themselves have little thyroid hormone activity. The "coupling" of two such compounds, however, results in the final synthesis of the hormones, two molecules of DIT to form thyroxine (T_4), and DIT and MIT to form triiodothyronine (T_3). It is also possible that T_3 is formed in the gland by deiodination of T_4, a process that involves thyroxine 5'-deiodinase, which also occurs in peripheral tissues, especially liver and kidney.

D. Control of synthesis

As described earlier (Section 4.3), the thyroid gland is principally under the control of thyroid-stimulating hormone, TSH, which is secreted by the anterior lobe of the pituitary gland. This trophic hormone stimulates all aspects of thyroid gland function, including I^- uptake, intermediary metabolism, protein and phospholipid synthesis, and hormone synthesis and release. Internal, autonomous mechanisms in the thyroid gland itself, however, also contribute to the control of hormonal synthesis. The activity of the I^- pump is influenced by the quantities of available I^- in the plasma. For instance, when a great excess of I^- is presented to the gland, it is initially accumulated in

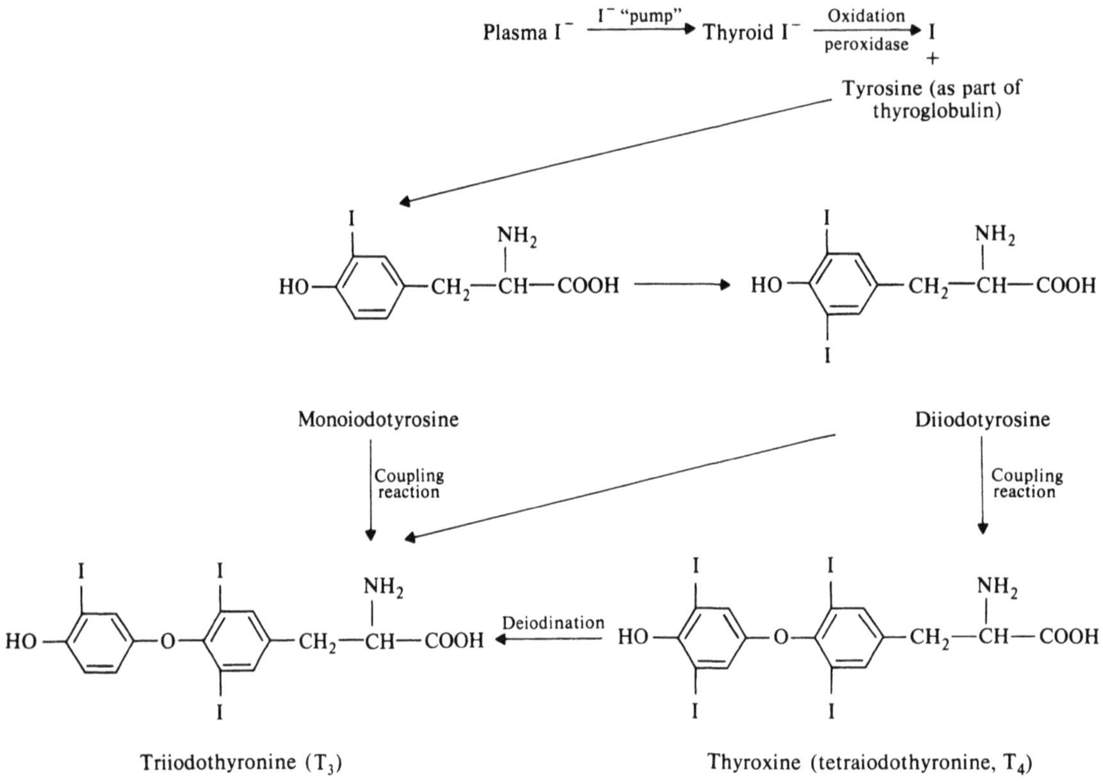

Fig. 4.4 Biosynthesis of the thyroid hormones.

large amounts. The presence of such an excess of I^- (Wolff and Chaikoff, 1948) reduces thyroid hormone formation and release (*Wolff–Chaikoff effect*). Subsequently, over a period of a few days, an "escape" from the Wolff–Chaikoff effect occurs, probably mainly as a result of a decline in the rate of I^- uptake, although a decline in the inhibitory effect on the mechanism of the organic binding of the I may also be occurring. A lack of iodide has the opposite effect and results in an increased ability to accumulate this anion. In I^- deficiency the synthesis of T_3 may also be increased relative to that of T_4, which as T_3 is more active, provides a mechanism for maintaining the hormonal activity with less expenditure of iodine.

E. Release of thyroid hormones

The secretion of thyroid hormones (see Greer and Haibach, 1974) immediately ceases following hypophysectomy because of the absence of pituitary thyrotropic hormone. The mechanism of action of this hormone (see also Section 4.3D) on the thyroid gland involves an activation of membrane adenyl cyclase and the formation of cyclic AMP. The latter nucleotide can mimic the action of TSH under *in vitro* conditions. The trophic action of TSH increases the endocytotic processes involved in mobilization of the intrafollicular thyroglobulin stores. The formation and metabolism of cyclic AMP is influenced by a number of conditions, and it has been shown that drugs such as theophylline, which inhibits phosphodiesterase and thus the destruction of cyclic AMP, may potentiate the effects of TSH. Lithium (see Section 4.9) has been shown to inhibit release of thyroid hormones and it has been suggested that it may act by inhibiting adenyl cyclase. Iodide can also block the release of thyroid hormones, and its effects may also be mediated by an inhibition of the action of TSH on adenyl cyclase (Rapport, West, and Ingbar, 1976).

A combination of physiological, biochemical, and histological studies have provided a description of the process of thyroid hormone secretion, although the nature of the coupling of this process to cyclic AMP is unknown. Colloid from the lumen of the follicles is taken up across the apical side of the thyroid cells by a pinocytotic process, which as it occurs in an inward direction is called endocytosis (see Figure 4.5). Once inside the cells, the colloid, and the thyroglobulin it contains, starts to migrate toward the basal region of the cell. Lysosomes containing proteolytic digestive enzymes

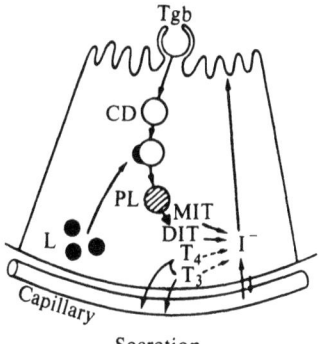

Secretion

Fig. 4.5. Diagrammatic representation of the process of secretion of thyroid hormones. Tgb, thyroglobulin; CD, colloid droplets; L, lysosomes; PL, phagolysosomes; MIT, monoiodotyrosine; DIT, diiodotyrosine. For a description of the events, see the text. (From Greer and Haibach, 1974.)

combine with these droplets to form phagolysosomes. The constituent iodoamino acids of the thyroglobulin are released and T_4 and T_3 pass across the basal membrane of the cell and into the blood. Some DIT and MIT are also released inside the cell, and this is deiodinated by a deiodinase. Most of the released I^- is conserved and passed back across the basal membrane into the follicular lumen. Some I^-, however, may escape into the circulation, the amount depending largely on the activity of the I^- pump. If this mechanism is inhibited, or if an excess of intracellular I is present, such a loss will be greater. Inhibitors of the I^- pump promote the loss of such intracellular I^-, an effect that forms the basis of the "perchlorate discharge test" (see Section 4.9B) for helping diagnose defects in the organic binding of I^- by the thyroid.

A summary of the physiological processes and drugs that may influence the synthesis and release of thyroid hormones is provided in Table 4.1. The pharmacological responses are described in more detail later (see Section 4.9).

4.5. Transport of thyroid hormones in the blood: binding to plasma proteins

More than 99 percent of the thyroid hormones in the plasma are bound to proteins. The total concentration of T_4 in the plasma in man is 5 to 11 μg/100 ml, while that of T_3 is only 100 to 150 ng/100 ml, a ratio of about 60:1. The unbound fraction of these hormones represents the biologically active portion and only amounts to about 0.1 percent of the total present. It should, however, be remembered that the bound and unbound forms are in a ready exchange equilibrium with each other, so that as the biological active, unbound fraction is used up it is replaced from the protein bound stores. About 99.95 percent of the thyroxine in the blood is bound compared to only 99.5 percent of the T_3; thus, the proportion of unbound T_3 is 10 times greater than that of T_4. There is about 6 times as much free T_4 as T_3 in the plasma.

The thyroid hormones are bound to three types of plasma proteins. The most important of these is *thyroid binding globulin* (TBG). This protein is only present in low concentration in the plasma, but it so avidly binds thyroid hormones that about 60 percent of the total is associated with it. About 30 percent of the T_4 is bound to a prealbumin (*thyroid-binding prealbumin*, TBPA) and 10 percent to albumin. TBG has not been identified in non-

Table 4.1. *Physiological and pharmacological conditions that influence the release of thyroid hormones*

	Response	Site of mechanism
Cold exposure (neonatal in man)	↑	Hypothalamus
Stress, acute	↑ or ↓	Hypothalamus?
Stress, chronic	↑	Hypothalamus?
Diurnal rhythms	↑ or ↓	Hypothalamus?
T_3 and T_4	↓	Pituitary (inhibit TSH)
TRH	↑	Pituitary (increases TSH release)
Somatostatin	↓	Pituitary (inhibits response to TRH)
Diphenylhydantoin	↓	Hypothalamus and/or pituitary
TSH, hCG, and LATS	↑	Thyroid
I^- excess (short-term)	↓	Thyroid
Li^+	↓	Thyroid (adenyl cyclase?)

Note: Other ions and drugs, such as SCN^-, ClO_4^-, propylthiouracil, and thioamides, block synthesis and so indirectly influence release.

mammals and thus appears to have been evolved specifically in relation to mammalian thyroid hormone function. It is synthesized by the liver, a process that is accelerated in pregnancy and in response to *oral contraceptives* and *estrogens,* but it is reduced by *androgens, anabolic steroids,* and *corticosteroids;* in some diseases, such as acromegaly and nephrosis; and in cases of malnutrition, as well as in some of the elderly. The binding sites on TBG and TBPA are accessible, respectively, to certain drugs such as *diphenylhydantoin* (which is used to treat epilepsy), *phenylbutazone* (an antiinflammatory drug), and *salicylates,* which may thus displace the hormones.

As most of the thyroid hormones are bound to proteins, and the residual binding of other iodine-containing molecules is small, the determination of protein-bound iodine (PBI) is a simple and convenient way of estimating effective circulating thyroid hormone levels. If, however, the TBG levels are affected by disease or pregnancy, or if other iodine compounds have been administered, such as in radiocontrast media or in the treatment of bronchial conditions, false indications of thyroid function may be obtained.

4.6. Metabolism

The secreted thyroid hormones persist for relatively long periods in the body prior to their metabolic destruction (see Van Middlesworth, 1974). The half-life of thyroxine is about 7 days, while that of triiodothyronine is 1.3 days. The disposal rate of the former hormone is thus about 10 percent a day compared to 50 percent for the T_3. The persistence of these hormones in the body largely reflects their binding to plasma proteins and the tissues, but also the fact that a substantial quantity is excreted in the bile and is partly reabsorbed from the intestine. About 60 percent of the thyroid hormones are nactivated by deiodination, a process that is ubiquitous among the bodily tissues but occurs especially in the liver and kidney. The enzyme concerned, thyroxine 5'-deiodinase, appears to be located in the cell membrane (Leonard and Rosenberg, 1978). The released iodine can be retrieved by the thyroid, but some appears in the urine. The amount of urinary excretion of iodide can be used to assess the amount of iodine available in the diet. The remainder of the thyroid hormones are decarboxylated, deaminated, and conjugated, mainly with glucuronate, but also with sulfate, in the liver. After excretion in the bile, about 40 to 60 percent of the hormones are reabsorbed from the intestine. This process is the result of the breakdown of the conjugated metabolites by bacterial action and the release of the hormones.

The metabolism of the thyroid hormones is influenced by a number of conditions, both physio-

logical and pharmacological. The requirements of old people are less than in the young, and athletes may utilize twice as much hormone. Exposure to cold also results in increased need for thyroid hormones. During fever their half-lives may decrease by 60 percent and the turnover rate may be increased threefold. Such observations are of practical significance when assessing the needs of patients for replacement doses of the hormones.

The chronic use of phenobarbital and diphenylhydantoin results in a decrease in the plasma levels of the thyroid hormones (Cavaliera, Sung, and Becker, 1973; Hansen et al., 1974). These effects appear to reflect an increased rate of hormone metabolism and fecal (biliary) excretion due to the induction of liver enzymes, which aid in the metabolism of the hormones. These observations may partly explain the old empirical observations that barbiturates may be useful in the treatment of thyrotoxicosis.

The process of absorption of thyroid hormones from the gut is of practical significance, as exogenous hormones are administered by this route. Triiodothyronine is absorbed more rapidly than thyroxine, which presumably contributes partly to the shorter period in the onset of its action. Absorption of thyroid hormones is influenced by the diet. When food is restricted, reabsorption of thyroxine from the gut is enhanced, while diets with a high fecal residue tend to impede absorption. It is also possible that certain dietary constituents may bind, and so reduce, absorption of thyroxine and could thus influence required dosages. Cholestyramine, which is used to lower plasma cholesterol levels, delays the absorption of thyroxine.

4.7. Mechanism of action

The precise mechanism of action of thyroid hormone is still, compared with other hormones, relatively controversial (see, e.g., Bernal and Refetoff, 1977; Schwartz and Oppenheimer, 1978). This problem may reflect the diversity of its physiological and biochemical effects on such processes as development, growth, and metabolism (Tables 4.2 and 4.3). The possibility of a unitary hypothesis for its various actions is attractive but may not apply. Until early in the 1960s it was thought that thyroid hormones may increase O_2 consumption by acting on the mitochondria to uncouple oxidative phosphorylation. This hypothesis was based on *in vitro* experiments in which mitochondria were exposed to high concentrations of thyroxine. It is now considered that these effects probably represented toxic responses of the hormone. It has, however, been shown recently that mitochondrial membranes also possess specific receptors for thyroid hormones (see Sterling, 1977), so that these

Table 4.2. *Multiple physiological actions of thyroid hormones*

Growth-promoting and developmental actions	Metabolic effects
Rate of growth of many mammalian and avian tissues	Regulation of basal metabolic rate
Maturation of central nervous system and bones	Regulation of water and ion transport
Obligatory requirement for all processes of amphibian metamorphosis	Calcium and phosphorus metabolism
Regulation of synthesis of some mitochondrial respiratory enzymes and structural elements	Regulation of cholesterol and fat metabolism
	Nitrogen (urea, creatine) metabolism

Source: Tata, 1974.

organelles are probably one of the sites for a direct action of these hormones. The various suggested mechanisms for the action of thyroid hormones are summarized in Table 4.4. It is considered that the final possibility listed, combinations of the mechanisms, is the most likely explanation.

Tata and his collaborators (Tata et al., 1963; Tata and Widnell, 1966; Tata, 1963, 1974) have shown that thyroid hormones increase the synthesis of proteins in the rat liver. Orotic acid was incorporated into newly synthesized RNA at an increased rate. RNA polymerase activity was also increased. The increased protein synthesis was followed by a stimulation of oxygen consumption. These effects can be blocked by actinomycin D as well as by inhibitors of protein synthesis, such as puromycin and cycloheximide. A variety of proteins are produced, including microsomal membrane enzymes

(as well as phospholipids) and mitochondrial respiratory enzymes (Table 4.3). These appear to mediate the various effects of the thyroid hormones. Genetic transcription and translation by the ribosomes are thus involved, such as are also seen in response to steroid hormones.

The thyroid hormones' effects on development and growth were studied by Tata in young thyroidectomized rats and tadpoles. The latter amphibians undergo a metamorphosis into adult frogs, and this transformation can be induced precociously by administering thyroid hormone. This effect, like those on metabolism, can also be blocked by actinomycin D and puromycin, and also involves *de novo* protein synthesis. In this instance, the adults proteins have been identified and include albumins, hemoglobin, the formerly deficient urea cycle enzymes, and hydrolytic enzymes which me-

Table 4.3. *Major biochemical actions of thyroid hormone in the liver of thyroidectomized rat and bullfrog tadpole liver*

Stimulation or increase in amount of:	Latent period (h)		Time of peak effect (h)	
	Rat	Tadpole	Rat	Tadpole
Synthesis *in vivo* of rapidly labelled nuclear RNA	4–6	25–30	22	50–60
RNA polymerase (ribosomal RNA product)	10–12	—	40	—
RNA polymerase (DNA-like RNA product)	18–20	—	50	—
Amino acid incorporation into protein by mitochondria and microsomes	18–24	—	40–45	—
Synthesis of microsomal phospholipids	12–16	32–36	40	70–80
Microsomal membrane enzymes	20–24	—	70–80	—
Mitochondrial respiratory enzymes	24–30	50–60	50–60	100
Serum albumin	—	90–100	—	250
Adult hemoglobin	—	80–90	—	250
Urea cycle enzymes	—	60–70	—	150

Note: A single injection of 15 to 25 μg of triiodothyronine was made to thyroidectomized rats or a dose of 0.5 to 1 μg to *Rana catesbeiana* tadpoles. The latent period and the time for peak stimulation of activity or increase in amount of a constituent are recorded.
Source: Tata, 1974.

Table 4.4. *Suggested mechanisms of action of thyroid hormone*

1. Nuclear transcription
2. Mitochondrial activation
3. Na,K-ATPase ("sodium pump")
4. Incorporation into tyrosine pathways
5. Adrenergic receptor sensitivity
6. Membrane action
7. Combinations of the mechanisms above

Source: Sterling, 1977.

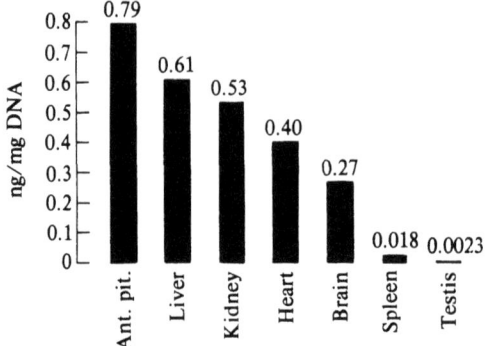

Fig. 4.6. Nuclear binding capacity for triiodothyronine (T_3) of various tissues in the laboratory rat. (From Oppenheimer, 1975. Reprinted by permission from *The New England Journal of Medicine 292,* 1065.)

diate such transformational processes as the resorption of the tail.

Direct evidence of a nuclear receptor for the thyroid hormones has been a relatively later development. Using ^{125}I- and ^{131}I-labelled triiodothyronine, such specific binding sites have now been identified in a variety of tissues (Samuels and Tsai, 1973; Surks et al., 1973; DeGroot et al., 1974; Oppenheimer, 1975). They bind (Table 4.5) T_3 about 10 times as strongly as T_4. A very active synthetic analogue, 3,5-diiodo-3′-isopropyl-L-thyronine (see Section 4.8), has a relative binding affinity as high as T_3. It seems likely that T_4 is normally converted to T_3 before it acts in the nucleus. Unlike the steroid hormones, T_3 can interact directly with its nuclear receptor; cytoplasmic receptors are apparently not directly involved in the activation of the nuclear receptor.

The number of thyroid hormone-binding sites has been measured in a variety of tissues (Figure 4.6). The pituitary has the greatest number, and these presumably contribute to the negative-feedback inhibition of the action of TSH by thyroid hormones. Metabolically active tissues such as the liver and heart have a high nuclear-binding capacity, while in unresponsive tissues such as the spleen and testes it is low. The brain, which lacks the anabolic response, has an intermediate number of receptors.

The nuclear receptors are associated with the chromatin. They have been prepared from rat liver nuclei in a soluble form (Macleod and Baxter, 1976; Latham, Ring, and Baxter, 1976). These receptors are nonhistone proteins with a molecular weight of about 50,000. The K_D for binding of T_3 is about 10^{-9} M. Other analogues could also bind to these receptors with affinities that reflect their relative biological potencies. The receptors can bind to DNA (Macleod and Baxter, 1975), which appears to account for their presence in the chromatin.

Table 4.5. *Relative nuclear binding affinity of thyroid-hormone analogues compared to T_3*

Analogue	Relative binding affinity ($T_3 = 1$)	
	In vitro	In vivo
L-T_3	1.0	1.0
D-T_3	0.6	0.7
Triiodothyroacetic acid (triac)	1.6	1.0
Isopropyl T_2	1.0	1.0
L-T_4	0.1	0.1
Tetraiodothyroacetic acid (tetrac)	0.16	0.05
3,3′,5′-T_3 (reverse T_3)	0.001	0
Monoiodotyrosine	0	0
Diiodotyrosine	0	0

Source: Oppenheimer, 1975. Adapted by permission from *The New England Journal of Medicine 292,* 1066.

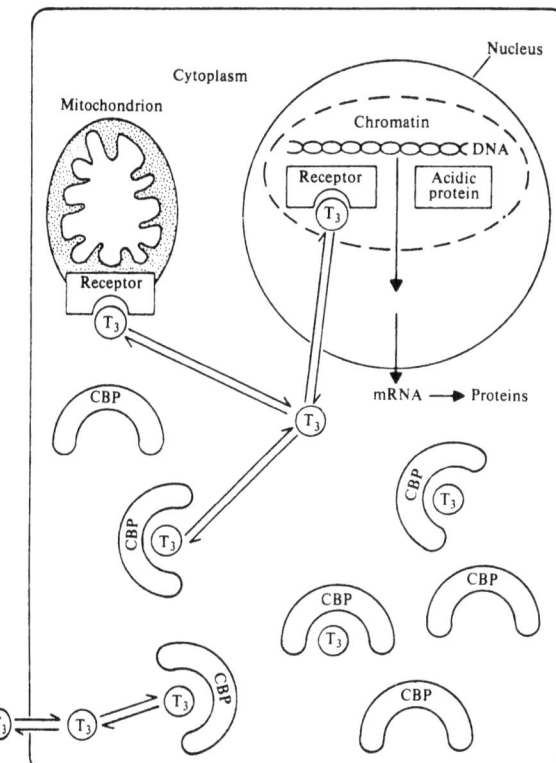

Fig. 4.7. Model showing the sequence of events that occurs when thyroid hormone enters its target cell. The unbound triiodothyronine (T$_3$) enters the cell by diffusion and binds to the cytosol-binding protein (CBP). The T$_3$–CBP complex remains in a reversible equilibrium with a very small amount of free T$_3$ in the cytoplasm. This unbound T$_3$ may interact with receptors in the nuclei or mitochondria, where it can trigger a response. (From Sterling, 1977.)

Free T$_3$ does not combine with the DNA, so that the presence of these receptor proteins accounts for the observed association of T$_3$ with cell nuclei.

Thyroid hormones can be shown to bind to these nuclear receptors in isolated chromatin preparations. The number of nuclear receptors is independent of the levels of thyroid hormone in the body (Spindler et al., 1975; Bernal, Coleoni, and DeGroot, 1978). However, fluctuations have been observed, such as the decline in hepatic nuclear receptors during starvation (Schussler and Orlando, 1978).

It has recently become apparent that specific thyroid hormone receptors are not confined to the nucleus (see Tata, 1975; Sterling, 1977). Thus, cell components that exhibit a high affinity, are saturable but have a low capacity for thyroid hormones, have also been identified in the cytoplasm, inner mitochondrial membranes, and plasma mem-

branes. Tata (1975) has suggested that: "One has, therefore, to consider the possibility that the final expression of a growth and developmental hormone may result from a cooperative action generated by simultaneous or sequential but independent interactions with more than one nuclear and extranuclear sites." Sterling has provided a summary (Figure 4.7) which involves the binding of T$_3$ to a cytosol-binding protein (CBP) from which it can dissociate and combine with receptors in the mitochondria and the nucleus. It has been estimated that there are about 2000 such mitochondrial receptors in each liver cell compared to 20 million cytosol sites. The mitochondrial membrane receptors are lipoproteins (Sterling et al., 1978) and can be chemically distinguished from the nuclear receptors. The former have a molecular weight of about 150,000. They have been identified in most tissues but are, notably, absent from the brain of adult rats yet are present in the brains of rats 12 days old or younger, where they presumably are involved in early development. A role for specific mitochondrial receptors for thyroid hormone has not yet, however, achieved consensus.

The nature of the thyroid hormone-receptor response is unknown, but a model has been proposed (Blake and Oatley, 1977). This hypothesis is based on the observation that the prealbumin molecule, which binds thyroid hormones in the plasma, also has a structural component that is complementary to double-helical DNA. Thus, part of the molecule may function like the DNA binding site of the nuclear thyroid hormone receptor. The prealbumin has a molecular weight of 55,000 and is a tetramer consisting of four identical subunits, each containing 127 amino acids. It is thus also comparable in size to the nuclear receptor. High-resolution X-ray analysis indicates that the putative DNA-binding site on the prealbumin consists of a deep semicylindrical groove. There are two thyroid hormone-binding sites, which are present in a cylindrical channel that runs right through the molecule. The nature of this channel can be related to the structure of the thyroid hormones and their tight binding to the receptors.

The natures of the responses that follow the interaction of thyroid hormones and their receptors are quite diverse. The nuclear effects appear to involve the formation of mRNA and an increased synthesis of proteins by the ribosomes. These proteins may be enzymes or structural components of the cell. The nature of the mitochondrial response is unknown, but it is thought to be direct and involve regulation of mitochondrial energy metabolism.

The anabolic effects of thyroid hormones presumably depend ultimately on the mitochondrial enzymes, but it has also been suggested that Na,K-activated ATPase is involved (Edelman,

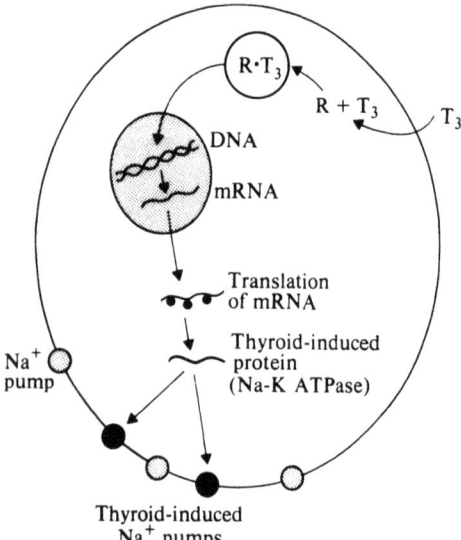

Fig. 4.8. Model to explain the effects of thyroid hormones on calorigenesis. For details, see the text. (From Edelman, 1974. Reprinted by permission from *The New England Journal of Medicine 290*, 1307.)

1974, 1976). A high proportion of the metabolic activity of many cells is known to be associated with their ion-regulating activity, especially the active transport of Na out of the cell. This process is related to the activity of Na,K-activated ATPase, which, in the presence of Na and K, splits ATP to ADP and inorganic P. The accumulated ADP determines the oxidative activity and production of energy, including heat, by the mitochondria. Ismail-Beigi and Edelman in 1970 suggested that this Na transport and its associated Na,K-ATPase may make a substantial contribution to the calorigenesis which is seen in response to thyroid hormones (Figure 4.8). They have observed that the levels of this enzyme (the V_{max} but not the K_m) are elevated in kidney and skeletal muscle in response to the presence of thyroid hormones (Lo and Edelman, 1976; Lo et al., 1976; Asano, Liberman, and Edelman, 1976). Such an effect on calorigenesis may involve both genetic transcriptional processes and direct effects on the mitochondria. Whether sufficient oxygen consumption is actually channelled *in vivo* via such a Na-dependent system is still, however, being discussed.

4.8. Thyroid hormones and their analogues: structure–activity relationships

By 1955, about 150 structural analogues of L-thyroxine (L-3,5,3',5'-tetraiodo-L-thyronine) had been synthesized, and the biological activities of many had been evaluated (Selenkow and Asper, 1955). Biology, however, triumphed over chemistry, for until the discovery of the naturally occurring analogue L-3,5,3'-triiodo-L-thyronine by Gross and Pitt-Rivers in 1953, no related chemical compound with a higher biological activity than L-thyroxine had been discovered. Indeed, despite further efforts, this record has not yet (with a single exception) been bettered. Another search has been made for chemical analogues that may have an antagonistic activity to T_4 and T_3, and although several interesting compounds (Figure 4.9), such as 3,3',5'-triiodo-L-thyronine (T_3' or "reverse" T_3), have been shown to exhibit such actions, the extremely high concentrations that must be used (100 to 200 times greater than the natural hormones) have generally precluded their therapeutic usefulness. A third line of research has been an attempt to dissociate the cholesterol-lowering effects of the hormones from their calorigenic and cardiac effects. Such an analogue could conceivably be used to lower blood cholesterol levels and possibly limit atherosclerotic changes in the vascular system. The optical isomer D-thyroxine was synthesized with such an effect in view. A recent controlled study for testing this compound in the United States indeed had to be abandoned because of the high incidence of cardiac problems and even deaths. Therefore, we may conclude that at this time the therapeutic applications of chemical analogues of T_3 and T_4 has not been widespread or especially successful.

Such studies are, however, of basic pharmacological interest, as they have helped delineate the requirements that these hormones have with respect to the interactions that they have with their receptors (see Pittman and Pittman, 1974). It may seem somewhat facile to note that the steric and electronic requirements for the activity of these molecules are very precise; this fact was, however, not always so apparent as it is today. The thyroid hormones are diphenyl ether compounds (Figure 4.9) with a hydroxyl group on one side and an alanine side chain on the other. They can be substituted with iodine at three or four positions in the molecule, 3',5' and 3,5 in thyroxine, and 3' and 3,5 in triiodothyronine. Seemingly quite modest changes can have dramatic and usually adverse effects on the molecule's activity, a notable exception being the deletion of the 5' iodine in T_4 to produce T_3. Substitution of the ether O between the phenyl groups with a thioether S reduces activity about tenfold, while a shortening of the bond by deleting the O results in a complete loss of activity. The phenolic OH group at the terminal 4' position is similarly important, although not vital; but a change at this site, such as *O*-methylation, results in a substantial decline in activity. Changes in the L-alanine side chain also reduces but does not

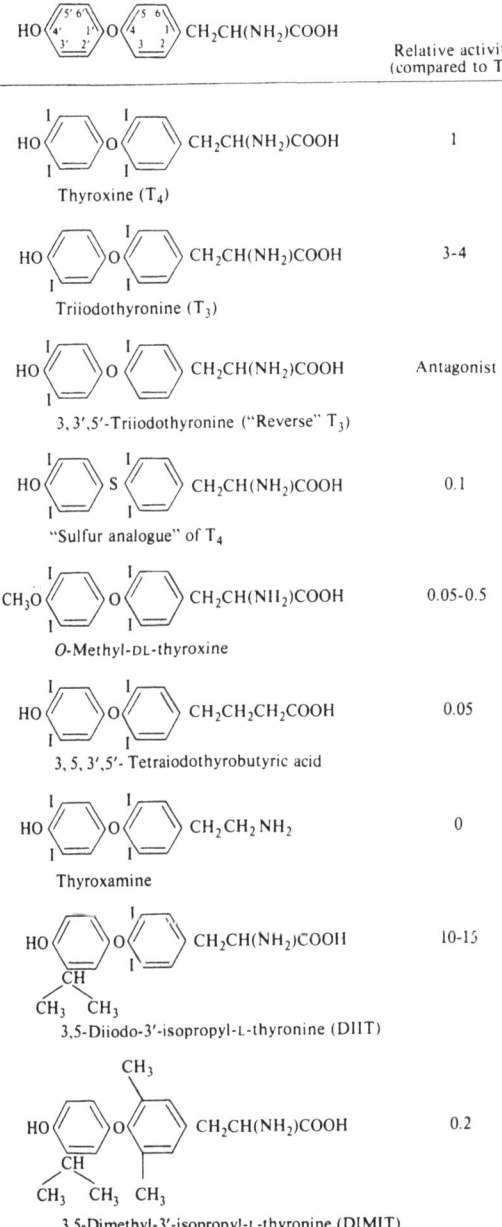

Fig. 4.9. Chemical structures and relative biological activities of some analogues of the thyroid hormones.

necessarily abolish activity. The amino group is not essential, but the terminal carboxyl is vital.

The role of iodine in the thyroid hormones has caused a great deal of speculation, and indeed it was once considered possible that the amino acid was merely acting as a "carrier" for iodine. Other suggestions are that the iodine atoms are essential for the maintenance of the three-dimensional con-

formation of the molecule. They could also be involved in electronic interactions and charge transfer in the combining of the hormone and the receptor. Substitutions at positions 3 and 5 are essential, but although iodine is most effective at these sites, other halogens, or even alkyl or aryl groups, result in a partial retention of activity. An isopropyl group (Figure 4.9) at the 3' position (3,5-diiodo-3'-isopropyl-L-thyronine, isopropyl T_2) gives a highly effective compound (10 times as active as T_4). However, the addition of a second isopropyl group at the 5' position, like the 5' iodine, greatly reduces activity. Analogues that completely lack iodine, such as 3,5,-dimethyl-3'-isopropyl-L-thyronine, still retain activity. An iodine-free analogue of triiodothyronine, 4'-methoxy-3,5,3'-trimethyl-thyronine-N-acetyl ethyl ester, has been prepared which has a similar conformation to the parent hormone (Cody, 1978). However, as the presence of the iodine enhances the molecule's biological activity, it is inferred that the iodine is also involved in electronic related processes concerned with the hormone's interaction with the receptor. Thus, iodine's role in maintaining the conformation of the hormone molecule is apparently not its only function.

The information derived from studies of the structure–activity relationships of such analogues of the thyroid hormones has provided a steric model that can be related to the interaction of T_3 with its receptors. The two phenyl rings appear to lie, through the ether oxygen, at a critical angle of 120° to each other (Figure 4.10). They are also in opposite planes which are perpendicular to each other, and if they come to lie in the same plane, activity is lost. Rotation of the inner ring about the ether O is, however, restricted by the I in positions 3 and 5. The I in the outer ring must be in the distal 3' position. Analogues in which rotation is prevented so that this I lies permanently in the proximal position are inactive. A model has been proposed by E. Jorgenson (Figure 4.10) in which the molecule of T_3 is thought to consist of two main parts: a "binding" region corresponding to the inner phenylalanine ring, and an outer "functional" region. These two hypothetical subunits appear to contribute, respectively, the pharmacological properties of "affinity" and "intrinsic activity" to the hormonal molecule.

4.9. Antithyroid chemicals and drugs

The possibility that thyroid function could be influenced by a variety of chemical compounds, unrelated to the hormones themselves, first became apparent when in 1928, Chesney, Clawson, and Webster showed that goiters occurred in rabbits fed on cabbage. This interesting observation led to the examination of the antithyroid effects of

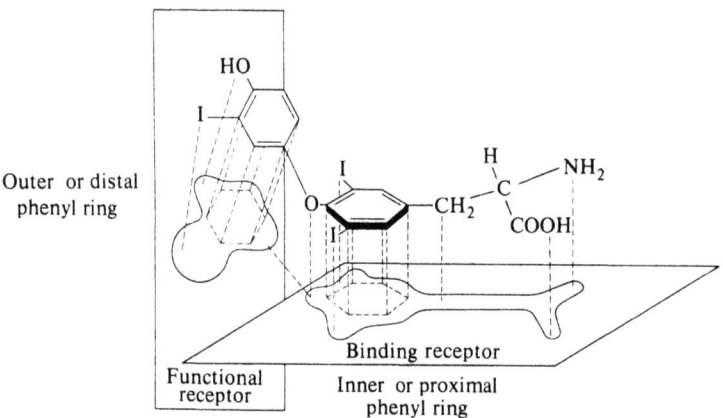

Fig. 4.10. Hypothetical model of triiodothyronine in relation to its possible interaction with the receptor. The two phenyl rings lie in opposite planes and at an angle of 120° to each other. (From Jorgensen et al., 1962.)

other plant foods and the isolation of L-5-vinyl-2-thiooxazolidone (Figure 4.11) from the Swedish turnip or rutabaga (*Brassica napobrassica*) (Astwood, Greer, and Ettlinger, 1949). This compound has been called *goitrin* and, along with some of its chemical analogues, has been identified in a number of plant materials from the *Brassica* family. The occurrence of such natural goitrogens and their possible role in the incidence of endemic goiter is discussed in Section 4.10A. In 1936, Barker made the astute observation that the use of potassium thiocyanate to lower the blood pressure in hypertensive patients was often accompanied by a considerable enlargement of the thyroid gland, as well as a small reduction in the basal metabolic rate. This effect, which is shared by a number of other anions, has subsequently been shown to reflect an inhibition of the iodide uptake mechanism of the thyroid follicular cells.

In 1941, MacKenzie, MacKenzie, and McCollum found that sulfaguanidine (Figure 4.11), which is used to treat intestinal infections, produced a hyperplasia of the thyroid gland in rats, and this effect was accompanied by a considerable reduction in the basal metabolic rate. Shortly after this observation, Kennedy (1942) found that allylthiourea had a goitrogenic effect in rats, and Richter and Clisby (1942) described a similar response to feeding them with small doses of phenylthiourea. These discoveries led to two broader surveys (MacKenzie and MacKenzie, 1943; Astwood, 1943) involving the description of the antithyroid actions of more than 100 such compounds. A variety of thiourea (or thioureylene) derivatives (Figure 4.11) were found to have considerable antithyroid activity. The *thioamide* grouping (—N-H—CS—NH—) is essential, and the effect is lost if the S is excluded. Thiouracil, on the basis of its

high activity and relatively low toxicity, was identified as the compound of "greatest interest clinically" (Astwood, 1943), a prophetic observation that led to the introduction of propylthiouracil for the treatment of hyperthyroidism. The activity of the sulfonamide compounds, which is generally less than the thioamides, was found to depend on the presence of an aniline moiety (NH_2—C_6H_5) in the molecule. When the amino group is substituted, activity is lost.

The antithyroid effects of *thiocyanate* can be readily overcome by additional dietary iodide, but this antagonism was not seen acutely with respect to the action of the aniline or thioamide compounds. This observation suggested a basically different mechanism of action. Thiocyanate prevents iodide "trapping" by the thyroid.

The thioamide or thioureylene antithyroid drugs prevent hormone synthesis by inhibiting the thyroid peroxidase-catalyzed reaction. Normally, the oxidized iodide that results from this process reacts with tyrosine. There are several possible sites where these drugs could interfere with this process, the precise one being somewhat controversial. It is considered that the sulfur atom of the thioamide is reactive and can exist as a sulfhydryl group, which has a reducing action and can form disulfides. In 1966, Morris and Hager suggested that the thioamides competed with the tryosine residues for the oxidized iodide. The peroxidase–iodinium complex (EI^+) formed can thus interact with either tyrosine or thioamide. These drugs have, however, also been shown to inhibit the enzyme itself in the absence of iodide, and this reaction may be irreversible, especially when the levels of iodide in the reaction mixture are low (Taurog, 1976; Nagasaka and Hidaka, 1976). It was suggested that such drugs directly interact with the peroxidase and are

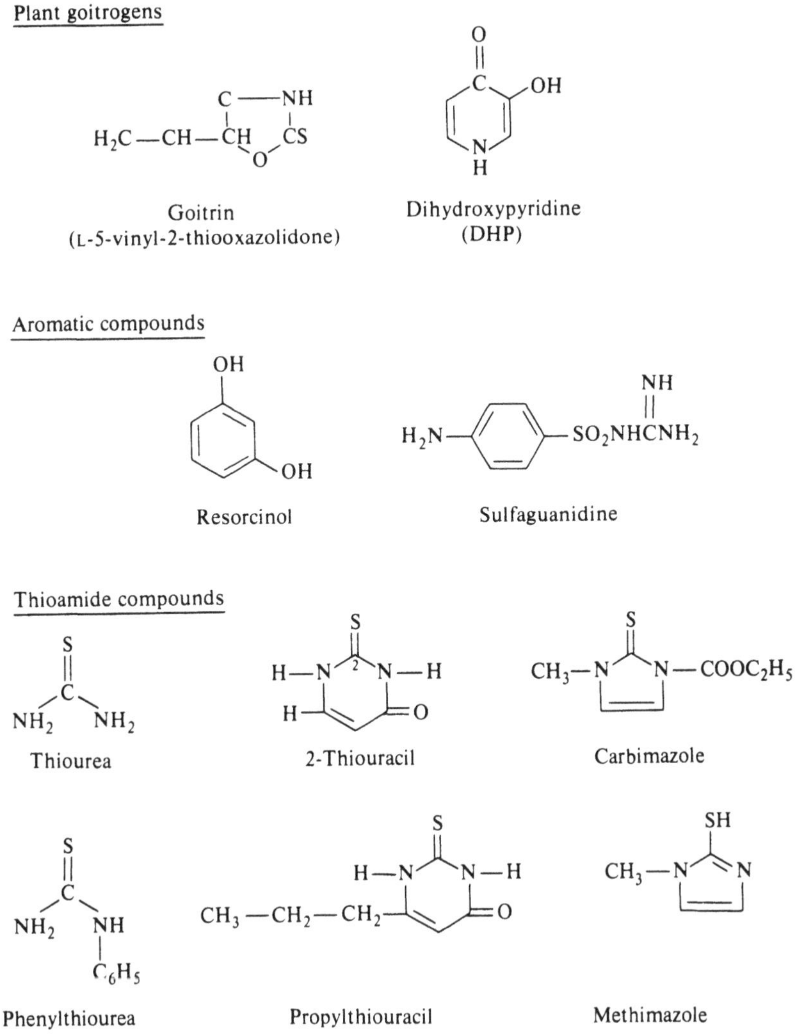

Plant goitrogens

Goitrin
(L-5-vinyl-2-thiooxazolidone)

Dihydroxypyridine
(DHP)

Aromatic compounds

Resorcinol

Sulfaguanidine

Thioamide compounds

Thiourea

2-Thiouracil

Carbimazole

Phenylthiourea

Propylthiouracil

Methimazole

Fig. 4.11. Chemical structures of some antithyroid chemicals and drugs.

competitive with iodide, with which they possibly share a common site. They thus have been described as acting as "general inhibitors."

An alternative scheme suggested by Davidson and his collaborators (1978) is basically similar to the original one of Morris and Hager. It is known that the presence of iodide can protect the enzyme from irreversible inactivation, as probably occurs under normal *in vivo* conditions. It was proposed that in the absence of iodide an oxidation product of the drug is formed as a result of the action of the peroxidase (scheme 4.1), and it is this substance which, indirectly, accounts for the irreversible inhibitory effect on the enzyme. In the presence of iodide, however, the oxidized or "active iodide" interacts with the thioamide group to form a

drug–sulfenyl iodide compound (scheme 4.2). The anion is thus not available to the tyrosine residue.

The drug–sulfenyl iodide can combine with another thioamide group to form a harmless disulfide and the reduced iodide is released. There will no doubt be addenda to this controversial subject.

Propylthiouracil has been shown (Jones and Van Middlesworth, 1960) to also act peripherally and reduce the deiodination of the thyroid hormones. It is difficult to predict the overall effect of such an action in the body, which has not yet been demonstrated *in vivo* in man, but it could contribute to the antithyroid effect if it reduced conversion of T_4 to the more active hormone T_3.

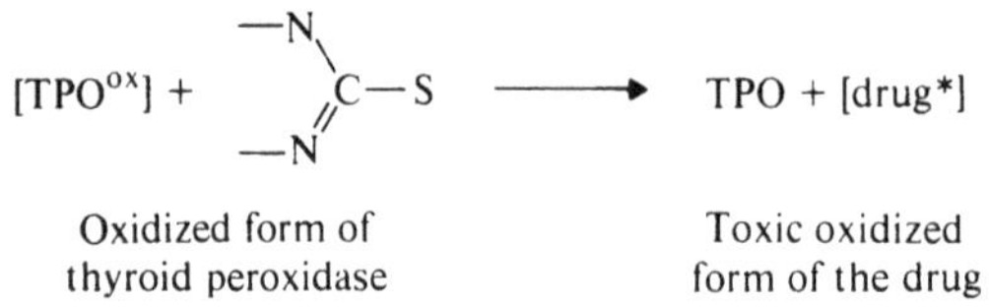

$$[\text{TPO}^{\text{ox}}] + \underset{-\text{N}}{\overset{-\text{N}}{>}}\text{C}-\text{S} \longrightarrow \text{TPO} + [\text{drug*}]$$

Oxidized form of   Toxic oxidized
thyroid peroxidase   form of the drug

Scheme 4.1.

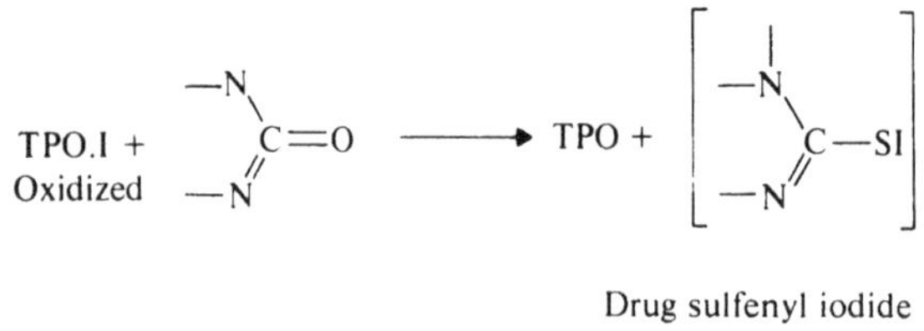

$$\underset{\text{Oxidized}}{\text{TPO.I} +} \ \underset{-\text{N}}{\overset{-\text{N}}{>}}\text{C}=\text{O} \longrightarrow \text{TPO} + \left[\underset{-\text{N}}{\overset{\overset{\displaystyle|}{\text{I}}}{\underset{}{\text{N}-}}}\text{C}-\text{SI} \right]$$

Drug sulfenyl iodide

Scheme 4.2.

An up-to-date comprehensive account of the basic and clinical pharmacology of the thioamide antithyroid drugs has been provided by Marchant, Lees, and Alexander (1978).

A variety of other drugs have been shown to have antithyroid side effects. The use of a *resorcinol* ointment to treat varicose ulcers on the legs has been found to result in hypothyroidism, an action that was reversed when such treatment was ceased (Bull and Frazer, 1950). This effect, and that of a number of related compounds, was confirmed in rats (Arnott and Doniach, 1951) and appears to reflect an inhibition of the organic binding of iodine in the gland. The use of *cobaltous chloride* to treat anemia has been shown to have a hypothyroid effect in man (Kriss, Carnes, and Gross, 1955).

More recently, the chronic use of *lithium* carbonate to treat manic depression has been shown to produce goiters and hypothyroidism in man (Schou et al., 1968; Candy, 1972). The mechanism of this effect is uncertain, but it may prevent the release of thyroid hormones by inhibiting adenyl cyclase (this response has been demonstrated *in vitro*) and so block the effect of TSH. The therapeutic use of lithium as an antithyroid drug has been investigated with differing conclusions. The central neural actions of this cation in such circumstances are classified as "undesirable side effects" and were found to be unacceptable in one study (Kristensen, Andersen, and Pallisgaard, 1976), although it had formerly been suggested that it "could be useful in the management of hyperthyroidism in some cases" (Lazarus et al., 1974), especially those that could not tolerate thioamides. Lithium reduces the turnover rate of iodine by the thyroid gland, and it may be "a useful adjunct" to help increase the effectiveness of ^{131}I in the treatment of hyperthyroidism, especially when it is de-

sirable to keep the dose of the radioactive iodine low (Turner, Brownlie, and Rogers, 1976).

Other drugs that exhibit antithyroid side effects include *carbutamide* (related to the sulfonamides), which is used as an oral hypoglycemic drug in Europe (but not in the United States), and *sodium nitroprusside*, which is used to lower blood pressure in hypertensive emergencies. The action of the latter probably depends on its metabolism to cyanates, which inhibit iodide "trapping" by the thyroid (Nourok et al., 1964). The diuretic *acetazolamide* (Diamox), which inhibits carbonic anhydrase, has a similar effect on I uptake. Such compounds, with a sulfamyl group, $-\text{SO}_2\text{NH}_2$, are known to inhibit the transport of halogen anions in a number of tissues.

Rats fed a soya bean diet have been shown to develop goiters and hypothyroidism (Van Middlesworth, 1957), an effect that is apparently related to a decreased intestinal reabsorption of the thyroxine that is excreted in the bile. A similar effect has been described in infants in which thyroxine is administered orally. This effect of a soya bean diet may reflect a general response due to the binding of thyroxine in the gut, as seen in diets with a high fecal residue content.

Cattle that graze on a tropical legume, *Leucaena leucocephala*, in various parts of northern tropical Australia and New Guinea have been observed to develop enlarged thyroid glands (Hegarty et al., 1976). In addition, both sheep and cattle that feed on this plant may give birth to goitrous young. The leucaena plants contain a nonprotein amino acid called *mimosine* (β-[N-(3-hydroxy-4-oxopyridyl)]-α-amino propionic acid), which has an inhibitory effect on the activity of the thyroid gland. However mimosine does not act directly; it, for instance, does not have a goitrous effect when fed to rats and mice. It is converted in the rumen of sheep and cattle, as a result of their digestive processes, to the

Table 4.6. *Anions that may inhibit the thyroid gland: relationship of uptake (K_m) and inhibition of iodide accumulation (K_i) to anion size (Φ_0)*

Anion	K_m(M)	K_i(M)	Partial molal volume, Φ_0, at 25°C (cm^3/mol)[a]
Bromide (Br$^-$)		$\sim 2 \times 10^{-2}$	25.1
Cyanate (OCN$^-$)		$1-2 \times 10^{-2}$	26.7
Nitrite (NO$_2^-$)		4×10^{-3}	25 (20°C)
Nitrate (NO$_3^-$)		$1-2 \times 10^{-3}$	29.4
Iodide (I$^-$)	3×10^{-5}, $2-6 \times 10^{-5}$		36.7
Astatide (At$^-$)			40.2
Thiocyanate (SCN$^-$)		$2-3 \times 10^{-5}$	40.6
Monofluorosulfonate (SO$_3$F$^-$)		$1-2 \times 10^{-5}$	47.8
Selenocyanate (SeCN$^-$)		1×10^{-5}	50.3
Tetrafluoroborate (BF$_4^-$)		3×10^{-6}	44.0
Perrhenate (ReO$_4^-$)	1×10^{-6}	1×10^{-6}	48.7
Perchlorate (ClO$_4^-$)		4×10^{-7}	44.5
Pertechnetate (TcO$_4^-$)	$3-5 \times 10^{-7}$		46.0

[a] Defined as volume of anion occupied in solution, extrapolated to zero concentration.
Source: Wolff, 1964.

active substance *3-hydroxy-4(1H)-pyridine* or *DHP* (Figure 4.11). The DHP appears to inhibit the activity of thyroid peroxidase and so, like the thioamides and resorcinol, it blocks the incorporation of iodide into the active hormone. The action of DHP on the young of ruminants suggests that it can cross the placenta.

The clinically useful antithyroid compounds are (a) the *thioamides,* (b) *anion inhibitors* of the iodide trapping mechanism, and (c) *iodide* in large doses.

A. The thioamides
As just described (see also Marchant, Lees, and Alexander, 1978), these compounds are derivatives of thiourea. The chemical structures of several of these is shown in Figure 4.11. The choice of those drugs used therapeutically has been made on the basis of the incidence of their toxic effects, which initially were quite high. Chemical modification so as to increase their potency without a directly related change in toxicity has resulted in the introduction of several compounds. *Propylthiouracil* and *methimazole* are generally used in the United States, but *methylthiouracil* is also available, while *carbimazole* is widely used in the United Kingdom. The latter is converted to methimazole in the body. Methimazole is about 10 times as active as propylthiouracil. Only oral preparations are commercially available. The drugs are rapidly absorbed from the gut; the half-lives are about 2 hours for propylthiouracil and 12 hours for methimazole. They cross the placenta, but this does not exclude their use in pregnancy, although it is a consideration when choosing dosage. Breast feeding is not possible during therapy, as the drugs readily pass into the milk. The response to the thioamides is

delayed if the patient has previously been receiving iodide, apparently as a result of increased residual stores of the hormone. Addition of iodide to the diet increases the chance of a relapse of the hyperthyroidism as the thioamides become less effective. The reason for this effect is not clear.

The most common *side effects* are skin rashes, but joint and muscle pain, headache, drug fever, and cholestatic jaundice may also occur. If such effects are not too severe, they subsequently may disappear if the dosage is reduced or an alternative drug is used. Cross-sensitivity, however, may occur. The incidence of such side effects is about 3 percent for propylthiouracil and 7 percent for methimazole. The most dangerous side effect is agranulocytosis, which is life-threatening, especially if not detected early. It can develop quite suddenly and the patient is usually warned to discontinue the drug and seek aid if he develops a severe sore throat. The incidence of this condition is about 0.4 percent for propylthiouracil and 0.1 percent for methimazole.

B. Anion inhibitors
The action of the thyroid iodide pump can be inhibited by a number of anions which appear to compete with iodide for sites on the "carrier" mechanism (Wolff, 1964). These ions must be univalent; divalent ones are ineffective. All univalent ions are, however, not inhibitory, as there are additional requirements that appear to be related to the size of the ions. The estimated size is found to be related to the ability of the thyroid to take up the anion (expressed as K_m, half-saturation) or inhibit iodide uptake (K_i, 50 percent inhibition) (Table 4.6). It can be seen that the most effective are the complex anions perchlorate (ClO$_4^-$) and pertech-

netate (TcO_4^-). Of the halogens, F^- is ineffective, while Br^- has a weak action and astatide (At^-) is as effective as I^-. All these anions are concentrated by the thyroid gland with the exception of SCN^-. The order of their K_m or K_i reflects their goitrogenic potencies.

Perchlorate is the only inhibitor of I^- uptake that has been utilized clinically. It is an effective hypothyroid drug, but as its use has been associated with a number of cases of fatal aplastic anemia, it is not generally used. However, it offers a temporary alternative in patients who are hypersensitive to the thioamides.

Such inhibitors of I^- transport not only block uptake but also promote the loss of I from the thyroid follicular cells. This effect forms the basis for the *"perchlorate discharge test,"* which is used to diagnose disorders in the organification of iodine in the thyroid and in order to distinguish Hashimoto's disease from simple goiter (Gray et al., 1974). Normally, there is little free iodine in the gland, but if its organification is compromised, this can be detected as an excessive release (using ^{131}I) in response to the administration of 0.5 to 1 g of perchlorate.

C. Iodide

The effects of iodide on the thyroid gland are complex and paradoxical, and depend on the dose as well as the condition of the subject. The ability of dietary iodide to inhibit the thyroid and produce "coast goiter" is described in Section 4.10A and its occasional thyrotoxic action (*Jod–Basedow effect*) is referred to in Sections 4.10A and E. Large doses of iodide, in excess of 5 to 10 mg/day (which is over 100 times the normal dietary allowance), inhibit thyroid function, and long before the discovery of modern antithyroid drugs it was used clinically to treat hyperthyroidism. The effect appears to involve at least three different mechanisms. There is a reduced thyroid hormone formation due to an inhibition of the "organification" of iodine: the Wolff–Chaikoff effect, which is described in Section 4.4D. It usually only lasts for a few days, as an "escape" occurs. Iodide can inhibit the release of thyroid hormones. This effect occurs dramatically and promptly within a few hours following its administration and is its most important current therapeutic use. It is used in conjunction with other drugs in acute situations to reduce hormone release in thyroid "storm" (see Section 4.10C). Its mechanism is not clear, but it may reduce the mobilization of the follicular colloid by inhibiting proteolytic enzymes or adenyl cyclase.

Iodide reduces the vascularity of the thyroid and it becomes firmer in its consistency. The mechanism for this effect is unknown. Iodide is often used in conjunction with thioamides in the preoperative preparation of patients for subtotal thyroidectomy, as this action reduces bleeding and apparently aids the surgeon's task.

Iodide is an old drug with well-characterized *side effects.* Its potential to produce hypothyroidism and thyrotoxicosis is described in Sections 4.10B and E. General effects sometimes referred to as *iodism* include swollen salivary glands, acneform rash (similar to that due to bromism), headache, fever, sore throat and sneezing (symptoms reminiscent of a "head" cold), and a metallic taste in the mouth. These effects subside when the iodide is withdrawn, but urinary excretion can, if necessary, be promoted with diuretics.

A comprehensive account of antithyroid drugs and goitrogens that occur in nature has been provided by Langer and Greer (1977).

4.10. Diseases of the thyroid gland and their pharmacological treatment

Thyroid disease is manifested as an excessive endocrine activity of the gland (hyperthyroidism, thyrotoxicosis), a decrease in activity (hypothyroidism), or structural or morphological changes that do not necessarily result in a change in thyroid hormone status (euthyroid). The causes of such diseases are diverse and often incompletely understood, so that the treatment of the primary disorder is mostly not possible, although the signs and symptoms can be ameliorated.

The effects of hyperthyroidism and hypothyroidism on various bodily functions are summarized in Table 4.7. It should, however, be remembered that function may be affected in various degrees and be related to the particular pathological situation, so that the effects described are not invariably apparent in all instances.

A. Goiter: endemic, sporadic, and chemically induced

A goiter (see Stanbury, 1968) is an enlargement of the thyroid gland which is most often of a diffuse nature but may take the form of discrete nodules. It should be distinguished from enlargement of the gland due to malignancy, although solitary thyroid nodules may subsequently become cancerous (5 to 30 percent). The enlargement of the thyroid gland can, in most instances, be viewed as a homeostatic mechanism to maintain thyroid hormone levels, and it occurs in response to elevated concentrations of thyrotropic hormone. If this compensation is successful, as it usually is, adequate thyroid hormone levels will be maintained and the person will be euthyroid. In some instances, however, hypothyroidism may be associated with such goiters, and this condition is most important in children, when normal growth and development may be compromised. Even euthyroid goiters can result in specific problems,

Table 4.7. *Some effects of hypothyroidism and hyperthyroidism on bodily functions*

	Hypothyroidism	Hyperthyroidism
Oxygen consumption	↓	↑
Heart rate	↓	↑ (arrythmias)
Intestinal motility	↓ (constipation)	↑ (diarrhea)
Body weight	↑	↓
Skin	Dry, cool	Warm, sweating, myxedema
Sympathetic nervous system activity	↓	↑
Central nervous system	Slowness in thought, lethargy, sleepiness	Hyperactivity, nervousness, tremor, irritability
Eyes	?	Lid retraction (exophthalmos)
Skeletal muscles	Slow reflexes	Fast reflexes
Blood lipid and cholesterol levels	↑	↓
Renal function	↓	—
Growth	↓	↓

Note: ↑, increase; ↓, decrease.

especially if they are so large as to constrict the neck and interfere with breathing and eating. Their cosmetic disadvantages may also be a problem.

Endemic goiter afflicts about 200 million people and occurs in certain specific, usually mountainous, parts of the world, mostly where iodine is deficient in the soil (Stanbury, 1968). Such regions include the Alps and the Pyrenees in Europe; the Andes in South America; the Himalayas in Asia; mountainous regions in New Guinea, New Zealand, and Tasmania; the Great Lakes region in North America; and parts of Africa, including Zaire. Other predisposing factors may, however, contribute to the disease, including genetic ones, the presence of antithyroid substances in the diet, and bacterial contamination of the drinking water. It may afflict as many as 90 percent of people living in such areas. Whatever the precise cause of the disease, it can be prevented and cured by the administration of iodine. An especially tragic association of endemic goiter is the occurrence of endemic cretinism. This disorder, which is apparent in early neonatal life, is not seen in all such regions (it is, for instance, rare in the Great Lakes basin), probably reflecting the degree of iodine deficiency. It may, however, affect an alarming proportion of children; an incidence in about 8 percent of births has, for instance, been observed in parts of New Guinea. The most important effect of such endemic cretinism is mental retardation and deaf-mutism, which, as the developmental pattern of the nervous system is determined early, *in utero*, is not responsive to subsequent postnatal hormonal replacement.

Goiter may also occur in less-well-defined situations (*sporadic goiter*) as a result of genetic and other causes which may influence the requirements and availability of iodide. This condition is also manifested in children, where it is referred to as "sporadic cretinism" or "congenital hypothyroidism."

1. Prophylaxis and treatment. Endemic goiter can be prevented by the administration of iodide. The daily dietary requirement of I, in order to compensate for urinary and fecal loss is about 100 μg, although as a dietary supplement a more generous 200 μg/day is usually recommended. In some areas, such as Tasmania, once or twice weekly oral doses of KI (10 mg) to schoolchildren have been used. In other more remote areas, such as in New Guinea (Butterfield et al., 1965), intramuscular injections of iodinated poppyseed oil (4 ml, providing a slowly absorbable depot of about 2000 mg) have been used. The latter treatment affords protection for up to 4 years. Plasma TSH levels are reduced by this treatment (Medeiros-Neto et al., 1975). Less formal, but also less controlled administration of I is brought about by supplementing the diet, usually with table salt (iodized salt) and, as in Tasmania, sometimes flour. The quantities of iodide added vary from country to country; 10 mg/kg salt is used in Switzerland, 20 to 200 mg/kg in the United States. Iodide is oxidized to iodine during storage, so that the more stable *iodate* is sometimes substituted for it. The quantity of iodide added to the salt has been partly limited by the fear of inducing hyperthyroidism ("Jod–Basedow effect"). This condition is fortunately rare but may occur, especially in people who have an incipient Graves' disease, which is normally limited by a low dietary iodide intake. With modern transportation, exchange of foods is more common, so that many people living in potential goiter belts now receive foods, such as sea fish, which have a high iodide content. An involuntary increase in I intake also occurs as a result of modern methods of breadmaking, which utilize added iodine compounds. However, there remain many less accessible areas of the world where dietary supplements remain essential but where facilities for their administration are inadequate.

The administration of additional iodine in public health programs has resulted in startling decreases in the incidence of endemic goiter, cretinism, and deaf-mutism. Indeed, in Europe most cretins are over the age of 60 years and the birth of such children in areas where the disease was formerly common is now almost nonexistent. The problem remains, however, in less accessible areas of Africa, Asia, and New Guinea.

When iodide is administered, the goiter ceases to enlarge and will often regress somewhat, but this treatment is not always successful. Thyroxine will often cause the goiter to disappear, but this response is not invariable. The question then arises as to whether surgical intervention (subtotal thyroidectomy) or the use of radioactive iodide (usually ^{131}I) is justified, such as for cosmetic reasons or to relieve undue pressure in the neck region. As such people are usually euthyroid, the risks are not always justified, especially as lifelong replacement therapy with thyroxine may be necessary.

Endemic cretinism can be treated with thyroxine and the sooner the disease is identified postnatally, the more successful it will be expected to be. As described earlier, however, the neural effects and mental retardation are determined prenatally and cannot be reversed. The continued administration of thyroxine may have adverse effects and turn a placid, easily manageable patient into an aggressive and difficult one, which may necessitate withdrawal of the hormone replacement in later years.

Sporadic or "simple" goiter occurs in regions where the dietary iodine is considered to be adequate in normal situations but for some reason the thyroid can only gain or utilize a limited amount of it. It can have several causes.

Coast goiter is a disease that occurs in certain parts of Japan and results from an excessive intake of iodide due to the eating of kelp (a seaweed) (Suzuki et al., 1965). It may result from the inhibition of the organification of iodine by excessive intrathyroidal iodine concentrations, such as exemplified by the Wolff–Chaikoff effect. However, in contrast to the latter, no "escape" mechanism limits the response, and it has been conjectured that this defect may have a familial basis. These iodide goiters usually regress when the dietary intake of iodine is restricted.

Naturally occurring goitrogens may be consumed in the diet and limit the production of thyroid hormones in various ways (see Clements, 1960). An interaction may occur in regions where there is endemic goiter and the availability of iodine is limited, so that a goitrous condition is precipitated more readily. The possibility that such dietary substances exist was first demonstrated by Chesney, Clawson, and Webster in 1928. They found that rabbits they were using in a study of syphilis developed enlarged thyroid glands which were likened to "endemic goiter." These animals were fed a diet of cabbage, which belongs to the *Brassica* group of plants, also including kale, cauliflower, and turnips. This type of goiter can be overcome by the administration of iodide and appears to result from the presence of thiocyanate (SCN$^-$), which prevents the trapping of I$^-$ by the thyroid or goitrins, which inhibits its organification. Thiocyanate has been identified in a number of plants and is formed during the preparation of cassava, which makes up a large portion of the diet in some parts of West Africa. It is considered unlikely, on a quantitative basis, that such diets are alone responsible for goiters in man. Thus, a daily consumption of about 10 kg of cabbage would be necessary to provide a minimally effective dose of SCN.

Antithyroid compounds that inhibit the organic binding of I in the thyroid gland have also been found in foodstuffs. In 1949, Astwood, Greer, and Ettlinger chemically identified the active antithyroid principle in Swedish turnips or rutabaga (*Brassica napobrassica*) as L-5-vinyl-2-thiooxzolidine. This substance and related compounds have been isolated from foodstuffs such as cabbage, kale, rape, white turnips, and many plants that occur in pastures. The latter observation is of importance, as such substances can pass into the milk of cows and may have been contributing to a goiter "epidemic" among schoolchildren in Tasmania in the 1950s (see Clements, 1955, 1960; Greene, Farran, and Glascock, 1958).

B. Hypothyroidism

Hypothyroidism was first described in 1874 by Gull and was shortly afterward called Gull's disease. As a dramatic accompaniment there was often a swelling of the skin and subcutaneous tissues, so it was also commonly referred to as a "myxoedema." The latter symptom does not, however, invariably occur, and other manifestations (Table 4.7) are more common. Hypothyroidism is now recognized to be a "graded" condition (Evered et al., 1973; Evered, 1976) and also has different manifestations, depending on whether it occurs in young children or adults. It can thus be divided into several categories.

1. Adult hypothyroidism. The incidence of this condition is about 0.2 percent of the community, but this estimate is almost certainly too low, especially as many nonovert cases of the disease go undiagnosed. The presence of autoimmune thyroid antibodies, which are thought to be a common cause of the disease, have, in a survey in the United Kingdom, been found to be "very high" in 4.6 percent of women and 1.6 percent of men, while those with levels above normal are 16 and 4 percent, respectively (Evered and Hall, 1972). Hypothyroidism is also commonly associated with overeffective treatment of hyperthyroidism due to surgery or irradiation of thyroid tissue. It may also result from antithyroid drugs (see Sections 4.9 and 4.10E) and naturally occurring dietary contaminants such as plant goitrogens and excess iodide (see Section 4.9C). The deficiency may also be a secondary response to pituitary disease (lack of TSH) or even a tertiary one related to inadequate functioning of the hypothalamus (lack of TRH). Genetic abnormalities occur but are rare, and on occasions the disease may be due to tissue damage following thyroiditis.

Hypothyroidism can be divided into four graded categories (Evered and Hall, 1972):

a. *Overt hypothyroidism* includes a wide range of physiological deficiencies, which are summarized in Table 4.7. The clinical findings rarely present any diagnostic problems. This condition always requires replacement therapy with thyroid hormones. If untreated, myxedema coma may eventually occur. The latter situation is an acute medical emergency in which the state of consciousness is impaired, the body temperature may decrease to 27°C, and there is hypoventilation and hypoxia, and cardiac arrythmias.

b. *Mild hypothyroidism.* This condition is more difficult to diagnose, as the symptoms are not always clear and usually include a general tiredness and lack of energy. It can be readily diagnosed by measuring plasma TSH levels (by radioimmunoassay), which are above normal. This condition is treated by thyroxine replacement.

c. *Preclinical hypothyroidism.* No symptoms are apparent in this condition, but it can be detected by measuring plasma TSH levels. An incipient thyroid deficiency is compensated for by increased stimulation by the TSH. It is not known whether such a condition may progress to mild or overt hypothyroidism, or whether it needs treatment. It has been suggested that preclinical hypothyroidism could result in an increased risk of coronary artery disease (see Evered and Hall, 1972).

d. *Autoimmune thyroid disease.* As described earlier, antithyroid antibodies occur in a much higher proportion of the population than those with diagnosed hypothyroidism. In most of these people, no deficiency in thyroid function can be detected, but some consider that they may be "at risk" for subsequent development of the disease.

Thyroid hormone replacement therapy in adult hypothyroidism and myxedema coma. Three hormone preparations are available for replacement therapy; synthetic Na-L-thyroxine and Na-L-triiodothyronine as well as desiccated thyroid gland obtained from animals. The latter contains variable amounts of T_3 and T_4 but some well-standardized preparations are available that give predictable results. Although this material is an "official" preparation and is sometimes still used, it has largely been replaced by the synthetic hormones. The standardization of the desiccated thyroid gland preparation is often based on its iodide content, which can be misleading and has even been artificially elevated by the unscrupulous addition of iodide.

L-Thyroxine is the preparation of choice in the treatment of hypothyroidism. L-Triiodothyronine is more potent but has a much shorter duration of action and so must be administered more often, which subjects the patients to larger oscillations, or surges, in the plasma hormone concentrations. A tablet combining T_4 and T_3 in the ratio 4:1 is available but confers no advantages over T_4 alone and it is suspected may be associated with an increased incidence of side effects (Smith, Taylor, and Massey, 1970). The dosages of T_4 that are used range from 100 to 200 μg/day for full replacement (about 2.25 μg/kg), the older estimates of 300 μg/day appear to be excessive (Stock, Surks, and Oppenheimer, 1974). The onset and duration of the effects of thyroid hormone preparations are given in Table 4.8. The initial dose is lower than the final maintenance dose, about 50 μg/day in adults over 40 years, although 100 μg may be used in younger people. Allowances in dosage should be made for the increased hormone demands that accompany exercise and exposure to cold.

Side effects may occur with overdosage, the most important being those related to the heart. Patients with ischemic heart disease and angina pectoris require special consideration, and concurrent administration of propranolol may allow the latter to tolerate larger doses of the hormone. In diabetes mellitus the need for insulin or oral antidiabetic drugs may be increased. The effect of coumarin anticoagulant drugs is enhanced and the prothrombin time is prolonged. Diphenylhydantoin can cause transient increases in serum thyroid hormone concentrations. The concurrent use of

Table 4.8. *Onset and duration of action of different preparations of thyroid hormones*

	Dose (mg)	Onset (days)	Duration (days)
Na-L-thyroxine	0.1–0.2	4	10
Na-L-triiodothyronine	0.06–0.12	6 h–3 days	5
Crude thyroid gland extract	60–180	4	10

cholestyramine delays absorption of the drug from the gut.

In *myxedema coma,* T_4 is also the drug of choice, even though the onset of action of T_3 is faster. The latter has a propensity to cause cardiac arrythmias, although it is used by some. It can be given by i.v. or i.m. injection. Thyroxine can be administered i.v. on the first day of treatment and can be maintained orally on succeeding days. The therapy of myxedema coma is somewhat contentious but also includes measures to support respiration and body temperature, the maintenance of the heart, and the i.v. administration of hydrocortisone Na succinate.

2. Childhood hypothyroidism (sporadic congenital cretinism). Hypothyroidism in children has special significance, as the thyroid hormones are necessary for optimal growth and development, including the central nervous system and the skeleton. The relationship to endemic goiter and the occurrence of endemic cretinism, and the role of iodide deficiency, are described in Section 4.10A. Sporadic congenital hypothyroidism may result from a variety of causes, including genetic defects in hormone synthesis, thyroid aplasia, and atrophy or the acquisition of antithyroid substances from the mother. It may also occur secondarily to defects in the pituitary and hypothalamus. As with endemic cretinism, the sooner the disease is recognized and treated the better, but even then prenatally determined mental changes are not reversible. Thyroxine is administered in the highest doses that can be tolerated without precipitating symptoms of hyperthyroidism, and this must be continued throughout the childhood growth period. If the condition has not been treated by the time that the child is about 10 years of age, the effects on development are irreversible and there is little point in starting therapy with T_4, as this will often only result in a formerly placid and easily manageable child becoming a difficult and unhappy one.

The possibility of administering thyroid hormone preparations to thyroid-deficient infants while they are still *in utero* is receiving active consideration (Comite, Burrow, and Jorgenson, 1978). Thus, it may be possible to avoid the early irreversible effects of hypothyroidism on the development of the nervous system. Thyroid hormones only cross the placenta with some difficulty and if administered to the mother can result in maternal hyperthyroidism. Several analogues of the thyroid hormones have been tested with a view to improving their rates of transfer across the placenta. These compounds include 3,5-diiodo-3'-isopropyl-L-thyronine (DIIIT) and 3,5-dimethyl-3'-isopropyl-L-thyronine (DIMIT) (Figure 4.9). These drugs could prevent fetal goiter, produced by propylthiouracil in rats, but had no hyperthyroid action on the mother. The favorable effects of these analogues may be the result of an increased lipid solubility and placental transfer and their lack of inactivation by deiodination.

C. Hyperthyroidism

This disease usually results from an insufficiently controlled release of thyroxine, although it has recently been recognized that in some instances triiodothyronine excess ("T_3 thyrotoxicosis") alone may be involved. The most common form is Graves' disease, which if untreated results in the death of about 20 percent of the patients within 3 years. About 25 percent recover spontaneously. The symptoms of thyrotoxicosis (see Table 4.7) include an increased metabolic rate, tachycardia and disorders in cardiac rhythym, diarrhea, and neural symptoms, including hyperirritability and nervousness. All of these changes are typical of hyperthyroidism but in Graves' disease a protrusion of the eyeballs (exophthalmos) also occurs (Harvard, 1972). This condition (see also Section 4.3E) may reflect a buildup of tissue at the back of the eye (infiltrative exophthalmos occurs in about 25 percent of cases) or a retraction of the eyelid (spastic exophthalmos), which may be unilateral, reflecting an increased sympathetic tone of the lid muscles (an incidence of about 50 percent). Exophthalmos is more often seen in the younger than the older hyperthyroid patient. It may also occur on other occasions unassociated with hyperthyroidism (euthyroid Graves' disease). In some cases of

Graves' disease, an infiltration of mucopolysaccharides occurs under the skin of the pretibial areas, the tops of the feet, and parts of the arms.

Graves' disease occurs most commonly in young women. Its causes are uncertain, but there is evidence to suggest a genetic relationship, as it is more prevalent in certain families and the coincidence is 30 to 60 percent in identical twins but only 3 to 9 percent in fraternal twins. It is not due to an increased TSH secretion, except in very rare occasions involving a malignancy of the pituitary. In at least 90 percent of instances the disease is associated with the presence of thyroid-stimulating antibodies (TSAb) in the serum. These are almost certainly autoantibodies to the TSH receptors that have the ability to stimulate the activity of the thyroid gland (see Sections 4.3 A and D). The disease may be precipitated in predisposed individuals in response to physical and emotional stress. In the absence of a known definitive cause, specific treatment of Graves' disease is not possible.

There are a number of less common and rare causes of hyperthyroidism. It may be congenital, or result in some individuals from a large iodide intake (Stewart and Vidor, 1976). In pregnancy, hyperthyroidism may be due to the formation of a TSH-like substance, either in the normal placenta or arising from choriocarcinoma or a hydatidiform mole (Ricketts, 1976). The active material appears to be human chorionic gonadotropin (hCG; see Section 7.3). Thyrotoxicosis may also result from a thyroxine "addiction" (Harvey, 1973) in people who surreptitiously take thyroxine tablets ("factitious hyperthyroidism"). Thyroid tissue disorders, including cancer, subacute thyroiditis, a solitary autonomous toxic adenoma, or multinodular goiters may result in hyperthyroidism. Changes in the control mechanisms for the thyroid occur rarely, but TSH-producing tumors of the pituitary and an excess of hypothalamic TRH have been described. The treatment of some of these hyperthyroid conditions can clearly be specifically directed to eliminating the cause of the disease.

An acute manifestation of hyperthyroidism is *thyroid "storm"* or "hyperthyroid crisis." This condition results from a massive release of thyroid hormones, which may be due to infection or surgery. It is a medical emergency which without treatment results in 100 percent mortality and even with treatment mortality may be 20 to 50 percent.

1. Uses of drugs in the treatment of hyperthyroidism. There are several possible ways to treat hyperthyroidism (see Kendall-Taylor, 1972; Evered, 1976; Irvine and Toft, 1976), the choice depending on such factors as age, pregnancy, and previous treatment. It should be recalled that in about 25 percent of cases of Graves' disease, spontaneous remission occurs. There are three main choices, all of which, at least partly, involve the use of drugs.

Surgery. Removal of a part of the thyroid gland (subtotal thyroidectomy) can be undertaken. Initially, surgical procedures were associated with a high mortality and other complications. The thyroid has an excellent blood supply, so that hemorrhage may occur. It is also closely associated with the laryngeal nerves, which if cut, may result in paralysis of the vocal cords (laryngeal palsy). The parathyroid glands also lie close by and accidental removal or damage to these tissues or their blood supply was formerly quite common and resulted in transient or even permanent hypocalcemia and tetany. The operation is not always completely successful, as insufficient glandular tissue may be removed, so that the hyperthyroidism may persist or there may be a resulting hypothyroidism, which has an incidence as high as 40 percent. The latter may reflect the formation of tissue antibodies.

The pharmacological contribution which aids the surgical treatment of hyperthyroidism may be substantial. If the surgery results in hypothyroidism, replacement with thyroxine can maintain the patient in essentially normal condition. Recurrence of hyperthyroidism can be treated with drugs or radioiodine.

It is vital that the patient be delivered to the surgeon in a euthyroid state so as to avoid the possibility of precipitating thyroid "storm" and to reduce the vascular supply to the gland. The patient is treated with a thioamide drug such as methimazole, for a period of up to 2 to 3 months prior to surgery. Two weeks before the operation thyroxine or iodide is also administered, to limit the stimulation of the gland by TSH. As described earlier, the iodide may also decrease the vascularity of the thyroid. The drugs are usually continued for a few days postoperatively. The use of propranolol has also been found, in one study, to provide adequate protection for patients undergoing subtotal thyroidectomy (Michie et al., 1974). The β-adrenergic blocking drug need only be administered for about 1 week preoperatively and is continued for 7 days following surgery. It is stated that the gland is soft and pliable and more easy to manipulate than following treatment with thioamides, and that less bleeding occurs. This procedure is, however, still a novel one, and thioamides and thyroxine are the most widely recognized preoperative treatment.

Radioiodine. The isotope ^{131}I is probably the most widely used and convenient method for treating hyperthyroidism. The radiation from the accumulated radioiodine reduces division of the follicular cells. One of the problems associated with this treatment is a high incidence of hypothyroidism, which develops at the rate of 3 to 6 percent annually and may be seen in 80 percent of cases after 15 years. An attempt to overcome this effect by using ^{125}I, which has less powerful radiation emissions, appears to have been unsuccessful (Bremner, McDougall, and Greig, 1973). In some clinics a relatively high dose of ^{131}I is administered, so that hypothyroidism is expected and replacement therapy, with thyroxine, is given as a standard procedure.

Toxic nodular goiter cannot be treated by drugs, and if radioiodine is used it is usually combined with triiodothyronine, which blocks the uptake of the isotope by the healthy thyroid tissue, and so protects it. Potassium perchlorate may be used if T_3 cannot be tolerated, such as if cardiac irregularities occur.

The use of radioiodine in the treatment of hyperthyroidism was introduced in 1942. The possibility of producing cancer of the thyroid, leukemia, and genetic damage to the gonadal germ cells was recognized, but thorough and prolonged follow-up studies have failed to show such effects. The use of isotopes is, however, usually still avoided in childhood, in young adults, and during pregnancy [see also Section 4.10C(2)]. In the latter condition the isotope could damage the developing tissues, especially the fetal thyroid.

Drugs. As just described, the use of drugs may have an important role in relation to the treatment of hyperthyroidism by subtotal thyroidectomy or radioiodine. In many instances, they may be the major way of controlling the disease. Drugs are especially important in treating hyperthyroidism in children, young adults, and during pregnancy and provide a form of treatment that can be used in most patients while awaiting a possible remission in the disease. Recurrence of hyperthyroidism may occur despite the use of drugs, and this may necessitate a resort to surgery or radioiodine. The antithyroid drugs and their side effects are described in detail in Section 4.9. *Propylthiouracil* and the thioamides *methimazole* and *carbimazole* are widely used. *Propranolol,* which blocks the sympathetic β-adrenergic effects seen in hyperthyroidism, has also been tested (McLarty et al., 1973; McDevitt, 1977) and alleviates the symptoms in some but not all patients. Its most important use is in the treatment of thyroid "storm." It may also be useful in neonatal thyrotoxicosis (Pearl and Chambers, 1977).

The thioamides are usually given for a period of 12 to 18 months. Beneficial effects are usually seen in 7 to 10 days, and the euthyroid condition is usually attained in 1 to 3 months. After the full course

of treatment, about half the patients remain euthyroid. In the remainder, the recurrence can be treated with a second course of drugs, but in this instance about 75 percent recontract the disease, so that the alternative of surgery or radioiodine must then be considered. Occasionally, patients have been maintained for periods as long as 20 years on antithyroid drugs, but this choice is not a usual one. It is generally considered that the antithyroid drugs do not cure the disease but merely control it until a natural remission occurs. However, the course of the disease may be influenced, as the antithyroid drugs are successful in about 50 percent of cases, while natural remission occurs in only about 25 percent of untreated patients. More recently, it has been advocated that therapy need only be continued until the patient becomes euthyroid (Greer, Kammer, and Bouma, 1977). The remission rate following such "short-term antithyroid drug therapy" was as good as that seen after the more prolonged use of the drugs.

Hyperthyroidism in *pregnancy* is usually treated with antithyroid drugs, but as these can cross the placenta, there is a danger that they may produce goiter and hypothyroidism in the fetus. Thyroxine can also cross the placenta, and it has been administered in conjunction with the antithyroid drugs in order to overcome this problem (Herbst and Selenkow, 1965), but this procedure is currently contentious.

Thyroid storm is a life-threatening condition and must be treated immediately (see Mackin, Canary, and Pittman, 1974). The details are controversial, but, apart from general procedures to lower the body temperature, maintain hydration, and control congestive heart failure, usually involve the prompt use of propranolol to limit the stimulation of β-adrenergic mechanisms, and antithyroid drugs to reduce plasma T_3 and T_4 levels. Methimazole, carbimazole, or propylthiouracil, usually the latter, are given initially to prevent further formation of the thyroid hormones. These preparations are usually only available for oral administration, although methimazole is soluble in water and has been given intravenously. Propylthiouracil may have an additional action by reducing the peripheral deiodination of T_4 to T_3. After about 1 hour iodide can be administered, either orally (as an iodine solution, Lugol's iodine) or intravenously (as Na iodide). The release of thyroid hormones is then promptly reduced. Propranolol (McDevitt, 1977) (i.v. or orally) rapidly controls the sympathetic effects, especially those on the heart, which are life-threatening. Adrenocortical steroids, such as hydrocortisone sodium succinate i.v., are also often administered.

The *exophthalmos* or proptosis often associated with Graves' disease is difficult to treat, but corticosteroids, either as eyedrops, systemically, or injected intraorbitally, have been used, although they are not successful in all instances (Havard, 1972). Diuretics, such as the thiazides, are also sometimes used in an attempt to reduce local edema. Lid retraction ceases when the hyperthyroidism is controlled but can in the meantime be treated by the local application of α-adrenergic blocking drugs such as phentolamine.

2. Choice of treatment for hyperthyroidism –
Summary. The most appropriate form of treatment for hyperthyroidism is somewhat controversial and depends to a considerable extent on the available medical facilities. Surgery has a number of serious risks which can only be avoided by considerable training and skill. It also involves the use of inpatient hospital facilities. It is thus the most expensive and difficult option and is usually reserved for cases where two recurrences of the disease occur following drug therapy. However, if a malignancy of the thyroid is suspected, it is the treatment of first choice. Radioiodine is a relatively inexpensive and effective treatment which does not require in-patient care. It is never used during pregnancy and is usually reserved for those who are over 25 years of age. As the risks of resulting malignancy and genetic damage appear to be almost nonexistent, this form of therapy has become a first choice in many clinics. Hypothyroidism is a common after effect, which can, however, be readily controlled by the administration of thyroxine. Toxic nodular goiters must be treated by surgery or large doses of ^{131}I, the choice being largely an arbitrary one of the particular clinic. Antithyroid drugs offer an inexpensive option which can be used in most instances, but are especially favored in young adults and children, during pregnancy, in preparation for surgery, and in thyroid crisis. Side effects (see Section 4.9A) can sometimes exclude their use, and the recurrence rate of the disease is high, which ultimately may require the use of radioiodine or surgery.

D. Thyroiditis
This term covers several diseases of the thyroid in which an inflammation of the gland occurs. Thyroiditis may be *acute* or *subacute* and be accompanied by malaise, fever, and an enlargement of the thyroid gland, which is usually painful. *Subacute* (or *acute*) *nonsuppurative thyroiditis* (De Quervain's thyroiditis) is one of the commonest forms and may be associated with viral infections, but its cause is uncertain. It is self-limiting and usually runs a natural course of a couple of weeks to several months. It is more common in women than in men (in the ratio of about 5:1). Treatment is not always necessary, but if so consists of antiinflammatory doses of aspirin (which may also exert an antipyretic and analgesic effect) or phenyl-

butazone. If these drugs are inadequate, corticosteroids, such as prednisone, may be used. Administered thyroid hormones have also been reported to have favorable effects, probably acting by reducing stimulation of the gland via a reduced release of TSH. The actions of the drugs are not curative but provide relief while the disease runs its natural course.

Acute suppurative thyroiditis is quite rare and results from infections which can usually be treated with antibiotics.

Chronic forms of thyroiditis also occur, the most common being *chronic lymphocytic thyroiditis* or *Hashimoto's disease*. This condition appears to be due to an autoimmune reaction of antibodies, including those to thyroglubulin, with the thyroid tissue. It is also much more common in women than in men (in the ratio of about 9:1). It may be accompanied by hypothyroidism, which results in tiredness and fatigue. Hashimoto's disease, however, sometimes coexists with Graves' disease (hyperthyroidism), and treatment of the latter is then the primary consideration. In other instances, thyroid hormones are administered to suppress the release of TSH, so that the goiter regresses and it also affords replacement therapy. Sometimes, no therapy is needed, but if necessary it is usually for life. Surgical treatment was once frequently advocated, but it is now controversial, except if it is considered likely that thyroid cancer may also be involved.

Other rare chronic forms of this disorder are *chronic nonsuppurative thyroiditis*, which may be associated with diseases such as tuberculosis and syphilis, and the *fibrous (Reidel) type*, which is of unknown cause. The treatment of the latter is usually surgical, following which replacement therapy with thyroid hormones may be necessary. Thyroid gland function may be reduced in any type of thyroiditis and necessitate replacement therapy, usually with thyroxine.

E. Iatrogenic causes of thyroid imbalance and disorders – summary
It is apparent from the various observations already described that the use of a number of drugs can have side effects related to dysfunctioning of the thyroid gland. A euthyroid condition may often be maintained, but the metabolic turnover and regulation of secretion of the thyroid hormones may be altered. In more extreme cases, manifestations of goiter, hypothyroidism, and even thyrotoxicosis may become apparent. Tests of thyroid function may also be adversely affected so that their interpretation becomes a problem (see Section 4.10F). Such drugs may act by influencing various aspects of thyroid function, including hormone synthesis, release, binding to plasma proteins, metabolism, and absorption from the gut. A

Table 4.9. *Summary of drug interactions with thyroid gland function*

Binding of T_3 and T_4 to plasma proteins
Displacement:
Salicylates
Phenylbutazone
Diphenylhydantoin
Sulfonylurea oral hypoglycemic drugs
Plasma-binding protein (TBG) levels
Increase:
Estrogens
Oral contraceptives
Decrease:
Androgens
Corticosteroids
Increase in thyroid hormone metabolism (enzyme induction)
Phenobarbital
Diphenylhydantoin
Inhibition of thyroid hormone synthesis
Inhibition of I uptake:
Thiocyanate
Na nitroprusside
Acetazolamide
Inhibits "organification" of I^-:
Sulfaguanidine
Carbutamide?
Aminoglutethimide?
Iodide (but note I^- thyrotoxicosis, Jod–Basedow phenomenon)
Inhibition of release of thyroid hormone
Lithium
Iodide
Inhibition of absorption of hormone from gut
Cholestyramine

summary of some such effects is given in Table 4.9.

Interactions between effects at different sites may occur. For instance, a displacement from binding to plasma proteins or a decrease in TBG can result in an increased metabolism of the hormone and an inhibition of release of TSH. Some drugs may also act in more than one way: diphenylhydantoin may displace thyroid hormones from binding to plasma proteins and also induce liver enzymes that enhance their metabolism. The prolonged use of diphenylhydantoin, however, does not usually result in any manifestations of hypothyroidism. The ultimate effect of such drugs can thus be rather difficult to predict. In many instances the use of such drugs, which influence some aspect(s) of thyroid function, results in a new hormonal kinetic equilibrium, so that there is little or no obvious sustained change, although transient side effects may occur. If such effects in thyroid hormone me-tabolism or binding to plasma proteins occurs in patients with thyroid disease, adjustment of the dosage of replacement hormones or antithyroid drugs may be necessary.

In some instances, the use of certain drugs can result in thyroid disorders, although such effects are quite rare and may occur more frequently in patients who are predisposed to such effects. An enlargement of the thyroid gland occurs in a number of patients being treated with $LiCO_3$ for psychiatric disorders, but goiter and hypothyroidism are rare (Candy, 1972). In one survey involving 86 patients on lithium therapy, 20 were observed to have elevated TSH levels (Transbol, Christiansen, and Baastrup, 1978). Irreversible myxedema has been observed (Perrild, Madsen, and Hansen, 1978). Prolonged use of the hypotensive drug Na nitroprusside invariably results in an accumulation of thiocyanate, which is one of its metabolites, which can reversibly inhibit the synthesis of thyroid hormones (Nourok et al., 1964). The diuretic acetazolamide can inhibit iodide trapping by the thyroid gland, but it is difficult to find authentic reports of hypothyroidism.

A number of other drugs have been shown to influence thyroid function. These include thiocyanate (once used to lower blood pressure), carbutamide (an early oral hypoglycemic drug), aminoglutethimide (an investigational drug used to block secretion of cortisol), and sulfaguanidine (an antimicrobial drug). Many old proprietary pharmaceutical preparations contained iodide. These were and, in some instances, still are available as blood "tonics," powders for the treatment of "rheumatism," expectorants, and for the treatment of asthma. In addition, excessive dietary supplements of this anion may be ingested. Such iodide intake can result in goiter and hypothyroidism (Paris et al., 1960; Murray and Stewart, 1967). These effects are rare and somewhat unexpected, as such individuals apparently do not utilize the normal "escape" mechanism from the inhibitory effect of iodide (Wolff–Chaikoff effect) on hormonal synthesis. Paradoxically, iodide can also induce thyrotoxicosis (Editorial, 1972b). This is called the Jod–Basedow phenomenon and is rare, apparently only occurring in those patients suffering from some underlying abnormality of thyroid function such as a nontoxic goiter containing autonomous glandular tissue. This condition is more commonly found in areas where goiter is endemic.

The absorption of exogenous thyroid hormones and its endogenous conjugated metabolites which are excreted in the bile are reduced in patients taking cholestyramine in attempts to lower plasma cholesterol and lipoprotein levels.

The most common cause of drug-induced

Table 4.10. *Drugs that influence tests of thyroid function*

	PBI[a]	T-3 resin uptake	[131]I uptake	Mechanism
Estrogens Phenothiazines (prolonged)	Increased	Decreased	Unchanged	Increase in TBG
Androgens and anabolic steroids Corticosteroids (large doses)	Decreased	Increased	Unchanged	Decrease in TBG
Iodine-containing compounds	Increased	Unchanged	Decreased	Large iodide pool
Sulfonylureas PAS Phenylbutazone and related drugs	Decreased	Increased	Decreased	Impaired synthesis of thyroxine
Cobalt salts Diphenylhydantoin Salicylates	Decreased	Increased	Unchanged	Competition for binding sites on TBG
Heavy metals (mercury, gold, silver)	Decreased	Unchanged	Unchanged	Interference with chemical analysis of PBI

[a] Plasma protein-bound iodine.
Source: Kendall-Taylor, 1972.

hypothyroidism is overdosage with antithyroid drugs which are being used to treat thyrotoxicosis.

F. Role of drugs in tests of thyroid function

The clinical diagnosis of thyroid disease (Davies, 1972; Rosenberg, 1972; Havard and Boss, 1974; Havard, 1974) is aided by a variety of *in vivo* and *in vitro* laboratory tests. These procedures involve the use of a number of pharmacological preparations (Table 4.10), and they are subject to interference, depending on the physiological condition of the patient and the simultaneous presence of certain drugs.

Probably the oldest and formerly very widely used test of thyroid function is the measurement of *protein-bound iodide* (PBI). It is assumed that most of the thyroid hormones are bound to plasma proteins and contribute a relatively constant proportion of the iodide bound at these sites. The plasma proteins are separated and their iodide content is chemically determined to give an estimate of the thyroid hormone levels. Interference in this test occurs in the presence of excessive amounts of inorganic iodide, such as can be obtained in the diet, which cause falsely high values. Organic iodides, such as in expectorants (cough mixtures) and radiocontrast media, have a similar effect. Changes in plasma protein levels, especially TBG, also influence the test; estrogens and oral contraceptive pills increase TBG and give a spuriously high PBI

level; and androgens and large doses of corticosteroids may lower the plasma protein concentrations. Drugs that also bind to plasma proteins, such as salicylates and diphenylhydantoin, displace bound hormone and so lead to falsely low values. The PBI test is now rarely used and has been supplemented by more accurate direct determinations of T_3 and T_4 using radioimmunoassay procedures.

The uptake of radioiodine (*Thyroid radioiodine uptake test*) [131]I by the thyroid is useful for the diagnosis of hyper- and hypothyroidism when the rate is, respectively, increased or decreased. The accumulation of the administered isotope is directly measured by placing a gamma counter over the thyroid. Like the PBI test, it can also be interfered with by the presence of excess iodides in the plasma. It has been observed that as the dietary intake of I has increased over recent years, the expected "normal" rates of ^{131}I uptake have declined. Euthyroid people show a low ^{131}I uptake (or response) when the plasma iodide levels are raised and a high one when they are low. Technetium-99[m], a short-life radioisotope with a weaker radiation emission, is now sometimes used as an alternative to ^{131}I in this test (Van't Hoff, Pover, and Eiser, 1972).

Thyroxine and triiodothyronine levels can now be measured directly using *competitive binding displacement assays*. These tests are influenced by the total amount of TBG, which, as described above, may be elevated by estrogens and the "pill" as well as conditions such as pregnancy and hepatitis. Under such conditions, false "high" values may be observed in euthyroid individuals. On the other hand, the TBG may be lowered in certain illnesses and in old age (Jefferys et al., 1972). Androgens and anabolic steroids, which lower TBG, and drugs that compete for the protein binding sites (salicylates and diphenylhydantoin), will produce false low values.

There are a number of additional tests which further aid the diagnosis of hypo- or hyperthyroidism by localizing the nature of the defect. The administration of triiodothyronine can be used to differentiate a raised ^{131}I uptake due to thyrotoxicosis or merely an I deficiency (T_3-*suppression test*). This hormone will suppress I uptake (via inhibition of TSH release) in the case of the latter individuals, as well as normal people, but has no effect in thyrotoxicosis. It is also ineffective in euthyroid (opthalmic) Graves' disease and toxic adenoma (a localized toxic nodule in the thyroid). Administration of T_3 can cause a transient hyperthyroidism, which can be a problem in old people and those with congestive heart failure.

Thyrotropin (TSH) preparations are obtained from cattle and are not used in the treatment of thyroid diseases but are utilized to distinguish hypothyroidism due to pituitary deficiency from that directly resulting from thyroid deficiency (respectively "secondary" and "primary" hypothyroidism). This *TSH stimulation test* results in a raised ^{131}I uptake by the thyroid if there is a pituitary deficiency but no change if the disease involves the thyroid gland itself.

Thyrotropin-releasing hormone, TRH, has been synthesized and is available for testing thyroid (*TRH stimulation test*) function (Shenkman et al., 1972; Hershman, 1974). It can be administered intravenously or intramuscularly and is even effective, although at much higher doses, orally. Side effects are transient and not serious and include lightheadedness, mild nausea, urinary urgency, and an odd taste in the mouth. Doses up to 1000 μg (i.v.) have been administered repeatedly without serious toxic effects. TRH can be used to distinguish hypothalamic deficiency ("tertiary" hypothyroidism) from that due to a reduced pituitary function. The response is usually measured by the immunoassay of TSH in the plasma (Azizi et al., 1975). If the defect is in the pituitary, TRH will be ineffective. TRH also promotes secretion of prolactin from the pituitary and so can be used as a more general test of pituitary function. It seems likely that it will replace the thyroid suppression test for diagnosing euthyroid Graves' disease. Most patients suffering from this condition do not exhibit an increased release of TSH in response to TRH. In thyrotoxicosis, TSH release in response to the injected TRH is suppressed, owing to the high circulating levels of thyroid hormones. On the other hand, in primary hypothyroidism the TSH release is considerably exaggerated, reflecting the ready availability of TSH.

5
Steroid hormones: introduction

Steroids are lipid compounds which are ubiquitous constituents of living organisms. Cholesterol is a predominant sterol in animals, where it can be totally synthesized from acetate and stored in droplets or incorporated into cell membranes. A number of metabolites of cholesterol are utilized as hormones, including the endocrine secretions of the adrenal cortex, the testes, and the ovaries. The chemical nature and properties of such compounds contrast markedly with those of other hormones, all of which are derived from amino acids. This dichotomy is reflected in the nature of their actions. The ability of the adrenocortical and gonadal tissues to synthesize such compounds is reflected in their common embryological origins from the urogenital ridge. These endocrine tissues have, in addition, similar histological, biochemical, and physiological characters, slight modifications of which confer on them a preferential ability to produce one type of steroid hormone over another. In musical terms these endocrine glands play slightly different tunes all based on the same theme. It is thus academically logical, and also economical, to deal with them as a group.

5.1. Chemical structure

The principal steroid hormones in man are cortisol and aldosterone from the adrenal cortex, testosterone from the testes, and estradiol and progesterone, which are produced by the ovaries. As will become apparent later, this is not an exclusive list but represents those hormones that are secreted in the highest active concentrations into the plasma. These hormones are chemically based on a structure called a cyclopentanophenanthrene nucleus (also called gonane or sterane), containing four carbon rings linked together to give a total of 17 carbon atoms. These carbons are numbered consecutively starting from the A-ring (Figure 5.1). This basic tetracyclic compound is initially synthesized as part of the cholesterol molecule, which has

a side chain, containing eight additional carbons, attached at the C-17 position. The active hormones are formed from the cholesterol by the cleavage of different sections of this side chain and the substitution or insertion of various chemical groups, such as methyl, hydroxyl, ketone, and even an aldehyde, at various positions, including C-11, C-13, C-17, and C-20. These modifications are mediated by specific enzymes which are predominantly present in the particular endocrine glands, but may also occur in peripheral tissues such as the liver. The changes result in alterations in the relative water and lipid solubility of the molecule and its steric configuration, thus conferring on it a degree of specificity which makes it a useful and distinct excitant. Such changes may not only influence the specific binding of the steroids to their receptors, but also their accessibility to these sites (which are intracellular), binding to plasma proteins, and their rates of inactivation and clearance from the circulation.

The steroid hormones are classified into three groups based on their chemical parent tetracyclic hydrocarbon compounds. These are (see Figure 5.1) (a) *pregnane*, which contains 21 carbon atoms (C_{21}) and includes the progestins and adrenocorticosteroids; (b) *androstane*, C_{19}, to which the male steroidal sex hormones or androgens belong; and (c) *estrane*, C_{18}, compounds that have the actions of the female steroidal sex hormones or estrogens. For a detailed description of the nomenclature for the natural and synthetic steroid hormones, a specialized volume should be consulted (e.g., Schulster, Burstein, and Cooke, 1976). In the present text the common or "trivial" names of these compounds will generally be used rather than the internationally agreed upon chemical terminology, which while more descriptive is less well known and longer. The numbering of the carbon atoms, given in Figure 5.1, proceeds in an orderly manner, starting in ring A and proceeding to C-17 in ring D. The position of a substituent group

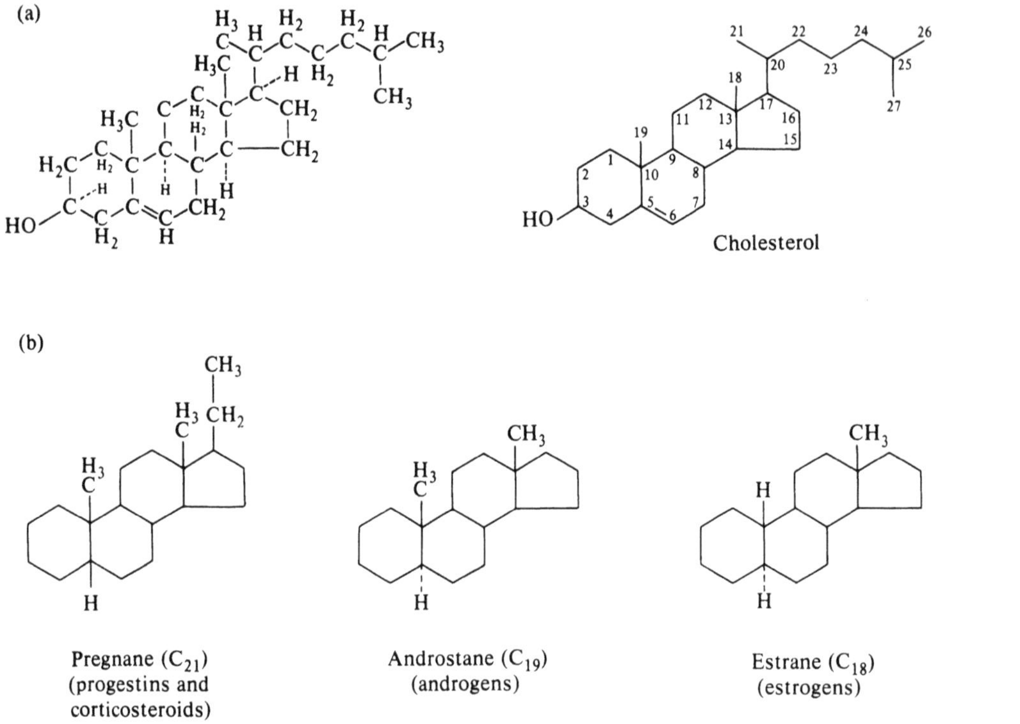

(a)

Cholesterol

(b)

Pregnane (C$_{21}$)
(progestins and
corticosteroids)

Androstane (C$_{19}$)
(androgens)

Estrane (C$_{18}$)
(estrogens)

Fig. 5.1. (a) Chemical structure of cholesterol and the conventional manner of numbering the carbon atoms. (b) Parent steroid compounds for the progestins and corticosteroids (C$_{21}$), androgens (C$_{19}$), and estrogens (C$_{18}$).

is identified by these numbers. Such groups may have different steric configurations: if their position is perpendicular to the rings, in a direction below the plane of the paper, it is referred to as an α (alpha) group (e.g., 17α-hydroxyprogesterone) and is diagrammatically denoted as a broken line, while if it is upwardly facing, it is a β (beta) group and is shown on the chemical structure as an unbroken line. Unsaturation, or double bonds, are commonly (but not officially) designated by the prefix Δ (delta) – for instance, that between carbons 4 and 5 being referred to as a Δ^4 function.

The following is a brief, incomplete summary of the principal prefixes and suffixes used in the "official" chemical names of steroid hormones. These are added to the name of the *parent* tetracyclic compound and numbered according to their position.

a. The insertion of a double bond is indicated by changing the suffix from "-ane" to "-ene." Two double bonds are "-adiene," three "-yne." The terminal "e" is sometimes omitted.
b. Ketone (or oxo-) functions (=O) are given by the prefix "oxo-" or the suffix "-one." Two such groups are "dione" and three "trione."
c. Alcohol or hydroxyl groups (—OH) are described by using the prefix "hydroxy-" or the suffix "-ol." If two hydroxyls are present, "-diol" is used; and so on.

d. Aldehydes (—CHO). The change from a methyl group, as in aldosterone, is indicated by the prefix "al-."

Two examples are: progesterone is described as 4-pregnene-3,30-dione; cortisol is referred to as 11β,17α,21-trihydroxy-4-pregnene-3,20-dione.

The chemical structures of the principal steroid hormones are shown in Figure 5.2. The similarities in their structures are much more apparent than the differences, so that it is probably less surprising to be told that there may be some crossover and interactions in their effects than that each can exert remarkably specific actions in the body. Consideration of the similarities in their structures, fortified by plenty of hindsight, suggest that they may share common metabolic pathways, both with respect to their synthesis and destruction.

5.2. Biosynthesis

All of the steroid hormones, whether they originate in the adrenal cortex, testes, ovaries, or placenta, share a basically common mechanism for their biological synthesis (see Figure 5.2). Thus, the adrenal cortex, apart from forming aldosterone, cortisol, and corticosterone, can also secrete small amounts of androgens, and even estrogens.

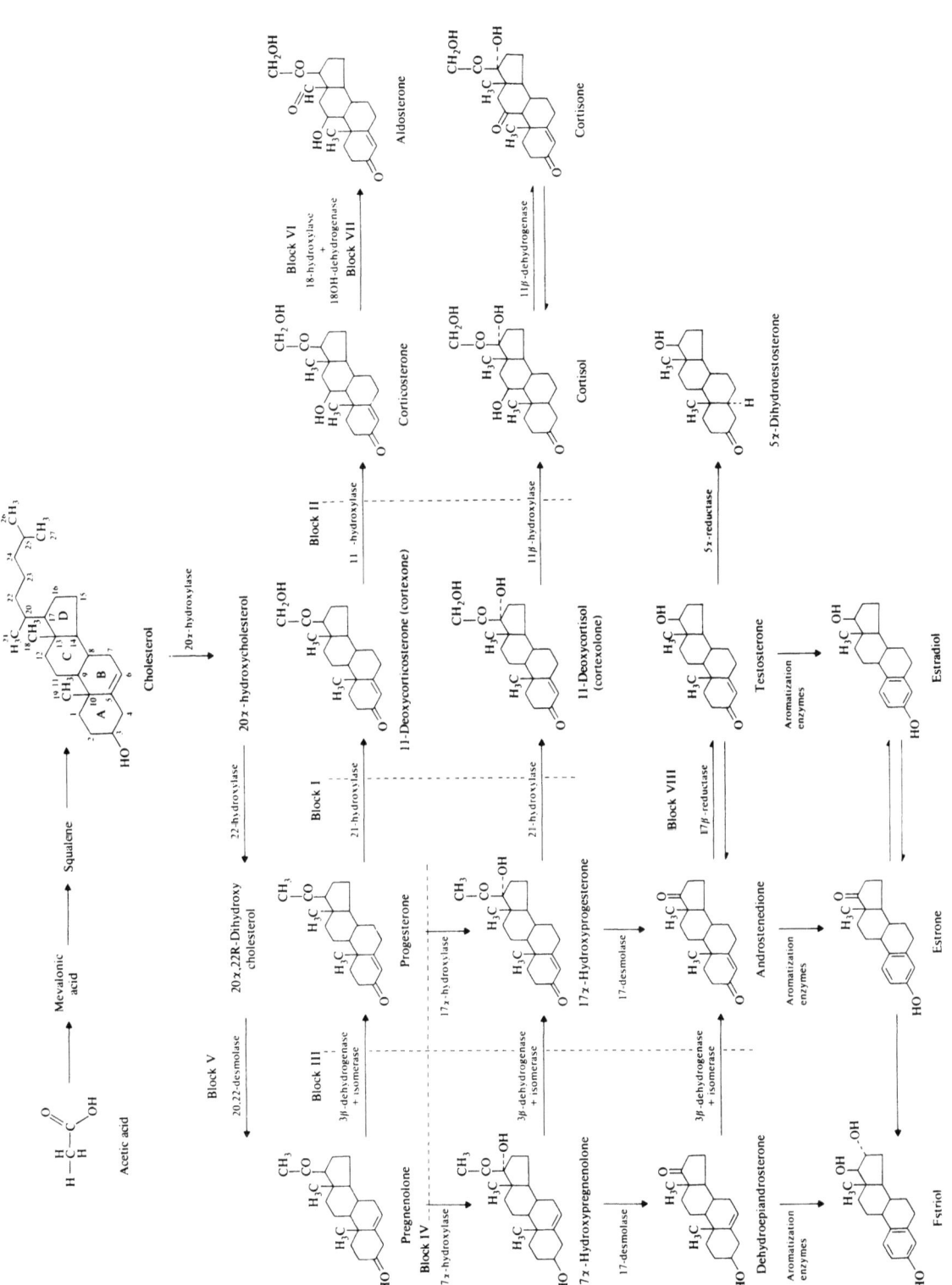

Fig. 5.2. Summary of the biosynthesis of the steroid hormones.

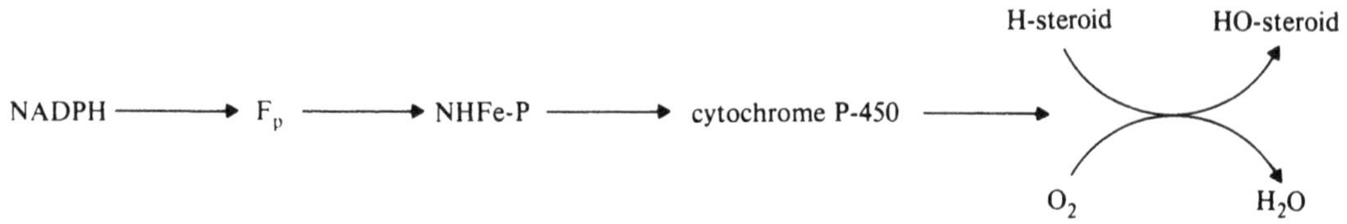

$$\text{NADPH} \longrightarrow F_p \longrightarrow \text{NHFe-P} \longrightarrow \text{cytochrome P-450} \longrightarrow$$

H-steroid   HO-steroid

O_2 H_2O

Scheme 5.1.

The predominant secretions of the ovary are progesterone and estradiol-17β, but they also release distinct amounts of androgens. The testes can form estrogens as well as androgens and indeed the urine of stallions is a rich source of the former more typically "female" sex hormones. When the steroidal metabolites in the different endocrine tissues are examined, such as during incubation under *in vitro* conditions, a whole plethora of common steroids can be identified, which reflects the similar pathways that are utilized to form the different hormones. These syntheses involve the same enzyme systems. The final differences in the actual hormones that are secreted appear to be largely dependent on the relative quantitative importance of each enzyme-mediated reaction, which determines the amounts of the final products. In at least one instance a unique enzyme system is present; this is seen in the region of the zona glomerulosa in the adrenal cortex and converts corticosterone to aldosterone. When certain enzyme reactions in the paths of normal syntheses are inhibited, such as may occur in genetic diseases or be due to the presence of certain drugs, there may be hyperplasia of the glandular tissue and a buildup of metabolites at certain sites in the biosynthetic chain. Such an effect may result in a diversion of synthesis to the formation of abnormal quantities of other hormones. Thus, in some circumstances the adrenal cortex can synthesize large quantities of androgens, which may have a virilizing action in women.

In general terms (see Figure 5.2), the biosynthesis of the steroid hormones occurs as follows. Cholesterol is the original steroidal substrate. It undergoes a series of reactions which involve mitochondrial, microsomal, and cytosol enzymes. Cleavage of the C-17 side chain by desmolases occurs both initially and in the later formation of the steroidal sex hormones. Oxidation reactions involving hydroxysteroid dehydrogenases occur at C-3, C-11, C-17, and C-18 and may utilize the nucleotides NAD$^+$ (nicotinamide adenine dinucleotide) and NADP$^+$ (nicotinamide adenine dinucleotide phosphate) as cofactors and proton acceptors. A series of hydroxylation reactions occurs at the C-21, C-11, C-17, and C-3 positions. The hydroxylase enzymes utilize NADPH and O_2 as cofac-

tors. Each type of hydroxylation at a particular carbon position was once thought to be probably due to a separate protein, but this concept appears to be an oversimplification. The multiple-enzyme cytochrome P-450 system appears to be involved, and its components include a flavoprotein dehydrogenase (F_p), a nonheme iron protein (NH Fe-P) called adrenodoxin, and a protoporphyrin hemoprotein called cytochrome P-450 [see Section 2.4A(e)]. The particular hydroxylation reaction initiated may depend on the specificity of the cytochrome P-450, which may exist in several forms. This component appears to be the terminal link in the reaction, which directly introduces the hydroxyl group into the steroid. The hydroxylation process can be summarized as shown in Scheme 5.1. The ability of a drug to specifically inhibit a hydroxylase could depend on its interaction with the particular type of cytochrome P-450 involved.

A. Conversion of acetate to cholesterol

Cholesterol is the common precursor of all the steroidal hormones. The origin of this complex molecule was initially in some doubt. Many were skeptical as to whether its complete synthesis could occur in the body. It was, however, clearly shown that it could be formed from acetate, which was labelled with ^{14}C. The synthesis is a complex one involving numerous two-carbon units derived from the acetate, and the formation of mevalonic acid, squalene, lanosterol, and desmosterol as precursors. Each carbon atom in the cholesterol can be related to either the labelled carbon in the methyl or carboxyl group of the acetate. The endocrine glands can utilize cholesterol from other sources in the body but, with the exception of the placenta, can synthesize their own requirements locally. External sources of cholesterol cannot be used with equal facility by all endocrine tissues; thus while the adrenal cortex can readily accumulate exogenous cholesterol, this ability is very limited in the testes.

Cholesterol synthesis can be prevented by several drugs, of which *triparanol* is the best known. It prevents the conversion of desmosterol to cholesterol. Triparanol has been shown to block synthesis of corticosteroids. Its clinical use is controversial.

B. Conversion of cholesterol to pregnenolone

The rate of this process appears to limit the production of the steroid hormones in each glandular tissue. The cholesterol is first hydroxylated at the C-20 and C-22 positions in the side chain. The latter is then split by a 20,22 desmolase, resulting in the formation of Δ^5-pregnenolone, which is a pregnane (C_{21}) compound. This transformation occurs in the mitochondria and is stimulated by corticotropin (ACTH), angiotensin II, and high K concentrations in the adrenal cortex, or by luteinizing hormone (LH) in the ovary and testis. The actions of ACTH and LH are mediated by cyclic AMP (see Sections 6.6A and 7.3D). If the pregnenolone accumulates in the mitochondria, its further formation is inhibited. Subsequent steroid hormone synthesis is dependent on it passing out across the mitochondrial membrane to gain access to the smooth endoplasmic reticulum and cytoplasm of the cells. The stimulatory effects of ACTH and LH may depend on their ability to increase the permeability of the mitochondrial membrane and so prevent the process of a negative-feedback inhibition due to a local buildup of pregnenolone concentration.

A congenital block has been observed in the 20,22 desmolase system, which limits the formation of pregnenolone. *Aminoglutethimide* inhibits the conversion of cholesterol to 20α-cholesterol and blocks the formation of pregnenolone. It thus prevents the formation of corticosteroids and has undergone clinical trials to reduce cortisol secretion in autonomous adrenal tumors and for the treatment of human breast cancer ("chemical adrenalectomy").

C. Conversion of pregnenolone to progesterone or 17α-hydroxypregnenolone

These processes occur outside the mitochondria, although in close association to it, and result in the formation of (a) progesterone, which is formed as a result of the actions of two enzymes, a 3β-dehydrogenase and an isomerase; and (b) 17α-hydroxypregnenolone, which is synthesized via the action of a 17α-hydroxylase. These two compounds are the intermediates for all the steroidal hormones; progesterone can be converted to corticosterone and aldosterone, while either steroid can be utilized to synthesize cortisol, androgens, or estrogens. Genetic deficiencies in both of these enzyme systems have been described. *Amphenone* and *SU-9055* (see Figure 6.5) inhibit 17 α-hydroxylase and so prevent steroid hormone formation. Amphenone B is the best known of these compounds but has limited therapeutic use because of its toxicity. It influences the adrenals and the testis and apart from 17α-hydroxylase also inhibits 11β- and 21-hydroxylase reactions, and possibly even the cholesterol 20α-hydroxylase. Considering its ubiquitous effects on such enzymes, its toxicity is perhaps not surprising.

D. Formation of corticosterone and aldosterone from progesterone

In the adrenal cortex progesterone can be converted to 11-deoxycorticosterone by 21-hydroxylase. The subsequent stages in the hormone syntheses occur in the mitochondria. The formation of corticosterone is mediated by 11β-hydroxylase. The zona glomerulosa tissue has the special ability to convert corticosterone to aldosterone, and this process involves two enzymes, an 18-hydroxylase and an 18-OH-dehydrogenase.

Genetic deficiencies have been described for 21-hydroxylase, 11β-hydroxylase, and 18-hydroxylase. *Metyrapone* (SU-4885) blocks 11β-hydroxylation and so increases the formation of 11-deoxycorticosterone by the adrenal cortex. This, combined with additional 11-deoxycortisol secretion, leads to hypertension and salt retention, as these compounds have mineralocorticoid activity. Cortisol secretion is blocked so that ACTH levels rise. The conversion of corticosterone to aldosterone can be blocked by SU-9055, which inhibits 18-hydroxylase. It is also interesting that heparin can inhibit the 18-hydroxylation *in vivo*, although the effect takes several days to develop.

E. Formation of cortisol from progesterone or 17α-hydroxypregnenolone, via 17α-hydroxyprogesterone

17α-Hydroxyprogesterone is formed from either 17α-hydroxypregnenolone under the influence of 3β-dehydrogenase and isomerase or from progesterone via 17α-hydroxylase. The product is converted to 11-deoxycortisol by 21-hydroxylase, and then enters the mitochondria, where it is converted to cortisol by an 11β-hydroxylase. The effects of drugs on these enzymes have been described in the preceding sections. Cortisol can be converted to cortisone, which has little biological activity, but this reaction is reversible, a response that is utilized when cortisone is used therapeutically.

F. Formation of androgens from pregnenolone and progesterone

The synthesis of androgens occurs principally in the testes but also in the adrenal cortex and ovary. In the testis it takes place in the Leydig cells in the interstitial tissue. In the ovary it is an intermediate in the process leading to the formation of estrogens. The ovarian interstitial or stromal tissue appears to be the principal source of the secreted androgens. Under the influence of 17α-hydroxylase either 17α-hydroxypregnenolone or

17α-hydroxyprogesterone may be formed, respectively, from pregnenolone and progesterone. These give rise, respectively, due to cleavage of the side chain by 17-desmolase (or C_{17}–C_{20} lysase), to dehydroepiandrosterone (which is predominant in the adrenal, where it is sulfated) or androstenedione. These reactions occur in the endoplasmic reticulum. Dehydroepiandrosterone can be converted to androstenedione, which, via a 17-reductase, can be reversibly transposed to testosterone. The more active form of this hormone is 5α-dihydrotestosterone, which is formed from testosterone by saturation of the Δ^4 double bond by 5α-reductase, an enzyme that is present in the testis as well as in peripheral target tissues, such as the prostate and also the liver.

G. Formation of estrogens from androstenedione and testosterone

The A ring of estrogens is an aromatic nucleus containing a phenolic hydroxyl group. This characteristic confers the typical estrogenic effects on the hormones, and a similar chemical function in nonsteroidal compounds may even confer such activity on them also. The latter property is thus seen in such drugs as diethylstilbestrol (DES), which is not a steroid. The possible ovarian synthesis of such benzenoid compounds from small acyclic precursors was initially considered unlikely. However, it was clearly shown that ^{14}C-acetate, when injected *in vivo* into pregnant mares, resulted in the recovery of ^{14}C-labelled estrogens from their urine. The intermediate roles of testosterone and androstenedione in the formation of estradiol-17β and estrone were also demonstrated in ovarian, as well as testicular, adrenocortical, and placental tissues *in vitro*. The estrogens thus share a common pathway (see Figure 5.2) with the other steroidal hormones, commencing with the synthesis of cholesterol from acetate and utilizing androgens as intermediates. The final aromatization process is incompletely understood but is enzymic and has been shown to occur in the ovary and placenta and to a lesser extent in the testis, adrenals, and liver, the usual hormonal products being estradiol-17β and estrone. An *in vitro* inhibitor of the aromatization process, *androst-4-ene-3,5,17-trione* (ADT), has been described (Schwartzel, Kruggel, and Brodie, 1973). It has also been used *in vivo* in rats (Booth, 1978). *Aminoglutethimide* also inhibits aromatization processes *in vitro* (Chakraborty, Hopkins, and Parke, 1972). In the ovary the estrogen synthetic activity occurs in the follicular cells. The separated granulosa and theca cells of the follicles can both mediate estrogen synthesis, but the former are probably more effective. Some estrogens are also formed by the corpus luteum and a little by the stroma. Estriol is commonly found in the urine and can be formed from estrone or dehydroepiandrosterone in the liver, especially that of the fetus.

5.3. Catabolism of steroid hormones

The steroid hormones usually have a longer half-life in the plasma than other types of hormones, but their survival is nevertheless finite. They are progressively metabolized and excreted so that their levels, if unreplenished, continue to fall, although their complete disappearance may take several days. This delay partly reflects the protection afforded them by binding to plasma proteins, their lipid solubility, which allows them to enter fat stores in the body, and the time taken for their catabolic conversion to products that are less active and more readily excreted in the urine and bile. Only small amounts of steroid hormones are excreted unchanged in the urine, although in the instance of their synthetic analogues, this avenue may be more substantial.

Literally scores of metabolites of the steroid hormones have been described in the urine, plasma, and feces, and during the incubation of the hormones with tissues *in vitro*. The relative physiological importance of most of these is unknown and doubts linger regarding the positive identification of some of them. Thus, while recognizing that a wide array of such compounds may be formed from steroid hormones, I will attempt to deal mainly with the underlying principles and the *major* products that are formed.

The catabolic metabolites usually have less biological activity than the parent molecule and are more water-soluble, which facilitates their excretion.

A. Nature of the catabolic chemical reactions

1. Conjugation in which the steroids are linked covalently to a nonsteroidal moiety. The latter are usually glucuronic acid and sulfuric acid, forming glucuronides and sulfates (Figure 5.3). These reactions occur principally in the liver, where the glucuronic acid is synthesized, and to a lesser extent in other tissues, including the kidneys. The linkage can be broken by hydrolase enzymes and this reaction commonly occurs in the intestine, allowing the steroids that are secreted as conjugated compounds in the bile to be reabsorbed in the blood.

The conjugation reactions occur with the hydroxyl groups in the steroid; C-3, C-16, and C-17 are more commonly involved. More than one site may sometimes be conjugated on the same molecule and the reaction can be "mixed" so that a combined sulfate and glucuronide may be formed.

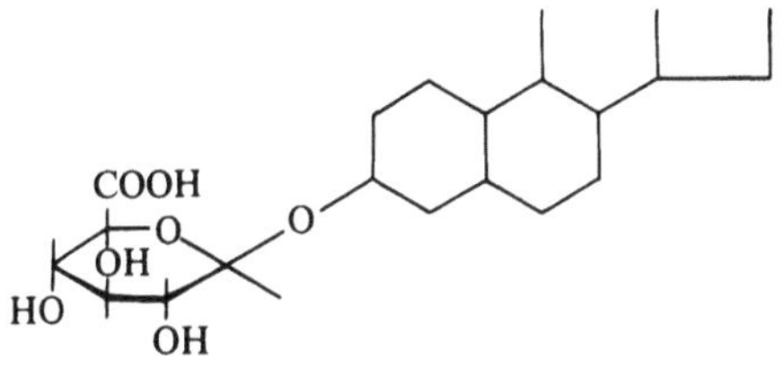

(a) Steroid glucuronide

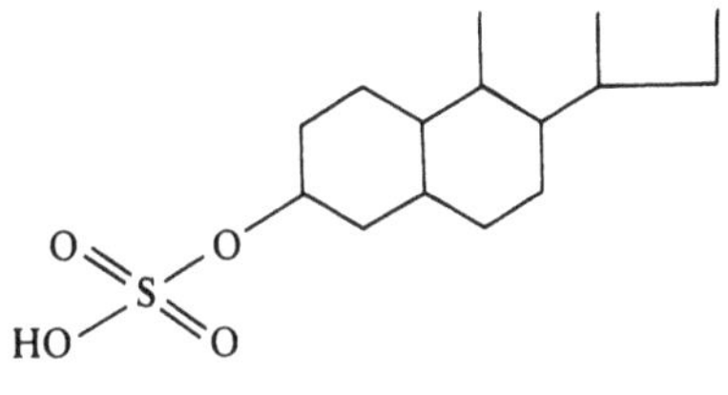

(b) Steroid sulfate

Fig. 5.3. Conjugated metabolites of steroid hormones.

2. Chemical modification of the steroid nucleus

a. The introduction of hydroxyl groups has been described at numerous carbon positions in the molecule; C-3, C-16, and C-17 are especially common.
b. There is a saturation of the Δ^4 (between C-4 and C-5) double bond in the A ring due to the action of a 5α- or 5β-reductase. These enzymes are present in high concentrations in the liver, the former is microsomal and the latter in the cytosol. They are also present in peripheral tissues.
c. A reduction of the 3-ketone group which prepares the way for subsequent conjugation (Figure 5.4a).
d. The 17α-hydroxylated side chain of the C_{21} steroids can be attacked. The C-20 ketone can be hydroxylated to a 20α or 20β group (Figure 5.4b). The side chain can undergo cleavage at C-17, leaving a C-17 hydroxyl which can be oxidized to a ketone group (Figure 5.4c). This reaction only occurs in compounds possessing a 17α-hydroxyl group.
e. Hydroxyl and ketone moieties can be oxidized or reduced, as already described, at the C-3 and C-17 positions as well as C-11. This latter reaction occurs in the reversible transformation between cortisol and cortisone.

B. Sites of steroid hormone catabolism

The liver is the principal site for the metabolism of the steroid hormones. This involves both conjugation reactions and the modification of the steroid nucleus. Some of the products may be secreted in the bile and thus enter the intestine, from which some may be reabsorbed due to hydrolysis and liberation of the free steroid. The fetal liver [see also Section 7.6H(3)] is also an important site for steroid hormone metabolism, but its function differs somewhat from that of the adult, depending on the particular stage of its development. Liver function can be affected by disease and the actions of drugs, which may reduce it or, owing to the induction of microsomal enzymes, increase its catabolic ability. Thus, the half-lives of steroid hormones can be changed.

The kidney has an important role in the clearance of hormones both indirectly as a result of its tissue metabolic abilities, and directly as a result of the filtration of water-soluble products of steroid catabolism. Other tissues take a smaller role in the metabolic conversion of steroid hormones, and these include the red cells, intestine, skin, and placenta.

C. Catabolism of corticosteroids and progestins (C_{21})

Cortisol exists in a reversible equilibrium with its inactive 11-keto form, cortisone. This reaction, which occurs in the liver and peripheral tissues, involves the action of an 11β-dehydrogenase. The levels of this enzyme can be increased by thyroid hormones. The catabolism of these corticosteroids principally occurs in the liver, about 50 percent undergoing reduction of the Δ^4 double bond and a hydroxylation of the C-3 keto group. As these reactions result in the addition of four hydrogens to the molecules, they are called tetrahydrocortisol and tetrahydrocortisone. Part of these compounds may be hydroxylated at C-20 to form cortol and cortolone (20 to 30 percent of the total metabolites). These products are conjugated principally with glucuronic acid prior to excretion in the urine. Little appears in the bile. A small quantity of the corticosteroids (about 10 percent) undergoes side-chain cleavage to form the 17-ketosteroid compound.

Corticosterone and *progesterone* lack the 17α-OH group and so cannot be converted to 17-ketosteroids, but as in cortisol, a reduction of the Δ^4 double bond and the C-3 and C-20 keto groups occurs. *Aldosterone* is metabolized similarly, but 10 to 15 percent apparently undergoes glucuronide conjugation at the C-18 position.

The *synthetic corticosteroids* which are used therapeutically have a much longer half-life than cortisol. For prednisolone and dexamethasone, this is about 190 minutes compared to 90 minutes for cortisol. As they are bound less readily to plasma proteins, the difference appears to reflect a reduced catabolism. The Δ^2 double bond which is inserted in ring A of such compounds appears to interfere with the saturation of the Δ^4 double bond

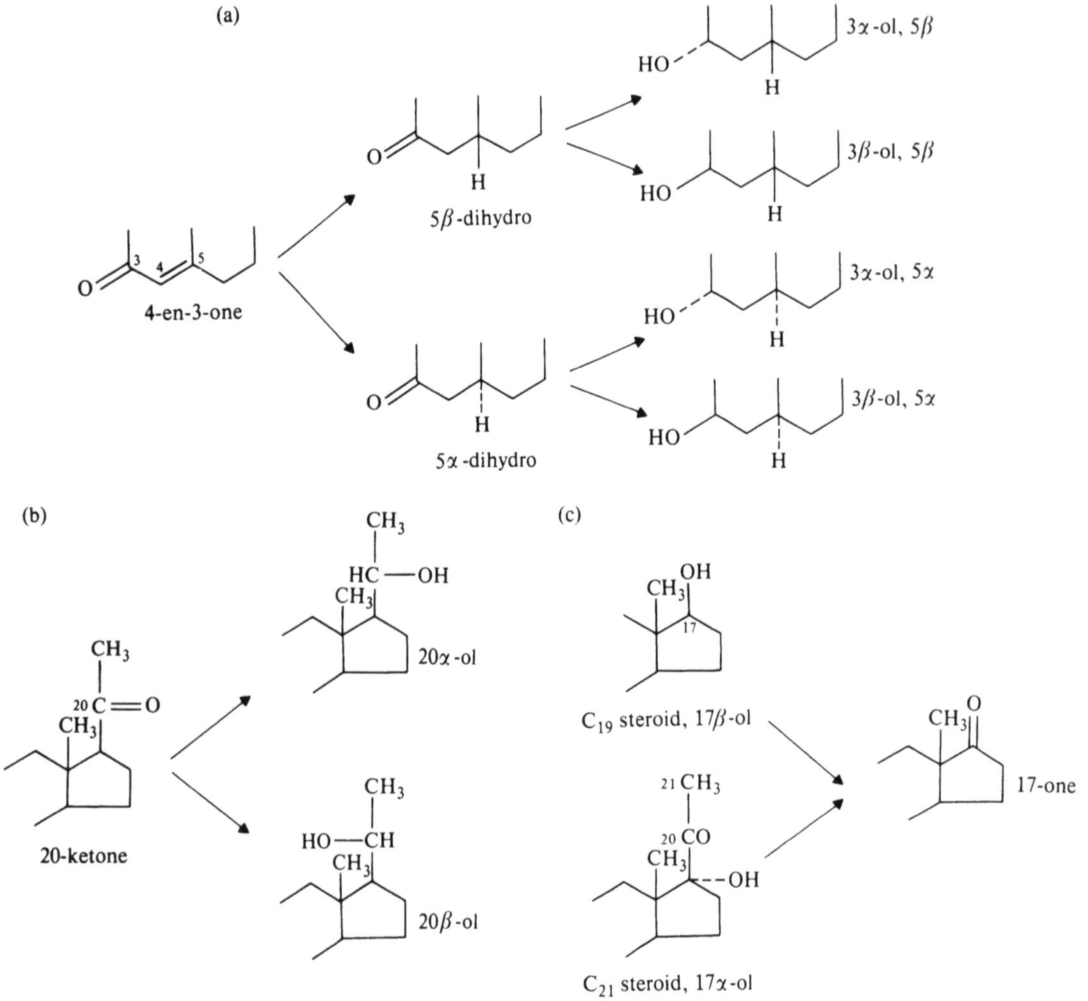

(a)

(b)   (c)

Fig. 5.4. Some of the chemical modifications that occur in the metabolism of the steroid hormones. (a) Reduction of the 4-en-3-one conjugated double-bond system to produce a 3-hydroxyl function. (b) Reduction of the 20-ketone function to a 20α- or 20β-hydroxyl group. (c) Generation of a ketone function at the C-17 position either by oxidation of the 17-hydroxyl group in C_{19} steroids, or by cleaving off the C-21,20 side chain of C_{21} steroids containing a 17α-hydroxyl group.

and the 3-keto group. Several of these compounds also contain a 9α-fluoro group which interferes with the reduction of the hydroxyl at C-11. The insertion of a 16α- or β-hydroxyl moiety protects the C-20 keto group. As a result of their reduced metabolism, more of these compounds are excreted unchanged in the urine, so that the kidney assumes greater importance for their excretion than is the case for the natural hormones.

D. Catabolism of the androgenic steroids (C_{19})

More than 60 metabolites of testosterone and androstenedione (which are interconvertible) have been identified.

Testosterone can undergo saturation of the Δ^4 double bond, in ring A, to 5α-dihydrotestosterone. This reaction can occur in the liver or peripheral tissues under the influence of 5α-reductase. Dihydrotestosterone has a higher androgenic activity than its precursor and is usually the most active form of the hormone. Androgens can also be transformed to estrogens by aromatization enzymes.

The inactivation of the androgens principally involves the Δ^4 saturation reaction (5α- or 5β-reductase) just described, which if accompanied by reduction of the 3-keto group results in the formation of 5α- and 5β-androstenediol from testosterone, and androsterone and aetiocholanolone from androstenedione. These compounds are mainly excreted in the urine as the glucuronides.

The adrenal also produces androgens, especially dehydroepiandrosterone, which can be conjugated with sulfate, a reaction that is especially important in the fetal liver. This compound can be converted to estrogens by the placenta [see Section 7.6H(3)].

E. Catabolism of the estrogens (C_{18})

More than 20 different metabolites of the estrogen hormones have been identified. These are principally excreted in the urine but also appear in the feces. Large amounts of estrogen metabolites appear in the bile but these are largely reabsorbed from the intestine, the so-called "enterohepatic circulation." The liver is the principal tissue involved in metabolism of the estrogens, but this process also occurs in peripheral tissues, including the kidney, skin, intestine, red blood cells, and the placenta.

Two types of chemical reactions are mainly involved: (a) hydroxylation has been described at most of the carbon positions, although more important sites are the C-16, C-2, and C-17; and (b) conjugation reactions, principally with sulfate and glucuronic acid but also with N-acetylglucosamine. These processes occur mainly in the liver but also in the intestine.

The three main estrogens in the body are estradiol-17β, estrone, and estriol. The estradiol-17β can be reversibly converted to estrone by a 17β-dehydrogenase. This enzyme has been identified in the red cells and the placenta, where it has been called estronase. Estriol is a major urinary metabolite which is formed from estrone. The estrone also appears in the urine, principally in the form of the sulfate, while the estriol is present as the glucuronide.

The enterohepatic circulation can "trap" significant quantities of estrogens. Conjugates of estrone and estradiol are mainly excreted in the bile, but it is a less important channel for estriol, which is predominant in the urine. The conjugates can be broken down in the intestine as a result of the action of tissue enzymes or bacteria. A more complete breakdown of the steroids can also occur at this site, but the details of this process are unknown. The freed unconjugated steroids can be reabsorbed and subsequently either be excreted in the urine or reenter the enterohepatic circulation.

During pregnancy the fetal liver and adrenals play an interesting and important metabolic role in conjunction with the placenta. The urinary excretion of estrogens rises markedly during pregnancy. This increase arises partly from synthesis by the placenta but also from the metabolism of fetal adrenocorticosteroids. The liver of the developing infant can convert the latter steroids to estriol, de-

hydroepiandrosterone being the substrate. This steroid is conjugated with sulfate, the fetal liver having a poor ability to form glucuronides, which is then passed to the mother. It has been proposed that the conjugation affords the fetus protection against the effects of the free steroids. During transit the placenta hydrolyses this steroid conjugate, as it possesses a sulfatase (but no β-glucuronidase), so that the free steroid enters the parental circulation, where it undergoes subsequent metabolism by the maternal tissues. Although estrogens can readily pass across the placenta from the fetus to the mother, there is little transfer in the opposite direction.

5.4. Binding of steroid hormones to plasma proteins

The steroid hormones can be bound to plasma proteins, and this can have important effects on their functions in the body. Such binding can delay their clearance from the circulation by interfering with their catabolic metabolism and glomerular filtration. The hormones' effects may also be modified, and it has been suggested that in certain conditions, such as pregnancy, they afford protection against elevated levels of testosterone and aldosterone. It has also been observed that corticosteroids have more side effects in patients who have low plasma albumin levels. Binding to plasma proteins may also influence a hormone's ability to gain access to certain tissue sites and even help them to cross some capillary beds. Circulating bound steroid hormones provide a readily accessible storage site, which helps to buffer and maintain the levels of the soluble effective forms of the hormones.

Binding of steroid hormones to plasma proteins is of practical interest, as the levels of such proteins can change in response to disease and the administration of drugs. As the binding is not completely specific for any single steroid, other chemical compounds may be able to occupy the same or adjacent sites, so that interactions between drugs and endogenous hormones may be expected to occur.

The steroid hormones that are bound to plasma proteins are in chemical equilibrium with those in solution. Thus, a decline in the soluble hormone will result in some leaving their binding sites, so that the effective level will be maintained. Conversely, as there is usually a reserve of unoccupied binding sites, a rise of hormone levels in solution will result in a movement back onto the plasma proteins. The forces involved in the binding are weak ones, such as hydrogen and ion bonds and van der Waals' forces. The attraction can nevertheless be quite specific, so that one type of steroid hormone may be much preferred compared to another. This selectivity is partly dictated by the

Table 5.1. *Apparent association constants for the interactions of some steroid hormones with cortisol binding globulin,* CBG

	Apparent association constant, K_A (liters/mol at $37°C$)
Progesterone	9×10^7
Corticosterone	3×10^7
Cortisol	2.4×10^7
Aldosterone	6.5×10^6
Testosterone	1.4×10^6
Estradiol-17β	2×10^4

Source: Based on data from Westphal, 1975.

chemical nature of the steroid as well as that of the protein. The addition of more polar groups tends to decrease binding affinity, and this is referred to as the "polarity rule." In the instance of cortisol-binding globulin, it has been suggested that it is the relatively smooth α-surface, as opposed to the knobby β-side of the hormone, which is important for binding. Modifications of the molecule which disturb the symmetry of its α-surface, such as seen in the more potent synthetic corticosteroid dexamethasone, are thought to reduce binding.

The plasma albumins have a propensity to bind many drugs and hormones, including the steroids. They are, however, not very selective and have a low affinity for individual molecules. Owing to their large mass, they have a "high capacity," but this deficiency is somewhat compensated for by the large quantities that are present in the plasma. Certain proteins have, however, been identified which have a much more specific ability to bind the steroid hormones (they have a high affinity for them), although, as they are present in much lower concentrations, their capacity is said to be low. These include *corticosteroid-binding globulin* (CBG), or *transcortin,* and *sex-steroid binding globulin* (SBG or SSBG).

CBG is an α-globulin with a molecular weight of about 52,000. It is synthesized in the liver and has a half-life of about 5 days. CBG binds steroid hormones on a molecule for molecule basis. About 94 percent of the cortisol in the plasma is bound, about 60 percent to the CBG. The remainder is nearly all attached to the plasma albumins. When it is recalled that the albumin concentration is about 5.5×10^{-4} M compared to 7×10^{-7} M for CBG, the differences in their binding abilities become clearly apparent. Other steroid hormones also bind to CBG, and their association constants are shown in Table 5.1. It can be seen that progesterone binds even more strongly than cortisol and corticosterone, but testosterone and estradiol are only weakly attached. Progesterone can competitively displace corticosteroids and about 50 percent of the total in the plasma is bound in this way to CBG. Only about 1 percent of progesterone is unbound, the remainder being attached to plasma albumin. Aldosterone is poorly bound (60 percent) to plasma proteins, and the albumins appear to be of greater relative importance than CBG.

The levels of CBG rise considerably during pregnancy and following the administration of estrogens to men or women. The latter response can occur due to the use of certain types of contraceptive pills. Pregnancy is also associated with large increases in progesterone levels, which can be bound to CBG and even displace cortisol. The elevated CBG thus may afford another example of the "protective" function of such plasma proteins. CBG levels fall in liver disease, presumably because of decreased synthesis and in nephrosis as a result of its excretion in the urine.

In man, a plasma α-globulin has been isolated which binds estradiol-17β and testosterone. This is the sex steroid-binding globulin (SSBG or SBG). It has a much lower affinity for estrone and estriol. Like CBG, the levels increase in pregnancy and in response to estrogen treatment.

6
The adrenal glands

The adrenal glands, as their name indicates, lie adjacent to each kidney. In man, they weigh about 3 to 5 g, but this can vary, especially in disease. Two endocrine tissues are present which in man have a distinct distribution; the chromaffin cells lie in a medullary position surrounded by the cortical (adrenocortical) tissue. In man, the latter makes up about 20 percent of the total tissue, but it shows considerable interspecific variability. Both types of tissue have an endocrine function: the medulla releases catecholamine hormones, principally epinephrine, and the cortex steroidal ones. The tissues have different embryological origins: the cortex is mesodermal and the medulla is neural tissue, analogous to that of sympathetic ganglia. The possible importance of the close morphological association of the chromaffin and adrenocortical tissues is not clear, and in nonmammals they often occur at separate sites. The blood supply of the two regions appears to be partly contiguous, so that their products may mix. The conversion of norepinephrine to epinephrine in the medulla occurs under the influence of phenylethanolamine-N-methyl transferase (PNMT), and the formation of this enzyme is induced by high concentrations of corticosteroids, such as could only occur at such a site (Pohorecky and Wurtman, 1971). This interesting observation provides one possible reason for the juxtaposition of these two endocrine tissues.

The adrenal cortex

6.1. Morphology

In mammals the cortical region of the adrenal gland can be seen to have three distinct layers or zones (Figure 6.1). The outer one, which abuts onto the capsule of the gland, is called the *zona glomerulosa* and consists of round cells with many mitochondria but relatively few lipids. Underlying this outer region is the *zona fasciculata*, which is the predominant mass of the adrenal cortex and consists of elongated "clear" cells which are rich in mitochondria and lipid droplets. Below this and abutting onto the chromaffin tissue lies the *zona reticularis*, which consists of flattened cells that are poor in lipids.

The morphological differences appear to reflect the respective functions of the different zones in the biosynthesis of the corticosteroid hormones. These observations are based on differences in the production of the steroid hormones in diseases where the various zones are hypo- or hypertrophied. In addition, the hormones have been localized in each type of tissue, and the ability of cultured cells from each region to synthesize hormones has been measured. The zona glomerulosa is the principal source of aldosterone and contains relatively little cortisol. Its hypertrophy is associated with increased production of aldosterone. The zona fasciculata is the predominant source of cortisol, but it also contains appreciable amounts of corticosterone. It seems likely that the zona reticularis is the principal source of the adrenal androgen hormones, as its activity appears to increase in the adrenogenital syndrome. The evidence is, however, equivocal, and androgenic steroids also appear to be synthesized in the zona fasciculata.

The cytological appearance of the cells of the adrenal cortex is in many respects unique. The mitochondria do not have the cristae that are so typical in other types of cells but have tubular infoldings which, when sectioned, can have the appearance of vesicles. There are numerous large droplets, or liposomes, which are the storage sites for cholesterol, which is either accumulated from the blood or synthesized in the cell. The endoplasmic reticulum is abundant and has a smooth appearance, as it is not studded with ribosomes. These cytological characteristics appear to be associated with the cell's special abilities to synthesize

(a)

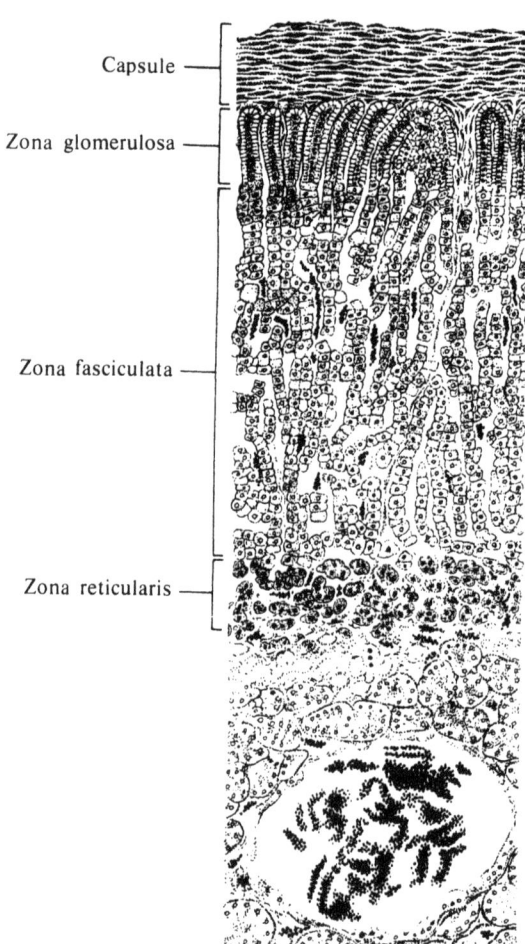

Capsule

Zona glomerulosa

Zona fasciculata

Zona reticularis

(b)

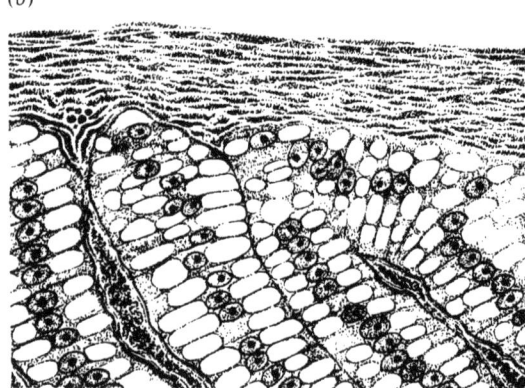

Fig. 6.1. (a) Histological section of a mammalian adrenal (a racoon), showing the zonation of the adrenal cortex. (b) Enlargement of the capsular glomerular zone. (From F. A. Hartman and K. A. Brownell, 1949. *The Adrenal Gland.* Philadelphia: Lea & Febiger.)

steroid hormones; a process that involves the lipid storage droplets, the mitochondria, the endoplasmic reticulum, and the cytoplasm, all of which contribute, via their enzymes, to the orderly synthesis of the corticosteroids (see Section 5.2). Structural changes can be observed which are associated with activity, and these include diminution of the liposomes, mitochondria, and the endoplasmic reticulum in the zona fasciculata when the gland is hypoactive. Hyperactivity has the opposite effects, except that the liposomes are depleted.

6.2. Functions and effects

The importance of the adrenal cortex for life was first recognized by the London physician Thomas Addison in 1855. He described the effects of hypofunction of the adrenal cortex, which, because of the contemporary prevalence of tuberculosis, was probably seen more often at that time. Addison's disease is associated with muscular weakness, low blood pressure, lack of appetite, decreased blood volume, a hypoglycemia, and a decline in renal function, including an inability to excrete water. There is a decrease in plasma Na^+ concentration and an elevation of K^+, reflecting an excessive loss of Na in the urine which is accompanied by an inability to excrete sufficient K. Death ultimately occurs. That the ubiquitous effects of hypoadrenocortisolism were due to an endocrine-type deficiency was not recognized until the 1920s, although Sir William Osler in 1896 had tested, with some limited success, the effects of adrenocortical extracts in Addison's disease. Later is was shown experimentally that adrenalectomy in cats produced similar deficiencies to those observed in Addison's disease, and extracts of the adrenal cortex maintain such animals in adequate health indefinitely. The first demonstration that such adrenal extracts may assure survival in man suffering from Addison's disease (adrenocortical insufficiency) was made at the Mayo Clinic in 1930. Unfortunately, the adrenal cortex stores little of its active hormonal products, so that glandular extracts were a poor, limited, and expensive source of such therapeutic material. Once the hormones had been chemically identified, however, it was only a matter of time before they were made synthetically. In the 1950s, cortisone, a precursor of cortisol, became available in the quantities necessary for its full-scale therapeutic use. Prior to this important technological advance, an active hormone analogue, deoxycorticosterone, was synthesized by Von Steiger and Reichstein in 1937 and, although lacking many of the effects of cortisol, could be used to treat adrenocortical insufficiency.

As was readily apparent to Thomas Addison, the adrenal cortex has ubiquitous physiological effects (Table 6.1). These are traditionally divided into

Table 6.1. *Summary of the major effects of corticosteroids*

Effects		Target organs
Mineralocorticoid	*(mainly aldosterone)*	
Na retention	↑	Kidney, intestine, salivary glands, sweat glands
K excretion	↑	Kidney
Glucocorticoid (cortisol)		
Gluconeogenesis	↑	Liver
Glycogen synthesis	↑	Liver
Glucose uptake by cells	↓	Peripheral tissues
Lipolysis	↑	Adipose tissue
Protein synthesis	↓	Muscle, skin, bones
Proteolysis	↑	As above
Growth	↓	
Lympholytic	↑	Lymphoid tissues, including thymus
Antiinflammatory	↑	Pharmacological effect at high dose
Androgenic (androstenedione, dehydroepiandrosterone)		In women (? pubic and axial hair, behavior)

Note: ↑, increase; ↓, decrease.

two groups: (a) *mineralocorticoid actions,* referring to the effects on sodium, potassium, and hydrogen ion; (b) *glucocorticoid effects,* all other actions, such as on intermediary metabolism of fats, carbohydrates, and proteins, on growth, and on muscular function. The therapeutic actions of the corticosteroids in suppressing inflammation are considered with glucocorticoid effects.

The foregoing classification has a real physiological basis, which reflects the particular function of each type of adrenocortical hormone. Aldosterone has a mineralocorticoid function and, except in excessive doses, negligible glucocorticoid effects. Cortisol and corticosterone have principally glucocorticoid effects, although they also possess small, but significant, mineralocorticoid actions.

A. Mineralocorticoid effects

Death usually occurs within a few days following adrenalectomy, and this is principally due to serious disturbances in electrolyte metabolism. There is a hyponatremia and a hyperkalemia which is accompanied by a metabolic acidosis. Dietary measures, such as the provision of additional NaCl and the restriction of the intake of K, can often prolong survival for many months. The administration of corticosteroids is a more effective treatment and can completely correct the electrolyte imbalance.

1. Kidney. The disturbance in electrolyte metabolism is primarily the result of an inability to conserve Na^+ and to excrete K^+ and H^+. These processes occur principally in the kidney, where excretion in the urine is affected, but the intestine and sweat glands are also involved.

Normally more than 99 percent of the Na that

appears in the renal glomerular filtrate is reabsorbed back into the blood as it passes along the renal tubule. A fine regulation of this process occurs in the distal part of the nephron, where it is under the control of the corticosteroids. Aldosterone is normally the main hormone involved, although cortisol and corticosterone have sufficient mineralocorticoid activity to maintain these processes and prevent excessive salt loss in the urine. The administration of exogenous aldosterone initially results in a retention of Na due to its renal action, but after a few days an "escape" occurs and the normal pattern of salt excretion is resumed. The reasons for this response are not clear but may involve renal circulatory adjustments. This effect accounts for the lack of excessive salt retention in hyperaldosteronism.

Potassium excretion in the urine depends on processes that involve its reabsorption from the glomerular filtrate as well as its secretion, into the urine, by the tubular cells. Aldosterone promotes the latter process in the distal segment of the renal tubules. While the effect is often described as a Na^+-K^+ exchange process, the mechanisms involved appear to be quite distinct. For instance, while Na^+ reabsorption in response to aldosterone can be prevented by actinomycin D, the secretion of K^+ is unaffected and there is no "escape" mechanism for the latter ion. Similarly, the ability to secrete H^+ is also unrelated to the antinatriuretic effects of corticosteroids. Administered corticosteroids thus produce a hypokalemia which is accompanied by a metabolic alkalosis.

2. Intestine. Clinical observations have clearly shown that in hyperaldosteronism, there is a decreased Na and an increased K level in the feces. It has even been suggested that the Na/K ratio in the feces may afford a useful clinical index of adrenocortical function (Charron et al., 1969). The change in fecal electrolytes appears to reflect a direct action of aldosterone on the processes of Na reabsorption and K secretion in the gut. *In vivo* perfusion and *in vitro* studies indicate that these hormone-sensitive processes occur in the colon (Wrong, 1968), although an action in more proximal segments of the intestine has not been completely excluded.

3. Salivary glands. The administration of aldosterone to man decreases the Na concentration in the saliva (Simpson and Tait, 1955). It has been observed that the Na/K ratio in this secretion may vary from 0.21 in hyperaldosteronism to 3.37 in normal subjects (Lauler, Hickler, and Thorn, 1962). Such changes have been used to help diagnose hyperaldosteronism. The effects of aldosterone on salivary secretion appear to involve the duct epithelium rather than changes in the rate of formation of the "primary" secretion (Mangos and McSherry, 1969). Its action is thus analogous to its renal effects.

4. Sweat glands. In man, the sweat glands play an important physiological role in providing water for evaporation, which aids temperature regulation. In hot climates considerable fluid loss may occur in this way. The sweat contains substantial amounts of solutes, especially NaCl. During acclimatization to work in hot environments, the Na content of the sweat declines over a period of several days as a result of the action of aldosterone (Conn, 1963). The reabsorption of Na from the exocrine ducts of the sweat glands increases. This effect is seen in the eccrine sweat glands in man. The salt content of the sweat has been observed to increase fifty- to sixtyfold in patients suffering from Addison's disease (Conn et al., 1947).

B. Glucocorticoid effects

Corticosteroids have widespread effects on the metabolic functioning of tissues. It is uncertain whether these involve a common mechanism, but such a hypothesis has the attraction of orderly simplicity. As will be described later, the glucocorticoid effects of the adrenocortical hormones appear to initially involve the process of genetic transcription and a *de novo* synthesis of proteins which may have enzymic functions. It seems likely that a generally common mechanism of action exists, but a variety of enzymes may be induced which mediate the diverse effects of the corticosteroids. For convenience we can classify such actions into several major types, although it will be apparent that there may be some crossover between them.

1. Intermediary metabolism. This process involves the transpositions that occur between fats, carbohydrates, and proteins, principally involving the liver, adipose tissue, and skeletal muscle. It should be recalled that these processes are influenced by several types of hormones, apart from corticosteroids, especially insulin but also growth hormone, glucagon, and epinephrine. These excitants may interact in various ways, either antagonizing or facilitating each other's effects. The predominant response will depend on the particular physiological situation, such as recent feeding or if fasting is occurring. In abnormal situations, such as disease, the responses may be altered. When attempting to relate experimental observations on animals to man, it should be remembered that species differences in response may occur and that for the purposes of scientific clarification, the animal's physiology may have undergone considerable technical modification. Thus, responses to corticosteroids that are observed in animals suffering from experimental diabetes mellitus may have little direct bearing on a response seen in a healthy man.

Corticosteroids have an especially important physiological role during fasting, as they promote the formation of glucose, a substrate that is vital for the metabolism of many cells, especially the brain. Many tissues have a limited ability to utilize fatty acids as a source of energy, and corticosteroids are also indirectly involved in the release of these substrates. Corticosteroids *promote glucose synthesis*, or *gluconeogenesis*, in the liver. Amino acids are broken down, and these come from the peripheral tissues, especially skeletal muscle, a process that is also promoted by the corticosteroids. *Mobilization of fatty acids and glycerol* (lipolysis) from triglycerides that are stored in fat cells is increased by hormones such as epinephrine, glucagon, and growth hormone, but the corticosteroids are also important and exert a "permissive" action which indirectly promotes their effects. The fatty acids are utilized for energy and the glycerol is converted to glucose by the liver. The entire process is thus a well-integrated one which results in the maintenance of a supply of energy substrates, especially glucose. The injection of corticosteroids into intact animals can even be shown to elevate blood glucose levels, but this effect is usually only a modest one which is rapidly compensated for by an increased release of insulin.

The *synthesis of glycogen* in the liver is increased by corticosteroids, and this effect appears to be associated with the formation of additional hepatic enzymes, especially glycogen synthetase.

Corticosteroids *decrease the peripheral utilization of glucose* and its uptake by these tissues. This response does not appear to play a direct role in maintaining blood glucose levels.

The actions of corticosteroids on fat cells are

complex and depend on the species and its physiological condition. In man, an excess of corticosteroids results in a redistribution of body fat; there are the characteristic local accumulations resulting in "moon face" and "buffalo hump" and a depletion in peripheral stores, such as in the limbs. As described above, corticosteroids, especially in conjunction with other types or hormones, promote mobilization of fatty acids. The effect can, however, be antagonized by insulin, which is released in response to the corticosteroid-induced hyperglycemia. The net result may depend on the sensitivity of the particular type of adipose tissue to each hormone; for instance, the fat cells in the limbs may have a low sensitivity to the insulin, so that the effects of the corticosteroids predominate; adipose tissue in the face and back may be more responsive to insulin, so that additional fat is laid down.

2. Muscle. The catabolic "wasting" that characteristically occurs with an excess of corticosteroids in skeletal muscles and collagenous tissues, such as skin, largely reflects the mobilization of amino acids from them. This effect is thought to principally involve an inhibition of protein synthesis rather than a promotion of its breakdown. However, other more subtle effects are apparent, such as a decrease in contractility which may be especially apparent in the heart and possibly vascular smooth muscle, where it may contribute to the hypotension observed in Addison's disease.

3. Growth. Pediatricians were one of the first to observe that the administration of corticosteroids to children can result in an inhibition of growth. This effect, which occurs at quite low doses, can be readily demonstrated in young rats and mice. The rate of cell proliferation can be measured in different tissues by recording the incorporation of thymidine into DNA. It can be seen in Figure 6.2 that in young rats this effect is seen in liver, heart, skeletal muscle, and kidney but not in the gut mucosa, brain, testis, or spleen. In man, the growth of cartilage is also inhibited. At higher doses an inhibiting effect on the gastric mucosa also becomes apparent, which could be contributing to the controversial effects of corticosteroids in precipitating peptic ulcer. The mechanism for the inhibition of growth is unknown but would appear to be consistent with the catabolic effects of these hormones. It has been suggested that the corticosteroids may interfere with the release of growth hormone, which has been observed under certain conditions, or an antagonism at the tissue level could be occurring. The administration of large doses of growth hormone, however, does not overcome the inhibition. When corticosteroid therapy is stopped, there is usually a "catch-up" in the growth. It is interesting that in children the administration of ACTH

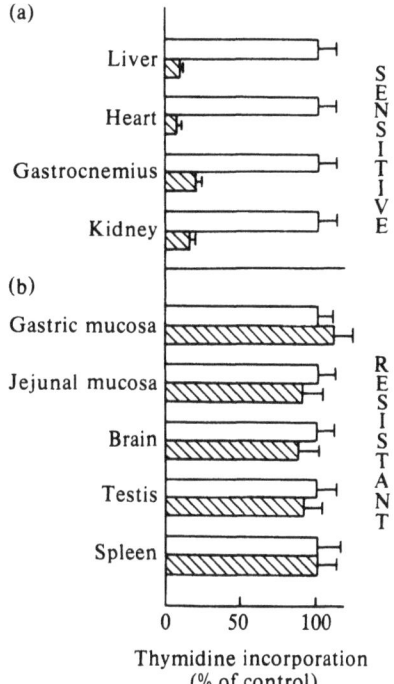

Fig. 6.2. Effect of a low dose of cortisone on the incorporation of thymidine into the DNA of different tissues of weanling rats (mean ± S.E.M.). Thymidine incorporation by the tissues in panel A is greatly suppressed by cortisone treatment, while incorporation by the tissues in panel B is unaffected. Open bars denote control animals; crosshatched bars, cortisone-treated animals. (From Loeb, 1976. Reprinted by permission from *The New England Journal of Medicine 295*, 549.)

does not have such severe effects on growth as do corticosteroids. It has also been found that when corticosteroids are administered on alternate days, the effects are not as severe. A normal physiological role of corticosteroids on the regulation of growth is doubtful.

4. Calcium metabolism. This process is also influenced by corticosteroids, and a common side effect of their therapeutic administration is osteoporosis. There are several reasons for this effect (see Section 9.5), probably the most important being an inhibition of the growth of the collagenous matrix of bone, which is another manifestation of their catabolic effects on tissues.

5. Lymphoid tissues. The thymus, 'spleen, lymph nodes, and circulating lymphocytes break down under the influence of corticosteroids (they have a lympholytic action). These effects have widespread application in the therapeutic uses of such compounds to combat inflammation, reduce antibody formation, and inhibit the immunological responses of tissues such as seen following surgical grafting and transplantation. There is considerable species variation in such responses: mice, rats, hamsters, and rabbits are "steroid-sensitive," while man, monkeys, and guinea pigs are "steroid-resistant." The latter may respond, but much higher doses of corticosteroids are usually needed to give a very much smaller response. This distinction in responsiveness should be kept in mind when considering the human relevance of experiments on certain animals. In man, corticosteroids do not produce a reduction in lymphocyte number, although this response may occur in leukemia cells. The thymus of infants is reduced in size in response to corticosteroids, but this effect is reversed when therapy ceases. The immunological response of tissues in man is decreased, which is the basis of the use of corticosteroids as immunosuppressive drugs. The mechanism of the lytic actions of corticosteroids on lymphoid tissue is controversial. Glucose uptake by the cell is inhibited, and this response can be blocked by actinomycin D. Genetic transcription and the synthesis of a protein that can inhibit glucose uptake thus appears to initiate the response, and this results in protein breakdown, the release of amino acids, and the death of the cell.

6. Antiinflammatory actions. The most widespread therapeutic uses of corticosteroids involve their antiinflammatory actions. It is doubtful if they have such an effect under normal physiological conditions, as very high dosages are required. They are used for this purpose in many inflammatory diseases (see Section 6.8B.) The inflammatory response is associated with redness, swelling, and pain, which reflects a local vasodilatation of blood vessels, an increase in capillary permeability, and a stimulation of sensory nerves. It is sometimes viewed as a protective response which leads to healing. Thus, it is not surprising to observe that administered corticosteroids can delay wound healing. The exact causes are controversial and are discussed in Section 6.8C(5).

7. Stress. As described in Section 6.6A, glucocorticoids are secreted in large amounts in response to a large variety of nonspecific stimuli, such as pain, infection, emotion, and severe exercise. These have been described as "stress." There has been a lot of speculation about the possible physiological significance and uses of the massive releases of glucocorticoids that occur under these conditions. The problem is reminiscent of the early considerations about the functions of released epinephrine in the "fright-and-flight" response. Little is known, however. It is notable that patients who have a hypofunctioning adrenal cortex respond poorly to stress and may rapidly die under

such conditions unless large doses of glucocorticoids are rapidly administered.

C. Androgenic effects
Significant amounts of androgens are secreted by the adrenal cortex, especially in the fetus and in the infant. Subsequently, at puberty this process is resumed. It appears that at the latter time they contribute to the growth of pubic and axial hair. It has been speculated that they may also influence the psychosexual behavior of women, but this aspect is controversial. Their potential anabolic effects in normal individuals is also uncertain. An excess adrenocortical androgen secretion can occur in certain diseases [see Section 6.8A(2)], where it can lead to virilism in both sexes and disorders in the maturation of the sexual apparatus. The androgens may be converted to estrogens in peripheral tissues, and this may be an important source of these hormones in postmenopausal women.

6.3. Corticosteroids and their synthetic analogues: structure–activity relationships

In man, the principal secreted corticosteroid hormones (Figure 6.3) are cortisol (also called hydrocortisone and originally compound F) and aldosterone. Only small amounts of corticosterone (compound B) are secreted in man, but in rats and mice this steroid is the principal glucocorticoid hormone. Cortisol is absent in these laboratory animals because of a deficiency of 17α-hydroxylase in the adrenal cortex. Cortisone is also present in the circulation, but this is an inactive corticosteroid that must be converted to cortisol. A variety of intermediates in the synthesis of these steroids exists in appreciable amounts in the adrenal cortex and exhibit activity; the best known of these is probably 11-deoxycorticosterone, the immediate precursor of corticosterone. This mineralocorticoid was the first to be chemically synthesized.

Androgens are also secreted by the adrenal cortex but usually in small amounts, although in disease, when genetic blockade of certain enzymic processes occurs [see Section 6.8A(2)], considerable quantities may appear in the plasma. The principal adrenal androgen to appear in the circulation is 11β-hydroxyandrostenedione. Dehydroepiandrosterone, which is secreted as the sulfate, appears to be an androgen which is unique to the adrenal cortex but may also be secreted by tumors of the testis. The adrenal cortex also secretes androstenedione. The structure of these compounds is shown in Figure 6.3.

A. Structure–activity relationships
This subject has been described in detail by Bush (1962). Changes in chemical structure can change

the biological "activity" of corticosteroids in several ways, apart from its intimate ability to react with its receptors.

a. *Absorption* may be affected. In orally administered preparations this is enhanced by lipid solubility, and this process tends to decrease as the molecules become more polar by the addition of hydroxyl groups. Conversely, it can be increased by nonpolar moieties such as methyl groups. Absorption across the buccal mucosa (and, following topical application, across the skin) tends to be inversely related to the polar nature of the molecules. In contrast, absorption from parenteral injection sites is facilitated in steroids of a more polar nature.

The solubility of many corticosteroid preparations is modified by the preparation of their esters, usually at the C-21 hydroxyl position. This modification is used to modify their solubility; thus, the phosphate and succinate esters are water-soluble and can be readily administered intravenously. The acetates are lipid-soluble and are used to provide "depot" preparations as pellets or in oil, from which absorption is slow but regular.

b. *Metabolic degradation* of the hormone may be changed and can be affected by several factors. When absorbed from the gut, the compounds will pass directly to the liver, which is the major site of their catabolism. This process will also be promoted in compounds that are less well bound to plasma proteins. The availability of hydroxyl groups where conjugation reactions occur may also promote catabolism. In synthetic corticosteroids the Δ^{1-2} unsaturation (double bond) in the A-ring, which is seen in such compounds as prednisolone and dexamethasone, delays metabolism. The 16α or 16β substitution of an hydroxyl or methyl group is thought to delay the degradation of the C-17 side chain. The widely used synthetic analogues with such substitutions are, largely for this reason, mainly excreted unchanged in the urine.

c. A facility by which *activation* of an otherwise inactive compound occurs. Cortisone and prednisone have an 11β-keto group which blocks the molecule's glucocorticoid activity. This moiety can, however, be readily hydroxylated in the body.

d. *Plasma binding* may be altered. If decreased, this change can increase activity due to higher free concentration in the plasma, although degradation can also be hastened. Lack of protein binding will also hasten excretion in the urine. As described earlier (Section 5.4) binding tends to follow the "polarity" rule and is decreased as the number of polar groups, such as hydroxyls, on the molecule increases. It also appears to be related to the relatively smooth α-surface of the corticosteroid, so that the addition of an irregularity, such as α-hydroxyl group, may decrease binding.

e. *Access across the cell membrane* can influence the accessibility to the receptors in the cytoplasm and nucleus. There appears to be little information about this process, but on theoretical grounds it would appear to be favored by the molecule's lipid solubility. The addition of methyl groups could conceivably facilitate this process but may also make the molecule more accessible to its catabolic sites.

f. *Specific interaction with its receptors.* The receptors appear to be protein moieties, the initial interaction taking place in the cytoplasm, from which the steroid is transferred to a nuclear site. Ideally, the characterization of this process will ultimately depend on an analysis of the corticosteroid's ability to interact with the "isolated receptors." Some progress has been made and perhaps, considering all the other factors that may influence the hormones action, a surprising degree of correlation between such binding and the hormone's potency has been observed. Little direct information is, however, available about the intimate type of interaction between corticosteroids and their receptors. Different processes appear to be involved from plasma protein binding; thus, dexamethasone is not significantly bound to plasma proteins but nevertheless has a very potent glucocorticoid action.

The forces that determine the hormone's interaction with its receptor are comparatively weak ones probably involving hydrogen bonding, such as that contributed by protons or hydroxyl groups and London forces. The correct steric configuration of the molecule is vital. The insertion of α-hydrogens at ring junctions, such as the C-5 position, completely alters the planes of alignment of the A- and B-rings and abolishes normal activity.

The *levo* forms of the corticosteroid hormones are inactive, and indeed even substitution of a differently oriented single group such as an 11α-hydroxyl in cortisol for the 11β-hydroxyl results in a complete loss of activity.

The Δ^{4-5}, the 3-keto group and the C-17 side chain appear to be vital for both mineralo- and glucocorticoid actions. The latter effects require the 21-hydroxyl and 20-keto group. Removal of the 21-OH (21-deoxy compounds) results in a complete lack of glucocorticoid activity, but the retention of a little mineralocorticoid action (but see the effects of a 9α-fluoro substitution, end of this section.)

The 11β-hydroxyl group is also essential for glucocorticoid activity. The 11β-keto compounds are inactive (but as described above can be converted metabolically to the hydroxyl). The 17α-hydroxyl group enhances but is not essential for glucocorticoid activity. This relationship is seen in corticosterone, which lacks the 17α-OH and principally exhibits mineralocorticoid effects but which also has a significant residual glucocorticoid activity.

Neither the 11β-hydroxyl or -keto group, or the

17α-hydroxyl are necessary for mineralocorticoid activity. 11-Deoxycorticosterone lacks both of these moieties, yet has substantial mineralocorticoid activity and exhibits no glucocorticoid effects. It is, however, notable that aldosterone, which is far more active, has an 11β-hydroxyl group (but no 17α-OH). The aldehyde group at C-18 plays an important role in the potent mineralocorticoid action of aldosterone, but the nature of this effect is unknown.

Substitution of a 9α-fluoro group, which is adjacent to the 11β-OH, increases the mineralocorticoid activity of cortisol 125 times but the glucocorticoid action only about 10 times. This substitution increases the electronegativity of the 11β-OH. The difference in the ratio of the activity suggests that this hydroxyl is important in the mineralocorticoid response. It is, however, interesting that the insertion of a 9α-fluoro group into deoxycorticosterone, which lacks an 11β-hydroxyl,

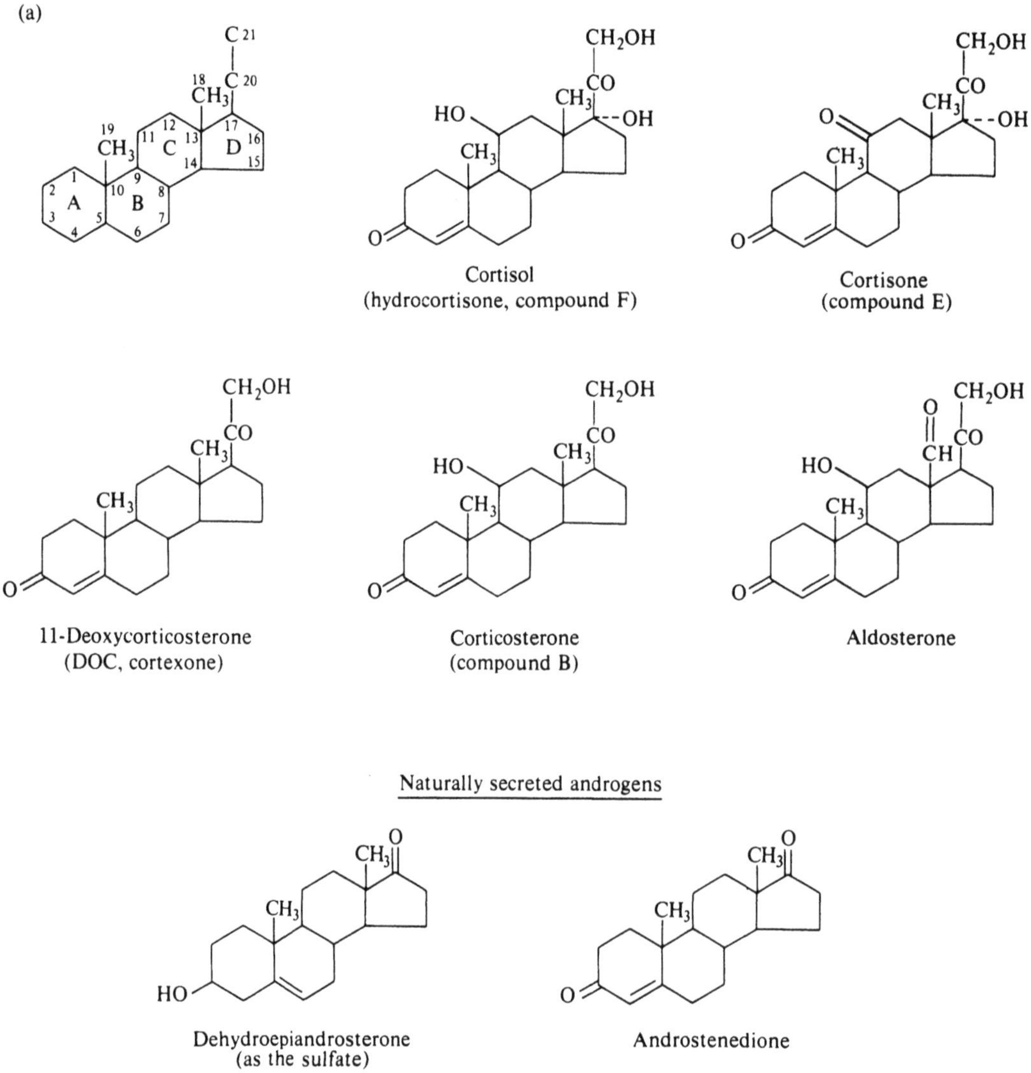

(a)

Cortisol
(hydrocortisone, compound F)

Cortisone
(compound E)

11-Deoxycorticosterone
(DOC, cortexone)

Corticosterone
(compound B)

Aldosterone

Naturally secreted androgens

Dehydroepiandrosterone
(as the sulfate)

Androstenedione

Fig. 6.3. The adrenocorticosteroids: natural and synthetic. (a) In man, cortisol and aldosterone are the principal adrenocortical hormones. However, the androgens androstenedione and dehydroepiandrosterone sulfate are also secreted and are an important source of androgens in women, and in certain diseases. Deoxycorticosterone and corticosterone are metabolic intermediates of the secreted hormones in man, but the latter is the major glucocorticoid in rodents. (b) Synthetic derivatives of the adrenocorticosteroid hormones which have modified activities, especially a reduced mineralocorticoid action, which make them useful therapeutically.

(b)   <u>Δ^1-Double bond derivatives</u>

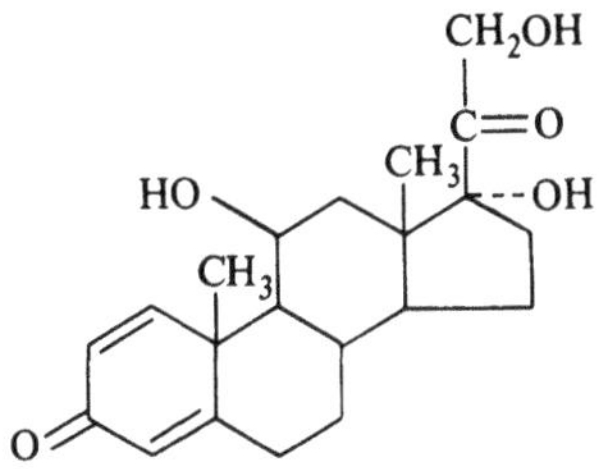

Prednisolone
(1,2-dehydrocortisol)

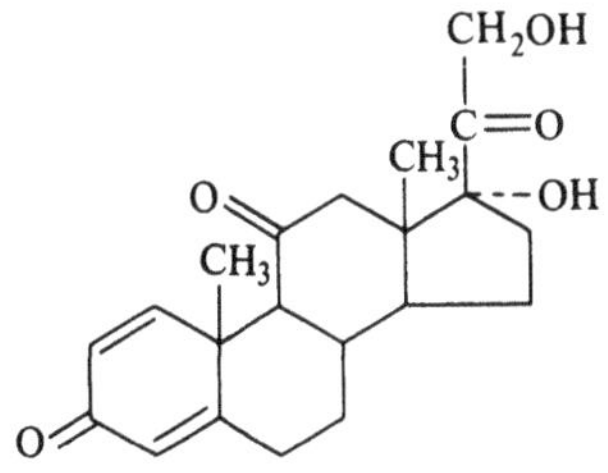

Prednisone
(1,2-dehydrocortisone)

<u>6α-Substitution</u>

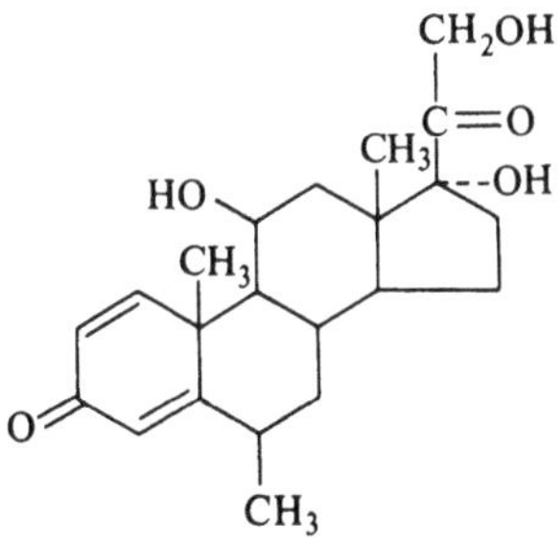

Methylprednisolone
(6α-methylprednisolone)

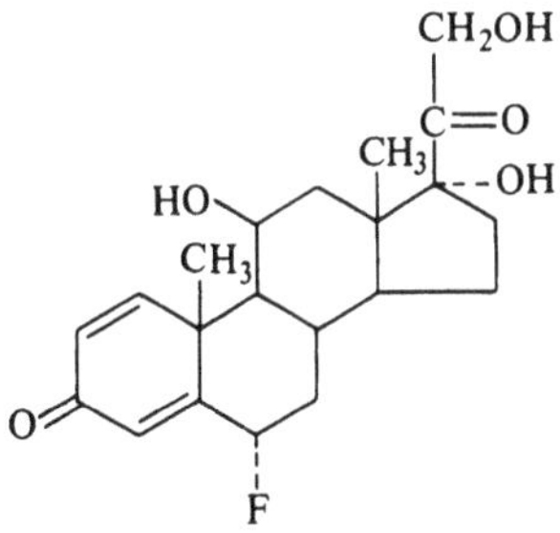

Paramethasone
(6α-fluoroprednisolone)

<u>9α-Fluoro derivatives</u>

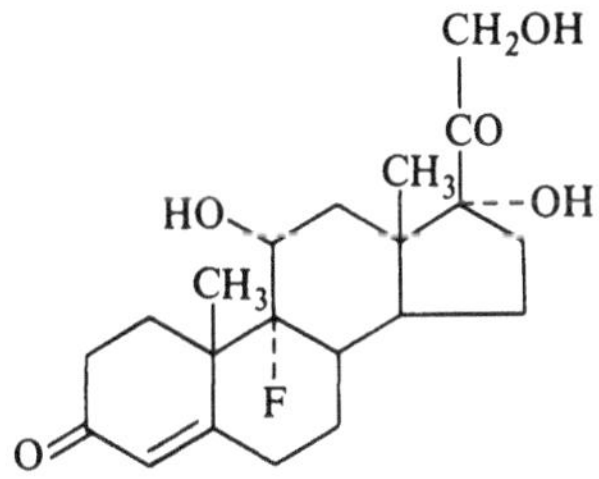

9α-Fluorocortisol

<u>16α- or β-Substitution and Δ^1 double bond</u>

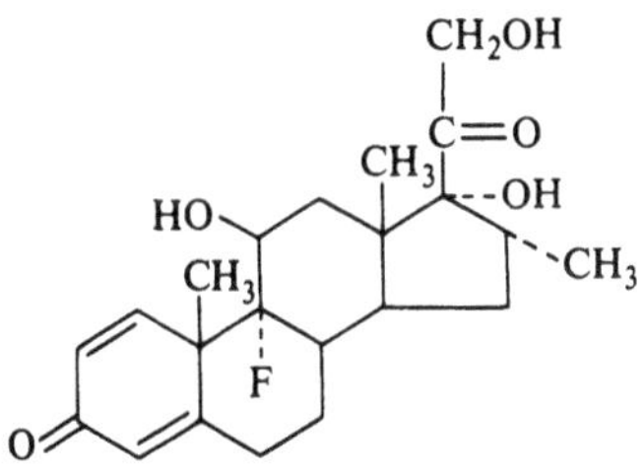

Dexamethasone
(9α-fluoro-16α-methylprednisolone)

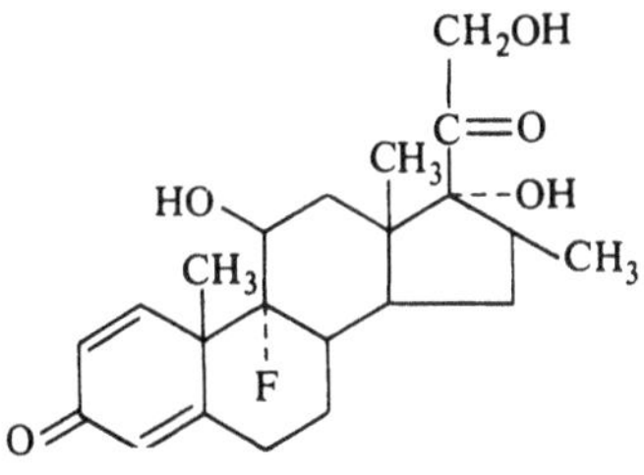

Betamethasone
(9α-fluoro-16β-methylprednisolone)

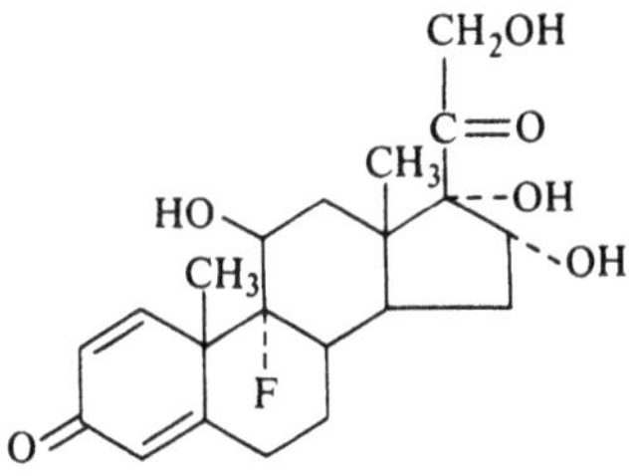

Triamcinolone
(9α-fluoro-16α-hydroxyprednisolone)

increases its mineralocorticoid activity 12 times. This observation indicates that the fluorine can have effects which are independent of interactions involving the 11β-hydroxyl.

B. Synthetic analogues of corticosteroids

The early successful use of corticosteroids in the treatment of inflammatory diseases, especially rheumatoid arthritis, resulted in considerable elation among patients, physicians, the general public, and the mass media. The self-congratulations were, however, short-lived, as the chronic use of the available preparations, at that time principally cortisone, was soon seen to result in a horrendous variety of side effects. These included mineralocorticoid effects resulting in edema and hypokalemia, as well as metabolic disturbances such as those involving fat metabolism ("moon face"), osteoporosis, gastric ulceration, inhibition of the adrenal cortex, delayed wound healing, an increased likelihood of infection, and even behavioral changes, including psychosis. (For a more detailed description, see Section 6.8C.) There was clearly a need for more specifically active compounds, a host of which were chemically synthesized with results of substantial practical, but still limited, importance.

From the practical standpoint three or four different types of modifications of the corticosteroid molecule are important (see Figure 6.3).

a. The introduction of the Δ^1 double bond in ring A of cortisol (to produce prednisolone) increases its glucocorticoid (or antiinflammatory) activity but not its mineralocorticoid actions. The ratio of the two actions is thus changed, allowing an adjustment of the dose with a resulting decrease in mineralocorticoid-related side effects. Although the rate of the compounds catabolism is slowed, the change in the ratio of the two types of effects suggests that interactions with receptors are also influenced.

b. 9α (or 12α)-Halogenation, usually with fluoride, markedly increases both the glucocorticoid and mineralocorticoid activities of the natural corticosteroids. This important discovery, made in 1954, was of considerable practical importance in the chemical design of new therapeutically useful corticosteroids. The 9α-fluoro analogue of cortisol has about 125 times the mineralocorticoid and 10 times the glucocorticoid activity of cortisol. It is thought that the halogen increases the electronegativity of the adjacent 11β-hydroxyl group and so enhances its ability to interact with its receptor. It is interesting that while removal of the C-21 hydroxyl (21-deoxycortisol) virtually abolishes the corticosteroids activity, the insertion of a 9α-fluoro moiety in such a molecule substantially restores its actions. This suggests that while the two hydroxyls are normally vital, if the propensity of one to inter-

act with its receptor is increased, a successful steroid–receptor interaction can occur. On the other hand, a 9α-methyl group reduces activity, presumably by reducing the electronegativity of the 11β-hydroxyl.

Apart from influencing the steroids interaction with its receptor, 9α-fluorination can also inhibit the metabolism at the 11β-site. Thus, 9α-fluorocortisone is inactive, as it cannot be hydroxylated to its active form. Conversely, however, it also limits the oxidation of the 11β-hydroxyl, so that the active cortisol form retains its effectiveness longer.

c. The introduction of methyl or hydroxyl groups at the C-16 position in ring D improves the ratio of glucocorticoid to mineralocorticoid effects. A 16α- or 16β-methyl group decreases the Na-retaining effects of cortisol but increases the glucocorticoid effects. Similarly, a 16α-hydroxyl decreases mineralocorticoid actions but has little effect on its antiinflammatory action. The insertion of a 16α-methyl group into deoxycorticosterone lowers its activity, showing that this substitution has a specific antimineralocorticoid action.

d. Substitution at the 6α position in ring B has rather unpredictable but some useful effects. In cortisol or prednisolone, a 6α-methyl group increases glucocorticoid and decreases mineralocorticoid activity. Halogenation, with fluoride or chloride, has a similar type of effect.

A number of useful analogues (Table 6.2) of the corticosteroids have been made available for therapeutic use which combine the various changes just described and with minimal, and sometimes virtually no, mineralocorticoid-related side effects. An example of the latter is dexamethasone (Figure 6.2), which incorporates the Δ^{1-2} double bond, the 9α-fluoro group, and the 16α-methyl to produce a compound with high antiinflammatory activity but no Na-retaining effects. It is, however, disappointing that while mineralocorticoid-type side effects have been eliminated, the others remain. It appears that these actions and the antiinflammatory effects of the corticosteroids are firmly linked, and it is unlikely that it will be possible to dissociate them by modification of the steroid's chemical structure.

C. Summary of the nature of the corticosteroid–receptor interaction

The relationships of structure to activity discussed above have provided some insight into the nature of the combination reaction of the corticosteroids and their receptors. In the instance of the glucocorticoids, the 11β-hydroxyl appears to be essential, although the 21-OH and the 17α-OH can also contribute. Neither the 11β- nor 17α-OH groups are vital for the action of mineralocorticoids, but removal of the remaining 21-OH

Table 6.2. *Relative potencies of some natural and synthetic corticosteroids*

	Anti-inflammatory[a] effects	Na retention
Cortisol	1	1
9α-Fluorocortisol	10	125
Cortisone	0.8	0.8
(must be 11β-hydroxylated to act)		
Corticosterone	0.35	1.5
11-Deoxycorticosterone	0	10
Aldosterone	0 (0.35)[b]	300
Δ¹-*Double bond derivatives*		
Prednisolone	4	0.8
Prednisone	4	0.8
(must be 11β-hydroxylated to act)		
6α-Methylprednisolone	5	0.5
6- and 9α-Fluorinated, 16α- or β-substituted prednisolone		
Paramethasone	10	0
Triamcinolone	5	0
Dexamethasone	30	0
Betamethasone	25	0

[a] Approximately equal   [b]glucocorticoid effects.

in deoxycorticosterone (to produce progesterone) nearly abolishes its activity. Progesterone, however, acts as a competitive antagonist to DOC, suggesting that it can still react with the receptor. It is interesting that the position of the hydroxyl group which promotes mineralocorticoid activity does not appear to be fixed, thus, 9α-fluoro-11β-hydroxy-21-deoxycorticosterone is as active as DOC.

The interaction of the corticosteroid and its receptor is thought to involve "physical" rather than a chemical type of reaction. The initial attachment may be due to hydrogen bonding, which in glucocorticoids involves the 11β-OH. Once attached, the molecule can then complete its alignment with the receptor by rotating around this primary bond. The latter process is thought to involve short-range London forces. In the instance of the mineralocorticoids such a hydroxyl may also be important, but its preferred site may be more flexible and can include the 21-hydroxyl, as in 11-deoxycorticosterone.

D. Mineralocorticoid antagonists

The specific ability to antagonize the effects of aldosterone is of both theoretical interest and therapeutic importance. Removal of the 11β- and 18-aldehyde groups of aldosterone results in a decline in its mineralocorticoid activity, but the resulting compound 11-deoxycorticosterone (DOC) still has a potent and very specific effect on electrolyte metabolism (see Table 6.2). Removal of the re-

maining 21-OH, however, almost abolishes such effects, except at very high doses. The resulting compound is progesterone, which acts as an antagonist to the mineralocorticoid effects of DOC and aldosterone. The antagonistic properties of this modification of the 17β side chain led to the synthesis by Searle and Co. of a number of similar steroids containing a γ-lactone ring at C-17. These compounds (Figure 6.2) are 17-spirolactosteroids (a family to which the cardiac glycosides also belong) which are commonly called spirolactones. The lactone ring needs to be closed for optimal activity. Although some active compounds possess an open ring, it is thought that this must be closed before the antagonist can act. The spirolactones only act in the presence of endogenous or exogenous mineralocorticoids such as aldosterone or DOC (Kagawa, Cella, and van Arman, 1957; Liddle, 1957). In high doses some of them exhibit some agonist actions. They behave as classical reversible competitive antagonists and can inhibit all the actions of aldosterone, on Na+, K+, and H+. They can be persuaded to act *in vitro*, where they inhibit the effects of mineralocorticoids on model membranes such as the toad's urinary bladder (see Sakauye and Feldman, 1976). *In vivo* they exert a similar effect on the kidney, where they act on the distal region of the nephron. The result is an enhanced excretion of Na in the urine accompanied by a K retention. *In vitro* they can also be shown to bind to aldosterone receptors isolated from the cytoplasm of kidney cells (Herman, Fimognari, and Edelman, 1968; Marver et al., 1974). Spirolactones exclude mineralocorticoids from binding to these receptors. At a concentration ratio of about $10^4:1$ (spirolactone/aldosterone), binding of aldosterone is reduced by about 70 percent. The cytoplasmic steroid–receptor complexes of the antagonist, in contrast to steroid agonist–receptor complexes, cannot bind to the nuclear chromatin acceptor sites. This presumably accounts for their lack of activity and the inhibition of Na transport that results.

In terms of the allosteric–steroid receptor model, the action of a spirolactone can be viewed in either of two ways (Marver et al., 1974):

a. It combines with the receptor but, unlike an agonist, no conformational change, such as is necessary for it to bind in the nucleus, occurs.
b. If an active and an inactive form of the receptor exist in equilibrium, the antagonist binds preferentially to the latter.

A number of different spirolactones have been tested for pharmacological activity. The first two compounds were SC-5233, which has a 19-methyl group, and SC-8109, which is the 19-nor compound (Figure 6.4). These drugs were quite active parenterally, but very large doses had to be administered orally. The addition of a 7α-acetylthio group improved the oral activity, and this compound (SC-9420) is used therapeutically as a diuretic under the generic name *spironolactone*.

The *structural requirements for antagonist activity* are (Sprague, 1968; Funder et al., 1974):

a. A 3-keto and a Δ^{4-5} double bond in the A-ring. The insertion of an additional Δ^{1-2} double bond increases activity even more. (This is reminiscent of the increased effectiveness of prednisolone.)
b. Opening of the lactone ring usually reduces activity drastically. However, some compounds, for instance SC-14266 and canrenoate-K (Figure 6.4), are active, but the lactone ring must be reconstituted before it can act. γ-Lactone unsaturation also decreases activity.
c. Esterification or thioesterification at the 7α position increases activity and binding to renal receptors. This group is stereospecific, as 7β-substituents are inactive. The relative effects *in vivo* on the kidney, *in vitro* on the toad's urinary bladder, and binding to isolated kidney receptors usually parallel each other. This 7α moiety is, however, removed (Figure 6.4) during metabolism of spironolactone, the predominant active metabolite in man being canrenone (Ramsay et al., 1977).
d. Some of the main side effects of spironolactone are related to its antiandrogenic effects. These actions are apparently due to its ability to combine with androgen receptors thus antagonizing the effects of the natural steroid hormones. Attempts have been made to prepare analogues in which this antiandrogenic activity is reduced (Cutler et al., 1978). One of these steroids, SC-25152, has a $-CO_2CH_3$ group at the 7α position (Figure 6.4). This compound retains its ability to combine with renal mineralocorticoid receptors (*in vitro*), but it has a reduced affinity (3 to 4 times) for androgen receptors from human and rat prostate. This drug could thus provide a more specifically acting mineralocorticoid antagonist than spironolactone, but reports of its effects *in vivo* in man are not yet available.

6.4. Adrenostatic drugs

A number of drugs (Figure 6.5 and Table 6.3) are available which can inhibit the biosynthesis of the adrenocorticosteroid hormones (see Chart et al., 1962; Lipsett, 1968; Samuels and Nelson, 1975). They have limited therapeutic use, although one of them, metyrapone, is widely used to diagnose adrenocortical disorders [see Section 6.8A(3)]. The experimental investigation of adrenocortical function may also be facilitated by the use of these drugs.

As the biosynthesis of the corticosteroid hormones involves numerous metabolic transformations, it is perhaps not surprising to observe that synthesis can be inhibited by a variety of drugs that act in different ways. The discovery of the adrenostatic actions of many of these drugs resulted from observations of their clinical side effects. Thus, an insecticide DDD ([1,1-dichloro-2,2-bis(P-chlorophenyl)ethane]), a relative of DDT, was found in toxic quantities to produce an adrenal necrosis in dogs. Aminoglutethimide was originally used as an antiepileptic drug. One of its side effects is a masculinizing action in women, which appears to be the result of a derangement in synthesis of corticosteroids. Triparanol is a hypolipidemic drug which was once used in attempts to reduce blood cholesterol levels (it has now been withdrawn from use), and it also prevents the synthesis of corticosteroids.

Mitotane (o,p'-DDD, o,p'-dichlorophenyldichloroethane) (Figure 6.5) is an isomer of the insecticide DDD. It has a cytotoxic effect on adrenocortical cells, apparently causing a degeneration of the mitochondria, so that the secretion of corticosteroids is inhibited. It is active in normal and cancerous tissue and is used to treat Cushing's syndrome (Schteingart, 1978), especially when it is due to an inoperable carcinoma when metastases have occurred.

Aminoglutethimide decreases corticosteroid secretion by autonomous adrenal tumors. It is usually ineffective in normal tissue or in Cushing's syn-

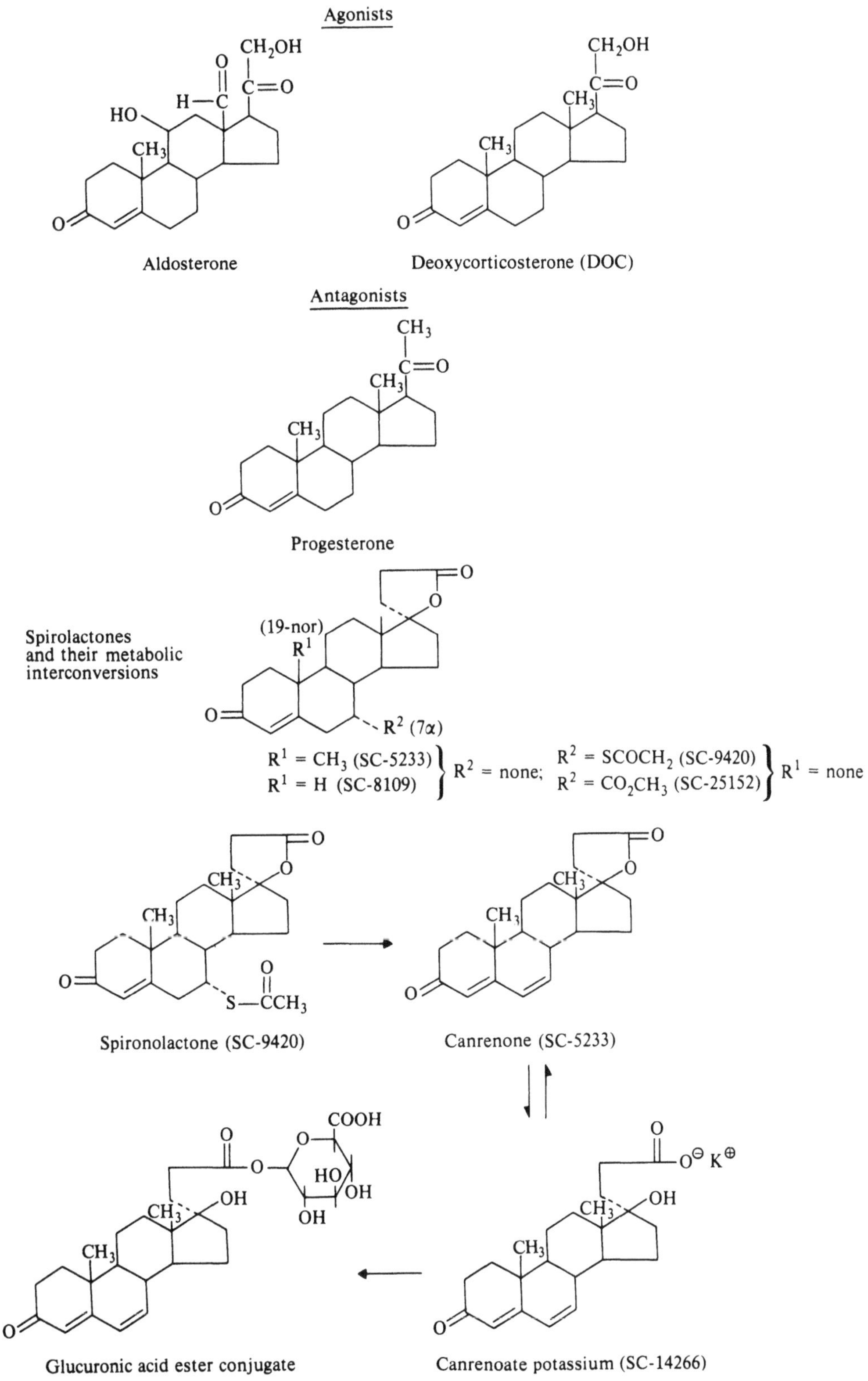

Agonists

CH₂OH
O
H—C
HO
C=O
CH₃

Aldosterone

CH₂OH
C=O
CH₃
CH₃
O

Deoxycorticosterone (DOC)

Antagonists

CH₃
C=O
CH₃
CH₃
O

Progesterone

Spirolactones
and their metabolic
interconversions

O
O
(19-nor)
R¹
CH₃
O
R² (7α)

R¹ = CH₃ (SC-5233) } R² = none;   R² = SCOCH₂ (SC-9420) } R¹ = none
R¹ = H (SC-8109)   R² = CO₂CH₃ (SC-25152)

O
O
CH₃
CH₃
O
O
S—CCH₃

Spironolactone (SC-9420)

→

O
O
CH₃
CH₃
O

Canrenone (SC-5233)

↓↑

COOH
O
O
HO
OH
OH
OH
CH₃
CH₃
O

Glucuronic acid ester conjugate

←

O
O⁻ K⁺
OH
CH₃
CH₃
O

Canrenoate potassium (SC-14266)

Fig. 6.4. Molecular structures of agonists and antagonists of the mineralocorticoid effects of the adrenal steroid hormones.

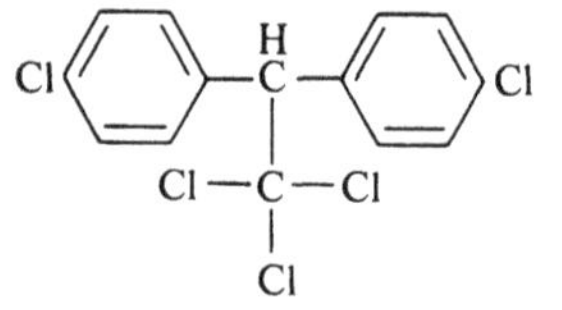

DDT

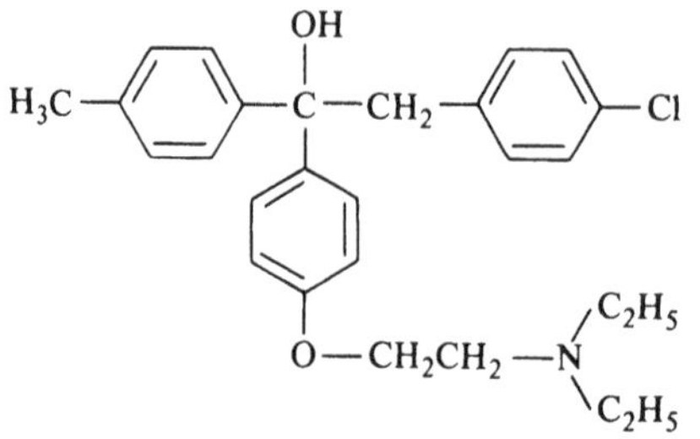

Triparanol

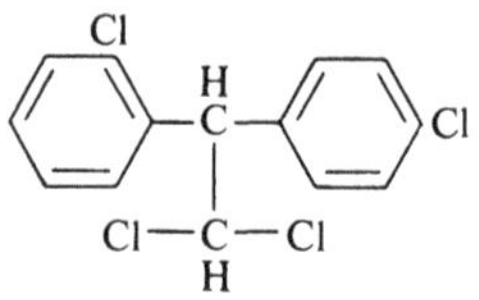

(o, p'-DDD)
Mitotane

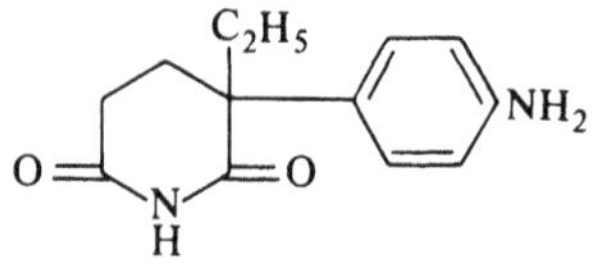

Aminoglutethimide

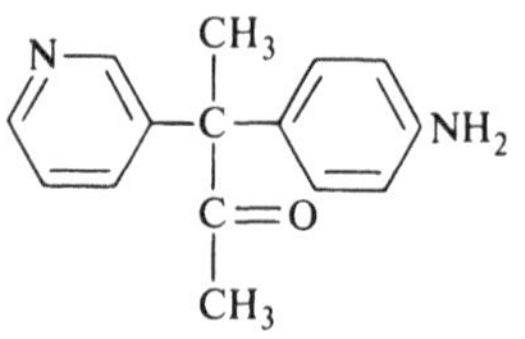

Amphenone

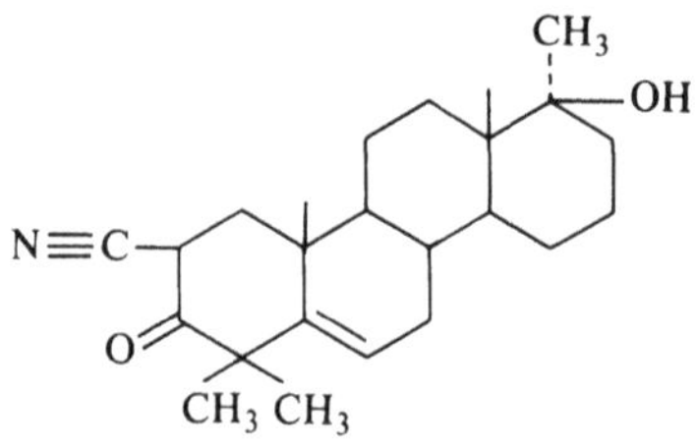

Cyanotrimethylandrostenolone

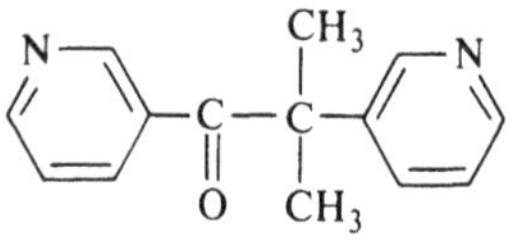

Metyrapone

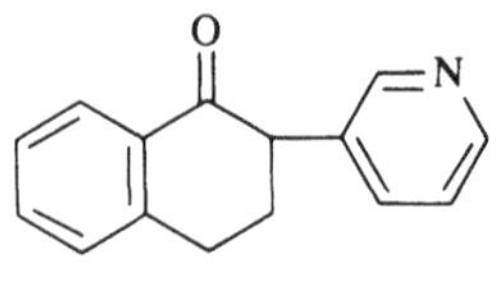

SU-9055

Fig. 6.5. Structures of some adrenostatic drugs.

drome, as its inhibitory effects are compensated for by an increased ACTH secretion (see Zachmann et al., 1977). This effect, which produces adrenal hyperplasia, could be the reason for the masculinizing effect of this drug. The response could be increasing the secretion of adrenal androgens. Aminoglutethimide inhibits the conversion of cholesterol to 20α-hydroxycholesterol, so that the former substrate accumulates in the adrenal cortex. The adrenal hyperplasia can be prevented by the administration of dexamethasone or cortisol, which inhibits the release of ACTH (Santen, Lipton, and Kendall, 1974, Santen et al., 1977).

Amphenone B and metyrapone are chemically related to DDD (see Figure 6.5) and were specifically synthesized for their ability to inhibit the adrenal cortex. They probably react with specific cytochrome P-450 components of the hydrolase reac-

Table 6.3. *Actions of adrenostatic drugs*

Drug	Biosynthetic reaction inhibited	Enzyme inhibited
Triparanol	Desmosterol → cholesterol	
Aminoglutethimide	Cholesterol → 20-hydroxycholesterol	20α-Hydroxylase?
Amphenone B	Acts at several sites	11β-, 17α-, and 21-Hydroxylases; 3β-dehydrogenase
Metyrapone	DOC → corticosterone 11-Deoxycortisol → cortisol	11β-Hydroxylase; at higher levels also 18- and 21-hydroxylases
Cyanotrimethylandrostenedione	Pregnenolone → progesterone	3β-Dehydrogenase
SU-9055	Progesterone → 17α-hydroxyprogesterone and corticosterone → aldosterone	17α- and 18-Hydroxylases
Mitotane (o,p'-DDD)	Degeneration of adrenal mitochondria	

tions. Amphenone B acts at several such enzymic sites (Table 6.2). In contrast to mitotane, which has a general cytotoxic action, amphenone B causes an adrenal hyperplasia. This effect is the result of a decline in cortisol secretion, so that ACTH release is stimulated. This drug has many side effects and is not used clinically.

Metyrapone is a chemical derivative of amphenone B but has a more specific action. It inhibits the 11β-hydroxylase reaction so that there is a decline in cortisol secretion but an increase in DOC and 11-deoxycortisol. In high doses it can inhibit 18- and 21-hydroxylases. It also appears to inhibit the conversion of cholesterol to pregnenolone (Carballeira, Fishman, and Jacobi, 1976). Metyrapone is widely used to test adrenocortical function [Section 6.8A(3)]. It is often ineffective as a therapeutic drug for hypercortisolism, as a compensatory adrenal hyperplasia, which is a result of increased ACTH secretion, occurs which negates its effects. It has been used with some success, however (Jeffcoate et al., 1977b).

Spirolactones. It is known that the presence of certain steroids may also inhibit synthesis of the adrenocortical hormones. Such effects appear to be due to a competition, with normal steroid intermediates, for biosynthetic intermediates and biosynthetic sites, especially cytochrome P-450 (see Greiner et al., 1978). It has recently been shown that spirolactones may also exert their effect by inhibiting aldosterone secretion by competing with the normal substrates in the 11β- and 18-hydroxylation reactions (Cheng et al., 1976).

6.5. Mechanisms of action of corticosteroids

The steroid hormones all appear to have a mechanism of action that involves the response of the cell nucleus, which results in the process of nuclear transcription of messenger RNA and the synthesis of new proteins which mediate their effects. The initial evidence for such a mechanism of action was

of a largely pharmacological nature. It generally involved the observations that the responses to the steroid hormones could be inhibited by actinomycin D or puromycin. Actinomycin D inhibits RNA depolymerase and hence mRNA formation, while puromycin blocks the synthesis of new proteins by the ribosomes. Such evidence, however, is equivocal, as such substances have widespread effects on cells and they can inhibit other nuclear and ribosomal processes. Hence, additional information needs to be sought such as the site and nature of the receptors, direct evidence that mRNA and a *de novo* synthesis of proteins are occurring, and the precise nature of the effector process.

A. Nature of the corticosteroid receptors

The original observations on the nature of the steroid hormone receptors were made in studies of the interactions of estradiol-17β and the uterus (see Section 7.4). The processes, however, appear to be remarkably similar for all steroid hormones. The differences probably reflect the particular specificity of each receptor for a particular family of steroids.

Corticosteroid receptors have been isolated from the cytoplasm of a number of effector organs, including liver, lymphoid tissue, kidney, intestine, adipose tissue, and salivary gland. They are identified in whole tissues or homogenates by:

a. Their ability to bind preferentially a particular type of steroid. This is measured by incubating them with a labelled compound, such as [³H]aldosterone and measuring the relative abilities of related, unlabelled, steroids, including antagonists, and the "cold" agonist to displace it (see Section 2.11D.).

b. Such receptors may also display a saturability to the hormone at concentrations that are consistent with levels of the hormone that elicit their responses.

c. The bound steroids can be released by the actions of proteolytic enzymes.

Table 6.4. *Properties of corticosteroid receptors which have been identified in the rat kidney*

	Receptors		
	Type I	Type II	Type III
Optimal steroid	Aldosterone	Dexamethasone	Corticosterone
K_D	5×10^{-10} M (37°C)	5×10^{-9} M (25°C)	3×10^{-9} M (25°C)
Percent relative potency[a]			
Aldosterone	100	20	<1
DOC	85	20	25
Corticosterone	2	40	100
Dexamethasone	2	100	<1

[a] Determined by competitive binding assay against tritiated optimal steroid. The potency of unlabelled optimal steroid is taken as 100 percent.
Source: Feldman, Funder, and Edelman, 1972.

The receptors behave like proteins which can be reproducibly fractionated and characterized by density-gradient centrifugation. They display stereospecific binding to corticosteroid hormones. Two (and possibly three) types of corticosteroid receptors have been identified. In rat kidney tissue, these have been separated on the basis of their relative abilities to bind different types of corticosteroids (Table 6.4). Type I has a high affinity for mineralocorticoids, especially aldosterone, but a low affinity for glucocorticoids. Conversely, type II binds glucocorticoids, such as dexamethasone, most strongly. Type III has a low affinity for both aldosterone and dexamethasone but binds corticosterone and thus behaves like plasma corticosteroid-binding globulin. Types I and II receptors have also been identified in human kidney (Fuller and Funder, 1976). Liver hepatoma cells contain only a single class of cytoplasmic receptors and these appear to be like the type II variety in the kidney (Rousseau et al., 1972). They bind both dexamethasone and aldosterone, but have a much higher affinity for the former, the ratio of the values being equivalent to the ability of each steroid to induce tyrosine aminotransferase.

The mineralocorticoid (type I) and glucocorticoid receptors (type II) exist in several forms. A separate type of receptor has been isolated from the cell nuclei (Fanestil and Edelman, 1966; Swanek, Chu, and Edelman, 1970) which behaves differently from the cytoplasmic receptors and requires the presence of intact DNA for the binding of the steroid to occur.

The sequence of events and changes involving the steroid–receptor interactions and its transfer to the nucleus are summarized in Figure 6.6. The cytoplasmic receptor is thought to be an allosteric protein that exists in two forms, an inactive and an active form which are in equilibrium. The steroids that have a high intrinsic activity interact with the "active" form, while those with a low activity, such as antagonists, interact with the "inactive" type. Depending on the ratio of the two, the equilibrium will shift. Thus, if the active form interacts with steroids, they no longer are involved in the chemical equilibrium, so that a transposition from inactive to active forms will occur. The "active" steroid–receptor complex undergoes a conformational change and in a temperature-dependent reaction enters the nucleus, where it is converted to the 3S form and then to the 4S type, which binds to the chromatin.

This final interaction appears to be proximate to the process initiating genetic transcription and the formation of messenger RNA. The manner by which this process is accomplished is uncertain, but two hypotheses are provided in Figure 6.7.

B. Cellular processes involved in the actions of corticosteroids

1. Aldosterone

Effects on Na (antinatriuretic action) (see Bentley and Scott, 1978). The initial suggestion that a process of genetic transcription may be mediating the antinatriuretic effects of aldosterone in the kidney was made by H. E. Williamson in 1963. It was shown that actinomycin D inhibited the renal antinatriuretic response to aldosterone. These experiments were extended to a "model" preparation, the toad's urinary bladder *in vitro*, where a similar inhibition of aldosterone's effect on transmural Na transport was demonstrated. Puromycin, which inhibits translation by the ribosomes, also reduced the response (Edelman, Bogoroch, and Porter, 1963; Crabbé and De Weer, 1964). A more direct confirmation of a nuclear site of action was the

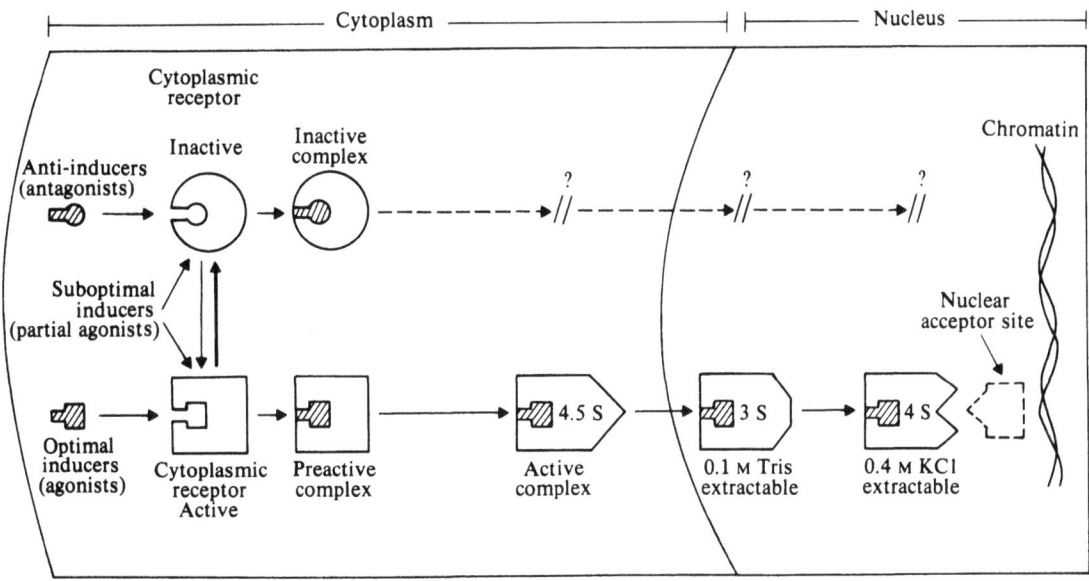

Fig. 6.6. Composite model to describe the steroid hormones (such as an adrenocorticosteroid) interaction with its target cell and its transfer to the site of its ultimate action in the nucleus. For a detailed description, see the text. (From Feldman, Funder, and Edelman, 1972.)

increased incorporation of [³H]uridine into RNA extracted from the epithelial cells and its radioautographic localization in the cell nuclei (Porter, Bogoroch, and Edelman, 1964). An increased incorporation of [³H]leucine into cellular protein has also been observed. Thus, the basic framework of the aldosterone response on the toad's urinary bladder involves a nuclear accumulation of the steroid hormone, a DNA-dependent synthesis of RNA, and the formation of new proteins. Such a mechanism is consistent with the observed delay or "lag" in the response and, most probably, reflects a mechanism common to the action of aldosterone at other sites in the body, including the kidney.

The presence of specific receptors for aldosterone was first described by Sharp, Komack, and Leaf (1966). They showed that [³H]aldosterone was bound to the tissue and that this could be displaced in different degrees by other steroids, the most effective being "cold" aldosterone itself, followed by deoxycorticosterone, cortisol, and spironolactone, which reflects the relative decreasing activities of these compounds in stimulating Na transport. Nuclear-binding sites with similar properties were identified by Ausiello and Sharp (1968). It seems likely that, as observed in the kidney and other steroid hormone-sensitive tissues, cytoplasmic receptors are also present.

The precise nature of the protein(s) induced by aldosterone has been elusive. However, newly synthesized proteins have been isolated from toad

urinary bladder preparations exposed to aldosterone (Benjamin and Singer, 1974; Scott and Sapirstein, 1975). The latter found that they had a molecular weight of 17,000 to 38,000.

The nature of the effect of the induced protein is controversial. Current evidence suggests that Na transport across such epithelial cells may be a two-step process, an influx, possibly passive, along existing electrochemical gradients, across the mucosal, or luminal border of the cells, followed by an extrusion across the inner serosal, or basal side. The latter process is thought to occur against an electrochemical gradient and is synonymous with the Na "pump." Such pumps are associated with the enzyme Na,K-activated ATPase, but its activity in the toad bladder has not been shown to change in the presence of aldosterone (Bonting and Canady, 1964; Hill, Cortas, and Walser, 1973). We are thus left with two (or three) general possible types of action (Figure 6.8).

An *increase in the permeability of the mucosal border of the cell to Na.* It has been suggested that the induced protein may function like a "permease" (Sharp and Leaf, 1964) and increase the permeability of the cell to sodium. The evidence for such an effect was originally largely based on changes in the size of the tissue "Na pool"; an increase was assumed to reflect an accelerated mucosal entry, while a decrease would be expected if an initial stimulation of the "pump" occurred. The validity of this model and hence the assumption based on it have, however, been seriously questioned (Leaf

(a)

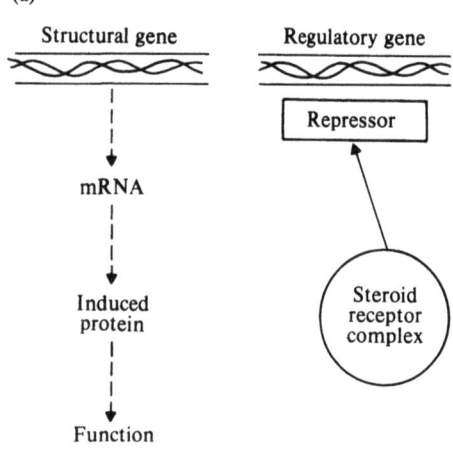

(b)

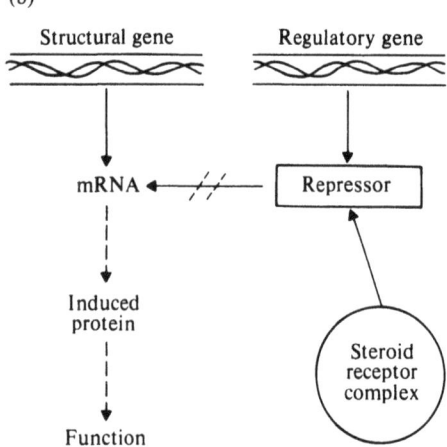

Fig. 6.7. Possible genetic mechanisms by which a steroid hormone–receptor complex may be able to act in the nucleus to influence the induction of proteins that mediate the hormones' effect. (a) One possible mechanism for transcriptional control. In the absence of the hormone, the structural genes coding for steroid-induced protein are not expressed because of the presence of a repressor at the regulatory gene site. With the introduction of the steroid–receptor complex, the repressor is inactivated, resulting in derepression of the genome. As indicated by the dotted lines, DNA transcription and RNA translation then proceed. (b) One possible mechanism for post-transcriptional control. In the absence of the hormone, the genes coding for steroid-induced protein are transcribed into mRNA; however, translation of this mRNA is prevented by a repressor that is synthesized concomitantly. With the introduction of the steroid–receptor complex, the repressor is inactivated. As indicated by the dotted lines, the mRNA can then be translated to yield induced protein. (From Feldman, Funder, and Edelman, 1972; based on Tomkins et al., 1969.)

and MacKnight, 1972), although the site of the effect may still be correct.

A *stimulation of the pump due to an increased supply of a metabolic intermediate(s)* via an action on the tricarboxylic acid cycle *or to a direct stimulating effect* may occur (Edelman, Bogoroch, and Porter, 1963; Fimognari, Porter, and Edelman, 1967).

Amiloride is a diuretic drug which inhibits Na transport across a variety of epithelial membranes, including the toad's urinary bladder (Bentley, 1968; Ehrlich and Crabbé, 1968). It appears to act on the mucosal side of the cell to block Na channels which are thought to be present there. Cuthbert and Shum (1976) have noted that amiloride acts on many tissues which are also responsive to aldosterone: the distal renal tubule, the colon, salivary gland, as well as amphibian skin and urinary bladder. They have found that aldosterone increases the number of amiloride-binding sites in toad bladder epithelial cells. This would be consistent with a mucosal site of action for the hormone.

A number of comparable, but not usually such detailed, studies have been made on the kidney (Fimognari, Fanestil, and Edelman, 1967; Fanestil and Edelman, 1966; Funder, Feldman, and Edelman, 1972), while sporadic observations are available on the salivary glands (Funder, Feldman, and Edelman, 1972), sweat glands (Dixon and Schwarz, 1969), and colon (Edmonds and Marriott, 1967; Pressley and Funder, 1975). The results tend to confirm the similar nature of the Na response to aldosterone.

Effects on K^+ (kaliuretic response). Actinomycin D inhibits the renal antinatriuretic response to aldosterone but has no effect on the observed increase in urinary K^+ or H^+ loss. This evidence, and other information described earlier, suggests that aldosterone acts on these processes in a fundamentally different manner from that on Na. Studies on perfused renal tubules suggest that this corticosteroid can increase the permeability of the apical side of the lining epithelial cells to K^+ (Wiederholt et al., 1973) but no precise information as to the mechanism involved is available. It has been suggested that metabolites of aldosterone may be mediating the kaliuretic response (Morris and Davis, 1974).

2. Glucocorticoids. The glucocorticoids have diverse effects on different types of tissues, which often exhibit multiple responses and interactions with several hormones. It is therefore not surprising to find that although a basic framework of their mechanisms of action are known, the evidence for the processes is often equivocal and invariably incomplete.

The effector tissues in which the molecular mechanism of glucocorticoid actions have been

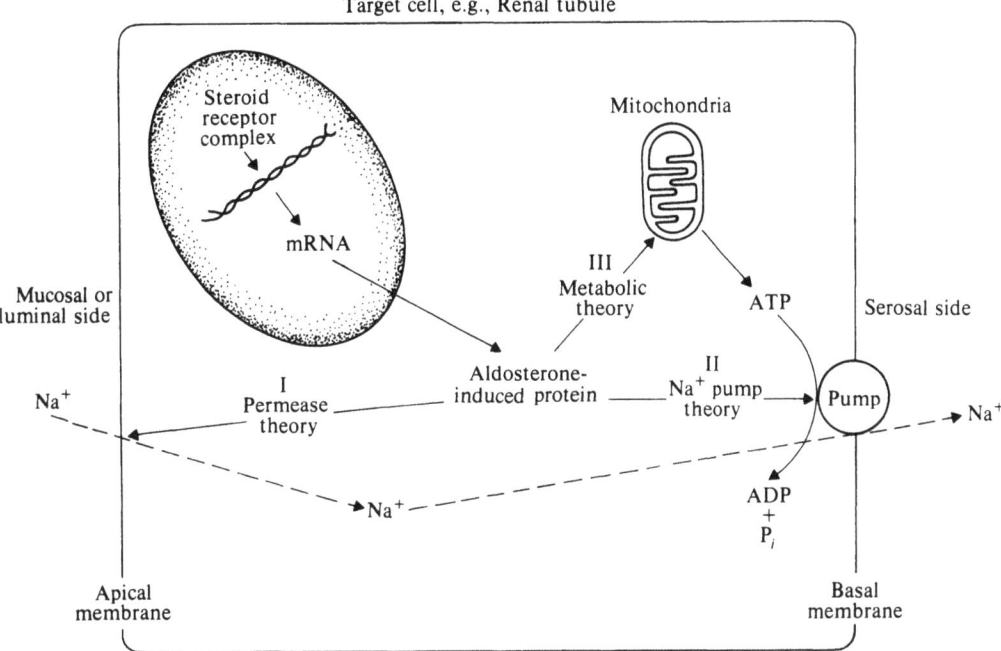

Fig. 6.8. Three possible mechanisms by which a protein induced by aldosterone may act on epithelia, such as the renal tubule, to increase transepithelial Na transport: (I) at the apical Na^+ entry step (*permease theory*), (II) directly on the Na,K-ATPase (*pump theory*), or (III) on the oxidative pathway generating high-energy intermediates to fuel the Na^+ pump (*metabolic theory*). (From Feldman, Funder, and Edelman, 1972.)

studied in some detail include the liver, lymphoid tissues, and fat cells.

Liver. The hepatic responses to glucocorticoids include (see Figure 6.9):

a. *Gluconeogenesis,* a process that involves the deamination and transamination of glucogenic amino acids, as well as fatty acids and glycerol, which are supplied locally by the liver and also by the peripheral skeletal, lymphoid, and adipose tissues.
b. The *synthesis of glycogen,* in which glycogen synthetase plays a determining role, although there is no evidence that this enzyme is directly induced by glucocorticoids.
c. *Protein synthesis,* which not only includes plasma proteins but also the many enzymes necessary for the metabolic functioning of the liver.

Elucidation of the glucocorticoid mechanisms in the liver are complicated, as these steroids can simultaneously influence several related and unrelated processes. In addition, this organ is the site of catabolism of steroid hormones, and the plasma proteins that can bind them are also formed there.

Experiments on liver tissues can be carried out *in vivo* or *in vitro,* and an especially favored tool is a line of hepatoma cells (HTC) which are grown and studied in tissue culture.

Proteins that specifically bind glucocorticoids

have been identified in the cytosol, and these receptor–hormone complexes are concentrated in the nucleus, where they appear to stimulate the formation of messenger RNA (see, e.g., Litwack et al., 1973; Higgins et al., 1973; Feigelson et al., 1975; Simons, et al., 1976; Wrange and Gustafsson, 1978). The activities of a host of enzymes that influence amino acid, glycolytic, and urea cycle metabolism can be increased by glucocorticoids. However, most of these only appear after 2 or more days, and so are probably secondary parts of the responses. Two amino acid-metabolizing enzymes appear within a few hours of exposure to such steroids, at a time which precedes the onset of gluconeogenesis. These are *tyrosine aminotransferase* and *tryptophan pyrolase* (or tryptophan oxygenase) (Lee, Reel, and Kenney, 1970; Feigelson et al., 1975). The possibility that such enzyme synthesis can be controlled by post-transcriptional effects of glucocorticoids has been considered but not substantiated (Kenney et al., 1973; Feigelson et al., 1975). The process appears to involve the formation of messenger RNA specifically coded for these particular enzymes. Other enzymes may also be involved, of course.

The functional steps in the glucocorticoid response in liver cells are summarized in Figure 6.10. After entering the cell, the steroid is bound to a

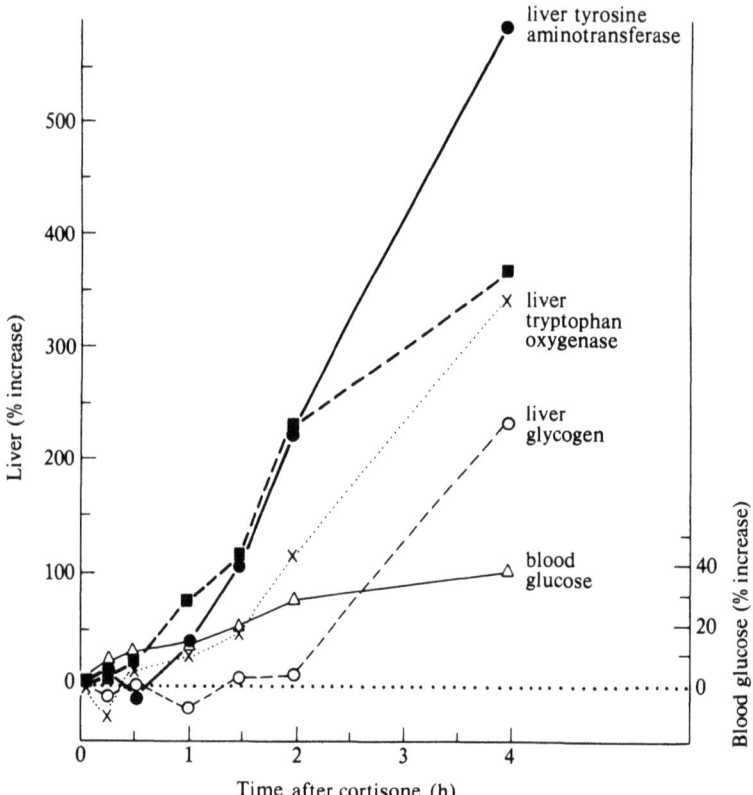

Fig. 6.9. Some responses of the liver to the administration of cortisone in rats. (From Feigelson and Feigelson, 1966.)

cytoplasmic receptor, which is a protein that then undergoes a conformational change, allowing it to enter the nucleus. It then combines with a nuclear receptor (or acceptor). Messenger RNA is transcribed and the enzyme(s), including tyrosine aminotransferase, is induced, which catabolizes the amino acid(s) and leads to glucose synthesis.

Lymphoid tissues. Glucocorticoids stimulate the destruction of lymphoid tissues. This lympholytic effect results from an inhibition of protein synthesis, the final result being lysis of the cells and the release of amino acids.

The mechanism of this catabolic effect of glucocorticoids has been investigated principally by A. Munck (Munck, 1971; Munck and Young, 1975). The lymphoid tissues that have been studied, usually *in vitro*, include thymus cells (or thymocytes), lymphocytes, and cells from lymphoid tumors. The lympholytic effects of glucocorticoids are associated with several metabolic changes. *A decreased uptake of glucose occurs.* The levels of this energy substrate inside the cell become very low despite high external concentra-

tions. This effect is thought to reflect an inhibition of the glucose transport system in the cell membrane. *Amino acid loss is increased* while uptake is decreased. There is a *decreased synthesis of DNA, RNA, and protein synthesis.* The RNA polymerase levels drop.

There are two possible mechanisms by which glucocorticoids could be exerting these effects. The observed decrease in RNA and protein synthesis and the lysis, which occurs after about 1 hour, could be reflecting a *direct* inhibitory effect of the steroids on these processes. However, on the basis of the mechanism of action of steroid hormones in other tissues, where they invariably *stimulate* RNA formation and protein synthesis, such a process would be somewhat unexpected.

Alternatively, the changes that are observed could be *secondary*, gross manifestations of an earlier response of the more conventional type. The evidence presented by Munck suggests that this latter alternative is correct, although all the details are not yet clear. The catabolic effect may thus be a manifestation of an earlier anabolic one.

The proposed mechanism of action on thymus

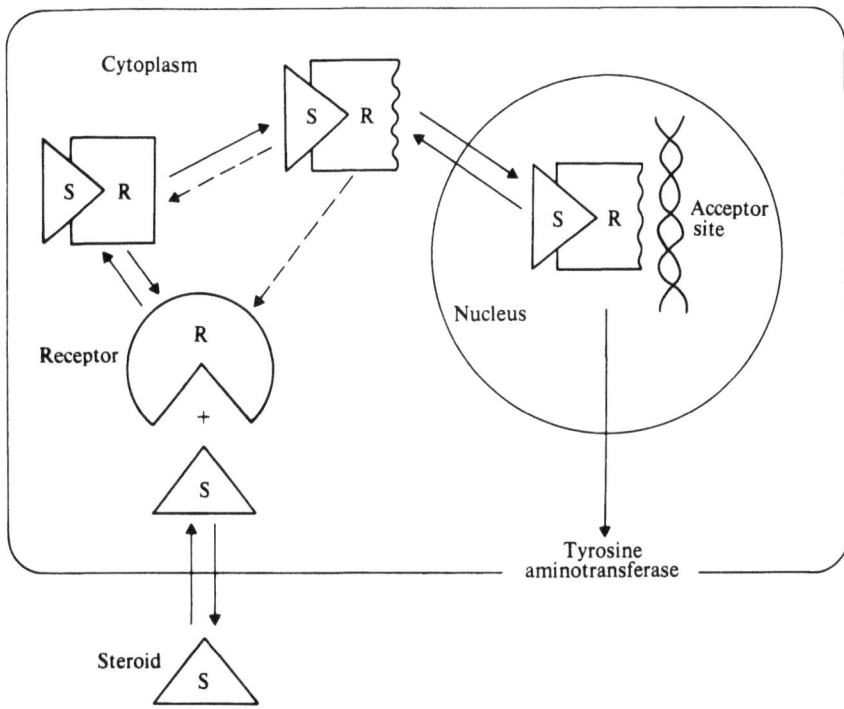

Fig. 6.10. Schematic description of the mechanism of action of a glucocorticoid on the liver cell. (From Higgins et al, 1973.)

cells is summarized in Figure 6.11. Cytoplasmic protein receptors have been identified (Turnell et al., 1974). The steroid–receptor complex enters the nucleus, where it initiates transcription of messenger RNA. This process, which can be inhibited by actinomycin D, occurs within the first 10 to 15 minutes' exposure to the glucocorticoid. After about 20 minutes there is a decline in glucose transport into the cell, and this response is the first to be observed. Munck has suggested that this primary effect is due to the inhibitory action of an induced protein. The glucose response can be inhibited by cycloheximide, which is consistent with this proposal. The mechanism by which the lack of available glucose leads to the catabolic part of the response, the decreased RNA and protein synthesis, in the lymphoid cells is not clear but may be due to a lack of glucose-dependent ATP. Munck proposed that this type of mechanism is also seen in skin and fat cells but probably not in skeletal muscle.

Fat cells. The ability of glucocorticoids to mobilize fatty acids from adipose tissue is a complex effect that is influenced by the presence of other hormones, especially insulin. Normally, insulin antagonizes its action so that it is seen most prominently in diabetic animals. Epinephrine and glucagon increase lipolysis via an activation of adenyl cyclase which results in the formation of cyclic AMP. This nucleotide activates a protein kinase responsible for the conversion of a lipase into its active form.

Thus, there are several possible mechanisms by which glucocorticoids could promote lipolysis (Fain and Czech, 1975; Livingston and Lockwood, 1975). As in lymphoid tissue, the initial response of adipose tissue to such steroids is an inhibition of their glucose uptake (Figure 6.12). It is thought that this response contributes to lipolysis by preventing reesterification of fatty acid, because of a lack of α-glycerophosphate. Glucocorticoids also appear to promote lipolysis more directly, but the precise nature of this effect is unknown. It apparently does not involve changes in cyclic AMP levels but may involve an increased sensitivity of the lipolytic process to stimulation. The response of adipose tissue to glucocorticoids shows a definite "lag" period and it can be blocked by inhibitors of RNA and protein synthesis, suggesting that like their other effects, a process of genetic transcription is probably involved. Specific glucocorticoid

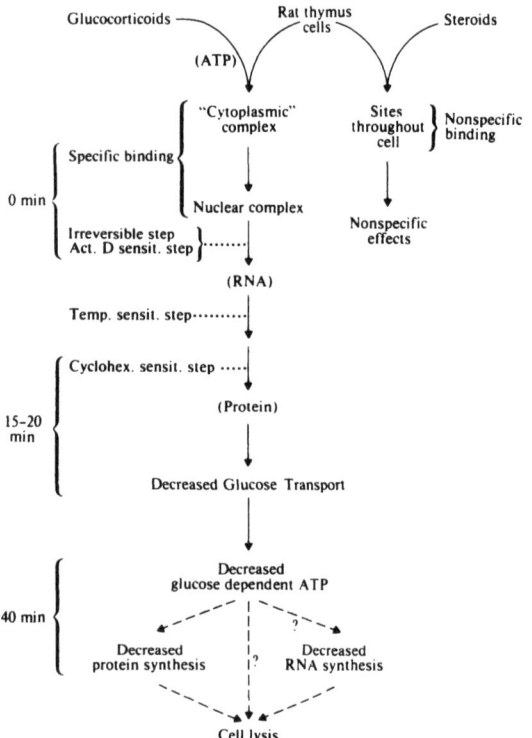

Fig. 6.11. Principal events that have been identified in the lympholytic effects of glucocorticoids on rat thymus cells *in vitro.* The approximate time sequence of the events is shown on the left side of the diagram. For a detailed description, see the text. (From Munck, 1971. Copyright 1971 by The University of Chicago Press.)

receptors have been identified in epididymal fat pad tissue from rats (Feldman and Loose, 1977). The K_D for dexamethasone was about 6×10^{-9} M.

Mechanism of antiinflammatory effects. The mechanism of the antiinflammatory actions of corticosteroids is unknown but has been the subject of continuing investigation and speculation. Inflammation can have a variety of internal and external causes: physical, mechanical, chemical, immunological, and microbiological. There are several prominent effects on the tissue, including vasodilatation, increased capillary permeability and edema, stimulation of sensory nerves resulting in pain, and ultimately, the initiation of tissue repair. These effects may be initiated by a variety of locally released substances, including kinins, histamine, 5-hydroxytryptamine (serotonin), certain enzymes (such as lipases, hydrolases, and proteases), and prostaglandins. An integral part of the inflammatory response includes the migration of white blood cells, especially phagocytic monocytes and granulocytes, to the site of the damage, and even-

tually an increased activity of fibroblast cells. There are thus numerous possible loci where corticosteroids may act to reduce the inflammatory response. They are usually, however, effective in all forms of inflammation, suggesting that they exert a quite basic type of action. Whether or not this involves a single site is not clear, although a unitary hypothesis certainly has its attractions.

Studies on structure–activity relationships of different corticosteroids indicates that the antiinflammatory effects exhibit a parallelism to glucocorticoid-type effects. This observation suggests that they may share a similar type of mechanism, but it is disturbing to recall that relatively high, nonphysiological concentrations of corticosteroids are needed to produce antiinflammatory actions.

The emphasis of recent research into the mechanism of the antiinflammatory effects of corticosteroids has tended to involve the role of prostaglandins. However, earlier hypotheses suggested other mechanisms, some of which could also involve prostaglandins.

The mechanisms include:

a. Stabilization of lysosomal membranes, thus inhibiting the release of their enzyme contents (Weissman and Lewis, 1964).

b. Inhibition of the migration of white blood cells, especially mononuclear phagocytes, to the sites of tissue damage (Thompson and van Furth, 1970).

c. A reduction of the adherent properties (or "stickiness") of granulocytes to the damaged vascular endothelium (McGregor, Spagnuolo, and Lentnek, 1974).

d. An inhibition of the activity of fibroblast cells (Pratt and Aronow, 1966). Glucocorticoid-type receptors have recently been characterized in such cells (Aronow, 1978).

e. Stimulation of the formation of a peptide by monocytes and macrophages, which, possibly via a stimulation of formation of cyclic AMP, may promote the migration of polymorphs away from the site of injury (Stevenson, 1977).

f. Miscellaneous effects, such as an inhibition of the formation of kinins and the antigen–antibody reaction, have also been suggested, but these do not offer the possibility of an action on a common pathway.

There is clear evidence that prostaglandins may be involved in the tissue inflammatory response (see Vane, 1976). Following the demonstration that nonsteroidal antiinflammatory drugs, such as aspirin, phenylbutazone, and indomethacin, act by inhibiting the formation of prostaglandins (Vane, 1971), there was speculation that corticosteroids may act in a comparable manner. The evidence has been contentious. However, in some model tissue systems corticosteroids have been shown to interfere with the metabolism of prostaglandins. This may involve the production of the prostaglandins (Tashjian et al., 1975; Krantowitz et al., 1975). This

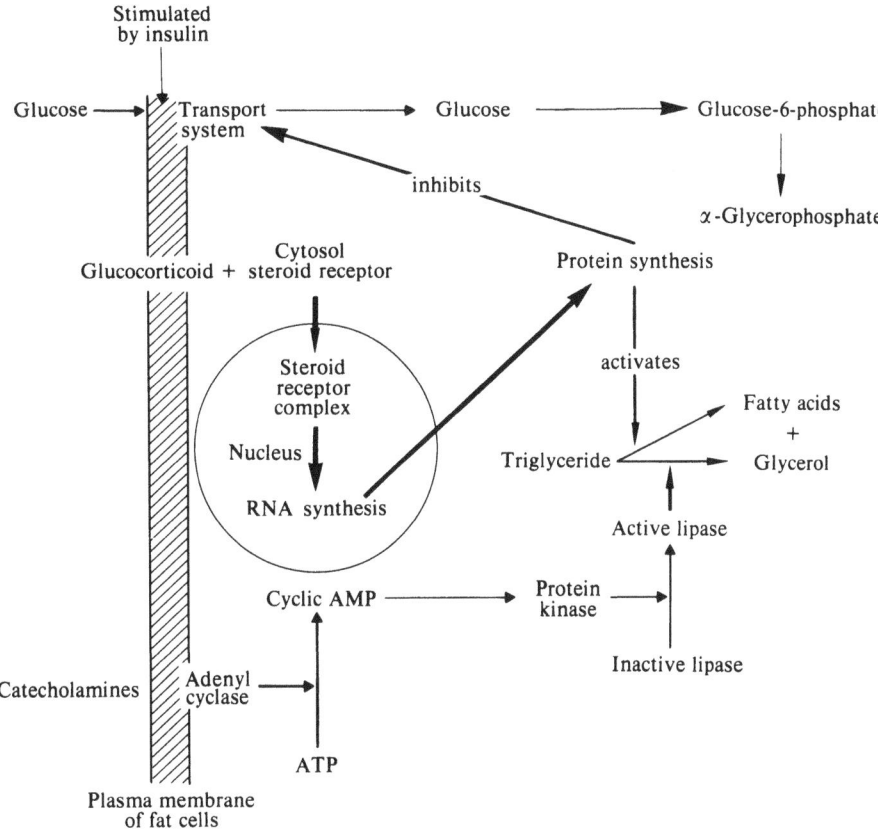

Fig. 6.12. Model for the action of glucocorticoids on adipose tissue and its relationship to the effects of insulin and catecholamines. The glucocorticoids are postulated to act through a mechanism that is dependent on RNA and protein synthesis. Proteins made under the influence of the hormone activate lipolysis and inhibit glucose uptake. Inhibition of glucose transport reduces the amount of α-glycerophosphate available for fatty acid reesterification. Thus, glucocorticoids increase net fatty acid release by inhibiting reesterification and enhance lipolysis through unelucidated mechanisms. (J. N. Fain, personal communication; based on Fain and Czech, 1975).

effect appears to differ from that of the nonsteroidal antiinflammatory drugs, as it involves and interference with the supply of the substrate arachidonic acid rather than inhibition of the prostaglandin synthetase system (Gryglewski, 1976; Floman and Zor, 1976; Hong and Levine, 1976). The actions of different corticosteroids on the production of prostaglandins *in vitro* is paralleled by their relative antiinflammatory potencies *in vivo* (Tam, Hong, and Levine, 1977). Phospholipase A_2 mobilizes arachidonic acid from cell membrane phospholipids, and corticosteroids may inhibit the action of this enzyme (Hammatström et al., 1977; Chandrabose et al., 1978).

Consensus about the manner of the antiinflammatory action of corticosteroids has not yet been achieved. It is, for instance, also considered likely that corticosteroids block the release or transport (rather than the production) of the prostaglandins

out of the cells (Gryglewski, 1976; Chang, Lewis, and Piper, 1977). It is also possible that they may inhibit the response to the prostaglandins by direct competition for tissue sites (De Asua et al., 1977).

The locus of the antiinflammatory effects of corticosteroids in the cell is unknown. It is not even clear whether or not they have to enter the cells or act directly on its surface. In 1961, Willmer proposed that steroidal hormones can directly interact with the plasma membrane, and the relatively high concentration of corticosteroids that are needed to initiate an antiinflammatory response would be consistent with such a physicochemical interaction. However, steroids can initiate a variety of metabolic effects in cells, and these are classically mediated by receptors in the cytosol and nuclear "acceptor" sites. That this well-known type of effect, due to an interaction with intracellular sites, plays some role thus seems quite likely. In one model system, the rat renal papilla, it has been shown that cortisol inhibits prostaglandin synthesis, and this effect (like more classical glucocorticoid responses) depends on the initiation of RNA and protein synthesis (Danon and Assouline, 1978).

6.6. Secretion of corticosteroids

The release of the glucocorticoids cortisol and corticosterone from the adrenal cortex involves a generally distinct mechanism from that of aldosterone, which is consistent with their different physiological roles in the body. There is little storage of corticosteroids in the adrenal cortex, so that the process of their secretion or release into the blood is intimately related to their biosynthesis (see Section 5.2). It is, however, uncertain whether simple diffusion or a specific membrane-related process is involved in the extrusion of the steroids from the cell.

A. Glucocorticoids

Corticosteroids are not maintained at uniform levels throughout the day, but they are released episodically at discrete times. There is a circadian rhythm; the concentrations in the plasma and urine start to rise in the morning shortly before waking at 3 to 4 A.M., reach a peak at about 8 A.M., and then gradually decline to a minimum at about midnight. This cycle appears to be related to waking activity, as it is unaffected by light or whether the person is recumbent or not. In diurnal animals such as the laboratory rat, which are active at night, the pattern is reversed. In man, working at night and sleeping in the day results in a similar reversal but one that may take several weeks for completion. Infants do not exhibit such a circadian rhythm in glucocorticoid release until they reach the age of 2 to 3 years.

Sudden, often massive, secretion of glucocorticoid hormones can occur in response to a variety of nonspecific "stressing" stimuli. Injury, pain, violent exercise, and infection, as well as psychogenic stimuli such as fright and emotion, can all have such an effect.

In 1930, P. E. Smith showed that removal of the pituitary resulted in a degeneration of the adrenal cortex and that the administration of pituitary extracts could antagonize this effect, suggesting the presence of a trophic hormone. This is now known to be the polypeptide hormone corticotropin (or ACTH), which is secreted by the adenohypopysis (see Section 3.2). The normal physiological control of glucocorticoid synthesis and secretion appears to be totally under the control of the pituitary due to the release of corticotropin. In its absence, both the circadian rhythm and the response to stress are abolished.

Corticotropin stimulates the release and biosyn-

thesis of cortisol, corticosterone, and androgens but has little effect on aldosterone. It also stimulates the growth and maturation of the adrenal cortex. In the absence of the pituitary, or corticotropin secretion, the weight of the gland declines dramatically. The tissue degeneration is, however, largely confined to the zona reticularis and zona fasciculata, which produce the androgens and glucocorticoids.

As described in Section 3.1C, the adenohypophysis is connected by a blood portal system to the median eminence and hypothalamus. The release of ACTH is promoted by a releasing hormone, corticotropin-releasing hormone (or factor), which appears to be a neurosecretory product of the hypothalamus. Its precise chemical identity is still unknown, but it appears to be a peptide [see Section 3.2E(3)]. The neurons that secrete CRF receive information via nervous pathways from diverse, but as yet not precisely defined sites through the autonomic nervous system and the brain, including its "higher" centers (see Redgate, 1976).

The normal release of ACTH is controlled by a negative-feedback system involving glucocorticoids. Increased circulating levels of these hormones or their administered analogues, such as dexamethasone, inhibit release of ACTH. Stimuli related to stress can, however, overcome this inhibition. The site or sites of the negative-feedback inhibition are still equivocal. The results of *in vitro* experiments indicate that the steroids act directly on the adenohypophysis. Other information, including the effects of implantation of small pellets of steroids into the region of the median eminence, suggests that they also act in this part of the brain. These results have, however, been criticized, owing to the difficulties in precisely localizing the implanted hormone's distribution. The conclusions may nevertheless be correct.

1. Role of corticotropin (ACTH). The pituitary adrenocorticotropic hormone is a linear polypeptide containing 39 amino acids. The sequence is shown in Figure 6.13. The animal hormones differ slightly from the human hormone. Porcine ACTH has leucine instead of serine at position 31, while the bovine hormone has glutamine replacing glutamic acid at the 33 position. These single amino acid substitutions are sufficient to make the animal hormone immunogenic when administered to man, a property of practical significance, as they are used therapeutically.

Corticotropin may exist in more than one form. Early observations on this hormone suggested that it was much larger and had a molecular weight of about 20,000 (the Li–Sayers ACTH), while more recent immunochemical observations have shown that an even larger form called "big ACTH"

Human:

H-Ser -Tyr -Ser -Met -Glu -His -Phe -Arg -Trp -Gly-
  1   2   3   4   5   6   7   8   9   10

Lys -Pro -Val -Gly -Lys -Lys -Arg -Arg -Pro -Val-
11   12   13   14   15   16   17   18   19   20

Lys -Val -Tyr -Pro -Asn -Gly -Ala -Glu -Asp -Glu-
21   22   23   24   25   26   27   28   29   30

Ser -Ala -Glu -Ala -Phe -Pro -Leu -Glu -Phe -OH
31   32   33   34   35   36   37   38   39

Porcine:

- Asn -Gly-Ala-Glu-Asp-Glu- Leu -
  25   31

Bovine:

- Asn -Gly-Ala-Glu-Asp-Glu-Ser-Ala- Gln
  25   33

Ovine:

-Asp -Gly -Ala -Glu -Asp -Glu -Ser -Ala -Gln
  25   26   27   28   29   30   31   32   33

Fig. 6.13. Amino acid sequence of corticotropin (ACTH). The differences between the structure of the human hormone and those from animals is shown. It can be seen that these differences occur in the 25–33 segments; the different amino acids are underlined.

(Yalow and Berson, 1973) exists in plasma and pituitary and tumor extracts. These molecules are glycoproteins (Eipper, Mains, and Guenzi, 1976). Normal ACTH can be split off big ACTH by trypsin. The functional significance of these forms of the hormone are not clear, but they probably represent precursors (see Section 3.2C). ACTH-like material is also secreted by nonpituitary tumors which can be present in a variety of tissues. The human placenta also contains ACTH-like activity, and when incubated *in vitro* the trophoblast cells appear to be able to synthesize this peptide (Liotta et al., 1977). The precise chemical identity(s) of the ACTH from these ectopic sources is unknown, but it behaves like pituitary ACTH and in many instances it has been shown to stimulate the adrenal cortex to produce cortisol. It can also be a pathological cause of Cushing's syndrome.

Assay. Corticotropin has a number of effects, some of which have been used to quantitatively standardize its activity. Its most prominent effect is to stimulate the secretion of glucocorticoids and androgens from the adrenal cortex. This steroidogenic response has been used both *in vivo* and *in vitro* as a bioassay procedure. ACTH also causes a depletion of adrenocortical ascorbic acid, and this test (the Sayers assay), using hypophysectomized rats, has been used very widely. It has a lipolytic action and in large doses stimulates melanin synthesis in melanocytes and the dispersal of

this pigment in the melanophores in the skin of frogs and toads, and so causes their skin to darken. The latter response has also been used to characterize the biological activities of different analogues of the hormone. Radioimmunoassay procedures are now available for measuring ACTH concentrations, and these are sensitive enough to allow almost routine measurements of its levels in the plasma.

Terminology. The naming of the different natural and synthetic corticotropin molecules and their fragments is unfortunately not uniform and so can be unnecessarily confusing. Initially, more than one type of molecule that displayed corticotropic activity was isolated from the pituitaries of several species. Li proposed that the first identified in a particular species be called α-ACTH (or α-corticotropin), the second β-ACTH, and so on. The species involved was indicated by a subscript: thus, α_h-ACTH was from human pituitaries and α_p-ACTH from porcine glands. The fragments that were then prepared by enzymic breakdown of the parent hormones were identified by adding a superscript that corresponded to the amino acids present, counting from the N-terminus. Thus, α_p^{29-35}-ACTH refers to the C-terminal segment of the porcine hormone and α_h^{1-24}-ACTH the N-terminal segment of human corticotropin. This terminology is still used. However, as pointed out by Hofmann (1974), this system is unnecessarily complex when referring to the N-terminal 1–24 fragments, which are identical in all species. He proposed omitting the prefix and identifying amino acids present (counted from the N-terminus) by a subscript suffix, for example corticotropin$_{1-24}$. An even shorter alternative is ACTH$_{1-24}$. A more recent, simple, uniform, and, it appears, acceptable method is to omit the super- or subscripts and insert the numbering of the amino acids in parentheses at the same level: for example, ACTH-(1–24). Amino acid substitution can be indicated by appropriate numbering in parentheses. Thus, [1-D-serine-17,18,-dilysine]corticotropin (1–18) amide indicates that the N-terminal 1–18 fragment has been substituted with D-serine at position 1 and lysine at both positions 17 and 18, while it has an amide group at its C-terminus. Final judgment on which is the best and most useful terminology must await the formation and decision of an international committee.

Structure–activity relationships. An understanding of the structure–activity relationships of ACTH has several attractions. This hormone is used for therapeutic and diagnostic purposes, but there are several associated problems. As it is a polypeptide, it must be given by injection, and since it is readily destroyed by proteolytic enzymes, its period of action is relatively short. While the animal hormones offer a readily available source of biologically active material, they are difficult to purify and are immunogenic. From the more theoretical standpoint, an appreciation of the roles of certain amino acid configurations in the hormone's biological effects may provide information about its mechanism of action, particularly with respect to hormone–receptor interactions.

An enlightened account of the structure–activity relationships of ACTH has been provided by Hofmann (1974). The activity of the molecule survives considerable amino acid substitution. Many of the positions appear to be relatively unimportant, but a change in some, for instance the arginine at the 8 position, is profoundly important. Such moieties have been dubbed, respectively, "filler" and "functional" amino acid residues. Certain larger segments of the ACTH molecule seem to be of relatively specific importance for certain aspects of its biological activity.

The first 24 amino acids at the N-terminal of ACTH are identical in the hormones from man, pig, ox, and sheep (Figure 6.13). The interspecific differences occur in the 25–33 segment, while the 34–39 are also identical in all those species. It is thus probably not surprising to observe that the entire molecule is not necessary for its biological activity and that the segment containing the first 24 amino acids, at the N-terminus ACTH-(1–24), is highly active (Figure 6.14). The ACTH-(25–39) segment is, on the other hand, inactive.

ACTH-(1–24) has an activity (in the rat adrenal ascorbic acid depletion test) of 300 IU/μmol, compared to about 500 IU/μmol for the intact hormone (on a unit-weight basis they have similar activities). Removal of amino acids from the C-terminal end of the ACTH-(1–24) fragment progressively reduced activity (see Figure 6.14), especially in segments smaller than ACTH-(1–18). These effects appear to result from an increased susceptibility to proteolysis, as the presence of a "protective" C-terminal amide or substitution of lysine for arginine increases the activity of such analogues. Thus, a substituted 17 amino acid fragment of ACTH, [Ala^1,Lys^{17}]ACTH-(1–17)-4-amino-*n*-butylamide, has been shown to have considerable activity in increasing plasma cortisol levels in man (Vierhapper and Waldhäus, 1978). The Lys-Lys-Arg-Arg sequence of amino acids, at positions 15 to 18, are, however, important "attachment" sites (see later) which also undoubtedly influence the activity of such fragments of the hormone. The tridecapeptide ACTH-(1–13) (= α-MSH) and decapeptide ACTH-(1–10) also retain some steroidogenic activity, although the fragments ACTH-(11–24) and ACTH-(7–23) amide are inactive. *In vitro* tests

using the rat adrenal cells show that even ACTH-(4–10) and ACTH-(6–10) are steroidogenic, although very high concentrations are necessary. Replacement of the histidine at position 6, however, abolishes activity. Thus, the segment Hist-Phen-Arg-Trp-Gly (in the 6–10 position) appears to be vital, and Hofmann considers it to be the "active center" (or "active core") of the molecule. This segment is also present in the α- and β-MSH and β-lipotropin, suggesting that it may be of general importance for the manifestation of biological responses. In pharmacological parlance it would appear to contribute to the "intrinsic activity" of such molecules. Others consider the slightly larger fragment ACTH$_{4-10}$ to be the active site or "message sequence" (see Lang et al., 1976).

The active center is very sensitive to amino acid substitutions. Replacement of the arginine at position 8 by lysine nearly abolishes activity, while the adjacent tryptophan is also highly sensitive to manipulation. Substitution of D-isomers for the usual L-amino acids in positions 5–9 abolishes activity, although analogues in which the first pair of amino acids, between the active center and the N-terminus, are changed in this way retain activity. The amino acids in the 5–9 segment of the ACTH molecule may be closely involved in functional aspects of the hormones effects which occur after it is bound to the receptor.

The parts of the ACTH molecule which are adjacent to the active center appear to be important for its binding to the receptor. The Lys-Lys-Arg-Arg sequence (in positions 15–18) is especially important, and masking of the ε-amino groups with formyl drastically reduces its effectiveness.

ACTH-(1–24) does not evoke an allergic response in patients who are sensitized to intact ACTH. While the amino acids in positions 25–39 appear to be unimportant for the hormone's mechanism of action, they act as an immunogenic "tail" to the molecule and it is possible that they may also protect it somewhat from inactivation. Hofmann (Figure 6.15) has thus proposed that the molecule's functions are distributed in three general segments: the 1–10, which contains the active center; the 11–24, which is important for binding to the receptor ("attachment site" or "address sequence"); and the remainder, which is immunogenic and could contribute other functions which are not yet clear.

It has been shown that some of the different effects of ACTH can be dissociated and depend on the structure of the analogue used. Thus, ACTH-(7–24) can stimulate adenyl cyclase activity in fat cell membrane preparations but lacks the usual lipolytic action (Lang et al., 1976). Similarly, differences have been observed in the relative steroidogenic and lipolytic effects of ACTH

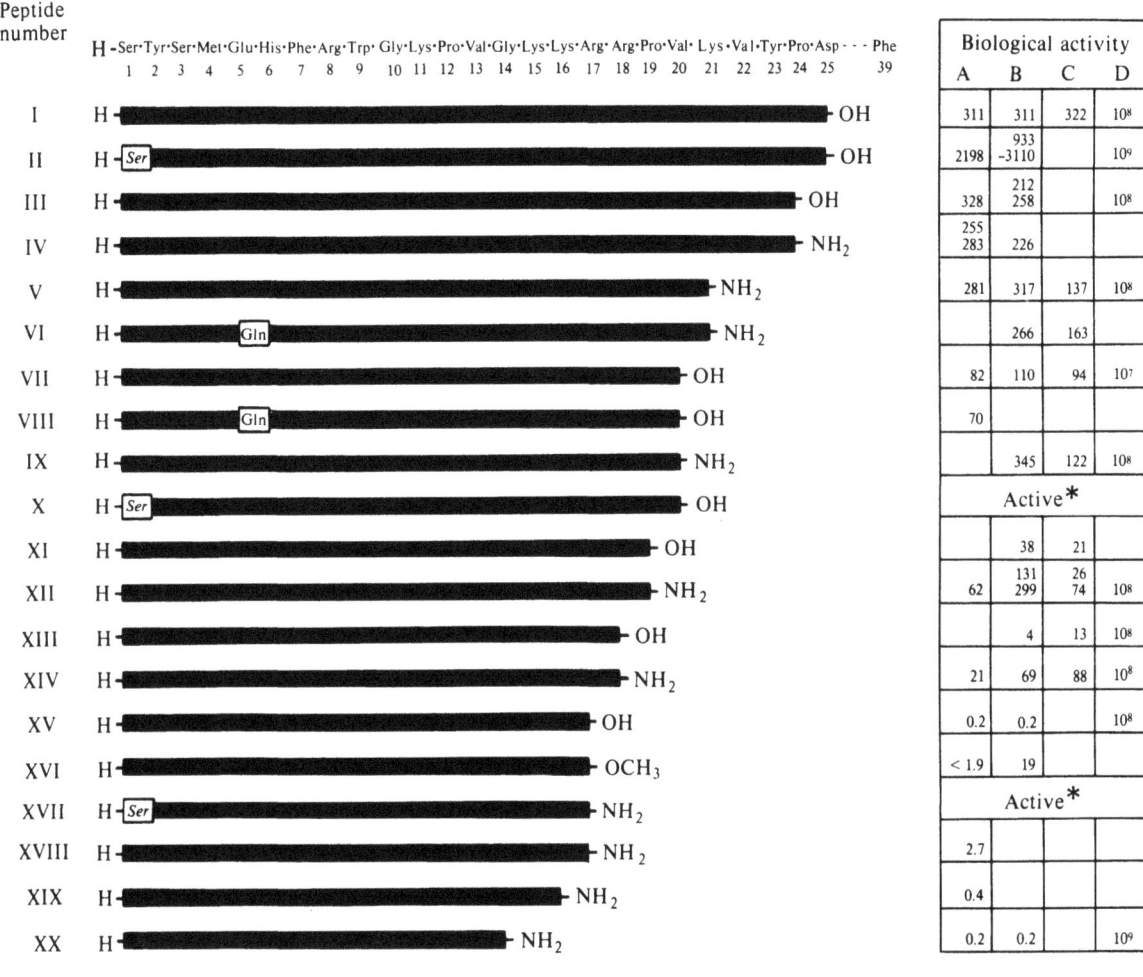

Fig. 6.14. Biological activities of fragments of ACTH. Amino acid substitutions are shown in boxes. D-Amino acid residues appear in italic type. A, adrenal ascorbic acid-depleting activity (units/μmol); B, *in vivo* steroidogenic activity (units/μmol); C, *in vitro* steroidogenic activity (units/μmol); D, *in vitro* melanocyte-stimulating activity (units/g). *Activities not recorded in conventional units. (From Hofmann, 1974.)

analogues in which the arginine is substituted in positions 3 and 5 of ACTH-(3–10) (Jean-Baptiste, Draper, and Rizack, 1977).

A peptide has been identified in human pituitary extracts which can inhibit the steroidogenic action of ACTH on the rat adrenal cortex (Li et al., 1978). This molecule has been called corticotropin-inhibiting peptide, or CIP, and has 32 amino acids, corresponding to α-ACTH-(738). It completely lacks endogenous ACTH steroidogenic agonist activity and could be acting as a competitive antagonist.

The destruction of the active fragments of ACTH can be slowed and its duration of effect prolonged by replacing the L-serine at the N-terminus with D-serine (see, e.g., Jeffcoate et al., 1977a). As described earlier, the C-terminal amides are also generally more effective for the same reason. These observations are of practical significance in the chemical design of therapeutically useful analogues.

Metabolism of ACTH. Estimates of the half-lives of endogenous and exogenous ACTH, including ACTH-(1–24), in man vary from 1 to 20 minutes (see Nicholson et al., 1978). Studies with C^3H_3-

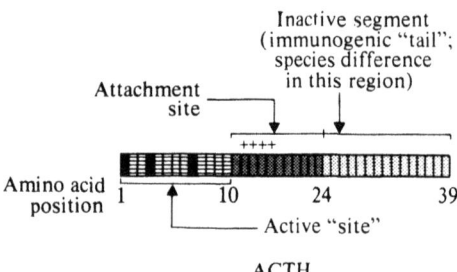

Fig. 6.15. Hofmann (1974) model of the ACTH molecule, showing the three functional parts: the active site, which includes amino acids 1–10; the attachment site, 11–24: and the "inactive immunogenic" tail, 25–39. Each box represent a single amino acid.

methylated ACTH indicate that most of the polypeptide is destroyed in the kidneys and liver, but other tissues may also be involved. Although the biological activity of the ACTH in the plasma declines rapidly, the immunoreactivity is changed much less, suggesting that the initial structural changes which occur in the molecule are quite modest ones.

Preparations of ACTH. Corticotropin is extracted from the pituitary glands of pigs, cattle, and sheep. Highly purified preparations which are suitable for intravenous injection are available. These hormones do, however, differ structurally from human ACTH and are immunogenic. Nevertheless, allergic reactions are rare, although they do occur both locally and systemically and have resulted in fatalities. The prevalence of such hypersensitivity reactions is related to the purity of the preparations used and so not only reflect the specific structural differences in the hormones but also the presence of associated contaminants. As corticotropin is a peptide, it is readily destroyed in the body, so that the duration of its action is short. Depot preparations in which it is adsorbed onto gelatine, carboxymethyl cellulose, or zinc hydroxide are available. When given i.m., absorption is slow, so that they only need to be administered once a day.

The complete human ACTH molecule has been prepared by chemical synthesis, but this is an expensive process and it is not available in sufficient quantities for general use. Synthetic fragments of the molecules are (as described above) also active and easier to prepare. Corticotropin-(1–24) is available for diagnostic tests. It is called *tetracosactrin.* Its action is, however, too brief for general therapeutic purposes, but a preparation in which it is combined with zinc phosphate (Zn-tetracosactrin) is available which when administered i.m. is effective for up to 48 hours. In this

form, however, it is immunogenic. This effect is probably the result of its prolonged exposure to tissue proteins with which it combines. A substituted fragment, [1-D-serine-17,18-dilysine] corticotropin-(1–18) amide, has also been tested. When given i.m. it has 10 times the activity of corticotropin-(1–24). It can also be given subcutaneously, which has the advantage that it can be self-administered and is rarely immunogenic. When given in this way it is effective for 7 to 19 hours (Irvine, Wilson, and Toft, 1973) and so needs to be administered only once a day. It can also be absorbed as snuff (see also Jeffcoate et al., 1977a), but uncertainties as to the size of the absorbed doses do not recommend this route at this time.

Mechanism of the steroidogenic effects of ACTH. The manner by which ACTH stimulates synthesis and release of glucocorticoids has been the subject of intensive investigation. The secretory response is rapid; it takes only about 2 minutes, which places a number of restrictions on its experimental study. The initial interaction of the hormone occurs on the surface of the adrenocortical cells: when it is bound to larger carrier materials, such as polyacrilamide beads, which should prevent its entrance into the cell, the tissue is still stimulated. A rapid change in the properties of the cell membrane occurs following its interaction with the ACTH (Rowlands and Allen-Rowlands, 1978). Changes in the conformation of the membrane proteins could be detected using electron spin resonance (ESR) techniques, and crystalline patches of phospholipids appeared on the surface. These changes could be contributing to the activation of the membrane adenyl cyclase.

Cyclic AMP can mimic the effects of ACTH *in vitro,* and this hormone stimulates the tissue levels of the nucleotide. The changes occur rapidly, owing to activation of adenyl cyclase, and precede the corticosteroid release. Cyclic AMP appears to act as it does at the other sites of its effects in hormone-responsive tissues, by binding to an intracellular protein which results in activation of a protein kinase (Gill, 1972). The latter can phosphorylate proteins, which presumably mediates the effect of ACTH, although the exact nature of this process is unknown. There are, nevertheless, plenty of suggestions about it (see Figure 6.16).

The precise role of cyclic AMP, especially as a sole second messenger in the steroidogenic response, is uncertain (see Halkerston, 1975; Neher and Milani, 1976; Perchellet, Shanker, and Sharma, 1978). Steroidogenesis can be initiated using concentrations of ACTH which are too low to produce a change in tissue cyclic AMP levels. Whether this ability is due to inadequate techniques for measuring small changes in the levels of the

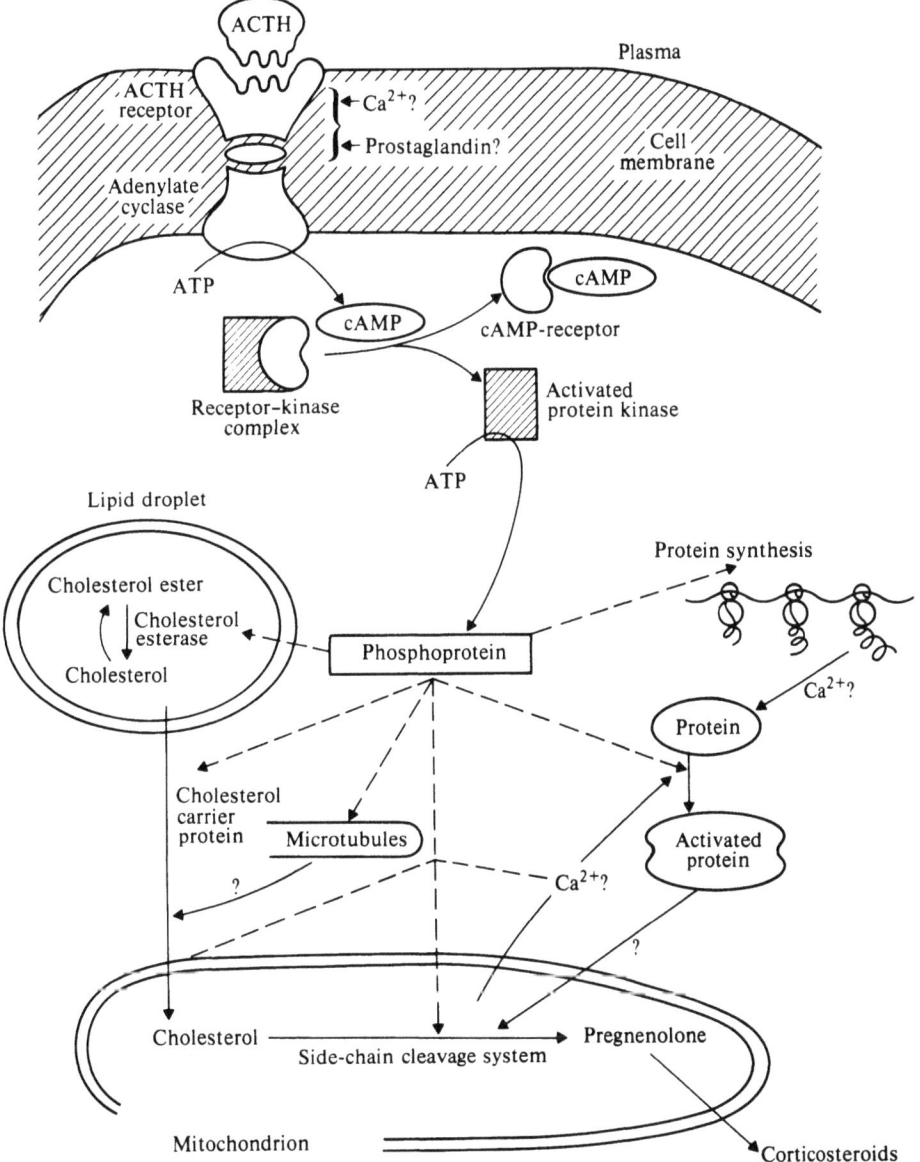

Fig. 6.16. Model provided by Schulster, Burstein, and Cooke (1976) to describe the various ways that ACTH may promote steroidogenesis in adrenocortical cells. The role of the phosphorylated protein, which is thought to be formed in the cells as a result of the activation of a protein kinase, is not known, but some possibilities are shown by the dashed lines. (From Schulster, Burstein, and Cooke, 1976. Reproduced with permission from *Molecular Endocrinology of the Steroid Hormones,* John Wiley & Sons Ltd., copyright 1976.)

nucleotide or whether the cyclic AMP does not have a unique role in initiating steroidogenesis is unknown. Other "messengers" may, however, be involved, including Ca²⁺ (Neher and Milani, 1976) and cyclic GMP (Perchellet, 1978).

Both the initiation of synthesis and the release of corticosteroids by ACTH are dependent on the ele-

vation of free intracellular Ca²⁺ concentrations, and this divalent ion, may, like in many other excitable tissues, act as a messenger and activate the steroidogenic machinery (see Neher and Milani, 1976). The calcium appears to play a permissive role at the level of the mitochondria, where it amplifies the action of the ACTH (Farese and

Prudente, 1978). The increase in Ca^{2+} levels in response to ACTH may reflect a conformational change in the plasma membrane so that external Ca^{2+} enters the cells, and/or a release of the ion could occur from internal storage sites. Further binding of ACTH to its receptors may then initiate an activation of adenyl cyclase (or guanyl cyclase) and a rise in cell cyclic AMP (or GMP), which could also play a secondary role in helping to maintain maximum steroidogenesis. Calcium could also be involved in the coupling of the hormone-receptor response to the adenyl or guanyl cyclase enzymes.

The precise nature of the manner in which steroidogenesis is initiated is also enigmatic. A limiting reaction in the synthesis of corticosteroids is the side-chain cleavage of cholesterol to form pregnenolone, a process that occurs in the mitochondria. A possible increase in the availability of energy substrates and cofactors, such as NADPH, has been considered but discarded (Haynes, 1975). When isolated mitochondrial preparations are incubated under conditions that promote their swelling, such as in the presence of Ca^{2+}, detergents, fatty acids, and proteolytic enzymes, the synthesis of pregnenolone is increased (Koritz, 1968). This observation suggested that the permeability of the mitochondrial membrane may be important, as increase could either promote an uptake of cholesterol or facilitate the loss of the pregnenolone, the product, which may exert a feedback inhibition of its further synthesis. Such an ingenious scheme has, however, insufficient evidence for its total acceptance. Cholesterol is stored mainly in an esterified form in the liposomes and is released by the action of the enzyme cholesterol esterase. It has been suggested that one of the actions of a newly phosphorylated protein may be to activate this enzyme (Trzeciak and Boyd, 1973). The released cholesterol is believed to be transported to the mitochondria bound to a carrier protein. The movement of this complex across the mitochondrial membrane could also be facilitated as a result of the action of the phosphorylated protein. However, the possible sites of its action even at this early stage of hormone synthesis are numerous, and the complete unequivocal description of these sites is not yet possible.

A summary of the possible mechanisms of action of ACTH have been provided by Schulster and his collaborators (Figure 6.16).

B. Aldosterone

1. Role of the renin–angiotensin system. Aldosterone also shows a daily rhythm in its plasma concentration. It rises in the morning, but this change is not seen when the person stays recumbent. It thus appears to depend on postural

changes, which can influence the blood flow to various organs. Thus, the blood supply to the liver is decreased on standing, which could indirectly increase circulating aldosterone levels due to a decline in its catabolism. Assuming an erect posture, however, has a direct effect via renin release (see Section 11.5). Aldosterone levels also change in response to a variety of stimuli, including dietary sodium depletion and an elevation of the plasma K levels. These responses are consistent with the hormone's physiological role. Dramatic changes in aldosterone concentrations also occur in disease or can be induced by experimental procedures. Acute hemorrhage and constriction of the inferior vena cava or the renal artery elicit release. Increased plasma concentrations are seen in congestive heart failure, renal artery stenosis, the nephrotic syndrome, cirrhosis of the liver and ascites, and in some patients suffering from hypertension. Diuretic drugs increase the release of aldosterone, and as these drugs are commonly used in many of the above mentioned diseases, it is sometimes difficult to identify the primary response.

The hormonal role of aldosterone was discovered much later than the other corticosteroid hormones, so that information about the mechanism that controls its secretion was available much later. The importance of ACTH in regulating adrenocortical function had already been recognized, although the relatively independent integrity of the zona glomerulosa had already been observed. In the absence of ACTH, aldosterone secretion usually declines somewhat but not nearly as dramatically as that of cortisol, which is virtually abolished. Optimum secretion of aldosterone appears to depend on ACTH, but its role is more of a "supportive" nature. Chronic administration of ACTH results in an initial rise in aldosterone secretion, but this returns to normal in a few days. Other factors are clearly involved.

In 1960, J. O. Davis (Davis, 1961) found that extracts of kidney tissue when injected into dogs brought about an increase in aldosterone release. He called this substance adrenal-stimulating hormone (or ASH). It was also found that while acute hemorrhage is a potent stimulus for release of aldosterone, this effect is not seen following removal of the kidneys. Further evidence that a bloodborne "hormone" was involved came from crosscirculation experiments in dogs and sheep (see Davis and Freeman, 1976). When blood from a donor animal that had been stimulated by hemorrhage was perfused through the adrenals of a normal recipient, it promoted the release of aldosterone from the latter.

An important clue to the nature of this "hormone" was provided by Laragh and his collaborators (1960). They found that the i.v. injection of angiotensin II could bring about aldosterone release in man. When injected into the adrenal artery, this peptide was very effective. It produced these effects at dosages so low that the potent hypertensive effects of angiotensin II were absent. The possibility that the ASH from the kidney was synonymous with renin was clear and proved to be correct.

The renin–angiotensin system thus exerts a controlling effect on aldosterone secretion; it has no effect on the release of glucocorticoids. A detailed description of the origins and actions of renin and angiotensin are given elsewhere (Chapter 11). It is relevant, however, to observe that the renin content of the kidney increases during salt depletion and declines when mineralocorticoids or salt are administered.

The release of renin is a complex process that can involve several types of stimuli and receptors, which are discussed in Section 11.5.

2. Effects of increases in K and declines in Na. Several other factors apart from the renin–angiotensin system contribute to controlling the secretion of aldosterone. The most notable of these is an increased K concentration, an effect that was first observed by Laragh and Stoerk (1957) in man but which is also seen *in vitro*. It is thought that this change results in an elevation of K levels in the cells of the zona glomerulosa. Experimentally, Cs^+, Rb^+, and NH_4^+ have a similar effect (Bartter et al., 1964). The action of the K^+ may be mediated by increases in intracellular Ca^{2+} levels (Mackie, Warren, and Simpson, 1978). A decline in Na concentration also appears to exert a direct stimulating effect on the adrenal cortex, and may increase sensitivity to angiotensin. 5-Hydroxytryptamine (Müller, 1971) is also effective but appears to be only active *in vitro*, so that its physiological significance is obscure.

3. Mechanism of stimulation of aldosterone secretion. Increased secretion of aldosterone involves an acceleration of its biosynthesis (see Peach, 1977). The effects of angiotensin II, elevated K, and serotonin (as well as ACTH) appear to involve an early step in this synthesis, the conversion of cholesterol to pregnenolone. When other radioactively labelled substrates are supplied, there is usually no change in the incorporation of the label into aldosterone. A single exception exists: [³H]corticosterone is converted more rapidly to the hormone in the presence of high K^+ concentrations. Thus, the effects of these trophic substances on the zona glomerulosa involve both an early and also a late stage in the hormone synthesis (see McKenna et al., 1978). Ca^{2+} is necessary for the response, *in vitro*, and as with other effects

of angiotensin, it may play a primary role in the effect. Cyclic AMP is not involved in the actions of angiotensin II or potassium on the zona glomerulosa but the effect of ACTH there appears to be similar to its action on the zona fasciculata, where it activates adenyl cyclase.

Angiotensin II can also exert a trophic effect (see Section 11.3B) on the zona glomerulosa cells. It increases both cell proliferation (Gill, Ill, and Simonian, 1977) and the number of its own receptor sites (Hauger, Aguilera, and Catt, 1978).

6.7. Metabolism and transport

In man, injected cortisol has a half-life of about 90 minutes in the circulation while that of aldosterone is 15 to 35 minutes. The metabolism and plasma binding of the corticosteroids are described in detail in Section 5.4.

Metabolism occurs principally in the liver and mainly involves hydroxylation of the C-3 and C-20 keto groups, reduction of the Δ^4 double bond in the A-ring, cleavage of the side chain on ring D, and conjugation, usually with glucuronide, at the C-3 position. The metabolites are mainly excreted in the urine. It is possible that some of these, such as 5α-dihydroaldosterone, may retain significant biological activity (Sekihara, Island, and Liddle, 1978). As will be described later, many synthetic analogues of the corticosteroids are not as readily metabolized, so that they are mainly excreted in the urine.

About 94 percent of cortisol and 60 percent of aldosterone in the circulation is bound to the plasma proteins. Cortisol binding globulin (CBG) provides a site with a high affinity but low capacity, while plasma albumin has a high capacity but low affinity. The levels of CBG rise during pregnancy and may be influenced by disease. They are increased by oral contraceptives and estrogens. These changes can influence the half-lives and responses to corticosteroids.

The rate of clearance of cortisol from the circulation is increased by about 20 percent in human patients being treated with the anticonvulsant drug diphenylhydantoin (Choi et al., 1971). This effect appears to be due to the induction of liver enzymes concerned with the hormones metabolism. Binding to plasma proteins is not influenced by the drug. Normal plasma levels of cortisol are maintained so that the turnover rate of the steroid is increased. The synthetic glucocorticoid dexamethasone is not metabolized in the same way as cortisol and undergoes a 6-hydroxylation reaction in the liver. The clearance of injected dexamethasone increases by 140 percent in patients being treated with diphenylhydantoin (Haque et al., 1972). Phenobarbital and aminoglutethimide have comparable effects on the metabolism of such synthetic glucocorticoids (Santen, Lipton, and Kendall, 1974). Aminoglutethimide, however, does not influence the metabolism of cortisol. This difference forms the basis for the use of cortisol, rather than dexamethasone, to suppress endogenous ACTH levels in patients undergoing "medical adrenalectomy" for the treatment of breast cancer [Section 7.6K(1); Santen et al., 1977].

6.8. Corticosteroids and disease

Corticosteroids enjoy a more ubiquitous use than any other group of drugs (see Labhart, 1974; Myles and Daly, 1974; Melby, 1977; Swartz and Dluhy, 1978). They were originally classified among the "wonder drugs," a verdict that has subsequently received some reconsideration, especially in view of their frequent and often alarming side effects. The therapeutic uses of corticosteroids fall into two general categories: (a) *specific* diseases involving the adrenal cortex, and (b) *nonspecific* use in a wide variety of diseases, the causes of which are not directly related to adrenocortical function. The latter constitutes the greatest and most widespread use of these steroids. It is also the most controversial application and most commonly results in side effects which basically reflect a physiological overdosage phenomenon.

The specific adrenocortical related uses of the corticosteroid drugs involve their use in diseases of hypo- and hyperactivity of the adrenal cortex. This can involve three types of steroids: the glucocorticoids, principally cortisol (or hydrocortisone); aldosterone; and the adrenal androgens. Dysfunction of the adrenal cortex may be direct, a "primary" disorder of the gland itself, or be the "secondary" consequence of external changes that influence its activity.

The nonspecific uses of corticosteroid drugs were first heralded by the demonstration by P. Hench and his collaborators in 1949 that the administration of cortisone brought about a dramatic regression in the symptoms of patients suffering from rheumatoid arthritis. This effect reflects their antiinflammatory properties, which have been subsequently utilized for the treatment of many other diseases. Shortly after their introduction a pervasive feeling appeared to develop which suggested that the corticosteroids may be effective in almost any disease for which, at that time, no effective treatment was available. They thus appear to have been used at one time or another to treat most such diseases, usually with inconclusive or controversial results. Possibly as a result of governmental regulations, the prospect of litigation, and more informed medical education, their usage has now become more circumspect.

A. Abnormalities of adrenocortical function and the role of drugs in their treatment

1. Corticosteroid hormone deficiencies

Primary deficiencies. Addison's disease is the result of a primary deficiency in the functioning of the adrenal cortex. It usually results in lack of all three types of corticosteroids: cortisol, aldosterone, and androgens. Addison's disease usually has a prolonged onset and is not manifested clinically until about 90 percent of the glandular tissue is destroyed.

Its *incidence* is low; in England it has been estimated to be about 0.04 per thousand population. In Thomas Addison's time the principal cause appeared to be tuberculosis of the adrenals, but today this is only apparent in less than one-third of cases. The majority are the result of a cytotoxic response of the tissue, the causes of which are uncertain (idiopathic adrenocortical atrophy) but may involve an autoimmune response to adrenal tissue antibodies. More rarely, it can result from fungal infections and even parasites. Carcinoma can also produce a destruction of the adrenal, although more commonly hyperproduction of steroids results from such tumors.

Other forms of primary adrenocortical insufficiency are becoming more frequent. With readily available corticosteroid replacement preparations, surgical adrenalectomy is performed more commonly, such as in attempts to correct a hyperfunctioning adrenal cortex and mammary gland and prostate carcinoma. Intravascular hemorrhage into the adrenal has been described especially in patients undergoing anticoagulant treatment.

Congenital disorders that result in hypofunction of the adrenal cortex are rare but are of considerable interest, as the effects can often be localized to correspond with the deficiency of a specific enzyme in the biosynthetic pathway. This can occur at several sites (see Figure 6.17). T most common such deficiency is of the 21-hydroxylase enzyme, which converts progesterone or 17α-hydroxyprogesterone to, respectively, 11-deoxycorticosterone or 11-deoxycortisol, so that the formation of cortisol and aldosterone is limited. There is a back buildup of precursors and a hyperplasia of the adrenal cortex which can lead to other deficiencies, such as a virilization due to overproduction of adrenal androgens ["the adrenogenital syndrome"; see Section 6.8A(2)], hypertension, and "salt wasting." The latter is the result of a deficiency of aldosterone and probably also the antagonistic effects of accumulated steroid precursors on the mineralocorticoid response of the kidney. Infants can die within a few days if this condition is not correctly diagnosed. Lack of cor-

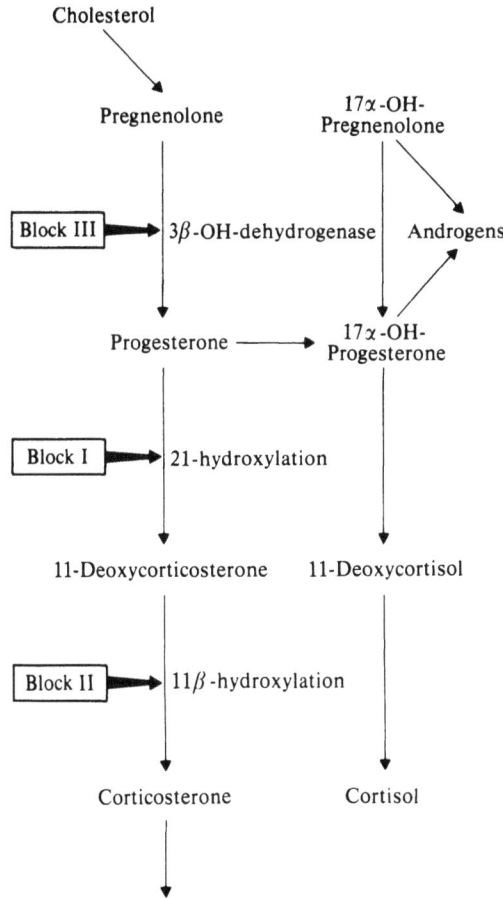

Fig. 6.17. Sites of congenital deficiencies in adrenocortical biosynthetic enzymes which can result in adrenogenital syndromes. The prevention of the synthesis of such steroids as aldosterone and cortisol results in a buildup of metabolic intermediates, so that there is an increased production of androgens. The steroids can result in virilization. (From Schulster, Burstein, and Cooke, 1976. Reproduced with permission from *Molecular Endocrinology of the Steroid Hormones,* John Wiley & Sons Ltd., copyright 1976.)

tisol results in an excessive secretion of ACTH so that a characteristic adrenal hyperplasia (but associated with a hypofunction) occurs.

Primary adrenocortical insufficiency has a variety of important effects in the body, which can be seen suddenly in a "crisis" situation or may develop and persist on a chronic basis over a period of many months or even years. If, however, the disease is untreated, it is inevitably fatal. Fatigue is an early and persistent symptom which appears to reflect several aspects of the deficiency in cortico-

steroids. This is associated with muscular weakness, hypotension, mental changes, and even, rarely, psychosis. Disturbances in intermediary metabolism result in a hypoglycemia, which is compounded by anorexia, which can extend to gastrointestinal disturbances, nausea, and vomiting. There is thus usually a weight loss. Kidney function is also disturbed, and this includes difficulty in excreting a water load due to cortisol deficiency. There is also an excessive urinary Na loss and K retention, leading to hyponatremia and hyperkalemia, which reflect the absence of mineralocorticoids. The hypotension is the result of both a decline in extracellular fluid volume and probably a reduced peripheral vascular tone. In women a lack of androgens can result in a lack of pubic and axial hair and there may be a loss in libido, but this effect is controversial. Probably, the most clear external physical sign of Addison's disease is an increase in the pigmentation of the skin, which results from a high rate of secretion of pituitary ACTH and possibly MSH, due to the lack of negative-feedback control via cortisol.

Secondary insufficiency. This condition is always due to a deficiency of pituitary ACTH. This situation contrasts with a primary adrenocortical insufficiency where there are high circulating levels of ACTH. The consequences are that increases in pigmentation do not occur in secondary adrenocortical insufficiency and aldosterone levels are not seriously compromised. There are two principal causes of this condition:

a. *Hypopituitarism,* which can have diverse causes and can be congenital. Usually, this condition is associated with a deficiency of other pituitary hormones, which can result in dwarfism, hypogonadism, hypothyroidism, and diabetes insipidus (panhypopituitarism). Adrenocortical insufficiency is thus only one aspect of a serious disease. The hypopituitarism may involve inadequacies in the hypothalamic control of the gland.

b. *Pituitary suppression due to corticosteroid therapy,* specifically a depression of ACTH secretion, which deprives the adrenal cortex of its trophic response. Short-term use of such steroids, while acutely decreasing ACTH secretion, does usually result in an atrophy of the gland sufficient to compromise normal corticosteroid secretion. The pituitary–adrenal axis retains sufficient resilience to bounce back rapidly into a normal pattern of activity. If, however, therapy extends to periods of 6 months to 1 year, sudden withdrawal of the administered steroids can result in adrenocortical insufficiency and even an acute "crisis" situation. This condition is reversible, but it may take several months. The particular effect and the time of recovery are related to the dosage used and the period of the treatment. The administration of ACTH can elicit adrenocortical responses in such instances, but as the released cortisol will further inhibit the pituitary, such a course of action makes little therapeutic sense. Withdrawal of corticosteroids thus should be performed gradually, and care needs to be exercised to guard against, or deal with, any stressing situations, such as those associated with infection, accidents, or surgery, that bring about an increased physiological need for the steroid hormones.

More recently, it has been found that if corticosteroids are administered every other day, suppression of the pituitary ACTH is less likely to occur. A "2-day dose" is given in the morning, when physiological secretion is normally high, on alternate days.

Endocrine treatment of adrenocortical insufficiencies. Prior to the identification of the adrenocortical steroid hormones and the chemical syntheses that made them readily available, more than 60 percent of people suffering from Addison's disease died within 1.5 years of its diagnosis. The use of crude adrenocortical extracts plus the provision of additional dietary salt improved this condition somewhat, but it was not until synthetic deoxycorticosterone became available in the late 1930s that prolonged survival was possible. With the subsequent preparation of glucocorticoids, initially cortisone, specific and highly effective treatment became possible. The particular strategy for replacement of the hormones depends on the nature of the insufficiency.

a. *Acute adrenal insufficiency (Addisonian "crisis").* This condition may be the result of the patient not taking his replacement corticosteroids or an exacerbation of the disease due to a sudden stress. It can also be the initial manifestation of a latent or subacute form of the disease. It is important to identify the precise nature of the crisis in treated Addisonian patients, as an overdosage phenomenon in patients being treated with a mineralocorticoid, such as DOCA, may be manifested as fluid retention and high blood pressure, which is opposite to the dehydration and hypotension seen in the absolute steroid deficiency. Thus, while NaCl and fluid replacement are important in the latter condition, they are contraindicated if an overdosage of DOCA has occurred.

Addisonian crisis is a life-threatening emergency necessitating the rapid elevation of corticosteroid levels in the body. Ideally, this is performed by the i.v. infusion of water-soluble preparations such as hydrocortisone (cortisol) hemisuccinate. This steroid contributes both mineralo- and glucocorticoid effects but, if unavailable, preparations such as prednisolone phosphate or hemisuccinate can be used. If i.v. administration is not possible, they can be given i.m. An intramuscular reserve of cortisone acetate is also often given routinely, but the effects of this do not become apparent for some time. The blood pressure can drop to very low levels, a systolic pressure of less than 70 Torr, and may necessitate other supportive measures, such as norepinephrine and plasma or blood transfusion.

b. *Primary adrenocortical insufficiency.* All three types of corticosteroids may be deficient and require replacement. Oral administration of cortisone acetate is usually preferred to the directly acting hormone cortisol, as it is absorbed less rapidly and only has to be administered every 12 hours instead of every 6 hours. This steroid is usually accompanied by a mineralocorticoid, and 9α-fluorocortisol acetate is now usually administered, although DOCA, which is less potent, is also used. Overdosage of the mineralocorticoid can produce fluid retention, and in patients with heart failure it is sometimes omitted. It is unnecessary to use the newer synthetic glucocorticoids, such as prednisolone and prednisone, for replacement therapy, but they are used nevertheless. Such steroids were primarily developed for use in other conditions when adrenocortical deficiency is not present. As they lack significant mineralocorticoid activity, their use may necessitate higher doses of DOCA or 9α-fluorocortisol.

In women it may be necessary to institute androgen replacement, and fluoxymesterone is then used.

c. *Secondary adrenocortical insufficiency.* Although the primary cause of the disorder is a lack of ACTH, this hormone is not used, as it must be administered by injection. Replacement is the same as for the primary deficiency, but as there is not usually a lack of aldosterone, the mineralocorticoid is unnecessary. Androgens may also be administered. The precautions necessary to avoid steroid-induced secondary adrenocortical insufficiency are discussed in Section 6.8D.

The physiological requirements for corticosteroids can vary quite dramatically and, as described earlier, increase during stress, such as that due to trauma, surgery, and infections. The dosages of the corticosteroids thus must be adjusted accordingly to avoid an Addisonian crisis. Patients may be advised to carry suitable identification as to the nature of their condition and even "kits" containing preparations that can be administered parenterally in an emergency. Precautions are taken if surgery is to be performed on such patients. An additional oral dose of cortisone acetate may be given before minor surgery, such as the extraction of a tooth. Intravenous infusions of cortisol hemisuccinate are given during more serious surgical procedures, including adrenalectomy, and this treatment is continued postoperatively.

2. Excess of corticosteroids. The hyperactrivity of the adrenal cortex can involve any of the three types of hormones that it secretes: cortisol, aldosterone, or androgens.

Hypercortisolism. Historically, the first descriptions of the effects of hypersecretion of cortisol by the adrenal cortex were made by Harvey Cushing in 1932 (see Besser and Jeffcoate, 1976; Schteingart, 1978). He ascribed the disease to a pituitary abnormality, although this view was controversial, as many considered it always the result of a primary disorder of the adrenal cortex. With the aid of modern diagnostic procedures and an ability to measure hormone levels in the plasma, it is now known to be mostly due to a hyperplasia of the adrenal which is secondary to an overproduction, or inadequate control, of ACTH release by the pituitary.

There are several types of hypercortisolism:

a. *Primary hypercortisolism,* in which the hypersecretion of cortisol is autonomous and generally independent of ACTH. It is due to an adenoma or, more rarely, a carcinoma of the adrenal cortex. The circulating plasma cortisol inhibits pituitary ACTH release, so that the levels of the trophic hormone are low. The tumor is usually unilateral and the contralateral gland atrophies due to the lack of ACTH. While adrenal adenomas may respond to administered ACTH, a carcinoma does not.

b. *Secondary hypercortisolism* results in a hyperplasia of the gland which is dependent on pituitary or ectopic ACTH secretion. This pituitary condition is the classical "Cushing's syndrome" and is associated with normal to high-normal levels of ACTH in the plasma. The cause of this is uncertain but may in some instances be related to a small adenoma in the adenohypophysis (about 15 percent of cases, "Nelson's syndrome") or a derangement in the mechanism controlling ACTH release, possibly also involving hypothalamic CRF and the transmitters that control it. The ACTH production is not autonomous, however, as it can still be inhibited by *high* doses of the potent glucocorticoid dexamethasone. About 65 percent of cases of Cushing's syndrome are due to excess pituitary ACTH.

ACTH or ACTH-like material can be secreted by ectopic (nonpituitary) tissues. A variety of tumors, especially of the bronchus, thymus, and pancreas, secrete ACTH. The latter is not suppressed by high doses of dexamethasone (although a few exceptions involving "oat cell" carcinoma of the bronchus have been recorded).

c. *Iatrogenic hypercortisolism* is a result of the therapeutic administration of glucocorticoids either due to replacement therapy or their use in nonadrenocortical disorders.

A summary of the nature of the disorders result-

ing in hypercortisolism in relation to plasma levels of ACTH, cortisol secretion (as 17-hydroxycorticosteroids in the urine), and diagnostic-tests is shown in Figure 6.18.

The *incidence* of Cushing's disease is fortunately rare; a variety of statistics suggest it is about 0.05 case per thousand people. The most common cause of hypercortisolism is the therapeutic use of corticosteroids.

When it occurs, Cushing's disease is a very distressing condition, as it results in physical disfigurement. Life expectancy may be from 6 months to several years. Most of the signs are predictable on the basis of the excessive levels of cortisol. Disturbances in carbohydrate metabolism, and an increase in appetite, result in obesity, the lipids tending to be deposited in certain areas such as the face ("moon face") and in the abdominal and truncal region beneath the clavicles, and lower cervical region of the spine ("buffalo hump"). The excessive production of glucose can precipitate diabetes mellitus, often referred to as "steroid" diabetes, which differs from that due to a lack of insulin as peripheral utilization of the substrate is not reduced, so that fat can be readily laid down. There are prominent catabolic effects, including muscle wasting and osteoporosis. The skin also atrophies and becomes thin and stretched, which results in the characteristic red face and the purple striae. Growth in children is retarded. Hypogonadism occurs, which may be manifested as impotence or amenorrhea. Hirsutism occurs in women, and this is probably related to an associated excess of adrenal androgens. Hypertension is common; its cause is uncertain but may be related to Na retention, possibly due to DOC secretion. Mental changes are frequent. Primary hypercortisolism exhibits a similar picture, but virilism is more common and a hypokalemic alkalosis frequently occurs. In Cushing's disease, which is due to an excessive ectopic production of ACTH from nonpituitary tumors, life expectancy is not usually long, so that the full effects of the excess of corticosteroids are not always manifested. It is accompanied by a hypokalemic alkalosis and pigmentation due to the greater amounts of circulating ACTH. Hypercortisolism due to administration of corticosteroids (their side effects) can result in most of the symptoms described for Cushing's syndrome. A notable exception is peptic ulcer, which is not associated with Cushing's syndrome; however, its relationship to corticosteroid therapy has recently been questioned.

Treatment of hypercortisolism is predominantly by irradiation, but where available surgical procedures can be effective. In the instance of a unilateral adenoma or a carcinoma, the adrenal is removed and as the contralateral gland is invariably atrophied, replacement therapy such as with cor-

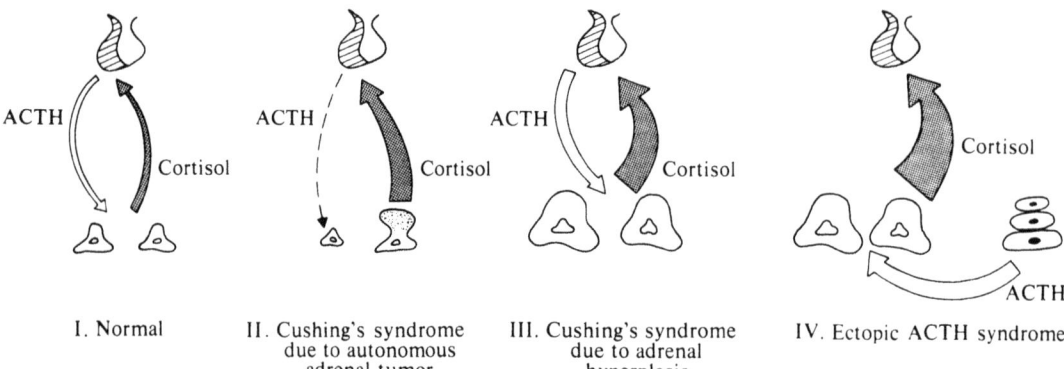

I. Normal   II. Cushing's syndrome due to autonomous adrenal tumor   III. Cushing's syndrome due to adrenal hyperplasia   IV. Ectopic ACTH syndrome

	I. Normal	II. Adrenal tumor	III. Cushing's disease	IV. Ectopic ACTH syndrome
Plasma cortisol	10–25 μg% rhythmic	High, no rhythm	High, no rhythm	High, no rhythm
Plasma ACTH	0.1–0.4 mU%	Low	High	High
Urinary 17-OHCS				
Response to ACTH	3–5 fold rise	+.0	+	+.0
Response to metyrapone	2–4 fold rise	0	+	+.0
Response to dexamethasone	0–3 mg/day	No fall	Partial fall	No fall
Plasma ACTH after adrenalectomy, on normal cortisol	Normal	Low	High	High

Fig. 6.18. Pituitary–adrenal function in some adrenocortical disorders that result in hypercortisolism. The profile of plasma levels of cortisol and ACTH and urinary 17-hydroxycorticosteroids (17-OHCS) in normal conditions and following the administration of exogenous ACTH, dexamethasone, and the metyrapone test. (From Liddle, 1972.)

tisol acetate and 9α-fluorocortisol needs to be instituted to maintain the patient postoperatively. This treatment is gradually withdrawn so that ACTH levels and the function of the remaining adrenal slowly return. In instances of bilateral hyperplasia, in the absence of an obvious adrenal tumor both adrenals may be removed and replacement therapy is given for life.

Chemical treatment of hypercortisolism may be used as an alternative to irradiation, although more usually it is utilized in conjunction with this therapy. Adrenostatic drugs (Section 6.4) may also be administered prior to surgery. The use of such drugs has involved several types of approaches. The release of hypothalamic CRF may be partly under the control of 5-hydroxytryptamine (serotonin), and *cyproheptadine*, which is an antagonist of this hypothalamic neurotransmitter, has been used successfully to control the disease (Krieger, Amorosa, and Linick, 1975). It is, however, not effective in all cases, possibly being most useful in those instances where a specific hypothalamic disorder is involved (Allgrove, Husband, and Brook, 1977; Schteingart, 1978). *Mitotane*

(*o,p'*-DDD) has been used with some success in a trial in the United States (Schteingart, 1978). Therapy was continued for up to 1 year or more. Some side effects, usually anorexia, nausea, and diarrhea, were observed. Cortisol replacement was occasionally necessary. *Metyrapone* has been used in the United Kingdom (Jeffcoate et al., 1977b) and resulted in an improvement of the condition. Treatment was prolonged, for up to 5.5 years. Virilization was observed in some women, but other side effects were rare and not a problem. It should be emphasized that both of these trials involved concurrent radiation treatment.

Aminoglutethimide also has an adrenostatic action (Section 6.4) but is more effective in hypercortisolism due to autonomous adrenal tumors than in adrenal hyperplasia, and its action is short-lived. These effects are probably due to the compensation for its effects by additional ACTH secretion in adrenal hyperplasia (see Zachmann et al., 1977). Aminoglutethimide is also used for "medical adrenalectomy" for the treatment of breast cancer [Section 7.6K(1)].

Adrenal surgery must be carried out under an

effective pre- and postoperative cover of administered corticosteroids utilizing i.v. cortisol hemisuccinate (see Figure 6.20).

Hyperaldosteronism. Primary hyperaldosteronism ("*Conn's syndrome*") was described by J. W. Conn in 1954 (Conn, 1955) and is usually the result of an autonomous adenoma or carcinoma of the adrenal cortex which results in an excessive secretion of aldosterone. The remaining cortical tissue does not hypertrophy. In 20 to 30 percent of cases there is a hyperplasia but no tumor can be identified (idiopathic hyperaldosteronism).

Secondary hyperaldosteronism can occur due to changes in its control mechanism and it has been associated with high levels of circulating renin and angiotensin. Small renin-secreting tumors (Editorial, 1973b) have recently been identified in the kidneys. Release of renin is increased in a variety of diseases, including edematous conditions such as congestive heart failure, cirrhosis of the liver, and the nephrotic syndrome. Sodium depletion, such as that due to the therapeutic use of diuretic drugs, also activates the renin–angiotensin–aldosterone axis.

The *incidence* of hyperaldosteronism is controversial, but in the United States about 1000 cases of Conn's syndrome have been diagnosed. It is, however, in the view of some, more widespread, and it was formerly implicated in many cases of "essential" hypertension. This suggestion has, however, not been substantiated.

Generally, hyperaldosteronism is associated with a hypokalemia, due to excessive urinary K^+ excretion, and a metabolic alkalosis. The latter can result in paresthesias and even tetany. Hypokalemia has widespread effects on many tissues, including skeletal muscle, the heart, and kidney. In the latter it is manifested as a polyuria. Hypertension is invariably present, but this is benign and not progressive, although diastolic pressures as high as 160 Torr have been recorded. This may be related to a high incidence of headache.

When an aldosterone-secreting tumor of the adrenal cortex is identified, it is usually removed surgically. It is in 98 percent of instances unilateral, so that replacement therapy is unnecessary. Specific aldosterone-blocking drugs, called spirolactones (see Section 6.3D) are utilized prior to surgery, especially as this aids the restoration of plasma K levels to normal. If surgery is contraindicated or if the problem appears to be due to a bilateral hyperplasia, spironolactone can be administered on a chronic basis, especially if it is well tolerated. Its use is, however, often associated with gastrointestinal problems, which can be avoided to some extent by their administration at mealtimes. Spironolactone also has some effects on sex hormone metabolism, as it can result in breast tenderness and gynecomastia, loss of libido and impotence in men, and menstrual irregularities in women. The most serious problem encountered in its use is excessive K retention and hyperkalemia, which if undetected in time can have serious results, especially on the heart [Section 6.8B(4)].

If spironolactone cannot be tolerated, an alternative form of drug therapy is to administer the "potassium-sparing" diuretic *amiloride* (Kremer et al., 1977). This drug promotes urinary sodium excretion but reduces potassium loss. Its effects are not related to any direct antagonism of the actions of aldosterone. In a clinical trial involving 19 patients with primary hyperaldosteronism, amiloride resulted in a significant decrease in blood pressure and the return of body sodium and potassium levels toward normal values.

Adrenogenital syndrome (an excess of adrenal androgens). The adrenal cortex normally secretes a small amount of androgens, especially androstenedione and dehydroepiandrosterone sulfate. The androgenic effects of the latter are weak. An excessive secretion of such steroids can have important effects, especially during development of the female fetus and child. This hypersecretion of adrenocortical androgen can be due to congenital hereditary defects in the biosynthesis of the corticosteroids, a general hyperplasia of the adrenal cortex, or the presence of tumors of the adrenal gland.

Congenital hereditary adrenocortical enzyme deficiencies (congenital adrenocortical hyperplasia) are due to a deficiency in the activities of certain enzymically mediated processes of corticosteroid biosynthesis. They most commonly involve the 21-hydroxylase system but also 11β-hydroxylase and 3β-hydroxysteroid dehydrogenase.

The effects depend on the particular site in the biosynthetic pathway (see Figure 6.17) which is affected. Thus, 21-hydroxylase deficiency blocks the formation of 11-deoxycorticosterone and 11-deoxycortisol and the secreted hormones cortisol and aldosterone. In the instance of 11β-hydroxylase deficiency, the two precursors accumulate, but the final synthesis of the hormones is reduced. Changes in the relative quantities of the hormones and accumulations of their known steroid precursors can be used to diagnose the particular site of the enzymic defect.

The low plasma cortisol levels that result from such congenital defects fail to activate the negative-feedback control mechanism for ACTH release, so that its levels rise. A compensatory hyperplasia of the adrenal cortex thus occurs, so that in some instances the gland may be 10 times its normal size. There is rarely a complete absence of cortisol and aldosterone, the reason apparently being that the necessary enzymes, while compromised, are not entirely absent and the hypertrophic response often results in sufficient restoration for reasonable physiological function. The accumulation of high levels of androgens appears to be due to the general hyperplasia and the backup of precursors so that the biosynthetic pathways are diverted.

The adrenogenital syndrome evidently has a genetic basis; it is more common in siblings and familial relationships involving a specific type of defect have been described. Thus, it is not surprising to find that its incidence varies; it is 1 : 500 among Eskimos and a recent survey in Switzerland showed it to be 1 : 5000, but other surveys show a 1 : 62,500 ratio.

The effects of hyperandrogen secretion are more dramatic in the female. Virilization occurs, which starts to affect the infant *in utero*. The internal female sex characters are retained, but the secondary sex characters tend to become male and the clitoris hypertrophies. A condition of pseudohermaphrodism thus may occur. Hair distribution on the body and face becomes like that in the male. In the male, little obvious change may occur, but the penis hypertrophies. The testes, however, remain infantile due to the androgens inhibiting the release of pituitary gonadotropins. There is an acceleration in growth and bone maturation. In the female, amenorrhea commonly results. "Salt wasting" may occur due to an inadequate secretion of mineralocorticoids and probably an associated accumulation of steroid metabolites, such as progesterone, which antagonize their effects on the kidney. Hypertension, probably the result of the action of accumulated 11-deoxycorticosterone, also occurs. The incidence of these particular disorders varies and depends principally on the nature of the enzyme defect.

Related disorders include a 17α-hydroxylase and 20,22-desmolase deficiency, which do not result in virilization. They also affect the gonads, however, so that sexual maturation does not occur. There is an excessive accumulation of DOC, which results in hypertension and hyperkalemia, in the 17α-hydroxylase deficiency. The 21-hydroxylase defect shows a strong virilizing action which may or may not be accompanied by salt wasting. The 11β-hydroxylase deficiency, however, results in an accumulation of DOC, so that salt wasting does not occur but there is hypertension. When 3β-dehydrogenase is compromised, the formation of many steroids is affected, even androgens. Dehydroepiandrosterone is formed, but this steroid only has a weak action so that virilization is slight. Salt loss is severe, however, so that most infants with this condition die within the first few months of life.

The *treatment* of the hereditary adrenogenital syndrome depends on the administration of cortisone. This serves to both inhibit the release of ACTH and so reduce the adrenal hyperplasia and also provides replacement therapy. The dose should be kept as low as possible so as to avoid side effects, especially the reduction of somatic growth in these children. The possibility that the requirements may increase during stress should be considered. If salt wastage remains a problem, DOCA or 9α-fluorocortisol can also be administered. Surgical correction of the altered external genitalia is important in young girls, and this is usually performed within the first 18 months so that they may grow up in the normal way.

The 3β-dehydrogenase or 17α-hydroxylase deficiency has widespread endocrine effects, so that the formation of androgens and estrogens are prevented and these must be replaced at the time of puberty.

Adrenal tumors and hyperplasia. The adrenogenital syndrome can also initially occur ("acquired") in the adult. In these instances it is due to an adrenal tumor or hyperplasia which may be congenital but slow to manifest itself. Such a tumor is usually a carcinoma and can be distinguished from hyperplasia by the ability of cortisone or dexamethasone to suppress the latter, which is ACTH-dependent. The clitoris enlarges but the other genitalia are unaffected, although hirsutism in the male pattern and acne occur. In the male, early puberty occurs and growth is accelerated. The testes, however, do not mature, because of inhibition of gonadotropin release by the adrenal androgens. Treatment is surgical removal of the carcinoma. Metastases can be treated with *o,p'*-DDD, the use of which may, however, require corticosteroid replacement therapy. Hyperplasia of the adrenal cortex can be treated, like the condition in children, with cortisone. Problems may, however, arise with the selection of a dose which, while being effective, does not result in Cushingoid-like effects.

3. Pharmacological tests of adrenocortical function. The diagnosis of the nature of adrenocortical diseases, such as the involvement of the hypothalamic–pituitary–adrenal axis and tumors, utilizes a considerable variety of laboratory tests, including suppressive and provocative tests using drug preparations.

Such procedures basically depend on the identification and measurement of plasma and urine levels of corticosteroids and their metabolites. The methods vary in their precision with respect to particular steroids and involve chemical, competitive-protein binding, and radioimmunoassay procedures. ACTH can also be measured by radioimmunoassay.

Several methods are used to estimate urine and plasma corticosteroids and their metabolites. In the United States 17-hydroxycorticosteroids (17-OHCS) are often measured by the Porter–Silber methods (in which phenylhydrazine interacts with the steroids which have a 17β,21-dihydroxy-20-ketone side chain). These values give a good estimate of cortisol secretion but also include metabolites such as cortisone and tetrahydrocortisone. An important precursor, 11-deoxycortisol (cortexolone, compound S), is also included. The levels of this compound are usually low but may rise considerably as a result of the administration of metyrapone and afford a basis for the metyrapone test. In England a slightly different procedure for estimating cortisol secretion is used. This method involves a periodate oxidation of 17-hydroxylated steroids, which includes a larger number of steroid hormones still including 11-deoxycortisol. The British method is said to measure *17-ketogenic steroids* (17-KGS, sometimes referred to as "total 17-hydroxycorticosteroids"). For practical purposes the results of the two methods are similar but not strictly identical. A more specific method for measuring cortisol involves the use of a fluorometric procedure. This does not include 11-deoxycortisol and so cannot be used during a metyrapone test. Some drugs fluoresce and can interfere with the determination; especially notable ones are the spirolactones, whose effects may persist for many days. Competitive-protein binding and radioimmunoassays also exist for cortisol (and aldosterone). The competitive-protein binding assay also measures 11-deoxycortisol.

Another important determination for the diagnosis of the adrenogenital syndrome is that of 17-ketosteroids (17-KS), which (utilizing the Zimmerman reaction) reflects androgen secretion. In women these steroids are derived entirely from the adrenal cortex, while in men about 40 percent are of testicular origin.

ACTH tests of adrenocortical functions. The functional integrity of the adrenal cortex can be assessed by measuring its ability to respond to injected ACTH (see, e.g., Tyler and West, 1972). The response can be gauged in terms of plasma or urine cortisol or hydroxycorticosteroid (17-OHCS) levels. There are a number of variants of the test which depend on the type of end response being measured and the suspected nature of the deficiency. The ACTH preparation can be injected i.m. or infused intravenously.

The most popular screening test for assessing whether the adrenal cortex is responsive or not involves the i.v. or i.m. injection of ACTH and the measurement of the change in plasma corticosteroids 30 minutes later (Kehlet et al., 1976; Lindholm et al., 1978). Procedures involving the collection of urine take longer. An inadequate response indicates adrenocortical insufficiency but

does not distinguish the cause. If the adrenal cortex is merely suppressed as a result of a low endogenous ACTH secretion or corticosteroid therapy, it can usually be persuaded to respond to longer-term stimulation, such as repeated i.m. injections or i.v. infusion carried out on 2 or 3 successive days. If a response still fails to occur, it is suspected that the primary problem lies in the adrenal gland itself. This test is considered to be as reliable as the provocative insulin hypoglycemia test (see later in this section).

The ACTH preparations originally used were obtained from animals and, apart from having a different amino acid constitution from the human hormone, these preparations often contained protein contaminants. In addition, in order to prolong their effects, the preparations for i.m. injection contained additives, such as gelatine, to slow absorption. Hypersensitivity reactions and local irritation have been recorded, and these may be exacerbated by an associated adrenocortical insufficiency. Prophylaxis against an allergic response often involves the prior administration of 0.5 to 1 mg of dexamethasone. The dosage of this glucocorticoid is too low to affect the corticosteroid measurements. Human ACTH has been synthesized and the segment containing the N-terminal 1–24 amino acids has the full hormonal activity. This pure polypeptide preparation of ACTH is called *tetracosactrin* and is administered i.m. or i.v. for diagnostic purposes.

Metyrapone test of pituitary–adrenal function. Metyrapone is related to amphenone and reduces adrenocortical function. It inhibits 11β-hydroxylase and so prevents the conversion of 11-deoxycortisol (compound S) to cortisol (and 11-deoxycorticosterone to corticosterone). This effect is not its sole effect on adrenocortical enzymes, but it appears to be the one most utilized for the purpose of the diagnostic test. Normally, plasma cortisol levels decline in the presence of metyrapone, but these are replaced by 11-deoxycortisol, which is also measured by the Porter–Silber reaction or the competitive-protein binding assay. When pituitary function is adequate, the decline in plasma cortisol results in a release of ACTH, which if the adrenal is normal will, in turn, produce an increased secretion of 11-deoxycortisol.

The metyrapone test (see Figure 6.19) has several uses:

a. It can be used as a general screening test for adrenocortical insufficiency. A failure of corticosteroid levels to rise in response to metyrapone indicates that function is inadequate but does not provide information about the nature of the deficiency, that is, if the pituitary or adrenal is at fault.

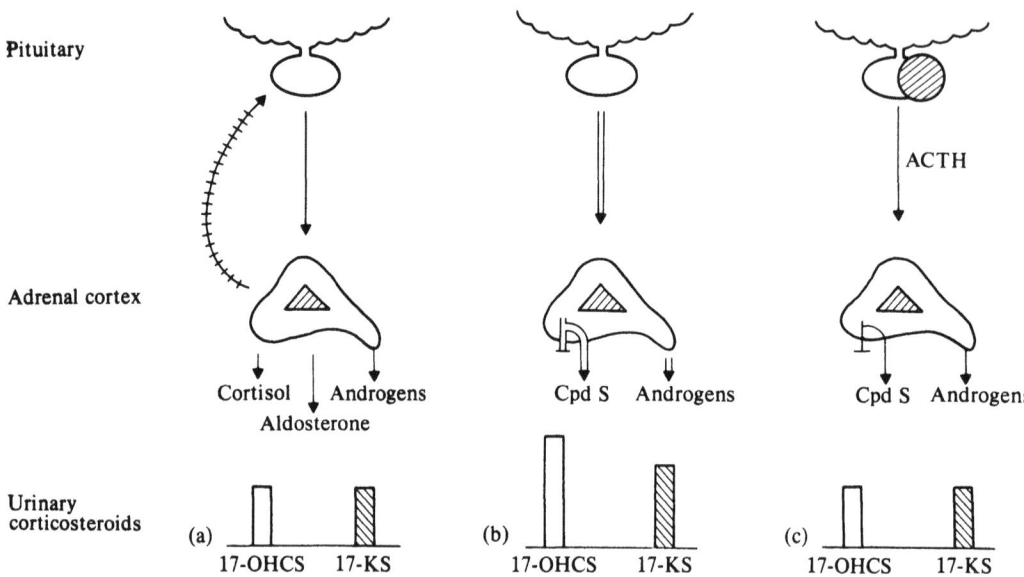

Pituitary

Adrenal cortex

ACTH

Cortisol   Androgens
Aldosterone

Cpd S   Androgens

Cpd S   Androgens

Urinary
corticosteroids   (a)   (b)   (c)

17-OHCS   17-KS   17-OHCS   17-KS   17-OHCS   17-KS

Fig. 6.19. *Metyrapone test* for adrenocortical and pituitary (ACTH) function. (a) Normal person, showing the functioning of the adrenal cortex–pituitary ACTH axis. The urinary excretion of 17-hydroxycorticosteroids (17-OHCS, reflecting cortisol secretion) and 17-ketosteroids (17-KS, reflecting adrenal androgens) is shown. (b) Normal person treated with metyrapone. The blockade of 11-hydroxylase inhibits the formation of cortisol, but the intermediate steroid cortexolone (compound S, CpdS) is formed. The measured urinary 17-OHCS includes the latter steroid. However, cortexolone cannot act on the pituitary to control release of ACTH, which then increases and stimulates the adrenal cortex, so that there is an increased urinary excretion of 17-OHCS (in this instance, compound S) and 17-KS. (c) Patients with partial pituitary deficiency do not respond to metyrapone, as the reserve of ACTH is decreased. (Based on Labhart, 1976.)

b. If other tests indicate that the adrenal cortex itself is normal, then a failure to respond to metyrapone suggests that pituitary function is inadequate, as ACTH secretion cannot be adequately stimulated (see Figure 6.19).

c. This test is also useful for identifying the cause of Cushing's syndrome (Figure 6.18). If an adrenal tumor is present, the pituitary is chronically suppressed by the continual cortisol release, and metyrapone fails to influence ACTH release. On the other hand, if the disease is the result of an adrenal hyperplasia, due to an abnormal pituitary ACTH mechanism, then additional release of this hormone will occur and corticosteroid levels will rise. It will be noted that the test does not specifically diagnose Cushing's syndrome but only helps in the identification of its cause.

Metyrapone is usually given orally as successive doses at 4-hour intervals for 24 hours. It frequently causes nausea and even vomiting. Intravenous administration is also sometimes used. As cortisol secretion is reduced, an Addisonian crisis or insufficiency can be provoked but is uncommon.

Provocation tests of the pituitary–adrenal axis using insulin and vasopressin. The integrity of the hypothalamic–pituitary system controlling the adrenal cortex can be tested in various ways which appear essentially to invoke its response to stress, which normally should induce a substantial rise in urinary corticosteroid levels.

The most frequently used is the *insulin hypoglycemia test.* Hypoglycemia is brought about by an intravenous dose of insulin. The discomforts and hazards of this procedure are obvious but appear to be unavoidable for the "test." Severe untoward reactions such as a hypoglycemic coma are, however, rare.

Vasopressin was initially used in this type of provocative test, as it was thought to be synonymous or closely related to the evasive CRF. Like CRF, it acts directly on the pituitary gland to stimulate release of ACTH (Yasuda et al., 1978). It was thus hoped that the effect of the vasopressin would be quite specific, but this prediction has not been realized. This neurohypophysial peptide (usually as the synthetic lysine-vasopressin) is administered i.m. or i.v. and results in a peripheral vasoconstriction and palor, and frequently abdominal cramps, headache, and other discomforts. As it constricts the coronary vessels, it offers the prospect of even greater potential hazards. If the metyrapone and insulin tests suggest an abnormality and the vasopressin test is normal, it has been suggested that the problem may lie in the hypothalamus.

Dexamethasone suppression of ACTH. This test is used to differentiate the causes of hypercortisolism, whether it is due to an overproduction of ACTH (Cushing's syndrome) or an autonomous adrenocortical tumor. Small oral doses (1 mg) of the potent glucocorticoid dexamethasone depress cortisol secretion in normal people via a direct inhibitory effect on the release of ACTH from the pituitary (see Yasuda et al., 1978). As the concentration of the administered glucocorticoid is so low, it does not interfere with the steroid determinations. In hypercortisolism, no matter what the cause, *low doses* (2 mg/day) fail to have such an effect. *High doses* (8 mg/day) of dexamethasone, however, can "break through" and inhibit ACTH release and so lower cortisol secretion. If the disease is due to a tumor which is not dependent on endogenous ACTH, the administered glucocorticoid will be ineffective. Ectopic sources of ACTH are also usually unresponsive, although there are some exceptions, such as bronchial adenomas. The low dose of dexamethasone can also be used to differentiate the cause, in adrenogenital syndrome, whether it is due to an ACTH-mediated hyperplasia or an autonomous tumor. In this latter test 17-ketosteroid levels are used as indicators of the response.

B. Use of corticosteroids in nonadrenocortical diseases

With the possible exception of antimicrobial agents, the corticosteroids have probably been used more widely in different diseases than any group of drugs in history. As described earlier in the late 1940s, cortisone was shown to have very dramatic effects in the treatment of rheumatoid arthritis, an action that appears to principally reflect its antiinflammatory properties. The dosage used to treat this condition is 5 to 10 times greater than that which is necessary for effective replacement therapy in Addison's disease. It is therefore not especially surprising to observe that the initial optimism about the potential uses of such steroids was considerably dampened when patients given long-term therapy with them invariably developed the symptoms of hypercortisolism or Cushing's syndrome. As described earlier [Section 6.8A(2)], this is a serious situation which in many instances was fatal. Rheumatoid arthritis is a crippling disease but rarely has a direct effect on mortality, so that the choice of whether to use the corticosteroids in this condition became a difficult one. Subsequently, there was a reversion to the use of other antiinflammatory drugs, of which aspirin is

the most widely used. Today, corticosteroids are still used in the treatment of rheumatoid conditions but form a second line of defense for use when other, more conservative, measures are inadequate.

The early qualified success in the use of corticosteroids led to the testing of their actions in a wide range of other related and unrelated diseases. It is probably true to say that they have been tried in most diseases but are only effective in some. Nevertheless, compared to other nonantimicrobial drugs, they have a very wide spectrum of action, which principally reflects their general antiinflammatory properties and also possibly their immunosuppressive actions. As a large number of diseases have inflammatory manifestations and many are the result of autoimmune responses, the widespread use of corticosteroids is not surprising. Such diseases or their symptoms do not reflect inadequate adrenocortical function; the therapeutically effective doses are enormous compared to normal corticosteroid secretion. These drugs do not effect a "cure" but exert symptomatic relief while awaiting the favorable effects of other treatment or a spontaneous remission of the disease.

1. Diseases where corticosteroids are used. Much of the initial enthusiasm for the use of corticosteroids in diseases where there is no known effective treatment has waned. Their use at one time appeared to be virtually indiscriminate, but this situation has been generally corrected due to a wider appreciation of their toxic, and even fatal, effects. If improperly used, corticosteroids are dangerous drugs!

A current list of the uses of corticosteroids is given in Table 6.5. It includes about 20 types of conditions and areas of use, including some in which, as indicated, their efficacy is doubtful. The summary given is not a complete one nor are the verdicts necessarily final. Properly conducted and statistically analyzable programs for testing the effects of such drugs are not often available. It seems likely that in the future more specific information, on which one can base a reasonable judgment, will come to hand, but this is usually a slow process. Comprehensive reviews are presented by Azarnoff (1973) and Myles and Daly (1974).

Musculoskeletal diseases. As described in Section 6.8D, the dramatic effects of cortisone in alleviating the symptoms of *rheumatoid arthritis* were the first demonstration that such steroids may be effective in the treatment of nonadrenocortical diseases. Currently, the use of corticosteroids is reserved for those situations when other forms of therapy are inadequate (Gumpel, 1978). The course of the dis-

Table 6.5. *Summary of the principal nonendocrine uses of corticosteroids*

Musculoskeletal diseases
Rheumatoid arthritis[a]
Rheumatoid fever with carditis[a]?
Osteoarthritis
Systemic lupus erythematosus
Periarteritis nodosa
Cranial ("giant cell") arteritis
Polymyositis and dermatomyositis
Myasthenia gravis[c]

Respiratory diseases
Asthma[a]
Status asthmaticus
Respiratory distress syndrome in infants?[b]
Hay fever[a]
Tuberculosis[a]
Fibrosing alveolitis[c]

Skin diseases
As an antiinflammatory and antiallergic drug

Eye diseases
As an antiinflammatory and antiallergic drug

Gastrointestinal diseases
Ulcerative colitis
Idiopathic steatorrhea[a]
Regional enteritis (Crohn's disease)
Hepatitis, chronic (and acute)[b]

Kidney disease
Nephrotic syndrome in adults,[c] and children

Malignant diseases
Leukemias and lymphomas
Acute lymphocytic leukemia
Chronic lymphocytic leukemia
Hodgkin's disease
Acute and chronic myelocytic leukemia[c]
Carcinoma of the breast and prostate (especially with metastases)[c]

Blood diseases
Hemolytic anemia
Idiopathic thrombocytopenic purpura
Agranulocytosis[b]

Miscellaneous
Hypercalcemia
Postmyocardial infarction syndrome[b]
Organ transplantation (inhibition of rejection reaction)
Anaphylactic shock
Cardiogenic, hermorrhagic, and septic shock[b]

Note: For reference, see Myles and Daly, 1974.
[a] Usually only if other treatment ineffective.
[b] Doubtful efficacy.
[c] Appears to be effective but only in minority of cases.

ease is unchanged by corticosteroids; it may even be worsened, and there are the inevitable side effects. The dosage is thus kept as low as possible and periodic attempts to reduce it are made. Prednisone is a favored preparation.

While there is no evidence to indicate that corticosteroids alter the course of *rheumatic fever,* including the incidence of cardiac lesions, they do exert prominent acute effects in suppressing the manifestations of the disease. They can thus be life-saving, especially in the interim period before treatment with aspirin becomes effective. Some patients cannot tolerate such high doses of aspirin, so that the steroids may offer a reasonable alternative.

A variety of *collagen diseases* (which are probably mainly due to autoimmune responses) respond favorably to treatment with corticosteroids. These include the muscular weakness seen in *polymyositis* and *myasthenia gravis* (Grob, 1976; Dahl, 1976). In the latter condition they are used in cases that are resistant to anticholinesterase drugs. There may be an initial period of increased muscular weakness, although if the dosage of the steroids is increased gradually on alternate days, this problem is reduced (Seybold and Drachman, 1974). *Periarteritis nodosa* is a rare disease that affects blood vessels, and corticosteroids are the only effective treatment. They prolong survival and give symptomatic relief. In arteritis of the ophthalmic artery (*cranial "giant cell" arteritis*), corticosteroid treatment may prevent blindness. Some of the manifestations of *systemic lupus erythematosus* respond favorably to corticosteroids; these include articular and dermatological complications as well as renal and central nervous sytem involvement.

Respiratory diseases. Chronic asthma (see Editorial 1975a) is usually a childhood disease. Corticosteroids are effective in preventing this allergic response, but their side effects, especially inhibition of growth, make it desirable first to exhaust all other forms of therapy, such as β-adrenergic drugs and Na cromoglycate. If it should be necessary, the use of bronchial aerosol sprays of *beclomethasone dipropionate* are available (see Figure 6.20), which minimize the possibility of systemic absorption and side effects.

Status asthmaticus is a medical emergency and treatment includes very large doses of corticosteroids (Franklin, 1974). Bronchial sprays are obviously of no use in this situation, as the airways are not sufficiently patent.

Hay fever can also be prevented by corticosteroids systemically or sometimes as nasal sprays (Mygind, 1973), but they are usually only used in extreme situations.

Premature infants may have an inability to produce phospholipids and other substances that normally line the lung and have a surfactant effect.

Some surveys (Editorial, 1976a) have shown that the administration of corticosteroids to the mother 24 hours before birth may decrease infant deaths from this *respiratory distress syndrome.*

Fibrosing alveolitis is a collagen disease that affects the lungs, and in some instances corticosteroids may dramatically reduce the fibrosis.

Corticosteroids were at one time used in conjunction with antitubercular drugs in the treatment of *tuberculosis.* Clinical trials showed that only when a pleural effusion was present was there any improvement. The use of corticosteroids in patients who have formerly had tuberculosis exposes them to the risk of reactivation of the disease.

Skin diseases. The corticosteroids revolutionized the practice of dermatology, as it became possible to produce a prompt remission of the effects of many skin diseases, including eczema, dermatitis, and psoriasis. These responses utilize the steroids antiinflammatory effects, and as with all such actions provide symptomatic relief which can be associated with the usual side effects and the masking of infection (Editorial, 1973b, 1977a, 1977b). The corticosteroids thus need to be used, as always, with care and prognostic circumspection in treating skin conditions. They can be used systemically to treat dermatological conditions but are more often applied locally. They can be absorbed in sufficient quantities from such surface sites to produce their classic side effects, including Cushing's syndrome (Staughton and August, 1975). Special related problems may arise as corticosteroids have been shown on occasions to produce an allergic contact dermatitis. In addition, when applied chronically to certain areas, such as on the face, for prolonged periods they can result in a thinning of the skin which when the steroid is withdrawn results in red rosacea-like eruptions. This can be treated by the substitution of a weaker corticosteroid such as hydrocortisone or lower concentrations of the synthetic analogues (Cunliffe, 1976), which are gradually withdrawn.

Adverse reactions resulting from systemic absorption through the skin have been described in workers involved in the manufacture of glucocorticoids (Newton et al., 1978). Despite precautions to limit contact with the drugs, facial plethora developed in many, and a depression of adrenocortical function was observed.

The indiscriminate use of corticosteroids for the treatment of skin diseases has been deplored (Hunter, 1973), as it can make a correct diagnosis difficult and has "become a cloak for diagnostic ignorance."

Eye diseases. The eye is composed of very delicate tissues whose functions include the ability to transmit light. Widespread or persistent inflam-matory reactions (Levine and Leopold, 1973; Editorial, 1975b) and a disorganization of the processes of repair can readily result in changes in these tissues so that they lose their structural integrity and can no longer function and transmit light properly. Deficiencies in vision and even blindness can result. The corticosteroids thus play a very important role in the treatment of many ocular diseases and disorganizations of the processes of repair. These are generally of the type that are *not* caused by infection and include general inflammatory and allergic responses. They are also important in inhibiting the formation of scar tissue during wound healing. The latter effect is especially useful when the wounds are extensive, such as in chemical or thermal burns of the cornea. They should not, however, be used in treating minor corneal abrasions. Particular care must be taken in using corticosteroids in ocular infections and then only in the presence of an effective antibiotic. They are strongly contraindicated in herpes simplex. Failure to take these precautions has led to permanent damage to the cornea and blindness. Corticosteroids are of no use in degenerative diseases of the eye.

Gastrointestinal diseases. The antiinflammatory effects of corticosteroids are effectively utilized in some intestinal diseases.

In *ulcerative colitis* there is an inflammation of the mucosal cells lining the colon and rectum. This can lead to bleeding and perforation and the disease has a significant mortality. Corticosteroids are used to control the initial acute attacks and can produce a remission in a large number of cases. They, however, do not effect a cure and their use is probably not warranted in long-term treatment of the condition (Wall, 1973). If possible, local application, such as suppositories and retention enemas, may be used. Systemic absorption of the steroids is, however, still usually significant, so that the classic side effects are commonly observed. Corticosteroids provide symptomatic relief in *regional enteritis* (Crohn's disease). They are also used in the treatment of *idiopathic steatorrhea* (celiac sprue). About 70 percent of cases respond favorably to the withdrawal of gluten from the diet and the steroids are only used in the remaining unresponsive patients, and they are not effective in all of these.

Liver diseases. The corticosteroids have been widely used in the treatment of chronic and acute liver disease. These habits appear to have been perpetuated from the early indiscriminate use of corticosteroids, especially in serious diseases where there appeared little that could be done. Despite their extensive use in hepatic diseases, reliable statistics about possible beneficial effects are rare. Two controlled studies (Cook, Mulligan, and Sherlock, 1971; Juhl et al., 1974) showed that corticosteroid treatment reduced mortality in nonalcoholic chronic active hepatitis. Such treatment appears to be more effective in women than in men. Corticosteroids are also used in acute emergency (*not* chronic) situations in patients who are severely ill with alcoholic hepatitis. They are contraindicated in the presence of ascites and other types of liver disease, as the rate of survival appears to be decreased.

Kidney disease. The use of corticosteroids has increased the 5-year survival time of children suffering from the nephrotic syndrome disease by about 70 percent. The response is not as favorable in adults and most do not respond. They are ineffective in acute or chronic glomerulonephritis.

Malignant diseases. Following the initial observed "miraculous" effects of corticosteroids in the treatment of rheumatoid arthritis, they were soon utilized for the treatment of various malignant neoplasms, with mixed success (see De Vita and Schein, 1973). The aims are twofold: to secure a remission of the disease as a result of an inhibition of the activity of the tumorous tissue, and to treat associated problems, such as hypercalcemia, cerebral edema, and thrombocytopenia. When used for the treatment of such conditions, large amounts of the corticosteroids are usually given in divided doses (alternate-day therapy is *not* used).

The lympholytic effects of corticosteroids are utilized in the treatment of the leukemias and lymphomas. The steroids are seldom used alone but in combination with cytotoxic drugs, including vincristine, methotrexate, and 6-mercaptopurine (Cline, 1973). In acute lymphocytic leukemia, such treatment initially results in a high (>70 percent) rate of remission, although subsequently this declines. They are also effective in chronic lymphocytic leukemia and lymphoma, including Hodgkin's disease. It is now generally accepted that corticosteroids are not effective enough to justify their routine use in acute or chronic myelocytic leukemia.

Corticosteroids are also used in the treatment of carcinoma of the breast, especially in the presence of metastases. There is some controversy as to their effectiveness, estimates of tumor regression varying from 6 to 48 percent, with 15 percent considered a realistic assessment (Brennan, 1973). Corticosteroids may also have beneficial effects in the treatment of carcinoma of the prostate following castration. The mechanisms of these effects are uncertain but appear to be related to the inhibition of secretion of pituitary hormones, and hence the adrenal androgens and estrogens, which support the growth of the tumors.

Surgical adrenalectomy is sometimes used for the treatment of breast cancer. "Medical adrenalectomy" using aminoglutethimide offers an alternative method for inhibiting the synthesis of adrenocortical steroids (Santen, Lipton, and Kendall, 1974; Lipton and Santen, 1974; Smith et al., 1978). The primary aim is to reduce the secretion of adrenal androgens, which act as precursors for estrogens. These latter steroids can be formed in peripheral tissues. Normally, aminoglutethimide only has a relatively brief action when it is used alone, as there is a compensatory release of ACTH from the pituitary so that steroidogenesis is restored. To overcome this effect and to provide corticosteroid replacement, dexamethasone or, preferably, cortisol is also administered to inhibit the release of the ACTH.

Corticosteroids are used to reduce *cerebral edema* associated with neoplasms in the brain and *hypercalcemia,* which often occurs in such diseases.

Blood diseases. Apart from malignant diseases, such as the leukemias, corticosteroids are also widely used in the treatment of other blood diseases. They are the treatment choice in *idiopathic-thrombocytopenic purpura,* where they are effective in about 60 percent of adults and 80 percent of children. This treatment is often combined with immunosuppressive therapy, usually with azathioprine (Gerber and Steinberg, 1976), but the relative effectiveness of the combined therapy versus steroids alone is not clear. The corticosteroids are also effective in many types of *hemolytic anemias.* Corticosteroids may be effective in some cases of *agranulocytosis,* especially if leucocyte antibodies are present, although the additional risks of infection really do not justify their use.

Miscellaneous uses. Although corticosteroids have been widely used in the treatment of *shock,* their effectiveness, except in Addisonian shock due to a deficiency of corticosteroids, is in doubt and their use has been described as a type of "medical last rites." They are clearly ineffective in cardiogenic shock (such as due to myocardial infarction) or hemorrhagic shock, although the question of their effectiveness in anaphylactic and septic shock is still open (Reichgott and Melmon, 1973).

Corticosteroids in combination with cytotoxic drugs are a standard procedure in *organ transplantation* (Gerber and Steinberg, 1976). Their effectiveness appears to be dependent on an inhibition of the inflammatory response, once an antigenic effect is initiated, rather than any direct immunosuppressive action. The dosage is thus usually increased when there are signs of tissue rejection.

The effectiveness of corticosteroids in treating patients following *myocardial infarction* has been controversial (Marx, 1976a); some claim a decrease in the size of the infarct, while others found that the steroid increased the incidence of cardiac arrythmias and the infarct.

Corticosteroids may also have a beneficial effect (Adour et al., 1972) in reducing complications and hastening recovery in idiopathic facial paralysis (*Bell's palsy*). A swelling of the facial nerve may play an important role in the early stages of this disease and corticosteroids appear to reduce this edema.

C. Side effects of corticosteroids
In a review in 1968, Sawin estimated that "as many as 6 percent of patients may die as a result of corticoid treatment." They are probably the cause of more iatrogenic (drug-induced) disease than any other group of drugs. In England it was estimated that in the 1960s they accounted for about 20 percent of deaths due to drugs. These are alarming statistics but serve to reemphasize the problems associated with the therapeutic use of corticosteroids. I have been unable to find more recent estimates of the incidence of their fatal effects, but as no new toxic effects have become known since then, and we have no reason to suspect medical education has advanced in the interim, the current situation is probably similar.

An awareness of the possible toxic effects of corticosteroids has undoubtedly helped to lower their incidence, but it has also created some problems with respect to the decision that a physician must make. The drugs unquestionably may alleviate the symptoms of many painful and distressing diseases especially in short-term treatment, although if they are used for long-term therapy it may merely amount to the substitution of one problem for another. An evaluation of the lesser evil may then be necessary. In other situations, such as rheumatoid arthritis, mortality may actually be increased by corticosteroids. In potential terminal illnesses such as cancer of the breast and prostate, acute hepatic failure, or the survival of a life-saving organ transplant, a more straightforward decision may be possible.

The corticosteroids are usually administered in supraphysiological amounts so that their side effects are to a large extent similar to those seen in Cushing's syndrome [Section 6.8A(2)]. It would, however, be untrue to say that they are generally an extension of their physiological effects, for as described earlier, such effects do not always occur in hypercortisolism. One, however, suspects that the side effects reflect the general basic mechanisms of the steroids' actions, especially their catabolic effects on tissues. Such problems are clearly related to the dose and the period of administration. A daily dosage of less than about 10 mg of prednisolone is unlikely to result in side effects that are a serious problem.

1. Electrolyte imbalance. The retention of Na and excretion of K, resulting in a hypokalemic alkalosis, was one of the first side effects to be observed from corticosteroid therapy. This action is an extension of the physiological effects of the corticosteroids on salt metabolism and may be associated with hypertension. The syntheses of chemical analogues with high glucocorticoid activity and extremely low mineralocorticoid effects have virtually abolished this problem, except when very high doses are used.

2. Catabolic effects. The abilities of corticosteroids to inhibit protein synthesis and facilitate the mobilization of amino acids, which is associated with a negative nitrogen balance, contribute to a variety of their most troublesome side effects. It has not been possible to dissociate these, or indeed any, nonelectrolyte effects from their desired therapeutic actions. Attempts to antagonize them by the concurrent administration of anabolic, androgen-like steroids have not met with a generally accepted success.

a. *Muscle wasting (myopathy)* may occur. The identification of this condition can be difficult in patients who are receiving corticosteroids for arthritic conditions and who are thus often relatively immobile. Apart from the effects on protein catabolism, an associated hypokalemia can also manifest itself as muscular weakness. The condition is not usually seen when "low doses" of steroids are used and the synthetic fluorinated steroids are allegedly more of a problem. The condition of the muscles is restored when steroid treatment is ceased.

b. The *skin* may become thinner due to wasting of the supporting collagen. It thus tends to become transparent so that the blood vessels may be visible, giving it a red or purple appearance. This condition is also seen locally following prolonged local application for dermatological conditions. A surgical problem involving wound closing can sometimes result. Bruising may occur relatively easily, and acne and the growth of hair on the trunk and face may be stimulated. The latter effects appear to be related to interactions with androgens.

c. The *bones* respond to corticosteroids in various ways. *Osteoporosis* is a common accompaniment of long-term therapy but may not become apparent for a prolonged time. It occurs in children and adults but is most common in women over 50 years of age. Vetebral collapse can occur, but other bones, such as the neck of the femur and the ribs, may also be affected so that fractures are more likely to occur. The main cause of the condition is a reduction in the bone remodelling process and the failure to lay down a collagenous matrix which is adequate for rebuilding. There may also be an increased resorption of bone, but this is uncertain. The problem can be compounded by the ability of the corticosteroids to reduce absorption of dietary calcium across the intestine and from the renal glomerular filtrate. In patients who are immobilized, bone remodelling is also reduced. Vitamin D supplements (more recently $1\alpha,25$-$(OH)_2$ vitamin D_3) and physiotherapy are sometimes given to limit the problem. The damage to the bones is generally repaired when the steroid administration is stopped.

Another osseous problem which is seen in adults is most commonly associated with a necrosis of the head of the femur, but other bones may also be involved, as well as the knee and ankle joints. This complication may have a relatively rapid onset and occur within a month with high dosage therapy. It has thus often been observed during immunosuppressive therapy for renal transplants. It is called *osseous aseptic necrosis* and its cause is unknown.

d. *Inhibition of somatic growth* is a problem commonly arising in children. It is mainly reversible when therapy ceases and there is a "catch-up" of growth. The effects appear to be general ones involving various tissues, including bone and connective tissue. It can be reduced by alternate-day therapy or the use of ACTH or, in the cases of chronic asthma, the use of bronchial aerosol sprays so that the dose is localized and small [see Section 6.8D(1)]. A reduced birthweight has also been observed in the infants of mothers treated with prednisone (Reinisch et al., 1978). The frequency of this occurrence and its significance is not at present clear.

3. Glucogenic effects. Corticosteroid therapy commonly results in the accumulation of fat in the trunk, abdomen, and around the face ("moon face"), but there is a loss from the limbs. A rise in blood glucose may also occur, which can result in steroid diabetes. There is often an increase in appetite, which contributes to these problems, but they are also the result of the gluconeogenic effects of the corticosteroids. An increased insulin release normally compensates for the increased production of glucose, which is laid down as fat but only in certain localized areas. If inadequate insulin secretion occurs, in predisposed individuals diabetes may be precipitated. This condition is, however, not difficult to control and usually disappears when therapy ceases. The use of corticosteroids in diabetes mellitus is not contraindicated but usually requires the use of a higher dose of insulin.

4. Suppression of the hypothalamic–pituitary–adrenal axis. Any therapeutically significant dose of glucocorticoids will normally suppress the release of pituitary ACTH. This response reflects the activation of the physiological negative-feedback control mechanism that regulates the release of cortisol. The chronic use of corticosteroids can produce a persistent depression of endogenous ACTH release as well as an atrophy of the adrenal cortex (Editorial, 1975c), which loses its normal ability to respond. If corticosteroid therapy is ex-

tended for periods of 6 months to 1 year, restoration of normal pituitary and adrenal function may only return after several months, although following shorter periods of treatment it may only take a few days. The use of alternate-day therapy reduces the suppression of the pituitary–adrenal axis (see Section 6.8D). It is important to withdraw corticosteroid therapy gradually so that the risk of precipitating an Addisonian crisis during a period of sudden stress is avoided. The integrity of the pituitary–adrenal axis can be tested by the insulin-hypoglycemia test [Section 6.8A(3)]. If this function is found to be deficient, attempts to hasten its recovery, following corticosteroid withdrawal, are sometimes made by intermittent administration of ACTH every few days until a normal response to insulin hypoglycemia occurs. This procedure is, however, considered to be debatable, as the most serious problem is usually hypothalamic–pituitary depression, and the administration of ACTH may exacerbate this problem.

5. Miscellaneous effects. Corticosteroid therapy is associated with a variety of undesirable responses which it is difficult to clearly put into any single one of the former categories.

a. *Increased sensitivity to infection* is a major problem associated with the use of corticosteroids. These effects are not due to any immunosuppressive actions, which are not generally of practical significance in man, but to a suppression of the inflammatory response. Thus, an important normal defense mechanism is inhibited. This problem may be further increased, as the signs of an infection may be masked. A general feeling of well-being which is sometimes seen in patients treated with corticosteroids may also contribute to obscuring the effects of such an infection, which may thus spread alarmingly before it becomes readily visible.

Infections due to bacteria, viruses, and fungi are all more likely to occur during corticosteroid therapy, and latent infections may suddenly become apparent. The latter has been a problem with respect to tuberculosis, so that the dosage of antitubercular drugs is increased. The use of corticosteroids in infection is thus, although not completely contraindicated, a situation where special care and precautions are necessary. They are sometimes used in conjunction with effective antibiotic therapy. Some infectious diseases may, however, be so debilitating as to be acutely life-threatening, in which instances steroids may be used in the initial period before the antibiotics have had time to produce their effects.

b. *Delayed wound healing* is also largely the result of the antiinflammatory response, as well as a reduced activity of fibroblast cells, and can be intimately related to a delayed response to infection.

As referred to earlier, the thinning of the skin can also contribute to a slow closing of wounds, and this type of effect is apparently also seen in patients with ulcerative colitis who undergo surgery of the colon. Such effects may be of importance when performing surgical procedures under "cover" of a high concentration of corticosteroids.

c. The *activity of the central nervous system* is also affected by corticosteroids, but the causes of such effects are unknown. There may be a euphoria which can result in addiction to the steroids (Morgan, Boulnois, and Burns-Cox, 1973). Insomnia can also occur. In some instances, especially in predisposed individuals, the patient may go into a state of depression. This effect can be serious, although if necessary it may be possible to treat it with drugs.

An organic response of the central nervous system is the occurrence of intracranial hypertension during corticosteroid *withdrawal*. It is apparently due to insufficient corticosteroids, but the precise nature of the change is unknown. It is most common in children, where it results in headache and vomiting and may become especially alarming if vision is affected. It can be corrected by increasing the dosage of the corticosteroids.

d. The *eye* also may respond adversely to corticosteroids during their use to treat either ocular or nonocular diseases. The precautions already described with respect to responses to infection are important with respect to the eye; the topical use of corticosteroids in eye infections has resulted in perforation of the cornea.

The prolonged systemic use of corticosteroids over periods of several years has been shown to result in cataracts (Oglesby et al., 1961). Small opaque areas appear beneath the posterior capsule of the lens. They are usually not clinically important, but if they should become so can be treated surgically. They do not appear to be reversible when corticosteroids are withdrawn.

Topical use of corticosteroids, usually over a period of several weeks, can induce a rise in intraocular pressure, a condition called *corticosteroid-induced glaucoma* (Laval and Collier, 1955; Goldmann, 1962). This condition has also been described in patients receiving corticosteroid nasal drops for the treatment of hay fever or asthma. The ocular hypertension is usually reversible when the steroid treatment is ceased. The response is greater in patients suffering from chronic simple glaucoma, so that special precautions need to be taken in such a situation. This glaucomatous sensitivity to local, topically applied corticosteroids is inherited (Armaly, 1966) and is apparently due to a dominant gene which is present in one-third of the population. Individuals who are homozygous suffer from chronic simple glaucoma. The response is thus relatively easy to predict in the relatives of people suffering from this disease. The cause of the ocular hypertension in response to corticosteroids is uncertain but appears to involve a reduction in the outflow of the aqueous fluid from the eye. Ocular side effects of corticosteroids may be considerably exaggerated in people wearing contact lenses (Burde and Becker, 1970), presumably because of an enhancement of local absorption.

e. An increased occurrence of *peptic ulcers* has been widely noted in patients treated with corticosteroids. Their possible causative role has, however, recently been strongly challenged (Conn and Blitzer, 1976) on the basis that such patients suffer innately from such gastric problems anyway. However, there remain some unaccounted for aspects, such as an increased incidence in children, who normally very rarely suffer from peptic ulcer. From the practical standpoint the physician should be aware, whatever the cause, that patients being treated with corticosteroids have an increased likelihood of developing peptic ulcers.

f. *Allergy* to intravenous injections of hydrocortisone and methylprednisolone have been described in patients who had been receiving these drugs for the treatment of asthma (Editorial, 1974a). Contact allergies to topically applied dermatological preparations of hydrocortisone have also been described (Bacon and Spencer, 1973). These observations are rather surprising, as such drugs are often used to combat anaphylactic responses.

D. Administration of corticosteroids

As corticosteroids are potentially dangerous drugs, the decision to use them needs to be made after serious consideration of their disadvantages and side effects and only after alternative, more direct ways of combating the disease have failed (see Thorn and Lauler, 1972; Myles and Daly, 1974; Streeten, 1975; Swartz and Dluhy, 1978). Corticosteroids do not constitute a cure, nor do they usually appear even to alter the natural history of a disease. Their role is generally to provide symptomatic relief. They clearly need to be used with circumspection in patients suffering or predisposed to such conditions as osteoporosis, peptic ulcer, congestive heart failure, diabetes mellitus, glaucoma, psychosis, or underlying infections.

An important consideration is the choice of dose and whether or not long-term therapy will be necessary. The occurrence of side effects is related to both of these factors. Thus, a minimal dose for the shortest possible period of time is a prime aim. As the required dosage is usually difficult to predict, it often requires experimental adjustment. This process can be aided by the availability of tablets containing small amounts of the steroid; thus, prednisolone can be obtained in 1-mg tablets, which allow a fine gradation of the dose. Local application which limits the systemic dispersal of the drugs may be made by topical application, the use of bronchial aerosol sprays, intraarticular injection (Yates, 1977), and rectal suppositories.

The administration of corticosteroids on *alternate days* (see Streeten, 1975) can play an important role in reducing side effects during long-term treatment, especially the suppression of the pituitary–adrenal axis and the retardation of growth in children. A single dose is given in the morning, usually before breakfast, so that the levels rise early in the day and thus follow the normal circadian rhythm of their secretion. It is also important to use short- or intermediate-acting preparations (Table 6.6), as the longer-acting ones, such as dexamethasone, persist for too long in the circulation and tend to negate the intentions of the schedule.

The general principles in initiating administration of corticosteroid therapy in nonadrenal related conditions have been described by Thorn and Lauler (1972).

a. The disease should be initially brought under control by round-the-clock high-dose therapy.
b. Change to a single daily dose as soon as possible using a short- or intermediate-acting steroid preparation.
c. Subsequently, shift to an alternate-day dosage schedule (Table 6.7).
d. Initiate supplementary therapy designed to reduce side effects of the steroids (Table 6.8).

1. Preparations. A large number of different preparations of corticosteroids are available for therapeutic use. For adrenocortical substitution therapy, some mineralocorticoid actions are desirable, but such effects usually constitute an undesirable side effect in the use of corticosteroids in nonadrenocortical diseases. The synthetic analogues with negligible mineralocorticoid effects are usually used when systemic administration on a chronic basis is desired. However, cortisol is widely used topically and when an acute need arises, such as in shock and allergic reactions.

The time of onset and duration of the effects of corticosteroids depend on several factors. A classification into short-, intermediate-, and long-acting preparations is given in Table 6.6. In these instances, the differences depend principally on the rates of their metabolism. Another important factor is the relative water and lipid solubility of the preparations. The corticosteroids exist as free alcohols that are poorly soluble in water. Many corticosteroids which are esterified at either, or both, the C-21 and C-17 positions are available which have higher water or lipid solubilities than do the parent free alcohols (Figure 6.20). The hemisuccinates and disodium phosphate esters are very water soluble and are used for intravenous administration or intramuscular injection when rapid absorption is desired. The acetates have a high lipid

solubility and are thus absorbed only slowly into the circulation when injected. They are thus useful for the establishment of a slowly absorbable depot or the local injection into joints and tendons where it is desirable to maintain a high local concentration. The free alcohol compounds are also used but their absorption is more rapid. The butylacetate ester or the hexacetonide are very water insoluble. Some preparations of corticosteroids are available as crystalline suspensions which are slowly absorbed and often used for intraarticular injection. Absorption from the gastrointestinal tract is favored by lipid solubility, so that the free alcohols and acetates are commonly used. Inactivation of corticosteroids occurs mainly in the liver, and compounds that are absorbed across the intestine enter the hepatic portal circulation and are carried directly to that organ. Many of the synthetic analogues are not readily metabolizable, while cortisol is bound to plasma proteins, which afford it some protection against catabolism. Aldosterone, however, is poorly bound to plasma proteins and is thus rapidly inactivated by the liver and so is relatively ineffective when administered orally.

It should be recalled that the 11-keto compounds cortisone and prednisone are biologically inactive until they undergo 11-hydroxylation. This process occurs mainly in the liver. When used topically or injected into local sites, such compounds are likely to be less effective than 11-hydroxylated compounds and may even be ineffective, as in the skin. If liver function is compromised, their actions may also be reduced.

In order to be effective on the *skin*, the corticosteroids must be absorbed into the cells across lipid barriers, although they must initially pass through a film of water on the cell surface. Lipid-soluble compounds are most effective therapeutically, but some degree of water solubility is also needed, so that *highly* nonpolar compounds may be less effective. Absorption into the cells is slow and takes several hours. It is facilitated by the presence of a covering sheet of thin plastic over the area being treated. This procedure increases the response about tenfold. Commercial tape impregnated with corticosteroids is also available. The absorption of highly potent glucocorticoids across the skin into the systemic circulation can be a problem if large surfaces of skin are involved, and this process is also increased by a plastic covering. Beclomethasone dipropionate (Figure 6.20) is poorly absorbed across the skin and other membranes.

Topical application of corticosteroids on the *eye* utilize both water-soluble Na phosphate compounds and lipid-soluble free alcohols and acetate esters in ointments and suspension. The free alcohols and acetate esters are more readily absorbed into the eye and pass into the systemic circulation.

Table 6.6. *Adrenal corticosteroid preparations*

Drug	Anti-inflammatory potency[a]	Equivalent potency[a] (mg)	Sodium-retaining potency	Daily dose (mg) above which HPA[b] axis suppression possible[c]		Plasma half-life (min)	Biological half-life (h)
				Males	Females		
Short-acting							
Cortisol (hydrocortisone)	1	20	2+	20–30	15–25	90	8–12
Cortisone	0.8	25	2+	25–35	20–30	90	8–12
Intermediate-acting							
Prednisone	3.5	5	1+	7.5–10	7.5	200 or >	18–36
Prednisolone	4	5	1+	7.5–10	7.5	200 or >	18–36
Methylprednisolone	5	4	0	7.5–10	7.5	200 or >	18–36
Triamcinolone	5	4	0	7.5–10	7.5	200 or >	18–36
Long-acting							
Paramethasone	10	2	0	2.5–5	2.5–5	300 or >	36–54
Betamethasone	25	0.6	0	1–1.5	1–1.5	300 or >	36–54
Dexamethasone	30	0.75	0	1–1.5	1–1.5	300 or >	36–54

[a] Potency is defined as a milligram-for-milligram equivalence with hydrocortisone.
[b] Hypothalamo–pituitary–adrenal axis.
[c] Intended as a guide only. The dose in an individual depends on total body surface area. The figures quoted are those which apply in general.
Source: Swartz and Dluhy, 1978.

Table 6.7. *Suggested program for accomplishing the transition from a dosage of 50 mg of prednisone each day to an every-other-day schedule*

	Day	Prednisone dose (mg)
	1	60
	2	40
	3	70
	4	30
	5	80
	6	20
	7	90
	8	10
	9	95
	10	5
Then one could begin an overall reduction as follows		
	11	90
	12	5
	13	90
	14	5
	15	90
	16	5
Then		
	17	85
	18	5
	19	85
	20	5
	21	85
	22	5
Then		
	23	80
	24	5

Source: Thorn and Lauler, 1972.

If the topical concentrations are high enough and therapy is long-term, systemic side effects may become apparent. Cortisone can be activated by C-11 hydroxylation once it is absorbed into the eye but appears to be less effective than hydrocortisone on the surface. The water-soluble forms are less well absorbed, although apparently can still enter the eye, where they can raise intraocular pressure as well as exert their desired antiinflammatory actions. Hydrocortisone preparations are often preferred to treat surface problems, as they are less likely to cause rises in intraocular pressure than are the more potent synthetic fluorinated steroids. Attempts have been made to dissociate the antiinflammatory and ocular hypertensive effects of the corticosteroids. Dilution of the topically applied solutions reduces absorption but also their therapeutic actions. Medrysone (11β-6α-methyl-progesterone) enters the eye and has a weak, but useful antiinflammatory effect and apparently lacks a glaucomatous action (Podos, Kolker, and Becker, 1970). The absence of the latter effect may reflect its general low potency Another analogue, tetrahydrotriamcinolone acetonide, on the other hand, has a potent antiinflammatory action but little glaucomatous effect, which apparently reflects its poor penetration (Levine and Leopold, 1973).

Table 6.8 *Supplementary measures that can be considered to alleviate the possible side effects of long-term therapy with corticosteroids*

1. A diet low in sodium chloride and high in potassium
2. A diet restricted in calories to prevent weight gain
3. Antacids – three or four times daily
4. Supplementary vitamin D – calciferol 50,000 units, serum calcium to be monitored
5. Fluoride – tablets twice daily
6. Androgen therapy for female patients – fluoxymesterone
7. Physiotherapy – increased active muscular exercise as tolerated

Source: Thorn and Lauler, 1972.

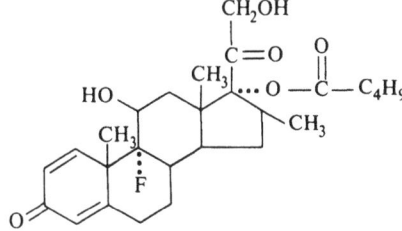

Fig. 6.20. Some clinically useful corticosteroids that are esterified at the C-17 or C-21 position.

In the treatment of *ulcerative colitis,* corticosteroids can be applied locally in the form of a retention enema or rectal suppository. A low rate of absorption is desirable, so that water-soluble compounds, such as the hemisuccinate and disodium phosphates, are used. Betamethasone valerate (Figure 6.20) is especially useful in this respect.

Attacks of *asthma* in children can be prevented by corticosteroids. Systemic administration is effective but results in the usual side effects, including inhibition of growth. It has been found that two compounds, beclomethasone dipropionate and betamethasone valerate, are effective when applied directly to the bronchial tract with the aid of an aerosol spray (Brown, Storey, and George, 1972; Campbell, 1975). These compounds are very poorly absorbed across the pulmonary membranes, so that it is possible to deliver effectively high local concentrations without such risk of systemic absorption.

Corticosteroids can be expensive, so that when choosing a preparation fiscal considerations may be appropriate. If long-term therapy is used, this can be at considerable financial cost to the patient or the social program that supports him. Considerable differences exist between the cost of preparations which have therapeutic equivalence.

2. Withdrawal of corticosteroid therapy. A sudden cessation of the administration of corticosteroids can result in a variety of problems and in some situations can even be fatal. The response depends on several factors, including the underlying disease which is being treated, the dosage and its schedule (the administration as divided doses or on alternate days), and the duration of the therapy.

The higher the dose of the corticosteroids and the longer the duration of their administration, the greater will be the problem of their withdrawal. The brief administration of moderate amounts (e.g., <40 mg prednisolone/day) of corticosteroids, for a few days, is unlikely to result in problems. However, these steroids are often used chronically for many months, often extending into years, and under these conditions even the use of low doses (e.g., <10 mg prednisolone/day) can result in problems when the dosage is reduced.

Detailed descriptions of the procedures, problems, and recommended dosage schedules during withdrawal of corticosteroids have been made several times (Sawin, 1968; Myles and Daly, 1974; Byyny, 1976). The general strategy involves avoidance of sudden precipitous changes in dose so that the drugs are withdrawn gradually. Concurrent testing of the integrity of the pituitary–adrenal axis is often necessary, and precautions must be taken to avoid or promptly deal with a stress situation that results in an Addisonian crisis due to adrenal insufficiency.

The problems that are frequently experienced can be divided into three categories:

a. *A flare-up of the underlying disease* is usually the most immediate response. A slow, cautious reduction of the dosage may help, but if the disease has not undergone a spontaneous remission, it may be necessary to maintain corticosteroid therapy, although it may be found that a lower dosage will suffice. It is this problem which most frequently limits the withdrawal of the drugs.

b. *Withdrawal symptoms* may occur. These may be of a rather general nature, such as a loss of a feeling of well-being and a reduction of appetite. The latter may result in a loss in weight accompanied by a redistribution of fat in a more normal manner; for instance a loss of "moon face." Conjunctivitis, rhinitis, and itchy nodules (panniculitis), may occur but these problems usually persist for only a few days. The patient may suffer from tiredness, muscular weakness, anorexia, and nausea, and these symptoms have been compared to adrenal insufficiency (*corticoid withdrawal syndrome*). It appears, however, that this condition is due to a sudden decline in plasma corticosteroid levels and may reflect an habituation of the tissues to abnormally high steroid concentrations. This condition is also usually only transitory. The occasional sudden increase in intracranial pressure has been described elsewhere [Section 6.8C(5c).

c. *Suppression of the pituitary–adrenal axis* was described earlier [Section 6.8C(4); Editorial, 1975c]. Its integrity can be tested using the insulin hypoglycemia and ACTH stimulation tests [Section 6.8A(3)]. These tests are especially useful in predicting the recovery of the patient's ability to respond to sudden stress. Recovery from the sup-

pression of the pituitary and adrenal almost always occurs, but it can take many months.

A proposed schedule for withdrawal of corticosteroids is given in Table 6.9.

3. Use of ACTH; comparison to corticosteroids. When the therapeutic benefits of corticosteroid therapy were first recognized, the steroids were expensive and available only in very limited quantities. The possibility of stimulating the patient's own adrenal to produce more corticosteroids was apparent, and as ACTH preparations were readily available afforded an alternative method for instituting such therapy. The disadvantages are obvious, and include the necessity for parenteral administration, which is inconvenient and often painful, and, as animal hormone preparations are used, allergic reactions may occur. It is also dependent on the functional activity of that particular patient's adrenal cortex. This indirect mechanism of action also imposes limitations with respect to the circulating corticosteroid levels which can be achieved and makes it difficult to predict what the precise response may be to a given dose of ACTH.

The administration of ACTH has some effects which are different from those of corticosteroids and which in some situations may be advantageous. Adrenocortical function is not suppressed by ACTH, and even the pituitary remains more responsive. Thus, withdrawal of ACTH is not likely to be as difficult as that of corticosteroids. There is also less risk of impairing growth in children. The reason for this effect is not clear, but it has been suggested that it may reflect a more adequate release of pituitary growth hormone. A direct effect of ACTH on secretion of growth hormone appears unlikely (Lee et al., 1973) and the effect cannot be combated by the administration of this latter hormone. There have also been claims that the use of ACTH results in less osteoporosis, fewer peptic ulcers, and less atrophy of the skin. It is difficult to evaluate such observations, which could be due to lower plasma levels of corticosteroids. On the other hand, ACTH also stimulates the release of adrenal androgens, which have anabolic effects, and it has been suggested, but not confirmed, that these steroids could be contributing to the amelioration of such side effects.

The administration of ACTH has some side effects that are more prominent than those seen with corticosteroid therapy. Acne and hirsutism are more common in women, probably reflecting the release of androgens. The retention of Na and hypokalemia are also more of a problem; cortisol, the endogenous hormone, unlike many of its synthetic relatives, has substantial mineralocorticoid activity. ACTH has a melanocyte-stimulating action and can increase pigmentation of the skin, an

Table 6.9. *Suggested protocol for withdrawal of administered glucocorticoids*

Step	Interval	Observation	Result	Glucocorticoid and dose
I	Variable	Underlying disease	Worsening of underlying disease	Variable; gradual decrements in dose to biologic equivalent of hydrocortisone, 20 mg/day
			Symptoms and signs of steroid withdrawal	Raise dosage for flare-up of disease; continue if disease quiescent; supplement for stress
II	4 weeks	8 A.M., plasma cortisol	Plasma cortisol: When <10 μg/100 ml	Begin hydrocortisone, 20 mg/day, then taper by 2.5 mg/once a day/week to 10 mg each morning and continue this dosage; supplement for stress
			When >10 μg/100 ml	Stop maintenance hydrocortisone; supplement for stress
III	4 weeks, indefinite	8 A.M., 250 μg i.m., ACTH[a] test	When plasma cortisol increment <6 μg/100 ml or maximum <20 μg/100 ml (or both)	Supplement for stress
IV	4 weeks, indefinite	8 A.M., 250 μg i.m., ACTH test	When plasma cortisol increment >6 μg/100 ml and maximum >20 μg/100 ml	Stop supplementation for stress
V	Indefinite	Routine		As indicated

[a] Synthetic ACTH-(1–24).
Source: Byyny, 1976. Adapted by permission from *The New England Journal of Medicine 293*, 31.

effect that is absent during corticosteroid therapy, when the secretion of ACTH is inhibited. Myles and Daly (1974) have commented: "It would be unwise to use ACTH in white-skinned people from a country where excessive melanin is regarded as a disqualification from normal and political life!"

The therapeutic use of ACTH is clearly limited and usually reserved for patients where the dose of corticosteroids does not need to be large but is long-term: for instance, rheumatoid arthritis and in children for the prevention of attacks of asthma (Malone et al., 1972). It is, however, clearly more inconvenient to use. When it is decided to change from corticosteroid to ACTH therapy, this is done with great caution, as it is impossible to predict what the equivalent doses may be. In the interim period both drugs may be used, but such a duality is not recommended as a standard practice.

4. Use of spironolactone in the treatment of edema and hypertension. The aldosterone antagonist spironolactone was initially developed in an attempt to produce a diuretic drug that while promoting urinary Na loss did not also increase K excretion (a "K-sparing" diuretic). This latter problem, which can produce hypokalemia, is one of the major side effects experienced in the use of most diuretic drugs, such as the thiazides, ethacrynic acid, and furosemide. Patients who take these diuretics on a chronic long-term basis, such as in the treatment of congestive heart failure, cirrhosis, and hypertension, are advised to choose foods containing large amounts of K, such as orange juice, nuts, and bananas, or they are given supplements of KCl.

Aldosterone promotes K loss so that the antagonism of its effects would be expected to produce a retention of this ion. Spironolactone has this predicted effect. It is, however, only a weak diuretic drug, as it can only promote the excretion of less than 2 percent of the Na filtered at the glomerulus (compared to more than 20 percent for ethacrynic acid and furosemide). It is thus not very useful as a diuretic drug when used alone but is often combined with more active ones. The aim and hope is that its K-sparing effects will compensate for the kaliuretic effects of the other diuretic. The responses to such "cocktails" are difficult to predict and their use is frowned upon by some, but they are nevertheless still used.

Spironolactone is also used in certain cases where the mobilization of edema fluid is relatively unresponsive (refractory edema) to other diuretics.

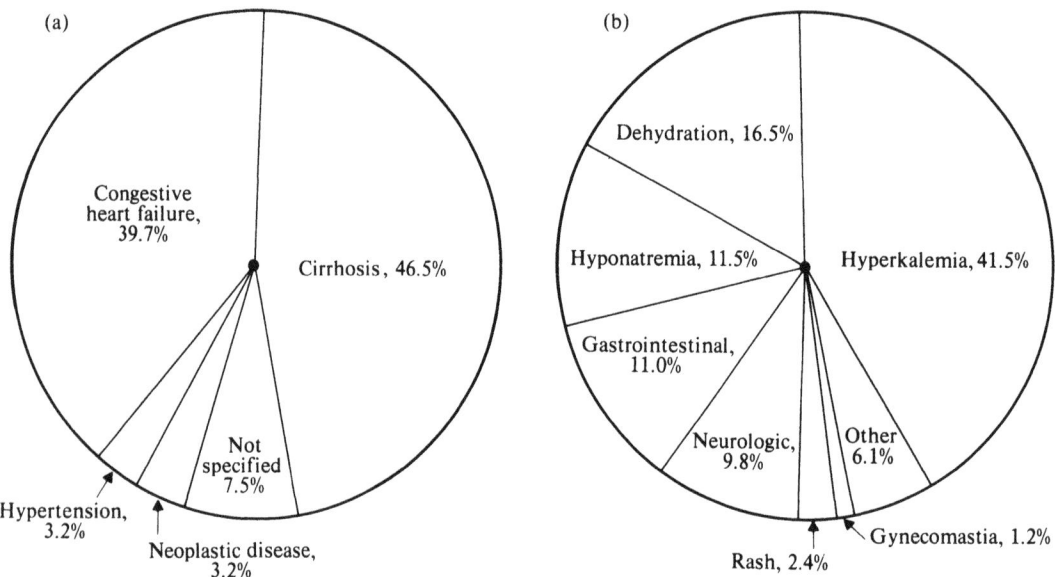

Fig. 6.21. Use of spironolactone. (a) Reasons for using spironolactone in a group of 788 patients. (b) Distribution of side effects and adverse reactions in a group of 164 affected patients. (From Greenblatt and Koch-Weser, 1973.)

This may occur when plasma aldosterone levels are high, such as in cirrhosis of the liver. Its possible effectiveness in such conditions is unpredictable.

The use of spironolactone is associated with a number of side effects (Figure 6.21). Some of these can be serious and others rather interesting. An extensive international survey of patients taking this drug (Greenblatt and Koch-Weser, 1973) showed that 21 percent of patients had adverse side effects and *hyperkalemia* was the most predominant problem. In some patients it may have resulted in death. The hyperkalemic effect was three times more common in those people who were also receiving K supplements.

A small number of men and women suffered from sexually related disorders; a loss of libido, amenorrhea, and hirsutism in women and impotence and gynecomastia in men. These effects appear to be related to disorders in steroid hormone metabolism, but the locus of such effects has been difficult to detect (Stripp et al., 1975). It may act by excluding androgens from their receptors (Pita et al., 1975; Cutler et al., 1978). The original drug trials showed that spironolactone could produce tumors in rats, although these were apparently not discovered until after the drug was released for use. In the United States the FDA has suggested that spironolactone should therefore only be used when other measures have failed.

The use of the steroidal compound *carbenoxolone sodium* for the treatment of gastric ulcer is discussed elsewhere [Section 10.6A(2)]. This com-

pound has some troublesome side effects, which include Na retention, hypokalemia, and hypertension. These responses are reminiscent of the actions of aldosterone (see Lewis, 1974; Langman, 1976). It is interesting that while the side effects of carbenoxolone can be antagonized by spironolactone, the healing effects of the carbenoxolone are also blocked (Doll, Langman, and Shawdon, 1968). The precise nature of these interactions is uncertain but the compounds have some structural similarities (Figure 6.22). Carbenoxolone is bound to plasma proteins and so could be displacing aldosterone and hence increase its effective concentration. This possibility, however, appears to have been excluded, and it is considered more likely that it reacts directly with aldosterone receptors or interferes with the hormone's metabolism. The antagonistic effect of spironolactone on its desired therapeutic effect is consistent with the possibility of such an interaction. Carbenoxolone is a derivative of licorice and it is interesting that patients suffering from Addison's disease have been observed to maintain themselves by self-administration of licorice (Cotterill and Cunliffe, 1973). From the therapeutic standpoint, the aldosterone-like effects of carbenoxolone can be satisfactorily combated by the judicious use of other diuretics combined with K supplements. The chronic taking of licorice can also result in hypokalemia and it has been observed to lead to cardiac arrest (Bannister, Ginsburg, and Shneerson, 1977). Its action appears to be a direct one, as

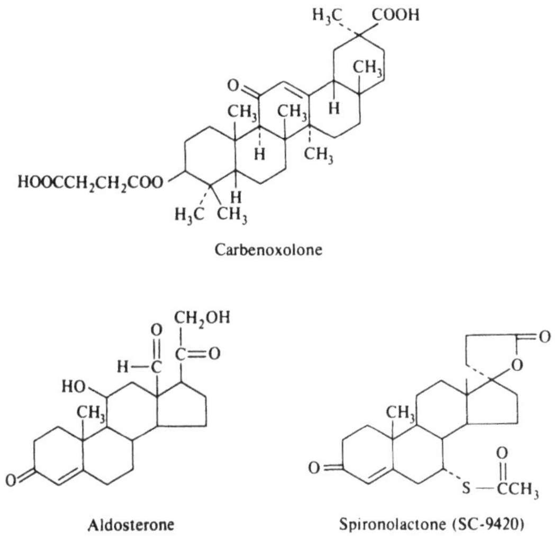

Fig. 6.22. Chemical structure of carbenoxolone compared to that of aldosterone and spironolactone.

no changes in rates of secretion of cortisol or urinary steroid metabolites are apparent (Epstein et al., 1978).

The adrenal medulla

The hormones of the adrenal medulla were among the first to be identified; their vasoactive properties were described by Oliver and Schäfer in 1895. They were soon shown to be catecholamines and in 1904 adrenaline, or epinephrine, was synthesized. This hormone was the first to be prepared in this way. The important functions of catecholamines as neurotransmitters in the peripheral sympathetic nervous system and in the central nervous system were recognized later. The roles of norepinephrine (or noradrenaline) and dopamine in neural transmission have since been the predominant focus of research on the physiology and pharmacology of catecholamines. The endocrine role of epinephrine, which is the principal secretion of the adrenal medulla, is clearly of less physiological importance, and indeed the absence of this glandular tissue does not compromise human activities in any notable way. There are, however, many functional analogies between neuronal and adrenal medullary catecholamine metabolism. The morphological and embryological similarities between the adrenal medulla with its cholinergic preganglionic nerve fibers and sympathetic ganglia were first noted more than 60 years ago. It is therefore not surprising to observe that information regarding synthesis, metabolism, and release of catecholamines has been obtained from studies on

both types of tissues. The information obtained for one tissue is very often applicable to the other.

The principal interest of pharmacologists in catecholamines has been related to their roles as neurotransmitters. Many important bodily functions, including the regulation of blood pressure and behavior, may be altered by drugs that mimic or alter the actions and metabolism of neuronal catecholamines. As these effects bear little or no relationship to the endocrine functions of the adrenal medulla, I will not deal with them in detail.

A comprehensive account of the first 75 years of research on the adrenal medulla has been provided under the editorial supervision of Blaschko, Sayers, and Smith (1972). More current information on the subject is available in the published proceedings of the Third International Catecholamine Symposium (Usdin and Snyder, 1973). A fourth such volume is iminent. The major advances since that time are mainly concerned with information about the mechanisms of action of the catecholamine hormones, especially the identification, roles, and functioning of their receptors.

6.9. Morphology

The adrenal medulla consists of epitheloid cells derived embryonically from neuroectodermal cells [they are APUD cells; see Section 3.1B(c)] which invade the adrenal cortex. They differentiate into two main types of cells, the medullary or chromaffin tissue, which is predominant, and a few scattered sympathetic ganglion cells. The chromaffin cells are so named because, histologically, they stain a dark brown color with chromium salts. They are rich in cell organelles, especially granules or vesicles, which contain the stainable material. These cells possess a prominent Golgi apparatus and a rough endoplasmic reticulum, which reflect their hormone-secreting abilities. They are also quite rich in mitochondria and lysosomes. There are two main types of chromaffin cells:

a. A darkly staining variety ("dark" cells), which make norepinephrine. These cells are autofluorescent and can be stained with silver salts.
b. Most (about 80 percent) of the cells present, which form epinephrine, stain lightly ("clear" cells). They have smaller, less electron dense granules and are not autofluorescent.

The chromaffin cells are separately innervated by fibers that are carried in the splanchnic nerve, which arises in the celiac ganglion and the spinal cord. They are preganglionic cholinergic fibers whose action is pharmacologically classified as "nicotinic." Neural control of the epinephrine- and norepinephrine-secreting cells appears to be distinct.

The blood supply to the medullary tissue comes predominantly from the adrenocortical venous sinuses and constitutes a portal system. These vessels are thought to supply the epinephrine-secreting cells, which require high concentrations of adrenocorticosteroids for their functioning. There is also an arterial blood supply to the medulla which appears to be directed principally to the norepinephrine-secreting cells.

6.10. Functions and actions

The adrenal medulla is not essential for life and, indeed, especially in man, its physiological significance is not very clear, except possibly in conditions of extreme stress. The principal secretion is epinephrine, which is known to exert many effects in the body, especially when administered in pharmacological doses. Under these conditions it shares many effects with the neurotransmitter norepinephrine, so that its actions can often be considered to mimic stimulation of the sympathetic nervous system ("sympathomimetic"). Epinephrine is, however, released from the adrenal medulla both chronically and acutely in response to specific stimuli and may contribute to homeostasis in the body. In some instances, such as in the presence of catecholamine-secreting tumors, it can exert prominent toxic effects. The original suggestion of Cannon that the release of epinephrine is concerned with the mobilization of the body's physiological resources in response to physical and emotional stresses ("fright-or-flight" reactions) is still valid, although its possible importance and utility in modern man is uncertain and controversial. This hormone may, however, be of more physiological significance in some animals where it is involved, apart from fright-or-flight responses, in temperature regulation, arousal from hibernation, and, under certain conditions, the regulation of intermediary metabolism. Some pharmacological responses to epinephrine are utilized therapeutically, and drugs that mimic or antagonize certain of its actions also have widespread uses.

The effects of catecholamines in the body are ubiquitous, mainly reflecting the distribution of neuronal adrenergic processes which are situated both in the peripheral and central nervous systems. Some of these effects are summarized in Table 6.10.

The first effect of adrenal medullary hormones, which was described by G. Oliver and E. A. Schäfer, was a prominent, but evanescent, *increase in blood pressure*. Except in extraordinary circumstances, such as a prominent hypotension and the presence of catecholamine-secreting tumors, circulating epinephrine does not appear to play a significant role in the regulation of blood pressure. This particular response is also related to the method of administration and the dose of the drug. Low doses may even result in a decline in blood pressure. The activity of compensatory cardiac and vascular reflexes may be altered in the presence of other drugs, and the particular experimental preparation used can influence the type of effect. There are several components of the response of blood pressure to catecholamine hormones. Epinephrine constricts many peripheral arterioles and venules, especially in the skin and viscera, but vessels in skeletal muscle dilate so that the diastolic pressure may drop. The threshold for these two responses differs; higher concentrations are needed for the constrictor effects. There may be reflex slowing of the heart in response to the increased peripheral vasoconstriction, but epinephrine also has a direct effect on both the heart rate and the contractility of the cardiac muscle (speeds relaxation and exerts a positive inotropic effect) which can contribute to the rise in blood pressure. The effects on the heart also include a dilatation of the coronary vessels. The net effect of epinephrine when given by rapid intravenous injection in relatively high doses is a considerable increase in blood pressure, but if given more slowly, or subcutaneously, the rise is not as great.

Catecholamines have prominent effects on the *contractility of nonvascular types of smooth muscle*. They decrease the motility of muscle in the intestine and stomach by inducing a relaxation, but they contract most of the sphincters present in the gastrointestinal tract. The same types of effects occur in the urinary bladder. The response of uterine smooth muscle depends on the animal species, the stage of the reproductive cycle, and pregnancy. Usually, relaxation occurs in the nonpregnant condition and contraction takes place during pregnancy. In the eye, the radial muscle of the iris contracts so that mydriasis results while the ciliary muscle relaxes and there is accommodation, as for distant vision. The bronchial muscle of the pulmonary tract is relaxed.

The *contractility of skeletal muscle* is usually enhanced by epinephrine, and this effect is partly due to a facilitation of neuromuscular transmission.

The *secretion from various exocrine* and *endocrine glands* can be altered by epinephrine. Some effects, such as an ability to increase sweat gland secretion, are dependent on the species, and in man this response is confined to the glands on the palms of the hands and soles of the feet. The secretion of insulin can be either increased or decreased, the latter usually being more prominent. The decreased secretion of mucus in the nasal passages and the decline in gastrointestinal secretions appear mainly to reflect decreases in the blood supply to these tissues, but there may also be direct effects.

Epinephrine can have prominent *effects on metabolism*. Oxygen consumption may be stimulated, es-

Table 6.10. *Some effects of catecholamines in the body*

Tissue	Response	Type of receptor
Vasculature (arterioles)		
Skin	Contract	α
Visceral	Contract	α
Skeletal muscle	Dilate, contract	β_2, α
Coronary	Dilate	β
Heart		
Rate	Increase	β_1
Contractility	Increased force	β_1
Skeletal muscle		
Contraction	Augmented	β
Neurotransmission	Augmented	β
Gut		
Sphincters	Contract (except pyloric)	α
Motility	Decrease	α, β
Bladder		
Sphincter	Contract	α
Motility	Decrease	β
Uterus		
Motility	Decrease, increase in pregnancy	β, α
Bronchi	Dilate	β_2
Eye		
Iris	Radial muscle contracts (mydriasis)	α
Ciliary muscle	Relax (accommodation for distant vision)	β
Glands		
Salivary secretion	Increased volume	α
Islets of Langerhans	Inhibit or increase insulin release	α, β
Metabolism		
Calorigenic	Increased O_2 consumption	β
Glycolysis	Blood glucose increased	β
Lipolysis	Plasma fatty acids increased	β

Note: These effects do not necessarily reflect the endocrine role of the adrenal medulla and are usually sympathomimetic.

pecially in some animals which possess "brown" fat, and blood glucose rises as a result of an increased glycogenolysis, reflecting an activation of glycogen phosphorylase. The action of glycogen synthetase is concurrently inhibited. Physiologically, the most important site of its action on glycogenolysis is muscle, but at relatively high doses the liver is also involved. Fatty acids and glycerol are mobilized from adipose tissue (lipolytic effect) due to an activation of a lipase enzyme.

Catecholamines can promote the *aggregation of blood platelets* (see Alexander, Cooper, and Handin, 1978). Platelet aggregation is inhibited by prostaglandin E_1 and this response is associated with an increase in the formation of cyclic AMP. Epinephrine inhibits this response to prostaglandin E_1.

Epinephrine also produces changes in the *central nervous system*, including sensations of apprehension and satiety. As circulating epinephrine does not cross the blood-brain barrier, the mechanism of these effects is not clear.

Catecholamines, including epinephrine, have a wide variety of effects which can be broadly divided into stimulatory and inhibitory types of responses. In 1948, R. P. Ahlquist suggested that two basic types of adrenergic receptors were present which principally mediated each type of response. The stimulatory ones he referred to as *alpha*-type (α) and the inhibitory ones as *beta*-type (β). In addition, exceptions were proposed to account for the stimulatory effects on the heart, which were also classified as β-responses. Although there were some imperfections in this original classification, subsequent studies have shown that it is correct and reflects real differences in the types of receptors involved. Thus, drugs have subsequently been prepared which specifically mimic or antagonize each type of response, and some of the specific receptors have even been isolated. It has also been shown that β-adrenergic responses are related to the activity of adenyl cyclase and the formation of cyclic AMP. The types of receptors present in each tissue are shown in Table 6.10. The β-adrenergic receptors were further differentiated by A. M. Lands and his collaborators in 1967 into two types, β_1 and β_2. This distinction largely depended on the further synthesis of specific adrenergic agonists and antagonists. The contractility and rate of the heart beat is a β_1 response that can be blocked by practolol. Vasodilatation, relaxation of the bronchi, and probably most other responses are β_2 effects, which can be mimicked by salbutamol and blocked by butoxamine. This subclassification compensates for some of the inadequacies of the original one provided by Ahlquist.

Dopamine is considered to interact with a different specific type of receptor (dopaminergic receptors), and this response can be blocked by pimozide and haloperidol and mimicked by apomorphine, bromocriptine, and lergotrile mesylate. Dopamine from the adrenal medulla is of little or no importance, but in pharmacological doses it can dilate the blood vessels in the kidneys. Its principal physiological role appears to be as a central neurotransmitter, where it can influence the release of some hypophysiotropic hormones and, most probably, has a direct effect in inhibiting the release of prolactin [see Sections 3.2E(4) and 7.6G(3)].

6.11. Adrenergic hormones; structure–activity relationships and synthetic agonists and antagonists

Epinephrine, like norepinephrine and dopamine, is a catecholamine. It consists (Figure 6.23) of an aromatic benzene ring which is hydroxylated at the 3 and 4 positions and has a side chain (ethylamine) which consists of two carbons (α and β) which terminate in an amino group. The hydroxylated benzene ring compound o-dihydroxybenzene is known as catechol, hence the naming of the catecholamines. The levorotatory isomers (−) are much more active than the dextrorotatory forms (+).

Epinephrine differs from norepinephrine in having a methyl group at the terminal amino position. Bulky constituents in this position appear to favor both β_1- and β_2-type activity, and this can be enhanced in synthetic analogues such as isoproterenol, which has an isopropyl group in this position. This analogue has negligible α-adrenergic

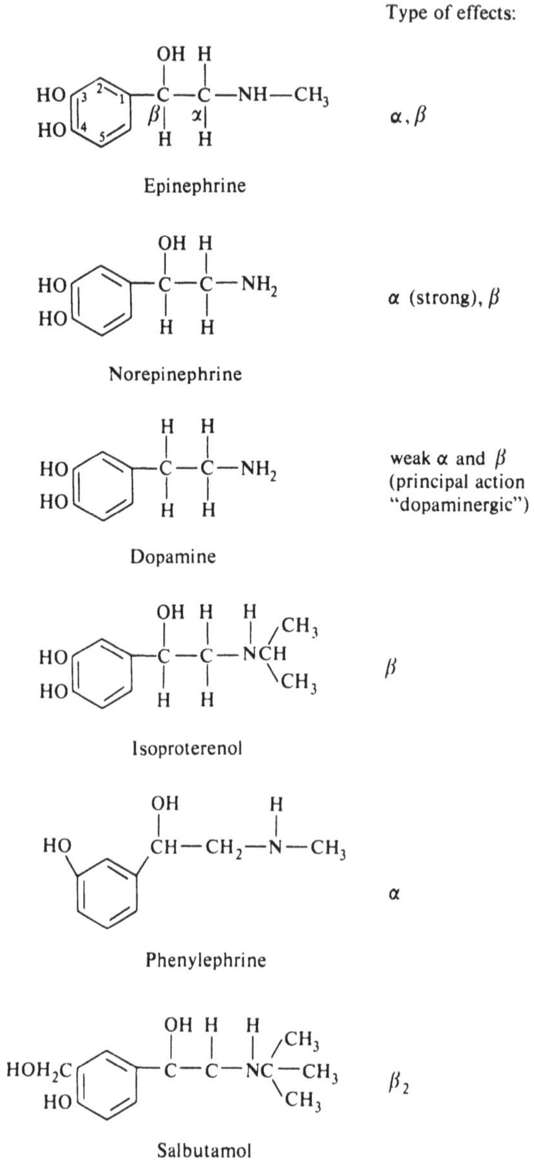

Type of effects:

Epinephrine α, β

Norepinephrine α (strong), β

Dopamine   weak α and β (principal action "dopaminergic")

Isoproterenol β

Phenylephrine α

Salbutamol β_2

Fig. 6.23. Structure of epinephrine and related catecholamines.

activity and is the prototype of a pure β-adrenergic agonist. The amino terminal appears to be important for the interaction (or affinity) of the catecholamines with their receptors. It should be noted that α-adrenergic activity is also increased by methylation of the amino terminal, so that epinephrine has greater α-adrenergic activity than norepinephrine; but as it also has β-type activity, it is not as selective in its actions.

The length of the side chain of the catecholamines is also important, and two carbons appear to be optimal. Norepinephrine only differs from dopamine, which is its immediate metabolic precursor, by the presence of a hydroxyl on the β-carbon. Epinephrine is also hydroxylated in this position, so that this group appears to be important for both α- and β-type activity. Artificial substitutions have been made on the α-carbon which interfere with the metabolism of the compound by monoamine oxidase and so prolong its period of action.

The aromatic ring appears to be of principal importance in determining the molecule's intrinsic activity. Substitution for the hydroxyl groups in the 3 and 4 positions or the insertion of other moieties results in profound changes in the molecule's activity. Deletion of the 4-OH reduces both α- and β-adrenergic effects and in norepinephrine results in the formation of the analogue phenylephrine. This change virtually abolishes β-adrenergic activity, but some α-activity remains so that this compound is a pure α-adrenergic agonist, for which it is the prototype. The $β_2$ agonist salbutamol has a CH_2OH group in position 3 of the aromatic ring, which, combined with the three methyl groups at the amino terminus, confers on it this special selective action.

The β-adrenergic antagonists have the same bulky substituents at the amino terminus as the β-agonists, but the aromatic ring is either altered or replaced by another type of structure. The first synthetic compound to exhibit activity as a β-antagonist had Cl substituted for the three and four hydroxyls in the ring structure. However, this compound, *dichlorisoproterenol* (DCI), had substantial residual β-agonist activity, and so it was not useful clinically. In 1964, J. W. Black developed a pure β-adrenergic antagonist called *propranolol* (Black et al., 1964), which retains the favorable bulky configuration at the amino terminus, but the aromatic ring portion of the molecule has undergone considerable change (Figure 6.24). Propranolol is the prototype of the β-adrenergic antagonist. Other modifications, also principally in the aromatic ring portion, have resulted in the selective $β_1$-antagonist *practolol* as well as the $β_2$-antagonist *butoxamine*.

The first α-adrenergic blocking agent to be described was *ergot*. In 1906, H. Dale showed that extracts of this natural product (which is a mixture of alkaloids) blocked the increase in blood pressure which is normally seen when epinephrine is injected into an anesthetized cat. Instead of the usual rise, a lowering of the blood pressure was observed which for the first time displayed the β-adrenergic vasodilatatory effects of this hormone. Ergot has a wide spectrum of pharmacological effects, including a direct vasoconstrictor action, and it promotes emesis, so that it is not a clinically useful α-adrenergic antagonist. Two more useful groups

of α-adrenergic antagonists have been developed (Figure 6.24). These are the β-haloalkylamines, of which *phenoxybenzamine* is the most useful, and some imidazoline derivatives, notably *phentolamine*. These are competitive antagonists and appear to be more effective against circulating catecholamines than those released at nerve endings. Phenoxybenzamine has a prolonged effect, as it forms a covalent bond with the α-adrenergic receptor. The action of phentolamine is quite brief. Both types of antagonists have other actions apart from their blockade of α-adrenergic receptors.

The hydroxyl groups on the catecholamines (apart from contributing the molecule's affinity and intrinsic activity), because of their polar nature, also appear to be able to influence its absorption and distribution. The hydrophilic nature of these molecules appears to be largely responsible for their inability to be absorbed from the gastrointestinal tract or to cross the blood-brain barrier. Thus, adrenergic agonist drugs such as ephedrine and amphetamine, which lack one or both of the hydroxyl groups on the aromatic ring, are effective orally and exert prominent effects in the central nervous system.

The adrenal medullary catecholamines, epinephrine and norephinephrine, have both α- and β-adrenergic effects. Epinephrine is the most active with respect to $β_1$- and $β_2$-type effects. Norepinephrine has a more selective action, as it displays α-adrenergic responses but its β effects are relatively less. Its vasodilatatory, $β_2$, activity is small, although its cardiac effects, $β_1$, remain substantial. The metabolic effects of epinephrine are much greater than those of norepinephrine.

6.12. Synthesis and release

The catecholamine hormones (and neurotransmitters) are made from phenylalanine or tyrosine. The pathway of their biosynthesis, which occurs in the adrenal medullary chromaffin cells and sympathetic neurons, was first proposed by H. Blaschko in 1939. It has subsequently been amply confirmed and the enzymes and intermediate products involved have been identified and isolated from certain components of the cells (see, e.g., Goldstein et al., 1973; Fuller, 1973). In many instances, drugs are available which can specifically inhibit certain parts of the biosynthetic process (Figure 6.25). The enzymes themselves, apart from phenylethanolamine-N-methyltransferase (PNMT), are not especially specific for the synthesis of catecholamines but can act on many other types of compounds with similar chemical groups. Drugs that influence these enzymes may thus have other effects in the body.

After the precursor amino acids are taken up by the cell, the tyrosine (see Figure 6.26) is hydroxyl-

α-Adrenergic antagonists

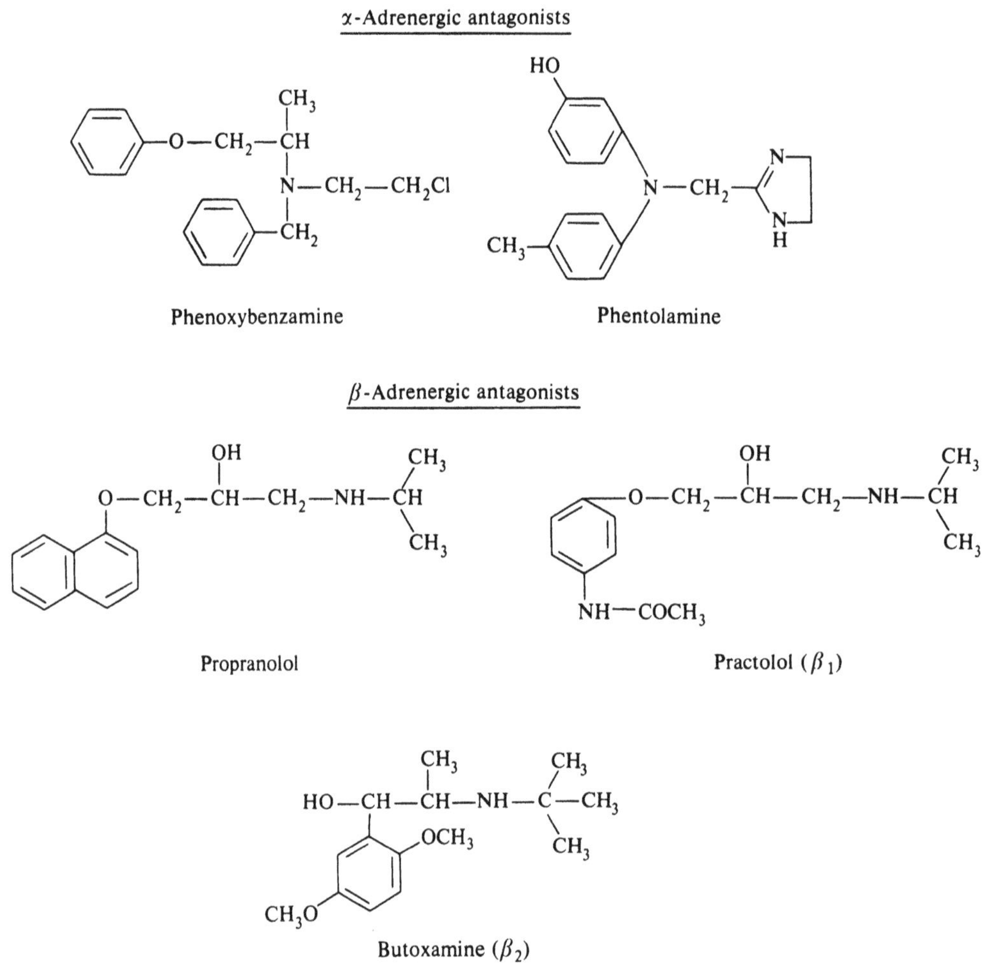

Phenoxybenzamine   Phentolamine

β-Adrenergic antagonists

Propranolol   Practolol (β_1)

Butoxamine (β_2)

Fig. 6.24. Structures of some adrenergic blocking drugs.

ated by tyrosine hydroxylase to dopa. This enzyme is soluble and the process occurs in the cytoplasm. This step is rate-limiting for the formation of norepinephrine and epinephrine and the enzyme can be inhibited by *α-methyl-p-tyrosine*. This drug can thus be used to reduce the formation of these catecholamines. Dopamine is formed from dopa under the influence of dopa decarboxylase and is then taken up from the cytoplasm into the storage granules. This uptake can be blocked by *reserpine*. The dopamine is converted to norepinephrine by dopamine β-hydroxylase, which can be inhibited by *disulfiram*. The norepinephrine can be stored for release in the sympathetic neurons and the dark cells of the adrenal medulla, or it can be converted to epinephrine as occurs in the medullary clear cells. This process consists of N-methylation of the terminal amino group and occurs under the influence of phenylethanolamine-*N*-methyltransferase

(PNMT). This interesting enzyme is principally found in the adrenal medulla and its formation is induced by cortisol, which passes directly into this tissue from the adrenal cortex. The steroid is thus present in high concentrations and its action can be mimicked by ACTH or high doses of synthetic glucocorticoids like dexamethasone. The norepinephrine is released from the storage granules into the cytoplasm, where the PNMT promotes its methylation to epinephrine. This hormone is then again taken up and stored, probably in another type of granule.

Catecholamines can be taken up by storage granules against concentration gradients as high as 100 : 1 (see Pleitscher et al., 1973) in a process that requires the utilization of energy provided by ATP. The uptake can be inhibited by reserpine and inhibitors of Mg^{2+}-dependent ATPase, such as *N*-ethylmaleimide. The catecholamines are stored

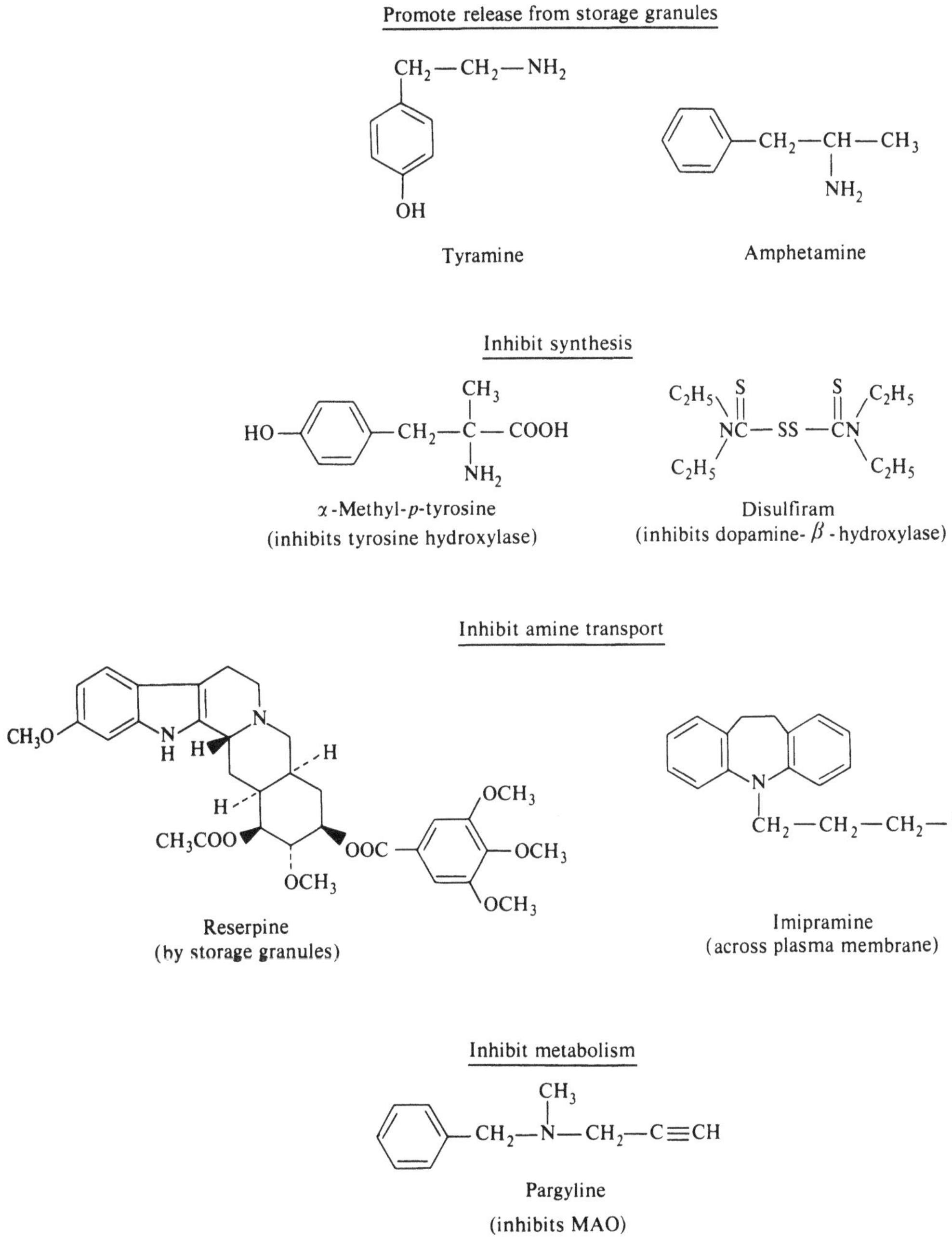

Promote release from storage granules

Tyramine

Amphetamine

Inhibit synthesis

α-Methyl-*p*-tyrosine
(inhibits tyrosine hydroxylase)

Disulfiram
(inhibits dopamine-β-hydroxylase)

Inhibit amine transport

Reserpine
(by storage granules)

Imipramine
(across plasma membrane)

Inhibit metabolism

Pargyline
(inhibits MAO)

Fig. 6.25. Some drugs that influence the synthesis, release, and metabolism of the catecholamine hormones.

principally in a complex with ATP, in the ratio of four molecules of the catecholamine to one of the adenine nucleotide. The granules, or vesicles, also contain dopamine decarboxylase and soluble proteins called chromogranins, and proteins and lipids which are parts of the vesicular membrane. Metal ions, especially Ca^{2+}, are also present and

appear to play a role in the formation of the storage complex. The stored catecholamines make up about 20 percent of the dry weight of the granules.

The physiological *release of catecholamines* from the adrenal medulla occurs in response to nerve stimulation (see, e.g., Douglas, 1972; Kirshner and Viveros, 1972). This is a cholinergic response

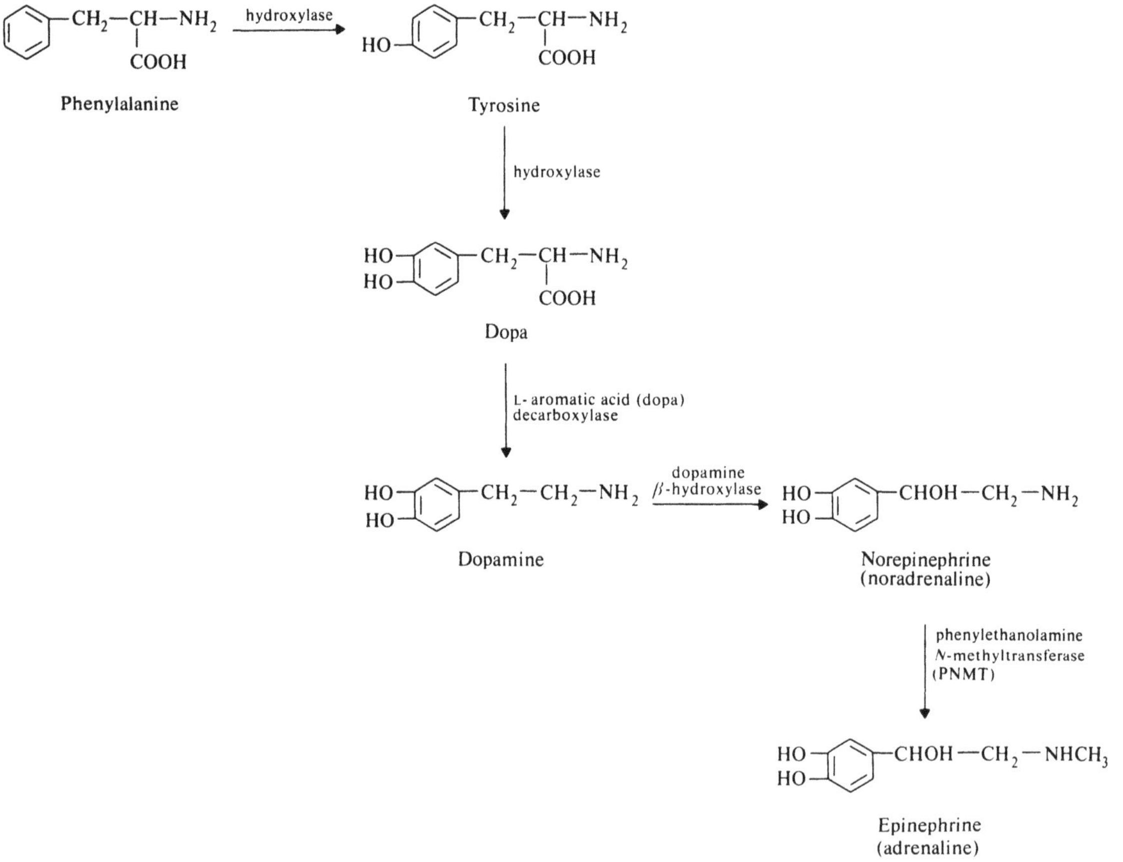

Fig. 6.26. Biosynthesis of dopamine, norepinephrine, and epinephrine from phenylalanine and tyrosine.

which can be mimicked by nicotine, but muscarinic drugs, such as pilocarpine, are also effective. Physiological stimuli (Table 6.11) include physical and emotional stress, exercise, hypotension, and hypoglycemia. The stimuli appear to trigger receptors in the brain which are present in regions of the hypothalamus and brain stem. The sites that are involved in sympathetic neuronal discharge also appear to initiate release of adrenal medullary

Table 6.11 *Stimuli and drugs that can promote the release of catecholamines from the adrenal medulla*

Physiological stimuli	Drugs
Emotional stress	Nicotine
Physical stress	Pilocarpine
Exercise	Histamine
Hypoglycemia	Kinins
Hypotension	Angiotensin I and II
Cold environment	Tyramine

catecholamines. A number of drugs may have more direct effects on the release of the adrenal medullary catecholamines, including tyramine, which displaces them from their storage granules. The actions of substances such as histamine and kinins on epinephrine release may involve vasodilatatory or other stressful effects, so that their actions could be reflex ones, but direct actions on the glands may also be involved.

The process of release of catecholamines from their storage granules was first described by Douglas and Rubin in 1961 and served largely as a model for the release of other types of hormones, which are also stored in this manner. The events involved appear to be as follows: acetylcholine released from the adjacent nerve endings results in a depolarization of the glandular cells and an accompanying influx of Ca^{2+}. Neither acetylcholine nor K^+-induced depolarization are effective stimuli for the release of catecholamines in the absence of external Ca^{2+}: The effect can, however, be mimicked by the Ca ionophore A23187, which increases Ca^{2+} uptake by the cells (Cochrane et al., 1975). A small increase in intracellular ionized Ca^{2+} concentration then appears to mediate the discharge of the contents of the storage granules across the plasma membrane by exocytosis. The precise role of the Ca^{2+} in this process is not clear, but it may promote the movements of the granules toward the plasma membrane and the fusion of the latter with the granule membrane. The catecholamines, ATP, and protein contents of the granules are all discharged in a quantal, all-or-none, manner. It is uncertain whether the remnants of granules can be reconstituted and used again or if a completely new synthesis must always take place.

The process of the release of adrenal medullary catecholamines has been studied *in vitro* in pure preparations of isolated bovine chromaffin cells (Schneider, Herz, and Rosenheck, 1977). Under these conditions, stimulation can be promoted by acetylcholine. It can be blocked by tetracaine, which reduces the ionic permeability of the cell membranes, and vinblastine, which disrupts microtubules in the cell. Cytochalasin B, which breaks the microfilament system near the cell membrane, is without effect. It was interesting that the vinblastine did not alter K-induced release of catecholamines, so that it appears that the microtubules are either important at a very early stage in the action of acetylcholine or the K-induced release reflects a different type of mechanism.

Following acute or chronic stimulation of the adrenal medulla, the levels of catecholamines are restored over a period of 2 to 4 days (see, e.g., Kirshner and Slotkin, 1973; Fuller, 1973). This synthesis is controlled in two ways.

a. The activity of the enzymes, especially the rate-limiting tyrosine hydroxylase, and PNMT are increased. The effect may be the result of a change in negative-feedback inhibition due to a decreased level of their respective final products norepinephrine, dopamine, and epinephrine.

b. There may be a stimulation in the *amounts* of the biosynthetic enzymes, especially following chronic stimulation. This enzyme induction appears to be under neuronal control for tyrosine hydroxylase and dopamine-β-hydroxylase. In the instance of PNMT, the response is the result of stimulation by adrenocorticosteroids. All these effects can be mimicked by butyryl cyclic AMP and ACTH, which activates adrenocortical adenyl cyclase. They have, however, been shown to have effects that appear to be independent of increases in corticosteroid secretion. The precise mechanisms involved are thus not clear.

A summary of the possible processes by which cholinergic stimulation results in the secretion and increased synthesis of catecholamines by the adrenal medulla is shown in Figure 6.27. This scheme illustrates the interaction between nerve stimulation, changes in intracellular free Ca^{2+}, and the possible role of cyclic AMP in these processes.

6.13. Metabolism

Catecholamines are rapidly cleared from the plasma, where they have a half-life of about 10 minutes. Only a small proportion, about 3 to 6 percent, appears unchanged in the urine. Initially, they are removed from the circulation and extracellular fluids as a result of their uptake by sympathetic neuronal tissue. Norepinephrine, which is released from nerve endings, is reaccumulated by these tissues as a result of the activity of a neuronal membrane amine pump. This is called the "uptake$_1$," process and can be blocked competitively by other amine compounds. Uptake$_1$, however, does not appear to be of major significance with respect to catecholamines that are released from the adrenal medulla and which are present in the general circulation. In this instance, a less rapid accumulation occurs by the nonneuronal tissues and is called "uptake$_2$." Inactivation of catecholamines mainly occurs as a result of the actions of two enzymes, monoamine oxidase (MAO) and catechol-O-methyltransferase (COMT). In the instance of the circulating compounds, these processes take place mainly in the liver and kidney. The MAO is associated with the mitochondria and is also present in sympathetic nerve endings, where it plays an integral role in the metabolism of norepinephrine. The COMT is a soluble enzyme.

The COMT methylates epinephrine and norepinephrine, at the 3-hydroxyl position in the aromatic ring to form metanephrine (Figure 6.28). Some of this metabolite may then be conjugated at position 4 in the aromatic ring, usually with sulfate in man but also with glucuronic acid. Alternatively, it may undergo oxidative deamination under the influence of MAO to form vanillylmandelic acid or VMA. (This compound is more correctly named 4-hydroxy-3-methoxymandelic acid.) About 40 percent of the metabolites appear in the urine as VMA and another 40 percent as metanephrine and its conjugates. About 7 percent of the metanephrine is converted to 4-hydroxy-3-methoxyphenylglycol (see Figure 6.28). Norepinephrine is metabolized in a similar manner to epinephrine, the primary methylated metabolite in this instance being normetanephrine.

Measurements of the levels of catecholamine metabolites in the urine are useful for the diagnosis of adrenal medullary disorders, especially pheochromocytomas, where there is a hypersecretion of adrenal medullary hormones.

The levels of tissue and circulating cate-

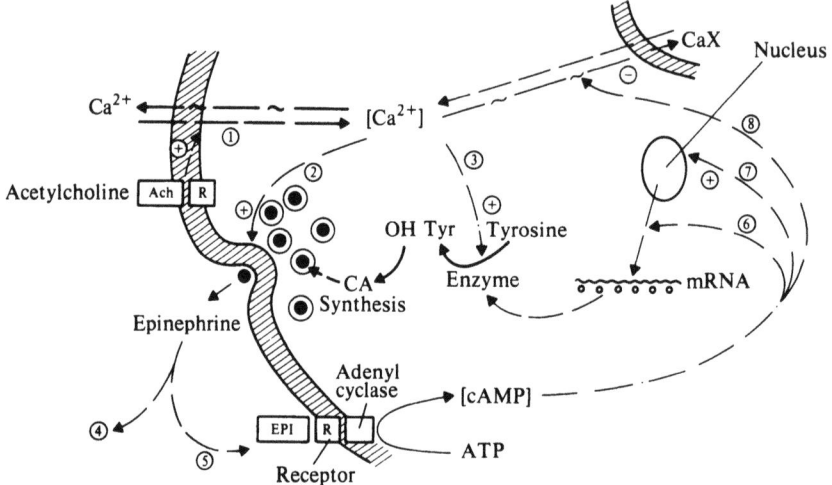

Fig. 6.27. Secretion of catecholamines from the adrenal medulla in response to neural cholinergic stimulation. The possible interactions of acetylcholine, cyclic AMP, and Ca^{2+}. When acetylcholine (ACh) acts, it causes an increased entry of Ca^{2+} (1) into the cell and raises the calcium concentration in the cell cytosol ($[Ca^{2+}]$). The rise in $[Ca^{2+}]$ is the activator of catecholamine secretion (2) and also activates one or more key enzymes (e.g., tryosine hydroxylase) in catecholamine (CA) synthesis (3). The released catecholamine, epinephrine (EPI), acts systemically (4), but also acts locally (5) to stimulate adenylate cyclase activity in the parent cell. The resulting rise in [cAMP] leads to (a) an increase in the synthesis of key enzymes [e.g., tyrosine hydroxylase (6) involved in catecholamine (CA) synthesis] and (b) cell growth and cell division (7). It also may stimulate the mobilization of Ca (8) from an intracellular source (CaX), thereby enhancing the strength of the original secretory stimulus. (From Rasmussen and Goodman, 1977.)

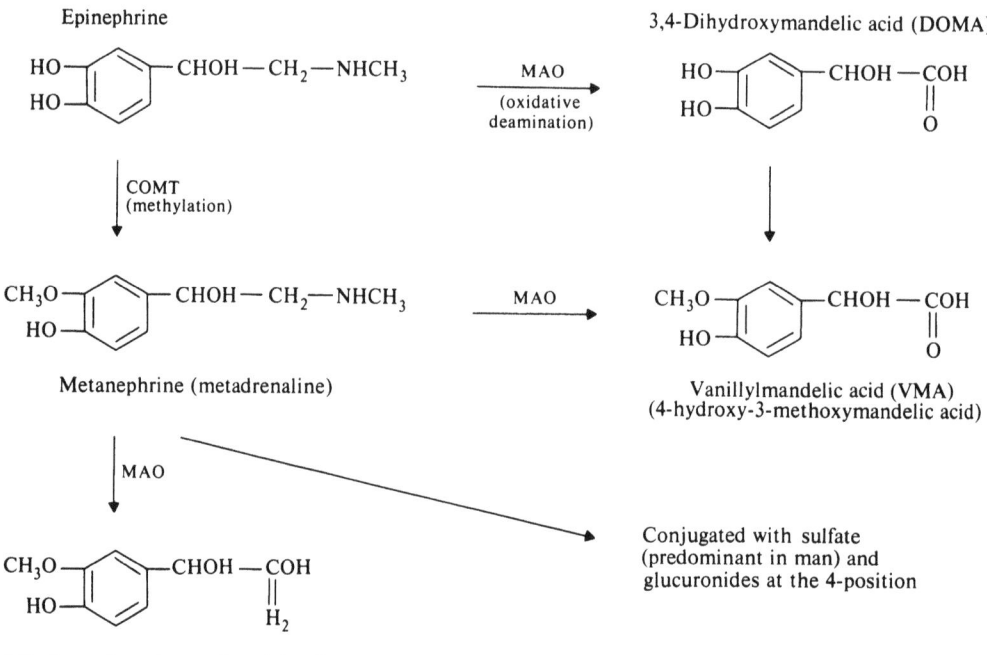

Fig. 6.28. Principal pathways for the metabolism of catecholamine hormones. (Note: VMA and free and conjugated metanephrine are the major metabolites (~80%) that appear in the urine. Norepinephrine is metabolized in a parallel manner. MAO, monoamine oxidase; COMT, catechol-O-methyltransferase.

cholamines can be altered by a number of drugs that influence their uptake and metabolism. Uptake$_1$, into sympathetic nerves, is inhibited by competitors for the neuronal membrane "amine pump," including cocaine, tranquilizers such as chlorpromazine, and tricyclic antidepressant drugs such as imipramine. This effect may result in drug interactions with catecholamines, whose actions can be enhanced. Uptake$_2$, into nonneuronal tissues, can be inhibited by phenoxybenzamine, but as this drug is also an α-adrenergic antagonist, it will not increase α-adrenergic effects. Its action with respect to β-adrenergic responses is not clear, but it could facilitate them, which may contribute to the frequent necessity for the administration of propranolol with α-adrenergic blockers in treating pheochromocytoma. Some "antidepressant" drugs used to treat psychiatric disorders inhibit the enzyme MAO and so can prolong the half-life of circulating catecholamines. Tyramine is present in many foods, including some cheeses and wines, and is normally destroyed by MAO. When this enzyme is inhibited, tyramine accumulates and can promote a massive release of catecholamines, which may have disastrous cardiovascular consequences.

6.14. Mechanism of action

Probably the first major insight into the molecular mechanism of action of hormones was provided by E. W. Sutherland in 1958, when he showed that epinephrine (and glucagon) acted on the liver to promote the formation of cyclic AMP (see Sutherland, 1972). This nucleotide was shown to be formed as the result of the activation of adenyl cyclase, which is present in cell membranes. A phosphorylase enzyme, which initiates the breakdown of liver glycogen to glucose, was activated by the cyclic AMP, which was thus shown to act as a "second messenger," epinephrine, the hormone, being the "first messenger." Sutherland and his collaborators subsequently demonstrated that catecholamines also increased cyclic AMP levels in other epinephrine-sensitive tissues, such as cardiac muscle, adipose tissue, and uterine smooth muscle. It is thus apparent that cyclic AMP can mediate a number of the effects of epinephrine, including the mobilization of glucose and fatty acids, contractility of heart muscle, and the relaxation of smooth muscle. These are all β-adrenergic responses. Information on the mechanisms of α-adrenergic responses has been much more evasive.

A. Receptors

An excellent account of the roles of adrenergic receptors in the actions of epinephrine and norepinephrine has been provided by Williams and Lefkowitz (1978) (see also Lefkowitz, 1978).

The initial step in the action of catecholamines on cells involves an interaction with specific receptors which are present in the cell membrne. Catecholamines apparently do not have to enter the cell in order to act as shown by their retention of activity when convalently linked with large structures such as synthetic polymers (Verlander et al., 1976).

Despite the early advances that Sutherland made into understanding the mechanism of action of catecholamines, direct studies of the properties of the α- and β-adrenergic-type receptors have, compared with steroid and peptide hormones, been relatively slow in their progress. Attempts were made to characterize such receptors by the use of radioactively labelled hormones and their analogues, including [^{3}H]epinephrine, [^{3}H]norepinephrine, and [^{3}H]isoproterenol. Although these agonists could be shown to bind, in relatively large amounts, to the target tissues, they displayed a number of disturbing characteristics which indicated that the interaction was not specifically related to their hormonal functions (see Cuatrecasas et al., 1975; Lefkowitz, 1976; Wolfe, Harden, and Molinoff, 1977). For instance, the binding was not readily reversible, it was not stereospecific [the $l(-)$ and $d(+)$ forms bound with similar facility], and a large number of catechol-containing compounds which lacked any agonist or antagonist activity whatsoever could inhibit binding. A surprisingly large number of "receptors" appeared to exist in relation to those which needed to be occupied to produce a response, so that it was necessary to postulate the presence of a vast number of "spare" receptors. Some of these problems appear to be related to the ease with which these agonist molecules are oxidized to compounds that also can be bound and also to their relatively low radiolabelled specific activities. Many antagonists of adrenergic activity have a much higher affinity for the receptors than agonists, and they are not as readily oxidized. These can be labelled with tritium or ^{125}I to give molecules with a high specific activity. Such compounds as [^{3}H]propranolol (Figure 6.29), [^{3}H]dihydroalprenolol (DHA), and [^{125}I]-hydroxybenzylpindolol (IHYP) have been used to characterize β-adrenergic receptors (Levitzki, Atlas, and Steer, 1974; Mukherjee et al., 1975; Aurbach et al., 1974; Brown et al., 1976a). The α-adrenergic antagonist [^{3}H]dihydroergocryptine (DHE) has been used to study α-adrenergic receptors (Williams, Millikan, and Lefkowitz, 1976). The binding of these compounds can be inhibited stereospecifically by both adrenergic agonists and antagonists, and the ability to do so, and the K_D, usually closely parallels their relative biological activities (see Lefkowitz et al., 1976; Maguire et al., 1977).

Adrenergic receptors have been studied in a var-

α-Adrenergic antagonist

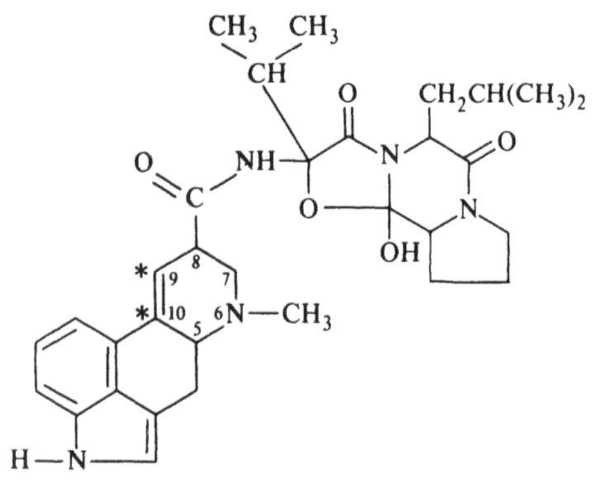

Ergocryptine

β-Adrenergic antagonists

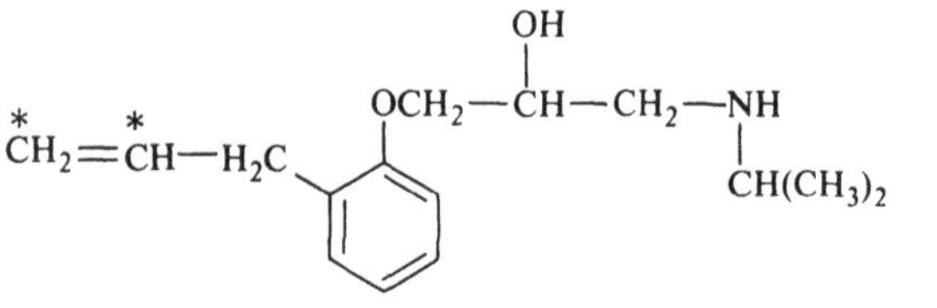

(−) Alprenolol (= dihydroalprenolol, DHA)

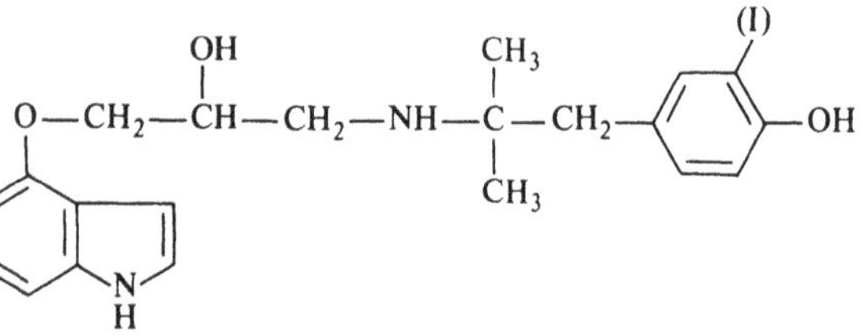

Iodohydroxybenzylpindolol (I-HYP)

Fig. 6.29. Chemical structures of α- and β-adrenergic antagonists used to identify α- and β-adrenergic receptors. [³H]Dihydroergocryptine (DHE) was prepared by catalytic reduction of ergocryptine at the double bond in position 9,10. Possible sites of tritiation are indicated by the asterisks.

iety of tissues, including membrane preparations made from liver, cardiac muscle, and uterine myometrium. Blood cells, especially erythrocytes from turkeys and frogs as well as human lymphocytes, are often favored because of their simplicity. These blood cells can be studied intact, although usually red cell "ghost" and membrane preparations are made from them. A soluble preparation of β-adrenergic receptors has been made from frog erythrocytes (Caron and Lefkowitz, 1976).

The numbers of β-adrenergic receptors on cells varies with the particular tissue and its condition, including previous exposure to catecholamines and other hormones, such as thyroid hormones, glucocorticoids, estrogens, and progesterone. There appear to be about 1000 such β-receptor sites in a single turkey or frog erythrocyte and as many as 80,000 in a cultured line of rat skeletal muscle myoblasts (Atlas, Hanski, and Levitzki, 1977). The latter number is extraordinarily high and it was suggested that such muscle myoblast cells may provide a good source of β-adrenergic receptors for experimental studies. The numbers of β_2-receptors in uterine myometrium is relatively low (Roberts et al., 1977a).

Binding of iodohydroxybenzylpindolol to turkey erythrocytes (Brown et al., 1976b) or dihydroalprenolol to frog erythrocytes (Mukherjee et al., 1975) both display a dissociation constant which usually corresponds to that calculated from the concentration necessary to inhibit (by 50 percent) isoproterenol-activated adenyl cyclase (apparent K_D or K_i apparent) prepared from the same tissue (see Table 6.12). There appears to be a one-to-one relationship between the binding to the receptor-adenyl cyclase complex and the inhibition of the enzyme's activation. This close relationship is, however, not invariably seen, so that the K_D calculated in each way can differ depending on the particular agonist or antagonist and the concentrations of guanine nucleotides present (see Section 6.13B). Binding of dihydroalprenolol rapidly reaches an equilibrium, in about 5 minutes, and is readily dissociable (Figure 6.30). The half-time for the dissociation is about 30 seconds. Competitive displacement of dihydroalprenolol by both agonists and antagonists is in the order of their biological potencies.

The iodohydroxybenzylpindolol binding sites in membrane preparations from rat cardiac muscle have also been studied (Harden, Wolfe, and Molinoff, 1976). They displayed stereospecificity, and the drug could be displaced by a series of agonists and antagonists in an order which reflected their potency as β-adrenergic drugs. There were differences, however, from the behavior of turkey erythrocytes and uterine myometrium. Equilibration was only achieved relatively slowly and dissociation from binding was not as rapid (Table 6.13). The K_D for the drug was also quite high compared to the other tissues.

Iodohydroxybenzylpindolol has also been used to characterize the β_2-receptors in rabbit uterine myometrium membranes (Roberts et al., 1977b). Binding is stereospecific and can be competitively inhibited by other antagonists and agonists in an order which reflects the latter's ability to relax this muscle. Equilibrium is attained in about 10 minutes and binding is reversible with a half-time of 5

Table 6.12. *Comparison of K_D for binding of β-adrenergic antagonists to receptors in turkey erythrocytes with K_i apparent for inhibition of tissue adenyl cyclase activity*

Antagonist	K_D, binding (M)	K_i, cyclase (M)
(−)-I-HYP[a]	2.5×10^{-11}	7.5×10^{-11}
(±)-HBP[b]	3.6×10^{-10}	1.0×10^{-9}
(−)-Propranolol	1.1×10^{-9}	2.1×10^{-9}
(+)-Propranolol	2.7×10^{-7}	3.7×10^{-7}
(−)-Alprenolol	3.2×10^{-9}	1.6×10^{-9}
(+)-Alprenolol	3.7×10^{-7}	2.3×10^{-7}
(−)-Oxprenolol	2.3×10^{-9}	1.4×10^{-9}
(±)-Oxprenolol	3.7×10^{-9}	3.3×10^{-9}
(±)-Pindolol	1.9×10^{-9}	1.7×10^{-9}
(±)-Dichloroisoproterenol	1.3×10^{-7}	2.3×10^{-7}
(±)-Pronethalol	1.4×10^{-7}	2.1×10^{-7}
(±)-p-Chloroisoproterenol	1.8×10^{-7}	3.0×10^{-7}
(±)-Nylidrin	6.0×10^{-7}	3.6×10^{-7}
(±)-Sotalol	9.3×10^{-7}	1.1×10^{-6}
(−)-Isoprophenamine	2.8×10^{-6}	5.3×10^{-6}
(±)-"Reverse" soterenol	1.3×10^{-5}	2.9×10^{-5}

Notes: The K_D binding was calculated by measuring the ability of the drugs to displace ^{125}I-hydroxybenzylpindolol (I-HYP) from specific binding sites. The adenyl cyclase was activated with isoproterenol.
[a] One-half of the added I-HYP was assumed to be (−)-I-HYP.
[b] Hydroxybenzylpropranolol.
Source: Brown et al., 1976b.

minutes. The K_D is higher than in turkey erythrocytes.

The β-adrenergic receptors, characterized by binding to dihydroalprenolol, have been prepared in a solubilized form from frog erythrocytes with the aid of the plant glycoside detergent digitonin (Caron and Lefkowitz, 1976). These receptors

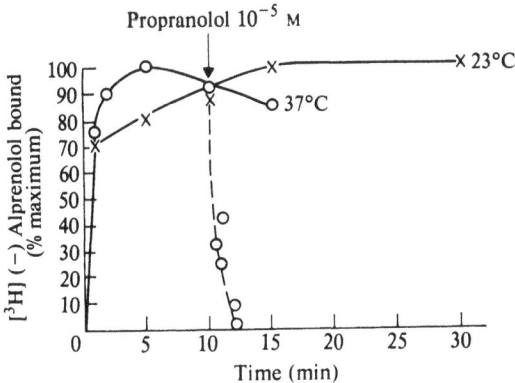

Fig. 6.30. Time course of binding of (−)-[³H]alprenolol to frog erythrocyte membranes. Maximum binding refers to the amount of (-)-[³H]alprenolol bound at equilibrium, which was 0.07 pmol. After equilibrium binding was reached, 10 μM (±)-propanolol was added and the amount of (−)-[³H]alprenolol bound was determined thereafter at 30-second intervals. (From Mukherjee et al., 1975.)

show remarkable similarities to the receptor–adenyl cyclase membrane preparations with respect to their ability to combine with different adrenergic agonists (Table 6.14) and antagonists. They have a molecular weight of about 130,000 to 150,000 and appear to be proteins with a lipid component. They can be degraded by trypsin and phospholipase A. Free lysine, tryptophan, serine, and sulfhydryl groups may play a role in their interaction with adrenergic hormones and drugs. The size of the solubilized β-receptor increases when it is exposed to a β-adrenergic agonist prior to its isolation in solution (Limbird and Lefkowitz, 1978). The increase in apparent receptor size is not seen following exposure to β-adrenergic antagonists. The mechanism of the change in the character of the receptor or its possible physiological significance is unknown.

The *α-adrenergic receptor sites* in rabbit uterine myometrium have also been characterized by measuring the binding of [³H]dihydroergocryptine (Williams, Mullikan, and Lefkowitz, 1976). This α-adrenergic antagonist could be displaced from its receptor site by (−)-epinephrine, which displayed a K_D of 0.23 μM. The (+)-epinephrine was more than 20 times less effective. Other α-adrenergic agonists and antagonists (Table 6.15) also competed for binding, and the order of their effectiveness paralleled their potency in stimulating or inhibiting α-adrenergic responses. The α-adrenergic receptors were about 20 times

Table 6.13. *Characteristics of β-adrenergic receptors, as measured with* [125]*I-hydroxybenzylpindolol in membrane preparations from different tissues*

Tissue	K_D (nM)	Equilibrium time (min)	Dissociation half-time t_{50} (min)	Number of binding sites (pmol/mg protein)
Turkey erythrocytes[a]	0.25	10	2.5	0.25
Rat heart muscle[b]	1.4	40 to 60	15	0.16
Uterine myometrium[c]	0.12	10	5	0.015

[a] Brown et al., 1976b; Brown et al., 1976a.
[b] Harden, Wolfe, and Molinoff, 1976.
[c] Roberts et al., 1977b.

more numerous than the β-adrenergic receptors (Roberts et al., 1977a).

Human blood platelets also contain α-adrenergic receptors, and these have also been characterized using [³H]dihydroergocryptine (Alexander, Cooper, and Handin, 1978). The K_D for the interaction with epinephrine was 0.34×10^{-6} M. There are about 100 such receptor sites on each platelet. The amount of binding was correlated with the degree of platelet aggregation or inhibition of prostaglandin E_1 stimulated adenyl cyclase which could be elicited by epinephrine.

1. Regulation of adrenergic receptors. The numbers of α- and β-adrenergic receptors in a tissue have been shown to vary under a number of conditions (see Kunos, 1978; Williams and Lefkowitz, 1978). Prolonged exposure of membrane preparations to the β-adrenergic agonist isoproterenol results in a progressive decline in the synthesis of cyclic AMP. This "desensitization" has been observed in frog erythrocyte membranes and a cultured line of mouse lymphoma cells (Mickey, Tate, and Lefkowitz, 1975; Shear et al., 1976). Although the preparations are refractory to stimula-

Table 6.14. *Comparison of the dissociation constants, K_D, for β-adrenergic agonists in membrane-bound and solubilized receptors*

Compound	4	3	R	Membrane-bound receptors, K_D (μM)	Solubilized receptors, K_D (μM)
(−)-Norepinephrine	OH	OH	H	49 ± 2	41 ± 14
(+)-Norepinephrine	OH	OH	H	200	132 ± 28
(−)-Epinephrine	OH	OH	CH_3	4.6 ± 0.2	37 ± 0.7
(+)-Epinephrine	OH	OH	CH_3	137 ± 4	28 ± 14
(−)-Isoproterenol	OH	OH	$-HC<^{CH_3}_{CH_3}$	0.40 ± 0.005	0.21 ± 0.05
(+)-Isoproterenol	OH	OH	$-HC<^{CH_3}_{CH_3}$	183 ± 0.6	68 ± 22
(±)-Soterenol	OH	CH_3SO_2NH	$-CH<^{CH_3}_{CH_3}$	2.4 ± 0.46	0.084 ± 0.01
(±)-Cc34	OH	OH	$-\underset{CH_3}{\overset{CH_3}{C}}-CH_2-\bigcirc-OH$	0.080 ± 0.02	0.012 ± 0.001

Note: The receptor preparations are from frog erythrocytes and the K_D was measured by the ability of the agonists to displace (−)[³H]dihydroalprenolol from binding sites.
Source: Caron and Lefkowitz, 1976.

Table 6.15. *Inhibition of* [³H]*dihydroergocryptine binding to membranes prepared from rabbit uterine smooth muscle*

Adrenergic antagonist	Inhibition of [³H]DHE[a] binding K_D (nM)
Dihydroergocryptine	10
Dihydroergotamine	15
Phentolamine	15
Phenoxybenzamine	18
Ergotamine	21
Dihydroergocristine	42
Dihydroergocornine	60
Ergocornine	100
Ergocryptine	120
Dibozane	130
Ergocristine	140
Yohimbine	220
Ergonovine	450
Azapetine	650
Chlorpromazine	650
Trifluoperazine	1,250
Tolazoline	2,100
Ergotaminine	≫1,000
(±)-Dichlorisoproterenol	27,000
(±)-Propranol	27,000
Lysergic acid	>10⁵
Practolol	>10⁶

[a] [³H]DHE, [³H]dihydroergocryptine.
Source: Williams, Mullikan, and Lefkowitz, 1976.

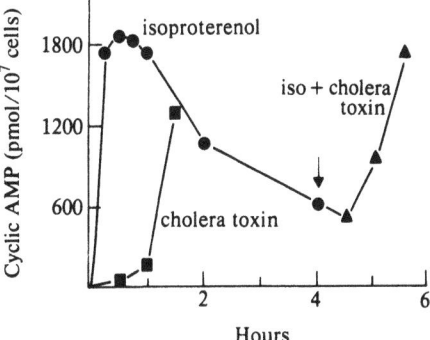

Fig. 6.31. Decline is the responsiveness ("fade") of adenyl cyclase in mouse lymphoma cells following continual exposure to isoproterenol (iso). The stimulating effect of cholera toxin is seen to be undiminished. (From Shear et al., 1976.)

tion by β-adrenergic agonists, they can still respond to other types of substances, such as prostaglandins and cholera toxin (Figure 6.31). Such desensitization is not seen when the membrane preparations are exposed to β-adrenergic antagonists, and it is dependent on the activity of adenyl cyclase. The effect is accompanied by substantial loss in β-adrenergic receptors, which, once the agonist is removed, are gradually restored. There is no change in the affinity of these β-receptors. Lefkowitz has presented a "model" to describe these effects which is summarized in Figure 6.32.

Other types of hormones have also been shown to alter or regulate the numbers of adrenergic receptors, and this may account for the associated changes in sensitivity which are sometimes observed *in vivo*. Treatment of rabbits with estrogens markedly increases the numbers of α-adrenergic receptors, so that they are 3 times more frequent than seen in progesterone-treated animals (Roberts et al., 1977a). The number of β-adrenergic receptors was unchanged. Thyroid hormones are known to increase the sensitivity of the heart to catecholamines. It has been shown that rats treated with thyroid hormones have more than twice the number of dihydroalprenolol-binding sites in their heart muscle than seen in normal con-

trol rats (Williams et al., 1977). The dissociation constant K_D was unchanged by the thyroid hormones.

B. Relationship of the adrenergic receptors to adenyl cyclase

The β-adrenergic receptor appears to exist as a distinct and separate molecule from the adenyl cyclase with which it can, however, become associated. Several observations make such a separation of the receptor and enzyme apparent. The solubilization of the receptor is consistent with such an arrangement. Furthermore, the adenyl cyclase activity, minus the receptor, can also be isolated (Limbird and Lefkowitz, 1977). The quantities of the two components can be shown to vary independently of each other, such as in two genetic strains of mouse lymphoma cells and in reticulocyte-rich erythrocyte preparations from rats treated with phenylhydrazine (Insel et al., 1976; Charness et al., 1976). It has been suggested that the adenyl cyclase and the receptors are under separate genetic control.

An especially fascinating illustration and application of knowledge about the distinct nature of the receptor and adenyl cyclase has been provided by Orly and Schramm (1976) (Figure 6.33). β-Adrenergic receptors obtained from turkey erythrocytes, in which the adenyl cyclase has been inactivated by treatment with N-ethylmaleimide (NEM), were "grafted" or coupled to a line of cultured mouse Friend erythroleukemia cells (F cells). This procedure was performed with the aid of Sendai virus. The F cells possess adenyl cyclase, but it normally cannot be activated by β-adrenergic agonists. Following the artificial coupling, however, isoproterenol can activate this enzyme. Such a procedure has been repeated with other types of cells

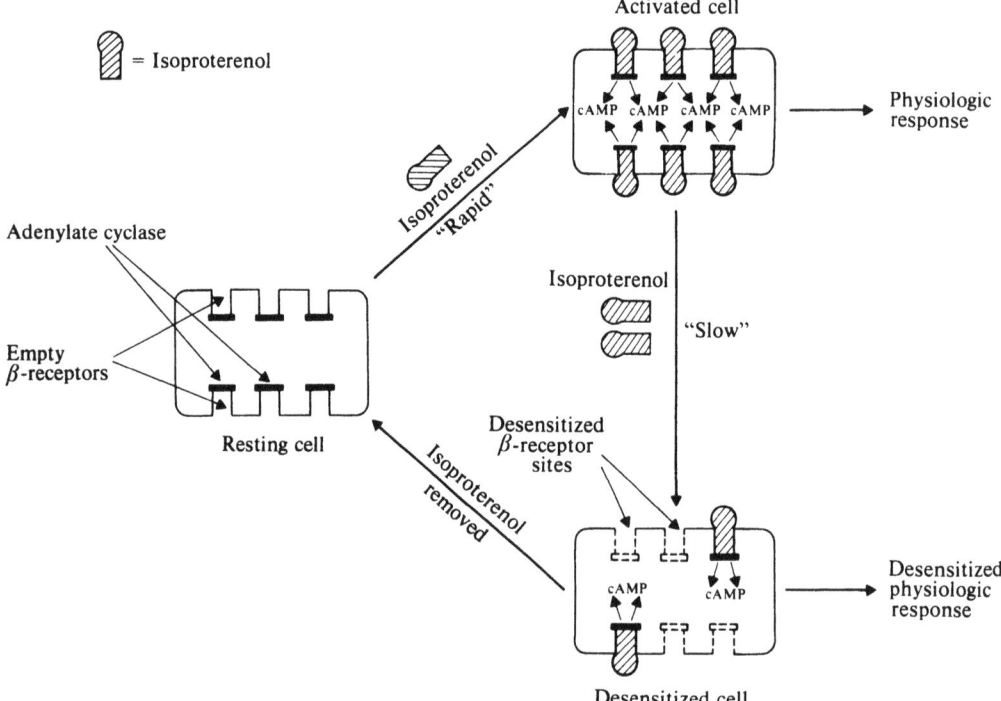

Fig. 6.32. Model to describe the "activation" and "inactivation" of β-adrenergic receptors in the presence of β-adrenergic agonists such as isoproterenol. "Rapid" activation responses occur in seconds, while subsequent desensitization and reactivation are "slow" and take minutes. (From Lefkowitz, 1976. Reprinted by permission from *The New England Journal of Medicine* 295, 327.)

and receptors (Schramm et al., 1977; Schulster et al., 1978). It appears that receptors that can activate adenyl cyclase have common properties which only differ in their propensity to interact with a particular hormone. The nature of the linkage of the receptor to the adenyl cyclase is unknown, but it appears to be quite flexible, as suggested in the "mobile receptor" hypothesis of P. Cuatrecasas (see Figure 6.33b,ii). An even more indirect relationship has been suggested by M. Sonnenberg and A. S. Schreider (Figure 6.33c). They have proposed that the receptor may mediate changes in the electrical potential or ionic permeability of the cell membrane which activates the enzyme. Whatever the nature of the linkage between the receptor and the enzyme, it can, apparently, only take place when an agonist, not an antagonist, is bound to the receptor. It has been suggested that membrane lipids may contribute to the linkage (Maguire et al., 1977).

In order to provide a more complete picture of the nature of the β-adrenergic drug–receptor complex and the manner in which it may activate adenyl cyclase, it is necessary to consider the role of purine nucleotides on the response of this enzyme to hormones and their analogues. The important

role of these compounds in hormonal responses has principally been due to the investigations of M. Rodbell (1975). It has been observed that certain purine nucleotides, especially GTP and ITP, can synergistically increase the responsiveness of adenyl cyclase to stimulation by several types of hormones, including glucagon and catecholamines (Figure 6.34). When the GTP analogue, 5′-guanylylimidodiphosphate [Gpp(NH)P], which is not readily hydrolyzed, is used, the increase in adenyl cyclase activity is even greater than with GTP and, in this instance, occurs in the absence of the hormone (Figure 6.35) or even the presence of receptors. Binding of the nucleotides appears to occur at a specific site on the adenyl cyclase (Salomon and Rodbell, 1975; Pfeuffer and Helmreich, 1975). Thus, there are three sites present in the adenyl cyclase system: a "catalytic site" for the substrate, a "receptor site" for the hormone, and a "regulatory site" for the GTP. The latter nucleotide is thought to play an important role in the functioning of the adenyl cyclase system, the activity of which is modulated by hormones.

A more direct interaction between the regulatory and receptor sites also appears to occur. It has

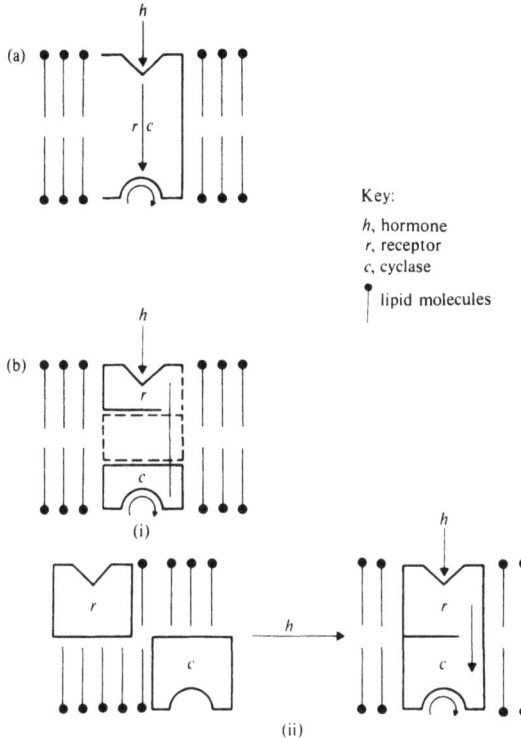

Key:

h, hormone
r, receptor
c, cyclase

● lipid molecules

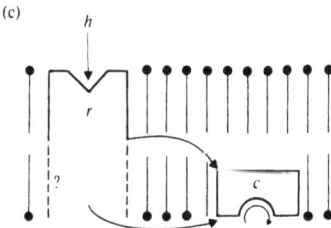

Fig. 6.33. Models of possible physical relationships between hormone receptors and adenylate cyclase. (a) Single molecule. (b) Physical coupling models: (i) stable association – tandem or triplet schemes; (ii) mobile receptor hypothesis. (c) Indirect interaction models. (From Greaves, 1977.)

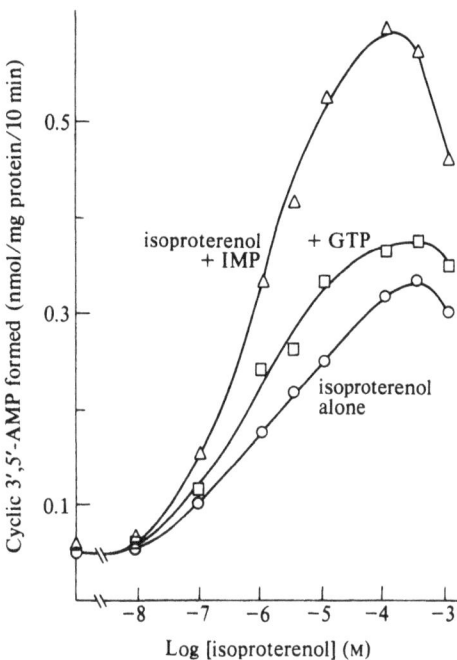

Fig. 6.34. Influence of purine nucleotides on the response of adenyl cyclase (from turkey erythrocytes) to isoproterenol. The concentration of GTP and IMP was 0.2 mM. (From Bilezikian and Aurbach, 1974.)

been shown that purine nucleotides, when combined with the adenyl cyclase, can induce changes in the receptor (Lefkowitz, Mullikan, and Caron, 1976). These changes, which are only seen when the receptor is occupied by an agonist, and so is associated with the enzyme system, decrease its affinity for agonists but also increase the efficacy by which it can activate the adenyl cyclase system.

Rodbell (see Rodbell et al., 1975) has constructed a kinetic model of the adenyl cyclase system (Figure 6.36). This was originally applied to

the action of glucagon on adenyl cyclase, but it also applies to catecholamines. The adenyl cyclase system can exist in three states, between which, as a result of allosteric effects, transformations can occur. These changes alter its properties with respect to its ability to form cyclic AMP from its substrate ATP (catalytic activity). The transformations can occur spontaneously but only very slowly, and they are enhanced by the presence of GTP and hormones. These ligands each bind to distinct but interacting sites, the regulatory site and receptor site. The enzyme system exists in a basal or *ground state* designated E, which has a low V_{max}. When GTP binds to the regulatory site, an allosteric change occurs so that the enzyme is transformed to a *second state* called E_N^1, which has a high V_{max}. (The GTP exists in two forms, an unchelated protonated one, which is the activator, as opposed to the inactive Mg^{2+} chelated nucleotide.) The E_N^1 adenyl cyclase system, while having a high V_{max}, can, however, be readily inhibited by the protonated form of its substrate $HATP^{3-}$ and so is said to have a low K_i. The active form of the substrate is a chelated form, $MgATP^{2-}$, and its interaction with the catalytic site is competitively inhibited by the $HATP^{3-}$. The E_N^1 form of the enzyme system can, however, be spontaneously transformed to a *third state*, E_N^{11}, which

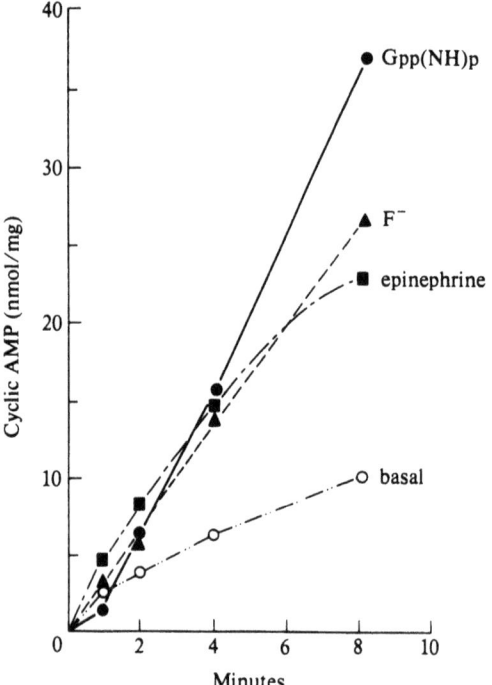

Fig. 6.35. Effects of epinephrine (0.35 mM), fluoride (F⁻, 10 mM), and Gpp(NH)p (1 μM) on adenyl cyclase activity from rat fat cells. (From Londos et al., 1974.)

retains the high V_{max} but also has a high K_i, so that the inhibition by $HATP^{2-}$ does not occur as readily. The attainment of the E_N^{11} state is facilitated by the presence of the hormone at the receptor site.

It has been proposed that the catecholamines activate adenyl cyclase by inducing an "opening" of the regulatory site so that tightly bound GDP can be displaced by the free, activating GTP (Cassel and Selinger, 1978). The response is terminated by

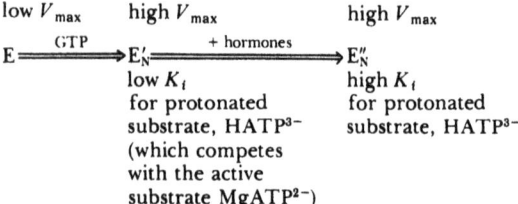

Fig. 6.36. Model proposed by M. Rodbell (1975) to describe and account for the effects of purine nucleotides on the actions of glucagon and catecholamines on the adenyl cyclase system. E, different states of the adenyl cyclase system. K_m for the active substrate, $MgATP^{2-}$, is the same in all three states. GTP is only effective in its nonchelated protonated form (it can exist chelated with Mg).

the conversion of the GTP to GDP, by GTPase which is associated with this site.

C. Function of the adenyl cyclase system in adrenergic responses

"Activation" of the adenyl cyclase system, under the influence of GTP and catecholamines, allows this enzyme to promote formation of cyclic adenosine 3′,5′-monophosphate (cyclic AMP) from ATP. As described earlier, this cyclic nucleotide mediates β-adrenergic responses to catecholamines and in many instances addition of exogenous cyclic AMP to tissues can mimic such responses. The cyclic adenosine 3′,5′-monophosphate is rapidly converted in the cell to 5′-AMP, which is inactive. This transformation occurs under the influence of a soluble enzyme(s) phosphodiesterase, which can be inhibited by compounds such as theophylline, caffeine, and papaverine. Such drugs can thus maintain cellular levels of cyclic AMP and so mimic, and increase, many of its actions as well as those of the catecholamines.

As we have seen, β-adrenergic responses to catecholamines are quite diverse in their nature and occur in a variety of tissues and organs. They include effects on the intermediary metabolism of carbohydrates and fats, ion movements and relaxation of muscle, and secretion by glands. Quite a lot of information is available about the cellular events that mediate such responses, but it is only within the last 10 years that the nature of the common linkage of these processes to cyclic AMP has become apparent.

In 1968, E. G. Krebs and his collaborators (Walsh, Perkins, and Krebs, 1968) isolated a *protein kinase* from rabbit skeletal muscle, which could phosphorylate casein and protamine. This enzyme was activated by cyclic AMP. They stated that "It is attractive to think that this enzyme serves as a link between the epinephrine stimulation of adenyl cyclase and the activation of phosphorylase kinase . . . ," which is the process that results in glycogenolysis. This important observation resulted in the forging of another link between the hormone–receptor interaction and the final response, as it indicated the nature of the immediate effect of cyclic AMP in the cell. The role of protein kinases in cells has been reviewed by Krebs (1972) and Walsh and Ashby (1973).

The presence or protein kinase enzymes in animal tissues has been known for about 25 years. They promote the transfer of a phosphoryl group from nucleotides, such as ATP, to proteins. The recipients are serine and threonine residues. The process requires Mg^{2+}, but other metals, such as Mn^{2+}, are also effective. The opposite reaction also occurs, so that such phosphorylated proteins can be dephosphorylated under the influence of a phosphoryl protein phosphatase. The two reactions can be summarized as follows:

$$\text{protein} + n\text{ATP} \xrightarrow{\text{protein kinase}} \text{protein} - P_n + n\text{ADP}$$
$$\text{protein} - P_n + n\text{H}_2\text{O} \xrightarrow[\text{phosphatase}]{\text{phosphoryl protein}} \text{protein} + n P_i$$

The important contribution of E. G. Krebs and his collaborators was the recognition that some such protein kinases can be activated by cyclic AMP. It has been shown (Kuo and Greengard, 1969) that such protein kinases are widespread in the animal kingdom, where they occur in many tissues and can even be activated by other types of cyclic nucleotides, apart from cyclic AMP. In catecholamine-sensitive tissues, however, the latter is more effective. It was suggested that such cyclic AMP-dependent protein kinases may mediate *all* hormonal responses that utilize cyclic AMP as a "second messenger." The activation of cell protein kinases has been shown to occur at physiological levels of cyclic AMP (Beavo, Bechtel, and Krebs, 1974).

Protein kinase enzymes can transfer a phosphoryl group to many types of proteins, including those in cell components, such as enzymes, plasma membranes, nuclei, mitochondria, and the endoplasmic and sarcoplasmic reticulum. The phosphorylation of such structures may alter their conformation and activity so as to produce a response such as that seen to a hormone (Figure 6.37). Cyclic AMP-dependent kinases have molecular weights of 120,000 to 280,000 and can be quite specific in their actions. In the presence of cyclic AMP they dissociate into two subunits (see Brostrom et al., 1971): a *regulatory subunit* (R) to which the cyclic AMP is bound, and the active *catalytic subunit* (C). The intact holoenzyme is inactive. The process can be summarized thus:

R · C   + cyclic AMP → R · cyclic AMP   +   C
inactive   regulatory   active catalytic
holoenzyme   subunit   subunit

Calcium plays a ubiquitous role in the functioning and effects of adenyl cyclase and cyclic AMP. Its role is not doubted, but it is a complex one which is quite controversial (see, e.g., Steer, Atlas, and Levitzki, 1975; Rasmussen and Goodman, 1977). It has been variously shown and suggested that Ca^{2+} inhibits or activates adenyl cyclase and phosphodiesterase. Such effects could be involved in the initiation of positive- or negative-feedback mechanisms in the responses of the adenyl cyclase system to catecholamines.

1. Metabolic effects. Catecholamines promote the conversion of glycogen to glucose, inhibit the synthesis of glycogen, and increase the conversion of triglycerides to free fatty acids and glycerol. These effects occur in a cascade of reactions and

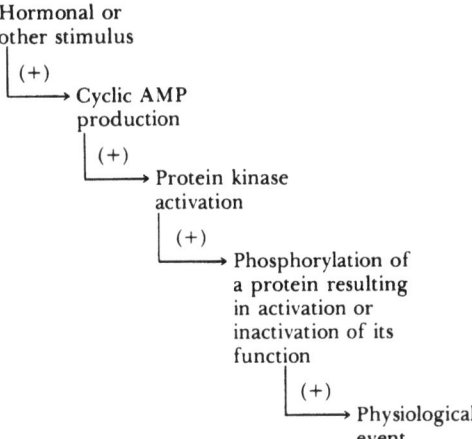

Hormonal or other stimulus

Fig. 6.37. Model to account for the role of cyclic AMP in regulating cell functions through activation of a protein kinase. (From Krebs, 1972.)

they are all ultimately the result of changes in the phosphorylation of enzymes, under the influence of a cyclic AMP-dependent protein kinase.

The conversion of glycogen to glucose in the liver and in muscle is under the control of a phosphorylase which is converted from its inactive *b* form to its active *a* type. This change occurs due to the effects of a phosphorylase kinase, which, in its turn, also exists in an inactive nonphosphorylated and an active phosphorylated form. The transfers of the phosphoryl group from ATP take place under the influence of a cyclic AMP-dependent protein kinase (Figure 6.38). It has thus also been called phosphorylase kinase kinase (or sometimes kinase kinase or kinase II). This enzyme also promotes the transformation of glycogen synthetase from its active (I) to its inactive (D) state. The ratio of the amounts of phosphorylase/phosphorylase kinase/protein kinase in the cells has been estimated as 800:30:1 (Krebs, 1972).

The mobilization of fatty acids and glycerol from triglycerides follows a similar sequence (Figure 6.39). The inactive form of the lipase enzyme (*b*) is converted to the active (*a*) type as a result of its phosphorylation by a cyclic AMP-dependent kinase.

2. Effects on muscle. Catecholamines, either in their roles as neurotransmitters or hormones, have a variety of effects on the contractility of different types of muscle. Some, but *not all*, of these may be related to changes in cellular levels of cyclic AMP and the phosphorylation of cell proteins. Consensus regarding possible mechanisms has, however, not been arrived at (see Osnes and Øye, 1975; An-

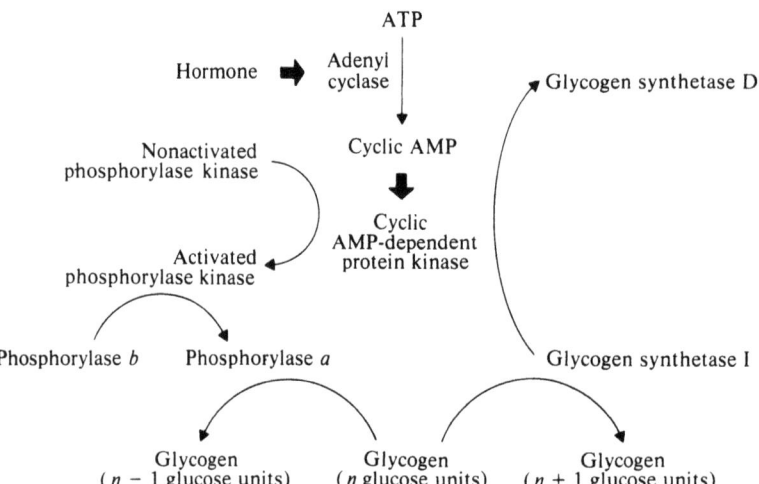

Fig. 6.38. Dual role of cyclic AMP-dependent protein kinase in regulating glycogen metabolism in muscle. Active phosphorylase *a* is formed under the influence of the protein kinase, so that glycogen is broken down to glucose. At the same time, glycogen synthetase is converted to its inactive D-form, so that the synthesis of glycogen is reduced. (From Soderling et al., 1970.)

dersson et al., 1975; Katz, Tada, and Kirchberger, 1975; Krause et al., 1975; Tsien, 1977).

It will be recalled that catecholamines can either facilitate a relaxation or a contraction of muscle. In

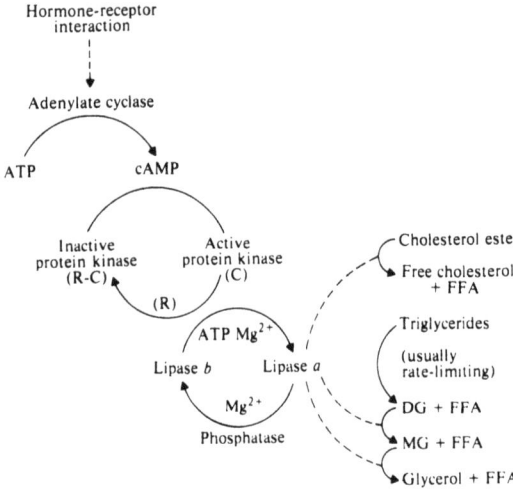

Fig. 6.39. Schematic diagram showing the role of hormones, including catecholamines, in the mobilization of glycerol and fatty acids in adipose tissue. A cyclic AMP-dependent protein kinase converts the inactive lipase *b* to active lipase *a*, resulting in the conversion of triglycerides to diglycerides (DG), monoglycerides (MG), free fatty acids (FFA), and glycerol. R and C refer to the regulatory and catalytic subunits of the protein kinase system. (From Steinberg and Khoo, 1977.)

the heart they increase pacemaker activity, speed relaxation, and have a prominent positive inotropic effect. These responses are usually considered to be mainly β-adrenergic types of responses. However, as the inotropic effect is also seen in the presence of β-adrenergic antagonists, it also appears to have an α-adrenergic component. Smooth muscle may be contracted (= α-adrenergic) or relaxed (= β-adrenergic) by catecholamines. Most peripheral blood vessels contract, but some, especially those in skeletal muscle, relax. Muscles in the gut may either contract (usually the sphincters) or relax. The uterus behaves similarly, but in this instance the same type of muscle can respond in either way, depending on the species and whether it is dominated by the effects of estrogens or progesterone. The bronchi relax. There have been many attempts to find a common mechanism for these various actions, but this search has been only partly successful. Many of the results and their interpretation remain contentious. It is generally agreed that α-adrenergic effects do not involve increases in cyclic AMP levels, although a decrease, or a rise in cyclic GMP, has been suggested. The β-adrenergic responses are invariably associated with a rise in cyclic AMP levels, but whether this always has a cause-and-effect relationship to the observed changes in contractility is uncertain. Some of the relationships seem to be paradoxical. Thus, a rise in cyclic AMP is associated with an increased contractility of heart muscle but a relaxation of smooth muscle. The differences, however, appear to reflect their patterns of contractility and the more rapid rhythmical activity of the heart.

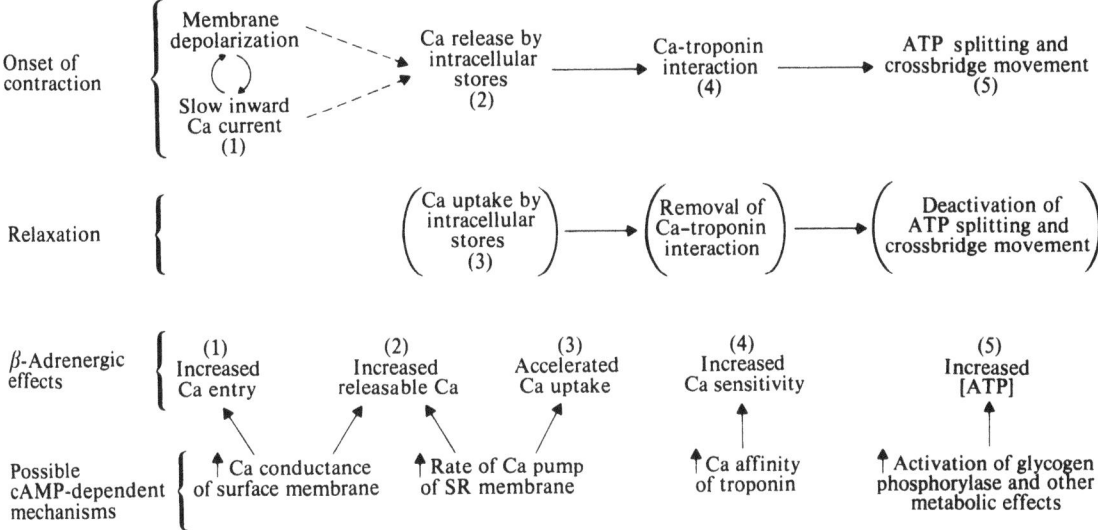

Fig. 6.40. Multiple possible sites of action of catecholamine-dependent cyclic AMP on excitation–contraction coupling in heart muscle. (From Tsien, 1977. Cyclic AMP and contractile activity in heart. In *Advances in Cyclic Nucleotide Research*, Vol. 8, pp. 364–420 (edited by P. Greengard and G. A. Robinson). New York: Raven Press.)

The mechanisms may be basically the same in each instance (see later).

The contraction of muscle is intimately associated with changes in free ionized Ca^{2+} in the cell. This divalent ion plays an important role in the excitation–contraction coupling mechanism. When it combines with the protein troponin the inhibitory effect of tropomyosin on the contractile actinomyosin system is relieved. It is thus suspected that catecholamines, through their product cyclic AMP, may act by changing the levels of free Ca^{2+} in the cell (see, e.g., Triggle, 1972). The effects and roles of calcium in cells are, however, ubiquitous, which makes it difficult to disentangle the precise events that lead to and result from changes that occur in the presence of catecholamines. There seems little doubt that catecholamines influence calcium metabolism in muscle cells via changes in the formation of cyclic AMP, but other mechanisms for this probably also exist, especially in relation to α-adrenergic activity. It is difficult to define the exact relationship of the cyclic AMP to contractile events, because, like Ca^{2+}, it has other roles in the cell. A brief recapitulation of the processes that control calcium levels in the cell and contraction is necessary to appreciate the possible mechanisms of action of catecholamines on muscle tone.

Only a very small fraction of the calcium content of the cell is in a free, active, ionized state. The normal concentration is about 10^{-7} M, but small changes can have considerable effects on such processes as contractility. Most of the calcium in muscle is segregated in the mitochondria and the vesicles of the sarcoplasmic reticulum (or its equivalent

in smooth muscle). It is also present in the plasma membrane. Most calcium exists in either a bound state or is precipitated in the cell organelles, but there also remains a free Ca^{2+} "pool." Accumulation of Ca^{2+} by the mitochondria and sarcoplasmic reticulum is thought to involve the action of a "Ca-pump," which in the latter instance is associated with a Ca^{2+}-activated ATPase. Changes in free ionized Ca^{2+} in the cytoplasm are the result of an exchange with the stores in these organelles and also changes in Ca^{2+} movements from the extracellular fluids across the plasma membrane into the cell.

Contraction of muscle follows an electrical depolarization of the plasma membrane, which usually reflects an increase in Na^+ movement into the cell. In cardiac muscles this event is followed by a "slow inward current" which is carried by Ca^{2+}. This ion movement results in an increase in Ca^{2+} stores in the plasma membrane and free Ca^{2+} inside the cell. The latter, in addition, appears to trigger a "calcium-dependent" release of Ca^{2+} from the sarcoplasmic reticulum; contraction is thus effected. Relaxation follows a decline in cytoplasmic Ca^{2+} due to its reaccumulation by the cell organelles and its extrusion from the cell.

The catecholamines could be acting at several stages of the process. In cardiac muscle there is evidence that their effect may be a multiple one mediated by changes in cyclic AMP (see Figure 6.40). There is an increase in the initial rate of Ca^{2+} entry into the cell. Catecholamines also stimulate uptake of calcium by the sarcoplasmic reticulum, so that with the continued stimulation there is sub-

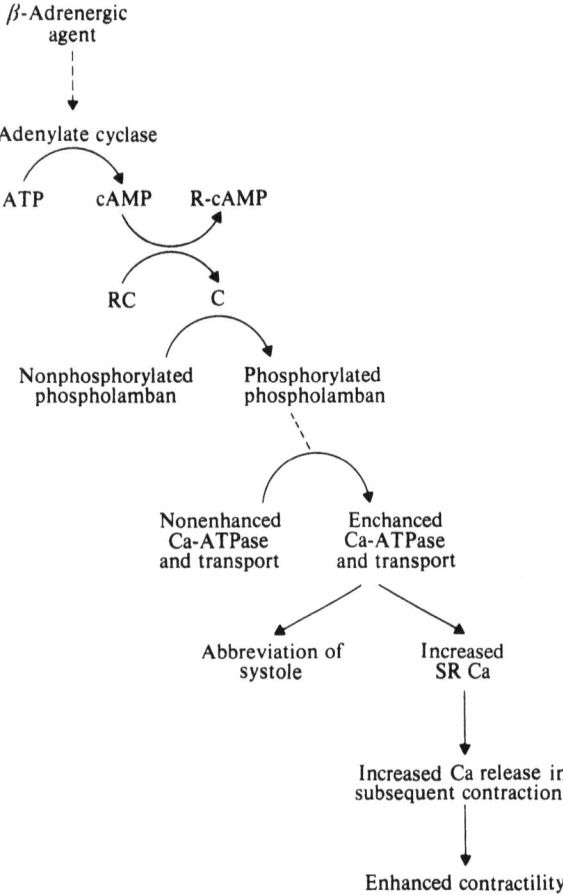

Fig. 6.41. Possible mechanism for β-adrenergic effects on contractile activity, mediated by enhanced Ca uptake by the sarcoplasmic reticulum (SR). RC is protein kinase in soluble form or in association with SR membranes. R is the regulatory subunit; C is the catalytic subunit. Phospholamban denotes a 20,000- to 22,000-dalton protein in the SR membrane, which is the substrate for cAMP-dependent phosphorylation. (From Tsien, 1977. Cyclic AMP and contractile activity in heart. In *Advances in Cyclic Nucleotide Research,* Vol. 8, pp. 364–420 (edited by P. Greengard and G. A. Robinson). New York: Raven Press. Based on Katz, Tada, and Kirchbergerger, 1975.)

sequently more releasable Ca^{2+} available in these organelles, which increases subsequent responses. Both of these effects have been related to changes in the phosphorylation of constituent proteins which are mediated by the cyclic AMP-dependent protein kinase system (see, e.g., Figure 6.41). There may also be effects on the contractile proteins themselves, which may involve a phosphorylation and/or be the result of the simultaneous metabolic effects of the catecholamines on tissue glycogen and lipid metabolism.

Rises in cyclic AMP levels in cardiac muscle have been shown to precede changes in contractility.

The metabolic effects of catecholamines on glycogen metabolism have a different and slower time course from the effects on contraction, so that the two types of effects are not directly related. The α-adrenergic component of the positive inotropic effect of catecholamines in heart muscle does not appear to be related to cyclic AMP metabolism.

The mechanism of action of catecholamines on smooth muscle appears to be more complex than in heart muscle. Smooth muscle is more diverse, as it is present at many sites in the body, where it has different functions. Catecholamines may elicit either a contraction or relaxation and in some instances, such as the uterus, either effect may be seen to depend on the physiological circumstances. It is usually agreed that both types of effects reflect, as appropriate, increases or decreases in the free Ca^{2+} concentration in the cells. The problematical question is if, and how, such effects are related to changes in cyclic nucleotide levels. The experiments and reviews by Andersson and his collaborators (1975) highlight some of the difficulties. Observed changes in cyclic AMP (and cyclic GMP) levels in smooth muscle are variable and depend on the particular muscle and agonist used, the time following stimulation when the measurements were made, and the *in vitro* conditions, such as the external Ca^{2+} concentrations. The relaxing effect of catecholamines on smooth muscle seems to be generally associated, temporally and quantitatively, with a rise in the levels of tissue cyclic AMP. It appears that, as in heart muscle, this stimulates an uptake of cytoplasmic Ca^{2+} into the cell organelles. Precise and consistent information about this process is not plentiful and an increased efflux of Ca from the cells could also be involved. The relationship of cyclic AMP to contractile, including α-adrenergic, effects are complex, as increases and decreases in the nucleotide levels are observed. Elevated cyclic AMP concentrations commonly occur during the course of contraction just prior to relaxation and are also seen in response to nonadrenergic agents such as carbachol. This effect could be part of a negative-feedback system mediating a decline in cell Ca^{2+} levels prior to relaxation. Andersson and his collaborators observed an initial decline in cyclic AMP levels in intestinal vascular and bronchial smooth muscle following α-adrenergic stimulation and carbachol. The change was not dependent on Ca^{2+}, which suggests that it may be a primary response. The subsequent rise in cyclic AMP levels was, however, Ca-dependent and could be inhibited by indomethacin, which blocks the synthesis of prostaglandins. No consistent changes in cyclic GMP were observed.

3. Effects on glandular secretions. Catecholamines may increase or decrease the rates of secretion of certain exocrine and endocrine glands, including the salivary glands, mucous glands, and the insulin-secreting B-cells of the islets of Langerhans. In some instances, such as the B-cells, secretion may be either increased or decreased, which corresponds, respectively, to β-, or the usually predominant, α-adrenergic effects. Many such effects of catecholamines are indirect, as they may involve changes in the blood supply to the glands. Direct effects, however, may also be involved.

In rat parotid gland, epinephrine increases the rate of fluid secretion, which is associated, *in vitro*, with a release of K^+. This α-adrenergic response is dependent on the presence of external Ca^{2+} (Selinger, Eimerl, and Schramm, 1974). The Ca ionophore A23187 can mimic the effect of epinephrine in this system. It was suggested that the α-adrenergic receptor may function like an ionophore. Such a mechanism could be mediating these effects of epinephrine in other responsive glands (e.g., sweat glands) and tissues.

Beta-adrenergic effects on glandular secretion are rare, but when they do occur, such as in the B-cells of the islets of Langerhans, they involve increased formation of cyclic AMP and a lowering of free Ca^{2+} in the cells.

6.15. Uses of catecholamine hormones and related drugs

Adrenergic agonists and antagonists are used clinically to treat a large variety of disorders. Most of these are related to the roles of the catecholamines as neurotransmitters in such processes as regulation of blood pressure, contraction of heart muscle, and behavior. Thus, propranolol, a β-adrenergic antagonist, is used to treat cardiac arrythmias, angina pectoris, and hypertension. They are also utilized for their effects on the central nervous system, as antidepressants, analeptics, and to promote weight loss (anorectics). Such applications of sympathomimetic drugs will not concern us here. The use of adrenergic types of drugs, including those that mimic and block the effects of dopamine in order to influence the release of anterior pituitary hormones, have already been described elsewhere [Sections 3.2F(3) and 7.6G(3)].

The vasoconstrictor properties of epinephrine are utilized as a *local hemostatic agent* during certain types of surgery, such as that involving the nasal passages, and in patients with bleeding disorders. It is also often added to solutions of local anesthetics so as to delay their absorption and so prolong their actions and limit systemic toxicity. The wisdom of using epinephrine in heart block and cardiac arrest is questionable and may result in ventricular fibrillation, but in extreme situations it may be justified. Mechanical and electrical measures are favored for initial treatment and electrical pacemakers for chronic use.

Injected epinephrine may be lifesaving in instances of severe *allergic and hypersensitive reactions*, in which it limits tissue swelling. This effect is of special importance if it involves the pulmonary tract. Epinephrine also relaxes the bronchi and so is of special use in status asthmaticus, where a decrease in pulmonary edema is also a useful effect. Adrenergic drugs are a mainstay in the treatment of chronic bronchial asthma. The more specific B_2 agonists are now more favored, as they have fewer side effects.

The use of catecholamines to treat *shock* and hypotension such as that which follows traumatic injury, massive bacterial sepsis, and hemorrhage is controversial. In such shock conditions adrenergic mechanisms are usually already acting and limiting the blood flow to certain tissues. The wisdom of increasing this effect further is in doubt, and indeed it has been suggested that adrenergic blocking drugs may even be more appropriate. However, the maintenance of blood pressure may be of primary importance on some occasions, such as maintaining the blood supply to the heart and brain. The use of norepinephrine, rather than epinephrine, is then more appropriate. Such measures are, however, only of temporary use, restoration of blood volume being of primary importance. The risks of decreasing the perfusion rate of the tissues and even the blood volume further, promoting cardiac arrythmias and obscuring the true nature of the situation, are factors that need to be considered in individual cases.

Sympathomimetic drugs, including epinephrine, are used on the *eye*. They offer one possible way of dilating the pupil to facilitate ophthalmic examinations. They can also be used for the treatment of primary open-angle glaucoma. Solutions of epinephrine can be applied topically to the eye and result in a decrease in the formation of aqueous fluid together with an increased outflow of this fluid, so that the intraocular pressure declines. Catecholamines should not be used in patients with narrow-angle glaucoma as they can sometimes increase intraocular pressure. The use of epinephrine on the eye is not free of possible side effects, which include irritation, blurred vision, headache, allergic conjunctivitis, and, with chronic use, deposits of pigment on the conjunctiva and cornea. Appreciable systemic absorption can occur, especially if the cornea is damaged.

Epinephrine has been used to treat *hypoglycemia*, including that induced by insulin, but glucose and glucagon are currently preferred, as they have fewer side effects.

Epinephrine is usually injected s.c. or i.m. Intravenous administration is often used in the treatment of shock but can result in dramatic effects on the heart and has resulted in death, usually due to cardiac arrythmias but also to cerebral hemorrhage and respiratory arrest. In asthmatics it can be given chronically by inhalation. Oral and sublingual preparations of some analogues are available. Analogues of epinephrine and norepinephrine are widely used. Thus *phenylephrine* (α) is administered to reduce nasal congestion; *salbutamol* (β_2) has recently been introduced to relax the bronchi in chronic asthmatics and those suffering from other pulmonary diseases, such as emphysema. It can be given as an aerosol or orally. The specific β_2 actions of salbutamol exclude side effects related to the actions of adrenergic-type drugs on the heart. Other drugs which are used for this purpose but which exhibit both β_1 and β_2 effects are *isoproterenol* (given parenterally or as an aerosol) and *ephedrine* (active orally but with side effects on the central nervous system). The use of the β-adrenergic drugs, especially of the β_2 type, in relaxing the uterus and delaying labor is discussed in Section 7.6H(4).

Hypersecretion of catecholamines from tumors, such as pheochromocytomas, can be treated with specific adrenergic antagonists. The α-adrenergic effects are best blocked chronically by *phenoxybenzamine*, but the rapid, though brief, action of *phentolamine* may be utilized in acute emergencies. If β-adrenergic effects are prominent, *propranolol* is also administered. The synthesis of catecholamines can be inhibited by α-methyl-*p*-tyrosine, but this drug produces a troublesome crystallinuria. The use of drugs in the diagnosis of pheochromocytoma is discussed in Section 6.16.

The effects of circulating catecholamines are potentiated by thyroid hormones and are the major problem when acute increases in the secretion of these hormones occur in hyperthyroidism. This emergency situation is called "thyroid storm," and propranolol plays a major role in its treatment (Section 4.10C).

A. Side effects and interactions with other drugs

The injection of epinephrine is frequently associated with subjective feelings of apprehension, fear, and anxiety, but these usually pass quickly. The major side effects are related to its actions on the heart (sinus tachycardia and ventricular fibrillation) and increases in blood pressure that are especially likely to occur during intravenous administration. The cardiac effects occur as the result of a direct action on the heart, but indirect reflex responses involved in the adjustment of the blood pressure can contribute.

Peripheral vasoconstriction may also result in problems due to a decreased blood supply to certain organs such as the kidney, toes, and fingers. They may also promote a reduction in plasma volume. Injections into the fingers or toes or repeated chronic injections at single sites can result in tissue necrosis.

There are a number of clinical situations when epinephrine may be expected to have toxic side effects, especially in relation to the genesis of cardiac arrythmias. Clearly, their use in patients with cardiac disorders and peripheral vascular disease can be problematical. Some anesthetics, such as *cyclopropane, trichlorethylene,* and *halothane,* potentiate the effects of endogenous and exogenous catecholamines so that the use of epinephrine under such conditions needs to be undertaken with care. As described above, thyroid hormones also increase the actions of catecholamines, so that their use in patients with hyperthyroidism or in those being treated with *thyroid hormones* also needs circumspection. The concurrent use of digitalis *cardiac glycoside* compounds for the treatment of heart disorders also increases the likelihood of epinephrine precipitating cardiac arrythmias. *Mercurial diuretics,* which are rarely used today, may also have such effects in conjunction with epinephrine.

The clearance of catecholamines from the circulation can be delayed by several types of drugs. $Uptake_1$ by the sympathetic neurons is slowed by substances that compete with it for the "amine pump" in the plasma membrane. These include *cocaine, amphetamine, imipramine, chlorpromazine,* and *guanethidine. Phenoxybenzamine* decreases $uptake_2$ into nonneuronal tissues.

Many of the potential side effects of the catecholamines can be avoided by using more selective varieties of the drugs. Thus norepinephrine has relatively greater effects on the peripheral circulation than on the heart. It is now possible to use agonists with specific α (e.g, phenylephrine) or β_2 (e.g., salbutamol) activity, so that adverse β_1 effects on the heart are not a problem. The general β-adrenergic antagonists, such as propranolol, antagonize both β_1 and β_2 effects and so when used to treat cardiovascular disorders may result in bronchoconstriction. The β_1 antagonists, such as practolol, are alternative drugs which lack this effect on the pulmonary tract but can have other types of side effects.

6.16. Diseases of the adrenal medulla

No well-defined clinical conditions have been described which are due to a hyposecretion of adrenal medullary hormones, although hypersecretion due to the presence of tumors can occur (see Besser and Jeffcoate, 1976; Fernandes and Bellini, 1977). Pheochromocytomas, which involve the chromaffin cells, are the most common of these, but neuroblastomas, which are derived from the sympathoblast cells in childhood, may also secrete catecholamines. They commonly accompany medullary carcinoma of the thyroid gland.

Pheochromocytomas are, in 80 percent of cases, present in the adrenal medulla but may also be located at other sites, especially the sympathetic ganglia and urinary bladder. Only about 10 percent of pheochromocytomas are malignant. The incidence is difficult to define, but some retrospective postmortem studies have suggested that they may occur in 0.05 to 0.1 percent of people. It has been estimated that about 1000 people die each year in the United States as a result of undiagnosed pheochromocytomas.

These tumors can secrete large amounts of epinephrine and norepinephrine and their diagnosis is best based on measurements of the excretion of the hormone metabolites, such as VMA, in the urine. In 60 percent of instances there is a sustained increase in blood pressure, but in the remaining 40 percent the hormones may be released periodically (paroxysmal). This secretion may occur spontaneously or in response to stimuli such as exercise, smoking, or emotional stress. The blood pressure may rise dramatically and attain a systemic level of 200 to 300 Torr. This may be accompanied by headache, cardiac arrythmias, palor, sweating, anxiety, nausea, and vomiting. Patients suffering from pheochromocytomas are usually underweight and may exhibit a glucosuria. These characteristics of the disease can readily be related to the effects of excessive amounts of catecholamines.

Effective treatment of pheochromocytoma is surgical, but drugs also play an important role in presurgical control, as an adjunct to surgery, and, formerly, in the diagnosis of the disease. The latter procedure is now mainly dependent on quantitative measurements of urinary catecholamines. However, tests based on provocation of release or blockade of the peripheral effects of the circulating catecholamines have been used. Thus, *histamine* injection can promote a dramatic increase in blood pressure due to a stimulation of secretion, while *phentolamine* can cause an equally impressive decline in blood pressure due to its α-adrenergic blocking action. Such uncontrolled oscillations in blood pressure can be dangerous, and even fatal, so that these procedures are not generally approved of at this time.

Drug treatment of pheochromocytoma is based on the use of α- and β-adrenergic blocking drugs. In hypertensive crises due to high circulating catecholamines, i.v. *phentolamine* can be used to lower blood pressure, although better control is possible with i.v. *Na nitroprusside* (Lipson et al., 1978; Daggett, Verner, and Carruthers, 1978). This latter drug has a direct peripheral vasodilatatory effect and can be used in conjunction with propranolol and phenoxybenzamine. The brief action of phentolamine makes it unsuitable for chronic use, so that *phenoxybenzamine* is routinely used. Propranolol is given to control cardiac arrythmias. However, it should only be used following α-adrenergic blockade or pulmonary edema may result due to the inability of the heart to compensate for the peripheral vasoconstriction (Wark and Larkins, 1978). The blockade of the synthesis of catecholamines by α-methyl-p-tyrosine has also been used successfully but is associated with the deposition of crystals in the kidney tubules. The chronic use of such drugs is usually reserved for treatment prior to surgery or if, for some reason such as malignancy, surgery is not possible.

Exploratory manipulations aimed at locating the sites of the tumors, surgery, and labor may initiate a sudden release of catecholamines from pheochromocytomas. To guard against this possibility and as a preparation for surgery, phenoxybenzamine and propranolol are given for several days preoperatively. Because of the persistent peripheral vasoconstriction, the blood volume may be reduced, so that sudden removal of such a tumor may result in a profound hypotension. Pretreatment with the blocking drugs thus not only antagonizes the effects of suddenly released catecholamines but allows the blood volume to attain more normal levels before the tumor is removed. Plasma transfusions are, however, usually also required. Diuretics should be avoided preoperatively in order to help maintain the volume of the extracellular fluids.

7

The endocrinology and pharmacology of reproduction

The reproductive apparatus is necessary for the perpetuation of the species but is not essential for the life of the individual. It does, however, make life more interesting and plays a considerable role in the social and economic framework of our lives. Hormones, produced principally by the pituitary, testes, and ovaries, play a vital role in the many processes involved in reproduction. These include conception, embryonic differentiation, birth, lactation, development at puberty, and the desire and ability to procreate. A number of related factors have entered into the reproductive pattern of man, mainly as a result of his intellectual abilities, which have allowed him a measure of control over his environment. Life expectancy has increased considerably over a relatively short period of less than 100 years. Reproduction has become more effective as a result of the dramatic decline in infant mortality and prolonged life span. The age of the menarche has declined and the menopause comes later, so that the reproductive life of women has been extended. There has thus been a real and potential increase in fecundity which has become a nutritional and economic threat. The menopause in women occurs at about 50 years of age. At the turn of the century, this age corresponded approximately to life expectancy. Today, the average woman subsequently survives for more than another 25 years. Women are thus in the unique position in the animal kingdom of living about one-third of their lives after they have lost the ability to reproduce. This period is accompanied by a profound change in female endocrine function.

Although the male remains relatively fertile throughout his life, an increased life expectancy also creates associated problems, including a gradual decline in the endocrine activity of the testes and an increased incidence of diseases, such as enlargement and even carcinoma of the prostate. In addition, the social adjustments that need to be made to such changes should not be under-estimated. The endocrine-related modifications and treatment of the human reproductive system is thus an important branch of medicine, which has not yet reached its zenith.

7.1. Steroidal sex hormones

The primary hormones that control reproductive functions in the male and female are steroids (Figure 7.1). The general chemical nature of these compounds has been described in Section 5.1. In the male, these hormones are called *androgens*, the principal one being *testosterone*, but some *androstenedione* is also formed. Testosterone is metabolized to a more active form, *5α-dihydrotestosterone*. In the female there are two main types of steroidal sex hormones, (a) the *estrogens*, of which the most active is estradiol-17β, but *estrone* and *estriol* are also formed; and (b) the *progestins*, of which the most active is *progesterone*, but *17α-hydroxyprogesterone*, which has only a weak action, is also present in substantial amounts in the circulation. Some *androgens* are also formed in women and are important physiologically.

The sources of these steroidal hormones are the testes in the male and the ovaries and placenta in the female. The adrenal cortex also makes a significant contribution, especially in women, as it secretes androstenedione, which can be converted peripherally to estrogens and testosterone.

7.2. Embryonic development of the reproductive apparatus

In its early stages of development, the embryo, whether genetically male or female, lacks any special morphological sex characters and it is said to be sexually "indifferent." The gonocytes, which give rise to the germ cells, can be identified during the fourth week of gestation in man, and several hundred of these migrate to their eventual resting

228

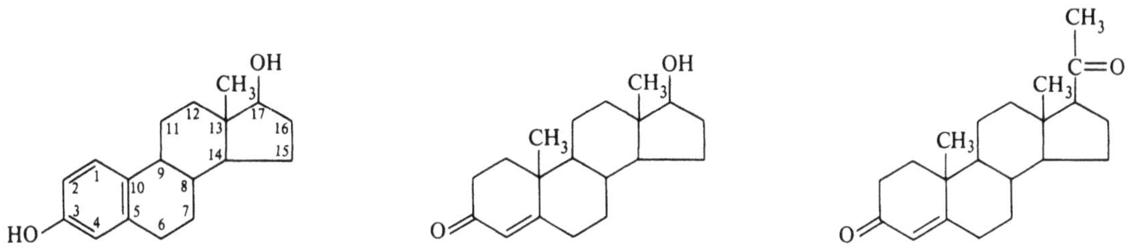

Fig. 7.1. Principal steroidal sex hormones.

place in the genital ridge. Here they lie in tissue derived from the coelomic epithelium, which is mesodermal in origin. The primordial gonads start to differentiate in the seventh week of gestation and, depending on the genetic sex, either the cortex will form an ovary or the medulla a testis. It is unknown what controls this development, but it does not appear to be due to the steroid sex hormones. The gonads at this stage are associated with several structures, the Wolffian ducts, the Mullerian ducts, the genital tubercle, and the urinogenital sinus. These structures are predestined to become the adult sexual apparatus.

The embryonic testis develops an ability to produce steroidal sex hormones and respond to the trophic effects of pituitary gonadotropins. In the third month the sexual apparatus differentiates (Figure 7.2).

In the male, under the influence of the testicular androgenic hormones, the Wolffian ducts become the epididymis, vas deferens, and seminal vesicles, and the genital tubercle forms the external sexual apparatus, including the penis. The urinogenital sinus gives rise to the prostate and urethra. The Mullerian ducts become vestigial, a process that does not appear to be dependent on the actions of hormones.

In the genetic female, differentiation does not appear to be dependent on ovarian hormones; the Mullerian duct forms the oviduct, uterus, and upper part of the vagina, and the clitoris arises from the genital tubercle. The Wolffian ducts degenerate in the female. During this vital third month the fetus is especially sensitive to the actions of any abnormal levels of androgenic hormones. These may be present in concentrations that do not affect the mother but after crossing the placenta can cause profound disturbances in sexual differentiation. Abnormal levels of such steroid sex hormones may be derived from tumors, but this problem has occurred more frequently as a result of misplaced therapy with hormone preparations. Female fetuses may be virilized as a result of the presence of male androgenic steroids, so that the external sex organs have a male appearance. Even

after the third month development may still be influenced but not quite so dramatically. Similarly, the use of antiandrogens may influence the development of a male fetus, as has been observed experimentally in animals.

The appropriate patterns of male or female sexual behavior are set neonatally, and possibly even prenatally, under the influence of the sex hormones. If newborn female rats are injected with male androgenic hormones, they do not develop the normal adult pattern of cyclical sexual behavior. Conversely, if newborn male rats are castrated, they exhibit female patterns of behavior. This process involves differentiation of the brain, especially the hypothalamus.

Following the differentiation of the ovary or the testis, the gonocytes multiply. In the male they are dispersed in the developing seminiferous tubules and subsequently form the sperm. In the female the oogonia multiply and form oocytes, which mature by about the twentieth week of gestation. They are then surrounded by a layer of cuboidal granulosa cells. At birth there are a total of about 600,000 such primary oocytes in the ovaries. Most of these subsequently degenerate, but in some, further maturation occurs and the granulosa cells proliferate to form tertiary follicles, which periodically mature to form Graafian follicles. This process commences at the menarche, at about 12 to 14 years of age, and ceases about 35 years later at the menopause. Spermatogenesis occurs throughout mans adult life, although it may decline somewhat, but it has even been observed to be active in men aged up to 90 years.

7.3. Gonadotropins: the control of testicular and ovarian function

A. Functions

Deficiencies in pituitary function, either as a result of hypophysectomy or due to disease, result in disturbances in reproductive function of the male and the female. In the prepuberal condition there may be a failure to attain sexual maturity; if such problems arise in the adult, a hypogonadal condi-

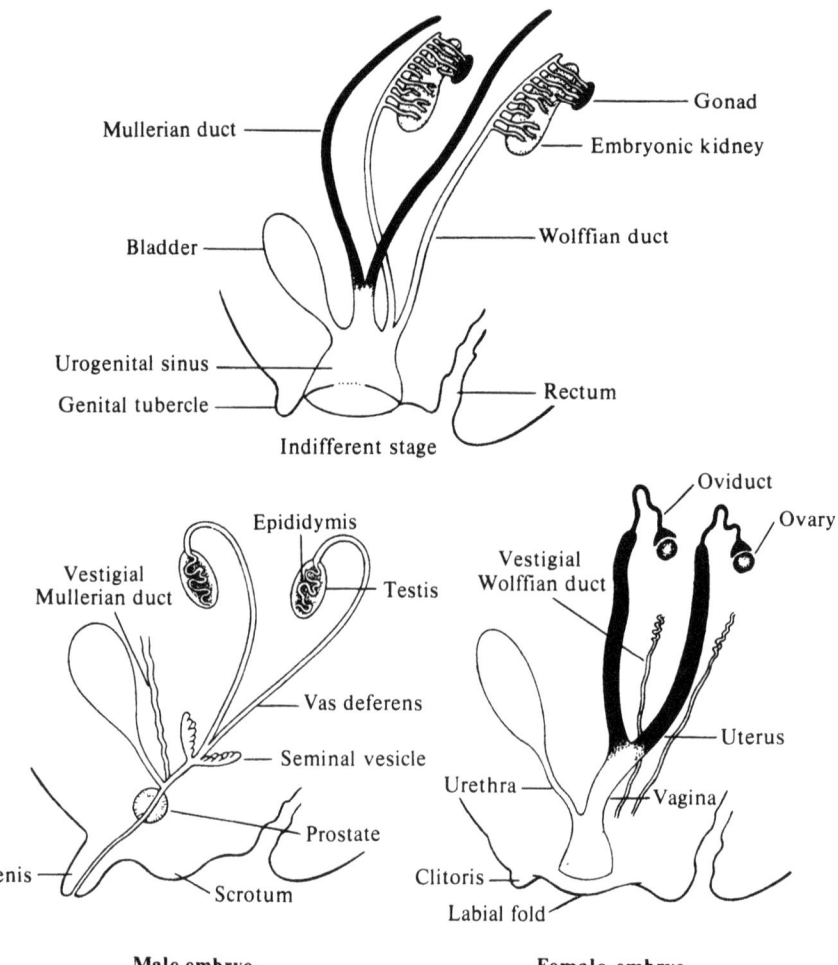

Fig. 7.2. Differentiation of the sexual organs in an embryonic mammal. The appropriate development of the genetic male is dependent on the presence of androgens. This pattern can be changed by removal or antagonism of the natural hormones or by injecting hormones more typical of the opposite sex. (From Frye, 1967. By permission of Macmillan Publishing Co., Inc., New York. Copyright 1967.)

tion and infertility may result in either the male or female. Gonadotropic hormones were identified in the pituitary more than 50 years ago. Such substances were also identified in the urine and formed the basis for early pregnancy tests. The urine of pregnant women is very rich in a gonadotropin, and this has been shown to originate principally as a result of synthesis by the placenta and is thus called human chorionic gonadotropin (hCG). The pituitary and placental gonadotropins are closely related chemically.

There are two distinct gonadotropins present in the pituitary of both men and women. Follicle-stimulating hormone (FSH) controls the development of the ovarian follicles, and this process eventually results in the maturation of the mature

ovum, which is made available for fertilization on about the fifteenth day of the menstrual cycle. It also stimulates the formation of estrogens by the follicles. In the male, FSH acts on the seminiferous tubules in the testis to promote the formation of the sperm. In the female, luteinizing hormone (LH) is released in a sudden surge just prior to ovulation and is thought to trigger the process. It subsequently controls the development of the corpus luteum, the growth of the lutein cells, and their production of progesterone. LH also, in conjunction with FSH, influences steroidogenesis by the ovarian follicles. In the male, LH controls androgen synthesis by the Leydig cells in the testicular interstitial tissue and is thus also called interstitial cell-stimulating hormone (ICSH). The latter name has, however, been generally discarded. Human chorionic gonadotropin principally exhibits an LH-like activity and contributes to the control of the activity of the corpus luteum during pregnancy. It is, however, possible that it has other effects, such as influencing the descent of the testes in the fetus. The hCG has an intrinsic FSH-like activity, but using a radioligand-receptor assay its affinity for its receptor was found to be less than 1 percent that of FSH (Siris et al., 1978). The production of a chorionic gonadotropin during pregnancy appears to be confined to primates, and in other animals the pituitary may be the predominant source of gonadotropins at this time. The blood of the pregnant horse has high concentrations of gonadotropins during pregnancy, and pregnant mares' serum (PMS) is a useful source of this hormone. In contrast to hCG, PMS exhibits mainly an FSH-like action. Another pituitary hormone, prolactin, may also exhibit gonadotropic activity, but this action, which is on the corpus luteum (it is luteotropic), is not, apparently, seen in man.

B. Structure

The gonadotropins (see Sairam and Papkoff, 1974) are glycoproteins and bear a close chemical similarity to thyrotropin (TSH). They have a molecular weight of about 26,000 to 30,000 and, depending on the species, contain variable amounts of carbohydrates, including hexose, hexosamine, fucose, and sialic acid. In man these carbohydrate moieties make up about 12 to 17 percent of the weight of the molecule. The gonadotropins, including human chorionic gonadotropin (and also TSH), consist of two major subunits which can be readily dissociated and subsequently recombined chemically. These are called the α-subunit (or CI), which contains about 90 amino acids, and the β-subunit (or CII), which has about 120 amino acids. The carbohydrate moieties are present on both sections of the molecule, but in general much more appears to be attached to the α-subunit. They are usually linked to asparagine and serine residues. There are interspecific differences in the precise amino acid composition of each part of the molecule which results in them having antigenic properties when they are administered to another species. Thus, the animal hormones are of limited therapeutic use in man.

The amino acid composition of the α- and β-subunits of the gonadotropins have recently been elucidated in several animals and in man (Figure 7.3). Human FSH-α contains 92 amino acids and hFSH-β contains 118 of these residues (Sairam and Papkoff, 1974; Rathnam and Saxena, 1975; Saxena and Rathnam, 1976). The α-subunit of hCG has an identical structure to the hFSH-α (Morgan, Birken, and Canfield, 1975) and that of human LH-α, except that the latter lacks the

N-terminal residue sequence Ala-Pro-Asp (TSH-α is also similar). The similarities in the α-subunits of these glycoprotein hormones are rather remarkable and suggest that this part of the molecule lacks a specific role in determining the particular action of each hormone.

The β-subunits of the gonadotropins are the sections of the molecule that confer their special activities, and they are thus called the "hormone-specific" sections. It is thus, for example, possible to combine the α-subunit of TSH with the β-section of LH so that one gets normal LH, but not TSH, activity. Thus, it is to be expected that differences exist in the β-subunits of these hormones, although there remain considerable homologies.

Human FSH-β contains 118 amino acids and at the N-terminal it shows considerable similarity to hTSH, hLH, and hCG (Figure 7.3). The hCG-β has 145 amino acids and hLH-β has 115 such residues (Morgan, Birken, and Canfield, 1975). As described above, LH and hCG both have LH activity, so that it is interesting to observe that about 80 percent of the amino acids in hLH-β are shared by hCG-β. The latter, however, possesses an additional 30 amino acids at the C-terminus, and these residues are unique to this hormone. Thus, there are considerable homologies and differences in the structure of the β-subunits of the glycoprotein hormones which contribute to the similarities and specificity of their effects.

Precise information about the structure–activity relationships of the gonadotropins is not available. When the molecules are dissociated into their separate subunits, these fragments display little or no biological activity. Thus, although the α-subunit has little specificity, its presence is vital for the activity of the hormone. This observation, combined with its remarkable structural uniformity, indicates that it has an important function, but at present this is unknown. An N-terminal Ala-Pro-Asn-sequence is lacking in hLH-α, and it is interesting that when this subunit is combined with FSH-β, the molecule has less FSH-activity than when hCG-α or FSH-α is, instead, combined with FSH-β. This portion of the molecule thus appears to be important for the full expression of FSH activity (see Rathnam and Saxena, 1975).

There has been speculation about the function of the carbohydrate moieties in the gonadotropins (see Bahl, 1977). Removal of these sections of the molecule can result in a considerable reduction in activity *in vivo*, and it has been shown to reduce its half-life in the circulation. It thus would appear that the molecule is somewhat protected against inactivation by the carbohydrates. Removal of carbohydrate from hCG only results in a small reduction in its ability to promote steroidogenesis *in vitro*, but there is a marked decline in cAMP production

Human chorionic gonadotropin and hFSH α-subunit:

```
                                                      ┌→ hLH-α
                                                      │
                         1                            │
                         Ala - Pro - Asp - Val - Gln - Asp - Cys - Pro - Glu -
   10                                              20
 - Cys - Thr - Leu - Gln - Glu - Asp - Pro - Phe - Phe - Ser - Gln - Pro - Gly - Ala - Pro - Ile - Leu - Gln - Cys -
       30                                         40
 Met - Gly - Cys - Cys - Phe - Ser - Arg - Ala - Tyr - Pro - Thr - Pro - Leu - Arg - Ser - Lys - Lys - Thr - Met -
           50                                           60
 Leu - Val - Gln - Lys - Asn - Val - Thr - Ser - Glu - Ser - Thr - Cys - Cys - Val - Ala - Lys - Ser - Tyr - Asn -
               70                                           80
 Arg - Val - Thr - Val - Met - Gly - Gly - Phe - Lys - Val - Glu - Asn - His - Thr - Ala - Cys - His - Cys - Ser -
                   90
 Thr - Cys - Tyr - Tyr - His - Lys - Ser.
```

β-Subunit hCG:

```
                         1                                                              10
                         Ser - Lys - Glu - Pro - Leu - Arg - Pro - Arg - Cys - Arg - Pro - Ile -
                             20                                                     30
 Asn - Ala - Thr - Leu - Ala - Val - Glu - Lys - Glu - Gly - Cys - Pro - Val - Cys - Ile - Thr - Val - Asn - Thr -
                                 40                                                     50
 Thr - Ile - Cys - Ala - Gly - Tyr - Cys - Pro - Thr - Met - Thr - Arg - Val - Leu - Gln - Gly - Val - Leu - Pro -
                                     60
 Ala - Leu - Pro - Gln - Val - Val - Cys - Asn - Tyr - Arg - Asp - Val - Arg - Phe - Glu - Ser - Ile - Arg - Leu -
 70                                                     80
 Pro - Gly - Cys - Pro - Arg - Gly - Val - Asn - Pro - Val - Val - Ser - Tyr - Ala - Val - Ala - Leu - Ser - Cys -
     90                                                 100
 Gln - Cys - Ala - Leu - Cys - Arg - Arg - Ser - Thr - Thr - Asp - Cys - Gly - Gly - Pro - Lys - Asp - His - Pro -
         110                                                120
 Leu - Thr - Cys - Asp - Asp - Pro - Arg - Phe - Gln - Asp - Ser - Ser - Ser - Ser - Lys - Ala - Pro - Pro -
             130                                            140
 Pro - Ser - Leu - Pro - Ser - Pro - Ser - Arg - Leu - Pro - Gly - Pro - Ser - Asp - Thr - Pro - Ile - Leu - Pro - Gln.
```

β-Subunit hFSH:

```
 1                                                         10
 Asn - Ser - Cys - Glu - Leu - Thr - Asn(CHO) - Ileu - Thr - Ileu - Ala - Ileu - Glu - Lys - Glu - Glu - Cys -
     20                                                    30
 Arg - Phe - Cys - Leu - Thr - Ileu - Asn(CHO) - Thr - Thr - Trp - Cys - Ala - Gly - Tyr - Cys - Tyr - Thr -
         40                                                50
 Arg - Asp - Leu - Val - Tyr - Lys - Asn - Pro - Ala - Arg - Pro - Lys - Ileu - Gln - Lys - Thr - Cys - Thr -
                 60                                                    70
 Phe - Lys - Glu - Leu - Val - Tyr - Glu - Thr - Val - Arg - Val - Pro - Gly - Cys - Ala - His - His - Ala -
                     80
 Asp - Ser - Leu - Tyr - Thr - Tyr - Pro - Val - Ala - Thr - Gln - Cys - His - Cys - Gly - Lys - Cys - Asp -
     90                                          100
 Ser - Asp - Ser - Thr - Asp - Cys - Thr - Val - Arg - Gly - Leu - Gly - Pro - Ser - Tyr - Cys - Ser - Phe -
         110                                        118
 Gly - Glu - Met - Lys - Gln - Tyr - Pro - Thr - Ala - Leu - Ser - Tyr
```

Fig. 7.3. Amino acid sequences of the α- and β-subunits of human pituitary and chorionic gonadotropin. The α-subunits are remarkably similar in structure, both to each other and to TSH-α. The differences in the activity of the molecules appear to be largely carried by the β-subunits. Human β-chorionic gonadotropin and hLH-β are quite similar and have similar amino acids at about 80 percent of sites. (Based on Morgan, Birken, and Canfield, 1975, and Saxena and Rathnam, 1976.)

(Moyle, Bahl, and März, 1975). The hCG with carbohydrate removed inhibited the actions of the intact molecule on the accumulation of cyclic AMP, suggesting that binding to the receptor was still occurring (Bahl, 1977; Channing, Sakai, and Bahl, 1978). It was suggested that the carbohydrate moiety may principally increase the intrinsic activity of the molecule with respect to a "cyclic AMP receptor." While the hormones potency with respect to steroidogenesis may decline somewhat following removal of the carbohydrate, a maximal response can still be elicited.

C. Release

The control of release of the adenohypophysial hormones is described in Section 3.2E. FSH and LH are under the control of the hypothalamic peptide hormone LHRH. A negative-feedback system is present in which the gonadal hormones, estrogens, progestins, and androgens, influence the release of LHRH in the hypothalamus and possibly its action on the pituitary gland. LH is also under positive-feedback control. Prior to ovulation, rising plasma estrogen levels stimulate its release. Various external and internal processes, rhythms, and drugs impinge on this process to influence the release of the gonadotropins. These will be described in detail later in this chapter.

It has been known for many years that a material can be isolated from testes, semen, and cultures of seminiferous tubules which can inhibit the release of FSH but usually not LH. It has been called *inhibin* and it is a polypeptide (Franchimont et al., 1975; Baker et al., 1976; Eddie et al., 1978). It has been suggested that it may have a physiological role in the regulation of testicular function. A comparable substance has also been isolated from the ovarian granulosa cells of female rats (Erickson and Hsueh, 1978).

D. Mechanism of action

The receptors for the gonadotropins are present on the plasma membrane and the hormones do not appear to have to enter the cell in order to act. Receptor preparations, either present on particles of membranes or as "soluble" complexes with the hormones, have been prepared from isolated components of ovaries and testes (see, e.g., Mendelson, Dufau, and Catt, 1975; Bellisario and Bahl, 1975; Azhar and Menon, 1976; Dufau et al., 1975).

The receptors obtained from each of the effector tissues display considerable specificity for FSH or LH and hCG (see, e.g., Bellisario and Bahl, 1975; Means, 1975). Membrane preparations from other organs, such as liver, do not specifically bind these hormones. In addition, there is no crossover in the interactions, so that, while FSH will bind to receptors prepared from seminiferous tubules, LH will not. LH will, however, interact with receptors from

Leydig cells or the corpus luteum. Maximal interactions occur at a concentration of about 10^{-8} M while the K_D is 10^{-10} M. These values are consistent with the normal levels of the hormones *in vivo*. The separated α- and β-subunits do not bind to the receptors. The molecular weight of solubilized gonadotropin receptors from rat testes has been estimated as about 200,000 (Dufau et al., 1975).

The nature of the gonadotropin receptors has been explored with the aid of proteolytic and lipolytic enzymes. Their activity is destroyed by the action of trypsin, indicating their protein nature. If a trypsin inhibitor is added, the receptors regenerate in a few hours, indicating that they are constantly being formed (see Means, 1975). Phospholipids also appear to be intimately involved in the receptor function, as treatment of preparations with phospholipases decreases the number of binding sites, although not the affinity of each site for the hormone (in this case, hCG) (Azhar and Menon, 1976). This effect may, however, be indirect, owing to a general action of the enzyme which influences the membrane environment and inhibitory effects of the accumulated products of the digestion (Azhar, Hajra and Menon, 1976). The gonadotropin receptors, from ovaries of rats, can be labelled with N-acetyl-D[C^{14}] glucosamine, indicating that they contain associated carbohydrate moieties and may be glycoproteins (Bahl, 1977).

The various actions of LH, hCG, and FSH may involve activation of adenyl cyclase and the formation of cyclic AMP. This effect has been demonstrated for the actions of FSH on the seminiferous tubules and LH on the Leydig cells, the corpus luteum, and the granulosa cells of the Graafian follicle (see Marsh, 1975; Means, 1975; Mendelson, Dufau, and Catt, 1975; Ling and Marsh, 1977). These effects on cyclic AMP precede the responses, which include steroidogenesis. The cyclic AMP response is potentiated by the presence of theophylline, which is a phosphodiesterase inhibitor, and so favors the accumulation of the nucleotide. This response is accompanied by an activation of protein kinase so that the effect could be of the classical type, where cyclic AMP acts as a "second messenger." A discrepancy has, however, been observed, as the doses of gonadotropins that are necessary to elicit a measurable change in cyclic AMP content of the effector tissues are usually much greater than those that are observed to stimulate steroidogenesis. Thus, in isolated Leydig cells hCG stimulates steroidogenesis at concentrations where no change in cyclic AMP can be detected (Mendelson, Dufau, and Catt, 1975). A maximum steroidogenesis is, in fact, seen when only 1 percent of the total receptor sites are occupied. When the concentration of the hCG is increased and receptor occupancy rises, cyclic AMP also increases, but this has no obvious relationship to the observed rise in steroidogenesis. The two responses, accumulation of cyclic AMP and steroidogenesis, can also apparently be dissociated when using preparations of hCG in which the carbohydrate moieties have been depleted (Bahl, 1977). These components may serve to increase cyclic AMP accumulation but are not essential for an effect on steroidogenesis. Similar observations have been made with respect to the action of LH on the corpus luteum, but it has recently been shown that, by using a sensitive assay for cyclic AMP-dependent protein kinase, a relationship between this response and the steroidogenesis still seems to exist (Ling and Marsh, 1977). It is doubtful, however, if the final word has yet been said regarding this interesting problem.

The active form of cyclic AMP-dependent protein kinase can initiate a variety of responses in different effector tissues. The relationship of this to the actions of the gonadotropins is not completely clear. The responses are of a dual nature, (a) under the influence of FSH, there may be an increased spermatogenesis and proliferation of follicular granulosa cells, and (b) increased steroidogenesis by the Leydig cells, Graafian follicles, and corpus luteum occurs in response to LH or hCG.

The actions of FSH on the ovarian and testicular germinal epithelium are probably mediated, respectively, by the granulosa cells and Sertoli cells. However, other types of cells may also respond. Means (1975) has provided a scheme to account for the effects on the testes and this is summarized in Figure 7.4. Following the activation of the protein kinase, a general increase in protein synthesis is observed and this is mediated by increased genetic transcription. Mitotic activity is then increased and the cells multiply. Increased protein synthesis has also been observed in the follicular granulosa cells, but a detailed analysis of this response in relation to a possible intermediary role of cyclic AMP does not appear to have been made at this time.

The steroidogenic effect of LH on the testes and ovaries is rapid and not accompanied by a general increase in protein synthesis. There is an increased conversion of cholesterol to Δ^5-pregnenolone due to stimulated cleavage of the side chain. It has also been proposed that cholesterol is mobilized from its storage droplets via activation of cholesterol esterase (Moyle, Jungas, and Greep, 1973a,b).

7.4. Mechanisms of action of the gonadal steroid sex hormones

Our understanding of the molecular mechanism of action of the steroid sex hormones arose from the pioneer observations of J. Monod and F. Jacob in elucidating the process of genetic transcription

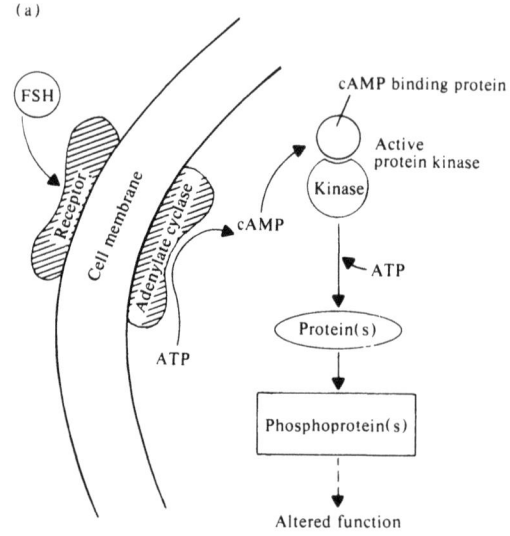

(b)

Event	Time
Binding to plasma membrane	<5 min
Activation of adenylate cyclase	<5 min
Intracellular accumulation of cyclic AMP	5 min
Activation of protein kinase	5 min
Phosphorylation of proteins	15–30 min
Transcription	
Rapidly labelled nuclear RNA	15–30 min
Stimulation of RNA polymerase II	15–30 min
Increased chromatin template activity	30 min
Stimulation of RNA polymerase I	60 min
Protein synthesis	60–120 min
Wet weight	2–4 h
Cytoplasmic accumulation of ribosomal RNA	3–6 h
Dry weight	3–6 h
Mitotic activity	6–9 h
Decreased spermatogonial degeneration	9–12 h

Fig. 7.4. Mechanism of action of FSH on the seminiferous tubules as proposed by Means (1975). (a) The events following the interaction of FSH with the membrane receptor, the activation of adenyl cyclase, the formation of cyclic AMP, and the active protein kinase are shown. The phosphorylation of cell proteins may then result in an increased formation of proteins via a nuclear transcription process. (b) The sequence of events and their timing are summarized. (From Means, 1975.)

in bacteria. Shortly after these discoveries, P. Karlson pointed out the analogy between the action of inducers, which regulate genetic transcription in bacteria, and the possible effects of steroid hormones on their target cells. This brilliant insight has been amply confirmed. Useful references include those of Jensen and DeSombre (1973); Chan and O'Malley (1976a,b,c); Minguell and

Sierralta (1975); O'Malley and Schrader (1976); Schrader and O'Malley (1978); and Mainwaring (1977).

The initial process concerned with the action of steroid hormones on cells appears to be basically similar for estrogens, androgens, progestins, and even adrenocortical steroids. Differences are due to the specificity of particular target organs, the subsequent metabolism of the steroid within its effector tissue, and the particular genetic processes that are stimulated.

The steroid hormones must enter their target cells in order to act, which contrasts with the action of most protein hormones, which appear to act at the cell surface. Their ability to enter the cell will depend on such factors as their external concentration, binding to plasma proteins, and lipid solubility. It has been suggested that they may cross the cell membrane by a process of facilitated diffusion, but this interesting possibility does not appear to have been extensively studied.

The availability of labelled, tritiated steroid hormones, initially estradiol-17β, with a high specific activity, made it possible to trace the pathways followed by such hormones, not only within the body but also in specific subcellular organelles. These studies on the steroidal sex hormones have been mainly performed in the laboratories of Jensen, Hamilton, Gorski, O'Malley, Liao, and Mainwaring.

Early observations by Jensen showed that when tritiated estradiol-17β was injected into rats, there was a preferential uptake, accumulation, and retention in its specific target organs, especially the uterus. This special affinity of sex steroids for their particular effector tissues is a general phenomenon. When the uteri were homogenized and their components separated, it was found that most of the labelled estradiol-17β was localized in the nuclei but that some also remained in the cytoplasm. The cytoplasm was then fractionated by sucrose density centrifugation and the hormone was found to be concentrated in a region with a sedimentation coefficient of 8S. The binding material was found to be a protein with a molecular weight of about 200,000 to which two molecules of the steroid are bound strongly, but not covalently, with a K_D of about 10^{-10}M. This cytoplasmic protein from the uterus will not easily bind nonestrogenic steroids, although more closely related compounds can be bound and may interfere with the uptake of estradiol-17β. The cytoplasmic-binding protein thus has a high affinity, but a low capacity, for specific types of steroids and it is thought to be the "receptor." Binding usually parallels the biological activities of the different steroids and their analogues, but there are differences that reflect other properties of the molecules. In the case of estrogens, these receptors are called *estrophiles*,

and there are about 10,000 to 100,000 present in each cell. When these cytoplasmic receptors are exposed to a high salt concentration *in vitro*, they reversibly dissociate into two subunits, A and B.

The ultimate site of action of the steroid hormone is the nucleus. This locus has not only been demonstrated using cell fractionation procedures but also by radioautography. Jensen and Gorski separately proposed that such steroid hormones work in a "two-stage" process and that following combination with the cytoplasmic receptor they are "activated" and move into the nucleus. This process is temperature-dependent; it is not seen at 0°C but occurs rapidly when the tissues are incubated at 37°C. Following this transfer the labelled hormone can be shown to be bound to a nuclear protein with a sedimentation coefficient of 5 S, and is called the "acceptor" or sometimes the "nuclear receptor." There are about 5000 such acceptor sites in each cell.

The acceptor is present in the chromatin. Binding of the steroid to this site will not occur in the absence of the cytoplasmic receptor, but if this hormone–receptor complex is present, combination can be shown to occur *in vitro*, in cell-free preparations. The acceptor is a nonhistone acidic protein which has been isolated in the so-called AP$_3$ fraction of the chromatin. It has special characteristics related to the specificity of its target cell. Removal of this protein abolishes the ability of the hormone to combine with the chromatin from its target organ. However, if it is transplanted to a normally nontarget type of chromatin, the hormone can still bind to it.

There has been considerable speculation regarding the precise nature of the effect of sex steroid hormones in initiating genetic transcription. It appears that unlike the prediction based on observations in bacteria, they do not negate the effect of a repressor protein, but act directly to initiate a response at a particular gene site. In a brilliant series of experiments involving the action of progesterone on the chick oviduct, O'Malley and his collaborators (Schwartz et al., 1976) have shown that there is an increase in the number of "initiation" sites where RNA polymerase can act on the DNA template.

Subsequent events are a *de novo* synthesis of cell proteins following the effects of the messenger RNA on the ribosomes. The particular response will, of course, depend on the nature of these proteins. The events are summarized in Figure 7.5.

The various sites of the steroids action in the cell provide known and potential loci where drugs may exert their effects to block or even potentiate the actions of the hormones. These are summarized in Figure 7.6. The transfer of the steroids across the cell membrane can be affected by altering the concentration of free unbound hormones in the

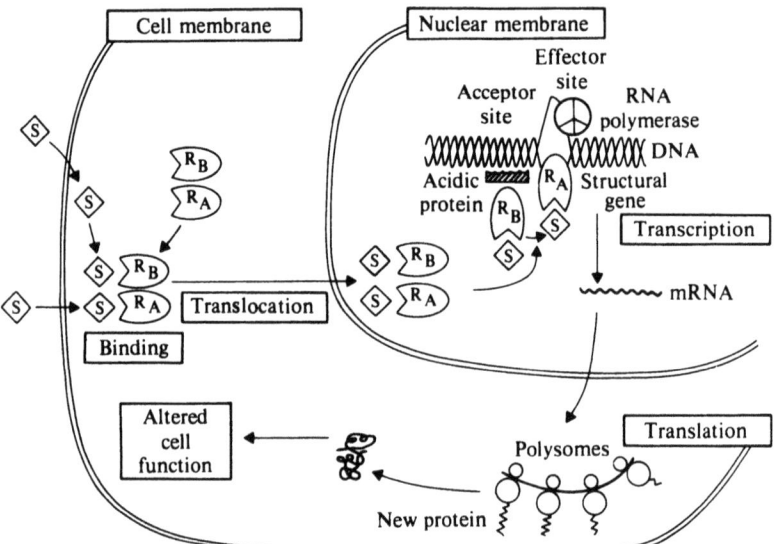

Fig. 7.5. Genberal mechanism of the action of sex steroid hormones as constructed by Chan and O'Malley, 1976a. This scheme was composed principally on information gathered from studies on the progesterone receptor in the chick oviduct. S, steroid hormone; R_A and R_B, subunits of the steroid hormone receptor. (From Chan and O'Malley, 1976a. Reprinted by permission from *The New England Journal of Medicine 294*, 1322.)

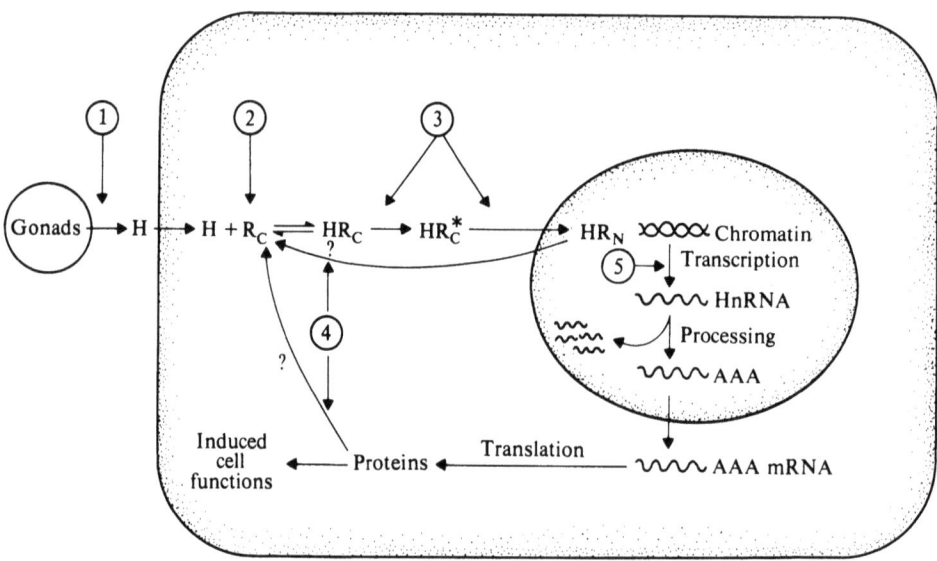

Fig. 7.6. Major potential sites for the actions of antagonists of the steroid sex hormones. (1) Depletion of the endogenous plasma steroid-hormone levels; (2) depletion of the specific steroid-hormone receptor; (3) inhibition of the nuclear translocation of the cytoplasmic steroid–hormone–receptor complex; (4) perturbation of the receptor cycle; (5) inhibition of steroid hormone-induced gene transcription. H, sex steroid hormone; R_C, cytoplasmic receptor; R_C^*, activated cytoplasmic receptor; R_N, nuclear receptor; HnRNA, heterogeneous RNA; AAA, polyadenylate. (From Chan and O'Malley, 1976c. Reprinted by permission from *The New England Journal of Medicine 294*, 1431.)

plasma. Such changes can be brought about by altering synthesis or release of the steroids and the formation of the proteins that bind them, especially SHBG. It is also possible that the process of their transfer across the cell membrane could be influenced especially if, as some suspect, an interaction that involves facilitated diffusion occurs. The interaction with the cytoplasmic receptor may be blocked by compounds that compete with the hormone. Such complexes may be unable to enter the nucleus or, if they do, may fail to interact as effectively. The number of receptors present in the cytoplasm is also important, and these are known to vary under different conditions, owing to disease and the presence of other hormones or to drugs. Thus, progesterone and antiestrogens produce a decline in estrogen receptors. Conversely, estrogens can stimulate the formation of progesterone receptors. The final effect, genetic transcription, can be prevented by drugs such as actinomycin D, which inhibits RNA depolymerase.

Finally, a note of warning has been sounded (Spaziani, 1975; Shields, 1978). Many of the effects of the steroid sex hormones, such as increases in local blood flow and changes in distribution of water and electrolytes, occur very rapidly and it may be difficult to account for these via a synthesis of messenger RNA. General trophic (pleiotypic) effects on tissues also do not always appear to depend directly on genetic transcription. Therefore, the possibility of extragenomic effects of the sex steroids should not be excluded. Increases in protein synthesis that involve changes in translation, rather than transcription, are also considered to be likely.

Steroid hormones have been shown to interact with lysosomes from target-specific tissues, so that the permeability of these organelles increases and the hydrolase enzymes they contain are released (Szego, 1974). Such lysosomes, when combined with certain hormones, have been observed to enter the nucleus, where it has been proposed that their released enzymes may contribute to the response. They may, for instance, interact with components of the chromatin and/or assist in the development of the effects of the cytosol receptor–steroid complex. Anabolic and mitogenic effects of steroid hormones often cannot be readily explained by simple transcriptional changes, and Szego has suggested that the lysosomes may contribute to such effects.

Steroids can directly interact with plasma membranes, and it was formerly considered likely that this may be a mechanism for the actions of such hormones on their target tissues (see Willmer, 1961). This interesting possibility has been neglected in the wake of the dramatic demonstrations that the steroid hormones can initiate genetic transcription. Baulieu and his collaborators (1978)

have reinvestigated such a phenomenon and have shown that progesterone can induce meiotic division in amphibian oocytes. This effect is initiated at the cell surface and appears to be related to an increase in free intracellular Ca^{2+}. This ion may act as a "second messenger" and induce protein synthesis and meiotic division.

7.5. Male reproductive system

Sexual dimorphism and the production of viable germ cells in man and other animals is largely the result of the actions of hormones produced by the pituitary gland and either the ovaries or the testes. In the male, the testes have the dual function of producing spermatozoa and the hormones that facilitate the maturation and delivery of these cells in the process of reproduction. The physiological role of the testes has been recognized for a longer period than has that of any other endocrine gland. The effects of castration in man and animals has been known since early historic times. In man, it was practiced for a variety of practical, social, and religious reasons, including the taming of slaves, enforced celibacy, and the maintenance of the boyish soprano voice for choirs. The taming effects of castration in domestic animals, such as horses, cattle, and chickens, have been utilized in farming for hundreds of years. It is, however, less than 50 years since the androgenic (or male) hormones were isolated from the testes and shown to be steroids. Their uses are still being studied but at present they are used to replace hormones that may be absent or deficient, to treat breast cancer and certain types of anemia, and to promote the growth of muscle in athletes. Hopes that they may be utilized as a male contraceptive pill have not yet been realized, but further chemical modifications of the hormones offer some prospect of success.

A. Male reproductive apparatus

The primary genetic effector in the male is the testis, which produces sperm and androgenic hormones that control the embryonic and puberal development of the primary and secondary sexual characters. The spermatozoa pass from the testes via a series of short ducts (vasa efferentia) into the epididymis, which lies adjacent to the testis. The sperm are stored and undergo maturational changes in this organ and can be ejected into the *vas deferens,* which is contiguous with the urethra. A series of glands open into this duct and contribute secretions which make up the major volume of the semen and promote the viability and reproductive success of the sperm. These *accessory sex glands* are the *seminal vesicles* (or vesicular glands), the *prostate,* and the *bulbo-urethral* (or Cowper's) *glands.* The development and maintenance of these *primary sex characters,* which also include the penis and scrotum, are under the control of the androgenic steroid hormones.

A variety of *secondary sex characters* are similarly influenced by the testicular hormones. They include aspects of general morphology and behavior which are associated with "maleness." These characters include the skin – such factors as its pigmentation, the distribution of hair, and the development of the sebaceous glands – and the depth of the voice. The general morphological stature of the body, including size, height, the breadth of the shoulders and the pelvis, and the development of the skeletal muscles are also under the control of the endocrine secretions of the testis.

The testis has a dual morphology (Figure 7.7) which is related to its ability to form spermatozoa and androgenic steroid hormones. The bulk of this gland (about 65 percent in man) consists of the seminiferous tubules. These are convoluted structures lined by the germinal epithelium and the Sertoli cells.

The *germinal epithelium* continually undergoes a series of divisions and maturational events which result in the formation of the sperm. This process of *spermatogenesis* initially involves the formation of spermatocytes from the spermatogonia, which lie in the basement membrane of the tubules. The haploid spermatids are formed as a result of meiotic division in two stages. There is subsequently a third stage of maturation, called spermiogenesis, where sperm ready for transfer to the epididymis are formed. Successful spermatogenesis only occurs under the influence of the pituitary gonadotropic and testicular hormones.

The role of the *Sertoli cells* in the spermatogenic process is not clear. They envelop the developing sperm and it has often been considered that they play a vital role in their nutrition – hence their other name, sustentacular cells. They are the site of secretion of fluids that accumulate in the seminiferous tubules. It has been suggested that under the influence of FSH, they may synthesize androgenic steroids specifically related to the needs of the germinal epithelium. The synthesis of an *androgen-binding protein* (ABP) also occurs, via released testosterone (Tindall and Means, 1976). This protein may facilitate transport of androgens to the developing sperm and the epididymis. The Sertoli cells are potential sites for the action of drugs that may have a contraceptive effect in the male.

The seminiferous tubules lie in a matrix called

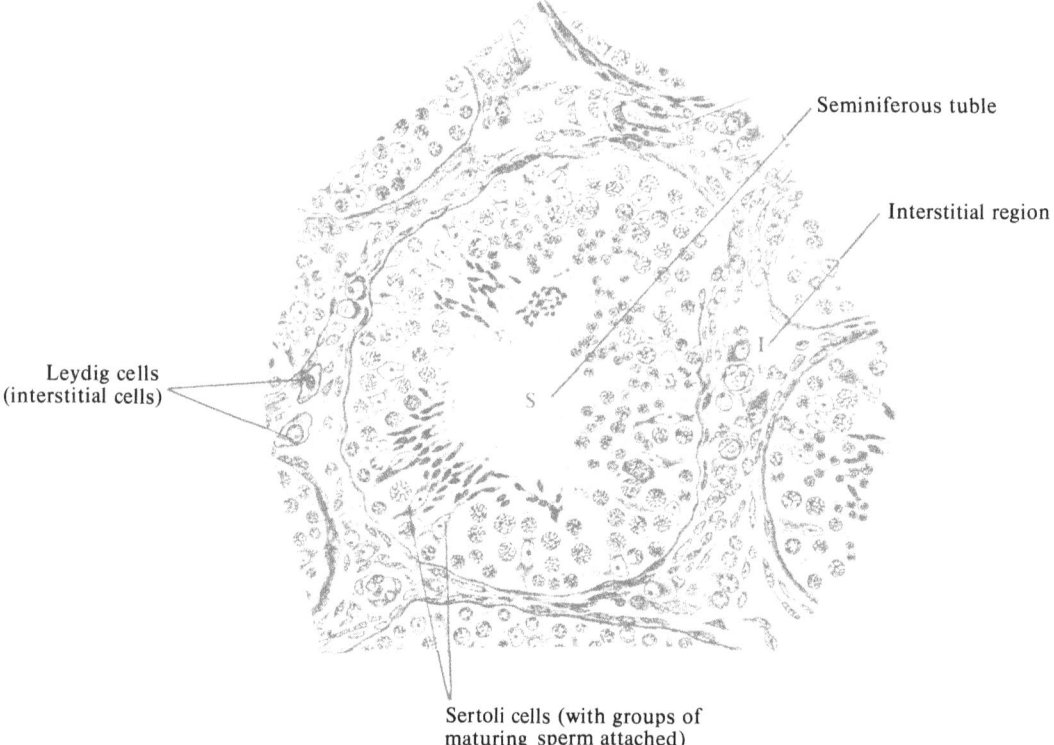

Leydig cells
(interstitial cells)

Seminiferous tuble

Interstitial region

Sertoli cells (with groups of maturing sperm attached)

Fig. 7.7. Morphology of the testes, showing the seminiferous tubules, the Sertoli cells, and the Leydig (or interstitial) cells.

the *interstitial tissue*. This region of the testis is well vascularized and contains connective tissue, fibroblasts, mast cells, and the *Leydig cells* (also called the *interstitial cells*). Androgenic steroids are synthesized by the latter cells (which make up about 12 percent of the testicular volume), and these either pass via the extracellular fluid to the seminiferous tubules or into the blood for dissemination to other androgen-dependent effector tissues. The Leydig cells contain abundant smooth endoplasmic reticulum, which is thought to play an important role in the synthesis of such steroid hormones. The enzyme 3β-hydroxysteroid dehydrogenase, which mediates the formation of androstenedione, has been localized histochemically in these cells. The Leydig cells first appear in the human embryo at about 7 weeks and can make up about 50 percent of the total testicular volume. They start to invo-

lute at about 18 weeks and substantially disappear before birth. They are absent in children but undergo a second period of development during puberty, usually starting at about 13 years of age.

Hypophysectomy results in a regression of male characteristics and sterility reflecting the controlling role of the pituitary gland in male reproduction. Its role in the production of the gonadotropins FSH and LH was described in Section 7.3.

B. Androgenic steroid hormones

The general chemistry and biosynthesis of the androgens was described earlier (Section 5.2). They are C_{19} steroids, which are produced principally by the testis but also by the adrenal cortex and ovary. Several such steroids may be present in the circulation. These secretions (Figure 7.8) include androstenedione and dehydroepiandro-

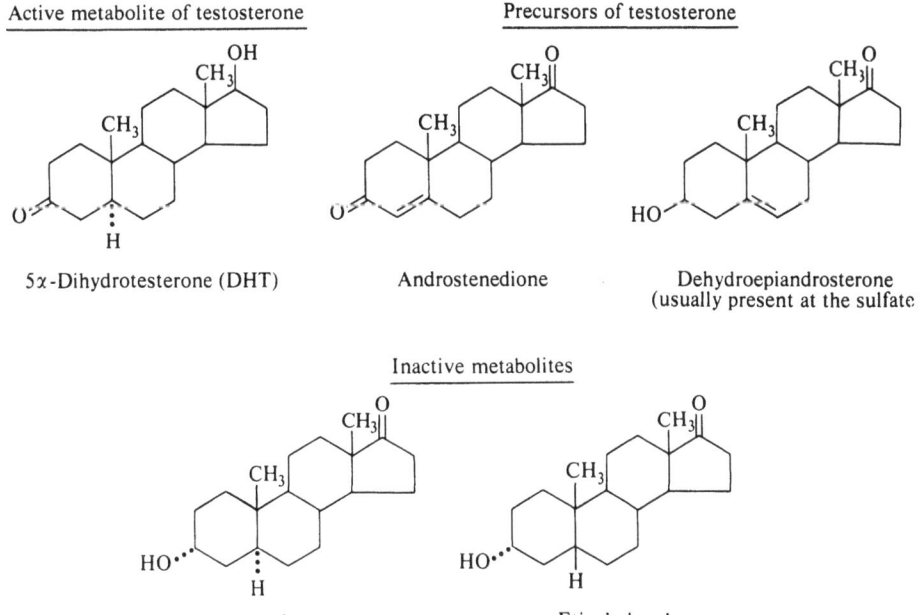

Fig. 7.8. Principal androgenic steroids and their metabolites. Testosterone is the main steroid to be secreted by the testis, but it is metabolized to a more active form, 5α-dihydrotestosterone (DHT), at peripheral sites. Dehydroepiandrosterone and androstenedione can be converted to testosterone but have little androgenic activity themselves. Etiocholanolone and androsterone are inactive metabolites.

sterone (mostly as the sulfate conjugate), which have little androgenic activity but can be converted to more active compounds or even to estrogens. Testosterone is the main androgen to be secreted by the testis, but it is converted peripherally, especially in many of the androgen effector tissues, to the more active metabolite 5α-dihydrotestosterone (DHT). This conversion takes place under the influence of a 5α-reductase, which saturates the Δ⁴ double bond. The reaction is irreversible and this compound, unlike the other androgens, cannot undergo peripheral aromatization to estrogens. In most tissues, DHT is the most active form of the hormone and it is this compound that combines with the receptors in the cytoplasm of the cell. The 5α-reductase enzyme appears to be present in most of its effector tissues, where it has been found in the endoplasmic reticulum and the nucleus. Its tissue levels, however, vary considerably and it is not at all clear whether dihydrotestosterone is the active form of the hormone in *all* tissues, especially in the brain, skeletal muscle, and bone marrow. The testis principally secretes testosterone, but some DHT and estrogens are also formed.

C. Functions and actions

Although the most prominent effects of the androgens are observed in the male genitalia and the secondary and accessory sex glands and tissues, they also influence the activity of other organs. Their effects are indeed ubiquitous, as they have actions on skeletal muscle, cardiac muscle, the kidney, the liver, the bones (including the bone marrow), and the brain. Although it would be attractive to postulate some common underlying mechanism for all these effects, conclusive evidence that this is so is lacking. Several types of processes may be involved, and it is difficult to even relate these to a common type of effect at the level of genetic transcription. Some effects may, indeed, be extragenomic (Spaziani, 1975). Some biochemical effects of the androgens on different tissues are summarized in Table 7.1.

Embryonic differentiation. Embryonic differentiation in the male fetus is dependent on the actions of androgens. These are produced from the embryonic testis, starting in man at about the seventh week of gestation. Under their influence the Wolffian ducts differentiate into the vas deferens, epididymis, and seminal vesicles; the genital tubercle forms the external genitalia; and the prostate develops. The development of the noncyclical tonic pattern of male sexual behavior (see McEwen, 1976) also results from the action of androgens on the developing brain. In rats, the patterns of metabolism of steroid hormones in the liver appear to be determined by androgens in the neonatal period (Gustafsson and Stenberg, 1976).

Changes at puberty. At puberty there is a second activation of the male sexual characters, which results in reproductive maturity. The effects include further growth and maturation of the genitalia and secondary sex glands, the production of mature spermatozoa by the testes, and the development of an appropriate pattern of behavior.

In addition, a number of general morphological changes occur. Under the influence of androgens the *skin* thickens (in castrates, the skin has a thin, white, waxy appearance), and there is a characteristic development of hair on the face and in the genital regions. The hairline on the head takes a characteristic appearance and balding (which is not seen in castrates) becomes a future possibility. Pigmentation is promoted so that the skin can more readily tan in response to exposure to sunlight. The sebaceous glands develop. These cutaneous developments are often accompanied by the outbreak of acne. In women, hirsutism, in the male pattern, may develop due to an excessive production of androgens by the ovary or adrenal cortex. A recent study showed that androstenedione is the principal such steroid produced in such women (Kuttenn et al., 1977).

Growth is initially promoted but subsequently ceases as a result of the closure of the epiphyses. Early sexual maturation may thus result in an early cessation of growth. (In castrates, the epiphyses of the long bones may still be open at 40 years of age. The effect of this is to produce a disproportionate length of the lower portion of the body.) The *muscles develop* into the adult form. (Even with continual exercise such development does not occur in castrates.) The *thymus* involutes under the influence of androgens.

The *larynx* grows and the "Adam's apple" appears. This results in a deepening of the voice. [Boys were castrated (castrati) as late as the nineteenth century in order to preserve their soprano voices for cathedral choirs.]

There is an *activation and maintenance* of the male sexual characters. When castration occurs after puberty, the principal morphological characters are maintained but there is a regression in their size and activity. There is a decline in libido and a change in the psyche. The lack of aggression and the submissiveness of the eunuch are well known and were used to break the spirit and control captives of war and slaves. Such characteristics can, however, be maintained or restored by the administration of androgens.

Anabolic effect. The androgens exert a general *anabolic effect* and increase protein synthesis in the body. This response is reflected by a decreased nitrogen excretion and the retention of salts, including calcium, sulfate, potassium, chloride, and sodium. The Na retention is probably due to the mineralocorticoid-like effects of the androgenic

Table 7.1. *Some biochemical effects produced by testosterone in different tissues*

Species	Tissue	Effect
Rat	Prostate	Increased synthesis of cyclic AMP
		Increased enzyme activity of folate coenzymes
		Decreased ribonuclease activity
		Decreased ATP concentration
		Increased Na^+,K^+-ATPase activity
		Stimulated synthesis of rRNA
		Enhanced DNA polymerase activity
		Increased turnover of endoplasmic reticulum membranes
	Liver	Promoted formation of rough endoplasmic reticulum
		Enhanced oxidase activity
		Induced aryl hydrocarbon hydroxylase enzymes
	Muscle	Increased total protein synthesis
		Increased chromatin template activity
	Bone marrow	Stimulated RNA synthesis
		Enhanced RNAase activity
	Preputial gland	Promoted lysosomal labilization
		Regulated total RNA content
	Hepatoma	Inhibited induction of TAT by cortisol
		Modified cell adhesiveness
Mouse	Kidney	Stimulated pyrimidine salvage pathways
		Stimulated erythropoietin production
		Stimulated leucine incorporation into proteins
	Submaxillary gland	Induced esteropeptidase enzymes
		Stimulated nuclear RNA synthesis
		Modulated genesis of new Golgi apparatus
Chick	Oviduct	Regulated protein biosynthesis

Source: Minguell and Sierralta, 1975.

steroids. These hormones increase the size and strength of the skeletal muscles, an attribute that has been utilized in competitive athletics by both men and women [see Section 7.5I(8)]. Androgens also have a "renotrophic" effect, the size of the kidneys decreases following castration. The growth and maintenance of cartilage is promoted; lack of androgens can result in "knock knees" and flat feet. The red cell and hemoglobin levels are characteristically somewhat higher in the male than the female which appears to be related to an "erythropoietic effect" of androgens [see Section 7.5I(8)].

Spermatogenesis. Spermatogenesis cannot proceed in the absence of the pituitary because of resultant absence of the gonadotropins FSH and LH. The effects of LH are indirect, owing to their ability to stimulate the formation of androgens by the Leydig cells. FSH appears to have a more direct effect on the seminiferous tubules, possibly involv-

ing the Sertoli cells. The precise role of androgens in spermatogenesis is uncertain and exhibits variation between species. The early embryonic differentiation of the gonocytes may be androgen-dependent as may also be the final stages of the meiotic division that result in the formation of the spermatids. The final maturation of the spermatids appears to depend on FSH and also, possibly, on androgens. Androgen receptors have been identified in the seminiferous tubules, but their precise location is unknown. It is uncertain whether the effect of androgens in promoting spermatogenesis is due to a direct action on the germ cells or whether it involves other types of cells. High doses of androgens may have a paradoxical action and result in an inhibition of spermatogenesis. This effect appears to be due to an inhibition of pituitary gonadotropin release. In animals, testosterone has been shown to be able to support spermatogenesis even in the absence of FSH and LH, but this effect has not been demonstrated in man.

Release of the pituitary gonadotropins. FSH and LH are under the negative-feedback control of testicular androgens. Both FSH and LH levels rise following castration, but their ratio changes from (FSH/LH) 1:1 to 3:1. The reason for this alteration is uncertain, especially as exogenous testosterone is more effective in blocking LH release. The site of the feedback inhibition appears to be in the hypothalamus, but it remains possible that the adenohypophysis may also be directly involved. These inhibitory effects of androgens on the release of gonadotropins offer the prospect of a practical male antifertility agent.

Androgens have characteristic effects in women. The effects of androgens in women can be seen when androgens are administered in the treatment of cancer of the breast, as anabolic agents, or result from excessive endogenous production by the adrenal cortex (see Section 6.8A). Depending on the dose, they may have a virilizing effect, which is manifested as an excessive growth of hair on the body and face, a deepening of the voice, an atrophy of the vaginal epithelium, and an enlargement of the clitoris. At high doses there may be an abnormal increase in libido. Androgens can delay menstruation and have a virilizing effect on a female fetus. They can be used to suppress lactation.

D. Synthesis and release

As described earlier, the principal source of androgens in the male are the Leydig cells in the testis. The adrenal cortex also produces significant quantities of such steroids and these may be especially important in the female and in the fetus [see Section 7.6H(3)]. The biosynthesis of androgens (Section 5.2) is under the control of LH from the anterior lobe of the pituitary gland. As release of this gonadotropin is subject to negative-feedback control, it responds not only to circulating endogenous androgens but also to administered androgens, estrogens, and progestins. Administered gonadotropins, including hCG and PMSG, cause a rapid rise in plasma androgen levels due to an increase in synthesis. Preparations of hypothalamic LHRH can also initiate release via LH.

Androgens are not stored in appreciable amounts in the Leydig cells and are apparently released fairly promptly following their biosynthesis. The precise nature of the release mechanism is, however, unknown.

Changes in the rate of synthesis and release of androgens appear to be related to hypothalamic–pituitary function and LH release. However, an independent release of testosterone occurs during sleep and exercise (see Sutton et al., 1973). Catecholamines, prostaglandins, and increases in testicular blood flow may also promote release, at least under experimental conditions.

Levels of androgens in the blood may, however, rise for reasons apart from changes in rate of their secretion, including changes in their rates of metabolic clearance and the quantity of their binding proteins, especially SHBG.

E. Transport and metabolism

The androgens, principally (about 90 percent) testosterone, are released into the spermatic venous blood. There is a relatively rapid "clearance" in the first hour, with a half-life of about 10 minutes, but subsequently the decline in plasma levels is much slower (half-life of about 100 minutes). Only about 1 to 3 percent of the androgens in the plasma are in a free, biologically active form, the remainder being bound to plasma proteins. These include the globulins, SHBG, and transcortin (CBG), which have a high affinity for such sex steroids but a limited capacity. The more plentiful albumins bind these steroids less strongly but have a greater total capacity. The binding of different steroid hormones to such proteins is not specific, although there are differences in their relative binding affinities. Thus, 5α-DHT is bound three times more strongly to SHBG than is testosterone. Estradiol-17β also binds to SHBG but somewhat less strongly than does testosterone. Such steroids may compete with each other for the same binding sites, so that they can reciprocally influence each other's activity.

SHBG is synthesized in the liver, and this process is increased by estrogens and reduced by androgens. In late pregnancy the levels rise markedly, and this response results in an increased binding of plasma androgens. It has been surmised that this decline in free androgen levels may help protect the fetus against their masculinizing effects.

The general metabolism (see Wilson, 1975) of the steroid hormones is described in Section 5.3. About 5 to 8 percent of the testosterone is converted peripherally by target organs and the liver to the more active metabolite 5α-DHT. Some estrogens are also formed by aromatization reactions in the liver, hypothalamus, adrenal cortex, and adipose tissue. Less than 0.1 percent of testosterone is excreted unchanged in the urine. Most of the androgens are metabolized in the liver, in reactions that involve dehydrogenation of the 17β-OH group, reduction of the 3-keto, and saturation of the double bond in the A-ring. These metabolites (mainly epiandrosterone, etiocholanolone, androsterone, and 5α- and 5β-androstanediol) are nearly all conjugated with sulfate or glucuronide and are excreted in the urine or enter the bile and hence the enterohepatic cycle.

F. Analogues and structure–activity relationships

The synthesis of analogues of the androgenic steroids has been mainly prompted by the desire to obtain compounds that (a) can be administered

Table 7.2 *Relative androgenic activities (RA) and relative competition (for receptor binding) indices (RCI) of various androgens*

Steroid	RA[b]	RCI[a] Nuclear retention	RCI[a] Receptor binding
Testosterone (T)	0.4	0.7	<0.1
7α-CH₃-T	0.4	0.2	0.2
7β-CH₃-T	0.1	<0.1	<0.1
17α-CH₃-T	0.4	0.6	0.1
7α,17α-(CH₃)₂-T	0.6	0.3	0.2
7β,17α-(CH₃)₂-T	0.1	<0.1	<0.1
5α-Dihydrotestosterone (5α-DHT)	*1.0*	*1.0*	*1.0*
7α-CH₃-5α-DHT	1.2	0.4	0.4
17α-CH₃-5α-DHT	0.8	1.2	1.1
7α,17α-(CH₃)₂-5α-DHT	1.5	0.7	0.6
7β,17α-(CH₃)₂-5α-DHT	0.0	<0.1	<0.1
19-Nortestosterone (NorT)	0.2	0.7	0.0
7α-CH₃-NorT	2.6	1.8	2.6
17α-CH₃-NorT	0.3	0.9	1.2
7α,17α-(CH₃)₂-NorT (DMNT)	5.7	3.6	3.5
19-Nor-DHT (Nor-5α-DHT)	0.1	0.6	0.5
7α-CH₃-Nor-5α-DHT	0.3	0.4	0.6
7α,17α-(CH₃)₂-Nor-5α-DHT	0.3	0.4	1.0

[a] Relative competition index. The competition index (CI or a) for 5α-DHT is theoretically 1 for the ability of the "cold" unlabelled steroid to displace labelled 5α-[³H] DHT from the ventral prostate of rats. Because of the unavoidable presence of endogenous 5α-DHT in the tissue preparations, the value must be adapted relative to displacement by "cold" DHT, as determined experimentally at the same time so that

$$RCI = \frac{a \text{ for the analogue}}{a \text{ for DHT}}$$

[b] Relative androgenic activity measured by the abilities of the steroids to maintain the growth of the ventral prostate in castrated rats.
Source: Liao et al., 1973.

orally; (b) if given parenterally, have a prolonged period of action; and (c) have a selective anabolic action, that is, a high ratio of anabolic/androgenic activity. The uses of androgenic steroids for treatment of such conditions as breast cancer, osteoporosis, or for their anabolic and erythropoietic effects may be associated, especially in women and children, with an undesirable virilizing action, so that the testosterone molecule has been modified in order to change the ratio of its activities and so reduce the undesired androgenic side effects. This subject has been extensively reviewed by Krüskemper (1968) and Liao and Fang (1969). The relative activities of a variety of such steroids are summarized in Table 7.2.

The response to administered androgens depends on several factors, so that it is difficult to pinpoint the precise molecular requirements of its effector tissues and organs. These steroids may undergo considerable structural modification as a result of their metabolism in the body, and this may not only be reflected as differences in the rate of their clearance from the circulation but also in

the nature of the metabolites themselves. These products may have considerable endogenous activity, which, as in the case of 5α-dihydrotestosterone, may even exceed that of the parent compound. In some instances, other types of steroids with different activities may be formed. Thus, as a result of the action of peripheral aromatization enzymes, estrogens may be synthesized from androgens, and these steroids can interact with the androgens and influence the final response. Thus, estrogens will usually have a stronger effect in inhibiting pituitary gonadotropin release than the androgens from which they are formed.

Androgens have widespread actions in the body, and a variety of these responses have been used in attempts to quantify and compare their activities. These "assays" include the response of accessory sex glands, such as the prostate and seminal vesicles of castrated animals, usually rats and guinea pigs. However, other species, such as chickens, have also been widely used, and the response of the comb of roosters was the first bioassay to be utilized for the identification of androgens. The anabolic effects of androgens are usually measured by their ability to restore the weight of the levator ani muscle in castrated guinea pigs or rats. There are, not surprisingly perhaps, considerable differences in the responses of the different tissues, organs, and species, and this variation has made it difficult to obtain uniform standards for comparison. Some human tests are available, and these include measurement of the volume and fructose concentration in the seminal ejaculate, and the nitrogen balance of the body. Extrapolation from animal experiments to man, although useful, can be rather tenuous. Precise definition of structure–activity relationships at the purely receptor–effector level is thus rather difficult and requires an *in vitro* system where metabolism of the steroids is not significant. As the facility for preparing isolated androgen receptors increases, this system would appear to offer the brightest prospects for an accurate analysis of the structure–activity relations of these steroids, and it has been utilized by Liao and his collaborators (1973) and Skinner et al. (1975).

The activity of metabolites and precursors of testosterone indicate that several positions in the molecule are important for its activity (see, e.g., Figure 7.7 and Table 7.2). The 17β-hydroxyl group appears to be of primary importance, as when this moiety is oxidized to a 17-keto group there is a considerable loss in activity, as seen in the precursor androstenedione. 17α-OH compounds are not as active as the 17β compounds. Reduction of the 3-keto group to a 3-hydroxyl also lowers activity, but the effect of this change may be partly indirect and not influence interaction with the receptor. Thus conjugation reactions, with SO_4 and glucuronide, occur at the 3-hydroxyl site, and these make the molecule more polar and thus less able to cross the cell membrane and enter the cell. Androgenic steroids lacking the 3-keto group but retaining some activity are known.

Saturation of the Δ^4 double bond to form *5α-dihydrotestosterone* results in an increase in the activity of the molecule at many, but not all, of its receptor sites. This configuration in the A-ring may influence the shape of the molecule. 5α-DHT thus has a flatter, more planar, configuration, which could facilitate its access or interaction with its receptor.

As in the corticosteroids, the presence of a fluoride may enhance the activity of androgens. Thus, the fluoride at C-9 in *fluoxymesterone* (Figure 7.9) considerably enhances its activity, probably by increasing the reactiveness of the C-11 hydroxyl group which is present in this compound.

The substitution of the C-19 methyl group by a hydrogen, to form *19-nortestosterone*, reduces androgenic activity. However, it is interesting that the anabolic action of this compound is essentially un-changed. This observation led to the development of a series of analogues with preferential anabolic, as compared to androgenic, effects.

The nature of the A-ring of the androgenic steroids has an important effect on its action. If the degree of unsaturation of this part of the molecule is increased, or even if the Δ^4 is shifted to Δ^1 position, the molecule gains in estrogenic activity. On the other hand, saturation of the Δ^4 double bond, to produce 5α-dihydrotestosterone and related compounds, blocks possible conversion to estrogens. Substitutions at the C-1, C-2, C-4, and C-7 positions in the A-ring enhances the ratio of anabolic/androgenic activity. The compound with the highest such ratio is *stanozolol*, where the A-ring is condensed with hydrazine. A steroid that is widely used to treat breast cancer is *calusterone*, which is identical to 17α-methyltestosterone but has a β-CH$_3$ group at C-7 in the B-ring. It should be recalled that initial estimates of the relative activities of the various analogues is based on animal experiments. Clinical studies in man have, however, not always confirmed these preliminary results, so that their usage is ultimately dictated by clinical experience. Danazol is another compound with a modified A-ring which, however, only exhibits weak androgenic activity (Potts, 1977). However it has a relatively selective action in inhibiting the release of pituitary gonadotropins.

The activity of exogenous androgens can be increased indirectly by reducing the rate of their catabolism in the body. When testosterone is given orally, it enters the hepatic-portal circulation and is rapidly inactivated by the liver. Alkylation at the 17α-position, as *17α-methyltestosterone* (Figure 7.9), prevents this process by stabilizing the 17β-hydroxyl moiety and preventing its oxidation to a 17-keto group. The presence of the Δ^4 double bond or methyl groups at C-1, C-2, and C-6 also enhances the stability of the 17β-hydroxyl. It is also possible that methylation, which increases lipid solubility, promotes the accumulation of the molecules in the cell and so enhances their interactions with the receptors that are present there (Wilson and French, 1976).

The effects of androgens can be promoted by protracting the period of their effects. This can be achieved by prolonging the duration of their absorption. Esterification at the 17β-position (Figure 7.9) is commonly employed in steroid hormone preparations (injected in oil or aqueous suspensions) to enhance their lipid solubility and create local depots from which they can be absorbed over a prolonged period of time. In order to act, the free alcohol form of the steroid must also be released as a result of the action of a steroid esterase. The more bulky the ester group, the more slowly it is absorbed. The action of the esterase is also reduced. Thus, *testosterone propionate*, when injected

Oral or buccal administration

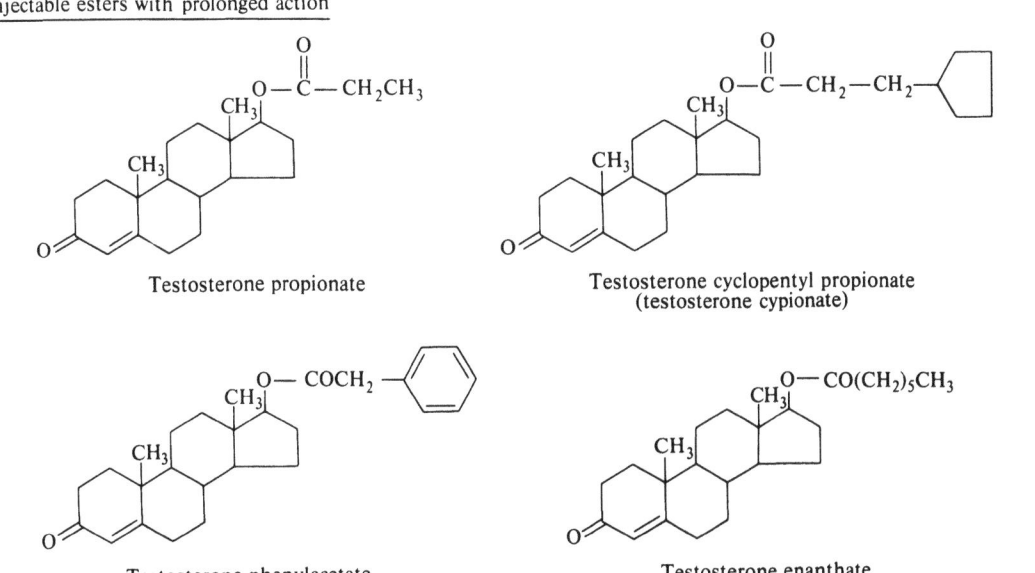

17α-Methyltestosterone

Fluoxymesterone
(9α-fluoro-11β-hydroxy-
17-methyltestosterone)

Danazol
(17α-pregna-3,4-20-yno
(2,3-d)-isoxazol-17ol)

Injectable esters with prolonged action

Testosterone propionate

Testosterone cyclopentyl propionate
(testosterone cypionate)

Testosterone phenylacetate

Testosterone enanthate

Fig. 7.9. Preparations of androgens that are used therapeutically. The oral or buccal preparations are given on a daily basis. The esters are injected intramuscularly, usually in oil (testosterone phenylacetate is prepared in aqueous solution), and have a prolonged period of action. The propionate is given 3 or 4 times a week, but the larger esters may act for 2 to 4 weeks. Pellets of testosterone are also available which when implanted under skin remain an active site for absorbable androgens for 4 to 6 months.

in an oil vehicle, needs to be administered every 2 to 3 days, but the larger *testosterone cypionate* is effective for 2 to 4 weeks.

The binding of numerous analogues of testosterone to its cytoplasmic receptors (β-protein) has been measured by Liao et al. (1973) and Skinner et al. (1975). Such measurements involve binding of tritiated 5α-dihydrotestosterone and recording the ability of unlabelled "cold" analogues to displace it. The values are expressed as the relative competition index (RCI), which is essentially a reflection of the relative binding abilities of the analogues compared to 5α-DHT, which is taken as 1. Some of these measurements, as well as the ability of the steroids to be retained in the nucleus, are shown in

Table 7.2. Generally, there is a good correlation between such binding and androgenic activity.

G. Antiandrogens

Compounds with antiandrogenic actions are of considerable theoretical interest and they are also of potentially important therapeutic use. This use has been considered in such androgen-dependent conditions as acne, hirsutism in women, enlargement and tumors of the prostate, baldness, satyriasis, and as a male contraceptive pill. The first such compounds to be used for these purposes were the estrogens, including estradiol-17β and stilbestrol. The actions of androgens may be prevented in several ways:

a. Inhibition of secretion of gonadotropins. The testicular atrophy observed under the influence of administered estrogens appears to be mainly a result of this response.
b. Metabolism of the androgens to more active metabolites, especially 5α-dihydrotestosterone, may be inhibited. Such compounds are mainly hypothetical at this time, but efforts are being made to develop them. The aim is to specifically inhibit 5α-reductase.
c. Synthesis of androgens may be blocked by appropriate inhibitors of different stages in steroidal biosynthesis. Such compounds, which are unfortunately not specific to androgens alone, are described in Section 5.2.
d. A specific competitive antagonism may occur which limits the interaction of androgens at their effector sites. *Antiandrogens* are often considered specifically to be this type of compound. In the development of such antagonists, the aim is to provide compounds with a minimum of intrinsic androgenic activity but a high affinity for the androgen receptors.

If such a drug is to be useful therapeutically, it should have a minimum of other actions. It has been recognized for many years that other steroids, including progestins and possibly estrogens, may have a direct antagonistic action on responses to androgens. This property led to the development of several types of antiandrogens (Figure 7.10): *cyproterone,* which in structure is progestin-like; *estracyt,* which is related to estrogen; and *BOMT,* which is related to 5α-DHT. It is also interesting that the competitive aldosterone blocker *spironolactone* also competes with androgens, which may account for its side effect in promoting gynecomastia (Corvol et al., 1975).

Cyproterone and *cyproterone acetate* are potent antiandrogens which compete with 5α-DHT for binding to the cytoplasmic receptor (Fang and Liao, 1969). They inhibit a variety of androgenic responses, including the embryonic differentiation of the Wolffian ducts and the prostate and the seminal vesicles (Elger, Neumann, and Von Berswordt-Wallrabe, 1971). Cyproterone also appears to block the effects of testosterone on the hypothalamic negative-feedback control system, so that gonadotropin release is increased. This response, by stimulating testosterone secretion, will tend to oppose the direct antagonistic actions of the drug. The ester cyproterone acetate, however, retains substantial progestin-like activity, so that gonadotropin release is inhibited while its other antiandrogenic effects are unimpaired. This drug causes testicular atrophy, inhibits spermatogenesis, and causes degeneration of the accessory sex glands in rats (Neumann and Von Berswordt-Wallrabe, 1966). Clinical trials in man indicate that it has similar effects, but it is not in general therapeutic use. Cyproterone acetate has been shown to have favorable effects in the treatment of male sexual deviants (Murray et al., 1975). The decline

in libido is, however, unacceptable in most other normal men. There is a decline in the plasma testosterone levels, so that the effect may at least partly reflect its action in inhibiting the release of gonadotropin. Cyproterone acetate has also been used, with a measure of success, to treat hirsutism in women (Ismail et al., 1974; Thomas et al., 1977). When used in these women, it was given with ethinyl estradiol to avoid menstrual abnormalities and pregnancy. In view of the effects of cyproterone on sexual differentiation in male rats, its use in human pregnancy would appear to be potentially hazardous. During more than a year of treatment for female hirsutism, no adverse side effects were noted except possibly a decline in libido.

Estracyt (Figure 7.8) has been tested clinically in patients with carcinoma of the prostate (Müntzing et al., 1974). Four of 11 patients experienced an objective improvement. Side effects were not a problem except for gastrointestinal symptoms. This result confirmed earlier European trials. *BOMT* (Figure 7.8) is another steroidal antiandrogen which competes with 5α-DHT for sites on the cytoplasmic receptors (Mangan and Mainwaring, 1972). In animal tests, it has been shown to lack progestin activity but exhibits some antigonadotropic effects, which may also contribute to its actions (Boris, DeMartino, and Trmal, 1970).

The most widely studied nonsteroidal antiandrogen is *flutamide* (Figure 7.8). Despite its apparent lack of chemical similarity to androgens, it inhibits the binding of 5α-DHT to its cytoplasmic receptor (Liao, Howell, and Chang, 1974). Its side chains appear to be flexible and it has been suggested that it can assume the necessary geometrical planar configuration to interact with the androgen receptor. It is metabolized to a more active form. Like cyproterone (*not* CP acetate), it stimulates release of gonadotropins and circulating testosterone levels increase (Södersten et al., 1975). *DIMP* is another nonsteroidal antiandrogen which acts peripherally (Boris et al., 1973).

H. Mechanisms of action of androgens
The principal effects of androgens are mediated following their interaction with specific cytosol receptors, and the response usually appears to involve the formation of messenger RNA. Other effects, however, may occur. There is no immediate rapid response to androgens such as the almost instantaneous increase in uterine blood supply which is seen in the presence of estrogens. It also seems unlikely that there are direct membrane effects, such as result in the release of lysosomal enzymes (Sepsenwol and Hechter, 1976). However, one such nongenomic action has been described, as androgens have been shown to produce an increased accumulation of cyclic AMP in the prostate

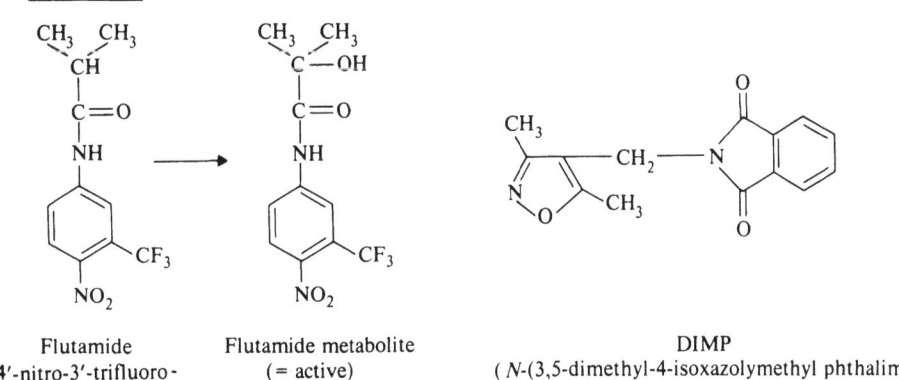

(a) Steroidal

Cyproterone acetate

BOMT
(6α-bromo-17β-hydroxy-17α-methyl-
4-oxa-5α-androstan-3-one)

Estracyt
[estradiol-3-N-(bis (2-chlorethyl))-
carbamate-17-phosphate, estramustine phosphate]

Spirolactone (SC-9420)

(b) Nonsteroidal

Flutamide
(4′-nitro-3′-trifluoro-
methylisobutrylanilide)

Flutamide metabolite
(= active)

DIMP
(N-(3,5-dimethyl-4-isoxazolymethyl phthalimide))

Fig. 7.10. Steroidal and nonsteroidal drugs with antiandrogen effects.

Table 7.3. *"Early" biochemical events that can be stimulated by androgens*

Induction of enzymes
Cytochrome oxidase
Malate dehydrogenase
Citrate-condensing enzyme
Glycosidases
Aldolase
Aldose and ketose reductase
Adenosinetriphosphatase (Na,K-activated)
RNA polymerase

Synthesis of constituents of low molecular weight
Citric acid
Polyamines (spermidine and spermine)

Synthesis of macromolecular constituents
Polyribosomes
Messenger RNA
Endoplasmic reticulum (membranes)
Nuclear proteins (nonhistones)
Nuclear membranes

Source: Mainwaring, 1977.

Table 7.4. *Some properties of cytoplasmic androgen receptor (β-protein) in the rat ventral prostate gland*

Molecular weight	276,000
Einstein–Stokes radius (Å)	84
Sedimentation coefficient (S)	8.0
S value in 0.5 M KCl	4.5
Fractional ratio f/f_0	1.96
Isoelectric point (pI)	5.8
Requirement for —SH groups	Yes
Thermal stability	Extremely labile
Dissociation constant (K_D)	$2.4–4.0 \times 10^{-9}$ M
Effect of cyproterone acetate	Active competitor
Relative binding of estradiol	Low
Transfer of 5α-dihydro-testosterone into chromatin and DNA	Yes
Effect of warming at 20°C	Modified form; 4.5 S, pI 6.5

Source: Mainwaring, 1977.

(see Mainwaring, 1977). This effect is related to the activity of the pentose phosphate "shunt" cycle. This response is not inhibited by antiandrogens, so does not appear to be mediated via the usual cytosol receptors. Cyclic AMP also cannot mimic the effects of androgens on growth of the prostate. The significance of this effect of androgens on nucleotide metabolism is not clear.

Androgens are relatively slow to act and the responses have distinctive time sequences. On this basis, Mainwaring (1977) has classified the receptor-mediated effects of androgens into three categories:

a. *Initial,* which involves the entry of the hormone into the cell, its metabolism, and binding to the cytosol receptors. There is a stimulation of ribosomal (extranuclear) RNA. The events usually occur within the first 30 minutes of exposure to the androgens.
b. *Early,* which involves the formation of messenger RNA and the synthesis of cell proteins, phospholipids, and membranes (Table 7.3). These changes are seen after about 16 to 24 hours.
c. *Late,* referring to effects that take several days to develop and involve cell replication ("mitogenic") and increased synthesis of nuclear constituents such as DNA and histones. Repeated administration of androgens is needed to elicit these effects experimentally or therapeutically. Human prostatic hyperplasia is an example of such a response.

The general nature of androgenic responses is also classified into two types:

a. *Switch,* which is a qualitative change involving the initiation of an event such as embryonic differentiation.

b. *Amplification* (or modulation), which involves quantitative changes, such as in the production of enzymes or structural components of the cells. This type of effect is the principal one elicited by androgens and is concerned with the maintenance and growth of sexually related glands and tissues.

The general mechanism of action, involving genetic transcription, of the androgens conforms to the description given earlier (Section 7.4), but there are some special features that deserve comment (see especially Minguell and Sierralta, 1975; Mainwaring, 1977; Liao and Fang, 1969). The anabolic effects, for instance, appear to involve a different type of mechanism.

The androgens may be unique in that, in order to act, they are usually converted at many (but not all) of their target sites from the secreted "prohormone" testosterone, by 5α-reductase, to 5α-dihydrotestosterone. This transformation usually occurs locally, after the steroid has entered its target cell. It then combines in the usual way with its cytoplasmic receptor. This protein (β-protein) has been isolated and some of its properties are given in Table 7.4. Other proteins that bind androgens have been identified in the cytoplasm of effector tissues; one of these, α-protein, has a high affinity for 5-DHT, but it does not enter the nucleus and thus may, by sequestering the steroid, antagonize its effects (see Mainwaring, 1977).

It has been proposed that the androgenic steroids react by apposing their planar α- or β-faces, rather than their periphery, to their receptors ("receptor-face" hypothesis). This could involve their ability to enter a narrow cleft in the receptor. Steric, rather than electronic, factors are thought to be more important in the interaction of the androgens with their receptors. Models of testosterone, 5-DHT, and two analogues which have a high endogenous ability to bind to androgen receptors are compared in Figure 7.11. It can be seen that testosterone, which has a very low receptor binding ability, is distorted on its α-face because of bending of the A-ring, which may interfere with its approach to the active site in the receptor. The other molecules have a sufficiently elongated planar α-face to make such interaction possible. It is, however, likely that such models provide an oversimplified interpretation of the hormone–receptor interaction.

High-affinity receptor proteins have not been equivocally demonstrated in all androgen target tissues. They are notably deficient in bone marrow, although in this tissue high-affinity nuclear receptor sites for testosterone are present. This situation may be unique for the steroid hormones. There has been some doubt as to whether cytosol receptors for androgens were present in skeletel muscle, but they have now been identified (Krieg and Voigl, 1977). They are, however, present at relatively low concentrations compared to other target organs, such as the prostate (70 times less). The precise nature of receptors in the brain is somewhat elusive. Androgen receptors may also be deficient as a result of genetic diseases in man and animals (testicular feminization syndromes).

Not all androgen effector tissues have the ability to transform testosterone to 5-DHT; it is, for instance, absent in most skeletal muscles, bone mar-

row, and during embryonic development in the Wolffian duct (where testosterone, but not 5-DHT, mediates embryonic differentiation).

The nature of the final effect of androgens varies with the target organ. The physiological or pharmacological responses usually appear to result ultimately from the formation of specific proteins by the ribosomes. These may be structural components of the cell, or enzymes. Some of these are listed in Table 7.3. The particular response may vary in different species, so one must take care in extrapolating the results too freely. In the instance of the enzyme prostate aldolase in rats, a specific messenger RNA which mediates its formation has been identified. This type of specific transcription no doubt applies also to other constituents of the cells which respond to the presence of androgens. Dihydrotestosterone maintains adenyl cyclase levels in the rat prostate, but it only influences one form (type I, not type II) of the enzyme (Fuller, Byus, and Russell, 1978).

The "late" *mitogenic effects* of androgens involve cell replication and are seen in organs such as the prostate, whose innate integrity is related to the presence of androgens ("androgen-dependent") and other tissues, such as the kidney, which under normal physiological conditions are relatively unaffected but respond to pharmacological doses ("androgen-sensitive"). When testosterone is administered to castrated rats (see Minguell and Sierralta, 1975; Mainwaring, 1977), there is a latent

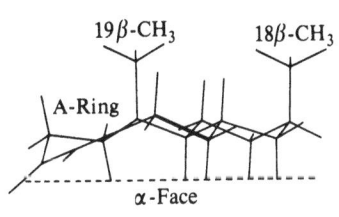

Testosterone

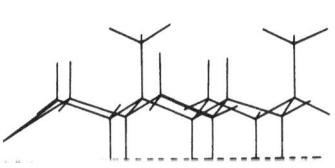

5α-Dihydrotestostosterone (DHT)

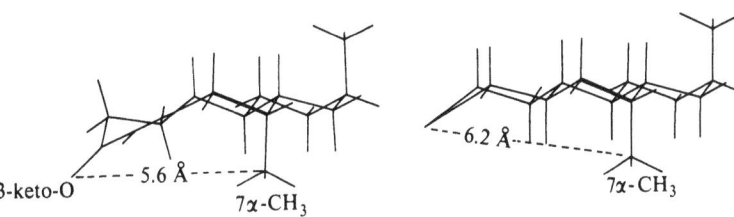

7α,17α,DICH₃-19-nortestosterone (DNMT)   7α,17α,DiCH₃-19-nor-5α-DHT

Fig. 7.11. Dreidings models of testosterone, DHT, and two analogues which have a high ability to bind to the receptors. A planar face is thought to be important, but in testosterone this is broken by the alignment of the A-ring. The presence of the 7α-CH₃ group enhances the molecules' ability to interact with the receptor. The distance in the plane between this group and the 3-keto-oxygen appears to be important. (From Liao et al., 1973.)

period of about 2 days followed 3 to 4 days later by a dramatic increase in the growth of the prostate. This response involves cell replication and an increase in DNA. The effect follows an interaction of 5-DHT and the receptors during the latent period, which can be inhibited by the antiandrogen cyproterone acetate and metabolic inhibitors. Subsequently, there is an increase in many cell constituents, notably DNA polymerase and histones, which are a vital constituent of the chromatin. It is not known if these changes involve the direct transcription by the androgen of the appropriate messenger RNAs. It is possible that some other component stimulates cell division or that the changes are a response to the generally increased activity of the cells.

I. Disorders of testicular function and clinical uses of androgenic anabolic steroids

Androgens have several types of uses:

a. Those related to *disturbances of testicular function,* which include hypogonadal conditions that result in sterility and/or disturbance in sexual development and function.
b. *Utilization of the "anabolic effects" of these steroids* in conditions that are not the result of any known physiological androgen deficiency. These include their action on muscles and bone, and erythropoiesis.
c. *The treatment of cancer,* especially that of the breast and its metastases.

1. Testicular dysfunctions. The field of testicular dysfunctions has been recently reviewed by Odell and Swerdloff (1978).

Hypogonadal conditions arise from a variety of causes, which can be broadly divided into two groups: (a) *primary,* due to a basic disturbance arising in the testes; and (b) *secondary,* due to pituitary or hypothalamic failure which results in inadequate gonadotrophic stimulation of the testes.

The effects of both types of conditions may be similar and be displayed as inadequate function of the seminiferous tubules and/or deficient activity of the Leydig cells in the interstitial tissue, so that androgen production is inadequate. The result can be sterility and/or failure of sexual differentiation, development, and function.

a. *Primary hypogonadism* is most explicitly displayed in the absence of the testes. The effects of castration were discussed earlier (Section 7.5C). This operation is rare today as a mutilation procedure, but it occurs as a result of accidents and in therapy for cancer of the testicles and prostate. Its use to control certain forms of sexual deviation is now generally frowned upon. The testes may also fail to develop ("congenital anorchia") or degenerate due to disease later in life.

A primary deficiency of testosterone can be corrected by administration of the hormone. Oral preparations of the free alcohol form of testosterone are not often used, as they are considered to be relatively ineffective as a result of their rapid inactivation by the liver. However, it has been shown that high doses, 100 mg four times a day, were able to maintain adequate plasma levels in eunuchs (Johnsen, Bennett, and Jensen, 1974). Prolonged oral therapy with the 17α-alkylated androgens, such as 17α-methyltestosterone, is usually avoided because of their propensity to cause hepatic disorders. Parenterally administered preparations with prolonged actions are more favored, although by their use one sacrifices some measure of controlling the dose. Intramuscular injections of such compounds as testosterone propionate can be given at 2- or 3-day intervals, while testosterone cypionate need only be given every 2 to 4 weeks. Aqueous suspensions of testosterone can be injected intramuscularly every 2 to 3 days. Pellets of testosterone are often implanted subcutaneously and can provide sufficient circulating testosterone for 4 to 8 months. They may cause some local irritation. The response to such treatment is rapid, within a few days, and the development of the secondary sex characters becomes complete in less than 4 months.

Less acute changes in testicular androgen production and responses include *Klinefelter's syndrome,* which is a chromosomal abnormality where there is an additional X chromosome so that the genetic male has the XXY, instead of the XY constitution. Sexual development often proceeds normally at first, but function may start to decline shortly after puberty. Although this condition invariably results in sterility, sexual function may initially be adequate but is not sustained. Testosterone levels are low and can be restored by the administration of the steroid, usually as periodic injections of slowly absorbed esters.

Primary examples of *male pseudohermaphrodism* are the testicular feminization syndrome, the 5α-reductase deficiency, and the Reifenstein syndrome.

The *testicular feminization syndrome* is congenital and results in a failure of appropriate sexual differentiation and development. In this condition there is a normal male XY chromosomal karyotype, but the testes and male accessory sex organs fail to develop properly. Instead, the external genitalia have the female form, but there is no uterus or ovaries. Many of the tissues are unresponsive to androgens, which may be due to a lack of androgen receptors (see Williams-Ashman, 1975). Such patients do not respond to testosterone or 5α-DHT, so there is nothing that can be done pharmacologically.

The syndrome of 5α-reductase deficiency (sometimes classified as "incomplete testicular feminization") is also manifested as a development of female-type external genitalia but in a male genotype. There appears to be a genetic deficiency of 5α-reductase, so that testosterone cannot be converted to its active metabolite 5α-DHT. Testosterone-mediated responses such as differentiation of the Wolffian ducts occur. Some male development occurs at puberty and, in contrast to the testicular feminization syndrome, there is no development of the breasts. Possible uses of steroids in such a condition have not been defined.

The *Reifenstein syndrome* ("partial androgen sensitivity") is characterized by incomplete development at puberty accompanied by infertility and gynecomastia (Amrhein et al., 1977). The production of androgens is normal, 5α-reductase is present, and the levels of LH are elevated. It appears that this congenital condition, which may or may not be familial, is due to a defect in the ability of the target end organs to respond to androgens. In one of the patients described by Amrheim et al., there was a deficiency in the cytoplasmic binding of DHT, but this receptor function was normal in four other people. Large doses of testosterone enanthate reduced the gonadotropin levels and had a mild virilizing effect. The routine long-term use of such high doses of androgens to treat this condition was, however, not recommended.

The "male climacteric" is a subject of much popular debate. Unlike women, men usually maintain their fertility (at least theoretically) throughout their lives, and adequate spermatogenesis has been described in very elderly men. Effective testosterone levels, however, appear to decline with age (Editorial, 1975d), and this change is due to both a decreased production and an increased level of sex hormone-binding globulin (SHBG), so that the free unbound level of testosterone is reduced. Pituitary gonadotropin levels increase. This process of decline in testicular function may be accelerated, a condition sometimes called the "male climacteric." The use of androgens in old age has been the subject of much debate and speculation. In 1889, Brown-Sequard injected himself with extracts of animals testes and reported favorable results. Subsequent measurements, however, showed that the extracts that he used were undoubtedly devoid of androgenic activity. Injections of testosterone esters have been shown to overcome the symptoms of the male climacteric, but whether such therapy is routinely desirable in aging men is questionable. High testosterone levels may be associated with prostatic hyperplasia.

The *seminiferous tubules* are more prone to damage than is the interstitial tissue. This probably reflects their high rate of activity and cell division. *Tubular failure* results in infertility but does not alter other manifestations of male sexual function. Testosterone is necessary for maturation of the sperm, so that tubular function is not completely independent of Leydig cell activity. Conditions that primarily damage the seminiferous tubules include various diseases, such as mumps, which result in testicular inflammation. Spermatogenesis requires a temperature that is 1 to 2°C lower than the normal body temperature, so that tubular damage occurs when the testes are retained in the body cavity or inguinal canal (*cryptorchidism*), or if the body temperature rises excessively, as in fever. The alleged contraceptive value of taking a hot bath before copulating has been extolled. Low oxygen tensions such as those experienced at high altitudes may reduce spermatogenesis. The deleterious effects of ionizing radiation are well known. A number of chemical agents also influence the tubules; these will be described in a later section related to chemical contraception in the male.

b. *Secondary causes of hypogonadism* in the male result from deficiencies in the anterior lobe of the pituitary gland and the hypothalamus, which controls its activity. The problem may be of a general nature, such as following hypophysectomy, or in panhypopituitarism. More rarely, a specific LH and/or FSH deficiency has been observed. As the release of these gonadotropins are controlled by LHRH from the hypothalamus, a dysfunction of this region of the brain can also result in male hypogonadism. The respective roles of the pituitary and hypothalamus can be assessed by the injection of LHRH (see Besser et al., 1972). If the problem is hypothalamic, gonadotropin and androgen release will occur, but if pituitary function is inadequate, this response will not be seen.

Deficiencies in the activity of the hypothalamus–pituitary system can result in a lack of activity of the seminiferous tubules or the Leydig cells or, most often, both.

Secondary hypogonadism may also result from the administration of estrogens or androgens, which may inhibit release of gonadotropin. Androgens can also be converted to estrogens, which exert an even stronger inhibitory effect. An excessive release of androgens in the adrenogenital syndrome may have a similar effect. Phenothiazines, which are used as tranquilizers, can decrease gonadotropin release and this action has been related to a decrease in the weight of the testes.

The widespread abuse of drugs such as alcohol, marihuana, and heroin has also been associated with changes in testicular function, which may at least partly be related to their action on the hypothalamus and pituitary. The mechanism of their effects is uncertain but may involve these endocrine tissues.

These "secondary" effects on testicular function offer the prospect of the development of a male contraceptive pill which could act via the hypothalamus or pituitary. In some instances, infertility in the male may be treated with gonadotropins or LHRH.

Cryptorchidism is another endocrine-related deficiency in male gonadal function. The testes normally descend into the scrotum in fetal life but on some occasions (2 to 3 percent) one or both of these organs may be retained in the abdomen or inguinal canal. Descent may then occur shortly after birth or at puberty. However, in about 0.2 to 0.8 percent of males, this event does not occur spontaneously. A late puberal descent of the testes can leave the legacy of a deficient tubular function and sterility. This effect appears to be due to prolonged exposure to the unphysiologically high temperature of the abdominal cavity. Leydig cell function is not appreciably affected. The incidence of cancer of the testicles is very much higher in cases of cryptorchism and has promoted some to recommend routine castration.

The physiological processes that control the descent of the testes are not understood but appear to involve the maternal chorionic gonadotropin. Late descent in puberty probably reflects a rise in the circulating levels of pituitary gonadotropins.

The most successful treatment is surgical, but hormonal therapy with hCG is often tried initially. To avoid permanent damage to the testes, it should be started as early as possible after the fifth year. Intramuscular injections of the hCG (prepared from the urine of pregnant women) are given at 2- to 3-day intervals for a period of up to about 8 weeks. There is a moderate rate of success (about 40 percent), especially if the condition is bilateral and the testes are retained in the inguinal canal.

Synthetic LHRH has also been shown to have a favorable effect in cryptorchidism. In a recent clinical trial it was tested in 84 boys. Complete testicular descent or substantial improvement occurred in a large number of instances (Illig et al., 1977). A synthetic preparation of the hormone was given intranasally six times a day for 4 weeks. No untoward side effects, such as could result from increases in plasma androgen levels, were observed. This form of therapy has the advantage of convenience, but the preparation is expensive and relatively large doses must be given when using this route of administration. Others prefer to be more frugal and inject LHRH, which has the added advantage that absorption is more predictable. Whether LHRH is more effective than hCG has been questioned, and if one resorts to parenteral administration the advantages of the former are in doubt.

2. Benign prostatic hyperplasia. In fetal life and at puberty the development of the prostate is under the control of 5α-dihydrotestosterone, but it ceases to grow after this time. In about 70 percent of men, there is a second growth spurt of this gland in middle age. The enlargement can cause urinary retention due to a mechanical resistance to micturition. It has been suggested (Wilson, 1972) that this disease is a result of an excessive conversion of testosterone to 5α-DHT in the gland due to a high 5α-reductase level. The growth thus appears to be androgen-dependent and can be antagonized by the estrogen stilbestrol. This treatment (see Section 7.6N) can result in a variety of undesirable side effects, including loss of libido, gynecomastia, and an increased death rate from cardiovascular and thrombotic accidents. The antiandrogen cyproterone acetate has been considered as an alternative form of endocrine therapy, but its use has not yet been established.

3. Gynecomastia. In response to certain stimuli, the breasts of the male may undergo varying degrees of development. This is called gynecomastia, and it can be a painful as well as a disfiguring condition. Usually, only a development of the mammary ducts occurs, but occasionally the secretory alveoli may also grow and milk can even be expressed. Gynecomastia can occur in response to therapy with a variety of drugs. It is common in hypogonadal diseases and commonly occurs, but subsequently regresses, at puberty. It may be associated with high levels of plasma gonadotropins and prolactin.

The endocrine control of the mammary glands and lactation is described in detail elsewhere [Sections 7.6G and H(5)] Several types of hormones are involved, including estrogens and progestins, thyroid hormones, glucocorticoids, growth hormone, and prolactin. There are thus multiple possible causes of gynecomastia, although the evidence about the mechanism of action of each is usually equivocal.

One cause of the disorder is a relative or absolute excess of estrogens. This situation may be due to a high rate of endogenous formation of the steroids, estrogen therapy, and probably the conversion of other steroids to such compounds. Thus, the use of the steroidal compounds deoxycorticosterone, spironolactone, and digitalis is sometimes associated with gynecomastia. Competition by such steroids for endogenous androgens may also be involved. Stimulation of steroid hormone release, such as during treatment of cryptorchidism with hCG, may result in this condition.

Disturbances in liver function may be associated with gynecomastia. Alcoholics sometimes experience this condition, probably as a result of derangements in liver enzymes which either fail to inactivate endogenous estrogens or facilitate the conversion of androgens to estrogens.

The role of pituitary hormones in producing gynecomastia is probably reflected in the rare occurrence of this condition during the use of the hypotensive drugs reserpine and α-methyldopa and the tranquilizer chlorpromazine (Robinson, 1957). These compounds influence hypothalamic function and, probably via a depletion of dopamine, stimulate an excessive secretion of prolactin. However, other factors and hormones could be involved. Treatment involves withdrawal of the drugs, although this may not always be possible. Radiotherapy prior to using the drugs may prevent the effects of estrogens. Danazol (Figure 7.9) is sometimes effective and appears to act by decreasing the release of pituitary gonadotropin (Buckle, 1977).

4. Galactorrhea. Although secretion by the mammary glands is expected in postpartum women, it is unexpected in men. This condition may occur, however (see, e.g., Kleinberg, Noel, and Frantz, 1977), and is associated with high plasma levels of prolactin. The high rate of secretion of this hormone may be due to the presence of pituitary tumors, a malfunctioning of the hypothalamic–pituitary axis, or the use of certain drugs (see previous section). This condition is discussed in more detail later [Section 7.6H(6)]. It can be treated with bromocriptine, which is a dopaminergic drug that inhibits the release of prolactin.

5. Control of fertility in the male. About 10 percent of marriages are childless, and in about half of these the problem reflects infertility in the male. There are a variety of possible causes (see London, 1972) and many of them are not due to endocrine disturbances. For example, there may be a blockage of the sperm ducts, which can be congenital or result from venereal diseases or tuberculosis. A circulatory disturbance, varicocele, of the testes is a common cause of infertility, and psychic problems that result in impotence can also result in this condition. The production of sperm by the seminiferous tubules may be inadequate due to cryptorchidism or such diseases as gonorrhea or mumps orchiditis. The production of sperm is, however, also dependent on pituitary FSH and LH, and androgens. As described earlier, there is a deficiency of these hormones in primary hypogonadism or secondary hypogonadotropic hypogonadism. Endocrine therapy does not restore fertility in primary hypogonadism but may be effective if the deficiency is due to a hypothalamic or pituitary disorder. The prognosis is, however, not very favorable. Attempts can be made to replace the deficient gonadotropins, either directly or by administration of LHRH (see Mortimer et al., 1974a). Normal fertility occurs when the sperm count is about 60 million/ml of ejaculate, but below 20 million/ml, infertility is likely to occur. Human

menopausal gonadotropins (HMG), which contain a predominance of FSH, have been used to promote spermatogenesis (see Labhart, 1976). If there is a concurrent LH deficiency, which results in inadequate function of the Leydig cells, then hCG may be administered concurrently. The sperm take about 3 months to mature so that therapy is prolonged, and as it requires repetitive parenteral (i.m.) administration accompanied by frequent checks of the sperm count, it is expensive. Also, the rate of success is not high. Synthetic LHRH has also been used to treat secondary hypogonadotropic hypogonadism (Mortimer, et al., 1974a). It was injected subcutaneously at 8-hour intervals for periods of up to a year or more. Spermatogenesis was initiated in four patients.

Impotence (see Lording, 1978) is usually a psychosexual disorder, although it may also accompany the use of some drugs, such as guanethidine, clonidine, reserpine, and the ganglionic blockers, which are used to treat hypertension. Drugs of addiction, such as opiates and ethanol, and psychoactive compounds such as MAO inhibitors and tricyclic antidepressants, may also precipitate such problems. Androgenic steroids have no established place in the treatment of this disorder. Hypogonadism, however, is usually accompanied by a lack of libido, including impotence, and replacement therapy with testosterone usually corrects this condition and can even result in a high incidence of painful priapism. Hyperprolactinemia has also been identified as a cause of impotence and may be associated with hypogonadism, infertility, and galactorrhea. The dopaminergic agonist bromocriptine is now widely used to treat hyperprolactinemic conditions in women and it may also be effective in men. In a group of 25 men, this drug restored potency in 19, although the response was not always retained when the drug was withdrawn (Thorner and Besser, 1978). Bromocriptine has also been shown to restore testicle endocrine function and increase sperm levels. In one study, fertility was restored in several oligospermic men (Saidi, Wenn, and Sharif, 1977). The favorable effects of bromocriptine on potency are not invariable, however, as in a group of impotent patients with similar hormonal levels to those of controls, no difference in effect was observed when the action of bromocriptine was compared to that of a placebo (Ambrosi et al., 1977).

The *pharmacological induction of infertility* in the male is of considerable socioeconomic importance but, despite numerous attempts, has not met with the practical success of the female contraceptive pill (see Van der Vies, 1977; Ewing and Robaire, 1978). Women usually produce only a single ovum, which matures rapidly and is delivered at a predictable time, every 28 days. The male, however, continuously produces millions of sperm, and these take 3 months to mature and retain their viability for about 12 days during maturation and storage in the epididymis. Several types of mechanisms, endocrine and nonendocrine, may be utilized to control the production of sperm. The loci of a drug's action may involve sperm formation in the seminiferous tubules or spermaticidal effects in the epididymis. The latter is of considerable theoretical interest but is not yet practical.

Endocrine interruption of spermatogenesis depends on an ability to interfere with the release of FSH from the pituitary. The production of sperm is also dependent on androgens, so that blockade of the release of LH would also be effective but results in hypogonadism, with all its unacceptable consequences. A selective blockage of FSH is therefore desirable. The administration of a variety of steroids can block the release of gonadotropins, but they usually also influence androgen levels. Steroids that have been shown to be effective include small doses of estrogens, androgens, and progestins either alone or sometimes in combination, such as androgens and progestins, and androgens plus estrogens (Heller et al., 1959; Jackson and Jones, 1972; Patanelli, 1975; Ewing and Robaire, 1978). Estrogens are effective inhibitors of spermatogenesis in man, but the accompanying loss of libido and gynecomastia are unacceptable. Injections of testosterone propionate or enanthate (the latter once a week) inhibits spermatogenesis, but it takes 2 to 3 months to be effective. The orally active 17α-alkylated androgens are not generally of practical long-term use, as, in the high doses that are necessary, they adversely influence liver function.

Danazol (see Figure 7.9) is a 17α-alkylated derivative of testosterone which has only low androgenic activity but which selectively inhibits the release of pituitary gonadotropins. It has no detectable estrogen or progestin activities. When administered orally to men, it reduces the sperm count to levels below those necessary for fertility (Ulstein et al., 1975). At the doses used it had no reported adverse side effects on liver function. Its action on the sperm count is slow, taking about 12 weeks, and is reversible when the drug is discontinued. Androgen levels in the plasma decline and this effect may be associated with a decrease in libido, which, however, was compensated for by the concurrent intramuscular injection of testosterone enanthate once a month.

The antiandrogen cyproterone acetate and the antiestrogen 19-norspiroxone also induce infertility in the male (Neumann and Von Berswordt-Wallrabe, 1966; Goldman, Shapiro, and Root, 1976). Their effects appear to be mediated by an inhibition of the release of gonadotropins. Cyproterone may also have a direct effect on spermatogenesis and inhibit the functioning of the

epididymis. These drugs have not yet been shown to be of practical use in controlling fertility in men.

In rats, the administration of a potent analogue of hypothalamic LHRH, [D-Ala^6, des-Gly-NH$_2$^{10}]LHRH, results in a decline in testicular weight and an inhibition of spermatogenesis (Auclair et al., 1977; Pelletier et al., 1978). This effect is associated with a rise in the plasma levels of LH and FSH and a reduction ("down regulation") in the numbers of testicular LH receptors. Such receptor depletion and desensitization in response to high doses of LH and hCG have been described previously, and result in impaired steroidogenic response of the Leydig cells (Cigorraga, Dufau, and Catt, 1978). These observations have resulted in speculation regarding the possible utilization of such responses to LHRH analogues in order to block spermatogenesis in man (Ewing and Robaire, 1978). It seems likely, however, that a deficiency in androgen could result in undesirable side effects in man, although it may be possible to replace this steroid.

Nonendocrine mechanisms have been reviewed by Patanelli (1975). Various chemical and cytotoxic drugs have antispermatogenic actions. These include heterocyclic compounds such as the nitrofurans, thiophenes, diamines, and dinitropyrroles. Alkylating drugs such as the nitrogen mustards have a similar action. Cadmium has a well-known toxic effect on the seminiferous tubules, and this response appears to reflect a degeneration of the vasculature of the testes. Such drugs may affect other cells in the body, so that their actions are not considered to be specific enough to use in men. Their possible mutagenic effects must also be rigorously excluded before the adoption of such drugs is feasible.

6. Endocrine management of sexual deviation in man. The treatment of male sexual deviation has been reviewed by Brandon (1975). He states: "Hormone treatment has followed castration into oblivion as a form of treatment." Estrogens are effective and have been used for 20 years, but they have feminizing actions, including gynecomastia. The tranquilizer benperidol is also effective (see Murray et al., 1975) in the control of such conditions. Exhibitionism is the commonest form of sexual deviation and cyproterone acetate has been used to treat this disorder, but it undoubtedly reduces libido and sexual drive. It has, however, been questioned (Brandon, 1975) whether it is wise to treat a disorder that may reflect anxiety about impotence with a drug that can promote this condition.

7. Effects of drug abuse on endocrine aspects of sexual function. The effects of common drugs of abuse, such as alcohol, marihuana, and heroin, on sexual activity is largely a matter of legend and hearsay. The acute effects, such as dictated by changes in mood and behavior, do not appear to be related to the endocrine system and so will not concern us here. The long-term chronic use of such drugs, however, can have endocrine-related effects on the male reproductive system.

It is well known that severe *alcoholism*, such as results in cirrhosis of the liver, is associated with the occurrence of gynecomastia, reduced sexual activity, and atrophy of the testes (Adlercreutz, 1970).

The mechanisms of the effects of alcohol on sexual function in the male are complex and appear to differ, depending on the conditions, such as the period of imbibition and whether or not cirrhosis has developed. Alcohol when administered chronically to normal subjects causes a decline in plasma testosterone levels (Gordon et al., 1976). This effect appears to be related to a decreased hormone synthesis by the gonads, which is mediated by a direct testicular effect and also, possibly, indirectly as a result of a centrally mediated inhibition of gonadotropin release (see Van Thiel and Lester, 1974). Clearance of testosterone is increased, probably as a result of the induction of liver enzymes that metabolize the steroids and a decline in SHBG. In later stages of the disease, when cirrhosis develops, SHBG concentrations increase and the clearance of testosterone declines. There is an increased conversion of androgens to estrogens (Adlercreutz, 1974). Testosterone binds preferentially to the SHBG, so that the free active levels of estrogens undergo a further increase relative to androgens.

In acute experiments, ethanol (as 100-proof vodka in a dose of 2.4 ml/kg body weight consumed within 15 minutes) was also observed to bring about a reduction in plasma testosterone levels in normal men (Mendelson, Mello, and Ellingboe, 1977). This effect was, however, accompanied by a rise in plasma LH, so that the changes in the androgen levels appear to be of peripheral origin. It was suggested that a direct effect on its metabolism in the liver or an increase in the blood supply to that organ could be occurring. The rise in LH was especially interesting and there was speculation regarding its possible effects on sexual behavior. It has, for instance, been reported that plasma LH levels rise in men viewing erotic movies, and this response may be contributing to the concurrent sexual arousal.

The acute use of *marihuana* (hashish or cannabis) is alleged to result in increased sexual desires and activity, but the long-term use of this drug may result in hypogonadism. Chronic "intensive" use has been associated with the occurrence of gynecomastia and reduced circulating levels of testosterone and inhibition of spermatogenesis (Harmon and Aliapoulios, 1972; Kolodny et al., 1974). The cause of these responses is uncertain but may

involve the hypothalamic–pituitary system or "higher" centers in the brain. Cannabinol has recently been shown to reduce testosterone secretion from slices of mouse testis *in vitro*, so a direct action on the gonads may be occurring (Dalterio, Bartke, and Burstein, 1977). The results of Kolodny et al. were not confirmed by Mendelson and his collaborators (1974), and the subject continues to be a contentious one (see Maugh, 1975a).

Heroin and *morphine* addiction are often associated with sexual difficulties. These drugs are central depressants, so there are a variety of possible mechanisms for their effects. They depress the hypothalamus and this action is reflected in changes in pituitary function, including a decreased release of gonadotropins. *Methadone,* which is a chemical analogue of these drugs, is now widely used to maintain people that are addicted so as to prevent withdrawal symptoms. This drug has also been shown to influence male gonadal function. Heroin decreases plasma testosterone levels, although the results are contentious (Mendelson, Mendelson, and Patch, 1974, 1975; Azizi et al., 1973; Cushman, 1973). Methadone also causes a substantial drop in plasma testosterone (Cicero et al., 1975; Mendelson, Mendelson, and Patch, 1975) and this response is associated with a decreased activity of the secondary sex glands, a decline in sperm motility, and, apparently, decreased fertility.

8. Anabolic steroids and their uses. All androgenic steroids have anabolic effects and can be administered for this purpose. Compounds such as testosterone and its esters, however, although ideal for hormone replacement therapy, are not always suitable when it is desired to produce a specific anabolic action. Androgens, especially in high doses, may have virilizing effects in children and women, and may even have undesirable effects in men. Hence, a variety of alternative steroids have been developed in which the ratio of anabolic to androgenic activity has been increased. As described earlier, however, these compounds have largely been screened using animal tests, and their relative effectiveness in man does not always correspond to that originally predicted. They are, however, used quite widely for a variety of purposes.

The chemical nature of the anabolic steroids was described in Section 7.6F, and the structures of some more useful ones are given in Figure 7.12. *Norethandrolone* was one of the earlier prototypes, although it has not been released in the United States. *Calusterone* is relatively new and under active investigation. Several esters, such as *nandrolone phenyl propionate,* are available for parenteral use. *Stanozolol* and *testolactone* have, allegedly, the highest ratio of anabolic/androgenic activity. *Mesterolone* appears to be devoid of antigonadotropic actions.

Like the antiinflammatory glucocorticoids, the

Testosterone derivatives

Testosterone

Calusterone

Testolactone

Methandrostenolone

19-Nortestosterone derivatives

19-Nortestosterone

Norethandrolone

Nandrolone phenyl propionate

Dihydrotestosterone derivatives

5 α -Dihydrotestosterone
(stanolone)

Androstalone

Oxandrolone

Modification of the A-Ring

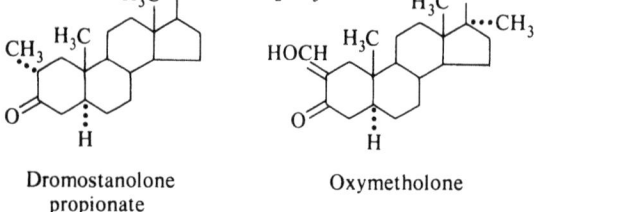

Dromostanolone
propionate

Oxymetholone

Stanozolol

Fig. 7.12. Chemical structures of some anabolic steroid compounds.

anabolic steroids are not necessarily administered for conditions that are known to be due to a specific hormone deficiency. The primary aim is to increase protein synthesis, but when this process is inadequate it is not usually associated with an androgen deficiency.

There are several *precautions and prerequisites* needed for the use of the anabolic steroids and certain conditions when special caution is needed. To be effective, there must be an adequate dietary protein intake. They should not be used in pregnancy, as they can have a virilizing effect on the fetus. They should also be avoided in cases of enlargement of the prostate, as they will exacerbate this condition. Certain of these compounds may also cause biliary and liver problems, so they should be avoided in patients with impaired hepatic function.

A wide variety of uses has been proposed for the anabolic steroids, but all of these have not met with unmitigated success or approval.

a. *To induce growth, development, and even hypertrophy of skeletal muscles.* Muscular wasting may result from malnutrition, debilitating illnesses, malabsorption of proteins from the gut, anorexia nervosa, hyperthyroidism, and the use of antiinflammatory corticosteroids. The induction of muscular hypertrophy in some athletes is also sometimes considered to be desirable. The use of anabolic steroids is often associated with a feeling of well-being and increased appetite. Generally speaking, however, anabolic steroids have not been highly successful or met with wide acceptance when used to promote anabolic effects. They also do not appear to be useful for those athletes participating in events that require stamina. The practice of taking these drugs has not been sanctioned by the organizations controlling athletic competitions or the medical profession (hence the sources are "nonprescription" ones). It is thought, however, to be practiced by up to 80 or 90 percent of male competitors in national and international sports (see Wade, 1972). Anabolic steroids are also used by "body builders" and are popularly thought to contribute to the desired excessive development of the muscles. Although their use is mainly confined to men, they have apparently also been used by women, in whom their virilizing effects may become apparent.

Despite the myotrophic effects of anabolic steroids on the levator ani muscle of rats and guinea pigs, a general anabolic effect has not been conclusively demonstrated in animals. In man, their effects are contentious. It seems likely, however, that in individuals on an adequate protein diet, who are also undergoing intensive athletic training, they do increase the lean weight of the body (Freed et al., 1975; Hervey et al., 1976). It is not clear whether this response is associated with increased strength, and it may even be partly due to fluid retention. The effects wear off within a few weeks of discontinuing the drugs.

The anabolic steroids that are popular among athletes are methandrostenolone (methandienone), stanozolol, and oxandrolone. The accounts of their use are largely anecdotal, but it is said that they are used in amounts up to 300 mg/day. In controlled trials, smaller doses of 10 to 100 mg/day have been used. The anabolic steroids appear to be freely available through "underground" sources.

Side effects are frequent but are usually reversible when the drug is discontinued. Acne and headache are common, and blood pressure may increase. Circulating testosterone levels decrease, apparently as a result of an inhibition of release of gonadotropins. Libido has been reported to increase or decrease. Hepatic disorders and jaundice have been described.

b. *Osteoporosis.* This bone disease is an especially common condition in postmenopausal women (Section 9.4A,2), where it is often considered to reflect a decline in steroid hormone levels. Anabolic steroids have been used to treat this condition but without conspicuous success, except possibly over brief periods of time. Their virilizing action is also undesirable.

c. *To treat delayed growth in children.* Delayed growth in children has been treated with anabolic steroids. In some instances this is due to a pituitary deficiency, and general replacement therapy is needed. Large doses will have virilizing effects and disturb sexual maturation in boys and girls, but small doses have been reported to be successful and hasten the attainment of the normal adult height (see Sobel, 1968). These steroids also hasten closure of the epiphyses, thus limiting growth, so they need to be used with considerable caution. They are usually given for intermittent periods of 6 weeks to 6 months so that constant reassessment can be made. [The use of growth hormone in pituitary dwarfism is described in Section 3.2F(6)].

d. *To stimulate erythropoiesis in several types of anemia.* Experiments on animals have shown that androgens can stimulate erythropoiesis, either in the female, or in the male following castration. In women, it was observed that the use of high doses of androgens to treat carcinoma of the breast was accompanied by an increase in the number of red cells. Hypogonadism in men may also be associated with anemia, which can be corrected by androgens. These observations have led to the use of anabolic steroids for the treatment of blood diseases associated with deficiencies of the bone marrow, and constitute one of the most important uses for these drugs (Shahidi, 1973). The conditions include acquired and constitutional aplastic anemia, Fanconi's anemia, and anemia associated with renal failure (Hendler et al., 1974).

The erythropoietic effects of androgens appear to involve two mechanisms:

(i) *Increased production of erythropoietin.* This response may reflect an action on the kidney, possibly the well-known renotrophic effects of these steroids. However, in man, erythropoietin is also produced at extrarenal sites, and this may account for the effectiveness of anabolic steroids in anephric patients. Erythropoietin is classified as a "growth factor" which is androgen-dependent.

(ii) *Direct effect on the erythropoietic stem cells.* Androgens may also have a direct effect on the erythropoietic stem cells, resulting in a larger population of tissue on which the erythropoietin can act.

The use of androgens and anabolic steroids to treat anemias requires relatively high doses. They may be administered intermittently. When a remission occurs, this is often sustained for many months, although in some conditions, such as constitutional aplastic anemia, continual therapy may be needed. As such high doses are used, side effects are common, including acne and a virilizing effect in women and children. Liver function may be adversely affected, especially by 17α-alkylated androgens. More rarely, hepatocellular carcinoma has been described following long-term treatment (Johnson et al., 1972; Farrell et al., 1975).

e. *Therapy of advanced metastatic breast cancer.* Many breast tumors appear to be hormone-dependent, and in premenopausal women a remission may be observed following removal of the ovaries. This subject is described more fully in a succeeding section (Section 7.6K). Another form of therapy, especially in postmenopausal women, is the use of androgens or, for preference, as they have a reduced virilizing effect, the anabolic steroids. Calusterone has undergone clinical trials (Goldenberg et al., 1973; Gordan, Wessler, and Avioli, 1972) and in a selected group of women was shown to produce remission in about 30 percent of cases. Women chosen for such therapy are usually postmenopausal and have widespread metastases with secondary tumors in the bones. Some doubt has been expressed as to whether such treatment is as effective as the use of cytotoxic drugs (Priestman et al., 1977). The mechanism of action of the androgenic anabolic steroids in this condition is not clear, but they are known to inhibit uptake of amino acids by the tumor tissue *in vitro*. It has been suggested that they exert a direct antiestrogenic effect, and calusterone has been shown to initiate a decline in the synthesis of estrogens (Fishman and Hellman, 1976; Fukushima et al., 1976). Hypercalcemia occurs in 10 to 20 percent of cases treated with such drugs but can be combated with diuretics such as furosemide, and with glucocorticoids.

9. Summary of the side effects on androgenic–anabolic steroids.

a. *Virilizing (or masculinizing) effects.* These effects are seen in women and children and may take the form of accelerated sexual maturation in boys. In girls and women, there may be a masculine-like growth of hair on the face and body. The voice may deepen. The clitoris enlarges (this effect may also be seen in female infants when such drugs are taken during pregnancy). If taken before puberty, the morphology of the body of girls may start to assume the masculine form. The sebaceous glands develop and "oily" skin and acne may occur. Libido may be increased in women, but in men the effect varies. The "anabolic" steroids were developed to reduce these side effects, but they are still present.

b. *Growth and skeletal maturation.* In children, the puberal spurt in growth occurs but the epiphyses tend to close, so that the final height of the individual may be reduced.

c. *Water and salt retention.* Possibly owing to an endogenous aldosterone-like effect or displacement of corticosteroids from plasma binding sites, edema due to salt retention may occur. This effect can be combated by a low-salt diet and/or the use of diuretic drugs.

d. *Liver problems.* Liver damage, including jaundice and cholestasis, may occur. This effect is seen most often with the 17α-alkylated analogues (see Westaby et al., 1977) such as 17α-methyltestosterone, and it is usually reversible when the drug is discontinued. Hepatic adenocarcinomas occur with protracted high doses. The tumors regress in size when the steroid is withdrawn.

e. *Antigonadotropic effect on the pituitary.* Androgens can inhibit gonadotropin release, and this can result in hypogonadism, gynecomastia, and a reduction in the number of viable sperm. Menstrual disorders may occur in women. This effect may be enhanced by conversion of the steroid to estrogen. Mesterlone is an anabolic steroid that cannot be converted to estrogens and is said to have no antigonadotropic activity.

f. *Hypercalcemia.* Hypercalcemia can occur when the anabolic steroids are used to treat advanced carcinoma of the breast. This effect can be combated with diuretics such as furosemide, and glucocorticoids.

7.6. Female reproductive system

Like the testes in the male, the ovaries of the female have a dual function, the formation of mature germ cells, the ova, and the production of hormones. In this instance there are two major types produced, the estrogens and progestins, which serve to coordinate the development and maintenance of the normal female sexual characters, as well as the processes associated with fertilization of the ovum, pregnancy, parturition, and lactation. The female contribution to reproduction is somewhat more complicated and committed than that of the male, and this is reflected in the endocrinology of the reproductive system. In the human male, sperm is produced continuously, but ripe ova suitable for fertilization mature only once about every 28 days. Usually, only one egg is produced at a time and it is only available for fertilization for about 1 day or less. Even if fertilized, the ovum will not continue to develop or survive for long if it is not quickly provided with an optimal morphological, physiological, and biochemical resting place in the oviduct and uterus. It is clear that to ensure even such initial success, a far greater degree of physiological coordination is needed in the female than in the male.

This complex process is reflected in the human menstrual cycle, during which, over a period of about 28 days, the ovum matures, is extruded from the ovary, and enters the oviduct, where it is available for fertilization. This event occurs in about the middle of the cycle, when the uterus is in a suitably primed condition to receive and nurture the blastocyst. And this is only the beginning!

The orderly succession of physiological events that must occur during this preparatory process are programmed by a timed release of the appropriate hormones. As in the male, the hypothalamus and the anterior lobe of the pituitary gland are of primary importance. These endocrine tissues can respond to both internal and external environmental "cues," resulting in the release of the gonadotropins FSH and LH, which act on the ovary to influence the development of the ovum and the biosynthesis of its steroidal hormone secretion. The hypothalamus–pituitary system in turn can respond, via a negative- or a positive-feedback mechanism, to the released estrogens and progesterone. The basic cyclical nature of the release of the gonadotropins develops in the brain after birth, in girls probably in their first 7 or 8 years. In animals, this development can be overcome by administration of androgens, or it can be promoted to develop in genetic males by neonatal castration.

These endocrine complexities can be utilized pharmacologically to control fertility in a manner that is not possible in the male. An intimate understanding of the physiological processes involved provided the theoretical basis for the development of the oral contraceptive pill and the use of drugs to treat infertility.

The reproductive endocrinology of women is also complicated, compared with that of men, by a relatively shorter period of fertility in the life span. Puberty, or the menarche, in girls occurs at about the age of 12 to 14 years and the last menstruation or menopause, at about 50 years. The latter event appears to reflect a senescence of the ovary and its inability to respond to stimulation by the gonadotropins. There is thus not only a failure to produce ova but also a decline in the production of ovarian steroid hormones. The immediate pre- and postmenopausal period, or climacteric, is thus associated with some quite dramatic physiological changes. These include some loss of the secondary sexual characters, as well as more general changes such as an increased propensity to suffer from osteoporosis, hypertension, and obesity. Despite this senescence, women on the average live for more than 25 years after the menopause and their life expectancy is several years longer than that of men. Sex hormones are clearly not everything!

A. Female reproductive apparatus

The primary female organs of reproduction are the two ovaries that lie in the pelvic region of the body cavity. Their embryonic origin has been described earlier. In general terms related to their origins, the ovaries consist of two types of tissue: germinal tissue, originating from the gonocytes, which are embedded in a matrix consisting of interstitial tissue; and batches of hilar or stromal cells, which are homologous to the testicular Leydig cells. The ovary's general appearance is not very uniform, as it is in a continual oscillating state of growth, maturation, and senescence. A layer of germinal epithelial cells (Figure 7.13) surrounds each ovary and underlying this lies a mass of partly developed ova or *primordial follicles*. At the menarche there are about 200,000 of these structures in each ovary. As described earlier (Section 7.2), some of these develop through stages of primary, secondary, and tertiary follicles. During reproductive life, usually one of these follicles matures about every 28 days and delivers a ripe ovum into the body cavity, from which it passes into the oviduct. This maturing follicle is called the *Graafian follicle*. The primordial follicle consists of the oocyte and a single surrounding layer of granulosa cells. The latter are thought to be derived from the germinal epithelium. The growth of the follicles is associated with an increase in the number of layers of granulosa cells and the acquisition of two outer "shells" derived from the ovarian stroma. Immediately adjacent to the granulosa cells is the *theca interna*, a well-vascularized layer of connective tissue separated from the granulosa by a basement membrane. The outer layer of the follicle consists of fibrous tissue called the *theca externa*. The granulosa cells, the theca interna, and the ovarian stroma can synthesize steroid hormones: estrogens, progesterone, and androgens.

The Graafian follicle (Figure 7.13) is formed as a

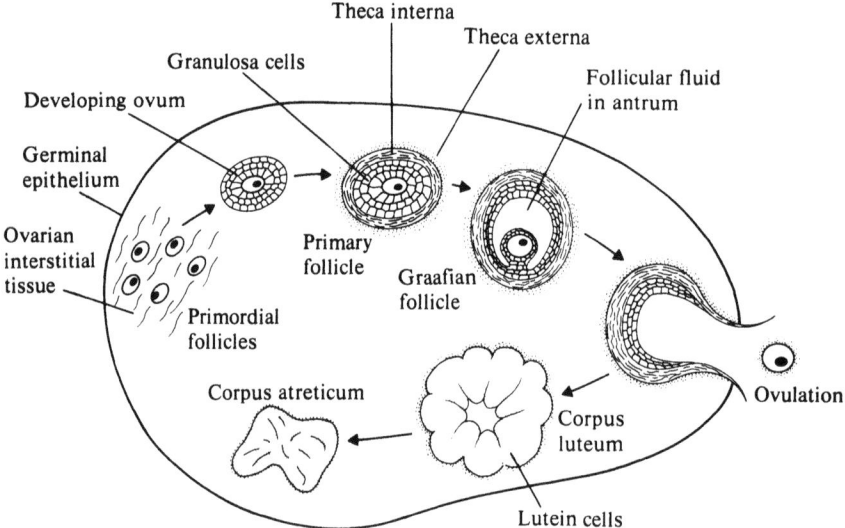

Fig. 7.13. Structure of the ovary.

result of a proliferation of a tertiary follicle under the influence of pituitary FSH. LH is concerned with the regulation of the secretion of its hormones. The structure expands, mainly as a result of the accumulation of the follicular fluid, and it takes up a position at the surface of the ovary, preparatory to ovulation. This process, by which the ovum is extruded into the body cavity, appears to be due to several factors, including a buildup in the fluid content and the action of enzymes in the region of an area of weakness, the stigmata, in the follicles' envelope. This process occurs in response to a surge in release of LH, but FSH may also be involved.

Following ovulation, the corpus luteum is formed from the remnants of the Graafian follicle, minus its ovum. The granulosa cells proliferate and the basement membrane between these cells and the theca interna breaks down. Prior to this change the granulosa cells lack a direct blood supply, but they now become vascularized and even intermixed with the theca interna cells. The cells of the corpus luteum have a bright yellow appearance due to accumulated lipids which form the substrates for the synthesis of the steroid hormones. The cells are then said to be luteinized and are called the granulosa *lutein cells.* Luteinization takes place under the influence of pituitary LH. The corpus luteum persists if pregnancy occurs, but if fertilization does not take place, it degenerates in about 10 days.

The Mullerian duct differentiates in the fetus into a continuous canal consisting of the oviduct, uterus, cervix, and upper vagina. These structures are vital secondary sex organs in the female. This tube is surrounded by smooth muscle, is lined by epithelial cells, and its various segments are separated by muscular and glandular sphincters. The degree of development and activity of these structures follows a cyclical pattern during the menstrual cycle which is influenced by ovarian hormones.

The following description of the female reproductive tract follows the natural pathway that the ovum takes in its quest for sperm and a safe resting place where it may develop.

The *oviduct,* or *Fallopian tubes,* opens into the abdominal cavity, from which it collects the ovum. It is separated from the uterus by the *uterotubal junction.* The oviduct has two segments, the *ampulla* and the *isthmus,* which are separated by the *ampullary–isthmic junction.* It is lined with secretory epithelia and cilia. The ovum passes rapidly down the ampulla, assisted by the action of the cilia and muscular contractions, but is delayed for 1 or 2 days at the ampullary–isthmic junction. Fertilization occurs in the ampulla and the egg then passes down the isthmus and on about day 3 into the uterus, where it implants about 8 to 13 days after fertilization. The role of the oviduct in aiding the transportation and viability of the sperm is poorly understood. The activity of the oviduct, including the rate of egg transport, is influenced by hormones. Estrogens appear to promote the transport of the ovum whereas progesterone delays it. The oviduct is the true "marriage bed" and reproductive success can be determined here. Thus, it is a logical site to consider for the actions of drugs that may promote or inhibit fertility.

The *uterus* is the largest segment of the female

reproductive canal. It undergoes considerable cyclical changes in its activity, which are associated with the menstrual cycle and pregnancy. The encompassing layers of smooth muscle, or *myometrium,* can hypertrophy, as seen in pregnancy, and exhibit contrasting patterns in their activity. Contractility is promoted by estrogens and is usually inhibited by progesterone. The epithelium that lines the lumen of the uterus, the endometrium, undergoes cyclic changes; there is a buildup of cells and glands under the influence of estrogen prior to ovulation. It is "primed" in readiness to receive the fertilized egg. Subsequently, progesterone, from the corpus luteum, promotes an increased differentiation and secretion of the glandular cells. The endometrium becomes receptive for the implantation of the blastocyst and, largely under the influence of progesterone, stays in a suitably secretory condition for the duration of the pregnancy.

The uterus and the vagina are separated by the *cervix.* This structure plays a very important role in several aspects of human reproduction, including protection of the uterus from external invasion, promotion of the transport of sperm at the time of ovulation, the maintenance of pregnancy, and parturition. In women, it contains a short canal (2 to 3 cm long), called the *endocervical canal,* which is lined with epithelial cells. Some of these have a secretory function and form the *cervical mucus.* This fluid (see Moghissi, 1973) displays a host of interesting and physiologically important physicochemical properties which change under the influence of the ovarian hormones. It can, for instance inhibit or promote the transfer of the sperm into the uterus and, not surprisingly, it has the latter effect about the time of ovulation. This property can, however, be prevented by small doses of progestins, and this effect forms the basis for the actions of some types of contraceptive pills (the "minipill"). During pregnancy the mucus forms a seal or plug which isolates the fetus from external invasion. The cervix is mainly composed of connective tissue, which has the ability to relax at the time of parturition.

The *vagina* is a muscular canal lined by epithelial cells which undergo cyclic cornification under the influence of estrogens. In women, it is the site for the collection of the sperm, although in some animals ejaculation occurs directly into the uterus.

The *external genitalia* in women consist of the clitoris and labia, which are the embryological homologues of the penis and scrotum. There are, in addition, a number of secretory glands which form mucus and aid copulation. These tissues and glands are dependent on ovarian hormone activity.

The *secondary sex characters* in women also include the distribution of fat; the pitch of the voice; the distribution of hair; the general morphology of the

skeleton, including the pelvis; the musculature; and, of course, female behavior. The mammary glands are discussed later [Section 7.6H(5)].

B. Ovarian steroid hormones: estrogens and progestins

Three types of steroid hormones that are concerned with the reproductive function are produced in women: estrogens (C_{18}), progestins (C_{21}), and the androgens (C_{19}) (Figure 7.14). The chemistry of these compounds was described earlier (Section 5.1). The most prolific source of these steroids are the ovaries, but the adrenal cortex can also be important, and during pregnancy the fetus and the placenta also contribute. It should be recalled that peripheral transformation of steroids, such as androgens to estrogens, may occur in such organs as the liver, brain, skin, and adipose tissue.

a. The predominant and most active estrogen secreted by the ovaries is estradiol-17β. Some estrone (about 10 times less) is also produced. These steroids are formed by the theca cells of the follicles and also, possibly, by the corpus luteum. In the latter, they are probably formed by the luteinized theca cells. The hormones accumulate in high concentration in the follicular fluid. Estriol is also present in the circulation, but this compound is formed peripherally as a result of metabolism of the ovarian estrogens. During pregnancy it is synthesized by the placenta from the fetal corticosteroid dehydroepiandrosterone. This androgen must first undergo 16-hydroxylation in the fetal liver. In postmenopausal women, a significant quantity of estrogens is derived from the peripheral conversion of adrenal androgens.

b. The principal source of progesterone is the corpus luteum. 17α-Hydroxyprogesterone is also secreted by the ovary. This steroid has little biological activity and its function is unknown. As it is an intermediary in the biosynthetic chain, its occurrence may reflect an "overflow." It may be formed by the theca cells. During pregnancy progesterone is also secreted by the placenta, where it is formed from exogenous cholesterol. Synthesis is thus "incomplete," as this tissue cannot form cholesterol from acetate. The placental progesterone can enter the fetal circulation, where it can be utilized by the adrenal cortex.

c. Androgens, principally androstenedione, but also dehydroepiandosterone and testosterone, can be produced by the stromal (or hilar) cells of the ovary. The other important source of androgens in women is the adrenal cortex.

C. Functions and effects

1. Estrogens. These steroids have a general tonic effect on most tissues in the body, but they have more specifically directed actions on the de-

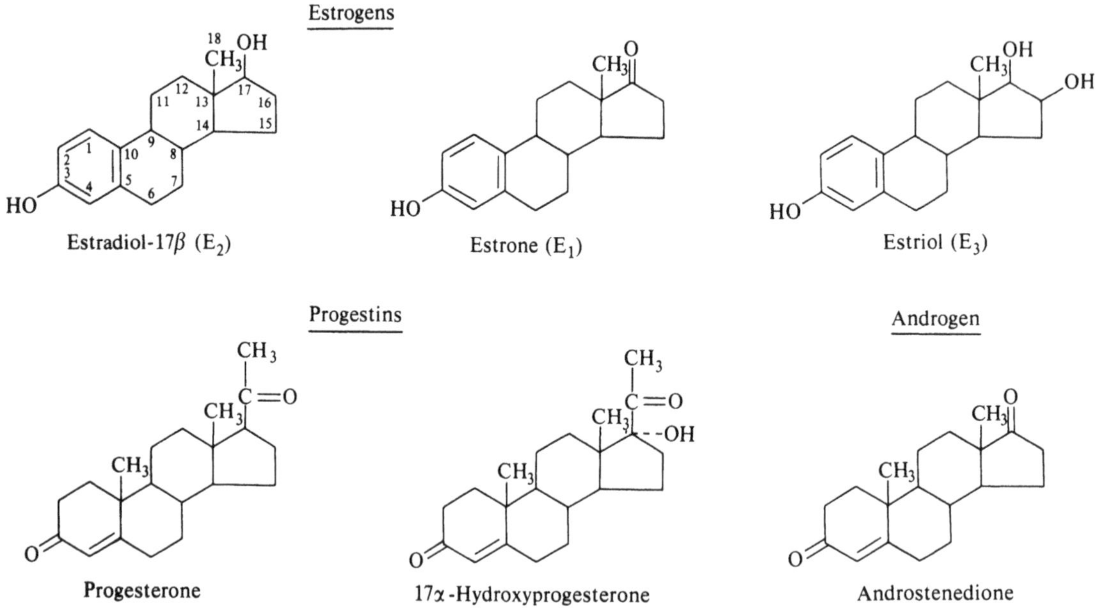

Fig. 7.14. Principal steroid sex hormones in women. Estradiol-17β is the main estrogen produced by the ovary, but some estrone is also formed. Estriol is a metabolite found in the peripheral circulation. The androgen androstenedione is formed in the ovary, and some testosterone and dehydroepiandrosterone is also produced. The adrenal cortex is also an important source of androgens in women.

velopment and function of the female reproductive organs. This effect may also be of a tonic nature, reflecting a continual basal level of the hormones, or it may take part in the cyclical changes associated with ovulation and pregnancy. The effects of estrogens are of a general stimulatory nature associated with such changes as hypertrophy and hyperplasia, increased blood supply, the retention of water and electrolytes, the accumulation of amino acids, and the synthesis of proteins. They thus have an anabolic action but, unlike the androgenic–anabolic steroids, their effects are more specifically directed, in this instance to the female reproductive system. It should, however, be remembered that estrogens usually act in conjunction with other hormones, especially progesterone.

Uterus and oviduct. Estrogens induce a hyperemia of the uterus, and this increase in the blood supply appears to be mediated by a release of histamine. The growth and activity of the myometrium are also increased, so that it contracts more readily and strongly and increases in size. The endometrium proliferates, as seen in the preovulatory stage of the human menstrual cycle. The epithelium multiplies and the granular cells grow and differentiate.

Implantation of the blastocyst in the endometrium depends on the presence of estrogens in conjunction with progesterone. Rabbit blastocysts have

been observed to produce estrogens on the seventh day, and this could be directly contributing to implantation (George and Wilson, 1978).

Estrogens stimulate the growth and development of the uterus prior to sexual maturity.

Cervix. The cyclical activity of the cervical epithelium is influenced by estrogens. Especially notable is the increased activity of the mucous secretory cells. The mucus, under the influence of estrogens, has characteristic physicochemical properties which aid the successful transfer of the sperm into the uterus. These properties include its volume, pH, viscosity, and elasticity. When suitably prepared on a microscope slide, the cervical mucus at this stage dries in fern-shaped patterns, which is the basis for the *fern test* to determine the time of ovulation. Its transparency and volume reaches a high peak and its viscosity is low. It can then be drawn out into long, cohesive threads (Spinnbarkeit).

Vagina. Estrogens have very specific effects on the vagina. They promote the development and increase in the size of this organ. The vaginal epithelium and its secretions undergo cyclical changes during the menstrual cycle and the estrogens play a predominant role in the increased glycogen storage and the proliferation and cornification that occur at about the time of ovulation, and they control its secretions. The properties of the latter are especially important in influencing the activity of microorganisms. Vaginal infections are thus more likely to occur in the absence of estrogens. The changes in the appearance of the vaginal epithelial cells are quite characteristic at different times of the menstrual cycle, and they are used to assess the activity of estrogens in the body.

Mammary glands. The control of differentiation, growth, and secretion by the mammary glands depends on a host of hormones working in conjunction with one another [see Section 7.6H(5)]. There are, however, considerable interspecific differences. Estrogens are necessary for proliferation of both the ducts of the mammary glands and the secretory alveoli, although with respect to the latter their action is probably only secondary to that of progesterone. The secretion of milk does not appear to be dependent on estrogens, and indeed these steroids have been administered to women in order to "dry up" the milk.

Brain. Estrogens have been identified by autoradiography in various regions of the brain (see McEwen, 1976), including the hypothalamus. Specific estrogen receptors have also been isolated (e.g., Ginsburg et al., 1977). These observations are consistent with physiological observations suggesting that these steroids have direct effects on this tissue. They influence the activity of the hypothalamic–pituitary axis, which controls the release of gonadotropins. As described earlier, they are responsible for the preovulatory "surge" in the release of LH. On the other hand, estrogens can also, paradoxically, inhibit the secretion of gonadotropins.

In animal experiments it is well known that high endogenous levels of estrogens or the administration of this steroid may result in "heat" and an increased receptivity and even aggressive invitations to the male (see Michael, 1973). These effects can be seen when small amounts of estrogens are directly implanted into the hypothalamus. Other factors and hormones may, however, also play a role, although there appears to be quite a lot of species' variability. Thus, progesterone, in addition to estrogens, may be necessary for the appropriate behavior of rats at estrus. When androgens are administered to postmenopausal women, there is often an increase in libido, and this effect is also seen when endogenous production of these steroids occurs, as seen in some disorders of the adrenal cortex. The antiandrogen cyproterone acetate has been reported to decrease libido in some women. Thus, although estrogens clearly do act on the brain and influence behavior, it is doubtful if they alone are responsible for what is

euphemistically called "female behavior." As many have suspected, this subject is a complex business.

Connective tissue and bone. Estrogens promote the closing of the epiphyses and the deposition of calcium in bone. They thus appear to limit growth at puberty and their lack contributes to osteoporosis after the menopause. The skin is also influenced, and their effects may take the form of increased formation of collagen and mucopolysaccharides as well as local edema.

Cardiovascular system. The coagulability of the blood is increased by estrogens, and this effect appears to be due to an increase in the levels of prothrombin and factor V. The fragility of the capillaries is decreased and the blood vascular system appears to be more flexible, which may explain the increased ability of women to withstand chronic hypertension. There is an increased incidence of hypertension following the menopause, and it is paradoxical to observe that the administration of estrogens in oral contraceptive pills may also be associated with increases in the blood pressure. Plasma cholesterol levels are reduced under the influence of estrogens, apparently due to an increased accumulation by the liver and an associated enhanced excretion and breakdown. Estrogens have been used in attempts to lower blood cholesterol levels in men and so, hopefully, to reduce arteriosclerotic deposition of this sterol. Somewhat unexpectedly, this treatment was associated with an increased incidence of myocardial infarction, which is consistent with the observation that men with high plasma estrogen levels are more prone to this disease (Phillips, 1976).

Liver. The most prominent anabolic effect of estrogens outside the reproductive system is on the liver. It increases its size, and protein synthesis is promoted. The increased levels of transcortin and SHBG have already been described, and angiotensinogen synthesis may also be promoted. The latter may contribute to the effects of estrogens on blood pressure.

Cancers of the female reproductive organs. It has been recognized for many years that tumors of the female reproductive organs, especially of the breast and endometrium, may be related to the presence of exogenous or endogenous estrogens. Thus, in many premenopausal women, ovariectomy can produce a regression in breast cancers and the administration of androgens and antiestrogens may do the same. Conversely, the therapeutic administration of estrogens is apparently associated with an increased incidence of cancer of the endometrium. Prenatal exposure of female fetuses to estrogens, in attempts to prevent abortion, are associated with the occurrence of "clear-cell" adenocarcinoma of the vagina in the late teenage years. Thus, there is little doubt that estrogens can support and increase the risk of induction of some cancers in women. Those that appear to be estrogen-dependent and responsive to some form of hormone therapy have been shown to contain relatively high concentrations of cytoplasmic estrogen receptors (see Sections 7.4 and 7.6K). Estrogens are natural hormones and cannot really be considered carcinogens. However, they support and promote the growth and activity of the female reproductive organs and so may increase the likelihood of the success of a carcinogenic agent on such tissue. The activity of tumors from such tissues may also, not surprisingly, be promoted by estrogens. Paradoxically, estrogens in certain situations may also cause a regression of tumors associated with mammary gland carcinoma (Section 7.6K).

2. Progestins. The physiological effects of the progestins generally do not appear to be elicited alone, but are superimposed or are dependent on the priming effects of estrogens. The latter may promote the formation of progesterone receptors. Both usually act on the same organs and tissues. Some of the actions of the progestins appear to be diametrically opposite to those of estrogens, but in other important respects they complement each other.

Uterus and oviduct. Progesterone mediates the postovulatory maturation of the secretory glands and spiral arteries of the uterine endometrium and the lining of the oviduct. The spontaneous activity of the myometrium is depressed so that it contracts less readily and frequently. Progesterone is especially noted for being the hormone associated with pregnancy. In its absence, abortion results. In the early part of pregnancy in women, progesterone is obtained from the corpus luteum, but subsequently the placenta is the main physiological source of this hormone. As described in the last section, progesterone and estrogens are necessary for the successful implantation of the blastocyst and they appear to contribute to the regulation of the delivery of the fertilized egg through the oviduct. Their role in maintaining a quiescent myometrium appears to be a very important part of their role in pregnancy. A decrease in the concentration of progesterone in the circulation may influence parturition, but this role is controversial.

Cervix. Progesterone promotes the secretion of a form of mucus by the cervix whose composition and properties inhibit the transport of sperm. It has a small volume and is very viscous. This effect is the basis for a contraceptive action of progestins (the "minipill").

Vagina. Progesterone opposes the effects of estrogens in inducing cornification of the vaginal epithelium. Following ovulation it promotes the desquamation of the epithelial cells, which take on the appearance of irregular "cracker" biscuits in which the nucleus is less discrete and obvious.

Mammary glands. In conjunction with estrogens, the differentiation and growth of the secretory alveoli are promoted by progesterone, especially during pregnancy, when lactation is imminent.

Brain. In some animals progesterone may be necessary for the promotion of mating behavior at the time of ovulation. Its normal role in the behavior of women is not clear, but the use of progestin contraceptive pills has occasionally been reported to reduce libido. Progesterone, by its action on the hypothalamus, can inhibit the release of gonadotropins, an action that is utilized for its contraceptive effect. Progestin receptors have been identified in the brain and some of these are induced by estrogens (MacLuskey and McEwen, 1978).

Other actions. Apart from its effect on the reproductive organs, progesterone has a catabolic action and promotes a negative nitrogen balance. It can promote salt excretion by the kidney, an effect that may be mediated by a competitive antagonism to the action of aldosterone. The plasma levels of this corticosteroid increase considerably during pregnancy, and this response may reflect an increased need for this hormone due to the antagonistic effect of the progesterone. In contrast to estrogen, progesterone has a thermogenic effect, which is reflected in the rise in body temperature of about 0.3 to 0.6°C which follows the estrogen-dependent decline that occurs at ovulation in women. This effect may be due to an action on the hypothalamic temperature-regulating center. Progesterone has an interesting effect on pulmonary respiration, as it lowers the partial pressure of CO_2 in the alveoli in both women and men. The possible use of this steroid in the treatment of respiratory disease has been considered. Like its action on body temperature, its effect appears to be on the brain, via the respiratory center.

D. Synthesis and release
There is little storage of ovarian steroidal hormones, so that the release and synthesis of either estrogens or progesterone are closely associated. As described earlier, this is principally under the control of LH, which appears to act by modulating the levels of cyclic AMP. LH secretion is inhibited by certain drugs that act on the hypothalamus, in-

cluding the tranquilizer chlorpromazine. Administration of LH or hCG stimulates the release of ovarian steroids. Estrogens and LH are involved in an interesting positive-feedback system in which LH stimulates the secretion of estrogens and when the circulating concentration attains a critical level, the steroid triggers a massive release or "surge" of LH which promotes ovulation. Subsequently, the negative-feedback control appears to predominate and LH levels decline. Release of estrogens and progesterone by the ovary may occur in a cyclical manner, which is controlled by the brain, as it does during the menstrual cycle. Release may also occur tonically, as in pregnancy, although even here the levels change over the period of gestation.

E. Transport and metabolism
Estimates of the half-lives of the ovarian steroids in the circulation vary, but in women they appear to be about 6 minutes for the estrogens and 2 minutes for progesterone. Disappearance may, however, follow a different time course following administration and be more rapid in the first hour than subsequently. The disappearance of these steroids from the plasma principally reflects their uptake by the tissues and catabolism and excretion.

Binding occurs to plasma proteins. Like other steroid hormones, they are loosely bound ("low affinity and high capacity") to plasma albumin. Progesterone also binds with a high affinity to transcortin (CBG), while estrogens are firmly bound to SHBG. The synthesis of these globulins by the liver is promoted by estrogens. The binding to plasma proteins reduces the rate of the catabolism of the steroid hormones.

Estrogens and progesterone are mainly metabolized in the liver. Progesterone is hydroxylated at the C-20 position to give 20α- or 20β-hydroxypregn-4-ene-3-one and also at the C-3 to give pregnanediol (pregnane-3α,20α-diol), which may function as an active metabolite. These compounds are conjugated, mainly as glucoronides, and are excreted in the urine and bile.

A vast number (see Adlercreutz, 1970) of metabolites of the estrogens have been isolated, usually during *in vitro* studies, and it is not clear how many of these are significant in quantity under normal *in vivo* conditions. Oxidation of the C-17 hydroxyl moiety of estradiol-17β occurs in the liver to produce estrone, which is less active. Estriol, on the other hand, is the result of additional α-hydroxylation of C-16. These hydroxyl groups provide sites for the conjugation of the estrogens to form glucuronides and, to a lesser extent, sulfates. Such polar compounds are excreted in the urine or bile. A high proportion may be retained by the enterohepatic circulation. The conjugates appear to be broken down in the intestine by the action of endogenous or bacterial enzymes,

which increases their lipid solubility so that they are reabsorbed.

Disease and drugs can have dramatic effects on the hepatic metabolism of estrogens (see Adlercreutz, 1970).

F. Ovarian steroid hormones: sources, structure–activity relationships, and mechanisms of action
The estrogens and progestins that occur naturally in the body are of rather limited use as a source of therapeutically active materials. One of the reasons for this situation is that relatively little of the active hormonal material is stored in the ovaries, so that unlike many protein hormones, animal glands cannot be used as a commercial source of material. Although these natural steroid hormones can now be readily made in the laboratory, they have a rather low activity when administered. This problem is the result of their rapid inactivation by the liver. Orally administered preparations are especially vulnerable, as they are carried directly to the liver in the hepatic–portal veins. Some commercial preparations from natural sources are available, but these are confined to conjugated estrogens which are obtained from the urine of pregnant mares (hence the trade name Premarin) and women. Large quantities of such compounds are excreted in the urine and they are effective when given orally. Substances that have estrogenic activity also occur in plants. These compounds (see Farnsworth et al., 1975a,b) may be steroids, including estrone and estriol, but a number of nonsteroidal compounds (Figure 7.15) which have estrogenic activity have also been identified in plants. Subterranean clover, for instance, contains an isoflavone called genistein, which many years ago was shown to cause infertility in sheep in Australia (Bennetts, Underwood, and Shier, 1946; Wong and Flux, 1962). Quail living in desert conditions in California display irregular breeding cycles associated with the growth of forage. In dry years the desert plants on which they feed have been shown to contain increased amounts of such phytoestrogens (Leopold et al., 1976). They thus seem to provide a natural form of contraception, which is an interesting but, no doubt, fortuitous relationship between animals and plants. It is possible that such substances may contribute to the alleged folkloric contraceptive actions of various herbal extracts (see Farnsworth et al., 1975a,b). Such substances have not been used as a natural source of estrogens; however, plants containing steroidal materials are commonly used as a starting point for the partial synthesis of steroid hormones. As already described in the instances of the corticosteroids and androgens, the natural estrogens and progestins have been subjected to considerable chemical modification in order to provide useful

(a) Stilbenes

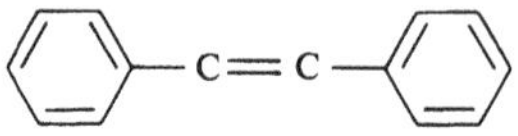

Stilbene nucleus

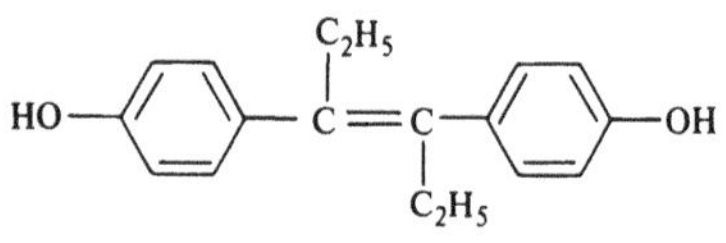

Diethylstilbestrol (DES)

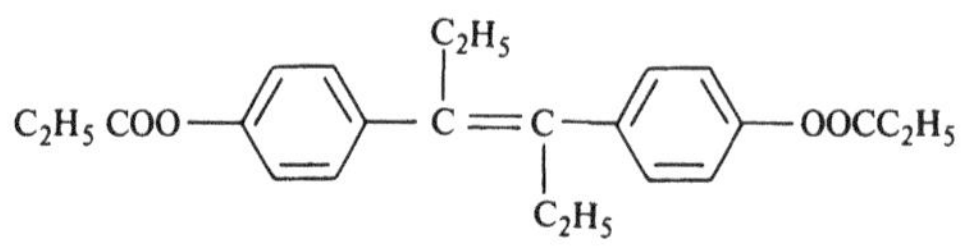

Diethylstilbestrol dipropionate

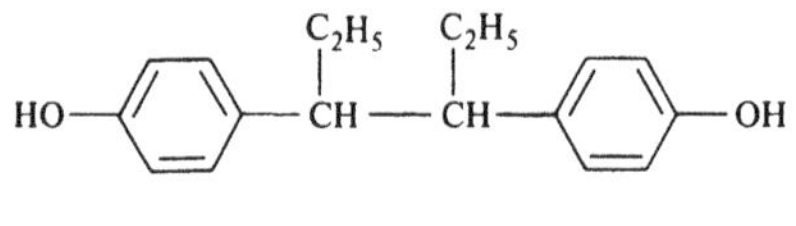

Hexestrol

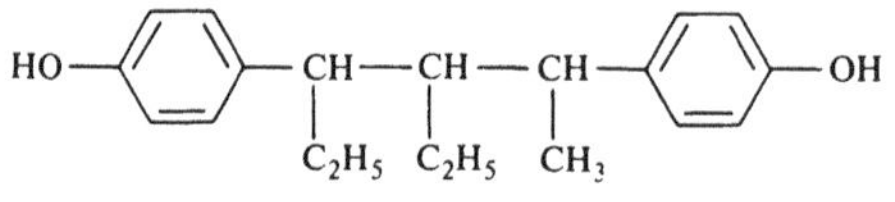

Benzestrol

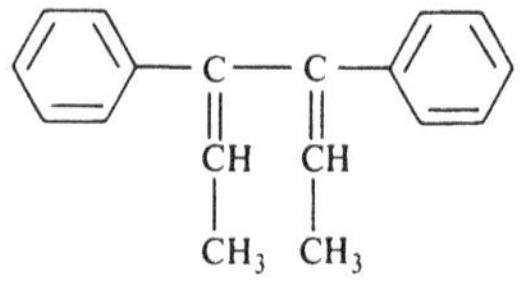

Dienestrol

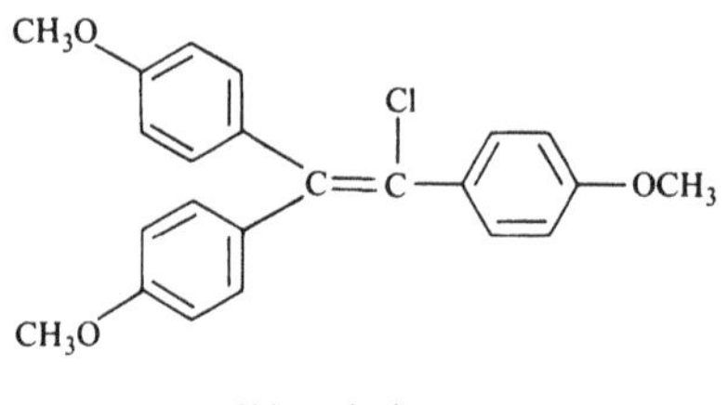

Chlorotrianisene

(b) Isoflavonoids

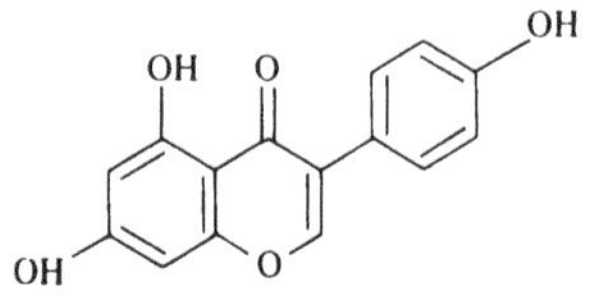

Genistein

(c) Resorcylic acid lactones, RALs (mycotoxins)

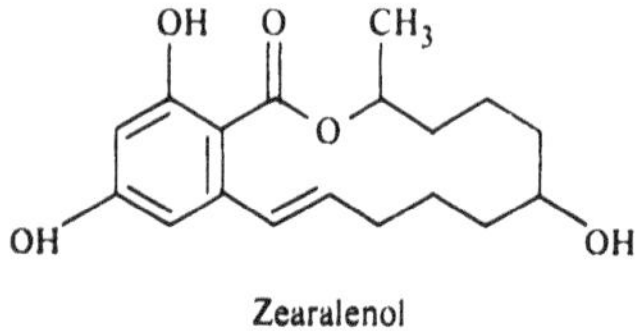

Zearalenol

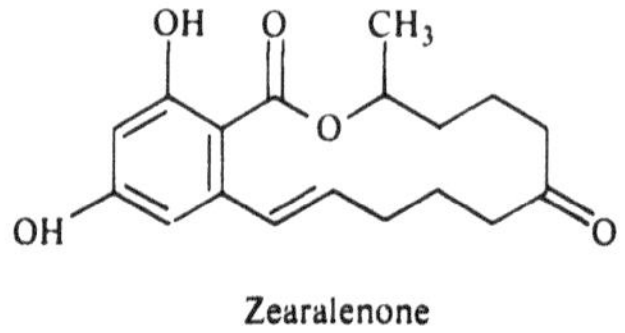

Zearalenone

(d) Coumestans

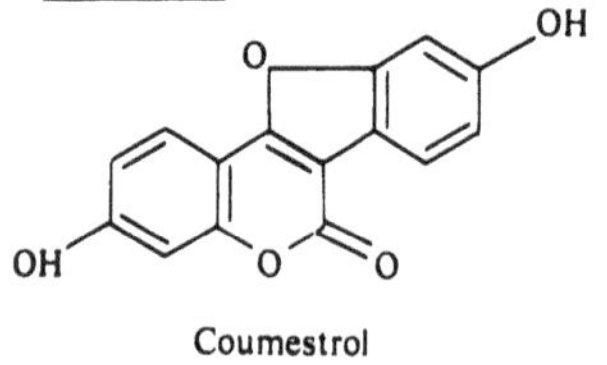

Coumestrol

Fig. 7.15. Some nonsteroidal estrogens. (a) Derivatives of stilbene. These are compounds prepared by chemical synthesis. (b)–(d) Naturally occurring phytoestrogens that have been isolated from plants. The isoflavanoids and coumestans occur in higher plants, such as clover, which may be used as forage by animals. The mycotoxins have been isolated from the fungus *Fusarium graminearum*, which grows on cereals such as wheat and barley and may also be ingested by animals. Zearalenone has been used as an anabolic agent in animals.

therapeutic preparations. Such changes influence the duration of their actions, the route by which they may be administered, their potency, and the spectrum of their activities. In the instance of the estrogens, it is especially interesting to observe that activity is not confined to steroidal compounds but can be displayed by other types of chemicals, such as the stilbenes, which contain appropriately spaced benzene rings and phenolic hydroxyl groups. These structures provide them with estrogenic activity.

A host of synthetic preparations of estrogens and progestins have been made, and many of these are available commercially. It should be remembered that these are often subject to patent rights, and this fact, rather than any real therapeutic advantage, explains part of the plethora presented for choice by the physician.

1. Estrogens

Sources. Estradiol-17β is the most active of the natural estrogens, followed by estriol and estrone. A conjugate of estrone, *Na estrone sulfate,* is used commercially in a mixture which also contains *Na equilin sulfate* and *Na equilenin sulfate* (Figure 7.16). This potion is obtained from the urine of pregnant mares who are members of the genus *Equus*, hence the naming of equilin and equilenin. This mixture (Premarin) is given orally and has minimal effects on the uterine endometrium. They have therefore been dubbed "soft" estrogens. Their relatively low potency and somewhat exotic sources provide only a limited use for them, mainly for the treatment of menopausal disorders.

By far the most commonly used estrogens are prepared synthetically. Probably the most dramatic breakthrough in this area was the discovery by E. C. Dodds in 1937 (Dodds et al., 1938) that derivatives of a group of chemical compounds called *stilbenes* (Figure 7.15) had a high estrogenic activity and were effective when given orally. They are not readily inactivated by the liver. The most famous and widely used prototype is *diethylstilbestrol* (DES). These estrogens can be produced cheaply and have been used in enormous quantities both in human therapy and to promote the

growth and edible qualities of farm animals. The dipropionate ester provides an injectable depot preparation. Pellets, from which absorption is slow and prolonged, can also be implanted. The derivative *chlorotrianisene* (TACE) is an interesting compound, as it is given orally but has a very prolonged period of action. It is rapidly taken up into body fat, presumably helped by the presence of its three methoxy groups, from which it is slowly released. It would appear that in order to exert an estrogenic effect, demethylation would need to occur (see later).

Although estradiol-17β is relatively ineffective therapeutically, especially orally, the molecule can be modified to enhance its stability (Figure 7.17). Esters such as *estradiol cypionate* and the benzoate are sequestered in adipose tissue and thus are protected, owing to their slow release. The most effective chemical modification is the insertion of an ethinyl moiety at the 17α position to form *ethinyl estradiol*. This group protects the molecule from inactivation by the liver. It also enhances interaction with the receptor, as it has also been observed to do in estriol, which normally acts only very briefly (Lan and Katzenellenbogen, 1976). *Mestranol* has a methoxy group substituted at the C-3 position and is a widely used derivative with a similar but more prolonged activity. In order to act, however, the phenolic hydroxyl group must be restored by metabolic cleavage of the methoxy moiety. The 3-methoxy group also decreases plasma binding (Raynaud et al., 1973). The insertion of cyclopentylenol groups enhances the lipid solubilities of such steroids and so prolongs their effects, and *quinestrol* is such a derivative of ethinyl estradiol. It is given orally and is effective for several weeks. The cyclopentylenol group is at the C-3 position and must also be subjected to metabolic cleavage before the molecule can become active. It thus acts as a type of "prohormone," although only in a pharmacological sense.

The *relationship of structure to activity* of the estrogens is a complex problem. Thus, when they are administered *in vivo*, a variety of factors can influence the final response, especially their absorption, plasma protein binding, metabolism, and uptake by the target tissue (see Raynaud et al., 1973). Ide-

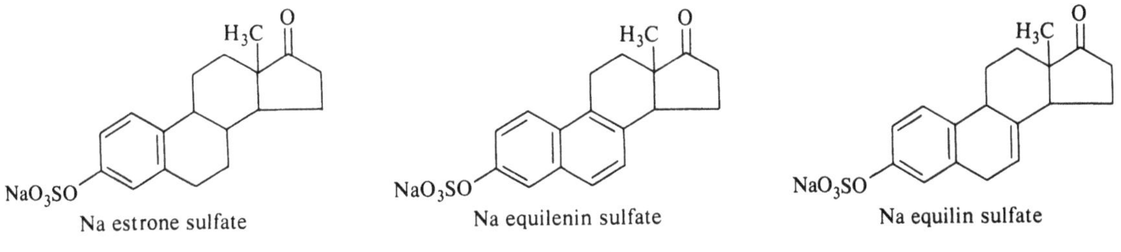

Fig. 7.16. Natural conjugated ("soft") estrogens that are used therapeutically.

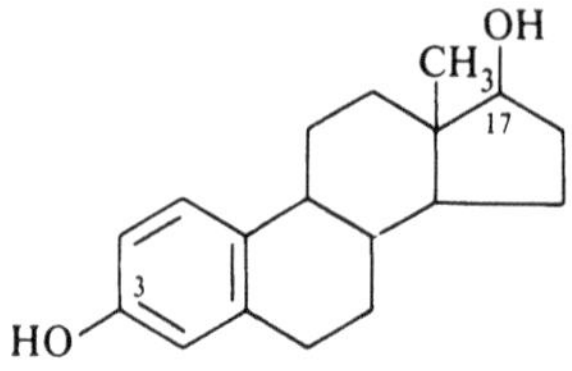

Estradiol-17β

Esters for parenteral use (long-acting)

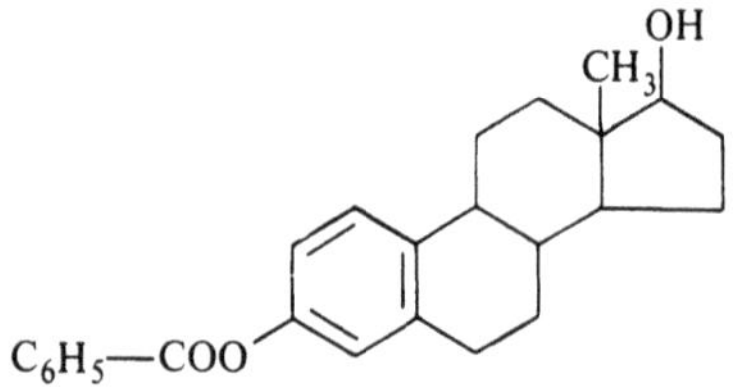

Estradiol benzoate

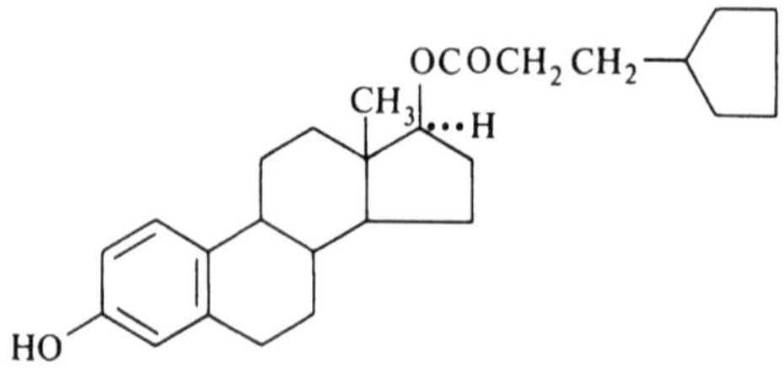

Estradiol 17β-cypionate

17α-Derivatives (oral use)

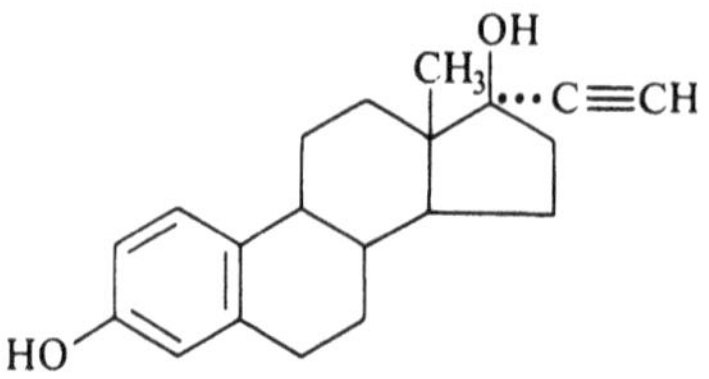

Ethinyl estradiol

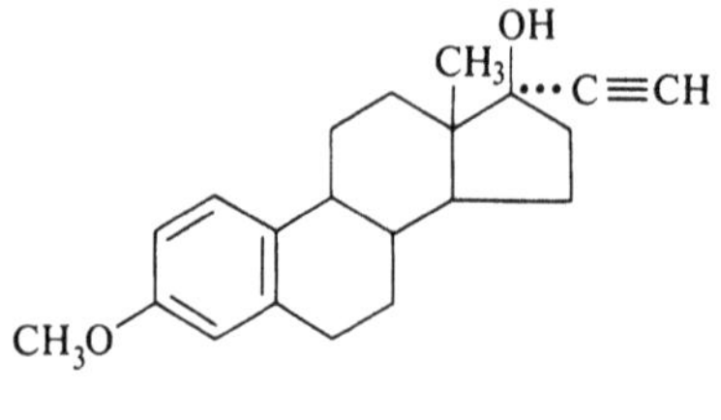

Mestranol

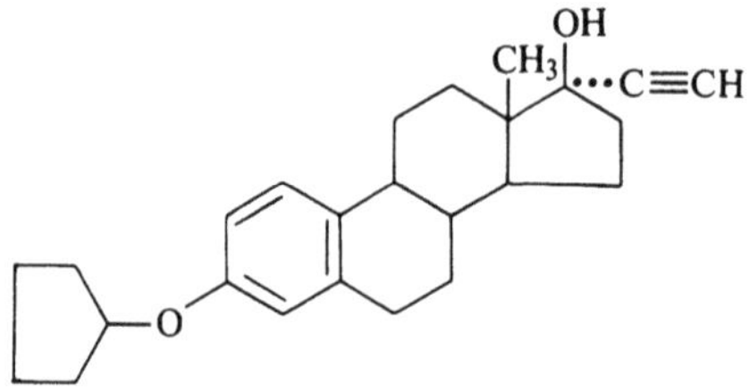

Quinestrol

Fig. 7.17. Therapeutically useful steroidal estrogen drugs that are derived chemically from estradiol-17β.

ally, although perhaps not always in the practical sense, the pharmacologist is interested in how effective such drugs are locally at the effector site. When estrogen receptors were identified and it became possible to isolate them, a more specific analysis of structure–activity relationships became possible. Such procedures involve isolation of the cytosol receptors, usually from the uterus, and measurement of the ability of various analogues to displace tritiated estradiol-17β from binding to

them. Such measurements do not, of course, tell the whole story regarding relative activities; they do not, for instance, distinguish an agonist from a competitive antagonist. Two substantial surveys appear to have been made. Korenman (1969) used cytosol from rabbit uterus and Hähnel, Twaddle, and Ratajczak (1973a) used the human uterus. The present account is based mainly on the latter observations, but generally the two surveys were in agreement.

"Cold" estradiol-17β could completely displace the tritiated form of the hormone, but other types of steroid hormones, including progestins, androgens, and corticosteroids, were without effect.

The primary importance of the unsaturated benzene *A-ring* was confirmed. If it was saturated, activity was abolished. The hydroxyl group could not be moved to any other adjacent position, such as C-1 or C-2. Substitution of the C-3 phenolic hydroxyl (such as with a 3-methoxy, 3-sulfate, or 3-glucosiduronate) virtually abolished affinity for the receptor. The insertion of methyl groups at C-1 or C-4 (but not C-2) led to a loss of effectiveness.

The other major feature of the molecule which had a very important influence on its affinity for the receptors is the alcoholic β-hydroxyl at C-17. This is not quite as important as the C-3 hydroxyl, as an oxo group (as in estrone) or a 17α-OH or 16α-OH is also effective, although they result in a reduction of activity.

17α-Ethinyl estradiol, as well as the nonsteroidal estrogens diethylstilbestrol and hexestrol, had an affinity for the receptors similar to that of the natural hormone estradiol-17β.

It was proposed that the interaction of estrogens and their receptors involve two active sites. The initial attachment involves the C-3 hydroxyl, which apparently must have no steric hindrance and be unimpeded by any adjacent bulky groups. Following this combination, there is a steric change in the receptor which facilitates a second combination with the β-hydroxyl group at C-17. The distance between the two hydroxyls appears to be about 11 Å, and this may determine the success or otherwise of related compounds. The phenolic hydroxyl groups in the active stilbenes derivatives appear to be separated by about this distance. The relationships of the affinities of the different estrogens for the isolated receptors are reasonably consistent with their relative activities determined *in vivo*.

Mechanism of action of estrogens. The earliest effect of injected estradiol-17β is a considerable increase in the blood supply to the uterus. This effect can be seen with the naked eye and occurs within 1 minute. There is also a change in membrane permeability and an uptake of fluid by the cells. These effects take place too early to be accounted for by changes in genetic transcription (see Szego, 1965; Szego and Davis, 1967; Warren and Crist, 1973; Spaziani, 1975). This response could, however, enhance such effects by increasing the rate of delivery of the steroid hormone to its cytosol receptors.

The mechanism of this *very* early effect of estrogens appears to involve the membrane and may be mediated by a release of lysosomal enzymes. It has been suggested that a local release of histamine could be involved. Prostaglandins are another possibility. A transient increase in cyclic AMP has been observed in the uterus; it can be seen 15 seconds after exposure to estradiol-17β and starts to decline after 5 minutes. This effect can be blocked by the β-adrenergic inhibitor propranolol, suggesting that it reflects a β-adrenergic response. The role of the cyclic AMP is not clear, although it has been suggested that it may contribute to the anabolic effects of the steroid. These fascinating first metabolic events in the mechanism of action of estrogens appear to have been unjustly neglected, owing probably to the dramatic advances that have been made regarding their effects on genetic transcription.

The binding of estrogens to cytosol receptors, their passage into the nucleus and combination with nuclear "acceptor" sites, the synthesis of mRNA, and the *de novo* synthesis of cell proteins were described earlier (Section 7.4). The precise nature of the induced proteins is not clear; however, in the chick oviduct, ovalbumin, programmed by a specific mRNA, has been identified, and in the rat an "induced protein" with a molecular weight of 45,000 has been found (see Chan and O'Malley, 1976a,b,c). It seems likely that more than one protein may be induced by such steroid hormones which promote their ultimate effects, but the precise way in which they mediate the responses is unknown. Estrogens specifically induce the formation of a peroxidase in growth-responsive tissues such as the uterus, vagina, and mammary gland tumors (Jellinck and Newcombe, 1977; Lyttle and DeSombre, 1977). It has been suggested that this general effect may serve as a useful marker and indicator for responses and responsiveness to estrogens. The complete role of the induced peroxidase is not clear, but in the presence of H_2O_2, it catalyzes the binding of estradiol to cell proteins.

a. The effects of estrogens on the uterus appear to occur as follows: A very "early" response involving hyperemia and changes in membrane permeability occurs within minutes.
b. A formation of messenger RNA is seen after about 30 minutes.
c. A *de novo* synthesis of proteins commencing at about 3 hours, which appears to initiate many biochemical and morphological changes in the uterus.
d. An increase in DNA synthesis at 18 to 24 hours, which is associated with cell multiplication.

A dramatic instance of the specific role of RNA in the action of estrogens on the uterus was provided by Segal, Davidson, and Wada (1965). After injecting estradiol-17β into rats, they extracted the RNA from the uterus and found that when they placed it in the lumen of uteri of ovariectomized rats, it induced an estrogen-like response.

2. Antiestrogens. There has been a considerable amount of interest in compounds that may antagonize the effects of estrogens. Conventionally, these do not include "physiological" antagonists, such as androgens or progestins or substances that may block secretion or have a general toxic action on the target cells. They refer to substances that act specifically to block the effects of the hormones at their effector sites. These are of considerable potential therapeutic use for treatment of infertility, hormone-dependent cancers, and as antifertility drugs (see Emmens, 1970; Lunan and Klopper, 1975).

A number of compounds have been prepared which exhibit antiestrogen activities (Figure 7.18). All of them (except it appears Mer-25) may also exhibit some estrogenic activity and are thus "weak" or "partial" agonists. As the concentrations required to block a response are usually relatively high (twentyfold) compared to that of the estrogen, they could also be termed "weak antagonists." Before applying information from animal experiments to man, it is well to remember that their effects, as agonists and antagonists, vary a great deal and depend on the particular assay preparation, such as the vagina, uterus, or a release of gonadotropins, and whether the tests involved local or systemic administration of the drugs. Rats, mice, and men (as well as women) have been used, and there can be considerable interspecific differences between their responses. It has also been found that the relative activities may differ, depending on the proportions of the *trans* or *cis* isomers in the available preparations. A considerable amount of information regarding their possible mechanisms of action have been obtained from measurements of their abilities to displace [³H]estradiol-17β from binding sites. Such observations do not, of course, clearly indicate whether the compound will act as an antagonist or agonist.

Chemically, the antiestrogens belong to three broad groups: (a) *estrogenic steroids,* such as estriol; (b) *diphenolic compounds* related to diethylstilbestrol; and (c) *derivatives of polycyclic phenols* containing three benzene rings (Figure 7.18).

Functionally, they are divided into two groups, the polycyclic phenols and the *impeded estrogens.* The latter compounds can interact with the receptor but dissociate from it too rapidly to exert a strong estrogenic effect. However, if present at very high *local* concentrations, such as can be applied to the vagina, they block or "impede" the access of estradiol-17β to its site of action. They are ineffective if administered systemically, as they cannot then maintain a sufficiently high concentration in the region of the target organ. They include the steroidal and diphenolic estrogens. The *polycyclic phenols* display relatively strong and persistent

binding to the receptors, which is in contrast to the impeded estrogens.

The effects of a series of different types of antiestrogens on the binding of estradiol-17β to mouse uterus, *in vitro,* is shown in Figure 7.19a. The potencies have a wide range, over four orders of magnitude. These compounds also exert a distinct estrogenic effect, as shown by their ability to stimulate development of the mouse uterus *in vivo.* They, however, displayed different types of responses (Figure 7.19b). Nafoxidine (compound 10) was, for instance, only a partial agonist (Figure 7.19bii) while the impeded estrogens (Figure 7.19biii) had a very shallow dose–response curve. Most of the polycyclic phenols exhibited a bell-shaped curve, more like a "normal" estrogen. They do not exhibit their antagonistic actions at doses lower than their maximum uterotrophic effect. This relationship is often seen in antagonists that are also "partial agonists." It is the compounds within this group that have been found to have greatest practical applications. Thus, *clomiphene* is used to treat infertility [see Section 7.6H(7)] by stimulating the release of gonadotropins, while *tamoxifen* and *nafoxidine* are used for the treatment of estrogen-dependent breast cancer (see Section 7.6K). Other possible uses that are being considered are the treatment of endometrial disorders and infertility in the male (see Lunan and Klopper, 1975).

The *mechanism of action of the antiestrogens* has been investigated using *in vitro* procedures to measure the ability of the compounds to displace tritiated estradiol-17β. These have involved the uterus and vagina of the mouse (Terenius, 1970, 1971) and cytoplasmic receptors prepared from human uterus and breast carcinoma tissue (Hähnel, Twaddle, and Ratajczak, 1973b).

There are several possible ways that the antiestrogens could be acting:

a. They could bind to the same receptor site as estradiol-17β. Such an interaction could result in a competitive antagonism and the formation of a hormone–receptor complex that is either unable to enter the nucleus and trigger the response (see Ruh and Ruh, 1974) or, if it did, it may ineffective there (Katzenellenbogen and Ferguson, 1975). Neither explanation, however, appears to be completely satisfactory in the case of the polycyclic phenols. The drug–receptor complex has been observed to enter the nucleus and can trigger a response, hence the agonistic effects of such compounds. This movement of receptors into the nucleus results in a *depletion of cytoplasmic receptors* and it has been suggested that, unlike in the presence of estradiol, these are apparently not readily replaced (Clark, Peck, and Anderson, 1974; Ferguson and Katzenellenbogen, 1977). The tissue then becomes unresponsive to the effects of additional

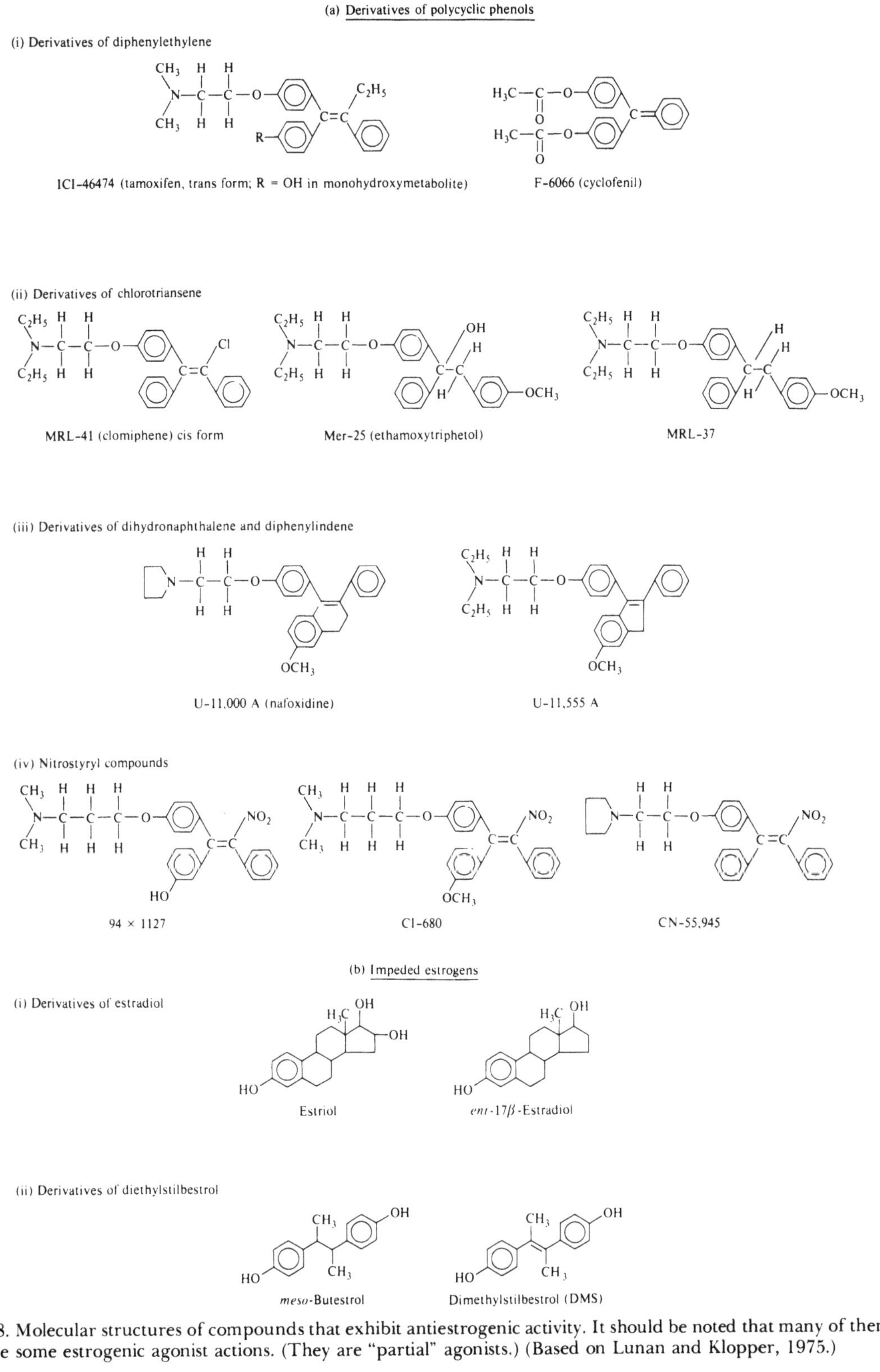

(a) Derivatives of polycyclic phenols

(i) Derivatives of diphenylethylene

ICI-46474 (tamoxifen, trans form; R = OH in monohydroxymetabolite)

F-6066 (cyclofenil)

(ii) Derivatives of chlorotriansene

MRL-41 (clomiphene) cis form

Mer-25 (ethamoxytriphetol)

MRL-37

(iii) Derivatives of dihydronaphthalene and diphenylindene

U-11,000 A (nafoxidine)

U-11,555 A

(iv) Nitrostyryl compounds

94 × 1127

Cl-680

CN-55,945

(b) Impeded estrogens

(i) Derivatives of estradiol

Estriol

ent-17β-Estradiol

(ii) Derivatives of diethylstilbestrol

meso-Butestrol

Dimethylstilbestrol (DMS)

Fig. 7.18. Molecular structures of compounds that exhibit antiestrogenic activity. It should be noted that many of them also have some estrogenic agonist actions. (They are "partial" agonists.) (Based on Lunan and Klopper, 1975.)

(a)

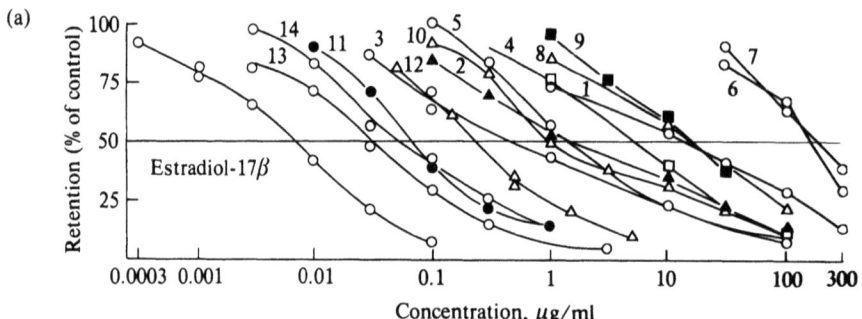

(b)

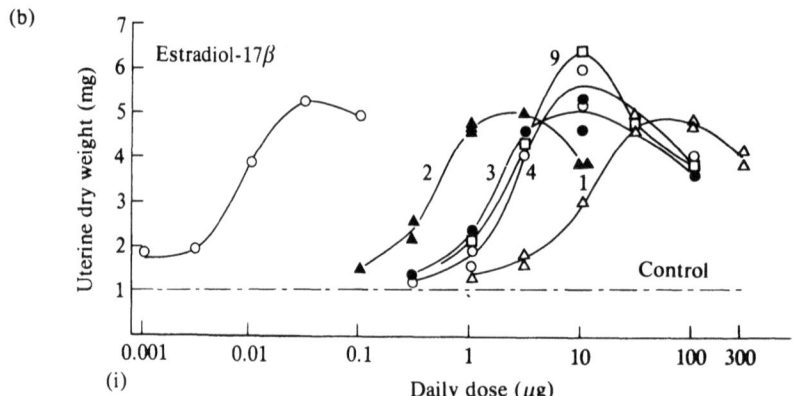

(i)

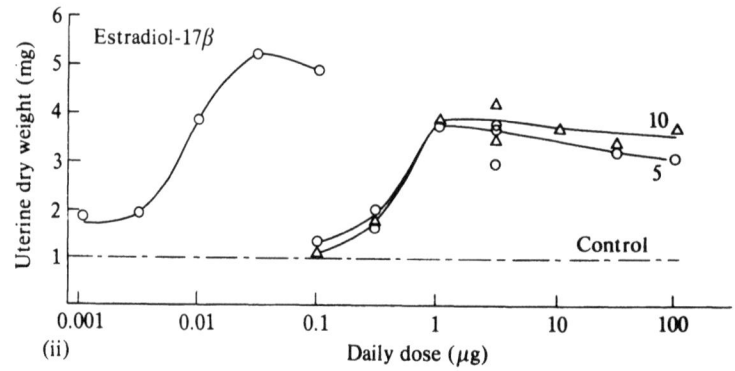

(ii)

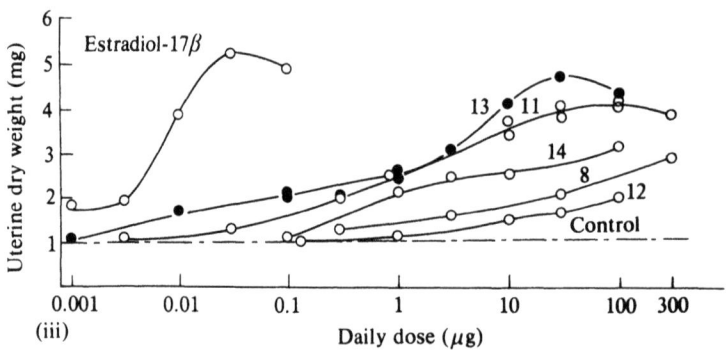

(iii)

estrogens. This hypothesis has, however, not been accepted by all. It has, for instance, been shown that cytoplasmic receptor levels in the rat uterus can be restored in the presence of tamoxifen. The defect may then be due to a failure of its dissociation from binding in the nucleus (Koseki et al., 1977). In addition (Sutherland, Mešter, and Baulieu, 1977), failure of replenishment of receptors in the cytosol cannot account for the pure and rapid antagonist effects of tamoxifen in the chick oviduct. In this avian tissue the antagonist receptor complex is transferred to the nucleus, where it is unable to elicit a response.

b. In the instance of the "impeded" estrogens, an interaction occurs with the receptor, but it is of such brief duration that it fails to trigger a response, although if present at high enough local concentrations, it can exclude the hormone.

c. It has also been proposed that the polycyclic phenols exclude estrogens from their receptor sites by exerting an allosteric effect on the receptor (Hühnel, Twaddle, and Ratajczak, 1973b). The drug thus may interact with a site that is adjacent to the receptor and induce a conformational change which adversely influences the binding of the hormone to it.

d. Tamoxifen has been observed to increase the numbers of progesterone receptors in the rat uterus (Jordan and Prestwick, 1978). Such an action could result in an increased sensitivity of tissues to the effects of progesterone. It was suggested that such an effect could be important in sensitizing endometrial cancer cells to the therapeutic effects of progesterone.

The possible *structure–activity relationships* of the antiestrogens (see Figure 7.18) have been described by Terenius (1971), Hähnel, Twaddle, and Ratajczak (1973b), and Ferguson and Katzenellenbogen (1977). The impeded estrogens apparently can interact with the receptors in the same way as the hormones: possibly a two-stage interaction involving the two hydroxyl groups that are present in all these compounds. The polycyclic phenols usually contain a *N*-ethylether moiety and also often an aromatic methoxy group. These

groups may provide points for the attachment of the drug to its receptor. The interaction is stereochemically specific, as illustrated by the difference observed between *cis* and *trans* isomers. The 3-methoxy moiety is considered to be in an analogous position to the 3-hydroxyl group in estradiol. It is notable that the 3-methoxy compounds take longer to act but are more persistent in their effects than those with a hydroxyl group replacing the methoxy one. It appears that methoxy compounds, before acting, must be metabolized to their active hydroxylated form. Thus, the receptor binding, relative to estradiol-17β, of 94X1127 (hydroxy) is 222 percent, but it is only 34 percent in CI-680 (methoxy).

Tamoxifen is metabolized to its monohydroxylated and dihydroxylated derivatives (Jordan et al., 1977). These metabolites also exhibit antiestrogenic activity, and monohydroxytamoxifen is even more active than its parental compound. It seems likely that the *para* hydroxyl groups at the phenolic site in tamoxifen and the C-3 phenyl group in estradiol are interacting with the same site on the estrogen receptor.

3. Progestins. The two natural hormonal progestins are progesterone and 17α-hydroxyprogesterone. Although the latter appears in substantial amounts in the circulation, it only has a weak action. Progesterone is effective but cannot be given orally because of its rapid inactivation by the liver. It can be injected i.m. in oil, but this procedure is inconvenient and, because of its irritant effects, results in the formation of sterile ulcers. Thus, chemical modifications of the natural hormones were necessary to produce therapeutically useful compounds.

The progestins exhibit a variety of responses both quantitatively and qualitatively. It is often difficult to compare their potencies, as animal experiments do not accurately predict their responses in man. The progestins display a range of physiological and pharmacological effects, but there are often considerable differences in the spectrum of action of the different compounds. They thus can exhibit

Fig. 7.19. Properties of compounds that exhibit antiestrogenic actions. (a) Relative abilities of antiestrogen compounds to displace [^{3}H]estradiol-17β from binding to receptor sites in the mouse uterus *in vitro*. The key to the compounds is given below. (b) Estrogenic uterotrophic effects, as tested on the mouse *in vivo*, of the different types of antiestrogen compounds. (i) The actions of compounds that exhibit a "normal" type of estrogenic response can be seen, compared to the effect of estradiol-17β. However, relatively high doses are required. (ii) These compounds, which include nafoxidine, exhibit only partial agonist activity and cannot achieve the same maximal response exhibited by estradiol-17β. (iii) "Impeded" estrogens, in which the dose-response curves do not parallel those of estradiol-17β; there is only a very gradual increase in the response. Key: 1, ICI-47,699 (*cis*); 2, tamoxifen (*trans*); 3, clomiphene (*cis*); 4, clomiphene (*trans*); 5, CN-55, 945; 6, WSM-4513; 7, ethamoxytriphetol (MER-25); 8, MRL-37; 9, U-11, 555A; 10, nafoxidine (U-11, 100A); 11, *meso*-butestrol; 12, *ent*-17-estradiol; 13, estriol; 14, dimethylstilbestrol (DMS). The structures of these compounds are shown in Figure 7.18. (From Terenius, 1971.)

some estrogenic, androgenic, and anabolic properties, but different analogues characteristically differ in these respects.

The initial primary aim in synthesizing new progestins was to produce compounds that were active when taken orally. Four general groups of such steroids have been produced:

a. *Derivatives of progesterone (Figure 7.20a).* These are limited in number and include *dydrogesterone* and *quingestrone.* The latter contains a 3-cyclopentylenol ether group and appears to be sequestered in fat, from which it is slowly released.

b. *Derivatives of 17α-hydroxyprogesterone (Figure 7.20b).* The 17α-acetoxy analogue can be given orally but enormous doses are needed for an effect. It is more effective i.m., but for this purpose *17α-hydroxyprogesterone caproate* is more useful. This analogue is more soluble in oil and so can be administered in smaller volumes and is active for 7 to 10 days. It is sometimes used in the treatment of endometrial cancer.

Substitution of 17α-hydroxyprogesterone acetate at the C-6 position results in a series of compounds which are effective orally (and also by injection) and have a higher activity. The best known is *chlormanidone acetate*, which has a Cl at C-6 and a Δ⁶ double bond. Methyl groups at C-6, with or without the adjacent unsaturated moiety, are also effective, as in *megestrol acetate* and *medroxyprogesterone acetate.* [Chlormanidone and megestrol acetate were withdrawn from use in the United Kingdom when long-term administration of high dosages was found to produce tumors, apparently nonmalignant, in the breasts of beagle bitches (Editorial, 1975e).] Medroxyprogesterone acetate has been used when given i.m. as a long-acting contraceptive preparation and for the treatment of endometrial cancer. These changes in the molecule appear to act by stabilizing it and slowing the rate of its inactivation.

The previous two groups of progestins have no significant androgenic or anabolic actions, which is in contrast to groups (c) and (d).

c. *Derivatives of testosterone* (Figure 7.21). Androgens have some endogenous progestational effects, and early attempts to utilize these resulted in the synthesis of ethisterone. Substitution of testosterone at the 17α-position enhances its oral effectiveness by inhibiting the inactivation of the molecule. 17α-Alkylation with a methyl group is used, but such compounds retain high androgenic activity. If, instead, a 17α-ethinyl moiety is substituted, the compound is stabilized and is also active orally, but its androgenic and anabolic effects are reduced. *Ethisterone* is such a compound and was first synthesized in 1937. However, it still retains significant androgenic activity, which makes it unacceptable, for most purposes, as a progestational drug. *Dimethisterone* (which is methylated at C-6) is

not very active but is used at relatively high doses in at least one sequential oral contraceptive preparation.

d. *Derivatives of 19-nortestosterone* (Figure 7.21b). Testosterone has a methyl group (C-19) attached at the C-10 position in the molecule. When this moiety is removed, 19-*nor*testosterone is formed, which has a substantially reduced androgenic activity, although it still retains a substantial anabolic action. This compound has provided the basis for a number of progestins which are especially useful in oral contraceptive preparations.

When 17α-substitution is performed, the nortestosterone derivates become effective orally. The 17α-ethyl analogue is *norethandrolone*, which has progestin activity but has an anabolic action and is used mainly for the latter purpose. However, as in the testosterone derivatives, a 17α-ethinyl group suppresses the anabolic effects and results in orally effective compounds which at normal dosages do not have significant androgenic effects. The most widely used of these steroids are norethynodrel and *norethindrone (norethisterone)*. An esterified form of the latter, *norethisterone enanthate*, when injected i.m. is active for 2 to 3 months and has been used as a contraceptive in women. Substitution of an ethyl (for a methyl) group at C-13 in norethindrone results in a compound that has a much higher activity, *norgestrol*, and is used in low-dose oral contraceptive preparations and the "minipill" (see Section 7.6I). Another interesting modification is reduction of the C-3 keto moiety to produce *ethynodiol*, which is also highly active. *Norethynodrel* has a similar action to norethindrone (the two compounds were prepared at about the same time), but the Δ⁴ double bond in the A ring is displaced to the Δ⁵ position. It should be noted that the compounds derived from testosterone (groups c and d above) exhibit some estrogenic activity due to partial metabolism to estrogens.

The *relationship of chemical structure to the activity* of progestins is, as with the other steroid hormones, complicated by such factors as their absorption, protein binding, and metabolism. Interactions of different analogues of progesterone with its receptors have been measured. The original and basic work on the cytosol progesterone receptors was performed on the chick oviduct, but some observations have been made in mammals, including the human uterus (Smith et al., 1974). There are similarities in the abilities of the chick and human progesterone receptors to interact with different progestins, but they are not identical. The following account refers mainly to studies with the human uterine receptors.

The chemical nature of the binding of progesterone to its cytosol receptors is unknown. However, in contrast to many other types of steroids, it appears to be relatively strong. Thus, only 50 per-

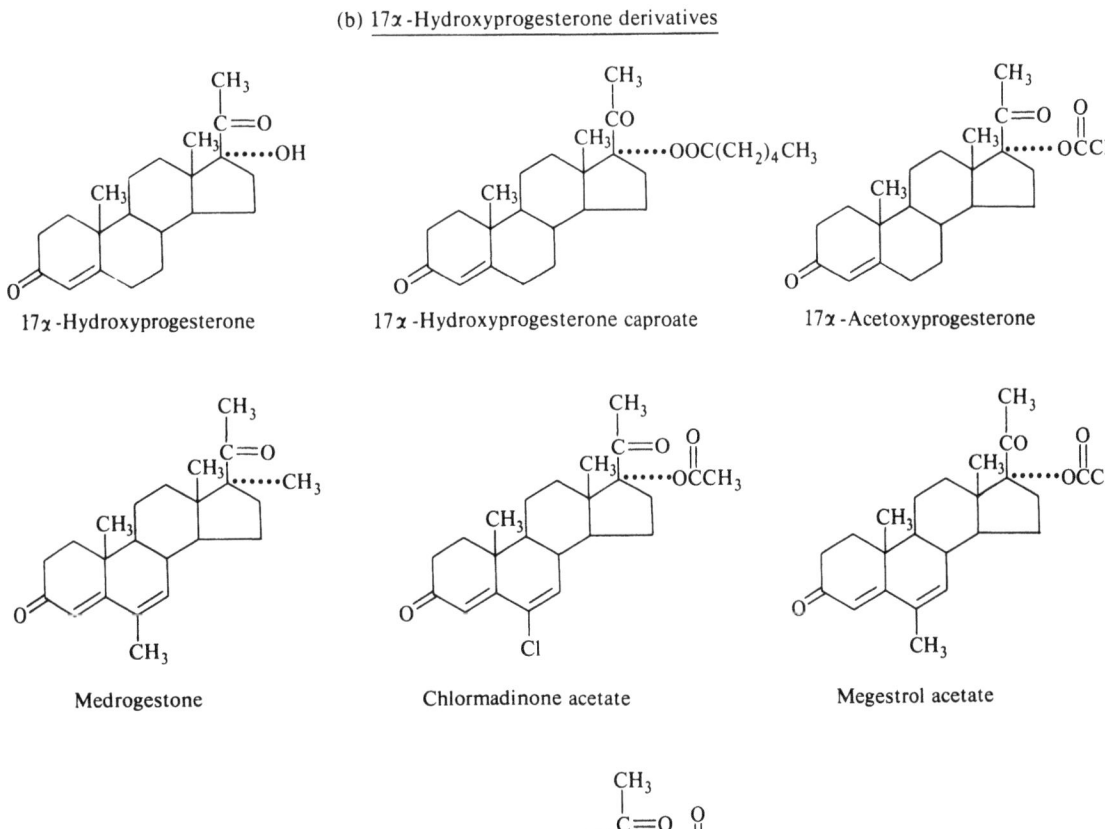

(a) Progesterone derivatives

Progesterone   Quingesterone   Dydrogesterone

(b) 17α-Hydroxyprogesterone derivatives

17α-Hydroxyprogesterone   17α-Hydroxyprogesterone caproate   17α-Acetoxyprogesterone

Medrogestone   Chlormadinone acetate   Megestrol acetate

Medroxyprogesterone acetate

Fig. 7.20. Progestin compounds derived chemically from progesterone and 17α-hydroxyprogesterone.

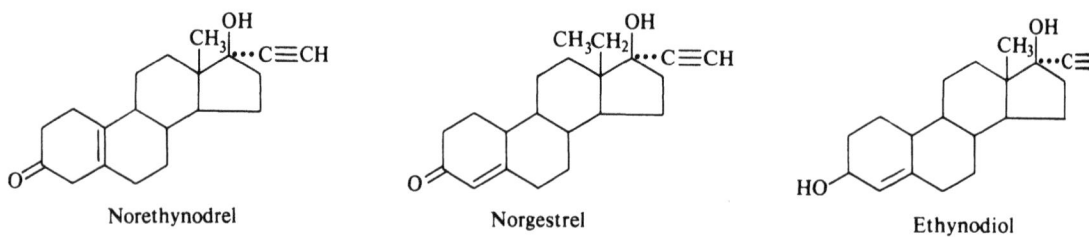

Fig. 7.21. Progestin compounds derived from testosterone and 19-nortestosterone.

Table 7.5. *Relative competitive binding of progestational steroids to the cytosol receptor from the human uterus*

	Relative receptor binding	Biological activity: Clauberg assay[a]
Progesterone (standard)	100	100
19-Norprogesterone	168	≃600
Additional —OH groups		
Deoxycorticosterone	28	20
Cortisol	<1	—
17α-OH derivatives		
17α-Hydroxyprogesterone	3	1
17α-Acetoxyprogesterone	40	500
C-6 substitution		
Medroxyprogesterone acetate	90	2000
Chlormanidone acetate	50	5500
Testosterone derivatives		
Testosterone	2	—
19-Nortestosterone	22	Inactive
17α-Methyl nortestosterone	99	≃50
17α-Ethinyl nortestosterone, norethindrone	150	≃1000
17α-Methyltestosterone	7	Weak
17α-Ethinyl testosterone, ethisterone	44	30

[a] Transformation of endometrium in young rabbits.
Source: Based on Smith et al., 1974.

cent of bound tritiated progesterone in the presence of 100-fold excess of "cold" progesterone is eluted from the receptors in 12 hours (Kuhn et al., 1975). Progesterone lacks hydroxyl groups, which appear to be important foci for the interactions of other steroid hormones – the estrogens, androgens, and corticosteroids – with their receptors. This difference may contribute to the rather special properties of the progesterone–receptor association. It is thus notable that deoxycorticosterone, which has a C-21 hydroxyl group but is otherwise identical to progesterone, is active and combines with the progesterone receptor, but it is only about one-fifth as effective (Table 7.5). The addition of hydroxyl groups at C-11 and the C-17 positions (as in cortisol) abolishes receptor binding and activity. Likewise, 17α-hydroxyprogesterone binds poorly and has only a very weak action. Hydroxylation at C-20 also reduces activity. Thus, the presence of hydroxyl groups appears, if anything, to be detrimental to the activity of progesterone.

With a single exception, any structural change in the progesterone molecule appears to reduce its ability to bind to its receptor (Table 7.5). Removal of the C-19 methyl groups to form 19-nor-progesterone *increases* both binding and activity. This effect is a consistent one, which is seen in other molecules that exhibit progestational actions.

Thus, 17α-methyltestosterone has weak activity, but its 19-nor derivative, 17α-methylnortestosterone, is nearly as effective as progesterone. Substitution of Cl or a methyl group at the C-6 position in 17α-acetoxyprogesterone, to produce chlormanidone acetate or medroxyprogesterone acetate, increases binding, but the biological activity is enhanced far more. The biological effectiveness of such analogues thus appears to reflect other factors, probably a slower rate of their metabolism.

The progestational effects of the testosterone derivatives appear to be related to two main factors. Testosterone has a small but significant ability to combine with the progesterone receptor, and this property is enhanced by certain 17α-substitutions. Thus, 17α-methyltestosterone binds more strongly than testosterone, while a 17-ethinyl group, as in ethisterone, considerably increases both binding and activity. Similarly, a 17α-ethinyl group increases these parameters in nortestosterone derivatives, such as norethindrone. These observations suggest that in this region of the molecule, the α-face is important for interaction with the receptor.

The basic interest of such observations is related to the possibility of constructing a molecular model which may indicate something about the nature of the interaction of the hormone with its receptor. Unlike in other types of steroid hormones, one cannot relate binding to the receptor to a predominant type of interaction, such as those involving either the α or β face of the molecule. Overall, steric factors are undoubtedly important, as indicated by the detrimental effects of changes in the A ring or hydroxylation at the peripheral C-11.

The A-ring appears to be of primary importance for the actions of the progesterone molecule, especially its C-3 keto substituent. Changes in the steric configuration of this ring reduces, but does not abolish, the ability of the molecule to interact with its receptor. Thus, saturation of the Δ^4 double bond to produce 5α- or 5β-dihydroprogesterone reduces both binding and activity considerably. Likewise, flattening the A ring, by the insertion of another double bond at Δ^1 or Δ^6, results in a 50 percent decline in binding. It has been suggested that the β-face of the A-ring is important. The reactivity of the C-3 keto appears to involve an α alignment as the binding is decreased when it is tilted upward as a result of steric changes in the A-ring.

The D-ring appears to be more flexible, and a number of substitutions can be made, but the molecule retains its capacity to bind to its receptor. For example, the 17β side chain can be substituted by an OH group, provided that there is a concurrent addition at the 17α-position, for instance an acetoxy, methyl, or ethinyl group. Thus, in the substituted testosterones with progestational activity, the successful 17α-substitution suggests an interaction at the α face, combined with a peripheral one due to the 17β-hydroxyl.

The interaction of progesterone and its analogues with its cytosol receptor can be seen to involve both faces and the periphery of the steroid, depending on which part of the molecule is involved. Smith et al. (1974) conclude: "Our picture of the receptor active site implies envelopment of the steroid by the receptor and hence extensive conformational changes in the protein." The intimate nature of this association may contribute to the relatively strong binding of progesterone to its receptors.

The ability to bind to the receptor correlates reasonably well with the biological activity *in vivo* of the progestational steroids. There are, however, differences which appear to be dictated by other properties of the molecules.

Mechanism of action of progesterone. Progesterone displays an unusual variety of different types of effects, such as promotion of the secretory phase of uterine growth, an increase in body temperature, stimulation of the respiratory center, an antagonism to aldosterone, and an ability to inhibit the electrical activity and contractility of the myometrium. Whether a unitary theory, related to its effects on genetic transcription, is sufficient to account for all these effects is unknown.

Progesterone has been shown to display local effects *in vitro*, and it is possible that some of these may reflect direct actions on membranes. Thus, this steroid has been shown to increase cyclic AMP levels in chick oviduct (Rosenfeld and O'Malley, 1970), an effect that is usually mediated by activation of membrane adenyl cyclase. This response did not appear to be related to increased protein synthesis, although it occurred simultaneously. The release of rat liver lysosomal enzymes is increased by progesterone in high concentrations (5×10^{-4}M) (Badenok-Jones and Baum, 1973). In high concentrations, progesterone also appears to be able to inhibit the action of aldosterone, and this response is probably a direct action due to exclusion of the mineralocorticoid from its receptor site. The application of progesterone, *in vitro*, to the rat uterus has been shown to abolish the action of oxytocin after only 30 minutes of exposure (Marshall and Csapo, 1961). The response returned after washing the progesterone from the tissue.

It appears, however, that the most important aspect of the mechanism of action of progesterone is related to an initiation of genetic transcription. This process has been studied extensively by O'Malley and his collaborators using the chick oviduct. This nonmammalian tissue has provided a considerable amount of important information about the general actions of such steroid hormones (see Section 7.4). The progesterone receptors can be induced by the presence of estrogens, which may account for the interrelated and synergistic effects of the two hormones, progesterone usually succeeding that of the estrogens. In the chick oviduct, progesterone induces the synthesis of a specific protein, avidin, but at present there is no information about the nature of such substances in mammals or how they may be acting to mediate the numerous and varied progestagenic effects.

The *progesterone receptor* has been especially elusive compared to other steroid hormone receptors. This problem has been largely the result of the propensity of progesterone to bind to other components found in cell cytosol preparations, such as glucocorticoid and androgen receptors and a cortisol-binding globulin (CBG)-like protein. The latter may be a contaminant from the plasma. There is, in addition, considerable nonspecific binding material in such preparations. In 1973, Philibert and Raynaud used a synthetic progestin, *R 5020* (17,21-dimethyl-19-nor-4,9-pregnadienne-3,20-dione), to "tag" and isolate specific progesterone receptors. This steroid was labelled with tritium and was shown to display an affinity for such receptors which was four times greater than that of progesterone. (In animal experiments it can exhibit a biological activity which is about 100 times greater than that of progesterone.) Such progesterone receptors were initially identified in the cell cytosol of the uteri of mice and rats. They have also been found in guinea pig and rabbit uteri (Philibert and Raynaud, 1974). The bound tritiated R 5020 could be displaced by the "cold" steroid, as well as by progesterone and norgestrol but not by cortisol or estradiol. The cytosol receptors, which exhibit a high affinity but low capacity for progestins, can be separated in the "7–8 S" fraction of a sucrose gradient. The K_D for R 5020 is about 0.6×10^{-9}M, while that for progesterone is 2 to 3×10^{-9}M. Another substance found in the cytosol, a "4–5 S" component, can bind progesterone, but it also binds cortisol and exhibits no specific binding for R 5020.

Using labelled R 5020, progesterone receptors have been measured in human breast cancers (Horwitz and McGuire, 1975) and in dimethylbenzanthracene (DMBA)-induced mammary tumors in rats (Asselin et al., 1976). This technique may be useful in predicting the responses of human breast cancer to hormonal therapy (see Section 7.6K).

Nuclear receptor (or "acceptor") sites for progesterone have been identified using R 5020 (Vu Hai and Milgrom, 1978). The cytosol receptors move into the nucleus when they are bound to R 5020 or progesterone but not when they are bound to cortisol or testosterone.

The levels of progesterone receptors in the cytosol and nucleus have been observed to change during the estrous cycle and pregnancy in the rat (Vu Hai, Logeat, and Milgrom, 1978). The number of nuclear receptors, for instance, peaked in midpregnancy (on day 15) but then declined and virtually disappeared prior to parturition. This cycle appears to be related to the relative activities of progesterone at these times. The cytosol receptors, on the other hand, were low in early pregnancy, but rose progressively between days 15 and 22. The change appears to correspond to a decrease in plasma progesterone levels and could represent a response which helps to maintain the sensitivity of the tissue to this steroid hormone.

G. Prolactin: the lactogenic hormone

In man, the adenohypophysial hormone prolactin has a unique and essential role in reproduction, as it is involved in regulating lactation and is sometimes called "lactogenic hormone" (see also Section 3.2). It has, however, also been dubbed the "Jack-of-all-trades" of the pituitary gland (Nicoll and Bern, 1972). At the time of that review, 82 different actions (Table 7.6) of this hormone had been reported throughout the vertebrates, from fishes to man. These effects can be classified into several groups, including reproduction, growth, water and ion metabolism, the integument, and a synergism with the actions of steroid hormones. The physiological significance of most of these reported actions have not been established and many are undoubtedly only of pharmacological interest. Contamination of the preparations of prolactin with other substances and hormones may have contributed to some of the observations. However, it is interesting that this protein hormone can exhibit such a wide spectrum of activity and it appears to be a very biologically reactive molecule. In man, prolactin has only been conclusively shown to have a single action: *its lactogenic effect on the mammary glands.* It may also have an antidiuretic effect, but this is contentious because of the demonstration that prolactin preparations are usually contaminated with small, but significant, amounts of antidiuretic hormone. In some rodents it has been shown to stimulate the corpus luteum and is thus luteotrophic. It has thus formerly also been called luteotrophic hormone or LTH. Prolactin can stimulate growth in some animals. This somatotropic action appears to reflect its chemical similarity to growth hormone, so that it may not represent a distinctive physiological effect.

1. Structure. Prolactin (PRL) is a protein containing about 200 amino acids. It exists in other vertebrates, including mammals, but its precise structure appears to have undergone considerable evolutionary change, so that the hormone isolated from one species is not always effective in another. The different types of prolactin also exhibit antigenic properties in different species, and this limits their possible uses in man. Like growth hormone, the ideal preparation for human experimentation and use would be human prolactin. The principal current source of the hormone for research purposes is that obtained from sheep, ovine prolactin (oPRL).

Prolactin shows a considerable degree of general structural similarity to growth hormone and human placental lactogen and it shares some of the actions of both of these hormones. Human growth hormone indeed has a similar (50 to 80 percent) lactogenic activity to prolactin and it was at one time doubtful if the two distinct hormones actually existed together in man. The evidence of comparative bioassays and immunoassays, as well as the purification of a hPRL, now show that they do have distinct identities (see Niall et al., 1973).

The structural homologies between fragments of human prolactin, hGH, and human placental lactogen are illustrated in Figure 7.22. These results, as well as interspecific comparisons of such hormones, are summarized in Table 7.7. It can be seen that there are close homologies between hPRL and oPRL, but that hPRL and hGH, while exhibiting considerable similarities, are also quite distinct.

Table 7.6. *Comparative endocrinology of prolactin (unabridged list of the manifold actions claimed for prolactin, May 1971)*

Cyclostomes
Electrolyte metabolism in hagfish (ACTH-like)

Teleosts
Osmoregulatory actions, including:
1. Survival of hypophysectomized euryhaline fresh-water species
2. Restoration of water turnover in hypophysectomized *Fundulus kansae*
3. Restoration of plasma Na$^+$ and Ca^{2+} in hypophysectomized eels when given cortisol
4. Skin, buccal, and gill mucus secretion
5. Reduced gill Na$^+$ efflux (reduced permeability)
6. Reduced gill permeability to water
7. Inhibition of gill Na$^+$, K$^+$-ATPase
8. Renotropic (glomerular and tubular changes)
9. Increased urinary water elimination and decreased salt excretion
10. Stimulation of renal Na$^+$, K$^+$-ATPase
11. Decreased water absorption and increased Na$^+$ absorption in flounder bladder
12. Decreased salt and water absorption from eel gut
Adrenocorticotropic
Resistance to high-temperature stress
Dispersion of pigment in xanthophores of *Gillichthys mirabilis, Arothron hispidus,* and *Gobius minutus*
Melanogenesis and proliferation of melanocytes (synergism with MSH)
Thyroid (TSH?) stimulation
Lipid metabolism
Growth and secretion of catfish seminal vesicles
Reduction of toxic effects of oestrogen
Parental behaviour (nest building, fin fanning, buccal incubation of eggs)
Maintenance of brood pouch in male seahorse
Gonadotropic (increased 3β-ol dehydrogenase in cichlid ovaries)

Amphibians
Water drive, including skin and tail changes
Larval growth (especially tail, including collagen synthesis, and gills)
Somatotropic in postmetamorphic anurans, including tail growth in urodeles
Lipid metabolism in anurans
Peripheral thyroxin antagonism (antimetamorphic)
Goitrogenic–thyrotropic (TSH release)
Growth of brain in frog tadpoles
Limb regeneration
Decreased urea excretion in some anuran tadpoles
Increased hepatic arginase activity in other anuran tadpoles secondary to thyroid activation
Hyperglycaemic–diabetogenic
Proliferation of melanophores
Skin yellowing in frogs
Restoration of plasma Na$^+$ levels in hypophysectomized newts
Na$^+$ and water transport across toad bladder

Secretion of oviductal jelly
Antispermatogenic
Cloacal gland development (urodele)
Stimulation of ultimobranchial and possible hypercalcaemia (toads)

Reptiles
Somatic growth
Tail regeneration
Hyperphagia
Epidermal sloughing
Lipid metabolism
Antigonadotropic
Restoration of plasma sodium levels in hypophysectomized lizard

Birds
Production of crop "milk" (columbids)
Formation of brood patch
Stimulation of feather growth
Somatic growth (including splanchnomegaly)
Lipid metabolism
Hyperglycaemic–diabetogenic
Stimulation of nasal (orbital) salt gland secretion
Antigonadal (antigonadotropic)
Synergism with steroids on female reproductive tract
Parental behavior
Suppression of sexual phase of reproductive cycle (including calling and mating in quail)
Premigratory restlessness (*Zugunruhe*)

Mammals
Stimulation of mammary growth and development
Stimulation of milk secretion
Stimulation of sebaceous gland size and activity (including preputial glands)
Hair maturation
Stimulation of somatic growth (including splanchnomegaly)
Lipid metabolism
Hyperglycemic–diabetogenic; increased insulin secretion
Erythropoietic
Renotropic, including Na$^+$ retention
Increased fertility in male and female dwarf mice
Actions on male reproductive organs, including:
1. Synergism with androgens on male sex accessory glands
2. Increased androgen binding in human prostate
3. Increased cholesterol levels in mouse testes
4. Stimulation of β-glucuronidase activity in rodent testes
Decreased copulatory activity in male rabbits
Advancement of puberty in rats
Progesterone secretion by mouse ovaries: possible synergism in other species (luteotropic)
Luteolytic action in rats
Parental behavior (retrieval of young by laboratory rats)
Vaginal mucification in rats

Source: Nicoll and Bern, 1972.

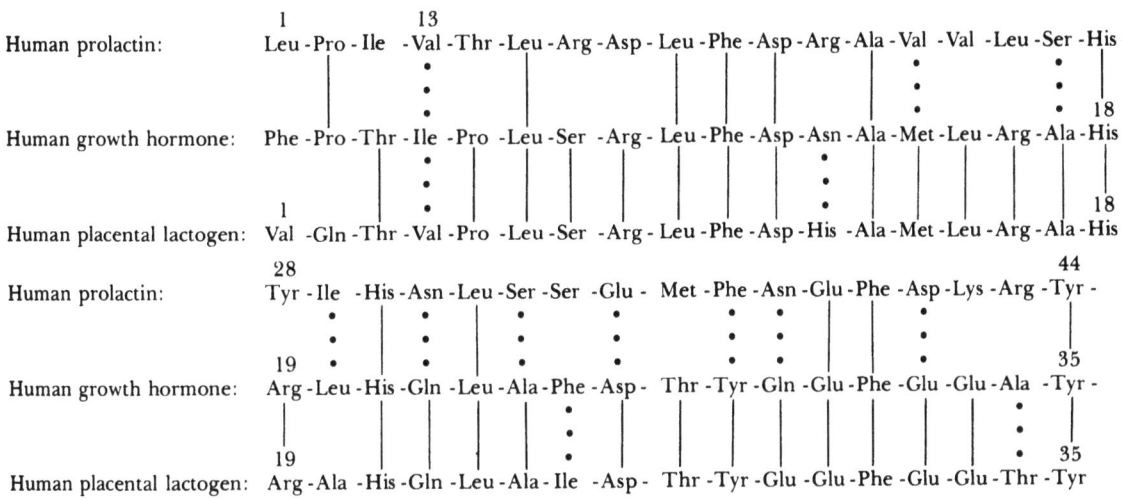

Fig. 7.22. Structural homologies among sections of human prolactin, human growth hormone, and human placental lactogen. The residues in the 4–12 sequence of human prolactin have been omitted to improve the homology. The vertical lines indicate identical amino acids; the three dots, "acceptable" homologies. (From Niall et al., 1973.)

2. Functions and effects. Prolactin has a lactogenic effect in mammals, including man, and promotes the maturation and maintenance of the secretory activity of the mammary glands. It is luteotrophic in some rodents, but this effect is not apparent in man. The injection of a preparation of the ovine prolactin into man had an antidiuretic effect and promoted retention of Na and K (Horrobin et al., 1971). High endogenous levels of prolactin in man have also been related to an antidiuresis (Buckman and Peake, 1973). The effect on water and ion metabolism is reminiscent of some of its actions on fishes and suggested that it could be involved in the physiological control of osmoregulation in man. This proposal is, however, contentious, and recent studies have failed to confirm such a function (Auty et al., 1976; Baumann

Table 7.7. *Sequence homology of the N-terminal regions of human growth hormone (hGH), human placental lactogen (HPL), human prolactin (hPRL), ovine prolactin (oPRL), bovine growth hormone (bGH), and monkey placental lactogen (MPL)*

Comparison[a]	% Identity
hGH-HPL (1–50)	76
hGH-hPRL (1–50)	24
hPRL-oPRL (1–50)	78
hGH-bGH (1–50)	58
hGH-HPL (1–24)	75
hGH-MPL (3–26)	83

[a] N-terminal amino acids are shown in parentheses.
Source: Niall et al., 1973.

and Loriaux, 1976). Ovine prolactin preparations appear to be contaminated with ADH, which could be contributing to some of these effects (see, e.g., Bond et al., 1976). In rats, ovine prolactin has also been shown to increase fluid transfer across the jejunum (Mainoya, 1975) and to have an ADH-like effect on the renal collecting tubule (Wallin and Lee, 1976). The possible physiological significance of these effects are not clear.

Prolactin may play a role in mammary gland carcinoma in man (see Smithline, Sherman, and Kolodny, 1975). Experiments on mice and rats have shown that the development of mammary carcinoma is at least partly under the control of prolactin; however, once the tumors have developed, they are independent of the action of this hormone. Mammary cancer in these rodent models arises from alveolar tissue, whereas most mammary cancer in women is due to proliferation of the duct tissues; hence, the application of these results to the human situation is uncertain. Measurements of plasma prolactin levels have failed to demonstrate a difference between women suffering from mammary carcinoma and control subjects, but this does not exclude the possibility that the hormone is involved. Culture of human mammary tumor tissue *in vitro* indicates that in about one-third of cases, the tissue is dependent on the presence of prolactin and that when none of the hormone is present, a regression of the tumor tissue occurs (Salih et al., 1972). Such results have given rise to speculation that in some instances it may be possible to control breast cancer with drugs, such as the ergot alkaloids [see Section 7.6G(3)], which block the release of prolactin. These drugs, however, do not completely abolish secretion, which may contribute to the disappointing results in clinical trials.

3. Release. Prolactin levels in the circulation are normally similar in men and women and follow a diurnal pattern, being greatest during the hours of sleep (Figure 7.23). Release of this hormone also occurs in response to a variety of physiological, pathological, and pharmacological stimuli (Tables 7.8 and 7.9). The principal normal stimulus for release is the suckling of the infant (Figure 7.24), which is a reflex the afferent part of which is along neural pathways to the brain. It has also been reported that an increased osmotic concentration of the plasma increases release of prolactin, whereas plasma hypotonicity decreases it, but the latter response has not been confirmed. The release of prolactin may also rise in such conditions as pituitary tumors, pregnancy, galactorrhea (sometimes), insulin hypoglycemia, severe physical exercise, coitus in women, surgical section of the pituitary stalk, and acute stress, such as that associated with surgery (see Table 7.8).

A number of drugs can either inhibit or stimulate the secretion of prolactin from the pituitary gland. To understand their probable mechanisms of action, it is necessary to recall (see Section 3.2E) the nature of the mechanisms controlling the release of prolactin. Like other adenohypophysial

Table 7.8. *Bodily conditions associated with release of prolactin in man*

Suckling	↑
Plasma hypertonicity[a]	↑
Plasma hypotonicity[a,b]	↓
Pituitary tumors	↑
Galactorrhea	↑↓
Exercise (strenuous)	↑
Stress	↑
Insulin hypoglycemia	↑
Gynecomastia	o
Pituitary stalk section	↑
Pregnancy	↑
Hypothyroidism	↑
Sexual intercourse (women)	↑

Note: ↑, increase; ↓, decrease.
[a] Buckman and Peake, 1973.
[b] Contentious remainder from Frantz, Kleinberg, and Noel, 1972, and Frantz and Kleinberg, 1978.

hormones, this process is mediated by the hypothalamus, which releases a local hormone (or hormones) that passes in the hypothalamic–hypophysial portal vessels to the adenohypophysis. The mechanism is, however, distinctive from that concerned with regulation of other adenohypophysial hormones; when the portal system is interrupted or the pituitary is transplanted so that it is not under the influence of the hypothalamus, there is an *increase* in secretion of prolactin. The dominant hypothalamic control therefore appears to be of an *inhibitory* nature. A factor or hormone has been identified in the hypothalamus which can inhibit the release of prolactin, and this is called the *prolactin release-inhibitory hormone* (P-R-IH). It has not yet been successfully isolated nor is its chemical nature quite clear (see Blackwell and Guillemin, 1973; de Wied and de Jong, 1974; Vale, Rivier, and Brown, 1977). It is widely suspected that, like other hypothalamic hormones that control adenohypophysial function, it is a peptide. However, the catecholamine dopamine has similar effects and is present in adequate concentrations in the hypothalamus. It has been difficult to separate their effects, but some feel that dopaminergic neurons may be acting to influence release of P-R-IH. Nevertheless, dopamine has been shown, *in vitro,* to act directly on the pituitary and inhibit the release of prolactin (MacLeod and Lehmeyer, 1974). It is possible that it can be carried from the hypothalamus in the portal circulation to the pituitary. Such an effect does not exclude an action in the hypothalamus also. The release of prolactin can be stimulated by another "factor" which has been isolated from the hypothalamus, called *prolactin-releasing factor* (PRF). The thyrotropin-releasing hormone (TRH) also stimulates release of prolactin, and it is possi-

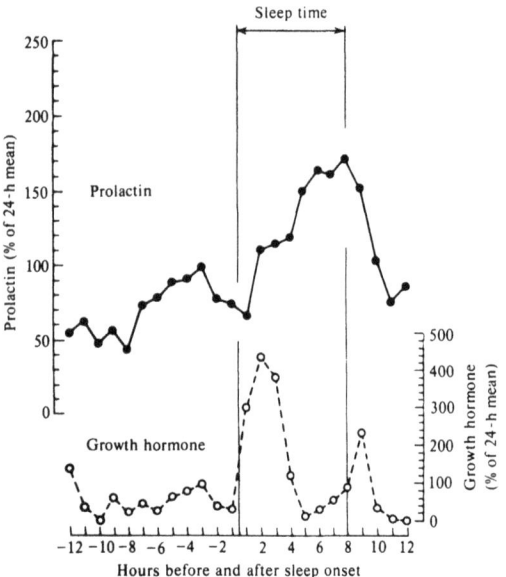

Fig. 7.23. Diurnal pattern in the release of prolactin and growth hormone in man. It can be seen that the release of both these hormones was accentuated during sleep. (From Sassin et al., 1972. Copyright 1972 by the American Association for the Advancement of Science.)

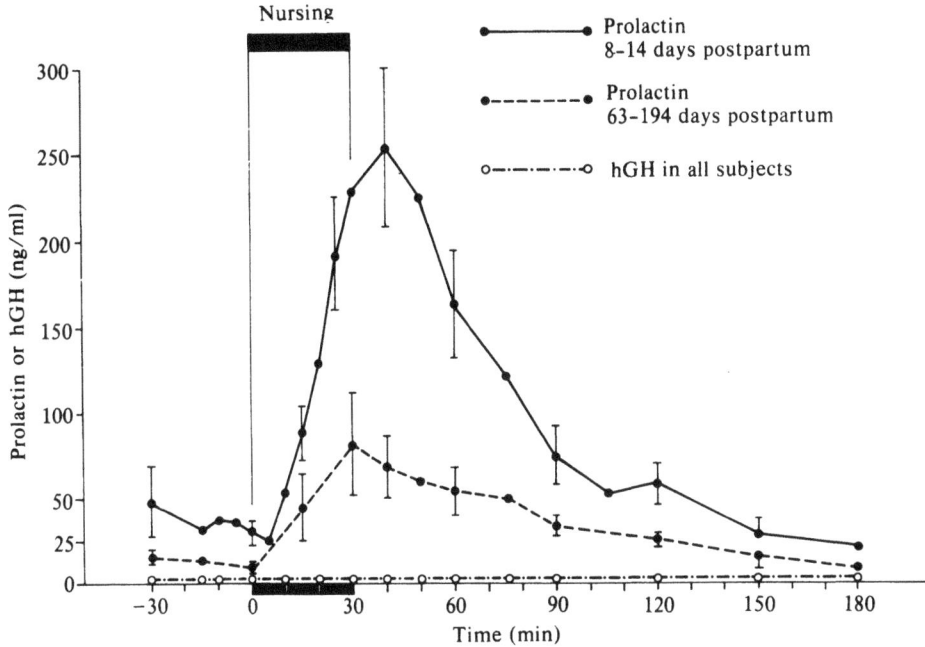

Fig. 7.24. Plasma prolactin values (determined by radioimmunoassay) before, during, and after nursing in 8 women 8 to 41 days postpartum, and in 6 women 63 to 194 days postpartum. It can be seen that suckling did not influence the release of pituitary growth hormone. (From Noel, Suh, and Frantz, 1974.)

ble that the two substances are identical. It is, however, also possible that they are chemically rather similar and that TRH has some PRF activity.

There are several possible mechanisms that can be utilized to increase or decrease the secretion of prolactin. The i.v. injection of *TRH* produces a rapid increase in prolactin levels in the plasma (Figure 7.25) and provides an experimental method for studying this process (Baumann and Loriaux, 1976). It may also be useful for clinically testing the integrity of the hypothalamic–pituitary system. *Morphine* stimulates release of prolactin in man (Table 7.9). It is especially interesting that the peptide *β-endorphin*, which has been isolated from the brain and has morphine-like analgesic properties (see Section 3.2A), also has this action (Rivier et al., 1977). These effects can be blocked by morphine antagonists such as *naloxone*. The effects of morphine are not seen *in vitro* and appear to reflect an inhibition of hypothalamic function, possibly involving a dopamine antagonism there. Estrogen in high doses can also increase prolactin release (Figure 7.26).

A variety of drugs (Table 7.9 and Figure 7.27) which interfere with adrenergic function, especially those with dopamine-blocking actions, have been shown to promote the release of prolactin. In the clinical situation this sometimes results in galactorrhea and, it has been suspected, may even play a

role in the genesis of breast cancer. Thus, *reserpine*, which is a psychotropic drug but which has been principally used to control hypertension, results in elevated plasma prolactin levels. Several studies have suggested that its use may result in an in-

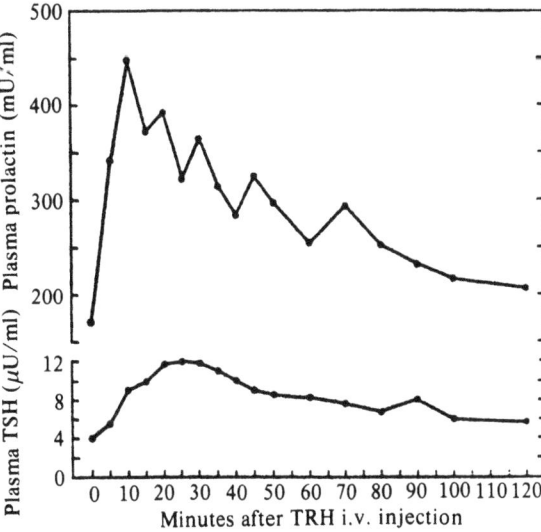

Fig. 7.25. Effects of injected TRH (200 μg i.v.) on plasma prolactin concentrations (mean values in four human subjects). (From L'Hermité et al., 1972.)

Table 7.9. *Effects of various drugs on the release of prolactin*

		Reference
L-Dopa	↓	Frantz, Kleinberg, and Noel (1972)
Dopamine[a]	↓	See de Wied and de Jong (1974)
Ergot derivatives (e.g., α-bromocriptine, lergotrile)	↓	See Clemens et al. (1974)
Apomorphine	↓	Martin et al. (1974)
TRH	↑	L'Hermité et al. (1972)
Phenothiazines (e.g., chlorpromazine)	↑	See de Wied and de Jong (1974)
Tricyclic antidepressants	↑	See de Wied and de Jong (1974)
Reserpine	↑	See de Wied and de Jong (1974)
α-Methyldopa	↑	See de Wied and de Jong (1974)
Metoclopramide	↑	Judd, Lazarus, and Smythe (1976)
Estrogens[b]	↑	Frantz, Kleinberg, and Noel (1972)
Morphine[c]	↑	Tolis, Hickey, and Guyda (1975)
β-Endorphin[c]	↑	Rivier et al. (1977)

Note: ↑, increase; ↓, decrease.
[a] Only *in vitro*.
[b] Antagonized by tamoxifen in rats (Jordan and Koerner, 1976).
[c] Antagonized by naloxone.

creased incidence of breast cancer (Boston Collaborative Drug Surveillance Program, 1974; Armstrong, Steven, and Doll, 1974; Armstrong et al., 1976), although this relationship has not been confirmed by all (O'Fallon, Labarthe, and Kurland, 1975; Mack et al., 1975). Reserpine depletes stores of catecholamines at adrenergic nerve junctions which could be contributing to a reduced inhibitory action of dopamine. Another antihypertensive drug, *α-methyldopa*, also increases plasma prolactin concentrations, but its use has not been related to a higher incidence of breast cancer. This

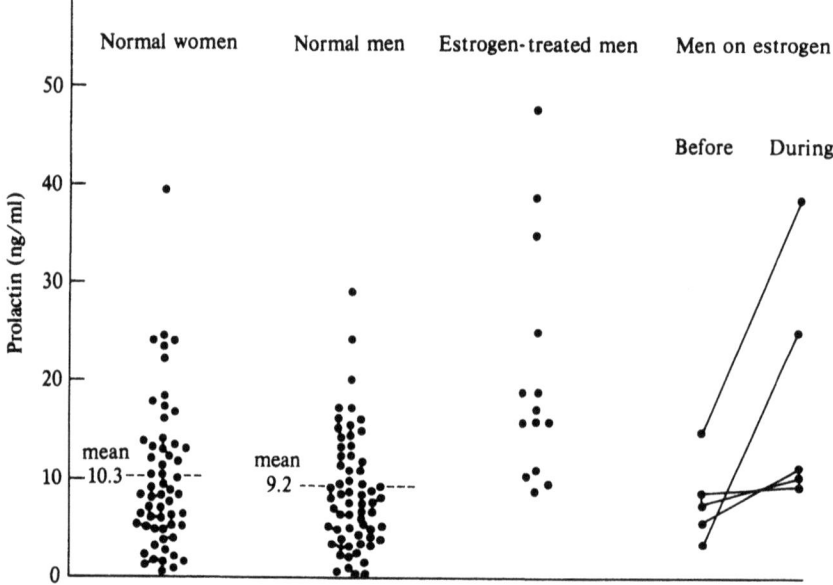

Fig. 7.26. Plasma prolactin (determined by radioimmunoassay) in men on chronic estrogen therapy (5 to 50 mg diethylstilbestrol/day). The right-hand column represents prolactin in 5 normal males before and during 1 week of diethylstilbestrol administration, 15 to 50 mg/day. (From Frantz, Kleinberg, and Noel, 1972.)

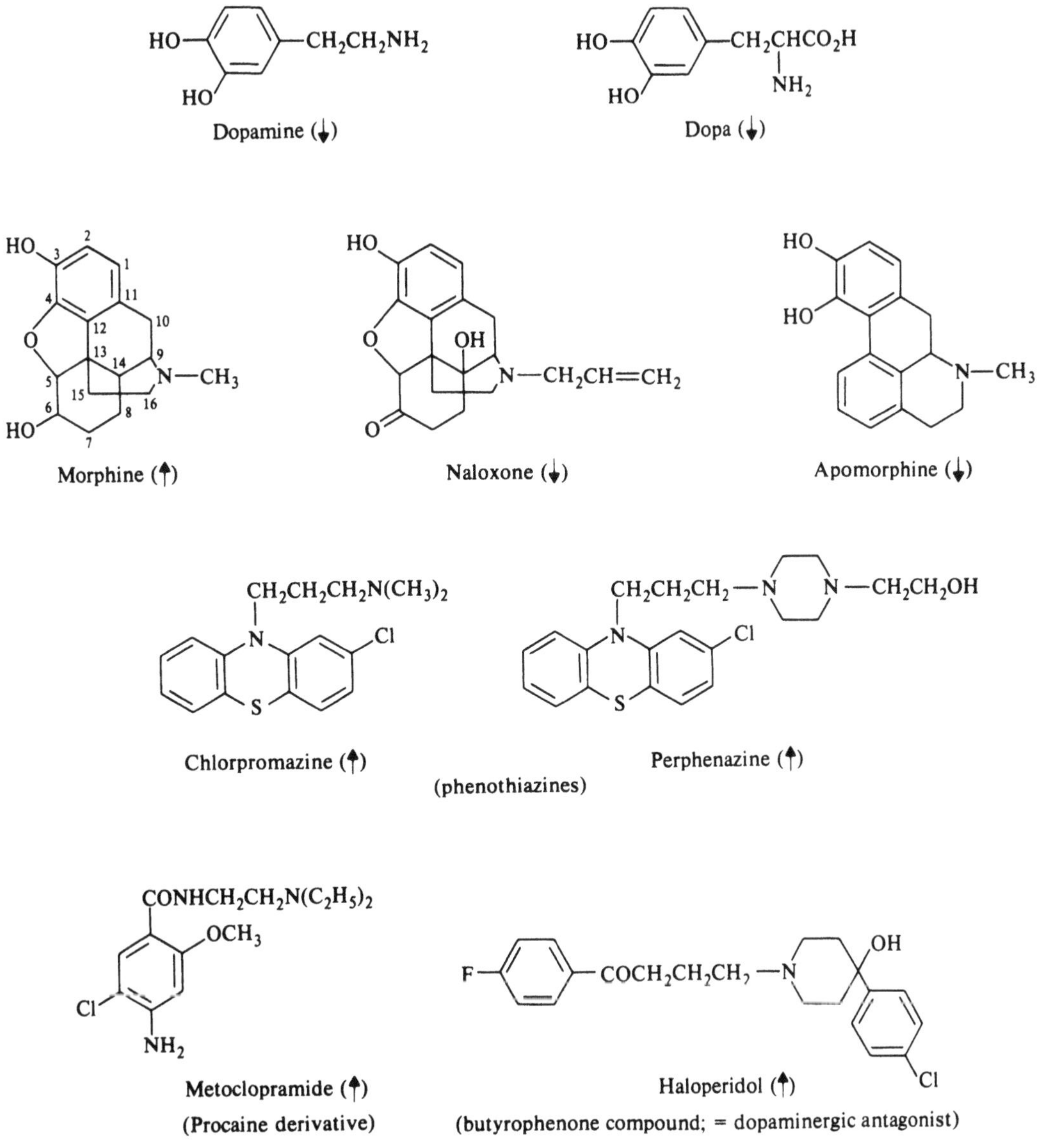

HO—⬡—CH$_2$CH$_2$NH$_2$ (HO)

Dopamine (↓)

HO—⬡—CH$_2$CHCO$_2$H (HO) | NH$_2$

Dopa (↓)

Morphine (↑)

Naloxone (↓)

Apomorphine (↓)

CH$_2$CH$_2$CH$_2$N(CH$_3$)$_2$

Chlorpromazine (↑)

CH$_2$CH$_2$CH$_2$—N⬡N—CH$_2$CH$_2$OH

Perphenazine (↑)

(phenothiazines)

CONHCH$_2$CH$_2$N(C$_2$H$_5$)$_2$
OCH$_3$
Cl
NH$_2$

Metoclopramide (↑)

(Procaine derivative)

F—⬡—COCH$_2$CH$_2$CH$_2$—N⬡OH ... Cl

Haloperidol (↑)

(butyrophenone compound; = dopaminergic antagonist)

Fig. 7.27. Some drugs that can increase (↑) or decrease (↓) the release of prolactin from the pituitary gland. Reserpine, α-methyldopa, TRH, and β-endorphin also increase release.

drug has a number of effects, including an inhibition of dopa decarboxylase, so that it reduces the formation of dopamine, which is a possible mechanism for its action on the release of prolactin. A number of types of drugs used as tranquilizers and antidepressants also promote the release of prolactin. The *phenothiazines chlorpromazine* and *perphenazine,* and the butyrophenone drug *haloperidol,* have dopamine blocking actions which appear to mediate their effects. *Tricyclic antidepressant drugs,* such as *imipramine,* may also cause galactorrhea (Klein, Segal, and Warner, 1964) due to increased

secretion of prolactin. *Metoclopramide* is an antiemetic drug which is a derivative of procainamide, and it also has a dopamine blocking action and stimulates secretion of prolactin. 5-Hydroxytryptophan, a precursor of 5-hydroxytryptamine (5-HT), can stimulate release of prolactin, and this effect can be antagonized by ergocornine (Larson, Sinha, and Vanderlaan, 1977). Histamine is present in the hypothalamus and i.v. or intracisternal injections of this amine also increases the secretion of prolactin in rats (Rivier and Vale, 1977). Its effects can be blocked by specific histamine antagonists and also by phenoxybenzamine (an α-adrenergic antagonist) and γ-aminobutyric acid (GABA). It is likely that the effects of the drugs described in this paragraph reflect the specific involvement with the hypothalamic mechanisms that control the release of prolactin.

There is more current therapeutic interest in drugs that can *inhibit* the release of prolactin than in those which can stimulate it. The latter response may contribute to side effects of drugs, but an ability to block release offers the prospect of controlling undesired effects of hyperprolactinemia, which can result in several disorders, including galactorrhea and infertility. They have also been promoted for the treatment of certain types of breast cancer and for suppressing normal lactation. *Dopamine* is effective *in vitro* but not *in vivo*, as it apparently cannot cross the blood-brain barrier in sufficient amounts to exert its inhibitory effect. *Levodopa* (L-*dopa*) is the metabolic precursor of dopamine and can enter the brain, where it is apparently decarboxylated to dopamine, which mediates an inhibition of the release of prolactin (Figure 7.28). *Apomorphine*, which has a dopaminomimetic effect, has a similar action. This drug also has an emetic effect and only a brief action, so it is not useful therapeutically. Two longer-acting dopamine agonists have been tested, *lergotrile* and *bromocriptine*, and the latter has been

introduced into clinical practice for the treatment of hyperprolactinemic conditions. Bromocriptine and lergotrile are chemical derivatives of *ergot* (see Lemberger, 1978). They appear to derive their origins from the speculation of Lindner and Shelesnyak (1967), who suggested that the ergotoxine group of alkaloids may influence endocrine function by inhibiting the release of prolactin. These alkaloids are pharmacologically an especially interesting group of substances, with a variety of effects, many of which are toxic and unpleasant. They occur naturally in certain plants, including a fungus, *Claviceps purpurea*, which contaminates the cereal rye. In the Middle Ages this periodically resulted in epidemics with many deaths (ergotism, "St. Anthony's Fire") due to eating contaminated bread. It causes a peripheral vasoconstriction, gangrene, abortion, vomiting, convulsions, and madness. It has been suggested that ergotism may have contributed to the strange patterns of behavior which led to the Salem witchcraft trials in 1692 (Caporael, 1976). A number of useful drugs and pharmacological discoveries have resulted from the study of the ergot alkaloids, including the phenomenon of α-adrenergic inhibition by Henry Dale. They have been formerly used to contract the uterus and assist labor and they can effectively prevent postpartum hemorrhage [see Section 7.6H(4)]. Ergot drugs are also used to treat migraine headaches.

In view of the colorful and sordid history of the ergot alkaloids, it is not surprising to observe that they commonly exert undesirable side effects which have resulted in attempts to chemically modify their structures so as to improve their therapeutic usefulness. Systematic studies have been made in attempts to find more effective compounds to inhibit the release of prolactin but with a minimum of side effects. One such survey (Clemens et al., 1974) involved an examination of about 300 derivatives of ergot and resulted in the clinical testing of lergotrile. Another group, led by E. Flückiger (see Flückinger, 1978), developed the use of bromocriptine. The mechanism of action of ergot alkaloids in inhibiting the release of prolactin reflects their dopaminergic effects. They are effective in both men and women. An inhibition of prolactin secretion has been demonstrated *in vitro*, and they inhibit the effects of TRH so that a direct action on the pituitary seems likely. However, an additional effect in the brain, such as a stimulation of dopaminergic neurons in the hypothalamus, is also considered possible. Ergot alkaloids are derivatives of lysergic acid and exist in two general forms:

a. *Amino acid alkaloids*, in which the lysergic acid moiety is combined with a polypeptide (Figure 7.29). These substances include the naturally occurring compounds *ergocornine, ergocristine,* and *ergocryptine*, which before their separation were

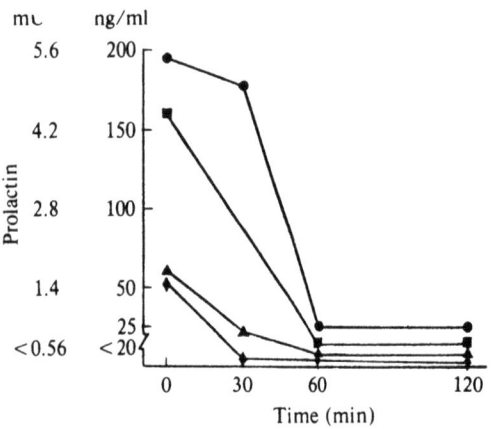

Fig. 7.28. Prolactin (determined by bioassay) following L-dopa, 500 mg orally, at zero time in 4 subjects with initially elevated plasma prolactin: ■, 50-year-old male tested several weeks after incomplete surgical removal of a prolactin-secreting chromophobe adenoma; ●, 25-year-old female with persistent galactorrhea following withdrawal of oral contraceptives; ▲, 29-year-old female with galactorrhea possibly related to chlordiazepoxide administration; ◆, 39-year-old female with galactorrhea beginning after oral contraceptive administration. (From Frantz, Kleinberg, and Noel, 1972.)

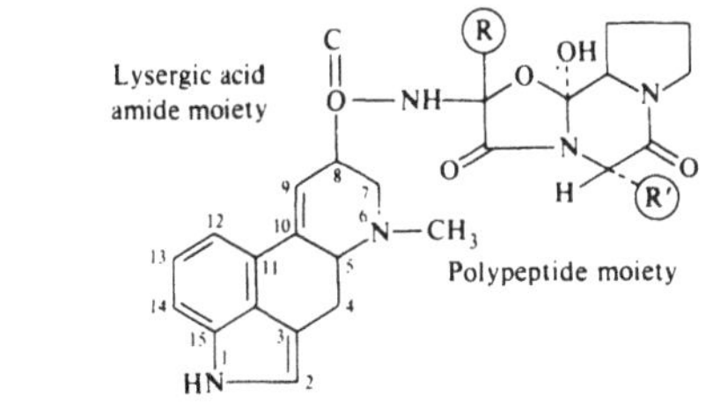

R = $CH(CH_3)_2$	$CH(CH_3)_2$	$CH(CH_3)_2$	CH_3
R′ = $CH(CH_3)_2$	$CH_2—C_6H_5$	$CH_2—CH(CH_3)_2$	$CH_2—C_6H_5$
Ergocornine	Ergocristine	Ergocryptine	Ergotamine

Effects of various peptide containing alkaloids on prolactin release from male rats treated with reserpine

Treatment[a]	Serum prolactin levels (ng/ml)	
Non-reserpine-treated controls	21.2 ± 2.1	
Reserpine controls	31.4 ± 3.4	
Ergocornine	5.1 ± 0.3	(<0.001)
Ergocorninine	21.5 ± 2.9	(N.S.)
Dihydroergocornine	11.3 ± 1.0	(<0.001)
Ergocryptine	8.5 ± 0.4	(<0.001)
Dihydroergocryptine	5.6 ± 0.5	(<0.001)
Ergotamine	17.5 ± 0.9	(<0.05)
Ergocristine	6.8 ± 0.5	(<0.001)

[a] Each ergot was administered at a dose of 50 μg/kg to 10 rats.

Fig. 7.29. Peptide-containing ergot alkaloids (amino acid alkaloids) that can inhibit the release of prolactin. Bromocriptine has a bromine substituted at position 2 in the lysergic acid amide moiety of ergocryptine and is thus called 2-bromo-α-ergocryptine. (From Clemens et al., 1974.)

jointly called *ergotoxine*. Such alkaloids can be effective inhibitors of the release of prolactin. When tested in rats treated with reserpine to give standard elevated baseline levels of the hormone, ergocornine, ergocryptine, and ergocristine lowered plasma prolactin levels (Figure 7.29). An isomer of ergocornine, *ergocorninine* (where the bond angle at position 8 linking the LSD amide to the polypeptide is changed), was ineffective. However, when the double bond between C-9 and C-10 was saturated to give the dihydrogenated derivatives, these were as effective as the parent compounds. This double bond appears to be responsible for the vasoconstrictor effects of the molecule and so offers the prospect of a possibly advantageous compound. The insertion of bromine in position 2 in ergocryptine to give *2-brom-α-ergocryptine*, or *bromocriptine*, results in a compound that is used clinically to treat hyperprolactinemic conditions such as galactorrhea (Thorner et al., 1974). Another interesting response (or side effect) to this drug is its ability to decrease blood pressure and produce postural hypotension, especially in patients with hypertension who also have elevated plasma prolactin levels (Stumpe et al., 1977). It was suggested that a disorder of the hypothalamic dopaminergic system may be mediating both the hyperprolactinemia and the hypertension. Bromocriptine mesylate may have a wide range of therapeutic uses, including, apart from those related to the release of prolactin, parkinsonism, acromegaly, and senility. It has been described as "an important new advance in therapeutics" (Judd, 1978).

b. *Amine alkaloids* or *ergoline derivatives* lack the peptide moiety and are thus easier to prepare and do not exhibit as many toxic effects as do the amino acid alkaloids. For instance, they lack α-adrenergic inhibitory actions and have little vasoconstrictor effect. They do, however, retain their uterotonic actions and can be potent inhibitors of prolactin secretion (Figure 7.30). Clemens et al. (1974) investigated the structure–activity relationships of such substances in inhibiting release of prolactin. The parent compound lysergic acid was devoid of activity. Substitution, other than a hydrogen, at the C-1 or C-2 position invariably had adverse effects on activity. At C-2 this could be related to the size, and ability to accept electrons, of the substituent group. A Cl retained more activity than an I at this position, while Br was intermediate between these. One of these ergoline derivatives, 2-chloro-6-methylergoline-8β-acetonitrile (Compound 83636 or *lergotrile mesylate*) has been tested clinically (Lemberger et al., 1974; Frantz and Kleinberg, 1978). In men and women it produced a 70 to 80 percent decline in baseline levels of prolactin in the plasma of patients suffering from various disorders. It also markedly reduced the increase seen in response to perphenazine. Lergotrile has also formerly been used to treat acromegaly [see Section 3.2F(6)]. However, because of observed toxicities in animals and man, it is not now recommended for clinical use.

4. Mechanism of action of prolactin on mammary glands. Studies on the mechanism of action of prolactin on the mammary gland have been principally performed using mice. Under *in vitro* conditions, this tissue has been shown to require the presence of insulin and cortisol for full expression of the effects of prolactin (see Lockwood, Stockdale, and Topper, 1967). The three hormones act synergistically. The composition of such "lactogenic complexes" varies, depending on the species, so that one must be careful in extrapolating the results to man. Such hormonal interactions may also contribute to difficulties in a clear interpretation of the effects of a single hormone. Pro-

| | | Substituents | | % Inhibition |
Generalized structure	Compound	R	R'	of prolactin
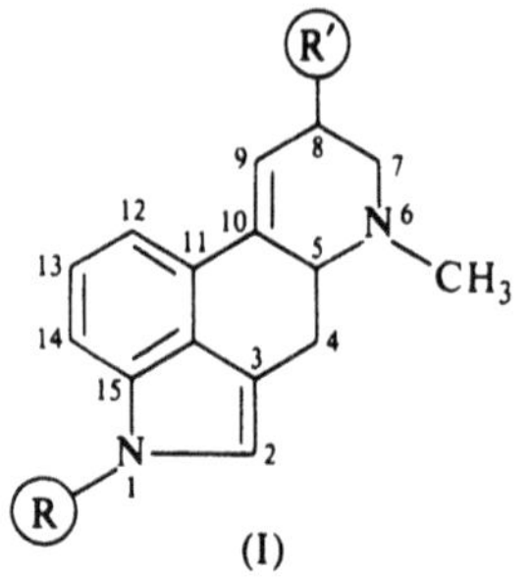 (I)	Lysergic acid	H	C—OH (with =O above)	0
	Methergine	H	C—NH—CH—CH₂—CH₂ (with O and CH₂OH)	47
	Lysergol	H	CH_2OH	56
	d-9,10-Didehydro-6-methyl-ergoline-8β-acetonitrile	H	CH_2CN	61
	dl,N-(9,10-Didehydro-6-methyl-8α-ergolinyl) formamide	H	NH—C—H (with =O)	76
Lysergic acid derivatives	dl,N-(9,10-Didehydro-6-methyl-8α-ergolinyl) acetamide	H	NH—C—CH₃ (with =O)	77

Lysergic acid derivatives — basic structure (I): ring positions labeled 1, 2, 3, 4, 5, 6, 7, 8, 9, 10, 11, 12, 13, 14, 15; N6—CH₃; R on N1; R' on position 8.

| | | Substituents | | % Inhibition |
Generalized structure	Compound	R	R'	of prolactin
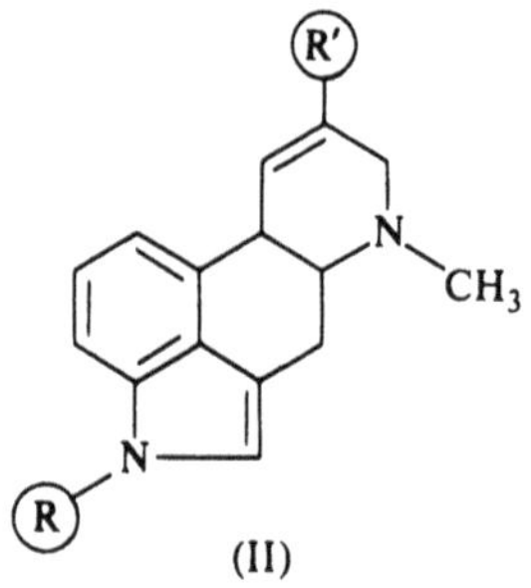 (II)	d-8,9-Didehydro-6,8-dimethyl-ergoline propionitrile	CN—(CH₂)₂	CH_3	0
	d-8,9-Didehydro-6-methyl-8-(piperidino-methyl) ergoline	H	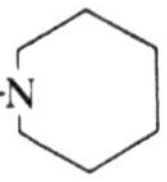 CH₂—N (piperidine ring)	34
	Agroclavine	H	CH_3	45
		H	CH₂—C—CH₃ (with =O)	57
		H	CH_2OH	71
Clavine derivatives	Elymoclavine-o-acetate			
	Elymoclavine			

| | | Substituents | | | % Inhibition |
Generalized structure	Compound	R	R'	R''	of prolactin
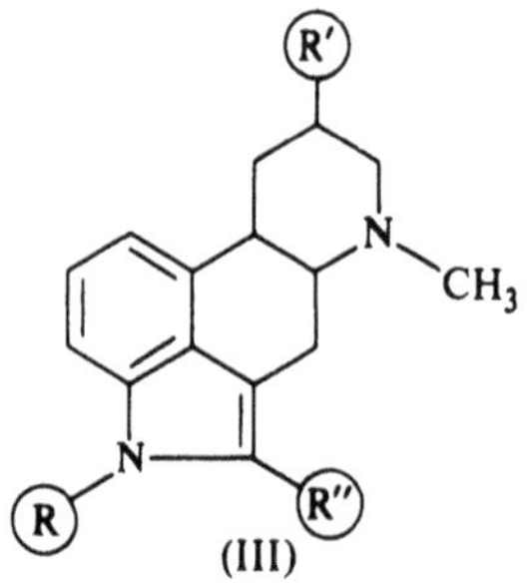 (III)	dl-6-dimethylergoline-8β-acetonitrile	CH_3	CH_2CN	H	38
	2-Iodo-6-methylergoline-8β-acetonitrile	H	CH_2CN	I	44
	8β(Chloromethyl)-6-methylergoline	H	CH_2Cl	H	45
	d-2-Chloro-1,6-dimethylergoline-8β-acetonitrile	CH_3	CH_2CN	Cl	47
	2-Bromo-6-methylergoline-8β-acetonitrile	H	CH_2CN	Br	53
Dihydrolysergic acid derivative	2-Chloro-6-methylergoline-8β-acetonitrile Lilly 83636 (Lergotrile)	H	CH_2CN	Cl	63
	1-Formyl-6-methylergoline-8β-acetonitrile	H—C— (with =O)	CH_2CN	H	74
	6-Methylergoline-8β-acetonitrile	H	CH_2CN	H	85

Fig. 7.30. Ergoline derivatives (amine alkaloids; basic structures I, II, and III) that inhibit the release of prolactin in reserpine-treated male rats. (From Clemens et al., 1974.)

lactin has been shown *in vitro* to stimulate the formation of casein and β-lactalbumin, and these effects appear to be preceded by the synthesis of RNA (Juergens et al., 1965; Green, Bunting, and Peacock, 1971). Prolactin has also been shown to stimulate the formation of the enzymes lipoprotein lipase and lactic synthetase (Zinder et al., 1974; Turkington et al., 1968), and it increases the uptake of polyamines such as spermidine (Kano and Oka, 1976). The latter may play an important role in the synthesis of milk proteins.

The specific nature of the prolactin receptors and the process by which they stimulate RNA synthesis is not yet clear. There are, however, some especially intriguing suggestions. Cell fractionation studies and the ability of prolactin bound to Sepharose beads to act suggests that the prolactin receptors may be on the outer membrane of the cells (Turkington, 1970). A prolactin receptor has been solubilized from particulate membrane fractions of rabbit mammary gland (Shiu and Friesen, 1974). It was shown to have a molecular weight of about 220,000 and, apart from prolactin, can also bind human placental lactogen and human growth hormone, both of which exhibit lactogenic activity. Such membrane-bound receptors would be consistent with observations regarding the site of receptors for other protein hormones. There are, however, some disquieting observations. Prolactin has been observed to increase RNA synthesis by *isolated nuclei* from mammary gland epithelia (Chomozynski and Topper, 1974). This effect did not require the presence of other hormones; it was not seen in other tissues, such as liver or kidney; and it could not be elicited by insulin. Furthermore, using immunohistochemical procedures it has been shown that endogenous prolactin is present *inside* the alveolar cells of the rat mammary gland (Nolin and Witorsch, 1976). It was found near the apical membrane, which is distal to the blood supply. Thus, it is possible that while prolactin binds to membranes, it may be doing this from the inside. This is a novel, but potentially important, observation with respect to the possible mechanisms of action of some protein hormones.

Estrogens appear to induce the formation of additional prolactin receptors in the liver (Posner, Kelly, and Friesen, 1975). It would be especially interesting to know if this effect is also seen in the mammary gland, as it could be an important feature of prelactational development of the secretory tissue.

The nature of the link between the interaction with the receptors and the synthesis of RNA in mammary glands is uncertain. It appears that cyclic AMP does not act as a "second messenger," although it could still modulate the effects of prolactin. Some interesting observations that may implicate prostaglandins and phospholipase A have been made (Rillema, 1976; Rillema and Wild, 1977). Indomethacin, which inhibits prostaglandin synthetase, blocks the effects of prolactin on RNA synthesis. In addition, certain prostaglandins – B_2, E_2, and $F_{2\alpha}$ – can imitate the effects of prolactin, but their actions are not additive to it, nor were they seen in tissues that were not suitably primed with cortisol and insulin. It was suggested that certain prostaglandins may act as a messenger to stimulate RNA synthesis. Prostaglandins are formed from arachidonic acid, which is released in the cell under the influence of phospholipase A. Prolactin was shown to activate membrane-bound preparations of this enzyme.

H. Roles of hormones and effects of drugs

1. Menstrual cycle. Women produce a "ripe" ovum, which is available for fertilization, about every 28 days. This cycle, in the absence of pregnancy, is normally repeated consecutively throughout the year. As it corresponds closely to a lunar month, it is called the menstrual cycle (mensis = month). It differs from the reproductive cycle in most animals, with the exception of some monkeys, as a vaginal discharge containing endometrial debris and some blood (the menses) occurs between the successive ovulations. The menstrual cycle has a distinct pattern that reflects changing levels of pituitary gonadotropins, estrogens, and progesterone, which control morphological and physiological changes in the reproductive organs, especially the ovary, uterus, and vagina. The cyclical rhythm is ultimately controlled from the brain, especially the hypothalamus, which by a programmed release of LH/FSH-RH regulates the secretion of gonadotropins (see Section 7.3). In many animals there are prominent external environmental signals that influence the hypothalamus and the reproductive cycle, but such effects are rare, or more subtle, in women, although they may contribute to chronological variation and irregularities.

The first menstrual cycle is called the *menarche* and usually occurs at about 13 years of age. The time of this event is, however, far from constant, and it can occur as early as 9 years or as late as 18 years. If it occurs outside these limits, it is considered to be abnormal. The time of puberty, and its climax the menarche, appears to be mainly controlled genetically, but it is also influenced by other factors, especially nutrition. The age of the menarche has been (progressively) receding in Western countries over the last 100 years, with temporary lapses in wartime Europe.

Although the average length of the menstrual cycle is 28 days, there is considerable variation, and periods ranging from 21 to 35 days are not unusual. Such variation can occur in the same woman, either throughout the year or at different times of her life. The cycle tends to get shorter as the menopause approaches but then it gets longer immediately prior to this event.

For the purposes of scientific convenience, the first day of the menstrual cycle is taken as the day when the menses (or menstruation) commences and the last day is that which precedes the next menses, about 28 days later. Menstruation lasts for about 3 to 8 days, usually 5 days. This event is precipitated by low plasma levels of estrogens and progesterone. In the face of the withdrawal of these hormones, the uterine endometrium cannot sustain itself and starts to break down. Small spiral blood vessels in this tissue break so that blood leaks out amidst the tissue and a massive sloughing occurs, which is discharged through the vagina. Usually, about 15 to 50 ml of blood is lost during each menstrual period. It is, in effect, a hormone-"withdrawal" bleeding. It can be imitated by the appropriate administration and withdrawal of estrogen and progesterone. *Withdrawal bleeding* differs from *"breakthrough" bleeding,* which may occur on different occasions, as the blood does not clot, because of the presence of fibrinolytic enzymes released from the endometrial cells. Breakthrough bleeding occurs by leakage through capillaries and is not due to breakage of vessels. The spiral arteries, which develop under the influence of progesterone, are not involved. Excessive blood loss is called *menorrhagia.*

The climax of the menstrual cycle occurs on about day 16, when ovulation takes place (see Figure 7.31). As described in Section 7.3A, the Graafian follicle, under the influence of FSH, matures prior to this event. There is a concomitant secretion of estradiol-17β, which results in a proliferation of the endometrium. This hormone also initiates a sudden massive increased release of LH, which triggers the ovulation. In women, this event is said to be spontaneous, as, unlike in some animals, it is not usually considered to be directly related to sexual stimuli. There is, in addition, a spurt in the secretion of FSH, which also contributes to ovulation. The ovum enters the oviduct to await iminent fertilization and the corpus luteum starts to form under the influence of LH. The ovum survives for 1 to 2 days if not fertilized. Progesterone secretion increases and this is associated with a rise in body temperature. This hormone also initiates the transformation of the endometrial glands into their secretory or postovulatory phase in preparation for possible implantation of a fertilized ovum. (17α-Hydroxyprogesterone levels rise prior to ovulation, but the role of this weak progestin is not known.) The activity of the corpus luteum is maintained for up to about 8 days after ovulation (day 24 of the cycle) and then, if implantation does not

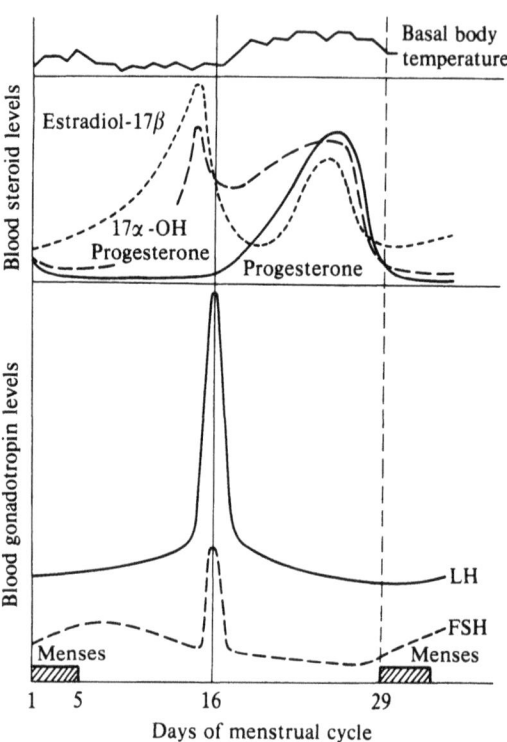

Fig. 7.31. Changes in the hormone levels in the blood during the human menstrual cycle. (From Short, 1972.)

occur, starts to regress so that progesterone secretion drops over the next 4 days. After a brief rise following ovulation, estrogen levels also decline and the next menses results.

Ovulation does not occur in all cycles and in about 5 percent of women it never occurs. Even in normal women it appears that ovulation fails to occur in 3 to 10 percent of the cycles. Anovulatory cycles become more common prior to the menopause and they also occur frequently in adolescence just after the menarche and after parturition. This latter period of relative infertility may be related to lactation. The disorders of the menstrual cycle and the use of exogenous hormones in their treatment are discussed in Section 7.6H(7).

2. Pregnancy. Pregnancy is the period from the time of fertilization until the birth of the young, which in women is about 260 days. Fertilization of the ovum takes place in the oviduct, and on about day 3 following ovulation the egg enters the uterus. Differentiation to a blastocyst takes place and implantation occurs on days 8 to 13. This first direct physical association of mother and embryo (it has been compared to that of host and parasite) is a precisely timed event, and both the maternal and embryonic tissues must be in a receptive condition. If conditions are unsuitable, then implantation will not occur and in women the blastocyst will die. (In some animals there will be a delay and implantation may occur much later.) The uterus must be optimally primed with progesterone, and both this hormone and estrogens are necessary for success. This receptive period when implantation is possible lasts for only a few hours. Thus, this stage provides a critical period for possible pharmacological interference with fertility.

Implantation involves changes in the endometrium, called *decidualization*, when the decidua is formed. This provides a cushion on which the blastocyst rests and under which the maternal capillaries undergo an increase in permeability. The blastocyst differentiates into the embryoblast, which forms the fetus, and the trophoblast, which grows and interdigitates with the endometrial cells to forge the connection with the maternal circulation and form the placenta. This tissue is mainly derived from the ovum and provides the structure whereby exchanges of respiratory gases and nutrients can occur between the mother and fetus. In women it also has an important endocrine role and can form several protein and steroid hormones, including hCG, human placental lactogen (HPL, also called human chorionic somatomammotropin, HCS), estrogens, and progesterone.

The uterus hypertrophies during pregnancy, increasing in weight from less than 100 g to about 1000 g, and its volume expands more than 800 times. This growth appears to occur under the influence of estrogens, but as hypertrophy also occurs as a simple result of physical expansion, the precise contribution of the steroid is uncertain. Estrogen levels rise throughout pregnancy (Figure 7.32), and these come from both the ovary and placenta.

Progesterone levels in the plasma rise considerably during gestation, and it is thought that their principal function is concerned with maintaining the quiescence of the myometrium and inhibiting the normal cyclic release of pituitary gonadotropins. Progesterone initially is formed by the corpus luteum, which is principally under the control of hCG. The corpus luteum starts to degenerate in the third month of pregnancy and the placenta remains as the principal physiological source of progesterone.

During pregnancy, estrogens and progesterone also function to promote the development and preparation of the ducts and secretory alveoli of the mammary glands for future lactation.

Other changes that occur in pregnancy and appear to involve hormones are a relaxation of ligaments in the pelvic region and an increased elasticity of the vagina, both of which may be related to future parturition. Areas of pigmentation occur

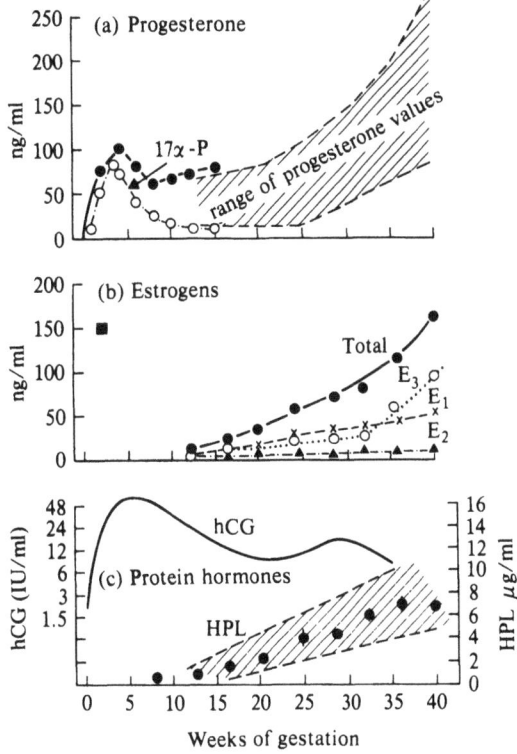

Fig. 7.32. Hormone levels in the peripheral blood during pregnancy in women. 17α-P, 17α-hydroxyprogesterone; E_1, estrone; E_2, estradiol-17β; E_3, estriol; hCG, human chorionic gonadotropin; HPL, human placental lactogen. (Based on Heap, Perry, and Challis, 1973.)

on various parts of the body and may be related to increased secretion of estrogens and pituitary ACTH or its fragments of melanocyte-stimulating hormone.

3. Endocrine function of the placenta; the "feto-placental unit." The placental protein and steroid hormones are thought to be formed principally by the cells of the syncytiotrophoblast tissues, which are on the chorionic villi in contact with the maternal blood. The fully developed human placenta weighs about 500 g and produces enormous quantities of hormones during the period of pregnancy, commencing, in the instance of human chorionic gonadotropin, about 12 to 14 days after fertilization has occurred.

The principal *steroid hormones* that are released by the placenta (Figure 7.33) are progesterone and estriol, but small amounts of estradiol-17β and estrone are also secreted. The placenta, however, has only a limited capacity to synthesize these steroids. Thus, the substrate for progesterone is maternal cholesterol, and not acetate, as in the ovaries. Es-

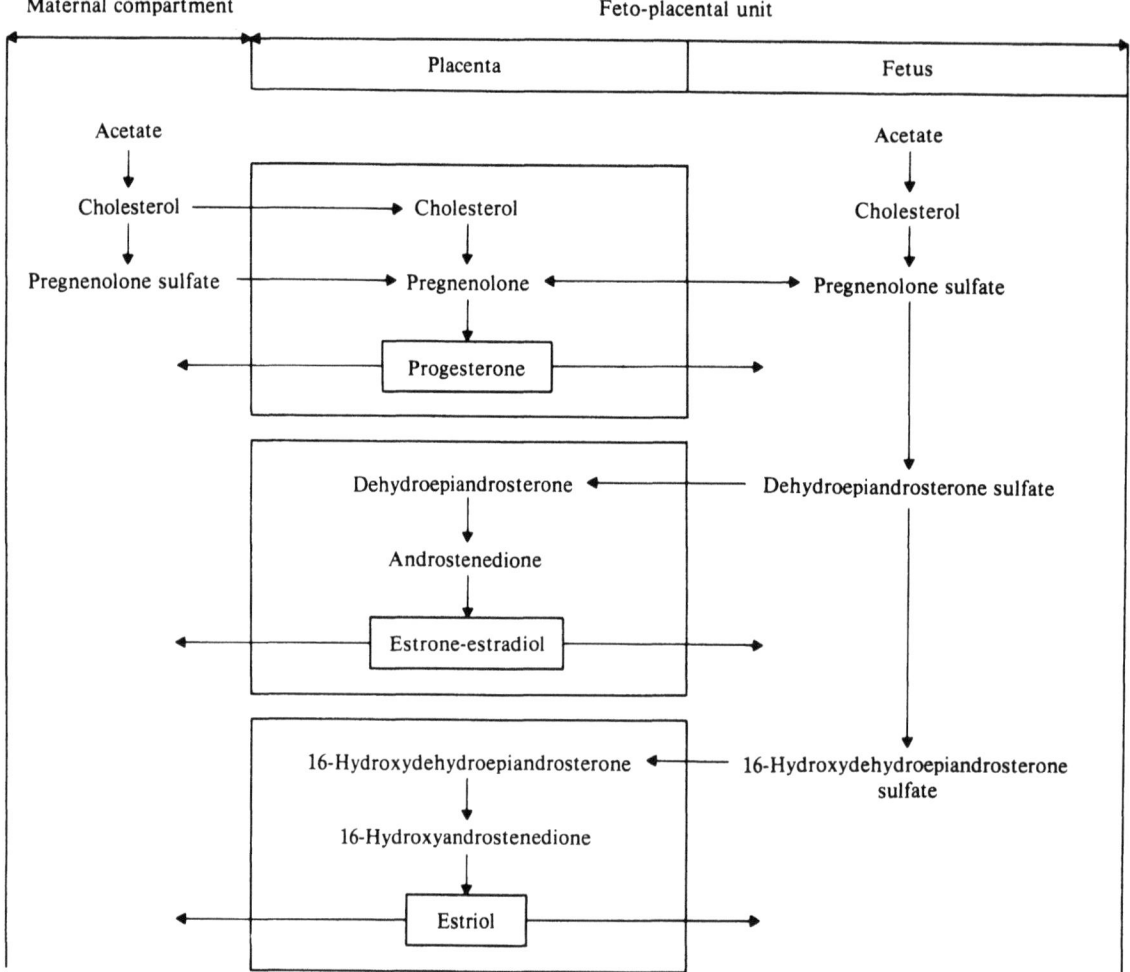

Fig. 7.33. Summary of the steroid hormone metabolism by the human feto-placental unit and its relationship to the maternal circulation. For a more detailed description, see the text. (Based on Ryan, 1969, and 1973.)

triol is formed by placental aromatizing enzymes from 16-hydroxydehydroepiandrosterone. This substrate is formed from dehydroepiandrosterone sulfate (DHAS) from the fetal adrenal cortex and it is 16-hydroxylated by the liver of the fetus. DHA from the infant can also be converted, via androstenedione, to estradiol-17β and estrone. The steroid hormones that are formed by partial synthesis in the placenta can pass into the circulation of the mother and fetus. In the parent they play an important role in maintaining pregnancy and the development of the mammary glands, so that ovariectomy after about the second to third month does not usually result in abortion. The fetus and placenta thus make an important contribution to the maintenance of pregnancy which is independent of the maternal tissues.

A variety of proteins that exhibit hormone-like

activity and appear in the circulation were described earlier (Section 7.3); these include human placental lactogen and possibly also a placental thyrotropin, which is probably identical to hCG (see Section 4.3A). Human chorion-decidual tissue has also been shown to synthesize and secrete a prolactin that cannot be distinguished from that produced by the pituitary (Golander et al., 1978).

Human placental lactogen is present in high concentrations in the placenta, where it is formed by the syncytiotrophoblast. It can be identified in the tissue about 3 to 4 weeks after fertilization and appears in the plasma after 7 to 8 weeks. Its concentration rises throughout pregnancy and the rate of its secretion may be as high as 1 g/day. This interesting material was first observed in the human placenta in 1962 by Josimovich and MacLaren, who noted that it had a strong lactogenic effect and cross-reacted immunologically with antibodies to human growth hormone. In addition to its effects in stimulating the mammary glands, it also has a growth hormone-like (somatotrophic) effect, but it has only 0.1 to 1 percent of the activity of hGH. In view of these effects, on the mammary glands and growth, it has also been called human somatomammotropin.

The amino acid sequence of HPL has been described (Niall et al., 1971; Li, Dixon, and Chung, 1971; Sherwood et al., 1971). As shown in Figure 7.34, like hGH, HPL has 190 amino acid residues; 160 of these are identical in the two hormones, and even in the remaining 30, only 7 are considered to be "nonhomologous." These similarities extend to all three molecules, as illustrated in Figure 7.35. There are similar amino acid sequences within each hormone, which suggests that they have a common ancestry and evolved from shorter peptides by a "tandem-like duplication" (Niall et al., 1971).

Some studies have been made on the structure-activity relations of HPL in comparison to hGH (Handwerger et al., 1972; Aloj et al., 1972). Although there is a considerable difference in the somatrophic activity of hGH and HPL, their lactogenic activity is similar, suggesting that there may be two overlapping active sites in the molecules. Disruption of the disulfide bridges has little effect on the lactogenic potency of either molecule but profoundly decreases their immunological activities. Such changes result in considerable changes in the native tertiary structure of the molecules. It was suggested that while such three-dimensional factors may be important for immunological activity, they are not vital for biological actions. Smaller polypeptide fragments may be responsible for the latter, and these may still retain a tertiary structure which is essential for biological activity.

The function of HPL is not known; it cannot readily enter the fetal circulation, so that its actions may be entirely on the mother. It may be concerned with the development of the mammary glands during pregnancy, but this is uncertain. HPL can be readily assayed in the circulation using an immunological procedure, and it has been shown that its levels are relatively low in many, but not all, instances of fetal distress, threatened abortion, and "high-risk" pregnancies (Niven, Landon, and Chard, 1972; England et al., 1974; Letchworth, Slattery, and Dennis, 1978). It has thus been suggested that such measurements could provide a useful test to predict the outcome of a threatened abortion, but the false-positive rate is high. Attempts to utilize the somatotrophic effects of this protein in hypopituitarism have not met with success. It appears to be about 500 times less active than human growth hormone. However, like

growth hormone, ovine placental lactogen can stimulate the formation of somatomedin in rats, and it has been suggested that it could thus contribute to the control of growth in the fetus (Hurley et al., 1977).

The placenta (see Josimovich, 1973) is the barrier across which various molecules can pass between the maternal and fetal circulation. Apart from nutrients, metabolites, and respiratory gases, hormones and drugs can also cross the placenta. As just described, protein hormones pass with difficulty, but an interchange of steroids can occur. Factors that influence the effectiveness of such exchanges, apart from permeability, are the binding to plasma proteins and metabolism by the placenta and the fetus itself. Thus, the possibility that administered exogenous hormone preparations can enter the fetal circulation and influence the development of the young needs to be considered. The use of progestins, which have androgenic activity, in pregnant women can have a masculinizing effect on the female fetus. As already described, estrogens administered to the mother may have a long-term action and subsequently result in clear-cell adenocarcinoma in teenage girls. Sulfonylurea oral hypoglycemic drugs can result in hypertrophy of the B-cells of the islets of Langerhans and hypoglycemia in the newborn. Antithyroid drugs, such as propylthiouracil, can also cross the placenta and exert a goitrogenic action in the fetus. Such an effect may, however, be antagonized by thyroxine or triiodothyronine, which also cross the placental barrier in small but physiologically significant quantities. Insulin appears to cross the placenta from mother to fetus but is rapidly degraded there.

4. Parturition and labor. Labor and delivery of the infant occurs as a result of relaxation of puberal ligaments and the dilation of the cervical canal in conjunction with rhythmical contractions of the uterus. The physiological events that determine the initiation and progress of parturition are uncertain, especially as animal experiments have merely emphasized that considerable interspecific variability can occur.

Parturition may be assisted directly by two hormones, *oxytocin* (see Section 3.3) and *relaxin*. The latter is a peptide with a molecular weight of about 6000 (see James et al., 1977; Schwabe and McDonald, 1977). It was originally identified over 50 years ago by F. L. Hisaw (1926). He isolated it from corpora lutea of pregnant rabbits and guinea pigs, and in these species it was found to have a potent effect in relaxing the ligaments of the pubic symphysis. It also enhances relaxation of the cervix and vagina and inhibits contractions of the uterus. Relaxin has since been identified in other species, including the human corpus luteum of pregnancy

HPL:	Val	-Gln	-Thr	-Val	-Pro	-Leu	-Ser	-Arg	-Leu	-Phe	-Asp	-His	-Ala	-Met	-Leu	-Gln	-Ala	-His	-Arg	-Ala	-His	-Gln	-Leu	-Ala	-Ile -
	0	*	I	*	I	I	I	I	I	I	I	*	I	I	I	*	I	I	I	0	I	I	I	I	*
hGH:	Phe	-Pro	-Thr	-Ile	-Pro	-Leu	-Ser	-Arg	-Leu	-Phe	-Asp	-Asn	-Ala	-Met	-Leu	-Arg	-Ala	-His	-Arg	-Leu	-His	-Gln	-Leu	-Ala	-Phe -
	1									10										20					25
HPL:	Asp	-Thr	-Tyr	-Gln	-Glu	-Phe	-Glu	-Glu	-Thr	-Tyr	-Ile	-Pro	-Lys	-Asp	-Gln	-Lys	-Tyr	-Ser	-Phe	-Leu	-His	-Asp	-Ser	-Glx	-Thr -
	I	I	I	*	I	I	I	I	*	I	I	I	I	*	I	I	I	I	I	I	*	I	*	I	I
hGH:	Asp	-Thr	-Tyr	-Glu	-Glu	-Phe	-Glu	-Glu	-Ala	-Tyr	-Ile	-Pro	-Lys	-Glu	-Gln	-Lys	-Tyr	-Ser	-Phe	-Leu	-Gln	-Asp	-Pro	-Glu	-Thr -
	26									35										45					50
HPL:	Ser	-Phe	-Cys	-Phe	-Ser	-Asx	-Ser	-Thr	-Pro	-Thr	-Pro	-Ser	-Asx	-Met	-Glx	-Glx	-Thr	-Glx	-Lys	-Ser	-Asx	-Leu	-Glx	-Leu	-Leu -
	I	*	I	I	I	*	I	*	I	I	I	I	I	*	I	I	I	I	I	I	I	I	I	I	I
hGH:	Ser	-Leu	-Cys	-Phe	-Ser	-Glu	-Ser	-Ile	-Pro	-Thr	-Pro	-Ser	-Asn	-Arg	-Glu	-Glu	-Thr	-Gln	-Lys	-Ser	-Asn	-Leu	-Gln	-Leu	-Leu -
	51									60										70					75
HPL:	Arg	-Ile	-Ser	-Leu	-Leu	-Leu	-Ile	-Glx	-Ser	-Trp	-Leu	-Glx	-Pro	-Val	-Arg	-Phe	-Leu	-Arg	-Ser	-Met	-Phe	-Ala	-Asx	-Asx	-Leu -
	I	I	I	I	I	I	I	I	I	I	I	I	I	I	*	I	I	I	I	*	I	I	I	*	I
hGH:	Arg	-Ile	-Ser	-Leu	-Leu	-Leu	-Ile	-Gln	-Ser	-Trp	-Leu	-Glu	-Pro	-Val	-Gln	-Phe	-Leu	-Arg	-Ser	-Val	-Phe	-Ala	-Asn	-Ser	-Leu -
	76									85										95					100
HPL:	Val	-Tyr	-Asx	-Thr	-Ser	-Asx	-Asx	-Asx	-Ser	-Tyr	-His	-Leu	-Leu	-Lys	-Asx	-Leu	-Glx	-Glx	-Gly	-Ile	-Glx	-Thr	-Leu	-Met	-Gly -
	I	I	†	*	I	I	*	I	†	I	†	I	I	I	I	I	I	I	I	I	I	I	I	I	I
hGH:	Val	-Tyr	-Gly	-Ala	-Ser	-Asn	-Ser	-Asp	-Val	-Tyr	-Asp	-Leu	-Leu	-Lys	-Asp	-Leu	-Glu	-Glu	-Gly	-Ile	-Glu	-Thr	-Leu	-Met	-Gly -
	101									110										120					125
HPL:	Arg	-Leu	-Glx	-Asx	-Gly	-Ser -		-Arg	-Thr	-Gly	-Glx	-Ile	-Leu	-Lys	-Glx	-Thr	-Tyr	-Ser	-Lys	-Phe	-Asx	-Thr	-Asx	-Ser	-His -
	I	I	I	I	I	I		I	I	I	I	I	*	I	I	I	I	I	I	I	I	I	I	I	I
hGH:	Arg	-Leu	-Glu	-Asp	-Gly	-Ser	-Pro	-Arg	-Thr	-Gly	-Gln	-Ile	-Phe	-Lys	-Gln	-Thr	-Tyr	-Ser	-Lys	-Phe	-Asp	-Thr	-Asn	-Ser	-His -
	126									135										145					150
HPL:	Asx	-His	-Asx	-Ala	-Leu	-Leu	-Lys	-Asx	-Tyr	-Gly	-Leu	-Leu	-Tyr	-Cys	-Phe	-Arg	-Lys	-Asx	-Met	-Asx	-Lys	-Val	-Glx	-Thr	-Phe -
	I	†	I	I	I	I	I	I	I	I	I	I	I	I	I	I	I	I	I	I	I	I	I	I	I
hGH:	Asn	-Asp	-Asp	-Ala	-Leu	-Leu	-Lys	-Asn	-Tyr	-Gly	-Leu	-Leu	-Tyr	-Cys	-Phe	-Arg	-Lys	-Asp	-Met	-Asp	-Lys	-Val	-Glu	-Thr	-Phe -
	151									160										170					175
HPL:	Leu	-Arg	-Met	-Val	-Gln	-Cys	-Arg	-Ser	-Val	-Glu	-Gly	-Ser	-Cys	-Gly	-Phe -OH										
	I	I	*	I	I	I	I	I	I	I	I	I	I	I	I										
hGH:	Leu	-Arg	-Ile	-Val	-Gln	-Cys	-Arg	-Ser	-Val	-Glu	-Gly	-Ser	-Cys	-Gly	-Phe -OH										
	176									185					190										

Fig. 7.34. Structure of human placental lactogen (HPL) and a comparison with the structure of human growth hormone (hGH). Glx and Asx indicate that the presence of glutamic acid versus glutamine and of aspartic acid versus asparagine has not yet been determined. I indicates identical residues in the two hormones. *, Highly favored substitutions; †, acceptable substitutions; 0, unfavored substitutions. (From Sherwood et al., 1971.)

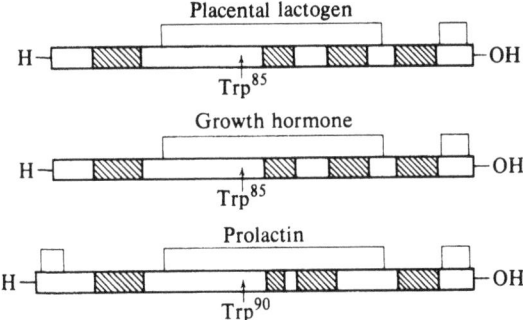

Placental lactogen

H—[ ]—OH

Trp^{85}

Growth hormone

H—[ ]—OH

Trp^{85}

Prolactin

H—[ ]—OH

Trp^{90}

Fig. 7.35. Diagrammatic representation of the structures of placental lactogen and growth hormone from man, and prolactin from sheep. The crosshatched areas represent regions of internal homology in the sequence of the amino acids. Other similarities can be seen in the presence of disulfide bridges (narrow lines) and the tryptophan residues at position 85 in placental lactogen and growth hormone and 90 in prolactin. (From Niall et al., 1971.)

(Weiss, O'Byrne, and Steinetz, 1976). The precise role of relaxin in parturition in women has not, however, been described. It is interesting that its chemical structure has many similarities to that of insulin (Figure 7.36) as well as the somatomedins, although it lacks the biological activity of insulin and its antigenic properties are different.

Theories regarding the initiation of parturition include a simple maternal uterine response to the size of its contents, a decline in progesterone, and/or increased estrogen levels and an increased release of oxytocin from the neurohypophysis. Experiments on sheep have suggested that fetal adrenal corticosteroids may mediate a release of prostaglandins, and this model currently has its devotees. None of these theories, however, adequately accounts for parturition in women. Thus, progesterone does not fall or estrogen rise significantly at this time, and hypophysectomized women, who lack oxytocin, can still give birth in a normal manner. Oxytocin is, however, released during parturition and appears to enhance uterine contractions. It is interesting that this hormone is also released in response to suckling by the infant. An old practice by midwives was to place the baby on the breast as soon as possible after birth. It has been conjectured that the additional release of oxytocin may then aid the expulsion of the placenta and help guard against postpartum hemorrhage.

The human gestation period is normally 259 days (37 weeks) from the first day of the last menstrual period. There is, of course, some variation of this time, but if labor occurs prior to 34 weeks it is termed *preterm labor,* and if it is over 41 weeks it is considered to be a *prolonged pregnancy.*

Both conditions have their dangers, the former especially in relation to the fetus, and the latter can also signify problems for the mother. With special care the fetus is viable after 26 weeks of gestation, the difficulties being due to a lack of maturity of the organ systems, especially a failure of the lungs to adopt their normal respiratory function. The latter is often due to the inadequate secretion of a surfactant which inhibits the frothing of the pulmonary fluids. This problem can be exaggerated in some infants, when it is known as hyaline membrane disease ("respiratory distress syndrome"). It can be treated by administering betamethasone to the mother about 24 to 48 hours prior to delivery [see Section 6.8A(1)]. Prolonged pregnancy suggests possible problems for a normal delivery and has been related, by some, to an increased perinatal mortality ("mature unexplained").

Accelerated and augmented labor. The induction of labor may be medically necessary in a number of situations, especially prolonged pregnancy and hypertension such as that associated with pre-eclamptic toxemia. Labor can be initiated surgically by amniotomy (cutting the membranes) and even without further aid may then progress normally. This does not, however, always occur in a reasonable time, which creates a number of problems, especially the risk of uterine infection. Labor may be accelerated following amniotomy by the infusion of the neurohypophysial peptide hormone oxytocin. An excellent account of this procedure is given by MacVicar (1973).

Oxytocin (see also Section 3.3) was first used to assist labor more than 70 years ago, less than 10 years after the discovery, by Henry Dale, of its uterine-contracting effect. Its use has had a somewhat stormy history, as it has resulted in a number of fetal deaths, mainly due to asphyxia, and to maternal mortality as a result of rupture of the uterus. By the 1930s oxytocin had fallen from grace. The problems associated with its use were mainly related to the inappropriate methods involved: i.m. and s.c. injections and excessively high doses. It has also been administered by nasal spray and by the buccal route. Absorption is somewhat unpredictable from such sites, so that contractions of the uterus are often too strong and too frequent. In 1948, Theobald et al. introduced the method of intravenous infusion, in which the dose of oxytocin could be adequately controlled, and low, more physiological, levels could be maintained in the circulation. Since that time some modifications of the i.v. method have been introduced, but it has become widely used in obstetrics, not only to induce labor but also to assist or augment normal labor. In the United Kingdom and the United States, oxytocin is used in about 25 to 50 percent of deliveries (see Barber, Graber, and Orlando, 1972; Steer et

Relaxin

A-Chain: H-Arg-Met-Thr-Leu-Ser-Glu-Lys-Cys-Cys-Glu-Val-Gly-Cys-Ile-Arg-Lys-Asp-Ile-Ala-Arg-Leu-Cys-OH

B-Chain: <Glu-Ser-Thr-Asn-Asp-Phe-Ile-Lys-Ala-Cys-Gly-Arg-Glu-Leu-Val-Arg-Leu-Trp-Val-Glu-Ile-Cys-Gly-Val-Trp-Ser-OH

Insulin

A-Chain: H-Gly-Ile-Val-Glu-Gln-Cys-Cys-Thr-Ser-Ile-Cys-Ser-Leu-Tyr-Glu-Leu-Glu-Asn-Tyr-Cys-Asn-OH

B-Chain: H-Phe-Val-Asn-Gln-His-Leu-Cys-Gly-Ser-His-Leu-Val-Glu-Ala-Leu-Tyr-Leu-Val-Cys-Gly-Glu-Arg-Gly-Phe-Phe-Tyr-Thr-Pro-Lys-Ala-OH

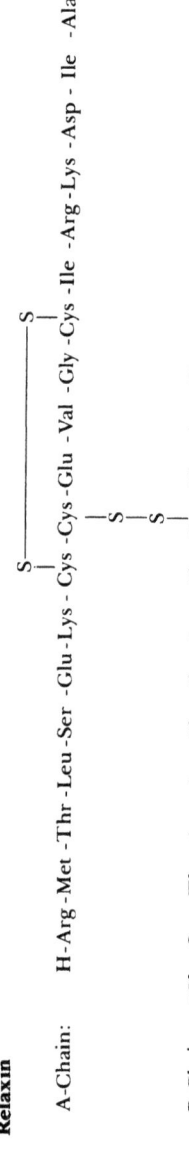

Fig. 7.36. Primary structures of porcine relaxin and insulin with their cysteine residues aligned. In the insulin structure, the underlined residues signify those which are either homologous or conservatively replaced with reference to the respective chains in relaxin. It can be seen that the disposition of the disulfide bridges is the same in both relaxin and insulin. They have also been found to have a similar tertiary conformation (see Bedarkar et al., 1977, and Schwabe and Harmon, 1978). (From Schwabe and McDonald, 1977. Copyright 1977 by the American Association for the Advancement of Science.)

al., 1975; Chalmers et al., 1976). It has also become a quite common practice to utilize these procedures to initiate what may otherwise be a spontaneous delivery – the so-called *"elective" induction of labor*.

The recognized practice for accelerated labor and elective induction is to start an i.v. infusion of oxytocin immediately following amniotomy, not alone (Lilienthal and Ward, 1971), and delivery usually then occurs within 12 hours. It can be given as a "drip" in a dilute solution or in a small volume of fluid controlled by an infusion pump. The dosage is usually "titrated" in relation to the uterine response. This can be done with the aid of devices to monitor the rate of contraction and the intrauterine pressure. A "Cardiff" pump is available which does this automatically in response to the strength of the recorded contractions. The same system is used for the augmentation of spontaneous labor, such as when uterine contractions are considered to be uncoordinated or are inadequate in frequency or pattern. The infusion is tapered off following delivery and terminated about 1 hour postpartum.

There are a number of problems associated with this use of oxytocin, but they can be avoided by appropriate precautions. *Contraindications* have been firmly laid down (see Barber, Grober, and Orlando, 1972) and include malposition; fetal–maternal disproportion; multiparity; damage and scars to the vagina, cervix, and uterus; fibroids; and severe preeclamptic toxemia. Excessive contractions can result in maternal death due to rupture of the uterus. There is an increased incidence of neonatal jaundice (Chalmers, Campbell, and Turnbull, 1975) and postpartum hemorrhage (Brinsden and Clark, 1978). Fetal distress and death can occur following asphyxia due to a reduced blood supply or compression of the head (see, e.g., Liston and Campbell, 1974). These problems are now uncommon due to careful monitoring of the uterine contractions, the fetal heart rate, and the pH of the blood.

Oxytocin is prepared either from animal glands (such as Pitocin) or by chemical synthesis (Syntocinon). Early preparations were contaminated with vasopressin (antidiuretic hormone, ADH), and oxytocin itself has a small, but significant, antidiuretic activity. When given in large doses, this can result in excessive water retention, especially when the administration is by the drip method and substantial volumes of fluid are concurrently administered. Such fluid retention can be a special problem in cases of preeclamptic toxemia. Oxytocin preparations contain 0.5 percent chlorbutanol as a preservative. This compound has a hypnotic effect and causes vasodilatation. If excessive amounts of oxytocin are administered, the chlorbutanol present can have significant toxic effects.

The induction of labor using oxytocin is now a common practice, but it has become a controversial one, especially in the United Kingdom. It is now possible to program delivery to occur on a particular day and even, with some certainty, within a period of several hours. Labor and delivery can thus be an elective procedure which may be used for "social or medical convenience." The controversy arose following a television program called "A time to be born" (see Editorial, 1976b), in which the question arose: "Elected by who, patient or the obstetrician?" Claims that the use of oxytocin to augment or accelerate labor reduces perinatal mortality are controversial or inconclusive (see Barber, Graber, and Orlando, 1972; Cole, Howie, and Macnaughton, 1975; Chalmers et al., 1976; Fedrick and Yudkin, 1976). When used for "elective induction" the mean duration of labor is similar to controls. There are clearly logistic advantages in being able to time delivery. This procedure helps the patient in her domestic arrangements and ensures that delivery occurs at a time of the day when the various hospital services are most readily and efficiently available. It has, however, been reported to result in excessive crowding in certain hospital shifts. A relatively precise timing of the delivery, especially in the daytime, can of course also be convenient for the obstetrician. While the latter possibility appears to annoy some of the general public, the most important question is whether there is any evidence of additional danger to the mother or fetus and whether the mother may find a normal delivery "as satisfying" (see Editorial, 1976b). At present there appears to be no evidence that the child suffers any permanent disadvantages. It was suggested that "our patients are made to feel that they have taken part in the decisions that affect their lives."

While oxytocin is preeminent in its use to assist uterine contractions in labor, other drugs have been used (Figure 7.37). The *ergot alkaloids* [see also Section 7.6G(3)] are potent oxytocic drugs and in the early part of the last century were used, especially in the United States, to assist labor. They have had an illegal use until even more recently to induce abortion, but they are not ideal for this purpose. In large doses the ergot alkaloids can cause sustained contractions of the uterus and have resulted in many fetal and also maternal deaths. They can cause violent vomiting. Their use in obstetric practice to promote labor was thus abandoned a long time ago. Two ergot derivatives, *ergonovine (ergometrine)* and *methylergonovine* (Figure 7.38), are, however, frequently used to control and prevent *postpartum hemorrhage*. These can be given i.m., when they act more rapidly than oxytocin, or i.v. When given in the latter way, they are effective in less than 0.5 minute.

Sparteine (Figure 7.37) has in the past been used

Oxytocic

Cys-Tyr-Ile-Gln-Asn-Cys-Pro-Leu-GlyNH$_2$

Oxytocin

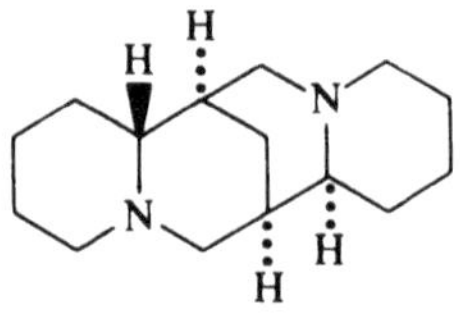

Sparteine

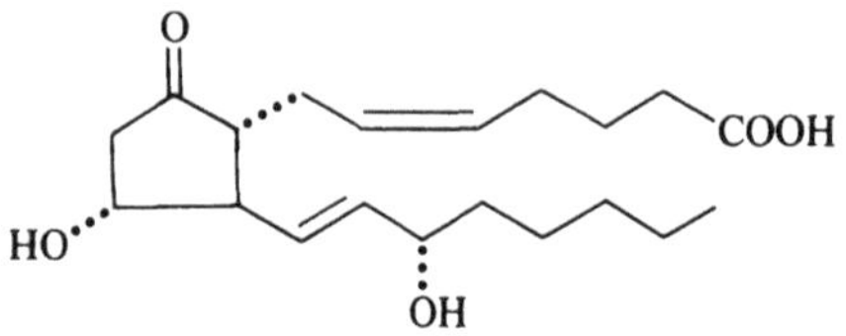

Prostaglandin E$_2$

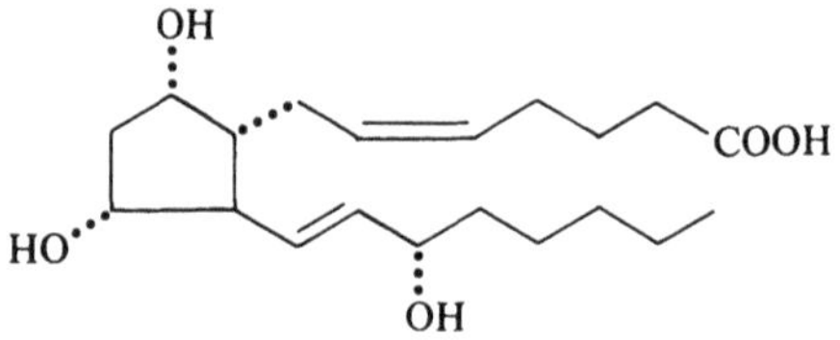

Prostaglandin F$_{20}$

Antioxytocinergic (tocolytic)

1. β_2-Adrenergic drugs

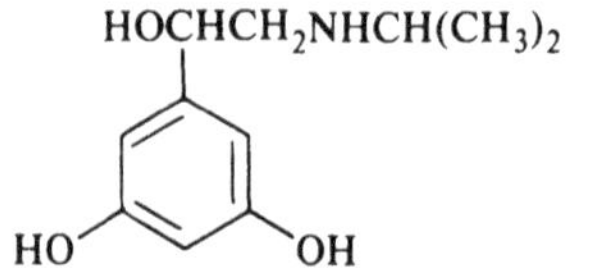

HOCHCH$_2$NHCH(CH$_3$)$_2$

Metaproterenol
(= orciprenaline)

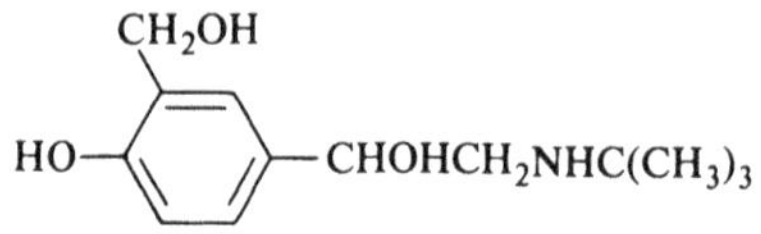

CH$_2$OH

CHOHCH$_2$NHC(CH$_3$)$_3$

Salbutamol

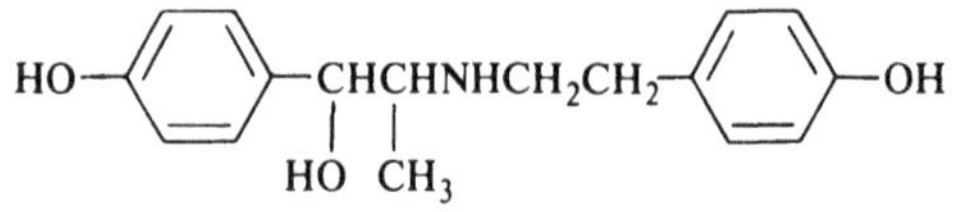

CHCHNHCH$_2$CH$_2$

Ritodrine

CHCHNHCHCH$_2$O

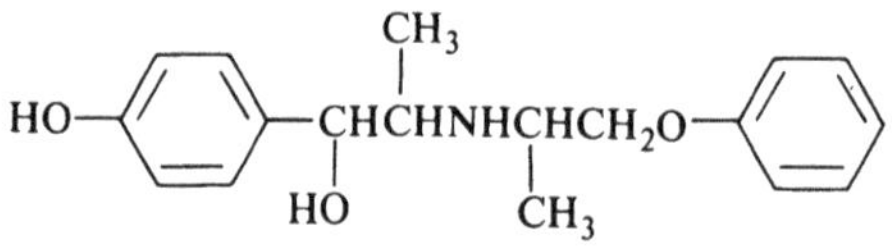

Isoxsuprine
(also α-adrenergic blocker)

2. Peripheral muscle relaxant (direct action)

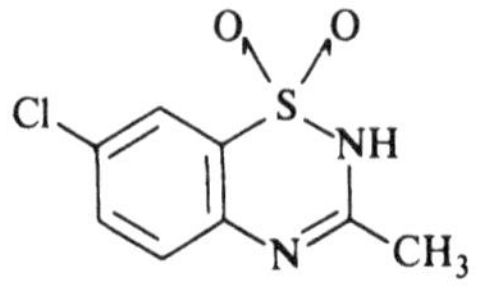

Diazoxide

Fig. 7.37. Oxytocic and antioxytocic drugs that have been used therapeutically to induce or hasten labor or to delay it.

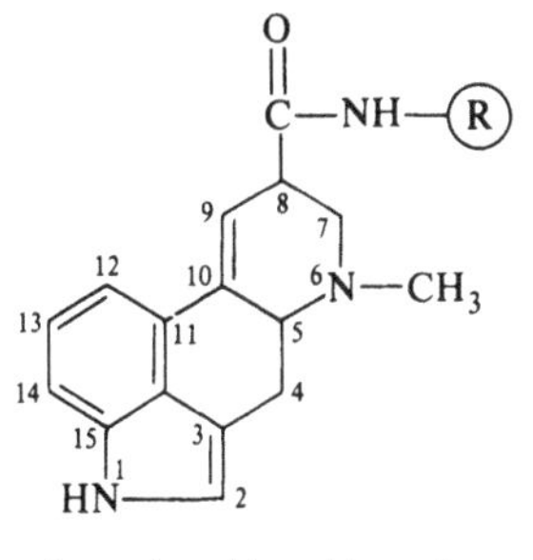

Lysergic acid amide moiety

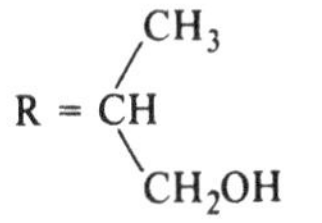

$$R = \overset{CH_3}{\underset{CH_2OH}{CH}}$$

Ergonovine (ergometrine)

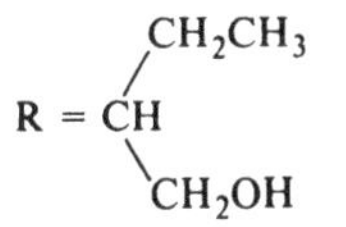

$$R = \overset{CH_2CH_3}{\underset{CH_2OH}{CH}}$$

Methylergonovine
(methylergometrine)

Fig. 7.38. Derivatives of ergot that are used to prevent or control postpartum hemorrhage.

sporadically to enhance contractions of the uterus. About 20 years ago it had a period of popularity but was withdrawn from the market following a series of episodes involving fetal asphyxia and rupture of the uterus. It was given i.m., a procedure that results in problems similar to those seen in the comparable use of oxytocin. It may thus have been unjustifiably villified due to an inappropriate method of administration.

The *prostaglandins* (Figure 7.37) are a family of fatty acids, derived from prostanoic acid, which have a vast range of biological activities and are ubiquitous in their distribution in the body. It is interesting that they were first identified in semen and shown to contract the human uterus. Currently, their possible physiological roles and practical applications as drugs are being investigated in a wide range of medical conditions. They have, however, yet to gain an accepted role in therapy in man, but the use of their ability to contract the uterus has probably come closer to this possibility than any of their other actions. Oxytocin can only contract the "prepared" uterus such as at term in women, although it can be primed with estrogens, and will then respond. Prostaglandins, especially the forms E_2 and $F_{2\alpha}$, ripen the cervix and are oxytocic at any stage of its cycle and may even be involved in the normal process of parturition. Furthermore, preparations that are active by oral, nasal, vaginal, and parenteral administration are available. They do, however, have a wide range of side effects, including nausea, vomiting, diarrhea, and a reduction in blood pressure, which is consistent with their ubiquitous activities in the body. Oral prostaglandin E_2 (Prostin E_2, dinoprostone)

given in incremental doses at regular intervals has also been used for the *induction of labor* (Miller, Welply, and Elstein, 1975). Oxytocin, however, was more successful, but the prostaglandin, because of its greater simplicity of oral administration, was preferred by the nursing staff. While neonatal hyperbilirubinemia is often seen after the induction of labor with oxytocin, it does not seem to occur following the use of prostaglandin E_2 for this purpose (Chew, 1977). Subsequent to its use, no significant decline in fertility has been observed, although there is a slight increase in the incidence of spontaneous abortions (MacKenzie and Hillier, 1977). Intravaginal administration of prostaglandin E_2 has been used for the induction of labor.

Prostaglandins E_2 and $F_{2\alpha}$ have been used *to induce abortion*, especially during the midtrimester, using i.v., oral, intravaginal, and intrauterine (in the extraamniotic space) methods of administration (Karim, 1971; Karim and Sharma, 1971a,b; Miller, Calder, and Macnaughton, 1972).

Prostaglandins have also been used to *induce early abortions* (MacKenzie et al., 1978). In a survey involving 309 women at the John Radcliffe Hospital in Oxford, a total of 229 such pregnancies were successfully terminated. The 16:16 dimethyl prostaglandin E_2 analogue, used as a vaginal suppository, was found to be most effective and resulted in fewest side effects. The procedure is best done at the earliest possible time following the first missed menstrual period, and it then resulted in the highest rate of success. It was considered advisable to remove IUDs before using the drug, to avoid possible pelvic sepsis. The procedure was found to be safe and convenient, and resulted in a minimum of emotional distress. Self-administration was considered to be a feasible possibility.

The use of prostaglandin E_2, in a vaginal suppository, was approved in the United States in 1978 for the induction of abortions after the twelfth week of pregnancy. It can also be used for uterine evacuation in missed abortion or in cases of fetal death up to 28 weeks of gestation. The prostaglandin is contraindicated in acute pelvic inflammatory disease and should be used with care in other infections of the reproductive tract. Prostaglandins have not yet received unequivocal acceptance in obstetrics, especially as alternative proven, more expensive, methods are available. On occasions, however, the uterus is unresponsive to oxytocin and prostaglandins may then emerge preeminent.

Suppression of labor. Premature labor and spontaneous abortion occur with varying frequencies, the causes of which are often not clear. In women over 45 years of age, one pregnancy in three aborts. Fetal wastage can occur for a variety of reasons, and in some women habitual abortion may occur so that they are effectively infertile.

The use of progesterone and estrogen replacement in attempts to prolong and save pregnancies is described in Section 7.6H(7). Their efficacy is in serious doubt, but it is possible that in some instances progesterone replacement may be effective (see, e.g., Csapo, Pohanka, and Kaihola, 1974; Johnson et al., 1975).

Premature labor and abortion are most commonly related to an antepartum hemorrhage, but a variety of factors, such as RH blood group incompatibility, fetal malformations or death, and pre-eclampsia may also be involved. Certain drugs can be used to suppress labor, and their use may be justified in situations where the fetus is too young to survive (26 weeks) or if there are unfavorable circumstances which make it desirable to delay delivery. There are two approaches:

a. To inhibit the release of oxytocin by the i.v. infusion of *ethanol*. This effect is only seen when the blood levels of alcohol are *rising*. It can be effective, but alcohol has other effects and it is uncertain whether its action is solely due to blocking the release of the neurohypophysial hormone.

b. Relaxation of the myometrium can be promoted by the direct actions of several types of drugs. The uterus has a cholinergic and adrenergic nerve supply, but the role of these in influencing contractility is not clear. A contraction or relaxation may occur in response to adrenergic drugs, but this often varies with the condition of the uterus, the stage of pregnancy, and the species. The human uterus near term relaxes in response to adrenergic drugs with β-adrenergic activity. The β_2 subgroup is most desirable, as it has a more selective effect. Thus, these drugs can relax uterine smooth muscle (*tocolytic effects*), as well as the bronchus and the blood vessels in the skeletal muscles, but they have little or no direct effect on the heart. Several of these β_2-sympathomimetic drugs have been used (Figure 7.37), including *ritodrine, salbutamol, metaproterenol, fenoterol,* and *isoxuprine*. The latter also has an α-adrenergic blocking action, so a decline in blood pressure may occur. These drugs are given by i.v. infusion and their side effects include nervousness, tremor, palpitations, and nausea. No one of them has gained universal acceptance, but ritodrine has probably been used most often. Fenoterol is widely used in West Germany. Corticosteroids, such as betamethasone, are often used in conjunction with these drugs in an attempt to hasten the maturity of the surfactant system in the fetal lung [see Section 6.8A(1)]. Fluid retention and pulmonary edema have been reported when using such combinations of drugs (Elliott, Abdulla, and Hayes, 1978). *Diazoxide* has a direct relaxant effect on smooth muscle and is widely used to lower the blood pressure in hypertensive emergencies. It also

relaxes the human uterus (Landesman, Coutinho, and Wilson, 1968) and has been gaining favor for delaying labor. Its main side effects are hypotension, hyperglycemia [see Section 8.5C(1)], and salt and water retention. The anesthetic gas *halothane* also relaxes the uterus, and its use has accidentally resulted in postpartum hemorrhage. *Magnesium* (as magnesium sulfate) is used to reduce uterine contractions. *Inhibitors of prostaglandin synthetase,* such as indomethacin, can delay labor but their clinical safety and effectiveness are still under investigation. The administration of sedatives, especially barbiturates, and morphine has been utilized for delaying labor. It must be remembered that many such drugs can cross the placenta and may have adverse effects on the fetus. Also, they cannot be administered on a chronic basis for extended periods of time; thus they may delay delivery for a few hours to a few days in some cases. One hopes that pregnancy may proceed normally after they have been withdrawn.

The effectiveness of such pharmacological procedures for delaying labor is at present not clear nor agreed upon.

5. Lactation. Before the relatively recent practice of artificially feeding human infants ("bottle feeding"), the secretion of milk by women during the puerperium was just as essential for reproduction as such processes as ovulation and fertilization. The milk of mammals differs considerably in its composition, which is related to the needs of each species, and it was rather serendipitous that cow's milk should support the survival of human infants. Today, the human mammary glands are more medically important in relation to cosmetic effects and their propensity to be the site of tumors than to the survival of the infant. Some, however, may question this statement because of doubts about whether artificial feeding provides an optimal nutritional and social basis for the development of the infant.

The mammary glands are a mammalian prerogative. Their development and function go through a number of phases, which are under the control of several hormones (sometimes referred to as the "lactogenic complex"), including estrogens, progesterone, thyroid hormones, corticosteroids, insulin, possibly parathyroid hormone, growth hormone, prolactin, and oxytocin. The precise needs depend on the particular stage of the development and function of the mammary glands. The experimental animals that have been examined display considerable interspecific differences. The methods used to study the role of the hormones involve such procedures as hypophysectomy, adrenalectomy, ovariectomy (often all in the same animal), and tissue culture. In view of all these observations, and the experimental limitations of women, one needs to be a little circumspect in relating the observations on animals to man.

In the fetus, the mammary glands develop from ectodermal tissue and they appear to have analogues to the various glands of the skin, such as sebaceous glands. Initial development in the human fetus involves the formation of a rather rudimentary system of ducts and the nipples. Subsequently, they grow and differentiate further at puberty. The human mammary glands each consist of 15 to 20 lobules arranged radially around the nipple. They consist of numerous alveoli, which have a secretory function and form the milk. These are small sac-shaped structures lined by epithelial cells containing a prominent Golgi apparatus. The outer surface of the alveoli are covered with a network of contractile myoepithelial cells, which have an important function in promoting the expulsion of the secreted milk from the lumen of the alveolus. The ducts that lead from the alveoli eventually coalesce and expand at their base to form a series of sinuses where the milk can be stored prior to its release. A mass of stromal tissue is normally interspersed between the lobules of the mammary glands, but this tissue regresses and is replaced by expanded alveoli when lactation is imminent.

The role of hormones in the development and function of the mammary glands of the *rat* is summarized in Figure 7.39. As described earlier, there is considerable interspecific variation, but it is clear that the processes are complex and involve several hormones. Women appear to be no exception to this observation, so it is not surprising to observe that disturbances in the function of the human mammary glands can arise from several endocrine foci.

The development and function of the mammary glands occurs in several stages:

a. In the fetus, the formation of the rudimentary duct system occurs in both the female and male. In some species, such as the mouse, sex differences appear late in pregnancy and appear to be related to the presence of sex steroids. This apparently does not occur in man.

b. The mammary glands start to develop during the early stages of puberty in girls. The ducts grow, mainly under the influence of estrogens. Growth hormone and corticosteroids are involved in animals, and it seems likely that they are also important in women. An excessive development of the duct system sometimes also occurs in boys (pubertal gynecomastia), where it may reflect an increased production of estrogens by the testis or their peripheral formation from androgens.

The secretory alveoli develop under the influence of both progesterone and estrogens, but corticosteroids as well as prolactin and growth hormone also may be important.

c. In women, the mammary glands increase in size during pregnancy and there is a considerable proliferation of both the ducts and the alveoli in preparation for lactation. The placental hormones, including human placental lactogen, contribute to these developments. These processes include not only multiplication of the cells but also the differentiation of the alveolar epithelium so that it acquires the complement of enzymes which are necessary to secrete the milk. Milk, apart from water and salts, contains lactose, triglycerides, and proteins, which are principally derived by the synthetic activities of the alveoli. Thus, prolactin has been shown to promote the synthesis of lipoprotein lipase (Zinder et al., 1974), which is concerned with triglyceride synthesis by the mammary gland.

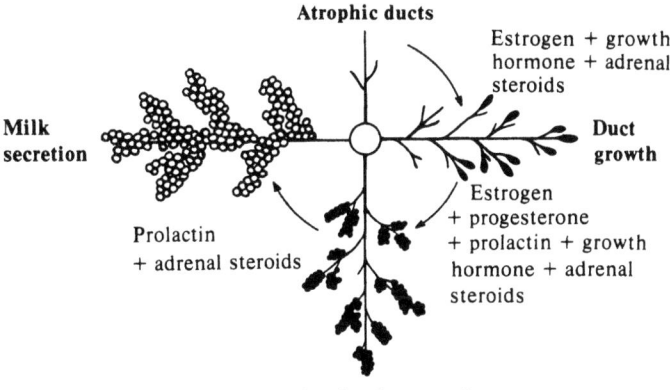

Atrophic ducts

Estrogen + growth hormone + adrenal steroids

Milk secretion

Duct growth

Prolactin + adrenal steroids

Estrogen + progesterone + prolactin + growth hormone + adrenal steroids

Lobulo-alveolar growth

Fig. 7.39. Role of hormones in the growth of the mammary gland and the secretion of milk. In this instance, the example is the laboratory rat, but in other species different combinations of hormones (called the "lactogenic complex") may be required. (Based on Lyons, 1958; from Cowie, 1972.)

d. The secretion of milk or *lactogenesis* occurs in the first few days following parturition. The hormonal events that are concerned with the initiation of this process are not clear in man, but prolactin appears to be important. During pregnancy, its lactogenic effects may be held in check by the high circulating levels of estrogens and progesterone, and the decline in these steroids following parturition may allow prolactin to fully express its actions. Lactation can be maintained for many months following the birth of the young, and this process is called *galactopoiesis*. This process does not depend on estrogens or progesterone but requires prolactin and is usually depressed if the adrenal cortex or thyroid is removed. The maintenance of optimal prolactin levels during lactation involves its periodic release as a result of the suckling stimulus.

e. *Milk ejection* normally occurs as a result of the suckling of the infant on the nipple. Evacuation of the milk is partly a physical process due to the effect of suction, but it mainly involves a neurohumoral reflex (the milk-ejection reflex). As a result of the suckling stimulus, the release of oxytocin, via a neural mechanism in the brain, occurs from the neurohypophysis. This hormone (see Section 3.3) contracts the myoepithelial cells surrounding the secretory alveoli, so that the pressure in the ducts increases and a "draught" of milk flows. Other stimuli, such as copulation, can also initiate the release of oxytocin and the sudden ejection of milk.

6. Endocrine disorders of the mammary glands. These conditions are almost all associated with high plasma levels of prolactin and often amenorrhea and infertility. Such *hyperprolactinemia* (see Frantz and Kleinberg, 1978) may have several general causes:

a. Tumors, which may either be sources of the hormone itself or interfere with the hypothalamic control mechanism, such as may be due to mechanical processes such as compression of the pituitary stalk.
b. A disorder of the hypothalamic regulatory mechanism, which can be primary (or "functional") or the result of the actions of administered drugs.
c. Miscellaneous causes, such as those seen occasionally in hypothyroidism, Addison's disease, and polycystic ovarian disease.

Lactation may follow parturition in a physiologically orderly manner and subside when the infant is weaned. The *syndrome of inappropriate lactation* may, however, occur in a number of circumstances and it is then called *galactorrhea* (see Kleinberg, Noel, and Frantz, 1977). Following parturition and weaning, lactation may continue and this is associated with two types of conditions, both of which appear to be due to hypothalamic disorders. In the

Chiari–Frommel syndrome, normal menstrual cycles are not resumed and amenorrhea accompanies the galactorrhea. This condition is usually associated with a high plasma prolactin level and a decline in gonadotropins. The high prolactin levels may also inhibit the actions of the gonadotropins, so that their correction may result in a return of normal menstrual cycles (Thorner et al., 1974). In other instances, lactation may persist in the presence of normal menstrual cycles and is called *idiopathic galactorrhea,* as the cause is unknown. It also occurs in men.

Pituitary tumors, usually chromophobe adenomas which stain histologically in a characteristic way, can also result in galactorrhea. This is called the *Forbes–Albright syndrome* and is characterized by an excessive secretion of prolactin and low gonadotropin levels. It is not associated with pregnancy and is accompanied by amenorrhea, obesity, and hirsutism.

Hypothyroidism is occasionally associated with galactorrhea and it has been suggested (L'Hermité et al., 1972) that this very rare condition is due to an excessive secretion of TRH, which stimulates the release of prolactin.

A number of *drugs can induce galactorrhea,* especially those that act on areas of the brain associated with the hypothalamus and so stimulate the release of prolactin [Section 7.6G(3)]. Tranquilizers such as the phenothiazines and haloperidol, as well as tricyclic antidepressants, such as imipramine, may nave this effect. Antihypertensive drugs, such as reserpine and α-methyldopa (Vaidya et al., 1970), also occasionally elicit this condition. Galactorrhea can also result from administration of estrogens and progesterone and has occurred due to use of the contraceptive pill or following the withdrawal of this treatment.

Gynecomastia in the male was discussed earlier [Section 7.6I(3)].

The most serious and common disease of the mammary glands is breast cancer. This often fatal disease has reached alarming proportions. Its possible relationship to prolactin and estrogen and the presence of steroid hormone receptors is described elsewhere (Section 7.6K).

Summary of the use·of drugs that influence mammary gland function. Galactorrhea in women and men is probably currently the most valid clinical reason for using drugs that suppress lactation. The ergot derivatives, especially *bromocriptine,* have revolutionized the treatment of galactorrhea (Friesen and Tolis, 1977). They can be taken orally over extended periods of time, apparently without serious adverse side effects. Vertigo, nausea, vomiting, constipation, headache, and postural hypotension can occur. Their action appears to be specifically related to the inhibition of release of prolactin and, apart from growth hormone, they do not generally influence other adenohypophysial hormones. The onset of action of *lergotrile mesylate* is shorter, but its duration of action is more brief than that of bromocriptine and the incidence of side effects may be greater (Thorner et al., 1978). In addition, in animal studies (see Kleinberg, Noel, and Frantz, 1977) its use has been associated with the occurrence of small tumors in the reproductive tract, and the purveyor (Eli Lilly) initially limited its use to the treatment of acromegaly, Parkinson's disease, and breast cancer. A high incidence of liver damage in humans has since led to its withdrawal.

The *suppression of normal puerperal lactation* has, in the past, utilized various regimens, including drugs. *Estrogens* have been used successfully (although they fail in 10 to 20 percent of cases), but they have widespread effects in the body, especially breakthrough bleeding due to their effects on the uterus, and a "rebound" lactation is common. Other more cogent risks include various thromboses. They are not now recommended for the routine suppression of lactation (Editorial, 1977c). If a drug is required, bromocriptine is the choice (Walker et al., 1975; MacLeod et al., 1977; Rolland and Schellekens, 1978; Editorial, 1977c).

Levodopa is widely used in the treatment of parkinsonism. However, its action in inhibiting the release of prolactin is brief and it must be used in large doses, which result in side effects that are unacceptable in the treatment of galactorrhea.

The possible use of antiprolactin drugs for the treatment of breast cancer has been considered, but at present their efficacy has not been proved in man (see Smithline, Sherman, and Kolodny, 1975).

Attempts *to promote lactation* in women by the use of prolactin have not been successful, probably partly as a result of the necessity to use animal hormones, which promote the formation of antibodies. Oxytocin has been administered by the nasal route as a spray in order to facilitate milk ejection and relieve breast engorgement (Thornton, 1961), but this is not a common practice. Thyroid hormones have been used to promote lactation in cows.

Local applications of estrogen creams have been recommended for *cosmetic reasons* in attempts to increase the size of the breasts. They produce a proliferation of the ducts but usually without growth of the stroma, so the overall effect is often not dramatic. A bizarre side effect was the report of the development of gynecomastia in a lady's husband, as a result of thoracic contact. The rumor is that he strongly advocated his wife's treatment, but the twist of fate is something that he could scarcely have imagined.

7. Disorders of the menstrual cycle

Amenorrhea. This condition refers to a failure of the occurrence of the normal menstrual cycle. It may be *primary* or *secondary*, depending on whether the menarche has in fact taken place or if the condition arises following years of normal menstrual cycles. Secondary amenorrhea occurs normally in pregnancy, in the puerperal period, and often during lactation.

Primary amenorrhea is usually diagnosed if menstruation has not occurred by the nineteenth year. The menarche normally takes place in about the fourteenth year, and it is said to be *delayed* if it occurs between 16 and 18 years. There are a variety of causes of primary amenorrhea, such as inadequate nutrition, debilitating diseases, chromosomal disorders (Turner's syndrome), psychological factors, and certain endocrine diseases. The disorder is usually specifically related to the pituitary–gonadal axis and may involve the activity of the hypothalamus, pituitary, or ovaries. A more direct action on the uterus, such as the effects of tuberculosis, can also occur. Extragenital endocrine disorders involving the adrenal cortex (adrenogenital syndrome, Cushing's disease), the thyroid (hypo- or hyperthyroidism), and the pancreas (diabetes mellitus) can also result in primary amenorrhea.

Secondary amenorrhea is usually defined as a failure to menstruate for a period of at least several months (a year is often taken as the arbitrary time) provided that pregnancy can be excluded and lactation is not occurring. Like primary amenorrhea, it can have various causes, which influence the activity of the hypothalamus, pituitary, or gonads. It may result from a general illness or nutritional disturbances, including abnormal weight, psychological problems, and endocrine disorders. It commonly accompanies hyperprolactinemia. Secondary amenorrhea may also be *drug-induced* as a result of the use of tranquilizers such as the phenothiazines and hypotensive drugs, such as reserpine and α-methyldopa, which lower gonadotropin secretion and enhance the release of prolactin.

Hyperprolactinemia is often associated with amenorrhea and infertility (see Bergh, Nillius, and Wide, 1978). It may or may not occur in conjunction with galactorrhea (Seppälä et al., 1975). High prolactin levels have also been observed in infertile women who may (Lenton, Sobowale, and Cooke, 1977) or may not (Pepperell et al., 1977) ovulate. The possible causes of the high rate of secretion of prolactin were described in Section 7.6G(3). The reasons for the effects of the prolactin is uncertain, but it may involve an inhibitory action on the production of progesterone by the corpus luteum (McNatty, Sawers, and McNeilly, 1974). Serum estrogen levels are also low (Bergh, Nillius, and Wide, 1978). In male rats, there is a decline in plasma FSH and LH which appears to reflect an increased sensitivity of the hypothalamic and/or pituitary feedback mechanism to gonadal steroids (McNeilly et al., 1978).

Ovarian disorders that result in amenorrhea include cystic ovaries, where ovulation fails and there is a prolonged persistent secretion of estrogens; hypoplastic ovaries, where hormone secretion is irreversibly too low; and various degrees of deficiency in ovarian hormone secretion. The latter may occur postpartum when normal menstrual cycles fail to be restored. Amenorrhea also occurs in the Stein–Leventhal syndrome, where there are polycystic ovaries and an associated high secretion of androgens.

Gonadotropin secretion may be inadequate due to pituitary or hypothalamic disorders. These problems can be due to tumors and necrotic degeneration. It is suspected that emotional disturbances can influence the activity of the hypothalamus. Local effects involving the *uterus* may also be involved in the genesis of secondary amenorrhea.

Hormone preparations play a role in both the *diagnosis* and *treatment* of amenorrhea. The adequacy of ovarian function can be assessed by the *gestagen test*. An oral progesterone preparation is administered for 5 days and then withdrawn. If there has previously been a normal proliferative buildup of the endometrium under the influence of ovarian estrogen, withdrawal bleeding should occur 4 to 5 days later. Such a positive result suggests that there is a failure of both ovulation and growth of the corpus luteum. However, there is presumably some secretion of estrogen, so that, at least, the hypothalamus and pituitary are intact. The failure is probably then due to an inadequate hypothalamic feedback mechanism. If the test is negative and bleeding does not occur, a hypoplasia of the ovaries may be present or the uterus may be unresponsive to the hormones. In this instance, an *estrogen test* is performed. An oral preparation, such as ethinyl estradiol, is given for a couple of weeks and then withdrawn. (This treatment is sometimes combined with use of a progestin.) If bleeding occurs within 7 or 8 days, it can be assumed (when the gestagen test is negative) that the endometrium is responsive, but there is a deficiency of estrogens. No response indicates that the uterus is unresponsive. A deficiency in estrogen may be caused by a primary ovarian deficiency or be secondary to inadequate stimulation by the hypothalamus or pituitary. High circulating gonadotropins suggest ovarian failure, while low levels can be due to deficiency of the hypothalamus or pituitary. A deficiency in hypothalamic and pituitary function can then be tested for by the use of synthetic preparations of LHRH (Besser et al., 1972; Newton and Collins, 1972). This hypothalamic hormone is given i.v. and changes in LH, FSH, and estradiol-17β can be measured. A rise in the gonadotropin levels suggests that the disorder is either hypothalamic or pituitary in origin, while a failure to respond indicates that the pituitary is involved (provided that it has been ascertained that there is ovarian function). The antiestrogen clomiphene also normally stimulates release of pituitary gonadotropins and can be used to further identify the site of the problem (Ginsburg et al., 1975). Clomiphene is effective in hypothalamic feedback failure but not in primary pituitary or hypothalamic disorders. A flow sheet summarizing the nature of amenorrheic disorders is given in Figure 7.40. To this could be added, where available, measurements of plasma prolactin concentrations.

Endocrine treatment of amenorrhea must be appropriate to the site and nature of the disorder. If the cause is extragenital, attempts should be made to correct the problem. In cases of ovarian or hypothalamic pituitary disorders, treatment is mainly indicated to correct the accompanying infertility, if children are desired (see later this section). However, in some instances the problem can still need treatment because of a feeling of inadequate femininity. Sequential administration of estrogens and progesterone, such as an oral contraceptive pill, can be used to simulate a normal menstrual cycle. Such treatment is, however, only of a "cosmetic" nature and will not restore fertility. Indeed, if pregnancy is likely to be desired in the future, such a course of action should be discouraged, as suppression of the hypothalamic-pituitary axis will occur. If development of the primary and secondary sex characters is inadequate, substitution therapy with estrogens can be undertaken.

If the disorder is related to an excessive secretion of prolactin, bromocriptine can be administered (Seppälä, Hirvonen, and Ranta, 1976; Pepperell et al., 1977; Friesen and Tolis, 1977; Bergh, Nillius, and Wide, 1978). It has even been found to be effective in instances of secondary amenorrhea which are not associated with hyperprolactinemia. It is important to determine the cause of the high prolactin levels, because if this is secondary to such conditions as tumors, drugs, or hypothyroidism, it is more rational to attempt to treat the primary disorder. Menstruation usually returned within 8 weeks of commencing treatment, but, unfortunately, in most instances, this response did not persist for long following the withdrawal of the drug. Such a drug-induced remission may, however, be adequate in duration to result in a pregnancy. When the latter occurs, the drug is immediately withdrawn. Bromocriptine is expensive and its

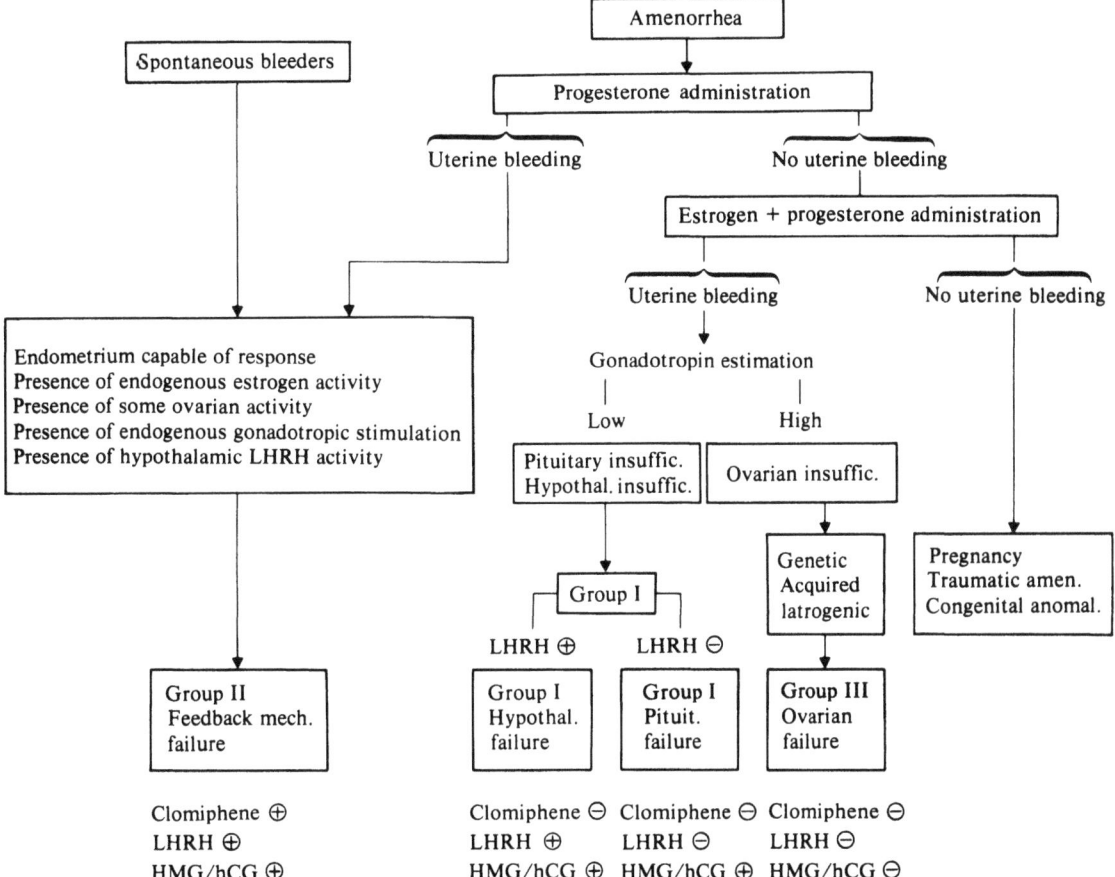

Fig. 7.40 "Flowsheet" of responses to pharmacological tests used to classify and diagnose the reasons for amenorrheic states in women. ⊕, ⊖, positive or negative responses to a designated stimulation to an administered hormone preparation or drug (e.g., LHRH⊕ = positive response to LHRH). Group I: pituitary–hypothalamic failure, reflected in low gonadotropin levels and low estrogen levels (< 10 μg/day). The latter is due to a failure of ovarian stimulation. Group II: anovulatory women who have normal gonadotropin; urinary estrogen excretion on the low side (but >10 μg/day) (feedback mechanism failure). Group III: very low estrogen levels in the urine due to ovarian failure, with extremely high gonadotropin levels due to lack of negative-feedback inhibition (ovarian failure). HMG, human menopausal gonadotropin; hCG, human chorionic gonadotropin. (From Lunenfeld and Insler, 1974.)

long-term use just to assure regular menstruation, rather than a pregnancy, is questionable.

Amenorrhea itself is not harmful, but it signifies a number of disorders, including sterility. The treatment of this latter condition will be described later (see later this section).

Uterine bleeding. Uterine bleeding that is not associated with the normal menstrual cycle can occur in a variety of circumstances, many of which are related to disturbances of ovarian function. Other causes of abnormal bleeding that may occur from various parts of the reproductive tract include the presence of tumors, lesions, inflammation, spontaneous or incomplete abortion, or an ectopic pregnancy. The problem may be cyclical, acyclical,

midcycle, or excessive menstrual bleeding (menorrhagia). It is frequently acyclical but can be a recurrent problem (metrorrhagia). The length of the menstrual cycle may be irregular and if it is between 35 to 90 days, it is usually classified as oligomenorrhea; or if less than 21 days, polymenorrhea. Such bleeding disorders can occur at any time of life, starting in the newborn. They are, however, more common during puberty and in the period of the menopause.

The usual basic problem is due to a persistent and prolonged action of estrogens on the uterine endometrium. The primary problem can, however, arise in any part of the hypothalamic-pituitary–ovarian axis. Under such conditions the normal proliferative phase of the menstrual cycle persists and there is no conversion, under the influence of progesterone to the secretory phase. The latter mediates the normal differentiation of the endometrial glands and the spiral arteries of the vasculature, so that normal menstrual degeneration and bleeding occurs. In the presence of adequate estrogens, the proliferative phase of the endometrium can be maintained for extended periods of time and it becomes highly developed and may result in glandular cystic hyperplasia. If, however, the levels of the hormone decline for any reason or they become insufficient to maintain the tissue, *"breakthrough" bleeding* will occur. The estrogen levels may thus be variable or steady. This bleeding differs from normal menstrual bleeding, as the blood is coagulable and it involves increases in capillary permeability, rather than breakage and contraction of the endometrial spiral arteries. It is not analogous to a normal "period" which follows progestational stimuli to the uterus.

Bleeding sometimes occurs from the reproductive tract of newborn infant girls. This is thought to be due to a relatively high level of estrogens in the circulation, which fall following birth, thus precipitating a type of withdrawal bleeding. A failure of ovulation in women results in the lack of development of a corpus luteum, and this can produce irregular and severe uterine bleeding. Disorders in hypothalamic-pituitary function may result in inadequate development of the follicles and a failure to ovulate. The latter may occur as a result of the absence of the midcycle "surge" in the release of LH. Polycystic ovaries may thus occur, and this has been observed in 1 to 2 percent of women (during laparotomy). The follicles persist and continually secrete estrogen. In a rare type of this condition, the *Stein–Leventhal syndrome*, there is an excessive secretion of androgens from the ovary, apparently as a result of a lack or aromatizing enzymes. This condition is thus usually associated with hirsutism. Amenorrhea is common, but dysfunctional bleeding can also occur and be severe.

Endocrine preparations are used in the *treatment* of abnormal uterine bleeding. A correct diagnosis is, however, initially important so as to exclude nonendocrine causes of the condition. Curettage and other surgical procedures are usually used to aid diagnosis and in attempts to correct the condition. Endocrine therapy, however, has a distinct role, especially in juvenile bleeding, if there is a future desire to have children and if curettage does not correct the condition. A rational basis for the use of endocrine preparations can be advocated on the known bases of the disorders. However, not all of these are really practical. For instance, one can attempt to restore a normal rhythm by the administration of gonadotropins. Such treatment, however, requires continual careful monitoring to assess the correct dose and avoid hyperstimulation of the ovaries, and is inconvenient, as it involves parenteral administration. It is thus expensive and time-consuming and is only really justified in order to restore fertility (see later this section). The antiestrogen clomiphene has also been used to promote the release of gonadotropins. Substitution with ovarian steroids is the most practical method of drug treatment for such bleeding disorders. The use of estrogens alone to maintain the proliferative endometrium has been advocated but may only ultimately increase the problem. Progesterone can transform the endometrium into its secretory "postovulatory" phase, and upon withdrawal will precipitate a more normal controlled menstrual type of bleeding. Currently, the use of combinations of estrogens and progesterone, such as the "pill," is generally favored. The dosages are usually higher than those used for contraceptive purposes. They are given daily for 24 days, and an orderly withdrawal bleeding usually follows in 48 to 72 hours. This cycle of treatment can be repeated on two or three occasions and then is usually stopped to see if normal cycles recur. Excessive menstrual bleeding can also be controlled by such estrogen-progesterone preparations, but they are usually only administered for a few days. Relatively high initial doses are used, and these usually stop the bleeding within 24 to 36 hours. Withdrawal bleeding follows cessation of the use of the drugs.

Dysmenorrhea. This condition refers to painful menstruation due to cramplike pains in the lower abdomen (see Bender, 1977). It can be quite severe and debilitating. Dysmenorrhea occurs just before or at the beginning of menstruation and usually lasts a few hours or, at the most, 2 days. Primary dysmenorrhea usually arises soon after the menarche (teenage dysmenorrhea) and often disappears following the birth of the first child. A "secondary" form may arise later in life and be due to mechanical problems such as fibroids and endometriosis. The cause of primary dysmenorrhea is unknown, but it may be associated with an excessive contractility of the myometrium, which possibly reduces the blood supply to the uterus. It has been suggested that prostaglandins may be involved, which could contribute to the effectiveness of *mefenamic acid, ibuprofen,* and *indomethacin* (which inhibit prostaglandin synthetase) in relieving the pain (Editorial, 1975f). It is suspected that an imbalance of hormones associated with an ovulatory cycle may occur, but its precise nature has not been specified. Analgesic and spasmolytic drugs, or even a placebo, are quite effective. The suppression of ovulation by the use of a *contraceptive pill* nearly always results in a painless "period" and is used in severe cases of the disorder. Sometimes it is thought advisable that a teenager or her mother should not be informed about the nature of the medication, so the pharmacist is asked to remove the labelling and packing. This treatment with oral contraceptives can have other problems, especially if the woman wishes to become pregnant. Secondary dysmenorrhea requires closer investigation.

Premenstrual tension syndrome. The changes in mood, irritability, anxiety, headaches, breast discomfort, and the sensations of edema that may commence 7 to 14 days prior to menstruation are well known. They may occur in about 40 percent of women and can have serious consequences on family life. Their causes are uncertain and could be multiple. In 1931, R. T. Frank suggested that this premenstrual tension syndrome was caused by hormonal imbalance during the luteal phase of the menstrual cycle. This explanation has been accepted by many but information about its precise nature has been elusive. Modern methods for the determination of hormone levels in the plasma have, however, allowed a classification of the sufferers into three groups (Kerr, 1977):

a. Those with notably low progesterone levels during the luteal phase of the cycle (Taylor, 1977). Fifty such women were treated with the progestin *dydrogesterone.* (This drug is active when administered orally, which is a considerable advantage over other available progestins that often must be injected or given as rectal or vaginal suppositories.) A marked improvement was reported in about 70 percent of the women in this group. Most symptoms were reduced, though not always abolished, an exception being breast discomfort. No side effects were apparent.

b. A small number of women having observably high levels of plasma prolactin. This condition offers the prospect of specific treatment with *bromocriptine,* which blocks the release of this hormone. This dopaminergic drug has, however, been tested less specifically and, in one clinical trial involving 40 women, it was found that the administration of bromocriptine had favorable effects on the mental symptoms (Andersch et al., 1978). However, it was concluded that the response reflected a direct central dopaminergic response rather than an action on the release of prolactin. Another study using a lower dose of bromocriptine to avoid common side effects such as nausea, diarrhea, and vertigo failed to find a favorable response (Ghose and Coppen, 1977).

c. The vast majority of women with no currently detectable abnormalities in hormone levels. More arbitrary therapy may thus be justified. Treatment of premenstrual tension syndrome has, and does, include psychotherapy, mild sedation, tranquilizers, placebos, and diuretics. An interesting more rational approach has involved the administration of *pyridoxine (vitamin B_6).* As described later, the use of oral contraceptives is often associated with emotional and physiological changes which are similar to those seen in the premenstrual syndrome. It was found that these may sometimes be overcome by the administration of pyridoxine, reflecting a disturbance in tryptophan metabolism and possibly reduced levels of 5-hydroxytryptamine levels in the brain. These observations led Kerr (1977) to test the effects of pyridoxine in 70 women who had no notable abnormalities in their plasma estrogen, progesterone, or prolactin levels. Substantial relief of many of their symptoms was reported by 50–60 percent of these women. No side effects were observed.

At present there is no universally effective or accepted treatment for premenstrual tension syndrome but advances in the ability to classify the condition in each patient offer considerable hope for the future.

Female infertility. The general problem of infertility and the contribution of the male to this disorder has been described earlier [Section 7.6I(5)]. Infertility is sometimes classified into two categories: (a) *absolute,* when pregnancy has never occurred, also referred to as *primary sterility;* and (b) *relative,* when pregnancy has formerly taken place, either unsuccessfully or successfully, but the woman has not subsequently been able to become pregnant. Sometimes a fetus may develop but does not attain the stage where it is viable. If this occurs on several occasions, it is called *habitual* or *recurrent abortion.* Endocrine therapy is often successful in correcting such disorders, although in instances of habitual or even threatened abortion, the administration of hormones is now generally unacceptable.

Women are most fertile between the ages of about 15 and 25 years; there is a modest decline up to 35 years which subsequently becomes more rapid. Infertility has a number of causes:

a. *Anatomical problems and lesions* associated with the vagina, cervix, uterus, and oviducts. These may be gross malformations or disturbances in the lining of the endometrium or vagina which create an inhospitable environment for transport of the sperm, ovum, or blastocyst. Such disorders may be congenital or the result of disease, such as tuberculosis, or physical trauma due to damage during curettage or the performance of an abortion. The most common such disorder, which accounts for about 30 percent of all cases of infertility, is occlusion of the oviducts.

b. The *physiological environment of the reproductive tract* may not be consistent with adequate survival of the sperm. This can result from vaginitis, when the presence of leucocytes damages the sperm and results in secretion of a cervical mucus which is inadequate in quantity or quality to assist the passage of the sperm toward the oviduct. Estrogens, such as ethinyl estradiol, are sometimes administered for several days prior to ovulation in an effort to improve the secretion of cervical mucus. Immunological or allergic problems occasionally arise when antibodies to the sperm are formed which incapacitate them.

c. *Psychological factors* may contribute to a failure to court the presence of sperm frequently enough or on the most appropriate occasions. They can also result in vaginismus. The release of gonadotropins can be influenced by the activity of the central nervous system, and it is thought that this type of effect may sometimes contribute to sterility. However, precise information about this possibility is lacking. Psychogenic spasm of the oviduct or an adverse pattern of uterine contractility have also been considered likely.

d. *Severe undernutrition,* such as seen in anorexia nervosa (see Boyar et al., 1974), frequently results in menstrual disorders and sterility. These effects appear to be mediated via the hypothalamus and the central nervous system.

e. *Specific endocrine disorders* can result in sterility. These may be direct, on the hypothalamic–pituitary–ovarian axis, including hyperprolactinemia, or be the indirect result of other hormonal problems. The latter include adrenocortical disorders, such as the adrenogenital syndrome, Cushing's disease, and Addison's disease, as well as diabetes mellitus and hypothyroidism. Thyroid hormones have in the past been administered to women in attempts to improve fertility, but without notable success. Thyroid preparations should only be used in specific instances of hypothyroidism.

Deficient function of the corpus luteum has been considered a possible cause of sterility, resulting in a truncation of the secretory phase of the menstrual cycle so that an inadequate environment for implantation occurs. In addition, if the corpus luteum fails to function properly early in pregnancy, the embryo may fail to survive. The importance of this disorder is, however, controversial. Progestational steroids have thus been administered following ovulation on days 15 to 25 of the cycle in an attempt to improve the uterine environment. They have also been used in cases of habitual or threatened abortion, where it has been speculated that the problem is due to insufficient production of progesterone. Normally, the placenta produces sufficient progesterone to maintain pregnancy, and indeed women can be ovariectomized after about the 6 or 7 weeks of pregnancy without loss of the fetus. The administration of progestins, which also have androgenic activity, can result in a virilization of female children, especially if used in the early stages of pregnancy (Wilkins, 1960). Furthermore, clinical trials (see Todays

Drugs, 1972a) have failed to demonstrate a favorable effect of the administered progestins in preventing habitual or threatened abortions. The possibility that such treatment has teratogenic actions has been raised (Editorial, 1974b,c). It also appears that they may have long-term effects on personality and temperament (Reinisch, 1977).

In the past, estrogens have also been in vogue, especially in the United States but also in other parts of the world, including the United Kingdom, for the treatment of habitual and threatened abortion. The rationale for their use was never clear but presumably rested on the unproved assumption that a substantial relative deficiency of these steroids existed. The efficacy of this treatment was never demonstrated. It has, however, been conclusively related to the occurrence of clear-cell adenocarcinoma of the cervix and vagina in the teenage daughters of mothers who were treated with diethylstilbestrol (DES) (Herbst et al., 1972, 1975).

Failure to ovulate is the principal endocrine-related cause of infertility. As a result of recent advances in knowledge of the hormonal control of this process, rational therapy is available. Moreover, it is often successful, but can be time-consuming, expensive, and even dangerous. Therefore, before embarking on such a program it is important to eliminate other possible causes of infertility, including that of the male partner and anatomical problems, especially occlusion of the oviducts. The precise locus of the endocrine disorder should then be isolated, as this will determine the nature of the therapy that is likely to be the most successful. Anovulation is, of course, related to amenorrhea but can also occur in apparently normal menstrual cycles. The incidence of the latter increases with age. The classification and causes of amenorrhea have been described earlier in this section and are summarized in Figure 7.40. Such a series of endocrine diagnostic tests (see beginning of this section), including the progesterone test, estrogen test, or "estrogen + progesterone test," combined with measurements of endogenous levels of gonadotropins and estrogens, serves to localize the cause of the disorder. It can be primarily due to dysfunction of the uterus, ovary, or the hypothalamus–pituitary axis. The integrity of the adenohypophysis can now be tested by the administration of synthetic LHRH (Besser et al., 1972) or its more active analogues (such as [D-Leu[6],des-Gly[10]]LHRH ethylamide or [D-Leu[6]]LHRH ethylamide). If the hypothalamus has a lesion that makes it nonfunctional, then clomiphene, an antiestrogen, will fail to elicit a pituitary response. If the cause of the infertility is related to primary failure at the level of the reproductive tract or the ovary, ovulatory therapy will not be effective.

Clomiphene. Most commonly the uterus, ovaries, pituitary, and hypothalamus are capable of functioning. However, there is a lack of an adequate and properly timed release of the gonadotropins, so an orderly successive development of the follicles, ovulation, and luteinization does not occur. This disorder is thought to be due to failure or overactivity of the feedback mechanism that controls the release of hypothalamic LHRH. Some antiestrogens can unmask this mechanism and promote the release of gonadotropins. The drug most widely used (since 1961) to promote ovulation is clomiphene citrate. It can be administered orally and when given to an appropriate group of women can induce ovulation in about 70 percent of cases and pregnancy in about half of these (see Lunenfeld and Insler, 1974). In the view of some, this rate of success may be a little optimistic. Clomiphene is usually given in doses of 50 to 100 mg/day for 5 days, starting, in women who have a menstrual cycle, on the fifth day. (This drug is, however, also effective in women experiencing amenorrhea.) Sometimes it is given earlier, starting with the first day of the cycle. If it is ineffective, the dose can be gradually increased up to 250 mg/day. This cyclical treatment is repeated about six times before it is deemed to be unsuccessful.

The possible *side effects* of clomiphene, especially in doses over 100 mg/day, include the risk of hyperstimulation of the ovaries and the formation of ovarian cysts. Hot flushes occur and, rarely, loss of hair, galactorrhea, and blurring of vision. Multiple pregnancies occur in about 6 percent of cases and about 20 percent of the fetuses are lost. The infants are normal in weight and do not have a different incidence of birth defects.

Gonadotropins. If clomiphene fails to induce ovulation, or if it is known that a hypothalamic or pituitary deficiency exists, one can utilize therapy with gonadotropins to induce ovulation. This procedure is an old and well-tried one; the first induction of pregnancy using human pituitary extract was described by Gemzell, Diczfalusy, and Tillinger in 1958. Since that time, quite a few technical refinements have been developed, and the procedure, although still a specialized one, is almost routine. It can, nevertheless, be hazardous and as a result of uncontrolled hyperstimulation of the ovaries has even resulted in death.

Animal preparations of gonadotropins are ineffective, while the human pituitary gland hormone, which must be obtained from cadavers, is not readily available. Two useful human preparations are, however, extracted from the urine of women. *Human menopausal gonadotropin* (menotropin, HMG) which is a mixture of both FSH and LH, is excreted in large amounts by postmenopausal women. The commercial preparation (Pergonal) contains 75 IU of FSH and 75 IU of LH in each ampule. Successive i.m. doses of this preparation are used to stimulate the development of the follicles, and if ovulation does not occur, this may be induced, by an appropriately timed dose of *human chorionic gonadotropin* (hCG, from the urine of pregnant women). This preparation has a predominance of LH activity and so mimics the preovulatory "surge" in the release of pituitary LH. Various dosages and schedules are advocated (see Todays Drugs, 1972a; Lunenfeld and Insler, 1974). Provided that the ovary is functional and steroids are secreted in response to the injected gonadotropin, ovulation is relatively frequent. The estimates vary, but indicate that about 30 to 50 percent of women have a successful pregnancy. The infants are normal, but there is a pregnancy wastage of about 20 percent, and multiple ovulations, which may result in abortion or multiple births, are common (about 20 percent).

The *side effects* of treatment of gonadotropins can be quite dangerous and are mainly the result of hyperstimulation of the ovary and an excessive production of estrogens. It is best if the plasma levels of these steroids can be monitored during treatment, and they should not be allowed to exceed more than 150 μg/day. Mild hyperstimulation can be detected by careful palpation of the ovaries. If signs of hyperstimulation occur, therapy is stopped until the ovaries return to their normal size. The condition of ovarian hyperstimulation results in abdominal tenderness and pain, progressing to nausea and vomiting, accumulation of fluid in the peritoneal (ascites) and pleural cavities, and contraction of the plasma volume. If an ovarian cyst bursts, it may result in serious intraabdominal hemorrhage requiring laparotomy. This condition can be very serious and has resulted in deaths, but it can be avoided by careful monitoring to avoid overdosage with the gonadotropins. The "vital statistics" of gonadotropin therapy are summarized in Table 7.10.

LHRH (LH/FSH-RH, Gn-RH). The early hopes that the hypothalamic gonadotropin-releasing hormone, LHRH, may provide a superior method to the use of clomiphene or the HMG/hCG regimen have not been fulfilled (see Lunenfeld and Insler, 1974). This decapeptide hormone has been synthesized and a number of analogues with more prolonged and stronger activity are readily available [see Section 3.2E(2)]. They are usually given parenterally, i.v., i.m., sometimes s.c., or even as a nasal spray. Early experiments (Keller, 1972; Akande et al., 1972, Besser et al., 1972) on women showed that they could stimulate the release of LH (Figure 7.41) and, to a somewhat lesser extent,

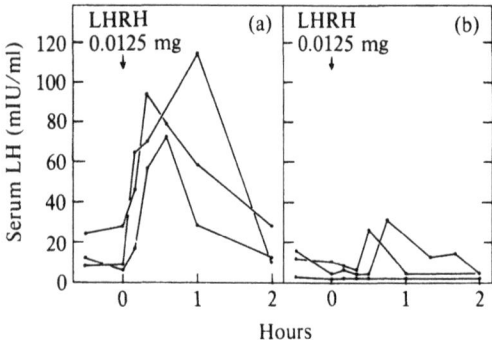

Fig. 7.41. Effects of injected LHRH into women suffering from amenorrhea. (a) Good response, suggesting that pituitary function is adequate and that the fault *may* lie at the level of the hypothalamus. (b) Poor response, suggesting that pituitary function may be compromised. (From Keller, 1972.)

FSH. They do not, however, induce ovulation unless the follicle is in a suitably prepared condition, such as following preliminary treatment with clomiphene. A clinical trial on the use of LHRH in women who were made anovulatory by the administration of a contraceptive preparation was not uniformly successful (Zañartu et al., 1974). Successive doses of LHRH were given once or twice a day for periods ranging from 7 to 20 days and resulted in some ovarian response and follicular development, but ovulation occurred in only 2 of 16 women. Gonadotropin-releasing hormone has also been used in an attempt to "program" ovulation in women (Zañartu et al., 1975). Its effects were reinforced by the administration of ethinyl estradiol on days 9 to 11 of the cycle, and then on day 13 a long-acting preparation of LHRH ([D-Leu^6]LHRH

ethylamide) was administered. A high proportion of the women ovulated on days 15 and 16. The number who would have ovulated spontaneously at this time was, however, unknown. If such a technique can be perfected, it will have certain practical applications, such as in the timing of artificial insemination and the practice of the rhythm method for birth control.

It will be recalled [Section 7.5I(5)] that LHRH may reduce fertility in the male possibly as a result of "down regulation" of LH receptors. A reduction in fertility in female animals has also been described which may have a similar basis or a decline in LHRH receptors. Thus LHRH also has the paradoxical potential of being used as an antifertility drug in women.

Bromocriptine may be effective in facilitating pregnancy in women who have hypogonadal conditions associated with high plasma prolactin levels (Thorner et al., 1975; Lenton, Sobowale, and Cooke, 1977; Friesen and Tolis, 1977; Pepperell et al., 1977; Thorner and Besser, 1978). It has been shown to restore ovulation, but it is even effective in some women who ovulate but are infertile. Bromocriptine has also been found to restore fertility in some women with normal plasma prolactin levels. Such treatment is usually embarked upon when clomiphene has been found ineffective. It has also been used subsequent to gonadotropins, but there is some question as to which order may be more appropriate. The drug is withdrawn as soon as pregnancy occurs. There have been 342 pregnancies reported, worldwide, following the use of bromocriptine, but no increased incidence of birth defects has been observed (Thorner and Besser, 1978). In many instances such infertility is associated with the presence of pituitary prolactin-secreting tumors, and these may some-

Table 7.10. *Results and complications of therapy with gonadotropins*

	Group I	Group II	Total
Number of treatment cycles	621	784	1405
Number of pregnancies	185	127	312
Number of mild hyperstimulations	34	85	119
Mild hyperstimulation rate (%)	5.5	10.8	8.4
Number of severe hyperstimulations	4	8	12
Severe hyperstimulation rate (%)	0.6	1.0	0.8
Number of abortions, first trimester	20	22	42
Abortion rate (%), first trimester	10.8	17.3	13.5
Number of abortions, second trimester	12	4	16
Abortion rate (%), second trimester	6.5	3.2	5.1
Number of multiple pregnancies	44	18	62
Multiple pregnancy rate (%)	23.8	14.3	19.9

Note: For a description of the two types of patients, group I and group II, see Figure 7.40.
Source: Lunenfeld and Insler, 1974.

times enlarge in pregnancy, owing to the rise in estrogen levels. As a precautionary measure, such tumors, when known, have been treated by irradiation prior to attempting pregnancy (Thorner and Besser, 1978). The rate of success of pregnancy depends on the manner that the group is selected, but it is as high as about 65 percent. A favorable outcome appears to be especially likely in women with high plasma prolactin concentrations due to pituitary tumors or functional disorders of the hypothalamic–pituitary axis. Success may also be predicted by a decline in plasma prolactin levels below normal, following administration of the bromocriptine.

I. Oral contraceptives: pharmacological control of fertility in women

The greater physiological complexities of the reproductive process in women enhance the prospects for successful interference with fertility by drugs. As described earlier, pharmacological control of fertility in the male, although feasible, is not yet practical. The gloomy Malthusian predictions of world overpopulation and resulting starvation currently appear to be nearing their fruition. Thus, the search for simple efficient and cheap methods of birth control has been considerably hastened in the last 20 years. A number of methods are available, any of which, if widely used, could adequately limit world population growth. These include surgical procedures, such as cutting the oviducts or the vas deferens; mechanical ones, such as diaphragms, intrauterine devices (IUDs), sheaths; and drugs. The current problem is no longer a serious technological one, but more a matter of overcoming cultural and religious customs and prejudices. The procedure ideally should, however, not be accompanied by any undue discomfort or have adverse effects on health. It is also more likely to be acceptable if its effects are reversible. Economic factors are important especially in poorer countries, so the cost should be low. In the context of all these requirements, the role of drugs as contraceptives is uncertain, but it has been estimated that about 15 million women in the United States use them and world consumption is upward of 60 million.

Useful steroidal drugs that can be used as oral contraceptives were developed in the 1950s, largely by the endocrinologist Gregory Pincus, prompted by the feminist and educator Margaret Sanger. An entertaining factual account of the development of these drugs has been given by Goldzieher and Rudel (1974). The basic physiological information that made this project feasible had been available for many years. Pregnant women do not ovulate and it had been recognized that infertility in farm animals may be associated with a persistent corpus luteum. Thus, the possibility that progesterone may be interfering with ovulation was clear and had been confirmed in animal experiments. More recent developments in the chemical synthesis of orally active analogues of the steroid sex hormones, which could be produced cheaply and in large quantities were, however, vital to the success of this project. The first clinical trials were commenced in 1955 in Puerto Rico using a synthetic progestational drug called norethynodrel plus the estrogen mestranol. Although early work emphasized the importance of progestins, it subsequently became apparent that estrogens may be even more effective, although the two types of steroids together generally have a happy synergism.

1. Mechanism of action. There are several ways that drugs can act to reduce fertility (see Emmens, 1970; Odell and Molitch, 1974; Van der Vies, 1977).

a. *Inhibition of ovulation.* As already described, ovulation involves the development of the follicles under the influence of FSH and the extrusion of the ripe ovum as a result of a "surge" in release of LH. The secretion of gonadotropins is controlled by both a negative-feedback system influencing FSH and LH and a positive-feedback system for LH. Estrogens and progesterone appear to act synergistically to control the negative feedback and estrogens the positive feedback. High doses of progestins plus smaller amounts of estrogens abolish development of the follicles and ovulation. The endogenous secretion of estrogens and progesterone is therefore decreased. The possibility that the exogenous steroids also have direct effects on the ovaries has not been excluded.

The first contraceptive pills to be used, and they are still available today (although smaller doses, especially of the estrogen component, are used), were mixtures of an orally active progestin and an estrogen. Examples of such combinations are *norethynodrel* (5 mg) + *mestranol* (0.075 mg), and *norethindrone* (1 mg) + *mestranol* (0.1 to 0.05 mg) or *norethindrone* (1 mg) + *ethinyl estradiol* (0.05 to 0.02 mg). These pills are usually given daily for 21 days. When withdrawn, menstruation (really "withdrawal bleeding") occurs. The next sequence is started 7 days later. To be effective, administration must commence on the fifth day of the cycle.

High doses of progestins can also inhibit ovulation when given alone, and in women they appear to synergize with endogenous estrogens. Their contraceptive action is highly effective. *Medroxyprogesterone acetate* has been used, especially in Southeast Asia; it is given i.m. once every 3 to 6 months (McDaniel and Pardthaisong, 1974; Parveen, Chowdhury, and Chowdhury, 1977). However, menstruation is irregular and uterine bleeding is frequent and unpredictable. The additional presence of estrogens in the combination preparations may help stabilize the endometrium, but this is not clear. It has been variously reported that the milk yield in lactating mothers is decreased (Parveen, Chowdhury, and Chowdhury, 1977) or increased (Chaudhury et al., 1977). The latter observations was related to an increased release of prolactin. *Norethisterone enanthate* is another injectable progestin which prevents conception, and it has recently undergone trials in England (Giwa-Osagie, Savage, and Newton, 1978). It is administered initially at intervals of 8 weeks and later every 12 weeks and was found to be highly effective. Some menstrual irregularities and amenorrhea occasionally occurred but no serious side effects were observed.

Estrogens alone also block ovulation, primarily by inhibiting the release of FSH. LH behaves in an unpredictable fashion and may decline, or suddenly be released. In the absence of a mature follicle, however, ovulation will not result. Long-acting preparations ("once-a-month" pill) have been tested. These contain *quinestrol* as the active ingredient. This estrogen has a cyclopentylenol moiety which enhances its storage in the body fat, from which it is slowly released over a period of several days and is converted to the active form, ethinyl estradiol. It is combined with a short-acting progestin, the declining levels of which initiate "menstruation" after about 4 days. The use of estrogens alone, however, has not gained favor, principally because of irregular breakthrough bleeding and more recently as a result of an increased awareness of their possible side effects (see Section 7.6N).

In an attempt to more closely mimic events that occur in the normal menstrual cycle, estrogens and progestins can be given separately or *sequentially* (also called *biphasic*). Either mestranol (0.08 mg) or ethinyl estradiol (0.1 mg) is given for 21 days and norethindrone (2 mg) or *dimethisterone* (25 mg) is added for the last 5 days. Withdrawal bleeding follows in 2 to 3 days, and the sequence is recommenced in about 7 days. This method is not quite as effective as the "combination" pill and the predominance of estrogen has been related to a higher incidence of major side effects. The sequential drugs have been withdrawn from use in the United States.

The doses of progestins and estrogens that were used in the initial trials were clearly more than required and were in the category of an "overkill." In view of the various side effects that may be encountered, especially those related to the presence of estrogens, the dosage has been progressively decreased. As little as 20 μg of ethinyl estradiol combined with 100 μg of norethindrone has been reported to be effective. Usually, it is recommended that no more than 50 μg of the estrogen be used,

but preparations with larger doses are still available in the United States. We are now moving into the era of the "minipill" and even "micropill," but, as described later, the mechanism of their actions may not always be related to an inhibition of ovulation.

b. *Local interference with the normal development of the endometrium* may also contribute to the antifertility effects of steroidal drugs. Indeed, the combination pill inhibits the normal development of the endometrium and in their presence it never attains its full potential. Progestins alone result in endometrial atrophy. Such conditions are not suitable for implantation.

c. *Implantation* depends on a correctly timed series of events, including the emergence of the blastocyst from the oviduct, a suitably primed endometrium, and optimal amounts of estrogens and progesterone to promote decidualization. Veterinarians have for many years utilized large doses of diethylstilbestrol to prevent pregnancy following what is deemed to be an undesirable mating ("misalliance") by pedigreed bitches or racing greyhounds. The mechanism is uncertain, but it could interfere with any of the foregoing events; possibly it facilitates the premature expulsion of the fertilized ovum from the oviduct. The use of this procedure to prevent pregnancy (postcoital contraception) following rape in women was first used tentatively about 40 years ago. It has, however, been more recently rediscovered and given much publicity as the "morning-after pill." A large dose of an estrogen, such as diethylstilbestrol (25 to 50 mg) or ethinyl estradiol (2 to 5 mg), can be given on 5 successive days following copulation. It has been especially recommended for use following rape, incest, or other such brief encounters. In the latter circumstances, it is said to have been widely used on college campuses. In view of possible serious side effects (see later), it is not recommended for more stable liasons when it could become the "daily" rather than the "morning-after" pill. In view of the effects of estrogens in inducing cancer of the vagina in the children of mothers treated with estrogens during pregnancy, a therapeutic abortion is usually recommended if this pharmacological procedure should fail.

d. Under the influence of estrogen, the *cervical mucus* acquires an optimal physical consistency and composition at the time of ovulation, and these properties aid the migration of the sperm toward the oviduct. Progestins result in the secretion of a cervical mucus which is thick and unfavorable to the passage of the sperm. This response is seen at extremely low doses and is the basis for the minipill, such as norethindrone (0.35 mg) or norgestrel (75 μg) (Eckstein et al., 1972). Even lower doses may be effective. These drugs are taken every day and in these doses do not appear to interfere with normal ovulation, but irregular bleeding occurs. The minipill is not quite as effective as "combination" or "sequential" formulations, but it is less likely to have undesirable side effects.

e. An increase in female-initiated sexual activity has been shown to occur at the time of ovulation in women (Adams, Gold, and Burt, 1978). It was found that this behavior was suppressed by the use of oral contraceptives, probably reflecting a failure of the increases in hormone secretion which accompany ovulation.

2. Side effects. The use of oral contraceptives is a unique pharmacological situation, as it involves the administration of drugs to such a vast number of normally healthy women, allegedly in the prime of their lives. They are undoubtedly the most "surveyed" group of drugs in history. Such studies of their side effects have provided a remarkable record of the diversity of the human response to the acute and chronic long-term use of exogenous chemicals. Side effects are not a problem with respect to most users of the pill, but in some they can result in discomfort or illness, and occasionally their effects may be fatal. The timely recognition of the more serious toxicities can contribute a great deal toward avoiding these problems.

a. *Minor side effects* are not infrequent but often disappear with continued use of the drugs. They can, however, result in sufficient discomfort to merit a change in the preparation or even the method of birth control. The newer formulations contain lower doses of the steroids, which appear to reduce the incidence of such side effects.

Nausea, and even vomiting, are not uncommon, especially in the first cycle, but usually this effect disappears by the second or third cycle. It has been likened to "morning sickness" in early pregnancy. There may be a substantial *weight gain,* 5 kg or even more. This effect is thought to be due to increased appetite and water and salt retention. A sudden incidence of *headaches,* and even migraine, may occur. Slight *breakthrough bleeding* can take place in midcycle and *amenorrhea* is quite common (9 percent). In the latter instance, pregnancy must be excluded. All these effects appear to be due predominantly to the estrogen present.

The progestin is thought to contribute to a feeling of *fullness in the breasts.* Some women report a feeling of *depression.* Either reduction or increase in *libido* may occur. The objective cause of these effects is not clear, but progestins have been more often implicated. It has been suggested that the depression is due to a vitamin B_6 (pyridoxine hydrochloride) deficiency, and one trial (Adams et al., 1973) showed that the administration of this metabolite had a favorable effect.

b. *Major side effects* are less frequent but of great concern. Epidemiological surveys of the health of

women taking contraceptive pills have been common, especially in Europe, the United Kingdom, and the United States. Many of these studies have been specifically related to suspected abnormalities, while others have had a broader, more general perspective. Reports of two long-term surveys involving about 63,000 English women have been recently published (Vessey et al., 1976; Vessey, McPherson, and Johnson, 1977; Beral, 1977). The findings generally confirm the existence of several problems which, although rare, can be serious, including an increase in the rate of overall mortality. A general review of the adverse effects of steroid contraceptives has been provided by McQueen (1978).

Venous thrombosis and embolism. There has been considerable argument and controversy about the existence and incidence of blood clotting disorders in women taking the pill. Initial studies carried out about 15 years ago in the United Kingdom, Sweden, and the United States suggested an increased occurrence of thrombophlebitis; thromboembolism, especially in the cerebral and pulmonary vessels; and deep vein thrombosis. The alleged relationship and risks were subsequently denied in several surveys, including those by the United States FDA in 1963 and 1968. The British were, however, more circumspect and recommended that the estrogen content of the pill should not exceed 50 μg. A more recent survey in the United States (Boston Collaborative Drug Surveillance Program, 1973) has confirmed the earlier association. They estimated an increased *risk factor* of 11 times, which is similar to the British conclusions, while in Sweden it is somewhat less (4.5). The incidence, or attack rate, in users of the pill is estimated at 1 in 1500 to 1 in 2000 per year. Disorders in the clotting mechanisms may explain the effects that have been described (see, e.g., Pilgeram, Ellison, and Von Dem Bussche, 1974; Sagar et al., 1976). Both the sequential and combination types of pills appear to be involved. An early report suggested that progestins are probably not involved (Poller, Thomson, and Thomas, 1972), but this conclusion is now in doubt (Meade, et al., 1977).

Cardiovascular diseases. The use of the contraceptive pill has been associated (see, e.g., Beral, 1976; Stern et al., 1976) with an increased incidence of *hypertension, stroke,* and *myocardial infarction.* An international survey indicated that the relative risk of death from all such cardiovascular disorders was increased about three times.

A small increase in systolic and diastolic blood pressure occurs quite commonly in women who use the pill, but this does not usually constitute a medical problem. However, in some instances, which have been estimated as about 4 percent of cases, the increase may necessitate cessation of this therapy. Usually, the blood pressure returns to normal levels within a few months (see Editorials, 1973c; 1976c). The effect may be related to salt retention and an increased level of angiotensinogen and renin in the plasma (Beckerhoff et al., 1973). It is currently recommended that a periodic record and check of the blood pressure should be made in women taking the pill. Vessey et al. (1976) noted a highly significant relationship to the incidence of stroke. An American study involving nearly 18,000 women (Pettiti and Wingerd, 1978) found that subarachnoid hemorrhage was 6.5 times as frequent in pill users than in nonusers. In women who smoked cigarettes and who also took such contraceptives, the increased risk of this disorder was 22 times as great as in women who indulged in neither of these practices.

Myocardial infarction is also more common in women taking contraceptive pills (see Hennekens and MacMahon, 1977). This increase was particularly apparent in women in the older age group of 40 to 44 years, and it may be synergistic with other disorders, such as cigarette smoking, hypertension, and diabetes (see Radford and Oliver, 1973; Mann and Inman, 1975; Mann et al., 1975). The relative increased risk factor appeared to be 4.5 in all users (Shapiro, 1975).

Plasma triglycerides consistently rise, although not usually above the normal range, in women taking the pill (Stern et al., 1976; Wallace et al., 1977). This effect may be related to the induction of hepatic microsomal enzymes for triglyceride synthesis (Martin, Martin, and Goldberg, 1976). The possible relationship of these changes to cardiovascular disease is uncertain.

Liver diseases. The sex steroids act on the liver and are metabolized by it, so it is perhaps not surprising to observe that they may sometimes have adverse effects on that organ. The hepatic handling of bromosulfthalein (a test of liver function) is adversely affected in a high proportion of women taking the pill (Editorial, 1974d; Adlercreutz and Tenhunen, 1970). The lipase activities in the liver are decreased in rats given ethinyl estradiol, and such an effect may contribute to the elevated concentrations of triglycerides observed in women taking the pill (Valette et al., 1977). Benign cholestatic jaundice (Drill, 1974) and hepatic porphyria occur in response to both the estrogens and progestins (especially the 19-nortestosterone derivatives) and they may have a synergistic effect. The incidence of gall bladder disease appears to be about doubled in users of the pill (Boston Collaborative Drug Surveillance Program, 1973; Ingelfinger, 1974). The latter effect may be related to an increase in the level of cholesterol and decrease in the proportion

of chenodeoxycholic acid in the bile (Bennion et al., 1976).

An increased incidence of benign liver cell adenomas appears to occur in users of contraceptive pills (Editorial, 1976d; 1977d). These can give rise to internal bleeding, and one report showed the malignant transformation of such a tumor (Davis et al., 1975). The relationship to oral contraceptives is, however, not proved and in any case the incidence is very low; less than 100 cases have been reported.

A survey in Boston has suggested that the possibility of admission to the hospital for an attack of acute hepatitis is 3.3 times greater than normal for women who use the pill (Morrison, Jick, and Ory, 1977).

Glucose tolerance. Many users of contraceptive pills show a deterioration in their glucose tolerance. Oral contraceptives are not contraindicated in diabetes mellitus, but an adjustment of the dosage of insulin or oral hypoglycemic drugs may be necessary. The statistical significance of these effects as compared to those normally seen in nontakers of the pill is not clear at this time.

Cancer. There has been considerable speculation as to whether oral contraceptives may promote the formation of malignant tumors, especially of the breast and cervix. Most evidence indicates that such a relationship is unlikely (Vessey, Doll, and Jones, 1975; Drill, 1975). These drugs appear to *reduce* the incidence of benign breast tumors (Ory et al., 1976). The prognosis in patients with breast cancer who have been taking oral contraceptives appears to be unchanged (Spencer, Millis, and Hayward, 1978). A recent discordant note, however, has been the observation that the severity and progression of cervical dysplasia to cancer is enhanced in women using a combination-type contraceptive pill (1 mg of ethynodiol + 0.1 mg of mestranol) (Stern et al., 1977). Thus, it would appear inadvisable for women exhibiting cervical dysplasia to take the pill. Sequential oral contraceptives have also not yet received a "clean bill of health," especially with respect to endometrial cancer (see Marx, 1976b).

Fertility after ceasing to take the pill. There has been some concern about the future prospects of a woman being able to become pregnant after ceasing to take oral contraceptives (see Editorial, 1972c). About 2 to 3 percent of women subsequently suffer from amenorrhea, but this form of infertility, if persistent, can usually be successfully treated with clomiphene [see Section 7.6H(7)]. The relationship to use of the pill is, however, questionable (Jacobs et al., 1977). Vessey et al. (1976, 1978) confirmed the observations of earlier surveys that users of the pill become pregnant less readily after they cease taking it than women who have used other methods of birth control. This is more common in nulliparous and parous subjects, and the relative infertility tends to decline with time. After 42 months, however, the nulliparous oral contraceptive users failed to display any differences from normal, and in the multiparous women this interval was only 30 months. Impairment of fertility is thus only temporary. It was not apparent in past users of an IUD. Pregnancies following oral contraception did not show a difference in the rates of multiple births, miscarriages, ectopic gestation, birth weight, or congenital malformations. Contrary to earlier reports suggesting a predominance of female offspring, no difference in sex ratio was observed (see also Rothman and Liess, 1976).

Birth defects. The question of whether oral contraceptives are teratogenic has been considered (see "Are sex hormones teratogenic," Editorial, 1974b). The administration of estrogens and progestins during pregnancy, such as in attempts to prevent abortion, is associated with disorders such as clear–cell adenocarcinoma of the vagina and virilization. Congenital limb defects have also been reported (Janerich, Piper, and Glebatis, 1974).

In a recent survey carried out on nearly 8000 infants born in the United States, no significant relationship was observed between taking contraceptive pills and malformations at birth (Rothman and Louik, 1978). There appears to be a slightly higher incidence of minor defects in infants whose mothers conceived shortly after taking the pill or who continued to use it during early pregnancy. The statistical significance of this increase is, however, not clear. The use of pregnancy tests that involve administration and subsequent withdrawal of progestins is, under the circumstances, probably unwise, especially as it is unnecessary. A precautionary measure may also be to try and delay pregnancy for 2 to 3 months after ceasing to take oral contraceptives.

Effects on mortality. The results of two long-term surveys for assessing changes in mortality that may occur in women taking contraceptive pills have recently become available (Beral, 1977; Vessey, McPherson, and Johnson, 1977). The conclusions have been summarized in an editorial in *The Lancet* (Editorial, 1977e). The surveys involved one group of 46,000 women (Royal College of General Practitioners or RCGP survey) and another of 17,000, both in the United Kingdom. There was a definite rise in mortality in women who use or have ever used the pill. This increase was about 40 percent. The cause of the increased mortality was cardio-vascular diseases (1 in 5000 ever users per year). Deaths tended to be concentrated in certain groups of women and were eight times greater in the 35- to 44-year-old group than 25- to 34-year-olds. Excesses in mortality increased with age, cigarette smoking, and the period of use of the pill. The British Royal College of General Practitioners issued recommendations based on these findings (Kuenssberg et al., 1977).

a. In women *under 30 years of age* no changes were recommended, but the possible advantages of stopping cigarette smoking should be pointed out.
b. Women who are *30 to 35 years of age* are in a period of increasing risk, especially if they have used oral contraceptives for 5 years or more. The possibility of changing the method of contraception should be considered, especially if they are smokers.
c. All women *over 35 years of age* should definitely reconsider their method for birth control. This decision is especially important if they smoke and/or have been using oral contraceptives for 5 years or more.

The increased risk of mortality was described as "relatively small" and "carries a minute risk of premature death." In an attempt to put it into perspective, *The Lancet* editorial compared the risk to deaths in traffic accidents. In England and Wales, this is a uniform 4 to 6 deaths per 100,000 women per year between the ages of 25 and 44 years. The excess mortality associated with the use of oral contraceptives in the RCGP survey was 4.4 deaths per 100,000 in the 25- to 34-year-old group per year and 33 deaths per 100,000 women per year in the 35- to 44-year-old group. The overall increased risk for all age groups was about 10 times greater than that of the excess death rate due to the extra pregnancies in the control group.

3. Contraindications. In view of the side effects that have been described, there are a number of contraindications to observe when considering using oral contraceptives. These are rare and may be based on such factors as genetic differences, current health, prior diseases, and age.

They can be summarized as follows, (a) hypertension, (b) a history of thrombotic episodes and recent surgery, (c) heart disease, (d) cerebrovascular disease, (3) liver disease, (f) hormone-dependent tumors, (g) the development of serious headaches and migraine, and (h) hyperlipidemia. Extra care should be taken in epilepsy, diabetes, bronchial asthma, obesity, and women over 30 years of age, especially if they are cigarette smokers.

4. Conclusions. Oral contraceptives offer the most efficient method of birth control available. When properly used, the combination and sequential pills are virtually 100 percent effective, 0.1 and

0.5 pregnancies/100 women per year, respectively. The minipill is reportedly less safe: about 1.3 pregnancies per 100 women-years. A summary of unplanned pregnancies and their outcome, using different methods of contraception, is shown in Table 7.11. The relative values for pregnancy per 100 woman-years are 0.15 for all oral contraceptives, 2.0 for the IUD, and 2.4 for the diaphragm. Other surveys indicate that the latter can be much less efficient, the difference possibly reflecting a better local education in its use. The value for the occurrence of pregnancy using no contraceptive precautions would be expected to be about 100 per 100 woman-years.

Although oral contraceptives are known to be associated with an increase in death from certain diseases described earlier, an estimate of relative mortality has only recently become available in one country (the United Kingdom). It has been said that in the United Kingdom (admittedly a rather law-abiding country) the risk of death from taking oral contraceptives is about the same as being murdered. It seems, however, that these estimates were somewhat low, especially in women over 35 years of age or in younger women who are smokers. With the current reduction in dosages of the constituent steroids and reservations about using the sequential type of oral contraceptives, a decline, in what is already a minimal risk, can be expected. An awareness of the possible problems that allow a precise definition of the contraindications can also reduce the magnitude of any such problems. In terms of public health the use of oral contraceptive pills appears to be justified, although in the case of each individual, some informed circumspection is necessary. The use of the pill does appear to need closer medical supervision than had been originally envisaged and should not cover the entire reproductive span of a woman's life.

However, the risk needs geographical assessment in relation to overall mortality rates, especially during pregnancy, as this can be much higher in certain parts of the world than others. The adverse cardiovascular effects of oral contraceptives have recently been the subject of a successful damages suit against a drug company in the United States.

J. Menopause (or the "climacteric syndrome")

The menopause is the last menstruation, in the retrospect of an absence of this event for 12 months (see Schreiner, 1976). It usually occurs at about 50 years of age and is preceded by a period when ovarian function declines, the premenopause, and is followed by the postmenopause, when this process is brought to completion. The entire phase, which lasts several years, is called the climacteric and is followed by the life stage of senility. The climacteric results from a decline in the ability of the ovary and its follicles to respond to pituitary gonadotropins. The causes are unknown, but it has been speculated that it may be due to ovarian vascular failure. The onset of the climacteric is signaled by a decline in estrogen levels and, as a result of the failure of the negative-feedback control, gonadotropin levels rise. The normal urinary excretion of estrogens is about 30 μg/day, at the age of 51 to 55 years it is only 12 μg/day, and in the senile phase, after 65 years, it is only 5 μg/day. The estrogens excreted following the menopause are not from the ovary but are derived as a result of peripheral aromatization of androstenedione in the liver, skin, and adipose tissue. This androgen is secreted by the adrenal cortex and possibly also the ovarian stroma.

Apart from infertility, the menopause results in a wide variety of physiological and behavioral changes. Probably the most prominent is a vascular

Table 7.11. *Frequency and outcome of unplanned pregnancies among women using different methods of contraception*

Outcome of unplanned pregnancies[a]	Rates per 100,000 per year among women using		
	Oral contraceptives	Diaphragm	IUD
Term birth	91	1462	497
Miscarriage	19	306	776
Ectopic gestation	1	10	121
Termination	39	622	606
Total	150	2400	2000

[a] The data on outcome for the oral contraceptive and diaphragm groups are based on the pooled figures for methods of contraception other than the IUD.
Source: Vessey et al., 1976. Reproduced with permission from *Journal of Biosocial Science.*

lability which is manifested as "hot flushes" followed by a period of sweating. This distressing problem can be quite infrequent or occur as often as 15-minute intervals throughout the day. Its causes are unknown, but it probably involves changes in the central nervous system rather than any direct effects on the blood vessels. Other vascular problems include hypertension and changes in the ECG ("palpitations") and giddiness. The other important change is related to the nervous system and results in periods of depression, irritability, insomnia, headaches, and lack of libido. Ovulation does not occur, so progesterone is not formed, although certainly in the early stages sufficient estrogens to stimulate endometrial proliferation may be present. Thus, dysfunctional uterine bleeding and menorrhagia commonly occur. In the absence of sufficient estrogens there is an atrophy of the reproductive tract, and this degeneration is manifested as atrophic vaginitis and loss of myometrial tone. The latter can result in a prolapse of the uterus. Other metabolic changes include a propensity to gain weight and an increased resorption of calcium from bone, which in the absence of compensating accretion leads to osteoporosis [see Section 9.4A(2)]. Some hirsutism may develop as a result of the action of androgens, which are now only inadequately opposed by estrogens. The texture and elasticity of the skin also changes. There is an increased likelihood of developing diseases, such as peptic ulcer, cardiovascular diseases, and certain cancers. The physiological changes associated with the menopause can all be mimicked by oophorectomy and appear to be the result of a decline in the levels of estrogen. The elevation in gonadotropin secretion does not appear to be involved in the disorders although it is the subject of some speculation.

1. Treatment. It is possible to treat many of the disorders associated with the menopause by the administration of estrogens. This is a type of replacement or substitution therapy, but it is, of course, rather inexact and incomplete by physiological standards. The administration of estrogens does not constitute what has been improperly described as a pathway to "eternal youth." Estrogens during the climacteric and even after do, however, alleviate many of the distressing associated disorders. It has been estimated that in the United States about 50 percent of women have used estrogens during the postmenopausal period (Weiss, 1975). The oral preparations used include *diethylstilbestrol, ethinyl estradiol,* and, most frequently, a mixture of conjugated natural estrogens (Premarin) consisting of *estrone sulfate* and *equilin sulfate.* The latter are called "soft estrogens" and are thought to have less effect on the endometrium. Stimulation of the latter results in a proliferative

buildup which can result in bleeding disorders. *Estriol* is also a weak estrogen with little effect on the endometrium but which can restore vasomotor stability (Tzingounis, Aksu, and Greenblatt, 1978). To avoid endometrial problems, estrogens have been administered in cycles alternated with progestins, a process that mimics the more orderly menstrual cycle.

Intramuscular depot preparation of estrogens and pellets implanted s.c. are sometimes used to treat menopausal disorders, but oral administration is preferred. Local application of estrogens in the form of suppositories or creams, usually containing *dienestrol,* can be used to treat vaginal disorders associated with the menopause and have the advantage that while some absorption occurs, systemic levels are more limited. At one time many creams were freely available which contained estrogens. They were marketed with the purported aim of maintaining a youthful texture of the skin. In view of the possible absorption of these steroids through the skin, and the undesirable effects that they may then have, they are generally no longer available. Androgens, such as methyltestosterone or testosterone propionate, can also be effective and increase libido and have a general anabolic effect (see Sections 7.5B and C). Some virilization may occur, however.

Favorable placebo effects following history taking, injections of saline, or the administration of "dummy" pills can be quite prominent in menopausal women (see Thompson and Oswald, 1977). Depression, anxiety, and even hot flushes may decline by 50 percent or more and cannot always be readily distinguished from the actions of administered estrogens.

The role and uses of estrogens in the treatment of postmenopausal osteoporosis are described later [Section 9.4A(2)].

Side effects. The relatively free use of estrogens to treat disorders in women during the climacteric and after is currently very controversial. This reassessment has largely been the result of the recognition that estrogens are not harmless drugs and can cause a number of serious, indeed fatal, disorders. These include [see also Section 7.6N(1)] such conditions as thromboembolic disorders, disturbances in liver function such as cholestasis and gall bladder disease, and the risk of cancer. There appears to be a relationship between the incidence of endometrial cancer and the use of estrogens (see Lipsett, 1977; Editorial, 1977f). Estimation of the increased risk varies, but it is normally about 1/1000 women per year and with estrogen treatment may increase to as much as 7.6/1000 per year. The average increase or "risk factor" appears to be about fivefold and is related to the length of the period of administration of the drugs. Such a causal relationship has been contested, however, on the basis that there may have been a preferential selection for the studies of women who are being treated with estrogens (Horwitz and Feinstein, 1978). This proposal is also contentious (Hutchison and Rothman, 1978). It is also considered possible that there may be a relationship between use of estrogens and an increase in the rate of breast cancer, but at present the evidence is equivocal and generally does not appear to indicate that this is so.

The use and abuse of estrogens in the treatment of menopausal disorders is thus controversial (see Studd, 1976; Philipp, 1976; Lipsett, 1977). The chronic risks need to be weighed against the acute benefits. Their use is certainly *contraindicated* in the presence of recent thromboembolic diseases, hepatic disorders, endometrial problems such as endometriosis or fibroids, and, depending on the type (see Section 7.6K), mammary gland tumors. The future possible problem of endometrial cancer can be resolved by prophylactic hysterectomy, but such a cavalier surgical solution is not acceptable to all, especially to patients and internists. Undoubtedly, estrogens will now be used more circumspectly. It has been estimated that they are probably only really justified in about 10 percent of women (Philipp, 1976) in *minimal doses* and for the *shortest possible periods of time*. The conclusions are as follows: "These are difficult judgements and must be made by the *adequately informed patient*" (Lipsett, 1977, the italics are mine). "Above all, it is necessary to have the courage to withhold therapy when it is not needed" (Philipp, 1976).

K. Hormone-related cancers and their endocrine therapy

Malignant tumors that are related to the presence of hormones occur in the mammary glands, uterus, cervix, and the vagina in women, and in the prostate and mammary tissue in men. These cancers usually arise spontaneously, but in some instances their occurrence can be related to the use of hormone preparations, especially estrogens. The role of endogenous hormones is shown by the observation that cancer of the prostate does not occur in eunuchs and the incidence of breast and uterine cancer is low in women who are deficient in sex steroid hormones, either as a result of ovariectomy or disease. Conversely, such tumors often regress following castration. The administration of steroidal sex hormones is also associated with an increased frequency of cancers in animals and women. This observation has resulted in the suggestion that they are carcinogens, an allegation that requires some definition, as it does not appear to be strictly correct. The steroidal sex hormones maintain the integrity and stimulate the growth, differentiation, and multiplication of a variety of epithelial tissues, especially those associated with reproduction. Derangements and disorganization of these patterns of cell activity can occur in response to various types of stimuli, genetic and environmental, including toxic chemicals and radiation, which result in cancers. The hormones are considered to provide the necessary metabolic background that allows such carcinogenic responses to occur and, once a tumor is present, may support its growth (see Lipsett, 1977). Chemical carcinogens such as dimethylbenanthracene can induce mammary tumors in rats, but this effect is inhibited by ovariectomy or hypophysectomy. The activity of such tumors, when developed, is also inhibited by such glandular extirpations (see Smithline, Sherman, and Kolodny, 1975). Many tumors have also been shown to contain specific receptors for steroid hormones, especially estrogens. The pituitary hormone prolactin (Section 7.6G) also appears to be involved in supporting the activity of mammary gland tumors, although as yet this has only been conclusively demonstrated in animal models.

The possibility of a malignant tumor developing not only depends on the presence of such hormones but also on the period of exposure. Therefore, the length of administration of exogenous steroid hormones that may support the activity of tissues prone to the actions of carcinogens increases the risk of malignant transformation occurring. Disorders in endogenous metabolism of steroid hormones may also contribute to a more favorable environment for cancer to occur. Thus, in postmenopausal women, who are especially at risk, an increased-peripheral conversion of adrenocortical androgens to estrogens may take place. It has been observed that cancer of the uterus is more frequent in obese women, and this has been related to the propensity of adipose tissue to affect this aromatization of androgens to estrogens (see Lipsett, 1977).

1. Carcinoma of the breast. Breast carcinoma is the most frequent cause of death from cancer. It has a morbidity of 3 percent, but it has been projected that in the United States about 10 percent of women born today will develop it and, with the present lack of success in treatment, about half of these will die of it (Gordan, Wessler, and Avioli, 1972). More than 50 percent of mammary tumors appear to be hormone-dependent. While prolactin and estrogens can be shown to induce this condition in animals, exogenous steroids, including oral contraceptives, have not been implicated in women (Drill, 1975; Smithline, Sherman, and Kolodny, 1975).

Hormonal preparations play a distinct role in the treatment of breast cancer. The estrogen dependence of the tumor and its metastases can be predicted as a result of a diagnostic test involving determination of the presence of cytosol estrogen receptors (ER). This interesting application of basic medical research was developed by E. V. Jensen (Jensen, DeSombre, and Jungblut, 1967) and L. Terenius (1968) and has been confirmed by many others (e.g., Englesman et al., 1973; McGuire, 1975; Leclercq et al., 1975). It has been found that the presence of estrogen receptors provides a good indication as to the potential effectiveness of endocrine-related therapy, including ablation of the ovaries and the administration of hormone preparations. A review and summary have been provided by Lippman and Allegra (1978).

Therapy usually follows several lines:

a. Surgical or, sometimes, radiological treatment of the primary tumor and the investigation of possible adjacent metastases in the lymph nodes. The latter can be treated surgically, radiologically, or with cytotoxic drugs.

b. If metastases subsequently develop, several possible courses of action may be taken. In premenopausal women, ovariectomy to remove the source of endogenous estrogen, either surgically or by radiation treatment, results in a regression of the tumor in about one-third of all cases. The rate of success can, however, be predicted in about 55 percent of women on the basis of identification of estrogen receptors in the primary or metastatic tissue (McGuire, 1975). If receptors are absent, the rate of success is only 8 percent. If ovariectomy results in regression of the tumors but later on more metastases develop, then further ablative treatment involving adrenalectomy [see Section 6.8A(1)] and hypophysectomy, to further limit endogenous estrogen synthesis, can be considered. In instances of mammary carcinoma in men, castration may be effective.

c. In postmenopausal women, or if ablative treatment is unsuccessful, cytotoxic drugs or steroid hormone preparations ("additive" therapy) can be used. The choice is somewhat controversial (see Priestman et al., 1977). A low estrogen receptor value has been related to the success rate of cytotoxic therapy (Lippman et al., 1978).

Hormone treatment can involve the use of estrogens, androgens [see Section 7.5I(8)], progesterone, and corticosteroids. The latter are not very successful but have been used especially when metastases occur in the liver (For further discussion, see Section 6.8B.) Only one such type of preparation is used at a time and, before changing to another type, therapy is ceased for a period during which a withdrawal remission may occur. The use of estrogens, such as *diethylstilbestrol* and *ethinyl estradiol*, is more frequent, especially if the metastases are confined to the "soft" tissues. If they occur in the bones ("hard" tissues), then *androgens* are used more often [Section 7.5I(8)]. High doses

of estrogens are used, about 10 times those utilized for the treatment of gynecological disorders. Such hormone additive treatment is successful in about one-third of women, and the likelihood of this success can be predicted in about 60 percent of cases in which estrogen receptors have been identified in the tumor tissue.

The reasons for the failure of many tumors that contain estrogen receptors to respond to hormonal therapy are not known. It has been suggested (McGuire et al., 1977), however, that it is due to a failure of the hormone effector mechanism at a site which is distal to that of the estrogen receptors. The numbers of progesterone receptors in many tissues are increased in response to estrogens, so their numbers could serve as a basis for further predicting the likelihood of a favorable response. It was found (McGuire et al., 1977) that a higher proportion of the responsive tumors had both estrogen and progesterone receptors than estrogen receptors alone. The method appeared to be especially useful in predicting hormonal responses to metastatic tumors, where an 81 percent response was seen in those containing both types of receptors compared to only a 41 percent response in those containing estrogen receptors alone. Another study carried out in Europe by Leclercq and Heuson (1977) failed to show a relationship between the presence and concentration of progesterone receptors and responses to hormonal therapy.

It is paradoxical that while breast tumors may be estrogen-dependent, the administration of high doses of exogenous estrogens may induce them to regress. There is no adequate explanation for this discrepancy (see Stoll, 1973), but it has been suggested that estrogen may antagonize the effects of prolactin or a type of dose-threshold phenomenon may be involved.

Antiestrogens, especially *nafoxidine* and *tamoxifen*, and clomiphene have been used in the treatment of breast cancer (Ward, 1973; Bloom and Boesen, 1974; Heuson et al., 1975; Leclercq and Heuson, 1977). The patients were postmenopausal and remissions were observed in 20 to 40 percent of cases. Nafoxidine appeared to be more effective in metastatic tumors of soft tissue. One study (Sasaki, Leung, and Fletcher, 1976) found that only tumors with estrogen receptors responded. In one trial (Heuson et al., 1975) nafoxidine was found to be more effective than ethinyl estradiol. Several "life-threatening" (cardiovascular) complications occurred with the estrogen, but nafoxidine had no serious effects, although it did cause some problems related to the skin, including hair loss, excessive dryness (ichthyosis), and phototoxicity. Tamoxifen appears to have even fewer side effects.

The mechanism of action of the antiestrogens

appears to be a direct one, probably involving estrogen receptors in the tumor. Tamoxifen can reduce prolactin secretion but not to a degree that suggests it is working via such a mechanism (Jordan and Koerner, 1976). These drugs have been observed to directly interact with estrogen receptors from breast tumor and to displace estrogens (Hähnel, Twaddle, and Ratajczak, 1973a; Jordan and Jaspan, 1976).

2. Cancer of the uterine endometrium. Uterine endometrium cancer has been associated with the administration of estrogens for the treatment of menopausal disorders (see Smith et al., 1975; Mack et al., 1976). This "side effect" of estrogens was described earlier (Section 7.6J). The increase in the risk due to the use of estrogens may be as high as elevenfold. Treatment is primarily surgical and radiological, but it may be necessary to use steroids. Long-acting preparations of progestins, such as medroxyprogesterone acetate or 17-hydroxyprogesterone caproate, are administered i.m. (Richardson, 1972), and these suppress the activity of the endometrium. There appear to be no serious side effects.

3. Cervical cancer. Cervical cancer is the next most common cause of death after breast cancer in women of reproductive age. Its occurrence has been related to herpes, early marriage, and promiscuity, including that of the sexual partner. It is the commonest of gynecological cancers, but there are considerable geographical differences in its incidence. Thus, it ranges from 1/1000 women-years in Colombia to 0.06/1000 women-years in Israel. An early predictive sign of the disease is dysplasia of the cervical epithelium, which subsequently may regress or progress to cancer. It has been shown that estrogens can induce cervical cancer in mice, but no consistent relationship to the use of estrogens or oral contraceptives has been shown in women (Drill, 1975). A recent study (Stern et al., 1977) carried out in California, however, suggested that use of an oral contraceptive preparation (1 mg of ethynodiol and 0.1 mg of mestranol) increased the progression of cervical dysplasia. The relative promiscuity of the two experimental groups was not compared but both "were . . . presumably sexually active."

4. Clear-cell adenocarcinoma. Clear-cell carcinoma of the cervix and vagina is normally a very rare disease. It has, however, been identified in the daughters of women who were given diethylstilbestrol to prevent abortion in the first trimester of pregnancy [see Section 7.6H(7)]. The condition usually occurs between the ages of 15 and 22 years with an incidence of less than 1 per 1000. Herbst and his collaborators extended their original study (Herbst et al., 1975) and found a very high incidence of morphological and histological malformations in the vagina. The cause appears to be failure of proper differentiation of the genital tract in the upper region of the vagina, where the Mullerian duct epithelium is normally replaced by that of the urinogenital sinus. Small cysts (vaginal adenosis) were present in 74 percent of women whose mothers were given diethylstilbestrol in the first 2 months of pregnancy. Malignant transformation of the vaginal adenosis was not observed in this survey, but the adenosis invariably accompanies the clear-cell adenocarcinoma. The use of oral contraceptives was associated with less adenosis and a preliminary report suggested that progesterone-theobroma oil suppositories may promote healing.

L. Miscellaneous uses of estrogens

1. Inhibition of linear growth. The administration of estrogens to prepuberal girls hastens the onset of the menarche and results in closure of the epiphyses, thus halting linear growth. This response has been utilized by pediatric endocrinologists for about 30 years to limit the height of girls who may become excessively tall (see Greenblatt, McDonough, and Mahesh, 1968). It is considered by some that a height in excess of 6 feet is a social and even possibly an economic disadvantage to a woman. Projected growth is based on family history and current bone age, from X-ray examination and recent growth trends. Treatment in appropriate cases consists of the administration of large doses of estrogens. These can be given as implants of *estradiol* about once every 6 months, or frequent oral doses of *conjugated estrogens* (Premarin) in a dosage of 10 to 15 mg/day. With such treatment breakthrough bleeding occurs, but this is limited by the periodic administration of a progestin. Thus, a usual regimen would be 21 days of estrogen and 5 days of progestin, which is followed by withdrawal bleeding. Treatment continues while the epiphyses remain open and usually lasts about 2 years. It has been estimated that there is an average inhibition of about 3.25 to 6.5 cm in linear growth. Apart from withdrawal bleeding, side effects are typical of estrogen treatment and those reported include nausea, weight gain, and hypertension. Normal menses have been shown to follow cessation of treatment. The recent association of the use of estrogens and endometrial cancer has, however, provoked some reservation about such treatment in young girls (Brogan, 1976). This use of estrogens has not been approved by the U.S. Food and Drug Administration.

2. Endometriosis. This is an interesting condition in which extant endometrial tissue, including glandular cells, grows in various parts of the body, usually the pelvic regions and including the ovaries. The cause is uncertain, but it has been suggested that it may result from a reflux of endometrial cells along the oviducts into the body cavity. The condition is usually cured by pregnancy, a form of therapy that may not be feasible or acceptable to all. This cure, however, afforded an important clue regarding therapy, as it has been found that progestins, such as medroxyprogesterone, dydrogesterone, and norgestrol can suppress the growth of the tissue. Long-term use is required, but in cyclical patterns, such as for 21 days with 5 to 7 days off. A progestin can be given alone, but more often a combination with an estrogen (e.g., ethinyl estradiol), such as in oral contraceptive pills, is used. This regimen has been called the "pseudopregnancy treatment." The induction of an analogous anovulatory condition can also be produced by the selective pituitary gonadotropin inhibitor danazol (see Section 7.5F). This weak androgenic steroid, which has no estrogen or progestin activity, has undergone successful clinical trials for the treatment of endometriosis in the United Kingdom (Chalmers, 1977) and United States (Young and Blackmore, 1977). Side effects appeared to be less than those occurring with the use of progestins and estrogens. This regimen has been dubbed the "pseudomenopause treatment."

3. Uses of estrogen in the male. Estrogens have been used for the treatment of a number of conditions in men. Probably the only truly recognized beneficial one is *cancer of the prostate*, which is an androgen-dependent condition. The estrogen appears to act by suppressing the secretion of androgen. *Diethylstilbestrol* is usually used, but there has been considerable confusion as to the appropriate dosage. Enormous doses, up to 1000 mg/day, have been used. A U.S. Veterans Administration survey of treated patients showed that many of them did not die from cancer but from cardiovascular and thrombotic complications due to the drug. It has been recommended that the dosage of diethylstilbestrol should be 1 mg/day (Editorial, 1974e; Shahmanesh et al., 1973). The overall place of estrogens in the treatment of cancer of the prostate is, however, not well defined (Editorial, 1977g). Indeed, there remain some questions as to whether it is even useful to treat such tumors at all in asymptomatic patients. Other methods of treatment, including orchidectomy, chemotherapy with cytotoxic drugs, and radiotherapy, are used. Most but not all prostatic tumors respond to hormonal treatment. The reason for the failures is unknown, but the measurement of androgen receptors in the tumors possibly could provide a way of more reliably predicting the response.

Two interesting, indeed bizarre, uses of estrogens in men have been for the prevention of myocardial infarction and the treatment of peptic ulcer. The basis for such uses appears to be epidemiological, as women, at least certainly in the past, suffer from these conditions less than men. Such uses for estrogens have not been acceptable to most men, as they may have a feminizing effect and cause a lack in libido (although one reviewer questioned the latter and thought they may be better off without it!). Other unacceptable side effects included an *increased* incidence of myocardial problems which resulted in the cessation of one extensive trial.

M. Estrogens in food products

Diethylstilbestrol is widely used in the meat industry to improve the quality and foster a more efficient weight gain in cattle, sheep, and chickens. In chickens, pellets may be implanted in the neck just below the head, while in cattle they are placed at the base of the ear. It is important that these parts of the animal be discarded following slaughtering. A more widely used procedure is the addition of diethylstilbestrol to the feed of steers. This improves the efficiency of the conversion of their food intake to meat, and there is an increased weight gain of 10 to 25 percent. The mechanism appears to involve an anabolic effect. Such feed is withdrawn from the animals at least 48 hours prior to slaughtering, when the levels of the drug are usually undetectable in most edible tissues. However, it has been found in some instances, at greater than acceptable levels, in the liver. Public discomfort about the possible carcinogenic effects of such residues of estrogens led to the banning of this practice in many countries, including the United Kingdom, Canada, Australia, and West Germany. An attempt by the U.S. FDA to prevent its use in the United States was overturned in court but it is still being challenged. Such a ban obviously would have considerable economic consequences, especially if the cattle are being fed on grain. What will happen in the future is uncertain, and whatever the decision, the association with cancer in man will probably remain equivocal.

As described earlier (Section 7.6F and Figure 7.15), a number of substances that have estrogenic activity occur in many common plants (phytoestrogens) which may be consumed in large quantities by domestic animals. These natural hormone analogues may diminish or increase fertility and they may also have anabolic effects. The increase in the growth of domestic animals that is associated with the "spring flush" in plant growth is well known in folklore and may be partly related to changing levels of such substances. Clearly, phytoestrogens may be obtained in the diet of man, but there appears to be little information about the magnitude of such effects. The potential actions of such estrogens are uncertain but as in animals, they could be influencing normal physiological processes. Several phytoestrogens have also been shown to be able to combine with estrogen receptors isolated from human breast cancer cells (Martin et al., 1978), but the possible significance of such effects, for good or evil, is unknown. Regulation by governmental decree of the type of grass on which animals may be fed is somewhat novel to consider but bureaucratically feasible.

N. Summary of the side effects of estrogens and progestins

1. Estrogens. There appears to be little to choose between preparations of estrogens when considering their possible side effects. Some drugs have been reported to have different effects; for instance, the conjugated, so-called soft estrogens may have fewer effects on the endometrium than other "hard" estrogens. Diethylstilbestrol (DES), in particular, has been involved in a number of controversies. There is, however, no compelling reason to believe that any one estrogenic compound has any special innate properties that induce certain types of side effects. The only exception may be the propensity of the synthetic 17α-alkylated estrogens to cause hepatic disorders. The differences that are observed appear to mainly reflect the relative potencies of the preparations, the doses that are usually given, and their routes of administration. Certain estrogens are favored for particular conditions. For instance, currently DES is used to treat cancer of the prostate and conjugated estrogens for menopausal disorders, so that, when side effects are uncovered, the particular drug preparation being used may, unjustifiably, bear the brunt of the criticism.

The incidence and severity of the side effects will of course vary with the dose, the route of administration, the duration of therapy, the sex of the patient, age, and underlying differences, whether due to genetic factors, disease, or the concurrent administration of other drugs.

It is often difficult to disentangle the effects of estrogens and progestins, as they are frequently given at the same time.

a. Minor side effects include *nausea*, which is quite common and occurs initially in about one-third of patients, a feeling of *fullness in the breasts, weight gain* due to the *retention of fluids,* and *headaches.*
b. Disturbances in blood-clotting mechanisms that can result in *thrombophlebitis, deep-vein thrombosis,* and *thromboembolism.*
c. *Hypertension,* which is usually reversible when the drug is discontinued.

d. *Cerebrovascular disease* and an increased incidence of stroke.
e. An increase in *myocardial infarction* which is seen in both sexes. In women, it is more common in those over 40 years of age.
f. Metabolic changes, including a rise in *plasma triglycerides* and a lowering of *glucose tolerance.* The former condition has been associated (rarely) with the occurrence of pancreatitis.
g. *Hepatic disorders* are quite common and include benign cholestatic jaundice, precipitation of porphyria, and benign adenomas. There is an increased incidence of gall bladder disease.
h. *Proliferation of the endometrium* associated with menstrual disorders and breakthrough bleeding.
i. *Cancer of the endometrium* associated with long-term use for the treatment of menopausal disorders. Despite suspicions based on animal experiments, no relationship to cancer of the cervix or breast has been found. Estrogens may, however, attenuate an underlying hormone-dependent tumor of the breast, especially in premenopausal women.
j. Teratogenic effects when used in pregnancy, including adenosis and clear-cell adenocarcinoma of the vagina and, possibly, limb defects.
k. Estrogens will *hasten puberty* in girls and result in a closure of the epiphyses so that linear *growth is halted.*
l. When administered to inhibit lactation, a "rebound" lactation may occur a few days after ceasing administration.
m. In *men,* they have a *feminizing effect* and result in a decline in libido, impotence, gynecomastia, and atrophy of the testes.

2. Progestins. These preparations have fewer serious side effects than do estrogens. There may, however, be differences in the side effects, depending on the particular preparation used. Thus, some preparations, such as the testosterone derivatives, have higher androgenic effects and can be metabolized to estrogens. They also may exhibit a significant anabolic action.

a. Progestins have, like estrogens, been implicated in the occurrence of *cholestatic jaundice.*
b. When administered in early pregnancy, progestins may result in pseudohermaphrodism in female infants and even later in pregnancy can produce virilization. The principal defect is a masculinization of the external genitalia. Other deformities are controversial.
c. Hair loss and oily skin, but this is rare. It may reflect the androgenic effects of the preparations.
d. They may produce uterine bleeding, amenorrhea, and occasionally heavy bleeding.
e. The progestin component of oral combination contraceptives has been implicated in *hypertension* (Royal College of General Practitioners Oral Contraception Study, 1977).
f. No relation to cancer has been observed, although in some animal studies using medroxyprogesterone (in beagle dogs), increases in breast tumors occur.

8
The islets of Langerhans

The islets of Langerhans, which are embedded in the exocrine pancreatic tissue, are known to secrete three hormones: insulin, glucagon, and somatostatin. Whereas the former two excitants appear in the general systemic circulation, pancreatic somatostatin appears to be mainly confined locally within the endocrine tissue, where it exerts a regulatory influence on the secretion of the other two hormones.

8.1. Historical background

Diabetes mellitus is the most common endocrine disease and affects about 2 percent of the population in Western countries. It has been recognized in various societies for many hundreds of years, where its cardinal signs were known to be the formation of large volumes of urine, which was, rather curiously, recognized to have a sweet, sugary flavour. This disorder is only one facet of the disease, which has many other effects in the body, but it reflects the central metabolic problem, an inability to regulate the levels of glucose, with the result that some may spill over into urine and produce an osmotic diuresis. Survival was once limited to a few years, the sufferers often dying in a diabetic coma. The role of the islets of Langerhans in diabetes mellitus was suspected in 1909 by J. de Meyer, who named the hypothetical secretion insulin, but this was not proved until 1921, when F. G. Banting and C. H. Best successfully extracted insulin from the pancreas and were able to sustain the survival of a pancreatectomized dog indefinitely. The lifesaving capacity of this hormone was soon demonstrated in man, initially using insulin extracted from the pancreas of fetal calves. This discovery is considered one of the first and most influential advances that heralded the subsequent era of modern medicine. More than 50 years later our knowledge of diabetes mellitus, its causes and treatment, is, however, still incomplete. This ignorance is a source of

chagrin, as the incidence of the disease is increasing, and with early death from diabetic coma now rare, the disastrous long-term vascular effects of the disease have become apparent. There is, however, considerable information available about many aspects of islet cell function. The amino acid sequence of insulin was determined by F. Sanger in 1955, (see Sanger, 1959) and its precise three-dimensional structure was described by D. C. Hodgkin (see Hodgkin, 1972) and her collaborators in 1969. Insulin has also been prepared by chemical synthesis. Its radioimmunoassay was developed by S. Berson and R. Yalow (Yalow and Berson, 1960) so that its concentration can be readily determined in small volumes of plasma *in vivo* or under experimental conditions *in vitro*. Considerable advances have been made in understanding its mechanism of action, especially the work of P. Cuatrecasas and J. Roth, in relation to its interaction with receptors. Such studies on insulin have had a wide impact on biology generally, particularly in relation to protein chemistry, the extensive application of the principles of radioimmunoassay to all aspects of endocrinology, and a more general understanding of the nature of the elusive pharmacological concept of the "receptor." More detailed information about the relationship of function to cell ultrastructure is probably available for the B-cell than for any other type of endocrine tissue.

The presence of the other islet hormone, glucagon, which has an opposite, hyperglycemic, effect to that of insulin was demonstrated shortly after the isolation of the latter, in 1923 by C. P. Kimball and J. R. Murlin. It was not, however, chemically characterized until 1957, by W. W. Bromer et al. Glucagon has a less dramatic medical history than insulin, but it is of undoubted physiological importance. Its role in disease states does not appear to be common, although it has recently been suggested that its unopposed actions may influence,

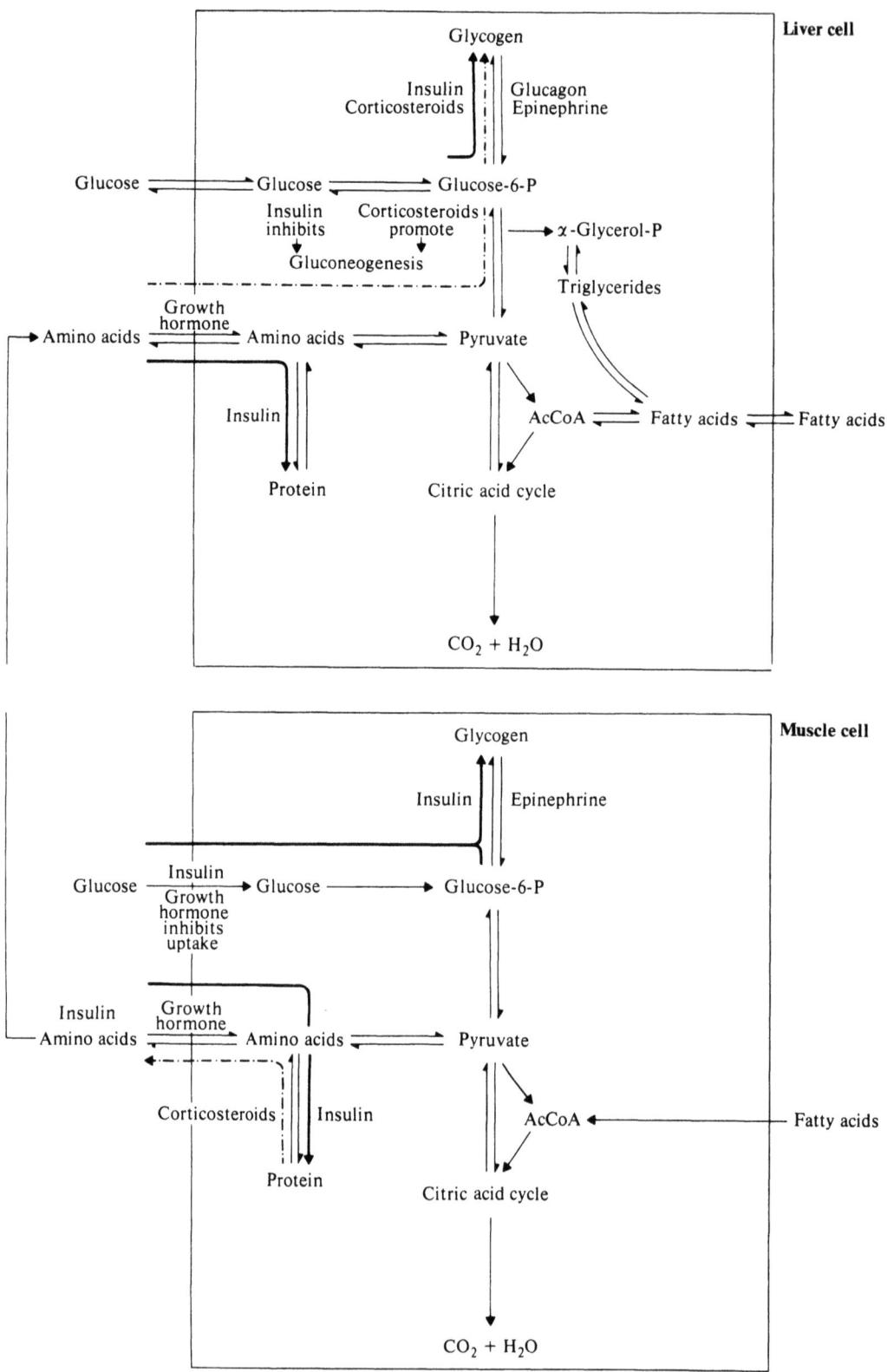

Fig. 8.1. Summary of the effects of hormones on the transfer of metabolites across the cell membranes and their metabolic transformation within liver cells, muscle cells, and fat cells. The principal metabolic transformations that are influenced by insulin are indicated by the solid lines and those by cortiocosteroids by dashed lines. (From Bentley, 1976.)

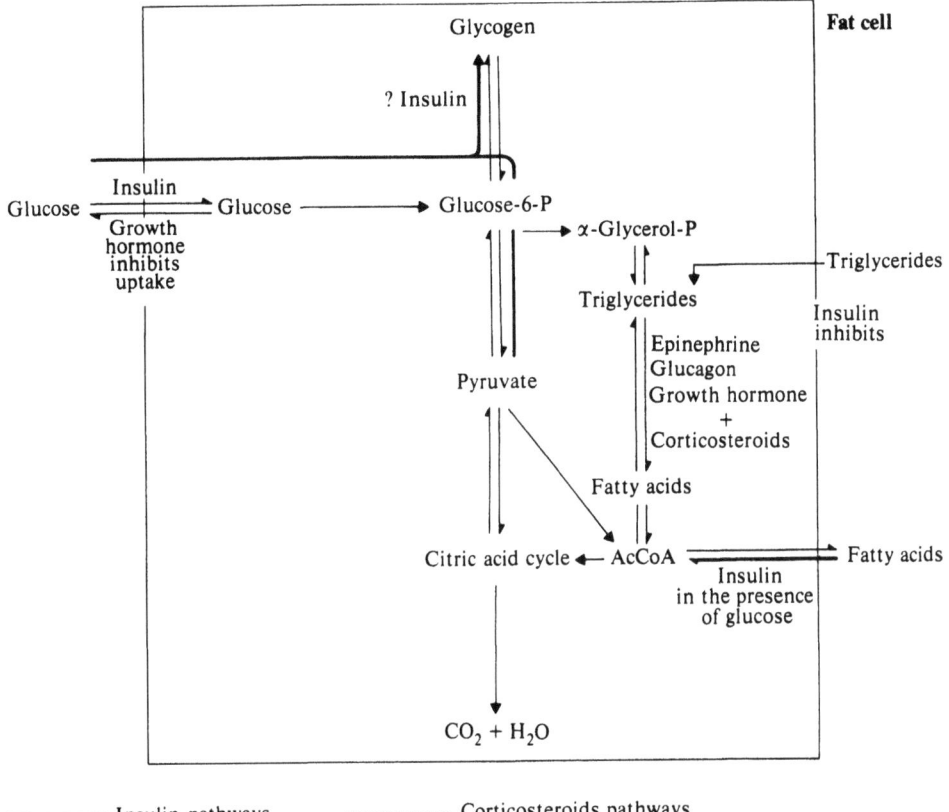

Insulin pathways   ·—··—··—··—· Corticosteroids pathways

Fig. 8.1. (*cont.*)

and exacerbate, the pathological difficulties associated with a lack of insulin and so contribute to diabetes mellitus. Glucagon has an interesting historic role in endocrinology, as it was the investigation of its mechanism of action in conjunction with that of epinephrine by Earl Sutherland (see Sutherland, 1972) that resulted in the discovery of the role of cyclic AMP in hormone action.

A. Role of hormones in intermediary metabolism

The most ubiquitous effects of hormones in the body concern their actions in regulating intermediary metabolism. These processes involve regulation of the various interconversions that occur among proteins, fats, and carbohydrates. The effects may be anabolic, such as the synthesis of glycogen, fats, or proteins, or catabolic due to mobilization and utilization of energy substrates, including glucose and fatty acids. Such processes occur in all cells in the body, but certain tissues, such as the liver, muscle, and adipose tissue, may have special roles. These are summarized in Figure 8.1.

Numerous hormones can influence intermediary metabolism, including insulin, adrenocorticosteroids, catecholamines, glucagon, growth hormone, estrogens, androgens, and thyroid hormones. Insulin appears to play a broad pivotal role, primarily by virtue of its ability to control glucose concentrations in the body fluids. However, it also has important actions on fat and protein synthesis. Corticosteroids have effects that tend to oppose those of insulin; they promote gluconeogenesis and the mobilization of amino acids and fatty acids. The role of the thyroid hormones appears to be the regulation of the oxidative breakdown of substrates for the supply of energy.

Other hormones, such as catecholamines, glucagon, and growth hormone, have more specialized roles which appear to be important in more special and unique circumstances, such as during maturation of the young and the needs that are related to reproduction. The principal effects of estrogens, androgens, and growth hormone are anabolic ones.

It is important to remember that no hormone

can exert its normal effects when acting completely alone. The control of intermediary metabolism is a truly multihormonal discipline. Thus, although the principal functions of each hormone are discussed separately in this book, such an arrangement is merely a convenient teaching device which is especially useful within a pharmacological framework.

B. Morphology of the islets of Langerhans

Paul Langerhans, in 1869, was the first to describe the islets of pancreatic tissue that bear his name, but it took about another 50 years before their endocrine role was established. The islets of Langerhans are scattered throughout the pancreas, but they are more common in the region of the tail than in the head of the gland. In normal humans they make up about 2 percent of the total pancreatic parenchyma, but this is reduced to less than 1 percent in most forms of diabetes. Embryologically, these tissues are derived from the epithelium of the exocrine ducts and they are endodermal. Adrenergic and cholinergic nerves of the autonomic nervous system are proximate to the endocrine tissues and even form junctions with them. The islets have an abundant blood supply. There are about 2 million islets of such endocrine tissue in the human pancreas, each with a diameter of 100 to 200 μm.
200 μm.

Each islet is composed of several types of cells, and the role of all of these is not yet clear. The three types of cells that secrete the known hormones have been positively identified, most recently using immunocytochemical techniques which utilize antibodies with radio or fluorescent labels to each hormone. Insulin originates in the B-cells (also called beta or β-cells), glucagon in the A-cells (α_2-cells), and somatostatin in the D-cells (delta or A_1-cells). Early experiments suggested that gastrin may also be present in the D-cells, but it is not yet clear if this is so.

The arrangement of the three types of cells in each islet show a definite pattern which can vary in different species. In man (Figure 8.2), there is a central core of B-cells surrounded on the periphery by A-cells which are closely associated with D-cells. This consistent and orderly arrangement may be related to the functioning of the tissue (Orci and Unger, 1975). Somatostatin from the D-cells may control the secretion of glucagon from the adjoining A-cells, and both may influence the B-cells near them. It has been suggested that the cells at the periphery function as a unit that rapidly responds to changes in the physiological needs for insulin and glucagon while the B-cells at the core provide the basic background need for insulin.

The cells are separated by narrow intercellular spaces, but they are also connected by "gap" junctions through which direct transmission of infor-

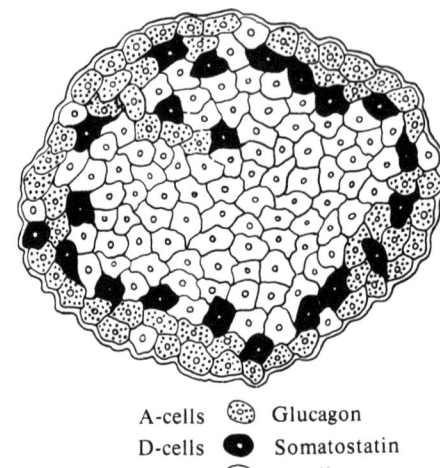

A-cells ⊛ Glucagon
D-cells ● Somatostatin
B-cells ⊙ Insulin

Fig. 8.2. Schematic representation of an islet of Langerhans, showing the distribution of glucagon, somatostatin, and insulin-containing cells. Islet cell types for which a function has not yet been positively established are omitted. (From Orci and Unger, 1975.)

mation, in the form of metabolites, may occur. There are also "tight" junctions between the cells, and it is thought that these may limit the free movement of substances through the intercellular spaces within the islets and so allow a local buildup and prevent an undesirable dispersion of regulators such as somatostatin.

The A-, B-, and D-cells of the pancreatic islets were originally distinguished from each other on the basis of their histochemical staining properties. Under the electron microscope they are readily distinguishable due to differences in the hormone storage granules that they contain. In the absence of such granules, however, they would be ultrastructurally indistinguishable from each other or from many other types of unrelated cells. The granules of the A-cells have a high electron density which is a little reduced toward the border of their peripheral bounding membranes. B-cell granules have an electron-dense core which often has an irregular appearance and a crystal-like structure which is similar to that of crystalline insulin. The peripheral regions of these granules are usually quite clear and are thought to contain the soluble "C"-peptide (see Section 8.3C) remnant of the insulin precursor molecule. The granules in the D-cells are slightly larger than in the other types of cells, are less electron-dense, and have a uniform texture. The remainder of the cells' structure is not remarkable; they contain rough endoplasmic reticulum, usually on one side of the nucleus, and a Golgi apparatus. These organelles are important for the synthesis of the hormones and their storage granules.

A microtubular system has been identified in the

B-cells (Lacy and Malaisse, 1973). These structures, which have a diameter of about 200 Å, are made up of units of a protein called "tubulin," which is like actin and may have contractile properties. They can be broken by a number of drugs, including the vinca alkaloids, vincristine and vinblastine, and colchicine. The structure can be stabilized by deuterium (D₂O), hexylene glycol, and ethanol. Such treatment has been found to prevent the release of insulin under *in vitro* conditions. The granules appear to be associated with the microtubules, which may contribute to their peripheral migration, preparatory to insulin release, in the B-cell. The antimicrotubular drugs appear to be quite specific in their effects on release, as they do not influence Ca uptake in response to glucose stimulation or the biosynthesis of the hormone. However, they also block release in response to other stimuli, such as leucine and sulfonylurea drugs.

A web of microfilaments has been identified at the periphery of the B-cell just below the cell membrane (Orci, Gabbay, and Malaisse, 1972). This region is 500 to 3000 Å wide and the individual filaments have a diameter of about 60 Å. The filamentous network is disrupted by cytochalasin B. This drug increases the secretion of insulin by cells which are stimulated by high external glucose concentrations, although it has no effect on the associated calcium uptake or insulin biosynthesis. It has been suggested that the microfilament web limits the fusion of the B-granules with the cell membrane so that, when destroyed by cytochalasin B, the process increases. However, the specificity of the action of cytochalasin B is at present uncertain, as it also appears to act on the plasma membrane, so proof of this theory is lacking.

The reason for the fragmentation of the endocrine tissue of the islets of Langerhans into so many small pieces of tissue which are dispersed about the pancreatic exocrine tissue is not known. Henderson (1969) has pointed out this interesting problem: "Why are the islets of Langerhans?" He has noted that this dispersal is not seen as clearly in all vertebrates, but it is very prominent in birds and mammals. It is also intriguing that certain gut hormones, such as secretin, gastrin, and cholecystokinin, stimulate secretion of *both* the endocrine and exocrine pancreatic tissues. Henderson suggested that the endocrine tissue may play a role in controlling the exocrine function of the pancreas, which could be facilitated by its dispersal into small units with their relatively large surfaces. Experimental support for this interesting hypothesis is lacking, but it was pointed out that the pancreas of alloxan-treated dogs (alloxan has a cytotoxic effect on B-cells) shows abnormalities in the exocrine duct cells. It is also well known that the pancreas of patients suffering from diabetes mellitus may be atrophied and fibrotic and is smaller than normal. A common explanation in textbooks is "that it may be an expression of the hereditary diabetic trait," but Henderson suggests that it reflects a lack of insulin.

Physiological activity, disease, and drugs may affect the islet tissue both qualitatively and quantitatively. Intense activity, such as its response to changes in plasma glucose levels or inadequate insulin formation in early juvenile diabetics, results in a degranulation of the cells. Certain drugs such as alloxan and streptozocin (see Section 8.7) are selectively accumulated by the B-cells and have a cytotoxic action on them. Normally in man, the islet tissue makes up about 2 percent of the total pancreatic parenchymous tissue, but it is less than half this in adult-onset diabetics and about 0.5 percent in juvenile onset diabetics. About 80 percent of the islet cells are insulin-secreting B-cells, but in diabetes mellitus the number declines markedly. In contrast, children of diabetic mothers have been observed to have a much larger mass (there is an increase in number and size) of B-cells, which may be as great as 10 percent of the total pancreatic parenchyma. The islet cells replicate themselves, but this appears to be a slow process about which there is little information in man (Logothetopoulos, 1972). The possibility of formation of new islet cells (neogenesis) has been explored. Persistent compensatory increases in response to surgical removal of part of the pancreas or the administration of alloxan have not been observed. A B-cytotrophic effect of the administration of sulfonylurea drugs or continual administration of glucose has been described in animals but not in man. The potential ability of the human islet tissue to morphologically repair its deficiencies thus does not seem to be promising.

8.2. Somatostatin

This tetradecapeptide hormone was first identified in the hypothalamus, where it inhibits the release of growth hormone [Section 3.2E(5)]. Somatostatin has, however, subsequently been identified at other sites in the body, including the gut and the islets of Langerhans, where it appears to act as a "paracellular" hormone. It has been localized in granules in the D-cells (Polak et al., 1975; Goldsmith et al., 1975). Prior to this observation, injected somatostatin was seen to inhibit the release of insulin and glucagon (Alberti et al., 1973; Koerker et al., 1974; Mortimer et al., 1974b; Gerich et al., 1974). These effects were seen in the presence of various stimuli that increased the release of these hormones, including glucose and arginine infusions, and tolbutamide. It was thus suggested that it may also play a role in regulating the release of hormones from the islets of Langerhans. As the D-cells are in

Table 8.1. *Summary of the general metabolic actions of insulin*

Glucose transport (cardiac and skeletal muscle, fat cells)	↑
Glycogen synthesis	↑
Gluconeogenesis	↓
Glycogenolysis	↓
Amino acid transport	↑
Protein synthesis	↑
Proteolysis	↓
Fatty acid transport	↑
Lipid synthesis	↑
Lipolysis	↓
K^+ accumulation in tissues	↑
Mg^{2+} accumulation in tissues	↑
Polarization of cell membranes	↑

close proximity to the A-cells in these tissues, somatostatin secreted by the former may regulate glucagon release by the latter. Direct measurements showing the inhibitory effects of glucagon on the release of somatostatin (Patton et al., 1977) and a stimulation of glucagon release by antisomatostatin serum (Barden et al., 1977) indicate that such a local control mechanism is physiologically functional. There appears to be preferential effect of somatostatin on the release of glucagon (Gerich, Lovinger, and Grodsky, 1975), but somatostatin released in response to elevated glucagon levels may also inhibit release of insulin from the B-cells (Patton et al., 1977).

Somatostatin appears to act on the A- and B-cells by blocking the entry of Ca^{2+} into the cells, thus preventing the extrusion of the hormones from their storage granules (Fujimoto and Ensinck, 1976; Basabe, Cresto, and Aparicio, 1977; Wollheim et al., 1977). The mechanism of release of somatostatin from the D-cells is unknown.

A somatostatin-secreting tumor has recently been identified in the pancreas (pancreatic somatostatinoma) (Larsson et al., 1977). It consisted of D-cells and was accompanied by a diabetic-type glucose tolerance as well as a hypochlorhydria (via its gastric effects), which is consistent with an excessive systemic release of somatostatin. Somatostatin, because of its diverse effects and short half-life in the circulation, has limited therapeutic application. It has, however, been shown to suppress insulin secretion in a pancreatic islet cell carcinoma (Curnow et al., 1975). Attempts to produce analogues with more prolonged and selective activities have met with some success [see Section 3.2E(5)].

8.3. Insulin

A. Functions and effects

Although the most acute and dramatic effect of insulin is to maintain plasma glucose at normal concentrations, it has equally important effects in promoting and maintaining cell lipids and proteins. In the absence of insulin, children fail to grow properly and tissue wasting occurs. Even when insulin is administered chronically to replace a deficiency of the hormone, long-term degenerative changes usually occur which especially affect the blood vessels. The causes of these latter changes are uncertain, although it is suspected that they may result from the difficulty in maintaining administered insulin at its appropriate physiological levels.

Insulin has the following major effects (see Table 8.1.):

1. Transport of metabolites. The *transport of glucose* into skeletal and cardiac muscle and fat cells (but not liver, kidney, or brain) is increased. This process is a passive one involving facilitated diffusion (which is "carrier-mediated") and can be shown to occur in isolated vesicles prepared from the plasma membranes of fat cells which have been treated with insulin (Carter, Avruch, and Martin, 1972).

This process can be inhibited by phlorizin and SH-blocking drugs such as N-ethylmaleimide. Other sugars, such as D-galactose, D-3-O-methyl glucose and 2-deoxyglucose, are also readily transported. The process is stereospecific, as while D-glucose is transported, L-glucose is not.

The *uptake of fatty acids* into fat cells is increased but only in the presence of glucose.

The *transport of amino acids* into liver, adipose tissue, and skeletal and cardiac muscle is increased. This effect is even seen for the nonmetabolizable amino acid, α-aminoisobutyric acid, indicating that the effect is independent of the intracellular changes in protein metabolism, which also occur in response to insulin. This process of amino acid transport occurs against a concentration gradient and is therefore *active*. It is thus dependent on a supply of metabolic energy and for certain amino acids is related to the presence of Na transport (cotransport). Amino acid transport occurs in the absence of glucose transport.

The *transport of ions* can be influenced by insulin. Possibly the earliest indication that insulin may influence transport was related to early observations that it produced a decline in plasma K^+ concentrations, reflecting the ions accumulation in the cells. This well-established effect has been utilized clinically for the treatment of hyperkalemia, such as may occur in renal failure. It has been demonstrated in muscle, liver, and fat cells. The reasons for the effect are controversial (see Zierler, 1972), but the response is accompanied by a hyperpolarization of the cells. There appears to be a decrease in the permeability of the cell membrane to K^+, but the efflux is reduced more than the influx. Potassium influx into the cell may be increased due to a stimulation of Mg-dependent Na,K-activated ATPase (see Hougen, Hopkins, and Smith, 1978). Insulin also increases the accumulation of Mg by cells (via an ATPase effect?) and it can stimulate Na transport across some epithelia. It has been suggested that insulin may contribute to the normal physiological regulation of intracellular K^+ levels (Cox, Sterns, and Singer, 1978).

2. Intracellular biosynthetic processes. While insulin's effects on transport of glucose, amino acids, and fatty acids across cell membranes contribute to the intracellular synthesis of glycogen, proteins, and triglycerides, the hormone may also have quite independent effects on the synthesis of these substrates. Insulin exerts separate and specific effects on these processes by modulating the activity of certain cellular enzymes. These effects of insulin are directed toward a buildup of glycogen, proteins, and lipids, which is accomplished not only by a stimulation of their synthesis but also by an inhibition of their degradation.

The *formation of glycogen* from glucose is promoted. This process, which occurs principally in the liver and muscle, is dependent on the activity of *glycogen synthetase* (or synthase). This enzyme exists as two types, an inactive *D form*, which, under the influence of insulin, can be converted to the active *I form*. Phosphorylation of the I type, under the influence of cyclic AMP, via a protein kinase, leads to the formation of the inactive D form. Thus, a decline in cellular cyclic AMP, which, it has been proposed, is mediated by insulin, favors the active I form (cyclic AMP *I*ndependent) of glycogen synthetase. Hormones that increase cyclic AMP levels, such as glucagon and epinephrine, will thus favor the phosphorylation of the enzyme to its inactive D form.

Gluconeogenesis, the synthesis of glucose from amino acids, is inhibited by insulin. This process, which occurs principally in the liver, is promoted by

corticosteroids and glucagon and involves the *de novo* synthesis or activation of certain hepatic enzymes which are opposed by the effects of insulin. Insulin also, less directly, reduces gluconeogenesis by inhibiting the mobilization of amino acids from muscle. The breakdown of glycogen to glucose by glucagon and epinephrine (glycogenolytic action) depends on the activation of a phosphorylase enzyme (via an increase in cyclic AMP), and this process may also be opposed by insulin.

The *synthesis of triglycerides* (*lipogenesis*) from glucose in adipose tissue is promoted. This process appears to involve a stimulation in the activity of pyruvate dehydrogenase. This enzyme exists in inactive phosphorylated and active nonphosphorylated forms, the formation of the latter being favored by insulin. Insulin also exerts an antilipolytic effect on adipose tissue, inhibiting the mobilization of fatty acids by glucagon and epinephrine. The actions of the latter hormones depend on the activation of a lipase enzyme via an increase in the cellular level of cyclic AMP. It is considered likely that insulin has the opposite effect on the nucleotide concentration and so antagonizes this response. In the absence of insulin the increased release of fatty acids results in depression of glucose metabolism, so that the lipogenic reesterification of cellular fatty acids declines further.

Insulin has important effects on *protein metabolism*, as reflected by the muscle wasting and negative nitrogen balance which is observed in its absence. It appears to both stimulate protein synthesis and inhibit proteolysis. These actions have been observed in muscle, adipose tissue, and liver. There is a general effect on the formation of all cellular proteins, but in some instances, such as in the liver, specific enzymes may be formed. There appears to be an increased formation of those enzymes concerned with glycogen synthesis and a decrease of those involved in gluconeogenesis. The mechanisms of these effects are poorly understood but appear to be independent of the actions of insulin in stimulating amino acid transport. It seems that there is stimulation of the translation process by the ribosomes. It is unlikely that genetic transcription via the formation of messenger RNA is involved, although there have been some reports that show an inhibition of insulin-stimulated protein synthesis by actinomycin D.

B. Structure and its relationship to activity
Insulin is a protein, which accounts for the early difficulties in its isolation from the pancreas, the exocrine tissue of which secretes proteolytic enzymes (see Humbel, Bosshard, and Zahn, 1972). Crystals of insulin were first prepared by J. J. Abel in 1926 and were shown to have a rhombohedral form. Their molecular weight is about 36,000, and they are hexamers consisting of three dimers ar-

Table 8.2. *Amino acid sequence in vertebrate insulins*

a. Amino acid sequences of insulin A-chains

Type of insulin	1	2	3	4	5	6	7	8	9	10	11	12	13	14
Human[a]	Gly	Ile	Val	Glu	Gln	Cys	Cys	Thr	Ser	Ile	Cys	Ser	Leu	Tyr
Sei whale	Gly	Ile	Val	Glu	Gln	Cys	Cys	*Ala*	Ser	*Thr*	Cys	Ser	Leu	Tyr
Horse	Gly	Ile	Val	Glu	Gln	Cys	Cys	Thr	*Gly*	Ile	Cys	Ser	Leu	Tyr
Cattle	Gly	Ile	Val	Glu	Gln	Cys	Cys	*Ala*	Ser	*Val*	Cys	Ser	Leu	Tyr
Sheep, goat	Gly	Ile	Val	Glu	Gln	Cys	Cys	*Ala*	*Gly*	*Val*	Cys	Ser	Leu	Tyr
Elephant	Gly	Ile	Val	Glu	Gln	Cys	Cys	Thr	*Gly*	*Val*	Cys	Ser	Leu	Tyr
Rat, mouse (I and II)	Gly	Ile	Val	*Asp*	Gln	Cys	Cys	Thr	Ser	Ile	Cys	Ser	Leu	Tyr
Guinea pig	Gly	Ile	Val	*Asp*	Gln	Cys	Cys	Thr	*Gly*	*Thr*	Cys	*Thr*	*Arg*	*His*
Chicken, turkey	Gly	Ile	Val	Glu	Gln	Cys	Cys	*His*	*Asn*	*Thr*	Cys	Ser	Leu	Tyr
Cod	Gly	Ile	Val	*Asp*	Gln	Cys	Cys	*His*	*Arg*	*Pro*	Cys	*Asp*	*Ile*	*Phe*
Tuna (II)	Gly	Ile	Val	Glu	Gln	Cys	Cys	*His*	*Lys*	*Pro*	Cys	*Asn*	*Ile*	*Phe*
Angler fish	Gly	Ile	Val	Glu	Gln	Cys	Cys	*His*	*Arg*	*Pro*	Cys	*Asn*	*Ile*	*Phe*
Toadfish (I)	Gly	Ile	Val	Glu	Gln	Cys	Cys	*His*	*Arg*	*Pro*	Cys	*Asp*	*Ile*	*Phe*
Toadfish (II)	Gly	Ile	Val	Glu	Gln	Cys	Cys	*His*	*Arg*	*Pro*	Cys	*Asp*	*Lys*	*Phe*

b. Amino acid sequences of insulin B-chains

Type of insulin	−1	1	2	3	4	5	6	7	8	9	10	11	12	13
Pig[b]		Phe	Val	Asn	Gln	His	Leu	Cys	Gly	Ser	His	Leu	Val	Glu
Man, elephant		Phe	Val	Asn	Gln	His	Leu	Cys	Gly	Ser	His	Leu	Val	Glu
Rabbit		Phe	Val	Asn	Gln	His	Leu	Cys	Gly	Ser	His	Leu	Val	Glu
Rat, mouse (I)		Phe	Val	*Lys*	Gln	His	Leu	Cys	Gly	*Pro*	His	Leu	Val	Glu
Rat, mouse (II)		Phe	Val	*Lys*	Gln	His	Leu	Cys	Gly	Ser	His	Leu	Val	Glu
Guinea pig		Phe	Val	*Ser*	*Arg*	His	Leu	Cys	Gly	Ser	*Asn*	Leu	Val	Glu
Chicken		*Ala*	*Ala*	*Asn*	Gln	His	Leu	Cys	Gly	Ser	His	Leu	Val	Glu
Cod	Met	*Ala*	*Pro*	*Pro*	Gln	His	Leu	Cys	Gly	Ser	His	Leu	Val	*Asp*
Tuna (II)	*Val*	*Ala*	*Pro*	*Pro*	Gln	His	Leu	Cys	Gly	Ser	His	Leu	Val	*Asp*
Angler fish	*Val*	*Ala*	*Pro*	*Ala*	Gln	His	Leu	Cys	Gly	Ser	His	Leu	Val	*Asp*
Toadfish (I)	Met	*Ala*	*Pro*	*Pro*	Gln	His	Leu	Cys	Gly	Ser	His	Leu	Val	*Asp*
Toadfish (II)	Met	*Ala*	*Pro*	*Pro*	Gln	His	Leu	Cys	Gly	Ser	His	Leu	Val	*Asp*

Note: The italicized amino acids indicate the principal differences.
[a] Sequence is identical in man, rabbit, dog, pig, and sperm whale.
[b] Sequence is identical in pig, horse, ox, dog, sheep, sperm whale, and sei whale.
Source: Humbel, Bosshard, and Zahn, 1972.

ranged around two atoms of zinc. The latter metal is necessary for crystallization and is present in the storage granules of the B-cells. Two monomers of insulin aggregate to form a dimer. The latter has little activity, the monomeric protein, with a molecular weight of 6000, being the active hormone.

F. Sanger in 1955 described the amino acid sequence of insulin. It consists of two adjacent chains of amino acids, an A-chain containing 21 residues and a B-chain with 30 (Figure 8.3). These are joined by two disulfide bridges provided by cystine residues.

There is considerable variability between insulins from different species of animals. More than 20 varieties have been identified, with substitutions being recorded in 29 of the 51 positions in the molecule. Some of these are shown in Table 8.2. It can be seen that insulin from guinea pigs differs from human insulin by 17 substitutions. The more commercially and medically useful pig and cattle insulins only differ, respectively, from the human variety by one and three amino acid replacements. There is surprisingly little difference in the biological potencies of the different animal insulins. However, when the amino acid disparities are large, as in guinea pig insulin, a difference in biological potency is seen. When assayed in rats, the latter hormone's ability to reduce glucose levels is only about 20 percent as great as that of bovine insulin (Zimmerman, Moule, and Yip, 1974). Animal insulins are the only commercial source of the hormone, so that the similarities in their activity

15	16	17	18	19	20	21
Gln	Leu	Glu	Asn	Tyr	Cys	Asn
Gln	Leu	Glu	Asn	Tyr	Cys	Asn
Gln	Leu	Glu	Asn	Tyr	Cys	Asn
Gln	Leu	Glu	Asn	Tyr	Cys	Asn
Gln	Leu	Glu	Asn	Tyr	Cys	Asn
Gln	Leu	Glu	Asn	Tyr	Cys	Asn
Gln	Leu	Glu	Asn	Tyr	Cys	Asn
Gln	Leu	Glu	*Ser*	Tyr	Cys	Asn
Gln	Leu	Glu	Asn	Tyr	Cys	Asn
Asp	Leu	*Gln*	Asn	Tyr	Cys	Asn
Asp	Leu	*Gln*	Asn	Tyr	Cys	Asn
Asp	Leu	*Gln*	Asn	Tyr	Cys	Asn
Asp	Leu	*Gln*	*Ser*	Tyr	Cys	Asn
Asp	Leu	*Gln*	*Ser*	Tyr	Cys	Asn

14	15	16	17	18	19	20	21	22	23	24	25	26	27	28	29	30
Ala	Leu	Tyr	Leu	Val	Cys	Gly	Glu	Arg	Gly	Phe	Phe	Tyr	Thr	Pro	Lys	Ala
Ala	Leu	Tyr	Leu	Val	Cys	Gly	Glu	Arg	Gly	Phe	Phe	Tyr	Thr	Pro	Lys	*Thr*
Ala	Leu	Tyr	Leu	Val	Cys	Gly	Glu	Arg	Gly	Phe	Phe	Tyr	Thr	Pro	Lys	*Ser*
Ala	Leu	Tyr	Leu	Val	Cys	Gly	Glu	Arg	Gly	Phe	Phe	Tyr	Thr	Pro	Lys	*Ser*
Ala	Leu	Tyr	Leu	Val	Cys	Gly	Glu	Arg	Gly	Phe	Phe	Tyr	Thr	Pro	*Met*	*Ser*
Thr	Leu	Tyr	*Ser*	Val	Cys	*(Gln*	*Asp*	*Asp)*	Gly	Phe	Phe	Tyr	*Ile*	Pro	Lys	*Asp*
Ala	Leu	Tyr	Leu	Val	Cys	Gly	Glu	Arg	Gly	Phe	Phe	Tyr	*Ser*	Pro	Lys	Ala
Ala	Leu	Tyr	Leu	Val	Cys	Gly	*Asp*	Arg	Gly	Phe	Phe	Tyr	*Asn*	Pro	Lys	
Ala	Leu	Tyr	Leu	Val	Cys	Gly	*Asp*	Arg	Gly	Phe	Phe	Tyr	*Asn*	Pro	Lys	
Ala	Leu	Tyr	Leu	Val	Cys	Gly	*Asp*	Arg	Gly	Phe	Phe	Tyr	*Asn*	Pro	Lys	
Ala	Leu	Tyr	Leu	Val	Cys	Gly	*Asp*	Arg	Gly	Phe	Phe	Tyr	*Asn*	*Ser*		

are of considerable practical importance. The differences can, however, confer immunogenic properties on the molecules. The natural variation in the structure of animal insulins has, in addition, provided considerable information of theoretical interest with respect to the relation of its molecular structure to its biological activity.

The three-dimensional structure of insulin has been determined (Blundell et al., 1971, 1972; Hodgkin, 1972) by X-ray analysis. A diagrammatic representation of the monomer is shown in Figure 8.4. The maintenance of its conformation is vital for its biological activity and depends on several factors, including the cystine disulfide bridges and a central hydrophobic core which includes a coiled α-helix between the 8 and 20 glycine residues in the B-chain (= B8 and B20). The nonpolar or hydrophobic groups face the inside of this core and stabilize the molecule in aqueous solutions. Other amino acids, however, also contribute to maintaining the correct three-dimensional structure and if a substitution produces a drop in activity, there is invariably an associated conformational change. The A-chain lies between the free terminal arms of the B-chain and above its central α-helix.

The relative importance of different parts of the molecule for its biological activity can be assessed by considering those residues that do not vary between species. These are shown in Figure 8.3. Substitution either in nature or in the chemist's laboratory show that the precise amino acid make up of

Table 8.3. *Effects of some changes in the structure of insulin on its biological activity and immunoreactivity*

Analogue[a]	% Biological activity		% Immuno-reactivity
	Fat cell	*In vivo*	
Des-Phe$_{B1}$		90[b]	103
Des-Gly$_{A1}$	2–10	10[b]	20–30
Des-Gly$_{A1}$-des-Phe$_{B1}$	1.6	7[b]	15
Des-(Phe$_{B1}$-Val$_{B2}$)	0.2		2
Des-(Gly$_{A1}$-Ile$_{A2}$)			
Desamino$_{A1}$	15		40
Acetyl$_{A1}$	40	100[b]	47
Acetyl$_{B29}$	75	100[b]	85
Diacetyl$_{B1.B29}$	85	100[b]	30
[*p*-Iodophenylalanine$_{B1}$]	60–70		85
[Di(*S*-carboxymethylcysteine)$_{A7B7}$]		100[b]	
		40[c]	
[Di(*S*-sulfocysteine)$_{A7B7}$]		15[c]	
Arg-Gly$_{A1}$	68	59[c]	40
Lys-Arg-Gly$_{A1}$	20		34

[a] A and B refer to the particular insulin chain.
[b] Blood sugar depression (rat). [c] Mouse convulsion assay.
Source: Zahn, Brandenburg, and Gattner, 1972.

the extreme terminal portions of the B-chain are of little functional importance.

The terminal invariant residues at both ends of the A-chain, however, are of considerable significance, so that even the addition of an acetyl or the deletion of an amino group in the Al glycine results in a substantial loss of biological and immunological activity. Deletion of this amino acid or the substitution of a more bulky residue can virtually abolish activity. Splitting the A7–B7 or A6–11 disulfide bridges have surprisingly little effect on the activity of the molecule. The effects of some such changes are summarized in Table 8.3.

Removal of the C-terminal octapeptide from the B-chain abolishes activity. This portion of the molecule, especially the B22–26, together with the B12 valine and B16 tyrosine, functions as a hydrogen-bonding system which is important for the aggregation to form the dimer and also for the monomers interaction with the receptor. There are several invariant amino acids in the A-chain – A1 glycine, A5 glutamine, A19 tyrosine, and A21 asparagine – which are exposed on the surface of the molecule and which also appear to be vital for

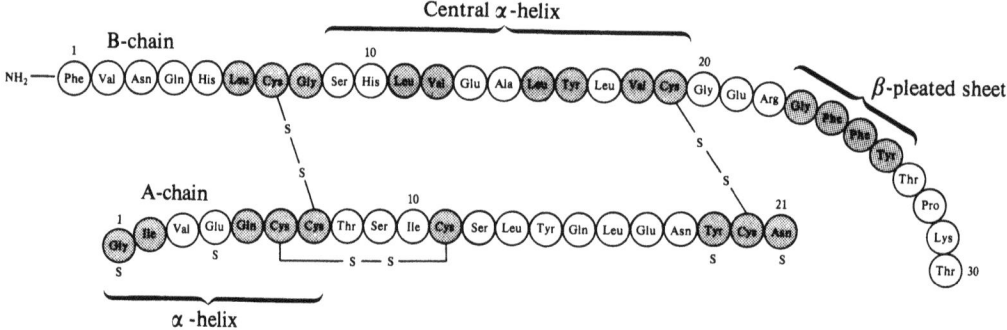

Fig. 8.3. Amino acid sequence of insulin, illustrating some of the features that may be important in determining the molecular aggregation of the hormone into dimers and hexamers and the interaction of the monomer with its receptors. Shaded circles indicate important invariant residues (there are some exceptions, as in coypu insulin); S, surface residues not involved in aggregation of insulin molecules but probably with the receptor interaction. The β-pleated sheet (nonpolar) is a hydrogen-bonding system which, with B$_{12}$ and B$_{16}$, is important for molecular aggregation. This is part of the hydrophobic "central core," which may also be involved in the interaction of the monomer with its receptor.

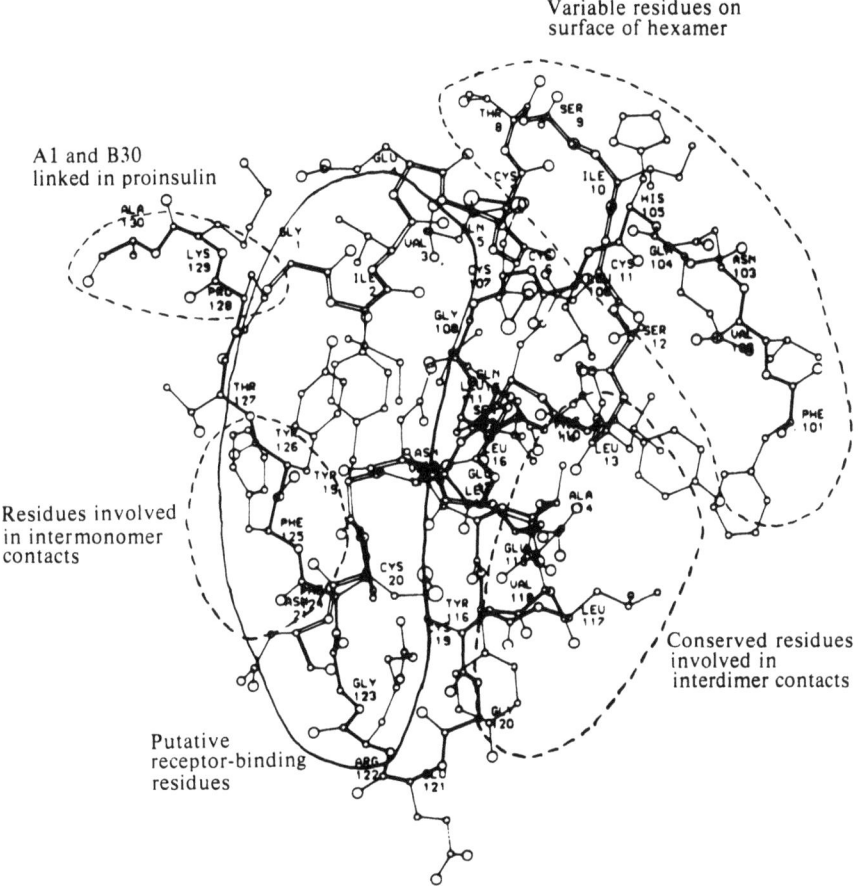

Variable residues on
surface of hexamer

A1 and B30
linked in proinsulin

Residues involved
in intermonomer
contacts

Conserved residues
involved in
interdimer contacts

Putative
receptor-binding
residues

Fig. 8.4. View of the insulin monomer, indicating how the surface may be described in terms of various areas involved in activation from the prohormone, binding to the receptor, and self-association. (From Blundell and Wood, 1975.)

the receptor interaction. These two major components together appear to function as the *receptor-binding region* for insulin, which involves hydrophobic interactions and the formation of hydrogen bonds (Pullen et al., 1976). The negative-cooperativity (see Section 2.11G) properties of insulin with respect to its receptor appear to depend on the C-terminal octapeptide in the B-chain and the A21 asparagine (De Meyts et al., 1978). The probable roles of the various parts of the molecules, in different aspects of its functions and the maintenance of its integrity, are shown in Figures 8.3 and 8.4.

Naturally occurring and synthetic analogues of insulin display differences in their affinities for the receptors, but they usually exhibit full intrinsic activity. Thus, although their biological potency may differ, it does so in parallel to the molecule's ability to bind to its receptor and, if enough of the polypeptide is present, a normal maximal response can usually be attained. In contrast to many other

hormones and biological excitants, the properties of affinity and intrinsic activity do not appear to reside in distinct and separate parts of the molecule.

The C-terminal part of the B-chain exhibits some properties of special interest, which have been explored by Weitzel and his colleagues (see Weitzel et al., 1973). In 1952, F. Sanger suggested that a part of the C-terminal region of the B-chain (B24–26, -Phe-Phe-Tyr-) may play a special role in determining the biological activity of the molecule. Weitzel found that when the terminal octapeptide (B22–30; -Arg-Gly-Phe-Phe-Tyr-Thr-Pro-Lys-Ala) was removed, the activity of the molecule was negligible. Removal of the terminal four amino acids, 27–30, made no appreciable difference to the biological activity. However, progressive removal of amino acids from positions 26 to 22 resulted in a gradual decline in activity which almost disappeared when the arginine at position 22 was deleted. It was also found that fragments containing the amino acids in the B22–26 region exhibited insulin-like activity (*in vivo* and *in vitro*) in rat diaphragm and adipose tissue. The smallest effective sequence was Arg-Gly-Phe, while Arg-Gly-Phe-Phe-NH₂ (B22–25) exhibited almost full agonist activity. Thus, the aromatic amino acids and the proximal arginine in this region appear to have an important role in the action of the hormone. It was very interesting to observe that inactivation of the insulin receptors in the tissues by incubating them with trypsin did not abolish the effects of the peptide fragments, suggesting that they can exert direct actions on the effector mechanisms involved. A peptide incorporating the B22–26 sequence, β-Ala-Arg-Gly-Phe-Phe-Tyr-NH₂ (DP-432), has been synthesized (Fujino et al., 1977) which exhibits a relatively high insulin-like activity in mouse diaphragm and adipose tissue, although not on liver, or plasma glucose levels.

Human insulin has been synthesized, but as the molecule is relatively large and complex, this process is at present too expensive to provide an alternative source to the hormones prepared from pig and cattle pancreases. The commercial supply of insulin is, however, not large, so that a more ready and abundant source of an active hormone preparation would be decidedly welcome. Thus, the design and preparation of suitable analogues that could be readily prepared by chemical synthesis or the modification of natural animal insulins is the ultimate aim of such studies of the structure–activity relationships.

C. Synthesis and release

Insulin is synthesized (see Steiner et al., 1974; Rubenstein, Horwitz, and Steiner, 1975; Steiner, 1977) in the B-cells and stored in granules from which it is released across the cell membrane by a process of emiocytosis (exocytosis). The mechanisms involved have been studied in considerable detail and involve a complex series of reactions which are incompletely understood. These are summarized in Figure 8.5.

The *rat* insulin gene has been isolated (Ullrich et al., 1977). The messenger RNA from the B-cells of rats and cattle has been translated in a cell-free (wheat germ) system (Chan, Keim and Steiner, 1976; Lomedico et al., 1977). The protein so formed has a molecular weight of about 11,500 (compared to about 6000 for insulin). It is a linear protein which has been called *preproinsulin* and normally only appears to exist intact in the B-cell for a very transient period of time. Preproinsulin has a "tail" of 23 amino acids at the N-terminus by which the ribosome, on which it is formed, is attached to the endoplasmic reticulum. A linear chain containing 86 amino acids separates from this molecule, and this contains the A- and B-chains of insulin joined by a segment containing

35 amino acids, including the "C" (*for "connecting"*)-*peptide*. This protein, which is called *proinsulin*, passes to the Golgi apparatus, where it is cleaved to insulin and packaged into storage granules.

Insulin is released from proinsulin by the action of two enzymes which act like trypsin and chymotrypsin, splitting the parent molecule at the 31 and 32 positions (where there are two arginine residues) and the 64 and 65 positions (a lysine and arginine) (Figure 8.6). These reactions release the three arginine and the lysine residues, leaving insulin and the C-peptide. This final process of activation of the hormone appears to begin in the Golgi apparatus but probably continues after the protein is packaged into its storage granules. The C-peptide is thought to be important in determining the correct pattern for folding and the formation of the disulfide bridges in the hormone.

The intact proinsulin molecule has little insulin-like activity, usually only about 5 percent of that of the hormone; however, it reacts with insulin antibodies. The C-peptide exhibits considerable interspecific variation in its amino acid sequence (Figure 8.7), and although this is of little or no significance with respect to the hormonal action of proinsulin, it confers immunogenic activity on the molecule. It is stored in the clear peripheral region of the granules and is released with the insulin. Commercial insulin preparations made from animal glands usually contain small quantities of contaminants, including proinsulin, which contributes to the formation of insulin antibodies when they are administered to another species. More highly purified preparations of insulin are available which have, for practical purposes, negligible antigenic properties [see Deckert, Andersen, and Poulsen, 1974; Oakley, 1976; and Section 8.5A(5)].

The C-peptide is released into the circulation and its concentrations in the plasma can be measured by radioimmunoassay. The antibodies are prepared using a synthetic human C-peptide as an antigen. The plasma levels reflect the insulin secretory activity of the B-cells, so this assay can be used to assess this function in patients suffering from diabetes mellitus. This test is especially useful, as determination of endogenous insulin may be impractical owing to the necessity to constantly administer exogenous hormone.

Many storage granules have been observed to contain a rhombohedral structure like that of insulin crystals prepared in the laboratory. They also contain zinc so that it appears that the insulin may be stored in the hexameric crystalline form. Following their formation, the storage granules migrate toward the periphery of B-cells preparatory to the release of their contents. This migration appears to be related to the activity of the microtubular system and can be prevented by such drugs as colchicine and vincristine (Lacy and Malaisse, 1973).

Beta Granule Formation

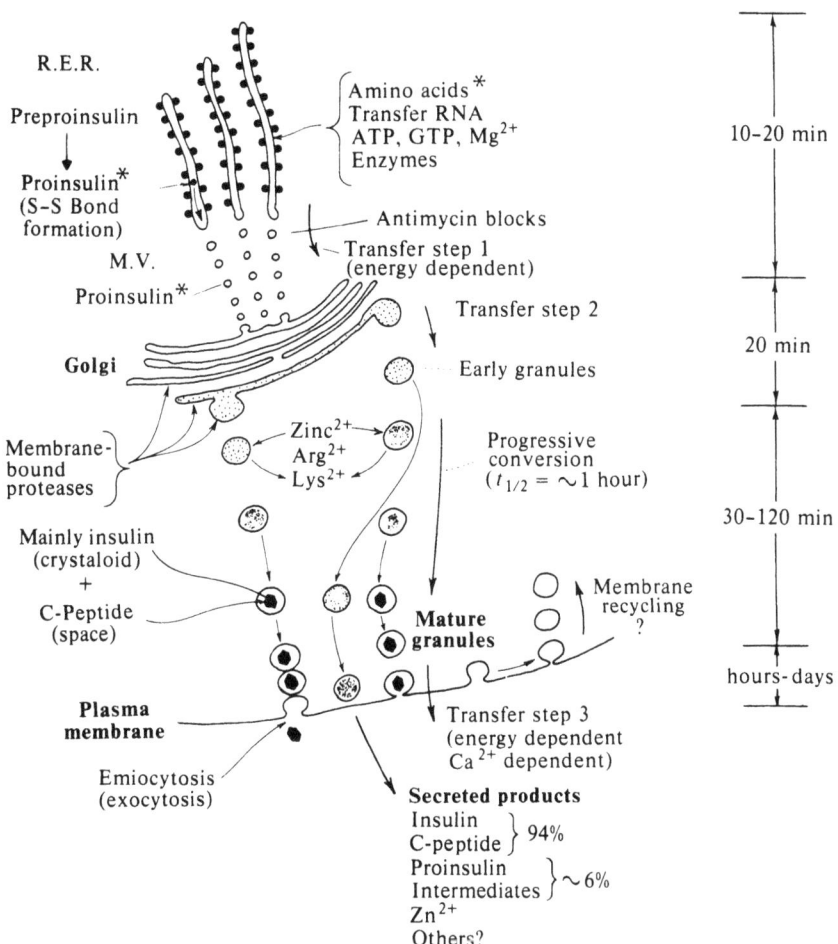

Fig. 8.5. Schematic summary of the insulin biosynthetic mechanism of the pancreatic B-cells. See the text for a discussion of this process. The time scale on the right side of the figure indicates the time required for each of the major stages in the biosynthetic process. R.E.R., rough endoplasmic reticulum; M.V., microvesicles. (From Steiner et al., 1974.)

Secretion of insulin occurs in response to a great variety of physiological and pharmacological stimuli (Table 8.4). In man, the most important physiological stimulus is an elevated glucose concentration in the plasma, but amino acid levels, especially leucine, are also important. The islets of Langerhans have a nerve supply (from the vagus) which can modify the process of release (Smith and Porte, 1976). There are cholinergic nerves, which have a stimulatory action, and adrenergic nerves, which can have dual effects: a release in response to β-adrenergic activity and an inhibition by an α-adrenergic action. A number of hormones, including glucagon, catecholamines, and gut hormones such as gastrin, secretin, enteroglucagon, and gastric inhibitory polypeptide, stimulate insu-

lin release. Such responses to these hormones could constitute part of the feeding response whereby insulin is mobilized in preparation for the needs that will follow the absorption of nutrients. Under the influence of these stimuli, the insulin storage granules are observed to fuse with the cell membrane and their contents pass to the outside of the cell. This process is called emiocytosis. It appears that there are specialized areas of the cell membrane where this process may occur and the granules can be seen to line up in queues awaiting the discharge of their contents.

The more intricate molecular details of insulin secretion by the B-cells have been studied. Release involves an interaction of excitants with receptors in the cell membrane and the initiation of a "cou-

Table 8.4. *Summary of stimuli that can influence the release of insulin* (in vivo *or* in vitro)

Increase
Glucose
Mannose
Glucosamine
Glyceraldehyde
Leucine
Epinephrine (β-adrenergic)
Prostaglandins
Glucagon
Gut hormones
Neural, vagal stimulation (cholinergic stimuli)
Sulfonylurea drugs (tolbutamide, chlorpropamide, etc.)
Ca ionophore (A23187)
PCMB
Ca^{2+}
K^{+} } *in vitro*
Ba^{2+}

Decrease
Epinephrine (α-adrenergic)
Somatostatin
Diazoxide
Thiazide diuretics
Mannoheptulose
Absence of Ca^{2+}
Prostaglandins

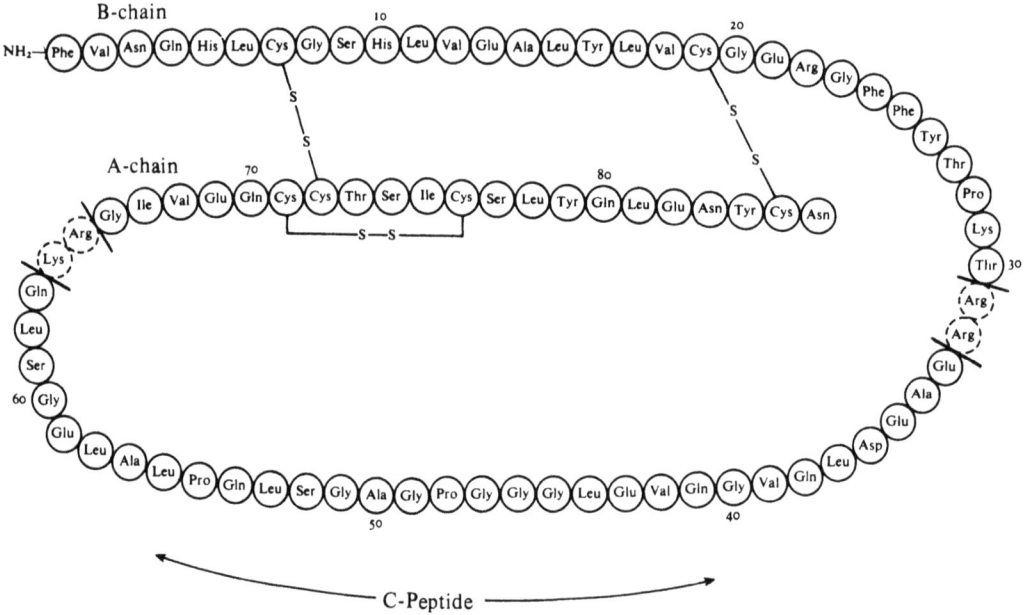

Fig. 8.6. Amino acid sequence of human proinsulin. The points of enzymic cleavage (trypsin and chymotrypsin-like) which result in the formation of insulin are indicated by the bars.

	1	2	3	4	5	6	7	8	9	10	11	12	13	14	15	16	17	18	19	20	21	22	23	24	25	26	27	28	29	30	31
Man	Glu	Ala	Glu	Asp	Leu	Gln	Val	Gly	Gln	Val	Glu	Leu	Gly	Gly	Gly	Pro	Gly	Ala	Gly	Ser	Leu	Gln	Pro	Leu	Ala	Leu	Glu	Gly	Ser	Leu	Gln
Monkey	Glu	Ala	Glu	Asp	Pro	Gln	Val	Gly	Gln	Val	Glu	Leu	Gly	Gly	Gly	Pro	Gly	Ala	Gly	Ser	Leu	Gln	Pro	Leu	Ala	Leu	Glu	Gly	Ser	Leu	Gln
Horse	Glu	Ala	Glu	Asp	Pro	Gln	Val	Gly	Glu	Val	Glu	Leu	Gly	Gly	Gly	Pro	Gly	Leu	Gly	Gly	Leu	Gln	Pro	Leu	Ala	Leu	Ala	Gly	Pro	Gln	Gln
Rat I	Glu	Val	Glu	Asp	Pro	Gln	Val	Pro	Gln	Leu	Glu	Leu	Gly	Gly	Gly	Pro	Glu	Ala	Gly	Asp	Leu	Gln	Thr	Leu	Ala	Leu	Glu	Val	Ala	Arg	Gln
Rat II	Glu	Val	Glu	Asp	Pro	Gln	Val	Ala	Gln	Leu	Glu	Leu	Gly	Gly	Gly	Pro	Gly	Ala	Gly	Asp	Leu	Gln	Thr	Leu	Ala	Leu	Glu	Val	Ala	Arg	Gln
Pig	Glu	Ala	Glu	Asn	Pro	Gln	Ala	Gly	Ala	Val	Glu	Leu	Gly	Gly	Gly	Leu	Gly	Gly	(-)	Gly	Leu	Ala	(-)	Leu	Ala	Leu	Glu	Gly	Pro	Pro	Gln
Cow, lamb	Glu	Val	Glu	Gly	Pro	Gln	Val	Gly	Ala	Leu	Glu	Leu	Ala	Gly	Gly	Pro	Gly	Ala	Gly	Gly	Leu	(-)	(-)	(-)	(-)	(-)	Glu	Gly	Pro	Pro	Gln
Dog	Asp	Val	Glu									Leu	Ala	Gly	Ala	Pro	Gly	Glu	Gln	Gly	Gly	(-)	Glu	Gln	Pro	Leu	Ala	Leu	Glu	Gly	Gln
Duck	Asp	Val	Glu	Gln	Pro	Leu	Val	Asn	Gly	Pro	(-)	Leu	His	Gly	Glu	Val	Gly	Glu	(-)	(-)	Leu	(-)	Pro	Phe	Gln	His	Glu	Glu	Tyr	(-)	Gln

Fig. 8.7. Comparison of the amino acid sequences of proinsulin C-peptides from man and several other mammals, and a bird. The solid bars indicate the residues that are identical in all species. (From Steiner et al., 1974.)

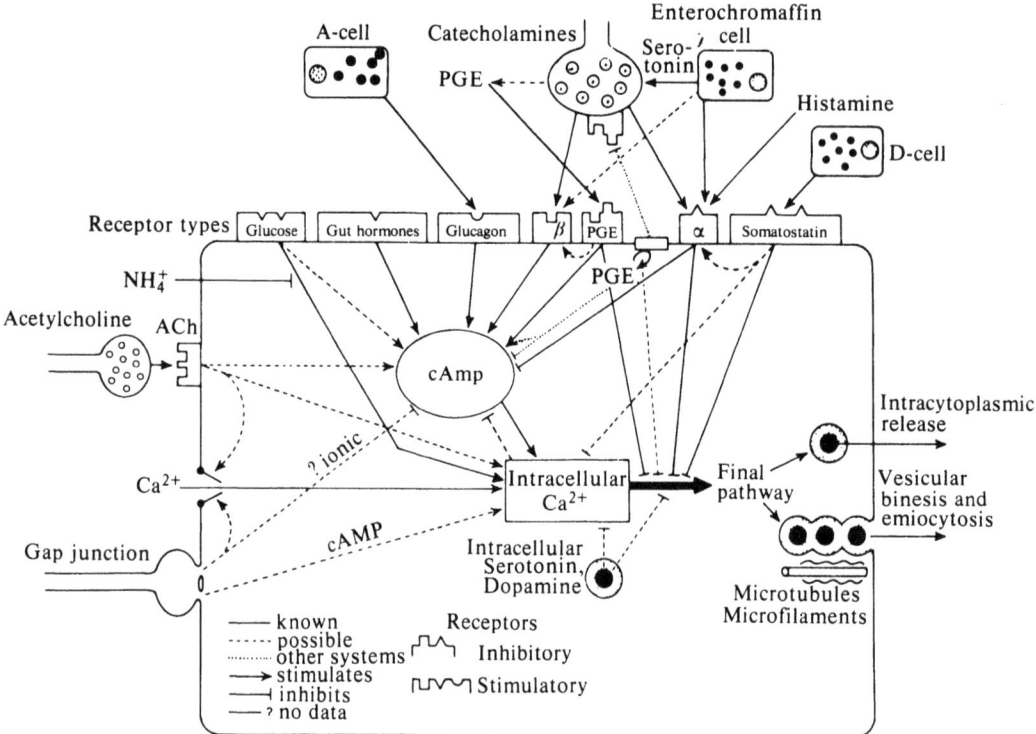

Fig. 8.8. Hypothetical model describing the various interacting systems that influence the release of insulin from the B-cell. (From Smith and Porte, 1976. Reproduced, with permission, from the *Annual Review of Pharmacology and Toxicology, 16;* copyright 1976 by Annual Reviews, Inc.)

pling reaction" which is dependent on the presence of Ca^{2+}. The response can be modulated and appears to be related to cyclic AMP levels in the cell. A hypothetical summary of the various processes is given in Figure 8.8.

Two broad categories of receptors are thought to exist on the cell membrane of the B-cell: *hormone receptors* such as respond to glucagon and gut hormones, and *glucoreceptors,* which are undoubtedly more important. Their presence is inferred from a variety of indirect evidence, including the observation that the B-cell responds by releasing insulin at concentrations just above the normal fasting glucose levels and that this response pattern follows that of a sigmoidal curve.

The precise nature of the signal and the *response of the glucoreceptor* is controversial. D-Glucose is its principal agonist, but mannose and glucosamine are also effective, as well as glyceraldehyde and leucine. These substances have a direct action and are called *initiators* (see Figure 8.9). There are, in addition, other compounds, such as fructose and *N*-acetylglucosamine, which alone are ineffective but which in the presence of low concentrations of glucose can increase the response of the B-cell and are thus called *potentiators.* The sulfonylurea drugs

(e.g., tolbutamide and chlorpropamide), theophylline, caffeine, certain amino acids, and vagal stimulation also act as potentiators.

There are two main models (Figure 8.9) for the glucoreceptors (see Ashcroft and Randle, 1975; Editorial, 1975g).

a. The *regulator-site model,* in which the agonist is thought to interact with the membrane receptor and promote a conformational change in the membrane resulting in an increased influx of Ca^{2+}. The evidence for such a mechanism depends on the positive identification of a response that does not involve metabolism of the agonist. Such evidence is controversial, although it has been observed for nonmetabolizable sugars, such as phlorizin, which substitutes for glucose in many of its carrier systems (Permutt and Kipnis, 1975). It has also been observed that the isomers of D-glucose differ in their ability to stimulate insulin release and this is not paralleled by their respective rates of metabolism. Thus, α-D-glucose is more effective than β-D-glucose (Niki et al., 1974).

b. The *substrate-site model,* in which the metabolism of the glucose is thought to be an indispensable part of the reaction, resulting in the formation of a metabolite that initiates the response, includ-

(a) Regulator-site model

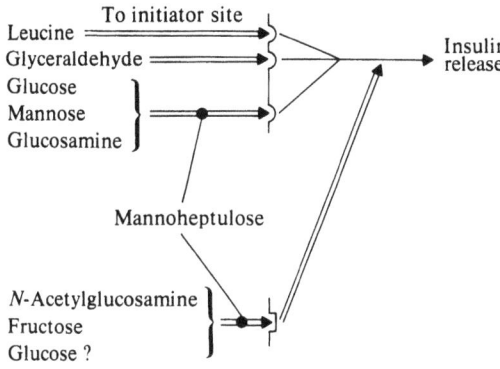

(b) Substrate-site model

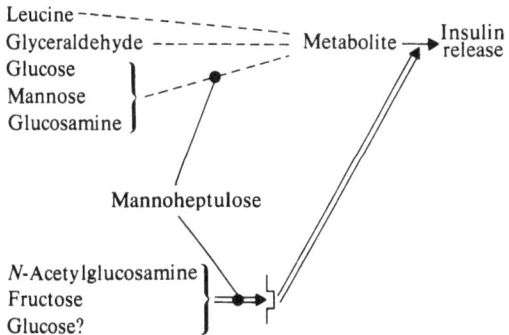

Fig. 8.9. Two models of the glucoreceptor, which controls the release of insulin from the pancreatic B-cell. (a) Regulator-site model. The B-cell is depicted as possessing initiator sites ( ) capable of eliciting insulin release when activated, and a potentiator site ( ), activation of which potentiates the effects of activation of the initiator site. (b) Substrate-site model, in which a trigger metabolite is formed whose action may be potentiated by activation of the potentiator site. The inhibitory effects of mannoheptulose are shown as filled circles. (From Ashcroft and Randle, 1975.)

ing the entry of Ca^{2+} into the cell. The only effective agonists are thus thought to be ones that can also be metabolized. Despite apparent exceptions to this rule (including those listed in the section above), this thesis has its adherents.

Each of these models is thought to have two receptor sites (Figure 8.9), a *primary* one that responds to initiators, such as glucose, and a *secondary* one, where potentiators, which modulate these effects, can act. The precise nature of the receptors is unknown, but they are thought to be present on the surface of the cell and they may be associated with membrane-SH groups (insulin can be released in response to stimulation by thiol-reactive compounds such as p-chloromercuribenzoate). In

the substrate-site model it has been suggested that there may be an enzyme associated with the receptor which controls the metabolism of glucose. The insulin-releasing effects of sulfonylurea drugs may also be related to the activity of membrane thiol groups (Hellman et al., 1976).

The response of the B-cell to increased glucose concentrations in the bathing media is dependent of the presence of external Ca^{2+} and can be prevented by verapamil, which can block membrane permeability to Ca^{2+} (Devis et al., 1975). The Ca ionophore A23187, which increases the permeability of membranes to Ca^{2+}, stimulates insulin release even in the absence of glucose (Sharp et al., 1975). Strontium and barium, but not beryllium, can substitute for the external Ca^{2+}. Glucose initiates action potentials and lowers the membrane potential of B-cells but only in the presence of Ca^{2+}. Calcium thus appears to be important for stimulus–response coupling as it is in many other tissues. Its precise role is unknown, but it could accelerate the movement of the granules to the cell membrane, possibly by an action on the microtubular or microfilament systems, or promote the fusion of the membranes of the granules and cell membranes. It is interesting that the changes in plasma Ca^{2+} levels that occur in parathyroid disease may be sufficient to influence release of insulin in man. Thus, insulin release appears to be reduced in hypoparathyroidism and increased in primary hyperparathyroidism (Yasuda et al., 1975).

The B-cells have an adenyl cyclase/phosphodiesterase system which mediates formation and metabolism of cyclic AMP (see Montague and Howell, 1975). The role, however, of this nucleotide in the control of insulin release is uncertain and controversial. Many of the reported experiments involving a possible role of cyclic AMP have not been confirmed or their interpretation has been criticized. Cyclic AMP does not appear to have a major direct role in influencing the secretion of insulin in response to glucose or leucine, but it may modulate this process (Ashcroft and Randle, 1975).

The action of a number of hormones and drugs that potentiate the release of insulin in the presence of glucose has also been shown to increase B-cell cyclic AMP levels. These include glucagon, epinephrine, prostaglandins, secretin, cholecystokinin, theophylline, and possibly sulfonylurea drugs such as tolbutamide (see Sharp et al., 1975). The latter drugs have been reported to act by stimulating adenyl cyclase and also by noncompetitively inhibiting B-cell phosphodiesterase. The actions of the hormones appear to be the result of an action on adenyl cyclase. The mechanism of the potentiating effect of cyclic AMP on glucose-mediated insulin release is uncertain. It was suggested that it may act by mobilizing intracellular Ca^{2+}, but this possibility is controversial (Sharp et al., 1975). The classical mechanism of action of cyclic AMP is an activation of a protein kinase which results in the phosphorylation and dephosphorylation, by a phosphatase, of cell proteins. Such kinases have been found in the B-cell, but the nature of their effects is unknown.

The *release of insulin can be inhibited* by catecholamines (an α-adrenergic effect) and somatostatin, as well as drugs such as the thiazide diuretics and their chemical relative diazoxide. α-Adrenergic stimuli have been reported to reduce cyclic AMP in the B-cell, but it is thought unlikely that such a response mediates their effect (see Smith and Porte, 1976). Changes in intracellular Ca^{2+} concentrations may be involved in these inhibitory responses. Epinephrine and diazoxide have been shown to stimulate Ca^{2+} efflux from B-cells (Brisson and Malaisse, 1973; Malaisse, Pipeleers, and Mahy, 1973), while somatostatin may inhibit the influx of this ion (Fujimoto and Ensinck, 1976).

Vagal stimulation can potentiate the release of insulin and this effect can be mimicked *in vitro* by cholinergic drugs such as acetylcholine, methacholine, and carbamylcholine (see Gagerman et al., 1978). In isolated mouse B-cells this effect appears to result from a depolarization of the cells and it does not involve cyclic AMP or cyclic GMP.

Secreted insulin is replaced as a result of its increased biosynthesis. This process is closely related to glucose stimulation of the B-cell glucoreceptor (see Permutt and Kipnis, 1975; Rubinstein, Horwitz, and Steiner, 1975) and does not automatically follow release in response to other stimulants. Thus, sulfonylureas stimulate secretion of insulin but do not enhance its biosynthesis (Steiner et al., 1972). Glucose fails to initiate insulin release when the external Ca^{2+} concentration is low but the biosynthetic response persists. Thus, although biosynthesis and secretion may both utilize the same glucoreceptor, the resulting responses appear to be distinct. Stimulation of insulin biosynthesis takes place rapidly, within a few minutes, in response to increased extracellular glucose concentration. This response is not affected by actinomycin D, indicating that a post-transcriptional process is involved. However, subsequent biosynthetic processes that occur about 40 minutes later can be partly inhibited by actinomycin D, suggesting that genetic transcription and the formation of messenger RNA could be involved.

D. Transport and metabolism

The normal concentration of insulin in human plasma is about 10^{-10} M. It has a half-life of about 5 minutes. Its nature in the blood has formerly been subject to considerable debate (see Berson and Yalow, 1966), but it appears that, contrary to earlier evidence, it is not significantly bound to plasma proteins. If antibodies are present, however, these will combine with it and so reduce its physiologically effective concentrations and also reduce its rate of inactivation. Insulin is probably present in the circulation as dimers or tetramers which can dissociate to the monomer, which is the active form.

Secreted insulin passes directly through the liver, where 40 to 50 percent is removed in a single passage (Field, 1972). Thus, the liver is not only an important receptor site for the hormone but is also the main site of its degradation. It is considered likely that specific binding of insulin to the plasma membrane must precede its inactivation (Terris and Steiner, 1975). Entry of the hormone, following such binding, into the peripheral regions of liver cells has been demonstrated (Gorden et al., 1978). The metabolic destruction of insulin (See Narahara, 1972) involves three types of reactions:

a. A nonenzymic disulfide–sulfhydryl reduction in the presence of thiols, such as cysteine and reduced glutathione.
b. An enzymic reduction of the disulfide bridges by an enzyme which in the presence of glutathione accelerates this process. Such an enzyme, glutathione-insulin transhydrogenase, has been isolated from liver and kidney.
c. Insulin can be inactivated by proteolytic enzymes which are ubiquitous in the body.

The latter process appears to be the most important and the principal site of this reaction is the liver (where a membrane-bound insulin protease is involved), although it also occurs in kidney and muscle. It seems that insulin can be filtered across the renal glomerulus, but little (about 1 percent) appears unchanged in the urine due to its reabsorption across the renal tubule.

As described in the next section, it is possible that some of the metabolites of insulin possess biological activity within the cell.

E. Mechanisms of action

Insulin appears to act, at least initially, on the outside of the cell membrane. When this hormone is covalently bound to materials such as sepharose beads, which are clearly too large to enter the cell, it still exerts its physiological action. Whether or not the insulin may leave such binding sites in sufficient amounts to have an effect has, however, been questioned (see Pilkis and Park, 1974). Some may also enter the cell.

The combination with the cell membrane triggers the definitive response of the cell to the insulin and involves a series of events which are only partly understood. The initial interaction occurs at specific receptor sites, and the complex that is formed appears to engender enough instability in the system to trigger the response.

1. Insulin receptor. Pharmacological inferences about the existence of receptors has in the last decade received dramatic confirmation from the study of the mechanisms of hormone action. These discoveries and their potential practical significance are especially notable with respect to the action of insulin. Two groups of endocrinologists led by P. Cuatrecasas and J. Roth have been especially active, and their results have been summarized in several reviews (Cuatrecasas et al., 1975; Hollenberg and Cuatrecasas, 1975a,b; Roth et al., 1975; Cuatrecasas, 1972a,b; Freychet, Roth, and Neville, 1971).

One of the basic tools for such studies is the preparation of a highly purified insulin which is labeled with ^{125}I and has a high specific activity. The binding of this ^{125}I-insulin to effector tissues such as liver and fat cells and lymphocytes can then be measured. Such molecules may bind to a variety of sites which are not specifically related to the tissues hormonal response. "Specific" binding of the insulin molecule to its effector tissues can, however, be demonstrated according to several criteria, including: (a) a saturability at physiological concentrations (10^{-10} to 10^{-11} M); and (b) displacement by "cold" unlabelled hormone and analogues of insulin, but not other polypeptide hormones.

The isolation of cell membranes from the effector tissues was found to result in a fiftyfold increase in the concentration of the specifically bound hormone. An especially dramatic discovery was the isolation of soluble insulin receptors (Cuatrecasas, 1972a,b) from liver and fat cell membranes with the aid of the detergents Triton X-100 and Lubrol-WX. These receptors have been concentrated and display similar properties to those in the intact cell membranes.

Such studies have provided a wealth of information about the nature and properties of the insulin receptor.

a. It is a glycoprotein with a molecular weight of about 300,000.
b. Incubation with enzymes show that it is partially buried in the cell membranes and faces in an outward direction. Its activity can be destroyed by trypsin.
c. Insulin combines with it reversibly and it is not directly changed or destroyed following such a combination. The dissociation constant, K_D, is about 10^{-10} M.
d. The response appears to be maximal when only about 5 percent of the receptors are occupied. This phenomenon has been described in many pharmacological preparations and appears to reflect a reserve of "spare" receptors.
e. There is an interaction (De Meyts, Bianco, and Roth, 1976) between the receptors so that dissociation of the hormone from them is enhanced as their occupation

increases (at physiological levels where 1 to 5 percent are occupied). This phenomenon is called "negative cooperativity" and is thought to be due to a specific region on the hormone called the "cooperative site" (De Meyts et al., 1978).
f. Their distribution in the cell may be influenced by the microfilament system, as binding can be reduced by cytochalasin A, B, and D (Van Obberghen, De Meyts, and Roth, 1976).
g. The insulin receptors can interact with other types of molecules, including the plant lectin protein concanavalin A and wheat germ agglutin (Cuatrecasas and Tell, 1973), somatomedins, epidermal growth factor, and nerve growth factor (see Megyesi et al., 1975, Hollenberg and Cuatrecasas, 1975a). These substances can displace ^{125}I-insulin in proportion to their endogenous insulin-like activity. They may, however, also have their own more specific receptors which reflect the physiological actions they may have in the body.
h. The continued presence of high concentrations of insulin reduces ("down regulation") the number of insulin receptors. There is some conflict as to the possible relevance of this observation to human disease, and it has been suggested that this phenomenon reflects a "proteolytic" activity of insulin (Huang and Cuatrecasas, 1975). It may result from a change in the affinity of the receptor for the hormone as a result of the hormone–receptor interaction. The response could reflect a regulatory mechanism for controlling the effects of insulin (Blackard, Guzelian, and Small, 1978).
i. Antibodies to purified insulin receptors have been prepared (Jacobs, Chang, and Cuatrecasas, 1978). It is interesting that they not only can bind to the receptors but can also elicit an insulin-like response. This observation confirms that such isolated receptors are the true physiological ones. The antibodies to insulin receptors do not displace bound insulin, so that each agonist thus appears to bind to a different site on the receptor.

The use of radioautography for the identification of ^{125}I-insulin on the plasma membrane of liver cells appears to reflect the presence of specific receptors at such sites (Bergeron et al., 1977). However using similar methods, it has been shown that following a period of incubation, the labelled insulin (possibly along with its receptor) is translocated into a region inside the cell periphery (Gorden et al., 1978). Fluorescent analogues of insulin have also been prepared which have allowed direct observation of the movements of the polypeptide molecule in living fibroblast cells (Schlessinger et al., 1978). Initially, the hormone was seen to be associated with the plasma membrane in which it could move in a lateral plane. After about 30 minutes it became immobile and appeared in endocytotic vesicles in the cell cytoplasm.

Intracellular receptor sites for insulin have some theoretical attractions, especially as many of the hormones actions involve effects that take place inside the cell. Their presence could reflect the cell's ability to regulate its receptor numbers (see, e.g., Posner et al., 1978) and contribute to the inactivation of the hormone. Thus, it is of special interest that such "receptors" have been identified in broken-cell preparations of nuclei (Goldfine and Smith, 1976) and the membranes from the Golgi apparatus (Posner, Josefsberg, and Bergeron, 1978; Bergeron et al., 1978). Such experiments, must, however satisfy stringent criteria which exclude possible contamination of the intracellular preparations by the plasma membranes.

It is possible that quantitative and qualitative differences in insulin receptors could be important causes of certain forms of diabetes mellitus. Suggestions that there is a reduction in the number of insulin receptors on fat cells in mature-onset diabetic patients who are insulin-resistant are controversial (Roth et al., 1975; Lockwood, Livingston, and Amatruda, 1975), although this has been observed in animal models (Kahn, Neville, and Roth, 1973). An insulin receptor disorder has been described in humans suffering from the skin disease acanthosis nigricans (Kahn et al., 1976). These patients were insulin-resistant. The disorder appears to have two causes: type A, with a primary receptor defect, and type B, where autoantibodies to the insulin receptor, which, it was suggested, may interfere with receptor binding of insulin, were demonstrated.

2. Nature of the coupling process. Although a modest amount of information is available about the biochemical nature of the final responses to insulin (see Section 8.3A), there is little unequivocal data about the processes that couple these effects to the formation of the insulin–receptor complex. However, there has been considerable speculation which involves two general types of processes: a direct effect on the cell membrane, and the possible formation of a "second messenger" which mediates the responses.

a. A change in the *conformation of the cell membrane* (see Zierler, 1972) has been proposed in an attempt to formulate a unitary theory to account for insulin's action. This hormone has an impressive number of actions on cell permeability, including changes in transport of ions, glucose, amino acids, and probably fatty acids. A general change in permeability could conceivably influence all these processes, although it is now clear that the precise mechanisms involved in each transport process differ. Probably as a result of observations regarding the primary importance of ion movements in initiating such processes as muscle contraction and glandular secretion, it also seemed possible that changes in permeability to ions could be initiating the other, intracellular, responses to insulin. It is now known, however, that the intracellular changes enjoy a measure of independence

from these membrane-related events. The possibility that this membrane hypothesis is correct was, however, reinforced by the interesting observations of Rodbell and his associates (Rodbell, 1966) that phospholipase C and A, which would be expected to deform the cell membrane, could mimic some of the effects of insulin. A broader view of the possible effects of insulin on the cell membrane has been provided by Zierler (1972). "Such a propagated change along and through the membrane could result in refolding of specific portions of the membrane . . . increasing permeability to glucose, revealing carrier sites, activating or inactivating membrane-centered or membrane-bound enzyme systems."

b. The general concept of the role of a *second messenger*, formed in response to an initial agonist–receptor interaction, was initially proposed to account for excitation–effector coupling in a number of excitable tissues, including those that respond to hormones. Known second messengers include Ca^{2+} and the nucleotides cyclic AMP and cyclic GMP. It has also been suggested that fragments of the hormone may, when released, mediate the actions of insulin.

Insulin has been shown to *decrease cyclic AMP* levels in a number of tissues, but the effects are usually small in magnitude and can only be demonstrated under certain conditions, such as when the nucleotide levels are elevated by glucagon or epinephrine. It has been suggested that such small changes may reflect a sequestration of the nucleotides into separate "pools," not all of which are responsive to insulin. The decreased cyclic AMP levels have been related to a decrease in activation of adenyl cyclase and/or an activation of phosphodiesterase (see Loten and Sneyd, 1970; Hepp, 1977; Kono, Robinson, and Sarver, 1975). The actions of the cyclic AMP may, in turn, be mediated via changes in the activation of cell protein kinases (Walaas, Walaas, and Grønerød, 1974).

Insulin has also been shown to *increase cyclic GMP* levels in fat cells, but this response is small and often difficult to reproduce (Illiano et al., 1973). It has been suggested that this cyclic nucleotide may be increased and oppose the effect of cyclic AMP (Hollenberg and Cuatrecasas, 1975b). A unitary hypothesis has been proposed to account for the actions of hormones such as glucagon and epinephrine, which increase cyclic AMP, and insulin, which opposes this effect but increases cyclic GMP (Figure 8.10). A single cyclase enzyme may exist in the cell membrane which, depending on the type of hormone present, may favor, for instance, the formation of cyclic GMP, which would in turn have the associated effect of reducing the quantity of the ATP-preferring form and lowering cyclic AMP levels.

Calcium has also been considered as a possible

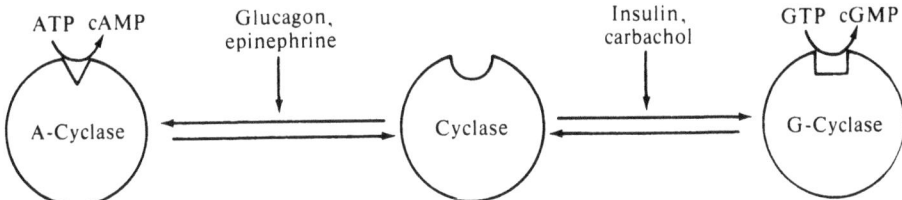

Fig. 8.10. Schematic representation of a hypothetical mechanism in which a single enzyme might catalyze the formation of either cyclic GMP or cyclic AMP, depending on the nature of the hormone agonist. (From Hollenberg and Cuatrecasas, 1975b.)

mediator of the effects of insulin (see Clausen, El-brink, and Martin, 1974; Kissebah et al., 1975). It has been proposed that insulin may release Ca^{2+} from a "pool" associated with the cell membrane. This could result in both a change in transport processes across the plasma membrane and, by increasing free intracellular Ca^{2+}, also initiate intracellular effects of the hormone. Insulin has been shown to have widespread actions on the calcium metabolism of cell organelles, including mitochondria, plasma membranes, and the endoplasmic reticulum (McDonald, Bruns, and Jarett, 1978). The latter authors suggested that Ca^{2+} is probably not, therefore, acting as a unique second messenger but could be "part of a tertiary messenger system concerned with the actions of the hormone."

It has been observed that certain fragments of the B-chain of insulin, especially the segment containing positions 22–26, exhibit some biological activity, even when the receptors are inactivated (see Section 8.3B). Steiner (1977) has suggested that such portions of the hormone may be split off the insulin molecule when, as described earlier, it enters the cell. Such fragments could be mediating some of the effects of the hormone, especially those that involve intracellular events. Steiner has shown that similar peptides may inhibit protein kinases *in vitro*. Thus, insulin fragments could be directly mediating the response. The normally low biological potency of the B22–26 fragments may reflect their difficulty in gaining access into the cell. This very interesting model has been called the *receptor/transducer/internalization model*. The hormone–receptor complex, once formed, allows the insulin to enter the cell where it is subject to degradation by proteolytic enzymes, and some of the peptide fragments may initiate the response.

8.4. Glucagon

The regulatory roles of insulin are intimately related to those of glucagon, which is secreted from closely associated anatomical sites in the A-cells of the islets of Langerhans. The presence of such a hormone with an opposing action to insulin was first suspected soon after the isolation of insulin. It was frequently observed that the administration of preparations of this hormone had a biphasic effect on blood glucose levels; the decline, which is the expected response to the insulin, was often preceded by an increase in the glucose concentration. It was subsequently shown that the latter effect was due to the presence of a distinct substance, which was called glucagon (referring to mobilization of glucose), which is synthesized and secreted by the A-cells. Glucagon is also secreted by A-type cells which are present in the gut (see Section 10.3E), and this hormone is called enteroglucagon. The role of glucagon in health and disease has been admirably reviewed by Unger and Orci (1976) and Unger et al. (1977).

A. Functions and actions

Glucagon has been shown to exert a variety of effects under experimental conditions *in vivo* and *in vitro* (Table 8.5). Many of these actions are of doubtful physiological significance and are only seen in the presence of excessive, unphysiological concentrations of the hormone. The most notable effects of glucagon are exerted on the liver, and this may be its only significant physiological site of action. This hormone, at concentrations of 10^{-10} to 10^{-11} M, rapidly mobilizes glucose from glycogen and stimulates gluconeogenesis, so that blood glucose concentrations increase. Hepatic glycogen synthesis is inhibited, which further contributes to the response. These effects are accompanied by increased formation of urea (ureogenesis) and later by the formation of ketone bodies. The latter effect, reflecting fatty acid metabolism, is thought to conserve amino acids, which, if used indiscriminately for the production of glucose, could contribute to a rapid depletion of the body protein reserves. Glucagon has thus been described as a hormone which is especially concerned with the supplying of energy substrates during starvation. Although it undoubtedly has such a role, it is more generally thought to be involved in less dramatic situations, such as the minute-to-minute regulation of hepatic glucose production and release, which it controls in conjunction with insulin. Glucagon acts on other tissues; it has a lipolytic effect on fat cells, it stimulates glycogenolysis and has a chronotropic and inotropic effect on the heart, and it causes a

Table 8.5. *Summary of the principal metabolic actions of glucagon*

Increased glyogenolysis
Increased gluconeogenesis
Increased lipolysis
Positive inotopic and chronotropic effects on cardiac muscle

release of K^+ and Ca^{2+} from the liver. These responses are probably not of physiological significance. Glucagon stimulates the release of insulin both by a direct action on the B-cells and indirectly as a result of the hyperglycemia it causes. The underlying reason for its many diverse actions is probably its ability to stimulate the formation of cyclic AMP, which normally mediates not only its particular effects but also has actions at many other sites not normally controlled by glucagon.

B. Structure and its relationship to activity

Glucagon is a polypeptide consisting of a linear chain of 29 amino acids (Figure 8.11) and has a molecular weight of about 3500 (Bromer et al., 1957). The sequence of the amino acids is remarkably uniform and has not been shown to differ in a variety of mammals, although bird glucagon may vary slightly from the mammalian variety. This conservatism may reflect precise structural requirements for its action. Indeed, analogues prepared by enzymic degradation or synthesis have little activity (Spiegel and Bitensky, 1969; Rodbell et al. 1971a). Even the removal of the histidine residue at the N-terminus abolishes biological activity, although this analogue can still bind to the receptor and act as an antagonist.

The tertiary structure of glucagon has been determined by X-ray analysis, and this is important in relation to its biological activity (Sasaki et al., 1975). In its active form it appears to have an α-helical arangement between amino acids 10 and 25. The formation of this helical type is favored in concentrated glucagon solutions, where it can form oligomers, and also in the vicinity of the receptor. Under these two sets of conditions, hydrophobic interactions are favored which result in the transformation of the hormone from one form to the other. These equilibria between the different types of glucagon are shown in Figure 8.12. The occurrence of the different structural varieties of glucagon may have important consequences with respect to its storage and inactivation, and its interaction with its receptors. In its storage granules it is

stabilized in its oligomeric crystalline form. In the circulation, where it is mostly present in its nonhelical configuration, it can be readily inactivated, which is necessary to allow it to function as an effective regulatory hormone. The binding of glucagon to its receptors appears to involve hydrophobic bonds (Rodbell et al. 1971a). Two such hydrophobic regions are present in the molecule, which with the acquisition of the α-helical arrangement, cluster in two groups at the surface of the molecule. These are about 20 Å apart and it has been suggested that each may react separately with different adjacent receptors.

C. Synthesis and release

There is little specific information available about the synthesis and storage of glucagon. However, it is known that this process occurs in the A-cells and that the hormone is packaged into granules. Like insulin, glucagon appears to be initially formed as part of a larger proglucagon precursor.

The principal stimuli controlling the release of glucagon from the A-cells is the concentration of glucose in the plasma and the local inhibitory effects of somatostatin released from the adjacent D-cells. Consistent with its action, glucagon secretion increases in response to hypoglycemia and is decreased by hyperglycemia. These effects are opposite in direction to the effects of plasma glucose on the release of insulin. Elevated fatty acid levels can also inhibit glucagon secretion, but this is probably of minor significance. Amino acid levels can stimulate release, and the infusion of arginine is a standard clinical test for A-cell function. Such aminergic stimuli are overridden by the inhibitory effects of elevated glucose levels if both should occur simultaneously. The glucagon is extruded from its storage granules by the process of emiocytosis, which appears to involve an increased influx of Ca^{2+} into the cells (see Wollheim et al., 1977).

Other hormones can influence glucagon release. This process appears to be inhibited by insulin and secretin. Other gut hormones, gastrin and cholecystokinin, and catecholamines with β-adrenergic effects, can stimulate the release of glucagon. Sympathetic nerve stimulation, such as during stress, can also initiate release of glucagon. The significance of these endocrine interactions is not clear, but it has been suggested that release of the gut hormones during feeding may contribute to changes in insulin and glucagon levels which are necessary during the process of absorption of nu-

Fig. 8.11. Amino acid sequence of glucagon.

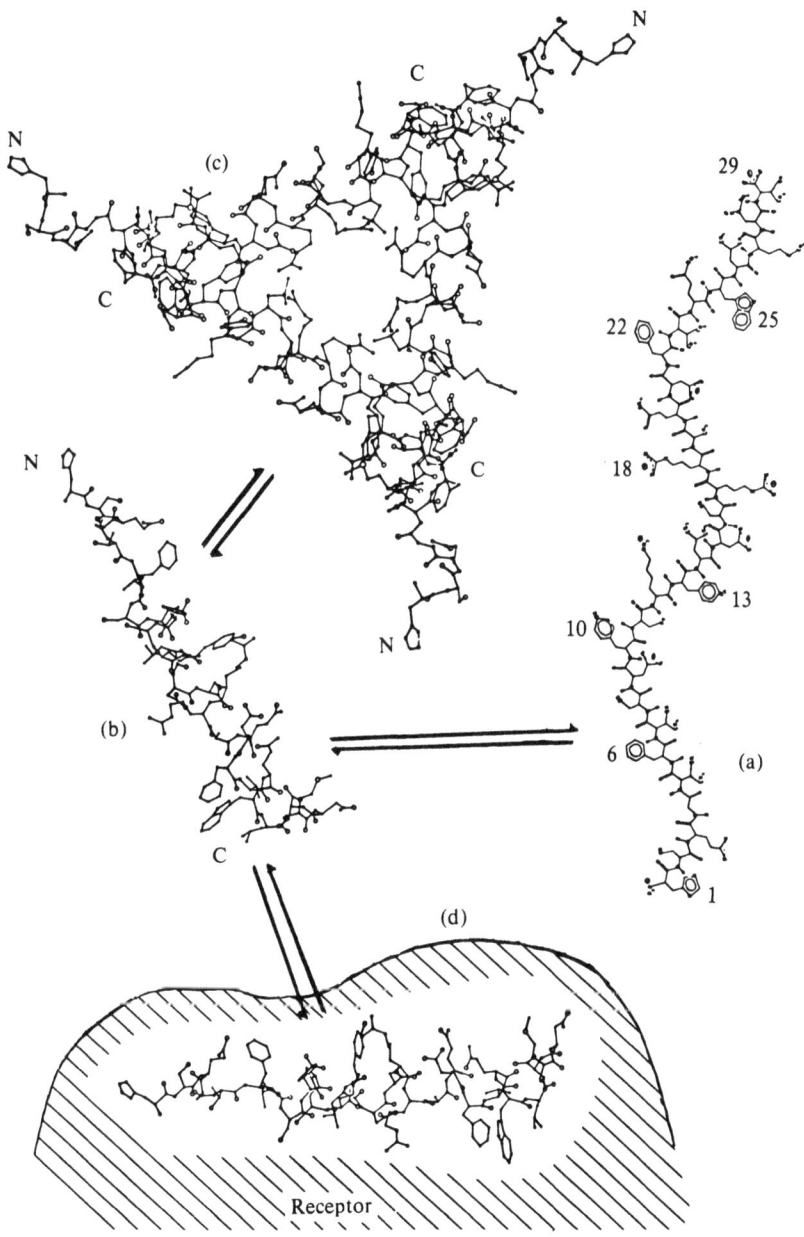

Fig. 8.12. Schematic representation of the equilibrium between different tertiary forms of glucagon. Hypothetical random coil structure (a) in equilibrium with a helical form (b). The helical form is stabilized either as trimers (c) or by association with a receptor (d) by hydrophobic interaction. (From Sasaki et al., 1975.)

trients. This response may be especially important during high-protein meals when the maintenance of optimal plasma glucose levels.may be necessary for the uptake of the absorbed amino acids and their incorporation into protein. The aminergic-stimulated release of glucagon may also be important on these occasions.

D. Transport and metabolism

Glucagon appears to circulate unbound in the plasma. Certain larger proteins with glucagon-immunoreactive activity, but lacking direct biological activity, have been identified in the plasma, but, despite earlier suggestions, these do not appear to represent bound forms of the hormone, although it is considered that one such substance ("large glucagon immunoreactivity" or LGI) could be a proglucagon. Glucagon probably retains little of its helical structure in solution in the plasma but attains its active tertiary structure under hydrophobic conditions, such as those that occur in the vicinity of the receptor.

Inactivation of glucagon occurs principally in the liver and kidney and is probably facilitated by the lack of its being in a tertiary structural form. The half-life of endogenous glucagon is about 3 minutes, but estimates based on the injected hormone are about double this value.

E. Mechanisms of action

1. Receptors. Information about the nature of the receptors for glucagon is principally due to the studies of Rodbell and his associates (see, e.g., Rodbell et al., 1971a,b, 1975). Glucagon labelled with ^{125}I and possessing high specific activity has been used in such studies, which have involved liver cells, fat cells, and cardiac muscle. This labelled hormone has been shown to be taken up and reversibly bound at specific sites (from which it can be displaced by unlabelled "cold" glucagon or 1-des-histidine-glucagon) on the cell membrane. As will be described in the next section, the action of glucagon is intimately related to its ability to activate adenyl cyclase. Isolated membrane preparations containing adenyl cyclase have been shown to take up the glucagon and exhibit parallel increases in the enzyme's activity. The receptor itself appears to be a lipoprotein and the interaction between it and the hormone seems to involve a lipophilic reaction. The relationship between the glucagon–receptor interaction and the activation of adenyl cyclase has been the subject of intensive investigations. These studies include the elucidation of the relationship between the receptor itself and the adenyl cyclase and the nature of the change that results in the increased activity of the enzyme. These problems have already been discussed in relation to the action of catecholamines on adenyl cyclase (Sections 6.14B and C).

Adenyl cyclase is a ubiquitous enzyme which is present in many cells and can be activated by a variety of substances. These include glucagon, ACTH, parathyroid hormone, thyrotropic hormone, luteinizing hormone, antidiuretic hormone, and the catecholamines, as well as such substances as the prostaglandins and cholera toxin. Although there is some specificity with respect to the sensitivity of adenyl cyclases from different organs to certain stimulants, the enzyme present in some tissues can respond to several such excitants (they are said to possess "multireceptors"). In fat cells for instance, activation of adenyl cyclase has been described in response to 5 or 6 different hormones,

including glucagon and epinephrine. The sensitivity to each excitant, however, varies and the actions of each hormone can be shown to be specifically blocked by certain drugs. For instance, propranolol inhibits the effect of epinephrine, 1-des-histidine-glucagon that of glucagon, and so on. The effects, however, are not additive, and a single adenyl cyclase appears to be involved which responds to the stimulation of a variety of distinct receptors (Birnbaumer and Rodbell, 1969).

The adenyl cyclase system for glucagon has been described as consisting of two associated parts: a "regulatory" subunit containing the receptor, and a "catalytic" subunit which mediates the formation of cyclic AMP. Combination of the hormone with the receptor could result in a conformational change so that the catalytic subunit is activated. A soluble preparation of adenyl cyclase has been isolated from cardiac muscle (Levey et al., 1974) and liver cells (Tomasi et al., 1970), and this has been shown to have a dissociable part which specifically binds glucagon. The total molecular weight of the enzyme preparation from cardiac muscle is 100,000 to 200,000, while that of the dissociable receptor region is about 26,000.

Thus, tissues such as fat cells possess a common adenyl cyclase which can be separately activated by a variety of hormones which act via distinct receptors. It is possible that each catalytic unit of the enzyme is in permanent physical contact with a series of separate hormone receptors, but such an arrangement appears to be rather clumsy and is not consistent with evidence which shows that the receptor and the adenyl cyclase may be present in different parts of the cell. Cuatrecasas and his collaborators (Bennett, O'Keefe, and Cuatrecasas, 1975; also De Haën, 1976) have presented an alternative hypothesis which they call "mobile receptor theory" (Figure 8.13). It is suggested that the cyclase exists as a distinct entity in the membrane but that it is only activated when it combines with a receptor, which exists separately. A variety of receptors, corresponding to distinct hormones, exist in the membrane, through which they can move. When combined with a hormone, such as glucagon, the receptors gain a specificity which allows them to interact with catalytic units of adenyl cyclase and so activate the cyclase.

The mobile receptor theory has received support from observations using a physical technique, radiation inactivation in an electron beam, to estimate the apparent target sizes of the glucagon–adenyl cyclase system. These measurements were made on plasma membrane preparations from rat liver cells in the presence and absence of stimulation by the hormone (Houslay et al., 1977). In the absence of stimulation by glucagon, the adenyl cyclase ("catalytic unit") had a target size of 160,000; when stimulated by the hormone, it was 380,000.

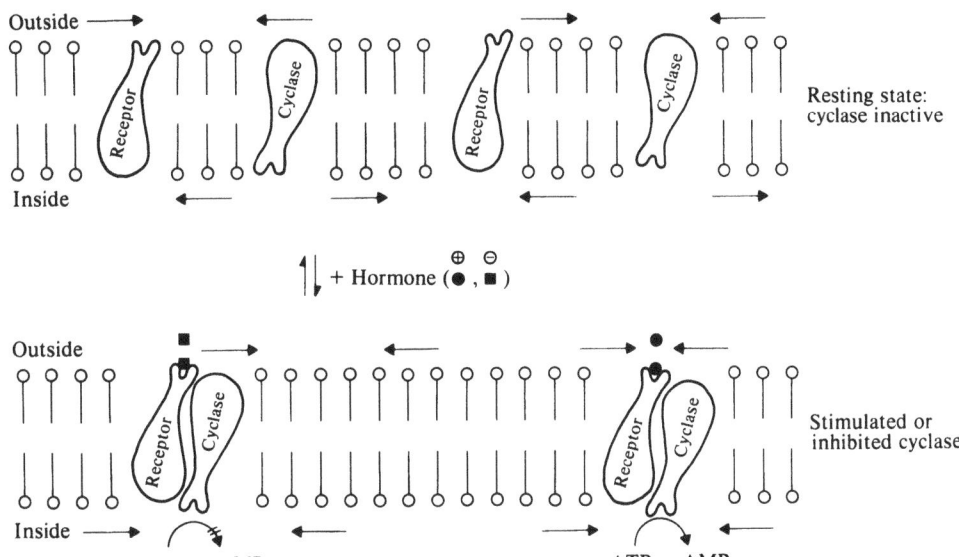

Outside →

Inside →

Resting state: cyclase inactive

↑↓ + Hormone (⊕●, ⊖■)

Outside →

Inside →

Stimulated or inhibited cyclase

ATP cAMP   ATP   cAMP

Fig. 8.13. Mobile receptor hypothesis for the mechanism of modulation of adenylate cyclase activity of cell membranes by hormones such as glucagon and epinephrine. The central feature is that the receptors and the enzyme are discrete and separate structures which acquire specificity and affinity for complex formation only after the receptor has been occupied by the hormone. These structures can combine after binding of the hormone because of the fluidity of cell membranes. The hormone binding sites of the receptor are on the external face, exposed in the aqueous medium, and the catalytic site of the enzyme is facing inward toward the cytoplasm of the cell. The activation of the cyclase may be either increased (⊕) or inhibited (⊖), depending on the nature of interaction with the receptor. (From Cuatrecasas et al., 1975.)

The size of the "uncoupled" glucagon receptor was found to be 217,000. It was concluded that the results were consistent with separate receptors and catalytic units being present in about equal numbers, and that in the presence of glucagon the two undergo a "locking interaction," so that a single larger target is observed.

The interaction of glucagon with its receptor and its ability to activate adenyl cyclase is related to the presence of the purine nucleotide GTP, which interacts with the regulatory subunit and modulates the action of the hormone. This system appears to be similar to that utilized when catecholamines react with the adenyl cyclase system. It is described in detail in Sections 6.14B and C.

Like insulin receptors, the numbers of glucagon receptors in tissues may be variable. It has been shown that in fat cells from rats injected with thyroid hormone (triiodothyronine), there is an increase in the number of glucagon receptors (Nistrup, Madsen, and Sonne, 1976). This change may account for the observed increase in the lipolytic response which is seen in such hyperthyroid rats.

2. Formation of cyclic AMP. Activation of adenyl cyclase results in the formation of cyclic AMP, which functions as a second messenger and initiates the responses to the glucagon. This process, which is now known to be involved in the actions of many hormones, was first described by Sutherland and his collaborators (see Sutherland, 1972) during their investigation of the mechanism of the glycogenolytic effects of glucagon and epinephrine on the liver. This action of glucagon is thus historically prototypical for the effects of several hormones.

Glycogenolysis under the influence of glucagon [and epinephrine; see Section 6.14C(1)], is mediated by a phosphorylase enzyme which converts glycogen to glucose-1-phosphate. It exists in two forms, inactive phosphorylase b and the active a variety. The process of glucose mobilization is also assisted by the inhibition of glycogen synthetase, which, under the influence of cyclic AMP, is phosphorylated from its I into its inactive D form.

Cyclic 3′,5′-AMP is formed from ATP under the influence of adenyl cyclase. It can be converted to 5′-AMP, which is inactive, by phosphodiesterase. This latter enzyme can be inhibited by methylxanthine drugs, such as theophylline and caffeine, which may thus mimic the peptide hormones actions. The cyclic AMP does not act directly on phosphorylase b but activates phosphorylase b kinase kinase (also called kinase II or kinase kinase). This enzyme consists of two parts, the "regulatory" and "catalytic" subunits; the holoenzyme being inactive. Cyclic AMP, by combining with the former, separates these segments and so, via the activated protein kinase, activates the enzyme phosphorylase b kinase, which, with the aid of ATP, phosphorylates phosphorylase b to a. This initiates the conversion of glycogen to glucose 1-phosphate and thus promotes the formation of glucose by glucagon.

8.5. Use of insulin in diseases associated with the islets of Langerhans

A. Diabetes mellitus

Diabetes mellitus is a disease that was known in ancient times and recognized by its association with the formation of large volumes of urine ("the pissing evil"), thirst, weight loss, and eventually coma and early death. The urine was found to have a sweet taste, due to the presence of glucose, which reflects the high concentrations of this sugar in the blood. Dehydration and thirst result from the osmotic diuresis, due to the renal excretion of the sugar. There are, however, many other notable facets of this disease. These include an excessive production of ketone bodies by the liver, a marked utilization of body proteins, and a predisposition to arteriosclerotic degeneration of blood vessels. Hyperglycemia occurs in a number of other endocrine diseases, such as pheochromocytoma, hypoglucagonemia, acromegaly, Cushing's disease, and hyperthyroidism.

Diabetes mellitus is, however, the most frequent cause of hyperglycemia and indeed the most common endocrine disease per se. It results from an insufficient production of insulin due to inadequate functioning of the B-cells of the islets of Langerhans. There may be an absolute deficiency of the hormone, or a relative one due to an increase in the normal needs in the body. The insufficiency of the endocrine tissue may be primary or secondary to some other condition.

Hypoglycemia associated with disease also occurs, but this is a rare condition. It is associated with a number of endocrine diseases, including hypothyroidism (cretinism), hypoadrenocorticolism (Addison's disease), hypopituitarism, and hyperinsulinism.

The most widely quoted figure for the overt occurrence of diabetes mellitus in Western countries is 2 percent of the population. Many of these people are unaware that they suffer from the condition. The estimated number can, however, vary considerably, depending on the precise geographic location, the ethnic group, the particular diagnostic procedures and criteria used, and the diligence of local health care programs. A recent report issued by the U.S. National Commission of Diabetes was far less conservative and indicated that in the United States the incidence was at least 5 percent of the population. About 35,000 deaths a year were directly attributed to diabetes, but probably more than 300,000 die as an indirect consequence of the disease. On this basis it is the third leading cause of death (behind cancer and heart disease) in the United States and the incidence is rising rapidly. The availability of insulin has reduced the death rate from diabetic coma to very low levels, but with the generally longer duration of life, problems associated with arteriosclerotic degeneration of the small blood vessels (microangiopathy) have become predominant. These changes occur in the kidney (nephropathy), the heart, the peripheral circulation, and the eye (retinopathy). With respect to the latter, diabetes is the second leading cause of blindness (some say the first). In diabetes mellitus there may also occur a degeneration of the nervous system (peripheral and autonomic neuropathy).

1. Classification. Diabetes mellitus is broadly classified into two types:

a. *Growth-onset (or juvenile)* diabetes, which makes its appearance early in life, usually before the age of 25 years, and includes about 10 percent of diabetics. In this condition there is a marked deficiency or an entire absence of insulin. This abnormality makes them prone to ketoacidosis, while growth is poor and there is muscle wasting. If untreated, such children die young, usually of infection or in a diabetic coma.

This category of diabetes has been subdivided (by the World Health Organization, WHO) into "infantile diabetes," with an onset before 15 years, and "juvenile-onset diabetes," which occurs between 15 and 24 years of age.

b. *Mature (or adult)-onset diabetes* is the predominant form of the disease. The vast majority of diabetics first contract the disease later in life; about 80 percent of diabetics are over 40 years of age. Such patients often have "normal" insulin levels or they may even be elevated. They, usually, however, show a "sluggish" response to the need for insulin, as there appears to be a delay in its secretion. The physiological problem is that the rates of secretion are no longer adequate for the patient's needs. Subsequently, degeneration of the B-cells may occur, but this is usually quite slow.

Mature-onset diabetics, unlike juvenile-onset ones, are not especially prone to ketosis and they are usually overweight. Reduction in weight will usually reduce their needs for insulin and somewhat ameliorate the disease. A reclassification into more functionally related categories, *insulin-dependent* and *noninsulin-dependent* diabetes, has recently been suggested.

2. Symptoms, effects, and prognosis. Before the discovery of insulin and its introduction for the treatment of diabetes mellitus, the average life expectancy following diagnosis of the disease was about 6 years. Death usually followed infection or diabetic coma, the latter being associated with severe dehydration, the accumulation of excess ketone bodies (ketosis), a metabolic acidosis, and a hypokalemia, following loss of potassium from the cells. The decline in the death rate, due to the availability of insulin, from this condition is shown in Figure 8.14. There has, however, been an alarming increase of morbidity from vascular diseases, especially those affecting the heart and kidneys. These changes are not initially apparent until about 10 years after the onset of the disease. The blood vessels of the retina are usually the first to suffer changes, which commonly lead to blindness. To add to this serious ocular problem, there is also an increased propensity to develop cataracts. It is widely suspected that a smooth efficient control of blood glucose levels with administered insulin may

delay the onset of such vascular problems, but definitive proof of this hypothesis is not available. It is, however, usually considered that early diagnosis and treatment of the disease will delay the onset of the vascular problems.

3. Causes. Complete information about the causes of diabetes mellitus could result in its prevention, more efficient treatment, and the application of measures calculated to reduce the problems of its long-term effects. Unfortunately, at this time there is no definitive information as to its causes, although there are many theories. Treatment is thus mainly pharmacotherapeutic, concentrating on the orderly replacement or supplementation of insulin and separate treatment of the associated disorders that accompany it.

Diabetes may have a number of possible causes which can be classified as follows (see for instance Craighead, 1978):

a. *Genetic influences* undoubtedly play a role, and it has been clearly demonstrated that a family history of the disease greatly increases the likelihood of its occurrence. It has been estimated that 20 to 30 percent of the population carry such "diabetic genes" but that this is only manifested in about 2 percent of the people. The detailed manner of inheritance and whether a single or several genes are involved is controversial. The site of such a genetic effect is also unknown, but it is thought to involve the B-cell, possibly via immune responses to toxic agents.

b. *Primary damage* to the islet cells is a possible cause of diabetes mellitus and could result in a failure of hormone synthesis or a failure of the secretory mechanism. Generally, however, the islet cells appear to be quite resistant to damage. In cases of pancreatic diseases, such as pancreatitis, B-cell function is rarely compromised, although a temporary decline in glucose tolerance may be observed. Physical trauma is also rarely a problem and a large proportion of the pancreas must be destroyed before a glycosuria is seen. Chemical damage is also rare, but the B-cell can be destroyed by some noxious chemicals such as alloxan and streptozocin. Degeneration of the B-cells as a result of autoimmune disease is considered likely, but a definitive description of its incidence is lacking. Damage due to infection, such as mumps and other viruses, is currently being investigated (see Maugh, 1975b), especially when it occurs in early postnatal or prenatal life.

c. The deficiency in B-cell function may not be a primary cause of diabetes mellitus. It is possible that degeneration of the insulin-secreting tissue occurs *secondarily* as a result of excessive stimulation from some other source. This could be hormonal or due to a metabolic defect which results in elevated blood glucose (see "The beta cell in

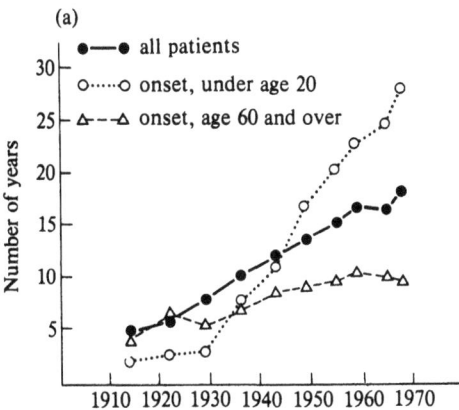

(a)

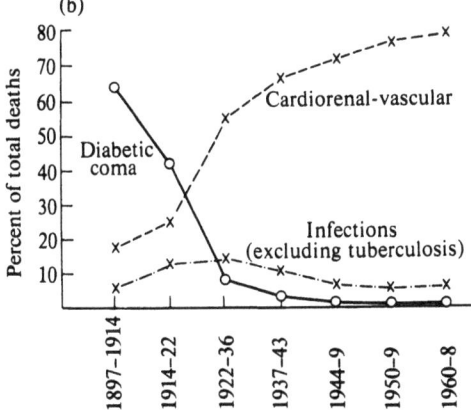

(b)

Fig. 8.14. Changes that have occurred in this century in the duration of life after the onset of diabetes mellitus (a) and the causes of death (b) in patients suffering from the disease. Insulin was introduced for the treatment of diabetes mellitus early in the 1920s. (From Marble, 1972.)

Diabetes – more sinned against than sinning?"
Editorial, 1977h). This would be consistent with
the "exhaustion theory" for the genesis of diabetes
mellitus.

d. An *increased "resistance"* to the action of insulin
has been demonstrated in a variety of conditions
involving other endocrine diseases. Some of these
are shown in Table 8.6, and include disorders of
the pituitary, adrenal cortex and medulla, pancreatic A-cells, thyroid, and gonads. The administration of corticosteroids and growth hormone can
result in diabetes. Such effects appear to be largely
the result of the metabolically antagonistic actions
of such hormones on glucose, fat, and protein metabolism. Changes in the responsiveness of the effector tissues, such as could involve changes in the
numbers and sensitivity of insulin receptors, may
also occur. Insulin-receptor antibodies have been
identified. The presence of circulating insulin antagonists has also been postulated, and the presence of insulin antibodies undoubtedly gives rise to
an increased need for insulin in those already
being treated for the disease. The occurrence of
insulins with an abnormal amino acid sequence is
also possible. Obesity is a well-recognized factor in
the manifestation of diabetes, which may (or may
not) be accompanied by a decline in insulin receptors [see Section 8.3E(1)]. Loss of weight undoubtedly reduces the need for insulin in mature-onset
diabetics. In the early stages of the disease, especially "chemical diabetes," when fasting glucose
levels are normal but glucose tolerance tests show
an abnormality, weight loss may result in a remission or even cure.

e. The A-cells of the islets of Langerhans often
proliferate in diabetes mellitus, and this condition
is invariably associated with a relative or absolute
excess of glucagon (Unger and Orci, 1976). There
has thus been considerable speculation as to the
role of this hormone in diabetes (see Unger et al.,
1977). In normal people or diabetics who are receiving insulin, the infusion of glucagon does not
cause glucose intolerance (Sherwin et al., 1976),
and its effects can apparently be compensated for
by insulin. Diabetics who are not receiving insulin,
however, show a very much exaggerated response
to glucagon, being 5 to 15 times more responsive.
It seems unlikely that elevated glucagon levels
play a primary role in the genesis of diabetes
mellitus, but the presence of this hormone may
exaggerate the hyperglycemic and ketotic effects of
insulin lack. This process has been called the
"bihormonal-abnormality hypothesis," in which it is
proposed that glucagon may play a role in the abnormal regulation of glucose in diabetes (see
Unger and Orci, 1976; Unger et al., 1977).

4. Prevention. Attempts to prevent diabetes
mellitus are relatively unsuccessful, partly reflecting a lack of precise information about its causes.

Table 8.6. *Association of insulin resistance with various diseases and altered metabolic conditions*

Hyperadrenocortisolism
  (Cushing's syndrome)
Growth hormone hypersecretion
  (acromegaly)
Catecholamine hypersecretion
  (pheochromocytoma)
Hyperthyroidism
Glucagon hypersecretion
  (pancreatic A-cell tumor)
Estrogen excess
  (pregnancy, oral contraceptives)
Diabetes mellitus
  (untreated, nonobese, nonketotic)
Streptozocin-induced diabetes
Starvation
Uremia
Myotonic dystrophy
Obesity

Source: Lockwood, Livingston, and Amatruda, 1975.

"Genetic counselling" and the discouragement of
marriage or childbearing in individuals who both
have a family history of the disease is considered
advisable, although it seems to be rarely adopted
or acted upon. The onset of the disease can often
be retarded in genetically predisposed individuals,
or in "latent" diabetics who show abnormal glucose
tolerance tests, or who have previously suffered diabetic "episodes." The measures that can be
adopted include discouragement of more than two
or three pregnancies and the avoidance of stressful
situations and obesity. Certain drugs may also precipitate diabetes in such individuals, and these include thiazide diuretics and corticosteroids. The
use of the oral antidiabetic drugs as a prophylactic
measure in early asymptomatic diabetes (as indicated by an abnormal glucose tolerance test) appears to be ineffective in preventing future progress of the disease (Shen and Bressler, 1977b).

**5. Treatment of diabetes mellitus: the role of
hormones and drugs.** The aims of treatment (see
Hockaday, 1974; Stowers, 1976a,b) in diabetes
mellitus are to alleviate the symptoms of the
hyperglycemia, forestall the development of
ketoacidosis and diabetic coma, and delay the onset
of the long-term vascular and neurological complications. These aims differ somewhat depending on
the type and stage of development of the
disease. For instance, "chemical diabetes" is usually
asymptomatic, but its recognition and treatment
are thought to be important in delaying the full
manifestations of the disease. The levels of glucose
in the plasma and the urine are the primary practical objective determinants used to assess the acute
success of treatment. Ultimately, however, the in-

cidence of degenerative changes in the tissues is of major importance.

The control of blood glucose concentrations in insulin deficiency can be influenced in several ways. While the administration of exogenous insulin is usually necessary, this treatment may be delayed or the amounts required can be reduced by modifying the patient's diet and manner of life. Stimulation of the sympathetic nervous system can inhibit release of insulin and promote glycogenolysis, so that unnecessary stress and excitement should be avoided. Similarly, infection, physical trauma, and pregnancy commonly increase the need for insulin, while exercise may decrease it. Nervous depression is diabetogenic, but some drugs used to treat it, such as the monoamine oxidase inhibitors, reduce the need for exogenous hypoglycemic drugs.

Dietary control. Dietary control is basic to all forms of treatment of diabetes. It has several aims:

a. The limitation of sudden rapid increases in carbohydrate intake, which increase the need for insulin. This is performed by reducing and more uniformly distributing the intake of refined carbohydrates, especially sucrose, which can be rapidly absorbed.

b. *Loss of weight* in mature-onset diabetics who are obese will reduce insulin "resistance" and the need for the hormone.

It has been suggested that arteriosclerotic changes may be associated with elevated blood cholesterol levels, part of which is acquired in the diet. Although the evidence for this theory is equivocal, it is felt by many that a *limitation of cholesterol intake,* and the use of unsaturated fatty acids, which limits cholesterol synthesis, may reduce the long-term vascular problems of the disease. In one study of borderline diabetics (Jarrett et al., 1977), it was found that progressive vascular changes were related to the blood pressure. Treatment with hypotensive drugs, combined when necessary with hypolipidemic treatment, was more effective than conventional antidiabetic therapy in preventing vascular degeneration.

Sometimes the diabetes can be controlled by dietary measures alone, but when this is not possible it is necessary to supplement them pharmacologically, and this can involve the administration of insulin. If there is functional B-cell tissue, drugs that stimulate its secretion or further reduce the need for the hormone may be considered.

Insulin. Insulin is always needed for the treatment of diabetic coma and those patients who are prone to ketosis. These people include all those suffering from growth-onset diabetes. Mature-onset diabetics usually have some functioning B-cell tissue and are not likely to become ketotic. They can thus be maintained, especially in the ear-

lier phases of the disease, with the aid of the sulfonylurea drugs, which stimulate B-cell function, or with biguanides, which reduce the need for insulin. The controversy about the use of these oral antidiabetic drugs as an alternative to insulin will be discussed in the next section (Section 8.6). It is, however, now widely considered that diet should be considered first, and if this alone is ineffective, insulin should also be the primary choice for the treatment of mature-onset diabetes.

As insulin is a polypeptide, it is destroyed in the gut and so must be administered by the subcutaneous route, although in emergencies, such as in diabetic coma or acute insulin resistance, it can be given intramuscularly or intravenously. It is rapidly destroyed when in the circulation, where it has a half-life of about 5 minutes. It is thus necessary to administer it frequently or create "depots" in the tissues from which it can be slowly absorbed. Preparations are available that need only be administered once or twice a day.

A variety of insulin preparations are provided commercially, but the use of many of these is no longer justified. Some of these are shown in Table 8.7. The number of such preparations reflects the various problems associated with the administration of exogenous insulin rather than commercialism. The aims, which have been largely achieved, are as follows:

To supply insulin in forms suitable for injection, mainly s.c. but also i.v. and i.m., which can be absorbed at different rates. They thus have different periods in the onset and duration in their actions and are classified as *short-, intermediate-,* and *long-acting.* The early preparations of insulin were rather crude pancreatic tissue extracts which were contaminated with a variety of proteins. When these were further purified, it was found that the hormone acted more promptly but that its duration of action was shorter. The contaminants apparently delayed its absorption from the sites of injection. Attempts were therefore made to reproduce this effect by adding proteins to the "purified" hormone, the principal ones being protamine and globin, which in the presence of zinc chloride form complexes with insulin. Depending on their concoction, they have different time courses of action. *Protamine zinc insulin* suspension (PZI) is long-acting, while *globin zinc insulin* injection and *isophane zinc insulin* suspension (or *NPH insulin* = neutral-protamine-Hagedorn, from whose laboratory it originated) are intermediate-acting. Soluble insulin preparations are absorbed rapidly and can be given i.v., so they have a rapid onset but a short duration of action. The first such preparations to be generally used contained some zinc, which is necessary for crystallization of the hormone and had a pH of 2.5 to 3.5. The acid nature of this preparation had several disadvantages, as it is more likely to cause irritation s.c.,

Table 8.7. Properties of various insulin preparations

Name	Duration of action (h)	Action (h), peak/onset	pH	Appearance	Source	Added protein	Special features
Soluble insulin							
Unmodified	Short ⎫	2–5 ⎫	Acid	Clear	Mainly beef	0	
Nuso	Short ⎬ 6–8	2–5 ⎬ 1	Neutral	Clear	Mainly beef	0	
Actrapid MC	Short ⎭	2–5 ⎭	Neutral	Clear	Pig	0	Virtually nonantigenic
Insulin zinc suspensions or lente group							
Amorphous or semilente (SL)	Intermediate, 14	3–6 1	Neutral	Cloudy	Pig	0	
Crystalline or ultralente (UL)	Long, 36	6–12 7	Neutral	Cloudy	Beef	0	
Lente 30% SL 70% UL	Intermediate-long, 24	3–8 2	Neutral	Cloudy	{30% pig 70% beef}	0	
Monocomponent insulin (MC)							
Actrapid MC (*see* Soluble insulin)							
Semitard MC	Intermediate	3–8 1	Neutral	Cloudy	Pig	0	Virtually nonantigenic
Monotard MC	Intermediate-long	6–10	Neutral	Cloudy	Pig	0	
Other insulin preparations							
Biphasic	Intermediate	2–5 1 / 6–13	Neutral	Cloudy	{Pig 25% Beef 75%}	0	Quick initial action
Globin zinc	Intermediate, 18	3–5 2	Neutral	Clear	Beef	x	
Isophane (NPH)	Intermediate, 24	3–6 2	Neutral	Cloudy	Beef	x	Contains no excess of protamine or zinc uncombined with insulin
Protamine zinc	Long, 36	5–14 7	Neutral	Cloudy	Beef	x	Contains excess of protamine and zinc

Source: Stowers, 1976.

and it cannot be premixed with PZI. It was, however, mixed with PZI immediately prior to injection, so that one could combine the early actions of one preparation and the prolonged effects of the other. This early onset of action was, however, somewhat modified due to the interaction of the soluble insulin with an excess of protamine in the PZI preparation. Some preferred to inject each preparation separately at distinct sites. Hallas-Møller and his group showed in the early 1950s that if insulin was precipitated or crystallized in the presence of certain amounts of zinc in differently buffered solutions, crystals of different size could be formed. Large crystals are absorbed from injection sites very slowly, so it is unnecessary to add proteins to prolong this process. *Ultralente insulin* (UL) has a very long action, while an amorphous precipitate of the hormone (*semilente insulin*, SL) has a much shorter period of action. More recent purified (monocomponent, Mc, see below) European equivalent products are called, respectively, *Ultratard* and *Semitard*. These may be mixed (e.g., 30 percent SL and 70 percent UL) to give a preparation which acts earlier but not for quite as long (*lente*, *Mixtard* and *Rapitard* MC).

The *injection of insulin* can result in several *localized and general reactions*, including allergy, irritation, and dystrophy of subcutaneous lipid at the injection site. These responses are due principally to the presence of contaminants and impurities. Such effects can now often be overcome by the use of more highly purified preparations of the hormone (Deckert, Andersen, and Poulsen, 1974; Oakley, 1976; Yue and Turtle, 1977; Tattersall, 1978).

Considerable advances were initially made in purifying insulin made from pig and beef pancreas, ultimately resulting in the general use of *"single-peak"* insulins. These preparations are, however, still only partially purified, and this can result in problems. For instance, contamination with proinsulin, and the C-peptide, which can be antigenic, still commonly occurs. A newer group of "highly" purified insulin preparations have been developed, principally in Denmark. These are called *single-component* or *monocomponent* (MC) insulins and are made, using anion-exchange chromatography, usually from porcine, but also bovine, insulin. The former hormone only differs from human insulin by a single amino acid substitution, compared to three for bovine insulin, and is thus less likely to be antigenic. When changing to the use of such insulin preparations, it may therefore be necessary to reduce the dose. The soluble monocomponent insulins (Actrapid) are stable at a neutral pH and have advantages compared to the former "regular" insulin, which is in an acidic solution. This neutral soluble insulin can be premixed with the other longer-acting, also

neutral, preparations so as to adjust the onset of the response. Other forms of regular insulin in solution at a neutral pH are also available. Longer-acting preparations, equivalent, respectively, to semilente and lente, for instance *Semitard* and *Monotard*, have also been made from monocomponent insulin. Unfortunately, at the time of writing the supplies of these highly purified insulin preparations are limited and expensive. Indeed, there never has been enough porcine insulin available to satisfy the world's demand – hence the wide use of bovine insulin. Human insulin has been synthesized, but this hormone is not available in commercial quantities. The proponents of "gene transplants" have suggested that it is conceivable that the large-scale production of human insulin could be made possible by transferring a replica of the human insulin gene to bacteria (see, e.g., Ullrich et al., 1977; Steiner, 1977). It appears that success in this area is quite imminent.

Occasionally, a more generalized allergic response may occur in response to insulin injection. Local reactions are often seen initially, especially following reinstitution of insulin therapy. They usually subside spontaneously. On occasion, antihistaminic drugs are administered orally or are mixed with the injected insulin solution. Low concentrations of antiinflammatory steroids, such as prednisolone, may also be given at the local site.

A *local atrophy* of the lipid tissue at the site of the injection may also occur, especially in women, and this can be unsightly. This effect is more common when using acidic preparations. It can be prevented by avoiding the same site for repeated injections, although as such sites tend to become anesthetic, it is tempting to use them. It has recently been shown that the mixing of dexamethasone with the insulin injection can result in a restoration of the depressed tissue (Whitely, Lawrence and Smith, 1976). This problem, like the allergic responses, is far less common when using the highly purified monocomponent insulins.

Several months after instituting insulin therapy, a *resistance* to the actions of the preparation may arise which can necessitate a considerable increase in the dosage of the hormone. There can be several reasons for this effect.

a. A *failure to absorb* the hormone from the s.c. injection site. The effect will not be seen if the preparation is tested i.v. (see Dandona et al., 1978).

b. In some instances, hormones with *opposing effects*, such as epinephrine, growth hormone, or corticosteroids, may be secreted.

c. The production of *insulin antibodies* may result in a binding of considerable amounts of the insulin. This bound hormone is usually released very slowly, although on occasion this process may occur quite suddenly and result in hypoglycemia.

Such patients may require several thousand units of insulin each day. The problem can be met in several ways, including desensitization of the patient, the administration of corticosteroids (the favorable effects may take several weeks to develop), or changing the insulin preparation. Porcine insulin will often be more effective, especially if patients have previously been using bovine insulin. Monocomponent porcine insulin has been shown to be effective in insulin-resistant individuals, so that the dosage can be gradually reduced (Oakley, 1976). Combination of insulin with the orally active sulfonylurea drugs has also been advocated in such difficult situations.

The *administration of insulin* is standardized in terms of *international units* (IU). This is based on a reference-standard crystalline insulin preparation which has a strength of about 24 IU/mg. This, in turn, is based on a bioassay procedure which measures the ability of the preparation to lower the blood glucose concentration of rabbits. Other bioassays are, however, available, such as the induction of hypoglycemic convulsions in mice. For most purposes, radioimmunoassays are now routinely used.

For normal use insulin is available at three concentrations: 40 IU/ml (U40), 80 IU/ml (U80), and 100 IU/ml (U100). It is intended that the latter "metrical" concentration, which was introduced into the United States in 1973, will eventually become international. It is hoped that many errors in administered dosage will be avoided by the availability of these standard preparations, and it should also facilitate the education of patients in the use of the hormone (Arky, 1973). Insulin is also available at a concentration of 500 IU/ml, but the uses of this are usually confined to cases of insulin resistance.

The decision to use insulin is based on many factors, which will not be discussed in detail here. Physicians have differing criteria and a consensus is unusual. It is always needed in juvenile-onset diabetes and in mature-onset diabetics suffering from ketosis, or if the fasting blood glucose concentration regularly exceeds about 250 mg/100 ml. The dosage is carefully adjusted and gradually increased, preferably in a hospital situation. It is frequently administered, subcutaneously, as a single daily injection, usually before breakfast; but in many instances, when a larger dose is needed and control is more difficult, a smaller dose is also given in the evening before dinner. Occasionally, a third dose may be given before lunch. By monitoring the appearance of glucose and ketone bodies in the urine and intelligently assessing his or her needs in different situations (e.g., rest, exercise, emotional stress, infection), the patient can adjust the dosage. Growing children have changing requirements. A finer control may be necessary during pregnancy, and then three injections a day are given.

Exercise is a reasonably common pursuit which may alter the need for insulin and can result in hypoglycemia. There are several possible reasons for this response, including increased utilization of glucose and a rise in the release of endogenous insulin. It has also been shown that the absorption of injected insulin, for instance from a site in the leg, is increased if the limb is exercised (Koivisto and Felig, 1978). It has therefore been suggested (see also Zinman et al., 1978) that a more distal, less active site, such as an arm, may be injected in such circumstances. It was also proposed that rather than decreasing the dose of insulin, additional readily absorbable carbohydrate should be taken before exercise.

Many patients will generally feel normal if the blood glucose levels are below about 180 mg/100 ml, although a persistent fasting level greater than 120 mg/100 ml is considered to be diabetic. It is not possible to use periodic insulin injections to maintain blood glucose concentrations at normal physiological levels and keep the normal daily fluctuation within the range of about 30 mg/100 ml. It is also difficult to keep the fasting values near the normal 100 mg/100 ml, because to do this one must risk the precipitation of hypoglycemia, the acute effects of which can be very damaging. Nevertheless, there is some latitude and the precise levels to be aimed at have resulted in considerable controversy. Those advocating *good control* ("tight" or "rigid") suggest that mean blood glucose levels between 110 mg/100 ml fasting and 150 mg/100 ml following a meal may be advantageous with respect to reducing the long-term vascular degeneration associated with the disease. Such guidelines have received the blessing of the American Diabetes Association (Cahill, Etzwiler, and Freinkel, 1976). Others feel that the benefits of such a program are unproven (see, e.g., Siperstein et al., 1977) and that it makes the life of the patient unnecessarily difficult and increases the risk of hypoglycemia. This school of *fair* (or "loose") *control* suggests that a range of blood glucose levels between 130 and 180 mg/100 ml is adequate.

The possible benefits of assuring a safe and effective "strict" control of blood glucose concentrations has recently led to the investigation and development of several systems for the accurate and appropriate administration of insulin. There are two basic types of such systems:

a. *Transplantation* of B-cell tissue, either as separated dispersed B-cells or portions of the pancreas. Although some success has been reported in animal experiments, the immunological problems that remain in man are considerable.

b. *Small "vest-pocket" mechanical and electronic devices* ("artificial pancreases"), sometimes incorporating a minicomputor, have been developed. There are two types. A "closed system" in which the infusion rate of the hormone is regulated directly in response to the blood glucose levels. The latter are continuously monitored, usually with a glucose electrode placed in a vein. The "open-loop" system is simpler and can be programmed according to the predicted need for insulin. Adjustments that are required, such as at mealtimes, can be made by the patient. These infusions are usually made subcutaneously which is much simpler than intravenously. Such a system has recently been used in man, with some success in the United States and United Kingdom.

The reasons for the degenerative vascular changes that accompany diabetes are unknown. It has, however, been shown (Halushka, Lurie, and Colwell, 1977) that the blood platelets from patients with diabetes mellitus aggregate more readily when they are exposed to arachidonic acid, which is a precursor of the prostaglandins. These platelets had an increased prostaglandin synthetase activity. It is considered possible that increased aggregation of platelets may be associated with, and contribute to, some vascular diseases, and this has led to clinical trials to test the ability of aspirin (which inhibits prostaglandin synthetase) to prevent or ameliorate such conditions (see Mustard and Packham, 1977). The final results are not yet available. It is, however, interesting that an early study (Powell and Field, 1964) showed that diabetic patients suffering from rheumatoid arthritis, for which they received large doses of salicylates, had a lower-than-expected incidence of diabetic retinopathy.

Special circumstances affecting insulin administration.
Diabetic coma caused death in 40 to 50 percent of the patients suffering from diabetes mellitus before the discovery of insulin. Today, this very serious situation is relatively rare as a result of the administration of insulin, but it still occurs with a death rate that has been estimated at 3 to 10 percent. Because of a lack of sufficient insulin, the patient suffers from hyperglycemia, resulting in a glycosuria and osmotic diuresis, which along with vomiting can lead to serious dehydration. Fatty acids are rapidly mobilized and are converted in the liver to acetylcoenzyme A. There is too much of this intermediate available to enter the tricarboxylic cycle, so ketone bodies are formed (acetoacetic acid, β-hydroxybutyric acid, and acetone). As the body has only a limited capacity to metabolize these substrates, they accumulate, resulting in a depletion of plasma bicarbonate and a metabolic acidosis. Acetone can be smelled on the breath. Cerebral edema may sometimes occur, and this condition is usually fatal.

The first aim of treatment is to restore insulin levels in the circulation. There are currently three, apparently equally effective, ways of doing this:

a. The administration i.v. of a bolus of regular insulin in a large dose [100 IU initially and 50 to 100 IU every 2 to 4 hours (Felig, 1974a)]. Absorption following s.c. or i.m. injection of insulin is considered to be too slow and unreliable, but a depot of insulin is sometimes created by simultaneous administration along with the i.v. hormone. In such circumstance and using these methods of administration, an overdose is not considered to be a serious threat; but it can result in complications (hypoglycemia, hypokalemia), which, nevertheless, can be treated.

b. Slow i.v. infusion of smaller amounts (2.4 to 8 IU/h) of regular insulin have been advocated (Page et al., 1974; Semple, White, and Manderson, 1974; Kidson et al., 1974; Felig, 1974b). This method ("low-dose insulin therapy") results in good control, as one only has to change the rate of the infusion to rapidly alter the levels of circulating insulin. Hypoglycemia is thus easier to avoid and control, while the lower levels of the hormone do not result in such serious hypokalemia. Such insulin infusions often contain human albumin, as this protein reduces the adsorption of the hormone to the walls of the receptacles and tubing; however, the necessity for this precaution has been questioned (Page et al., 1974).

c. A proposed modification of low-dose insulin therapy involves the initial rapid i.v. injection of a bolus of insulin (0.33 IU/kg), followed by an intramuscular injection (7 IV) at hourly intervals (Fisher, Shahshahani, and Kitabchi, 1977).

The use of insulin for the treatment of diabetic coma is accompanied by other measures, calculated to restore bodily hydration, pH, and potassium concentrations (Felig, 1974a,b).

The second islet hormone *glucagon* also influences intermediary metabolism and may play an important role in the genesis of diabetic ketoacidosis (Gerich et al., 1975). The dual, bihormonal role of the islets of Langerhans in controlling the levels of substrate metabolites such as glucose, fatty acids, ketone bodies, and amino acids is shown in Figure 8.15. Using volunteers who suffered from juvenile-onset diabetes, Gerich and his collaborators were able to show, following the withdrawal of insulin, that suppression of glucagon secretion by infused *somatostatin* delayed the onset of ketocidosis. These interesting observations emphasize the importance of glucagon in the manifestations of diabetes mellitus. A more recent clinical trial has, however, shown that the i.v. infusion of somatostatin did not correct diabetic ketoacidosis in juvenile diabetes following insulin withdrawal (Lundbaek et al., 1976). Somatostatin also has widespread effects in the body, so that a useful therapeutic role for it is at present uncertain (Editorial, 1975h).

Surgical procedures performed on patients suffering from diabetes mellitus are a special problem that necessitates careful consideration of the dosage of insulin and its method of administration.

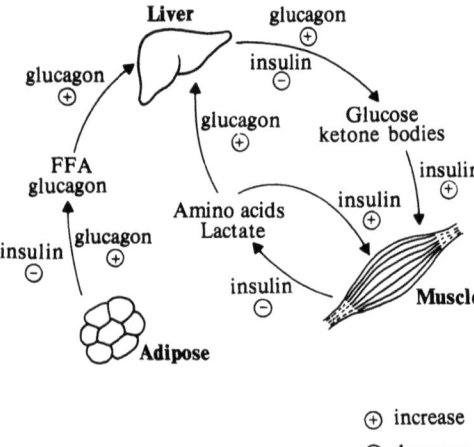

⊕ increase

⊖ decrease

Fig. 8.15. Schematic representation of the bihormonal control of substrate metabolism by glucagon and insulin. (From Gerich, Lovinger, and Grodsky, 1975. Reprinted by permission from *The New England Journal of Medicine 292*, 988.)

The complexities of this particular situation are the result of the necessity to fast the patient preoperatively, the frequent delay in his or her ability to take food postoperatively, the use of i.v. nutrients, and changes in insulin need as a result of the duration and severity of the surgical procedure, possible subsequent infection, and inactivity. For short procedures, with little after effect, few changes, except possibly a small reduction in dose of insulin, or appropriate timing of the operation in relation to this, may be necessary. In other instances, timed presurgical s.c. doses may be given combined with glucose infusion (it is of prime importance to avoid hypoglycemia) and a monitoring of the blood levels of this sugar. More recently, continuous i.v. infusion of small (1 to 2 IU/h) amounts of insulin (plus glucose) has been advocated (Taitelman, Reece, and Bessman, 1977).

B. Other uses of insulin
Insulin has received consideration and has been utilized clinically for a number of purposes other than hormone replacement therapy.

a. The ability of insulin to promote potassium uptake by the cells has been used to *treat hyperkalemia*, especially in crisis situations where a rapid decrease in plasma K levels is vital. It is administered in conjunction with glucose.

b. *Hypoglycemic convulsions* as an alternative to electroconvulsive treatment have been utilized in psychiatric institutions, but the considerable hazards associated with hypoglycemia (see the next section) have led to the general abandonment of this procedure.

c. Gastric acid secretion is increased in response

to insulin hypoglycemia, and this effect is largely mediated by the vagus. Insulin has thus been used to *test the completeness of surgical gastric denervation* (*Hollander test*). It has been suggested (Carter, Dozois, and Kirkpatrick, 1972) that the insulin is more safely administered as an i.v. infusion.

d. Growth hormone and ACTH is released in response to hypoglycemia, is used to *test pituitary function* (Wilson et al., 1972). The continuous i.v. infusion procedure was found to be preferable and safer.

e. Insulin appears to exert a direct effect *on the heart* and increases the rate of contraction and the cardiac output (Majid et al., 1972; Page, Smith, and Watkins, 1976). This effect is independent of any hypoglycemic action and may be the result of an improved access of substrates to the heart muscle. In one trial in which an infusion of glucose, insulin, and potassium was made in patients with ischemic heart failure (Majid et al., 1972), there was an improvement in cardiac output. Insulin has also been used in an attempt to overcome cardiac arrythmias following myocardial infarction (Mittra, 1965). Such uses for insulin do not appear to have gained general acceptance.

C. Side effects and drug interactions
The most common and serious side effect of insulin is *hypoglycemia*. Other undesirable actions, such as allergy and hypokalemia, are described elsewhere. Insulin hypoglycemia can be precipitated by inadequate carbohydrate intake, an overdose of the hormone which can be accidental, or as a result of changes in the need for insulin. A change in diet; reduced activity; increased sensitivity, such as in the first trimester of pregnancy; or an interaction with drugs that are administered concurrently may alter the requirements for insulin. Hypoglycemic symptoms arise when the blood glucose levels decline below about 30 to 40 mg/100 ml. They include tremor, palor, sweating, hypothermia, hunger, disorientation, convulsions, and coma.

Hypoglycemia may also occur in a variety of other circumstances, including hyperinsulinism due to disease of the pancreatic B-cells, insulin-secreting tumors (insulinoma), a deficiency of glucagon, or the overdosage of certain drugs (Marks, 1972; Fajans and Floyd, 1976). Hypoglycemia can result in death (insulin has been used to commit murder) or serious permanent damage to brain functions. Less startling fluctuations in plasma glucose levels associated with the therapeutic use of insulin are not infrequent but need to be corrected. If hypoglycemia occurs at night, during sleep, the patient may be unaware of it, and indeed when he wakes may be hyperglycemic due to an overcompensation by physiological mechanisms. If unaware of this sequence, he or she may feel it is

Table 8.8. *Drugs and drug combinations responsible for 473 episodes of hypoglycemia coma; comparative rates of mortality and sequelae*

	Number of patients			Percent of patients	
Drug or drug combination	Total	Se-quelae[a]	Deaths	Deaths	Deaths plus sequelae
Single hypoglycemic drug					
1. Alcohol	174	2	23	13.2	14.4
a. Adults	146	2	16		
b. Children	28	0	7		
2. Sulfonylurea	220	10	25	11.4	15.9
a. Carbutamide	21	1	2		
b. Tolbutamide	49	5	8		
c. Chlorpropamide	120	4	12		
d. Acetohexamide	9	0			
e. Tolazamide	6	0	0		
f. Glibenclamide	7	0	2		
g. Others	8	0	0		
3. Salicylate	15	1	5	33.3	40.0
a. Children	13	1	5		
b. Adults	2	0	0		
4. Phenformin	2	0	1		
Two hypoglycemic drugs	28	5	3	10.7	28.6
1. Insulin + sulfonylurea	11	1	1		
2. Insulin + alcohol	7	4	2		
3. Sulfonylurea + alcohol	4	0	0		
4. Sulfonylurea + salicylate	3	0	0		
5. Sulfonylurea + phenformin	3	0	0		
Hypoglycemic drug plus potentiating drug	28	2	3	10.7	17.9
1. Sulfonylurea plus:					
a. Sulfisoxazole, sulfaphenazole, or sulfadimidine	7	0	1		
b. Bishydroxycoumarin	5	2	1		
c. Phenylbutazone	4	0	1		
2. Insulin plus:					
a. Propranolol	5	0	0		
b. Oxytetracycline	3	0	0		
c. Ethylenediaminetetraacetic acid (EDT)	2	0	0		
d. Mebanazine (MAO inhibitor)	1	0	0		
e. Manganese	1	0	0		
Miscellaneous	6	0	0		
1. *para*-Aminobenzoic acid (PABA)	3	0	0		
2. Haloperidol	1	0	0		
3. Propoxyphene	1	0	0		
4. Chlorpromazine + orphenadrine	1	0	0		

[a] Permanent brain damage, other neurologic deficit, or acute myocardial infarction.
Source: Seltzer, 1972a.

appropriate to increase the dose of insulin. Such a response (the *Somogyi effect*) is, however, the result of an overdosage of the hormone, which precipitates a late hypoglycemia followed by the observed "rebound" hyperglycemia. The solution is either to *reduce* the morning dose of insulin or "split" it, so that less is given in the morning and the balance in the evening.

In toxic doses many drugs have a hypoglycemic effect (Table 8.8) and some, although they do not exhibit such an action when given alone, may interact with others to precipitate such a response. In the instance of insulin, the drinking of *ethanol* results in the highest incidence of such actions. This response is indirectly a result of inadequate nutrition but in chronic imbibers may result from hepatic problems and an inhibition of gluconeogenesis, which limits the body's ability to compensate for swings in blood glucose levels. *Propranolol* (Deacon and Barnett, 1976), a β-adrenergic blocking drug used to treat hypertension, angina pectoris, and cardiac arrythmias is contraindicated in diabetes, as it blocks the compensatory effects of epinephrine, which tend to overcome hypoglycemia. Many of the adrenergic symptoms (tremor, sweating, and hunger) are also hidden, thus complicating the diagnosis. Propranolol has, however, on occasion, also produced a hyperglycemia, which may reflect its ability to block β-adrenergic-controlled insulin release from the B-cells. It has been suggested that more selective $β_1$-antagonists would be preferable if they can be used, as they do not interfere with the responses and metabolic adjustments to hypoglycemia (Deacon, Karunanayake, and Barnett, 1977). Other interactions with insulin that have been reported to result in hypoglycemia include (Table 8.8), salicylates, sulfonylureas, oxytetracycline, and monoamine oxidase (MAO) inhibitors.

Some drugs may exacerbate or precipitate the *hyperglycemia* due to insulin deficiency. The possible effects of *corticosteroid* treatment have already been described (Section 6.8C). Several drugs used as tranquilizers, such as the *phenothiazines* and *chlordiazepoxide,* can have hyperglycemic effects which may aggravate diabetes mellitus (Zumoff and Hellman, 1977). The *thiazide diuretics* may reduce glucose tolerance, result in a rise in blood glucose levels, and even precipitate diabetes (see Amery et al., 1978). This side effect appears to be due to a direct effect on the B-cell, but other mechanisms, such as a hypokalemia and activation of peripheral adrenergic processes, may also be involved. This action of the thiazide diuretics is reversible. The more potent diuretics *furosemide* and *ethacrynic acid* (which are not thiazides) appear to have a similar effect (Cowley and Elkeles, 1978). *Diphenylhydantoin* has, under certain conditions, been observed to reduce the response to insulin, and it has been suggested that it should be used with caution in patients who "have a risk factor for diabetes" (Malherbe et al., 1972).

1. Treatment of insulin hypoglycemia. The principal aim of treatment is to restore the blood glucose level. In cases where the patient is still conscious, this can usually be done by oral administration of sugar. If he is uncooperative, an s.c. injection of glucagon may be effective. Its actions, however, are of relatively short duration. Epinephrine has also been used, but glucagon is now usually preferred, as it has fewer other effects, such as those that involve the cardiovascular system. Intravenous infusion of glucose may be necessary in more severe cases, especially if the patient is comatose. Glucagon can also be given by this route. In cases of persistent hypoglycemia, corticosteroids, such as

hydrocortisone Na succinate, may be added to the infusion.

Hypoglycemia may also result from an overproduction or release of insulin due, for instance, to an islet cell tumor. Treatment of this condition is usually surgical, but the hypoglycemia can often be controlled by the use of *diazoxide*. This drug is related to the thiazide diuretics (Figure 8.16). It has a wide range of actions, uses, and side effects (Todays Drugs, 1972b; Koch-Weser, 1976). The hyperglycemic effect is due to a blockade of insulin release and a sympathetic stimulation. This effect can be inhibited by α-adrenergic blocking drugs such as phentolamine. There is a decline in cyclic AMP levels in the B-cell which is consistent with such an α-adrenergic inhibition, and this effect may be due to a blockade of the Ca^{2+} influx, which normally mediates release of insulin (see Section 8.3C). The hyperglycemic effects of diazoxide are seen following oral administration; when given i.v. it has a potent hypotensive effect and is used to control blood pressure in hypertensive crises. On these occasions the hyperglycemia is classified as a "side effect" and can be antagonized with tolbutamide. Diazoxide also relaxes uterine smooth muscle and has been used in obstetrics to delay labor [see Section 7.6H(4)]. Although this drug is related chemically to the thiazide diuretics, it promotes *Na retention*. This action can be combated with diuretics. Diazoxide decreases uric acid excretion in the urine and so can result in hyperuricemia. Its chronic use can also result in hirsutism (lanugo-type hair on the face). There may also be gastric problems and anorexia, probably associated with its relaxant effect on stomach muscles. When diazoxide was administered, orally, to several normal and diabetic patients for about 1 week it was found that subsequent stimulation of the B-cells, with glucagon or tolbutamide, resulted in an enhanced release of the hormone (Greenwood, Mahler, and Hales, 1976). It was suggested that diazoxide may be useful in preserving and restoring insulin secretion in diabetes, but more extensive clinical trials are necessary to explore this interesting idea.

8.6. Oral hypoglycemic ("antidiabetic") drugs

Insulin, being a polypeptide, is destroyed in the gut, so it must be administered by injection. This pro-

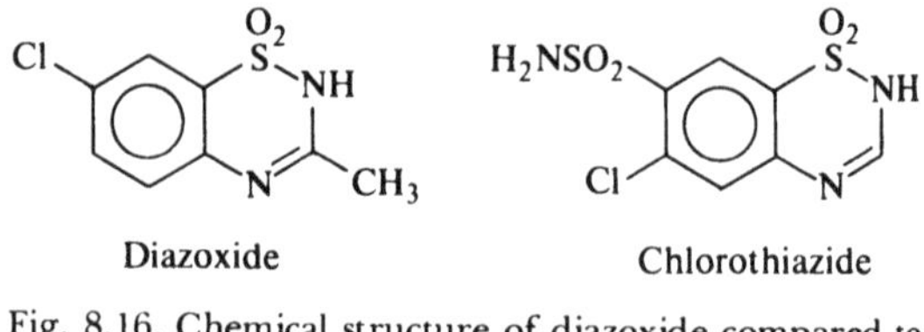

Fig. 8.16. Chemical structure of diazoxide compared to that of the thiazide diuretic chlorothiazide.

cedure has a number of disadvantages, including those of self-administration, which may be impractical in patients with poor sight, the elderly, those who suffer from tremorous conditions, and others. In the late 1950s, two groups of drugs were introduced which could be administered orally to lower blood glucose concentrations in diabetic patients. These compounds were the sulfonylureas and the biguanides (see, e.g., Stowers and Borthwick, 1977; Watkins, 1977), and they rapidly gained in popularity for the treatment of mature-onset diabetes. In 1970 in the United Kingdom, it was estimated that of the diabetics receiving medication (excluding those being treated by diet alone) about 60 to 70 percent were being treated with oral antidiabetic drugs. The status of such drugs in the treatment of diabetes mellitus is currently controversial. The reasons for this problem are described in more detail later in Sections 8.6D and E. They are partly the result of clinical trials, which strongly suggest that the use of either the sulfonylureas or biguanides results in an increased mortality due to cardiovascular diseases. The storm of controversy aroused by these results is probably without parallel in modern medicine (Editorial, 1975i). It has also been recently confirmed that the use of phenformin (a biguanide) is associated with lactic acidosis, a condition which, although relatively rare, has a very high rate of mortality. The use of such drugs as an alternative to diet or insulin is thus currently subject to considerable medical circumspection. Phenformin was recently withdrawn from use in the United States.

Guidelines have been provided for the use of oral antidiabetic drugs. For example, the FDA has issued a special warning (*FDA Drug Bulletin,* May 1972) regarding the labelling of *sulfonylureas:*

> *Diet and reduction of excess weight are the foundations of initial therapy of diabetes mellitus.* When the disease is adequately controlled by these measures, no hypoglycemic drug therapy is indicated. Because of the apparent increased cardiovascular hazard associated with oral hypoglycemic agents, they are indicated in adult-onset, non-ketotic diabetes mellitus only when the condition cannot be adequately controlled by diet and reduction of excess weight alone, and when, in the judgment of the physician, insulin cannot be employed because of patient unwillingness, poor adherence to injection regimen, physical disabilities such as poor vision and unsteady hands, insulin allergy, employment requirements, and other similar factors.

In the instance of *phenformin,* one manufacturer issued the following specific warning (USV Laboratories, February 25, 1977):

> There have been numerous reports of lactic acidosis associated with phenformin therapy. Because lactic acidosis is an often fatal metabolic acidosis, the following should be observed with phenformin:

> a. Use only in symptomatic adult-onset nonketotic diabetes mellitus.
> b. Diet and weight reduction should always be tried first. If diet therapy fails, a sulfonylurea or insulin should be used. If adequate control cannot be achieved by diet and sulfonylurea therapy or insulin therapy cannot be used, phenformin may be administered.
> c. Do not use in patients with conditions that increase the risk of lactic acidosis (such as kidney disease, cardiac insufficiency, or high ethanol intake).
> d. Advise patients to discontinue the drug immediately and notify the treating physician if nausea, vomiting, hyperventilation malaise, or abdominal pain occurs.
> e. Reevaluate patients frequently to ensure the absence of conditions that increase the risk of lactic acidosis. These usually occur more frequently in patients over 60 years of age.
> f. Reevaluate the need for continuing the drug at frequent intervals.
> g. Dosage should not exceed *100 mg/day.*

In 1977, phenformin was *withdrawn* from use in the United States on the order of the Secretary of the Department of Health, Education, and Welfare. (*FDA Drug Bulletin,* August 1977).

Despite these reservations, a large number of patients are being treated with oral antidiabetic drugs. Indeed, it has been pointed out that the world supply of insulin is insufficient to meet the sudden demand that would be necessary if insulin were completely substituted for these drugs.

A. Sulfonylureas

Chemically, the sulfonylureas are related to the antimicrobial sulfonamides and it was this similarity which led to their development. In 1942, M. Janbon and his collaborators, working at Montpellier in southern France, were testing the effectiveness of sulfonamide, "ITPD" (Figure 8.17), in the treatment of typhoid fever. They observed that a number of their patients developed a hypoglycemia, which in some instances was fatal. These effects were confirmed in animal experiments and provided the basis for the thesis investigations of A. Loubatières into its mechanism of action. He found that such compounds were ineffective in pancreatomized dogs, but that if 10 percent of this tissue was left intact, the drugs hypoglycemic action persisted. This observation suggested that the drug may be stimulating the release of insulin, and this was confirmed in cross-circulation experiments and histologically, when a degranulation of the B-cells was observed. In 1946, Loubatières (see Loubatières, 1972) suggested that this effect may be utilized for the treatment of diabetes in instances where there was a "sluggishness" in the response of the B-cell to secrete insulin. He also conjectured that it may be a useful diagnostic test for B-cell function. In 1957, carbutamide and tol-

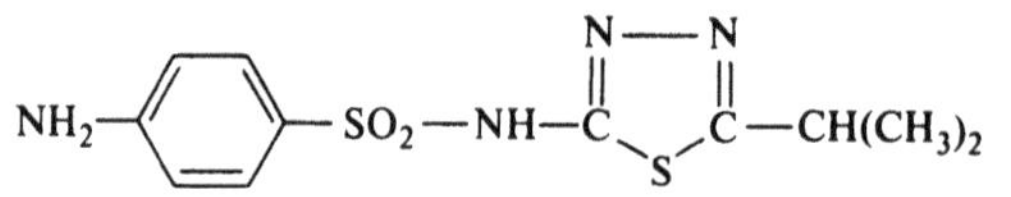

2-(*p*-Aminobenzoylsulfonamide)-5-isopropyl-1,3,4-thiodiazole
(IPTD, 2254RP)

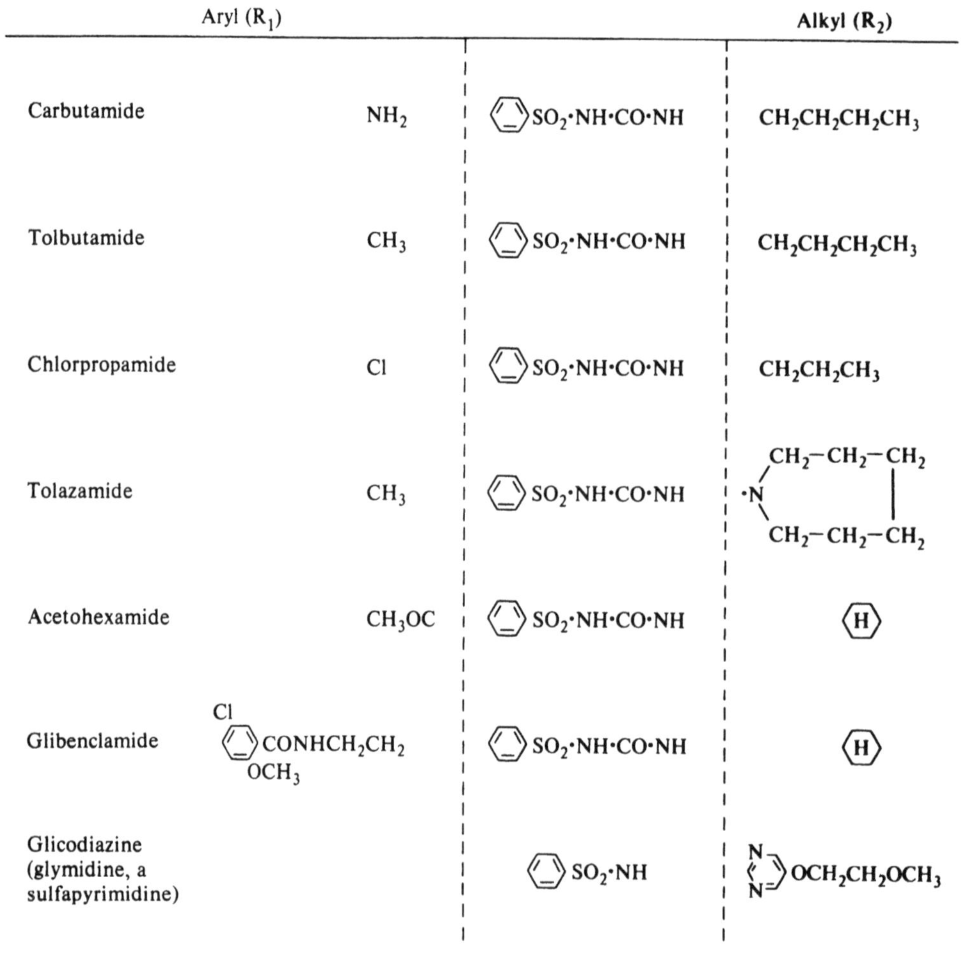

	Aryl (R$_1$)			Alkyl (R$_2$)
Carbutamide	NH$_2$		SO$_2$·NH·CO·NH	CH$_2$CH$_2$CH$_2$CH$_3$
Tolbutamide	CH$_3$		SO$_2$·NH·CO·NH	CH$_2$CH$_2$CH$_2$CH$_3$
Chlorpropamide	Cl		SO$_2$·NH·CO·NH	CH$_2$CH$_2$CH$_3$
Tolazamide	CH$_3$		SO$_2$·NH·CO·NH	·N⟨ CH$_2$—CH$_2$—CH$_2$ / CH$_2$—CH$_2$—CH$_2$ ⟩
Acetohexamide	CH$_3$OC		SO$_2$·NH·CO·NH	(H)
Glibenclamide	Cl ⟨ ⟩CONHCH$_2$CH$_2$ OCH$_3$		SO$_2$·NH·CO·NH	(H)
Glicodiazine (glymidine, a sulfapyrimidine)			SO$_2$·NH	N⟨ ⟩OCH$_2$CH$_2$OCH$_3$ N

Fig. 8.17. Structures of the sulfonylurea hypoglycemic (oral antidiabetic) drugs.

butamide were introduced for clinical use in the treatment of diabetes and a little later the *tolbutamide test*, to aid the diagnosis of the disease, was proposed.

The *chemical structure* of the sulfonylurea compounds, which have a hypoglycemic action, is shown in Figure 8.16. It can be divided into three parts: (a) an aryl (R$_1$), (b) a urea portion, and (c) an alkyl side chain or a cyclohexyl group (R$_2$). The activity of the molecule can be changed by substitution in the R$_1$ and R$_2$ sections. The first compound to be introduced clinically, *carbutamide*, has a *para*-amino group at R$_1$ which is responsible for a number of toxic effects, including agranulocytosis, hepatotoxicity, and an antithyroid action. The amino group is unnecessary for the molecule's hypoglycemic effect, and when it is substituted by a *para*-methyl group to form *tolbutamide*, these side effects are lacking. The compound nevertheless retains its hypoglycemic (but not antimicrobial) activity. Tolbutamide has a relatively short half-life and must be administered 2 to 3 times a day. When a Cl is substituted for the *para*-CH_3 group and the side chain (R_2) is shortened, the metabolism of the compound is reduced to insignificant levels, so that it has a long half-life. This drug, *chlorpropamide*, need only be administered once a day. *Tolazamide* and *acetohexamide* are also used clinically and mainly differ from the former compounds by the substitution of a ring structure at R_2. These have a half-life that is intermediate between tolbutamide and chlorpropamide. The principal reason for this difference is that acetohexamide is metabolized to hydroxyhexamide, which is more active (2.4 times) than its parent compound and is mainly excreted in the urine. Tolazamide has a slightly modified structure at the R_1 and R_2 positions and is absorbed more slowly from the gut. This appears to be an important factor in prolonging its half-life, but it also delays the onset of its effect. A "second generation" of hypoglycemic oral antidiabetic drugs was introduced clinically in the United Kingdom in 1967. *Glibenclamide* is the prototype; it has a cyclohexyl group at R_2, as in acetohexamide, but an alkene chain is introduced at R_1. This change increases its activity about 400-fold (on a molar basis) compared to tolbutamide. A related compound, which is available in Europe, is *glipizide*, which is even more potent. *Glymidine* is a sulfapyrimidine compound which has similar hypoglycemic properties but, because of its more distant chemical relationship to the sulfonylureas, does not display a cross-sensitivity reaction with them and so provides a possible substitute in cases where skin rashes occur.

The *metabolism* of the sulfonylurea hypoglycemic drugs is important in relation to the potency and duration of their effects, as well as their interactions with other drugs. Following absorption from the gut, tolbutamide, chlorpropamide, and glibenclamide are bound to plasma albumin (Crooks and Brown, 1974). The binding appears to be ionic in nature for the former two drugs but is nonionic for the latter. Only the unbound drug can act, and be metabolized and excreted, so that their effects are modified as a result of this binding. Displacement from plasma albumin can occur in the presence of other drugs, although this effect is probably noncompetitive in its nature (Brown and Crooks, 1976). Such a response can result in a sudden increase of the action of either drug and also foster its excretion and metabolism, so that a simple interaction of this nature would also be expected to decrease its duration of action. The latter

does not always occur, however, because of a concomitant inhibition of its metabolism. Tolbutamide is rapidly inactivated in the liver to inactive metabolites which are mainly excreted in the urine. Chlorpropamide, in contrast, is barely metabolized at all, so that the termination of its action depends principally on renal excretion. As it is also bound to plasma proteins, this process is very slow; it is found to persist in the body for more than 2 weeks and at therapeutically active levels for about 40 hours. As described above, acetohexamide is metabolized in the liver to hydroxyhexamide, which is *more* active, and this metabolite is excreted by the kidney. Thus, the effect of the sulfonylureas can be expected to change in liver and kidney diseases and in the presence of drugs that also interact with these organs. Special care must therefore be exercised in their use in such diseases.

The sulfonylureas' *mechanism of action* involves facilitating the release of insulin from the B-cells (see also Section 8.3C). They are thus effective only in mature-onset diabetes, where there are functional B-cells. Acutely, they bring about an elevated blood insulin level, although with more prolonged therapy this rise is not readily apparent. The B-cells, however, appear to respond more readily to stimulation by glucose (they are "potentiators"). Glibenclamide, on the other hand, appears to decrease the normal *fasting* level of glucose (Anderson et al., 1970). An increase in the number of B-cells (a B-cytotrophic effect) has been observed following chronic administration of sulfonylureas in animals, but this effect apparently does not occur in man. Chronic treatment with glipizide has been observed to increase the numbers of insulin receptors in the plasma membranes of liver cells from mice (Feinglos and Lebovitz, 1978). Sulfonylurea drugs could thus be acting in this way to increase the sensitivity of target tissues to insulin. Other actions of these drugs which may be expected to produce a hypoglycemia have been suggested, including an inhibition of an "insulinase" enzyme, which would prolong the half-life of the hormone, and an inhibition of glycogenolysis in the liver. These effects are, however, seen only under *in vitro* conditions with very high concentrations of the drugs and are not thought to reflect their therapeutic action.

The release of insulin involves rises in intracellular levels of free Ca^{2+} and cyclic AMP in the B-cells. The sulfonylureas could be acting by influencing such processes, but there is little definitive information about such possibilities. Chlorpropamide enhances the cyclic AMP-dependent antidiuretic response to vasopressin (ADH), and it has been shown that this effect is related to an inhibition of prostaglandin synthesis (Zusman, Keiser, and Handler, 1977a). It has been suggested that the action of the sulfonylureas on the B-cell also involves modulation of prostaglandin synthesis.

Some of the *precautions* to be observed in the therapeutic use of the sulfonylurea drugs have been described at the beginning of this section. They should not be used in patients whose diabetes can be *controlled by diet alone*, are *overweight*, or suffer from *ketosis*. They are usually ineffective in patients whose fasting blood glucose concentrations exceed 250 mg/100 ml. They are contraindicated in patients who have *renal, hepatic,* or *cardiovascular disease*. In view of the controversy as to their long-term effects on the cardiovascular system, it is considered by some that they should not be used at all, or used as an alternative to insulin, unless there are special circumstances such as serious allergy to insulin or in those patients where parenteral administration of the hormone is impractical. Such a view is still controversial, however.

Pregnancy may be associated with the precipitation or worsening of diabetes. Sometimes this can be controlled by diet, but low doses of sulfonylureas have been administered in attempts to avoid the use of insulin. The oral hypoglycemic drugs can, however, readily cross the placenta and so may affect the fetus. Infants born to mothers who have been treated with high doses of the sulfonylureas may suffer a hypertrophy of the B-cells and a hypoglycemia. A clinical trial on pregnant women suffering from "mild" diabetes showed that low doses of chlorpropamide (100 mg/day) could satisfactorily control the maternal blood glucose levels and did not adversely affect the outcome of the pregnancy (Sutherland et al., 1973). It has, however, been recommended that high blood sugar levels are better treated with insulin (Editorial, 1974f). The possibility of teratogenic responses to sulfonylurea drugs has not been excluded to the satisfaction of all and the use of these drugs during pregnancy remains controversial.

If one excludes the possible adverse cardiovascular consequences of using the sulfonylureas, the seriousness and incidence of their *side effects* is not great. In 1971, Harris estimated, on the basis of reported information, that an adverse reaction occurred only once in 15,000 patients per year. The most common reactions involve the *skin* (including photosensitivity): an incidence of about 1 percent for tolbutamide and 2 to 3 percent for chlorpropamide, and these usually occur within 3 weeks of commencing treatment. *Nausea* and *vomiting* may also occur, but this can often be controlled by reducing or dividing the dose. Blood reactions are even less common, but the early use of carbutamide in Europe resulted in 21 cases of *agranulocytosis*. Only 7 cases (all fatal) have been subsequently reported using the newer drugs. This problem is so rare that routine precautionary tests are no longer performed. Liver function may be

affected, and this includes *cholestatic jaundice* (usually transient in response to high doses of chlorpropamide).

The most common and generally serious side effect is *hypoglycemia*, which is an extension of the drug's therapeutic action. This response may be due to overdosage, a change in the need for the drug, such as inadequate carbohydrate intake; a failure to clear it from the body due to deficient liver or kidney function; or an interaction with other drugs. This response is not uncommon with chlorpropamide because of its prolonged time of action and its propensity to accumulate in the body. Treatment of the hypoglycemia, in turn, will also of necessity be prolonged, until enough of the drug is cleared from the body. The newer and more potent analogue, glibenclamide, is even more likely to produce hypoglycemia (Clarke and Campbell, 1975), and this reaction is also more severe.

An interesting side effect which is seen especially in response to chlorpropamide, but also tolbutamide, is a *hyponatremia* reminiscent of the condition of inappropriate secretion of antidiuretic hormone (ADH). This effect appears to be due to a potentiation of the effects of ADH in the body, possibly by an action on adenyl cyclase (Murase and Yoshida, 1973; Mendoza and Brown, 1974). Chlorpropamide may also inhibit the synthesis of prostaglandin E, which modulates the action of ADH (Zusman, Keiser, and Handler, 1977b). The effect is reversible and does not appear to occur in response to tolazamide or acetohexamide, which thus offers the possibility of an alternative drug. This side effect of chlorpropamide and tolbutamide has been utilized clinically for the treatment of diabetes insipidus. Provided that there is some remaining ADH, the urine flows of such patients can be controlled by these drugs. Although a hypertrophy of the B-cells may occur with prolonged treatment, this does not usually result in hypoglycemia, probably as a result of compensatory mechanisms in the body (Meinders, Cejka, and Bleijenberg, 1974).

Interactions may occur with a variety of other drugs and may precipitate hypoglycemia or hyperglycemia. They are reviewed by Hansen and Christensen (1977) and are summarized in Table 8.9.

Ethanol may interact with the sulfonylureas in several ways. A "disulfiram-like effect" may occur in which there is flushing, nausea, vomiting, and headache. The mechanism of this effect is unknown, but it has been likened to the action of disulfiram, which inhibits alcohol dehydrogenase and so promotes the accumulation of acetaldehyde. This effect, which may have a genetic basis, is seen most commonly with chlorpropamide, where it occurs in about 30 percent of patients, the

Table 8.9. *Some drug interactions with the oral antidiabetic sulfonylurea drugs*

Interactions that may produce hypoglycemia (and necessitate a decrease in dosage)[a]	
1. Tolbutamide	
Salicylate	Displaces from plasma protein binding, also direct hypoglycemic action via (?) insulin release
Phenylbutazone	Displaces from plasma protein binding and inhibits drug metabolism
Dicoumarol	As above
Sulfonamides (e.g., sulfaphenazole)	As above
Halofenate	Inhibits drug metabolism
Chloramphenicol	As above
2. Acetohexamide	
Phenylbutazone	Inhibits renal excretion of active metabolites
3. Chlorpropamide	
Dicoumarol, sulfonamides, and phenylbutazone	May displace from plasma protein binding and produce hypoglycemia (half-life increased), suggesting some effect on metabolism
4. Glibenclamide	
Few interactions reported	Drug bound to plasma proteins but difficult to displace (Brown and Crooks, 1976)
Interactions that may result in hyperglycemia	
Thiazide diuretics	Block insulin release (possibly also furosemide and ethacrynic acid)
Diazoxide	As above
Corticosteroids	Hyperglycemic
Oral contraceptives	?

[a] Additive effects with the drugs in this group may occur with MAO inhibitors, salicylate, and ethanol, resulting in hypoglycemia. Propranolol can inhibit compensatory hyperglycemic mechanisms.
References: Hansen and Christensen, 1977; Shen and Bressler, 1977a,b.

onset occurring about 10 minutes after taking the ethanol. Ethanol acutely may produce a hypoglycemia and so add to the drug's effect, but in chronic consumers it can increase the rate of metabolism of the drug via the induction of hepatic enzymes. In extreme cases of liver damage, however, metabolizable sulfonylureas will persist in the circulation for longer periods.

The interactions (see Table 8.9) of the oral hypoglycemic drugs with other therapeutic agents may result in inhibited or enhanced responses to either substance. Thus, the anticoagulant *dicoumarol* and tolbutamide displace each other from binding sites on plasma proteins and also mutually interfere with each other's metabolism in the liver, so that when both are present simultaneously, there may initially be an enhanced hypoglycemia as well as hemorrhage. Other drugs, such as *phenylbutazone, salicylates* and certain *sulfonamides,* interact with tolbutamide and chlorpropamide by such mutual displacement from plasma protein-binding sites. As with dicoumarol, there is also often a reduction in the metabolism of tolbutamide so that its half-life is increased. Such an effect is seen with *phenylbutazone, sulfonamides,* and *chloramphenicol.* Some drugs, such as ethanol, salicylates, and MAO inhibitors, have a hypoglycemic effect which may

be additive to that of the sulfonylureas. *Propranolol,* by inhibiting β-adrenergic responses such as glycogenolysis, may limit possible physiological compensatory mechanisms that occur in response to hypoglycemia and so facilitate such an adverse effect. Certain drugs block the actions of the sulfonylureas so that it may be necessary to increase the dosage. Diuretics, especially the *thiazides,* but also possibly *furosemide* and *ethacrynic* acid, can inhibit the release of insulin from the B-cells. *Corticosteroid* therapy, by promoting gluconeogensis, may also oppose their action. The *oral contraceptives* may increase the need for a higher dose of oral antidiabetic drugs, but the mechanism of their effect is not known.

B. Biguanides

In 1918, Watanabe showed that guanidine had a hypoglycemic action in animals. This compound is, however, toxic, so that attempts were made to produce chemical analogues which may have a clinically useful action. This resulted in the production of the diguanidines *synthalin A* and *B* (Figure 8.18), but these derivatives had toxic effects on the liver. It was not until 1959 that a useful chemically related compound was introduced for the treatment of diabetes mellitus. This was phenethylbiguanide

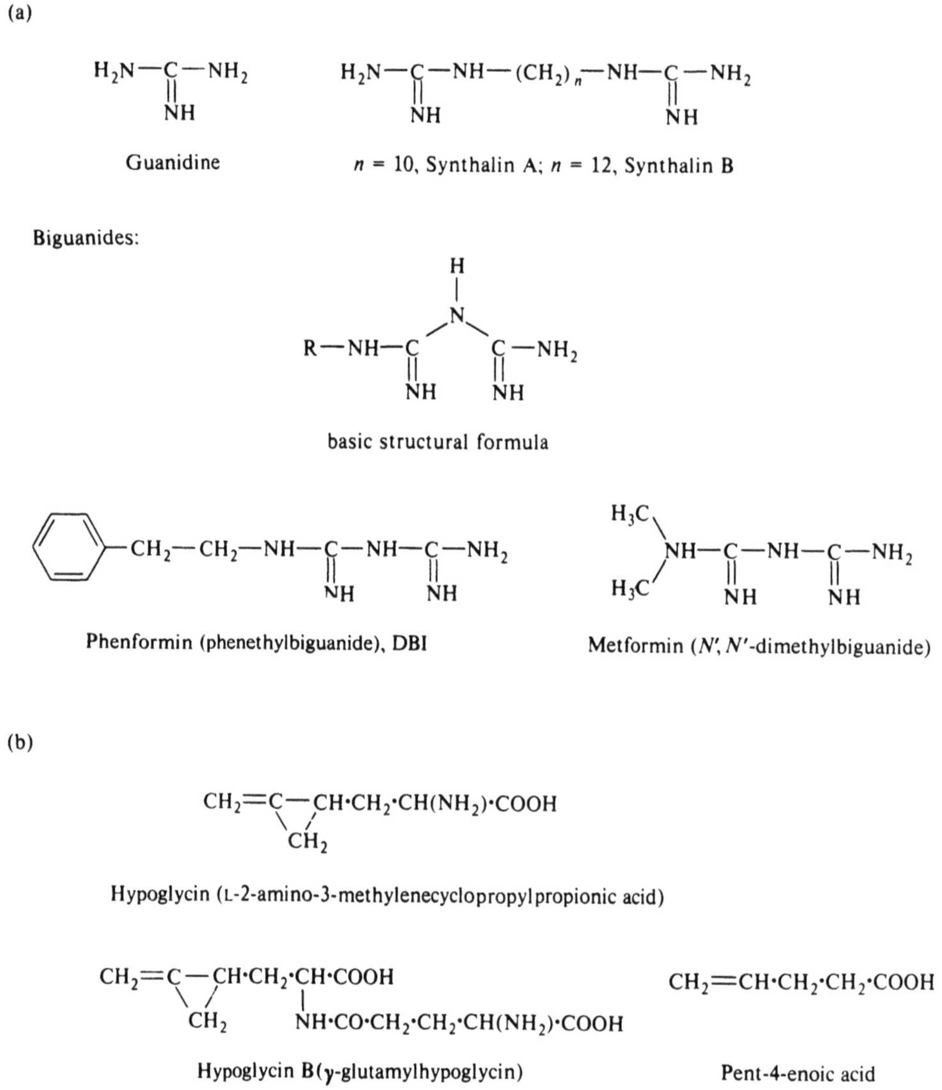

(a)

$$H_2N-\underset{\underset{NH}{\|}}{C}-NH_2$$

$$H_2N-\underset{\underset{NH}{\|}}{C}-NH-(CH_2)_n-NH-\underset{\underset{NH}{\|}}{C}-NH_2$$

Guanidine

$n = 10$, Synthalin A; $n = 12$, Synthalin B

Biguanides:

$$R-NH-\underset{\underset{NH}{\|}}{C}-\overset{\overset{H}{|}}{N}-\underset{\underset{NH}{\|}}{C}-NH_2$$

basic structural formula

$$\text{phenyl}-CH_2-CH_2-NH-\underset{\underset{NH}{\|}}{C}-NH-\underset{\underset{NH}{\|}}{C}-NH_2$$

Phenformin (phenethylbiguanide), DBI

$$\underset{H_3C}{\overset{H_3C}{>}}N-\underset{\underset{NH}{\|}}{C}-NH-\underset{\underset{NH}{\|}}{C}-NH_2$$

Metformin (N', N'-dimethylbiguanide)

(b)

$$CH_2{=}\underset{\underset{CH_2}{\diagdown\diagup}}{C}-CH{\cdot}CH_2{\cdot}CH(NH_2){\cdot}COOH$$

Hypoglycin (L-2-amino-3-methylenecyclopropylpropionic acid)

$$CH_2{=}\underset{\underset{CH_2}{\diagdown\diagup}}{C}-\underset{\underset{NH{\cdot}CO{\cdot}CH_2{\cdot}CH_2{\cdot}CH(NH_2){\cdot}COOH}{|}}{CH}{\cdot}CH_2{\cdot}CH{\cdot}COOH$$

$$CH_2{=}CH{\cdot}CH_2{\cdot}CH_2{\cdot}COOH$$

Hypoglycin B (γ-glutamylhypoglycin)

Pent-4-enoic acid

Fig. 8.18. (a) Structures of the biguanide hypoglycemic (oral antidiabetic) drugs. (b) Structures of hypoglycin and pent-4-enoic acid.

or *phenformin,* also known by the code name DBI. This drug was available until recently in the United States, but other derivatives – N',N'-dimethylbiguanide, or *metformin,* and N'-*n*-butylbiguanide, or *buformin* – are also used widely in Europe.

These drugs exist in their cationic form in the circulation and are poorly bound to plasma proteins. Phenformin is metabolized in the liver but about 70 percent of metformin is excreted unchanged in the urine. More phenformin than metformin thus accumulates in the liver, and it has been suggested that this difference may account for the higher incidence of lactic acidosis that is observed during the use of the former drug. These compounds have a relatively short duration of action, 5 to 7 hours, and so need to be given frequently. However, phenformin has been prepared in slow release (or timed-disintegration, TD) capsules, which prolong its action for about 14 hours.

The mechanism of action of the biguanides is poorly understood. They do not release insulin. They exert a hypoglycemic effect in diabetics but not in normal people, apparently due to the functioning of adequate compensatory processes in the latter. Several mechanisms of action have been proposed. Three of these appear to provide the most likely explanation of their actions.

a. There is a *delay in the absorption of carbohydrate from the gut.* This accounts for the observed improvement in the oral glucose tolerance test which is not seen when the glucose is administered intravenously. The absorption of other nutrients from the gut may also be slow, and prolonged use of these drugs has resulted in vitamin B_{12} and folic acid deficiency. This effect may also contribute to the observed weight loss in patients taking these drugs.

b. *Glucose uptake by peripheral tissues may be increased,* with a concomitant increase in anaerobic glycolysis which is reflected in the observed rise in plasma pyruvic and lactic acid levels. Such a process is consistent with the occurrence of lactic acidosis, which in certain clinical circumstances may occur when phenformin is being used.

c. It has also been suggested that these drugs produce a *decline in gluconeogenesis* from alanine, pyruvate, and lactate, which could also contribute to a lactic acidosis.

The biguanides have a ready capacity to bind to membranes, both on the cell surface and on subcellular particles, such as the mitochondria. This rather unspecific action may mediate their actions (Schäfer, 1976).

Some of the precautions and special circumstances that must be considered when using biguanides in the treatment of diabetes mellitus have been described at the beginning of Section 8.6. They have been used in conjunction with the sulfonylureas, especially when the diabetes is difficult to control with the former drugs alone. They have also been used in juvenile-onset diabetes in an attempt to reduce the dose of insulin and avoid possible hypoglycemia. In view of the predominance of the adverse effects of phenformin, there appears to be little to support this procedure. Phenformin alone is ineffective in juvenile diabetes, as it does not combat the ketosis, and for this reason is also not useful in ketosis-prone mature-onset diabetics. The clearance of the biguanides from the body is largely dependent on hepatic inactivation and urinary excretion, so that care must be exercised if hepatic or renal disorders accompany the diabetes. Its possible effects on the fetus have not been defined, so it has not been used during pregnancy. As phenformin promotes a loss in weight, it has been considered of special use in obese mature-onset diabetics whose weight is difficult to control. On the other hand, they have been avoided in thin mature-onset diabetics.

Troublesome minor *side effects* are infrequent when using biguanides. Skin rashes occur occasionally, there may be a metallic taste in the mouth, and there is anorexia. The latter appears to contribute to the observed weight loss. Nausea and vomiting sometimes occur, but this problem is usually associated with a too rapid increase of the size of the dose and so can be controlled. The possibility of vitamin B_{12} and folic acid depletion has been referred to above and it is recommended that patients on long-term therapy with biguanides be periodically checked for such deficiencies (Tomkin, 1973). There are two relatively rare but serious problems associated with the use of phenformin: the occurrence of lactic acidosis and an increased likelihood of death due to cardiovascular "incidents" (see later). The possible relationship of *lactic acidosis* to the administration of phenformin has been recognized for many years but was thought to be much more rare than now seems to be apparent (Gale and Tattersall, 1976; Wise et al., 1976; Conlay and Loewenstein, 1976; Luft, Schmülling, and Eggstein, 1978). Lactic acidosis is a serious condition and as many as two-thirds of patients who develop it may die of it. It occurs due to an imbalance in lactate metabolism – the anaerobic production of this metabolite exceeding the body's ability to oxidize and dispose of it. This can be precipitated by alcohol or by cardiovascular (leading to tissue anoxia), renal, and hepatic diseases. The latter conditions reduce metabolism and excretion of lactic acid, the ability to compensate for the acidosis, and may also result in an excessive accumulation of the drug and its metabolites. Phenformin itself may also reduce the kidneys' ability to produce ammonia, which limits acid excretion (Rooth and Bandman, 1973). The use of phenformin is thus not recommended in such diseases and the maximum dose should not exceed 100 mg/day.

Because of the incidence of lactic acidosis, phenformin is *no longer approved for use in the United States.* At this time, however, it is still available as an Investigatory New Drug (IND) and can be obtained by those willing to fill out the necessary forms required by the bureaucracy. Metformin has been widely used in Europe and lactic acidosis has only rarely been associated with its administration (see, e.g., Bergman, Boman, and Wiholm, 1978). It has therefore been suggested that metformin (where available for use) "is now the biguanide of choice" (Editorial, 1977i). Treatment of lactic acidosis (see Assan et al., 1975; Alberti and Nattrass, 1977; Luft, Schmülling and Eggstein, 1978) is somewhat controversial, as is always likely when there is a large rate of failure. Infusion of $NaHCO_3$ to correct the metabolic acidosis is the basic procedure. This may be accompanied by dialysis to aid excretion of the accumulated biguanides. The diuretic furosemide has also been recommended. The use of insulin is controversial. If substantial amounts of ketone bodies are present, which contribute to the metabolic acidosis, it may be useful. However, it also inhibits gluconeogenesis, which may further increase the accumulation of lactate. A novel, also controversial, proposal to also use dichloroacetate is discussed below.

C. Dichloroacetate

Sodium dichloroacetate (DCA) is an organic acid compound which can lower blood glucose, pyruvate, and lactate levels, as well as the concentrations of plasma cholesterol and triglycerides. These effects have been demonstrated in starved and diabetic animals (Lorini and Ciman, 1962), and it has also been tested experimentally in man (Vailati and Rabassini, 1962; Stacpoole, Moore, and Kornhauser, 1978). Dichloroacetate does not act by increasing the release of insulin. It appears to activate the enzyme pyruvate dehydrogenase (by inhibiting pyruvate dehydrogenase kinase, which converts pyruvate dehydrogenase from its active to its inactive form). Pyruvate is then utilized more readily, so there is a fall in the levels of pyruvate, glucose, and lactate. The mechanism of action of its effects on lipid metabolism is not known. The ability of DCA to lower plasma lactate levels offers the prospect of its use for the treatment of lactic acidosis (Editorial, 1978a). It has been used in experimental animals and in one instance clinically, but it was not effective. Future prospects for its success are controversial (Irsigler, Kaspar, and Kritz, 1977; Alberti, Nattrass, and Holloway, 1977).

D. Hypoglycin

An interesting naturally occurring hypoglycemic agent has been identified in a tropical fruit from Jamaica, the ackee *Blighia sapida* (Sherratt, 1969). It has been observed that when the unripe fruit is eaten it can result in a condition called "vomiting sickness," which may have caused as many as 5000 deaths. It is associated with a severe hypoglycemia, exhaustion of liver glycogen, fatty infiltration of the liver, and an increase in plasma free fatty acid levels. Two active compounds (Figure 8.18) were isolated in 1955; hypoglycin and γ-glutamylhypoglycin (respectively, hypoglycin A and B). Synthetic derivatives, including pent-4-enoic acid, have been prepared in exploring the possibility of utilizing their hypoglycemic actions in the treatment of diabetes mellitus. This application has not been successful, but the compounds and their mechanism of action remain of theoretical interest.

The observed elevation of plasma free fatty acids and the resulting ketosis contrast with the action of insulin and have provided clues as to the mechanism of action of hypoglycin and its derivatives. It appears that they may block the metabolism of long-chain fatty acids and gluconeogenesis, so that glucose is preferentially used and cannot be adequately replaced. The precise biochemical mechanism of action is controversial (see Billington, Osmundsen, and Sherratt, 1978) and includes a sequestration of cell CoA or inhibition of enzymes

responsible for the metabolism of free fatty acids, especially liver pyruvate carboxylase.

E. The UGDP controversy

In 1960, shortly after the introduction of the sulfonylureas and biguanides for the treatment of diabetes mellitus, a clinical trial was planned to test their long-term effects. This project involved 12 university hospital clinics and was called the UGDP program, which stands for "University Group Diabetes Program." The initial hope was that use of these drugs may reduce the long-term vascular complications of the disease (but the opposite appeared to occur). With the support of the U.S. National Institutes of Health, the trial commenced in 1961. It initially contained about 800 patients suffering from mild mature-onset, nonketotic diabetes (University Group Diabetes Program, 1970a,b, 1971). These people were admitted to the program on the basis of an abnormal glucose tolerance test. All were treated by diet and divided into four groups, each containing about 200 patients, which received different additional treatments. These were: (a) a placebo (i.e., diet alone); (b) a fixed dose of insulin (16 to 20 IU/day); (c) a variable dose of insulin; and (d) a fixed dose of tolbutamide (1.5 g/day).

About a year later a fifth group was added (University Group Diabetes Program, 1975), which received a fixed dose of phenformin (100 mg/day). In 1969, about 8.5 years after the trial began, it was observed that the death rate from cardiovascular causes in the tolbutamide-treated group was about double that in the other groups, and this treatment was terminated. The FDA and the press were notified and the storm broke. About a year later a similar increase in death rate was found in the phenformin-treated group, so this trial was also stopped. The FDA issued a "special warning" (beginning Section 8.6) which was to be inserted in the packages of the drugs. The experience of many physicians who had used these drugs extensively did not, however, appear to support the findings of the UGDP and they formed a group, the Committee on the Care of Diabetics, which obtained a legal injunction restraining the FDA from enforcing the new labelling of the drugs. A legal decision regarding the compulsory insertion of warning labels in preparations of the sulfonylureas is at present pending.

The criticisms of the drug trial program were extensive (see Seltzer, 1972b). These include:

a. The groups may not have been homogenous with respect to the baselines of their medical condition. It was, for instance, suspected that there was higher initial incidence of cardiovascular problems in the groups that reacted adversely.

b. The criteria for diagnosis of diabetes mellitus were not

considered to be adequate, and in fact a proportion of the patients would not by the usual standards be considered to have the disease at all.

c. The use of a fixed dose of drug is contrary to good medical practice: namely, this would be adjusted in relation to the responses and side effects of the patients.

d. The phenformin group was added later and did not constitute a part of the initial "pool" from which the controls were drawn.

e. The group of 1000 patients is small compared to the 1.5 million who were receiving the drug in the United States, and the statistical procedures were inadequate.

f. The results of other, smaller surveys did not confirm the UGDP results.

In order to obtain an expert opinion of the statistical evidence, the Biometric Society was asked to report on this aspect of the program (Gilbert et al., 1975). In general, they supported its validity but were somewhat equivocal. ("We consider the evidence of harmfulness moderately strong. We consider that in the light of the UGDP findings, it remains with the proponents of the hypoglycemics to conduct scientifically adequate studies to justify the continued use of such agents.") Two editorials from eminent Journals (Editorial, 1975i; Chalmers, 1975) provided reflected notes of caution regarding the use of these drugs, and Chalmers was especially critical of their continued use. A strong feeling of caution has certainly arisen and physicians have been reminded, as was always suggested, that in many instances diet alone may be sufficient to control the disease. The editorial in *Lancet* stated: "The UGDP war remains in the balance, and the combatants are obscured by increasingly heavy clouds of clinical, statistical and philosophical smoke." The only point of agreement seems to be that the patient should have the problem explained to him and give his "informed" (?) consent.

The reasons for the possible adverse cardiovascular effects of the oral hypoglycemics are not clear. Animal experiments have shown that tolbutamide may have an inotropic and chronotropic effect on the heart, probably by facilitating the activity of adenyl cyclase (Levey, Lasseter, and Palmer, 1974). However, these effects appear to be species-specific and their possible application to man is not clear. In the instance of phenformin, a rise in blood pressure and heart rate was observed in the UGDP trial, but the reasons for these effects or their significance are unknown. Certainly, the results of the UGDP have stimulated a lot of research into the basic mechanisms and other actions of the oral hypoglycemic drugs. It has probably not been adequately appreciated that therapeutic agents may have other effects in the body apart from those which justify their clinical use.

F. Efficacy of oral antidiabetic drugs

There are also conflicting reports as to the efficacy (in addition to the toxicity) of the oral antidiabetic drugs to control hyperglycemia in symptomatic and asymptomatic mature-onset diabetes mellitus. These have been reviewed and summarized by Shen and Bressler, 1977b. Primary failure (patients who never achieve satisfactory control with such drugs) occurs in about 40 percent of symptomatic diabetics, but this value depends on the criteria used for selection. Estimates of long-term secondary failure vary from 3 to 30 percent and are partly due to lack of compliance. There is considerable doubt as to whether people suffering from asymptomatic diabetes who take such drugs in the prophylactic sense, to prevent the further development of the disease, experience any benefit. Diet alone may be principally important, and it is often difficult to disentangle the benefits of this regimen from those of concurrently administered oral hypoglycemic drugs. Shen and Bressler (1977b) concluded, "At present oral hypoglycemic agents have not demonstrated a useful role in the management of maturity-onset diabetes mellitus, asymptomatic or symptomatic." There has been further heated discussion regarding this view, the evidence for which at present is equivocal (see Feldman, 1977; Gershberg, 1977).

8.7. B-cell cytotoxic chemicals

In 1943, Dunn and his collaborators demonstrated that alloxan (Figure 8.19) had a cytotoxic action on the islet B-cells of rabbits. This effect, which has been demonstrated in many other species, is remarkably specific; the only other tissue that appears to be affected is the kidney, where higher doses are required and the response is transient. The effect occurs within minutes following the injection of the chemical and results in a marked necrosis of the cells which is initially accompanied by a hyperglycemia, followed by a period of hypoglycemia when insulin is being released, and finally a chronic diabetic phase (see Rerup, 1970). In the latter stage the animals can be kept alive by the injection of insulin. This condition of experimental diabetes is often called "chemical diabetes," which is distinct from the term used clinically to describe early asymptomatic diabetes in man. The mechanism of action of alloxan is not known, but it can be combated with SH-active compounds such as glutathione, cysteine, and dimercaprol (BAL). It has thus been suggested that the alloxan may interact with membrane SH groups on the B-cells. It also has a capacity to bind heavy metals, which has led to the suggestion that it could act by binding zinc in the B-cell.

Streptozocin (also called streptozotocin) (Figure

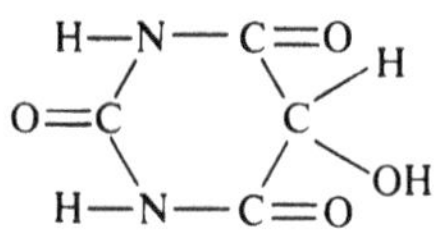

Alloxan

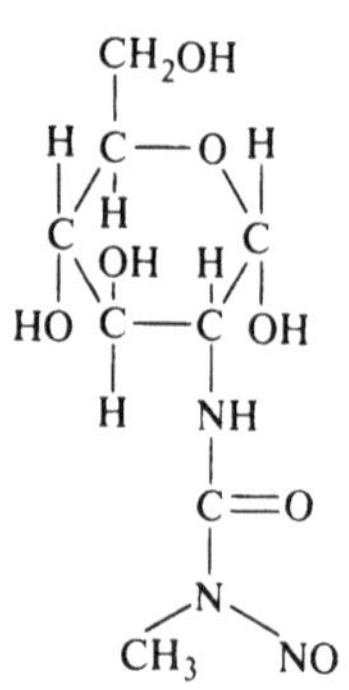

Streptozocin
(streptozotocin)

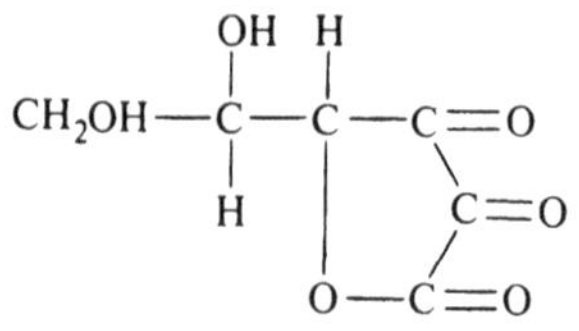

Dehydroascorbic acid

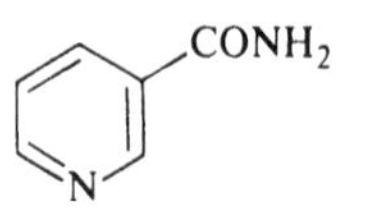

Nicotinamide 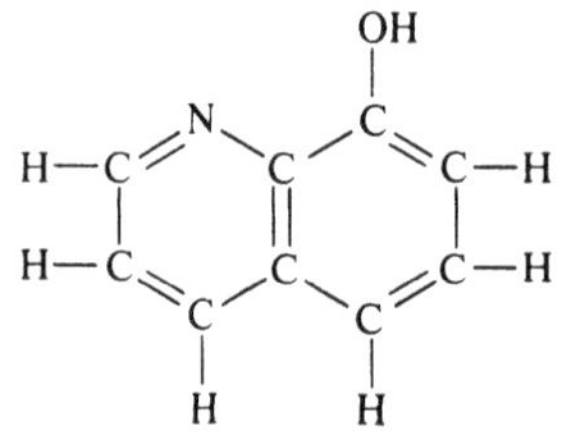   Oxine (8-hydroxyquinoline)

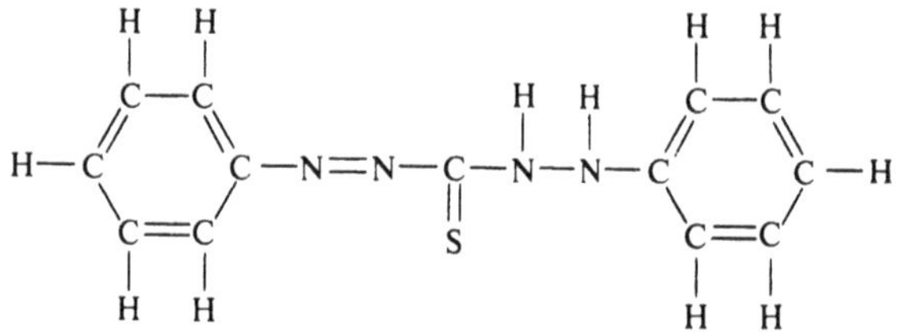

Diphenylethiocarbazide

Fig. 8.19. B-cell cytotoxic chemicals that can produce diabetes mellitus.

8.19) is a broad-spectrum antibiotic which is carcinogenic and also has a cytotoxic effect on the B-cell, producing diabetes mellitus (see Rerup, 1970). Its mechanism of action is not known and, in contrast to alloxan, SH-active chemicals do not afford protection against its effects. Glucose is a constituent part of its molecule, and this could afford it an ability to interact with glucoreceptor of the B-cell. Streptozocin has been used clinically in the treatment of hyperinsulinism.

A recently introduced rodenticide *N-3-pyridylmethyl N'-p-nitrophenyl urea* (*PNU*) has been observed to have a hyperglycemic action both in rats and in man when it has been accidentally or intentionally ingested (Prosser and Karam, 1978). In 12 of the surviving humans, ketotic diabetes mellitus, requiring the administration of insulin, developed. The poison appears to have a cytotoxic effect on the B-cells.

The B-cell cytotoxic actions of alloxan and streptozocin can be antagonized in animal experiments by the administration of *nicotinamide* (niacinamide), either before the administration of the alloxan or within 2 hours of giving streptozocin (Ganda, Rossini, and Like, 1976). It is possible that PNU may be acting as a nicotinamide antagonist (Prosser and Karam, 1978), so that the latter could also be an antidote to the action of the rat poison. The effect of alloxan can also be antagonized by D-*glucose,* and the nonmetabolizable sugar, *3-O-methyl glucose,* is even more effective (Rossini, Arcangeli, and Cahill, 1975). The latter sugar also protects against the effects of streptozocin (Ganda, Rossini, and Like, 1976).

8.8. Role of glucagon in disease and its therapeutic uses

At this time there are few clear instances which implicate excesses or deficiencies of glucagon in human disease. Glucagonoma (Unger and Orci, 1976), which results in a high rate of secretion of glucagon, is a very rare disease which does not cause a dramatic increase in blood glucose levels. Experimental hyperglucagonemia does not cause diabetes, provided that B-cell function is normal. A role for glucagon deficiency in hypoglycemia has not been described. In animal experiments it has not been possible to destroy or remove A-cells while leaving the B-cells intact, which makes an evaluation of their role difficult. The possible importance of glucagon for the manifestation of the full diabetic effects of a lack or insulin has been described earlier [Section 8.5A(3e)]. It has been suggested that a decreased secretion of glucagon may be associated with obesity. It has been reported that the glucagon secretory response of the A-cell to the infusion of alanine is reduced in obesity (Wise, Hendler, and Felig, 1973), but this observation was not confirmed in a subsequent study (Kalkoff, Gossain, and Matute, 1973). The reason for the discrepancy is not clear. If glucagon deficiency does have a role in obesity, it has been suggested that interest in the possible use of glucagon in the treatment of *obesity* may be revived (Editorial, 1973d). There do not appear to have been any subsequent reports about this interesting possibility. Glucagon has been used in the treatment of *acute pancreatitis,* and the preliminary results suggested that survival was improved (Condon, Knight, and Day, 1973). While glucagon has a positive inotropic and chronotropic effect on the heart of normal subjects, this effect is not seen in patients suffering from chronic heart failure, so that early expectations as to its possible use in this condition were not fulfilled. This change in sensitivity may be due to a decline in tissue adenyl cyclase (Gold et al., 1970). The only widely established therapeutic use for glucagon appears to be in the treatment of insulin hypoglycemia (Section 8.5C).

9

Calcemic hormones: parathyroid hormone, calcitonin, and vitamin D_3

Calcium metabolism is principally controlled by the actions of three hormones: parathyroid hormone, from the parathyroid glands; calcitonin, from the thyroid C (or parafollicular)-cells; and $1\alpha,25$-dihydroxycholecalciferol [$1\alpha,25$-$(OH)_2$ vitamin D_3], which is formed from vitamin D_3. The actions of these hormones are often dependent on each other: for instance, parathyroid hormone stimulates the final conversion of vitamin D_3 to $1\alpha,25$-$(OH)_2D_3$, while the presence of the latter may be necessary for parathyroid hormone to act on bone.

Calcium has a variety of different and vital functions in the body, so that when insufficient amounts are available profound physiological disturbances occur. Its most obvious contribution is toward the structure of the bones, where, principally in the form of insoluble phosphate salts, most of the body's calcium is sequestered. The structure of the "soft" tissues is, however, also dependent on calcium, and when not enough is present, cell membranes become more permeable and, as seen *in vitro*, the cells may even fall apart from each other. Such effects have led to its being described as a "stabilizer" of cell membranes. Many enzyme reactions depend on the presence of certain divalent ions, and among these processes Ca^{2+} is prominent, possibly even predominant. Its role in blood clotting, through the conversion of prothrombin to thrombin, is well known. Calcium also has a vital role in "coupling" various responses in tissues. The electrical depolarization of muscle cells leads to their contraction, and it is the increased availability of soluble Ca^{2+}, which reacts with the contractile protein system, that links these two processes. A similar role is played by Ca^{2+} in the processes that initiate secretion, including that of hormones, from cells. The immediate sources of such calcium are mainly cellular, as it can be mobilized from storage sites in the endoplasmic or sarcoplasmic reticulum, plasma membrane, and the mitochon-

dria, although an increased rate of accumulation from the extracellular fluids may also be important.

The ubiquitous importance of calcium in the body appears to be related to its ability to bind to proteins between which it can form stable cross-linkages. It thus influences their stability and configuration. This may be important in providing morphological cohesion as well as triggering changes in the functional role of proteins. Calcium also forms complexes with lipids and can change their surface tension. It has been suggested that if calcium had such effects at the cell surface, it could contribute to changes in the permeability of the plasma membrane. Calcium forms relatively insoluble salts, especially with phosphate and carbonate ions, and such complexes can assume crystalline forms that have considerable physical strength and which are utilized in the bony skeleton.

The calcium content of an adult man is about 1200 g, of which roughly 99 percent is present in the bones. It should be recalled that the bones are living tissues which, apart from growth in the young, are also, in adult life, being continuously "remodelled" or "rebuilt." There is a physicochemical exchange of calcium salts in the bones with those in the extracellular fluid, the Ca^{2+} and PO_4^{2-} in the latter tending to enter and precipitate in the bones more rapidly than they reenter the solution in the opposite direction. This process is compensated for by a "dynamic" physiological process in which bone Ca^{2+} and PO_4^{2-} is mobilized as a result of the activities of bone cells, and it is this type of process that is under hormonal control.

There are three main types of bone cells (Figure 9.1): the *osteoblasts*, which are subsequently transformed into *osteocytes*, and the *osteoclasts*. The osteoblasts lay down the collagen network, which is subsequently mineralized (accretion). They contain high levels of prolylhydroxylase and alkaline phosphatase. Hydroxyproline is a unique constitu-

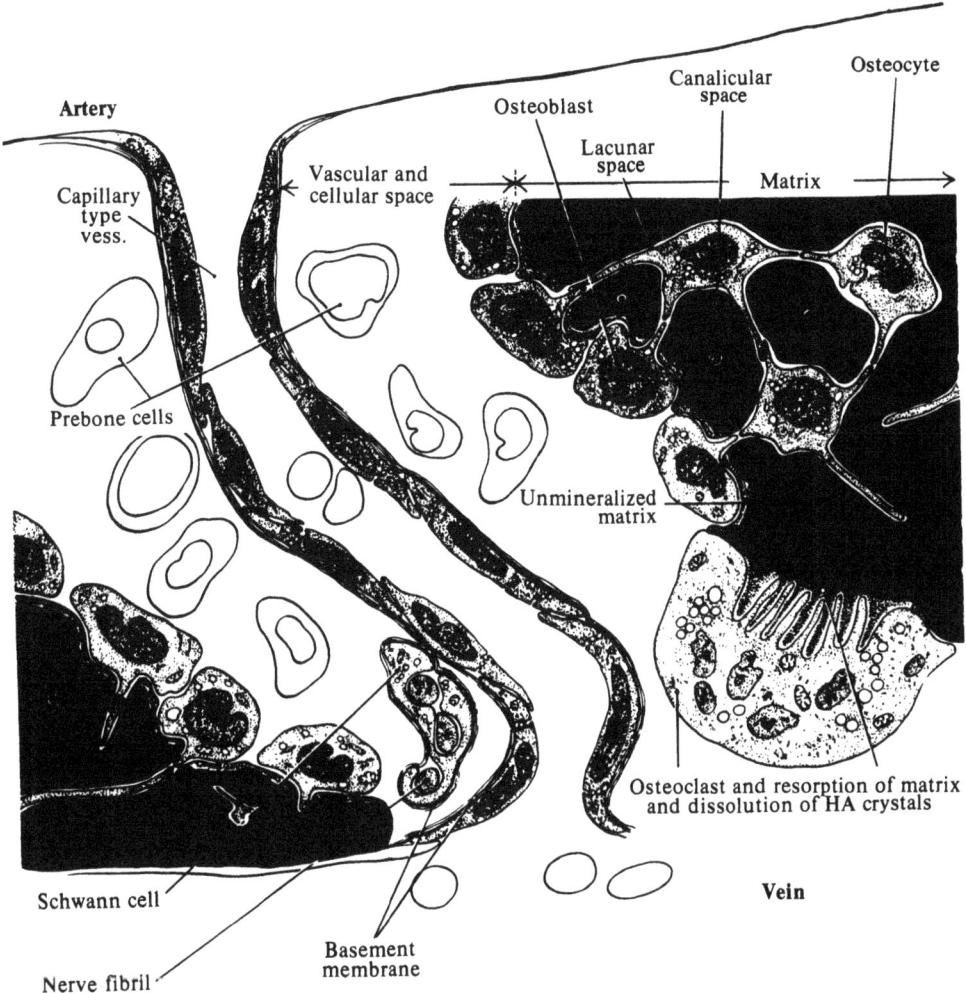

Fig. 9.1. Schematic representation of a physiological unit of bone tissue showing the osteocytes, osteoblasts, and osteoclasts. (From Doty, Robinson, and Schofield, 1976.)

ent of collagen, so that its levels in the blood and urine are taken as a reflection of bone formation; it is high when breakdown and calcium resorption are occurring. The osteocytes become enclosed in small bony chambers or lacunae and send out dendritic extensions through fine radiating channels, or canaliculi, which allow communication with the extracellular fluids. The osteocytes control local exchanges of calcium and thus have been referred to as "sentinels." The osteoclasts are large multinuclear cells whose principal role appears to be the mobilization or resorption of bone minerals. They appear to carry out this function by secreting enzymes, including an acid phosphatase, which mobilizes the precipitated salts.

There is a continual physiological turnover of calcium in the body which principally involves the gut and the kidney, but also the sweat glands and, during lactation, the mammary glands. Calcium is gained in the drinking water and food, from which it is absorbed across the wall of the intestine. This process is most rapid in the duodenum and jejunum, but quantitatively, probably reflecting its greater length, more occurs from the ileum and only little in the colon. The total absorbed each day, in man, is about 300 mg out of an intake of 1000 mg. Absorption occurs against an electrochemical gradient by a process of active transport which is principally controlled by vitamin D_3. Calcium is also secreted into the gut, principally in the bile. Calcium is excreted by the kidney, and in man about 200 mg is lost each day in the urine. The fraction of calcium that is not bound to the plasma proteins is filtered across the glomerulus.

About 95 to 99.5 percent of this calcium is reabsorbed across the renal tubules, and it appears to occur along its entire length: proximal, loop of Henle, and distal segments.

In man about 100 mg of calcium is lost each day in the sweat and this may be increased in hot conditions. Milk also contains calcium, so there are additional dietary requirements for this mineral during lactation. This amounts to 1.5 to 2 g/day, compared to the usual intake of about 1 g. Calcium also crosses the placenta in response to the needs of the fetus, which at birth contains about 30 g. This mineral is partly obtained in the maternal diet but also by depletion of calcium stored in the bones.

The metabolism of calcium is closely associated with that of phosphate, so it is sometimes difficult to dissociate the processes that influence each. The concentrations of Ca^{2+} and PO_4^{2-} in the extracellular fluid are related to their solubility product, so that when plasma PO_4^{2-} increases there is a decline in Ca^{2+} concentration, and vice versa. Thus, a loss of phosphate in the urine may lead to a hypophosphatemia and an elevation of plasma Ca^{2+} levels. Parathyroid hormone acts on the proximal renal tubule to inhibit phosphate reabsorption and promotes its urinary loss, and so indirectly acts to increase plasma Ca^{2+} concentration. Parathyroid hormone also has a separate and distinct effect in promoting Ca^{2+} reabsorption across the renal tubule. Active calcium transport across the intestine is apparently not related to the presence of phosphate; the anion passively follows the cation.

The normal Ca content of most cells is about 1 to 2 mM/kg water, but the concentration which is ionized in solution in the cell water is only about 10^{-7} M. The remainder is sequestered in two principal types of "compartments": an "inexchangeable" part, which constitutes that which is an integral part of the structural elements, and a "slowly" exchangeable portion, which exists in a type of equilibrium with that in solution but is sequestered in the mitochondria, the vesicles of the endoplasmic reticulum, and the plasma membrane. The slowly exchangeable calcium may be confined in the cell organelles due to its limited ability to cross their bounding membranes, or it may be in a precipitated crystalline form, which can nevertheless enter solution. Calcium is thought to be accumulated in the cell organelles by a process of active transport and tends to leave them by diffusion down electrochemical gradients. These processes may involve the action of cyclic AMP, which acts as a second messenger in many hormonal responses, including those of parathyroid hormone and calcitonin. There is also a continual exchange between Ca^{2+} in the cell water and that in the external bathing fluids, across the plasma membrane. It is thought that this process usually involves an active calcium extrusion and a passive entry. In transporting epithelia such as the intestine, these processes may be asymmetrical, with a passive influx across the mucosal surface followed by an active efflux across the serosa. Hormonal effects may be mediated at either site. Such a model is not entirely satisfactory (see, e.g., Borle, 1973) and it has been suggested that the intracellular exchanges may be the primary determinants of calcium movements within, and across, cells and that the hormonal responses may be primarily on the mitochondria and the endoplasmic reticulum.

In 1967 W. Y. Cheung (see Cheung, Lynch, and Wallace, 1978) isolated a Ca-binding protein from tissue homogenates containing cyclic nucleotide phosphodiesterase activity. Its presence was necessary for the activity of the latter enzyme. This protein has since been identified in a variety of tissues and animal species in which it appears to have ubiquitous functions. It has a molecular weight of 16,700 and has been called *calmodulin* (also modulator protein, Ca-regulated protein modulator). Calmodulin appears to mediate many effects of calcium by acting as an intracellular receptor for the metal. It assumes a helical configuration when combined with Ca and it can then combine with and activate its particular effector. Apart from phosphodiesterase, calmodulin has been shown to activate brain adenyl cyclase, red cell Ca-ATPase, and skeletal muscle phosphorylase kinase; it may be involved in events such as stimulus-secretion or -excitation coupling, release of neurotransmitters, and mitosis. Calmodulin clearly may have roles in hormone-mediated tissue responses and it is possible that its levels may be influenced by endocrine secretions.

9.1. Parathyroid glands

The parathyroid glands lie in the region of the neck proximal to the thyroid. In man, there are normally four present. They are derived embryonically from the III and IV branchial pouches. Occasionally, they may fail to develop (parathyroid aplasia), or in some instances more than four may be present, possibly representing the failure of another pair, derived from the II branchial pouch, to disappear during ontogeny.

The parathyroid glands were first described 100 years ago, in 1880, by I. V. Sandström, and although their possible relationship to calcium metabolism was suspected at the turn of the century, the presence of an active hormone secretion was not confirmed until 1925 by J. B. Collip. It was not chemically characterized until 1970. Subsequently, research on it has been accelerated by the application of radioimmunoassay procedures for its measurement in the blood.

A. Morphology

The parathyroids each weigh only about 40 mg. They lie near the posterior surface of the thyroid, and each receives a distinct blood supply from the carotid artery which, if compromised, can have serious effects on its function. These factors emphasize the problems that can arise due to damage to the parathyroids during thyroid surgery and the special difficulties associated with surgical procedures on these glands. Histologically, there are three main types of cells present in the parathyroids: fat cells; oxyphil cells, which appear to have no special endocrine function; and the chief cells. The latter synthesize and secrete the hormone and contain the Golgi apparatus and secretory granules concerned with the synthesis, storage, and secretion of the hormone.

B. Functions and actions

The knowledge that the parathyroid hormone was involved in the control of calcium metabolism resulted largely from early observations that skeletal diseases may be associated with pathological changes in the parathyroid glands. The primary experimental observation appears to be that of G. Vassale and F. Generali, who showed in 1900 that surgical removal of the parathyroid glands in dogs resulted in tetany. The recognition that tetany in man following thyroid surgery was associated with accidental damage to the parathyroids also provided an important impetus to research on the subject.

The secretion of parathyroid hormone is inversely related to the ionized calcium concentration in the blood that supplies the glands; it rises in response to hypocalcemia and is depressed by hypercalcemia. This effect can also be seen *in vitro* using pieces of glandular tissue incubated in media containing different concentrations of calcium.

The parathyroid hormone can be looked upon as maintaining plasma Ca^{2+} levels in opposition to the continual physicochemical tendency of Ca^{2+} to leave the extracellular fluids and enter the bones. It exerts this effect by acting on the bones, the kidneys, and, probably indirectly, on the intestines.

Parathyroid hormone promotes the resorption of calcium from the bones back into the extracellular fluid. This effect is apparently dependent on the presence of vitamin D_3 (see Section 9.3A), which may be responsible for the synthesis of an intermediary material that may act as a Ca "carrier." This effect of injected parathyroid hormone takes place in two stages: an initial one which takes several hours, and a longer-term effect which takes place over 2 or 3 days. The effector bone cells are the osteoclasts, osteocytes, and osteoblasts, but their precise contributions and the chronological order of their responses are not completely clear.

There is an increase in the number and activity of the osteoclasts, which may exert their phagocytotic activity by secreting proteases and hydrolase enzymes, including an acid phosphatase, as well as acid, which results in resorption of calcium. An activation or secretion of bone collagenase may also be involved in the Ca resorption (Sakamoto et al., 1975). Several enzymes, such as β-glucuronidase and N-acetyl-β-glucoaminidase, β-galactosidase, and cathepsin, may be released from lysosomes in the bone tissue under the influence of parathyroid hormone (Vaes, 1968a; Eilon and Raisz, 1978). This effect, which has been observed in bone tissue *in vitro*, may initiate bone resorption by an action on a protective "coat" of glycoproteins, allowing the subsequent action of the tissue collagenase. The initial faster effect of parathyroid hormone may involve the activity of the osteocytes as well as the osteoclasts. Osteocytes have been observed to become surrounded by a "halo" (Bélanger, 1969), in response to stimulation by parathyroid hormone, which is thought to reflect a release of calcium. The activity of the osteoblasts and their secretion of the collagen matrix for new bone formation is diminished by parathyroid hormone. In hens, it has also been observed that parathyroid hormone initially increases the blood supply to bone, which could also be an important factor in the initial "fast" response (Mueller et al., 1973).

More direct observations have been made on the effects of parathyroid hormone [as well as calcitonin and $1\alpha,25$-$(OH)_2D_3$] on separated cultures of "osteoblast-like" and "osteoclast-like" bone cells (Wong, Luben, and Cohen, 1977). The parathyroid hormone and $1\alpha,25$-$(OH)_2D_3$ both stimulated the enzymic activities of the osteoclasts and reduced those of the osteoblasts. The action of the parathyroid hormone, but not that of the $1\alpha,25$-$(OH)_2D_3$, was associated in both types of cells with an increase in the levels of cyclic AMP. Thus, although the two hormones act on the same type of cell, they apparently do so at different loci within the cells.

The kidney responds to injected parathyroid hormone much more rapidly than does bone; a response is seen in a few minutes. It has three actions on this organ:

a. There is an increased phosphate excretion, reflecting an inhibition of its reabsorption across the proximal segment of the renal tubule. This effect, by lowering the plasma phosphate concentration, will indirectly tend to increase the Ca^{2+} level.

b. Parathyroid hormone has a direct effect on Ca^{2+} by promoting its absorption from the glomerular filtrate, and this response appears to occur in both the proximal and distal tubules. Calcium absorption appears to be related to Na transport and it is difficult to dissociate the two processes. It has been suggested that parathyroid hormone may act in the kidney by initiating increases in Na reabsorption.

c. As will be described in detail later (Section 9.3), the activation of vitamin D_3 to its active form, $1\alpha,25$-$(OH)_2D_3$, depends on the action of 1α-hydroxylase in the kidney. This process is increased by parathyroid hormone, and its action at least partly reflects a decline on the cell phosphate concentrations, possibly related to its phosphaturic effect (Garabedian et al., 1974).

The role of the intestine in the hypercalcemic action of parathyroid hormone has been controversial. There is evidence that it can increase calcium transport across the wall of the intestine, but the observed effects were often inconsistent and small. It now appears that this action is an indirect one, reflecting increases in the "activation" of vitamin D_3 by the kidney.

Injected parathyroid hormone can increase the release of gastrin from the stomach of pigs (Bolman et al., 1977). This effect is not dependent on elevation of plasma Ca^{2+} levels and may contribute to the hypergastrinemia observed in humans suffering from hyperparathyroidism.

C. Parathyroid hormone

Purified parathyroid hormone (PTH) was first isolated in 1959 by H. Rasmussen and L. C. Craig and by G. D. Aurbach. It was shown to be a polypeptide with a molecular weight of about 10,000. The precise amino acid sequence of the bovine hormone was described in 1970 by H. B. Brewer and R. Ronan, and H. D. Niall and his collaborators, and shown to consist of a chain of 84 amino acids (Figure 9.2). There are differences in the amino acid sequences in the hormones from different species so that the animal hormones differ from those in man. There is, however, a considerable cross-responsiveness, and the bovine hormone is available for the treatment of human diseases.

The entire 84-amino acid single chain of parathyroid hormone is not essential for its activity. This discovery arose mainly from the observation that radioimmunoassay for circulating hormones yielded conflicting results, which suggested that the different antibodies that were used reacted in different ways with the plasma "hormones." It became apparent that several forms of parathyroid hormone were present in the circulation and that these had differing activities.

The parathyroid hormone molecule can be looked on as being in two sections: an N-terminal segment, usually taken to mean amino acids in the 1–34 positions; and a C-terminus. The N-terminal portion retains biological activity, but the C-terminus, for instance the 53–84 fragment, does not. The latter, however, remains immunologically active (Rosenblatt et al., 1978). It has been sug-

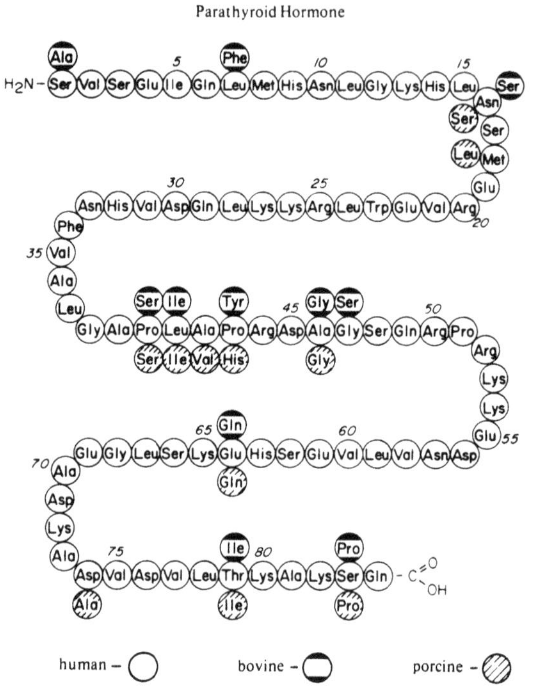

Fig. 9.2. Amino acid sequences of human, bovine, and porcine parathyroid hormone. (Reprinted with permission from H. T. Keutmann et al., 1978a. *Biochemistry* 17, 5723–9. Copyright by the American Chemical Society.)

gested that the C-terminal portion of the hormone protects the active N-terminus and slows its enzymic degradation (McIntosh and Hesch, 1976). These parts have been termed the "biologically active" and "immunologically active" portions of the hormone. It should be recalled that fragments of the N-terminus are also antigenic (Keutmann et al., 1978b). A general model of the three-dimensional conformation of the parathyroid hormone molecule has been proposed on the basis of its study by dark-field electron microscopy (Fiskin, Cohn, and Peterson, 1977). Its maximum dimension appears to be about 36Å and it contains two interconnected domains. These regions each have different masses and they probably correspond to the N-terminal amino acids 1–34 connected at positions 34–37 to the C-terminus.

The chemical synthesis of the N-terminal 1–34 sequence of bovine parathyroid hormone, PTH-(1–34), showed that it has an activity similar to that of the intact molecule (Table 9.1). In some other assay systems, however, it has been shown that the fragment has a lower activity, which appears to reflect an enhanced rate of degradation of the peptide (Di Bella, Arnaud, and Brewer, 1976). Modifications, including deletions and substitutions of amino acids and N- and C-terminal groups in

Table 9.1. *Biological activity of native and synthetic parathythroid peptides*

Polypeptide[a]	Rat adenylate cyclase *in vitro*		Chick hypercalcemia *in vivo*	
	Potency (units/mg)	Potency (%/mol)	Potency (units/mg)[b]	Potency (%/mol)
1–84	3000	100	2500	100
1–34	5400	77	7700	132
2–34	200	3	3800	64
3–34	<10	<0.3	<5	<0.2
1–26	<10	<0.3	<5	<0.2
1–27	200	2	<5	<0.2
1–28	440	5	<10	<0.3
1–31	740	10	4000	62
1–12 + 13–34	<10	<0.3	<5	<0.2

[a] Native bovine hormone 84 amino acids in length; other peptides are fragments.
[b] USP units based on MRC standard.
Source: Habener and Potts, 1976.

bovine PTH-(1–34), have given some insight into the nature of the structure–activity relationships of the hormone and has provided analogues with enhanced agonist or even antagonistic activities (see Habener and Potts, 1976; Goltzmann et al., 1975; Rosenblatt et al., 1976, 1977).

a. Removal of amino acids from the N-terminus results in a profound decline in the agonistic activity. Deletion of the alanine (position 1) produces a reduction, whereas if the valine (position 2) is also removed, agonist activity is abolished. These molecules, however, can act as antagonists (see below).
b. The removal of the seven amino acids, PTH-(1–27), at the C-terminus reduces, but does not completely abolish, activity. The PTH-(1–26), however, has negligible activity.
c. If the chain at the N-terminus is increased in length, activity declines.
d. Substitution of tyrosine for phenylalanine at the C-terminus (position 34) results in a 1.4-fold increase in agonistic activity, while if the carboxyl group at this end of the molecule is converted to a carboxamide (CONH$_2$), there is a threefold increase.
e. Deletion of the N-terminal amino group or substitution of serine or tyrosine for the alanine (position 1) reduces activity.
f. Substitution of the N-terminal L-alanine with its D-isomer increases activity, possibly due to a reduction in the rate of degradation of the hormone.
g. If the methionines at positions 8 and 18 are oxidized to the sulfoxides, activity is abolished.
h. However, replacement of these methionines by norleucine only results in a reduction of activity (by about 60 percent) so that the methionines are not essential.
i. The PTH-(1–34) has a single tryptophan residue at position 23, but attachment of an o-nitrophenylsulfenyl group through the indole nitrogen does not change activity. This group is very bulky, so it appears that although the tryptophan is present in the "active core"

of the molecule, there is considerable flexibility with respect to its configuration.

Two *competitive antagonists* have been prepared which can block the effects of parathyroid hormone, or its active 1–34 fragment, on the PTH-stimulated renal adenyl cyclase system (*in vitro*) in the rat (Goltzmann et al., 1975). These peptides were derived from the 1–34 fragment of the bovine hormone and deletions were made at the N-terminus. Removal of the NH$_2$ group to form [desamino-Ala^1]PTH-(1–34), or both the alanine and valine to give PTH-(3–34), results in a complete loss of agonist activity in this *in vitro* system. (The former analogue, however, retains some activity *in vivo* and thus may behave as a partial agonist.) These peptides, however, can competitively antagonize the effects of either intact parathyroid hormone or PTH-(1–34). The PTH-activated adenyl cyclase can be stimulated by fluoride, but the action of this anion is not reduced by the PTH antagonists. An adenyl cyclase which can be stimulated by salmon calcitonin (see Section 9.2F). is also present in rat kidney preparations, but this response is unchanged by the antagonists, showing that their effects are specific for parathyroid hormone.

These studies on the structure–activity relationships of analogues of parathyroid hormone suggest that the hormones ability to bind to the receptor (affinity) is a rather broad property of the molecule, as it can be seen in fragments encompassing residues 3–27. However, the biological response (intrinsic activity) has more stringent requirements, and the two amino acids at the N-terminus appear to be especially important in this respect.

The PTH-(3–34) is a relatively ineffective antagonist, as it must be present at a considerable excess, compared to the agonist, to inhibit the latter's effect. Further modifications of the molecule were therefore made (Rosenblatt et al., 1977) to increase its antagonistic potency. The changes were based on earlier observations of the effects of substitutions on the stability and activity of the agonist, bovine PTH-(1–34) (these have been described above). The two methionines (positions 8 and 18) were substituted for by norleucine, and the C-terminal phenylalanine was replaced by tyrosine and the COOH by a carboxamide. The resulting molecule, [Nle⁸,Nle¹⁸,Tyr³⁴]PTH-(3–34)amide, when present in equimolecular quantities to native parathyroid hormone, can inhibit the responses (*in vitro*, rat renal adenyl cyclase) by about 50 percent.

Information about the possible effects of such competitive antagonists on the action of parathyroid hormone *in vivo* does not appear to be available at present. They are, however, clearly of considerable potential experimental and possibly clinical use.

D. Synthesis and release

The synthesis of parathyroid hormone (Figure 9.3) appears to follow the same general pattern observed for other types of polypeptide hormones (see Habener, Potts, and Rich, 1976; Habener et al., 1977a,b; Habener and Potts, 1978a,b). A crude preparation of messenger RNA for parathyroid hormone has been translated in a cell-free, wheat germ system. DNA which is complementary to this mRNA has also been prepared and has been used

to direct the synthesis of parathyroid hormone (Kronenberg et al., 1977). The polypeptide precursor of parathyroid hormone is a single chain of 115 amino acids, which is 31 more than in the hormone. The additional amino acids are attached at the N-terminus of the hormone and contain many hydrophobic residues. This precursor is called *pre-proparathyroid hormone* and is formed by ribosomes that are attached to the rough endoplasmic reticulum. It exists intact for only a few seconds, passing across the membrane of the endoplasmic reticulum into the cisternal space, where an endopeptidase cleaves a 25-amino acid segment from the N-terminus to produce *proparathyroid hormone*. This polypeptide consists of parathyroid hormone plus a hexapeptide, which is also attached at the N-terminus. The proparathyroid hormone passes along the cisterna to the Golgi apparatus where, 12 to 15 minutes later, it is further cleaved by a specific membrane-associated enzyme (MacGregory, Chu, and Cohn, 1976) and is packaged into storage granules. The roles of the N-terminal peptide segments are uncertain. The 91–125 section probably serves to attach the ribosome to the endoplasmic reticulum and, because of its hydrophobic properties, facilitates the movement of the pre-proparathyroid hormone into the cisternal space. The N-terminal hexapeptide, positions 85–90, of proparathyroid hormone may assist the transfer to the Golgi apparatus and the packaging of the hormone into the storage granules. It could also mask the biological and immunological properties of the hormone.

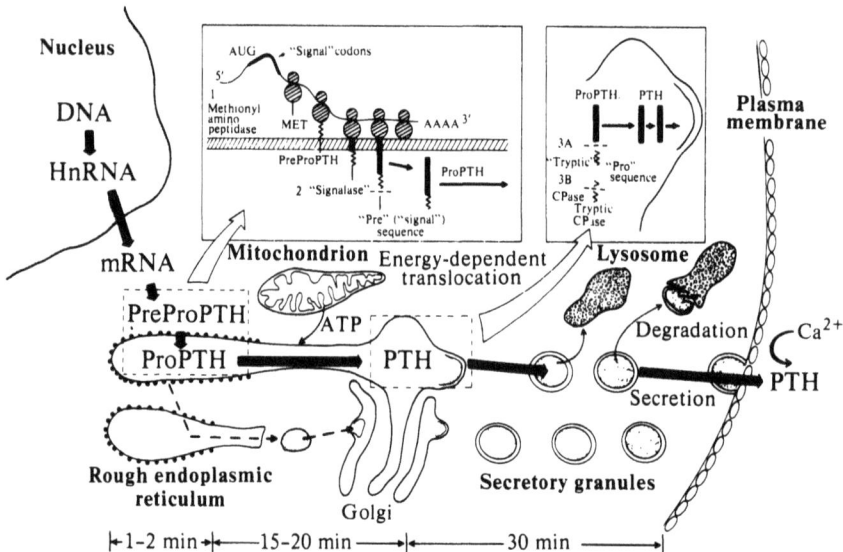

Fig. 9.3. Summary of the proposed pathway for the biosynthesis and secretion of parathyroid hormone from preproparathyroid hormone (115 amino acids) and proparathyroid hormone (90 amino acids). For details, see the text. (From Habener and Potts, 1978b. Reprinted by permission from *The New England Journal of Medicine 299*, 636.)

Parathyroid hormone is continually released into the circulation and this process is modulated by the concentration of Ca^{2+} in the bathing extracellular fluids. In contrast to other types of polypeptide hormones, the influx of Ca^{2+} into the glandular cells appears to *inhibit* secretion. It has been shown that the Ca ionophore A23187 can prevent the release of parathyroid hormone and also the conversion of proparathyroid hormone to parathyroid hormone (Habener et al., 1977b). Parathyroid hormone is released in response to stimulation by catecholamines and their effect can be blocked by propranolol, indicating that it is a β-adrenergic-type response. α-Adrenergic stimulation has been shown, *in vitro*, to inhibit the release of the hormone (Brown, Hurwitz, and Aurbach, 1978). Theophylline, which inhibits phosphodiesterase and the breakdown of cyclic AMP, can also elicit a release of parathyroid hormone. Adenyl cyclase has been identified in the cell membranes of the glandular cells, and tissue cyclic AMP levels rise when the tissue is exposed to low external Ca^{2+} concentrations. It thus seems likely that the release of parathyroid hormone involves the formation of cyclic AMP. The precise role of Ca^{2+} is uncertain, but it could be acting to inhibit the formation of the nucleotide.

Somatostatin has been observed to inhibit the release of parathyroid hormone (Hargis et al., 1978). However, it is not clear at this time whether or not this is a physiological or a pharmacological type of response.

The normal concentration of parathyroid hormone in the blood is about 10^{-10} M.

E. Metabolism

Parathyroid hormone levels rapidly decline in the circulation, the half-life being about 20 minutes, although the exact estimates depend on the type of assay used. The reason for this problem is that fragments of the hormone may retain biological and/or immunological activity. The precise functional significance of the metabolism is also not clear, as it is possible that it could involve a process of activation of the hormone molecule. A peptidase has been identified in the liver which can convert parathyroid hormone into fragments that retain their activity (Fischer et al., 1972). However, studies on membrane receptors (see below) indicate that intact parathyroid hormone *or* its fragments can effectively bind to the receptors, but the fragments apparently survive for a briefer period of time (McIntosh and Hesch, 1976; Di Bella, Arnaud, and Brewer, 1976). The kidney membrane receptor preparations also appear to possess an enzymic activity which can inactivate the parathyroid hormone or its active fragments (Zull, Malbon, and Chuang, 1977). It is not known if this process is dependent on binding to the receptors.

The importance of the kidney as a site for the inactivation of parathyroid hormone is sometimes seen in patients suffering from chronic renal failure who suffer from hyperparathyroidism, apparently due to an impaired metabolism of the hormone (Freitag et al., 1978).

F. Mechanism of action

Like several other hormones, including epinephrine, vasopressin (ADH), and ACTH, the actions of parathyroid hormone appear to be mediated by cyclic AMP. This mechanism is present in both bone and the kidneys (see Aurbach and Chase, 1976). There are several types of evidence which are consistent with this theory.

a. Parathyroid hormone increases cyclic AMP formation *in vitro* in kidney and bone and its time course closely follows that of the hormone's effects. In the kidney the response is greatest in the cortical tissue. *In vivo* substantial amounts of this nucleotide appear in the urine in response to the hormone.
b. Dibutryl 3′,5′-cyclic AMP *in vitro* or following its infusion *in vivo* mimics the effects of the hormone.
c. A membrane-bound adenylate cyclase which can be activated by the hormone has been isolated from bone and kidney cells.
d. Specific receptors for parathyroid hormone have been identified in the kidney.

Specific receptors for parathyroid hormone have been identified in plasma membrane preparations made from the kidneys of cattle, rats, and chicks. These receptor sites have been found in both the renal tubules (Sutcliffe et al., 1973; Di Bella et al., 1974; McIntosh and Hesch, 1976; Zull, Malbon, and Chuang, 1977) and glomeruli (in the rat) (Sraer et al., 1978). The native intact (84 amino acids) parathyroid hormone and its active N-terminal 1–34 fragment, from human and bovine parathyroid hormone, have been labelled with ^{125}I or tritium. These preparations of the hormone retain their biological activity and have been used to identify and characterize the receptors. Specifically bound hormone can be displaced by the native parathyroid hormone or the PTH-(1–34) fragment but not by the hormone with an oxidized methionine residue or by other polypeptide hormones. Binding is reduced by incubating the membranes with trypsin or phospholipase A or D. Calcium inhibits binding. The amount of binding could be correlated with the activation of adenyl cyclase in the preparations.

Bovine renal tubular receptors have been isolated and dispersed in detergent (Triton X-100) (Malbon and Zull, 1977). These preparations also display specific binding properties for the parathyroid hormone, but there is no associated adenyl cyclase activity present. Unlike the membrane preparations, these "solubilized" receptors are not sensitive to changes in the calcium content of the bathing medium. This suggests that associated membrane sites are influenced by the divalent ion.

The hormone affinity for the receptors is similar in both types of preparations, but it is relatively low (a K_D of about 10^{-7} M) compared to levels of the hormone which are active (10^{-10} to 10^{-11} M) *in vivo*. The reason for this disparity is unknown, but it is disturbing to some.

As described earlier, guanyl nucleotides, such as GTP, may increase the hormone-dependent responses of adenyl cyclase to glucagon and epinephrine. The activation of renal adenyl cyclase by parathyroid hormone can also be promoted by such nucleotides (Goltzman et al., 1978). This effect was especially apparent when testing the actions of analogues of the hormone with normally lower activities, such as [desamino-alanine-1]bPTH-(1–34) and bPTH-(1–26). The guanyl nucleotides may thus also act as important cofactors in the actions of parathyroid hormone.

The precise nature of the effect of the cyclic AMP is uncertain (see Aurbach and Chase, 1976), but in other hormone-responsive tissues it has been shown to activate protein kinases, which lead to a phosphorylation of cell components that initiate the response. The response differs, depending on the nature and the site of the hormone's action. In the case of ADH, cyclic AMP has been shown to increase the permeability of the cells to sodium, and it has been suggested that calcium is similarly affected by parathyroid hormone. This effect could involve the cell membrane and the distribution of calcium within the cells, such as by an action on the endoplasmic reticulum and mitochondria. It could, for instance, be triggering a release of ionized Ca^{2+} from the latter organelles.

It is not clear how cyclic AMP may mediate the longer-term effects that the hormone has on the number of bone cells or the increase in protein synthesis that occurs in these cells. An increased rate of synthesis and release of several hydrolase enzymes from the lysosomes in bone cells has been observed (*in vitro*) in response to parathyroid hormone (Vaes, 1968b; Eilon and Raisz, 1978). This response is accompanied by an increased rate of acid secretion. It has been suggested that the parathyroid hormone has a dual effect on the bone cells. By stimulating glycolysis, it increases acid secretion and it also initiates exocytosis of the contents of the lysosomes. The acid helps to solubilize the mineral content of the bone while the hydrolase enzymes foster the breakdown of the organic matrix. The release of the lysosomal enzymes can be mimicked by dibutryl cyclic AMP (Vaes, 1968b), suggesting that the cyclic AMP may mediate this response to the hormone. (The effects of the parathyroid hormone can be inhibited in these preparations by calcitonin, as well as cortisol and colchicine.)

9.2. Thyroid C-cells

A. Morphology

As described earlier, the mammalian thyroid gland contains cells that lie outside the follicles and are thus called parafollicular cells or now, more usually, as they are known to be the site of formation of calcitonin, C-cells. These cells were first described by E. C. Baber in 1876, but their possible endocrine function was not recognized for another 75 years. The C-cells contain large numbers of secretory granules and, apart from being present in the thyroid gland, they have been identified at a number of other sites, especially those tissues derived from the pharyngeal pouches. In lower vertebrates they are present in the ultimobranchial bodies, rather than the thyroid, although in the pigeon they are also found in the parathyroids and thyroid. It is now known that C-cells are not directly derived from the pharyngeal pouches but appear to be APUD cells originating in the neural crest, from which they migrate during ontogeny to find their various resting places.

B. Functions and actions

The discovery that the thyroid gland and the ultimobranchial bodies secreted a hormone that could lower the plasma calcium concentrations occurred in about 1960. D. H. Copp (see Copp et al., 1962) found that when he perfused the thyroid–parathyroid complex in dogs with solutions containing high concentrations of calcium, a substance was released into the effluent which, when perfused systemically back into the animal, exerted a hypocalcemic action. Removal of the parathyroids results in a lowering of plasma Ca^{2+} concentrations, but it was at first not certain whether this effect was merely a negative one, the absence of parathyroid hormone initiating the hypocalcemia, or whether there was also a positive hypocalcemic control mechanism present. Copp found that removal of the thyroid–parathyroid complex resulted in a much slower hypocalcemic effect than was seen when plasma from this gland complex, which had been perfused with high concentrations of calcium, was injected systemically. The active material was called calcitonin, and it was initially suggested that it came from the parathyroid glands. In a parallel series of observations, P. F. Hirsch and P. L. Munson (see Hirsch and Munson, 1969) found that the thyroid gland itself contained a hypocalcemic hormone which they called thyrocalcitonin. Calcitonin and thyrocalcitonin are now known to be identical. Studies with fluorescent antibodies to calcitonin show that in mammals it clearly originates from the thyroid C-cells. In nonmammals, including fishes, this hormone has been identified in extracts of the ultimobranchial bodies. It is also found in some human carcinoma cells. The precise structure of the hormone varies in different species and, as will be described later, this apparently rather academic observation has had important repercussions with respect to the therapeutic use of calcitonin in man.

Calcitonin has a hypocalcemic action in man, but as it is even more effective during hypercalcemia, its effect has, possibly more aptly, been described as antihypercalcemic. This response is principally due to an action on bone, but the kidneys, and possibly even the gut, are also involved. It has been shown, both *in vivo* and *in vitro*, that calcitonin inhibits resorption of calcium (and probably indirectly phosphate) from bone, by an action on the osteoclasts, exerting an antagonistic effect to that of parathyroid hormone (Wong, Luben, and Cohn, 1977). It also has a phosphaturic effect; renal phosphate excretion is increased, which is similar to the action of parathyroid hormone. It thus may produce a hypophosphatemia. Calcitonin promotes the renal loss of calcium in man, although this effect is not seen in all species. (It is, for instance, apparently absent in the laboratory rat.) Calcitonin may reduce the absorption of calcium across the intestine, but the evidence for this effect is equivocal. It is possible that it reflects an inhibition of the activation of vitamin D_3 in the kidney, but again the evidence is contentious.

Calcitonin has a variety of effects, especially in large doses, many of which are considered to be pharmacological (Munson, 1976). These include a diuretic action; reabsorption of urinary Na is inhibited so that a natriuresis results. Calcitonin also has effects on the gut; it can inhibit secretion of acid by the gastric mucosa and may affect motility. The effects on acid secretion appear to occur via an inhibition of release of gastrin (Bolman et al., 1977). In newborn sheep and rats, calcitonin has been shown to limit the absorption of lipids during feeding (Garel, Barlet, and Kervran, 1975). The responses to calcitonin shown considerable interspecific variability, so it is important to remember that one cannot necessarily extrapolate information about an effect from one species to another.

The normal physiological importance of calcitonin is not clear. If the thyroid glands are removed and thyroxine replacement therapy is instituted, no obvious debility, such as a persistent hypercalcemia, persists. However, it has been seen that calcitonin may limit rises in calcium concentrations that follow feeding, especially in young animals. The latter are more responsive to the hormone than are old animals, and this increased sensitivity is also seen during lactation and in Paget's disease. Bone metabolism is increased in all these circumstances. It is interesting that calcitonin is present in fishes, such as sharks, that lack a calcareous skeleton and where a mammalian-type hypocalcemic effect is not seen. It has been suggested that this hormone may have other physiological roles in such species, and it is possible that it may also have other effects in man.

C. Calcitonin

Calcitonin is a polypeptide consisting of a chain of 32 amino acids. Unlike parathyroid hormone, fragments of this molecule lack biological activity. However, it may endure considerable changes in its precise amino acid structure and retain its usual general actions. Such changes were initially identified in naturally occurring hormones, from nonmammals (Figure 9.4) which were shown to be associated with changes in the molecule's activity. The piscine calcitonins, from eels and salmon, exert a greater and more prolonged effect in man than the homologous, human, calcitonin. This difference appears to be the result of a higher affinity for its receptors and a greater persistence in the circulation. The differences in activity of the piscine calcitonins, compared to mammalian ones, is quite remarkable. When assayed *in vivo* in the rat, they display an activity of about 3000 IU/mg, compared to 50 to 200 IU/mg for calcitonin from cattle, pigs, sheep, or man (Potts and Aurbach, 1976). Salmon calcitonin thus provided the blueprint for the synthesis of an important therapeutic hormone preparation used to treat Paget's disease (see Section 9.4D).

Maier and his collaborators (Maier, Riniker, and Rittel, 1974; Maier et al., 1975, 1976) have studied the structure–activity relationships of calcitonin. Substitutions have been made based on the differences between salmon calcitonin and human calcitonin. Those analogues studied had changes in positions 8, 12, 16, 19, 22, 29, and 31. The C-terminal part of the polypeptide chain is very important, and removal of the terminal proline (position 32) or even its attached amide group results in a considerable decrease in biological activity. Salmon calcitonin has serine and threonine at positions 29 and 31, respectively, while in human calcitonin valine and alanine are present there. When those amino acids from the salmon hormone are substituted into the human one, the activity of the polypeptide increases five times. The opposite substitution, of valine and alanine into positions 29 and 31 in salmon calcitonin, resulted in a 50 percent reduction in its activity. Human calcitonin has the aromatic amino acids tyrosine, phenylalanine, and phenylalanine at positions 12, 16, and 19, respectively, while in salmon calcitonin leucine is present at these sites. When leucine was substituted at positions 12, 16, and 19 in human calcitonin, the biological activity was increased 10 times. The leucine at position 12 appears to be especially important. Substitution of valine and tyrosine, which are present at positions 8 and 22 in salmon calcitonin, into human calcitonin also increased the latters activity. The precise reasons for these changes in activity are unknown, but changes in the rate of

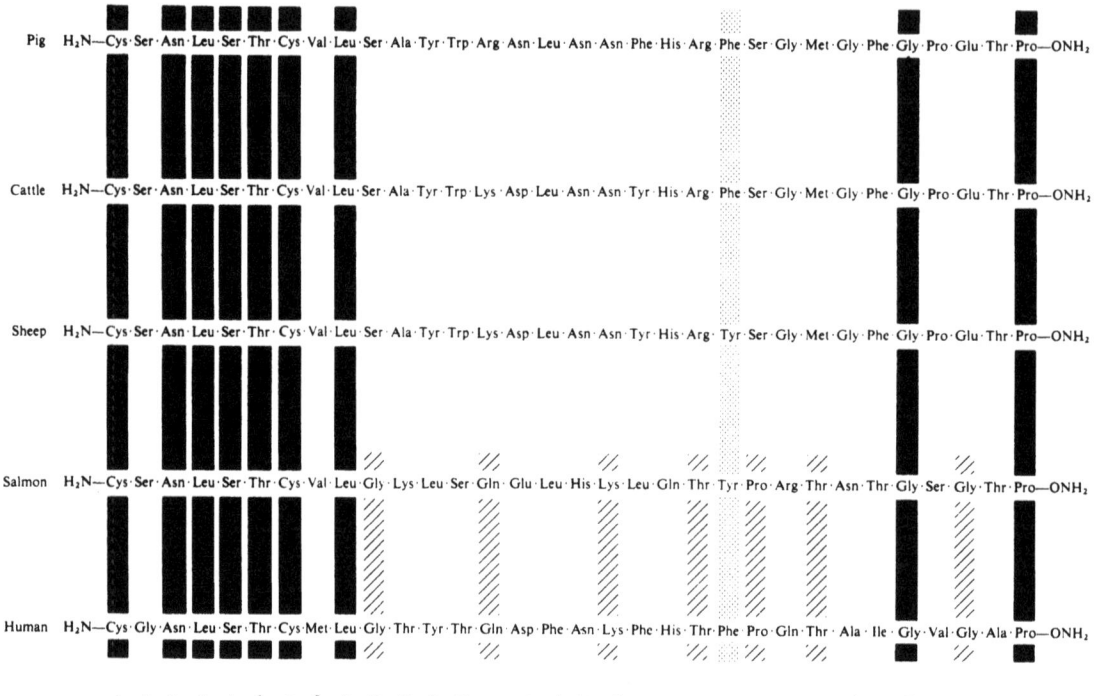

Fig. 9.4. Comparison of the amino acid sequence of calcitonin in four mammals (including man) and the salmon. The solid bars indicate the amino acids that are homologous in all species. It can be seen that extensive differences exist, especially in the central parts of the molecules. Crosshatched bars indicate homologies between salmon and man; stippled bar indicates comparable hydrophobic residues. (From Potts et al., 1972.)

metabolism, and hence the duration of action of the peptides appear to be involved. No one as yet appears to have ventured an analysis of possible effects with respect to the nature of interaction between the hormone and its receptor. It has been noted that salmon calcitonin is the most basic and hydrophilic of all calcitonins (see Potts and Aurbach, 1976).

D. Synthesis and release

The thyroid C-cells have prominent secretory granules which are the presumptive site for the storage of calcitonin. These have been observed to become depleted during hypercalcemia. The hormone is released in response to elevated Ca^{2+} levels in the plasma and is depressed by hypocalcemia in what appears to be a typical negative-feedback control mechanism. Glucagon also stimulates its secretion. The enteric hormones – gastrin, cholecystokinin, and secretin – can initiate calcitonin's release (Roos, 1977). Analogues of gastrin, such as pentagastrin, are also effective. Somatostatin, on the other hand, can inhibit release (Hargis et al., 1978). These observations have resulted in the suggestion that the secretion of calcitonin is part of the general hormone release that occurs in response to feeding and so may play an acute role in

limiting rapid excursions in blood calcium levels which could occur at this time. It is possible that release of calcitonin may involve the activation of adenyl cyclase, as cyclic AMP, theophylline, and adrenergic drugs can initiate its release, although the results vary somewhat depending on the species used.

E. Metabolism

Calcitonin is rapidly removed from the plasma, but species differences in the rate and site of metabolism exist. Salmon calcitonin persists much longer than the mammalian hormones. Porcine calcitonin is principally metabolized by the liver in the rat, while human and salmon calcitonin are mainly inactivated by the kidney (De Luise et al., 1972). Calcitonin is destroyed by the plasma, but this property is of little significance *in vivo*. Calcitonin apparently does not cross the placenta.

F. Mechanism of action

There is no consensus as to how calcitonin exerts its effects on the calcium metabolism of cells. The hormone has, like parathyroid hormone, been shown to elevate cyclic AMP levels in kidney and bone cells (Loreau, Lepreux, and Ardaillou, 1975; Aurbach and Chase, 1976). As the two hormones exert opposite effects on Ca metabolism, this observation may appear paradoxical. Their effects on cyclic AMP levels are, however, additive, indicating that different sites, possibly not even in the same cells, are involved. Unlike the effects of parathyroid hormone, calcitonin does not increase the excretion of cyclic AMP in the urine. Specific membrane receptor sites for calcitonin have been identified in bone cells and kidney tissue, and the degree of binding of the hormone from different species has been directly correlated with its ability to activate adenyl cyclase from rat kidney (Figure 9.5) (Marx and Aurbach, 1975). As in other aspects of calcitonin's effects, there are species differences in the tissue response; it activates adenyl cyclase from rat kidney, but that from man, cattle, and dogs is apparently unresponsive (Marx and Aurbach, 1975).

Guanyl nucleotides, such as GTP, appear to be involved in the interaction of the hormone and its receptor, as described previously for catecholamines (Section 6.14B) and glucagon. In rat kidney membranes the number of calcitonin receptor sites is diminished by such nucleotides, but their affinity for the hormone is increased considerably (Loreau et al., 1978). The K_D for the interaction is then about 60 pmol/liter, which is similar to the concentrations of the calcitonin observed in the plasma of rats (10 to 200 pmol/liter). The action of the guanyl nucleotides thus would appear to be essential for activation of the response under physiological conditions. It has been suggested by Loreau and her associates that the calcitonin–

adenyl cyclase system, like that for catecholamines and glucagon, consists of three components: a receptor, a nucleotide regulatory component, and a catalytic subunit.

The precise nature of the change in the reactive cells is not clear. DeLuca and his collaborators proposed a unified hypothesis to account for the actions of vitamin D₃, parathyroid hormone, and calcitonin in bone (Figure 9.6). In this simple model calcitonin is pictured as blocking calcium exit from the cells, in opposition to the effect of parathyroid hormone.

9.3. Vitamin D₃

Until about 10 years ago vitamin D₃ was classified as an "accessory food factor" which played some undetermined role in regulating the metabolism of calcium and phosphate in the body. A recent rapid series of scientific advances has now established that vitamin D₃ is actually a prohormone which is converted to a hormone by its hydroxylation in two stages, first in the liver and then in the kidney.

The discovery of the importance of vitamin D₃ in calcium and phosphate metabolism evolved out of observations of the disease called "rickets." This condition afflicts growing children (in adults the equivalent term is osteomalacia), who suffer gross deformities in their skeletal structure during growth. In modern times the condition was associated with life under crowded industrialized urban conditions, usually in northern latitudes, where exposure to sunlight was limited. The early history of vitamin D₃ has been reviewed by Nicolaysen and Eeg-Larsen (1953). The disease was described in the seventeenth century by the English physicians Glisson and Whistler. Two remedies were used to treat rickets (they have an "antirachitic activity"): cod-liver oil, the use of which apparently had its origins in folk medicine, and exposure to sunlight. Both treatments were subsequently shown to have a sound scientific basis.

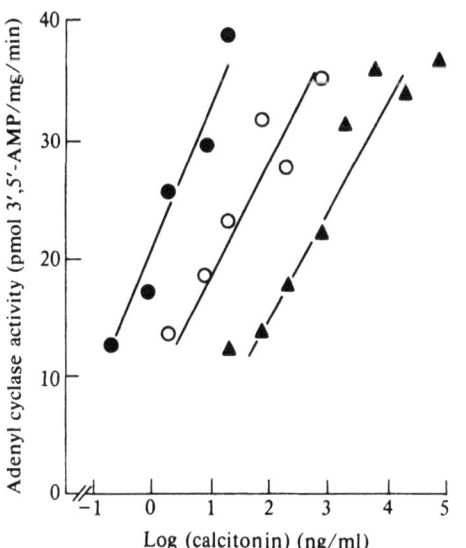

Fig. 9.5. Activation of adenyl cyclase from rat kidneys by salmon (●), porcine (○), and human calcitonins (▲). (From Marx and Aurbach, 1975.)

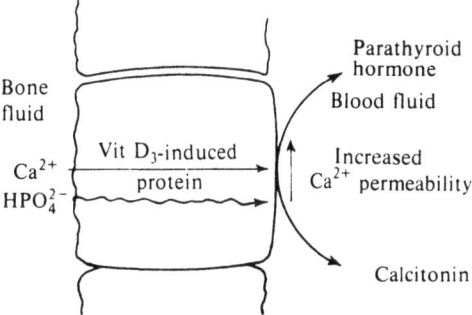

Fig. 9.6. Model for the endocrine control of calcium exchange in bone. (Modified from DeLuca, Morii, and Melancon, 1968.)

Cholecalciferol (vitamin D₃) is formed (in a photochemical reaction under the influence of ultraviolet light) from 7-dehydrocholesterol, which is synthesized in the skin (Holick and Clark, 1978). Sir Edward Mellanby in 1919 showed that experimentally induced rickets in puppies could be cured by a fat-soluble "dietary factor" in cod-liver oil and, shortly after, E. V. McCollum (McCollum et al., 1922) identified it as vitamin D₃ (Figure 9.7). In 1931, F. A. Askew and his collaborators isolated vitamin D₂ (ergocalciferol, Figure 9.8), which is not formed in animals but is obtained from plant materials. In man, it is an effective substitute for vitamin D₃. The structure of vitamin D₃ was described in 1938 by H. Brockmann and A. Buss. At this time it was also known that the defect resulting from the absence of vitamin D₃ was an inability to absorb adequate dietary calcium and phosphate from the intestine (Nicolaysen 1937a,b).

Thus, by the late 1930s this important nutritional factor had been identified, chemically characterized, and its principal locus of action pinpointed. The therapeutic importance of these ob-

servations was a dramatic decline in the incidence of what had been a common and serious disease, especially in children. The precise mechanism of action of vitamin D₃ was, however, unknown and information about these processes was not to become available for another 30 years, when modern methods for radioisotopically labelling the vitamin were developed.

The potential significance of isotopically labelling vitamin D₃ was recognized and principally used in three laboratories, that of Kodicek in England, and those of DeLuca and Norman in the United States (see Kodicek, 1974; Omdahl and DeLuca, 1973; DeLuca, 1974 and 1977; Norman and Henry, 1974). DeLuca and his collaborators succeeded in synthesizing vitamin D₃ with a specific activity sufficiently high to follow its metabolism in the body. This procedure initially led to the identification of 25-hydroxycholecalciferol (25-OHD₃, calcifediol), which is formed from the vitamin D₃ by hydroxylation in the liver (Figure 9.7). This compound was found to be about two to four times as active as the parent one in curing

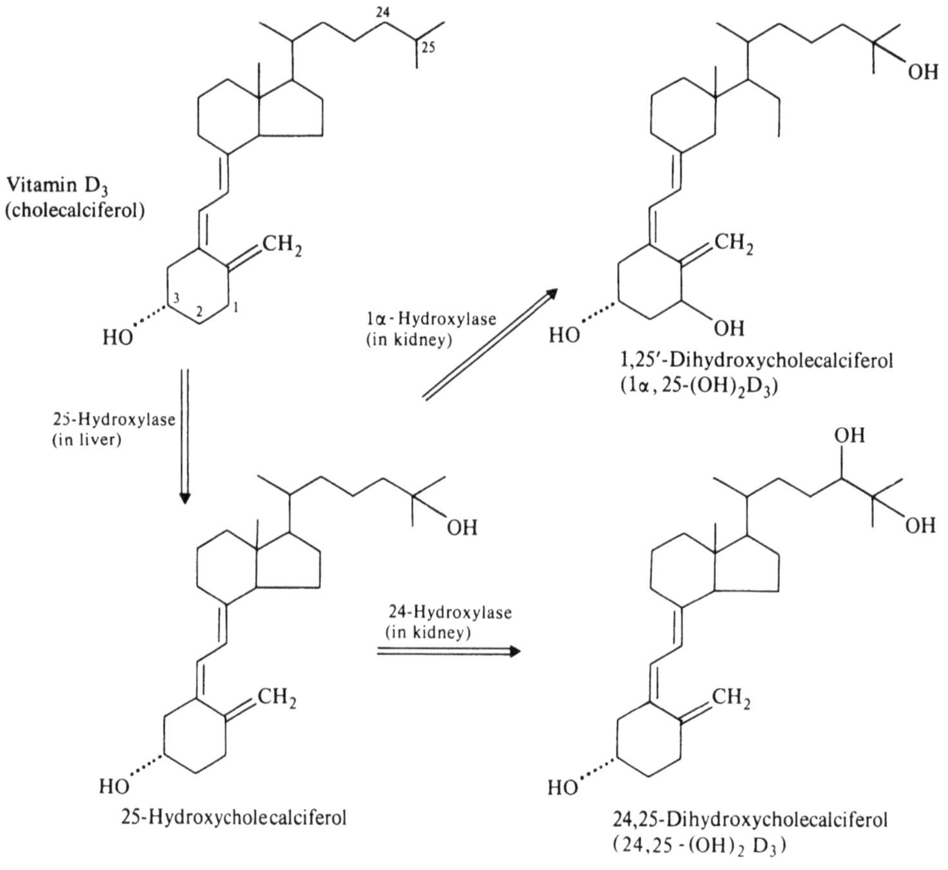

Fig. 9.7. Chemical structure of vitamin D₃, cholecalciferol and the pathways of formation of some of its active metabolites including 1α,25-dihydroxycholecalciferol (1α,25-(OH)₂ vitamin D₃).

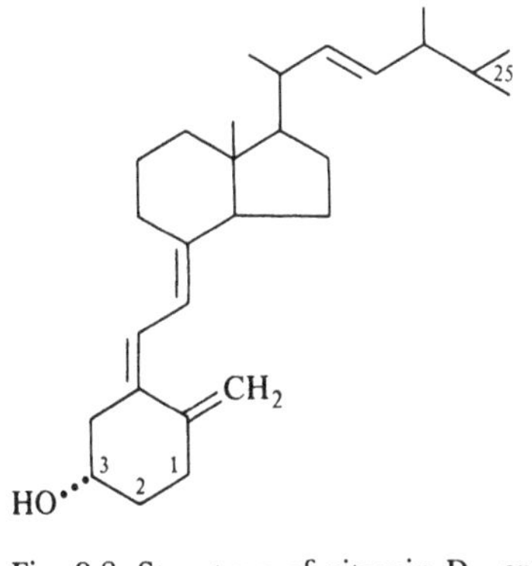

Fig. 9.8. Structure of vitamin D₂, ergocalciferol. This is not a natural hormone in mammals but is obtained from plant material, and when administered has vitamin D activity.

experimental vitamin D-deficient rickets. Subsequently, another even more active metabolite was identified, 1α,25-dihydroxycholecalciferol [1α,25-(OH)₂D₃]. This substance is the definitive active hormone and is formed by further hydroxylation of 25-OHD₃ in the kidney. Thus, vitamin D₃ is a prohormone which is converted to a hormone by two enzymes, one in the liver and the other in the kidney. The 1α,25(OH₂)D₃ is 15 times more active than its grandparental vitamin D₃.

There is a well-defined lag of about 8 hours between the administration of vitamin D₃ and the stimulation of calcium absorption across the intestine. This process of activation of the hormone partly accounts for the delay, the remainder being due to the induction of effector proteins in the target tissue.

A. Functions and actions
Vitamin D₃ deficiency results in abnormalities in the growth of bones; the collagenous matrix is laid down, but its mineralization is inadequate principally due to reduced absorption of calcium from

the intestine. Excessive amounts of calcium and phosphate are lost in the feces, reflecting the inadequate absorption. A hypocalcemia develops (which can result in tetany) due to this defect in absorption from the intestine, but also as a result of the inability of parathyroid hormone to stimulate resorption of calcium from bone. Vitamin D₃ may be responsible for the synthesis of an intermediary substance necessary for this action of parathyroid hormone, though this view has been challenged (Wong, Luben, and Cohn, 1977). It is also possible that 1α,25(OH)₂D₃ is necessary for the optimal effects of parathyroid hormone on the kidney (Puschett, Beck, and Jelonek, 1975).

B. Structure and analogues
Vitamin D₃ (Figure 9.7) or cholecalciferol (also referred to as calciferol) is the natural prohormone. It is a steroid and can be synthesized under the influence of ultraviolet light in the skin. Supplementary dietary sources are contained in foods of animal origin, including milk, butter, egg yolk, and fish oils, of which cod-liver oil is the best known. *Vitamin D₂*, or *ergocalciferol* (Figure 9.8), is not a natural hormone but is obtained from plant sources. Both vitamins have been artificially synthesized for clinical use. Vitamin D₂ has a similar activity to vitamin D₃ with respect to its antirachitic action, which principally depends on its effect on the intestine. However, it only has about half the activity in mobilizing calcium from bones and it has little effect at all in chicks and new world monkeys. As it is cheaper than vitamin D₃, it is probably used more often clinically in man.

The *dihydrotachysterols* (Figure 9.9) are interesting and important compounds that exhibit vitamin D-like activity. They are reduction products of vitamin D₂ (DHT₂) and vitamin D₃ (DHT₃). At low doses they have little activity in curing rickets; their potency is about 180 IU/mg compared to 40,000 IU/mg for vitamin D₂ and D₃. At high doses, however, they have often been found to be more effec-

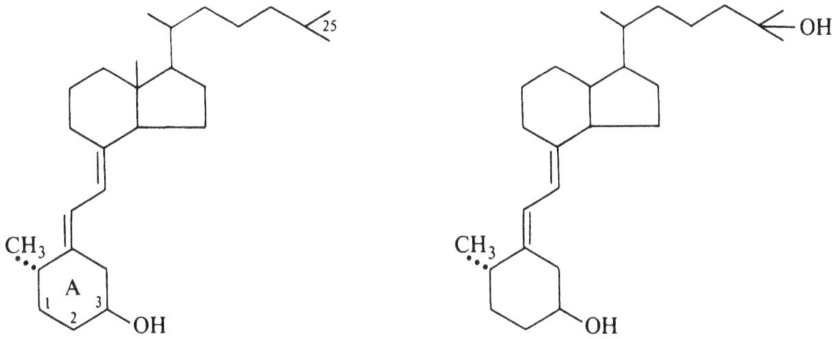

Dihydrotachysterol₃ (DHT₃)

25-Hydroxydihydrotachysterol₃ (active form)

Fig. 9.9. Structure of dihydrotachysterol₃ and its active metabolite.

tive than vitamin D and are thus useful for the treatment of hypoparathyroidism, vitamin D-resistant rickets, and renal osteodystrophy (see Section 9.4). Some of the differences in activity may reflect the route of dosage; for instance, DHT_3 is more effective orally than vitamin D_3 but less effective when given by intramuscular injection in oil. The principal reasons for the differences in activity at the different dose levels can, however, be related to chemical structure. It can be seen (Figure 9.9) that the lower ring, A, of the molecule of DHT is rotated by 180° as compared to vitamin D_3, so that the C-3 hydroxyl group lies sterically in a similar position to the C-1 in the active hormone $1\alpha,25\text{-}(OH)_2D_3$. In the latter compound this hydroxyl group is added to vitamin D_3 by the enzymic reaction in the kidney. In order to act, dihydrotachysterol must be hydroxylated at the C-25 position, and this occurs, like the similar hydroxylation of vitamin D_3, in the liver. Structurally, $25\text{-}OHDHT_3$ thus has many features of the active hormone, $1\alpha,25\text{-}(OH)_2D_3$, which can account for its activity, especially in disorders of calcium metabolism associated with renal disease, as the need for hydroxylation in the kidney is bypassed. Its very high activity in large doses, as compared to vitamin D_3, appears to be the result of a continual 25-hydroxylation in the liver, which, unlike this hydroxylation of vitamin D_3, is not subject to a negative-feedback inhibition by its products.

1α,25-Dihydroxycholecalciferol is highly active biologically. Various analogues have been chemically synthesized which demonstrate the importance of its structure in relation to its activity (Figure 9.10). Thus, the analogue 3-deoxy-1α-hydroxycholecalciferol ($3\text{-}D\text{-}1\alpha\text{-}OHD_3$) (Norman et al., 1975) has very little activity, about half that of vitamin D_3, indicating that the 3β-hydroxyl group is of considerable importance in the interaction of the hormone with its intestinal

receptors or its hydroxylation site in the liver. The other hydroxyl groups at C-1 and C-25 are equally essential, and compounds that lack these have little intrinsic activity alone but must be hydroxylated in the body before they can act.

1α-Hydroxycholecalciferol (Figure 9.10) is currently chemically synthesized (from cholesterol) more readily and cheaply than is $1\alpha,25\text{-}(OH)_2D_3$. It also has a high activity and is apparently even more effective than $1\alpha,25\text{-}(OH)_2D_3$ when given orally. Before it can act, it may possibly need to be hydroxylated at C-25 (Holick et al., 1975), but as this reaction can occur in the liver, a functioning kidney is not necessary. $1\alpha\text{-}OHD_3$ has a large number of clinical uses, which have been described in a recent symposium (Peacock, 1977). It has, for instance, been used for the treatment of osteoporosis, hypoparathyroidism, pseudohypoparathyroidism, and renal osteodystrophy. $1\alpha\text{-}OHD_2$ is also effective (Lam, Schnoes, and DeLuca, 1974), but it has no particular advantages over the vitamin D_3 derivative, either therapeutically or financially.

An interesting naturally occurring form of $1\alpha,25\text{-}(OH)_2D_3$ has been identified in the leaves of a South American plant *Solanum glaucophyllum* (also called *S. malacoxylin*). A calcinosis which can result in death has been observed in cattle grazing on this plant in Argentina. The vitamin D-like activity of extracts from the leaves of this plant was first described by Mautalen in 1972. Physiologically, it behaves in an almost identical manner to $1\alpha,25\text{-}(OH)_2D_3$, but it is water-soluble. It is active *in vivo* and *in vitro*, where it promotes the transport of calcium across the intestine (Lawson, Smith, and Wilson, 1974; Walling and Kimberg, 1975) and its resorption from bone. Synthesis of calcium-binding protein (CaBP) is promoted (Lawson, Smith, and Wilson, 1974; Wasserman, 1974) and it inhibits the activity of 25-hydroxycholecalciferol-

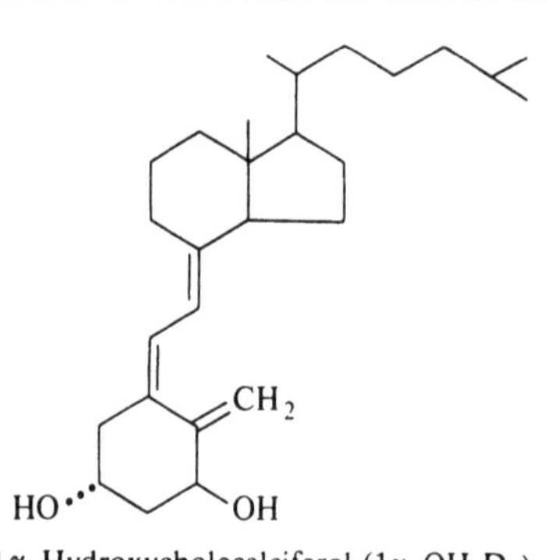

1α-Hydroxycholecalciferol (1α-OH-D₃)

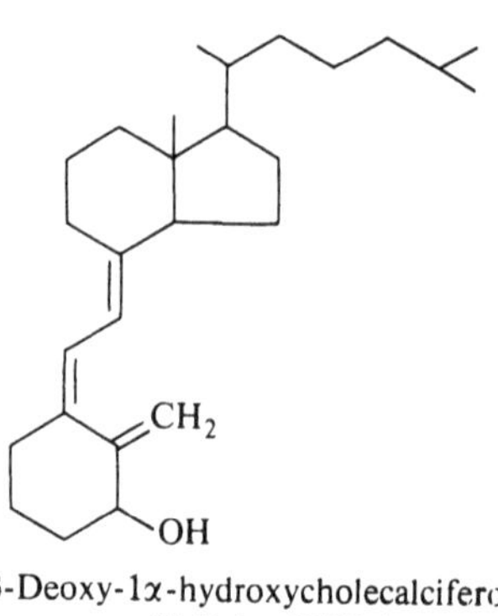

3-Deoxy-1α-hydroxycholecalciferol
(3-D-1α-OHD₃)

Fig. 9.10. Two synthetic analogues of vitamin D_3. The 3-deoxy-1α-hydroxycholecalciferol has little activity, reflecting the importance of the 3β-hydroxyl group. The 1α-hydroxycholecalciferol (1α-OH vitamin D_3) can be readily prepared from vitamin D_3 and is even more active orally than is $1\alpha,25\text{-}(OH)_2D_3$.

l-hydroxylase in the kidney. It competes with $1\alpha,25\text{-}(OH)_2D_3$ for receptor sites (Procsal et al., 1976). This plant extract behaves identically physiologically to the natural hormone, to which it presumably bears a close chemical resemblance. Analyses of its chemical nature indicate that it is a sterol-glycoside, the sterol presumably being identical to $1\alpha,25\text{-}(OH)_2D_3$ (Peterlik and Wasserman, 1975; Haussler et al., 1976; Wasserman et al., 1976; Napoli et al., 1977). A similar disease occurs in cattle that forage on *Cestrum diurnum* in the southeastern United States, Hawaii, and Jamaica. The active principle in this plant is also $1\alpha,25\text{-}(OH)_2D_3$-glycoside (Hughes et al., 1977).

The plant material is active in nephrectomized animals and in patients suffering from renal insufficiency (Von Herrath et al., 1974). It is possible that it may provide a useful source of $1\alpha,25\text{-}(OH)_2D_3$. These observations are of rather remarkable interest, in that a plant product should bear such a close similarity to a rather exotic animal hormone. However, the observation is not unique when one considers the similarities in the activities of vitamin D_2 and D_3, and the ability of licorice extract from the root of a plant to substitute for corticosteroids in Addison's disease.

C. Metabolism and transport

Vitamin D_3 is absorbed from the intestine and this process is assisted by bile salts. It is transported in the blood bound to a protein which in man appears to be an α-globulin, although there are some species differences in its nature. Its half-life in the circulation is about 12 hours; about 20 to 30 percent is excreted in the bile, 2 percent in the urine, and the remainder is either metabolized or stored in the fat depots of the body. The liver is a prominent storage site and it has been detected there many months after a single dose is administered. The active metabolites of vitamin D_3 and some analogues that are used therapeutically can also be absorbed from the intestine. $1\alpha,25\text{-}(OH)_2D_3$ is effective when given by this route, but $1\alpha\text{-}OHD_3$ is absorbed more readily.

The conversion of vitamin D_3 to $1\alpha,25\text{-}(OH)_2D_3$ has already been described (Figure 9.7). Vitamin D_3 is first converted to 25-hydroxycholecalciferol by a microsomal enzyme (25-hydroxylase) in the liver. The reaction is subject to negative-feedback inhibition due to the presence of $25\text{-}OHD_3$. This prohormone is transported, bound to an α-globulin, to the kidney where it is transformed, in the proximal tubule (Brunette et al., 1978) to $1\alpha,25$-dihydroxycholecalciferol by $25\text{-}OHD_3\text{-}1\alpha$-hydroxylase, which is a mitochondrial enzyme. This conversion is stimulated by low Ca^{2+} or PO_4^{2-} levels in the plasma. Hypocalcemia releases parathyroid hormone, which has a trophic effect on the renal tubules and initiates the effects of the

1α-hydroxylase. Low phosphate concentrations have a similar, but direct, effect. It is possible that the phosphaturic action of parathyroid hormone mediates a lowering of PO_4^{2-} concentrations in the renal target cells. This effect of parathyroid hormone appears to be initiated by cyclic AMP (Horiuchi et al., 1977). When there is sufficient $1\alpha,25\text{-}(OH)_2D_3$ present, the $25\text{-}OHD_3$ is converted to $24,25\text{-}(OH)_2D_3$, which is only half as active as vitamin D_3 (Tanaka and DeLuca, 1974). Thus, the formation of the active hormone is subject to a negative-feedback control. The 24-hydroxylase is present in the kidney and liver. The hormone $1\alpha,25\text{-}(OH)_2D_3$ is inactivated, probably partly in the liver, by conversion to a number of metabolites, one of which is $1,24,25\text{-}(OH)_3D_3$. The ability of certain drugs, such as barbiturates and diphenylhydantoin, to induce degradative microsomal enzymes in the liver can contribute to disorders in calcium regulation due to changes in the metabolism of vitamin D_3 (see Section 9.4A).

The vitamin D_3 metabolite $24,25\text{-}(OH)_2D_3$ has a very slow turnover rate in man; its half-life is several weeks compared to 1 to 3 days for $1\alpha,25\text{-}(OH)_2D_3$ (Kanis et al., 1978). It is present in the plasma at 100 times the concentration of the more active metabolite. The $24,25\text{-}(OH)_2D_3$ has been shown to produce a positive calcium balance when given orally to normal people or patients suffering from various disorders of calcium metabolism. It was active in anephric subjects, so that 1α-hydroxylation does not appear to be essential for its action. It appears that this metabolite of vitamin D_3 may also function as a hormone, possibly acting mainly on bone (Ornoy et al., 1978) and it may be useful therapeutically.

The "transport" protein to which vitamin D_3 and its metabolites bind in human plasma has been isolated (Haddad and Walgate, 1976). It is an α-globulin with a molecular weight of 59,000 and can bind either vitamin D_3, $1\alpha,25\text{-}(OH)_2D_3$, or $25\text{-}OHD_3$. The latter intermediary metabolite is bound most strongly with a K_D of about 6.4×10^{-8} M. There is one binding site on each molecule. The concentration of the binding protein in the plasma is about 10^{-5} M, so that in normal circumstances only about 5 percent of the total available sites are occupied by vitamin D_3 or its metabolites. The binding contributes to the relatively prolonged half-life of these sterols in the plasma, partly by restricting their excretion in the urine. The binding appears to be especially important with respect to $25\text{-}OHD_3$, as its integrity is preserved until it can be converted to the more active $1\alpha,25\text{-}(OH)_2D_3$ by the kidney.

D. Mechanism of action

Vitamin D_3, via its active metabolites, increases the concentrations of calcium and phosphate into the blood by an action at two sites, the intestine and the bones. The skeletal deficiencies seen in rickets are largely due to an inability to absorb sufficient calcium from the intestine; the formation of the collagen matrix of bone exceeds the ability of the body to mineralize it. Vitamin D_3 indeed promotes the resorption of calcium from bone and so, in conjunction with its effects on the intestine, aids the homeostasis of this mineral in the extracellular fluids. The movements of phosphate across the intestine appear to be secondary to vitamin D's effects on calcium, although whether or not it also has a direct effect on PO_4^{2-} metabolism is not clear.

Knowledge about the precise mechanism of action of vitamin D_3 on calcium transport is incomplete, and there are a number of conflicting reports. $1\alpha,25$-$(OH)_2D_3$ takes several hours to act, and its effects on bone and the intestine can be inhibited by actinomycin D. These observations suggest that it acts like steroid hormones via transcription of nuclear DNA to promote a *de novo* synthesis of proteins. As for other steroid hormones, cytoplasmic and nuclear receptors are present in intestine (DeLuca, 1977) that appear to mediate transfer of the hormone to the nucleus, where DNA is transcribed and messenger RNA is formed.

Little is known about this process in bone, but it is thought to involve principally the activity of both the osteoclasts, which increases (Baylink et al., 1973), and the osteoblasts, which decreases (Wong, Luben, and Cohn, 1977). It seems likely, from information on other tissues and types of such steroid hormones, that the action of $1\alpha,25$-$(OH)_2D_3$ may involve receptors in the cytoplasm and changes in nuclear transcription. A calcium-binding protein (CaBP) has been identified in the chick tibia, the levels of which can be increased up to 100 times by vitamin D_3 (Christakos and Norman, 1978). This protein could be involved in the mechanism of action of the sterol on bone. It has a higher molecular weight (34,000) than that previously identified [CaBP (see below)] in the intestine (28,000).

In the intestine $1\alpha,25$-$(OH)_2D_3$ seems to induce the formation of at least two proteins, a calcium-binding protein (CaBP) (Wasserman and Taylor, 1968) and a calcium-activated ATPase (Wong et al., 1970; Holdsworth, 1970). The precise role of these substances in the stimulation of calcium transport is contentious, partly reflecting uncertainties about the nature of the process of transmural calcium transport itself. Calcium can be transported across the intestinal epithelium against an electrochemical gradient and so requires energy, and the process is considered to be an example of active transport. It appears to occur in at least two steps: entry into the cell across the apical microvillar (or mucosal) border of the cell,

which appears to be the site of the hormones action (Wong and Norman, 1975), followed by the exit of the calcium from the cell across its basal (or serosal) surface. The calcium is thought to be transported from one side of the cell to the other within the mitochondria or it possibly may even be bound to the induced binding protein. The appearance of the induced proteins appears to correspond to changes in calcium transport and it is considered by some that the CaBP and the Ca-activated ATPase promote the entry of the calcium into the cell across its apical border. The CaBP may act as a "collecting" material on the microvillar membrane or, by having a higher affinity for calcium than the mitochondria, promote its release from these organelles. Two model processes for the action of vitamin D_3 are shown in Figure 9.11. The calcium, once released from the mitochondria near the basal side of the intestinal cells, is transported into the extracellular fluids by a mechanism that appears to involve a $Na^+ - Ca^{2+}$ exchange. This process can be inhibited by ouabain in the serosal fluid and so appears to involve the Na pump, whose role may be the active extrusion of Na from the cell, which restores and maintains intracellular Na levels and so establishes the electrochemical gradient that supplies the energy for the Ca transport.

A summary of the roles of parathyroid hormone, calcitonin, and vitamin D_3 in the regulation of calcium in the body is shown in Figure 9.12.

9.4. Disorders of calcium metabolism

The physiologically ubiquitous role of calcium in the body has been described at the beginning of this chapter. It is therefore not surprising to observe that disorders in calcium metabolism may have extremely diverse manifestations which contribute to the difficulties experienced in diagnosing the precise nature of the disease. As calcium phosphate salts make such a primary contribution to the structure and strength of the bones, changes in the skeleton have, not surprisingly, been a focus for attention in diseases of calcium metabolism. There is, however, a considerable array of other associated disturbances which may affect the nerves (depression, psychosis, and epilepsy), muscles (tetany and cardiac contractility), kidney (the formation of stones and polyuria), gut (gastric ulcers), and eyes (changes in the opacities of the cornea and lens). The potentially important role of calcium in many such disorders is only beginning to be appreciated. As just described, the control of calcium metabolism is an endocrine function primarily involving three hormones: $1\alpha,25$-dihydroxycholecalciferol, calcitonin, and parathyroid hormone. Other hormones may also have effects on calcium metabolism, including

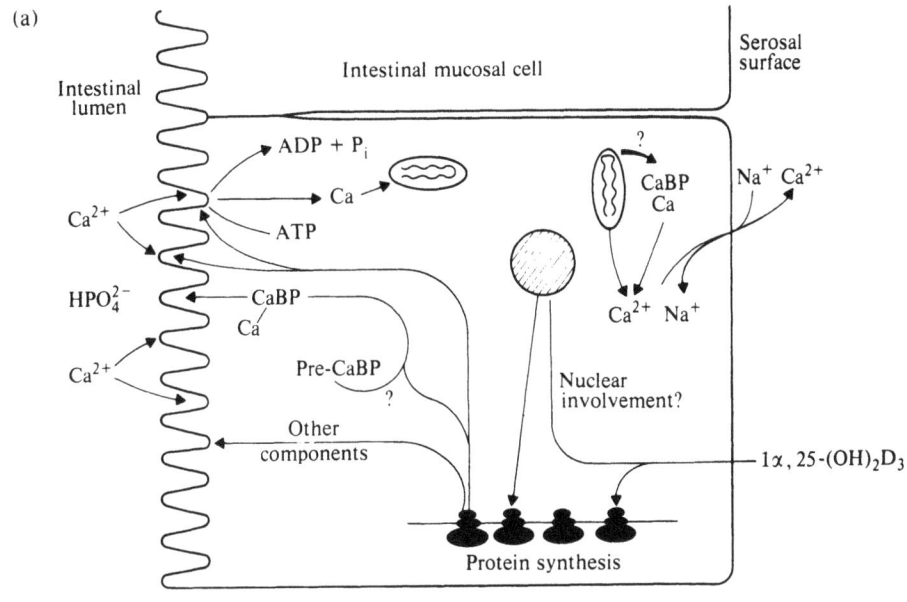

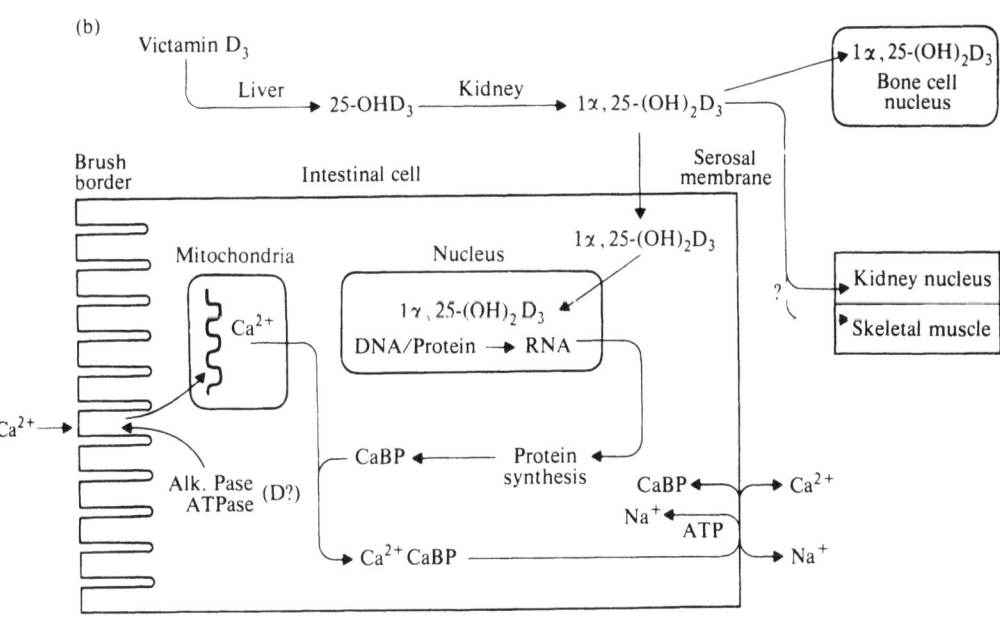

Fig. 9.11. Two models describing the effects of vitamin D₃ on the intestinal absorption of calcium. CaBP, calcium-binding protein. (a) from Omdahl and Deluca, 1973; (b) from Kodicek, 1974.

growth hormone, thyroxine, estrogens, and adrenocorticosteroids. The effects of the latter hormones are more prominently seen in endocrine diseases or when they are used therapeutically. A number of drugs also influence calcium metabolism, including diuretics, such as the thiazides, ethacrynic acid, and furosemide; antiepileptic agents, such as diphenylhydantoin and barbiturates; and even antacids, such as aluminum hydroxide. The diverse interrelations of human disease and calcium metabolism are shown in Figure 9.13.

Calcium homeostasis

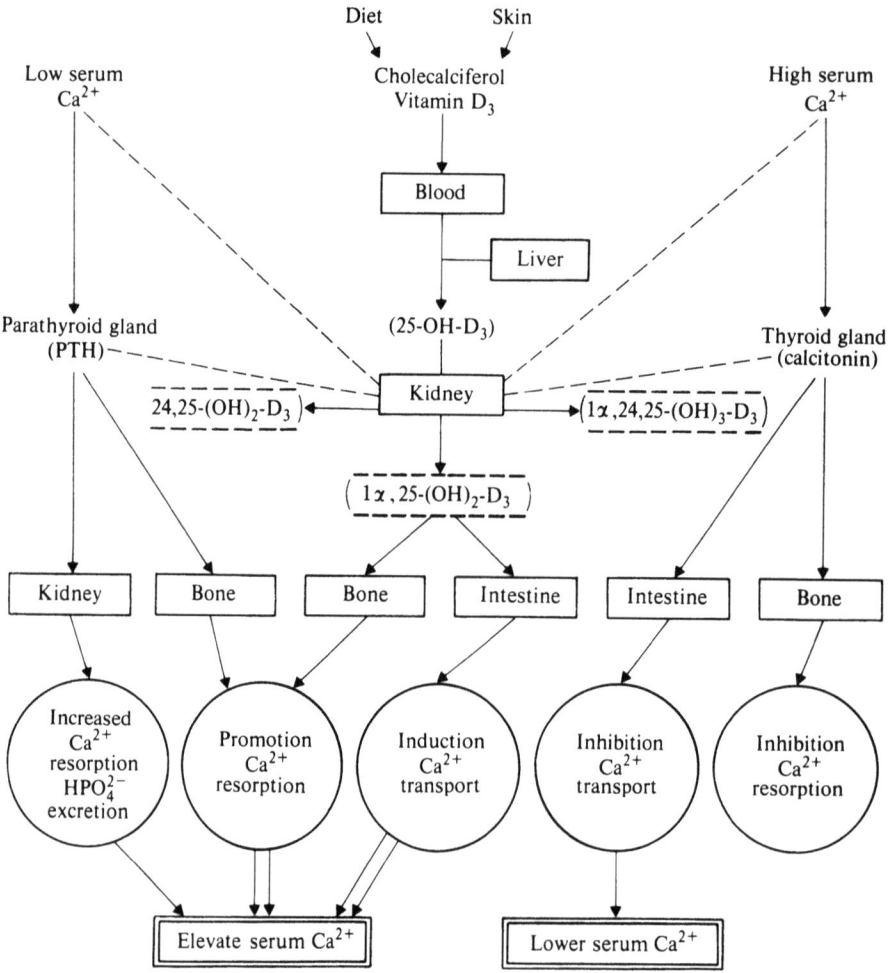

Fig. 9.12. Summary of calcium homeostasis, showing the roles of parathyroid hormone (PTH), vitamin D_3, and calcitonin. (From Norman and Henry, 1974.)

A. Metabolic bone diseases

1. Osteomalacia. This condition has a number of different causes, the primary defect being a lack of adequate mineralization of the collagenous matrix, or osteoid, which forms the basic framework of bone. It thus fails to gain its structural rigidity. The problem essentially arises due to a lack of sufficient calcium or, rarely, phosphate. This condition can result from inadequate dietary sources of the minerals, but this cause is usually only seen under experimental conditions. Insufficient absorption of calcium from the intestine is, however, quite common and can result primarily from reduced absorption, such as commonly follows gas-

trectomy, in steatorrhea, or secondarily due to insufficient vitamin D_3 or $1\alpha,25-(OH)_2D_3$. The latter hormonal deficiencies may not only be due to malnutrition but an inability to absorb vitamin D_3, due to gastrointestinal disorders, or a failure to activate it, such as in liver and kidney disease.

Osteomalacia is most commonly an age-related disease, as it occurs in growing children, when it is called rickets, and in the elderly. Childhood rickets was first identified in the seventeenth century (see Section 9.3). If people are exposed to sufficient sunlight, dietary supplements of vitamin D are probably unnecessary, but in many parts of the world, especially in high latitudes, children suffer an inadequate exposure to sunlight, so that addi-

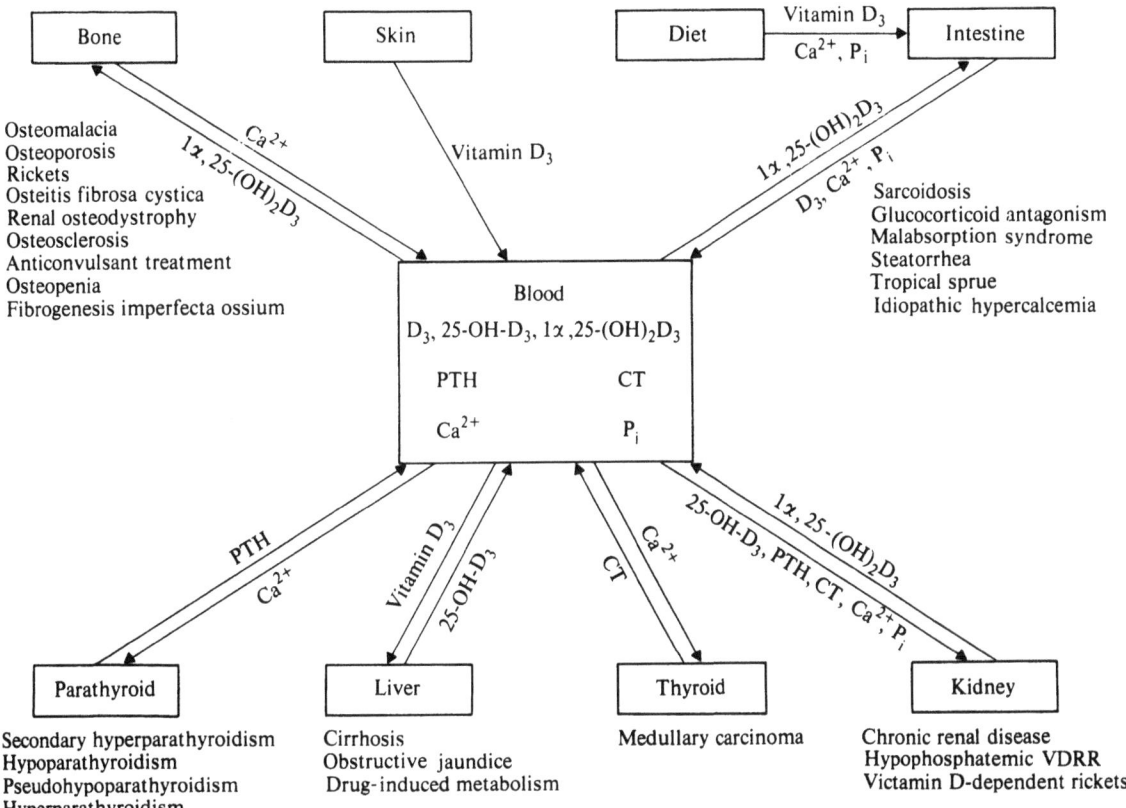

Fig. 9.13. Diseases that influence calcium metabolism due to interaction with the calcemic hormones. PTH, parathyroid hormone; CT, calcitonin; D_3, vitamin D_3. (From Norman and Henry, 1974.)

tional dietary vitamin D is essential. Vitamin D supplements, either as the D_2 or D_3 forms, are added to dairy products, such as butter and milk, in many Western countries, especially the United States. Schoolchildren are also often given regular doses of vitamin D, formerly in gastronomically not very acceptable forms, such as cod-liver oil. As a result of such programs, rickets was virtually wiped out in the United Kingdom in 1945. However, there has since been an increased incidence, especially among immigrant Asian and West Indian children (Editorial, 1973e), which has been related to a dietary vitamin D deficiency. An inadequate vitamin D intake may also contribute to the osteomalacia which is often observed in elderly people living under such unfavorable climatic conditions. The recommended daily intake of vitamin D is 400 IU (1 IU = 0.025 μg), but this may vary depending on such factors as diet and climate. For instance, people living in high latitudes are usually exposed to less sunlight, and this deficiency is greater in winter as compared to summer and in those who are confined indoors.

A hypocalcemia, reduced bone mineral content,

and osteomalacia have been described in patients being treated for epilepsy with diphenylhydantoin and barbiturates. The condition is called *anticonvulsant osteomalacia*. The problem was first reported in children in Germany and was quickly confirmed in independent surveys in other European countries and in the United States (see Editorial, 1972d). It has also been described in relation to the long-term use of hypnotics, including glutethimide (Greenwood, Prunty, and Silver, 1973). The disorder is related to the induction of liver enzymes, which include those important for the metabolism of vitamin D_3. The plasma levels of 25-hydroxyvitamin D_3 are reduced, probably reflecting the metabolism of vitamin D_3 to less active compounds (Hahn et al., 1972, 1975; Stamp et al., 1972). It has been suggested that patients who are being treated with such drugs receive a supplement of vitamin D, about 4000 IU/week in adults and 10,000 IU in children. Vitamin D_2 has been found to be more effective in restoring bone calcium than D_3, possibly reflecting the fact that D_2 is an artificial vitamin which may be less subject than the natural prohormone to degradation by induced liver enzymes (Christiansen et al., 1975). It has also been shown (one patient) that 1α-OHD$_3$ can provide effective replacement therapy, even while continuing to use the anticonvulsant (Juttman, Barth, and Birkenhäger, 1977). It is interesting that a "pilot study" showed that supplements of vitamin D$_2$ were found to reduce the number of seizures in patients being treated with anticonvulsant drugs (Christiansen, Rødbro, and Sjö, 1974).

It has been recognized since the latter part of the nineteenth century that renal disease may be associated with a deterioration of the bones. This osteomalacia is called *renal osteodystrophy.* It is highly resistant to treatment with vitamin D, although, as described earlier (Section 9.3B), large doses of this vitamin or dihydrotachysterol can be effective. The recently acquired knowledge that the active metabolite of vitamin D$_3$ is formed by the 25-hydroxyvitamin D$_3$-1α-hydroxylase enzyme, which is present in the kidney, has provided an explanation for this disease. If this enzyme is lacking, the formation and release of the active hormone is inadequate. Currently, renal osteodystrophy is commonly observed due to the prolonged survival of patients suffering from renal failure due to the use of blood dialysis machines. The logical replacement therapy is $1\alpha,25$-(OH)$_2$ vitamin D$_3$, and indeed this hormone (generic name *calcitriol*) has been found to be effective (Brickman, Coburn, and Norman, 1972; Chalmers et al., 1973; Henderson et al., 1974; Chesney et al., 1978b). 1α-Hydroxyvitamin D$_3$ is almost as effective as $1\alpha,25$-(OH)$_2$D$_3$, especially when given orally (Chalmers et al., 1973; Peacock, Gallagher, and Nordin, 1974; Catto et al., 1975; Chan et al., 1975).

A novel use of the β-adrenergic antagonist *propranolol* has been suggested in the treatment of renal osteodystrophy (Caro et al., 1978). It was observed that a group of patients maintained by hemodialysis and treated with propranolol (for angina pectoris or hypertension) showed less evidence of deterioration of the bones. The parathyroid hormone levels in the plasma were lower in patients treated with the β-adrenergic antagonist, which could thus be decreasing the rate of turnover of bone minerals.

Liver disease is also sometimes associated with osteomalacia. Levels of the vitamin D$_3$ metabolite, 25-OHD$_3$, which is formed in the liver, are, however, normal and the precise nature of the defect in bone metabolism is unknown. It was, nevertheless, found that $1\alpha,25$-(OH)$_2$D$_3$ (parenterally in an oil depot) had a beneficial effect in four patients suffering from this condition of *hepatic osteomalacia* (Long et al., 1978).

Osteomalacia and rickets may also have a hereditary basis. One form of this disease can be treated with enormous doses of vitamin D and is called *vitamin D-dependent rickets.* Small physiological doses of $1\alpha,25$-(OH)$_2$ vitamin D$_3$ can initiate bone healing, and it has been suggested (Fraser et al., 1973) that the hereditary defect is due to a deficiency in the renal 1α-hydroxylase system. Another form of the disease, *vitamin D-resistant rickets,* is associated with a hypophosphatemia which, with short-term treatment, cannot be corrected by $1\alpha,25$-(OH)$_2$D$_3$ (Brickman et al., 1973; Glorieux et al., 1973). However, in a single case treated with 25-OHD$_3$ for over a year, the healing of a previously resistant bone fracture was promoted (Manis and Norman, 1975).

2. Osteoporosis. This disease is a deficiency of the bone-remodelling possibly due to deficient synthesis of its collagenous protein matrix. The bone becomes porous due to a lack of calcification, but the primary defect is in the organic metabolism of the tissue; calcium is not deposited in the absence of sufficient of the matrix.

Osteoporosis is most commonly associated with old age, where it is sometimes given the prefix "senile" or, if in women it appears to be related to the menopause, "postmenopausal." The condition is usually slow to develop and it may take up to 20 years before serious symptoms appear. By this time 40 to 50 percent of the bone mass may have been lost. Osteoporosis can, however, occur more rapidly, such as that due to excess adrenal glucocorticoids, during periods of immobilization and if there is an inadequate protein intake. The periods of weightlessness experienced by astronauts also lead to an excessive loss of calcium, reflecting a reduction in the bone-remodelling process.

Various causes have been suggested for osteoporosis in the elderly. It may reflect declining levels of estrogens and androgens (Marshall, Crilly, and Nordin, 1977), and these hormones have been administered in attempts to correct the condition, but with controversial success. A dietary protein deficiency can contribute. Lack of exercise will also limit the remodelling process, while a propensity to stay indoors and thus be exposed to less sunlight can contribute to a vitamin D deficiency. It has also been suggested that a defect in the ability of the intestine to absorb calcium may develop, but it has now been clearly shown that this process is normally responsive to $1\alpha,25$-(OH)$_2$D$_3$ (Peacock, Gallagher, and Nordin, 1974). In Denmark a vitamin D deficiency appears unlikely, yet similar osseous problems arise in the aged which may reflect a slowing of the process of conversion of 25-OHD$_3$ to $1\alpha,25$-(OH)$_2$D$_3$ in the kidney (Lund, Sørensen, and Christensen, 1975).

Vitamin D and supplements of dietary calcium

have been used to treat osteoporosis. Calcium (usually given as the carbonate or gluconate) alone has been shown to reduce, but not abolish, bone calcium loss in postmenopausal women (Horsman et al., 1977). Vitamin D combined with calcium supplements has also been shown to be effective in osteoporosis (Riggs et al., 1976). Very large doses (50,000 IU twice weekly) of vitamin D, however, resulted in calciuria and were therefore not recommended for routine use. It was suggested that the oral calcium plus vitamin D reduced the levels of circulating parathyroid hormone, so that the rate of turnover of the bone is reduced. Supplements of vitamin D have been recommended in those of the elderly who are considered to be "at risk" (Editorial, 1973e).

Oral 1α-OHD_3 in a dose of 2 μg/day has been found to be effective in the treatment of osteoporosis in the elderly (Lund et al., 1975; Sørensen et al., 1977). However, in an earlier study it was found that prolonged treatment for 3 to 4 months with this drug could produce a hypercalcemia (Peacock, Gallagher, and Nordin, 1974). The daily dose was larger and considered to probably be excessive. Even with a dose of 2 μg/day, hypercalcemia can be a problem, and Lund et al. (1975) suggested that yet even smaller doses, in the absence of calcium supplements, may be preferable. The routine use of this highly active analogue thus should await the results of further trials. Other compounds, such as the hormone $1\alpha,25$-$(OH)_2D_3$ and 25-OHD_3 (Stamp, Haddad, and Twigg, 1977), may also be effective.

Several trials on the use of *low dosages of estrogen* have been performed. The administration of *mestranol* to a group of middle-aged ovariectomized women resulted in the maintenance of bone minerals (Aitken, Hart, and Lindsay, 1973; Lindsay et al., 1976). Similar results were found in another trial, which also included postmenopausal women (Horsman et al., 1977; Recker, Saville, and Heaney, 1977). It has been pointed out that vitamin D and its metabolites are probably only effective in postmenopausal osteoporotic patients who suffer from a deficiency in calcium absorption from the intestine (Marshall and Nordin, 1977). A lack of estrogens is probably also involved. It was found that a combination of therapy with ethinyl estradiol and 1α-OHD_3 was more uniformly effective than was either drug alone. While bone minerals can be conserved by estrogen treatment alone, it is not yet clear whether there is a corresponding reduction in the fracture rate of the bones. Anabolic steroids (see Section 7.5F) have also been used to treat osteoporosis, but as they have a virilizing effect in women, estrogens are usually preferred. It has been shown (Pike et al., 1978) that injected estrogens can modulate the levels of the

vitamin D hydroxylase enzymes in the kidney of chicks. A lack of such an effect if it occurred in women may contribute to osteoporosis in the elderly.

B. Hyperparathyroidism

Hyperparathyroidism may be primarily due to an oversecretion of parathyroid hormone, which occurs independently of the plasma calcium levels and so leads to a hypercalcemia. Alternatively, it is a secondary response to a hypocalcemia and merely represents a physiological effort to restore the plasma calcium levels to normal. This latter condition is more common, and clinically important, as it results in an excessive mobilization of calcium from bone.

1. Primary hyperparathyroidism. It has been estimated that this condition has an incidence of about 1 in 850, or 0.12 percent. In 80 to 90 percent of cases it results from a single adenoma of one of the parathyroid glands, although it can be multiple and occur in the others also. More rarely, the disease is the result of a more diffuse hyperplasia of the parathyroid tissue or cancer of the glands. The most basic change is a hypercalcemia which is the result of a rise in plasma parathyroid hormone levels, which can be measured by radioimmunoassay. These two parameters are important in diagnosis. There is usually a hypophosphatemia, reflecting the renal action of parathyroid hormone in promoting urinary loss of phosphate. The symptoms of the disease are, not surprisingly, diverse and involve the skeleton, which becomes deformed and may result in a shortening of the trunk and reduced height. Osteomalacia may ultimately occur. There are prominent effects on the kidney, including a high incidence (75 percent) of renal stones (nephrolithiasis) containing $CaPO_4$ and a calcification (nephrocalcinosis) which can ultimately be fatal. It has been estimated that 5 percent of patients with nephrolithiasis suffer from primary hyperparathyroidism. Hyperuricemia, which may be secondary to the renal changes, occurs in about 75 percent of cases. Gastric ulcer is also quite common (15 percent), and corneal calcification may occur as a result of the hypercalcemia. The hypercalcemia can be chronic or suddenly acutely worsen ("parathyroid crisis") and constitute a medical emergency. Hypercalcemia can, however, have a number of other causes, which are discussed in Section 9.4G. The pharmacological contribution to the treatment of primary hyperparathyroidism is only indirect. The disease can be cured surgically by removing the adenoma. Subsequently, calcium salts and vitamin D or dihydrotachysterol is administered to aid the process of recalcificaion of the bones and sometimes also to treat or guard against postoperative hypocalcemia and tetany. The latter course of treatment is undertaken with care, as hypercalcemia can be precipitated.

2. Secondary Hyperparathyroidism. This condition arises in response to reduced levels of plasma calcium, which may result from a vitamin D deficiency due to dietary, absorptive, or metabolic disorders and chronic renal failure (see Section 9.4A). It is distinguished from primary hyperparathyroidism by a lack of hypercalcemia. Rather surprisingly, an elevated plasma parathyroid hormone level cannot always be detected. There is usually a hyperphosphatemia. The pathological and clinical importance of secondary hyperparathyroidism is in dispute, but it may hasten the osteoporotic processes in bone, although it will be recalled that in severe vitamin D deficiency, parathyroid hormone has little effect on bone.

Therapy consists of treating the underlying condition, usually with vitamin D or its metabolites and analogues. Calcium salts may be administered to reduce the release of parathyroid hormone, while the high plasma phosphate concentrations can be reduced by the administration of aluminum hydroxide, which binds phosphate in the intestine and so prevents its absorption.

The hyperactivity of the glands in secondary hyperparathyroidism can sometimes result in an "escape" from normal physiological control by plasma calcium levels, so that they function autonomously. This condition has been called *tertiary hyperparathyroidism*, and its clinical effects are the same as those of primary hyperparathyroidism. It often follows renal transplantation. If it persists, it is treated surgically.

C. Hypoparathyroidism

Thyroid surgery is the most common cause of hypoparathyroidism occurs in 1 to 4 percent of cases following thyroid surgery. Hypoparathyroid glands or an obstruction of their arterial blood supply. The deficiency may be transient, recovery occurring due to regeneration of damaged glandular tissue, or it can be more permanent. It has been estimated that some degree of hypoparathyroidism occurs in 1 to 4 percent of cases following thyroid surgery. Hypoparathyroidism also occurs in infancy, where it may be due to hyperparathyroidism in the mother. It can also arise in the first 2 or 3 weeks of life, when it apparently reflects an increased need for the hormone at a time when the glands are insufficiently developed. Hypoparathyroidism, unassociated with thyroid surgery, can also occur in adult life, the causes being unknown (idiopathic hypoparathyroidism).

The basic effect of a deficiency of parathyroid hormone is a hypocalcemia which, characteristically, is accompanied by a hyperphosphatemia and a hypocalciuria. The plasma calcium levels can fall as low as 6 mg/100 ml or even less, and may result in an acute tetanic response. This overresponsiveness of the neuromuscular system not only can involve the skeletal muscles but also the autonomic nerves, so that widespread physiological problems, which include the cardiovascular system, can arise. It may also result in epileptic seizures. The more chronic manifestations of hypoparathyroidism include mental symptoms, such as psychosis and depression, loss of hair, a dry scaly skin, and degeneration of the teeth and nails. Hypoparathyroidism also commonly results in cataracts, and indeed the disease is often first identified by the ophthalmologist. The bones appear to be more dense. Small local calcium deposits may occur in parts of the brain.

Attempts to measure decreased levels of parathyroid hormone in the plasma have not been successful in this disease, so a direct endocrinological diagnostic test is not available. The hypocalcemia is, however, invariably accompanied by hyperphosphatemia, and these chemical determinations afford good evidence for the disease.

Pseudohypoparathyroidism is a condition where the end-organ responses (in the bones and the kidneys) to parathyroid hormone may be deficient and elevated circulating levels of the hormone (secondary hyperparathyroidism), which can be measured by radioimmunoassay, accompany the changes in plasma calcium and phosphate. This syndrome appears to be quite heterogeneous, so that deficiencies in the responses of the bones and the kidneys may occur independently. However, in one carefully studied patient (Metz et al., 1977), a deficiency of $1\alpha,25\text{-}(OH)_2D_3$ was also observed. It seems likely that the lack of this hormone may be the result of a primary inability of the kidney to respond to parathyroid hormone, so that the 1α-hydroxylation of $25\text{-}OHD_3$ is defective. The deficiency of the active sterol could result secondarily in a deficient response of bone to parathyroid hormone. The oral administration of $1\alpha,25\text{-}(OH)_2D_3$ has been shown to increase intestinal absorption of calcium and correct the hypocalcemia in patients suffering from pseudohypoparathyroidism (Bell and Stern, 1978). Hypercalcemia can, however, develop so that it is important to monitor the plasma calcium levels and adjust the dose of the hormone accordingly.

The administration of parathyroid hormones is not an effective way of treating hypoparathyroidism. Hormone preparations made from animal glands are available, but they are generally poorly standardized and their continual use leads to the formation of antibodies and resistance. A secondary response to hypoparathyroidism, and an important effect of the condition, may be a reduced activation of vitamin D_3 in the kidney, leading to a decreased intestinal absorption of calcium. Large doses of vitamin D and dihydrotachysterol (DHT) (Section 9.3B) are used to treat the chronic manifestations of the disease. Current experience indicates that vitamin D is as effective as DHT (also called A-T10, antitetanic factor 10), which is also quite expensive. The hyperphosphatemia is sometimes treated by administration of aluminum hydroxide to bind phosphate in the gut. If tetany occurs, the i.v. administration of calcium, as gluconate or chloride salt, is necessary. It has recently been shown that small doses of $1\alpha,25\text{-}(OH)_2D_3$ or $1\alpha\text{-}OHD_3$ (0.68 to 2.7 μg/day) produce a rapid and predictable rise in plasma calcium concentration, and it has been suggested that they may afford a more predictable and controlled form of therapy, both chronically and in a tetanic attack (Russell et al., 1974).

D. Paget's disease (osteitis deformans)

This bone disease arises due to a marked increase in the rate of metabolic turnover of bone. A process of rapid resorption is followed by the laying down of new tissue but in a disorganized manner, so that deformation of the bone occurs. The underlying causes of Paget's disease are unknown, but it has been estimated that the incidence is about 10 percent of people over the age of 80 years. The effects of the disturbance in the bone-remodelling process are widespread in the body and can result in considerable chronic pain. The bones become very susceptible to fracturing during the initial resorptive phase, while the disordered reconstitution can result in the compression of nerves which, apart from general pain, can produce paraplegia and deafness. Renal stones may be present. The blood supply to the bone increases, and this can be detected as a local hyperthermia. In extreme cases this change can result in a greatly reduced peripheral vascular resistance, leading to high-output congestive heart failure, which can be the ultimate cause of death. Hypercalcemia occurs, but it is not invariable and depends on the particular stage of the bone-remodelling cycle.

It is unknown whether the causes of Paget's disease are endocrine in their nature, but it has recently been shown that the administration of calcitonin has remarkable palliative effects and can even result in healing and restoration of the bone. Three types of calcitonin have been used: natural porcine calcitonin, synthetic salmon, and human calcitonin (Woodhouse et al., 1972; Editorial, 1973f; Rojanasathit, Rosenberg, and Haddad, 1974; Melick, Ebeling, and Hjorth, 1976). Salmon calcitonin is the most effective, but, like porcine calcitonin, a resistance to it may develop over a period of several months, possibly partly due to the

formation of antibodies. Synthetic human calcitonin has become available more recently, and resistance apparently does not arise as frequently with its use. The hormone preparations are given as single daily subcutaneous injections and result in a reduction in the pain, skin temperature, and bone turnover. There is an acute hypocalcemia and hypophosphatemia, but clinical effectiveness is not necessarily related to these responses. Long-term treatment also suggests that a healing of the bone lesions may occur. Even when administration of calcitonin ceases, symptomatic improvement may persist for up to a year, so that intermittent treatment with this hormone may be feasible (Avramides et al., 1976). Salmon and porcine calcitonin, have, however, been used continuously for several years and, although antibodies develop, their presence does not appear to be closely related to clinical resistance to the use of the drugs (Woodhouse et al., 1977). Side effects are minor and include nausea and local flushing and tingling at the site of the injection. Calcitonin has a diuretic action which does not appear to have been listed as a problem "side effect," but it can be prevented by indomethacin (Barnett, Edwards, and Smith, 1975). This effect on the kidney appears to be due to a renal vasodilatation resulting from a release of prostaglandins, the synthesis of which are prevented by the indomethacin.

Prior to the introduction of calcitonin, the treatment of Paget's disease was largely symptomatic and ineffective with respect to the course of the disease. The use of antiinflammatory drugs, such as the salicylates, phenylbutazone, and glucocorticoids, appeared to help, the former possibly because of their analgesic effects. Additional dietary calcium, vitamin D, anabolic steroids, and fluoride have been administered in attempts to promote healing and reduce the structural turnover of the bone. A high dietary phosphate intake has more recently been advocated with some evidence of improvement. Mithramycin provides an effective, but possibly more hazardous, though faster-acting, alternative to calcitonin for the treatment of Paget's disease (see Ryan, 1973; Aitken and Lindsay, 1973; Russell and Lentle, 1974). The drug is cytotoxic and its use requires care, but it has been shown to provide relief from pain, and remissions of a year or more have been reported. A safer alternative (Altman et al., 1973), which has the advantage over calcitonin of being active when given orally, is *diphosphonate* (ethane-1-hydroxy-1,1-diphosphonate, *EHDP*). This drug appears to inhibit the activity and reduce the numbers of the osteoclasts and possibly also of the osteoblasts. It is often preferred to calcitonin in the early, less active, stages of Paget's disease and especially when changes in the skull are involved. It can be used in conjunction with calcitonin. Other chemical derivatives such as dichloromethylene diphosphonate (Cl_2 *MPP*) and 3-amino-1-hydroxypropylidene-1,1-diphosphonate (*APD*) have also been clinically tested (Lemkes et al., 1978). The latter had a more potent action and a faster onset of effect than EHDP or calcitonin.

E. Medullary carcinoma of the thyroid

A hypersecretion of calcitonin characteristically occurs in the disease of medullary carcinoma of the thyroid gland (Tubiana et al., 1968; Tashjian and Melvin, 1968). These tumors do not involve the predominant thyroid tissue but the parafollicular or C-cells. In many instances it has a genetic basis and is probably an instance of an APUDoma (see Section 3.1B). A survey carried out over a period of 47 years at the Mayo Clinic in Minnesota (Chong et al., 1975) found that medullary carcinoma made up about 8 percent of all thyroid cancers, and about one-fifth of these were familial. The tissue may metastasize and can also be the site of formation of ectopic hormones such as ACTH (which can result in Cushing's syndrome) and 5-hydroxytryptamine, prostaglandins, and histaminase. The high levels of calcitonin do not appear to result in disturbances in calcium metabolism; plasma levels of calcium are normal, or a little low, and bone metabolism appears normal. A hyperplasia of the parathyroids commonly accompanies the condition, suggesting that a compensatory response may be occurring. Diarrhea occurs in about 30 percent of patients, and this may be due to the excessive formation of prostaglandins (Williams, Karim, and Sandler, 1968; Barrowman et al., 1975). In many instances the familial form of the disease is accompanied by a bilateral pheochromocytoma.

Diagnosis of the disease is most readily made by measuring plasma calcitonin levels, usually using a provocative test involving stimulation of the hormones release by the i.v. infusion of calcium or pentagastrin (Hennessy et al., 1973; Telenius-Berg et al., 1975; Ribeiro et al., 1976) Ethanol has also been reported to increase the release of calcitonin from such tumors and, as diarrhea and flushing also occurred, other products also appear to be released (Cohen et al., 1973). Pentagastrin-stimulated release of the peptide can be inhibited by i.v. infusion of somatostatin (Gordin et al., 1978). In one patient with diarrhea, it was found that the administration of nutmeg alleviated this condition (Barrowman et al., 1975). It was suggested that the effect was due to nutmeg's ability to inhibit prostaglandin synthesis, which is consistent with the observation that indomethacin (a prostaglandin synthetase inhibitor) was also effective.

Treatment is surgical, usually total thyroidectomy followed by replacement therapy with thyroxine. In the Mayo Clinic, 10-year survival following surgery was 86 percent if metastases had not occurred, but it was only about half this rate if they were present at the time of diagnosis.

F. Vitamin D intoxication

The toxic and even fatal effects of overdosage with vitamin D were recognized quite soon after its discovery. It produces a hypercalcemia due to an increase of calcium absorption from the intestine (its effects are thus greater on a high-calcium diet) and an increased resorption of mineral from bone. The latter effect may be more important in the response and can even lead to a significant loss of bone calcium. The hypercalcemia has a variety of effects, including, acutely, nausea, anorexia lethargy, mental depression, coma, muscle hypotonia, cardiac arrythmias, and renal failure. Chronically, the long-term effects may be more subtle and include the formation of renal stones and renal insufficiency. Vitamin D intoxication may occur due to overmedication either self-administered, accidentally ingested (especially in children), dietetically prescribed, or it may arise during the treatment of diseases such as osteoporosis or hypoparathyroidism. The effects may be prolonged for many months due to the storage of vitamin D in the fat deposits of the body.

It has recently been suggested on the basis of a Norwegian study that a chronic overdosage of vitamin D may result in myocardial infarction (Lindén, 1974). It may be coincidental, but the recent epidemic of myocardial infarction first became apparent about 10 years after the discovery of vitamin D by Mellanby in 1919. The incidence declined during World War II when cod liver was rationed. It is also interesting that vitamin D raises blood cholesterol levels. Lindén found in an epidemiological study that in Norway there was a statistically significant trend showing an increased probability of myocardial infarction related to the increased consumption of vitamin D. It will be interesting to see if a similar correlation is seen in other population groups.

The recommended dietary supplements of vitamin D have been criticized and adjusted over the years. The valetudinarianism of many people has led to an overconsumption of vitamin preparations, including vitamin D, so that it has been suggested that its availability be restricted. In 1963, The Committee on Nutrition of the American Academy of Pediatrics suggested that vitamin D enrichment be stopped in all products except milk and infant foods. A similar recommendation was made by the British Medical Association in 1950.

The average daily intake in the general population has been estimated as 12.5 μg. Lindén has suggested that it should not be greater than 30 μg/day, a level that is apparently exceeded by many American children. The official recommended intake is 10 μg (400 IU/day).

G. Treatment of hypercalcemia

Hypercalcemia has a number of endocrine-related causes, including vitamin D toxicity and hyperparathyroidism (Table 9.2). It may constitute a medical emergency which can be fatal: the beating of the ventricles terminating in their contracted state. The cause of the condition should be removed if possible (see Inesi, 1978). Urinary excretion can be promoted by inducing a saluresis by the administration of NaCl solutions and the diuretics ethacrynic acid or furosemide (not thiazides, which cause calcium retention). Phosphate solutions are often the initial choice; they can be initially administered i.v., but later are given orally. The PO_4^{2-} brings about a rapid decline in the plasma calcium concentration, but it can be hazardous, as calcium may be deposited in tissues, such as the kidney, which may already have undergone some calcification, especially in vitamin D intoxication. The effects of Na EDTA (i.v.) are transient and can also cause renal damage. Cortisol and synthetic corticosteroids, such as prednisolone, are useful, although their effects are several days in their onset. They appear to act by decreasing absorption of calcium from the renal tubules and the intestine. Calcitonin has also been used to treat hypercalcemia due to vitamin D intoxication and hyperparathyroidism; when given i.v., it produced a rapid decline in plasma calcium (Figure 9.14) which was sustained even after the treatment ceased (Buckle, Gamlen, and Pullen, 1972). The cytotoxic antibiotic mithramycin is effective, especially in Paget's disease, hyperparathyroidism, and hypercalcemia resulting from malignancies. It is, however, a potentially very toxic drug. Indomethacin and aspirin, in high doses, are effective in cases involving malignancies, but they are slow to act.

9.5. Adrenocorticosteroids, sex hormones, and calcium metabolism

The adrenocorticosteroids influence calcium and bone metabolism, and these effects can be important in diseases of the adrenal cortex, such as Addison's disease and Cushing's syndrome, and when glucocorticoids are administered for therapeutic reasons. The hypocalcemic effect of cortisol has already been described in the treatment of hypercalcemia. Enormous doses of related drugs, such as prednisolone and dexamethasone, are administered for their antiinflammatory effects and can cause osteoporosis. The latter is also seen in Cushing's syndrome. The osteoporotic effect of large amounts of corticosteroids could be due to an in-

Table 9.2. *Some endocrine-related causes of hypercalcemia and hypocalcemia*

Hypercalcemia
1. Hyperparathyroidism
2. Vitamin D intoxication: therapeutic, accidental, or self-administered
3. Adrenocortical insufficiency: Addison's disease, withdrawal of corticosteroid therapy, adrenalectomy
4. Hyperthyroidism (thyrotoxicosis)
5. Paget's disease
6. Tumors (e.g., breast cancer secreting parathyroid hormone)
7. Estrogen and androgen treatment of tumors

Hypocalcemia
1. Hypoparathyroidism
2. Vitamin D deficiency
   a. Dietary
   b. Relative, due to "resistance" or inability to activate vitamin D_3, "anticonvulsant" osteomalacia
3. Hyperadrenocorticoidism
   a. Cushing's disease
   b. Glucocorticoid therapy
4. Hypothyroidism
5. Excess calcitonin (e.g., medullary carcinoma of thyroid)
6. Renal insufficiencies
7. Metastatic skeletal carcinomas

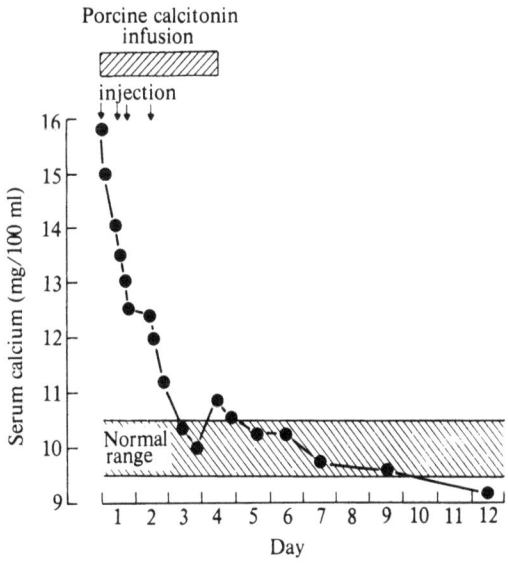

Fig. 9.14. Effects of i.v. infusion of porcine calcitonin on the serum calcium concentration in a patient who had mistakenly taken an overdose of vitamin D. (From Buckle, Gamlen, and Pullen, 1972.)

hibition of the synthesis of bone matrix proteins, but it also may involve an increased excretion of calcium in the urine, reduced absorption across the intestine, and an enhancement of the sensitivity of bone cells to parathyroid hormone and/or $1\alpha,25$-$(OH)_2$vitamin D_3. It has also been suggested that an increased secretion of parathyroid hormone may contribute to the osteoporosis, and this could be mediated by a hypocalcemia or, as shown recently *in vitro*, by a direct stimulating effect on the parathyroid glands (Au, 1976). In pigs, it has been shown that chronic administration of betamethasone decreases the absorption of calcium from the intestine (Fox, Care, and Marshall, 1978). This effect could be counteracted by the concurrent administration of 1α-OHD$_3$. The effect on intestinal Ca absorption could thus be due to a decrease in the 1α-hydroxylation of 25-OHD$_3$ in the kidney. However, measurements of the plasma levels of $1\alpha,25$-$(OH)_2D_3$ have not been made under these conditions, and a direct antagonism at the receptor level is also possible. However, it is interesting to observe that in children being treated with glucocorticoids, a reduction in the serum levels of $1\alpha,25$-$(OH)_2D_3$ has been observed (Chesney et al., 1978a). The osteoporosis associated with the administration of glucocorticoids can lead to "bone collapse," such as has been described in femur heads of patients undergoing corticosteroid therapy (Editorial, 1972e).

The administration of estrogens leads to a hypocalciuria and hypocalcemia, the mechanisms of which are unknown. To reconcile these two changes, it has been suggested that estrogens antagonize the effects of parathyroid on bone, but this hypothesis has not been substantiated. As described earlier [Section 9.4A(2)], the possible role of a lack of estrogens in causing postmenopausal osteoporosis is controversial (see Riggs et al., 1972), but their successful therapeutic use has been reported. Estrogens have also been used to reduce linear growth and hasten the closing of the epiphyses in girls who are growing too tall (see Greenblatt, McDonough, and Mahesh, 1968). The hormone induces premature puberty, with a resulting cessation of bone growth.

The regular therapeutic use of estrogens is at present controversial, especially as their use has been associated with endometrial cancer and cardiovascular disorders (see Sections 7.6J and N).

10

The gastrointestinal hormones

The gut is the most extensive and diverse endocrine tissue in the body. At the present time the number of accepted and putative hormones that have been identified in different parts of the mucosa of the gastrointestinal tract is at least 12, and it appears likely that others may yet be discovered. Gastrin, secretin, and cholecystokinin-pancreozymin (CCK or CCK-PZ) have clear status as endocrine hormones, and the evidence that the others have such a precise role is in various stages of equivocation. Several of the "excitants" have been chemically isolated and their structures are known, while others rest more cryptically in tissue extracts or their presence is at the moment only inferred from the results of physiological experiments. Some "hormones" have been identified in the blood using bioassay or radioimmunoassay techniques. The cells that synthesize them have in many instances been precisely pinpointed using immunohistochemical methods. These cells and their possible endocrine functions were originally described in 1938 by F. Feyrter, who called them "clear cells" (*helle Zellen*). Unlike other endocrine glands, these secretory cells do not form a mass of contiguous tissue, but they are distributed individually among the mucosal epithelial cells. They are thus in an admirable position to respond to stimulation by the contents of the gut. As they are also in intimate contact with a variety of neighboring mucosal cells, with different functions, they are also excellently situated to send local messages to their neighbors, a function that Feyrter called *paracrine*. Some of the active secretions have been located in association with nervous tissue in the gut and so may have roles as neurotransmitters. Thus, although excitant molecules may be judged as hormones, they do not necessarily act by being released into the general circulation, which is the criterion for an endocrine hormone.

Chemically, it appears that the gastrointestinal "hormones" are usually polypeptides, although some amines are present in the gut, but their role is uncertain. A recent exciting series of observations have shown that the peptides may be present not only in the gut but also in nervous tissue, especially the brain (Pearse, 1976). Substance "P" was probably the first such peptide to be identified at both sites, and it has recently been shown to have an analgesic action when injected (Stewart et al., 1976). Somatostatin (or growth hormone-release inhibitory hormone) was first identified in the hypothalamus [Section 3.2E(5)], but it is now known to be present at similar concentrations in the pancreas and the stomach, where it may have a paracrine function in inhibiting the release of insulin, glucagon, and gastrin. A gastrin-like activity, reacting with antigastrin antibodies, and vasoactive intestinal peptide (VIP) have also been found in the brain, following their initial identification in the gut.

The presence of cells that store or secrete the same peptides in the gut and in the brain may be explained in terms of Pearse's *APUD concept* (Pearse, 1974, 1976; also Section 3.1B). The APUD cells include the endocrine cells of the gut and are considered to have a common embryonic origin from neuroectodermal tissue. Some of these cells, although they are present at quite different sites, may still possess some common properties, including an ability to synthesize the same peptides. The skin glands in amphibians also contain cells of this type, and peptides have been extracted, such as caerulein, which have common amino acid sequences and activities to such mammalian hormones as gastrin, cholecystokinin, and secretin. It would thus appear that there is some embryological order and that certain groups of vertebrate cells have inherent abilities to synthesize certain types of reactive molecules which can be utilized as chemical messengers. Their particular functions apparently depend on the sites to which they migrate during development. The gastrointestinal tract has been blessed by an abundance of these interesting and useful cells, which it apparently utilizes to integrate its diverse digestive functions.

The gut is also unique as an endocrine organ, as it is the site of origin of the first hormone to be discovered. W. M. Bayliss and E. H. Starling, one afternoon in January 1902 at University College London, found that when the duodenum of a dog was exposed to acid, the secretion of pancreatic juices was stimulated even though all nervous connections between the two had been eliminated. They concluded that a "chemical reflex" was involved and suggested that the excitant (secretin) belonged to a class of substances which they called *chemical messengers*. Such excitants were more classically dubbed "hormones," from the Greek word meaning "I arouse to activity." Shortly after this unique endocrine event, J. S. Edkins in 1905 showed that extracts of the antrum of the stomach (but not the fundus), when injected into cats, stimulated the secretion of the oxyntic (or parietal), acid-forming cells in the fundus. He called this excitant gastrin, but its hormonal (and chemical) nature was to be disputed for another 40 years.

Cholecystokinin-pancreozymin also had a rather stormy history, as it was discovered twice and for many years, as its name implies, had a dual identity. A. C. Ivy and E. Oldberg in 1928 showed that an extract made from the "upper intestinal mucosa," when injected, caused the dog gallbladder to contract.' Using a cross-circulation procedure, they also demonstrated the presence of such a substance, in the blood, and called it cholecystokinin. In 1943, Harper and Raper showed that extracts made from the intestinal mucosa of pigs brought about an increased enzyme-rich pancreatic secretion in cats. They called this substance pancreozymin.

The chemical characterization of cholecystokinin and pancreozymin from extracts of pig intestinal mucosa was started in about 1960 by J. E. Jorpes and V. Mutt at the Karolinska Institute in Sweden (see Jorpes, 1968). It soon became apparent that the two biological activities showed a remarkably close parallelism in the successively more purified tissue extracts. It was finally shown that, in fact, only one active hormone was present which had a dual action in contracting the gall bladder and stimulating pancreatic secretion. It was named cholecystokinin-pancreozymin (CCK-PZ), although more recently a plea has been made to shorten its title to just cholecystokinin.

Gastrin, secretin, and cholecystokinin are clearly endocrine secretions. A number of other substances with distinct spectra of pharmacological activity have been isolated and in many instances chemically characterized, especially from the extracts of the upper intestinal mucosa from which

Jorpes and Mutt separated secretin and cholecystokinin. They include enteroglucagon, gastric inhibitory peptide (GIP) (Brown et al., 1969, 1975), and vasoactive intestinal peptide (VIP) (Said and Mutt, 1970). Other active materials include somatostatin, bulbogastrone, substance P, enterooxyntin, motilin, and chymodenin. The sites of their origins in the gastrointestinal tract and stimuli that initiate their release are shown in Tables 10.1, 10.2, and 10.3. As described earlier, the physiological functions of these excitants, and indeed even their discrete identity as chemical messengers is not always clear. They have thus been called "candidate hormones."

10.1. Gastrin

A. Functions and actions
Gastrin was primarily identified because of its ability to stimulate gastric acid secretion from the oxyntic or parietal cells in the fundus of the stomach. It has, however, a wide range of actions on the secretion of other types of glands: it influences membrane permeability and the contractility of smooth muscle, and it exerts a trophic effect on the growth of the gastrointestinal mucosa. Gastrin stimulates the secretion of enzymes and fluids from glands in the stomach, duodenum, and pancreas; it can inhibit the absorption of fluids from the ileum and contracts smooth muscle in the gut and gall bladder. Gastrin reduces the tone of sphincter of Oddi but increases the tone of the lower esophageal sphincter and the pyloric sphincter so that gastric emptying time is reduced. It increases the blood flow to the gastric mucosa. In larger quantities, gastrin stimulates release of insulin. Gastrin exerts a trophic effect on the growth of the oxyntic cells and other mucosal tissues, including small intestine and colon (see Johnson, 1977). These actions are quite distinct from those it has on secretion. Such effects of gastrin can be demonstrated when purified preparations of the hormone are injected, but they do not necessarily all reflect physiological roles.

B. Synthesis and release
Gastrin is synthesized in columnar epithelial cells, called "G" cells, which are found principally in the pyloric antrum but also in the duodenum and jejunum. They contain granules, the number of which seems to reflect the functional state of the tissue and which appear to be the intracellular storage sites for the hormone. Thus, in pernicious anemia, when gastrin secretion is increased, the cells are almost devoid of granules. The outer borders of the G cells have numerous microvilli, which appear to sense the changes in composition of the gastric contents that aid the stimulation or inhibition of the hormone's release.

Table 10.1. *Sites of origin of the gastrointestinal hormones and peptides*

Hormone or peptide	Site of origin
Gastrin[a]	G cells, gastric antrum, also duodenum and jejunum
Cholecystokinin (CCK)[a] (cholecystokinin-pancreozymin)	CCK cells (also called I cells), duodenum and jejunum
Secretin[a]	S cells, duodenum, jejunum, and gastric antrum
Gastric inhibitory peptide (GIP)[a]	K or GIP cells, duodenum, and jejunum
Vasoactive intestinal peptide (VIP)	H or VIP cells, throughout the gut
Enteroglucagon (glucagon-like immunoreactivity or GLI)	AL or A-like cells, stomach and small intestine
Motilin	EC or enterochromaffin cells of jejunum
Bulbogastrone	Upper duodenum
Substance P	EC cells
Somatostatin[a]	Somatostatin or D cells, gastric mucosa
Enterooxyntin	Jejunum?
Chymodenin	Duodenum
Urogastrone	Cells of Brunner's glands, upper duodenum

References: Barrington and Dockray, 1976; Bloom, 1976; Pearse, 1974, 1976; and Polak, 1976a,b. Also see the text.
[a] Hormonal role considered to be established (see Grossman, 1977). It seems likely that this is also the case for somatostatin (Gustavsson and Lundquist, 1978).

Table 10.2. *Principal effects of gastrointestinal peptides*

Gastrin	Stimulates gastric acid[a]; water and electrolyte secretion in stomach, pancreas, liver, and Brunner's glands (duodenum); enzyme secretion in stomach, intestine, and pancreas. Increases blood flow to gastric mucosa, small intestine, and pancreas. Stimulates smooth muscle of gut, including the lower esophageal sphincter and pyloric sphincter[a] (delays gastric emptying). Relaxes sphincter of Oddi. Exerts a trophic action on the gut mucosal cells. Stimulates release of insulin and calcitonin. Inhibits fluid resorption from ileum.
Secretin	Stimulates secretion of water and electrolytes from pancreas[a] and liver. Increases gastric pepsin secretion. Augments pancreatic enzyme secretion.[a] Releases insulin. Inhibits gastric emptying, gastric acid secretion, duodenal motility, and glucagon release.
Cholecystokinin, CCK	Stimulates pancreatic enzyme[a] and augments electrolyte secretion.[a] Contracts gallbladder[a] and inhibits stomach motility.[a] Stimulates smooth muscle of the intestine, gastric acid, and Brunner's glands. Relaxes lower esophageal sphincter and sphincter of Oddi.
Gastric inhibitory peptide, GIP	Inhibits gastric acid secretion[a] and release of insulin.[a] Stimulates secretion in jejunum and ileum.
Vasoactive intestinal peptide, VIP	Increases splanchnic and peripheral blood flow. It has secretin- and glucagon-like activity. Inhibits the release of gastrin.
Enteroglucagon	Stimulates insulin release and hepatic glycolysis. Inhibits gut motility and release of gastrin and pancreatic enzymes.
Motilin	Increases pepsin release and intestinal motility. Decreases gastric emptying. Inhibits the effects of pentagastrin and histamine on gastric acid secretion.
Bulbogastrone	Inhibits gastric acid secretion.
Enterooxyntin	Stimulates gastric acid secretion.
Chymodenin	Stimulates pancreatic chymotrypsinogen secretion.
Urogastrone	Inhibits gastric acid secretion. It has a trophic effect on gut epithelia.

Note: Many of the effects have only been demonstrated using pharmacological doses, but some of these may yet be shown to be of physiological significance or play a role in some diseases.
References: Andersson, 1973; Brown et al., 1975; H. Gregory, 1975; R. A. Gregory, 1974; Grossman, 1974; Rayford, Miller, and Thompson, 1976; and Walsh and Grossman, 1975.
[a] Physiological effects (see Grossman, 1977).

Table 10.3. *Conditions that increase the release of gastrointestinal peptides*

Gastrin	Digested protein and amino acids, distension; also Ca, epinephrine, and para-thyroid hormone (*inhibited* by gastric acid, secretin, glucagon, somatostatin, and calcitonin)
Cholecystokinin, CCK	Digested protein, fat, and gastric acid (small effect)
Secretin	Gastric acid in intestine (pH below about 4.5)
Gastric inhibitory peptide, GIP	Glucose and triglycerides
Enteroglucagon	Glucose
Motilin	Acid and fat
Bulbogastrone	Gastric acid
Somatostatin	Inhibited by decreased pH and amino acids
Enterooxyntin	Digested protein
Vasoactive intestinal peptide, VIP	?
Chymodenin	?

Note: For references, see the text.

The release of gastrin occurs in response to a number of stimuli. The most common physiological response is that due to the presence of food in the stomach, which results in a release in response to the local distension and presence of digested protein. These stimuli may exert a direct effect on the G cells or act via the intervention of a local nerve reflex. The vagus nerve has an important and ubiquitous role in the process of digestion, and one of its effects is to release gastrin. Inhibition of the hormone's release is due to a simple negative feedback due to acid conditions (a pH of less than about 2.5). Secretin can also inhibit the release of gastrin and reduces secretion by the oxyntic cells. The probable inhibitory role of somatostatin is discussed in Section 10.4.

Epinephrine and calcium stimulate the release of gastrin, and while it is unlikely that they have an action under normal conditions, their effects may sometimes be important clinically. The catecholamines released from pheochromocytoma tumors can produce elevated levels of gastrin in the plasma (Hayes et al., 1972). The concentrations return to normal after removal of the tumor or the administration of the α-adrenergic blocking drug phenoxybenzamine. The intravenous infusion of calcium is known to increase the release of gastrin, and a reflection of this effect is seen in hyperparathyroidism (see Section 9.4B) which is commonly associated with the presence of gastric ulceration. The parathyroid hormone also may act directly to release gastrin. Peptic ulcer disease can be treated with antacids, and it has been shown that calcium carbonate, which was formerly used quite commonly, results in a release of gastrin and a secretion of acid (Levant, Walsh, and Isenberg, 1973). Calcium carbonate is now used more judiciously.

C. Structure and its relationship to activity

Although gastrin was first identified in 1905, its discrete chemical identity remained an enigma until it was isolated and shown to be a peptide by R. A. Gregory and H. J. Tracy in 1962 (see Gregory, 1974). Prior to this time, it was suspected that it may be identical to histamine, which is a common contaminant of the gastric antral extracts. When injected, histamine stimulates gastric acid secretion, but it is now known to be about 1500 times less potent than purified gastrin.

Gregory and Tracy in their initial 2 years of work, extracted hormone material from the gastric antrums of 50,000 pigs! A chemical analysis of the purified hormone showed it to be a peptide containing a linear chain of 17 amino acids (G-17) (Figure 10.1). Two forms were found to be present; in one of these gastrin II, the single tyrosine residue, at position 12, is sulfated. The two forms of gastrin appear to be present in equal amounts in the G cells, and they have similar activities on acid secretion.

Several other forms of gastrin (see Gregory, 1974; Walsh, 1977) have, however, been identified in extracts of the normal gastric antrum, in serum, or in tumor tissue from patients suffering from the Zollinger–Ellison syndrome (see Section 10.6B). A larger molecule with gastrin activity, containing 34 amino acids (G-34), was found in serum, where it makes up about 50 percent of the total immunoreactive material, and also in antral tissue, where it is a minor component (5 to 10 percent). This gastrin, called "big gastrin" (Figure 10.1), exists in two forms, nonsulfated and sulfated, corresponding to gastrins I and II. It has a half-life in the circulation of dogs of about 15 minutes compared to 3 minutes for the heptadecapeptide, or "little gastrin." Big gastrins (BGI and II) are about five times less active in stimulating acid secretion than little gastrins. A similar pair of even smaller gastrins, containing 13 amino acids and called *minigastrins*, have also been identified in tumor tissue and normal serum. They have an even shorter half-life in the circulation than little gastrin and exhibit about half its biological activity. The serum

"Big" gastrin (G-34):

1	2	3	4	5	6	7	8	9	10	11	12	13	14	15	16	17
<Glu	-Leu	-Gly	-Pro	-Gln	-Gly	-His	-Pro	-Ser	-Leu	-Val	-Ala	-Asp	-Pro	-Ser	-Lys	-Lys

18	19	20	21	22	23	24	25	26	27	28	29	30	31	32	33	34
-Gln	-Gly	-Pro	-Trp	-Leu	-Glu	-Glu	-Glu	-Glu	-Glu	-Ala	-Tyr	-Gly	-Trp	-Met	-Asp	-Phe -NH_2

R

"Little" gastrin (G-17 = amino acid residues 18–34 of G-34):

1	*2*	*3*	*4*	*5*	*6*	*7*	*8*	*9*	*10*	*11*	*12*	*13*	*14*	*15*	*16*	*17*
(18	19	20	21	22	23	24	25	26	27	28	29	30	31	32	33	34)
<Glu	-Gly	-Pro	-Trp	-Leu	-Glu	-Glu	-Glu	-Glu	-Glu	-Ala	-Tyr	-Gly	-Trp	-Met	-Asp	-Phe -NH_2

R

"Mini" gastrin (G-13 = amino acid residues 22–34 of G-34):

1	*2*	*3*	*4*	*5*	*6*	*7*	*8*	*9*	*10*	*11*	*12*	*13*
(22	23	24	25	26	27	28	29	30	31	32	33	34)
Leu	-Glu	-Glu	-Glu	-Glu	-Glu	-Ala	-Tyr	-Gly	-Trp	-Met	-Asp	-Phe -NH_2

R

(Gastrin I, R = H; gastrin II, R = SO_3H.)

Fig. 10.1. Amino acid sequences of the human gastrins. <Glu = proglutamyl. (From Walsh and Grossman, 1975. Reprinted by permission from *The New England Journal of Medicine 292*, 1325.)

also contains another immunoreactive gastrin, which has a molecular weight of about 20,000 and has been called "big big gastrin" (Yalow and Berson, 1972). In fasting animals it is the major form of gastrin that is present, but in contrast to the other types its levels do not increase in response to feeding (Yalow and Wu, 1973). On the basis of the activity of chemical fragments of the molecule, the presence of yet other natural-related peptides has been predicted, perhaps "mini mini gastrins."

The structure–activity relationships of the natural gastrins and numerous artificial analogues have been studied. The C-terminal tetrapeptide of little gastrin, containing the 14 to 17 positions (Trp-Met-Asp-Phen-NH_2), has the full range of activities, although its potency is a little less, about $\frac{1}{10}$ that of the parent molecule. The N-terminal tridecapeptide (1–13) has no activity but its presence appears to protect the molecule against inactivation, especially in the liver, and this aids its transport to its active sites. The C-terminal tripeptide (Met-Asp-Phe-NH_2) retains some of the activities of the parent molecule, but it is only about 1/30,000 as active. An active compound based on the structure of the C-terminal tetrapeptide but with a β-alanine added, and the N-terminal blocked by a tertiary butyl oxycarbonyl group (*t*-Boc-β-Ala-Trp-Met-Asp-Phen-NH_2), has been produced commercially under the name of *pentagastrin*. It has a potency which is similar to the natural C-terminal pentapeptide and is used in experimental work and clinically to test the integrity of gastric acid function, where it is prefer-

able to the more traditional use of histamine, which has widespread side effects. While pentagastrin is used for this purpose in Europe, it is not yet available in the United States. It is also used as a provocative test to stimulate calcitonin release in cases of medullary carcinoma of the thyroid gland.

Morely (1968), in a study of 500 analogues of the C-terminal tetrapeptide, defined the structure–activity relationship of the molecule. These results can be summarized as follows:

a. Substitution of the tryptophan, methionine, or phenylalanine (positions 1, 2 or 4) is well tolerated and the molecule retains activity. These amino acids may be concerned with the binding of the molecule to its receptor.

b. The aspartic acid (position 3) is vital, and a wide range of substitutions resulted in a complete loss of activity. Its role thus appears to be a functional one. It has, however, been shown that when substitution of alanine for the aspartyl residue is made, the molecule retains some gastrin activity (Grossman, 1974).

c. The terminal amide is also necessary, although a limited replacement with a similar group such as methylamide or hydrazide can be tolerated.

d. The spatial arrangement of the side chains of the amino acids is important, for if these are disturbed, such as by inserting amino acyl residues between the amino acids, thus distorting the peptide backbone, then activity is lost.

Morely suggested that while the N-terminal portion of the gastrin is necessary for the transport of the molecule intact to its target tissue, the C-terminal tetrapeptide "stump" reacts with the receptors (see Figure 10.2). Binding is accomplished by the tryptophan, methionine, and phenylalanine residues, while the aspartic acid is attached to a substrate which is then linked to the receptor site. The latter can then act as an enzyme and initiate a response. Grossman (1974), on the basis of the more recent evidence that alanine can substitute for the aspartic acid, does not agree with this hypothesis and considers that no real distinction can be made between "binding" and "funtional" groups, and the only necessity is that a minimum number of binding groups are present.

D. Metabolism

As described above, in the dog the half-life in the circulation of big gastrin is about 15 minutes, but this decreases as the chain length diminishes, and it is only 3 minutes for little gastrin. Studies in man are more limited, but it has been estimated that survival is, respectively, about 5 minutes and 40 minutes for little gastrin and big gastrin. Big big gastrin has a much longer half-life. Most gastrin is inactivated in the kidney and small intestine, but the shorter-chain molecules pentagastrin and the C-terminal tetrapeptide can be destroyed by the liver. As the intact hormones must traverse the

liver immediately following their release from the G cells, their relative "protection" against destruction at this site seems to be functionally appropriate.

10.2. Cholecystokinin-pancreozymin

A. Functions and actions

The "primary" actions of cholecystokinin-pancreozymin (or as it will be referred to, cholecystokinin) are contraction of the gallbladder and a stimulation of exocrine pancreatic secretion, especially its enzymic component. While these effects were those that led to its discovery and may represent more important physiological actions, cholecystokinin, like gastrin, stimulates secretion of acid and pepsin by the stomach and decreases the emptying time of the stomach. It contracts the smooth muscle of the gastric antrum and the small intestine. The release of insulin and glucagon is increased. Cholecystokinin has a number of actions in common with gastrin, with which it also shares common chemical characteristics. These two hormones have been classified as members of the "gastrin family" of peptides, which also includes caerulein, an interesting natural analogue originally extracted from the skin of an Australian tree frog *Hyla caerulea* (Anastasi, Erspamer, and Endean, 1967).

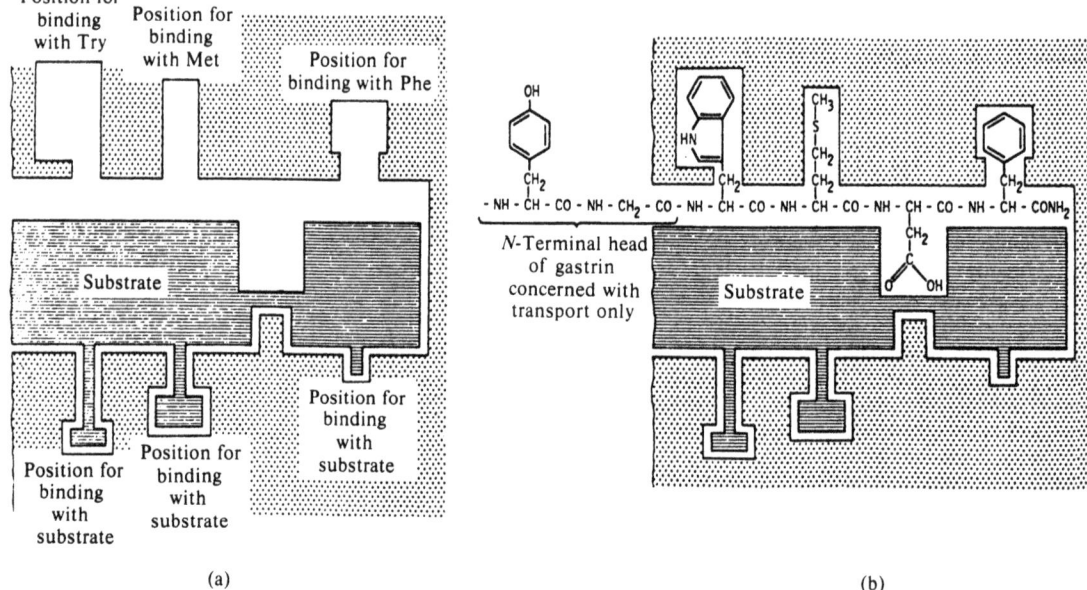

(a)   (b)

Fig. 10.2 Schematic model (based on studies of the structure–activity relationships of gastrin) proposed to account for the action of gastrin at its receptor site. (a) Cell receptor site equipped to receive a substrate but lacking an aspartyl group necessary for its function as an enzyme. (b) Missing carboxyl group supplied by the C-terminal sequence of a gastrin molecule at exactly the required position. The receptor site can now function as an enzyme. (From Morley, 1968.)

B. Synthesis and release

Cholecystokinin has been isolated in extracts of the duodenum and jejunum, where it was shown to originate in a specific cell type originally classified as I (for intermediate granule) cells but which now can be more specifically designated as CCK cells (Polak, 1976b).

The release of cholecystokinin occurs in response to the presence of the products of digestion in the duodenum. Fat is an especially effective stimulus, but digested protein and even the presence of acid are stimulatory. It has been suggested that saline, or osmotic, laxatives, such as Epsom salts ($MgSO_4$), may act by initiating a release of cholecystokinin with a resulting increased motility of the intestine (Harvey and Read, 1973).

C. Structure and its relationship to activity

Cholecystokinin is a linear polypeptide containing 33 amino acids (Figure 10.3), although another form containing 39 of these residues ("cholecystokinin variant") has also been identified. The C-terminal decapeptide is 10 to 15 times as active as the intact molecule, which suggests the possibility that the latter represents a prohormonal form of the secretion. The C-terminal octapeptide is also very active, about half that of the decapeptide, but it is still far more effective than is the parent molecule.

The C-terminal pentapeptide is identical to that of gastrin, so that this portion, as well as the tetrapeptide, has the same activities, which are gastrin-like. The more special actions of cholecystokinin on the gallbladder and pancreas depend on the sulfated tyrosine residue, which, counting back from the C-terminus, is at position 7, being separated from the glycine (position 5) by methionine (see Thompson, 1973). This particular arrangement is vital to the special activities of CCK; if the SO_3H is moved to a neighboring amino acid or even to another part of the tyrosine residue, this activity is lost. Even displacement of the sulfated tyrosine to the next position (to 6 from the C-terminus) abolishes the cholecystokinetic activity. As in gastrin, the penultimate (at the C-terminus) aspartic acid is vital for biological activity.

In *summary,* the sulfated tyrosine, the penultimate aspartic acid, and its neighboring methionine (positions 7, 2, and 3, respectively, counting back from the C-terminus) are vital. Basically, gastrin and CCK share their activities, but there are *relative* differences in their potency depending on the target organ. The sulfated tyrosine and its two neighboring amino acids, aspartic acid and methionine, appear to confer on CCK its ability to bind receptors in the gallbladder and pancreas.

Caerulein is a decapeptide and its C-terminal octapeptide is identical to that of cholecystokinin, except that the methionine which is adjacent to the sulfated tyrosine is replaced by threonine. This peptide also bears many similarities to gastrin (Figure 10.3). Qualitatively, it shares all the actions of cholecystokinin and gastrin. The importance of the sulfated tyrosine in such molecules is emphasized by the observation that this form of caerulein is 160 times more potent on the gallbladder than is the desulfated peptide. Caerulein is generally more potent than cholecystokinin and gastrin; on the gallbladder its activity is 16 times greater than CCK, while on pancreatic enzyme secretion it is 3 to 6 times more effective. It also has a greater

"Little" gastrin (G-17, gastrin II):

1   10   17
(Glu -Gly -Pro -Trp -Leu -Glu -Glu -Glu -Glu - Glu -Ala -Tyr -Gly -Trp -Met -Asp -Phe -NH₂
                                                                          |
                                                                       SO₃H

Cholecystokinin-pancreozymin:

1   10
Lys -Ala -Pro -Ser -Gly -Arg -Val -Ser -Met - Ile -Lys -Asn -Leu -Gln -Ser -Leu -

20   30   33
Asp -Pro -Ser - His -Arg -Ile -Ser -Asp -Arg -Asp -Tyr -Met -Gly -Trp -Met -Asp -Phe -NH₂
                                                                          |
                                                                       SO₃H

Caerulein:

1   10
<Glu -Gln -Asp -Tyr -Thr -Gly -Trp -Met -Asp - Phe -NH₂
                      |
                   SO₃H

Fig. 10.3. Amino acid sequence of cholecysytokinin (CCK), showing its relationship to gastrin II and caerulein (a peptide isolated from frog skin). <Glu = pyroglutamyl.

potency in stimulating pancreatic water and electrolyte secretion and in the contractility of gastrointestinal smooth muscle.

Cholecystokinin preparations have been used to stimulate contraction of the gallbladder, and it has been suggested that they could be useful for the treatment of paralytic ileus (see Rayford, Miller, and Thompson, 1976).

D. Metabolism

The half-life of cholecystokinin in the circulation is short, about 2 to 5 minutes. The kidney appears to be important for its inactivation, but other sites appear to exist.

10.3. Secretin and related peptides

A. Functions and actions

As described earlier, the word "hormone," in its physiological sense, was first used to describe secretin. It is indeed the prototype hormone, originally characterized by Bayliss and Starling in 1902 by its ability to stimulate the secretion of water and electrolytes by the pancreas. Secretin, however, like other gastrointestinal hormones, has quite ubiquitous effects (Table 10.1), as it can also stimulate the secretion of pancreatic and gastric enzymes, and bile. Insulin release is increased by pharmacological doses (Jensen et al., 1978a). It reduces gastric acid secretion and the tone of the lower esophageal sphincter. Gastric emptying is delayed and the motility of the duodenum is reduced. The effects of cholecystokinin on the gallbladder and pancreatic enzyme secretion are potentiated; indeed, it is doubtful if the plasma levels of CCK alone are sufficient to exert these effects.

B. Synthesis and release

Secretin originates in the "S" cells of the duodenum and jejunum (Polak, 1976b). This type of cell has also been identified in the antrum of the stomach and, in dogs, along the entire length of the small intestine (Chey and Escoffery, 1976). It

has characteristically small granules and it was for this reason that it received its name.

The release of secretin occurs in response to acid conditions, a pH below about 4.5, in the duodenum. If the pH rises above this value, its secretion ceases. Fatty acids have a weak releasing effect.

C. Structure and its relationship to activity

Although crude extracts of the duodenum were shown to contain the hormonal activity in 1902, it took another 60 years to chemically describe secretin (Jorpes et al., 1962). Some idea of the magnitude of the problem of isolation and purification, as well as the quantities of the hormone that are stored in the gut, is illustrated by the observation that 1000 pieces of pig duodenum, each 1 meter long, were necessary to produce 1 mg of the pure hormone. Secretin is a linear polypeptide containing 27 amino acids (Figure 10.4). In contrast to the hormones of the gastrin family, where fragments of the molecule may have high activity, the intact secretin molecule alone is effective. Even removal of the N-terminal histidine virtually abolishes its actions.

D. Metabolism

In dogs, the half-life of secretin is about 2.5 minutes (Rayford, Miller, and Thompson, 1976). *In vivo* studies indicate that the kidney is an important site for its inactivation while the liver is of little importance.

E. Other members of the secretin family

Secretin shares a close structural similarity to three other excitants that have been isolated from the gastrointestinal mucosa. Fourteen of its 27 amino acid residues are identical to those in glucagon, while "gastric inhibitory peptide," GIP, and vasoactive intestinal peptide, VIP, share, respectively, 8 and 9 residues with it (Figure 10.4). It seems likely that they may have common evolutionary origins, so the term "secretin family" is

	1	2	3	4	5	6	7	8	9	10	11	12	13	14
Secretin	His	Ser	Asp	Gly	Thr	Phe	Thr	Ser	Glu	Leu	Ser	Arg	Leu	Arg
VIP	His	Ser	Asp	Ala	Val	Phe	Thr	Asp	Asn	Tyr	Thr	Arg	Leu	Arg
Glucagon	His	Ser	Gln	Gly	Thr	Phe	Thr	Ser	Asp	Tyr	Thr	Lys	Tyr	Leu
GIP	Tyr	Ala	Glu	Gly	Thr	Phe	Ile	Ser	Asp	Tyr	Ser	Ile	Ala	Met

	15	16	17	18	19	20	21	22	23	24	25	26	27	28	29
Secretin	Asp	Ser	Ala	Arg	Leu	Gln	Arg	Leu	Leu	Gln	Gly	Leu	Val	NH$_2$	
VIP	Lys	Gln	Met	Ala	Val	Lys	Lys	Tyr	Leu	Asn	Ser	Ile	Leu	Asn	NH$_2$
Glucagon	Asp	Ser	Arg	Arg	Ala	Gln	Asp	Phe	Val	Gln	Trp	Leu	Met	Asp	Thr
GIP	Asp	Lys	Ile	Arg	Gln	Asp	Phe	Val	Asn	Trp	Leu	Ala	Gln	Gln	

Fig. 10.4 Comparison of the structures of porcine secretin, vasoactive intestinal peptide (VIP), glucagon, and gastric inhibitory peptide (GIP). The latter has 43 amino acids, but only the first 29 of these are shown. It can be seen that there are many similarities; hence, they have collectively been termed the "secretin family."

quite appropriate. Not unexpectedly, there is an overlap in their biological activities, such as an ability to release insulin, although each has specific individual properties. The hormonal status of glucagon is clear, but that of its intestinal analogue *entero*glucagon, as well as VIP and GIP, are equivocal and they have been put in a category called "candidate hormones."

Pancreatic glucagon is discussed in detail in Section 8.4. Principally on the basis of immunoreactivity, two forms of glucagon, one with a molecular weight of 3500 and the other 2900, have been isolated from the fundus of the stomach and the intestine, where they appear to originate, respectively, from the AL cells and the EG cells. The general terms "gut glucagon" or *"enteroglucagon"* are used to describe them. It seems likely that the type with the higher molecular weight is identical to pancreatic glucagon. The smaller form has been even less characterized and it has been called glucagon-like immunoreactivity (GLI). These hormones are released during feeding, in response to the presence of glucose, and like pancreatic glucagon, increase glycogenolysis in the liver and release insulin. It has been suggested that enteroglucagon may be an important initiator of the "feeding response" that follows the intake of food.

Gastric inhibitory peptide, or GIP, was isolated by Brown and Pederson in 1970 (see Brown et al., 1975) from CCK-enriched extracts of the duodenum. Its most interesting activity is an ability to inhibit gastric acid and pepsin secretion, which are reminiscent of the activities of an early elusive putative hormone called *enterogastrone.* The presence of such a hormone was suggested on physiological grounds by Kosaka and Lim in 1930. It is not clear whether GIP is indeed the long-time elusive enterogastrone. GIP has 43 amino acid residues and a molecular weight of about 5000 (Figure 10.4). It is released into the circulation in response to the presence of glucose and digested fat. GIP can also stimulate the release of insulin. It probably functions as a hormone (Grossman, 1977).

Vasoactive intestinal peptide, or VIP, was isolated by Said and Mutt in 1972. It contains 28 amino acids (Figure 10.4) and, as its name indicates, it has a powerful vasodilatatory action on both the peripheral and splanchnic blood vessels. This relaxing effect on smooth muscle extends to the trachea, gallbladder, and stomach, but it is rapidly destroyed *in vivo,* especially during transit through the liver. It is considered unlikely that these effects are physiological, although, as described in Section 10.6C, it may be active in the Verner–Morrison syndrome (pancreatic cholera). At such elevated levels it also inhibits gastric acid output and stimulates the flow of pancreatic electrolyte and intestinal secretions. It also stimulates lipolysis, glycogenolysis, and the

release of insulin and glucagon (Jensen et al., 1978b).

VIP and a series of fragments of the molecule have been synthesized (Bodanszky, Klausner, and Said, 1973). Only the intact molecule has significant biological activity, although the fragments do display some actions and their potency tends to increase as the C-terminal portion of the molecule is elongated.

VIP has recently also been identified in mast cell (Cutz et al., 1978) and in human brain and nerve fibers of the autonomic nervous system (Bryant et al., 1976). It is therefore possible that it acts as a local hormone and neurotransmitter, but whether these are its only roles or if it has a dual function, also as a gastrointestinal hormone, is not known.

10.4. Other putative hormones in the gastrointestinal tract

(For references see Andersson, 1973; Rayford, Miller, and Thompson, 1976.)

Motilin increases the contraction of the stomach, but gastric emptying is delayed. It also increases intestinal motility. It has been isolated from a crude commercial CCK preparation and is a peptide containing 22 amino acid residues (Figure 10.5). An immunoassay indicates that its levels are highest in the blood of man during fasting. In man, it is released in response to acid conditions and fat in the duodenum. Its physiological role is uncertain (see Editorial, 1977j).

Bulbogastrone has not been isolated chemically, but its presence is suggested by the observation that the acidification of the duodenal bulb results in a decreased gastric acid secretion. Its existence is considered to be equivocal and the responses observed could be due to a nervous reflex.

Chymodenin is an interesting substance, as it appears to specifically increase pancreatic chymotrypsin secretion. This observation suggests the possibility that a number of specific hormones, each able to separately influence the secretion of a certain enzyme, may exist. Chymodenin has been isolated from duodenal mucosal extracts and is a polypeptide with a molecular weight of about 5000.

Enterooxyntin may be released from the small intestine (hence it is also called intestinal-phase hormone) in response to the presence of protein, and it stimulates the secretion of acid by the oxyntic cells. The evidence for its presence is principally physiological, and it has not been chemically isolated.

Somatostatin was first isolated from the hypothalamus [see Section 3E(5)]. While its initial identification was the result of the demonstration of its ability to inhibit release of pituitary growth hormone, it has subsequently been shown to have a

Motilin:

```
1                                           10                                    20        22
Phe -Val -Pro -Ile -Phe -Thr -Tyr - Gly -Glu -Leu -Gln -Arg -Met -Gln - Glu -Lys -Glu -Arg -Asn -Lys -Gly -Gln
```

Fig. 10.5 Amino acid sequence of motilin. It contains 22 amino acids.

similar effect on the release of other hormones, including TSH, insulin, glucagon, parathyroid hormone, and calcitonin. In the gut (Rayford, Miller, and Thompson, 1976; Bloom, 1976) it not only reduces the release of gastrin but also the secretion of pepsin and acid from the gastric mucosa. It can block release of gastrin in response to stimulation by calcium, acetylcholine, and parathyroid hormone (Bolman, Cooper, and Wells, 1978).

Somatostatin thus has a very powerful antacid effect, even in patients suffering from the Zollinger–Ellison syndrome. Somatostatin also inhibits secretin and CCK (Schlegel et al., 1977) release and the secretion from the exocrine pancreas, as well as the zymic and cholecystokinetic effects of CCK. This peptide hormone has 14 amino acid residues and has been identified in other tissues, apart from the hypothalamus, and is present in high concentrations in the pancreatic D cells and a similar cell type in the upper gastrointestinal tract, including the stomach (Polak et al., 1975). It thus appears to have a general inhibitory effect on many secretory processes, and it may exert this effect by blocking release from secretory storage granules (Schofield, Mira, and Orci, 1974), possibly by inhibiting a membrane-mediated Ca transport (Fujimoto and Ensinck, 1976). In the stomach (Gustavsson and Lundquist, 1978) and pancreas its role may be that of a local hormone, a paracrine action, although as demonstrated by its effectiveness when administered parenterally, it could also be acting systemically like an endocrine hormone.

A detailed study has been made of the structure–activity relationships of somatostatin with pentagastrin-stimulated gastric acid secretion in cats (Brown et al., 1978). The linear somatostatin appeared to be less effective (Figure 10.6) in inhibiting the response than the cyclic form of the peptide. Deletion of the side chain at the N-terminus, as in the *N*-acetyl and *N*-benzoyl analogues, had little effect on the molecule's biological activity. Substitution of D-phenylalanine for the L-isomer in positions 6 and 7 caused a drastic decline in activity, probably reflecting a conformational change in the structure of the molecule. On the other hand, replacement of L-tryptophan by its D-isomer in position 8 increased activity. Somatostatin can be inactivated by cleavage at positions 8 and 9, and while it is possible that this reaction is slowed by this substitution, the results suggested that this is not the main reason for the increase in activity. Substitution of D- for L-cystine at position 14 increased

activity. The effects of the various changes on the molecule's activity may be additive. It can thus be seen that [Ala^2-D-Trp^8-D-Cys^{14}]somatostatin is more than 4 times as active in inhibiting gastric acid secretion as the native cyclic somatostatin.

Competitive antagonists of the action of somatostatin are of theoretical and possibly even practical interest. The analogues with low activity, however, failed to reduce the response to cyclic somatostatin. This observation suggests that their lack of agonistic effects results from a low affinity for their receptors. Thus, positions 6, 7, 10, and 12 may be important in determining the binding of the hormone to its receptors. Such effects could be direct or be due to an influence on the structural conformation of the molecule.

In 1938, Sandweiss identified a substance in human urine which could reduce gastric acid secretion when injected into dogs (Sandweiss, Saltzstein, and Farbman, 1938). The material was subsequently called *urogastrone* (Gray et al., 1940), but its site of origin and precise nature was elusive for more than 30 years. Sandweiss (1945) also reported that it promoted the healing of gastric ulcers by stimulating the growth of the epithelial cells of the gastrointestinal mucosa. Urogastrone has recently been prepared in pure form (Gregory, 1975) and has been shown to be a single-chain polypeptide containing 53 amino acids (Figure 10.7). Fluorescent-labelled antibodies have been prepared and used to trace its site of origin to the cells of Brunner's glands in the upper part of the duodenum (Elder et al., 1978). Urogastrone may thus be a gastrointestinal hormone and is another candidate for identity with enterogastrone. It was especially interesting to observe that the structure of urogastrone is remarkably similar to that of the *epidermal growth factor* isolated from the submaxillary gland of adult male mice (Savage, Inagami, and Cohen, 1972) and human urine (Cohen and Carpenter, 1975). Indeed, the latter appears to be identical to urogastrone (Gregory, 1975). Epidermal growth factor (EGF) is a powerful mitogen which can stimulate the growth of various epithelial tissues *in vitro* and *in vivo*. The action of urogastrone may thus not be confined to an action on gastric acid secretion, but, as originally proposed by Sandweiss, it may also have a role in promoting the growth of the epithelial cells lining the gut.

Gregory (1975) has speculated that urogastrone may have a use in the treatment of duodenal ulcer. The effects of urogastrone on gastric acid secretion have been tested in man (Elder et al., 1975). It was

Somatostatin analogue	Structure	Gastric acid secretion *in vivo*
	1 2 3 4 5 6 7 8 9 10 11 12 13 14	
Cyclic	Ala-Gly-Cys-Lys-Asn-Phe-Phe-Trp-Lys-Thr-Phe-Thr-Ser-Cys |——————— S – S ———————|	1.00
Linear	——————————————————————————	0.49
D-Cys^3	——D-Cys——————————————	0.75
Thr^5	—————— Thr ——————————	1.3
D-Phe^6	——————— D-Phe ———————	0.06^b
D-Phe^7	——————— D-Phe ———————	a
D-Trp^8	——————— D-Trp ———————	1.65
D-Thr^{10}	——————— D-Thr ———————	a
D-Thr^{12}	——————— D-Thr ———	a
D-Cys^{14}	——————————————— D-Cys	1.59^b
D-Ala^2-D-Trp^8	-D-Ala ——————— D-Trp ———————	3.31^b
Ala^2-D-Cys^{14}	— Ala —————————————— D-Cys	1.93^b
Ala^2-D-Trp^8-D-Cys^{14}	— Ala ——————— D-Trp ———————— D-Cys	4.30^b
D-Trp^8-D-Cys^{14}	——————— D-Trp ———————— D-Cys	2.87^b
D-Phe^6-D-Trp^8	——————— D-Phe — D-Trp ———————	a
N-acetyl	acetyl ——————————————	1.32
N-benzoyl	benzoyl ——————————————	1.07
5-F-Trp^8	——————— 5-F-Trp ———————	1.39
< Glu(Gly)$_2$-Trp^8	——————— D-Trp ———————	2.48^a

Fig. 10.6. Structure–activity relationships of some analogues of somatostatin. The response measured was inhibition of pentagastrin-stimulated gastric acid secretion in cats. The activities are relative to that of cyclic somatostatin (which is 1.0). ^{a} No significant inhibition at the dose range tested. ^{b} Potency significantly different from that of cyclic somatostatin. (Based on Brown et al., 1978.)

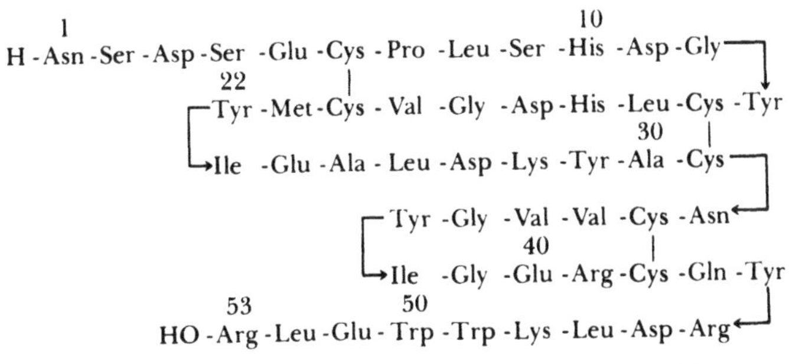

```
 1                                          10
H -Asn -Ser -Asp -Ser -Glu -Cys -Pro -Leu -Ser -His -Asp -Gly ┐
            22                   |                             ↓
          ┌ Tyr -Met -Cys -Val -Gly -Asp -His -Leu -Cys -Tyr
                            |                    30    |
          └→ Ile -Glu -Ala -Leu -Asp -Lys -Tyr -Ala -Cys ┐
                     ┌ Tyr -Gly -Val -Val -Cys -Asn ←┘
                                 40             |
                     └→ Ile -Gly -Glu -Arg -Cys -Gln -Tyr ┐
       53              50
      HO -Arg -Leu -Glu -Trp -Trp -Lys -Leu -Asp -Arg ←┘
```

Fig. 10.7. Amino acid sequence of β-urogastrone. (From Gregory, 1975.)

administered i.v. to a group of 12 volunteers and was found to reduce gastric acid secretion stimulated by pentagastrin, histamine, or insulin. Apart from headache, no adverse effects were observed in this limited "trial."

The secretion of gastric acid by the oxyntic cells can be inhibited by *prostaglandins,* especially those of the E series (see Way and Durbin, 1969; Nezamis, Robert, and Stowe, 1971; Robert et al., 1976). This effect has been observed *in vitro* on the amphibian gastric mucosa and *in vivo* in dogs, rats, and man. The acid content and the volume of the secretions are reduced and there is also an inhibition of secretion of pepsin. Prostaglandin E_1 can block the effects of stimulation of the gastric secretions in response to most stimuli, including pentagastrin, histamine, and vagal stimulation. There has been considerable speculation regarding the possible mechanism of action of prostaglandins in the gastric mucosa, and these include a reduction in the blood flow and an inhibition of release of gastrin. They have, however, been shown to act directly on the oxyntic cells. Prostaglandins of the E series can be formed by the gastric mucosa. The prostaglandins provide potential drugs for therapeutic use in modifying the process of gastric secretion in man [see Section 10.6A(b)].

10.5. Mechanisms of action of the gastrointestinal hormones

There is only limited information about how gastrointestinal hormones exert their actions, which is perhaps not surprising considering their number and the multitude of different effects that they have. The information that is available has resulted largely from studies of other peptide hormones that influence the cellular concentrations of "second messengers" such as cyclic AMP and Ca^{2+}. The effects of prostaglandins on the gut and their possible general function as local hormones that regulate cellular processes have also contributed. Even earlier information suggested that histamine could be a "final common mediator" of the hormonal effects on gastric acid secretion, but this has been strongly contested (see Johnson, 1977; Debas, 1977).

The mechanisms of action of the gastrointestinal hormones can be viewed at three levels: (a) the hormone–receptor interaction; (b) the effector response, such as a change in muscle tone or the secretion of acid, electrolytes, or enzymes; and (c) the coupling of these two processes through the mediation of a second messenger such as Ca^{2+} or cyclic AMP.

Few specific data are available about the nature of the interactions of the hormones with their receptors, although, as described earlier, the importance of certain of their chemical structural groups

has been assessed by the examination of the activities of fragments and analogues of the hormones.

The gastrointestinal hormones probably have a wider range of effects on more target tissues than any other group of hormones, and this property has led to some speculation about the possible common nature of the receptors. For instance, the gastric oxyntic cell can be stimulated by gastrin and CCK as well as by acetylcholine (released from the vagus) and histamine. Its secretion can be inhibited by secretin, prostaglandins, somatostatin, GIP, urogastrone, atropine, and the antihistaminic drugs, which "block" histamine's response on its H_2 receptors. The interactions of these excitants and blocking drugs can be rather complex.

To bring some order out of this potential chaos, and account for the pharmacological types of interactions observed in the oxyntic cell, Grossman (Grossman, 1970; Grossman and Konturek, 1974) has presented two interesting related hypotheses. He has suggested that the effects of some drugs and hormones on the oxyntic cell could be accounted for by their interaction with a single receptor with two active sites. In addition, other types of receptors could be present which interact with each other and the gastrin receptor. The combination of an excitant with one type of receptor may alter the properties of the other receptors.

Gastrin, CCK, secretin, and even glucagon could be acting on a single receptor on the oxyntic cell that possesses two active sites. Since gastrin and CCK have similar structures, they may be acting at the same receptor site (*site 1*). It is then possible to account for the antagonism that is observed when these two hormones are present together, as they have different intrinsic activities (or efficacies). (On the oxyntic cell, the ratio of the intrinsic activity of gastrin/CCK is 1 : 0.2.) Thus, each hormone alone is effective, but the maximal response to CCK will be less. At levels that produce submaximal responses, their stimulatory effects will be additive, but at maximally effective doses the less active CCK will tend to displace gastrin and so will competitively antagonize its action. Gastrin and CCK have a similar relationship on a number of other target sites, including the lower esophageal sphincter. This ring of smooth muscle prevents the reflux of the contents of the stomach into the esophagus. Gastrin is thought to normally maintain its tone and hence a resistance to high pressure. If this mechanism fails, however, a reflux of gastric acid and "heartburn" occur. This effect may be seen following a fatty meal, when it is thought that released CCK may compete with gastrin for a common receptor site, which results, as its intrinsic activity is less, in a relaxation of the sphincter muscle.

Secretin and gastric-inhibitory peptide (GIP) reduce acid secretion and are thought to act on another part of the oxyntic cell receptor called *site 2*. From here they could oppose, noncompetitively, the actions of gastrin and CCK.

Other distinct types of receptors for histamine and acetylcholine may also be present on the oxyntic cell, and Grossman and Konturek suggested that the blockade of one of these, for instance the histamine H_2-receptor by cimetidine or the cholinergic receptor by atropine, may change the properties of the gastrin receptor so that it responds less readily to the hormone. This antagonism could be due to a reduction in the receptor's affinity for the gastrin molecule or a reduced responsiveness of the cell to the effects of the hormone–receptor interaction. Prostaglandins (E_1, E_2, and A_1) also inhibit the effects of gastrin, histamine, and cholinergic stimuli on acid secretion, apparently also by a direct action on the oxyntic cell, and these could occupy an additional receptor site.

Histamine is a strong stimulant of gastric acid secretion which, as described earlier, resulted quite early in some uncertainty as to whether it was synonymous with gastrin. McIntosh in 1938 suggested that *histamine* may be the *final common mediator* or transmitter for the various stimulants, including nerves and hormones, of gastric acid secretion. This theory was also satisfactory to many, as it provided a long-sought-after physiological role for histamine. Histamine is found in the gas-

tric mucosa, but it was at that time disappointing to observe that the available histamine antagonist drugs did not block acid secretion. Subsequently, a new class of histamine "blockers," including *burinamide, metiamide*, and *cimetidine* (Figure 10.8), have been synthesized which prevent histamine-stimulated gastric acid secretion and have been classified as histamine H_2-receptor antagonists (Black, et al., 1972). The latter two drugs block the effects of all stimuli on acid secretion, including that of the vagus, gastrin and histamine, which would be consistent with histamine's action as a common transmitter substance. Such a role for histamine has been advocated by many (see, e.g., Code, 1965) but not by all.

Some (Gardner et al., 1978) have found that under certain conditions the effect of histamine on gastric acid secretion can be blocked by anticholinergic drugs such as atropine. Such an observation is difficult to reconcile with the amines' proposed role as a final common mediator of the response. (However, it should be pointed out that in the hands of many, atropine has not been found to block the actions of histamine.) Johnson (1978) has pointed out that while burinamide, a member of the family of histamine H_2 antagonists, can block the effects of histamine on acid secretion, it does not, in contrast to cimetidine, inhibit the effects of vagal stimulation. Thus, it seems possible

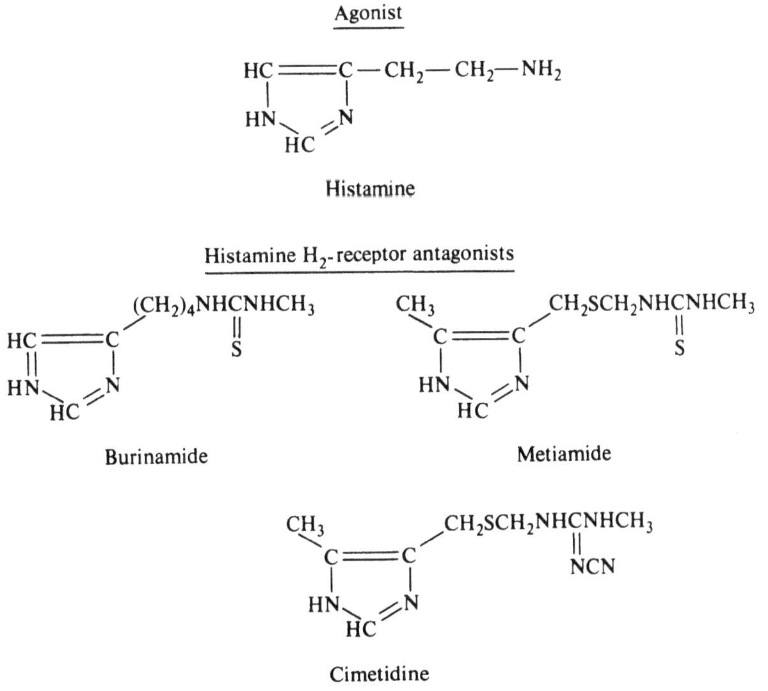

Fig. 10.8. Chemical structures of histamine and some histamine H_2-receptor antagonists that decrease gastric acid secretion.

that the cholinergic effects on gastric acid secretion can be dissociated from those related to histamine. The action of atropine described by Gardner could be consistent with acetylcholine being the common mediator, but the overall inhibitory effects of cimetidine, in turn, appear to exclude this possibility. There have, however, been questions raised regarding the specificity of the various types of antagonists at the concentrations used. The definitive statement about the nature of the "final common mediator" for neural and hormonal stimulation of gastric acid secretion, it appears, has not yet been made. Grossman and Konturek (1974) have provided an alternative explanation of the general action of the histamine H_2-receptor antagonists, suggesting that their interaction with one type of receptor on the oxyntic cell may influence the responsiveness of other adjacent types of receptors (see earlier in this section).

There is a considerable amount of information available about the nature of the *stimulus-secretion coupling* process for the various actions of the gastrointestinal hormones. Calcium and cyclic nucleotides play such roles in other types of tissue responses, so it is not surprising to find that they may also have a role with respect to the gut hormones.

Harris in 1965 (Alonso, Rynes, and Harris, 1965) proposed that the *adenyl cyclase system* may be involved in mediating the effects of secretatogues, including gastrin and histamine, on acid secretion by the oxyntic cells. Theophylline, which inhibits phosphodiesterase and so promotes the accumulation of cyclic AMP, stimulates acid secretion, while imidazole, which activates this enzyme, inhibits secretion. Exogenous cyclic AMP has also been shown, *in vitro*, to stimulate acid secretion. More direct evidence which consistently demonstrates an appropriate increase in tissue cyclic AMP levels following stimulation has been more difficult to obtain. Kimberg reviewed the subject in 1974 and then concluded that strong evidence regarding such a role for cyclic AMP was available only for the amphibian gastric mucosa. This view has even been challenged recently by Chew and Hersey (1978), who even failed to find a direct correlation between acid secretion and tissue cyclic AMP in the bullfrog gastric mucosa. In mammals, there have been numerous, often conflicting, reports regarding changes in gastric mucosal cyclic AMP levels in response to stimulation. There may be interspecific variation. If this is really the explanation, however, it could make it difficult to accept such a mechanism as a truly basic one. Apparent inconsistencies may arise due to experiments being carried out under different conditions, such as the presence or absence of phosphodiesterase and histamine H_2-receptor blocking drugs and the utilization of different time courses for making the measurements. In addition, it will be recalled that the stomach is quite a heterogeneous tissue, containing several types of cells which may obscure each other's actions and may even each respond differently. The gastric mucosa can be separated from its underlying muscle and connective tissue, but the relative purity of such preparations may vary. More recently, the mucosal cells have been disaggregated and the oxyntic cells have been separated (Major and Scholes, 1978). Isolated mucosal cell membrane preparations have also been made which contain adenyl cyclase activity (Thompson, Rosenfeld, and Jacobson, 1977).

Increases in gastric tissue cyclic AMP levels have been described following stimulation by pentagastrin and histamine in the rat, rabbit, guinea pig, and dog (see Kimberg, 1974; Case, 1976; Scholes et al., 1976; Keblad et al., 1978). However, as described above, there are also in many instances reports which fail to confirm these results. For instance, using membrane adenyl cyclase preparations made from gastric mucosal cells of rat, dog, rabbit, and guinea pig, Thompson, Rosenfeld, and Jacobson (1977) found that while they could all be stimulated by fluoride, only the guinea pig preparation was responsive to histamine (none responded to pentagastrin).

Scholes and his collaborators (Scholes et al., 1976; Major and Scholes, 1978) have used preparations of isolated oxyntic cells made from dog stomach to study the nature of the effects of histamine and prostaglandins on acid secretion. A histamine-sensitive adenyl cyclase was identified in these preparations. The H_2 agonist 4-methylhistamine was very active but 2-methylhistamine, an H_1 agonist, had only a weak effect. The accumulation of cyclic AMP in response to stimulation could be competitively blocked by the H_2-receptor antagonist drugs metiamide and burinamide. The H_1-receptor antagonists mepyramine and chlorpheniramine were, as predicted, ineffective. The histamine-sensitive adenyl cyclase system was found to be confined to oxyntic cell "rich" preparations of the gastric mucosal cells.

Prostaglandins, as described earlier, inhibit the effects of nearly all stimulants of gastric acid secretion, but not the response to exogenous cyclic AMP. They thus appear to be acting in quite a general manner on the oxyntic cell at a level which could be quite close to the acid secreting mechanism itself. In other tissues, prostaglandins have been shown to alter cyclic AMP levels, and in whole preparations of mixed gastric mucosal cells they have been observed, *like* gastrin and histamine, to increase tissue cyclic AMP levels (see Kimberg, 1974). Such effects even appear to be additive, and it is difficult to reconcile them with a prostaglandin adenyl cyclase-dependent antagonism. Using separated preparations of mucosal cells, Major and Scholes (1978) have, however, shown that the increase in cyclic AMP is only seen in the nonoxyntic cell fraction. The prostaglandins *reduce* cyclic AMP accumulation in response to histamine in oxyntic cell "rich" preparations. Therefore, it appears that the prostaglandins effects on acid secretion may be mediated by an inhibition of the adenyl cyclase system in the oxyntic cells. The effect is, notably, not competitive to that of histamine and is thus consistent with the presence of a receptor site quite distinct from that of the H_2 receptor.

The secretion of *fluid and electrolytes* by the pancreas in response to secretin and VIP appears to be mediated by cyclic AMP (Domschke et al., 1975; Case, 1976).

It does not seem that cyclic AMP is involved in the *enzyme-secreting response* of the pancreatic acinar cells to cholecystokinin (see Case, 1978). *Calcium*, however, seems to be involved in the process of stimulus-secretion coupling of this hormone on enzyme secretion by these cells (see Schreurs et al., 1976; Kondo and Schulz, 1976; Peterson and Ueda, 1976; Kanno, Saito, and Sato, 1977; Case, 1978). This response, and also that to acetylcholine, is dependent on the presence of external Ca^{2+} and can be imitated by the Ca ionophore A23187. It has, however, also been shown that the enzyme-secreting acinar cells become electrically depolarized and sodium needs to be present in the external solution (Dean and Matthews, 1972; Kanno, 1972; Matthews, Peterson, and Williams, 1973; Case and Clausen, 1973; Williams, 1975). Movements of Na across cell membranes and changes in intracellular Ca^{2+} levels have been associated in a number of different types of tissues. Kanno and his collaborators (1977) have summarized the two current hypotheses which have been proposed to account for this relationship in pancreatic acinar cells stimulated by cholecystokinin.

a. The hormone, after combining with its receptor, which is probably present on the cell membrane, may initiate an influx of Na^+ into the cell which results in a release of Ca^{2+} from intracellular storage sites (see Case, 1978) such as the endoplasmic reticulum and mitochondria. The Ca^{2+} then mediates the release of pancreatic enzymes, probably by exocytosis, from their storage granules.

b. This hypothesis differs from the other principally with respect to the proposed mechanism by which intracellular Ca^{2+} levels are increased. It is postulated (Kanno, Saito, and Sato, 1977) that this divalent ion mainly comes from outside the cell and is transported across the plasma membrane linked to a Na-dependent "carrier" or "complex." The principal evidence for such a "cooperative effect" is that the action of the Ca ionophore A23187 on Ca^{2+} influx is dependent on the presence of external Na^+. It has been calculated that the complex carries four Na molecules and one Ca. The

influx of the carrier appears to be a diffusional process which in order to be maintained requires the integrity of the cell Na-pump mechanism.

Similar changes in intracellular Ca^{2+} may mediate the contractile effects that these hormones often exert on *smooth muscle* (see Case, 1976). Conversely, reduced Ca^{2+} concentrations, possibly associated with increased cyclic AMP, could result in their effects in relaxing smooth muscle. There is, however, little information available specifically related to the mechanism of the effects of the gastrointestinal hormones on smooth muscle tone.

The *trophic effect of gastrin* on the growth of the gastric mucosa is distinct from its action on secretion and involves a different mechanism. The administration of gastrin has been shown to stimulate protein synthesis in the stomach as well as the pancreas and small intestine, and this is accompanied by increases in DNA and messenger RNA (see Walsh and Grossman, 1975; Enochs and Johnson, 1977; Johnson, 1977). Thus, gastrin appears to exert its trophic effects via nuclear transcription of DNA and the formation of messenger RNA. Hormones that stimulate glandular secretion, such as the action of CCK on the pancreas, resulting in increased enzyme levels, may have a similar type of action. It is, however, difficult to disentangle a primary effect on protein synthesis from one that is superimposed secondarily on a stimulation of secretion.

10.6. Gastrointestinal hormones and disease: pharmacological considerations

Unequivocal evidence of the role of gastrointestinal hormones in human disease has only become available recently when, as a result of the availability of pure hormones, their measurement in the plasma, using radioimmunoassays, has become possible (see Bloom, 1974).

A. Peptic ulcer

Peptic ulcer has a high incidence, and in the United States about 4 million people develop this disease each year. The possibility that it may result from excess gastric acid due to a high rate of gastrin secretion or an abnormal sensitivity to this hormone has been extensively explored but without finding a clear relationship. There has also been speculation that reflux esophagitis due to inadequate constriction of the lower esophageal sphincter may be the result of insufficient gastrin; but again the evidence is equivocal. It is, however, becoming increasingly apparent that the control of gastrointestinal function involves many more hormones than it has yet been possible to study exhaustively, so that the investigations as to their role in such diseases have certainly not been closed.

Although gastrin has not been specifically related to peptic ulcers, its normal physiological role in the control of gastric acid secretion may be considered in the treatment of the disease. Current therapy of peptic ulcer has several aims.

a. The neutralization of gastric acid by the use of antacids is used in almost all instances (Morrissey and Barreras, 1974). Such treatment alleviates the pain, although it remains controversial whether the healing process is promoted or even if the effect on pain is greater than that of a placebo. Calcium carbonate has been widely used as an antacid, but it has now been shown (Section 10.1B) that it promotes gastrin release, so it is not an ideal choice. As it also carries the hazards of possible hypercalcemia, especially if renal function is impaired, it is no longer an antacid of "first choice." Similarly, the use of milk has recently been shown to increase gastric acid secretion, and it has been suggested that its calcium and protein content may be the cause of the problem (Ippoliti, Maxwell, and Isenberg, 1976).

b. The blockade of gastric acid secretion can be brought about pharmacologically in a number of ways, although, with the probable exception of cimetidine, an ideal practical method has not yet been discovered or agreed upon. Blockade of vagal stimulation with atropine and related drugs has not really been successful, apparently due to the high dose threshold needed for an effect.

The histamine H_2-receptor blocking drugs *metiamide* and *cimetidine* (Figure 10.8) have recently been undergoing extensive clinical trials (see Finkelstein and Isselbacher, 1978). These drugs reduce resting acid secretion in patients with duodenal and gastric ulcer, block the effects of injected pentagastrin, and promote healing of the erosions. Cimetidine has been approved for clinical short-term use in the United Kingdom and the United States.

The earliest clinical trials (Haggie et al., 1976) were performed using metiamide, but it was not approved for clinical use. This drug was found to be associated with the occurrence of seven cases of agranulocytosis. This toxic effect was related to the presence of a thiourea moiety in the molecule, but when this was replaced by a cyanoguanidine, to produce cimetidine, this blood disorder did not occur. There is no doubt that the short-term use of cimetidine (up to about 8 weeks) is associated with an increased rate of healing of duodenal ulcer (Bodemar and Walan, 1976; Gray et al., 1977; Winship, 1978) and, most probably, gastric ulcer (Frost et al., 1977; Freston, 1978). When used for short-term therapy, cimetidine has not been shown to exhibit any serious toxic actions (Kruss and Littman, 1978). A total of 16 patients have been reported to develop gynecomastia; six of these had undergone longer-term treatment for the Zollinger–Ellison syndrome. There have been an increasing number of reports (see The Medical Letter, 1978) of the effects of cimetidine on the central nervous system, including confusion and sometimes slurred speech, delirium, and hallucinations. These side effects were usually seen in elderly patients receiving high doses of the drug, who often had an accompanying impairment of renal function. There have been isolated reports of fever, bradycardia, and diarrhea, but it is not always clear if the drug is directly implicated. Withdrawal of cimetidine unfortunately appears to be associated with a frequent recurrence of ulcers (Gudmand-Høyer et al., 1978; Editorial, 1978b). It therefore appears to be desirable to administer cimetidine in prophylactic or maintenance doses for extended periods of time. In a clinical trial involving 93 patients, cimetidine was used (in full doses) for up to a year and prevented the reoccurrence of peptic ulcers in most, but not all, patients (Cargill et al., 1978).

There are several other possible methods for blocking gastrin release and its effects on the oxyntic cell (see Burks, 1976). As described earlier, *somatostatin* inhibits the release of gastrin and secretion of acid and pepsin by the gastric mucosa. It has been described as an extremely effective antacid. As this "hormone" is a peptide, it needs to be given parenterally and has a relatively short half-life in the circulation. Analogues with a longer duration of action could overcome this problem. In one trial, intravenous infusions of somatostatin, over a period of 48 to 120 hours, have been found to be effective in the treatment of hemorrhage associated with peptic ulcer disease (Kayasseh et al., 1978). *Gastric inhibitory peptide* (GIP) is another candidate in the search for such therapeutically useful drugs. The possibility of synthesizing structural analogues of gastrin which act as competitive antagonists to the hormone has been considered but appears to have yielded no practical results. Some *prostaglandins,* including PGE_1, PGE_2, and PGA_1, suppress acid secretion and possibly also promote ulcer healing by a direct "cytoprotective" action in experimental animals and man (Konturek et al., 1976; Ippoliti et al., 1976; Gibiński et al., 1977; Carmichael, Nelson, and Russell, 1978). The methylation of such compounds at their 15 or 16 positions decreases their rates of degradation, so that they can then be administered orally. Side effects, related to the ability of the prostaglandins to alter the contractility of smooth muscle, such as diarrhea, vomiting, and abdominal cramps, are less of a problem with these methylated analogues.

c. The healing of peptic ulcers may be promoted by two types of steroidal compounds, estrogens and carbenoxolone.

1. Estrogens. Women usually (at least prior to Women's Liberation) have a lower incidence of peptic ulcer than men, which probably led to the use of estrogens in the treatment of this condition in the male. In one therapeutic trial, diethylstilbestrol was administered for 6 months (Truelove, 1960). The healing of duodenal ulcers (it is ineffective in gastric ulcer patients) appeared to be promoted and recurrence in a 5-year follow-up period was reduced. Diethylstilbestrol does not appear to influence gastric acid or pepsin secretion but promotes mucin formation and healing. The doses that are necessary, however, have strong feminizing effects, which generally make them unacceptable in the male. Such drugs have other side effects, including an increased risk of cardiovascular diseases. Their effects on the healing of duodenal ulcers thus appear to be more of academic interest than clinical use. It has been suggested that structural analogues with more specific actions could be developed, but they have not been forthcoming.

It is interesting that therapy with another type of steroid, the *adrenocorticosteroids,* which are administered for their antiinflammatory effects, has often been reported to produce peptic ulceration and the activation of dormant ulcers. Such an effect appears to be common to a lot of drugs with antiinflammatory actions. A recent careful analysis of hospital data, taken from 26 double-blind studies and 16 non-double blind investigations, however, completely failed to show a relationship between corticosteroid therapy and the incidence of peptic ulcer (Conn and Blitzer, 1976). It was concluded that "The steroid-ulcer myth arose from anecdotal reports. . . ." These conclusions need to be confirmed.

2. Carbenoxolone. Licorice, an extract from the plant *Glycyrrhiza glabra,* has a well-established use in folk medicine for the treatment of indigestion. Its effects were found to be associated with glycyrrhetic acid. The sodium salt of a succinic acid ester of this aglycone, carbenoxolone, has been synthesized and shown to have a healing effect in the treatment of gastric ulcers and possibly even duodenal ulcer. It has been used therapeutically for more than 10 years in the United Kingdom and is at present undergoing a duplicate trial in the United States. There seems to be little doubt about its therapeutic effectiveness, which appears to be related to an increased mucin formation, resulting in a decreased backflux of acid across the gastric mucosal epithelium and a more extended time of survival of these latter cells (see Lewis, 1974; Langman, 1976).

The side effects of carbenoxolone can be a problem and are of rather special endocrine interest. They include sodium retention and a loss of potassium in the urine, which can produce hypokalemia. The urinary Na retention can result in edema and hypertension. These effects are reminiscent of hyperaldosteronism and can be prevented by spironolactone, which is a specific antagonist of aldosterone. It is, however, unfortunate, but rather interesting, that spironolactone also prevents the ulcer-healing effects of carbenoxolone. Diuretics and potassium supplements or K-sparing diuretics, such as amiloride and triampterine, have been used to combat the side effects. Carbenoxolone has a steroid-like nucleus (Figure 10.9) and may thus have an aldosterone-like effect, which could be competing with binding sites on plasma proteins to increase the effective levels of the hormone, or it could even be acting on the kidney itself. The mechanism of the inhibitory effect of spironolactone on the ulcer healing action of carbenoxolone is unknown, but apparently it is not due to a direct antagonism in the gastric mucosa (Rees et al., 1975).

B. Zollinger–Ellison syndrome
This disease was first described in 1955 by the two surgeons for which it is named. The condition (see Walsh and Grossman, 1975) has a variety of characteristic but not unique symptoms, the most notable being a high rate of gastric acid secretion (hyperchlorhydria) and a mucosal tissue hyperplasia, especially involving the oxyntic cells. These effects usually result in ulcers that do not respond to standard therapy. In 75 percent of patients, erosions are present in the upper part of the duodenum. Diarrhea is present in about 30 percent of cases, apparently reflecting the large volumes of acid that are delivered to the intestine. The incidence of the disease is difficult to evaluate, but it has been suggested that about 10 percent of people suffering from severe peptic ulcer conditions may suffer from it. The ready availability of a radioimmunoassay to measure serum gastrin levels has recently facilitated its diagnosis. Intravenous infusion of Ca^{2+} (*calcium infusion test*) produces an exaggerated secretion of acid and very high gastrin levels. About 75 percent of patients suffering from the disease have a high gastrin response to the i.v. injection of secretin (*secretin infusion test*). The latter

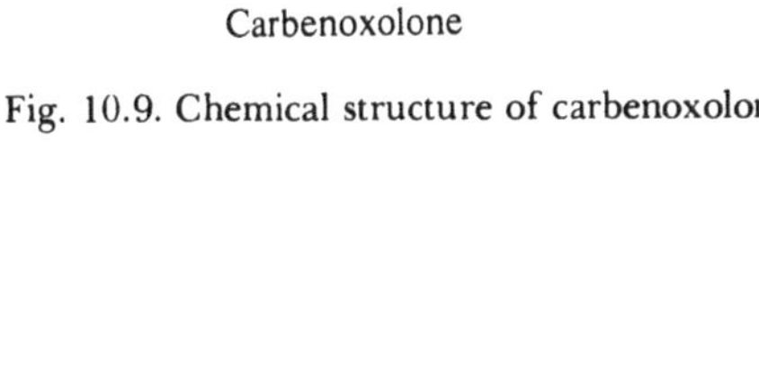

Fig. 10.9. Chemical structure of carbenoxolone.

hormone has little effect on ordinary ulcer patients, but those suffering from the Zollinger–Ellison syndrome show a rapid increase in serum gastrin in response to this provocation.

The Zollinger–Ellison syndrome is due to gastrin-secreting tumors, probably an APUDoma, usually in the pancreas, but sometimes they are present in the gastric antrum or the duodenal mucosa. They are called gastrinomas and are usually small, diffuse, and difficult to find; hence, their eradication is not usually feasible surgically, and while attempts have been made to eliminate them using chemotherapy with streptozocin, this has not been successful. Currently, treatment is removal of the stomach, which alleviates the principal problems and, as the tumors usually grow slowly, survival may be prolonged, for 1 to 5 years. Thus, at present the endocrine contribution is a diagnostic one, although there has been speculation about the potential use of gastrin antibodies or GIP to combat the disease.

An extensive survey (McCarthy, 1978) has shown that the histamine H_2-receptor blocking drug *cimetidine* may provide an alternative treatment to surgery in the Zollinger–Ellison syndrome. (Gastrectomy for this condition is associated with about a 20 percent mortality rate.) The symptoms of the disease are controlled, but the progression of the neoplasm is not retarded and is similar with the drug or surgery. The doses of cimetidine which are used are larger than those for peptic ulcer, but apart from six cases of gynecomastia, no serious side effects were observed when treatment was continued for a year or more.

Hypergastrinemia also occurs secondarily in gastritis when the antrum tissue is intact but secretion of gastric acid from the fundus is low. This lack of a negative-feedback inhibition results in hyperplasia of the G cells. This condition is commonly seen in cases of pernicious anemia. It also occurs in cancer of the body of the stomach if the antral tissue is relatively intact. Catecholamines released by pheochromocytoma cells result in a release of gastrin, and this effect can be prevented by the α-adrenergic blocking drug phenoxybenzamine.

C. Verner–Morrison syndrome

This disease is more descriptively called "pancreatic cholera" or "WDHA syndrome," which stands for water, diarrhea, hypokalemia, and achlorhydria. The diarrhea is unresponsive to the usual therapeutic measures and may amount to 10 to 20 liters/day. The effect is due to a flooding of the intestine with gastric, pancreatic, and possibly intestinal secretions. There is no deficiency in the intestine's ability to reabsorb electrolytes and fluids; they simply appear to exceed its normal capacity. The disorder has been discussed by Soergel (1975). The disease appears to result from an excess of one or more gastrointestinal hormones released from tumors usually present in the pancreas or bronchus. The precise nature of the excitant or excitants is uncertain, but GIP, which could account for the achlorhydria, and VIP are prime candidates. Very high serum levels of the latter have been detected, using a radioimmunoassay, but this of course does not exclude a multiple effect involving other substances. The tumors have been called VIPomas and probably represent another instance of an APUDoma. The Verner–Morrison syndrome can be readily diagnosed by a single fasting determination of serum VIP (Bloom, 1976). Removal of the tumor corrects the condition or, if this is not feasible, streptozocin may be effective, and this has been administered via the hepatic artery (Kahn et al., 1975). Corticosteroids, including prednisolone, are also effective in some, but not all, cases.

D. Ménétrier's disease (hypertrophic gastropathy)

Giant hypertrophy of the gastric mucosa, or Ménétrier's disease, is a rare condition (see Scharschmidt, 1977) in which there is a massive hypertrophy of the gastric mucosa. It is usually associated with a reduced secretion of gastric acid, a decline in motility, and a loss of protein from the stomach. The latter may lead to a hypoproteinemia. There is an increased risk of developing gastric carcinoma. The causes of this condition are unknown, but suggestions include excessive reflux of the contents of the intestine back into the stomach and an increased trophic action of gastrointestinal hormones, such as gastrin. Pharmacotherapeutic treatment has included the use of antacids and anticholinergic drugs but without notable success. Atropine, for instance, has been found to reduce the gastric protein loss but it is only effective when administered intravenously (Berenson, Sannella, and Freston, 1976). It was suggested that the histamine H_2-antagonist drugs, especially cimetidine, should be tested for the treatment of this disease. Potentially, antagonists of the gastrointestinal hormones could also repay investigation, especially those blocking the action of gastrin. In the meantime treatment is surgical, consisting of partial or total gastrectomy.

11

The renin–angiotensin system

One of the earliest descriptions of a putative hormone was made by R. Tigerstedt and P. G. Bergman in Sweden in 1898. They showed that a simple saline extract of the kidney, when injected into an anesthetized rabbit, elicited a considerable increase in blood pressure. They called this vasoactive material *renin*. Interest in the substance waned and it was not reinvestigated for another 35 years. In 1934, H. Goldblatt (Goldblatt et al., 1934) described a series of interesting observations in which he was able to produce experimental hypertension in dogs by constricting the renal artery of one kidney with a small silver clamp. These observations suggested that hypertension may be produced as a result of the release of a vasoactive material, reminiscent of renin, from the kidney. In 1940, the mechanism of action of this material was described simultaneously by two groups, that of E. Braun-Menendez (Braun-Menendez et al., 1940) in Argentina and that of I. A. Page (Page and Helmer, 1940) in the United States. They showed that the effect of renin was indirect and that it had an enzymic action on a plasma protein present in the α-2-globulin fraction. This resulted in the formation of a peptide, which was apparently the active material. Braun-Menendez called the latter *hypertensin,* while Page called it *angiotonin.* In an unusual instance of Pan-American cooperation, it was finally called *angiotensin.* Hypertension is a serious and increasingly common manifestation of human disease, so it was not surprising that these observations have resulted in a plethora of basic and clinical studies on the renin–angiotensin system.

Tigerstedt, who died about 1920, was an eminent physiologist, but his obituary contained no mention of his most famous discovery, renin. Nearly 80 years later we are still uncertain about the true roles of renin in normal physiological regulation and disease. Its effects can be mediated by several peptides – angiotensins I, II, and III –

which have diverse actions, not only on blood pressure, but also on other processes, such as the control of the secretion of aldosterone, the sensation of thirst, and possibly even some brain functions, such as learning. The existence of more than one form of angiotensin was first described by L. T. Skeggs and his collaborators (Skeggs et al., 1954). They have a quite remarkable propensity to elicit excitant effects on different tissues, so it is difficult to disentangle their putative physiological roles from their pharmacological actions.

Useful reviews of the subject include the volume *Angiotensin,* edited by Page and Bumpus (1974), and reviews by Oparil and Haber (1974a,b), Regoli, Park, and Rioux, (1974), Peart (1975, 1977), and Peach (1977).

11.1. Origins and nature of renin

Renin is a proteolytic enzyme, and renin-like activity has been identified at several sites in the body. They include, apart from the kidney, the uterus, the walls of blood vessels, the placenta, mouse salivary glands, and the brain (see Ganten et al., 1976). Apart from the mouse salivary gland, the kidney is by far the richest source of the enzyme. The exact relationships of the "renin" from these sites is often difficult to define, but in some instances it appears to constitute a series of isoenzymes (isorenins or pseudorenins). In the kidney the enzyme is formed in what appears to be modified smooth muscle cells in the afferent glomerular arteriole, adjacent to the glomerulus. These *juxtaglomerular cells* constitute a part of the *juxtaglomerular apparatus* (Figure 11.1), which makes up an integrated tissue complex concerned with regulation of the synthesis and release of renin. The juxtaglomerular apparatus in mammals, but not other vertebrates, also contains the *macula densa,* which is a thickened region of the distal renal

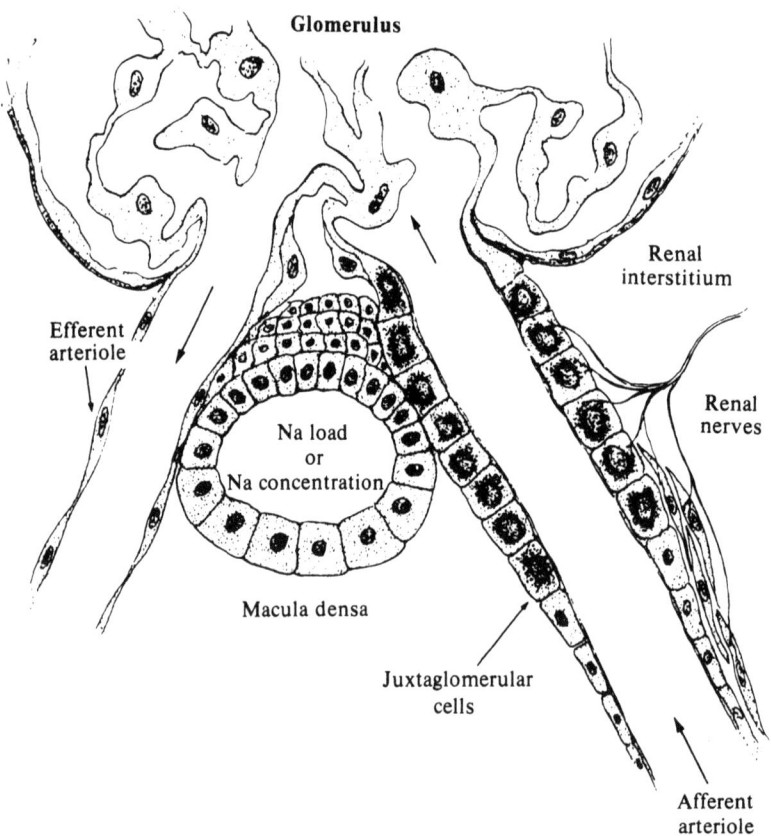

Glomerulus

Renal
interstitium

Efferent
arteriole

Renal
nerves

Na load
or
Na concentration

Macula densa

Juxtaglomerular
cells

Afferent
arteriole

Fig. 11.1. Juxtaglomerular apparatus of the kidney, showing its relationship to the afferent and efferent glomerular arterioles and the macula densa. The latter is a part of the distal renal tubule. The juxtaglomerular cells, which are in the afferent arteriole, are shown to contain granules (stainable by Bowie's method) which are thought to be the site of storage of renin. Both the juxtaglomerular cells and the smooth muscle cells are innervated by renal nerves. (From Davis, 1971. By permission of the American Heart Association, Inc.)

tubules, where it abuts onto a part of the glomerulus called the *polkissen* or *mesangium*.

The renin in the juxtaglomerular cells is associated with granules that can be characteristically stained (Bowie's granules). Several forms of renin (isorenins) have been isolated which have different molecular weights, the largest having been called such names as "prorenin" and "big renin." Two forms of renin have been isolated from mammalian kidneys. They are glycoproteins, with molecular weights of about 55,000 and 40,000, the smaller one being the more active in releasing angiotensin from its substrate in the plasma. The lower-molecular-weight form is apparently bound to an inhibitor protein, which if separated and re-added to the active form, will again block its effects (see Skeggs et al., 1977; Inagami and Murakami, 1977; Inagami et al., 1977). It has been suggested that the proteolytic enzyme kallikrein, which it appears can be associated with the juxtaglomerular apparatus,

may act as a "prorenin converting enzyme" and activate renin (Sealey et al., 1978).

Other forms of renin have been identified in the plasma (Skeggs et al., 1977; Atlas et al., 1977). The "renins" that have been found at other tissue sites may have roles to play in controlling local concentrations of the active peptides, but this possibility has not been proved.

11.2. Mechanism of action of renin

Renin does not have a direct excitant effect on peripheral tissues, but by its proteolytic action splits a decapeptide, angiotensin I, from a substrate molecule called *angiotensinogen* in the plasma. This is the result of the splitting of a leucyl–leucine bond that releases the decapeptide (Figure 11.2). The active site of renin contains a free carboxyl group and its action can be inhibited by pepstatin, which is an exogenous pepsin inhibitor (McKown, Workman,

Fig. 11.2. Formation and metabolism of the angiotensins from plasma "angiotensinogen" (R, an α-globulin). (From Peach, 1977.)

and Gregerman, 1974). Renin is thus classified as an acid protease. The plasma substrate, an α-globulin, angiotensinogen, is a glycoprotein formed in the liver. In man, it has a molecular weight of about 110,000 (see Skeggs et al., 1977). Its production is increased in pregnancy and under the influence of estrogens (including those present in oral contraceptives), glucocorticoids, and after nephrectomy (see Peach, 1977). It is decreased following adrenalectomy and hypophysectomy.

Angiotensin I itself usually has little activity but is converted to a more active octapeptide, angiotensin II. This reaction occurs under the influence of a *converting enzyme* (see Erdös, 1977), which is a peptidyl-dipeptidase that splits the histidine

and leucine residues from the C-terminus of angiotensin I. This enzyme is present at a number of tissue sites, especially the plasma membranes of the vascular endothelium of the lung and the kidney, and it is also present in plasma. It is a glycoprotein which contains zinc. Metal ions are needed as a cofactor in its action and it is activated by chloride. Angiotensin I is not a specific substrate for the converting enzyme. The enzyme is, for instance, synonymous with kininase II, which is responsible for the inactivation of kinins. This effect is of some theoretical and practical importance as, in contrast to angiotensin II, which has potent hypertensive effects, kinins have the opposite action and are very active hypotensive substances.

The activity of the converting enzyme can be

inhibited by EDTA, sulfydryl blocking compounds, and a number of peptides which can also act as its substrates. The prototype and best known such inhibitor of the converting enzyme is a nonapeptide which goes under the code name of *SQ 20881* (SQ = Squibb) or *teprotide*. This substance was originally identified in the venom of a poisonous South American snake *Bothrops jararaca* and was found to inhibit kininase II activity in plasma. It provided an important way for blocking the ultimate actions of renin. Other substances with a similar activity have been synthesized and may have a role in diagnosing and treating hypertensive diseases. This subject will be discussed later (Section 11.10).

Other endogenous proteolytic enzymes can act on the angiotensins to modify or destroy their activities. One notable change is the removal of the aspartic acid at the N-terminus to produce either des-Asp^1-angiotensin I or II. The latter heptapeptide has a much reduced activity on blood vessels but has been found (Blair-West et al., 1971; Freeman et al., 1977) to be as active as angiotensin II in promoting secretion of aldosterone. It has also been called *angiotensin III* and may possibly have a hormonal role, especially in regulating adrenocortical function.

11.3. Functions and actions

The angiotensins have ubiquitous effects in pharmacological situations, but their physiological roles are more restricted and some are still contentious. A summary is provided in Table 11.1. Angiotensin II has a powerful vasoconstrictor action, promotes sodium retention, principally by releasing aldosterone, and it can stimulate drinking. Its actions can thus be broadly characterized as the maintenance of the volume and hydrostatic pressure of the body fluids, notably that of the blood. This acton has been called the "vasoconstriction-volume hypothesis" (Laragh et al., 1977). It appears to do this in several quite different ways, some of which, such as its vasoconstrictor action, may only be manifested in rather unusual circumstances, such as hemorrhage or a severe depletion of the body's sodium.

A. Effects on vascular and other smooth muscle

The first effect of the renin-angiotensin system to be described was its ability to increase the blood pressure. On a molecular-weight basis it is more than 40 times as effective as norepinephrine and is considered to be the most active such substance known. This action principally reflects a constriction of arterioles in the peripheral vascular beds in the skin, splanchnic circulation, and the kidneys. It can also constrict the coronary vessels and reduce

Table 11.1. *Summary of some of the effects of angiotensin*

Increase in blood pressure (constricts arterioles)
Contracts smooth muscles of stomach, intestines, and
 uterus
Releases aldosterone
Releases adrenal medullary catecholamines
Releases antidiuretic hormone
Facilitates cholinergic and adrenergic transmission
Increases capillary permeability
Increases sodium transport (kidney, colon, and renal
 tubule)
Elicits drinking response
Inhibits learning behavior in rats

Note: It should be noted that the physiological significance of many of these has not been established.

the cerebral circulation, but it has little effect on the blood supply to skeletal muscles or the lungs. It has a powerful direct action on the vascular smooth muscle but can also exert indirect effects. Angiotensin II thus facilitates the activity of the sympathetic nervous system and can promote the release of epinephrine from the adrenal medulla. The principal site of this indirect effect appears to be in the area postramus of the fourth ventricle in the brain, but it has also been shown to have direct actions on the sympathetic ganglia and nerve terminals by promoting the synthesis, release, and inhibition of the re-uptake of norepinephrine.

One of the earliest effects of angiotensin to be reported was a brief antidiuretic action on the kidney. This response reflects changes in the renal blood supply, probably a constriction of the afferent glomerular arteriole. However, angiotensin has also been reported to stimulate release of ADH. It has been suggested that angiotensin may play a local role in the control of the regional distribution of blood, favoring the inner cortical and medullary regions of the kidney. Such a role is, however, not proved.

The possible physiological role of the renin-angiotensin system in the regulation of blood pressure is controversial. In normal circumstances it does not appear to be important, but it probably plays a role under abnormal conditions that results in a reduction in extracellular volume. Such circumstances may occur following hemorrhage or a depletion of body sodium, such as is associated with disease and the use of drugs, especially diuretics and vasodilators. Information regarding the diagnosis of such conditions has recently been aided by the availability of drugs that can block the release of renin, the conversion of angiotensin I to angiotensin II, and competitive antagonists of the latter peptide (see Section 11.7). The administration of such drugs to normal animals or man has little or no effect on blood pressure. If, however, there is an Na depletion, such as can be promoted by diuretics or a low-salt diet, blood pressure may decline when these inhibitors are administered. In normal human subjects the adjustment of the blood pressure when moving from the recumbent to an upright posture is unchanged by inhibitors of the converting enzyme. If, however, the subjects are first depleted of Na, they usually faint following the maneuver (Samuels et al., 1976). A comparable effect has been demonstrated in rats, where angiotensin inhibitors result in a decline in blood pressure of Na-depleted but not normal animals (Gavras et al., 1973). When hypertension was induced by constricting one of the renal arteries, as originally described by Goldblatt, a similar effect of angiotensin inhibitors was observed. The role of the renin-angiotensin system in disease will be discussed later (Section 11.9).

Apart from its effects on vascular smooth muscle, angiotensin can also have hemodynamic actions on the heart. In intact animals there is a bradycardia, which appears to be a reflex response to the elevation of the blood pressure. Following vagotomy or the administration of atropine, an increase in heart rate is usually observed. This effect is probably an indirect adrenergic one, due to an action on sympathetic ganglia and/or the adrenal medulla. Angiotensin II may also increase the force of contraction of heart muscle, and this positive inotropic effect is seen in isolated cardiac muscle preparations.

The permeability of capillaries is increased by angiotensin II so that fluid accumulates in the tissues. There has been speculation that this effect may be related to the local synthesis of prostaglandins (see Peach, 1977).

The smooth muscle of the uterus, stomach, and intestinal tract also contracts in response to angiotensins, especially angiotensin II. The response of the rat uterus and stomach strips and guinea pig ileum have provided useful bioassay procedures for such activity, although the blood pressure response of the rat is usually preferred. There is no evidence to indicate that such responses of nonvascular smooth muscle are of any physiological significance.

B. Stimulation of secretion of aldosterone

Aldosterone is released following hemorrhage or Na depletion. It was observed in animals that nephrectomy could abolish this response, suggesting the role of a renal "factor" (Davis et al., 1961; Mulrow and Ganong, 1961). As described in detail elsewhere (Section 6.6B), this essential link was renin and angiotensin II and III, which are potent stimulators of release of aldosterone from the zona glomerulosa cells of the adrenal cortex. The plasma levels of the angiotensins necessary to elicit this response are much lower than those that exert a vasoconstrictor effect.

Apart from an acute effect in regulating the synthesis and release of aldosterone, angiotensin II also appears to exert a trophic effect on the zona glomerulosa, and it increases its sensitivity to stimulation. This hormone has been shown to increase the proliferation of the glomerulosa cells in tissue culture preparations (Gill, Ill, and Simonian, 1977). An increased rate of incorporation of [^{3}H]thymidine into DNA was observed. The effect was quite specific; it could be blocked by the antagonist [Sar^1,Ile^8]angiotensin II and there was no response in six other types of cells. Angiotensin II, in addition, also increases the numbers of its receptors in these adrenocortical cells (Hauger, Aguilera, and Catt, 1978).

C. Effects on the central nervous system

Angiotensin has been shown to influence various activities of the central nervous system. These include an increase in blood pressure, thirst and drinking, salt appetite, the release of catecholamines and ADH, and a disruption of learning behavior. Such effects can be observed following peripheral administration of the peptide or renin, such as into the arterial supply to the brain, or after direct injection into the ventricles or localized areas of brain tissue. Such effects can be shown to be quite specific and can be inhibited by angiotensin antagonists or inhibitors of the converting enzyme. Nevertheless, it is still difficult to decide whether such observed effects are of physiological or merely pharmacological significance. Angiotensin can facilitate the release of neurotransmitters, especially norepinephrine (see Reit, 1972). It also can enhance catecholamine synthesis and inhibit re-uptake at nerve terminals. Such effects provide at least one rational basis for a central action of the peptide. Other possibilities, for which evidence exists, are a release of acetylcholine, local vasoconstrictor effects, and changes in cyclic nucleotide metabolism.

Circulating renin and angiotensin may gain access to certain circumventricular parts of the brain which lie outside the blood-brain barrier (see Ganong, 1977). One effect of this is seen in the area postramus and results in a sympathetic discharge and increase in blood pressure. Peripheral renin and angiotensin also appear to act on the subfornical organ and the organum vasculosum of the lamina terminalis, which lies in the wall of the third ventricle, to elicit a sensation of thirst and drinking. Renin and angiotensin I and II, as well as converting enzyme and angiotensinase, have also been identified in regions of the brain which are well sequestered from the peripheral circulation, behind the blood-brain barrier (see Ganten et al., 1971; Goldstein et al., 1972; Poth, Heath, and

Ward, 1975). The renin and renin substrate at such sites are independent of those present in the peripheral tissues. Their role is unknown. One study (Morgan and Routtenberg, 1977) has shown that when angiotensin II is injected into the brain nigro-neostriatal system of rats, they fail to retain a pattern of avoidance behavior that they learned 5 minutes earlier.

From the endocrine standpoint, the effects of peripheral renin and angiotensin on the brain are of special interest. Their centrally mediated action on blood pressure has already been described. *Thirst* in response to the decrease in the volume of the body fluids (hypovolemia) may be mediated by the peripheral renin–angiotensin system. The original observations on the role of the renin–angiotensin system in thirst were made by J. T. Fitzsimons (see Fitzsimons, 1972, 1976). It was shown in rats that drinking elicited by hemorrhage could be blocked if the animals were nephrectomized. Subsequently, it was demonstrated that angiotensin and renin, and substances that stimulate the release of the latter, can also elicit drinking behavior. Intracranial injections of renin and angiotensin I and II are also effective, but their action can be prevented by specific antagonists such as saralasin and SQ 20881 (see Section 11.7). These experiments have been extended to other animal species, including monkeys and dogs. In the latter, peripheral concentrations of angiotensin II that occur normally can stimulate drinking, and the results are consistent with a physiological response (Fitzsimons, Kucharczyk, and Richards, 1978).

The effects of renin or angiotensin on drinking can be reduced by vasodilator drugs or dopamine antagonists. The subfornical organ, and the organum vasculosum, as the latter's name indicates, are well-vascularized tissues. Fitzsimons has suggested that the thirst neurons become sensitized as the result of an altered input from their capacitance vessels due to the action of angiotensin. The effects of renin and angiotensin on thirst can be antagonized by vasodilatory drugs such as papaverine and Na nitroprusside, which is consistent with this possibility.

The dipsogenic effect of angiotensin III in dogs is much less than that of angiotenin II (Fitzsimons and Kucharczyk, 1978). This difference contrasts to their equipotent effects in adrenocortical steroidogenic responses but is similar to each of their actions on vascular smooth muscle. Thus, the receptors for the dipsogenic response may be similar to the myotropic receptors, which is consistent with the possibility that the drinking response involves local vascular changes.

It has also been suggested that renin may be concerned with controlling the *sodium appetite*. In one study (Chiaraviglio, 1976) it was shown that bilateral nephrectomy abolished Na appetite in rats and that intraperitoneally injected angiotensin II, or the administration of angiotensin I into the third ventricle, stimulated Na intake. The results have, however, not been consistent (Fitzsimons and Wirth, 1975, 1978) and their interpretation has been questioned.

D. Effects on Na transport

Angiotensin has been shown to increase the rate of Na transport across several types of epithelia under *in vitro* conditions. These effects have been observed in frog skin (McAfee and Locke, 1976); various segments of the intestine, including the colon of rats (Davies, Munday, and Parsons, 1970; Hornych, Meyer, and Milliez, 1973); and kidney slices, also from rats (Munday, Parsons, and Poat, 1972). These responses were observed at low concentrations of the peptides, and in some instances, high levels had inhibitory effects. It thus appears that angiotensin preparations may exert direct effects on Na transport in epithelia apart from the action that is mediated *in vivo* by the release of aldosterone. The possible physiological significance of such effects is obscure. The mechanism of the effect is also not clear but does not appear to involve the formation of cyclic AMP (Munday et al., 1976). As such responses can be inhibited by cycloheximide but not actinomycin D, it has been suggested that protein synthesis at the translational level may be involved (Bolton, Munday, and Parsons, 1977).

11.4. Angiotensins: structure–activity relationships

Angiotensin exists in three active forms: angiotensin I, II, and III (Figure 11.2). The first of these is a decapeptide which exhibits little activity. The deletion of the histidine–leucine residues from the C-terminus by the converting enzyme results in the octapeptide angiotensin II, which is the most active form of the hormone. An aminopeptidase, however, can also act on the opposite, N-terminus, to remove the aspartic acid. When angiotensin II is the substrate, the heptapeptide des-Asp^1 angiotensin II, or angiotensin III as it is also called, is formed. In man, this peptide (Kono et al., 1975) has only about 20 percent of the vasopressor activity of angiotensin II but is about equally effective in stimulating aldosterone secretion by the adrenal cortex and in blocking the release of renin from the juxtaglomerular cells. It is possible that angiotensin II and III have different mechanisms of action on vascular smooth muscle (Ackerly, Moore, and Peach, 1977). The hexapeptide, which includes amino acid residues 3 to 8, has only about 1 percent of the pressor activity of angiotensin II, and smaller fragments are devoid of any action.

Two natural forms of angiotensin II have been

identified. In man, as well as the horse, pig, and rat, there is isoleucine in position 5, while in cattle valine is present there. These analogues have similar biological activities. The structure of human angiotensin II is shown in Figure 11.3.

Two extensive reviews (Regoli, Park, and Rioux, 1974; Khosla, Smeby, and Bumpus, 1974) describing the structure–activity relationships of more than 300 analogues of angiotensin are available. These studies include the effects of these analogues on the blood pressure (*in vivo*) and isolated smooth muscle preparations (*in vitro*), including guinea pig ileum, rat stomach muscle, and rabbit aortic strips. All the amino acids appear to

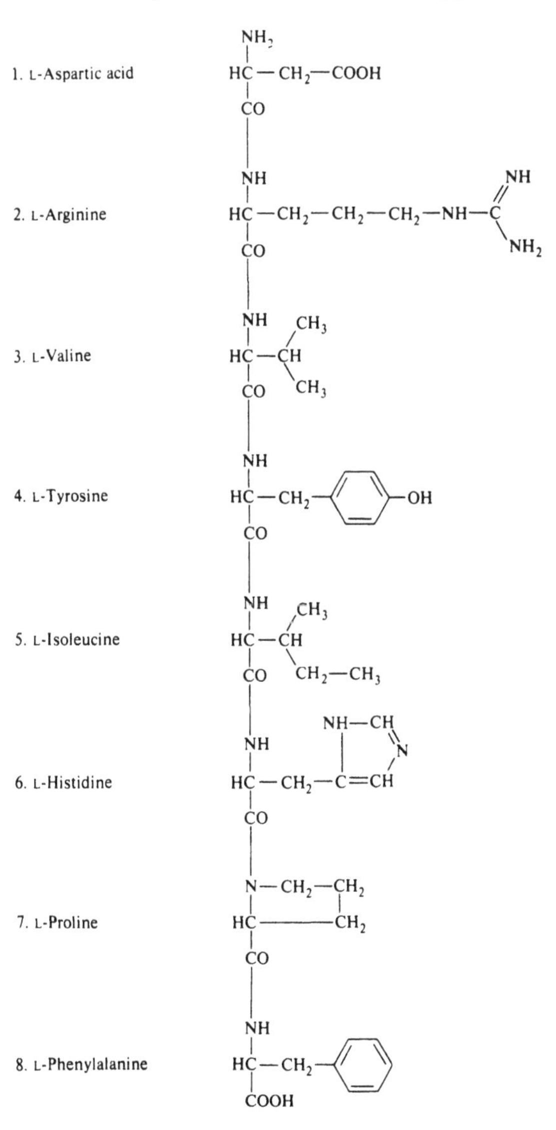

contribute in some way to the hormone's actions, whether it includes direct effects on binding to receptors or the expression of the molecule's intrinsic activity or the maintenance of its three-dimensional structure (which is usually also considered to be related to binding). The conclusions (see Figure 11.4) have been summarized by Regoli, Park, and Rioux (1974):

a. The minimum structure that retains activity is the hexapeptide corresponding to positions 3 to 8 of angiotensin II. The N-terminus is not essential.

b. However, removal of the two N-terminal amino acids, aspartic acid and arginine, results in a drastic reduction in activity, and it appears that both the N-terminal amino group and the guanido side chain of the 2-arginine normally make an important contribution to the binding of the molecule to its receptor. The carboxyl side chain of 1-aspartic acid seems to be of little importance. The N-terminal amino group plays an important role in the duration of the effect of the peptide, as it provides the focus of action of aminopeptidase enzymes.

c. Removal of either of the aromatic side groups contributed by the 4-tyrosine and 8-phenylalanine or the imidazole of the 6-histidine results in a 99 percent decline in activity. These rings all appear to contribute to receptor binding. The 8-phenylalanine, however, also plays a vital role in the ability of the molecule to elicit a response. Substitution of an aliphatic side chain in this position results in molecules such as [Ileu[8]]angiotensin II, which can still bind the receptor but has little intrinsic activity.

d. The neutral side chains contributed by the 3-valine, 5-isoleucine, and 7-proline appear to be important in maintaining the conformation of the molecule and the correct alignments of the amino acids at the 4, 6, and 8 positions.

e. The C-terminal carboxyl group appears to contribute to receptor binding. Elongation of the peptide chain at this end of the molecule, such as in angiotensin I, results in a marked decline in activity.

Such comprehensive surveys of the structure–activity relationships do not appear to have been carried out with respect to the action of angiotensin on the adrenal cortex. The observation (Blair-West et al., 1971) that des-Asp[1]-angiotensin II has undiminished effects on this target organ while the 3–8-hexapeptide is completely devoid of activity, however, indicates that the situation may differ from observations based on responses of smooth muscle preparations.

A number of attempts have been made to predict the natural *three-dimensional configuration* of angiotensin II. Such studies are important, as they may allow one to predict the particular topography

Fig. 11.3. Structure of angiotension II, showing the side chains of the constituent amino acids.

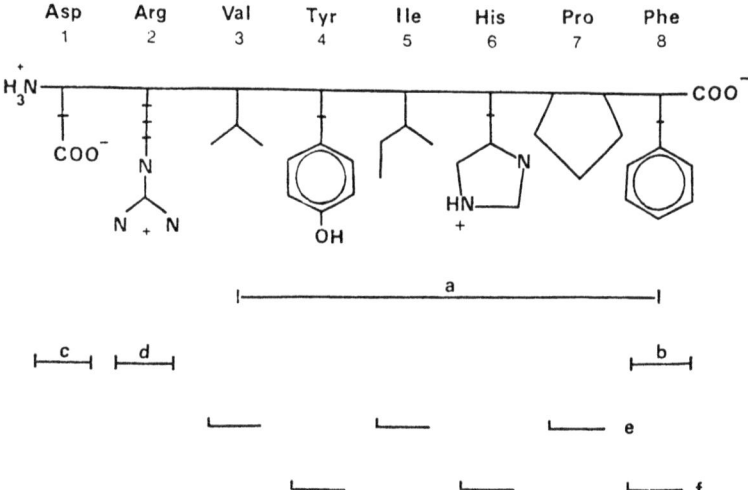

Fig. 11.4. Abbreviated chemical structure of [Asp^1,Ile^5]angiotensin II. a, Hexapeptide 3–8: minimum molecular length required for full biological activity; b, 8-Phe: hydrophobic ring essential for stimulation of receptors; c, N-terminal Asp: contributes to binding and influences duration of action; d, Arg residue: contributes to binding; e, hydrophobic side chains: stablize secondary structure; f, functional groups essential for binding to receptors (4-Tyr, 6-His) and for stimulation (8-Phe). (Copyright 1974, The Williams & Wilkins Co., Baltimore, Md.; from Regoli, Park, and Rioux, 1974.)

of the hormone and its receptor, which favors their interactions. These studies have involved a wide variety of different types of physical techniques, including optical rotatory dispersion (ORD), automated tritium-hydrogen exchange (ATHX), nuclear magnetic resonance (NMR), and circular-dichroism spectra measurements. It is not always practical to carry out such studies in physiological aqueous solutions, so nonaqueous solvents are often used. Because of this, the prediction from such measurements of the natural conformation of the molecule is uncertain or even controversial (see Khosla, Smeby, and Bumpus, 1974; Regoli, Park, and Rioux, 1974; Bumpus, 1977). Early optical rotatory dispersion studies by Bumpus and his collaborators suggested that angiotensin II does assume a stable conformational form. They suggested that it may have an α-helical twist, but this possibility now appears to have been excluded. The observations of Printz and his collaborators (Printz, Williams, and Craig, 1972; Printz, Némethy, and Bleich, 1972; Bleich, Galardy, and Printz, 1973) and Fermandjian and his group (Greff et al., 1976) indicate that intramolecular hydrogen and electrostatic (including head to tail) interactions confer an orderly shape to the molecule. The prediction of just what this conformation is likely to be, however, depends largely on the methods used. Printz, Williams, and Craig (1972), Printz, Némethy, and Bleich (1972) and Bleich, Galardy, and Printz (1973) have suggested β- and γ-turn models but prefer the latter. Greff et al. (1976) favor a β-type conformation in which the molecule has two turns, one near the N-terminus

and the other near the C-terminus. A diagram of such a model is shown in Figure 11.5. The aromatic groups provided by the tyrosine, histidine, and phenylalanine are conceived as being aligned so that they can enter a receptor site in the plasma membrane. Ionic bonding to the receptor may be provided by the C-terminus carboxyl group and the guanido side chain of the 2-arginine. It is also likely that hydrogen bonds are involved in the interaction. The observations of Printz, Williams, and Craig (1972) and Printz, Némethy, and Bleich (1972) led them to suggest that the receptor may possess a hydrophobic "receptor groove," which focuses its interactions with the amino acid side chains in the hormone.

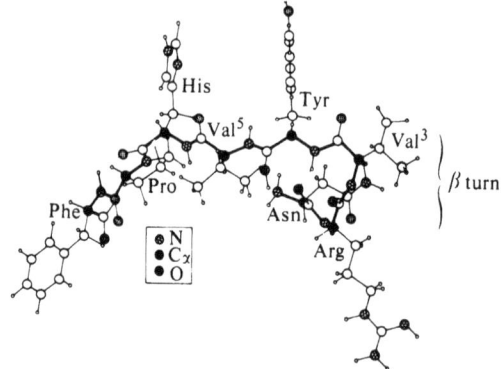

Fig. 11.5. Preliminary calculation of the conformation of bovine angiotensin II amide (valine is at position 5). (From Bumpus, 1977.)

11.5. Release of renin

The release of renin from its storage sites in the granules of the juxtaglomerular cells occurs in response to a large number of different stimuli: physiological, pathological, and pharmacological (Table 11.2). This subject has been extensively reviewed by Vander (1967), Davis and Freeman (1976), Johnston (1976), and Reid, Morris, and Ganong (1978). There appear to be several different types of mechanisms involved, although these may interact. The processes concerned include hemodynamic, neural, hormonal, and permeability effects.

The principal physiologically active stimuli that influence the release of renin appear to be Na depletion and changes in blood pressure such as may result from the assumption of an upright posture and physical exercise. A release of renin also occurs in more unusual circumstances, which include hemorrhage, cirrhosis of the liver, sometimes congestive heart failure, and hypocalcemia. The use of drugs in the treatment of diseases, especially those that affect the cardiovascular system, also influences the release of renin. Thus, stimulation occurs due to the use of vasodilator drugs such as diazoxide, hydralazine, and minoxidil, which are used to lower blood pressure. Diuretics, such as chlorothiazide, ethacrynic acid, and furosemide, are also effective. Other types of antihypertensive drugs, such as propranolol, reserpine, clonidine, and α-methyldopa, have the opposite action and block the release of renin. A summary of such drug-induced responses of the juxtaglomerular apparatus is shown in Table 11.2.

These responses of the juxtaglomerular apparatus are mediated in several ways:

a. There is a "vascular receptor" in the afferent glomerular arteriole. This appears to be a stretch-type baroreceptor which responds to decreases in the blood perfusion pressure to the kidneys. This response may be a direct one, such as can result from Na depletion, hemorrhage, or constriction of the renal artery. However, as the renal vessels also possess a sympathetic nerve supply and can respond to circulating hormones, changes in the tone of the renal vascular smooth muscle may also be involved. Such responses appear to contribute to the α-adrenergic inhibitory effects of catecholamines on the release of renin. Clonidine appears to act by stimulating such α-adrenergic receptors (Chevillard et al., 1978).

b. The renal *sympathetic nerves*, originating in the central nervous system, not only supply the muscular tonic elements of the glomerular blood vessels but directly innervate the juxtaglomerular cells and possibly even the macula densa. These nerve fibers contain β-adrenergic fibers, and their effect appears to be mediated by the formation of cyclic AMP. Stimulation of this nerve pathway results in

Table 11.2. *Physiological and pharmacological stimuli that influence the release of renin from the kidney*

Stimulate release
1. Physiological and disease processes
   Na depletion
   Physical exercise
   Hemorrhage
   Cirrhosis of the liver
   Renal artery stenosis
   Congestive heart failure (some cases)
   Sympathetic stimulation (direct or due to postural change in blood pressure, usually β-adrenergic)
   Renal vasodilatation
2. Drugs
   a. Vasodilators
      Na nitroprusside
      Hydralazine
      Minoxidil
      Bradykinin
      Theophylline
   b. α-Adrenergic antagonists
      Phentolamine
      Phenoxybenzamine
      Ergometrine
   c. β-Adrenergic agonist
      Isoproterenol
      Epinephrine
      Norepinephrine (low dose)
   d. Diuretics
      Furosemide
      Thiazides
      Ethacrynic acid
      Mercurials
      Spironolactone

Inhibit release
1. Physiological and disease processes
   Excess Na
   Hypokalemia
   Aldosterone
   Sympathetic nerve stimulation (α-adrenergic)
2. Drugs
   a. Vasopressors
      Angiotensin II
      Vasopressin
      Norepinephrine (high dose)
      Phenylephrine (α-adrenergic agonists)
   b. β-Adrenergic antagonists
      Propranolol
   c. Ganglionic blockers
   d. Centrally active antihypertensive drugs
      Clonidine
      Reserpine
      α-Methyldopa
   e. Miscellaneous
      L-Dopa
      Prednisone

Note: It should be noted that the responses can, however, vary depending on the physiological circumstances, such as basal blood pressure and the concurrent administration of other drugs.
Sources: Based on Johnston, 1976, and Peart, 1977.

the release of renin. Its effects can be blocked by β-adrenergic antagonists such as propranolol, and it is mimicked by agonists such as isoproterenol, glucagon, and theophylline, which can also promote the accumulation of cyclic AMP and may exert their stimulatory effects by means of this type of mechanism. Alpha-adrenergic nerve fibers may also contribute and either increase or decrease the secretion of renin. The response depends on the physiological and pharmacological conditions and the locus of the effects (for instance, vascular baroreceptors, renal blood vessels, or macula densa).

c. There has been considerable speculation that the *macula densa* cells of the distal renal tubule may also contribute to the control of the release of renin. It is thought that decreases in their permeability to Na and/or Cl, such as results from a decreased delivery of these ions along the renal tubule, may act as a stimulus which initiates the release of renin. It is also possible that diuretics, such as furosemide, which are known to decrease Cl permeability, may exert at least part of their effects by direct action on the macula densa cells.

d. Various *circulatory hormones* may also contribute to the control of the release of renin. Thus vasopressin and norepinephrine can inhibit, while epinephrine usually promotes, release of the enzyme. These hormones may act by altering the tone of blood vessels or exert β-adrenergic actions. Angiotensin II inhibits release, in a negative-feedback-type arc, possibly by virtue of its vasoconstrictor ability on the renal vasculature, although effects on the Ca^{2+} content of the juxtaglomerular cells may also be important.

The question immediately arises as to which particular mechanism mediates each type of stimulus that releases renin. It is usually difficult to implicate one mechanism to the exclusion of all the others. The response may indeed be due to more than one process. The effects of upright posture, exercise, mild hemorrhage, the cold pressor response, vasodilator drugs, and infused catecholamines appear to involve mainly the sympathetic β-adrenergic nerves (see Oparil and Haber, 1974a; Davis and Freeman, 1976; Davies and Slater, 1976). Studies on adrenalectomized animals with denervated kidneys and in which glomerular filtration is stopped indicate that conditions such as acute hemorrhage, Na depletion, and constriction of the renal blood supply can have a direct action on the vascular baroreceptors. However, a neural effect may still normally persist. Diuretic drugs and hyponatremia could be acting via the macula densa, but as they can also decrease the extracellular volume, such a mechanism need not be unique and baroreceptors could also be involved.

A. Process of the release of renin from the juxtaglomerular cells

The renin stored in the juxtaglomerular cells is associated with granules that can be characteristically stained by Bowie's method. The process of the release of the renin from these storage sites, in contrast to most other such storage mechanisms, apparently does not depend on increases in cytoplasmic Ca^{2+} levels and a resulting exocytosis. Van Dongen and Peart (1974) observed that in the perfused rat kidney the normal inhibitory effects of angiotensin II on the release of renin could be abolished if the external Ca^{2+} concentration in the perfusate was reduced. Basal levels of renin secretion increased when Ca^{2+}-free solutions were used. These observations have been substantially confirmed using isolated superfused rat glomeruli (Baumbach and Leyssac, 1977) and renal cortex slices from pigs (Park and Malvin, 1978). Peart (1977) has pointed out that the juxtaglomerular cells appear to be modified vascular smooth muscle cells and continue to respond similarly. Thus, angiotensin II increases free cytoplasmic Ca^{2+} levels and so initiates a contraction of vascular smooth muscle. It is suggested that Ca^{2+} continues to play a comparable role in the release of renin. Thus, a rise of this ion in the juxtaglomerular cells, in response to angiotensin II, results in a decline in the release of the enzyme. This mechanism has been called the *calcium flux hypothesis* (Figure 11.6). Baumbach and Leyssac have further suggested that the renin is released into the cell cytoplasm from the granules and can pass, in either its active or inactive form, across the cell membrane. The permeability of the latter increases when the Ca^{2+} concentrations are depressed.

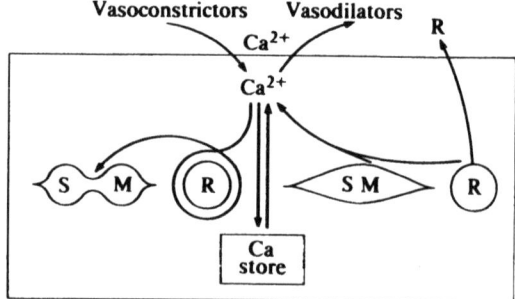

Fig. 11.6. Diagrammatic representation of the calcium flux hypothesis, relating renin release (R) and arteriolar smooth muscle contraction (SM). Left: influx of calcium leading to increased intracellular ionized calcium (Ca^{2+}) causes, on the one hand, smooth muscle contraction and, on the other, inhibition of renin release. Right: efflux of calcium leading to reduction of intracellular ionized calcium causes smooth muscle relaxation and increased release of renin. (From Peart, 1977.)

It is notable that β-adrenergic stimuli release renin. This sympathetic neural response is usually associated with increases in cyclic AMP levels in cells. The nucleotide has been shown to release renin. In smooth muscle, elevations of cyclic AMP in response to β-adrenergic stimuli have been related to a decrease in cytoplasmic Ca^{2+} levels. It is thus possible that the nucleotide is another link in the process controlling the release of renin from the juxtaglomerular cells. Direct measurements of its levels in the juxtaglomerular cells do not, however, appear to be available.

B. Negative-feedback control of renin release

Several parallel mechanisms appear to be involved in a negative-feedback control mechanism that limits the release of renin. Aldosterone, which is released as a result of the action of angiotensin II, stimulates Na reabsorption from the renal tubule. This response restores the extracellular volume and the renal perfusion pressure so that the renovascular receptor is no longer stimulated. This effect is called the "long-loop" or "volume-loop" feedback. Angiotensin II may also have a direct effect and act intrarenally to increase intracellular Ca^{2+} and constrict the afferent glomerular arteriole, and so restore the pressure on the vascular receptors. This mechanism is called the "short-loop" feedback. It can be inhibited by the angiotensin II antagonist saralasin. It is also possible that peripheral angiotensin II inhibits the central neural mechanism that influences the release of renin. The macula densa may be involved in the negative feedback response, but its role is uncertain. A hypernatremia could, for instance, increase the Na load to the macula densa and inhibit release of renin.

11.6. Metabolism of angiotensin

Angiotensin has only a short effect in the body, being metabolized to inactive fragments within one circulation time. As described earlier (see Figure 11.2), it is acted upon by a variety of enzymes, including the converting enzyme, which transforms the decapeptide angiotensin I to the active octapeptide angiotensin II. An aminopeptidase can also remove the aspartic acid at the N-terminus to produce the heptapeptide angiotensin III, which has reduced vasoconstrictor activity but retains its effectiveness on the adrenal cortex. Angiotensin II has a half-life of about 2 minutes in man, but the estimates are quite variable.

The clearance rates of various analogues of angiotensin II have been measured in the rat (Al-Merani et al., 1978). The half-lives of angiotensin II, des-Asp¹-angiotensin II and [Asn¹]angiotensin II were all about 15 seconds, suggesting that the presence of the aspartic acid or its amide in posi-

tion 1 does not slow metabolism. The antagonist [Sar¹,Ala⁸]angiotensin II, however, had a half-life of 6.4 minutes, confirming earlier observations that the presence of the sarcosine at this site considerably slows the degradation of the molecule. A number of peptidase enzymes that can hydrolyze, and finally inactivate, this peptide and angiotensin III have been identified in the plasma and in extracts of many tissues. These enzymes have been given the generic name *angiotensinases*, but no enzyme that specifically utilizes the angiotensins as a substrate has been identified. No useful inhibitors of these enzymes, such as could be used to prolong the action of angiotensin, are available.

11.7. Blockade of the renin–angiotensin system

There are several possible ways of inhibiting the renin–angiotensin system.

a. The *release* of *renin* can be blocked by *β-adrenergic antagonists* such as propranolol. This drug is used clinically to lower blood pressure but owes only part of its effect to this action on the release of renin.

b. The *enzymic action* of *renin* in splitting the decapeptide angiotensin I from its protein substrate can be inhibited (see Marshall, 1976). A naturally occurring pentapeptide called pepstatin (Figure 11.7), which was originally isolated from microorganisms, is effective *in vitro* and, in animals, *in vivo*. *Pepstatin*, as its name suggests, is not a specific renin inhibitor and was originally characterized by its ability to inhibit pepsin. Its action is noncompetitive. It inhibits 50 percent of renin activity at 1.3×10^{-10} M $(= K_i)$.

A series of *artificial substrates* for renin have been prepared (Figure 11.8) which can either competitively or noncompetitively inhibit the enzyme's activity. Trypsin splits a tetradecapeptide (A-O, Figure 11.8) from the N-terminal of angiotensinogen, and this fragment is used as a substrate for measuring the activity of renin. The renin splits the leucine–leucine bond between the amino acids at positions 10 and 11. Substitution of D-leucine, at either position, for the L-form blocks the action of

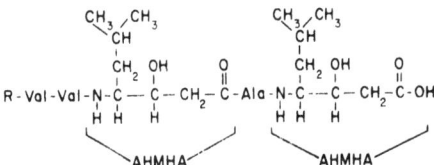

Fig. 11.7. Chemical structure of pepstatin, a peptidase enzyme originally isolated from microorganisms (*Actinomycetes*), which can inhibit the action of renin. AHMA, isovaleryl-L-valyl-L-valyl-4-amino-3-hydroxy-6-methylheptanoyl-L-alanyl-4-amino-3-hydroxy-6-methyl heptanoic acid. (From Marshall, 1976.)

Renin
↓

Tetradecapeptide substrate (A-O)   Asp¹ -Arg² -Val³ -Tyr⁴ -Ile⁵ -His⁶ - Pro⁷ -Phe⁸ -His⁹ -Leu¹⁰ —Leu¹¹ -Val¹² -Tyr¹³ -Ser¹⁴ = A-O

D-Leucine substitution (Competitive inhibitors)													
				His	- Pro	-Phe	-His	-Leu	—D-Leu	-Val	-Tyr	-Ser	
				His	- Pro	-Phe	-His	-D-Leu	—Leu	-Val	-Tyr	-Ser	
Arg	-Val	-Tyr	-Ile	-His	- Pro	-Phe	-His	-D-Leu	—Leu	-Val	-Tyr	-Ser	
Arg	-Val	-Tyr	-Ile	-His	- Pro	-Phe	-His	-Leu	—D-Leu	-Val	-Tyr	-Ser	
Arg	-Val	-Tyr	-Ile	-His	- Pro	-Phe	-His	-D-Leu	—D-Leu	-Val	-Tyr	-Ser	

Tetrapeptide fragments
(noncompetitive inhibitors)

His - Pro -Phe -His -Leu —Leu -Val -Tyr -Ser
Leu —Leu -Val -Tyr -OR (R = Me. Et)
Leu —Leu -Val -Phe -OR
Leu —Leu -Leu -Tyr -OMe
Leu —Leu -Leu -Phe -ol

Fig. 11.8. Artificial substrates for renin which can inhibit formation of angiotensin I. Note the substitution of D- for L-leucine to produce competitive inhibitors and the noncompetitive effects of certain tetrapeptides corresponding to positions 10 to 13 in the tetradecapeptide (A-O) substrate. (From Marshall, 1976.)

renin at this site. A number of fragments of the parent tetradecapeptide have been prepared which incorporate D-leucine and act as competitive inhibitors of renin. Compared with pepstatin they are, however, relatively ineffective, having a K_i of about 10^{-5} M. Several tetrapeptides, which include the amino acids and their substitutions in positions 10 to 13 of the parent substrate, can act as noncompetitive antagonists.

c. A number of naturally occurring *inhibitors* of the *converting enzyme* have been identified and have also been prepared by chemical synthesis. These substances (Figure 11.9) are peptides which were first identified by Ferreira, Bartelt, and Greene (1970) in the venom of a poisonous snake, the South American pit viper *Bothrops jaracara*. Similar compounds have been found in the venom of a Japanese snake, *Agistrodon halys blomhoffii* (Kato and Suzuki, 1971). These inhibitors were generally characterized by their ability to potentiate the effect of bradykinin due to inhibition of the enzyme kininase II, which inactivates the kinin, and which is identical to converting enzyme. The best known of these inhibitors, which is normally prepared by chemical synthesis, is a nonapeptide, *SQ 20881 or teprotide*. Its activity seems to reside principally in the C-terminal proline residues. The N-terminal

amino acids, however, also display inhibitory effects, as seen in the pentapeptide SQ 20475.

Several derivatives of proline have been prepared (Ondetti, Rubin, and Cushman, 1977) which inhibit converting enzyme. Unlike the venom peptides, these compounds are effective when taken orally. The manner in which they were discovered is an excellent example of drug design based on logical principles. As converting enzyme appears to be a carboxypeptidase which contains zinc, it was considered likely that it acts like carboxypeptidase A. The latter appears to possess three binding sites which are utilized in its interactions with its substrates. One of these sites contains the Zn^{2+} and another is a cationic site which can combine with the terminal COOH on its substrate. The converting enzyme differs from carboxypeptidase A as it splits two amino acids instead of one from the C-terminus of its substrate. It was thus thought that the distance between the cationic site at the active center of the enzyme and the Zn^{2+} must roughly correspond to the length of an additional amino acid. A model of this type of interaction is shown in Figure 11.10. As proline appears to be present at the C-terminal of all the venom inhibitors of the enzyme, it was chosen as the initial building block. With the optimal length require-

	< Glu¹	-Gly²	-Gly³	-Trp⁴	-Pro⁵	-Arg⁶ -	Pro⁷	-Gly⁸	-Pro⁹	-Glu¹⁰	-Ile¹¹	-Pro¹²	-Pro¹³	A
	< Glu	-	-	Trp	-Pro	-Arg -	Pro	-Thr	-Pro	-Gln	-Ile	-Pro	-Pro	A
	< Glu	-	Ser	-Trp	-Pro	-Gly	-	-	Pro	-Asn	-Ile	-Pro	-Pro	A
V-7:	< Glu	-	Asn	-Trp	-Pro	-His	-	-	Pro	-Gln	-Ile	-Pro	-Pro	A
SQ20881:	< Glu	-	-	Trp	-Pro	-Arg	-	-	Pro	-Gln	-Ile	-Pro	-Pro	A
	< Glu	-	Asn	-Trp	-Pro	-Arg	-	-	Pro	-Gln	-Ile	-Pro	-Pro	A
	< Glu	-	Gly	-Leu	-Pro	-Pro	-	Arg	-Pro	-Lys	-Ile	-Pro	-Pro	B
	< Glu	-	Gly	-Leu	-Pro	-Pro	-	Gly	-Pro	-Pro	-Ile	-Pro	-Pro	B
	< Glu	-	Lys	-Trp	-Asp	-Pro	-	Pro	-Pro	-Val	-Ser	-Pro	-Pro	B
SQ20475:	< Glu	-	Lys	-Trp	-Ala	-Pro	-	-	-	-	-	-	-	A

Fig. 11.9. Comparison of the amino acid sequences of several peptides, isolated from the venom of snakes, which can inhibit angiotensin I-converting enzyme. A = isolated from *Bothrops jaracara*; B = isolated from *Agistrodon halys blomhoffii*. (From Marshall, 1976.)

"Angiotensin-converting enzyme"

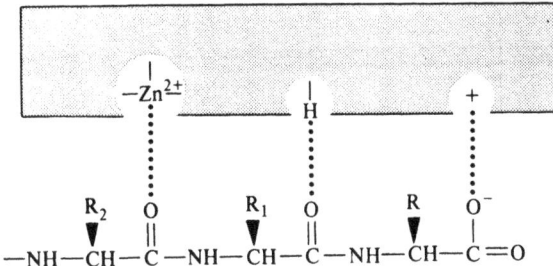

Prototype interaction with natural substrate

2-D-Methylsuccinyl-L-proline (SQ 13,297)

2-D-Methyl-3-mercaptopropanyl-L-proline (SQ 14,224)

Fig. 11.10. Schematic model illustrating the possible nature of binding of substrates and inhibitors to angiotensin-converting enzyme. (Based on Ondetti, Rubin, and Cushman, 1977.)

ments in mind, succinyl-L-proline was prepared and it was found to have weak but significant inhibitory effects. The insertion of a D-methyl group at position 2 in this molecule increased activity (Table 11.3), while the addition of another carbon, as in the 2-D-methylglutaryl-L-proline (SQ 13297), increased it even further. The 2-L-methyl analogues of these compounds were about 100 times less effective. In these proline derivatives the interaction involving the Zn^{2+} at the active site on the receptor appears to occur with a carboxyl group on the inhibitor (rather than with a carbonyl as in the natural substrate). It was considered likely that a mercapto moiety at this position might interact more readily with the Zn^{2+}, and when such a compound was prepared (3-mercaptopropanoyl-L-proline, SQ 13863) it was found to be much more active. The 2-D-methyl analogue (SQ 14225, D-3-mercapto-2-methylpropanoyl-L-proline, captopril) (see also Rubin et al., 1978) was about 10 times more active and on a molar basis is as effective as SQ 20881. This compound has been shown to lower the blood pressure of rats that are suffering from experimental renovascular hypertension (Laffan et al., 1978). It also has a hypotensive affect in man (see Section 11.10).

d. A number of *competitive antagonists* of *angiotensin II* have been prepared. The design of these drugs has been based mainly on knowledge about the structure–activity relations of angiotensin II and the processes by which it is metabolized (see Bumpus, 1977; Peach and Acherly, 1976; Marshall, 1976). The phenylalanine in position 8 at the C-terminal of the peptide is especially important and is a vital determinant of the molecule's intrinsic activity. Thus, modifications at this position such as replacement by amino acids lacking the aromatic ring or a modification of this moiety can reduce intrinsic activity while maintaining the peptide's ability to combine with its receptor. Such analogues can thus act as antagonists. The aspartic acid at the opposite N-terminal of angiotensin II (the position 1) is the locus for the action of an aminopeptidase, and modification of this position can reduce the rate of inactivation of the peptide. As this site may also be involved in attachment to the receptor, a facilitation of this process can also increase antagonist (and agonist) activity. Other parts of the molecule are important for its attachment to the receptor and for the maintenance of its optimal conformation. Thus, changes in these positions, while disturbing intrinsic activity, are also likely to reduce its ability to combine with receptors. Therefore, they would not be expected to be effective antagonists.

The antagonist activity of analogues of angiotensin II can be tested on a variety of systems *in vitro* and *in vivo*. These preparations include rat blood pressure, the release of catecholamines and aldosterone from adrenal gland preparations, and contraction (*in vitro*) of guinea pig ileum, rat uterus, and strips of aortic muscle obtained from rabbits. The structural requirements for antagonists, and agonists, may differ in these preparations. They can even vary depending on such conditions as the pH, the Mg^{2+} concentration, or whether the animal is Na-depleted. Differences in the antagonistic and agonistic properties of different analogues of angiotensin in three such assay systems are shown in Tables 11.4 and 11.5, respectively. It can be seen that one cannot necessarily extrapolate from an antagonistic activity tested on one system to its action in another, including clinical use in man. It will be observed that the available antiotensin II antagonists also exhibit appreciable agonist activity (they are partial agonists). Under some conditions the agonistic effect may be observed, so that, for instance, an increase in blood pressure can occur. The clinical aspects of this problem will be discussed later (Section 11.10).

The first analogue of angiotensin II, with antagonistic effects (see Bumpus, 1977), to be tested was [Ala⁸]angiotensin II. This lacks the aromatic ring normally contributed by phenylalanine and possesses only a methyl group in the side chain

Table 11.3. *Activities (determined* in vitro) *of various proline derivatives in inhibiting angiotensin-converting enzyme*

Structure[a]	Angiotensin-converting enzyme of rabbit lung (IC$_{50}$)[b]
1. <Glu-Trp-Pro-Arg-Pro-Gln-Ile-Pro-Pro (SQ 20881)	1.0
2. HO$_2$C—CH$_2$—CO—N⬠◀CO$_2$H	135
3. HO$_2$C—CH$_2$—CH—CO—N⬠◀CO$_2$H (SQ 13297) CH$_3$	12
4. HO$_2$C—CH$_2$—CH—CO—N⬠◀CO$_2$H CH$_3$	340
5. HO$_2$C—CH$_2$—CH$_2$—CH—CO—N⬠◀CO$_2$H CH$_3$	1.0
6. HO$_2$C—CH$_2$—CH$_2$—CH—CO—N⬠◀CO$_2$H CH$_3$	230
7. HS—CH$_2$—CH$_2$—CO—N⬠◀CO$_2$H (SQ 13863) CH$_3$	0.04
8. HS—CH$_2$—CH—CO—N⬠◀CO$_2$H (SQ 14225) CH$_3$	0.005
9. HS—CH$_2$—CH—CO—N⬠◀CO$_2$H	0.50

[a] *Key to compounds:* (2) succinyl-L-proline; (3) 2-D-methylsuccinyl-L-proline; (4) 2-L-methyl-succinyl-L-proline; (5) 2-D-methylglutaryl-L-proline; (6) 2-L-methylglutaryl-L-proline; (7) 3-mer-captopropanoyl-L-proline; (8) 2-D-methyl-3-mercaptopropanoyl-L-proline; (9) 2-L-methyl-3-mer-captopropanoyl-L-proline.

[b] Concentration (mg/ml) producing a 50% inhibition of enzyme activity.

Source: Ondetti, Rubin, and Cushman, 1977. Copyright 1977 by the American Association for the Advancement of Science.

(Table 11.6). This compound was followed by one in which the 4-tyrosine and the 8-phenylalanine were switched. The aromatic ring at position 8 thus had a hydroxyl present at the *para* position. This change appears to be adequate to disturb the aromatic group so that it loses the ability to mediate the response. Another series of compounds in which the amino group adjacent to the aromatic ring is methylated also have antagonistic effects (Pēna, Stewart, and Goodfriend, 1974). It seems that the methyl group prevents the aromatic ring from performing whatever function is necessary for the peptide to display its effect (intrinsic activity). The replacement of the aromatic ring in phenylalanine by an aliphatic side chain, such as in [Ile⁸]angiotensin II, also results in antagonistic activity. Angiotensin II and its antagonist and agonist analogues must be administered i.v. and they are rapidly destroyed by "angiotensinases." One of

these enzymes is an aminopeptidase which splits the aspartic acid from the N-terminal of the peptide. However, if the amino group is methylated, as when sarcosine is used to replace the aspartic acid, activity is much more prolonged. The compound [Sar¹,Ala⁸]angiotensin II, or *saralasin*, is the most widely used antagonist of this nature (see also Section 11.10).

Some new analogues of the compound [Sar¹]angiotensin II have been prepared which can covalently bind to the receptors in the target tissues and so have irreversible effects (Escher et al., 1978a,b). These compounds are photolabelled derivatives in which one of the aromatic amino acids has been substituted with the photolabel. One of these analogues [Sar¹,8-(4′-azidophenylalanine)]angiotensin II has been tested on the rabbit aorta *in vitro*. It was allowed to react with the tissue and was then exposed to the photolytic procedure, after

Table 11.4. *Comparison of the effects of various analogues of angiotensin II in inhibiting the action of* [Asp[1], Ile[5]]-*angiotensin II in three effector tissues*

Antagonist	Adrenal cortex	Adrenal medulla	Aortic strip
[Abu[8]]Ang II	No block	8.16	7.93
[Ala[8]]Ang II	8.53	No block	6.80
[Cha[8]]Ang II	7.33	8.70	Block at 10^{-6} M
[Ile[8]]Ang II	8.53	9.31	9.21
[Leu[8]]Ang II	7.00	8.53	8.26
[Val[8]]Ang II	7.00	8.80	8.61
[Sar[1],Ala[8]]Ang II	8.31	5.70	8.31
[Sar[1],Ile[8]]Ang II	8.31	9.53	9.33
[Sar[1],Leu[8]]Ang II	7.16	8.53	8.60
[Sar[1],Thr(Me)[8]]Ang II	8.16	7.00	8.76
[Sar[1],Thr[8]]Ang II	9.33	9.00	8.79
[MeIle[1],Ile[8]]Ang II	7.53	6.00	8.73
[Des-Asp[1],Ile[8]]Ang II	10.00	10.30	8.04
[Ile[8]]Ang I	—	8.53	8.00

Notes: Values are expressed as the pA_2 values. The pA_2 is indicative of the potency (see Section 2.11A) and is the negative logarithm of the molar concentration, which reduces the effects of a double concentration of the agonist to that of a single concentration.
Source: Peach and Ackerly, 1976.

Table 11.5. *Intrinsic agonist activities of various homologues and analogues of angiotensin II*

Peptide	Adrenal cortex	Adrenal medulla	Aortic strip
Ang I	2.0	100.0	<5.0
Ang II	*100.0*	*100.0*	*100.0*
Ang III	130.0	15.0	20.0
[Abu[8]]Ang II[a]	15.0	12.0	6.0
[Ala[8]]Ang II[b]	2.0	25.0	0.1
[Cha[8]]Ang II[a]	10.0	20.0	20.0
[Ile[8]]Ang II[a]	2.5	25.0	<5.0
[Leu[8]]Ang II[a]	1.0	10.0	10.0
[Val[8]]Ang II[b]	0.5	15.0	5.0
[Sar[1],Ala[8]]Ang II	0.5	3.0	0.5
[Sar[1],Ile[8]]Ang II	1.0	3.0	1.0
[Sar[1],Leu[8]]Ang II	1.5	3.0	0.5
[Sar[1],Thr(Me)[8]]Ang II	0.5	0.1	0.5
[Sar[1],Thr[8]]Ang II	1.0	0.1	0.5
[MeIle[1],Ile[8]]Ang II	0.5	1.0	0.5
[Des-Asp[1],Ile[8]]Ang II[b]	10.0	0	0.1
[Ile[8]]Ang I	N.D.[c]	0	<0.1

Notes: Many of these compounds also have antagonist activity. The results are expressed as percentages as compared to angiotensin II, which is given the value 100.
[a] Compounds with high intrinsic activity in all three tissues.
[b] Compounds with high intrinsic activity in at least one tissue.
[c] Not determined.
Source: Peach and Ackerly, 1976.

Table 11.6. *Structures of angiotensin II and some of its antagonists*

```
                  1    2     3     4     5     6     7     8
          NH₂- Asp - Arg - Val - Tyr - Ile - His - Pro - Phe -COOH
                          Angiotensin II
```

	Amino acid	
	Position 1	Position 8
Angiotensin II	NH₂ | HC—CH₂—COOH | CO |	NH | HC—CH₂—⬡ | COOH
[Ala⁸]Ang II	As above	NH | HC—CH₃ | COOH
[Try⁸,Phen⁴]Ang II	As above (Phen at position 4)	NH | HC—CH₂—⬡—OH | COOH
[Ile⁸]Ang II	As above	NH CH₃ | | HC—CH | COOHCH₂—CH₃
[NMePhe⁸]Ang II	As above	NCH₃ | HC—CH₂—⬡ | COOH
[Sar¹,Ala⁸]Ang II saralasin (Sar = *N*-methylglycine)	CH₃ | NH | CH₂ | CO |	NH | HC—CH₃ | COOH

which the aorta was unresponsive to angiotensin II. It could, however, still react to other agonists, such as epinephrine, histamine, and bradykinin. Such analogues are potentially useful in attempts to isolate the angiotensin II receptor.

11.8. Mechanism of action of angiotensin

It is apparent that the angiotensins can exert a wide variety of physiological and pharmacological effects. Many of these may involve indirect actions, such as changes in the blood supply to an organ or tissue; a release of neurotransmitters, especially norepinephrine from autonomic ganglia; the re-

lease of catecholamines from the adrenal medulla; and possibly even the release of local hormones such as prostaglandins and kinins. Primary effects of angiotensin on the tissue appear to be involved in contractile responses of smooth muscle and cardiac muscle, ion transport, and the increased synthesis of aldosterone. Both types of effects may, however, be involved; for instance in the heart, contractility can also be influenced by released catecholamines.

Extensive work has been performed in the search for a basic common cellular mechanism which mediates the actions of angiotensin. Many of these observations have been reviewed by Peach

(1977). Changes in cyclic AMP levels offered a likely prospective mechanism and, although the results are sometimes controversial, it does not appear that angiotensin acts primarily by changing the levels of this nucleotide in cells (see also Shima, Kawashima, and Hirai, 1978). The response of a number of target tissues, including the heart, ion transport in the colon and kidney (Munday, Parsons, and Poat, 1972; Munday et al., 1976; Bolton, Munday, and Parsons, 1977), and the zona glomerulosa of the adrenal cortex, can be blocked by cycloheximide and puromycin but not by actinomycin D. These observations suggest that angiotensin may stimulate protein synthesis by the ribosomes (translation) but does not directly affect genetic transcription.

Calcium appears to play an important role in responses to angiotensin, and it seems likely that an increase in free ionized Ca^{2+} may be an important initial part of the response. Angiotensin II can contract K^+-depolarized smooth muscle. In such muscle there appears to be an increased mobilization of the Ca^{2+} from intracellular stores (Baudouin et al., 1972). However, in some smooth muscle preparations, heart muscle, and the adrenal cortex, external Ca^{2+} is also necessary for a response, so that an increased influx of this ion into the cell may also be involved. The effects of angiotensin at such sites can often be inhibited by substances such as verapamil and Mn^{2+}, which can block Ca^{2+} movement across the cell membrane (Ackerly, Moore, and Peach, 1974).

The increased secretion of aldosterone that occurs from the adrenocortical zona glomerulosa cells in response to angiotensin II is not seen in Ca^{2+}-free media *in vitro* (see Peach, 1977). When the calcium concentration in the media bathing glomerulosa cells is increased, the synthesis of aldosterone rises (Shima, Kawashima, and Hirai, 1978). Verapamil, tetracaine, or lanthanum can block the entry of calcium into cells, and when they are added to the bathing media they block the steroidogenic response of the zona glomerulosa to angiotensin II. Thus, it seems likely that changes in cellular Ca^{2+} may also mediate the action of angiotensin on this effector. Adenyl cyclase does not appear to be involved. Inhibitors of protein synthesis, such as puromycin and cycloheximide (but not actinomycin D), can inhibit the response. Two enzymic reactions appear to be changed: there is an increase in the rate-limiting conversion of cholesterol to pregnenolone and also in the conversion of corticosterone to aldosterone. Thus, an increased synthesis of essential regulatory enzymes could be involved. However, the effect of angiotensin II on the secretion of aldosterone is seen in 2 to 5 minutes, while the observed increases in protein synthesis do not occur for 30 to 90 minutes. This effect on the process of translation thus does not

appear to be a primary one in the action of angiotensin on the adrenal cortex.

The initial interaction of angiotensins with their target tissues involves an interaction with their *receptors*. These are thought to be present on the cell membrane and were originally characterized by pharmacological analysis of the responses, especially in relation to the activity of different structural analogues of angiotensin. In 1970, Lin and Goodfriend prepared ^{125}I-labelled angiotensin II that had a high specific activity and retained about 80 percent of the biological potency of the parent peptide. They showed that the ^{125}I-labelled angiotensin II could bind to a variety of tissues, and that a proportion of this could be displaced in the presence of an excess of "cold" unlabelled angiotensin II. This fraction was considered to reflect "specific" binding, such as could occur to receptors. It was, however, noted that specific binding also occurred to nontarget tissues such as could be due to the presence of peptidase or "angiotensinase" enzymes, for which angiotensin II may be a substrate. Tritiated [3H]angiotensin II was prepared in several laboratories and retains 100 percent of its biological activity. This labelled hormone was shown to bind specifically to rabbit aorta and rat uterus, either the intact tissue or to subcellular membrane preparations made from it (Baudouin, Meyer, and Worcel, 1971; Baudouin et al., 1972; Devynck et al., 1973; Devynck and Meyer, 1978). The vascular receptors, which appear to be glycoproteins, have, with the aid of deoxycholate, been prepared and studied in a solubilized form (Devynck et al., 1974). Binding of the hormones is saturable and reaches equilibrium in about 5 minutes. The K_D is 2×10^{-8} M (compared to 2×10^{-9} M in plasma membrane preparations), and there are 2 pmol of specific binding sites per milligram of protein. The [3H]angiotensin II binding was also shown to be inhibited by a number of analogues of the peptide (Table 11.7).

Specific binding of labelled angiotensin II has also been demonstrated in isolated cells and membrane preparations of the adrenal cortex (Lin and Goodfriend, 1970; Brecher et al., 1973; Glossman, Baukal, and Catt, 1974). The zona glomerulosa cells possess the predominant number of such sites, and these cells have been studied following their dispersion with collagenase (Douglas et al., 1978). Increases in sodium concentration, dithiothreitol, and EDTA enhance the binding of angiotensin II for these receptor sites. The K_D for the interaction was about 5×10^{-10} M. The steroidogenic response of the cells could be correlated with the amount of binding. These dispersed zona glomerulosa cells were also responsive to ACTH.

Sodium deficiency increases the sensitivity of the adrenal cortex to the action of angiotensin II

Table 11.7. *Inhibitory effects of angiotensin II and some of its fragments and analogues on the binding of* [³H]*angiotensin II to specific angiotensin receptors in intact membrane preparations or when solubilized with the aid of deoxycholate*

	Membranes	Deoxycholate-solubilized material
1 2 3 4 5 6 7 8		
NH₂-Asp-Arg-Val-Tyr-Val-His-Pro-Phe-COOH	100	100
		45
		0
	81	100
	95	100
	39	30
	34	50
	4	35
Phe	92	85
Ile	43	90

Note: The concentrations of the [³H]angiotensin II and the other peptides were 2.5×10^{-7} M and 5×10^{-7} M, respectively.

Source: Devynck et al., 1974.

(Douglas and Catt, 1976). This effect, which is initiated by the angiotensin itself, reflects an increase in the interaction of the hormone with its receptors (Hauger, Aguilera, and Catt, 1978; Aguilera, Hauger, and Catt, 1978). Both the affinity of the receptors for the angiotensin II and the number of the receptors increase. In contrast, sodium deficiency results in a decline in the number of angiotensin II receptors in smooth muscle.

Studies on the structure–activity relationships of angiotensin suggest that there are considerable similarities in the structure of their receptors in different types of target tissues. However, it is also apparent that differences probably exist, but the precise nature of these must await a systematic comparison of the isolated receptors from each type of target tissue.

11.9. Disorders of the renin–angiotensin system

Hypertension is a common human disorder which, depending on the manometric standards used, may afflict 20 to 30 percent of the population. It is usually a chronic disorder which may be experienced over a period of many years. The ultimate ill effects include degeneration of the circulatory system, which becomes especially apparent in the kidneys, but deterioration can also be seen in other vascular beds, such as the retina and brain. The heart enlarges and there may be an increased likelihood of cardiac disorders, although this relationship is controversial. Ultimately, blood vessels may rupture, resulting in cardiovascular "accidents" or stroke. However, with the increased pharmacological ability to control blood pressure, this problem is not as common as it once was. Occasionally, the blood pressure may increase rapidly, over a period of a few weeks, and it is then called "malignant hypertension," which may be associated with traumatic accidents that may influence the physiological mechanisms of control, or diseases such as toxemia of pregnancy, pheochromocytoma, hyperaldosteronism, and kidney diseases. Sudden large increases in blood pressure can result in "hypertensive emergencies" and encephalopathy, necessitating the acute use of drugs to lower the blood pressure. The prevalence and effects of hypertension have prompted a vigorous search for its causes, which, initially, as a result of the experiments of H. Goldblatt, included the renin–angiotensin system. The causes in more than 90 percent of cases ("essential hypertension") remain unknown. An evaluation of the status and role of renin in hypertension may, however, be important. Renin can, for instance, sometimes play a primary role in the disease, but such information may also help in identifying other possible causes and in the consideration of the most appropriate drug therapy.

Angiotensin levels in the plasma were formerly measured using bioassays which utilized the ability of angiotensin II to contract such *in vitro* preparations as the rabbit aorta and the rat uterus and colon, or the response *in vivo* of the blood pressure of rats. Antibodies have now been prepared for this peptide, so that it can be measured more conveniently using radioimmunoassays. Renin levels in the plasma can be assayed by measuring its ability to form angiotensin I *in vitro*.

Many surveys have been made in an attempt to relate plasma renin levels to the blood pressure in normal and hypertensive people. The results have often been controversial, but it is now apparent that in the majority of instances there is no correlation that would suggest a causative relationship (see, e.g., Aurell, Petterson, and Berglund, 1975;

Lucas et al., 1974; Weir et al., 1973; Esler et al., 1977). Thus, plasma renin levels can be high with no manifestations of any hypertension, or they can be low and yet hypertension may be severe. Laragh and his associates (Brunner, Sealey, and Laragh, 1973) have classified patients suffering from hypertension into three groups: those with low, normal, and high plasma renin levels. The "low" category made up 20 to 30 percent of their group and the "high," 10 to 15 percent. Those in which renin may be a primary cause of hypertension (possibly 5 percent of the total) are expected to have high levels. It was also predicted that these patients, as well as the normals, should be most responsive to drugs such as propranolol, which can block the release of renin. The causes of hypertension in the low renin group may be associated with sodium retention and hypervolemia (which would reduce renin secretion), so that diuretic drugs were predicted to be more effective in these patients. At the time that this renin classification of hypertensive patients was made, it was suggested that the group with the low levels is less likely ultimately to suffer the vascular disorders that usually accompany the disease. However, this possibility has not generally been accepted (see Kaplan, 1975).

It will be recalled that many drugs, including those used to treat hypertension (see Johnston, 1976; Guthrie, Genest, and Kuchel, 1976), may either increase (e.g., diuretics) or decrease (e.g., propranolol, α-methyldopa, clonidine) the secretion of renin. Such changes may be seen for several weeks following the cessation of the drug therapy. Thus, the measurement of renin levels in patients who are currently being treated with such drugs may be difficult to interpret and may even be meaningless.

The hypersecretion of renin may be a primary cause of hypertension, although the highest estimates of the incidence of this disorder are only 5 percent of the total number who suffer from the disease. Sudden rises in blood pressure that occur during the malignant phase of the disease may also be related to a sudden increased release of renin. Hypertension that may result from the use of estrogens, including the contraceptive pill, are associated with increased levels of the renin substrate (angiotensinogen) in the plasma (see Stokes, 1976).

Primary hyperreninism (see Oparil and Haber, 1974b) can result from renovascular disorders, such as renal artery stenosis, which appear to stimulate renin secretion by decreasing the perfusion pressure to the receptors in the afferent glomerular arteriole. Primary renal diseases, such as pyelonephritis, may also cause an increased secretion of renin sufficient to elevate the blood pressure. Renin-secretory tumors, usually associated with the juxtaglomerular apparatus, have been consistently identified (see Editorial, 1973g;

Peart, 1975). These produce an increase in blood pressure which subsides following the removal of the tumor or its host kidney. They include hemoangiopericytomas, which are tumors associated with the blood vessels. Other types of tumors, including Wilms's tumor, are also associated with high renin secretion. The latter disorder occurs in children and the hypertension subsides when the offending kidney is removed. Such primary disorders of the secretion of renin are usually accompanied by secondary hyperaldosteronism (see Brown et al., 1972), which can result in severe hypokalemia. An intense thirst has also been described in some instances of hyperreninism, which may reflect the stimulation of the thirst center by angiotensin (see Fitzsimons, 1972, 1976).

Primary hyporeninism has also been described (see Editorial, 1973g). This condition can lead to hypoaldosteronism, which results in hyperkalemia, cardiac arrhythmias, and muscle weakness. With the availability of kidney transplants and dialysis procedures to maintain the composition of the body fluids, bilateral nephrectomy is not an uncommon operation. Such patients have little or no detectable renin in their circulation (Berman et al., 1972; Mitra et al., 1972). However, the regulation of aldosterone in such individuals does not appear to be compromised. "Secondary" hyporeninism occurs when there is an excess of aldosterone (hyperaldosteronism) (see Brown et al., 1972). Hypertension accompanies this condition, but it is clearly not directly related to the renin levels. An interesting example of secondary hyporeninism accompanied by secondary hypoaldosteronism has been described in relation to the chronic eating of licorice (Epstein et al., 1977). This confectionary has an aldosterone-like action and so can result in edema and hypokalemia, although there is no change in blood pressure. The condition is a pseudo-primary hyperaldosteronism in which *both* renin and aldosterone secretions decline. The levels of these hormones usually return to normal when the habit is stopped.

11.10. Experimental and clinical applications of blockade of the renin–angiotensin system

As already described, there are several possible ways of blocking the renin–angiotensin system. Not all of these are yet of practical importance in man. They include:

a. The use of the β-adrenergic antagonist *propranolol* to *block the release of renin*. This is the most practical and widely utilized procedure. Propranolol is currently one of the most useful drugs for the treatment of hypertension. As predicted by J. H. Laragh and his associates, it may be expected to be most effective in the small proportion of those patients that exhibit high renin levels (Buhler et al., 1972). This drug is, however, used chronically in the treatment of the many patients suffering from essential hypertension of which the cause is unknown. Although it can undoubtedly also inhibit renin secretion, in the large proportion of those patients that have normal plasma renin concentrations its hypertensive effects appear to be mediated by other mechanisms (see, e.g., Stokes, Weber, and Thornell, 1974). These involve an inhibition of vasoconstrictor activity. Propranolol may act in this way by controlling sympathetic discharge from the vasomotor center in the brain or it may "reset" the baroreceptors in the carotid sinus. A decline in the cardiac output could also contribute. Thus, the ability of propranolol to decrease renin release is probably only one facet of its antihypertensive action.

b. Direct *competitive antagonism* of the action of angiotensin II at its receptor site can sometimes initiate vasodilatation and a drop in blood pressure. The prototype and most widely used of these antagonists is *saralasin* ([Sar^1,Ala^8]angiotensin II) (see Pettinger and Mitchell, 1976). This peptide is not effective orally and must be administered i.v. either as a bolus or by continuous infusion, a procedure that limits its practical use. Saralasin is not yet approved by the FDA for any clinical use. It has been suggested that it may be useful in hypertensive crises (such as malignant hypertension) that result from a massive secretion of renin, but a number of better characterized and effective hypotensive drugs are available (see Streeten and Kaplan, 1977). Sodium nitroprusside, which has a direct relaxant effect on vascular smooth muscle, is currently preferred. It is given by i.v. infusion and has predictable and readily controlled effects. On the basis of experiments on animals it has been suggested that angiotensin II antagonists may be useful in the treatment of hemorrhagic shock, the effect of which can include acute renal failure (Needleman et al., 1976). A blockade of the action of circulating angiotensin may overcome vasoconstrictor effects and increase the blood supply to certain organs, including the kidney. Shock is, however, a complex condition, and its treatment is highly controversial.

The most widespread application of saralasin has been its use as a "screening" procedure to identify hypertension due to high renin levels (Brunner, Gavras, and Laragh, 1973). This condition may, for instance, be due to renovascular disease such as renal artery stenosis (Streeten et al., 1975; Streeten and Kaplan, 1977). In the latter type of condition, i.v. infusion of saralasin may produce a fall greater than about 10 Torr and 8 Torr, respectively, in the systolic and diastolic blood pressures. Such patients are then investigated more thoroughly by measuring renal vein renin levels and possibly taking a renal arteriogram. Saralasin is also a partial agonist, and occasionally the blood pressure may rise even to a level that results in hypertensive encephalopathy. This response appears to be associated with high body Na levels and can usually be avoided by placing the patient on a low-sodium diet for 2 to 3 days before the test or by administering a diuretic such as furosemide (see Ogihara et al., 1977). Excessive declines in blood pressure may also occur, but the blood pressure returns to normal when the infusion is stopped. Laragh (see "Second Opinion" in Streeten and Kaplan, 1977) questions the accuracy and utility of the use of angiotensin inhibitors as diagnostic drugs and suggests that the determination of plasma renin activity is more simple and preferable.

c. A number of *inhibitors of the converting enzyme* are available which inhibit the transformation of angiotensin I to its vasoactive form, angiotensin II. The best known is the nonapeptide SQ 20881 (teprotide). This must be given by i.v. injection, but inhibitors that are effective orally have been developed (Ondetti, Rubin, and Cushman, 1977). SQ 20881 has, like saralasin, been used to identify the role of renin in hypertension (Gavras et al., 1974; Case et al., 1976, 1977; Williams and Hollenberg, 1977; Williams, 1977). The use of this type of blocker avoids the problems due to the inherent agonist effects of saralasin. However, it will be recalled that converting enzyme is identical to kininase II, which inactivates kinins. Thus, the hypotensive effect of any bradykinin in the circulation will be increased, so that the specificity of the response is uncertain. It is generally considered that the effects of SQ 20881 on the response to bradykinin is small, as the circulating levels of such peptides are usually low.

The recently developed (see Section 11.7C) orally active angiotensin converting enzyme inhibitor SQ 14225 (captopril) has been tested in man (Gavras et al., 1978; Haber, 1978). It produced a substantial decline in the blood pressure of all 12 members of a group of hypertensive patients. Plasma aldosterone levels decreased and the activity of the converting enzyme was abolished. It was, however, unexpected, and surprising, to observe that the response bore no relationship to the initial levels of renin in the plasma. Thus, the mechanism of action of this drug is not clear. It could be exerting a nonspecific action, or its ability to inhibit the destruction of plasma kinins may be predominent.

d. Other possible ways of inhibiting the effects of renin are to block its action with *enzyme inhibitors* such as pepstatin, artificial renin substrates, or renin antibodies. Such methods are, however, not yet practicable in man.

References

Abe, K., Nicholson, W. F., Liddle, G. W., Orth, D. N. and Island, D. P. (1969). Normal and abnormal regulation of β-MSH in man. *J. Clin. Invest. 48*, 1580–5.

Acher, R., Chauvet, J. and Olivry, G. (1956). Sur l'existence éventuelle d'une hormone unique hypophysaire: I. Relations entre l'ocytocine, la vasopressine et la protéine de van Dyke extraites de la neurophypophysaire du boeuf. *Biochim. Biophys. Acta 22*, 421–7.

Ackerly, J. A., Moore, A. F. and Peach, M. J. (1977). Demonstration of different contractile mechanisms for angiotensin II and des-Asp¹-angiotensin II in rabbit aortic strips. *Proc. Natl. Acad. Sci. USA 74*, 5725–8.

Adams, D. B., Gold, A. R. and Burt, A. D. (1978). Rise in female sexual activity at ovulation blocked by oral contraceptives. *New Engl. J. Med. 229*, 1145–50.

Adams, D. D. and Purves, H. D. (1956). Abnormal responses in the assay of thyrotrophin. *Proc. University Otago Med. School. 34*, 11–12.

Adams, D. D., Kennedy, T. H. and Stewart, R. D. H. (1974). Correlation between long-acting thyroid stimulator protector level and thyroid ¹³¹I uptake in thyrotoxicosis. *Brit. Med. J. 2*, 199–201.

Adams, P. W., Wynn, V., Rose, D. P., Seed, M., Folkard, J. and Strong, R. (1973). Effect of pyridoxine hydrochloride (vitamin B₆) upon depression associated with oral contraception. *Lancet 1*, 897–904.

Adlercreutz, H. (1970). Oestrogen metabolism in liver disease. *J. Endocrinol. 46*, 129–63.

Adlercreutz, H. (1974). Hepatic metabolism of estrogens in health and disease. *New Engl. J. Med. 290*, 1081–3.

Adlercreutz, H. and Tenhunen, R. (1970). Some aspects of the interaction between natural and synthetic female sex hormones and the liver. *Amer. J. Med. 49*, 630–48.

Adour, K. K., Wingerd, J., Bell, D. N., Manning, J. J. and Hurley, J. P. (1972). Prednisone treatment of idiopathic facial paralysis (Bell's palsy). *New Engl. J. Med. 287*, 1268–72.

Aguilera, G., Hauger, R. L. and Catt, K. J. (1978). Control of aldosterone secretion during sodium restriction: adrenal receptor regulation and increased adrenal sensitivity. *Proc. Natl. Acad. Sci. USA 75*, 975–9.

Ahlquist, R. P. (1948). Study of adrenotropic receptors. *Amer. J. Physiol. 153*, 586–600.

Aitken, J. M. and Lindsay, R. (1973). Mithramycin in Paget's disease. *Lancet 1*, 1177–8.

Aitken, J. M., Hart, D. M. and Lindsay, R. (1973). Oestrogen replacement therapy for prevention of osteoporosis after oophorectomy. *Brit. Med. J. 3*, 515–18.

Akande, E. O., Bonnar, J., Carr, P. J., Corker, C. S., Dutton, A., Mackinnon, P. C. B. and Robinson, D. (1972). Effect of synthetic gonadotrophin-releasing hormone in secondary amenorrhea. *Lancet 2*, 112–16.

Alberti, K. G. M. M. and Nattrass, M. (1977). Lactic acidosis. *Lancet 2*, 25–9.

Alberti, K. G. M. M., Nattrass, M. and Holloway, P. A. H. (1977). Dichloroacetate in biguanide-induced lacticacidosis. *Lancet 2*, 1136.

Alberti, K. G. M. M., Christensen, N. J., Christensen, S. E., Prange Hansen, A. A., Iversen, J., Lundbaek, K., Seyer-Hansen, K. and Ørskov, H. (1973). Inhibition of insulin secretion by somatostatin. *Lancet 2*, 1299–301.

Alexander, R. W., Cooper, B. and Handin, R. I. (1978). Characterization of the human platelet α-adrenergic receptor. Correlation of [³H]dihydroergocryptine binding with aggregation and adenylate cyclase inhibition. *J. Clin. Invest. 61*, 1136–44.

Allgrove, J., Husband, P. and Brook, C. G. D. (1977). Cushing's disease. Failure of treatment with cyproheptadine. *Brit. Med. J. 1*, 686–7.

Al-Merani, S. A. M. A., Brooks, D. P., Chapman, B. J. and Munday, K. A. (1978). The half-lives of angiotensin II, angiotensin II-amide, angiotensin III, Sar¹-Ala⁸-angiotensin II and renin in the circulatory system of the rat. *J. Physiol. (Lond.) 278*, 471–90.

Aloj, S. M., Edelhoch, H., Handwerger, S. and Sherwood, L. M. (1972). Correlations in the structure and function of human placental lactogen and human growth hormone: II. The effects of disulfide bond modification on the conformation of human placental lactogen. *Endocrinology 91*, 728–37.

Alonso, D., Rynes, R. and Harris, J. B. (1965). Effect of imidazoles on active transport by gastric mucosa and urinary bladder. *Amer. J. Physiol. 208*, 1183–90.

Altman, R. D., Johnston, C. C., Khairi, M. R. A., Wellman, H., Serafini, A. N. and Sankey, R. R. (1973). Influence of disodium etidronate on clinical and laboratory manifestations of Paget's disease of bone (osteitis deformans). *New Engl. J. Med. 289*, 1379–84.

Altura, B. M. (1976a). DPAVP: a vasopressin analog with selective microvascular and RES actions for the treat-

435

ment of circulatory shock in rats. *Eur. J. Pharmacol. 37*, 155–67.

Altura, B. M. (1976b). Microcirculatory approach to the treatment of circulatory shock with a new analog of vasopressin, [2-phenylalanine,8-ornithine] vasopressin. *J. Pharmacol. Exp. Ther. 198*, 187–96.

Altura, B. M. and Altura, B. T. (1977). Vascular smooth muscle and neurohypophyseal hormones. *Fedn. Proc. 36*, 1853–60.

Amatruda, T. T., Mulrow, P. J., Gallagher, J. C. and Sawyer, W. H. (1963). Carcinoma of the lung with inappropriate antidiuresis. Demonstration of antidiuretic-hormone-like activity in tumor extract. *New Engl. J. Med. 269*, 544–9.

Ambrosi, B., Bara, R., Travaglini, P., Weber, G., Peccoz, P. B. Elli, R. and Faglia, G. (1977). Study of the effects of bromocriptine on sexual impotence. *Clin. Endocrinol. 7*, 417–21.

Amery, A. et al. (1978). Glucose intolerance during diuretic therapy. Results of a trial by the European Working Party on Hypertension in the Elderly. *Lancet 1*, 681–3.

Amir, S. M., Carraway, T. F., Kohn, L. D. and Winand, R. J. (1973). The binding of thyrotropin to isolated bovine thyroid plasma membranes. *J. Biol. Chem. 248*, 4092–100.

Amoss, M. S., Monahan, M. W. and Verlander, M. S. (1974). A long acting polymer-coupled LRF analog. *J. Clin. Endocrinol. Metab. 39*, 187–90.

Amrhein, J. A., Klingensmith, G. J., Walsh, P. C., McKusick, V. A. and Migeon, C. J. (1977). Partial androgen insensitivity. The Reifenstein syndrome revisited. *New Engl. J. Med. 297*, 350–6.

Anastasi, A., Erspamer, V. and Endean, R. (1967). Isolation and structure of caerulein, an active decapeptide from the skin of *Hyla caerulea*. *Experientia 23*, 699–700.

Andersch, B., Hahn, L., Wendestam, C., Öhman, R. and Abrahamsson, L. (1978). Treatment of premenstrual tension syndrome with bromocriptine. *Acta Endocrinol. 88*, Suppl. 216, 165–73.

Anderson, J., Coulson, R., Grassick, B. D. M., Morris, B. A., Thomas, W. D., Tomlinson, R. W. S. and Woodroffe, F. (1970). Clinical and metabolic study in diabetic patients treated with glibenclamide. *Brit. Med. J. 2*, 568–70.

Andersson, R., Nilsson, K., Wikberg, J., Johansson, S., Mohme-Lundholm, E. and Lundholm, L. (1975). Cyclic nucleotides and the contraction of muscles. In *Advances in Nucleotide Research*, Vol. 5, pp. 491–518 (edited by I. Drummond, P. Greengard, and G. A. Robison). New York: Raven Press.

Andersson, S. (1973). Secretion of gastrointestinal hormones. *Annu. Rev. Physiol. 35*, 431–52.

Anton, A. H. and Solomon, H. M., Eds. (1973). Drug–protein binding symposium. *Ann. NY Acad. Sci. 226*, 1–362.

Ariëns, E. J. and Simonis, A. M. (1964a). A molecular basis for drug action. *J. Pharm. Pharmacol. 16*, 137–57.

Ariëns, E. J. and Simonis, A. M. (1964b). A molecular basis for drug action. The interaction of one or more drugs with different receptors. *J. Pharm. Pharmacol. 16*, 289–312.

Arky, R. A. (1973). Medical metrication: U-100 insulin. *New Engl. J. Med. 28*, 580–1.

Armaly, M. F. (1966). The heritable nature of dexamethasone-induced ocular hypertension. *Arch. Ophthalmol. 75*, 32–6.

Armstrong, B., Steven, N. and Doll, R. (1974). Retrospective study of the association between use of rauwolfia alkaloids and breast cancer in English women. *Lancet 2*, 672–5.

Armstrong, B., Skegg, D., White, G. and Doll, R. (1976). Rauwolfia derivatives and breast cancer in hypertensive women. *Lancet 2*, 8–12.

Arnott, D. G. and Doniach, I. (1951). The effect of compounds allied to resorcinol upon the uptake of radioactive (^{131}I) by the thyroid of the rat. *Biochem. J. 50*, 473–9.

Aronow, L. (1978). The glucocorticoid receptor of mouse fibroblasts. *Fedn. Proc. 37*, 162–6.

Aronsen, K. F., Wettlerlin, S., Emås, S., Vojtišek, V., Mulder, J. L. and Cort, J. H. (1975). Die Wirkung von Triglycyl-Lysin-Vasopressin auf Kontrollpersonen und Patienten mit Blutungen des oberen Gastrointestinaltraktes. *Klin. Wochenschr. 53*, 747–53.

Asano, Y., Liberman, U. A. and Edelman, I. S. (1976). Relationships between Na$^+$-dependent respiration and Na$^+$-K$^+$-adenosine triphosphatase activity in skeletal muscle. *J. Clin. Invest. 57*, 368–79.

Ash, P. and Francis, M. J. O. (1975). Response of isolated rabbit articular and epiphyseal chondrocytes to rat liver somatomedin. *J. Endocrinol. 66*, 71–8.

Ashcroft, S. J. H. and Randle, P. J. (1975). The pancreas and insulin release. In *Diabetes; Its Physiological and Biochemical Basis* (edited by J. Vallance-Owen), pp. 31–62. Lancaster, England: MTP Press.

Ashton, M. G., Ball, S. G., Thomas, T. H. and Lee, M. R. (1977). Water intoxication associated with carbamazepine treatment. *Brit. Med. J. 1*, 1134–5.

Askew, F. A., Bourdillon, R. B., Bruce, H. M., Jenkins, R. G. C. and Webster, T. A. (1931). The distillation of vitamin D. *Proc. Roy. Soc. Lond. B 107*, 76–90.

Assan, R., Heuclin, C., Girard, J. R., LeMaire, F. and Attali, J. R. (1975). Phenformin-induced lactic acidosis in diabetic patients. *Diabetes 24*, 791–800.

Asselin, J., Labrie, F., Kelly, P. A., Philibert, D. and Raynaud, J-P. (1976). Specific progesterone receptors in dimethylbenzanthracene (DMBA)-induced mammary tumors. *Steroids 27*, 395–404.

Astwood, E. B. (1943). The chemical nature of compounds which inhibit the function of the thyroid gland. *J. Pharmacol. Exp. Ther. 78*, 79–89.

Astwood, E. B., Greer, M. A. and Ettlinger, M. G. (1949). The antithyroid factor of yellow turnip. *Science 109*, 631.

Atlas, D., Hanski, E. and Levitzki, A. (1977). Eighty thousand β-adrenoreceptors in a single cell. *Nature (Lond.) 268*, 144–6.

Atlas, S., Sealey, J. E., Laragh, J. H. and Moon, C. (1977). Plasma renin and "prorenin" in essential hypertension during sodium depletion, beta-blockade, and reduced arterial pressure. *Lancet 2*, 785–9.

Au, W. Y. W. (1976). Cortisol stimulation of parathyroid hormone secretion by rat parathyroid glands in organ culture. *Science 193*, 1015–17.

Auclair, C., Kelly, P. A., Coy, D. H., Schally, A. V. and Labrie, F. (1977). Potent inhibitory activity of [D-Leu^6,Des-Gly-NH$_2^{10}$]LHRH ethylamide on LH/hCG and PRL testicular receptor levels in the rat. *Endocrinology 101*, 1890–3.

Aurbach, G. D. (1959). Isolation of parathyroid hormone after extraction with phenol. *J. Biol. Chem. 234*, 3179–81.

Aurbach, G. D. and Chase, L. R. (1976). Cyclic nucleotides and biochemical actions of parathyroid hormone and calcitonin. *Handbook of Physiology:* Section 7, *Endocrinology*, Vol. 7, *Parathyroid Gland*, pp. 353–81. Washington, D. C.: American Physiological Society.

Aurbach, G. D., Fedak, S. A., Woodard, C. J., Palmer, J. S., Hauser, D. and Troxler, F. (1974). β-Adrenergic receptor: stereospecific interaction of iodinated β-blocking agent with high affinity site. *Science 186*, 1223–4.

Aurell, M., Petterson, M. and Berglund, G. (1975). Renin–angiotensin system in essential hypertension. *Lancet 2*, 342–5.

Ausiello, D. A. and Sharp, G. W. G. (1968). Localization of physiological receptor sites for aldosterone in the bladder of the toad *Bufo marinus*. *Endocrinology 82*, 1163–9.

Austen, B. M., Smyth, D. G. and Snell, C. R. (1977). Gamma endorphin, α endorphin and Met-enkephalin are formed extracellularly from lipotropin C fragment. *Nature (Lond.) 269*, 619–21.

Auty, R., Branch, R. A., Cole, E. N., Levine, D. and Ramsay, L. (1976). Prolactin, diuretics and urinary electrolytes in normal subjects. *J. Endocrinol. 70*, 173–81.

Avramides, A., Flores, A., DeRose, J. and Wallach, S. (1976). Paget's disease of the bone: observations after cessation of long-term synthetic salmon calcitonin treatment. *J. Clin. Endocrinol. Metab. 42*, 459–63.

Azarnoff, D. L. (1973). Symposium on steroid therapy. *Med. Clin. North Amer. 57*, 1155–359.

Azhar, S. and Menon, K. M. J. (1976). Gonadotropin receptors in plasma membranes of bovine corpus luteum: 1. Effect of phospholipases on the binding of ¹²⁵I-choriogonadotropin by membrane-associated and solubilized receptors. *J. Biol. Chem. 251*, 7398–404.

Azhar, S., Hajra, A. K. and Menon, K. M. J. (1976). Gonadotropin receptors in plasma membranes of bovine corpus luteum: II. Role of membrane phospholipids. *J. Biol. Chem. 251*, 7405–12.

Azizi, F., Vagenakis, A. G., Longcope, C., Ingbar, S. H. and Braverman, L. E. (1973). Decreased serum testosterone concentration in male heroin and methadone addicts. *Steroids 22*, 467–72.

Azizi, F., Vagenakis, A. G., Portnay, G. A., Rapoport, B., Ingbar, S. H. and Braverman, L. E. (1975). Pituitary–thyroid responsiveness to intramuscular thyrotropin-releasing hormone based on analyses of serum thyroxine, tri-iodothyronine and thyrotropin concentrations. *New Engl. J. Med. 292*, 273–7.

Azukizawa, M., Kurtzman, G., Pekary, A. E. and Hershman, J. M. (1977). Comparison of the binding characteristics of bovine thyrotropin and human chorionic gonadotropin to thyroid plasma membranes. *Endocrinology 101*, 1880–9.

Bacon, G. E. and Spencer, M. L. (1973). Pediatric uses of steroids. *Med. Clin. North Amer. 57*, 1265–76.

Badenoch-Jones, P. and Baum, H. (1973). Progesterone-induced permeability changes in rat liver lysosomes. *Nature New Biol. 242*, 123–4.

Bahl, O. P. (1977). Human gonadotropin, its receptor and mechanism of action. *Fedn. Proc. 36*, 2119–27.

Baker, B. L. (1974). Functional cytology of the hypophysial pars distalis and pars intermedia. *Handbook of Physiology:* Section 7, *Endocrinology*, Vol. 1, *The Pituitary Gland and Its Neuroendocrine Control*, Part 1, pp. 45–80. Washington, D. C.: American Physiological Society.

Baker, H. W. G., Bremner, W. J., Burger, H. G., de Kretser, D. M., Dulmanis, D. M., Eddie, L. W., Hudson, B., Keogh, E. J., Lee, V. W. K. and Rennie, G. C. (1976). Testicular control of follicle-stimulating hormone secretion. *Rec. Prog. Horm. Res. 32*, 429–76.

Bancroft, F. C., Wu, G-J and Zubay, G. (1973). Cell-free synthesis of rat growth hormone. *Proc. Natl. Acad. Sci. USA 70*, 3646–9.

Bannister, B., Ginsburg, R. and Shneerson, J. (1977). Cardiac arrest due to licorice-induced hypokalemia. *Brit. Med. J. 2*, 738–9.

Barber, H. R. K., Graber, E. A. and Orlando, A. (1972). Augmented labor. *Obstet. Gynecol. 39*, 933–41.

Barden, N., Lavoie, M., Dupont, A., Côté, J. and Côté, J-P. (1977). Stimulation of glucagon release by addition of antisomatostatin serum to islets of Langerhans *in vitro*. *Endocrinology 101*, 635–8.

Barer, R., Heller, H. and Lederis, K. (1963). The isolation, identification and properties of the hormonal granules of the neurohypophysis. *Proc. Roy. Soc. Lond. B 158*, 388–416.

Bargmann, W. (1968). Neurohypophysis, structure and function. In *Neurohypophysial Hormones and Similar Peptides* (edited by B. Berde), pp. 1–39. New York: Springer-Verlag.

Barker, M. H. (1936). The blood cyanates in the treatment of hypertension. *JAMA 106*, 762–7.

Barnett, D. B., Edwards, I. R. and Smith, A. J. (1975). Antagonism by indomethacin of diuretic response to calcitonin in man. *Brit. Med. J. 3*, 686.

Barrington, E. J. W. and Dockray, G. J. (1976). Gastrointestinal hormones. *J. Endocrinol. 69*, 299–325.

Barrowman, J. A., Bennett, A., Hillenbrand, P., Rolles, K., Pollock, D. J. and Wright, J. T. (1975). Diarrhoea in thyroid medullary carcinoma: role of prostaglandins and therapeutic effect of nutmeg. *Brit. Med. J. 3*, 11–12.

Bartter, F. C., Barbour, B. H., Carr, A. A. and Delea, C. S. (1964). On the role of potassium and of the central nervous system in the regulation of aldosterone secretion. In *Aldosterone* (edited by E. E. Baulieu and P. Robel), pp. 221–42. Oxford: Blackwell.

Basabe, J. C., Cresto, J. C. and Aparicio, N. (1977). Studies on the mode of action of somatostatin on insulin secretion. *Endocrinology 101*, 1436–43.

Baudouin, M., Meyer, P. and Worcel, M. (1971). Specific binding of ³H-angiotensin II in rabbit aorta. *Biochem. Biophys. Res. Commun. 42*, 434–40.

Baudouin, M., Meyer, P., Fermandjian, S. and Morgat, J.-L. (1972). Calcium release induced by interaction of angiotensin with its receptors in smooth muscle cell microsomes. *Nature (Lond.) 235*, 336–8.

Baulieu, E-E, Godeau, F., Schorderet, M. and Schorderet-Slatkine, S. (1978). Steroid-induced meiotic division in *Xenopus laevis* oocytes: surface and calcium. *Nature (Lond.) 275*, 593–8.

Baumann, G. and Loriaux, D. L. (1976). Failure of endogenous prolactin to alter renal salt and water excretion and adrenal function in man. *J. Clin. Endocrinol. Metab. 43*, 643–9.

Baumbach, L. and Leyssac, P. P. (1977). Studies on the mechanism of renin release from isolated superfused rat glomeruli: effects of calcium, calcium ionophore and lanthanum. *J. Physiol. (Lond.) 273*, 745–64.

Baxter, J. D. and Forsham, P. H. (1972). Tissue effects of glucocorticoids. *Amer. J. Med. 53*, 573–89.

Baylink, D., Sipe, J., Wergedal, J. and Whittemore, O. J. (1973). Vitamin D-enhanced osteocytic and osteoclastic bone resorption. *Amer. J. Physiol. 224*, 1345–57.

Bayliss, W. M. and Starling, E. H. (1902). The mechanism of pancreatic secretion. *J. Physiol. (Lond.) 28*, 325–53.

Beavo, J. A., Bechtel, P. J. and Krebs, E. G. (1974). Activation of protein kinase by physiological concentrations of cyclic AMP. *Proc. Natl. Acad. Sci. USA 71*, 3580–3.

Beckerhoff, R., Vetter, W., Armbruster, H., Leutscher, J. A. and Siegenthaler, W. (1973). Plasma aldosterone during oral-contraceptive therapy. *Lancet 1*, 1218–20.

Bedarkar, S., Turnell, W. G., Blundell, T. L. and Schwabe, C. (1977). Relaxin has conformational homology with insulin. *Nature (Lond.) 270*, 449–51.

Beitins, I. Z., Rattazzi, M. C. and MacGillivray, M. H. (1977). Conversion of radiolabeled human growth hormone into higher molecular weight moieties in human plasma *in vivo* and *in vitro*. *Endocrinology 101*, 350–9.

Bélanger, L. F. (1969). Osteocytic osteolysis. *Calcif. Tissue Res. 4*, 1–12.

Bell, N. H. and Stern, P. H. (1978). Hypercalcemia and increases in serum hormone value during prolonged administration of 1α,25-dihydroxyvitamin D. *New Engl. J. Med. 298*, 1241–3.

Bellisario, R. and Bahl, O. P. (1975). Human chorionic gonadotropin: V. Tissue specificity of binding and partial characterization of soluble human chorionic gonadotropin–receptor complexes. *J. Biol. Chem. 250*, 3837–44.

Bender, S. (1977). Treatment of dysmenorrhoea. *Prescr. J. 17*, 46–51.

Benjamin, W. B. and Singer, I. (1974). Aldosterone-induced protein in toad urinary bladder. *Science 186*, 269–71.

Bennett, V., O'Keefe, E. and Cuatrecasas, P. (1975). Mechanism of action of cholera toxin and the mobile receptor theory of hormone receptor–adenylate cyclase interactions. *Proc. Natl. Acad. Sci. USA 72*, 33–7.

Bennetts, H. W., Underwood, E. J. and Shier, F. L. (1946). A specific breeding problem of sheep on subterranean clover pastures in Western Australia. *Aust. Vet. J. 22*, 2–12.

Bennion, L. J., Ginsberg, R. L., Garnick, M. B. and Bennett, P. H. (1976). Effects of oral contraceptives on gallbladder bile of normal women. *New Engl. J. Med. 294*, 189–92.

Bentley, P. J. (1968). Amiloride: a potent inhibitor of sodium transport across the toad bladder. *J. Physiol. (Lond.) 195*, 317–30.

Bentley, P. J. (1976). *Comparative Vertebrate Endocrinology*, pp. 13, 25, 191–3. New York: Cambridge University Press.

Bentley, P. J. and Cobb, F. R. (1967). Neurohypophyseal-like activity of oat-cell carcinoma: actions on toad bladder. *J. Clin. Endocrinol. Metab. 27*, 1746–8.

Bentley, P. J. and Scott, W. N. (1978). Actions of aldosterone. In *General, Comparative and Clinical Endocrinology of the Adrenal Cortex* (edited by I. Chester Jones and I. W. Henderson), pp. 497–564. New York: Academic Press.

Benveniste, R., Stachura, M. E., Szabo, M. and Frohman, L. A. (1975). Big growth hormone (GH): conversion to small GH without peptide bond cleavage. *J. Clin. Endocrinol. Metab. 41*, 422–5.

Beral, V. (1976). Cardiovascular-disease mortality trends and oral-contraceptive use in young women. *Lancet 2*, 1047–52.

Beral, V. (1977). Mortality among oral-contraceptive users. Royal College of General Practioners oral contraceptive study. *Lancet 2*, 727–31.

Berde, B. and Boissonnas, R. A. (1968). Basic pharmacological properties of synthetic analogues and homologues of the neurohypophysial hormones. In *Neurohypophysial Hormones and Similar Peptides* (edited by B. Berde), pp. 802–70. New York: Springer-Verlag.

Berenson, M. M., Sannella, J. and Freston, M. D. (1976). Ménétrier's disease. Serial morphological, secretory and serological observations. *Gastroenterology 70*, 257–63.

Bergeron, J. J. M., Levine, G., Sikstrom, R., O'Shaughnessy, D., Kopriwa, B., Nadler, N. J. and Posner, B. I. (1977). Polypeptide hormone binding sites *in vivo*: initial localization of ¹²⁵I-insulin to hepatocyte plasmalemma as visualized by electron microscope radioautography. *Proc. Natl. Acad. Sci. USA 74*, 5051–5.

Bergeron, J. J. M., Posner, B. I., Josefsberg, Z. and Sikstrom, R. (1978). Intracellular polypeptide hormone receptors. The demonstration of specific binding sites for insulin and human growth hormone in Golgi fractions from the liver of female rats. *J. Biol. Chem. 253*, 4058–66.

Bergh, T., Nillius, S. J. and Wide, L. (1978). Hyperprolactinaemic amenorrhoea – Results of treatment with bromocriptine. *Acta Endocrinol. 85*, Suppl. 216, 147–61.

Bergland, R. M. and Page, R. B. (1978). Can the pituitary secrete directly to the brain? (Affirmative anatomical evidence). *Endocrinology 102*, 1325–8.

Bergman, U., Boman, G. and Wiholm, B-E. (1978). Epidemiology of adverse drug reactions to phenformin and metformin. *Brit. Med. J. 2*, 464–6.

Berl, T., Raz, A., Wald, H., Horowitz, J. and Czaczkes, W. (1977). Prostaglandin synthesis inhibition and the action of vasopressin: studies in man and rat. *Amer. J. Physiol. 232*, F529–37.

Berman, L. B., Vertes, V., Mitra, S. and Gould, A. B. (1972). Renin–angiotensin system in anephric patients. *New Engl. J. Med. 286*, 58–61.

Bernal, J. and Refetoff, S. (1977). The action of thyroid hormone. *Clin. Endocrinol. 6*, 227–49.

Bernal, J., Coleoni, A. H. and DeGroot, L. J. (1978). Thyroid hormone receptors from liver nuclei: characteristics of receptor from normal, thyroidectomized and triiodothyronine-treated rats; measurement of occupied and unoccupied receptors; and chromatin binding of receptors. *Endocrinology 103*, 403–13.

Berson, S. A. and Yalow, R. S. (1966). Insulin in blood and insulin antibodies. *Amer. J. Med. 40*, 676–90.

Besser, G. M. (1974a). Hypothalamus as an endocrine organ: I. *Brit. Med. J. 3*, 560–4.

Besser, G. M. (1974b). Hypothalamus as an endocrine organ: II. *Brit. Med. J. 3*, 613–15.

Besser, G. M. and Jeffcoate, W. J. (1976). Endocrine and metabolic diseases. Adrenal diseases. *Brit. Med. J. 1*, 448–51.

Besser, G. M., Wass, J. A. H. and Thorner, M. O. (1978). Acromegaly—results of long term treatment with bromocriptine. *Acta Endocrinol. 88*, 187–98.

Besser, G. M., McNeilly, A. S., Anderson, D. C., Marshall, J. C., Harsoulis, P., Hall, R., Ormston, B. J., Alexander, L. and Collins, W. P. (1972). Hormonal responses to synthetic luteinizing hormone and follicle stimulating hormone-releasing hormone in man. *Brit. Med. J. 3*, 267–71.

Bicknell, R. J. and Schofield, J. G. (1977). Mechanism of action of somatostatin: effect on $^{45}Ca^{2+}$ translocation and growth hormone release from dispersed bovine anterior pituitary cells. *J. Endocrinol. 72*, 31P.

Bilezikian, J. P. and Aurbach, G. D. (1974). The effects of nucleotides on the expression of β-adrenergic adenylate cyclase activity in membranes from turkey erythrocytes. *J. Biol. Chem. 249*, 157–61.

Billington, D., Osmundsen, H. and Sherratt, H. S. A. (1978). Mechanisms of the metabolic disturbances caused by hypoglycin and by pent-4-enoic acid *in vivo* studies. *Biochem. Pharmacol. 27*, 2891–900.

Birnbaumer, L. and Rodbell, M. (1969). Adenyl cyclase in fat cells. *J. Biol. Chem. 244*, 3477–82.

Bivens, C. H., Lebovitz, H. E. and Feldman, J. M. (1973). Inhibition of hypoglycemia-induced hormone secretion by the serotonin antagonists cyproheptadine and methysergide. *New Engl. J. Med. 289*, 236–9.

Black, J. W., Crowther, A. F., Shanks, R. G., Smith, L. H. and Dornhorst, A. C. (1964). A new adrenergic beta-receptor angaonist. *Lancet 1*, 1080–1.

Black, J. W., Duncan, W. A. M., Durrant, C. J., Ganellin, C. R. and Parsons, E. M. (1972). Definition and antagonism of histamine H_2-receptors. *Nature (Lond.) 236*, 385–90.

Blackard, W. G., Guzelian, P. S. and Small, M. E. (1978). Down regulation of insulin receptors in primary cultures of adult rat hepatocytes in monolayer. *Endocrinology 103*, 548–53.

Blackwell, R. E. and Guillemin, R. (1973). Hypothalamic control of adenohypophysial secretions. *Ann. Rev. Physiol. 35*, 357–90.

Blair-West, J. R., Coghlan, J. P., Denton, D. A., Funder, J. W., Scoggins, B. S. and Wright, R. D. (1971). The effect of the heptapeptide (2–8) and hexapeptide (3–8) fragments of angiotensin II on aldosterone secretion *J. Clin. Endocrinol. Metab. 32*, 575–8.

Blake, C. C. F. and Oatley, S. J. (1977). Protein–DNA and protein–hormone interactions in prealbumin: a model of the thyroid hormone nuclear receptor? *Nature (Lond.) 268*, 115–20.

Blaschko, H. (1939). The specific action of L-dopa carboxylase. *J. Physiol. (Lond.) 96*, 50–1P.

Blaschko, H., Sayers, G. and Smith, A. D., Ed. (1972). *Handbook of Physiology:* Section 7, *Endocrinology*, Vol. 6, *Adrenal Gland.* pp. 1–742. Washington, D. C.: American Physiological Society.

Bleich, H. E., Galardy, R. E. and Printz, M. P. (1973). Conformation of angiotensin II in aqueous solution. Evidence for the Y-turn model. *J. Amer. Chem. Soc. 95*, 2041–2.

Bloom, F., Segal, D., Ling, N. and Guillemin, R. (1976). Endorphins: profound behavioral effects in rats suggest new etiological factors in mental illness. *Science 194*, 630–2.

Bloom, H. J. G. and Boesen, E. (1974). Antioestrogens in treatment of breast cancer: value of nafoxidine in 52 advanced cases. *Brit. Med. J. 2*, 7–10.

Bloom, S. R. (1974). Hormones of the gastrointestinal tract. *Brit. Med. Bull. 30*, 62–7.

Bloom, S. R. (1976). Gastric inhibitory peptide, vasoactive intestinal peptide and motilin. *J. Endocrinol. 70*, 9P.

Blundell, T. L. and Wood, S. P. (1975). Is the evolution of insulin Darwinian or due to selectively neutral mutation? *Nature (Lond.) 257*, 197–203.

Blundell, T. L., Dodson, G. G., Dodson, E., Hodgkin, D. C. and Vijayan, M. (1971). X-ray analysis and the structure of insulin. *Rec. Prog. Horm. Res. 27*, 1–34.

Blundell, T. L., Cutfield, J. F., Cutfield, S. M., Dodson, E. J., Dodson, G. G., Hodgkin, D. C. and Mercola, D. A. (1972). Three-dimensional atomic structure of insulin and its relationship to activity. *Diabetes 21*, 492–505.

Bodanszky, M., Klausner, Y. S. and Said, S. I. (1973). Biological activities of synthetic peptides corresponding to fragments of and to the entire sequence of the vasoactive intestinal peptide. *Proc. Natl. Acad. Sci. USA 70*, 382–4.

Bodemar, G. and Walan, A. (1976). Cimetidine in the treatment of active duodenal and prepyloric ulcers. *Lancet 2*, 163–4.

Boden, G., Sivitz, M. C., Owen, O. E., Essa-Koumar, N. and Landor, J. H. (1975). Somatostatin suppresses secretin and pancreatic exocrine secretion. *Science 190*, 163–5.

Bolman, R. M., Cooper, C. W. and Wells, S. A. (1978). Somatostatin inhibition and reversal of parathyroid hormone-, calcium-, and acetylcholine-induced gastrin secretion. *Endocrinology 103*, 259–66.

Bolman, R. M., Cooper, C. W., Garner, S. C., Munson, P. L. and Wells, S. A. (1977). Stimulation of gastrin secretion in the pig by parathyroid hormone and its inhibition by thyrocalcitonin. *Endocrinology 100*, 1014–21.

Bolton, J. E., Munday, K. A. and Parsons, B. J. (1977). Effects of protein synthesis inhibitors on angiotensin-inhibited fluid transport by rat jejunum *in vivo*. *J. Endocrinol. 74*, 213–21.

Bond, G. C., Pasley, J. N., Koike, T. I. and Llerena, L. (1976). Contamination of an ovine prolactin preparation with antidiuretic hormone. *J. Endocrinol. 71*, 169–70.

Bonting, S. L. and Canady, M. L. (1964). Na-K activated adenosine triphosphatase and sodium transport in toad bladder. *Amer. J. Physiol. 207*, 1007–9.

Booth, J. E. (1978). Effects of aromatization inhibitor andros-4-ene-3,6,17-trione on sexual differentiation induced by testosterone in the neonatally castrated rat. *J. Endocrinol. 79*, 69–76.

Boris, A., DeMartino, L. and Trmal, T. (1970). Some endocrine studies on a new antiandrogen 6α-bromo-17β-hydroxy-17α-methyl-4-oxa-5α-androstan-3-one (BOMT). *Endocrinology 88*, 1086–91.

Boris, A., Scott, J. W., DeMartino, L. and Cox, D. C. (1973). Endocrine profile of a nonsteroidal antiandrogen N-(3,5-dimethyl-4-isoxazolylmethyl) phthalimide (DIMP). *Acta Endocrinol. 72*, 604–14.

Borle, A. B. (1973). Calcium metabolism at the cellular level. *Fedn. Proc. 32*, 1944–50.

Boston Collaborative Drug Surveillance Program (1973). Oral contraceptives and venous thrombo-embolic disease, surgically confirmed gallbladder disease, and breast tumours. *Lancet 1*, 1399–404.

Boston Collaborative Drug Surveillance Program (1974). Reserpine and breast cancer. *Lancet 2*, 669–71.

Botting, J. H., Linton, E. A. and Whitehead, S. A. (1977). Blockade of ovulation in the rat by a prostaglandin antagonist (N-0164). *J. Endocrinol. 75*, 335–6.

Bower, A., Hadley, M. E. and Hruby, V. J. (1974). Biogenic amines and control of melanophore stimulating hormone release. *Science 184*, 70–2.

Boyar, R. M., Katz, J., Finkelstein, J. W., Kapen, S., Weiner, H., Weitzman, E. D. and Hellman, L. (1974). Anorexia nervosa immaturity of the 24-hour luteinizing hormone secretory pattern. *New Engl. J. Med. 291*, 861–5.

Brandon, S. (1975). Management of sexual deviation. *Brit. Med. J. 3*, 149–51.

Braun-Menendez, E., Fasciolo, J. C., Leloir, L. F. and Muñoz, J. M. (1940). The substance causing renal hypertension. *J. Physiol. (London) 98*, 283–98.

Brecher, P., Tabacchi, M., Pyun, H. Y. and Chobanian, A. V. (1973). Angiotensin binding to rat adrenal capsular cell suspensions. *Biochem. Biophys. Commun. 54*, 1511–17.

Bremner, W. F., McDougall, I. R. and Greig, W. R. (1973). Results of treating 297 thyrotoxic patients with [125]I. *Lancet 2*, 281–2.

Brennan, M. J. (1973). Corticosteroids in the treatment of solid tumors. *Med. Clin. North Amer. 57*, 1225–39.

Breslow, E. and Walter, R. (1972). Binding properties of bovine neurophysins I and II: an equilibrium dialysis study. *Molec. Pharmacol. 8*, 75–81.

Brewer, H. B. and Ronan, R. (1970). Bovine parathyroid hormone amino acid sequence. *Proc. Natl. Acad. Sci. USA 67*, 1862–9.

Brickman, A. S., Coburn, J. W. and Norman, A. W. (1972). Action of 1,25-dihydroxycholecalciferol, a potent, kidney produced metabolite of vitamin D_3, in uremic man. *New Engl. J. Med. 287*, 891–5.

Brickman, A. S., Coburn, J. W., Kurokawa, K., Bethune, J. E., Harrison, H. E. and Norman, A. W. (1973). Actions of 1,25-dihydroxycholecalciferol in patients with hypophosphatemic, vitamin-D-resistant rickets. *New Engl. J. Med. 289*, 495–8.

Brinsden, P. R. S. and Clark, A. D. (1978). Postpartum haemorrhage after induced and spontaneous labour. *Brit. Med. J. 2*, 855–6.

Brisson, G. R. and Malaisse, W. (1973). The stimulus–secretion coupling of glucose induced insulin release: XI. Effects of theophylline and epinephrine on ^{45}Ca efflux from periperfused islets. *Metab. Clin. Exp. 22*, 455–65.

Brockmann, H. and Buss, A. (1938). Kristallisiertes Vitamin D aus Thunfischleberöl. *Naturwissenschaften 26*, 122–3.

Brodish, A. (1977). Tissue corticotropin releasing factors. *Fedn. Proc. 36*, 2088–93.

Brogan, N. (1976). The use of estrogen as a growth inhibitor in overtall girls is being questioned. *New York Times*, February 11, p. 55.

Bromer, W. W., Sinn, L. G., Staub, A., Behrens, O. K.,

Diller, E. R. and Bird, H. L. (1957). The amino acid sequence of glucagon: V. Location of amide groups, acid degradation studies, and summary of sequential evidence. *J. Amer. Chem. Soc. 79*, 2807–10.

Brostrom, C. O., Corbin, J. D., King, C. A. and Krebs, E. G. (1971). Interaction of the subunits of adenosine 3′ : 5′-cyclic monophosphate-dependent protein kinase. *Proc. Natl. Acad. Sci. USA 68*, 2444–7.

Brown, E. M., Hurwitz, S. H. and Aurbach, G. D. (1978). α-Adrenergic inhibition of adenosine 3′,5′-monophosphate accumulation and parathyroid hormone release from dispersed bovine parathyroid cells. *Endocrinology 103*, 893–9.

Brown, E. M., Aurbach, G. D., Hauser, D. and Troxler, F. (1976a). β-Adrenergic receptor interactions. Characterization of iodohydroxybenzylpindolol as a specific ligand. *J. Biol. Chem. 251*, 1232–8.

Brown, E. M., Fedak, S. A., Woodard, C. J., Aurbach, G. D. and Rodbard, D. (1976b). β-Adrenergic receptor interactions. Direct comparison of receptor interaction and biological activity. *J. Biol. Chem. 251*, 1239–46.

Brown, H. M., Storey, G. and George, W. H. S. (1972). Beclomethasone diproprionate: a new steroid aerosol for the treatment of allergic asthma. *Brit. Med. J. 1*, 585–90.

Brown, J. C., Pederson, R. A., Jorpes, E. and Mutt, V. (1969). Preparation of highly active enterogastrone. *Can. J. Physiol. Pharmacol. 47*, 113–14.

Brown, J. C., Dryburgh, J. R., Ross, S. A. and Dupré, J. (1975). Identification and actions of gastric inhibitory polypeptide. *Rec. Prog. Horm. Res. 31*, 487–526.

Brown, J. J., Fraser, R., Lever, A. F. and Robertson, J. I. S. (1972). Hypertension with aldosterone excess. *Brit. Med. J. 2*, 391–6.

Brown, K. F. and Crooks, M. J. (1976). Displacement of tolbutamide, glibenclamide and chlorpropamide from serum albumin by anionic drugs. *Biochem. Pharmacol. 25*, 1175–8.

Brown, M., Rivier, J. and Vale, W. (1977). Somatostatin: analogs with selected biological activities. *Science 196*, 1467–9.

Brown, M. P., Coy, D. H., Gomez-Pan, A., Hirst, B. H., Hunter, M., Meyers, C., Reed, J. D., Schally, A. V. and Shaw, B. (1978). Structure–activity relationships of eighteen somatostatin analogues on gastric secretion. *J. Physiol. (Lond.) 277*, 1–14.

Brown, W. A., Van Woert, M. H. and Ambani, L. M. (1973). Effect of apomorphine on growth hormone release in humans. *J. Clin. Endocrinol. Metab. 37*, 463–5.

Brownstein, M. (1977). Neurotransmitters and hypothalamic hormones in the central nervous system. *Fedn. Proc. 36*, 1960–3.

Brownstein, M., Arimura, A., Sato, H., Schally, A. V. and Kizer, J. S. (1975). The regional distribution of somatostatin in the rat brain. *Endocrinology 96*, 1456–61.

Brownstein, M. J., Palkovits, M., Saavedra, J. M., Bassiri, R. M. and Utiger, R. D. (1974). Thyrotropin-releasing hormone in specific nuclei of rat brain. *Science 185*, 267–9.

Brunette, M. G., Chan, M., Ferriere, C. and Roberts, K. D. (1978). Site of 1,25 $(OH)_2$ vitamin D_3 synthesis in the kidney. *Nature (Lond.) 276*, 287–9.

Brunner, H. R., Gavras, H. and Laragh, J. H. (1973). Angiotensin II blockade in man by Sar[1]-Ala[8]-

angiotensin II for understanding the treatment of high blood pressure. *Lancet 2*, 1045–8.

Brunner, H. R., Sealey, J. E. and Laragh, J. H. (1973). Renin subgroups in essential hypertension: further analysis of their pathophysiological and epidemiological characteristics. *Circ. Res. 32*, Suppl. 1, 99–109.

Bryant, M. G., Bloom, S. R., Pokla, J. M., Albuquerque, R. H., Modlin, I. and Pearse, A. G. E. (1976). Possible dual role for vasoactive intestinal peptide as gastrointestinal hormone and neurotransmitter substance. *Lancet 1*, 991–3.

Buckle, R. (1977). Studies on the treatment of gynaecomastia with Danazol (Danol). *J. Int. Med. Res. 5* Suppl. 3, 114–23.

Buckle, R. M., Gamlen, T. R. and Pullen, I. M. (1972). Vitamin D intoxication treated with porcine calcitonin. *Brit. Med. J. 3*, 205–7.

Buckman, M. T. and Peake, G. T. (1973). Osmolar control of prolactin secretion in man. *Science 181*, 755–7.

Buckman, M. T., Robertson, G. and Peake, G. T. (1974). Antidiuresis in patients with hyperprolactinemia. *Endocrinology 94*, A–140.

Buhler, F. R., Laragh, J. H., Baer, L., Vaughan, E. D. and Brunner, H. R. (1972). Propranolol inhibition of renin secretion. A specific approach to diagnosis and treatment of renin-dependent hypertensive diseases. *New Engl. J. Med. 287*, 1209–14.

Bull, G. M. and Frazer, R. (1950). Myxoedema from resorcinol ointment applied to leg ulcers. *Lancet 1*, 851–5.

Bumpus, F. M. (1977). Mechanisms and sites of action of newer angiotensin agonists and antagonists in terms of activity and receptor. *Fedn. Proc. 36*, 2128–32.

Burde, R. M. and Becker, B. (1970). Corticosteroid-induced glaucoma and cataracts in contact lens wearers. *JAMA 213*, 2075–7.

Burks, T. F. (1976). Gastrointestinal pharmacology. *Annu. Rev. Pharmacol. Toxicol. 16*, 15–31.

Bush, I. E. (1962). Chemical and biological factors in the activity of adrenocortical steroids. *Pharmacol. Rev. 14*, 317–445.

Buttfield, I. H., Black, M. L., Hoffmann, M. J., Mason, E. K. and Hetzel, B. S. (1965). Correction of iodine deficiency in New Guinea natives by iodized oil injection. *Lancet 2*, 767–9.

Byyny, R. L. (1976). Withdrawal from glucocorticoid therapy. *New Engl. J. Med. 295*, 30–2.

Cahill, G. F., Etzwiler, D. D. and Freinkel, N. (1976). "Control" and diabetes. *New Engl. J. Med. 294*, 1004.

Campbell, B. J., Woodward, G. and Borberg, V. (1972). Calcium-mediated interactions between the antidiuretic hormone and renal plasma membranes. *J. Biol. Chem. 247*, 6167–75.

Campbell, I. A. (1975). Inhaled corticosteroids compared with oral prednisone in patients starting long-term corticosteroid therapy for asthma. A controlled trial of the British Thoracic and Tuberculosis Association. *Lancet 2*, 469–73.

Candy, J. (1972). Severe hypothyroidism – an early complication of lithium therapy. *Brit. Med. J. 2*, 277.

Caporael, L. R. (1976). Ergotism: the satan loosed in Salem? Convulsive ergotism may have been a physiological basis for the Salem witchcraft crisis in 1692. *Science 192*, 21–6.

Carballeira, A., Fishman, L. M. and Jacobi, J. D. (1976). Dual sites of inhibition by metyrapone of human adrenal steroidogenesis: correlation of *in vivo* and *in vitro* studies. *J. Clin. Endocrinol. Metab. 42*, 687–95.

Cargill, J. M., Peden, N., Saunders, J. H. B. and Wormsley, K. G. (1978). Very long-term treatment of peptic ulcer with cimetidine. *Lancet 2*, 1113–15.

Carmichael, H. A., Nelson, L. M. and Russell, R. I. (1978). Cimetidine and prostaglandin: evidence for different modes of action on the rat gastric mucosa. *Gastroenterology 74*, 1229–32.

Caro, J. F., Besarab, A., Burke, J. F. and Glennon, J. A. (1978). A possible role for propranolol in the treatment of renal osteodystrophy. *Lancet 2*, 451–4.

Caron, M. G. and Lefkowitz, R. J. (1976). Solubilization and characterization of the β-adrenergic receptor binding sites of frog erythrocytes. *J. Biol. Chem. 251*, 2374–84.

Carr, D., Gomez-Pan, A., Weightman, D. R., Roy, V. C. M., Hall, R., Besser, G. M., Thorner, M. O., McNeilly, A. S., Schally, A. V., Kastin, A. J. and Coy, D. H. (1975). Growth hormone release inhibiting hormone: actions on thyrotrophin and prolactin secretion after thyrotrophin-releasing hormone. *Brit. Med. J. 3*, 67–9.

Carsten, M. E. and Miller, J. D. (1977). Effects of prostaglandins and oxytocin on calcium release from a uterine microsomal fraction. Hypothesis for ionophoretic action of prostaglandins. *J. Biol. Chem. 262*, 1576–81.

Carter, D. C., Dozois, R. R. and Kirkpatrick, J. R. (1972). Insulin infusion test of gastric acid secretion. *Brit. Med. J. 2*, 202–5.

Carter, J. R., Avruch, J. and Martin, D. B. (1972). Glucose transport in plasma membrane vesicles. *J. Biol. Chem. 247*, 2682–8.

Case, D. B., Wallace, J. M., Keim, H. J., Sealey, J. E. and Laragh, J. H. (1976). Usefulness and limitations of saralasin, a partial competitive agonist of angiotensin II, for evaluating the renin and sodium factors in hypertensive patients. *Amer. J. Med. 60*, 825–36.

Case, D. B., Wallace, J. M., Keim, H. J., Weber, M. A., Sealey, J. E. and Laragh, J. H. (1977). Possible role of renin in hypertension as suggested by renin-sodium profiling and inhibition of converting enzyme. *New Engl. J. Med. 296*, 641–6.

Case, R. M. (1976). Intracellular responses to gastrointestinal hormones. *J. Endocrinol. 70*, 10–11P.

Case, R. M. (1978). Synthesis, intracellular transport and discharge of exportable proteins in the pancreatic acinar cell and other cells. *Biol. Rev. 53*, 211–354.

Case, R. M. and Clausen, T. (1973). The relationship between calcium exchange and enzyme secretion in the isolated rat pancreas. *J. Physiol. (Lond.) 235*, 75–102.

Cassel, D. and Selinger, Z. (1978). Mechanism of adenylate cyclase activation through the β-adrenergic receptor: catecholamine-induced displacement of bound GDP by GTP. *Proc. Natl. Acad. Sci. USA 75*, 4155–9.

Catt, K. J. and Dufau, M. L. (1977). Peptide hormone receptors. *Annu. Rev. Physiol. 39*, 529–57.

Catto, G. R. D., MacLeod, M., Pelc, B. and Kodicek, E. (1975). 1α-Hydroxycholecalciferol: a treatment for renal bone disease. *Brit. Med. J. 1*, 12–14.

Cavalieri, R. R., Sung, L. C. and Becker, C. E. (1973). Effects of phenobarbital on thyroxine and triiodothyronine kinetics in Graves' disease. *J. Clin. Endocrinol. Metab. 37*, 308–16.

Celis, M. E., Taleisnik, S. and Walter, R. (1971). Regulation of formation and proposed structure of the factor inhibiting the release of melanocyte-stimulating hormone. *Proc. Natl. Acad. Sci. USA 68*, 1428–33.

Chakraborty, J., Hopkins, R. and Parke, D. V. (1972). Inhibition studies on the aromatization of androst-4-ene 3,17-dione by human placental microsomal preparations. *Biochem. J. 130*, 19–20P.

Chalmers, I., Campbell, H. and Turnbull, A. C. (1975). Use of oxytocin and incidence of neonatal jaundice. *Brit. Med. J. 2*, 116–18.

Chalmers, I., Zlosnik, J. E., Johns, K. A. and Campbell, H. (1976). Obstetrics practice and outcome of pregnancy in Cardiff residents 1965–73. *Brit. Med. J. 1*, 735–8.

Chalmers, J. A. (1977). Danazol treatment and follow-up of patients with endometriosis. *J. Int. Med. Res. 5*, Suppl. 3, 72–4.

Chalmers, T. C. (1975). Settling the UGDP controversy. *JAMA 231*, 624–5.

Chalmers, T. M., Davie, M. W., Hunter, J. O., Szaz, K. F., Pelc, B. and Kodicek, E. (1973). 1-α-Hydroxycholecalciferol as a substitute for the kidney hormone 1,25-dihydroxycholecalciferol in chronic renal failure *Lancet 2*, 696–9.

Chan, J. C. M., Oldham, S. B., Holick, M. F. and DeLuca, H. F. (1975). 1-α-Hydroxyvitamin D_3 in chronic renal failure. A potent analogue of the kidney hormone, 1,25-dihydroxycholecalciferol. *JAMA 234*, 47–52.

Chan, L. and O'Malley, B. W. (1976a). Mechanism of action of sex steroid hormones (part 1). *New Engl. J. Med. 294*, 1322–8.

Chan, L. and O'Malley, B. W. (1976b). Mechanism of action of sex steroid hormones (part 2). *New Engl. J. Med. 294*, 1372–81.

Chan, L. and O'Malley, B. W. (1976c). Mechanism of action of sex steroid hormones (part 3). *New Engl. J. Med. 294*, 1430–7.

Chan, S. J., Keim, P. and Steiner, D. F. (1976). Cell-free synthesis of rat preproinsulins: characterization and partial amino acid sequence determination. *Proc. Natl. Acad. Sci. USA 73*, 1964–8.

Chan, W. Y. (1976). An investigation of the natriuretic, antidiuretic and oxytocic actions of neurohypophysial hormones and related peptides: delineation of separate mechanisms of action and assessment of molecular requirements. *J. Pharmacol. Exp. Ther. 196*, 746–57.

Chan, W. Y. (1977). Relationship between the uterotonic action of oxytocin and prostaglandins: oxytocin action and release of PG-activity in solated nonpregnant and pregnant rat uteri. *Biol. Reprod. 17*, 541–8.

Chan, W. Y., Nestor, J. J., Ferger, M. F. and Du Vigneaud, V. (1974). Inhibition of oxytocic responses to oxytocin in pregnant rats by [1-L-penicillamine] oxytocin and [1-β-mercapto-β,β-diethylpropionic acid]oxytocin. *Proc. Soc. Exp. Biol. Med. NY 146*, 364–6.

Chandrabose, K. A., Lapetina, E. G., Schmitges, C. J., Siegel, M. I. and Cuatrecasas, P. (1978). Action of corticosteroids in regulation of prostaglandin biosynthesis in cultured fibroblasts. *Proc. Natl. Acad. Sci. USA 75*, 214–17.

Chang, J., Lewis, G. P. and Piper, P. J. (1977). Inhibition by glucocorticoids of prostaglandin release from adipose tissue *in vitro*. *Brit. J. Pharmacol. 59*, 429–32.

Changeux, J.-P., Thiéry, J., Tung, Y. and Kittel, C. (1967). On the cooperativity of biological membranes. *Proc. Natl. Acad. Sci. USA 57*, 335–41.

Channing, C. P., Sakai, C. N. and Bahl, O. P. (1978). Role of carbohydrate residues of human chorionic gonadotropin in binding and stimulation of adenosine 3′,5′-monophosphate accumulation by porcine granulosa cells. *Endocrinology 103*, 341–8.

Charness, M. E., Bylund, D. B., Beckman, B. S., Hollenberg, M. D. and Snyder, S. H. (1976). Independent variation of β-adrenergic receptor binding and catecholamine-stimulated adenyl cyclase activity in rat erythrocytes. *Life Sci. 19*, 243–50.

Charron, R. C., Leme, C. E., Wilson, D. R., Ing, T. S. and Wrong, O. M. (1969). The effect of adenal steroids on stool composition, as revealed by *in vivo* dialysis of faeces. *Clin. Sci. 37*, 151–67.

Chart, J. J., Sheppard, H., Mowles, T. and Howie, N. (1962). Inhibitors of adrenal corticosteroid 17α-hydroxylation. *Endocrinology 71*, 479–86.

Chaudhury, R. R., Chompootaweep, S., Dusitsin, N., Friesen, H. and Tankeyoon, M. (1977). The release of prolactin by medroxyprogesterone acetate in human subjects. *Brit. J. Pharmacol. 59*, 433–4.

Cheek, D. B. and Hill, D. E. (1974). Effect of growth hormone on cell and somatic growth. *Handbook of Physiology:* Section 7, *Endocrinology*, Vol. 2, *The Pituitary Gland and Its Neuroendocrine Control*, Part 2, pp. 159–85. Washington, D.C.: American Physiological Society.

Cheesman, D. W., Osland, R. B. and Forsham, P. H. (1977). Suppression of the preovulatory surge of luteinizing hormone and subsequent ovulation in the rat by arginine vasotocin. *Endocrinology 101*, 1194–202.

Cheng, S. C., Suzuki, K., Sadee, W. and Harding, B. W. (1976). Effects of spironolactone, canrenone and canrenoate-K on cytochrome P450, and 11β- and 18-hydroxylation in bovine and human adrenal cortical mitochondria. *Endocrinology 99*, 1097–106.

Chesney, A. M., Clawson, T. A. and Webster, B. (1928). Endemic goiter in rabbits: 1. Incidence and characteristics. *Bull. Johns Hopkins Hosp. 43*, 261–77.

Chesney, R. W., Mazess, R. B., Hamstra, A. J., DeLuca, H. F. and O'Reagan, S. (1978a). Reduction of serum-1-25-dihydroxyvitamin-D_3 in children receiving glucocorticoids. *Lancet 2*, 1123–5.

Chesney, R. W., Moorthy, A. V., Eisman, J. A., Jax, D. K., Mazess, R. B. and DeLuca, H. F. (1978b). Increased growth after long-term oral 1α,25-vitamin D_3 in childhood renal osteodystrophy. *New Engl. J. Med. 298*, 238–42.

Cheung, W. Y., Lynch, T. J. and Wallace, R. W. (1978). An endogenous Ca^{2+}-dependent activator protein of brain adenylate cyclase and cyclic nucleotide phosphodiesterase. In *Advances in Nucleotide Research* (edited by W. J. George and L. J. Ignarro) Vol. 9, pp. 233–51. New York: Raven Press.

Chevalier, J., Bourguet, J. and Hugon, J. S. (1974). Membrane associated particles: distribution in frog urinary bladder epithelium at rest and after oxytocin treatment. *Cell Tissue Res. 152*, 129–40.

Chevillard, C., Pasquier, R., Duchene, N. and Alexandre, J-M. (1978). Mechanism of inhibition of renin release by clonidine in rats. *Eur. J. Pharmacol. 48*, 451–4.

Chew, C. S. and Hersey, S. J. (1978). Dissociaton between oxyntic cell cAMP formation and HCl secretion in bullfrog gastric mucosa. *Amer. J. Physiol. 235,* E140–9.

Chew, W. C. (1977). Neonatal hyperbilirubinaemia: a comparison between prostaglandin E_2 and oxytocin inductions. *Brit. Med. J. 2,* 679–80.

Chey, W. Y. and Escoffery, R. (1976). Secretion cells in the gastrointestinal tract. *Endocrinology 98,* 1390–5.

Chiaraviglio, E. (1976). Effect of renin–angiotensin system on sodium intake. *J. Physiol. (Lond.) 255,* 57–66.

Chiodini, P. G., Liuzzi, A., Botalla, L., Cremascolli, G. and Silvestrini, F. (1974). Inhibitory effect of dopaminergic stimulation on GH release in acromegaly. *J. Clin. Endocrinol. Metab. 38,* 200–6.

Choi, Y., Thrasher, K., Werk, E. E., Sholiton, L. J. and Olinger, C. (1971). Effect of diphenylhydantoin on cortisol kinetics in humans. *J. Pharmacol. Exp. Ther. 176,* 27–34.

Chomozynski, P. and Topper, Y. J. (1974). A direct effect of prolactin and placental lactogen on mammary epithelial nuclei. *Biochem. Biophys. Res. Commun. 60,* 56–63.

Chong, G. C., Beahrs, O. H., Sizemore, G. W. and Woolner, L. H. (1975). Medullary carcinoma of the thyroid gland. *Cancer 35,* 695–704.

Christ, J. F. (1966). Nerve supply, blood supply and cytology of the neurohypophysis. In *The Pituitary Gland* (edited by G. W. Harris and B. T. Donovan), Vol. 3, *Pars Intermedia and Neurohypophysis,* pp. 62–130. Berkeley, Calif.: University of California Press.

Christakos, S. and Norman, A. W. (1978). Vitamin D_3-induced calcium binding protein in bone tissue. *Science 202,* 70–1.

Christiansen, C., Rødbro, P. and Sjö, O. (1974). "Anticonvulsant action" of vitamin D in epileptic patients? A controlled study. *Brit. Med. J. 2,* 258–9.

Christiansen, C., Rødbro, P., Munck, O. and Munck, O. (1975). Actions of vitamins D_2 and D_3 and 25-OHD₃ in anticonvulsant osteomalacia. *Brit. Med. J. 2,* 363–5.

Cicero, T. J., Bell, R. D., Wiest, W. G., Allison, J. H., Polakoski, K. and Robins, E. (1975). Function of the male sex organs in heroin and methadone users. *New Engl. J. Med. 292,* 882–7.

Cigorraga, S. B., Dufau, M. L. and Catt, K. J. (1978). Regulation of luteinizing hormone receptors and steroidogenesis in gonadotropin-desensitized Leydig cell. *J. Biol. Chem. 253,* 4297–304.

Clark, J. H., Peck, E. J. and Anderson, J. N. (1974). Oestrogens receptors and antagonism of steroid hormone action. *Nature (Lond.) 251,* 446–8.

Clarke, B. F. and Campbell, I. W. (1975). Long-term comparative trial of glibenclamide and chlorpropamide in diet-failed, maturity-onset diabetics. *Lancet 1,* 246–8.

Clausen, T., Elbrink, J. and Martin, B. R. (1974). Insulin controlling calcium distribution in muscle and fat cells. *Acta Endocrinol.* Suppl 191, 137–43.

Clemens, J. A., Shaar, C. J., Smalstig, E. B., Bach, N. J. and Kornfeld, E. C. (1974). Inhibition of prolactin secretion by ergolines. *Endocrinology 94,* 1171–6.

Clements, F. W. (1955). A thyroid blocking agent as a cause of endemic goitre in Tasmania: preliminary communication. *Med. J. Aust. 42*(2), 369–71.

Clements, F. W. (1960). Naturally occurring goitrogens. *Brit. Med. Bull. 16,* 133–7.

Cline, M. J. (1973). Adrenal steroids in the treatment of

malignant hematologic disease. *Med. Clin. North Amer. 57,* 1203–9.

Cobb, W. E., Spare, S. and Reichlin, S. (1978). Neurogenic diabetes insipidus: management with dDAVP (1-desamino-8-D-arginine vasopressin). *Ann. Intern. Med. 88,* 183–8.

Cochrane, D. E., Douglas, W. W., Mouri, T. and Nakazato, Y. (1975). Calcium and stimulus-secretion coupling in the adrenal medulla: contrasting stimulating effects of the ionophores X-537A and A23187 on catecholamine output. *J. Physiol. (Lond.) 252,* 363–78.

Code, C. F. (1965). Histamine and gastric acid secretion: a later look, 1955–65. *Fedn. Proc. 24,* 1311–21.

Cody, V. (1978). Molecular conformation of a halogen-free thyroxine analog: 4'-methoxy-3,5,3'-trimethyl-L-thyronine N-acetyl ethyl ester. *Science 201,* 1131–3.

Cohen, K. L. and Nissley, S. P. (1976). The serum half-life of somatomedin activity: evidence for growth hormone dependence. *Acta Endocrinol. 83,* 243–58.

Cohen, S. and Carpenter, G. (1975). Human epidermal growth factor: isolation and chemical and biological properties. *Proc. Natl. Acad. Sci. USA 72,* 1317–21.

Cohen, S. L., MacIntyre, I., Graham-Smith, D. and Walker, J. G. (1973). Alcohol-stimulated calcitonin release in medullary carcinoma of the thyroid. *Lancet 2,* 1172–4.

Cole, R. A., Howie, P. W. and Macnaughton, M. C. (1975). Elective induction of labour a randomized prospective trial. *Lancet 1,* 767–70.

Collip, J. B. (1925). Extraction of a parathyroid hormone which will prevent or control parathyroid tetany and which regulates the level of blood calcium. *J. Biol. Chem. 63,* 395–438.

Comite, F., Burrow, G. N. and Jorgenson, E. C. (1978). Thyroid hormone anaologs and fetal goiter. *Endocrinology 102,* 1670–4.

Condon, J. R., Knight, M. and Day, J. L. (1973). Glucagon therapy in acute pancreatitis. *Brit. J. Surg. 60,* 509–11.

Conlay, L. A. and Loewenstein, J. E. (1976). Phenformin and lactic acidosis. *JAMA 235,* 1575–8.

Conn, H. O. and Blitzer, B. L. (1976). Nonassociation of adrenocorticosteroid therapy and peptic ulcer. *New Engl. J. Med. 294,* 473–9.

Conn, J. W. (1955). Part 2. Primary aldosteronism, a new clinical syndrome. *J. Lab. Clin. Med. 45,* 6–17.

Conn, J. W. (1963). Aldosterone in man. Some clinical and climatological aspects (part 1). *JAMA 183,* 775–81.

Conn, J. W., Louis, L. H., Johnston, M. W. and Johnson, B. J. (1947). The electrolyte content of thermal sweat as an index of adrenalcortical function. *J. Clin. Invest. 27,* 529–30.

Cook, G. C., Mulligan, R. and Sherlock, S. (1971). Controlled prospective trial of corticosteroid therapy in active chronic hepatitis. *Quart. J. Med. 40,* 159–85.

Copp, D. H., Cameron, E. C., Cheney, B. A., Davidson, A. G. F. and Henze, K. G. (1962). Evidence for calcitonin – a new hormone from the parathyroid that lowers blood calcium. *Endocrinology 70,* 638–49.

Coppen, A., Montgomery, S., Peet, M., Bailey, J., Marks, V. and Woods, P. (1974). Thyrotrophin-releasing hormone in the treatment of depression. *Lancet 2,* 433–5.

Cornell, J. S. and Pierce, J. G. (1973). The subunits of human pituitary thyroid-stimulating hormone. Isolation, properties and composition. *J. Biol. Chem. 248*, 4327–33.

Cort, J. H., Strub, K. M., Häusler, G. and Rudinger, J. (1973). The natriuretic action of [4-leucine]-arginine-vasotocin. *Experientia 29*, 173–5.

Corvol, P., Michaud, A., Menard, J., Freifeld, M. and Mahoudeau, J. (1975). Antiandrogen effect of spirolactones: mechanism of action. *Endocrinology 97*, 52–8.

Cotterill, J. A. and Cunliffe, W. J. (1973). Self-medication with licorice in a patient with Addison's disease. *Lancet 1*, 294–5.

Cowie, A. T. (1972). Lactation and its hormonal control. In *Hormones in Reproduction* (edited by C. R. Austin and R. V. Short), Vol. 3, pp. 106–43. New York: Cambridge University Press.

Cowley, A. J. and Elkeles, R. S. (1978). Diabetes and therapy with potent diuretics. *Lancet 1*, 154.

Cox, M., Sterns, R. H. and Singer, I. (1978). The defense against hyperkalemia: the roles of insulin and aldosterone. *New Engl. J. Med. 299*, 525–32.

Crabbé, J. and De Weer, P. (1964). Action of aldosterone on the skin and bladder of the toad. *Nature (Lond.) 202*, 278–9.

Craighead, J. E. (1978). Current views on the etiology of insulin-dependent diabetes mellitus. *New Engl. J. Med. 299*, 1439–45.

Crine, P., Benjannet, S., Seidah, N. G., Lis, M. and Chrétien, M. (1977). *In vitro* biosynthesis of β-endorphin, α-lipotropin, and β-lipotropin by the pars intermedia of beef pituitary glands. *Proc. Natl. Acad. Sci. USA 74*, 4276–80.

Crooks, M. J. and Brown, K. F. (1974). The binding of sulphonylureas to serum albumin. *J. Pharm. Pharmacol. 26*, 304–11.

Csapo, A. I., Pohanka, O. and Kaihola, H. L. (1974). Progesterone deficiency and premature labour. *Brit. Med. J. 1*, 137–40.

Cuatrecasas, P. (1972a). Isolation of the insulin receptor of liver and fat-cell membranes. *Proc. Natl. Acad. Sci. USA 69*, 318–22.

Cuatrecasas, P. (1972b). Properties of the insulin receptor isolated from liver and fat cell membranes. *J. Biol. Chem. 247*, 1980–91.

Cuatrecasas, P. and Tell, G. P. E. (1973). Insulin-like activity of concanavalin A and wheat germ agglutinin-direct interactions with insulin receptors. *Proc. Natl. Acad. Sci. USA 70*, 485–9.

Cuatrecasas, P., Hollenberg, M. D., Chang, K-J. and Bennett, V. (1975). Hormone receptor complexes and their modulation of membrane function. *Rec. Prog. Horm. Res. 31*, 37–84.

Cuello, A. C. (1978). Endogenous opioid peptides in neurons of the human brain. *Lancet 2*, 291–3.

Cunliffe, W. J. (1976). Long-term treatment with 0.01% fluclorolone acetonid in children. *Brit. Med. J. 1*, 627.

Curnow, R. T., Carey, R. M., Taylor, A., Johanson, A. and Murad, F. (1975). Somatostatin inhibition of insulin and gastrin hypersecretion in pancreatic islet-cell carcinoma. *New Engl. J. Med. 292*, 1385–6.

Cusan, L., Dupont, A., Kledzik, G. S., LaBrie, F., Coy, D. H. and Schally, A. V. (1977). Potent prolactin and growth hormone releasing activity of more analogues of Met-enkephalin. *Nature (Lond.) 268*, 544–7.

Cushman, P. (1973). Plasma testosterone in narcotic addiction. *Amer. J. Med. 55*, 452–8.

Cuthbert, A. W. and Shum, W. K. (1976). Estimation of the lifespan of amiloride binding sites in the membranes of toad bladder epithelial cells. *J. Physiol. (Lond.) 255*, 605–18.

Cutler, G. B., Pita, J. C., Rifka, S. M., Menard, R. H., Sauer, M. A. and Loriaux, D. L. (1978). SC25152: a potent mineralocorticoid antagonist with reduced affinity for the 5α-dihydrotestosterone receptor of human and rat prostate. *J. Clin. Endocrinol. Metab. 47*, 171–5.

Cutz, E., Chan, W., Track, N. S., Goth, A. and Said, S. I. (1978). Release of vasoactive intestinal polypeptide in mast cells by histamine liberators. *Nature (Lond.) 275*, 661–2.

Daggett, P., Verner, I. and Carruthers, M. (1978). Intra-operative management of phaeochromocytoma with sodium nitroprusside. *Brit. Med. J. 2*, 311–13.

Dahl, D. S. (1976). The management of myasthenia gravis. *Drug Ther.* (October), 21–9.

Dale, H. (1906). On some physiological actions of ergot. *J. Physiol. (Lond.) 34*, 163–206.

Dalterio, S., Bartke, A. and Burstein, S. (1977). Cannabinoids inhibit testosterone secretion by mouse testes *in vitro. Science 196*, 1472–3.

Dandona, P., Foster, M., Healey, F., Greenbury, E., and Beckett, A. G. (1978). Low-insulin infusions in diabetic patients with high insulin requirements. *Lancet 2*, 283–5.

Danon, A. and Assouline, G. (1978). Inhibition of prostaglandin biosynthesis by corticosteroids requires RNA and protein synthesis. *Nature (Lond.) 273*, 552–4.

Datta, H. and Chaudhury, R. R. (1970). Further studies on the antidiuretic hormone blocking effect of [Asp⁴]-oxytocin. *J. Endocrinol. 46*, 117–18.

Daughaday, W. H. (1971). Sulfation factor regulation of skeletal growth. A stable mechanism dependent on intermittent growth hormone secretion. *Amer. J. Med. 50*, 277–80.

Daughaday, W. H., Salmon, W. D. and Alexander, F. (1959). Sulfation factor activity of sera from patients with pituitary disorders. *J. Clin. Endocrinol. Metab. 19*, 743–58.

Daughaday, W. H., Hall, K., Raben, M. S., Salmon, W. D., Van den Brande, J. L. and Van Wyke, J. J. (1972). Somatomedin: proposed designation for sulphation factor. *Nature (Lond.) 235*, 107.

Davidson, B., Soodak, M., Neary, J. T., Strout, H. V., Kieffer, J. D., Mover, H. and Maloof, F. (1978). The irreversible inactivation of thyroid peroxidase by methylmercaptoimidazole, thiouracil, and propyl-thiouracil *in vitro* and its relationship to *in vivo* findings. *Endocrinology 103*, 871–82.

Davies, A. G. (1972). Thyroid physiology. *Brit. Med. J. 2*, 206–9.

Davies, N. T., Munday, K. A. and Parsons, B. J. (1970). The effect of angiotensin on rat intestinal fluid transfer. *J. Endocrinol. 48*, 39–46.

Davies, R. and Slater, J. D. H. (1976). Is the adrenergic control of renin release dominant in man? *Lancet 2*, 594–6.

Davis, J. O. (1961). Mechanisms regulating the secretion and metabolism of aldosterone in experimental secondary hyperaldosteronism. *Rec. Prog. Horm. Res. 17*, 293–331.

Davis, J. O. (1971). What signals the kidney to release renin. *Circ. Res. 28*, 301–6.

Davis, J. O. and Freeman, R. H. (1976). Mechanisms regulating renin release. *Physiol. Rev. 56*, 1–56.

Davis, J. O., Carpenter, C. C. J., Ayers, C. R., Holman, J. E. and Bahn, R. C. (1961). Evidence for secretion of an aldosterone-stimulating hormone by the kidney. *J. Clin. Invest. 40*, 684–96.

Davis, M., Portmann, B., Searle, M., Wright, R. and Williams, R. (1975). Histological evidence of carcinoma in a hepatic tumour associated with oral contraceptives. *Brit. Med. J. 4*, 496–8.

Dawes, P. J. D., Petersen, V. B., Smith, B. R. and Hall, R. (1978). Solubilization and partial characterization of human and porcine thyrotrophin receptors. *J. Endocrinol. 78*, 89–102.

Dayton, P. G., Israili, Z. H. and Perel, J. M. (1973). Influence of binding on drug metabolism and distribution. *Ann. NY Acad. Sci. 226*, 172–94.

Deacon, S. P. and Barnett, D. (1976). Comparison of atenolol and propranolol during insulin-induced hypoglycemia. *Brit. Med. J. 2*, 272–3.

Deacon, S. P., Karunanayake, A. and Barnett, D. (1977). Acebutolol, atenolol, and propranolol and metabolic responses to acute hypoglycaemia in diabetics. *Brit. Med. J. 2*, 1255–7.

Dean, P. M. and Matthews, E. K. (1972). Pancreatic acinar cells: measurement of membrane potential and miniature depolarization potentials. *J. Physiol. (Lond.) 225*, 1–13.

De Asua, L. J., Carr, B., Clingan, D. and Rudland, P. (1977). Specific glucocorticoid inhibition of growth promoting effects of prostaglandin $F_{2\alpha}$ on 3T3 cells. *Nature (Lond.) 265*, 450–2.

Debas, H. T. (1977). Regulation of gastric secretion. *Fedn. Proc. 36*, 1933–7.

Deckert, T., Andersen, O. O. and Poulsen, J. E. (1974). The clinical significance of highly purified pig-insulin preparations. *Diabetologia 10*, 703–8.

DeGroot, L. J., Refetoff, S., Strausser, J. and Barsano, C. (1974). Nuclear triiodothyronine-binding protein: partial characterization and binding to chromatin. *Proc. Natl. Acad. Sci. USA 71*, 4042–6.

De Häen, C. (1976). The non-stoichiometric floating receptor model for hormone sensitive adenyl cyclase. *J. Theor. Biol. 58*, 383–400.

DeLorenzo, R. J., Walton, K. G., Curran, P. F. and Greengard, P. (1973). Regulation of phosphorylation of a specific protein in toad-bladder membrane by antidiuretic hormone and cyclic AMP, and its possible relationship to membrane permeability changes. *Proc. Natl. Acad. Sci. USA 70*, 880–4.

DeLuca, H. F. (1974). Vitamin D: the vitamin and the hormone. *Fedn. Proc. 33*, 2211–19.

DeLuca, H. F. (1977). Vitamin D metabolism. *Clin. Endocrinol. 7*, Suppl., 1S–17S.

DeLuca, H. F., Morii, H. and Melancon, M. J. (1968). The interaction of vitamin D, parathyroid hormone and thyrocalcitonin. In *Parathyroid Hormone and Thyrocalcitonin (Calcitonin)* (edited by R. V. Talmage and L. F. Belanger), pp. 448–54. International Congress Series 159. Amsterdam: Excerpta Medica Foundation.

De Luise, M., Martin, T. J., Greenberg, P. B. and Michelangeli, V. (1972). Metabolism of porcine,

human and salmon calcitonin in the rat. *J. Endocrinol. 53*, 475–82.

De Meyts, P., Bianco, A. R. and Roth, J. (1976). Site–site interactions among insulin receptors. Characterization of the negative cooperativity. *J. Biol. Chem. 251*, 1877–88.

De Meyts, P., Van Obberghen, E., Roth, J., Wollmer, A. and Brandenburg, D. (1978). Mapping of the residues responsible for the negative cooperativity of the receptor-binding region of insulin. *Nature (London) 273*, 504–9.

de Sousa, R. C., Berde, B. and Mach, R. S. (1965). Syndrome de sécrétion inappropriée d'hormone antidiurétique (syndrome de Schwartz–Bartter). Sécrétion intratumorale de l'arginine-vasopressine ou d'analogues des vasopressines? *Ann. Endocrinol. 25*, 756–63.

De Troyer, A. (1977). Demeclocycline treatment for syndrome of inappropriate antidiuretic hormone secretion. *JAMA 237*, 2723–6.

Devis, G., Somers, G., Van Obberghen, E. and Malaisse, W. J. (1975). Calcium antagonists and islet function: I. Inhibition of insulin release by verapamil. *Diabetes 24*, 547–51.

DeVita, V. T. and Schein, P. S. (1973). The use of drugs in combination for the treatment of cancer. *New Engl. J. Med. 288*, 998–1006.

Devynck, M-A. and Meyer, P. (1978). Angiotensin receptors. *Biochem. Pharmacol. 27*, 1–5.

Devynck, M-A., Pernollet, M.-G., Meyer, P., Fermandjian, S. and Fromageot, P. (1973). Angiotensin receptors in smooth muscle cell membranes. *Nature New Biol. 245*, 55–8.

Devynck, M-A., Pernollet, N-G., Meyer, P., Fermandjian, S., Fromageot, P. and Bumpus, F. M. (1974). Solubilisation of angiotensin II receptors in rabbit aortae membranes. *Nature (Lond.) 249*, 67–9.

de Wied, D. (1971). Long term effect of vasopressin on the maintenance of a conditioned avoidance response in rats. *Nature (Lond.) 232*, 58–60.

de Wied, D. (1976). Behavioral effects of intraventricularly administered vasopressin and vasopressin fragments. *Life Sci. 19*, 685–90.

de Wied, D. (1977a). Pituitary adrenal system hormones and behaviour. *Acta Endocrinol.* Suppl. 214, 9–18.

de Wied, D. (1977b). Peptides and behavior. *Life Sci. 20*, 195–204.

de Wied, D. and de Jong, W. (1974). Drug effects and hypothalamic–anterior pituitary function. *Annu. Rev. Pharmacol. 14*, 389–412.

Di Bella, F., Douša, T. P., Miller, S. S. and Arneaud, C. D. (1974). Parathyroid hormone receptors of renal cortex: specific binding of biologically active, ^{125}I-labeled hormone and relationship to adenylate cyclase activation. *Proc. Natl. Acad. Sci. USA 71*, 723–6.

Di Bella, F. P., Arnaud, C. D. and Brewer, H. B. (1976). Relative biologic activities of human and bovine parathyroid hormones and their synthetic, NH_2-terminal (1–34) peptides, as evaluated *in vitro* with renal cortical adenylate cyclase obtained from three different species. *Endocrinology 99*, 429–36.

Dixon, K. and Schwarz, V. (1969). Uptake of aldosterone by human skin and stimulation of RNA synthesis in the sweat gland *in vitro*. *J. Endocrinol. 45*, 231–44.

Dodds, E. C., Goldenberg, L., Lawson, W. and Robinson,

R. (1938). Oestrogenic activity of certain synthetic compounds. *Nature (Lond.) 141*, 247–8.

Doepfner, W. (1968). The influence of neurohypophysial polypeptides on adenohypophysial function. In *Neurohypophysial Hormones and Similar Peptides* (edited by B. Berde), pp. 625–54. New York: Springer-Verlag.

Doll, R., Langman, M. J. S. and Shawdon, H. H. (1968). Treatment of gastric ulcer with carbenoxolone: antagonistic effect of spironolactone. *Gut 9*, 42.

Domschke, S., Konturek, S. J., Domschke, W., Dembiński, A., Thor, P., Król, R. and Demling, L. (1975). Cyclic AMP and pancreatic bicarbonate secretion in response to secretion in dogs. *Proc. Soc. Exp. Biol. Med. 150*, 773–9.

Doneen, B. A., Bern, H. A. and Li, C. H. (1977). Biological actions of human somatotrophin and its derivatives on mouse mammary gland and teleost urinary bladder. *J. Endocrinol. 73*, 377–83.

Doty, S. B., Robinson, R. A. and Schofield, B. (1976). Morphology of bone and histochemical staining characteristics of bone cells. *Handbook of Physiology:* Section 7, *Endocrinology*, Vol. 7, *The Parathyroid Gland*, pp. 3–23. Washington, D.C.: American Physiological Society.

Douglas, J. and Catt, K. J. (1976). Angiotensin II receptors in the rat adrenal cortex: effects of dietary electrolyte changes. *J. Clin. Invest. 58*, 834–43.

Douglas, J., Aguilera, G., Kondo, T. and Catt, K. (1978). Angiotensin II receptors and aldosterone production in rat adrenal glomerulosa cells. *Endocrinology 102*, 685–96.

Douglas, W. W. (1972). Secretomotor control of adrenal medullary secretion synaptic, membrane, and ionic events in stimulus-secretion coupling. *Handbook of Physiology:* Section 7, *Endocrinology*, Vol. 6, *Adrenal Gland*, pp. 367–88. Washington, D.C.: American Physiological Society.

Douglas, W. W. (1973). How do neurons secrete peptides: Exocytosis and its consequences, including "synaptic vesicle" formation, in the hypothalamoneurohypophysial system. *Prog. Brain Res. 39*, 21–38.

Douglas, W. W. (1974). Mechanism of release of neurohypophysial hormones: stimulation-secretion coupling. *Handbook of Physiology:* Section 7, *Endocrinology*, Vol. 4, *The Pituitary Gland and Its Neuroendocrine Control*, Part 1, pp. 191–224. Washington, D.C.: American Physiological Society.

Douglas, W. W. and Rubin, R. P. (1961). The role of calcium in the secretory response of the adrenal medulla to acetylcholine. *J. Physiol. (Lond.) 159*, 40–57.

Dousa, T. P. (1977). Cyclic nucleotides in the cellular action of neurohypophyseal hormones. *Fedn. Proc. 36*, 1867–71.

Dousa, T. P. and Barnes, L. D. (1977). Regulation of protein kinase by vasopressin in renal medulla in situ. *Amer. J. Physiol. 232*, F50–7.

Dousa, T. P. and Valtin, H. (1976). Cellular actions of vasopressin in the mammalian kidney. *Kidney Int. 10*, 46–63.

Drill, V. A. (1974). Benign cholestatic jaundice of pregnancy and benign cholestatic jaundice from oral contraceptives. *Amer. J. Obstet. Gynecol. 119*, 165–74.

Drill, V. A. (1975). Oral contraceptives: relation to

mammary cancer, benign breast lesions, and cervical cancer. *Annu. Rev. Pharmacol. 15*, 367–85.

Dufau, M. L., Ryan, D. W., Baukal, A. J. and Catt, K. J. (1975). Gonadotropin receptors. Solubilization and purification by affinity chromatography. *J. Biol. Chem. 250*, 4822–4.

Dunn, J. S., Sheehan, H. L. and McLetchie, N. G. B. (1943). Necrosis of islets of Langerhans produced experimentally. *Lancet 1*, 484–7.

Dupont, A., Cusan, L., Garon, M., Labrie, F. and Li, C. H. (1977). β-Endorphin: stimulation of growth hormone release *in vivo. Proc. Natl. Acad. Sci. USA 74*, 358–9.

Du Vigneaud, V. and Tripett, S. (1953). The sequence of amino acids in oxytocin, with a proposal for the structure of oxytocin. *J. Biol. Chem. 205*, 949–57.

Du Vigneaud, V., Lawler, H. C. and Popenoe, E. A. (1953). Enzymic cleavage of glycinamide from vasopressin and a proposed structure for this pressor-antidiuretic hormone of the posterior pituitary. *J. Amer. Chem. Soc. 75*, 4880–1.

Ebling, F. J., Ebling, E., Randall, V. and Skinner, J. (1975). The synergistic action of α-melanocyte-stimulating hormone and testosterone on the sebaceous, prostate, preputial, Harderian and lachrymal glands, seminal vesicles and brown adipose tissue in the hypophysectomized-castrated rat. *J. Endocrinol. 66*, 407–12.

Eckstein, P., Whitby, M., Fotherby, K., Butler, C., Mukherjee, T. K., Burnett, J. B. C., Richards, D. J. and Whitehead, T. P. (1972). Clinical and laboratory findings in a trial of norgestrel, a low-dose progestogen only contraceptive. *Brit. Med. J. 3*, 195–200.

Eddie, L. W., Baker, H. W. G., Dulmanis, A., Higginson, R. E. and Hudson, B. (1978). Inhibin from cultures of rat seminiferous tubules. *J. Endocrinol. 78*, 217–24.

Edelman, I. S. (1974). Thyroid thermogenesis. *New Engl. J. Med. 290*, 1303–8.

Edelman, I. S. (1976). Transition from the poikilotherm to homeotherm: possible role of sodium transport and thyroid hormone. *Fedn. Proc. 35*, 2180–4.

Edelman, I. S., Bogoroch, R. and Porter, G. A. (1963). On the mechanism of action of aldosterone on sodium transport. The role of protein synthesis. *Proc. Natl. Acad. Sci. USA 50*, 1169–77.

Editorial (1972a). Lithium-induced diabetes insipidus. *Brit. Med. J. 2*, 726.

Editorial (1972b). Iodide-induced thyrotoxicosis. *Lancet 2*, 1072–3.

Editorial (1972c). Infertility after the pill. *Brit. Med. J. 4*, 59–60.

Editorial (1972d). Anticonvulsant osteomalacia. *Lancet 2*, 805–6.

Editorial (1972e). Corticosteroid-induced bone collapse. *Brit. Med. J. 1*, 581–2.

Editorial (1973a). Glucagon and growth hormone. *Brit. Med. J. 1*, 188–9.

Editorial (1973b). Hazards of potent topical corticosteroids. *Lancet 1*, 870–1.

Editorial (1973c). Blood Pressure and the pill. *Brit. Med. J. 1*, 693.

Editorial (1973d). Glucagon secretion in obesity. *Lancet 1*, 922–3.

Editorial (1973e). The need for vitamin-D supplements. *Lancet 1*, 1097–8.

Editorial (1973f). Treatment with calcitonin. *Brit. Med. J.* 1, 371–2.

Editorial (1973g). Primary excess and deficiency of renin. *Brit. Med. J.* 1, 627–9.

Editorial (1974a). Allergy to corticosteroids. *Brit. Med. J.* 4, 551–2.

Editorial (1974b). Are sex hormones teratogenic? *Lancet* 2, 1489–90.

Editorial (1974c). Synthetic sex hormones and infants. *Brit. Med. J.* 4, 485–6.

Editorial (1974d). Oral contraceptives and the liver. *Brit. Med. J.* 4, 430–1.

Editorial (1974e). Stilboestrol for prostatic cancer. *Brit. Med. J.* 2, 520.

Editorial (1974f). Chlorpropamide in diabetic pregnancy. *Lancet* 2, 32–3.

Editorial (1975a). Treatment of asthmatic children with steroids. *Brit. Med. J.* 1, 413–14.

Editorial (1975b). Steroids in the eye. *Brit. Med. J.* 1, 645–6.

Editorial (1975c). Steroid therapy and the adrenals. *Lancet* 2, 537–8.

Editorial (1975d). Hormones and elderly testes. *Brit. Med. J.* 3, 2–3.

Editorial (1975e). Cancer risks from hormone treatment. *Brit. Med. J.* 4, 608.

Editorial (1975f). A new line of dysmenorrhoea. *Brit. Med. J.* 2, 461–2.

Editorial (1975g). Glucoreceptors, insulin release and diabetes. *Lancet* 2, 646–7.

Editorial (1975h). Somatostatin and diabetes. *Lancet* 1, 1323–4.

Editorial (1975i). Oral hypoglycemics in diabetes mellitus. *Lancet* 2, 489–91.

Editorial (1976a). Corticosteroids and the fetus. *Lancet* 1, 74–5.

Editorial (1976b). Induction of labour. *Brit. Med. J.* 1, 729–30.

Editorial (1976c). The pill and raised blood pressure. *Brit. Med. J.* 1, 58–9.

Editorial (1976d). Oral contraceptives and liver nodules. *Lancet* 1, 843–4.

Editorial (1977a). The hazardous jungle of topical steroids. *Lancet* 2, 487–8.

Editorial (1977b). Topical steroids and relapses of psoriasis. *Brit. Med. J.* 1, 988.

Editorial (1977c). Suppressing lactation. *Brit. Med. J.* 1, 189.

Editorial (1977d). Liver tumours and the pill. *Brit. Med. J.* 2, 345–6.

Editorial (1977e). Mortality associated with the pill. *Lancet* 2, 747–8.

Editorial (1977f). Oestrogen therapy and endometrial cancer. *Brit. Med. J.* 2, 209–10.

Editorial (1977g). Primary treatment of prostatic cancer. *Brit. Med. J.* 2, 781–2.

Editorial (1977h). The beta cell in diabetes–more sinned against than sinning? *Lancet* 1, 177–8.

Editorial (1977i). Biguanides and lactic acidosis in diabetics. *Brit. Med. J.* 2, 1436.

Editorial (1977j). Motilin: actor in search of a play. *Brit. Med. J.* 1, 1372–3.

Editorial (1978a). Dichloroacetate. *Brit. Med. J.* 2, 674.

Editorial (1978b). Cimetidine for ever (and ever and ever . . .)? *Brit. Med. J.* 1, 1435–6.

Edkins, J. S. (1905). The chemical mechanism of gastric secretion. *J. Physiol. (Lond.)* 34, 133–44.

Edmonds, C. J. and Marriott, J. C. (1967). The effect of aldosterone and adrenalectomy on the electrical potential difference of rat colon on the transport of sodium, potassium, chloride and bicarbonate. *J. Endocrinol.* 39, 517–31.

Edwards, C. R. W., Kitau, M. J., Chard, T. and Besser G. M. (1973). Vasopressin analogue DDAVP in diabetes insipidus: clinical and laboratory studies. *Brit. Med. J.* 3, 375–8.

Ehrlich, E. N. and Crabbé, J. (1968). The mechanism of action of amipramazide. *Pfluegers Arch.* 302, 79–96.

Eilon, G. and Raisz, L. G. (1978). Comparison of the effects of stimulators and inhibitors of resorption on the release of lysozomal enzymes and radioactive calcium from fetal bone in organ culture. *Endocrinology* 103, 1969–75.

Eipper, B. A., Mains, R. E. and Guenzi, D. (1976). High molecular weight forms of adrenocorticotropic hormone are glycoproteins. *J. Biol. Chem.* 251, 4121–6.

Ekblad, E. B. M., Machen, T. E., Licko, V. and Rutten, M. J. (1978). Histamine, cyclic AMP and the secretory respones of piglet gastric mucosa. (Proc. Symp. Gastric Ion Transport Uppsala, 1977). *Acta Physiol. Scand.* Special Suppl., 69–80.

Elder, J. B., Ganguli, P. C., Gillespie, I. E., Gerring, E. L. and Gregory, H. (1975). Effect of urogastrone on gastric secretion and plasma gastrin levels in normal subjects. *Gut* 16, 887–93.

Elder, J. B., Williams, G., Lacey, E. and Gregory, H. (1978). Cellular localisation of human urogastrone/epidermal growth factor. *Nature (Lond.)* 271, 466–7.

Elger, W., Neumann, F. and Von Berswordt-Wallrabe, R. (1971). The influence of androgen antagonists and progestogens on the sex differentiation of different mammalian species. In *Hormones in Development* (edited by M. Hamburgh and E. J. W. Barrington), pp. 651–67. New York: Appleton-Century-Crofts.

Elliott, H. R., Abdulla, U. and Hayes, P. J. (1978). Pulmonary oedema associated with ritrodine infusion and betamethasone administration in premature labour. *Brit. Med. J.* 2, 799–800.

Emmens, C. W. (1970). Antifertility agents. *Annu. Rev. Pharmacol.* 10, 237–54.

Engelsman, E., Persijn, J. P., Korsten, C. B. and Cleton, F. J. (1973). Oestrogen receptor in human breast cancer tissue and response to endocrine therapy. *Brit. Med. J.* 2, 750–2.

England, P., Lorrimer, D., Fergusson, J. C., Moffatt, A. M. and Kelly, A. M. (1974). Human placental lactogen. The watchdog of fetal distress. *Lancet* 1, 5–7.

Enochs, M. R. and Johnson, L. R. (1977). Trophic effects of gastrointestinal hormones: physiological implications. *Fedn. Proc.* 36, 1942–7.

Epstein, M. T., Espiner, E. A., Donald, R. A. and Hughes, H. (1977). Effect of eating liquorice on the renin-angiotensin aldosterone axis in normal subjects. *Brit. Med. J.* 1, 488–90.

Epstein, M. T., Espiner, E. A., Donald, R. A., Hughes, H., Cowles, R. J. and Lun, S. (1978). Licorice raises urinary cortisol in man. *J. Clin. Endocrinol. Metab.* 47, 397–400.

Erdös, E. G. (1977). The angiotensin I converting enzyme. *Fedn. Proc.* 36, 1760–5.

Erickson, G. F. and Hsueh, A. J. W. (1978). Secretion of "inhibin" by rat granulosa cells *in vitro*. *Endocrinology* *103*, 1960–3.

Escher, E. H. F., Nguyen, T. M. D., Robert, H., St-Pierre, S. A. and Regoli, D. C. (1978a). Photaffinity labelling of the angiotensin II receptor: 1. Synthesis and biological activities of the labeling peptides. *J. Med. Chem. 21*, 860–4.

Escher, E. H. F., Nguyen, T. M. D., Guillemette, G. and Regoli, D. C. (1978b). Specific and irreversible block of the myotropic action of angiotensin II. *Nature (Lond.) 275*, 145–6.

Eskildsen, P. C., Svendsen, P. A., Vang, L. and Nerup, J. (1978). Long-term treatment of acromegaly with bromocriptine. *Acta Endocrinol. 87*, 687–700.

Esler, M., Julius, S., Zweifler, A., Randall, O., Harburg, E., Gardiner, H. and DeQuattro, V. (1977). Mild high-renin essential hypertension. Neurogenic human hypertension? *New Engl. J. Med. 296*, 405–11.

Evans, G. A., Hucko, J. and Rosenfeld, M. G. (1977). Preprolactin represents the initial product of prolactin mRNA translation. *Endocrinology 101*, 1807–14.

Evered, D. C. (1976). Endocrine metabolic diseases. Treatment of thyroid disease I. *Brit. Med. J. 1*, 264–6.

Evered, D. C. and Hall, R. (1972). Hypothyroidism. *Brit. Med. J. 1*, 290–3.

Evered, D. C., Ormston, B. J., Smith, P. A., Hall, R. and Bird, T. (1973). Grades of hypothyroidism. *Brit. Med. J. 1*, 657–62.

Ewing, L. L. and Robaire, B. (1978). Endogenous anti-spermatogenic agents: prospects for male contraception. *Annu. Rev. Pharmacol. Toxicol. 18*, 167–87.

Fain, J. N. and Czech, M. P. (1975). Glucocorticoid effects on lipid mobilization and adipose tissue metabolism. *Handbook of Physiology:* Section 7, *Endocrinology*, Vol. 6, *Adrenal Gland*, pp. 169–78. Washington, D.C.: American Physiological Society.

Fajans, S. S. and Floyd, J. C. (1976). Fasting hypoglycemia in adults. *New Engl. J. Med. 294*, 766–72.

Fanestil, D. D. and Edelman, I. S. (1966). Characteristics of the renal nuclear receptors for aldosterone. *Proc. Natl. Acad. Sci. USA 56*, 872–9.

Fang, S. and Liao, S. (1969). Antagonistic action of anti-androgens on the formation of a specific dihydro-testosterone–receptor protein complex in rat prostate. *Molec. Pharmacol. 5*, 420–31.

Farese, R. V. and Prudente, W. J. (1978). On the role of calcium in adrenocorticotropin-induced changes in mitochondrial pregnenolone synthesis. *Endocrinology 103*, 1264–71.

Farnsworth, N. R., Bingel, A. S., Cordell, G. A., Crane, F. A. and Fong, H. H. S. (1975a). Potential value of plants as sources of new antifertility agents I. *J. Pharm. Sci. 64*, 535–98.

Farnsworth, N. R., Bingel, A. S., Cordell, G. A., Crane, F. A. and Fong, H. H. S. (1975b). Potential value of plants as sources of new antifertility agents II. *J. Pharm. Sci. 64*, 717–54.

Farquhar, M. G., Reid, J. J. and Daniell, L. W. (1978). Intracellular transport and packaging of prolactin: a quantitative electron microscope autoradiographic study of mammotrophs dissociated from rat pituitaries. *Endocrinology 102*, 296–311.

Farrell, C. C., Joshua, D. E., Uren, R. F., Baird, P. J.,

Perkins, K. W. and Kronenberg, H. (1975). Androgen-induced hepatoma. *Lancet 1*, 430–2.

F. D. A. (1977). HEW secretary suspends general marketing of phenformin. *FDA Bull.* (August), 14–16.

Federick, J. and Yudkin, P. (1976). Obstetrics practice in the Oxford record linkage study area 1965–72. *Brit. Med. J. 1*, 738–40.

Feigelson, M. and Feigelson, P. (1966). Relationships between hepatic enzyme induction, glutamate formation, and purine nucleotide biosynthesis glucocorticoid action. *J. Biol. Chem. 241*, 5819–26.

Feigelson, P., Beato, M., Colman, P., Kalimi, M., Killewich, L. A. and Schutz, G. (1975). Studies on the hepatic glucocorticoid receptor and on the hormonal modulation of specific mRNA levels during enzyme induction. *Rec. Prog. Horm. Res. 31*, 213–39.

Feinglos, M. N. and Lebovitz, H. E. (1978). Sulphonylureas increase the number of insulin receptors. *Nature (Lond.) 276*, 184–5.

Fejes-Tóth, G., Magyar, A. and Walter, J. (1977). Renal response to vasopressin after inhibition of prostaglandin synthesis. *Ann. J. Physiol. 232*, F416–23.

Feldman, D. and Loose, D. (1977). Glucocorticoid receptors in adipose tissue. *Endocrinology 100*, 398–405.

Feldman, D., Funder, J. W. and Edelman, I. S. (1972). Subcellular mechanisms in the action of adrenal steroids. *Amer. J. Med. 53*, 545–60.

Feldman, R. (1977). Oral hypoglycemic agents. *New Engl. J. Med. 297*, 394.

Felig, P. (1974a). Diabetic ketoacidosis. *New Engl. J. Med. 290*, 1360–3.

Felig, P. (1974b). Insulin: rates and routes of delivery. *New Engl. J. Med. 291*, 1031–2.

Ferguson, D. R. and Twite, B. R. (1974). Effects of vasopressin on toad bladder membrane proteins: relationship to transport of sodium and water. *J. Endocrinol. 61*, 501–7.

Ferguson, E. R. and Katzenellenbogen, B. S. (1977). A comparative study of antiestrogen action: temporal patterns of antagonism of estrogen stimulated uterine growth and effects of estrogen receptor levels. *Endocrinology 100*, 1242–51.

Ferland, L., Labrie, F., Savary, M., Beaulieu, M., Coy, D. H., Coy, E. J. and Schally, A. V. (1976). Inhibitory activity of analogues of luteinizing hormone-releasing hormone (LH-RH) *in vitro* and *in vivo*. *Clin. Endocrinol. 5*, Suppl., 279S–89S.

Fernandes, M. and Bellini, G. (1977). Management of the patient with pheochromocytoma. *Drug Ther.* (May), 43–7.

Ferreira, S. H., Bartelt, D. C. and Greene, L. J. (1970). Isolation of bradykinin potentiating peptides from *Bothrops jararaca*. *Biochemistry 9*, 2583–93.

Feyrter, F. (1938). *Über diffuse endokrine Epitheliale Organe.* Leipzig: Barth.

Field, J. B. (1972). Insulin extraction by the liver. *Handbook of Physiology:* Section 7, *Endocrinology*, Vol. 1, *Endocrine Pancreas*, pp. 505–13. Washington, D.C.: American Physiological Society.

Fimognari, G. M., Fanestil, D. D. and Edelman, I. S. (1967). Induction of RNA and protein synthesis in the action of aldosterone in the rat. *Amer. J. Physiol. 213*, 954–62.

Fimognari, G. M., Porter, G. A. and Edelman, I. S. (1967). The role of the tricarboxylic acid cycle in the action of aldosterone on sodium transport. *Biochim. Biophys. Acta 135*, 89–99.

Finkelstein, W. and Isselbacher, K. J. (1978). Cimetidine. *New Engl. J. Med. 299*, 992–6.

Fischer, J. A., Oldham, S. B., Sizemore, G. W. and Arnaud, C. D. (1972). Calcium-regulated parathyroid hormone peptidase. *Proc. Natl. Acad. Sci. USA 69*, 2341–5.

Fisher, J. N., Shahshahani, M. N. and Kitabchi, A. E. (1977). Diabetic ketoacidosis: low-dose insulin therapy by various routes. *New Engl. J. Med. 297*, 238–41.

Fishman, J. and Hellman, L. (1976). 7β,17α-Dimethyltestosterone (calusterone)-induced changes in the metabolism, production rate, and excretion of estrogens in women with breast cancer: a possible mechanism of action. *J. Clin. Endocrinol. Metab. 42*, 365–9.

Fiskin, A. M., Cohn, D. V. and Peterson, G. S. (1977). A model for the structure of bovine parathormone derived by dark field electron microscopy. *J. Biol. Chem. 252*, 8261–8.

Fitzsimons, J. T. (1972). Thirst. *Physiol. Rev. 52*, 468–561.

Fitzsimons, J. T. (1976). The physiological basis of thirst. *Kidney Int. 10*, 3–11.

Fitzsimons, J. T. and Kucharczyk, J. (1978). Drinking and haemodynamic changes induced in the dog by intracranial injection of components of the renin-angiotensin system. *J. Physiol. (Lond.) 276*, 419–34.

Fitzsimons, J. T. and Wirth, J. B. (1975). The failure of peripheral–renin–angiotensin activation to induce sodium appetite. *J. Physiol. (Lond.) 251*, 31P–3P.

Fitzsimons, J. T. and Wirth, J. B. (1978). The renin-angiotensin system and sodium appetite. *J. Physiol. (Lond.) 274*, 63–80.

Fitzsimons, J. T., Kucharczyk, J. and Richards, G. (1978). Systemic angiotensin-induced drinking in the dog: a physiological phenomenon. *J. Physiol. (Lond.) 276*, 435–48.

Floman, Y. and Zor, U. (1976). Mechanism of steroid action in inflammation: inhibition of prostaglandin synthesis and release. *Prostaglandins 12*, 403–13.

Florsheim, W. H. (1974). Control of thyrotrophin secretion. *Handbook of Physiology:* Section 7, *Endocrinology*, Vol. 4, *The Pituitary Gland and Its Neuroendocrine Control*, Part 2, pp. 449–67. Washington, D.C.: American Physiological Society.

Flückiger, E. (1978). Effects of bromocriptine on the hypothalamo–pituitary axis. *Acta Endocrinol. 88*, Suppl. 216, 111–17.

Forrest, J. N. (1975). Lithium inhibition of cAMP mediated hormones: a caution. *New Engl. J. Med. 292*, 423–4.

Forrest, J. N. and Singer, I. (1977). Drug-induced interference with action of antidiuretic hormone. In *Disturbances in Body Fluid Osmolality* (edited by T. E. Andreoli, J. J. Grantham, and F. C. Rector), pp. 309–40. Bethesda, Md.: American Physiological Society.

Fox, J., Care, A. D. and Marshall, D. H. (1978). Reversal of betamethasone induced inhibition of intestinal calcium absorption by 1α-hydroxycholecalciferol. *J. Endocrinol. 78*, 187–94.

Franchimont, P., Chari, S., Schellen, A. M. C. M. and Demoulin, A. (1975). Relationship between gonadotrophins, spermatogenesis and seminal plasma. *J. Steroid Biochem. 6*, 1037–41.

Frank, R. T. (1931). The hormonal causes of premenstrual tension. *Arch. Neurol. Psychiatry 26*, 1053–7.

Franklin, W. (1974). Treatment of severe asthma. *New Engl. J. Med. 290*, 1469–72.

Frantz, A. G. and Kleinberg, D. L. (1978). The pathophysiology of hyperprolactinemic states and the role of newer ergot compounds in their treatment. *Fedn. Proc. 37*, 2192–6.

Frantz, A. G., Kleinberg, D. L. and Noel, G. L. (1972). Studies on prolactin in man. *Rec. Prog. Horm. Res. 28*, 527–73.

Fraser, D., Kooh, S. W., Kind, H. P., Holick, M., Tanaka, Y. and DeLuca, H. F. (1973). Pathogenesis of hereditary vitamin-D-dependent rickets. *New Engl. J. Med. 289*, 817–22.

Freed, D. L. J., Bank, A. J., Longson, D. and Burley, D. M. (1975). Anabolic steroids in athletics: crossover double-blind trial on weightlifters. *Brit. Med. J. 2*, 471–3.

Freeman, R. H., Davis, J. O., Lohmeier, T. E. and Spielman, W. S. (1977). [Des-Asp¹]Angiotensin II: mediator of the renin–angiotensin system? *Fedn. Proc. 36*, 1766–70.

Freitag, J., Martin, K. J., Hruska, K. A., Anderson, C., Conrades, M., Ladenson, J., Klahr, S. and Slatopolsky, E. (1978). Impaired parathyroid hormone metabolism in patients with chronic renal failure. *New Engl. J. Med. 298*, 29–32.

Freston, J. W. (1978). Cimetidine in the treatment of gastric ulcer. Review and commentary. *Gastroenterology 74*, 426–30.

Freychet, P., Roth, J. and Neville, D. M. (1971). Insulin receptors in the liver: specific binding of [¹²⁵I] insulin to the plasma membrane and its relation to insulin bioactivity. *Proc. Natl. Acad. Sci. USA 68*, 1833–7.

Friesen, H. G. and Tolis, G. (1977). The use of bromocriptine in the galactorrhea–amenorrhea syndromes: the Canadian Cooperative Study. *Clin. Endocrinol. 6*, Suppl., 91S–9S.

Froesch, E. R., Bürgi, H., Ramseier, E. B., Bally, P. and Labhart, A. (1963). Antibody-suppressible and nonsuppressible insulin-like activities in human serum and their physiologic significance. An insulin assay with adipose tissue of increased precision and specificity. *J. Clin. Invest. 42*, 1816–34.

Froesch, E. R., Burgi, H., Müller, W. A., Humbel, R. E., Jakob, A. and Labhart, A. (1967). Nonsuppressible insulin like activity in human serum: purification, physicochemical and biological properties and its relation to total serum ILA. *Rec. Prog. Horm. Res. 23*, 565–605.

Frost, F. et al. (1977). Cimetidine in patients with gastric ulcer: a multicentre controlled trial. *Brit. Med. J. 2*, 795–9.

Frye, B. E. (1967). *Hormonal Control in Vertebrates*, p. 104. New York: Macmillan.

Fujimoto, W. Y. and Ensinck, J. W. (1976). Somatostatin inhibition of insulin and glucagon secretion in rat islet culture: reversal by ionophore A23187. *Endocrinology 98*, 259–62.

Fujino, M., Wakimasu, M., Taketomi, S. and Iwatsuka, H. (1977). Insulin-like activities and insulin-potentiating actions of a modified insulin B₂₁₋₂₆ fragment: β-Ala-Arg-Gly-Phe-Phe-Tyr-NH₂. *Endocrinology 101*, 360–4.

Fukushima, D. K., Zumoff, B., Bulkin, W. and Hellman, L. (1976). Effect of 7β,17α-dimethyltestosterone (calusterone) on cortisol metabolism in women with advanced breast cancer. *J. Clin. Endocrinol. Metab. 43*, 38–45.

Fuller, D. J. M., Byus, C. V. and Russell, D. H. (1978). Specific regulation by steroid hormones of the amount of type I cyclic AMP-dependent protein kinase holoenzyme. *Proc. Natl. Acad. Sci. USA 75*, 223–7.

Fuller, P. J. and Funder, J. W. (1976). Mineralocorticoid and glucocorticoid receptors in human kidney. *Kidney Int. 10*, 154–7.

Fuller, R. W. (1973). Control of epinephrine synthesis and secretion. *Fedn. Proc. 32*, 1772–81.

Funder, J. W., Feldman, D. and Edelman, I. S. (1972). Specific aldosterone binding in rat kidney and parotoid. *J. Steroid Biochem. 3*, 209–18.

Funder, J. W., Feldman, D., Highland, E. and Edelman, I. S. (1974). Molecular modifications of antialdosterone compounds: effects on affinity of spirolactones for renal aldosterone receptors. *Biochem. Pharmacol. 23*, 1493–501.

Furchgott, R. F. (1978). Pharmacological characterization of receptors: its relation to radioligand-binding studies. *Fedn. Proc. 37*, 115–20.

Gagerman, E., Idahl, L-A., Meissner, H. P. and Taljedal, I-B. (1978). Insulin release, cGMP, cAMP, and membrane potential in acetylcholine-stimulated islets. *Amer. J. Physiol. 235*, E493–E500.

Gainer, H., Sarne, Y. and Brownstein, M. J. (1977). Biosynthesis and axonal transport of rat neurohypophysial proteins and peptides. *J. Cell. Biol. 73*, 366–81.

Gale, E. A. M. and Tattersall, R. B. (1976). Can phenformin-induced lactic acidosis be prevented? *Brit. Med. J. 2*, 972–5.

Ganda, O. P., Rossini, A. A. and Like, A. A. (1976). Studies on strepozotocin diabetes. *Diabetes 25*, 595–603.

Ganong, W. F. (1977). The renin–angiotensin system and the central nervous system. *Fedn. Proc. 36*, 1771–5.

Ganten, D., Marquez-Julio, A., Granger, P., Hayduk, K., Karsunky, K. P., Boucher, R. and Genest, J. (1971). Renin in dog brain. *Amer. J. Physiol. 221*, 1733–7.

Ganten, D., Schelling, P., Vecsei, P. and Ganten, U. (1976). Isorenin extrarenal origin. "The tissue angiotensinogenase" systems. *Amer. J. Med. 60*, 760–72.

Garabedian, M., Tanaka, Y., Holick, M. F. and DeLuca, H. F. (1974). Response of intestinal calcium transport and bone calcium mobilization to 1,25-dihydroxyvitamin D_3 in thyroparathyroidectomized rats. *Endocrinology 94*, 1022–7.

Gardner, J. D., Jackson, M. J., Batzri, S. and Jensen, R. T. (1978). Potential mechanisms of interaction among secratogues. *Gastroenterology 74*, 348–54.

Garel, J.-M., Barlet, J.-P. and Kervran, A. (1975). Metabolic effects of calcitonin in the newborn. *Amer. J. Physiol. 229*, 669–75.

Garland, J. T., Lottes, M. E., Kozak, S. and Daughaday, W. H. (1972). Stimulation of DNA synthesis in isolated chondrocytes by sulfation factor. *Endocrinology 90*, 1086–90.

Gavras, H., Brunner, H. R., Vaughan, E. D. and Laragh, J. H. (1973). Angiotensin-sodium interaction in blood pressure maintenance of renal hypertensive and normotensive rats. *Science 180*, 1369–72.

Gavras, H., Brunner, H. R., Laragh, J. H., Sealey, J. E., Gavras, I. and Vukovich, R. A. (1974). An angiotensin converting-enzyme inhibitor to identify and treat vasoconstrictor and volume factors in the hypertensive patients. *New Engl. J. Med. 291*, 817–21.

Gavras, H., Brunner, H. R., Turini, G. A., Kershaw, G. R., Tift, C. P., Cuttelod, S., Gavras, I., Vukovich, R. A. and McKinstry, D. N. (1978). Antihypertensive effect of the oral angiotensin converting-enzyme inhibitor SQ 14225 in man. *New Engl. J. Med. 298*, 991–5.

Gazis, D. and Sawyer, W. H. (1978). Elimination of infused arginine-vasopressin and its long-acting deaminated analogue in rats. *J. Endocrinol. 79*, 179–86.

Gemzell, C. A., Diczfalusy, E. and Tillinger, K. G. (1958). Clinical effect of human pituitary follicle stimulating hormone (FSH). *J. Clin. Endocrinol. Metab. 18*, 1333–48.

George, F. W. and Wilson, J. D. (1978). Estrogen formation in the early rabbit embryo. *Science 199*, 200–1.

Gerber, N. L. and Steinberg, A. D. (1976). Clinical use of immunosuppressive drugs (part II). *Drugs 11*, 90–112.

Gerich, J. E., Lovinger, R. and Grodsky, G. M. (1975). Inhibition by somatostatin of glucagon and insulin release from the perfused rat pancreas in response to arginine, isoproterenol and theophylline: evidence for a preferential effect on glucagon secretion. *Endocrinology 96*, 749–54.

Gerich, J. E., Lorenzi, M., Schneider, V. and Forsham P. H. (1974). Effect of somatostatin on plasma glucagon and tolbutamide in man. *J. Clin. Endocrinol. Metab. 39*, 1057–60.

Gerich, J. E., Lorenzi, M., Bier, D. M., Schneider, V., Tsalikian, E., Karam, J. H. and Forsham, P. H. (1975). Prevention of human diabetic ketoacidosis by somatostatin. Evidence for an essential role of glucagon. *New Engl. J. Med. 292*, 985–9.

Gershberg, H. (1977). Oral hypoglycemic agents. *New Engl. J. Med. 297*, 394.

Ghose, K. and Coppen, A. (1977). Bromocriptine and premenstrual syndrome: controlled study. *Brit. Med. J. 1*, 147–8.

Gibbs, D. M. and Neill, J. D. (1978). Dopamine levels in hypophysial stalk blood in the rat are sufficient to inhibit prolactin secretion *in vivo*. *Endocrinology 102*, 1895–900.

Gibiński, K., Rybicka, J., Mikoś, E. and Nowak, A. (1977). Double-blind clinical trial on gastroduodenal ulcer healing with prostaglandin E_2 analogues. *Gut 18*, 636–9.

Gilbert, J. P. et al. (1975). (Committee for the assessment of biometric aspects of controlled trials of hypoglycemic agents.) Report of the committee for the assessment of biometric aspects of controlled trials of hypoglycemic agents. *JAMA 231*, 583–600.

Gilkes, J. J. H., Rees, L. H. and Besser, G. M. (1977). Plasma immunoreactive corticotrophin and lipotrophin in Cushing's syndrome and Addison's disease. *Brit. Med. J. 1*, 996–8.

Gill, G. N. (1972). Mechanism of ACTH action. *Metabolism 21*, 571–88.

Gill, G. N., Ill, C. R. and Simonian, M. H. (1977). Angiotensin stimulation of bovine adrenocortical growth. *Proc. Natl. Acad. Sci. USA 74*, 5569–73.

Gillette, J. R. (1973). Overview of drug–protein binding. *Ann. NY Acad. Sci. 226*, 6–17.

Gillies, G., van Wimersma Greidanus, T. B. and Lowry,

P. J. (1978). Characterization of rat stalk median eminence vasopressin and its involvement in adrenocorticotropin release. *Endocrinology 103*, 528–34.

Ginsburg, J., Isaacs, A. J., Gore, M. B. R. and Harvard, C. W. H. (1975). Use of clomiphene and lutenizing hormone/follicle stimulating hormone-releasing hormone in investigation of ovulatory failure. *Brit. Med. J. 3*, 130–3.

Ginsburg, M. (1968). Production, release, transportation and elimination of neurohypophysial hormones. In *Neurohypophysial Hormones and Similar Peptides* (edited by B. Berde), pp. 286–371. New York: Springer-Verlag.

Ginsburg, M., Maclusky, N. J., Morris, I. D. and Thomas, P. J. (1977). The specificity of oestrogen receptor in brain, pituitary and uterus. *Brit. J. Pharmacol. 59*, 397–402.

Gispen, W. H. and Wiegant, V. M. (1976). Opiate antagonists suppress ACTH, $_{1-24}$-induced excessive grooming in the rat. *Neurosci. Lett. 2*, 159–64.

Gispen, W. H., Wiegant, V. M., Bradbury, A. F., Hulme, E. C., Smyth, D. G., Snell, C. R. and de Wied, D. (1976). Induction of excessive grooming in the rat by fragments of lipotropin. *Nature (Lond.) 264*, 794–5.

Giwa-Osagie, O. F., Savage, J. and Newton, J. R. (1978). Norethisterone oenanthate as an injectable contraceptive: use of a modified dose schedule. *Brit. Med. J. 1*, 1660–2.

Glorieux, F. H., Scriver, C. R., Holick, M. F. and DeLuca, H. F. (1973). X-linked hypophosphataemic rickets: inadequate response to 1,25-hydroxycholecalciferol. *Lancet 1*, 287–9.

Glossman, H., Baukal, A. and Catt, K. J. (1974). Cation dependence of high-affinity angiotensin II binding to adrenal cortex receptors. *Science 185*, 281–3.

Golander, A., Hurley, T., Barrett, J., Hizi, A. and Handwerger, S. (1978). Prolactin synthesis by human chorionic-decidual tissue: a possible source of prolactin in the amniotic fluid. *Science 202*, 311–13.

Gold, H. K., Prindle, K. H., Levey, G. S. and Epstein, S. E. (1970). Effects of experimental heart failure on the capacity of glucagon to augment myocardial contractility and activate adenyl cyclase. *J. Clin. Invest. 49*, 999–1006.

Gold, P. W., Goodwin, F. K. and Reus, V. I. (1978). Vasopressin in affective illness. *Lancet 1*, 1233–6.

Goldblatt, H., Lynch, J., Hanzal, R. F. and Summerville, W. W. (1934). Studies on experimental hypertension: I. The production of persistent elevation of systolic blood pressure by means of renal ischemia. *J. Exp. Med. 59*, 347–79.

Goldenberg, I. S., Waters, M. N., Ravdin, R. V., Ansfield, F. J. and Segaloff, A. (1973). Androgenic therapy for advanced breast cancer in women. *JAMA 223*, 1267–8.

Goldfine, I. D. and Smith, G. J. (1976). Binding of insulin to isolated nuclei. *Proc. Natl. Acad. Sci. USA 73*, 1427–31.

Goldman, A. S., Shapiro, B. H. and Root, A. W. (1976). Effects of new multi-site hormone blockers on the fertility of male rats. *J. Endocrinol. 69*, 11–21.

Goldmann, H. (1962). Cortisone glaucoma. *Arch. Ophthalmol. 68*, 621–6.

Goldsmith, P. C. (1977). Ultra-structural localization of some hypothalamic hormones. *Fedn. Proc. 36*, 1968–72.

Goldsmith, P. C., Rose, J. C., Arimura, A. and Ganong, W. F. (1975). Ultrastructural localization of somatostatin in pancreatic islets of the rat. *Endocrinology 97*, 1061–4.

Goldstein, A. (1976). Opioid peptides (endorphins) in pituitary and brain. *Science 193*, 1081–6.

Goldstein, D. J., Diaz, A., Finkielman, S., Nahmod, V. E. and Fischer-Ferraro, C. (1972). Angiotensinase activity in rat and dog brain. *J. Neurochem. 19*, 2451–2.

Goldstein, M., Anagnoste, B., Freedman, L. S., Roffman, M., Ebstein, R. P., Park, D. H., Fuxe, K. and Hökfelt, T. (1973). Characterization, localisation and regulation of catecholamine synthesising enzymes. In *Frontiers in Catecholamine Research* (edited by E. Usdin and S. H. Snyder), pp. 69–78. New York: Pergamon Press.

Goldzieher, J. W. and Rudel, H. W. (1974). How the oral contraceptives came to be developed. *JAMA 230*, 421–5.

Goltzman, D., Peytremann, A., Callahan, E., Tregear, G. W. and Potts, J. T. (1975). Analysis of the requirements for parathyroid hormone action in renal membranes with the use of inhibiting analogues. *J. Biol. Chem. 250*, 3199–203.

Goltzman, D., Callahan, E. N., Tregear, G. W. and Potts, J. T. (1978). Influence of guanyl nucleotides on parathyroid hormone-stimulated adenyl cyclase activity in renal cortical membranes. *Endocrinology 103*, 1352–60.

Goodman, R. L. (1978). The site of the positive feedback action of estradiol in the rat. *Endocrinology 102*, 151–9.

Gordan, G. S., Wessler, S. W. and Avioli, L. V. (1972). Calusterone in the therapy for advanced breast cancer. *JAMA 219*, 483–90.

Gorden, P., Lesniak, M. A., Hendricks, C. M. and Roth, J. (1973). "Big" growth hormone components from human plasma: decreased reactivity demonstrated by radioreceptor assay. *Science 182*, 829–31.

Gorden, P., Carpentier, J-L., Freychet, P., LeCam, A. and Orci, L. (1978). Intracellular translocation of iodine-125-labeled insulin: direct demonstration in isolated hepatocytes. *Science 200*, 782–5.

Gordin, A., Lamberg, B-A., Pelkonen, R. and Almquist, S. (1978). Somatostatin inhibits the pentagastrin-induced release of serum calcitonin in medullary carcinoma of the thyroid. *Clin. Endocrinol. 8*, 289–93.

Gordon, G. G., Altman, K., Southren, A. L., Rubin, E. and Lieber, C. S. (1976). Effect of alcohol (ethanol) administration on sex-hormone metabolism in normal men. *New Engl. J. Med. 295*, 793–7.

Graf, L., Szekely, J. I., Ronai, A. Z., Dunai-Kovacs, Z. and Bajusz, S. (1976). Comparative study on analgesic effect of Met^5-enkephalin and related lipotropin fragments. *Nature (Lond.) 263*, 240–2.

Gray, G. R., McKenzie, I., Smith, I. S., Crean, G. P. and Gillespie, G. (1977). Oral cimetidine in severe duodenal ulceration. A double-blind controlled study. *Lancet 1*, 4–7.

Gray, H. W., Greig, W. R., Thomson, J. A. and McLennan, I. (1974). Intravenous perchlorate test in the diagnosis of Hashimoto's disease. *Lancet 1*, 335–8.

Gray, J. S., Culmer, C. U., Wieczorowski, E. and Adkison, J. L. (1940). Preparation of pyrogen-free urogastrone. *Proc. Soc. Exp. Biol. Med. 43*, 225–8.

Greaves, M. F. (1977). Membrane receptor–adenylate cyclase relationships. *Nature (Lond.) 265*, 681–3.

Green, J. D. (1966). The comparative anatomy of the portal vascular system and the innervation of the hypophysis. In *The Pituitary Gland* (edited by G. W. Harris and B. T. Donovan), Vol. 1, *Anterior Pituitary*, pp. 127–46. Berkeley, Calif.: University of California Press.

Green, M. R., Bunting, S. L. and Peacock, A. C. (1971). Changes in labeling pattern of ribonucleic acid from mammary tissue as a result of hormone treatment. *Biochemistry 10*, 2366–71.

Greenberg, P. B., Beck, C., Martin, T. J. and Burger, H. G. (1972). Synthesis and release of human growth hormone from lung carcinoma in cell culture. *Lancet 1*, 350–2.

Greenblatt, D. J. and Koch-Weser, J. (1973). Adverse reactions to spironolactone. A report from the Boston Collaborative Drug Surveillance Program. *JAMA 225*, 40–3.

Greenblatt, D. J. and Koch-Weser, J. (1975a). Clinical pharmacokinetics (part 1). *New Engl. J. Med. 293*, 702–5.

Greenblatt, D. J. and Koch-Weser, J. (1975b). Clinical pharmacokinetics (part 2). *New Engl. J. Med. 293*, 964–70.

Greenblatt, R. B., McDonough, P. G. and Mahesh, V. B. (1968). Estrogen therapy for inhibition of linear growth. In *Clinical Endocrinology* (edited by E. B. Astwood and C. E. Cassidy), Vol. 2, pp. 117–31. New York: Grune & Stratton.

Greene, R., Farran, H. and Glascock, R. F. (1958). Goitrogens in milk. *J. Endocrinol. 17*, 272–9.

Greenwood, R. H., Mahler, R. F. and Hales, C. N. (1976). Improvement in insulin secretion in diabetes after diazoxide. *Lancet 1*, 444–7.

Greenwood, R. H., Prunty, F. T. G. and Silver, J. (1973). Osteomalacia after prolonged glutethimide administration. *Brit. Med. J. 1*, 643–5.

Greer, M. A. and Haibach, H. (1974). Thyroid secretion. *Handbook of Physiology:* Section 7, *Endocrinology*, Vol. 3, *Thyroid*, pp. 135–46. Washington, D.C.: American Physiological Society.

Greer, M. A., Kammer, H. and Bouma, D. J. (1977). Short-term antithyroid drug therapy for the thyrotoxicosis of Graves' disease. *New Engl. J. Med. 297*, 173–6.

Greff, D., Fermandjian, S., Fromageot, P., Khosla, M. C., Smeby, R. R. and Bumpus, F. M. (1976). Circular-dichroism spectra of truncated and other analogs of angiotensin II. *Eur. J. Biochem. 61*, 297–305.

Gregory, H. (1975). Isolation and structure of urogastrone and its relationship to epidermal growth factor. *Nature (Lond.) 257*, 325–7.

Gregory, R. A. (1974). The gastrointestinal hormones: a review of recent advances. *J. Physiol. (Lond.) 241*, 1–32.

Greiner, J. W., Rumbaugh, R. C., Kramer, R. E. and Colby, H. D. (1978). Relation of canrenone to the actions of spironolactone on adrenal cytochrome P-450-dependent enzymes. *Endocrinology 103*, 1313–20.

Grob, D. (1976). Use of drugs in myopathies. *Annu. Rev. Pharmacol. Toxicol. 16*, 215–29.

Gross, J. and Pitt-Rivers, R. (1953). 3,5,3'-Triiodothyronine: I. Isolation from thyroid gland and synthesis. *Biochem. J. 53*, 645–52.

Grossman, M. I. (1970). Gastrin, cholecystokinin and secretin act on one receptor. *Lancet 1*, 1088–9.

Grossman, M. I. (1974). Gastrointestinal hormones: spectrum of actions and structure–activity relations. In *Endocrinology of the Gut* (edited by W. Y. Chey and F. P. Brooks), pp. 65–75. Thorofare, N.J.: Charles B. Slack.

Grossman, M. I. (1977). Physiological effects of gastrointestinal hormones. *Fedn. Proc. 36*, 1930–2.

Grossman, M. I. and Konturek, S. J. (1974). Inhibition of acid secretion in dog by metiamide, a histamine antagonist acting on H_2 receptors. *Gastroenterology 66*, 517–21.

Gryglewski, R. J. (1976). Steroid hormones, anti-inflammatory steroids and prostaglandins. *Pharmacol. Res. Commun. 8*, 337–48.

Gudmand-Høyer, E., Jensen, K. B., Krag, E., Rask-Madsen, J., Rahbek, I., Rune, S. J. and Wulff, H. R. (1978). Prophylactic effect of cimetidine in duodenal ulcer disease. *Brit. Med. J. 1*, 1095–7.

Guillemin, R. (1976). Somatostatin inhibits the release of acetylcholine induced electrically in the myenteric plexus. *Endocrinology 99*, 1653–4.

Guillemin, R. (1977). The expanding significance of hypothalamic peptides, or, is endocrinology a branch of neuroendocrinology? *Rec. Prog. Horm. Res. 33*, 1–20.

Guillemin, R. (1978a). Peptides in the brain: the new endocrinology of the neuron. *Science 202*, 390–402.

Guillemin, R. (1978b). Control of adenohypophysial function by peptides of the central nervous system. *Harvey Lect. 71*, 71–131.

Guillemin, R. and Gerich, J. E. (1976). Somatostatin physiological and clinical significance. *Annu. Rev. Med. 27*, 379–88.

Guillemin, R., Vargo, T., Rossier, J., Minick, S., Ling, N., Rivier, C., Vale, W. and Bloom, F. (1977). β-Endorphin and adrenocorticotropin are secreted concomitantly by the pituitary gland. *Science 197*, 1367–9.

Gumpel, J. M. (1978). Rheumatoid arthritis. *Brit. Med. J. 2*, 1068–70.

Gustafsson, J-A. and Stenberg, A. (1976). Specificity of neonatal, androgen-induced imprinting of hepatic steroid metabolism in rats. *Science 191*, 203–4.

Gustavsson, S. and Lundquist, G. (1978). Participation of antral somatostatin in the local regulation of gastrin release. *Acta Endocrinol. 88*, 339–46.

Guthrie, G. P., Genest, J. and Kuchel, O. (1976). Renin and the therapy of hypertension. *Annu. Rev. Pharmacol. Toxicol. 16*, 287–308.

Habener, J. F. and Potts, J. T. (1976). Chemistry, biosynthesis, secretion, and metabolism of parathyroid hormone. *Handbook of Physiology:* Section 7, *Endocrinology*, Vol. 7, *Parathyroid Gland*, pp. 313–42. Washington, D.C.: American Physiological Society.

Habener, J. F. and Potts, J. T. (1978a). Biosynthesis of parathyroid hormone (part 1). *New Engl. J. Med. 299*, 580–5.

Habener, J. F. and Potts, J. T. (1978b). Biosynthesis of parathyroid hormone (part 2). *New Engl. J. Med. 299*, 635–44.

Habener, J. F., Potts, J. T. and Rich, A. (1976). Pre-proparathyroid hormone. Evidence for an early biosynthetic precursor of proparathyroid hormone. *J. Biol. Chem. 251*, 3893–9.

Habener, J. F., Kemper, B. W., Rich, A. and Potts, J. T. (1977a). Biosynthesis of parathyroid hormone. *Rec. Prog. Horm. Res. 33*, 249–99.

Habener, J. F., Stevens, T. D., Ravazzola, M., Orci, L. and Potts, J. T. (1977b). Effects of calcium ionophores on the synthesis and release of parathyroid hormone. *Endocrinology 101*, 1524–37.

Haber, E. (1978). Renin inhibitors. *New Engl. J. Med. 298*, 1023–5.

Haddad, J. G. and Walgate, J. (1976). 25-Hydroxy-vitamin D transport in human plasma. Isolation and partial characterization of calcifidiol-binding protein. *J. Biol. Chem. 251*, 4803–9.

Hadley, M. E., Hruby, V. J. and Bower, A. (1975). Cellular mechanisms controlling melanophore stimulating hormone (MSH) release. *Gen. Comp. Endocrinol. 26*, 24–35.

Haggie, S. J., Clark, C. G., Black, J. W. and Wyllie, J. H. (1976). Clinical experience with metiamide. *Fedn. Proc. 35*, 1948–52.

Hahn, T. J., Hendin, B. A., Scharp, C. R. and Haddad, J. G. (1972). Effect of chronic anticonvulsant therapy on serum 25-hydroxycalciferol levels in adults. *New Engl. J. Med. 287*, 900–4.

Hahn, T. J., Hendin, B. A., Scharp, C. R., Boisseau, V. C. and Haddad, J. G. (1975). Serum 25-hydroxycalciferal levels and bone mass in children on chronic anticonvulsant therapy. *New Engl. J. Med. 292*, 550–4.

Hähnel, R., Twaddle, E. and Ratajczak, T. (1973a). The specificity of the estrogen receptor of human uterus. *J. Steroid Biochem. 4*, 21–31.

Hähnel, R., Twaddle, E. and Ratajczak, T. (1973b). The influence of synthetic anti-estrogens on the binding of tritiated estradiol-17β by cytosols of human uterus and human breast carcinoma. *J. Steroid Biochem. 4*, 687–95.

Halkerston, I. D. K. (1975). Cyclic AMP and adrenocortical function. In *Advances in Cyclic Nucleotide Research*, Vol. 6, pp. 100–36 (edited by P. Greengard and G. A. Robison). New York: Raven Press.

Hall, K. (1972). Human somatomedin. Determination, occurrence, biological activity and purification. *Acta Endocrinol. Suppl. 163*, 1–52.

Hall, K. and Olin, P. (1972). Sulphation factor activity and growth rate during long-term treatment of patients with pituitary dwarfism with human growth hormone. *Acta Endocrinol. 69*, 417–33.

Hall, R., Ormston, B. J., Besser, G. M., Cryer, R. J. and McKendrick, M. (1972). The thyrotrophin-releasing hormone test in diseases of the pituitary and hypothalamus. *Lancet 1*, 759–63.

Hall, R., Besser, G. M., Schally, A. V., Coy, D. H., Evered, D., Goldie, D. J., Kastin, A. J., McNeilly, A. S., Mortimer, C. H., Phenekos, C., Tunbridge, W. M. G. and Weightman, D. (1973). Action of growth-hormone-release inhibitory hormone in healthy men and in acromegaly. *Lancet 2*, 581–4.

Halushka, P. V., Lurie, D. and Colwell, J. A. (1977). Increased synthesis of prostaglandin-E-like material by platelets from patients with diabetes mellitus. *New Engl. J. Med. 297*, 1306–10.

Hammarström, S., Hamberg, M., Duell, E. A., Stawiski, M. A., Anderson, T. F. and Vorhees, J. J. (1977). Glucocorticoid in inflammatory proliferative skin disease reduced arachidonic and hydroxyeicosatetraenoic acids. *Science 197*, 994–6.

Handler, J. S., Strewler, G. J. and Orloff, J. (1977). Role of protein synthesis and phosphoprotein metabolism in cellular response to vasopressin. In *Disturbances in Body Fluid Osmolality* (edited by T. E. Andreoli, J. J. Grantham, and F. C. Rector), pp. 85–95. Washington, D.C.: American Physiological Society.

Handwerger, S., Pang, E. C., Aloj, S. M. and Sherwood, L. M. (1972). Correlations in the structure and function of human placental lactogen and human growth hormone: I. Modification of the disulfide bonds. *Endocrinology 91*, 721–7.

Hansen, J. M. and Christensen, L. K. (1977). Drug interactions with oral sulphonylurea hypoglycemic drugs. *Drugs 13*, 24–34.

Hansen, J. M., Skovsted, L., Lauridsen, U. B., Kirkegaard, C. and Siersbaek-Nielsen, K. (1974). The effect of diphenylhydantoin on thyroid function. *J. Clin. Endocrinol. Metab. 39*, 785–9.

Haque, N., Thrasher, K., Werk, E. E., Knowles, H. C. and Sholiton, L. J. (1972). Studies on dexamethasone metabolism in man: effect of diphenylhydantoin. *J. Clin. Endocrinol. 34*, 44–50.

Harden, T. K., Wolfe, B. B. and Molinoff, P. B. (1976). Binding of iodinated beta-adrenergic antagonists to proteins derived from rat heart. *Molec. Pharmacol. 12*, 1–15.

Hargis, G. K., Williams, G. A., Reynolds, W. A., Chertow, B. S., Kukreja, S. C., Bowser, E. N. and Henderson, W. J. (1978). Effect of somatostatin on parathyroid hormone and calcitonin secretion. *Endocrinology 102*, 745–50.

Harmon, J. and Aliapoulios, M. A. (1972). Gynecomastia in marihuana users. *New Engl. J. Med. 287*, 936.

Harper, A. A. and Raper, H. S. (1943). Pancreozymin, a stimulant of the secretion of pancreatic enzymes in extracts of the small intestine. *J. Physiol. (Lond.) 102*, 115–25.

Harris, E. L. (1971). Adverse reactions to oral antidiabetic drugs. *Brit. Med. J. 3*, 29–30.

Hartman, F. A. and Brownell, K. A. (1949). *The Adrenal Gland*. Philadelphia: Lea & Febiger.

Harvey, R. F. (1973). Thyroxine 'addicts'. *Brit. Med. J. 2*, 35–6.

Harvey, R. F. and Read, A. E. (1973). Saline purgatives act by releasing cholecystokinin. *Lancet 2*, 185–7.

Hauger, R. L., Aguilera, G. and Catt, K. J. (1978). Angiotensin II regulates its receptor sites in the adrenal glomerulosa zone. *Nature (Lond.) 271*, 176–8.

Haussler, M. R., Wasserman, R. H., McCain, T. A., Peterlik, M., Bursac, K. M. and Hughes, M. R. (1976). 1,25-Dihydroxy-vitamin D_3-glycoside: identification of a calcigenic principle of *Solanum malacoxylon*. *Life Sci. 18*, 1049–56.

Havard, C. W. H. (1972). Endocrine exophthalmos. *Brit. Med. J. 1*, 360–3.

Havard, C. W. H. (1974). Which test of thyroid function. *Brit. Med. J. 1*, 553–6.

Havard, C. W. H. and Boss, M. (1974). *In vivo* tests of thyroid function. *Brit. Med. J. 3*, 678–81.

Hayashida, T., Farmer, S. W. and Papkoff, H. (1975). Pituitary growth hormones: further evidence for evolutionary conservatism based on immunochemical studies. *Proc. Natl. Acad. Sci. USA 72*, 4322–6.

Hayes, J. R., Ardill, J., Kennedy, T. L., Shanks, R. G. and Buchanan, K. D. (1972). Stimulation of gastrin release by catecholamines. *Lancet, 2*, 819–21.

Haynes, B. F. and Fauci, A. S. (1978). Diabetes insipidus

associated with Wegener's granulomatosis successfully treated with cyclophosphamide. *New Engl. J. Med. 299*, 764.

Haynes, R. C. (1975). Theories of the mode of action of ACTH in stimulating secretory activity of the adrenal cortex. *Handbook of Physiology:* Section 7, *Endocrinology*, Vol. 4, *Adrenal Gland*, pp. 69–76. Washington, D.C.: American Physiological Society.

Hays, R. M. (1976). Antidiuretic hormone. *New Engl. J. Med. 295*, 659–65.

Heap, R. B., Perry, J. S. and Challis, J. R. G. (1973). Hormonal maintenance of pregnancy. *Handbook of Physiology:* Section 7, *Endocrinology*, Vol. 2, *Female Reproductive System*, Part 2, pp. 217–60. Washington, D.C.: American Physiological Society.

Hechter, O., Kato, T., Nakagawa, S. H., Yang, F. and Flouret, G. (1975). Contribution of the peptide backbone to the action of oxytocin analogs. *Proc. Natl. Acad. Sci. USA 72*, 563–6.

Hegarty, M. P., Court, R. D., Christie, G. S. and Lee, C. P. (1976). Mimosine in *Leucaena leucocephala* is metabolized to a goitrogen in ruminants. *Aust. J. Vet. Sci. 52*, 490.

Heller, C. G., Moore, D. J., Paulsen, C. A., Nelson, W. O. and Liadlaw, W. M. (1959). Effects of progesterone and synthetic progestins on the reproductive physiology of normal men. *Fedn. Proc. 18*, 1057–64.

Hellman, B., Lernmark, Å., Sehlin, J., Söderberg, M. and Täljedal, I.-B. (1976). On the possible role of thiol groups in the insulin-releasing action of mercurials, organic disulfides, alkylating agents, and sulfonylureas. *Endocrinology 99*, 1398–406.

Hench, P. S., Kendall, E. C., Slocumb, C. H. and Polley, H. F. (1949). The effect of a hormone of the adrenal cortex [17-hydroxy-11-dehydrocorticosterone (compound E)] and of pituitary adrenocorticotropic hormone on rheumatoid arthritis. *Proc. Staff Meet. Mayo Clin. 24*, 181–97.

Henderson, J. R. (1969). Why are the islets of Langerhans? *Lancet 2*, 469–70.

Henderson, R. G., Russell, R. G. G., Ledingham, J. G. G., Smith, R., Oliver, D. O., Walton, R. J., Small, D. G., Preston, C., Warner, G. T. and Norman, A. W. (1974). Effects of 1,25-dihydroxycholecalciferol on calcium absorption, muscle weakness, and bone disease in chronic renal failure. *Lancet 1*, 379–84.

Hendler, E. D., Goffinet, J. A., Ross, S., Longnecker, R. E. and Bakovic, V. (1974). Controlled study of androgen therapy in anemia of patients on maintenance dialysis. *New Engl. J. Med. 291*, 1046–51.

Hennekens, C. H. and MacMahon, B. (1977). Oral contraceptives and myocardial infarction. *New Engl. J. Med. 296*, 1166–7.

Hennessy, J. F., Gray, T. K., Cooper, C. W. and Ontjes, D. A. (1973). Stimulation of thyrocalcitonin secretion by pentagastrin and calcium in two patients with medullary carcinoma of the thyroid. *J. Clin. Endocrinol. Metab. 36*, 200–3.

Hepp, K. D. (1977). Studies on the mechanism of insulin action: basic concepts and clinical implications. *Diabetologia 13*, 177–86.

Herbst, A. L. and Selenkow, H. A. (1965). Hyperthyroidism during pregnancy. *New Engl. J. Med. 273*, 627–33.

Herbst, A. L., Kurman, R. J., Scully, R. E. and Poskanzer, D. C. (1972). Clear-cell adenocarcinoma of the genital tract in young females. Registry report. *New Engl. J. Med. 287*, 1259–64.

Herbst, A. L., Poskanzer, D. C., Robboy, S. J., Friedlander, L. and Scully, R. E. (1975). Prenatal exposure to stilbestrol. A prospective comparison of exposed female offspring with unexposed controls. *New Engl. J. Med. 292*, 334–9.

Herington, A. C. and Veith, N. M. (1977). Solubilization of a growth hormone-specific receptor from rabbit liver. *Endocrinology 101*, 984–7.

Herington, A. C., Veith, N. and Burger, H. G. (1976). Characterization of the binding of human growth hormone to microsomal membranes from rat liver. *Biochem. J. 158*, 61–9.

Herman, T. S., Fimognari, G. M. and Edelman, I. S. (1968). Studies on renal aldosterone-binding proteins. *J. Biol. Chem. 243*, 3849–56.

Hershman, J. M. (1974). Clinical application of thyrotropin-releasing hormone. *New Engl. J. Med. 290*, 886–90.

Hervey, G. R., Hutchinson, I., Knibbs, A. V., Burkinshaw, L., Jones, P. R. M., Norgan, N. G. and Levell, M. J. (1976). "Anabolic" effects of methandienone in men undergoing athletic training. *Lancet 2*, 699–702.

Heuson, J. C., Engelsman, E., Blonk-van der Wijst, J., Maass, H., Drochmans, A., Michel, J., Nowakowski, H. and Gorins, A. (1975). Comparative trial of nafoxidine and ethinyloestradiol in advanced breast cancer: an E.O.R.T.C. study. *Brit. Med. J. 2*, 711–13.

Higgins, S. J., Rousseau, G. G., Baxter, J. D. and Tomkins, G. M. (1973). Early events in glucocorticoid action. *J. Biol. Chem. 248*, 5866–72.

Hill, J. H., Cortas, N. and Walser, M. (1973). Aldosterone action and sodium- and potassium-activated adenosine triphosphatase. *J. Clin. Invest. 52*, 185–9.

Hintz, R. L., Clemmons, D. R., Underwood, L. E. and van Wyk, J. J. (1972). Competitive binding of somatomedin to the insulin receptors of adipocytes, chondrocytes, and liver membranes. *Proc. Natl. Acad. Sci. USA 69*, 2351–3.

Hirsch, P. F. and Munson, P. L. (1969). Thyrocalcitonin. *Physiol. Rev. 49*, 548–622.

Hisaw, F. L. (1926). Experimental relaxation of the pubic ligament of the guinea pig. *Proc. Soc. Exp. Biol. Med. 23*, 661–3.

Hockaday, T. D. R. (1972). Diabetes insipidus. *Brit. Med. J. 2*, 210–13.

Hockaday, T. D. R. (1974). Diabetes mellitus. *Practitioner 213*, 535–51.

Hodgkin, D. C. (1972). The structure of insulin. *Diabetes 21*, 1131–50.

Hofmann, K. (1974). Relations between chemical structure and function of adrenocorticotropin and melanocyte-stimulating hormones. *Handbook of Physiology:* Section 7, *Endocrinology*, Vol. 4, *The Pituitary Gland and Its Neuroendocrine Control*, Part 2, pp. 29–58. Washington, D.C.: American Physiological Society.

Holaday, J. W., Wei, E., Loh, H. E. and Li, C. H. (1978). Endorphins may function in heat regulation. *Proc. Natl. Acad. Sci. USA 75*, 2923–7.

Holdsworth, E. S. (1970). The effects of vitamin D on enzymes' activity in the mucosal cells of the chick small intestine. *J. Membr. Biol. 3*, 43–53.

Holick, M. F. and Clark, M. B. (1978). The photobiogenesis and metabolism of vitamin D. *Fedn. Proc. 37*, 2567–74.

Holick, M. F., Holick, S. A., Tavela, T., Gallagher, B., Schnoes, H. K. and DeLuca, H. F. (1975). Synthesis of [6-³H]-1α-hydroxyvitamin D₃ and its metabolism *in vivo* to [³H]-1α,25-dihydroxyvitamin D₃. *Science 190*, 576–8.

Holladay, L. A. and Puett, D. (1976). Somatostatin conformation: evidence for a stable intramolecular structure from circular dichroism, diffusion, and sedimentation equilibrium. *Proc. Natl. Acad. Sci. USA 73*, 1199–202.

Holladay, L. A., Rivier, J. and Puett, J. (1977). Conformational studies on somatostatin and analogues. *Biochemistry 16*, 4895–900.

Hollenberg, M. D. and Cuatrecasas, P. (1975a). Insulin and epidermal growth factor. Human fibroblast receptors related to deoxyribonucleic acid synthesis and amino acid uptake. *J. Biol. Chem. 250*, 3845–53.

Hollenberg, M. D. and Cuatrecasas, P. (1975b). Insulin: interaction with membrane receptors and relationship to cyclic purine nucleotides and cell growth. *Fedn. Proc. 34*, 1556–63.

Hollenberg, M. D. and Cuatrecasas, P. (1978). Membrane receptors and hormone action: recent developments. *Prog. Neuro-Psychopharmacol. 2*, 287–302.

Hong, S-C. L. and Levine, L. (1976). Inhibition of arachidonic acid release from cells as the biochemical action of anti-inflammatory corticosteroids. *Proc. Natl. Acad. Sci. USA 73*, 1730–4.

Hope, D. B. and Pickup, J. C. (1974). Neurophysins. *Handbook of Physiology:* Section 7, *Endocrinology*, Vol. 4, *The Pituitary Gland and Its Neuroendocrine Control*, Part 1, pp. 173–89. Washington, D.C.: American Physiological Society.

Horiuchi, N., Suda, T., Takahashi, H., Shimazawa, E. and Ogata, E. (1977). *In vivo* evidence for the intermediary role of 3′,5′-cyclic AMP in parathyroid hormone-induced stimulation of 1α,25-dihydroxyvitamin D₃ synthesis in rats. *Endocrinology 101*, 969–74.

Hornych, A., Meyer, P. and Millicz, P. (1973). Angiotensin, vasopressin, and cyclic AMP: effects on sodium and water fluxes in rat colon. *Amer. J. Physiol. 224*, 1223–9.

Horrobin, D. F., Burstyn, P. G., Lloyd, I. J., Durkin, N., Lipton, A. and Muiruri, K. L. (1971). Actions of prolactin on human renal function. *Lancet 2*, 352–4.

Horsman, A., Gallagher, J. C., Simpson, M. and Nordin, B. E. C. (1977). Prospective trial of oestrogen and calcium in postmenopausal women. *Brit. Med. J. 2*, 789–92.

Horwitz, K. B. and McGuire, W. L. (1975). Specific progesterone receptors in human breast cancer. *Steroids 25*, 497–505.

Horwitz, R. I. and Feinstein, A. R. (1978). Alternative analytic methods for case-control studies of estrogens and endometrial cancer. *New Engl. J. Med. 299*, 1089–94.

Hospital Tribune (1972). Vasopressin analogue is used against enuresis, alcoholism. *Hosp. Trib.*, December 11.

Hougen, T. J., Hopkins, B. E. and Smith, T. W. (1978). Insulin effects on monovalent cation transport and Na-K-ATPase activity. *Amer. J. Physiol. 234*, C59–63.

Houslay, M. D., Ellory, J. C., Smith, G. A., Hesketh, T. R., Stein, J. M., Warren, G. B. and Metcalfe, J. C.

(1977). Exchange of partners in glucagon receptor-adenylate cyclase complexes. Physical evidence for the independent, mobile receptor model. *Biochim. Biophys. Acta 467*, 208–19.

Howe, A. (1973). The mammalian pars intermedia: a review of its structure and function. *J. Endocrinol. 59*, 385–409.

Huang, D. and Cuatrecasas, P. (1975). Insulin-induced reduction of membrane receptor concentrations in isolated fat cells and lymphocytes. Independence from receptor occupation and possible relation to proteolytic activity of insulin. *J. Biol. Chem. 250*, 8251–9.

Hughes, J., Smith, T. W., Kosterlitz, H. W., Fothergill, L. A., Morgan, B. A. and Morris, H. R. (1975). Identification of two related pentapeptides from the brain with potent opiate agonist activity. *Nature (Lond.) 258*, 577–9.

Hughes, M. R., McCain, T. A., Chang, S. Y., Haussler, M. R., Villareale, M. and Wasserman, R. H. (1977). Presence of 1,25-dihydroxyvitamin D₃-glycoside in the calcinogenic plant *Cestrum diurnum*. *Nature (Lond.) 268*, 347–9.

Humbel, R. E., Bosshard, H. R. and Zahn, H. (1972). Chemistry of insulin. *Handbook of Physiology:* Section 7, *Endocrinology*, Vol. 1, *Endocrine Pancreas*, pp. 111–32. Washington, D.C.: American Physiological Society.

Hunter, J. A. A. (1973). Diseases of the skin. The basis of skin therapy. *Brit. Med. J. 4*, 411–13.

Hurley, T. W., D'Ercole, A. J., Handwerger, S., Underwood, L. E., Furlanetto, R. W. and Fellows, R. E. (1977). Ovine placental lactogen induces somatomedin: a possible role in fetal growth. *Endocrinology 101*, 1635–8.

Hutchings, J. J., Escamilla, R. F., Deamer, W. C. and Li, C. H. (1959). Metabolic changes produced by human growth hormone (Li) in a pituitary dwarf. *J. Clin. Endocrinol. Metab. 19*, 759–69.

Hutchison, G. B. and Rothman, K. J. (1978). Correcting a bias? *New Engl. J. Med. 299*, 1129–30.

Illiano, G., Tell, G. P. E., Siegel, M. I. and Cuatrecasas, P. (1973). Guanosine 3′:5′-cyclic monophosphate and the action of insulin and acetylcholine. *Proc. Natl. Acad. Sci. USA 70*, 2443–7.

Illig, R., Exner, G. U., Kollmann, F., Kellerer, K., Borkenstein, M., Lunglmayr, L., Kuber, W. and Prader, A. (1977). Treatment of cryptorchidism by intranasal synthetic luteinizing-hormone releasing hormone. Results of a collaborative double-blind study. *Lancet 2*, 518–20.

Imura, H., Nakai, Y. and Yoshimi, T. (1973). Effect of 5-hydroxytryptophan (5-HTP) on growth hormone and ACTH release in man. *J. Clin. Endocrinol. Metab. 36*, 204–6.

Inagami, T. and Murakami, K. (1977). Pure renin. Isolation from hog kidney and characterization. *J. Biol. Chem. 252*, 2978–83.

Inagami, T., Hirose, S., Murakami, K. and Matoba, T. (1977). Native form of renin in the kidney. *J. Biol. Chem. 252*, 7733–7.

Inesi, G. (1978). Emergency management of hypercalcemia. *Drug Ther. Hosp. Ed.* (May), 14–22.

Ingelfinger, F. J. (1974). Gallstones and estrogens. *New Engl. J. Med. 290*, 51–2.

Insel, P. A., Maguire, M. E., Gilman, A. G., Bourne, H. R., Coffino, P. and Melmon, K. L. (1976). Beta adrenergic receptors and adenylate cyclase: products of separate genes? *Molec. Pharmacol. 12*, 1062–9.

Ippoliti, A. F., Maxwell, V. and Isenberg, J. I. (1976). The effect of various forms of milk on gastric-acid secretion. *Ann. Intern. Med. 84*, 286–9.

Ippoliti, A. F., Isenberg, J. I., Maxwell, V. and Walsh, J. D. (1976). The effect of 16,16-dimethyl prostaglandin E₂ on meal-stimulated gastric acid secretion and serum gastrin in duodenal ulcer patients. *Gastroenterology 70*, 488–91.

Irsigler, K., Kaspar, L. and Kritz, H. (1977). Dichloracetate in biguanide-induced lacticacidosis. *Lancet 2*, 1026–7.

Irvine, W. J. and Toft, A. D. (1976). The diagnosis and treatment of thyrotoxicosis. *Clin. Endocrinol. 5*, 687–707.

Irvine, W. J., Wilson, K. S. and Toft, A. D. (1973). Adrenocortical stimulation by substituted α1-18 corticotrophin. *Lancet 1*, 1417–19.

Ismail, A. A. A., Davidson, D. W., Souka, A. R., Barnes, E. W., Irvine, W. J., Kilimnik, H. and Vanderbeeken, Y. (1974). The evaluation of the role of androgens in hirsutism and the use of a new anti-androgen "cyproterone acetate" for therapy. *J. Clin. Endocrinol. Metab. 39*, 81–95.

Ismail-Beigi, F. and Edelman, I. S. (1970). Mechanism of thyroid calorigenesis: role of active sodium transport. *Proc. Natl. Acad. Sci. USA 67*, 1071–8.

Ivy, A. C. and Olberg, E. (1928). A hormone mechanism for gall bladder contraction and evacuation. *Amer. J. Physiol. 86*, 599–613.

Jackson, H. and Jones, A. R. (1972). The effects of steroids and their antagonists on spermatogenesis. *Adv. Steroid Biochem. Pharmacol. 3*, 167–92.

Jackson, I. M. D. and Reichlin, S. (1977a). Thyrotropin-releasing hormone: abundance in the skin of the frog, *Rana pipiens. Science 198*, 414–15.

Jackson, I. M. D. and Reichlin, S. (1977b). Brain thyrotrophin-releasing hormone is independent of the hypothalamus. *Nature (Lond.) 267*, 853–4.

Jacobs, H. S., Knuth, U. A., Hull, M. G. R. and Franks, S. (1977). Post- "pill" amenorrhoea—cause or coincidence? *Brit. Med. J. 2*, 940–2.

Jacobs, S., Chang, K-J. and Cuatrecasas, P. (1978). Antibodies to purified insulin receptor have insulin-like activity. *Science 200*, 1283–4.

Jacquet, Y. F. and Marks, N. (1976). The C-fragment of β-lipotropin: an endogenous neuroleptic or antipsychogen? *Science 194*, 632–5.

James, R., Niall, H., Kwok, S. and Bryant-Greenwood, G. (1977). Primary structure of porcine relaxin: homology with insulin and related growth factors. *Nature (Lond.) 267*, 544–6.

Janerich, D. T., Piper, J. M. and Glebatis, D. M. (1974). Oral contraceptives and congenital limb-reduction defects. *New Engl. J. Med. 291*, 697–700.

Jard, S. and Bockaert, J. (1975). Stimulus–response coupling in neurohypophysial peptide target cells. *Physiol. Rev. 55*, 489–536.

Jarrett, R. J., Keen, H., Fuller, J. H. and McCartney, M. (1977). Treatment of borderline diabetes: controlled trial using carbohydrate restriction and phenformin. *Brit. Med. J. 2*, 861–5.

Jean-Baptiste, E., Draper, M. W. and Rizack, M. A.

(1977). Steroidogenic activity of fragments of adrenocorticotropin with arginine in residue positions 3 and 5. *Proc. Natl. Acad. Sci. USA 74*, 4329–31.

Jeffcoate, W. J., Phenekos, C., Ratcliffe, J. G., Williams, S., Rees, L. and Besser, G. M. (1977a). Comparison of the pharmacokinetics in man of two synthetic ACTH analogues: α^{1-24} and substituted α^{1-18} ACTH. *Clin. Endocrinol. 7*, 1–11.

Jeffcoate, W. J., Rees, L. H., Tomlin, S., Jones, A. E., Edwards, C. R. W. and Besser, G. M. (1977b). Metyrapone in long-term management of Cushing's disease. *Brit. Med. J. 2*, 215–17.

Jefferys, P. M., Farran, H. E. A., Hoffenberg, R., Frazer, P. M. and Hodgkinson, H. M. (1972). Thyroid-function tests in the elderly. *Lancet 1*, 924–7.

Jellinck, P. H. and Newcombe, A-M. (1977). Induction of peroxidase in the rat uterus under various endocrine conditions. *J. Endocrinol. 74*, 147–8.

Jenkins, J. S. (1972). The hypothalamus. *Brit. Med. J. 2*, 99–102.

Jensen, E. V. and DeSombre, E. R. (1973). Estrogen-receptor interaction. *Science 182*, 126–34.

Jensen, E. V., DeSombre, E. R. and Jungblut, P. W. (1967). Estrogen receptors in hormone-responsive tissues and tumors. In *Endogeneous Factors Influencing Host-Tumor Balance* (edited by R. W. Wissler, T. L. Dao, and S. Wood), pp. 15–30. Chicago: University of Chicago Press.

Jensen, S. L., Fahrenkrug, J., Holst, J. J., Kuhl, C., Nielsen, O. V. and De Muckadell, O. B. S. (1978a). Secretory effects of secretin on isolated perfused porcine pancreas. *Amer. J. Physiol. 235*, E381–6.

Jensen, S. L., Fahrenkrug, J., Holst, J. J., Nielsen, O. V. and De Muckadell, O. B. S. (1978b). Secretory effects of VIP on isolated perfused porcine pancreas. *Amer. J. Physiol. 235*, E387–91.

Johnsen, S. G., Bennett, E. P. and Jensen, V. G. (1974). Therapeutic effectiveness of oral testosterone. *Lancet 2*, 1473–5.

Johnson, F. L., Feagler, J. R., Lerner, K. G., Majerus, P. W., Sigel, M., Hartmann, J. R. and Thomas, E. D. (1972). Association of androgenic–anabolic steroid therapy with development of hepatocellular carcinoma. *Lancet 2*, 1273–6.

Johnson, J. W. C., Austin, K. L., Jones, G. S., Davis, G. H. and King, T. M. (1975). Efficacy of 17α-hydroxy-progesterone caproate in the prevention of premature labor. *New Engl. J. Med. 293*, 675–80.

Johnson, L. R. (1977). Gastrointestinal hormones and their functions. *Annu. Rev. Physiol. 39*, 135–58.

Johnson, L. R. (1978). Histamine and gastric secretion. In *Histamine II and Anti-Histaminics Chemistry, Metabolism and Physiological and Pharmacological Actions* (edited by M. Roche e Silva), pp. 41–56. New York: Springer-Verlag.

Johnston, C. I. (1976). Effect of antihypertensive drugs on the renin-angiotensin system. *Drugs 12*, 274–91.

Jones, M. T., Gillham, B. and Hillhouse, E. W. (1977). The nature of corticotropin releasing factor from rat hypothalamus *in vitro. Fedn. Proc. 36*, 2104–9.

Jones, M. T., Hillhouse, E. W. and Burden, J. (1976). Effect of various putative neurotransmitters on the secretion of corticotrophin-releasing hormone from the rat hypothalamus *in vitro* – a model of the neurotransmitters involved. *J. Endocrinol. 69*, 1–10.

Jones, S. L. and Van Middlesworth, L. (1960). Normal I[131] L-thyroxine metabolism in the presence of potassium perchlorate and interrupted by propylthiouracil. *Endocrinology 67*, 855–61.

Jones, S. V. (1977). A compound for treating dwarfism (patents). *The New York Times*, p. 31. November, 5.

Jordan, V. C. and Jaspan, T. (1976). Tamoxifen as an anti-tumour agent: oestrogen binding as a predictive test for tumour response. *J. Endocrinol. 68*, 453–60.

Jordan, V. C. and Koerner, S. (1976). Tamoxifen as an anti-tumour agent: role of oestradiol and prolactin. *J. Endocrinol. 68*, 305–11.

Jordan, V. C. and Prestwich, G. (1978). Effect of non-steroidal anti-estrogens on the concentration of rat uterine progesterone. *J. Endocrinol. 76*, 363–4.

Jordan, V. C., Collins, M. M., Rowsby, L. and Prestwich, G. (1977). A monohydroxylated metabolite of tamoxifen with potent antioestrogenic activity. *J. Endocrinol. 75*, 305–16.

Jorgensen, E. C., Lehman, P. A., Greenberg, C. and Zenker, N. (1962). Thyroxine analogues: VII. Antigoitrogenic, calorigenic, and hypocholesteremic activities of some aliphatic, alicyclic, and aromatic ethers of 3,5-diiodotyrosine in the rat. *J. Biol. Chem. 237*, 3832–8.

Jorpes, J. E. (1968). Memorial lecture. The isolation and chemistry of secretin and cholecystokinin. *Gastroenterology 55*, 157–64.

Jorpes, J. E., Mutt, V., Magnusson, S. and Steele, B. B. (1962). Amino acid composition and N-terminal amino acid sequence of porcine secretin. *Biochem. Biophys. Res. Commun. 9*, 275–9.

Josimovich, J. B. (1973). Passage of hormones through the placenta, *Handbook of Physiology*: Section 7, *Endocrinology*, Vol. 2, Female Reproductive System, Part 2, pp. 277–84. Washington, D.C.: American Physiological Society.

Josimovich, J. B. and MacLaren, J. A. (1962). Presence in the human placenta of a highly lactogenic substance immunologically related to pituitary growth hormone. *Endocrinology 71*, 209–20.

Judd, S. J. (1978). Bromocriptine: a new advance. Rational therapy for some common disorders in obstetrics and gynaecology. *Drugs 16*, 167–70.

Judd, S. J., Lazarus, L. and Smythe, G. (1976). Prolactin secretion by metoclopramide in man. *J. Clin. Endocrinol. Metab. 43*, 313–17.

Juergens, W. G., Stockdale, F. E., Topper, Y. J. and Elias, J. J. (1965). Hormone-dependent differentiation of mammary gland *in vitro*. *Proc. Natl. Acad. Sci. USA 54*, 629–34.

Juhl, E. et al. (1974). Sex, ascites and alcoholism in survival of patients with cirrhosis. Effect of prednisone. A report from the Copenhagen study group for liver diseases. *New Engl. J. Med. 291*, 271–3.

Juttman, J. R., Barth, J. D. and Birkenhäger, J. C. (1977). Treatment of anticonvulsant osteomalacia with 1α-hydroxycholecalciferol. *Brit. Med. J. 1*, 551.

Kachadorian, W. A., Wade, J. B. and DiScala, V. A. (1975). Vasopressin: induced structural change in toad bladder luminal membrane. *Science 190*, 67–9.

Kagawa, C. M., Cella, J. A. and van Arman, C. G. (1957). Action of new steroids in blocking effects of aldosterone and deoxycorticosterone on salt. *Science 126*, 1015–16.

Kahn, C. R., Neville, D. M. and Roth, J. (1973). Insulin–receptor interaction in the obese-hyperglycemic mouse. A model of insulin resistance. *J. Biol. Chem. 248*, 244–50.

Kahn, C. R., Levy, A. G., Gardner, J. D., Miller, J. V., Gorden, P. and Schein, P. S. (1975). Pancreatic cholera: beneficial effects of treatment with streptozotocin. *New Engl. J. Med. 292*, 941–5.

Kahn, C. R., Flier, J. S., Bar, R. S., Archer, J. A., Gorden, P., Martin, M. M. and Roth, J. (1976). The syndromes of insulin resistance and acanthosis nigricans. Insulin receptor disorders in man. *New Engl. J. Med. 294*, 739–45.

Kalkoff, R. K., Gossain, V. V. and Matute, M. L. (1973). Plasma glucagon in obesity. Response to arginine, glucose and protein administration. *New Engl. J. Med. 289*, 465–7.

Kanis, J. A., Ledingham, J. G. G., Oliver, D. O., Walton, R. J. and Nairn, I. M. (1976). Effect of bromocriptine on plasma growth hormone and glucose tolerance in chronic renal failure. *Brit. Med. J. 1*, 879.

Kanis, J. A., Cundy, T., Bartlett, M., Smith, R., Heynen, G., Warner, G. T. and Russell, R. G. G. (1978). Is 24,25-dihydroxycholecalciferol a calcium-regulating hormone in man? *Brit. Med. J. 1*, 1382–6.

Kanno, T. (1972). Calcium-dependent amylase release and electrophysiological measurements in cells in the pancreas. *J. Physiol. (Lond.) 226*, 353–71.

Kanno, T., Saito, A. and Sato, Y. (1977). Stimulus–secretion coupling in pancreatic acinar cells: influences of external sodium and calcium on responses to cholecystokinin–pancreozymin and ionophore A23187. *J. Physiol. (Lond.) 270*, 9–28.

Kano, K. and Oka, T. (1976). Polyamine transport and metabolism in mouse mammary gland. General properties and hormonal regulation. *J. Biol. Chem. 251*, 2795–800.

Kaplan, N. M. (1975). The prognostic implication of plasma renin in essential hypertension. *JAMA 231*, 167–70.

Karim, S. M. M. (1971). Effects of oral administration of prostaglandins E₂ and F₂α on the human uterus. *J. Obstet. Gynecol. Brit. Commonw. 78*, 289–93.

Karim, S. M. M. and Sharma, S. D. (1971a). Therapeutic abortion and induction of labour by the intravaginal administration of prostaglandins E₂ and F₂α. *J. Obstet. Gynecol. Brit. Commonw. 78*, 294–300.

Karim, S. M. M. and Sharma, S. D. (1971b). Oral administration of prostaglandins for the inductions of labour. *Brit. Med. J. 1*, 260–2.

Karlin, A. (1967). On the application of "a plausible model" of allosteric proteins to the receptor for acetylcholine. *J. Theor. Biol. 16*, 306–20.

Kastin, A. J., Nissen, C., Nikolics, K., Medzihradszky, K., Coy, D. H., Teplan, I. and Schally, A. V. (1976). Distribution of ³H-α-MSH in rat brain. *Brain Res. Bull. 1*, 19–26.

Kato, H. and Suzuki, T. (1971). Bradykinin-potentiating peptides from the venom of *Agkistrodon halys blomhoffii*. Isolation of five bradykinin potentiators and the amino acid sequences of two of them, potentiators B and C. *Biochemistry 10*, 972–80.

Kato, J. and Onouchi, T. (1977). Specific progesterone receptors in the hypothalamus and anterior hypophysis of the rat. *Endocrinology 101*, 920–8.

Katz, A. M., Tada, M. and Kirchberger, M. A. (1975). Control of calcium transport in the myocardium by the cyclic AMP–protein kinase system. In *Advances in Cyclic Nucleotide Research*, Vol. 5, pp. 453–72 (edited by G. I. Drummond, P. Greengard, and G. A. Robison). New York: Raven Press.

Katzenellenbogen, B. S. and Ferguson, E. R. (1975). Antiestrogen action in the uterus: biological ineffectiveness of nuclear bound estradiol after antiestrogen. *Endocrinology 97*, 1–12.

Kayasseh, L., Gyr, K., Stalder, G. A. and Allgoewer, M. (1978). Somatostatin in acute gastrointestinal haemorrhage. *Lancet 2*, 833–4.

Kehlet, H., Blichert-Toft, M., Lindholm, J. and Rasmussen, P. (1976). Short ACTH test in assessing hypothalamic–pituitary–adrenocortical function. *Brit. Med. J. 1*, 249–51.

Keller, P. J. (1972). Induction of ovulation by synthetic luteinizing-hormone releasing factor in infertile women. *Lancet 2*, 570–2.

Kendall-Taylor, P. (1972). Hyperthyroidism. *Brit. Med. J. 2*, 337–41.

Kenimer, J. G., Hershman, J. M. and Higgins, H. P. (1975). The thyrotropin in hydatiform moles is human chorionic gonadotropin. *J. Clin. Endocrinol. Metab. 40*, 482–91.

Kennedy, T. H. (1942). Thio-ureas as goitrogenic substances. *Nature (Lond.) 150*, 233–4.

Kenney, F. T., Lee, K-L., Stiles, C. D. and Fritz, J. E. (1973). Further evidence against post-transcriptional control of inducible tyrosine aminotransferase synthesis in cultured hepatoma cells. *Nature New Biol. 246*, 208–10.

Kerr, G. D. (1977). The management of the premenstrual syndrome. *Curr. Med. Res. Opinion 4*, Suppl. 4, 29–34.

Keutmann, H. T., Sauer, M. M., Hendy, G. N., O'Riordan, J. L. H. and Potts, J. T. (1978a). Complete amino acid sequence of human parathyroid hormone. *Biochemistry 17*, 5723–9.

Keutmann, H. T., Hendy, G. N., Boehnert, M., O'Riordan, J. L. H. and Potts, J. T. (1978a). Complete amino human parathyroid hormone: recent studies and further observations. *J. Endocrinol. 78*, 49–58.

Khosla, M. C., Smeby, R. R. and Bumpus, F. M. (1974). Structure–activity relationship in angiotensin II analogs. *Handbook of Experimental Pharmacology*, Vol. 37, *Angiotensin*, pp. 124–61 (edited by I. H. Page and F. M. Bumpus). New York: Springer-Verlag.

Kidson, W., Casey, J., Kraegen, E. and Lazarus, L. (1974). Treatment of severe diabetes mellitus by insulin infusion. *Brit. Med. J. 2*, 691–4.

Kimberg, D. V. (1974). Cyclic nucleotides and their role in gastrointestinal secretion. *Gastroenterology 67*, 1023–64.

Kimura, T., Matsui, K., Sato, T. and Yoshinaga, K. (1974). Mechanism of carbamazepine (Tegretol)-induced antidiuresis: evidence for release of antidiuretic hormone and impaired excretion of a water load. *J. Clin. Endocrinol. Metab. 38*, 356–62.

Kirshner, N. and Slotkin, T. A. (1973). Secretion and recovery of catecholamines by adrenal cortex. In *Frontiers in Catecholamine Research* (edited by E. Usdin and S. H. Snyder), pp. 447–52. New York: Pergamon Press.

Kirshner, N. and Viveros, O. H. (1972). The secretory cycle in the adrenal medulla. *Physiol. Rev. 24*, 385–98.

Kissebah, A. H. Tulloch, B. R., Hope-Gill, H., Clarke, P. V., Vydelingum, N. and Fraser, T. R. (1975). Mode of insulin action. *Lancet 1*, 144–7.

Klein, J. J., Segal, R. L. and Warner, R. R. P. (1964). Galactorrhea due to imipramine: report of a case. *New Engl. J. Med. 271*, 510–12.

Kleinberg, D. L., Noel, G. L. and Frantz, A. G. (1977). Galactorrhea: a study of 235 cases, including 48 with pituitary tumors. *New Engl. J. Med. 296*, 589–600.

Kleinberg, D. L., Schaaf, M. and Frantz, A. G. (1978). Studies with lergotrile mesylate in acromegaly. *Fedn. Proc. 37*, 2198–201.

Knigge, K. M. and Silverman, A.-J. (1974). Anatomy of the endocrine hypothalamus. *Handbook of Physiology: Section 7, Endocrinology*, Vol. 4, *The Pituitary Gland and Its Neuroendocrine Control*, Part 1, pp. 1–32. Washington, D.C.: American Physiological Society.

Koch-Weser, J. (1974a). Bioavailability of drugs (part 1). *New Engl. J. Med. 291*, 233–7.

Koch-Weser, J. (1974b). Bioavailability of drugs (part 2). *New Engl. J. Med. 291*, 503–6.

Koch-Weser, J. (1976). Diazoxide. *New Engl. J. Med. 294*, 1271–4.

Koch-Weser, J. and Sellers, E. M. (1976a). Binding of drugs to serum albumin (part 1). *New Engl. J. Med. 294*, 311–16.

Koch-Weser, J. and Sellers, E. M. (1976b). Binding of drugs to serum albumin (part 2). *New Engl. J. Med. 294*, 526–31.

Kodicek, E. (1974). The story of vitamin D from vitamin to hormone. *Lancet 1*, 325–9.

Koerker, D. J., Ruch, W., Chideckel, E., Palmer, J., Goodner, C. J., Ensinck, J. and Gale, C. C. (1974). Somatostatin: hypothalamic inhibition of the endocrine pancreas. *Science 184*, 482–4.

Kohn, L. D. and Winand, R. J. (1971). Relationship of thyrotropin to exophthalmos-producing substance. Formation of a exophthalmos-producing substance by pepsin digestion of pituitary glycoproteins containing both thyrotropic and exophthalmogenic activity. *J. Biol. Chem. 246*, 6570–5.

Koivisto, V. A. and Felig, P. (1978). Effects of leg exercise on insulin absorption in diabetic patients. *New Engl. J. Med. 298*, 79–83.

Kolata, G. B. (1978). Polypeptide hormones: what are they doing in cells? *Science 201*, 805–7.

Kolodny, R. C., Masters, W. H., Kolodner, R. M. and Toro, G. (1974). Depression of plasma testosterone levels after chronic intensive marihuana use. *New Engl. J. Med. 290*, 872–4.

Kondo, S. and Schulz, I. (1976). Calcium uptake in isolated pancreas cells induced by secretogues. *Biochim. Biophys. Acta 419*, 76–92.

Kono, T., Robinson, F. W. and Sarver, J. A. (1975). Insulin sensitive phosphodiesterase. Its localization, hormonal stimulation and oxidative stabilization. *J. Biol. Chem. 250*, 7826–35.

Kono, T., Oseko, F., Shimpo, S., Nanno, M. and Endo, J. (1975). Biological activity of des-Asp[1]-angiotensin II (angiotensin III) in man. *J. Clin. Endocrinol. Metab. 41*, 1174–7.

Konturek, S., Kwiecień, N., Swierczek, J., Olesky, J., Sito, E. and Robert, A. (1976). Comparison of methylated prostaglandin E_2 analogues given orally in the inhibition of gastric responses to pentagastrin and peptone meal in man. *Gastroenterology 70*, 683–7.

Korenman, S. G. (1969). Comparative binding affinity of estrogens and its relation to estrogenic potency. *Steroids 13*, 163–77.

Koritz, S. B. (1968). On the regulation of pregnenolone synthesis. In *Functions of the Adrenal Cortex* (edited by K. W. McKerns), Vol. 1, pp. 27–48. New York: Appleton-Century-Crofts.

Kosaka, T. and Lim, R. K. S. (1930). On the mechanism of the inhibition of gastric secretion by fat. The role of bile and cystokinin. *Chin. J. Physiol. 4,* 213–20.

Koseki, Y. K., Zava, D. T., Chamness, C. C. and McGuire, W. L. (1977). Estrogen receptor translocation and replenishment by the antiestrogen tamoxifen. *Endocrinology 101*, 1104–10.

Kosman, M. E. (1978). Evaluation of a new antidiuretic agent desmopressin acetate (DDAVP). *JAMA 240,* 1896–7.

Kostyo, J. L. and Nutting, D. F. (1974). Growth hormone and protein metabolism. *Handbook of Physiology:* Section 7, *Endocrinology,* Vol. 4, *The Pituitary Gland and Its Neuroendocrine Control,* Part 2, pp. 187–210. Washington, D.C.: American Physiological Society.

Krantowitz, F., Robinson, D. R., McGuire, M. B. and Levine, L. (1975). Corticosteroids inhibit prostaglandin production by rheumatoid synovia. *Nature (Lond.) 258,* 737–9.

Krause, E. G., Will, H., Schirpke, B. and Wollenberger, A. (1975). Cyclic AMP-enhanced protein phosphorylation and calcium binding in a cell membrane-enriched fraction from myocardium. In *Advances in Cyclic Nucleotide Research,* Vol. 5, pp. 473–90 (edited by G. I. Drummond, P. Greengard, and G. A. Robison). New York: Raven Press.

Krebs, E. G. (1972). Protein kinases. *Curr. Top. Cell. Regul. 5,* 99–133.

Krejčí, I., Kupková, B. and Vávra, I. (1967). The effects of some 2-O-alkyltyrosine analogues of oxytocin and lysine vasopressin on the blood pressure of the rat, rabbit, and cat. *Brit. J. Pharmacol. 30,* 497–505.

Krejčí, I., Poláček, I. and Rudinger, J. (1967). The action of 2-O-methyltyrosine-oxytocin on the rat and rabbit uterus: effect of some experimental conditions on change from agonism to antagonism. *Brit. J. Pharmacol. 30,* 506–17.

Kremer, D., Boddy, K., Brown, J. J., Davies, D. L., Fraser, R., Lever, A. F., Morton, J. J. and Robertson, J. I. S. (1977). Amiloride in the treatment of primary hyperaldosteronism and essential hypertension. *Clin. Endocrinol. 7,* 151–7.

Krieg, M. and Voigl, K. D. (1977). Biochemical substrate of androgenic actions at a cellular level in prostate, bulbocavernosus/levator ani and in skeletal muscle. *Acta Endocrinol* Suppl. 214, 43–89.

Krieger, D. T. (1971). The hypothalamus and neuroendocrinology. *Hosp. Pract.* (September), 87–99.

Krieger, D. T. (1972). Circadian corticosteroid periodicity: critical period for abolition of neonatal injection of corticosteroid. *Science 178,* 1205–7.

Krieger, D. T. and Rizzo, F. (1969). Serotonin mediation of circadian periodicity of plasma 17-hydroxycorticosteroids. *Amer. J. Physiol. 217,* 1703–7.

Krieger, D. T., Amorosa, L. and Linick, F. (1975). Cyproheptadine-induced remission of Cushing's disease. *New Engl. J. Med. 293,* 893–6.

Krieger, D. T., Liotta, A. and Brownstein, M. J. (1977). Presence of corticotropin in brain of normal and hypophysectomized rats. *Prot. Natl. Acad. Sci. USA 74,* 648–52.

Kriss, J. P., Carnes, W. H. and Gross, R. T. (1955). Hypothyroidism and thyroid hyperplasia in patients treated with cobalt. *JAMA 157,* 117–21.

Kristensen, O., Andersen, H. H. and Pallisgaard, G. (1976). Lithium carbonate in the treatment of thyrotoxicosis. A controlled study. *Lancet 1,* 603–5.

Kronenberg, H. M., Roberts, B. E., Habener, J. F., Potts, J. T. and Rich, A. (1977). DNA complementary to parathyroid mRNA directs synthesis of pre-proparathyroid hormone in a linked transcription-translation system. *Nature (Lond.) 267,* 804–7.

Krulich, L., Quijada, M., Wheaton, J. E., Illner, P. and McCann, S. (1977). Localization of hypophysiotropic neurohormones by assay of sections from various brain areas. *Fedn. Proc. 36,* 1953–9.

Krüskemper, H. L. (1968). *Anabolic Steroids.* New York: Academic Press.

Kruss, D. M. and Littman, A. (1978). Safety of cimetidine. *Gastroenterology 74,* 478–83.

Kuenssberg, E. V., Kay, C. R., Dewhurst, J. and Booth, R. J. (1977). Recommendations from the findings by the RCGP oral contraceptive study on the mortality risks of oral contraceptive users. *Brit. Med. J. 2,* 947.

Kuhn, R. W., Schrader, W. T., Smith, R. G. and O'Malley, B. W. (1975). Progesterone binding components of chick oviduct: X. Purification by affinity chromatography. *J. Biol. Chem. 250,* 4220–8.

Kunos, G. (1978). Adrenoreceptors. *Annu. Rev. Pharmacol. Toxicol. 18,* 291–311.

Kuo, J. F. and Greengard, P. (1969). Cyclic nucleotide-dependent protein kinases: IV. Widespread occurrence of adenosine 3',5'-monophosphate-dependent protein kinase in various tissues and phyla of the animal kingdom. *Proc. Natl. Acad. Sci. USA 64,* 1349–55.

Kuttenn, F., Mowszowicz, I., Schaison, G. and Mauvais-Jarvis, P. (1977). Androgen production and skin metabolism in hirsutism. *J. Endocrinol. 75,* 83–91.

Labhart, A. (1976). *Clinical Endocrinology. Theory and Practice,* pp. 30, 388. New York, Heidelberg, Berlin: Springer-Verlag.

Lacy, P. E. and Malaisse, W. J. (1973). Microtubules and beta cell function. *Rec. Prog. Horm. Res. 29,* 199–221.

Laffan, R. J., Goldberg, M. E., High, J. P., Schaeffer, T. R., Waugh, M. H. and Rubin, B. (1978). Antihypertensive activity in rats of SQ 14225, an orally active inhibitor of angiotensin I converting enzyme. *J. Pharmacol. Exp. Ther. 204,* 281–8.

Lal, S., Tolis, G., Martin, J. B., Brown, G. M. and Guyda, H. (1975). Effect of clonidine on growth hormone, prolactin, luteinizing hormone, follicle-stimulating hormone, and thyroid-stimulating hormone in the serum of normal men. *J. Clin. Endocrinol. Metab. 41,* 827–32.

Lam, H-Y. P., Schnoes, H. K. and DeLuca, H. F. (1974). 1α-hydroxyvitamin D_2: a potent synthetic analog of vitamin D_2. *Science 186,* 1038–40.

Lan, N. C. and Katzenellenbogen, B. S. (1976). Temporal

relationships between hormone receptor binding and biological responses in the uterus: studies with short- and long-acting derivatives of estriol. *Endocrinology 98*, 220–7.

Landesman, R., Coutinho, E. M. and Wilson, K. H. (1968). The relaxant effect of diazoxide on non-gravid human myometrium *in vivo*. *Amer. J. Obstet. Gynecol. 102*, 1080–4.

Lands, A. M., Arnold, A., McAuliff, J. P., Luduena, F. P. and Brown, T. C. (1967). Differentiation of receptor systems activated by sympathomimetic amines. *Nature (Lond.) 214*, 597–8.

Lang, U., Fauchere, J-L., Pelican, G-M., Karlaganis, G. and Schwyzer, R. (1976). Hormone–receptor interactions. Adrenocorticotrophin-(17–24)-octadecapeptide stimulates adipocyte membrane adenylate cyclase without causing lipolysis in fat cells. *FEBS Lett. 66*, 246–9.

Langer, P. and Greer, M. A. (1977). *Antithyroid Substances and Naturally Occurring Goitrogens*. pp. 1–178. Basel: Karger.

Langman, M. J. S. (1976). Carbenoxolone sodium: its role in ulcer healing. *Drugs 11*, 241–4.

Laragh, J. H. and Stoerk, H. C. (1957). A study of the mechanism of secretion of the sodium retaining hormone (aldosterone). *J. Clin. Invest. 36*, 383–92.

Laragh, J. H., Angers, M., Kelly, W. G. and Lieberman, S. (1960). The effect of epinephrine, norepinephrine, angiotensin II, and others on the secretory rate of aldosterone in man. *JAMA 174*, 234–40.

Laragh, J. H., Case, D. B., Wallace, J. M. and Keim, H. (1977). Blockade of renin or angiotensin for understanding human hypertension: a comparison of propranolol, saralasin and converting enzyme blockade. *Fedn. Proc. 36*, 1781–7.

Larson, B. A., Sinha, Y. N. and Vanderlaan, W. P. (1977). Effect of 5-hydroxytryptophan on prolactin secretion in the mouse. *J. Endocrinol. 74*, 153–4.

Larsson, L-I., Hirsch, M. A., Holst, J. J., Ingemansson, S., Kühl, C., Lindkaer Jensen, S., Lundquist, G., Rehfeld, J. F. and Schwartz, T. W. (1977). Pancreatic somatostatinoma. Clinical features and physiological implications. *Lancet 1*, 666–8.

László, F. A., Kocsis, J., Manning, M. and Sawyer, W. H. (1975). Antidiuretic effect of 1-deamino-4-valine-8-D-arginine vasopressin in diabetes insipidus of various types. *Acta Endocrinol.* Suppl. 199, 151.

Latham, K. R., Ring, J. C. and Baxter, J. D. (1976). Solubilized nuclear "receptors" for thyroid hormones. Physical characteristics and binding properties, evidence for multiple form. *J. Biol. Chem. 251*, 7388–97.

Lauler, D. P., Hickler, R. B. and Thorn, G. W. (1962). The salivary sodium–potassium ratio. A useful "screening" test for aldosteronism in hypertensive subjects. *New Engl. J. Med. 267*, 1136–7.

Laval, J. and Collier, R. (1955). Elevation of intraocular pressure due to hormonal steroid therapy in uveitis. *Amer. J. Ophthalmol. 39*, 175–82.

Lawson, D. E. M., Smith, M. W. and Wilson, P. W. (1974). Relationship of calcium uptake and calcium-binding protein synthesis in chick and rat intestine in response to *Solanum malacoxylon*. *FEBS Lett. 45*, 122–5.

Lazarus, J. H., Richards, A. R., Addison, G. M. and

Owen, G. M. (1974). Treatment of thyrotoxicosis with lithium carbonate. *Lancet 2*, 1160–3.

Lazarus, L. H., Ling, N. and Guillemin, R. (1976). β-Lipotropin as a prohormone for the morphinomimetic peptides, endorphins and enkephalins. *Proc. Natl. Acad. Sci. USA 73*, 2156–9.

Leaf, A. and MacKnight, A. D. C. (1972). The site of the aldosterone induced stimulation of sodium transport. *J. Steroid Biochem. 3*, 237–45.

Leclercq, G. and Heuson, J. C. (1977). Therapeutic significance of sex-steroid receptors in the treatment of breast cancer. *Eur. J. Cancer 13*, 1205–15.

Leclercq, G., Heuson, J. C., Deboel, M. C. and Mattheiem, W. H. (1975). Oestrogen receptors in breast cancer: a changing concept. *Brit. Med. J. 1*, 185–9.

Lederis, K. (1974). Neurosecretion and the functional structure of the neurohypophysis. *Handbook of Physiology:* Section 7, *Endocrinology*, Vol. 4, *The Pituitary Gland and Its Neuroendocrine Control*, Part 1, pp. 81–102. Washington, D.C.: American Physiological Society.

Lee, G., Aloj, S. M., Beguinot, F. and Kohn, L. D. (1977). Existence of a soluble thyrotropin binding component in normal human sera. *J. Biol. Chem. 252*, 7967–70.

Lee, K-L., Reel, J. R. and Kenney, F. T. (1970). Regulation of tyrosine α-ketoglutarate transaminase in rat liver. *J. Biol. Chem. 245*, 5806–12.

Lee, P. A., Keenan, B. S., Migeon, C. J. and Blizzard, R. M. (1973). Effect of various ACTH preparations and of metyrapone on the secretion of growth hormone in normal subjects and in hypopituitary patients. *J. Clin. Endocrinol. Metab. 37*, 389–96.

Lee, T. H., Lee, M. S. and Lu, M-Y. (1972). Effect of α-MSH on melanogenesis and tyrosinase of B-16 melanoma. *Endocrinology 91*, 1180–8.

Lefkowitz, R. J. (1976). β-Adrenergic receptors: recognition and regulation. *New Engl. J. Med. 295*, 323–8.

Lefkowitz, R. J. (1978). Identification and regulation of alpha and beta-adrenergic receptors. *Fedn. Proc. 37*, 123–9.

Lefkowitz, R. J., Limbird, L. E., Mukherjee, C. and Caron, M. G. (1976a). The β-adrenergic receptor and adenylate cyclase. *Biochim. Biophys. Acta 457*, 1–39.

Lefkowitz, R. J., Mullikin, D. and Caron, M. G. (1976b). Regulation of β-adrenergic receptors by guanyl-5'-yl imidodiphosphate and other purine nucleotides. *J. Biol. Chem. 251*, 4686–92.

Legros, J., Gilot, P., Seron, X., Claessens, J., Adam, A., Moeglen, J. M., Audibert, A. and Berchier, P. (1978). Vasopressin in amnesia. *Lancet 1*, 41–2.

Lemberger, L. (1978). The pharmacology of ergots: past and present. *Fedn. Proc. 37*, 2176–80.

Lemberger, L., Crabtree, R., Clemens, J., Dyke, R. W. and Woodburn, R..T. (1974). The inhibitory effect of an ergoline derivative (Lergotrile, compound 83636) on prolactin secretion in man. *J. Clin. Endocrinol. Metab. 39*, 579–84.

Lemkes, H. H. P. J., Reitsma, P. H., Frijlink, W., Verlinden-Ooms, H. and Bijvoet, O. L. M. (1978). A new diphosphonate: dissociation between effects on cells and mineral in rats and a preliminary trial in Paget's disease. In *Homeostasis of Phosphate and Other Minerals* (edited by S. G. Massry, E. Ritz and A. Rapado). pp. 459–69. New York: Plenum Press.

Lenton, E. A., Sobowale, O. S. and Cooke, I. D. (1977). Prolactin concentrations in ovulatory but infertile women: treatment with bromocriptine. *Brit. Med. J. 2*, 1179–81.

Leonard, J. L. and Rosenberg, I. N. (1978). Subcellular distribution of thyroxine 5'-deiodinase in the rat kidney: a plasma membrane location. *Endocrinology 103*, 274–80.

Leopold, A. S., Erwin, M., Oh, J. and Browning, B. (1976). Phytoestrogens: adverse effects on reproduction in California quail. *Science 191*, 98–100.

Lerner, A. B. and McGuire, J. S. (1961). Effect of alpha and beta melanocyte stimulating hormones on the skin colour of man. *Nature (Lond.) 189*, 176–9.

Lesniak, M. A. and Roth, J. (1976). Regulation of receptor concentration by homologous hormone. Effect of human growth hormone on its receptor in IM-9 lymphocytes. *J. Biol. Chem. 251*, 3720–9.

Lesniak, M. A., Gorden, P., Roth, J. and Gavin, J. R. (1974). Binding of ^{125}I-human growth hormone to specific receptors in human cultured lymphocytes. *J. Biol. Chem. 249*, 1661–7.

Letchworth, A. T., Slattery, M. and Dennis, K. J. (1978). Clinical application of human-placental-lactogen values in late pregnancy. *Lancet 1*, 955–7.

Levant, J. A., Walsh, J. H. and Isenberg, J. I. (1973). Stimulation of gastric secretion and gastrin release by single oral doses of calcium carbonate in man. *New Engl. J. Med. 289*, 555–8.

Levey, G. S., Lasseter, K. C. and Palmer, R. F. (1974). Sulfonylureas and the heart. *Annu. Rev. Med. 25*, 69–74.

Levey, G. S., Fletcher, M. A., Klein, I., Ruiz, E. and Schenk, A. (1974). Characterization of ^{125}I-glucagon binding in a solubilized preparation of cat myocardial adenylate cyclase. *J. Biol. Chem. 249*, 2665–73.

Levine, J. D., Gordon, N. C. and Fields, H. L. (1978). The mechanism of placebo analgesia. *Lancet 2*, 654–7.

Levine, S. B. and Leopold, I. H. (1973). Advances in ocular corticosteroid therapy. *Med. Clin. North Amer. 57*, 1167–77.

Levitzki, A., Altas, D. and Steer, M. L. (1974). The binding characteristics and number of β-adrenergic receptors on the turkey erythrocyte. *Proc. Natl. Acad. Sci. USA 71*, 2773–6.

Lewis, J. R. (1974). Carbenoxolone sodium in the treatment of peptic ulcer. *JAMA 229*, 460–2. .

L'Hermité, M., Vanhaelst, L., Copinschi, G., Leclerq, R., Golstein, J., Bruno, O. D. and Robyn, C. (1972). Prolactin release after injection of thyrotrophin-releasing hormone in man. *Lancet 1*, 763–5.

Li, C. H. (1975). Human pituitary growth hormone: a biologically active hendekakaihekaton peptide fragment corresponding to amino-acid residues 15-125 in the hormone molecule. *Proc. Natl. Acad. Sci. USA 72*, 3878–82.

Li, C. H. and Chung, D. (1976). Primary structure of human β-lipotropin. *Nature (Lond.) 260*, 622–4.

Li, C. H., Dixon, J. S. and Chung, D. (1971). Primary structure of human chorionic somatomammotrophin (HCS) molecule. *Science 171*, 56–8.

Li, C. H. and Dixon, J. S. (1971). Human pituitary growth hormone: XXXII. The primary structure of the hormone: revision. *Arch. Biochem. Biophys. 146*, 233–6.

Li, C. H. and Gráf, L. (1974). Human pituitary growth hormone isolation and properties of two biologically

active fragments from plasmin digests. *Proc. Natl. Acad. Sci. USA 71*, 1197–201.

Li, C. H., Barnafi, L., Chrétien, M. and Chung, D. (1965). Isolation and amino-acid sequence of β-LPH from sheep pituitary glands. *Nature (Lond.) 208*, 1093–4.

Li, C. H., Yamashiro, D., Tseng, L-F. and Loh, H. H. (1977). Synthesis and analgesic activity of human β-endorphin. *J. Med. Chem. 20*, 325–8.

Li, C. H., Chung, D., Yamashiro, D. and Lee, C. Y. (1978). Isolation, characterization, and synthesis of a corticotropin-inhibiting peptide from human pituitary glands. *Proc. Natl. Acad. Sci. USA 75*, 4306–9.

Liao, S. and Fang, S. (1969). Receptor-proteins for androgens and the mode of action of androgens on gene transcription in ventral prostrate. *Vitam. Horm. 27*, 17–90.

Liao, S., Howell, D. K. and Chang, T-M. (1974). Action of a nonsteroidal antiandrogen, flutamide, on the receptor binding and nuclear retention of 5α-dihydrotesterone in rat ventral prostate. *Endocrinology 94*, 1205–9.

Liao, S., Liang, T., Fang, S., Castañeda, E. and Shao, T-C. (1973). Steroid structure and androgenic activity. Specificities involved in the receptor binding and nuclear retention of various androgens. *J. Biol. Chem. 248*, 6154–62.

Liao, T-H. and Pierce, J. G. (1971). The primary structure of bovine thyrotropin: II. The amino acid sequences of the reduced S-carboxymethyl and chains. *J. Biol. Chem. 246*, 850–65.

Liberti, J. P. and Miller, M. S. (1978). Somatomedin-like effects of biologically active bovine growth hormone fragments. *Endocrinology 103*, 29–34.

Liddle, G. W. (1957). Sodium diuresis induced by steroidal antagonists of aldosterone. *Science 126*, 1016–17.

Liddle, G. W. (1972). Pathogenesis of glucocorticoid disorders. *Amer. J. Med. 53*, 638–48.

Lilienthal, C. M. and Ward, J. P. (1971). Medical induction of labour. *J. Obstet. Gynecol. Brit. Commonw. 78*, 317–21.

Limbird, L. E. and Lefkowitz, R. J. (1977). Resolution of β-adrenergic receptor binding and adenylate cyclase activity by gel exclusion chromatography. *J. Biol. Chem. 252*, 799–802.

Limbird, L. E. and Lefkowitz, R. J. (1978). Agonist-induced increase in apparent β-adrenergic molecular size. *Proc. Natl. Acad. Sci. USA 75*, 228–32.

Lin, S-Y. and Goodfriend, T. L. (1970). Angiotensin receptors. *Amer. J. Physiol. 218*, 1319–28.

Lindén, V. (1974). Vitamin D and myocardial infarction. *Brit. Med. J. 3*, 647–50.

Lindholm, J., Kehlet, H., Blichert-Toft, M., Dinensen, B. and Riishede, J. (1978). Reliability of the 30-minute ACTH test in assessing hypothalamic–pituitary–adrenal function. *J. Clin. Endocrinol. Metab. 47*, 272–4.

Lindner, H. and Shelesnyak, M. (1967). Effect of ergocornine on ovarian synthesis of progesterone and 20α-hydroxy-pregn-4-en-3-one in the pseudo pregnant rat. *Acta Endocrinol. 56*, 27–34.

Lindsay, R., Hart, D. M., Aitken, J. M., MacDonald, E. B., Anderson, J. B. and Clarke, A. C. (1976). Long-term prevention of postmenopausal osteoporosis by oestrogens. Evidence for an increased bone mass after delayed onset of oestrogen treatment. *Lancet 1*, 1038–41.

Ling, N. and Guillemin, R. (1976). Morphinomimetic activity of synthetic fragments of β-lipotropin and analogs. *Proc. Natl. Acad. Sci. USA* 73, 3308–10.

Ling, W. Y. and Marsh, J. M. (1977). Reevaluation of the role of cyclic adenosine 3′,5′-monophosphate and protein kinase in the stimulation of steroidogenesis by luteinizing hormones in bovine corpus luteum slices. *Endocrinology* 100, 1571–8.

Liotta, A., Osathanondh, R., Ryan, K. J. and Krieger, D. T. (1977). Presence of corticotropin in human placenta: demonstration of *in vitro* synthesis. *Endocrinology* 101, 1552–8.

Liotta, A. S., Suda, T. and Krieger, D. T. (1978). β-Lipotropin is the major opioid-like peptide of human pituitary and rat pars distalis: lack of significant β-endorphin. *Proc. Natl. Acad. Sci. USA* 75, 2950–4.

Lippman, M. E. and Allegra, J. C. (1978). Current concepts: receptors in breast cancer. *New Engl. J. Med.* 299, 930–3.

Lippman, M. E., Allegra, J. C., Thompson, E. B., Simon, R., Barlock, A., Green, L., Huff, K. K., Do, H. M. T., Aitken, S. C. and Warren, R. (1978). The relation between estrogen receptors and response rate to cytotoxic chemotherapy in metastatic breast cancer. *New Engl. J. Med.* 298, 1223–8.

Lipsett, M. B. (1968). Rationale for chemotherapy of Cushing's syndrome. In *Clinical Endocrinology* (edited by E. B. Astwood and C. F. Cassidy), Vol. 2, pp. 489–98. New York: Grune & Stratton.

Lipsett, M. B. (1977). Estrogen use and cancer risk. *JAMA* 237, 1112–15.

Lipson, A., Hsu, T-H., Sherwin, B. and Geelhoed, G. W. (1978). Nitroprusside therapy for a patient with pheochromocytoma. *JAMA* 239, 427–8.

Lipton, A. and Santen, R. J. (1974). Medical adrenalectomy using aminoglutethimide and dexamethasone in advanced breast cancer. *Cancer* 33, 503–12.

Liston, W. A. and Campbell, A. J. (1974). Dangers of oxytocin-induced labour to fetuses. *Brit. Med. J.* 3, 606–7.

Litwack, G., Filler, R., Rosenfield, S. A., Lichtash, N., Wishman, C. A. and Singer, S. (1973). Liver cytosol corticosteroid binder II, a hormone receptor. *J. Biol. Chem.* 248, 7481–6.

Liuzzi, A., Chiodini, P. G., Botalla, L., Silvestrini, F. and Muller, E. E. (1974). Growth hormone (GH)-releasing activity of TRH and GH-lowering effect of dopaminergic drugs in acromegaly: homogeneity in the two responses. *J. Clin. Endocrinol. Metab.* 39, 871–6.

Livingston, J. N. and Lockwood, D. H. (1975). Effect of glucocorticoids on the glucose transport system of isolated fat cells. *J. Biol. Chem.* 250, 8853–60.

Lo, C-S. and Edelman, I. S. (1976). Effect of triiodothyronine on the synthesis and degradation of renal cortical ($Na^+ + K^+$)-adenosine triphosphatase. *J. Biol. Chem.* 251, 7834–40.

Lo, C-S., August, T. R., Liberman, U. A. and Edelman, I. S. (1976). Dependence of renal ($Na^+ + K^+$)-adenosine triphosphatase activity in thyroid status. *J. Biol. Chem.* 251, 7826–33.

Lockwood, D. H., Livingston, J. N. and Amatruda, J. M. (1975). Relation of insulin receptors to insulin resistance. *Fedn. Proc.* 34, 1564–9.

Lockwood, D. H., Stockdale, F. E. and Topper, Y. J. (1967). Hormone dependent differentiation of mammary gland: sequence of action of hormones in relation to cell cycle. *Science* 156, 945–6.

Loeb, J. N. (1976). Corticosteroids and growth. *New Engl. J. Med.* 295, 547–52.

Logothetopoulos, J. (1972). Islet cell regeneration and neogenesis. *Handbook of Physiology:* Section 7, *Endocrinology*, Vol. 1, *Endocrine Pancreas*, pp. 67–76. Washington, D.C.: American Physiological Society.

Lomedico, P. T., Chan, S. J., Steiner, D. F. and Saunders, G. F. (1977). Immunological and chemical characterization of bovine preproinsulin. *J. Biol. Chem.* 252, 7971–8.

London, D. R. (1972). Male infertility. *Brit. Med. J.* 1, 609–11.

Londos, C., Salomon, Y., Lin, M. C., Harwood, J. P., Schramm, M., Wolff, J. and Rodbell, M. (1974). 5′-Guanylimidophosphate, a potent activator of adenylate cyclase systems in eukaryotic cells. *Proc. Natl. Acad. Sci. USA* 71, 3087–90.

Long, R. G., Varghese, Z., Meinhard, E. A., Skinner, R. K., Wills, M. R. and Sherlock, S. (1978). Parenteral 1,25-dihydroxycholecalciferol in hepatic osteomalacia. *Brit. Med. J.* 1, 75–7.

Lord, J. A. H., Waterfield, A. A., Hughes, J. and Kosterlitz, H. W. (1977). Endogenous opioid peptides: multiple agonists and receptors. *Nature (Lond.)* 267, 495–9.

Lording, D. W. (1978). Impotence: role of drug and hormonal treatment. *Drugs* 15, 144–50.

Loreau, N., Lepreux, C. and Ardaillou, R. (1975). Calcitonin-sensitive adenylate cyclase in rat renal tubular enzymes. *Biochem. J.* 150, 305–14.

Loreau, N., Lajotte, C., Wahbe, F. and Ardaillou, R. (1978). Effects of guanyl nucleotides on calcitonin-sensitive adenylate cyclase and calcitonin binding in rat adrenal cortex. *J. Endocrinol.* 76, 533–45.

Lorini, M. and Ciman, M. (1962). Hypoglycaemic action of diisopropylammonium salts in experimental diabetes. *Biochem. Pharmacol.* 11, 823–7.

Loten, E. G. and Sneyd, J. G. T. (1970). An effect of insulin on adipose-tissue adenosine 3′,5′-cyclic monophosphate phosphodiesterase. *Biochem. J.* 120, 187–93.

Loubatières, A. (1972). Therapeutic modification of islet function. *Handbook of Physiology:* Section 7, *Endocrinology*, Vol. 1, *Endocrine Pancreas*, pp. 653–64. Washington, D.C.: American Physiological Society.

Lowry, P. J. and Scott, A. P. (1975). The evolution of vertebrate cortico trophin and melanocyte stimulating hormone. *Gen. Comp. Endocrinol.* 26, 16–23.

Lucas, G. P., Holzwarth, G. J., Ocobock, R. W., Sozen, T., Stern, M. P., Wood, P. D. S., Haskell, W. L. and Farquhar, J. W. (1974). Disturbed relationship of plasma-renin to blood-pressure in hypertension. *Lancet* 2, 1337–9.

Luft, D., Schmülling, M. and Eggstein, M. (1978). Lactic acidosis in biguanide-treated diabetics. A review of 330 cases. *Diabetologia* 14, 75–87.

Lunan, C. B. and Klopper, A. (1975). Antioestrogens: a review. *Clin. Endocrinol.* 4, 551–72.

Lund, B., Sørensen, O. H. and Christensen, A. B. (1975). 25-Hydroxycholecalciferol and fractures of the proximal femur. *Lancet* 2, 300–2.

Lund, B., Hjorth, L., Kjaer, I., Reimann, I., Friis, T., Andersen, R. B. and Sørensen, O. H. (1975). Treatment of osteoporosis of aging with 1α-hydroxy-cholecalciferol. *Lancet 2*, 1168–71.

Lundbaek, K., Christensen, S. E., Hansen, Aa. P., Iverson, J., Ørskov, H., Seyer-Hansen, K., Alberti, K. G. M. M. and Whitefoot, R. (1976). Failure of somatostatin to correct manifest diabetic ketoacidosis. *Lancet, 1*, 215–18.

Lundin, L., Ljunghall, S., Wide, L. and Bostrom, H. (1978). Bromocriptine therapy with acromegaly. *Acta Endocrinol. 88*, Suppl. 216, 207–16.

Lunenfeld, B. and Insler, V. (1974). Classification of amenorrhoeic states and their treatment by ovulation induction. *Clin. Endocrinol. 3*, 223–37.

Lyons, W. R. (1958). Hormonal synergism in mammary growth. *Proc. Roy. Soc. Lond. B 149*, 303–25.

Lyttle, C. R. and DeSombre, E. R. (1977). Generality of oestrogen stimulation of peroxidase activity in growth responsive tissues. *Nature (Lond.) 268*, 337–9.

Maany, I., Frazer, A. and Mendels, J. (1975). Apomorphine: effect on growth hormone. *J. Clin. Endocrinol. Metab. 40*, 162–3.

McAfee, R. D. and Locke, W. (1967). Effect of angiotensin amide on Na isotope flux and short circuit current of isolated frog skin. *Endocrinology 81*, 1301–5.

McCann, S. M. (1977). Luteinizing-hormone-releasing hormone. *New Engl. J. Med. 296*, 797–802.

McCann, S. M., Fawcett, C. P. and Krulich, L. (1974). Hypothalamic hypophysial releasing and inhibiting hormones. In *MIP International Review of Science Physiology*, Series One, Vol. 5, *Endocrine Physiology*, (edited by S. M. McCann), pp. 31–65. London: Butterworth.

McCarthy, D. M. (1978). Report of the United States experience with cimetidine in Zollinger–Ellison syndrome and other hypersecretory states. *Gastroenterology 74*, 453–8.

McCollum, E. V., Simmonds, N., Becker, J. E. and Shipley, P. G. (1922). Studies on experimental rickets: XXI. An experimental demonstration of the existence of a vitamin which promotes calcium deposition. *J. Biol. Chem. 53*, 293–312.

McConaghey, P. (1972). The production of "sulphation factor" by rat liver. *J. Endocrinol. 52*, 1–9.

McConaghey, P. and Dehnel, J. (1972). Preliminary studies of "sulphation factor" production by rat kidney. *J. Endocrinol. 52*, 587–8.

McDaniel, E. B. and Pardthaisong, T. (1974). Use-effectiveness of six-month injections of DMPA as a contraceptive. *Amer. J. Obstet. Gynecol. 119*, 175–80.

McDevitt, D. G. (1977). Management of hyperthyroidism with beta-blocking drugs. *Prescr. J. 17*, 143–8.

McDonald, J. M., Bruns, D. E. and Jarett, L. (1978). Ability of insulin to increase calcium uptake by adipocyte endoplasmic reticulum. *J. Biol. Chem. 253*, 3504–8.

McEwen, B. S. (1976). Interactions between hormones and nerve tissue. *Sci. Amer. 235*, 48–59.

McGarry, E. E. and Beck, J. C. (1972). Metabolic effects of human growth hormone. In *Human Growth Hormone* (edited by A. S. Mason), pp. 25–39. London: William Heinemann.

MacGregor, R. R., Chu, L. L. H. and Cohn, D. V. (1976). Conversion of proparathyroid hormone to parathyroid hormone by a particulate enzyme of the parathyroid gland. *J. Biol. Chem. 251*, 6711–16.

MacGregor, R. R., Spagnuolo, P. J. and Lentnek, A. L. (1974). Inhibition of granulocyte adherence by

ethanol, prednisone, and aspirin, measured with an assay system. *New Engl. J. Med. 291*, 642–6.

McIntosh, C. H. S. and Hesch, R. D. (1976). Characterization of the parathyrin receptor in renal plasma membranes by labelled hormone and labelled antibody binding techniques. *Biochim. Biophys. Acta 426*, 535–46.

MacIntosh, F. C. (1938). Histamine as a normal stimulant of gastric secretion. *Quart. J. Exp. Physiol. 28*, 87–98.

Mack, T. M., Henderson, B. E., Gerkins, V. R., Arthur, M., Baptista, J. and Pike, M. C. (1975). Reserpine and breast cancer in a retirement community. *New Engl. J. Med. 292*, 1366–71.

Mack, T. M., Pike, M. C., Henderson, B. E., Pfeffer, R. I., Gerkins, V. R., Arthur, M. and Brown, S. E. (1976). Estrogens and endometrial cancer in a retirement community. *New Engl. J. Med. 294*, 1262–7.

McKenna, T. J., Island, D. P., Nicholson, W. E. and Liddle, G. W. (1978). The effects of potassium on early and late steps in the aldosterone biosynthesis in cells of the zona glomerulosa. *Endocrinology 103*, 1411–16.

MacKenzie, C. G. and MacKenzie, J. B. (1943). Effect of sulfonamides and thioureas on the thyroid gland and basal metabolism. *Endocrinology 32*, 185–203.

MacKenzie, I. Z. and Hillier, K. (1977). Prostaglandin-induced abortion and outcome of subsequent pregnancies: a prospective controlled study. *Brit. Med. J. 2*, 1114–17.

MacKenzie, I. Z., Embrey, M. P., Davies, A. J. and Guillebaud, J. (1978). Very early abortion by prostaglandins. *Lancet 1*, 1223–6.

MacKenzie, J. B., MacKenzie, C. G. and McCollum, E. V. (1941). The effect of sulfanilylguanidine on the thyroid of the rat. *Science 94*, 518–19.

Mackie, C., Warren, R. L. and Simpson, E. R. (1978). Investigations into the role of calcium ions in the control of steroid production by isolated adrenal zona glomerulosa cells of the rat. *J. Endocrinol. 77*, 119–27.

Mackin, J. F., Canary, J. J. and Pittman, C. S. (1974). Thyroid storm and its management. *New Engl. J. Med. 291*, 1396–8.

McKown, M. M., Workman, R. J. and Gregerman, R. I. (1974). Pepstatin inhibition of human renin. Kinetic studies and estimation of enzyme purity. *J. Biol. Chem. 249*, 7770–4.

McLarty, D. G., Brownlie, B. E. W., Alexander, W. D., Papapetrou, P. D. and Horton, P. (1973). Remission of thyrotoxicosis during treatment with propranolol. *Brit. Med. J. 2*, 332–4.

MacLeod, K. M. and Baxter, J. D. (1975). DNA binding of thyroid hormone receptors. *Biochem. Biophys. Res. Commun. 62*, 577–83.

MacLeod, K. M. and Baxter, J. D. (1976). Chromatin receptors for thyroid hormones. Interactions of the solubilized proteins with DNA. *J. Biol. Chem. 251*, 7380–7.

MacLeod, R. M. and Lehmeyer, J. E. (1974). Studies on the mechanism of the dopamine-mediated inhibition of prolactin secretion. *Endocrinology 94*, 1077–85.

MacLeod, S. C., Scott, J., Lord, L., Brodie, G., Perlin, I. and Simpson, A. A. (1977). Prevention and suppression of post-partum lactation with 2-bromo-alpha-ergocrytine (CB-154). *Clin. Endocrinol. 6*, Suppl., 65s–70s.

MacLusky, N. J. and McEwen, B. S. (1978). Oestrogen modulates progestin receptor concentrations in some rat brain regions but not in others. *Nature (Lond.) 274*, 276–8.

McNatty, K. P., Sawers, R. S. and McNeilly, A. S. (1974). A possible role for prolactin in control of steroid secretion by the human Graafian follicle. *Nature (Lond.) 250*, 653–5.

McNeilly, A. S., Sharpe, R. M., Davidson, D. W. and Fraser, H. M. (1978). Inhibition of gonadotrophin secretion by induced hyperprolactinemia in the male rat. *J. Endocrinol. 79*, 59–68.

McQueen, E. G. (1978). Hormonal steroid contraceptives: a further review of adverse reactions. *Drugs 16*, 322–57.

MacVicar, J. (1973). Acceleration and augmentation of labour. *Scott. Med. J. 18*, 201–14.

Maffly, R. H. (1977). Diabetes insipidus. In *Disturbances in Body Fluid Osmolality* (edited by T. E. Andreoli, J. J. Grantham, and F. C. Rector), pp. 285–307. Bethesda, Md.: American Physiological Society.

Maguire, M. E., Ross, E. M. and Gilman, A. G. (1977). β-Adrenergic receptor: ligand binding properties and interaction with adenyl cyclase. In *Advances in Nucleotide Research*, Vol. 8, pp. 1–83 (edited by P. Greengard and G. A. Robison). New York: Raven Press.

McGuire, W. L. (1975). Endocrine therapy of breast cancer. *Annu. Rev. Med. 26*, 353–63.

McGuire, W. L., Horwitz, K. B., Pearson, O. H. and Segaloff, A. (1977). Current status of estrogen and progesterone receptors in breast cancer. *Cancer 39*, 2934–47.

Maier, R., Riniker, B. and Rittel, W. (1974). Analogues of human calcitonin: I. Influence of modifications in amino acid positions 29 and 31 on hypocalcaemic activity in the rat. *FEBS Lett. 48*, 68–71.

Maier, R., Kamber, B., Riniker, B. and Rittel, W. (1975). Analogues of human calcitonin: II. Influence of modifications in amino-acid positions 1, 8, and 22 on hypocalcaemic activity. *Horm. Metab. Res. 7*, 511–14.

Maier, R., Kamber, B., Riniker, B. and Rittel, W. (1976). Analogues of human calcitonin: IV. Influence of leucine substitutions in positions 12, 16 and 19 on hypocalcaemic activity in the rat. *Clin. Endocrinol. 5*, Suppl., 327S–32S.

Mainoya, J. R. (1975). Effect of bovine growth hormone, human placental lactogen and ovine prolactin on intestinal fluid and ion transport in the rat. *Endocrinology 96*, 1165–70.

Mains, R. E. and Eipper, B. A. (1978). Coordinate synthesis of corticotropins and endorphins by mouse pituitary tumor cells. *J. Biol. Chem. 253*, 651–5.

Mains, R. E., Eipper, B. A. and Ling, N. (1977). Common precursor to corticotropins and endophins. *Proc. Natl. Acad. Sci. USA 74*, 3014–18.

Mainwaring, W. I. P. (1977). *The Mechanism of Action of Androgens*. New York: Springer-Verlag.

Majid, P. A., Sharma, B., Meeran, M. K. M. and Taylor, S. H. (1972). Insulin and glucose in the treatment of heart-failure. *Lancet 2*, 937–41.

Major, J. S. and Scholes, P. (1978). The localization of a histamine H_2-receptor adenylate cyclase system in canine parietal cells and its inhibition by prostaglandins. *Agents Actions 8*, 324–31.

Malaisse, W. J., Pipeleers, D. G. and Mahy, M. (1973). The stimulus-secretion coupling of glucose-induced insulin release: XII. Effects of diazoxide and gliclazide upon ^{45}Ca efflux from perifused islets. *Diabetologia 9*, 1–5.

Malbon, C. C. and Zull, J. E. (1977). Studies of binding of parathyroid hormone to a detergent-dispersed preparation from bovine kidney cortex plasma membranes. *J. Biol. Chem. 252*, 1079–83.

Malherbe, C., Burrill, K. C., Levin, S. R., Karam, J. H. and Forsham, P. H. (1972). Effect of diphenylhydantoin on insulin secretion in man. *New Engl. J. Med. 286*, 339–42.

Malone, D. N. S., Drever, J. C., Grant, I. W. B. and Percy-Robb, I. W. (1972). Endocrine response to substitution of corticotrophin for oral prednisolone in asthmatic children. *Brit. Med. J. 3*, 202–5.

Mangan, F. R. and Mainwaring, W. I. P. (1972). An explanation of the anti-androgenic properties of 6α-bromo-17β-hydroxy-17α-methyl-4-oxa-5α-androstane-2-one. *Steroids 20*, 331–43.

Mangos, J. A. and McSherry, N. R. (1969). Micropuncture study of sodium and potassium excretion in rat parotid saliva: role of aldosterone. *Proc. Soc. Exp. Biol. Med. 132*, 797–801.

Manis, J. and Norman, A. (1975). Vitamin D-resistant rickets and 25-hydroxycholecalciferol. *Brit. Med. J. 2*, 478–9.

Mann, J. I. and Inman, W. H. W. (1975). Oral contraceptives and death from myocardial infarction. *Brit. Med. J. 2*, 245–8.

Mann, J. I., Vessey, M. P., Thorogood, M. and Doll, R. (1975). Myocardial infarction in young women with special reference to oral contraceptive practice. *Brit. Med. J. 2*, 241–5.

Manning, M., Balaspiri, L., Acosta, M. and Sawyer, W. H. (1973). Solid phase synthesis of [1-deamino,4-valine]-8-D-arginine-vasopressin (DVDAVP), a highly potent and specific antidiuretic agent possessing protracted effects. *J. Med. Chem. 16*, 975–8.

Manning, M., Lowbridge, J., Haldar, J. and Sawyer, W. H. (1977). Design of neurohypophyseal peptides that exhibit selective agonistic and antagonistic properties. *Fedn. Proc. 36*, 1848–52.

Mannucci, P. M., Ruggeri, Z. M., Pareti, F. I. and Capitanio, A. (1977). 1-Deamino-8-D-arginine vasopressin: a new pharmacological approach to the management of haemophilia and Von Willebrand's disease. *Lancet 1*, 869–72.

Marble, A. (1972). Insulin–clinical aspects: the first fifty years. *Diabetes 21*, Suppl. 2, 632–6.

Marchant, B., Lees, J. F. H. and Alexander, W. D. (1978). Antithyroid drugs. *Pharmacol. Ther. B 3*, 305–48.

Marks, V. (1972). Spontaneous hypoglycemia. *Brit. Med. J. 1*, 430–2.

Marsh, J. M. (1975). The role of cyclic AMP in gonadal function. In *Advances in Cyclic Nucleotide Research*, Vol. 6, pp. 137–99 (edited by P. Greengard and G. A. Robison). New York: Raven Press.

Marshall, D. H. and Nordin, B. E. C. (1977). The effect of 1α-hydroxyvitamin D with and without oestrogens on calcium balance in post-menopausal women. *Clin. Endocrinol. 7*, Suppl., 159S–68S.

Marshall, D. H., Crilly, R. G. and Nordin, B. E. C. (1977). Plasma androstenedione and oestrone levels in normal and osteoporotic women. *Brit. Med. J. 2*, 1177–9.

Marshall, G. R. (1976). Structure–activity relations of antagonists of the renin–angiotensin system. *Fedn. Proc.* *35*, 2491–501.

Marshall, J. M. and Csapo, A. I. (1961). Hormonal and ionic influences on the membrane activity of uterine smooth muscle cells. *Endocrinology 68*, 1026–35.

Martial, J. A., Baxter, J. D., Goodman, H. M. and Seeburg, P. H. (1977). Regulation of growth hormone messenger RNA by thyroid and glucocorticoid hormones. *Proc. Natl. Acad. Sci. USA 74*, 1816–20.

Martin, J. B., Lal, S., Tolis, G. and Friesen, H. G. (1974). Inhibition of apomorphine of prolactin secretion in patients with elevated serum prolactin. *J. Clin. Endocrinol. Metab. 39*, 180–2.

Martin, J. V., Martin, P. J. and Goldberg, D. M. (1976). Enzyme induction as a possible cause of increased serum-triglycerides after oral contraceptives. *Lancet 1*, 1107–8.

Martin, M. J. (1971). Demonstration of circulating antibodies to neurophysin in patients treated with pitressin. *J. Endocrinol. 49*, 553–4.

Martin, P. M., Horwitz, K. B., Ryan, D. S. and McGuire, W. L. (1978). Phytoestrogen interaction with estrogen receptors in human breast cancer cells. *Endocrinology 103*, 1860–7.

Martin, T. F. J., Cort, A. M. and Tashjian, A. H. (1978). Thyrotropin-releasing hormone modulation of uridine uptake in rat pituitary cells. *J. Biol. Chem. 253*, 99–105.

Marver, D., Stewart, J., Funder, J. W., Feldman, D. and Edelman, I. S. (1974). Renal aldosterone receptors: studies with [^{3}H]aldosterone and the antimineralocorticoid [^{3}H]spirolactone (SC-26304). *Proc. Natl. Acad. Sci. USA 71*, 1431–5.

Marx, J. L. (1976a). After the heart attack: limiting the damage. *Science 194*, 1147–50.

Marx, J. L. (1976b). Estrogen drugs: do they increase the risk of cancer? *Science 191*, 838–41.

Marx, S. J. and Aurbach, G. D. (1975). Renal receptors for calcitonin: coordinate occurrence with calcitonin-activated adenylate cyclase. *Endocrinology 97*, 448–53.

Mason, A. S. (1972). Short stature and its treatment. *Brit. Med. J. 2*, 519–22.

Matthews, E. K., Petersen, O. H. and Williams, J. A. (1973). Pancreatic acinar cells: acetylcholine-induced membrane depolarization, calcium efflux and amylase release. *J. Physiol. (Lond.) 234*, 689–701.

Maugh, T. H. (1975a). Marihuana: new support for immune and reproductive hazards. *Science 190*, 865–7.

Maugh, T. H. (1975b). Diabetes: epidemiology suggests a viral connection. *Science 188*, 347–51.

Maurer, R. A., Stone, R. and Gorski, J. (1976). Cell-free synthesis of a large translation product of prolactin messenger RNA. *J. Biol. Chem. 251*, 2801–7.

Mautalen, C. A. (1972). Mechanism of action of *Solanum malacoxylon* upon calcium and phosphate metabolism in the rabbit. *Endocrinology 90*, 563–7.

Meade, T. W., Chakrabarti, R., Haines, A. P., Howarth, D. J., North, W. R. S. and Stirling, Y. (1977). Haemostatic, lipid, and blood-pressure profiles of women on oral contraceptives containing 50 μg or 30 μg oestrogen. *Lancet 2*, 948–51.

Means, A. R. (1975). Biochemical effects of follicle-stimulating hormone on the testis. *Handbook of Physiology*: Section 7, *Endocrinology*, Vol. 5, *Male Reproductive System*, pp. 203–18. Washington, D.C.: American Physiological Society.

Medeiros-Neto, G. A., Penna, M., Monteiro, K., Kataoka, K., Imai, Y. and Hollander, C. (1975). The effect of iodized oil on the TSH response to TRH in endemic goiter patients. *J. Clin. Endocrinol. Metab. 51*, 504–10.

Megyesi, K., Kahn, C. R., Roth, J., Neville, D. M., Nissley, S. P., Humbel, R. E. and Froesch, E. R. (1975). The NSILA-s receptor in liver cell membranes. Characterization and comparison with the insulin receptor. *J. Biol. Chem. 250*, 8990–6.

Mehdi, S. Q., Badger, J. and Kriss, J. P. (1977). Thyrotropin binding and long-acting thyroid stimulator absorbing activities in subcellular fractions from isolated thyroid cells and thyroid homogenates. *Endocrinology 101*, 59–65.

Meinders, A. E., Cejka, V. and Bleijenberg, A. J. (1974). Insulin secretion in patients with diabetes insipidus during long-term treatment with chlorpropamide. *J. Clin. Endocrinol. Metab. 38*, 539–44.

Melby, J. C. (1977). Clinical pharmacology of systemic corticosteroids. *Annu. Rev. Pharmacol. Toxicol. 17*, 511–27.

Melick, R. A., Ebeling, P. and Hjorth, R. J. (1976). Improvement in paraplegia in vertebral Paget's disease treated with calcitonin. *Brit. Med. J. 1*, 627–8.

Mellanby, E. (1919). A further demonstration of the part played by accessory food factors in the etiology of rickets. *J. Physiol. (Lond.) 52*, Liii–Liv.

Mendelson, C., Dufau, M. and Catt, K. (1975). Gonadotropin binding and stimulation of cyclic adenosine 3′:5′-monophosphate and testosterone production in isolated Leydig cells. *J. Biol. Chem. 250*, 8818–23.

Mendelson, J. H., Mello, N. K. and Ellingboe, J. (1977). Effects of acute alcohol intake on pituitary-gonadal hormones in normal human males. *J. Pharmacol. Exp. Ther. 202*, 676–82.

Mendelson, J. H., Mendelson, J. E. and Patch, V. D. (1974). Effects of heroin and methadone on plasma testosterone in narcotic addicts. *Fedn. Proc. 33*, 232.

Mendelson, J. H., Mendelson, J. E. and Patch, V. D. (1975). Plasma testosterone levels in heroin addiction and during methadone maintenance. *J. Pharmacol. Exp. Ther. 192*, 211–17.

Mendelson, J. H., Kuehnle, J., Ellingboe, J. and Babor, T. F. (1974). Plasma testosterone levels before, during and after chronic marihuana smoking. *New Engl. J. Med. 291*, 1051–5.

Mendoza, S. A. and Brown, C. F. (1974). Effect of chlorpropamide on osmotic water flow across toad bladder and the response to vasopressin, theophylline and cyclic AMP. *J. Clin. Endocrinol. Metab. 38*, 883–9.

Metz, S. A., Baylink, D. J., Hughes, M. R., Haussler, M. R. and Robertson, R. P. (1977). Selective deficiency of 1,25-dihydroxycholecalciferol. A cause of isolated skeletal resistance to parathyroid hormone. *New Engl. J. Med. 297*, 1084–90.

Meyers, C., Arimura, A., Gordin, A., Fernandez-Durango, R., Coy, D. F., Schally, A. V., Drouin, J., Ferland, L., Beaulieu, M. and Labrie, F. (1977). Somatostatin analogs which inhibit glucagon and growth hormone more than insulin release. *Biochem. Biophys. Res. Commun. 74*, 630–6.

Michael, R. P. (1973). The effects of hormones on sexual behavior in female cat and rhesus monkey. *Handbook of Physiology*: Section 7, *Endocrinology*, Vol. 2, *Female Re-*

productive System, Part 1, pp. 187–221. Washington, D.C.: American Physiological Society.

Michie, W., Hamer-Hodges, D. W., Pegg, C. A. S., Orr, F. G. G. and Bewsher, P. D. (1974). Beta-blockade and partial thyroidectomy for thyrotoxicosis. *Lancet 1*, 1010–11.

Mickey, J., Tate, R. and Lefkowitz, R. J. (1975). Subsensitivity of adenylate cyclase and decreased β-adrenergic receptor binding after chronic exposure to (−)-isoproterenol *in vitro. J. Biol. Chem. 250*, 5727–9.

Miller, A. W. F., Calder, A. A. and Macnaughton, M. C. (1972). Termination of pregnancy by continuous intrauterine infusion of prostaglandins. *Lancet 2*, 5–7.

Miller, J. F., Welply, G. A. and Elstein, M. (1975). Prostaglandin E_2 tablets compared with intravenous oxytocin in induction of labour. *Brit. Med. J. 1*, 14–16.

Miller, M. and Moses, A. M. (1976). Drug-induced states of impaired water excretion. *Kidney Int. 10*, 96–103.

Minguell, J. J. and Sierralta, W. D. (1975). Molecular mechanism of action of the male sex hormones. *J. Endocrinol. 65*, 287–315.

Mirsky, I. A. (1968). Metabolic effects of neurohypophyseal hormones and related polypeptides. In *Neurohypophysial Hormones and Similar Peptides* (edited by B. Berde), pp. 613–24. New York: Springer-Verlag.

Mitra, S., Genuth, S. M., Berman, L. B. and Vertes, V. (1972). Aldosterone secretion in anephric patients. *New Engl. J. Med. 286*, 61–4.

Mittra, B. (1965). Potassium, glucose and insulin in treatment of myocardial infarction. *Lancet 2*, 607–9.

Miyai, K., Onishi, T., Hosokawa, M., Ishibash, K. and Kumahara, Y. (1974). Inhibition of thyrotropin and prolactin secretions in primary hypothyroidism by 2-Br-α-ergocryptine. *J. Clin. Endocrinol. Metab. 39*, 391–4.

Moghissi, K. S. (1973). Composition and function of cervical secreton. *Handbook of Physiology:* Section 7, *Endocrinology*, Vol. 2, *Female Reproductive System*, Part 2, pp. 25–48. Washington, D.C.: American Physiological Society.

Montague, W. and Howell, S. L. (1975). Cyclic AMP and the physiology of the islets of Langerhans. In *Advances in Cyclic Nucleotide Research*, Vol. 6, pp. 201–43 (edited by P. Greengard and G. A. Robison). New York: Raven Press.

Moore, W. V. and Wolff, J. (1974). Thyroid-stimulating hormone binding to beef thyroid membranes. Relation to adenylate cyclase. *J. Biol. Chem. 249*, 6255–63.

Morgan, F. J., Birken, S. and Canfield, R. E. (1975). The amino acid sequence of human chorionic gonadotropin. The α subunit and β subunit. *J. Biol. Chem. 250*, 5247–58.

Morgan, H. G., Boulnois, J. and Burns-Cox, C. (1973). Addiction to prednisone. *Brit. Med. J. 2*, 93–4.

Morgan, J. M. and Routtenberg, A. (1977). Angiotensin injected into the neostriatum after learning disrupts retention performance. *Science 196*, 87–9.

Morley, J. S. (1968). Structure–function relationships in gastrin-like peptides. *Proc. Roy. Soc. Lond. B 170*, 97–111.

Morris, D. J. and Davis, R. P. (1974). Aldosterone: current concepts. *Metabolism 23*, 473–94.

Morris, D. R. and Hager, L. P. (1966). Mechanism of the inhibition of enzymatic halogenation by antithyroid agents. *J. Biol. Chem. 241*, 3582–9.

Morrison, A. S., Jick, H. and Ory, H. W. (1977). Oral contraceptives and hepatitis. A report from the Boston Collaborative Drug Surveillance Program, Boston University Medical Center. *Lancet 1*, 1142–3.

Morrissey, J. F. and Barreras, R. F. (1974). Antacid therapy. *New Engl. J. Med. 290*, 550–4.

Mortimer, C. H., McNeilly, A. S., Fisher, R. A., Murray, M. A. F. and Besser, G. M. (1974a). Gonadotrophin releasing hormone therapy in hypogonadal males with hypothalamic or pituitary dysfunction. *Brit. Med. J. 4*, 617–21.

Mortimer, C. H., Tunbridge, W. M. G., Carr, D., Yoemans, L., Lind, T., Coy, D. H., Bloom, S. R., Kastin, A., Mallinson, C. N., Besser, G. M., Schally, A. V. and Hall, R. (1974b). Effects of growth-hormone release-inhibiting hormone on circulating glucagon, insulin, and growth hormone in normal, diabetic, acromegalic and hypopituitary patients. *Lancet 1*, 697–701.

Moses, A. C., Nissley, S. P., Cohen, K. L. and Rechler, M. M. (1976). Specific binding of a somatomedin-like polypeptide in rat serum depends on growth hormone. *Nature (Lond.) 263*, 137–40.

Moses, A. M. and Miller, M. (1974). Drug-induced dilutional hyponatremia. *New Engl. J. Med. 291*, 1234–9.

Mountjoy, C. Q., Price, J. S., Weller, M., Hunter, P., Hall, R. and Dewar, J. H. (1974). A double-blind crossover sequential trial of oral thyrotrophin-releasing hormone in depression. *Lancet 1*, 958–60.

Moyle, W. R., Bahl, O. P. and März, L. (1975). Role of the carbohydrate of human chorionic gonadotrophin in the mechanism of hormone action. *J. Biol. Chem. 250*, 9163–9.

Moyle, W. R., Jungas, R. L. and Greep, R. O. (1973a). Influence of luteinizing hormone and adenosine 3′-5′-monophosphate on the metabolism of free and esterified cholesterol in mouse Leydig-cell tumours. *Biochem. J. 134*, 407–13.

Moyle, W. R., Jungas, R. L. and Greep, R. O. (1973b). Metabolism of free and esterified cholesterol by Leydig cell tumour mitochondria. *Biochem. J. 134*, 415–24.

Mueller, W. J., Brubaker, R. L., Gay, C. V. and Boelkins, J. N. (1973). Mechanisms of bone resorption in laying hens. *Fedn. Proc. 32*, 1951–4.

Mukherjee, C., Caron, M. G., Coverstone, M. and Lefkowitz, R. J. (1975). Identification of adenylate cyclase-coupled β-adrenergic receptors in frog erythrocytes with (−)-[^{3}H]alprenolol. *J. Biol. Chem. 250*, 4869–76.

Müller, J. (1971). Regulation of aldosterone biosynthesis. In *Monographs in Endocrinology*, Vol. 5, p. 137. New York: Springer-Verlag.

Mulrow, P. J. and Ganong, W. F. (1961). The effect of hemorrhage upon aldosterone secretion in normal and hypophysectomized dogs. *J. Clin. Invest. 40*, 579–85.

Munck, A. (1971). Glucocorticoid inhibition of glucose uptake by peripheral tissues: old and new evidence, molecular mechanisms, and physiological significance. *Perspect. Biol. Med. 14*, 265–89.

Munck, A. and Young, D. A. (1975). Corticosteroids and lymphoid tissue. *Handbook of Physiology:* Section 7, *Endocrinology*, Vol. 6, *Adrenal Gland*, pp. 231–43. Washington, D.C.: American Physiological Society.

Munday, K. A., Parsons, B. J. and Poat, J. A. (1972). Studies on the mechanism of action of angiotensin on ion transport by kidney cortex slices. *J. Physiol. (Lond.)* *224*, 195–206.

Munday, K. A., Parsons, B. J., Poat, J. A., D'Auriac, G. A. and Meyer, P. (1976). The role of cyclic 3':5'-adenosine monophosphate in the responses of the intestine and kidney to angiotensin. *J. Endocrinol.* *69*, 297–8.

Munson, P. L. (1976). Physiology and Pharmacology of thyrocalcitonin. *Handbook of Physiology:* Section 7, *Endocrinology*, Vol. 7, *Parathyroid Gland*, pp. 443–64. Washington, D.C.: American Physiological Society.

Müntzing, J., Shukla, S. K., Chu, T. M., Mittelman, A. and Murphy, G. P. (1974). Pharmacoclinical study of oral estramustine phosphate (estracyt) in advanced carcinoma of the prostate. *Invest. Urol.* *12*, 65–8.

Murase, T. and Yoshida, S. (1973). Mechanism of chlorpropamide action in patients with diabetes insipidus. *J. Clin. Endocrinol. Metab.* *36*, 174–7.

Murray, I. P. C. and Stewart, R. D. H. (1967). Iodide goitre. *Lancet 1*, 922–6.

Murray, M. A. F., Bancroft, J. H. J., Anderson, D. C., Tennent, T. G. and Carr, P. J. (1975). Endocrine changes in male sexual deviants after treatment with anti-androgens, oestrogens or tranquillizers. *J. Endocrinol.* *67*, 179–88.

Mustard, J. F. and Packham, M. A. (1977). Platelets and diabetes mellitus. *New Engl. J. Med.* *297*, 1345–7.

Mygind, N. (1973). Local effect of intranasal beclomethasone diproprionate aerosol in hay fever. *Brit. Med. J.* *4*, 464–6.

Myles, A. B. and Daly, J. R. (1974). *Corticosteroid and ACTH Treatment: Principals and Problems*. Baltimore: Williams & Williams.

Nabarro, J. D. N. (1972). Pituitary tumours and hypopituitarism. *Brit. Med. J. 1*, 492–5.

Nagasaka, A. and Hidaka, H. (1976). Effect of antithyroid agents 6-propyl-2-thiouracil and 1-methyl-2-mercaptoimidazole on human thyroid iodide peroxidase. *J. Clin. Endocrinol. Metab.* *43*, 152–8.

Nair, R. M. G., Kastin, A. J. and Schally, A. V. (1971). Isolation and structure of hypothalamic MSH release-inhibiting hormone. *Biochim. Biophys. Res. Commun.* *43*, 1376–81.

Nair, R. M. G., DeVillier, C., Barnes, M., Antalis, J. and Wilbur, D. L. (1978). A bovine hypothalamic peptide possessing immunoreactive growth hormone-releasing activity. *Endocrinology 103*, 112–20.

Nakai, Y., Imura, H., Sakurai, H., Kurahachi, H. and Yoshimi, T. (1974). Effect of cyproheptadine on human growth hormone secretion. *J. Clin. Endocrinol. Metab. 38*, 446–9.

Nakano, H., Fawcett, C. P., Kimura, F. and McCann, S. M. (1978). Evidence for the involvement of guanosine 3',5'-cyclic monophosphate in the regulation of gonadotropin release. *Endocrinology 103*, 1527–33.

Nakano, J. (1974). Cardiovascular responses to neurohypophysial hormones. *Handbook of Physiology:* Section 7, *Endocrinology*, Vol. 4, *The Pituitary Gland and Its Neuroendocrine Control*, Part 1, pp. 395–442. Washington, D. C.: American Physiological Society.

Napoli, J. L., Reeve, L. E., Eisman, J. A., Schnoes, H. K. and DeLuca, H. F. (1977). *Solanum glaucophyllum* as a source of 1,25-dihydroxyvitamin D_3. *J. Biol. Chem. 252*, 2580–3.

Narahara, H. T. (1972). Binding of insulin to tissues in relation to biological action of the hormone. *Handbook of Physiology:* Section 7, *Endocrinology*, Vol. 1, *Endocrine Pancreas*, pp. 332–45. Washington, D.C.: American Physiological Society.

Needleman, P., Douglas, J. R., Jakschik, B. A., Blumberg, A. L., Isakson, P. C. and Marshall, G. R. (1976). Angiotensin antagonists as pharmacological tools. *Fedn. Proc. 35*, 2488–93.

Neher, R. and Milani, A. (1976). Mode of action of peptide hormones. *Clin. Endocrinol. Metab. 5*, Suppl. 29S–39S.

Nestor, J. J., Ferger, M. F. and Du Vigneaud, V. (1975). [1-β-Mercapto-β,β,-pentamethylenepropionic acid] oxytocin, a potent inhibitor of oxytocin. *J. Med. Chem. 18*, 284–7.

Neumann, F. and Von Berswordt-Wallrabe, R. (1966). Effects of the androgen antagonist cyprpterone acetate on the testicular structure, spermatogenesis and accessory sexual glands of testosterone-treated adult hypophysectomized rats. *J. Endocrinol. 35*, 363–71.

Newton, J. and Collins, W. P. (1972). Effect of synthetic luteinizing hormone releasing hormone (LH/FSH-RH) in women with menstrual disorders. *Brit. Med. J. 3*, 271–3.

Newton, R. W., Browning, M. C. K., Iqbal, J., Piercy, N. and Adamson, D. G. (1978). Adrenocortical suppression in workers manufacturing synthetic glucocorticoids. *Brit. Med. J. 1*, 73–4.

Nezamis, J. E., Robert, A. and Stowe, D. F. (1971). Inhibition by prostaglandin E_1 gastric secretion in the dog. *J. Physiol. (Lond.) 218*, 369–83.

Niall, H. D., Keutmann, H., Sauer, R., Hogan, M., Dawson, B., Aurbach, G. and Potts, J. (1970). The amino acid sequence of bovine parathyroid hormone. *Hoppe-Seyler's Z. Physiol. Chem. 35*, 1586–8.

Niall, H. D., Hogan, M. L., Sauer, R., Rosenblum, I. Y. and Greenwood, F. C. (1971). Sequences of pituitary and placental lactogenic and growth hormones: evolution from a primordial peptide by gene reduplication. *Proc. Natl. Acad. Sci. USA 68*, 866–9.

Niall, H. D., Hogan, M. L., Tregear, G. W., Segre, G. V., Hwang, P. and Friesen, H. (1973). The chemistry of growth hormone and lactogenic hormone. *Rec. Prog. Horm. Res. 29*, 387–404.

Nicholson, W. E., Liddle, R. A., Puett, D. and Liddle, G. W. (1978). Adrenocorticotropic hormone biotransformation, clearance, and catabolism. *Endocrinology 103*, 1344–51.

Nicolaysen, R. (1937a). Studies on the mode of action of vitamin D: II. The influence of vitamin D on the faecal output of endogenous calcium and phosphorus in the rat. *Biochem. J. 31*, 107–21.

Nicolaysen, R. (1937b). Studies on the mode of action of vitamin D: III. The influence of vitamin D on the absorption of calcium and phosphorus in the rat. *Biochem. J. 31*, 122–9.

Nicolaysen, R. and Eeg-Larsen, N. (1953). The biochemistry and physiology of vitamin D. *Vitam. Horm. 11*, 29–60.

Nicoll, C. S. and Bern, H. A. (1972). On the actions of prolactin among the vertebrates: is there a common denominator? In *Lactogenic Hormones* (edited by G. E. W. Wolstenholme and J. Knight), pp. 229–317. (A Ciba Foundation Symposium.) Edinburgh: Churchill Livingstone.

Niki, A., Niki, H., Miwa, I. and Okuda, J. (1974). Insulin secretion by anomers of D-glucose. *Science 186*, 150–51.

Nistrup Madsen, S. and Sonne, O. (1976). Increase of glucagon receptors in hyperthyroidism. *Nature (Lond.) 262*, 793–5.

Nisula, B. and Ketelslegers, J-M. (1974). Thyroid-stimulating activity and chorionic gonadotropin. *J. Clin. Invest. 54*, 494–9.

Niven, P. A. R., Landon, J. and Chard, T. (1972). Placental lactogen levels as guide to outcome of threatened abortion. *Brit. Med. J. 3*, 799–801.

Noel, G. L., Suh, H. K. and Frantz, A. G. (1974). Prolactin release during nursing and breast stimulation in postpartum and nonpostpartum subjects. *J. Clin. Endocrinol. Metab. 38*, 413–23.

Nolin, J. M. and Witorsch, R. J. (1976). Detection of endogenous immunoreactive prolactin in rat mammary epithelial cells during lactation. *Endocrinology 99*, 949–58.

Norman, A. W. and Henry, H. (1974). 1,25-Dihydroxycholecalciferol-A hormonally active form of vitamin D_3. *Rec. Prog. Horm. Res. 30*, 431–73.

Norman, A. W., Mitra, M. N., Okamura, W. H. and Wing, R. M. (1975). Vitamin D: 3-deoxy-1α-hydroxyvitamin D_3, biologically active analog of 1α,25-dihydroxyvitamin D_3. *Science 188*, 1013–15.

Nourok, D. S., Glassock, R. J., Solomon, D. H. and Maxwell, M. H. (1964). Hypothyroidism following prolonged sodium nitroprusside therapy. *Amer. J. Med. Sci. 248*, 129–38.

Oakley, N. W. (1976). Effect of "fractionated" insulins on the total plasma binding capacity and insulin requirement in severe diabetes. *Lancet 1*, 994–6.

Odell, W. D. and Molitch, M. E. (1974). The pharmacology of contraceptive agents. *Annu. Rev. Pharmacol. 14*, 413–34.

Odell, W. D. and Swerdloff, R. S. (1978). Abnormalities of gonadal function in men. *Clin. Endocrinol. 8*, 149–80.

O'Fallon, W. M., Labarthe, D. R. and Kurland, L. T. (1975). Rauwolfia derivatives and breast cancer. *Lancet 2*, 292–6.

Ogihara, T., Hata, T., Mikami, H., Mandai, T. and Kumahara, Y. (1977). Effects of two angiotensin II analogues on blood pressure in normal subjects with various sodium balances. *Life Sci. 20*, 1855–62.

Oglesby, R. B., Black, R. L., von Sallmann, L. and Bunim, J. J. (1961). Cataracts in patients with rheumatic diseases treated with corticosteroids. *Arch. Ophthalmol. 66*, 625–30.

Ojeda, S. R., Jameson, H. E. and McCann, S. M. (1977a). Hypothalamic areas involved in prostaglandin (PG)-induced gonadotropin release: I. Effects of PGE_2 and $PGF_{2α}$ implants on luteinizing hormone release. *Endocrinology 100*, 1585–94.

Ojeda, S. R., Jameson, H. E. and McCann, S. M. (1977b). Hypothalamic areas involved in prostaglandin (PG)-induced gonadotropin release: II. Effects of PGE_2 and $PGF_{2α}$ implants on luteinizing hormone release. *Endocrinology 100*, 1595–603.

Okada, Y., Onishi, T., Tanaka, K., Morimoto, S., Tsuji, M., Watanabe, K., Okazaki, A., Takeuchi, T. and Kumahara, Y. (1978). Prolactin and TSH responses to TRH, chlorpromazine, and L-dopa in children with human growth hormone deficiency. *Acta Endocrinol. 88*, 217–26.

Oliver, C. and Porter, J. C. (1978). Distribution and characterization of α-melanocyte-stimulating hormone in the rat brain. *Endocrinology 102*, 697–705.

Oliver, C., Mical, R. S. and Porter, J. C. (1977). Hypothalamic–pituitary vasculature: evidence for retrograde blood flow in the pituitary stalk. *Endocrinology 101*, 598–604.

Oliveros, J. C., Jandali, M. K., Timsit-Berthler, M., Remy, R., Benghezala, A., Audibert, A. and Moeglen, J. M. (1978). Influence of vasopressin on learning and memory. *Lancet 1*, 41–2.

O'Malley, B. W. and Schrader, W. T. (1976). The receptors of steroid hormones. *Sci. Amer. 234*, 32–43.

Omdahl, J. L. and DeLuca, H. F. (1973). Regulation of vitamin D metabolism and function. *Physiol. Rev. 53*, 327–72.

Ondetti, M. A., Rubin, B. and Cushman, D. W. (1977). Design of specific inhibitors of angiotensin-converting enzyme: new class of orally active antihypertensive agents. *Science 196*, 441–4.

Ondo, J. G. (1974). Gamma-aminobutyric acid effects on pituitary gonadotropin secretion. *Science 186*, 738–9.

Oparil, S. and Haber, E. (1974a). The renin–angiotensin system (part 1). *New Engl. J. Med. 291*, 389–401.

Oparil, S. and Haber, E. (1974b). The renin–angiotensin system (part 2). *New Engl. J. Med. 291*, 446–57.

Oppenheimer, J. H. (1973). Interaction of drugs with thyroid hormone binding sites. *Ann. NY Acad. Sci. 226*, 333–40.

Oppenheimer, J. H. (1975). Initiation of thyroid-hormone action. *New Engl. J. Med. 292*, 1063–8.

Orci, L. and Unger, R. H. (1975). Functional subdivision of islets of Langerhans and possible role of D cells. *Lancet 2*, 1243–4.

Orci, L., Gabbay, K. H. and Malaisse, W. J. (1972). Pancreatic beta-cell web: its possible role in insulin secretion. *Science 175*, 1128–30.

Orly, J. and Schramm, M. (1976). Coupling of catecholamine receptor from one cell with adenylate cyclase from another cell by cell fusion. *Proc. Natl. Acad. Sci. USA 73*, 4410–14.

Ornoy, A., Goodwin, D., Noff, D. and Edelstein, S. (1978). 24,25-Dihydroxyvitamin D is a metabolite of Vitamin D essential for bone formation. *Nature (Lond.) 276*, 517–19.

Ory, H., Cole, P. C., MacMahon, B. and Hoover, R. (1976). Oral contraceptives and reduced risk of benign breast diseases. *New Engl. J. Med. 294*, 419–22.

Osland, R. B., Cheesman, D. W. and Forsham, P. H. (1977). Studies on the mechanism of the suppression of the preovulatory surge of luteinizing hormone in the rat by arginine vasotocin. *Endocrinology 101*, 1203–9.

Osnes, J-B. and Øye, I. (1975). Relationship between cyclic AMP metabolism and inotropic response of perfused rat hearts to phenylephrine and other adrenergic amines. In *Advances in Cyclic Nucleotide Research*, Vol. 5, pp. 415–33 (edited by G. I. Drummond, P. Greengard, and G. A. Robison). New York: Raven Press.

Ott, I. and Scott, J. C. (1910). The action of infundibulin upon the mammary secretion. *Proc. Soc. Exp. Biol. Med. 8*, 48–9.

Page, I. H. and Bumpus, F., Eds. (1974). *Handbook of Pharmacology*, Vol. 37, 1–591. *Angiotensin*. New York: Springer-Verlag.

Page, I. H. and Helmer, O. M. (1940). Angiotonin-activator, renin- and angiotonin-inhibitor, and the mechanism of angiotonin tachyphylaxis in normal, hypertensive, and nephrectomized animals. *J. Exp. Med.* 71, 495–519.

Page, M. McB., Smith, R. B. W. and Watkins, P. J. (1976). Cardiovascular effects of insulin. *Brit. Med. J. 1*, 430–2.

Page, M., McB., Alberti, K. G. M. M., Greenwood, R., Gumaa, K. A., Hockaday, T. D. R., Lowy, C., Nabarro, J. D. N., Pyke, D. A., Sönksen, P. H., Watkins, P. J. and West, T. E. T. (1974). Treatment of diabetic coma with continuous low-dose infusion of insulin. *Brit. Med. J. 2*, 687–90.

Palade, G. (1975). Intracellular aspects of the process of protein synthesis. *Science 189*, 347–58.

Papkoff, H. (1972). Subunit interrelationships among the pituitary glycoprotein hormones. *Gen. Comp. Endocrinol.* Suppl. 3, 609–16.

Paris, J., McConahey, W. M., Owen, C. A., Woolner, L. B. and Bahn, R. C. (1960). Iodide goiter. *J. Clin. Endocrinol. Metab. 20*, 57–67.

Park, C. S. and Malvin, R. L. (1978). Calcium in the control of renin release. *Amer. J. Physiol. 235*, F22–5.

Parkin, J. M. (1976). Short stature. *Brit. Med. J. 1*, 1139–41.

Parveen, L., Chowdhury, A. G. and Chowdhury, Z. (1977). Injectable contraception (medroxyprogesterone acetate) in rural Bangladesh. *Lancet 2*, 946–8.

Patanelli, D. J. (1975). Suppression of fertility in the male. *Handbook of Physiology:* Section 7, *Endocrinology*, Vol. 5, *Male Reproductive System*, pp. 245–58. Washington, D.C.: American Physiological Society.

Patel, Y. C., Zingg, H. H. and Dreifuss, J. J. (1977). Calcium-dependent somatostatin secretion from rat neurohypophysis *in vitro. Nature (Lond.) 267*, 851–2.

Paton, W. D. M. (1961). A theory of drug action based on the rate of drug–receptor combination. *Proc. Roy. Soc. Lond. B 154*, 21–69.

Patton, G. S., Ipp, E., Dobbs, R. E., Orci, L., Vale, W. and Unger, R. H. (1977). Pancreatic immunoreactive somatostatin release. *Proc. Natl. Acad. Sci. USA 74*, 2140–3.

Paul, S. M. and Axelrod, J. (1977). Catecholestrogens: presence in brain and endocrine tissues. *Science 197*, 657–9.

Paul, S. M. and Skolnick, P. (1977). Catechol oestrogens inhibit oestrogen elicited accumulation of hypothalamic cyclic AMP suggesting role as endogenous anti-oestrogens. *Nature (Lond.) 266*, 559–61.

Paul, S. M., Axelrod, J. and Diliberto, J. J. (1977). Catechol estrogen-forming enzyme of brain: demonstration of a cytochrome P450 monooxygenase. *Endocrinology 101*, 1604–10.

Peach, M. J. (1977). Renin–angiotensin system: biochemistry and mechanisms of action. *Physiol. Rev. 57*, 313–70.

Peach, M. J. and Ackerly, J. A. (1976). Angiotensin antagonists and the adrenal cortex and medulla. *Fedn. Proc. 35*, 2502–7.

Peacock, M. (1977). Symposium: the clinical use of 1α-hydroxyvitamin D₃. *Clin. Endocrinol. 7*, Suppl., 1–246.

Peacock, M., Gallagher, J. C. and Nordin, B. E. C. (1974). Action of 1α-hydroxyvitamin D₃ on calcium absorption and bone resorption in man. *Lancet 1*, 385–9.

Pearl, K. N. and Chambers, T. L. (1977). Propranolol treatment of thyrotoxicosis in a premature infant. *Brit. Med. J. 2*, 738.

Pearse, A. G. E. (1968). Common cytochemical and ultrastructural characteristics of cells producing polypeptide hormones (the APUD series) and their relevance to thyroid and ultimobranchial C cells and calcitonin. *Proc. Roy. Soc. Lond. B 170*, 71–80.

Pearse, A. G. E. (1974). The endocrine cells of the GI tract: origins, morphology and functional relationships in health and disease. *Clin. Gastroenterol. 3*, 491–510.

Pearse, A. G. E. (1975). Neurocristopathy, neuroendocrine pathology and the APUD concept. *Z. Krebsforsch. 84*, 1–18.

Pearse, A. G. E. (1976). Peptides in brain and intestine. *Nature (Lond.) 262*, 92–4.

Pearse, A. G. E. and Takor, T. T. (1976). Neuroendocrine embryology and the APUD concept. *Clin. Endocrinol. 5*, Suppl., 229S–44S.

Peart, W. S. (1975). Renin–angiotensin system. *New Engl. J. Med. 292*, 302–6.

Peart, W. S. (1977). The kidney as an endocrine organ. *Lancet 2*, 543–8.

Pelletier, G., Dubé, D. and Puviani, R. (1977). Somatostatin: electron microscope immunohistochemical localization in secretory neurons of rat hypothalamus. *Science 196*, 1469–70.

Pelletier, G., Leclerc, R., Labrie, F., Cote, J., Chrétien, M. and Lis, M. (1977). Immunohistochemical localization of β-lipotropic hormone in the pituitary gland. *Endocrinology 100*, 770–6.

Pelletier, G., Cusan, L., Auclair, C., Kelly, P. A., Désy, L. and Labrie, F. (1978). Inhibition of spermatogenesis in the rat by treatment with [D-Ala⁶,Des-Gly-NH₂¹⁰] LHRH ethylamide. *Endocrinology 103*, 641–3.

Peña, C., Stewart, J. M. and Goodfriend, T. C. (1974). A new class of angiotensin inhibitors: *N*-methylphenylalanine analogs. *Life Sci. 14*, 1331–6.

Pepperell, R. J., Evans, J. H., Brown, J. B., Smith, M. A., Healy, D. and Burger, H. G. (1977). Serum prolactin levels and the value of bromocriptine in the treatment of anovulatory infertility. *Brit. J. Obstet. Gynecol. 84*, 58–66.

Perchellet, J.-P., Shanker, G. and Sharma, R. K. (1978). Regulatory role of guanosine 3′,5′-monophosphate in adrenocorticotropin hormone-induced steroidogenesis. *Science 199*, 311–12.

Permutt, M. A. and Kipnis, D. M. (1975). Insulin biosynthesis and secretion. *Fedn. Proc. 34*, 1549–55.

Perrild, H., Madsen, S. N. and Hansen, J. E. M. (1978). Irreversible myxoedema after lithium carbonate. *Brit. Med. J. 1*, 1108–9.

Peterlik, M. and Wasserman, R. H. (1975). 1,25-Dihydroxycholecalciferol-like activity in *Solanum malacoxylon:* purification and partial characterization. *FEBS Lett. 56*, 16–19.

Peterson, O. H. and Ueda, N. (1976). Pancreatic acinar cells: the role of calcium in stimulus–secretion coupling. *J. Physiol. (Lond.) 254*, 583–606.

Petitti, D. B. and Wingerd, J. (1978). Use of oral contraceptives, cigarette smoking, and risk of subarachnoid haemorrhage. *Lancet 2*, 234–6.

Pettinger, W. A. and Mitchell, H. C. (1976). Clinical pharmacology of angiotensin antagonists. *Fedn. Proc. 35*, 2521–5.

Pfeuffer, T. and Helmreich, E. J. M. (1975). Activation of

pigeon erythrocyte membrane adenylate cyclase by guanylnucleotide analogues and separation of a nucleotide binding protein. *J. Biol. Chem. 250,* 867–76.

Philibert, D. and Raynaud, J-P. (1973). Progesterone binding in the immature mouse and rat uterus. *Steroids 22,* 89–98.

Philibert, D. and Raynaud, J-P. (1974). Progesterone binding in the immature rabbit and guinea pig uterus. *Endocrinology 94,* 627–32.

Philipp, E. E. (1976). Management of the menopause: two views. Personal view II. *Prescr. J. 16,* 58–62.

Phillips, G. B. (1976). Evidence for hyperoestrogenaemia as a risk factor for myocardial infarction in men. *Lancet 2,* 14–18.

Pierce, J. G. (1974). Chemistry of thyroid-stimulating hormone. *Handbook of Physiology:* Section 7, *Endocrinology,* Vol. 4, *The Pituitary Gland and Its Neuroendocrine Control,* Part 2, pp. 79–101. Washington, D.C.: American Physiological Society.

Pietras, R. J., Seeler, B. J. and Szego, C. M. (1975). Influence of antidiuretic hormone on release of lysosomal hydrolase at mucosal surface of epithelial cells from urinary bladder. *Nature (Lond.) 257,* 493–5.

Pike, J. W., Spanos, E., Colston, K. W., MacIntyre, I. and Haussler, M. R. (1978). Influence of estrogen on renal vitamin D hydroxylases and serum $1\alpha,25$-$(OH)_2D_3$ in chicks. *Amer. J. Physiol. 235,* E338–43.

Pilgeram, L. O., Ellison, J. and Von Dem Bussche, G. (1974). Oral contraceptives and increased formation of soluble fibrin. *Brit. Med. J. 3,* 556–8.

Pilkis, S. J. and Park, C. R. (1974). Mechanism of action of insulin. *Annu. Rev. Pharmacol. 14,* 365–88.

Pita, J. C., Lippman, M. E., Thompson, E. B. and Loriaux, D. L. (1975). Interaction of spironolactone and digitalis with the 5α-dihydrotestosterone (DHT) receptor of rat ventral prostate. *Endocrinology 97,* 1521–7.

Pittman, C. S. and Pittman, J. A. (1974). Relation of chemical structure to the action and metabolism of thyroactive substances. *Handbook of Physiology:* Section 7, *Endocrinology,* Vol. 3, *Thyroid,* pp. 233–53. Washington, D.C.: American Physiological Society.

Pleitscher, A., Da Prada, M., Lütold, B., Berneis, K. H., Steffen, H. and Weder, H. G. (1973). Two mechanisms of catecholamine storage in adrenal chromaffin cells. In *Frontiers in Catecholamine Research* (edited by E. Usdin and S. H. Snyder), pp. 41–5. New York: Pergamon Press.

Pliška, V. and Rudinger, J. (1976). Modes of inactivation of neurohypophysial hormones: significance of plasma disappearance rate for their physiological responses. *Clin. Endocrinol. 5,* Suppl., 73S–84S.

Pliška, V., Chard, T., Rudinger, J. and Forsling, M. L. (1976). *In vivo* activation of synthetic hormonogens of lysine vasopressin: N^a-glycyl-glycyl-[8-lysine] vasopressin in the cat. *Acta Endocrinol. 81,* 474–81.

Podos, S. M., Kolker, A. E., and Becker, B. (1970). Topical corticosteroids: dissociation of effects. In *Ocular Anti-inflammatory Therapy* (edited by H. E. Kaufman), pp. 106–16. Springfield, Ill.: Charles C. Thomas.

Pohorecky, L. A. and Wurtman, R. J. (1971). Adrenocortical control of epinephrine synthesis. *Pharmacol. Rev. 23,* 1–35.

Polak, J. M. (1976a). Localization of gastric inhibitory peptide, vasoactive intestinal peptide and motilin. *J. Endocrinol. 70,* 8–9P.

Polak, J. M. (1976b). Localization of gastrin, secretin and cholecystokinin. *J. Endocrinol. 70,* 2–3P.

Polak, J. M., Pearse, A. G. E., Grimelius, L., Bloom, S. R. and Arimura, A. (1975). Growth-hormone release-inhibiting hormone in gastrointestinal and pancreatic D cells. *Lancet 1,* 1220–2.

Poller, L., Thomson, J. M. and Thomas, P. W. (1972). Effects of progestogen oral contraception with norethisterone on blood clotting and platelets. *Brit. Med. J. 4,* 391–3.

Porter, G. A., Bogoroch, R. and Edelman, I. S. (1964). On the mechanism of action of aldosterone on sodium transport. *Proc. Natl. Acad. Sci. USA 52,* 1326–33.

Porter, J. C., Ondo, J. G. and Cramer, O. M. (1974). Nervous and vascular supply of the pituitary gland. *Handbook of Physiology:* Section 7, *Endocrinology,* Vol. 4, *The Pituitary Gland and Its Neuroendocrine Control,* Part 1, pp. 33–43. Washington, D.C.: American Physiological Society.

Posner, B. I., Josefsberg, Z. and Bergeron, J. J. M. (1978). Intracellular polypeptide hormone receptors. Characterization of insulin binding sites in Golgi fractions from the liver of female rats. *J. Biol. Chem. 253,* 4067–73.

Posner, B. I., Kelly, P. A. and Friesen, H. G. (1975). Prolactin receptors in rat liver: possible induction by prolactin. *Science 188,* 57–9.

Posner, B. I., Kelly, P. A., Shiu, R. P. C. and Friesen, H. G. (1974). Studies of insulin, growth hormone and prolactin binding: tissue distribution, species variation and characterization. *Endocrinology 95,* 521–31.

Posner, B. I., Raquidan, D., Josefsberg, Z. and Bergeron, J. J. M. (1978). Different regulation of insulin receptors in intracellular (Golgi) and plasma membranes from livers of obese and lean mice. *Proc. Natl. Acad. Sci. USA 75,* 3302–6.

Poth, M. M., Heath, R. G. and Ward, M. (1975). Angiotensin-converting enzyme in human brain. *J. Neurochem. 25,* 83–5.

Potts, G. O. (1977). Pharmacology of Danazol. *J. Int. Med. Res. 5,* Suppl. 3, 1–14.

Potts, J. T. and Aurbach, G. D. (1976). Chemistry of calcitonins. *Handbook of Physiology:* Section 7, *Endocrinology,* Vol. 7, *Parathyroid Gland,* pp. 423–30. Washington, D.C.: American Physiological Society.

Potts, J. T., Keutmann, H. T., Niall, H. D., Habener, J. F. and Tregear, G. W. (1972). Comparative biochemistry of parathyroid hormone and calcitonin. *Gen. Comp. Endocrinol.* Suppl. 3, 405–10.

Powell, E. D. U. and Field, R. A. (1964). Diabetic retinopathy and rheumatoid arthritis. *Lancet 2,* 17–18.

Prange, A. J., Wilson, I. C., Lara, P. P., Alltop, L. B. and Breese, G. R. (1972). Effects of thyrotrophin-releasing hormone in depression. *Lancet 2,* 999–1002.

Pratt, W. B. and Aronow, L. (1966). The effect of glucocorticoids on protein and nucleic acid synthesis in mouse fibroblasts growing *in vitro. J. Biol. Chem. 241,* 5244–50.

Pressley, L. and Funder, J. W. (1975). Glucocorticoid and mineralocorticoid receptors in gut mucosa. *Endocrinology 97,* 588–96.

Priestman, T., Baum, M., Jones, V. and Forbes, J. (1977). Comparative trial of endocrine versus cytotoxic treatment in advanced breast cancer. *Brit. Med. J. 1*, 1248–50.

Printz, M. P., Némethy, G. and Bleich, H. (1972). Proposed models for angiotensin II in aqueous solution and conclusions about receptor topography. *Nature New Biol. 237*, 135–40.

Printz, M. P., Williams, H. P. and Craig, L. M. (1972). Evidence for the presence of hydrogen-bonded secondary structure in angiotensin II in aqueous solution. *Proc. Natl. Acad. Sci. USA 69*, 378–82.

Procsal, D. A., Henry, H. L., Hendrickson, T. and Norman, A. W. (1976). 1α,25-Dihydroxyvitamin D₃-like component present in plant *Solanum glaucophyllum*. *Endocrinology, 99*, 437–44.

Prosser, P. R. and Karam, J. H. (1978). Diabetes mellitus following rodenticide ingestion in man. *JAMA 239*, 1148–50.

Pullen, R. A., Lindsay, D. G., Wood, S. P., Tickle, I. J., Blundell, T. L., Wollmer, A., Krail, G., Brandenburg, D., Zahn, H., Gliemann, J. and Gammeltoft, S. (1976). Receptor-binding region of insulin. *Nature (Lond.) 259*, 369–73.

Purves, H. D. (1966). Cytology of the adenohypophysis. In *The Pituitary Gland* (edited by G. W. Harris and B. T. Donovan), Vol. 1, *Anterior Pituitary*, pp. 147–232. Berkeley, Calif.: University of California Press.

Puschett, J. B., Beck, W. S. and Jelonek, A. (1975). Parathyroid hormone and 25-hydroxyvitamin D. Synergistic and antagonistic effects on renal phosphate transport. *Science 190*, 473–5.

Radford, D. J. and Oliver, M. F. (1973). Oral contraceptives and myocardial infarction. *Brit. Med. J. 3*, 428–30.

Ramsay, L., Asbury, M., Shelton, J. and Harrison, I. (1977). Spironolactone and canrenoate-K: relative potency at steady state. *Clin. Pharmacol. Ther. 21*, 602–8.

Rang, H. P. (1971). Drug receptors and their function. *Nature (Lond.) 231*, 91–6.

Rapoport, B., West, M. N. and Ingbar, S. H. (1976). On the mechanism of inhibition of iodine of the thyroid adenylate cyclase response to thyrotropic hormone. *Endocrinology 99*, 11–22.

Rasmussen, H. and Craig, L. C. (1959). Purification of parathyroid hormone by use of counter current distribution. *J. Amer. Chem. Soc. 81*, 5003.

Rasmussen, H. and Goodman, D. B. P. (1977). Relationships between calcium and cyclic nucleotides in cell activation. *Physiol. Rev. 57*, 421–509.

Rathnam, P. and Saxena, B. B. (1975). Primary amino acid sequence of follicle-stimulating hormone from human pituitary glands. *J. Biol. Chem. 250*, 6735–46.

Rayford, P. L., Miller, T. A. and Thompson, J. C. (1976). Secretin, cholecystokinin and newer gastrointestinal hormones (parts I and II). *New Engl. J. Med. 294*, 1093–101, 1157–64.

Raynaud, J-P., Bouton, M-M., Gallet-Bourquin, D., Philibert, D., Tournemine, C. and Azadian-Boulanger, A. (1973). Comparative study of estrogen action. *Molec. Pharmacol. 9*, 520–33.

Reagan, C. R., Mills, J. B., Kostyo, J. L. and Wilhelmi, A. E. (1975a). Isolation and biological characterization of fragments of human growth hormone produced by digestion with plasmin. *Endocrinology 96*, 625–36.

Reagan,, C. R., Mills, J. B., Kostyo, J. L. and Wilhelmi, A. E. (1975b). Biological properties of plasmin digests of

S-carbamidomethylated human growth hormone. *Proc. Natl. Acad. Sci. USA 72*, 1684–6.

Reagan, C. R., Kostyo, J. L., Mills, J. B., Moseley, M. H. and Wilhelmi, A. E. (1978). Isolation and characterization of fragments of reduced and S-carbamidmethylated human growth hormone produced by plasmin digestion: II. Biological and immunological activities. *Endocrinology 102*, 1377–86.

Rechler, M. M. and Nissley, S. P. (1977). Somatomedins and related growth factors. *Nature (Lond.) 270*, 665–6.

Recker, R. R., Saville, P. D. and Heaney, R. P. (1977). Effect of estrogens and calcium carbonate on bone loss in postmenopausal women. *Ann. Intern. Med. 87*, 649–55.

Redgate, E. S. (1976). Central nervous system mediation of adrenal rhythmicity. *Life Sci. 19*, 137–46.

Rees, W. D. H., Rhodes, J., Cross, S. and Hale, D. (1975). The effect of an aldosterone antagonist on the protective action of carbenoxolone on the gastric mucosal barrier. *J. Pharm. Pharmacol. 27*, 903–6.

Rees Smith, B. and Hall, R. (1974). Thyroid-stimulating immunoglobulins in Grave's disease. *Lancet 2*, 427–31.

Rees Smith, B., Pyle, G. A., Petersen, V. B. and Hall, R. (1977). Interaction of thyroid-stimulating antibodies with human thyrotrophin receptor. *J. Endocrinol. 75*, 401–7.

Regoli, D., Park, W. K. and Rioux, F. (1974). Pharmacology of angiotensin. *Pharmacol. Rev. 26*, 69–123.

Reichgott, M. J. and Melmon, K. L. (1973). Should corticosteroids be used in shock? *Med. Clin. North Amer. 57*, 1211–21.

Reichlin, S. (1974). Regulation of somatotrophic hormone secretion. *Handbook of Physiology*: Section 7, *Endocrinology*, Vol. 4, *The Pituitary Gland and Its Neuroendocrine Control*, Part 2, pp. 405–47. Washington, D.C.: American Physiological Society.

Reid, I. A., Morris, B. J. and Ganong, W. F. (1978). The renin–angiotensin system. *Annu. Rev. Physiol. 40*, 377–410.

Reinisch, J. M. (1977). Prenatal exposure of human foetuses to synthetic progestin and oestrogen affects personality. *Nature (Lond.) 266*, 561–2.

Reinisch, J. M., Simon, N. G., Karow, W. G. and Gandelman, R. D. (1978). Prenatal exposure to prednisone in humans and animals retards intrauterine growth. *Science 202*, 436–8.

Reit, E. (1972). Interaction of angiotensin with the autonomic nervous system (Pharmacology Society Symposium). *Fedn. Proc. 31*, 1331–64.

Rerup, C. C. (1970). Drugs producing diabetes through damage of the insulin secreting cells. *Pharmacol. Rev. 22*, 485–518.

Ribeiro, F. M., Jullienne, A., Taboulet, J., Moukhtar, M. S. and Milhaud, G. (1976). Pentagastrin test and calcitonin heterogeneity. *Lancet 2*, 1017–18.

Richardson, G. S. (1972). Endometrial cancer as an estrogen–progesterone target. *New Engl. J. Med. 286*, 645–7.

Richter, C. P. and Clisby, K. H. (1942). Toxic effects of the bitter-tasting phenythiocarbamide. *Arch. Pathol. 33*, 46–57.

Ricketts, H. T. (1976). Long distance hyperthyroidism. *JAMA 235*, 287.

Riggs, B. L., Jowsey, J., Goldsmith, R. S., Kelly, P. J., Hoffman, D. L. and Arnaud, C. D. (1972). Short- and long-term effects of estrogen and synthetic anabolic hormone in postmenopausal osteoporosis. *J. Clin. Invest. 51*, 1659–63.

Riggs, B. L., Jowsey, J., Kelly, P. J., Hoffman, D. L. and Arnaud, C. D. (1976). Effects of oral therapy with calcium and vitamin D in primary osteoporosis. *J. Clin. Endocrinol. Metab. 42*, 1139–44.

Rillema, J. A. (1976). Effects of prostaglandins on RNA and casein synthesis in mammary gland explants of mice. *Endocrinology 99*, 490–5.

Rillema, J. A. and Wild, E. A. (1977). Prolactin activation of phospholipase A activity in membrane preparations from mammary glands. *Endocrinology 100*, 1219–22.

Rinderknecht, E. and Humbel, R. E. (1976a). Polypeptides with nonsuppressible insulin-like and cell-growth promoting activities in human serum: isolation, chemical characterization, and some biological properties of forms I and II. *Proc. Natl. Acad. Sci. USA 73*, 2365–9.

Rinderknecht, E. and Humbel, R. E. (1976b). Amino-terminal sequences of two polypeptides from human serum with nonsuppressible insulin-like and cell-growth-promoting activities: evidence for structural homology with insulin B chain. *Proc. Natl. Acad. Sci. USA 73*, 4379–81.

Rinderknecht, E. and Humbel, R. E. (1978). The amino acid sequence of human insulin-like growth factor I and its structural homology with proinsulin. *J. Biol. Chem. 253*, 2769–76.

Rivier, C. and Vale, W. (1977). Effects of γ-aminobutyric acid and histamine on prolactin secretion in the rat. *Endocrinology 101*, 506–11.

Rivier, C., Vale, W., Ling, N., Brown, M. and Guillemin, R. (1977). Stimulation *in vivo* of the secretion of prolactin and growth hormone by β-endorphin. *Endocrinology 100*, 238–41.

Robert, A., Schultz, J. R., Nezamis, J. E. and Lancaster, C. (1976). Gastric antisecretory and antiulcer properties of PGE$_2$, 15-methyl PGE$_2$, and 16,16-dimethyl PGE$_2$. Intravenous, oral and intrajejunal administration. *Gastroenterology 70*, 359–70.

Roberts, J. L. and Herbert, E. (1977). Characterization of a common precursor to corticotropin and β-lipotropin: cell-free synthesis of the precursor and identification of corticotropin peptides in the molecule. *Proc. Natl. Acad. Sci. USA 74*, 4826–30.

Roberts, J. M., Insel, P. A., Goldfien, R. D. and Goldfien, A. (1977a). α Adrenoreceptors but not β adrenoreceptors increased in rabbit uterus with oestrogen. *Nature (Lond.) 270*, 624–5.

Roberts, J. M., Insel, P. A., Goldfien, R. D. and Goldfien, A. (1977b). Identification of beta-adrenergic binding sites in rabbit myometrium. *Endocrinology 101*, 1839–43.

Roberts, J. S., McCracken, J. A., Gavagan, J. E. and Soloff, M. S. (1976). Oxytocin-stimulated release of prostaglandin F$_{2\alpha}$ from ovine endometrium *in vitro:* correlation with estrous cycle and oxytocin–receptor binding. *Endocrinology 99*, 1107–14.

Robinson, A. G. (1976). DDAVP in the treatment of central diabetes insipidus. *New Engl. J. Med. 294*, 507–11.

Robinson, B. (1957). Breast changes in the male and female with chlorpromazine or reserpine therapy. *Med. J. Aust. 2*, 239–41.

Rodbell, M. (1966). Metabolism of isolated fat cells: ii. The similar effects of phospholipase C (*Clostridium per-*

fringens α toxin) and of insulin on glucose and amino acid metabolism. *J. Biol. Chem. 241*, 130–9.

Rodbell, M. (1975). On the mechanism of activation of fat cell adenylate cyclase by guanine nucleotides. *J. Biol. Chem. 250*, 5826–34.

Rodbell, M., Birnheimer, L., Pohl, S. L. and Sunby, F. (1971a). The reaction of glucagon with its receptor: evidence for discrete regions of activity and binding in the glucagon molecule. *Proc. Natl. Acad. Sci. USA 68*, 909–13.

Rodbell, M., Krans, H. M., Pohl, S. L. and Birnbaumer, L. (1971b). The glucagon-sensitive adenyl cyclase system in plasma membranes of rat liver: II. Binding of glucagon: method of assay and specificity. *J. Biol. Chem. 246*, 1861–71.

Rodbell, M., Lin, M. C., Salomon, Y., Londos, C., Harwood, J. P., Martin, B. R., Rendell, M. and Berman, M. (1975). Role of adenine and guanine nucleotides in the activity and responses of adenylate cyclase systems to hormones: evidence for multisite transition states. In *Advances in Nucleotide Research*, Vol. 5, pp. 3–29 (edited by G. I. Drummond, P. Greengard, and G. A. Robison). New York: Raven Press.

Roemer, D., Buescher, H. H., Hill, R. C., Pless, J., Bauer, W., Cardinaux, F., Closse, A., Hauser, D. and Huguenin, R. (1977). A synthetic enkephalin analogue with prolonged parenteral and oral analgesic activity. *Nature (Lond.) 268*, 547–9.

Rojanasathit, S., Rosenberg, E. and Haddad, J. G. (1974). Paget's bone disease: response to human calcitonin in patients resistant to salmon calcitonin. *Lancet 2*, 1412–15.

Rolland, R. and Schellekens, L. A. (1978). Inhibition of puerperal lactation by bromocriptine. *Acta Endocrinol. 88*, Suppl. 216, 119–28.

Roos, B. A. (1977). Calcitonin secretion *in vitro:* III. Synergistic secretory effects of enteric polypeptide hormones. *Endocrinology 100*, 1679–83.

Rooth, G. and Bandman, U. (1973). Renal response to acid load after phenformin. *Brit. Med. J. 4*, 256–7.

Rosenberg, I. N. (1972). Evaluation of thyroid function. *New Engl. J. Med. 286*, 924–7.

Rosenblatt, M., Goltzman, D., Keutmann, H. T., Tregear, G. W. and Potts, J. T. (1976). Chemical and biological properties of synthetic, sulfur-free analogues of parathyroid hormone. *J. Biol. Chem. 251*, 159–64.

Rosenblatt, M., Callahan, E. N., Mahaffey, J. E., Pont, A. and Potts, J. T. (1977). Parathyroid hormone inhibitors. Design, synthesis, and biologic evaluation of hormone analogues. *J. Biol. Chem. 252*, 5847–51.

Rosenblatt, M., Segre, G. V., Tregear, G. W., Shepard, G. L., Tyler, G. A. and Potts, J. T. (1978). Human parathyroid hormone: synthesis and chemical, biological, and immunological evaluation of the carboxyl-terminal region. *Endocrinology 103*, 978–84.

Rosenfeld, M. G. and O'Malley, B. W. (1970). Steroid hormones: effects of adenyl cyclase activity and adenosine 3',5'-monophosphate in target tissues. *Science 168*, 253–5.

Rossini, A. A., Arcangeli, M. A. and Cahill, G. F. (1975). Studies on alloxan toxicity in the beta cell. *Diabetes 24*, 516–22.

Roth, J., Kahn, C. R., Lesniak, M. A., Gordern, P., De Meyts, P., Megyesi, K., Neville, D. M., Gavin, J. R., Soll, A. H., Freychet, P., Goldfine, I. D., Bar, R. S. and

Archer, J. A. (1975). Receptors for insulin, NSILA-s, and growth hormone: application to disease states in man. *Rec. Prog. Horm. Res. 31*, 95–126.

Rothman, K. and Liess, J. (1976). Gender of offspring after oral-contraceptive use. *New Engl. J. Med. 295*, 859–61.

Rothman, K. J. and Louik, C. (1978). Oral contraceptives and birth defects. *New Engl. J. Med. 299*, 522–4.

Rousseau, G., Baxter, J. D., Funder, J. W., Edelman, I. S. and Tomkins, G. M. (1972). Glucocorticoid and mineralocorticoid receptors for aldosterone. *J. Steroid. Biochem. 3*, 219–27.

Rowlands, J. R. and Allen-Rowlands, C. F. (1978). A spin label study of the interaction of ACTH in Y-1 adrenal cell membranes. *Molec. Cell Endocrinol. 10*, 63–80.

Roy, C., Barth, T. and Jard, S. (1975a). Vasopressin-sensitive kidney adenylate cyclase. Structural requirements for attachment to the receptor and enzyme activation: studies with vasopressin analogues. *J. Biol. Chem. 250*, 3149–56.

Roy, C., Barth, T. and Jard, S. (1975b). Vasopressin-sensitive kidney adenylate cyclase. Structural requirements for attachment to the receptor and enzyme activation: studies with oxytocin analogues. *J. Biol. Chem. 250*, 3157–68.

Roy, C., Rajerison, R., Bockaert, J. and Jard, S. (1975). Solubilization of the [8-lysine]vasopressin receptor and adenylate cyclase from pig kidney plasma membranes. *J. Biol. Chem. 250*, 7885–93.

Royal College of General Practioners Oral Contraception Study (1977). Effect of hypertension and benign breast disease of progestagen component in combined oral contraceptives. *Lancet 1*, 624.

Rubenstein, A. H., Horwitz, D. L. and Steiner, D. F. (1975). Proinsulin and insulin biosynthesis. In *Diabetes; Its Physiological and Biochemical Basis* (edited by J. Vallance-Owen), pp. 1–30. Baltimore, Md.: University Park Press.

Rubin, B., Laffan, R. J., Kotler, D. G., O'Keefe, E. H., Demaio, D. A. and Goldberg, M. E. (1978). SQ 14225 (D-3-mercapto-2-methylpropanoyl-L-proline), a novel orally active inhibitor of angiotensin I-converting enzyme. *J. Pharmacol. Exp. Ther. 204*, 271–80.

Rudinger, J. (1971). The design of peptide hormone analogs. In *Drug Design* (edited by E. J. Ariëns), pp. 319–419. New York: Academic Press.

Rudinger, J. and Krejči, I. (1968). Antagonists of the neurohypophysial hormones. In *Neurohypophysial Hormones and Similar Peptides* (edited by B. Berde), pp. 748–801. New York: Springer-Verlag.

Rudinger, J., Pliška V. and Krejči, I. (1972). Oxytocin analogs in the analysis of some phases of hormone action. *Rec. Prog. Horm. Res. 28*, 131–66.

Ruh, T. S. and Ruh, M. F. (1974). The effect of antiestrogens on the nuclear binding of the estrogen receptor. *Steroids 24*, 209–24.

Russell, A. S. and Lentle, B. C. (1974). Mithramycin therapy in Paget's disease. *Can. Med. Assoc. J. 110*, 397–400.

Russell, R. G. G., Smith, R., Walton, R. J., Preston, C., Basson, R., Henderson, R. G. and Norman, A. W. (1974). 1,25-Dihydroxycholecalciferol and 1α-hydroxycholecalciferol in hypoparathyroidism. *Lancet 2*, 14–17.

Ryan, K. J. (1969). Theoretical basis for endocrine control of gestation, a comparative approach. In *The Foeto-placental Unit* (edited by A. Pecilo and C. Finzi), pp. 120–31. Amsterdam: Excerpta Medica Foundation.

Ryan, K. J. (1973). Steroid hormones in mammalian pregnancy. *Handbook of Physiology:* Section 7, *Endocrinology*, Vol. 2, *Female Reproductive System*, Part 2, pp. 285–93. Washington, D.C.: American Physiological Society.

Ryan, W. G. (1973). Mithramycin in Paget's disease of bone. *Lancet 1*, 1319.

Saameli, K. (1968). The circulatory actions of the neurohypophysial hormones and similar polypeptides. In *Neurohypophysial Hormones and Similar Peptides* (edited by B. Berde), pp. 545–612. New York: Springer Verlag.

Sachs, H. and Takabatake, Y. (1964). Evidence for a precursor in vasopressin synthesis. *Endocrinology 75*, 943–8.

Sachs, H., Fawcett, P., Takabatake, Y. and Portanova, R. (1969). Biosynthesis and release of vasopressin and neurophysin. *Rec. Prog. Horm. Res. 25*, 447–84.

Sagar, S., Stamatakis, J. D., Thomas, D. P. and Kakkar, V. V. (1976). Oral contraceptives, antithrombin-III activity, and postoperative deep-vein thrombosis *Lancet 1*, 509–11.

Said, S. I. and Mutt, V. (1970). Polypeptide with broad biological activity: isolation from small intestine. *Science 169*, 1217–18.

Said, S. I. and Mutt, V. (1972). Isolation from porcine intestinal wall of vasoactive octacopeptide related to secretin and to glucagon. *Eur. J. Biochem. 28*, 199–204.

Saidi, K., Wenn, R. V. and Sharif, F. (1977). Bromocriptine for male infertility. *Lancet 1*, 250–1.

Sairam, M. R. and Li, C. H. (1973). Human pituitary thyrotropin: isolation and chemical characterization of its subunits. *Biochem. Biophys. Res. Commun. 51*, 336–42.

Sairam, M. R. and Papkoff, H. (1974). Chemistry of pituitary gonadotropins. *Handbook of Physiology:* Section 7, *Endocrinology*, Vol. 4, *The Pituitary Gland and Its Neuroendocrine Control*, Part 2, pp. 111–31. Washington, D.C.: American Physiological Society.

Sakamoto, S., Sakamoto, M., Goldhaber, P. and Glimcher, M. (1975). Collagenase and bone resorption: isolation of collagenase from culture medium containing serum after stimulation of bone resorption by addition of parathyroid hormone extract. *Biochem. Biophys. Res. Commun. 63*, 172–8.

Sakauye, C. and Feldman, D. (1976). Agonist and anti-mineralocorticoid activities of spirolactones. *Amer. J. Physiol. 231*, 93–7.

Salih, H., Flax, H., Brander, W. and Hobbs, J. R. (1972). Prolactin dependence in human breast cancers. *Lancet 2*, 1103–5.

Salmon, W. D. and Daughaday, W. H. (1957). A hormonally controlled serum factor which stimulates sulfate incorporation by cartilage *in vitro*. *J. Lab. Clin. Med. 49*, 825–36.

Salomon, Y. and Rodbell, M. (1975). Evidence for specific binding sites to guanine nucleotides in adipocyte and hepatocyte plasma membranes. *J. Biol. Chem. 250*, 7245–50.

Samuels, A. I., Miller, E. D., Fray, J. C. S., Haber, E. and Barger, A. C. (1976). Renin–angiotensin antagonists and the regulation of blood pressure. *Fedn. Proc. 35*, 2512–20.

Samuels, H. H. and Shapiro, L. E. (1976). Thyroid hormone stimulates *de novo* growth hormone synthesis in cultured GH_1 cells: evidence for the accumulation of a rate limiting RNA species in the induction process. *Proc. Natl. Acad. Sci. USA 73*, 3369–73.

Samuels, H. H. and Tsai, J. S. (1973). Thyroid hormone action in cell culture: demonstration of nuclear receptors in intact cells and isolated nuclei. *Proc. Natl. Acad. Sci. USA 70*, 3488–92.

Samuels, L. T. and Nelson, D. H. (1975). Biosynthesis of corticosteroids. *Handbook of Physiology:* Section 7, *Endocrinology*, Vol. 6, *Adrenal Gland*, pp. 55–68. Washington, D.C.: American Physiological Society.

Sandweiss, D. J. (1945). Enterogastrone, anthelone and urogastrone. *Gastroenterology 5*, 404–15.

Sandweiss, D. J., Saltzstein, H. C. and Farbman, A. (1938). The prevention or healing of experimental peptic ulcer in Mann–Williamson dogs with the anterior-pituitary-like hormone (antuitrin S). A preliminary report *Amer. J. Dig. Dis. 5*, 24–30.

Sanger, F. (1959). Chemistry of insulin. *Science 129*, 1340–4.

Santen, R. J., Lipton, A. and Kendall, J. (1974). Successful medical adrenalectomy with amino-glutethimide. Role of altered drug metabolism. *JAMA 230*, 1661–5.

Santen, R. J., Wells, S. A., Runić, S., Gupta, C., Kendall, J., Rudy, E. B. and Samojlik, E. (1977). Adrenal suppression with aminoglutethimide on glucocorticoid metabolism as a rationale for use of hydrocortisone. *Endocrinology 45*, 469–79.

Sara, V. R., Lazarus, L., Stuart, M. C. and King, T. (1974). Fetal brain growth: selective action by growth hormone. *Science 186*, 446–7.

Sarantakis, D., McKinley, W. A., Jaunakais, I., Clark, D. and Grant, N. H. (1976). Structure activity studies on somatostatin. *Clin. Endocrinol. 5*, Suppl. 275S–8S.

Sasaki, G. H., Leung, B. S. and Fletcher, W. S. (1976). Therapeutic value of nafoxidine hydrochloride in the treatment of advanced carcinoma of the human breast. *Surg. Gynecol. Obstet. 142*, 560–8.

Sasaki, K., Dockerill, S., Adamiak, D. A., Tickle, I. J. and Blundell, T. (1975). X-ray analysis of glucagon and its relationship to receptor binding. *Nature (Lond.) 257*, 751–7.

Sassin, J. F., Frantz, A. G., Weitzman, E. D. and Kapen, S. (1972). Human prolactin: 24-hour pattern with increased release during sleep. *Science 177*, 1205–7.

Savage, C. R., Inagami, T. and Cohen, S. (1972). The primary structure of epidermal growth factor. *J. Biol. Chem. 247*, 7612–21.

Sawin, C. T. (1968). Problems in the clinical use of corticoids. In *Clinical Endocrinology* (edited by E. B. Astwood and C. E. Cassidy), Vol. 2, pp. 499–518. New York: Grune & Stratton.

Sawyer, W. H. and Manning, M. (1971). 4-Threonine analogues of neurohypophysial hormones with selectively enhanced oxytocin-like activities. *J. Endocrinol. 49*, 151–65.

Sawyer, W. H. and Manning, M. (1973). Synthetic analogs of oxytocin and the vasopressins. *Annu. Rev. Pharmacol. 13*, 5–17.

Sawyer, W. H., Acosta, M., Balaspiri, L., Judd, J. and Manning, M. (1974). Structural changes in the ar-

ginine vasopressin molecule that enhance antidiuretic activity and specificity. *Endocrinology 94*, 1106–15.

Saxena, B. B. and Rathnam, P. (1976). Amino acid sequence of the β subunit of follicle-stimulating hormone from human pituitary glands. *J. Biol. Chem. 251*, 993–1005.

Scapagnini, U., Van Loon, G. R., Moberg, G. P. and Ganong, W. F. (1970). Effect of α-methyl-p-tyrosine on the circadian variation of plasma corticosterone in rats. *Eur. J. Pharmacol. 11*, 266–8.

Scatchard, G. (1949). The attractions of proteins for small molecules and ions. *Ann. NY Acad. Sci. 51*, 660–72.

Schäfer, G. (1976). On the mechanism of hypoglycemia-producing biguanides. A reevaluation and a molecular theory. *Biochem. Pharmacol. 25*, 2005–14.

Schally, A. V. (1978). Aspects of hypothalamic regulation of the pituitary gland. Its implications for the control of reproductive processes. *Science 202*, 18–28.

Schally, A. V., Arimura, A. and Kastin, A. J. (1973). Hypothalamic regulatory hormones. *Science 179*, 341–50.

Scharschmidt, B. F. (1977). The natural history of hypertrophic gastropathy (Ménétrier's disease). Report of a case with a 16 year follow up and review of 120 cases from the literature. *Amer. J. Med. 63*, 644–52.

Schild, H. O. (1957). Drug antagonism and pA_x. *Pharmacol. Rev. 9*, 242–6.

Schlegel, W., Raptis, S., Harvey, R. F., Oliver, J. M. and Pfeiffer, E. F. (1977). Inhibition of cholecystokinin-pancreozymin release by somatostatin. *Lancet 2*, 166–8.

Schlesinger, D. H., Frangione, B. and Walter, R. (1972). Covalent structure of bovine neurophysin: II. Localization of the disulfide bonds. *Proc. Natl. Acad. Sci. USA 69*, 3350–4.

Schlessinger, J., Schechter, Y., Willingham, M. C. and Pastin, I. (1978). Direct visualization of binding, aggregation, and internalization of insulin and epidermal growth factor on living fibroblast cells. *Proc. Natl. Acad. Sci. USA 75*, 2659–63.

Schlumpf, U., Heimann, R., Zapf, J. and Froesch, E. R. (1976). Nonsuppressible insulin-like activity and sulphation activity in serum extracts of normal subjects. acromegalics and pituitary dwarfs. *Acta Endocrinol. 81*, 28–42.

Schneider, A. S., Herz, R. and Rosenheck, K. (1977). Stimulus–secretion coupling in chromaffin cells isolated from bovine adrenal medulla. *Proc. Natl. Acad. Sci. USA 74*, 5036–40.

Schofield, Y. G., Mira, F. and Orci, L. (1974). Somatostatin and growth hormone secretion *in vitro:* a biochemical and morphological study. *Diabetologia 10*, 385–6.

Scholes, P., Cooper, A., Jones, D., Major, J., Walters, M. and Wilde, C. (1976). Characterization of an adenylate cyclase system sensitive to histamine H_2-receptor excitation in cells from dog gastric mucosa. *Agents Actions 6*, 677–82.

Schou, M., Amdisen, A., Jensen, S. E. and Olsen, T. (1968). Occurrence of goitre during lithium treatment. *Brit. Med. J. 3*, 710–13.

Schrader, W. T. and O'Malley, B. W. (1978). *Laboratory Methods Manual for Hormone Action and Molecular Endocrinology 1978*. Houston: Houston Biological Associates.

Schramm, M., Orly, J., Eimerl, S. and Korner, M. (1977). Coupling of hormone receptors to adenylate cyclase of different cells by cell fusion. *Nature (Lond.) 268*, 310–13.

Schreiner, W. E. (1976). The ovary. In *Clinical Endocrinology: Theory and Practice* (edited by A. Labhart), pp. 511–665. New York: Springer-Verlag.

Schreurs, V. V. A. M., Swarts, H. G. P., De Pont, J. J. H. H. M. and Bonting, S. L. (1976). Role of calcium in exocrine pancreas: II. Comparison of the effects of carbachol and the ionophore A-23187 on enzyme secretion and calcium movements in rabbit pancreas. *Biochim. Biophys. Acta 419*, 320–30.

Schteingart, D. E. (1978). Cushing's disease: an update. *Drug Ther. Hosp. Ed.* (February), 53–63.

Schueler, F. W. (1960). *Chemodynamics and Drug Design*, pp. 140–198. New York: McGraw-Hill.

Schulster, D., Burstein, S. and Cooke, B. A. (1976). *Molecular Endocrinology of the Steroid Hormones*, pp. 150–1, 113, and 200. New York: Wiley.

Schulster, D., Orly, J., Seidel, G. and Schramm, M. (1978). Intracellular cyclic AMP production enhanced by a hormone receptor transferred from a different cell. β-Adrenergic responses in cultured cells conferred by fusion with turkey erythrocytes. *J. Biol. Chem. 253*, 1201–6.

Schussler, G. C. and Orlando, J. (1978). Fasting decreases triiodothyronine receptor capacity. *Science 199*, 686–8.

Schwabe, C. and Harmon, S. J. (1978). A comparative circular dichroism study of relaxin and insulin. *Biochem. Biophys. Res. Commun. 84*, 374–80.

Schwabe, C. and McDonald, J. K. (1977). Relaxin: a disulfide homolog of insulin. *Science 197*, 914–15.

Schwalbe, S. L., Betts, P. R., Rayner, P. H. W. and Rudd, B. T. (1977). Somatomedin in growth disorders and chronic renal insufficiency in children. *Brit. Med. J. 1*, 679–82.

Schwartz, H. L. and Oppenheimer, J. H. (1978). Physiologic and biochemical actions of thyroid hormone. *Pharmacol. Ther. B. 3*, 349–76.

Schwartz, R. J., Kuhn, R. W., Buller, R. E., Schrader, W. T. and O'Malley, B. W. (1976). Progesterone-binding components of chick oviduct. *In vitro* effects of purified hormone receptor complexes on the initiation of RNA synthesis in chromatin. *J. Biol. Chem. 251*, 5166–77.

Schwartzel, W. C., Kruggel, W. G. and Brodie, H. J. (1973). Studies of the mechanism of estrogen biosynthesis: VIII. The development of inhibitors of the enzyme system in human placenta. *Endocrinology 92*, 866–80.

Scott, A. P. and Lowry, P. J. (1974). Adrenocorticotrophic and melanocyte-stimulating peptides in the human pituitary. *Biochem. J. 139*, 593–602.

Scott, A. P., Rees, L. H., Ratcliffe, J. G. and Besser, G. M. (1972). Corticotrophin-like peptide concentrations in the intermediate lobe of rat and guinea-pig pituitaries. *J. Endocrinol. 53*, xxxviii–xxxix.

Scott, A. P., Ratcliffe, J. G., Rees, L. H., Landon, J., Bennett, H. P. J., Lowry, P. J. and McMartin, C. (1973). Pituitary peptide. *Nature New Biol. 244*, 65–7.

Scott, W. N. and Sapirstein, V. S. (1975). Identification of aldosterone-induced protein in the toad's urinary bladder. *Proc. Natl. Acad. Sci. USA 72*, 4056–60.

Sealey, J. E., Atlas, S. A., Laragh, J. H., Oza, N. B. and Ryan, J. W. (1978). Human urinary kallikrein converts

inactive to active renin and is a possible physiological activator of renin. *Nature (Lond.) 275*, 144–5.

Seeburg, P. H., Shine, J., Martial, J. A., Baxter, J. D. and Goodman, H. M. (1977). Nucleotide sequence and amplification in bacteria of structural gene for rat growth hormone. *Nature (Lond.) 270*, 486–94.

Seeburg, P. H., Shine, J., Martial, J. A., Ivarie, R. D., Morris, J. A., Ullrich, A., Baxter, J. D. and Goodman, H. M. (1978). Synthesis of growth hormone by bacteria. *Nature (Lond.) 276*, 795–8.

Seelig, S. and Sayers, G. (1977). Bovine hypothalamic corticotropin releasing factor: chemical and biological characteristics. *Fedn. Proc. 36*, 2100–3.

Segal, S. J., Davidson, O. W. and Wada, K. (1965). Role of RNA in the regulatory action of estrogen. *Proc. Natl. Acad. Sci. USA 54*, 782–7.

Sekihara, H., Island, D. P. and Liddle, G. W. (1978). New mineral corticoids: 5α-dihydroaldosterone and 5α-dihydro-11-deoxycorticosterone. *Endocrinology 103*, 1450–2.

Selenkow, H. A. and Asper, S. P. (1955). Biological activity of compounds structurally related to thyroxine. *Physiol. Rev. 35*, 426–74.

Selinger, Z., Eimerl, S. and Schramm, M. (1974). A calcium ionophore simulating the action of epinephrine on the α adrenergic receptor. *Proc. Natl. Acad. Sci. USA 71*, 128–31.

Seltzer, H. S. (1972a). Drug-induced hypoglycemia. A review based on 473 cases. *Diabetes 21*, 955–66.

Seltzer, H. S. (1972b). A summary of criticisms of the findings and conclusions of the university group diabetes program (UGDP). *Diabetes 21*, 976–9.

Semple, P. F., White, C. and Manderson, W. G. (1974). Continuous intravenous infusion of small doses of insulin in treatment of diabetic keto acidosis. *Brit. Med. J. 2*, 694–8.

Seo, H., Vassart, G., Brocas, H. and Refetoff, S. (1977). Triiodthyronine stimulates specifically growth hormone mRNA in rat pituitary tumor cells. *Proc. Natl. Acad. Sci. USA 74*, 2054–8.

Seppälä, M., Hirvonen, E. and Ranta, T. (1976). Bromocriptine treatment of secondary amenorrhoea. *Lancet 1*, 1154–6.

Seppälä, M., Hirvonen, E., Ranta, T., Virkkunen, P. and Leppäluoto, J. (1975). Raised serum prolactin levels in amenorrhoea. *Brit. Med. J. 2*, 305–6.

Sepsenwol, S. and Hechter, O. (1976). Failure to observe testosterone induced nucleus-lysosome interactions in rat ventral prostate. *Molec. Cell. Endocrinol. 4*, 115–29.

Seybold, M. E. and Drachman, D. B. (1974). Gradually increasing doses of prednisone in myasthenia gravis. *New Engl. J. Med. 290*, 81–4.

Shahidi, N. (1973). Androgens and erythropoiesis. *New Engl. J. Med. 289*, 72–80.

Shahmanesh, M., Bolton, C. H., Feneley, R. C. L. and Hartog, M. (1973). Metabolic effects of oestrogen treatment in patients with carcinoma of prostate: a comparison of stilboestral and conjugate equine oestrogens. *Brit. Med. J. 2*, 512–14.

Shapiro, S. (1975). Oral contraceptives and myocardial infarction. *New Engl. J. Med. 293*, 195–6.

Share, L. (1974). Blood pressure, blood volume, and the release of vasopressin. *Handbook of Physiology:* Section 7, *Endocrinology*, Vol. 4, *The Pituitary Gland and Its*

Neuroendocrine Control, Part 1, pp. 243–55. Washington, D.C.: American Physiological Society.

Sharp, G. W. G. and Leaf, A. (1964). Biological action of aldosterone *in vitro*. *Nature (Lond.) 202*, 1185–8.

Sharp, G. W. G., Komack, C. L. and Leaf, A. (1966). Studies on the binding of aldosterone in the toad bladder. *J. Clin. Invest. 45*, 450–9.

Sharp, G. W. G., Wollheim, C., Muller, W. A., Gutzeit, A., Trueheart, P. A., Blondel, B., Orci, L. and Renold, A. E. (1975). Studies on the mechanism of insulin release. *Fedn. Proc. 34*, 1537–48.

Shear, M., Insel, P. A., Melmon, K. L. and Coffino, P. (1976). Agonist-specific refractoriness induced by isoproterenol. *J. Biol. Chem. 251*, 7572–6.

Shen, S-S. and Bressler, R. (1977a). Clinical pharmacology of oral antidiabetic agents (part 1). *New Engl. J. Med. 296*, 493–7.

Shen, S-S. and Bressler, R. (1977b). Clinical pharmacology of oral antidiabetic agents (part 2). *New Engl. J. Med. 296*, 787–93.

Shenkman, L., Mitsuma, T., Suphavai, A. and Hollander, C. S. (1972). Triiodothyronine and thyroid-stimulating hormone response to thyrotrophin-releasing hormone. A new test of thyroidal and pituitary reserve. *Lancet 1*, 111–13.

Sherratt, H. S. A. (1969). Hypoglycin and related hypoglycaemic compounds. *Brit. Med. Bull. 25*, 250–5.

Sherwin, R. S., Fisher, M., Hendler, R. and Felig, P. (1976). Hyperglucagonemia and blood glucose regulation in normal, obese and diabetic subjects. *New Engl. J. Med. 294*, 455–61.

Sherwood, L. M., Handwerger, S., McLaurin, W. D. and Lanner, M. (1971). Amino acid sequence of human placental lactogen. *Nature New Biol. 233*, 59–61.

Shields, R. (1977). Growth hormones and serum factors. *Nature (Lond.) 267*, 308–10.

Shields, R. (1978). Pleiotypic effects of hormone. *Nature (Lond.) 276*, 440–1.

Shima, S., Kawashima, Y. and Hirai, M. (1978). Studies on cyclic nucleotides in the adrenal gland: VIII. Effects of angiotensin on adenosine, 3′,5′-monophosphate and steroidogenesis in the adrenal cortex. *Endocrinology 103*, 1361–7.

Shiu, R. P. C. and Friesen, H. G. (1974). Solubilization and purification of a prolactin receptor from the rabbit mammary gland. *J. Biol. Chem. 249*, 7902–11.

Short, R. V. (1972). Role of hormones in reproductive cycles. In *Hormones in Reproduction* (edited by C. R. Austin and R. V. Short), Vol. 3, pp. 42–72. New York: Cambridge University Press.

Shuster, S. and Thody, A. J. (1974). The control and measurement of sebum secretion. *J. Invest. Dermatol. 62*, 172–90.

Shuster, S., Burton, J. L., Thody, A. J., Plummer, N., Goolamali, S. K. and Bates, D. (1973). Melanocyte-stimulating hormone and Parkinsonism. *Lancet 1*, 463–5.

Shuster, S., Smith, A., Plummer, N., Thody, A. and Clark, F. (1977). Immunoreactive beta-melanocyte-stimulating hormone in cerebrospinal fluid and plasma in hypopituitarism: evidence for an extrapituitary origin. *Brit. Med. J. 1*, 1318–19.

Silva, J. E. and Larsen, P. R. (1977). Pituitary nuclear 3,5,3′-triiodothyronine and thyrotropin secretion: an explanation for the effect of thyroxine. *Science 198*, 617–19.

Simantov, R. and Snyder, S. (1976). Isolation and structure identification of a morphine-like peptide "enkephalin" in bovine brain. *Life Sci. 18*, 781–8.

Simons, S. S., Martinez, H. M., Garcea, R. L., Baxter, J. D. and Tomkins, G. M. (1976). Interactions of glucocorticoid receptor steroid complexes with acceptor sites. *J. Biol. Chem. 251*, 334–43.

Simpson, S. A. and Tait, J. F. (1955). Recent progress in methods of isolation, chemistry and physiology of aldosterone. *Rec. Prog. Horm. Res. 11*, 183–219.

Singer, I. and Forrest, J. N. (1976). Drug-induced states of nephrogenic diabetes insipidus. *Kidney Int. 10*, 82–95.

Singh, R. N. P., Seavey, B. K., Rice, V. P., Lindsey, T. T. and Lewis, U. J. (1974). Modified forms of human growth hormone with increased biological activities. *Endocrinology 94*, 883–91.

Siperstein, M. D., Foster, D. W., Knowles, H. C., Levine, R., Madison, L. L. and Roth, J. (1977). Control of blood glucose and diabetic vascular disease. *New Engl. J. Med. 296*, 1060–3.

Siris, E. S., Nisula, B. C., Catt, K. J., Horner, K., Birken, S., Canfield, R. E. and Ross, G. T. (1978). New evidence for intrinsic follicle-stimulating hormone-like activity in human chorionic gonadotropin and luteinizing hormone. *Endocrinology 102*, 1356–61.

Skeggs, L. T., Marsh, W. H., Kahn, J. R. and Shumway, N. P. (1954). The existence of two forms of hypertensin. *J. Exp. Med. 99*, 275–82.

Skeggs, L. T., Levine, M., Lentz, K. E., Kahn, J. R. and Dorer, F. E. (1977). New developments in our knowledge of the chemistry of renin. *Fedn. Proc. 36*, 1755–9.

Skinner, R. W. S., Pozderac, R. V., Counsell, R. E. and Weinhold, P. A. (1975). The inhibitive effects of steroid analogues in the binding of tritiated 5α-dihydrotestosterone to receptor proteins from rat prostate tissue. *Steroids 25*, 189–202.

Smith, C. W. and Walter, R. (1978). Vasopressin analog with extraordinarily high antidiuretic potency: a study of conformation and activity. *Science 199*, 297–9.

Smith, D. C., Prentice, R., Thompson, D. J. and Herrmann, W. L. (1975). Association of exogenous estrogen and endometrial carcinoma. *New Engl. J. Med. 293*, 1164–70.

Smith, H. E., Smith, R. G., Toft, D. O., Neergaard, J. R., Burrows, E. P. and O'Malley, B. W. (1974). Binding of steroids to progesterone receptor proteins in chick oviduct and human uterus. *J. Biol. Chem. 249*, 5924–32.

Smith, I. E., Fitzharris, B. M., McKinna, J. A., Fahmy, D. R., Nash, A. G., Neville, A. M., Gazet, J.-C., Ford, H. T. and Powles, T. J. (1978). Aminoglutethimide in treatment of metastatic breast carcinoma. *Lancet 2*, 646–9.

Smith, P. E. (1930). Hypophysectomy and replacement therapy in the rat. *Amer. J. Anat. 45*, 205–73.

Smith, P. H. and Porte, D. (1976). Neuropharmacology of the pancreatic islets. *Annu. Rev. Pharmacol. Toxicol. 16*, 269–85.

Smith, R. N., Taylor, S. A. and Massey, J. C. (1970). Controlled clinical trial of combined triiodothyronine and thyroxine in the treatment of hypothyroidism. *Brit. Med. J. 4*, 145–8.

Smithline, F., Sherman, L. and Kolodny, H. D. (1975).

Prolactin and breast carcinoma. *New Engl. J. Med. 292,* 784–92.

Snyder, S. H. (1977). Opiate receptors and internal opiates. *Sci. Amer. 236,* 44–56.

Sobel, E. H. (1968). Anabolic steroids. In *Clinical Endocrinology* (edited by E. B. Astwood and C. E. Cassidy), Vol. 2, pp. 789–97. New York: Grune & Stratton.

Soderling, T. R., Hickenbottom, J. P., Reimann, E. M., Hunkeler, F. L., Walsh, D. A. and Krebs, E. G. (1970). Inactivation of glycogen synthetase and activation of phosphorylase kinase by muscle adenosine 3′,5′-monophosphate-dependent protein kinases. *J. Biol. Chem. 245,* 6317–28.

Södersten, P., Gray, G., Damassa, D. A., Smith, E. R. and Davidson, J. M. (1975). Effects of a non-steroidal antiandrogen on sexual behavior and pituitary–gonadal function in the male rat. *Endocrinology 97,* 1468–75.

Soergel, K. H. (1975). Hormonally mediated diarrhea. *New Engl. J. Med. 292,* 970–2.

Soloff, M. S. and Swartz, T. L. (1974). Characterization of a proposed oxytocin receptor in the uterus of the rat and sow. *J. Biol. Chem. 249,* 1376–81.

Soloff, M. S., Schroeder, B. T., Chakraborty, J. and Pearlmutter, A. F. (1977). Characterization of oxytocin receptors in the uterus and mammary gland. *Fedn. Proc. 36,* 1861–6.

Sørensen, O. H., Anderson, R. B., Christensen, M. S., Friis, T., Hjorth, L., Jørgensen, F. S., Lund, B., Melsen, F. and Mosekilde, L. (1977). Treatment of senile osteoporosis with 1α-hydroxyvitamin D₃. *Clin. Endocrinol. 7,* Suppl., 169S–75S.

Spaziani, E. (1975). Accessory reproductive organs in mammals: control of cell and tissue transport by sex hormones. *Pharmacol. Rev. 27,* 207–86.

Spencer, J. D., Millis, R. R. and Hayward, J. L. (1978). Contraceptive steroids and breast cancer. *Brit. Med. J. 1,* 1024–6.

Spiegel, A. M. and Bitensky, M. W. (1969). Effects of chemical and enzymatic modifications and glucagon on its activation of hepatic adenyl cyclase. *Endocrinology 85,* 638–43.

Spielman, L. L. and Bancroft, F. C. (1977). Pregrowth hormone: Evidence for conversion to growth hormone during synthesis on membrane-bound polysomes. *Endocrinology 101,* 651–8.

Spindler, B. J., MacLeod, K. M., Ring, J. and Baxter, J. D. (1975). Thyroid hormone receptors. *J. Biol. Chem. 250,* 4113–19.

Sprague, J. M. (1968). Duretics. *Top. Med. Chem. 2,* 1–63.

Sraer, J., Sraer, J. D., Chansel, D., Jueppner, H., Hesch, R. D. and Ardaillou, R. (1978). Evidence for glomerular receptors for parathyroid hormone. *Amer. J. Physiol. 235,* F96–103.

Stachura, M. E. and Frohman, L. A. (1975). Growth hormone: independent release of big and small forms from rat pituitary *in vitro. Science 187,* 447–9.

Stacpoole, P. W., Moore, G. W. and Kornhauser, D. M. (1978). Metabolic effects of dichloroacetate in patients with diabetes mellitus and hyperlipoproteinemia. *New Engl. J. Med. 298,* 526–30.

Stamp, T. C. B., Haddad, J. G. and Twigg, C. A. (1977). Comparison of oral 25-hydroxycholecalciferol, vitamin D, and ultra violet light as determinants of circulating 25-hydroxyvitamin D. *Lancet 1,* 1341–3.

Stamp, T. C. B., Round, J. M., Rowe, D. J. F. and Had-

dad, J. G. (1972). Plasma levels and therapeutic effect of 25-hydroxycholecalciferol in epileptic patients taking anticonvulsant drugs. *Brit. Med. J. 4,* 9–11.

Stanbury, J. B. (1968). Endemic goiter. In *Clinical Endocrinology* (edited by E. B. Astwood and C. E. Cassidy), Vol. 2, pp. 195–209. New York: Grune & Stratton.

Staughton, R. C. D. and August, P. J. (1975). Cushing's syndrome and pituitary–adrenal suppression due to clobetasol propionate. *Brit. Med. J. 2,* 419–21.

Steer, M. L., Atlas, D. and Levitzki, A. (1975). Interrelations between β-adrenergic receptors, adenylate cyclase and calcium. *New Engl. J. Med. 292,* 409–14.

Steer, P. J., Little, D. J., Lewis, N. L., Kelly, M. C. M. E. and Beard, R. W. (1975). Uterine activity in induced labour. *Brit. J. Obstet. Gynecol. 82,* 433–41.

Steinberg, D. and Khoo, J. C. (1977). Hormone-sensitive lipase of adipose tissue *Fedn. Proc. 36,* 1986–90.

Steiner, D. F. (1977). Insulin today. *Diabetes 26,* 322–40.

Steiner, D. F., Kemmler, W., Clark, J. L., Oyer, P. E. and Rubenstein, A. H. (1972). The biosynthesis of insulin. *Handbook of Physiology,* Section 7, *Endocrinology,* Vol. 1, *Endocrine Pancreas,* pp. 175–98. Washington, D.C.: American Physiological Society.

Steiner, D. F., Kemmler, W., Tager, H. S. and Peterson, J. D. (1974). Proteolytic processing in the biosynthesis of insulin and other proteins. *Fedn. Proc. 33,* 2105–15.

Steiner, R. A., Illner, P., Marques, P., Williams, D., Shen, L., Edwards, L. and Gale, C. C. (1977). Inhibition of dopamine-induced release of growth hormone by thyrotropin-releasing hormone. *Amer. J. Physiol. 233,* E430–3.

Stephens, W. P., Coe, J. Y. and Baylis, P. H. (1978). Plasma arginine vasopressin concentrations and antidiuretic action of carbamazepine. *Brit. Med. J. 1,* 1445–7.

Sterling, K. (1977). The mitochondrial route of thyroid hormone action. *Bull NY Acad. Med. 53,* 260–76.

Sterling, K. and Lazarus, J. H. (1977). The thyroid and its control. *Annu. Rev. Physiol. 39,* 349–71.

Sterling, K., Lazarus, J. H., Milch, P. O., Sakurada, T. and Brenner, M. A. (1978). Mitochondrial thyroid hormone receptor: localization and physiological significance. *Science 201,* 1126–9.

Stern, E., Forsythe, A. B., Youkeles, L. and Coffelt, C. F. (1977). Steroid contraceptive use and cervical dysplasia: increased risk of progression. *Science 196,* 1460–2.

Stern, M. P., Brown, B. W., Haskell, W. L., Farquhar, J. W., Wehrie, C. L. and Wood, P. D. S. (1976). Cardiovascular risk and use of estrogens or estrogen-progestagen combinations. Stanford three-community study. *JAMA 235,* 811–15.

Stevenson, R. D. (1977). Mechanism of anti-inflammatory action of glucocorticosteroids. *Lancet 1,* 225–6.

Stewart, J. C. and Vidor, G. I. (1976). Thyrotoxicosis-induced by iodine contamination of food—a common unrecognized condition? *Brit. Med. J. 1,* 372–5.

Stewart, J. M., Getto, C. J., Neldner, K., Reeve, E. B., Krivoy, W. A. and Zimmermann, E. (1976). Substance P and analgesia. *Nature (Lond.) 262,* 784–5.

Stock, J. M., Surks, M. I. and Oppenheimer, J. H. (1974). Replacement dosage of L-thyroxine in hypothyroidism. A re-evaluation. *New Engl. J. Med. 290,* 529–33.

Stokes, G. S. (1976). Drug-induced hypertension: pathogenesis and management. *Drugs 12*, 222–30.

Stokes, G. S., Weber, M. A. and Thornell, I. R. (1974). β-Blockers and plasma renin activity in hypertension. *Brit. Med. J. 1*, 60–2.

Stoll, B. A. (1973). Hypothesis: breast cancer regression under oestrogen therapy. *Brit. Med. J. 3*, 446–50.

Stowers, J. M. (1976a). Modern approaches to diabetes mellitus II. *Brit. Med. J. 1*, 573–4.

Stowers, J. M. (1976b). Modern approaches to diabetes mellitus I. *Brit. Med. J. 1*, 509–11.

Stowers, J. M. and Borthwick, L. J. (1977). Oral hypoglycaemic drugs: clinical pharmacology and therapeutic use. *Drugs 14*, 41–56.

Streeten, D. H. P. (1975). Corticosteroid therapy: 1. Pharmacological properties and principles of corticosteroid use. *JAMA 232*, 944–7.

Streeten, D. H. P. and Kaplan, N. M. (1977). A totally new approach to hypertension. *Curr. Prescr. 10/77*, 19–32.

Streeten, D. H. P., Anderson, G. H., Freiberg, J. M. and Dalakos, T. G. (1975). Use of an angiotensin II antagonist (Saralasin) in the recognition of "angiotensinogenic" hypertension. *New Engl. J. Med. 292*, 675–62.

Stripp, B., Taylor, A. A., Bartter, F. C., Gillette, J. R., Loriaux, D. L., Easley, R. and Menard, R. H. (1975). Effect of spironolactone on sex hormones in man. *J. Clin. Endocrinol. Metab. 41*, 777–81.

Strong, J. A. (1976). Endocrine and metabolic diseases. Pituitary diseases. *Brit. Med. J. 1*, 640–2.

Studd, J. (1976). Management of the menopause: two views. Personal view I. *Prescr. J. 16*, 51–8.

Stumpe, K. O., Kolloch, R., Higuchi, M., Krück, F. and Vetter, H. (1977). Hyperprolactinaemia and antihypertensive effect of bromocriptine in essential hypertension. Identification of abnormal central dopamine control. *Lancet 2*, 211–14.

Stumpf, W. E. and Sar, M. (1977). Steroid hormone target cells in the periventricular brain: relationship to peptide hormone producing cells. *Fedn. Proc. 36*, 1973–7.

Suda, T., Liotta, A. S. and Krieger, D. T. (1978). β-Endorphin is not detectable in plasma from normal human subjects. *Science 202*, 221–3.

Sulaiman, W. R. and Johnson, R. H. (1973). Effects of fenfluramine on human growth hormone release. *Brit. Med. J. 2*, 329–32.

Surks, M. I., Koerner, D., Dillman, W. and Oppenheimer, J. H. (1973). Limited capacity binding sites for L-triiodothyronine in rat liver nuclei. *J. Biol. Chem. 248*, 7066–72.

Sussman, P. M., Tushinski, R. J. and Bancroft, F. C. (1976). Pregrowth hormone: product of the translation *in vitro* of messenger RNA coding for growth hormone. *Proc. Natl. Acad. Sci. USA 73*, 29–33.

Sutcliffe, H. S., Martin, T. J., Eisman, J. A. and Pilczyk, R. (1973). Binding of parathyroid hormone to bovine kidney-cortex plasma membranes. *Biochem. J. 134*, 913–21.

Sutherland, E. W. (1972). Studies on the mechanism of hormone action. *Science 177*, 401–8.

Sutherland, H. W., Stowers, J. M., Cormack. J. D. and Bewsher, P. D. (1973). Evaluation of chlorpropamide in chemical diabetes diagnosed during pregnancy. *Brit. Med. J. 3*, 9–13.

Sutherland, R., Mešter, J. and Baulieu, E. M. (1977). Tamoxifen is a potent "pure" anti-oestrogen in chick oviduct. *Nature (Lond.) 267*, 434–5.

Sutton, J. R., Coleman, M. J., Casey, J. and Lazarus, L. (1973). Androgen responses during physical exercise. *Brit. Med. J. 1*, 520–2.

Suzuki, H., Higuchi, T., Sawa, K., Ohtaki, S. and Horiuchi, Y. (1965). Endemic coast goitre in Hokkaido, Japan. *Acta Endocrinol. 50*, 161–76.

Swabb, D. F., Nijveldt, F. and Pool, C. W. (1975). Distribution of oxytocin and vasopressin in the rat supraoptic and paraventricular nucleus. *J. Endocrinol. 67*, 461–2.

Swanek, G. E., Chu, L. L. H. and Edelman, I. S. (1970). Stereospecific binding of aldosterone to renal chromatin. *J. Biol. Chem. 245*, 5382–9.

Swartz, S. L. and Dluhy, R. G. (1978). Corticosteroids: clinical pharmacology and therapeutic use. *Drugs 16*, 238–55.

Szego, C. M. (1965). Role of histamine in mediation of hormone action. *Fedn. Proc. 24*, 1343–52.

Szego, C. M. (1974). The lysosome as a mediator of hormone action. *Rec. Prog. Horm. Res. 30*, 171–223.

Szego, C. M. and Davis, J. S. (1967). Adenosine 3′,5′-monophosphate in rat uterus: acute elevation by oestrogen. *Proc. Natl. Acad. Sci. USA 58*, 1711–18.

Taitelman, U., Reece, E. A. and Bessman, A. N. (1977). Insulin in the management of the diabetic surgical patient. Continuous intravenous infusion vs subcutaneous administration. *JAMA 237*, 658–60.

Talwar, G. P., Pandian, M. R., Kumar, N., Hanjan, S. N. S., Saxena, R. K., Krishnaraj, R. and Gupta, S. L. (1975). Mechanism of action of pituitary growth hormone. *Rec. Prog. Horm. Res. 31*, 141–70.

Tam, S., Hong, S-E. L. and Levine, L. (1977). Relationships, among the steroids, of anti-inflammatory properties an inhibition of prostaglandin production and arachidonic acid release by transformed mouse fibroblasts. *J. Pharmacol. Exp. Ther. 203*, 162–8.

Tamburrano, G., Tamburrano, S., Gambardella, S. and Andreani, D. (1976). Effects of alcohol on growth hormone secretion in acromegaly. *J. Clin. Endocrinol. Metab. 42*, 193–6.

Tanaka, M., de Kloet, E. R., de Wied, D. and Versteeg, D. H. G. (1977). Arginine-vasopressin affects catecholamine metabolism in specific brain nuclei. *Life Sci. 20*, 1799–808.

Tanaka, Y. and DeLuca, H. F. (1974). Stimulation of 24,25-dihydroxyvitamin D_3 production by 1,25-dihydroxyvitamin D_3. *Science 183*, 1198–200.

Tanner, J. M. (1972). Human growth hormone. *Nature (Lond.) 237*, 433–9.

Tanner, J. M., Whitehouse, R. H., Hughes, P. C. R. and Vince, F. P. (1971). Effect of human growth hormone treatment for 1 to 7 years on growth of 100 children, with growth hormone deficiency, low birth weight, inherited smallness, Turner's syndrome, and other complaints. *Arch. Dis. Child. 46*, 745–82.

Tashjian, A. H. and Melvin, K. E. W. (1968). Medullary carcinoma of the thyroid gland. Studies of thyrocalcitonin in plasma and tumor extracts. *New Engl. J. Med. 279*, 279–83.

Tashjian, A. H., Voelkel, E. F., McDonough, J. and Levine, L. (1975). Hydrocortisone inhibits prostaglandin production by mouse fibrosarcoma cells. *Nature (Lond.) 258,* 739–41.

Tasso, F., Picard, D. and Dreifuss, J. J. (1976). Ultrastructural identification of granules containing oxytocin and vasopressin. *Nature (Lond.) 260,* 621–2.

Tata, J. R. (1963). Inhibition of the biological action of thyroid hormones by actinomycin D and puromycin. *Nature (Lond.) 197,* 1167–8.

Tata, J. R. (1974). Growth and developmental action of thyroid hormones at the cellular level. *Handbook of Physiology:* Section 7, *Endocrinology,* Vol. 3, *Thyroid,* pp. 469–78. Washington, D.C.: American Physiological Society.

Tata, J. R. (1975). How specific are nuclear "receptors" for thyroid hormones? *Nature (Lond.) 257,* 18–23.

Tata, J. R. and Widnell, C. C. (1966). Ribonucleic acid synthesis during the early action of thyroid hormones. *Biochem. J. 98,* 604–20.

Tata, J. R., Ernster, L., Lindberg, O., Arrhenius, E., Pederson, S. and Hedman, R. (1963). The action of thyroid hormones at the cell level. *Biochem. J. 86,* 408–28.

Tate, R. L., Holmes, J. M., Kohn, L. D. and Winand, R. J. (1975a). Characteristics of a solubilized thyrotropin receptor from bovine thyroid plasma membranes. *J. Biol. Chem. 250,* 6527–33.

Tate, R. L., Schwartz, H. I., Holmes, J. M., Kohn, L. D. and Winand, R. J. (1975b). Thyrotropin receptors in thyroid plasma membranes. Characteristics of thyrotropin binding and solubilization of thyrotropin receptor activity by tryptic digestion. *J. Biol. Chem. 250,* 6509–15.

Tattersall, R. (1978). Highly purified insulins. *Prescr. J. 18,* 8–13.

Taurog, R. (1976). The mechanism of action of the thioureylene antithyroid drugs. *Endocrinology 98,* 1031–46.

Taylor, A. (1977). Role of microtubules and microfilaments in the action of vasopressin. In *Disturbances in Body Fluid Osmolality* (edited by T. W. Andreoli, J. J. Grantham, and F. C. Rector), pp. 97–124. Bethesda, Md.: American Physiological Society.

Taylor, A., Maffly, R., Wilson, L. and Reaven, E. (1975a). Evidence for involvement of microtubules in the action of vasopressin. *Ann. NY Acad. Sci. 253,* 723–37.

Taylor, A., Mamelak, M., Reaven, E. and Maffly, R. (1975b). Vasopressin: possible role of microtubules and microfilaments in its action. *Science 181,* 347–51.

Taylor, R. W. (1977). The treatment of premenstrual tension with dydrogesterone ('Duphaston'). *Curr. Med. Res. Opinion 4,* Suppl. 4, 35–40.

Telenius-Berg, M., Almquist, S., Hedner, P., Ingemansson, S., Tibblin, S. and Wästhed, B. (1975). Screening for medullary carcinoma of the thyroid. *Lancet 1,* 390–1.

Terenius, L. (1968). Selective retention of estrogen isomers in estrogen-dependent breast tumors of rats demonstrated by *in vitro* methods. *Cancer Res. 28,* 328–37.

Terenius, L. (1970). Two modes of interaction between oestrogen and anti-oestrogen. *Acta Endocrinol. 64,* 47–58.

Terenius, L. (1971). Structure-activity relationships of anti-oestrogens with regard to interaction with 17β-oestradiol in the mouse uterus and vagina. *Acta Endocrinol. 66,* 431–47.

Terenius, L. (1976). Somatostatin and ACTH are peptides with partial antagonist-like selectivity for opiate receptors. *Eur. J. Pharmacol. 38,* 211–13.

Terris, S. and Steiner, D. F. (1975). Binding and degradation of ^{125}I-insulin by rat hepatocytes. *J. Biol. Chem. 250,* 8389–98.

The Medical Letter (1978). Cimetidine (Tagamet): update on adverse effects. *Med. Lett. 20,* 77–8.

Theobald, G. W., Graham, A., Campbell, J., Gange, P. D. and Driscoll, W. J. (1948). The use of pituitary extracts in physiological amounts in obstetrics. *Brit. Med. J. 2,* 123–7.

Thody, A. J., Cooper, M. F., Bowden, P. E., Meddis, D. and Shuster, S. (1976). Effect of α-melanocyte stimulating hormone and testosterone on cutaneous and modified sebaceous glands in the rat. *J. Endocrinol. 71,* 279–88.

Thomas, A. K., Ebling, F. J., Cooke, I. D., Randall, V. A., Skinner, J. and Cawood, M. (1977). Treatment of hirsutism with cyproterone acetate. *J. Endocrinol. 75,* 28P–9P.

Thompson, J. and Oswald, I. (1977). Effect of oestrogen on the sleep, mood, and anxiety of menopausal women. *Brit. Med. J. 2,* 1317–19.

Thompson, J. and van Furth, R. (1970). The effects of glucocorticosteroids on the kinetics of mononuclear phagocytes. *J. Exp. Med. 131,* 429–42.

Thompson, J. C. (1973). Chemical structure and biological actions of gastrin, cholecystokinin, and related compounds. *International Encyclopedia of Pharmacology* (edited by P. Holton), Section 39A, pp. 261–86.

Thompson, W. J., Rosenfeld, G. C. and Jacobson, E. D. (1977). Adenylyl cyclase and gastric acid secretion. *Fedn. Proc. 36,* 1938–41.

Thorn, G. W. and Lauler, D. P. (1972). Clinical therapeutics of adrenal disorders. *Amer. J. Med. 53,* 673–84.

Thorner, M. O. and Besser, G. M. (1978). Bromocriptine treatment of hyperprolactinaemic hypogonadism. *Acta Endocrinol. 88,* Suppl. 216, 131–46.

Thorner, M. O., McNeilly, A. S., Hagen, C. and Besser, G. M. (1974). Long-term treatment of galactorrhoea and hypogonadism with bromocriptine. *Brit. Med. J. 2,* 419–22.

Thorner, M. O., Besser, G. M., Jones, A., Dacie, J. and Jones, A. E. (1975). Bromocriptine treatment of female infertility: report of 13 pregnancies. *Brit. Med. J. 4,* 694–7.

Thorner, M. O., Ryan, S. M., Wass, J. A. H., Jones, A., Bouloux, P., Williams, S. and Besser, G. M. (1978). Effect of dopamine agonist lergotrile mesylate, on circulating anterior pituitary hormones in man. *J. Clin. Endocrinol. Metab. 47,* 372–8.

Thornton, D. R. (1961). Oxytocin nasal spray in the treatment of breast engorgement. *Obstet. Gynecol. 18,* 701–3.

Thron, C. D. (1973). On the analysis of pharmacological experiments in terms of an allosteric receptor model. *Mol. Pharmacol. 9,* 1–9.

Tindall, D. J. and Means, A. R. (1976). Concerning the hormonal regulation of androgen binding protein in rat testis. *Endocrinology 99,* 809–18.

Tischler, A. S., Dichter, M. A., Biales, B. and Greene, L. A. (1977). Neuroendocrine neoplasms and their cells of origin. *New Engl. J. Med. 296,* 919–25.

Todays Drugs (1972a). Drugs in infertility. *Brit. Med. J. 4*, 167–70.

Todays Drugs (1972b). Diazoxide. *Brit. Med. J. 4*, 417–18.

Tolis, G., Hickey, J. and Guyda, H. (1975). Effects of morphine on serum growth hormone, cortisol, prolactin and thyroid stimulating hormone in man. *J. Clin. Endocrinol. Metab. 41*, 797–800.

Tomasi, V., Koretz, S., Ray, I. K., Dunnick, J. and Marinetti, G. V. (1970). Hormone action at the membrane level. The binding of epinephrine and glucagon to the rat liver plasma membrane. *Biochim. Biophys. Acta 211*, 31–42.

Tomkin, G. H. (1973). Malabsorption of vitamin B_{12} in diabetic patients treated with phenformin: a comparison with metformin. *Brit. Med. J. 3*, 673–5.

Tomkins, G. M., Gelehrter, T. D., Granner, D., Martin, D., Samuels, H. H. and Thompson, E. B. (1969). Control of specific gene expression in higher organisms. *Science 166*, 1474–80.

Transbol, I., Christiansen, C. and Baastrup, D. C. (1978). Endocrine effects of lithium: 1. Hypothyroidism, its prevalence in long-term treated patients. *Acta Endocrinol. 87*, 759–67.

Triggle, D. J. (1972). Adrenergic receptors. *Annu. Rev. Pharmacol. Toxicol. 12*, 185–96.

Truelove, S. C. (1960). Stilboestrol, phenobarbitone and diet in chronic duodenal ulcer. *Brit. Med. J. 2*, 559–66.

Trzeciak, W. H. and Boyd, G. S. (1973). The effect of stress induced by ether anesthesia on cholesterol content and cholesteryl-esterase activity in rat adrenal cortex. *Eur. J. Biochem. 37*, 327–33.

Tsien, R. W. (1977). Cyclic AMP and contractile activity in heart. In *Advances in Cyclic Nucleotide Research*, Vol. 8, pp. 363–420 (edited by P. Greengard and G. A. Robison). New York: Raven Press.

Tsushima, T. and Friesen, H. G. (1973). Radioreceptor assay for growth hormone. *J. Clin. Endocrinol. Metab. 37*, 334–7.

Tubiana, M., Milhaud, G., Coutris, G., Lacour, J., Parmentier, C. and Bok, B. (1968). Medullary carcinoma of the thyroid. *Brit. Med. J. 4*, 87–9.

Turkington, R. W. (1970). Stimulation of RNA synthesis in isolated mammary cells by insulin and prolactin bound to sepharose. *Biochem. Biophys. Res. Commun. 41*, 1362–7.

Turkington, R. W., Brew, K., Vanaman, T. C. and Hill, R. L. (1968). The hormonal control of lactose synthetase in the developing mouse mammary gland. *J. Biol. Chem. 243*, 3382–7.

Turnell, R. W., Kaiser, N., Milholland, R. J. and Rosen, F. (1974). Glucocorticoid receptors in rat thymocytes. *J. Biol. Chem. 249*, 1133–8.

Turner, J. G., Brownlie, B. E. W. and Rogers, T. G. H. (1976). Lithium as an adjunct to radioiodine therapy for thyrotoxicosis. *Lancet 1*, 614–15.

Tyler, F. H. and West, C. D. (1972). Laboratory evaluation of disorders of the adrenal cortex. *Amer. J. Med. 53*, 664–72.

Tzingounis, V. A., Aksu, M. F. and Greenblatt, R. B. (1978). Estriol in the management of the menopause. *JAMA 239*, 1638–41.

Ui, N. (1974). Synthesis and chemistry of iodoproteins. *Handbook of Physiology*: Section 7, *Endocrinology*, Vol. 3, *Thyroid*, pp. 55–79. Washington, D.C.: American Physiological Society.

Ullrich, A., Shine, J., Chirgwin, J., Pictet, R., Tischer, E., Rutter, W. J. and Goodman, H. M. (1977). Rat insulin genes: construction of plasmids containing the coding sequences. *Science 196*, 1313–19.

Ulstein, M., Netto, N., Leonard, J. and Paulsen, C. A. (1975). Changes in sperm morphology in normal men treated with Danazol and testosterone. *Contraception 12*, 437–44.

Unger, R. H. and Orci, L. (1976). Physiology and pathophysiology of glucagon. *Physiol. Rev. 56*, 778–826.

Unger, R. H., Raskin, P., Srikant, C. B. and Orci, L. (1977). Glucagon and the A cells. *Rec. Prog. Horm. Res. 33*, 477–511.

University Group Diabetes Program (1970a). A study of the effects of hypoglycemic agents on vascular complications in patients with adult-onset diabetes: I. design, methods and baseline results. *Diabetes 19*, 747–83.

University Group Diabetes Program (1970b). A study of hypoglycemic agents on vascular complications in patients with adult-onset diabetes: II. Mortality results. *Diabetes 19*, 789–830.

University Group Diabetes Program (1971). Effects of hypoglycemic agents on vascular complications in patients with adult-onset diabetes: III: Clinical results. *JAMA 218*, 1400–10.

University Group Diabetes Program (1975). A study of the effects of hypoglycemic agents on vascular complications in patients with adult-onset diabetes: V. Evaluation of phenformin therapy. *Diabetes 24*, Suppl. 1, 65–184.

Urry, D. W. and Walter, R. (1971). Proposed conformation of oxytocin in solution. *Proc. Natl. Acad. Sci. USA 68*, 956–8.

Usdin, E. and Snyder, S. H., Eds. (1973). *Frontiers in Catecholamine Research*. New York: Pergamon Press.

Uthne, K. and Uthne, T. (1972). Influence of liver resection and regeneration on somatomedin (sulphation factor) activity in sera from normal and hypophysectomized rat. *Acta Endocrinol. 71*, 255–64.

Vaes, G. (1968a). On the mechanisms of bone resorption. The action of parathyroid hormone on the excretion and synthesis of lysosomal enzymes and on the extracellular release of acid by bone cells. *J. Cell Biol. 39*, 676–97.

Vaes, G. (1968b). Parathyroid hormone-like action of N^6-$2'$-O-dibutyryl-adenosine-$3'5'$(cyclic)-monophosphate on bone explants in tissue culture. *Nature (Lond.) 219*, 939–40.

Vaidya, R. A., Vaidya, A. B., Van Woert, M. H. and Kase, N. G. (1970). Galactorrhea and Parkinsonism-like syndrome: an adverse effect of α-methyldopa. *Metabolism 19*, 1068–70.

Vailati, G. and Rabassini, A. (1962). Sull'azione impoglicemizzante del dicloroetanotato di diisopropilammonio Nell'uomo. *Riv. Crit. Clin. Med. (Firenze) 62*, 105–9.

Vale, W., Rivier, C. and Brown, M. (1977). Regulatory peptides of the hypothalamus. *Annu. Rev. Physiol. 39*, 473–527.

Vale, W., Brazeau, P., Rivier, C., Brown, M., Boss, B., Rivier, J., Burgus, R., Ling, N. and Guillemin, R. (1975). Somatostatin. *Rec. Prog. Horm. Res. 31*, 365–92.

Vale, W., Rivier, C., Brown, M., Leppaluoto, J., Ling, N., Monahan, M. and Rivier, J. (1976). Pharmacology of hypothalamic peptides. *Clin. Endocrinol 5*, Suppl., 261S–73S.

Vale, W., Rivier, C., Yang, L., Minick, S. and Guillemin, R. (1978). Effects of purified corticotropin-releasing factor and other substances on the secretion of adrenocorticotropin and β-endorphin-like immunoactivities *in vitro*. *Endocrinology 103*, 1910–15.

Valette, A., Vérine, A., Salers, P. and Boyer, J. (1977). Estrogen hormones and lipid metabolism. Effect of ethynyl-estradiol on liver lipases. *Endrocrinology 101*, 627–30.

Valtin, H., Stewart, J. and Sokol, A. W. (1974). Genetic control of the production of posterior pituitary principles. *Handbook of Physiology: Section 7, Endrocrinology*, Vol. 4, *The Pituitary Gland and Its Neuroendrocrine Control*, Part 1, pp. 131–71. Washington, D.C.: American Physiological Society.

Vander, A. J. (1967). Control of renin release. *Physiol. Rev. 47*, 359–82.

Van der Vies, J. (1977). Conception, contraception and misconception. *Acta Endocrinol.* Suppl. 214, 90–102.

Van Dongen, R. and Peart, W. S. (1974). Calcium dependence of the inhibitory effect of angiotensin on renin secretion in the isolated perfused kidney of the rat. *Brit. J. Pharmacol. 50*, 125–9.

Van Dyke, H. B., Chow, B. F., Greep, R. O. and Rothen, A. (1942). The isolation of a protein from the pars neuralis of the ox pituitary with constant oxytocic, pressor and diuresis inhibiting effect. *J. Pharmacol. Exp. Ther. 74*, 190–209.

Vane, J. R. (1971). Inhibition of prostaglandin synthesis as a mechanism of action of aspirin-like drugs. *Nature New Biol. 231*, 232–5.

Vane, J. R. (1976). The mode of action of aspirin and similar compounds. *Hosp. Formul. 10*, 618–29.

Van Middlesworth, L. (1957). Thyroxine excretion, a possible cause of goiter. *Endocrinology 61*, 570–3.

Van Middlesworth, L. (1974). Metabolism and excretion of thyroid hormones. *Handbook of Physiology: Section 7, Endrinology*, Vol. III, *Thyroid*, pp. 215–31. Washington, D.C.: American Physiological Society.

Van Obberghen, E., De Meyts, P. and Roth, J. (1976). Cell surface receptors for insulin and human growth hormone. Effects of microtubule and microfilament modifiers. *J. Biol. Chem. 251*, 6844–51.

Van Thiel, D. H. and Lester, R. (1974). Sex and alcohol. *New Engl. J. Med. 291*, 251–3.

Van't Hoff, W., Pover, G. G. and Eiser, N. M. (1972). Technetium-99^m in the diagnosis of thyrotoxicosis. *Brit. Med. J. 4*, 203–6.

Van Wyk, J. J., Underwood, L. E., Hintz, R. L., Clemmons, D. R., Voina, S. J. and Weaver, R. P. (1974). The somatomedins: a family of insulinlike hormones under growth hormone control. *Rec. Prog. Horm. Res. 30*, 259–95.

Vassale, G. and Generali, F. (1900). Fonction parathyroidienne et fonction thyroidienne. *Arch. Ital. Biol. 33*, 154–6.

Veber, D. F., Holly, F. W., Paleveda, W. J., Nutt, R. F., Bergstrand, S. J., Torchiana, M., Glitzer, M. S., Saperstein, R. and Hirschman, R. (1978). Conformationally restricted bicyclic analogs of somatostatin. *Proc. Natl. Acad. Sci. USA 75*, 2636–40.

Verlander, M. S., Venter, J. C., Goodman, M., Kaplan, N. O. and Saks, B. (1976). Biological activity of catecholamines covalently linked to synthetic polymers: proof of immobilized drug theory. *Proc. Natl. Acad. Sci. USA 72*, 1009–13.

Verney, E. B. (1947). The antidiuretic hormone and the factors which determine its release. *Proc. Roy. Soc. Lond. B 135*, 25–106.

Vessey, M., Doll, R. and Jones, K. (1975). Oral contraceptives and breast cancer. Progress report of an epidemiological study. *Lancet 1*, 941–4.

Vessey, M., Doll, R., Peto, R., Johnson, B. and Wiggins, P. (1976). A long-term follow-up study of women using different methods of contraception—an interim report. *J. Biosoc. Sci. 8*, 373–427.

Vessey, M. P., McPherson, K. and Johnson, B. (1977). Mortality among women participating in the Oxford/Family Planning Association contraceptive study. *Lancet 2*, 731–3.

Vessey, M. P., Wright, N. H., McPherson, K. and Wiggins, P. (1978). Fertility after stopping different methods of contraception. *Brit. Med. J. 1*, 265–7.

Vierhapper, H. and Waldhäus, W. (1978). Comparison of the corticotropic action of two synthetic, substituted analogues of ACTH$^{1–17}$ and ACTH$^{1–18}$. *J. Clin. Endocrinol. Metab. 47*, 208–11.

Von den Velden, R. (1913). Die Nierewirkung von Hypophysenextracten beim Menschen. *Berlin. Klin. Wochenschr. 50*, 2083–6.

Von Herrath, D., Kraft, D., Offerman, G. and Schaefer, K. (1974). *Solanum malacoxylon:* eine therapeutische Alternative fur 1,25-Dihydroxycholecalciferol bei urämischen Calciumstoffwechselstörungen. *Dtsch. Med. Wochenschr. 99*, 2407–9.

Von Steiger, M. and Reichstein, T. (1937). Desoxycorticosteron (21-Oxyprogesteron) aus A^5-Oxy-ätiocholensäure. *Helv. Chim. Acta 20*, 1164–79.

Vu Hai, M. T. and Milgrom, E. (1978). Characterization and assay of the progesterone receptor in rat uterine nuclei. *J. Endocrinol. 76*, 33–41.

Vu Hai, M. T., Logeat, F. and Milgrom, E. (1978). Progesterone receptors in the rat uterus: variations in cytosol and nuclei during the oestrous cycle and pregnancy. *J. Endocrinol. 76*, 43–8.

Wade, N. (1972). Anabolic steroids: doctors denounce them, but athletes aren't listening. *Science 176*, 1399–403.

Walaas, O., Walaas, E. and Grønnerød, O. (1974). Molecular events in the action of insulin on cell metabolism. The significance of cyclic AMP dependent protein kinases. *Acta Endocrinol.* Suppl. 191, 93–129.

Wales, J. K. (1975). Treatment of diabetes insipidus with carbamazepine. *Lancet 2*, 948–51.

Walker, S., Hibbard, B. M., Groom, G., Griffiths, K. and Davis, R. H. (1975). Controlled trial of bromocriptine quinoestrol, and placebo in suppression of peurperal lactation. *Lancet 2*, 842–5.

Wall, A. J. (1973). The use of glucocorticoids in intestinal disease. *Med. Clin. North Amer. 57*, 1241–52.

Wallace, R. B., Hoover, J., Sandler, D., Rifkind, B. M. and Tyroler, H. A. (1977). Altered plasma-lipids associated with oral contraceptives or oestrogen consumption. The lipid research clinic program. *Lancet 2*, 11–14.

Wallin, J. D. and Lee, P. A. (1976). Effect of prolactin on diluting and concentrating ability in the rat. *Amer. J. Physiol. 230*, 1524–30.

Walling, M. W. and Kimberg, D. V. (1975). Effects of 1α,25-dihydroxyvitamin D₃ and *Solanum glaucophyllum* on intestinal calcium and phosphate transport and on plasma Ca, Mg, and P levels in the rat. *Endocrinology 97*, 1567–75.

Walsh, D. A. and Ashby, C. D. (1973). Protein kinases: aspects of their regulation and diversity. *Rec. Prog. Horm. Res. 29*, 329–53.

Walsh, D. A., Perkins, J. P. and Krebs, E. G. (1968). An adenosine 3',5'-monophosphate-dependent protein kinase from rabbit skeletal muscle. *J. Biol. Chem. 243*, 3763–5.

Walsh, J. H. (1977). Gastrin heterogeneity: biological significance. *Fedn. Proc. 36*, 1948–51.

Walsh, J. H. and Grossman, M. I. (1975). Gastrin (parts I and II). *New Engl. J. Med. 292*, 1324–34, 1377–84.

Walter, R. (1973). The role of enzymes in the formation and inactivation of peptide hormones. In *Peptides 1972* (edited by H. Hanson and H-D. Jakubke), pp. 363–78. New York: Elsevier North-Holland.

Walter, R. (1977). Identification of sites in oxytocin involved in uterine receptor recognition and activation. *Fedn. Proc. 36*, 1872–8.

Walter, R. and Bowman, R. H. (1973). Mechanism of inactivation of vasopressin and oxytocin by the isolated perfused kidney. *Endocrinology 92*, 189–93.

Walter, R., Griffiths, E. C. and Hooper, K. C. (1973). Production of MSH-release-inhibiting hormone by a particulate fraction of hypothalami: mechanisms of oxytocin inactivation. *Brain Res. 60*, 449–57.

Walter, R., Van Ree, J. M. and de Wied, D. (1978). Modification of conditioned behavior of rats by neurohypophysial hormones and analogues. *Proc. Natl. Acad. Sci. USA 75*, 2493–6.

Walter, R., Yamanaka, T. and Sakakibara, S. (1974). A neurohypophyseal hormone analog with selective oxytocin-like activities and resistance to enzymatic inactivation: an approach to the design of peptide drugs. *Proc. Natl. Acad. Sci. USA 71*, 1901–5.

Walter, R., Schwartz, I. L., Darnell, J. H. and Urry, D. W. (1971). Relation of the conformation of oxytocin to the biology of neurohypophyseal hormones. *Proc. Natl. Acad. Sci. USA 68*, 1355–9.

Walter, R., Glickson, J. D., Schwartz, I. L., Havran, R. T., Meienhofer, J. and Urry, D. W. (1972). Conformation of lysine vasopressin: a comparison with oxytocin. *Proc. Natl. Acad. Sci. USA 69*, 1920–4.

Walter, R., Ballardin, A., Schwartz, I. L., Gibbons, W. A. and Wyssbrod, H. R. (1974). Conformational studies on arginine vasopressin and arginine vasotocin by proton magnetic resonance spectroscopy. *Proc. Natl. Acad. Sci. USA 71*, 4528–32.

Walter, R., Smith, C. W., Mehta, P. K., Boonjarern, S., Arruda, J. A. L. and Kurtzman, N. A. (1977). Conformational consideration of vasopressin as a guide to development of biological probes and therapeutic agents. In *Disturbances in Body Fluid Osmolality* (edited by T. E. Andreoli, J. J. Grantham, and F. C. Rector), pp. 1–36. Bethesda, Md.: American Physiological Society.

Ward, H. W. C. (1973). Anti-oestrogen therapy for breast cancer: a trial of tamoxifen at two dose levels. *Brit. Med. J. 1*, 13–14.

Wark, J. D. and Larkins, R. G. (1978). Pulmonary oedema after propranolol therapy in two cases of phaeochromocytoma. *Brit. Med. J. 1*, 1395–6.

Warren, J. C. and Crist, R. D. (1973). Effects of ovarian steroids on uterine metabolism. *Handbook of Physiology: Section 7, Endocrinology*, Vol. 2, *Female Reproductive System*, Part 2, pp. 49–67. Washington, D.C.: American Physiological Society.

Wass, J. A. H., Thorner, M. O., Morris, D. V., Rees, L. H., Mason, A. S., Jones, A. E. and Besser, G. M. (1977). Long-term treatment of acromegaly with bromocriptine. *Brit. Med. J. 1*, 875–8.

Wasserman, R. H. (1974). Calcium absorption and calcium-binding protein synthesis: *Solanum malacoxylon* reverses strontium inhibition. *Science 183*, 1092–4.

Wasserman, R. H. and Taylor, A. N. (1968). Vitamin D-dependent calcium-binding protein: response to some physiological variables. *J. Biol. Chem. 243*, 3987–93.

Wasserman, R. H., Henion, J. D., Haussler, M. R. and McCain, T. A. (1976). Calcinogenic factor in *Solanum malacoxylon*: evidence that it is 1,25-dihydroxyvitamin D₃-glucoside. *Science 194*, 853–5.

Watkins, P. J. (1977). Oral hypoglycaemic drugs. *Prescr. J. 17*, 76–83.

Watson, S. J., Barchas, J. D. and Li, C. H. (1977). β-Lipotropin: localization of cells and axons in rat brain by immunocytochemistry. *Proc. Natl. Acad. Sci. USA 74*, 5155–8.

Way, L. and Durbin, R. P. (1969). Inhibition of gastric secretion *in vitro* by prostaglandin E₁. *Nature (Lond.) 221*, 874–5.

Weir, R. J., Brown, J. J., Fraser, R., Kraszewski, A., Lever, A. F., McIlwaine, G. M., Morton, J. J., Robertson, J. I. S. and Tree, M. (1973). Plasma renin, renin substrate, angiotensin II, and aldosterone in hypertensive disease of pregnancy. *Lancet 1*, 291–4.

Weiss, G., O'Byrne, E. M. and Steinetz, B. G. (1976). Relaxin: a product of the human corpus luteum of pregnancy. *Science 194*, 948–9.

Weiss, N. S. (1975). Cancer risk and estrogen use in the menopause. *New Engl. J. Med. 293*, 1199–202.

Weissman, G. and Lewis, L. (1964). The effects of corticosteroids upon connective tissue and lysosomes. *Rec. Prog. Horm. Res. 20*, 215–39.

Weitzel, G., Eisele, K., Schulz, V. and Stock, W. (1973). Further studies on biologically active synthetic fragments of the B-chain. *Hoppe-Seyler's Z. Physiol. Chem. 354*, 321–30.

Weitzman, R. E., Fisher, D. A., Minick, S., Ling, N. and Guillemin, R. (1977). β-Endorphin stimulates secretion of arginine vasopressin *in vivo. Endocrinology 101*, 1643–6.

Westaby, D., Ogle, S. J., Paradinas, F. J., Randell, J. B. and Murray-Lyon, I. M. (1977). Liver damage from long-term methyl testosterone. *Lancet 2*, 261–3.

Westphal, U. (1975). Binding of corticosteroids by plasma proteins. *Handbook of Physiology: Section 7, Endocrinology*, Vol. 6, *Adrenal Gland*, pp. 117–25. Washington, D.C.: American Physiological Society.

White, M. G. and Fetner, C. D. (1975). Treatment of the syndrome of inappropriate secretion of antidiuretic hormone with lithium carbonate. *New Engl. J. Med. 292*, 390–2.

Whitley, T. H., Lawrence, P. A. and Smith, C. L. (1976). Amelioration of insulin lipoatrophy by dexamethasone injection. *JAMA 235*, 839–40.

Wiederholt, M., Schoormans, W., Fischer, F. and Behn, C.

(1973). Mechanism of action of aldosterone on potassium transfer in rat kidney. *Pfluegers Arch. 345*, 159–78.

Wilber, J. F. and Utiger, R. D. (1969). The effect of glucocorticoids on thyrotropin secretion. *J. Clin. Invest. 48*, 2096–103.

Wilhelmi, A. E. (1974). Chemistry of growth hormone. *Handbook of Physiology:* Section 7, *Endocrinology*, Vol. 4, *The Pituitary Gland and Its Neuroendocrine Control*, Part 2, pp. 59–78. Washington, D.C.: American Physiological Society.

Wilkins, L. (1960). Masculinization of female fetus due to use of orally given progestins. *JAMA 172*, 1028–32.

Williams, E. D., Karim, S. M. M. and Sandler, M. (1968). Prostaglandin secretion by medullary carcinoma of the thyroid. A possible cause of the associate diarrhoea. *Lancet 1*, 22–3.

Williams, G. H. (1977). Angiotensin-dependent hypertension–Potential pitfalls in definition. *New Engl. J. Med. 296*, 684–5.

Williams, G. H. and Hollenberg, N. K. (1977). Accentuated vascular and endocrine response to SQ 20881 in hypertension. *New Engl. J. Med. 297*, 184–8.

Williams, J. A. (1975). Na⁺ dependence of *in vitro* pancreatic amylase release. *Amer. J. Physiol. 229*, 1023–6.

Williams, L. T. and Lefkowitz, R. J. (1978). *Receptor Binding Studies in Adrenergic Pharmacology.* New York: Raven Press.

Williams, L. T., Mullikan, D. and Lefkowitz, R. J. (1976). Identification of α-adrenergic receptors in uterine smooth muscle membranes by [³H]α-dihydroergocryptine binding. *J. Biol. Chem. 251*, 6915–23.

Williams, L. T., Lefkowitz, R. J., Watanabe, A. M., Hathaway, D. R. and Besch, H. R. (1977). Thyroid hormone regulation of β-adrenergic receptor number. *J. Biol. Chem. 252*, 2787–9.

Williams-Ashman, H. G. (1975). Metabolic effects of testicular androgens. *Handbook of Physiology:* Section 7, *Endocrinology*, Vol. 5, *Male Reproductive System*, pp. 473–90. Washington, D.C.: American Physiological Society.

Williamson, H. E. (1963). Mechanism of the antinatriuretic action of aldosterone. *Biochem. Pharmacol. 12*, 1449–50.

Willmer, E. N. (1961). Steroids and cell surfaces. *Biol. Rev. 36*, 368–98.

Wilson, E. M. and French, F. S. (1976). Binding properties of androgen receptors. Evidence for identical receptors in rat testis, epididymis and prostate. *J. Biol. Chem. 251*, 5620–9.

Wilson, J. D. (1972). Recent studies on the mechanism of action of testosterone. *New Engl. J. Med. 287*, 1284–91.

Wilson, J. D. (1975). Metabolism of testicular androgens. *Handbook of Physiology:* Section 7, *Endocrinology*, Vol. 5, *Male Reproductive System*, pp. 491–509. Washington, D.C.: American Physiological Society.

Wilson, R. G., Singhal, V. K., Percy-Robb, I., Forrest, A. P. M., Cole, E. N., Boyns, A. R. and Griffiths, K. (1972). Response of plasma prolactin and growth hormone to insulin hypoglycaemia. *Lancet 2*, 1283–5.

Winand, R. J. and Kohn, L. D. (1975). Thyrotropin effects on thyroid cells in culture. Effects of trypsin on the thyrotropin receptor and on thyrotropin-mediated cyclic 3′ : 5′-AMP changes. *J. Biol. Chem. 250*, 6534–40.

Wingstrand, K. G. (1966a). Comparative anatomy and evolution of the hypophysis. In *The Pituitary Gland* (edited by G. W. Harris and B. T. Donovan), Vol. 1, *Anterior Pituitary*, pp. 58–126. Berkeley, Calif.: University of California Press.

Wingstrand, K. G. (1966b). Microscopic anatomy, nerve supply and blood supply of the pars intermedia. In *The Pituitary Gland* (edited by G. W. Harris and B. T. Donovan), Vol. 3, *Pars Intermedia and Neurohypophysis*, pp. 1–27. Berkeley, Calif.: University of California Press.

Winship, D. H. (1978). Cimetidine in the treatment of duodenal ulcer. Review and commentary. *Gastroenterology 74*, 402–6.

Wise, J. K., Hendler, R. and Felig, P. (1973). Evaluation of alpha-cell function by infusion of alanine in normal, diabetic and obese subjects. *New Engl. J. Med. 288*, 487–90.

Wise, P. H., Chapman, M., Thomas, D. W., Clarkson, A. R., Harding, P. E. and Edwards, J. B. (1976). Phenformin and lactic acidosis. *Brit. Med. J. 1*, 70–2.

Wolfe, B. B., Harden, T. K. and Molinoff, P. B. (1977). *In vitro* study of β-adrenergic receptors. *Annu. Rev. Pharmacol. Toxicol. 17*, 575–604.

Wolff, J. (1964). Transport of iodide and other anions in the thyroid gland. *Physiol. Rev. 44*, 45–90.

Wolff, J. and Chaikoff, I. L. (1948). Plasma inorganic iodide as a homeostatic regulator of thyroid function. *J. Biol. Chem. 174*, 555–64.

Wollheim, C. B., Blondel, B., Renold, A. E. and Sharp, G. W. G. (1977). Somatostatin inhibition of pancreatic glucagon release from monolayer cultures and interactions with calcium. *Endocrinology 101*, 911–19.

Wong, E. and Flux, D. S. (1962). The oestrogenic activity of red clover isoflavones and some of their degradation products. *J. Endocrinol. 24*, 341–8.

Wong, G. and Pawelek, J. (1973). Control of phenotypic expression of cultured melanoma cells by melanocyte stimulating hormones. *Nature New Biol. 241*, 213–15.

Wong, G. L., Luben, R. A. and Cohn, D. V. (1977). 1,25-Dihydroxycholecalciferol and parathormone: effects on isolated osteoclast-like and osteoblast-like cells. *Science 197*, 663–5.

Wong, R. G. and Norman, A. W. (1975). Studies on the mechanism of action of calciferol: VIII. The effects of dietary vitamin D and the polyene antibiotic, filipin, *in vitro*, on the intestinal cellular uptake of calcium. *J. Biol. Chem. 250*, 2411–19.

Wong, R. G., Adams, T. H., Roberts, P. A. and Norman, A. W. (1970). Studies on the mechanism of action of calciferol: IV. Interaction of the polyene antibiotic, filipin, with intestinal mucosal membranes from vitamin D-treated and vitamin D-deficient chicks. *Biochim. Biophys. Acta 219*, 61–72.

Woodhouse, N. J. Y., Joplin, G. F., MacIntyre, I. and Doyle, F. H. (1972). Radiological regression in Paget's disease treated by human calcitonin. *Lancet 2*, 992–4.

Woodhouse, N. J. Y., Mohamedally, S. M., Saed-Nejad, F. and Martin, T. J. (1977). Development and significance of antibodies to salmon calcitonin in patients with Paget's disease on long-term treatment. *Brit. Med. J. 2*, 927–9.

Wrange, Ö. and Gustafsson, J-Å. (1978). Separation of the hormone- and DNA-binding sites of the hepatic glucocorticoid receptor by means of proteolysis. *J. Biol. Chem. 253*, 856–65.

Wrong, O. (1968). Aldosterone and electrolyte movements in the colon. *Brit. Med. J. 1*, 379–80.

Yalow, R. S. and Berson, S. A. (1960). Immunoassay of endogenous plasma insulin in man. *J. Clin. Invest. 39*, 1157–75.

Yalow, R. S. and Berson, S. A. (1972). And now, "big big" gastrin. *Biochem. Biophys. Res. Commun. 48*, 391–5.

Yalow, R. S. and Berson, S. A. (1973). Characteristics of 'Big ACTH' in human plasma and pituitary extracts. *J. Clin. Endocrinol. Metab. 36*, 415–23.

Yalow, R. S. and Wu, N. (1973). Additional studies on the nature of big big gastrin. *Gastroenterology 65*, 19–27.

Yamasaki, N., Shimanaka, J. and Sonenberg, M. (1975). Studies on the common active site of growth hormone. Revision of the amino acid sequence of an active fragment of bovine growth hormone. *J. Biol. Chem. 250*, 2510–14.

Yamasaki, N., Kangawa, K., Kobayashi, S., Kikutani, M. and Sonenberg, M. (1972). Amino acid sequence of a biologically active fragment of bovine growth hormone. *J. Biol. Chem. 247*, 3874–80.

Yarborough, G. G. (1976). TRH potentiates excitatory actions of acetylcholine on cerebral cortical neurones. *Nature (Lond.) 263*, 523–4.

Yasuda, K., Hurukawa, Y., Okuyama, M., Kikuchi, M. and Yoshinaga, K. (1975). Glucose tolerance and insulin secretion in patients with parathyroid disorders. Effect of serum calcium on insulin release. *New Engl. J. Med. 292*, 501–4.

Yasuda, N., Greer, M. A., Greer, S. E. and Panton, P. (1977). Distribution of corticotrophin releasing factor activity within the hypothalamic-pituitary complex of rats and cattle. *J. Endocrinol. 75*, 293–303.

Yasuda, N., Greer, M. A., Greer, S. E. and Panton, P. (1978). Studies of the site of action of vasopressin in inducing adrenocorticotropin secretion. *Endocrinology 103*, 906–11.

Yates, D. A. H. (1977). Use of local steroid injections. *Brit. Med. J. 1*, 495–6.

Yen, S. S. C., Siler, T. M. and DeVane, G. W. (1974). Effect of somatostatin in patients with acromegaly. Suppression of growth hormone, prolactin, insulin and glucose levels. *New Engl. J. Med. 290*, 935–8.

Young, M. D. and Blackmore, W. P. (1977). The use of Danazol in the management of endometriosis. *J. Int. Med. Res. 5*, Suppl. 3, 86–91.

Yue, D. K. ad Turtle, J. R. (1977). New forms of insulin and their use in the treatment of diabetes. *Diabetes 26*, 341–5.

Zachmann, M., Gitzelmann, R. P., Zagalak, M. and Prader, A. (1977). Effect of aminoglutethimide on urinary cortisol metabolites in adolescents with Cushing's syndrome. *Clin. Endocrinol. 7*, 63–71.

Zahn, H., Brandenburg, D. and Gattner, H-G. (1972). Molecular basis of insulin action: contribution of chemical modifications and synthetic approaches. *Diabetes 21*, Suppl. 2, 468–75.

Zañartu, J., Dabancens, A., Rodriguez-Bravo, R. and Schally, A. V. (1974). Induction of ovulation with synthetic gonadotrophin-releasing hormone in women with constant anovulation induced by contraceptive steroids. *Brit. Med. J. 1*, 605–8.

Zañartu, J., Rosner, J. M., Guiloff, E., Ibarro-Polo, A. A., Croxatto, H. D., Croxatto, H. B., Aguilera, E., Coy, D. H. and Schally, A. V. (1975). Attempts to programme ovulation with exogenous oestrogens and LH-RH analogue. *Brit. Med. J. 2*, 527–9.

Zapf, J., Waldvogel, M. and Froesch, E. R. (1975). Binding of nonsuppressible insulin-like activity in human serum. Evidence of a carrier protein. *Arch. Biochem. Biophys. 168*, 638–45.

Zierler, K. L. (1972). Insulin, ions, and membrane potentials. *Handbook of Physiology:* Section 7, *Endocrinology,* Vol. 1, *Endocrine Pancreas,* pp. 347–68. Washington, D.C.: American Physiological Society.

Zimmerman, E. A. (1977). Localization of hormone secreting pathways in the brain by immunohistochemistry and light microscopy: a review. *Fedn. Proc. 36*, 1964–7.

Zimmerman, E. A. and Robinson, A. G. (1976). Hypothalamic neurons secreting vasopressin and neurophysin. *Kidney Int. 10*, 12–24.

Zimmerman, E. A., Moule, M. L. and Yip, C. C. (1974). Guinea pig insulin: II. Biological activity. *J. Biol. Chem. 249*, 4026–9.

Zimmerman, E. A., Robinson, A. G., Husain, M. K., Acosta, M., Frantz, A. G. and Sawyer, W. H. (1974). Neurohypophysial peptides in the bovine hypothalamus: the relationship of neurophysin I to oxytocin, and neurophysin II to vasopressin in supraoptic and paraventricular regions. *Endocrinology 95*, 931–6.

Zinder, O., Hamosh, M., Fleck, T. R. C. and Scow, R. O. (1974). Effect of prolactin on lipoprotein lipase in mammary gland and adipose tissue of rats. *Amer. J. Physiol. 226*, 744–8.

Zingg, A. E. and Froesch, E. R. (1973). Effects of partially purified preparations with nonsuppressible insulin-like activity (NSILA-S) on sulfate incorporation into rat and chicken cartilage. *Diabetologia 9*, 472–6.

Zinmam, B., Vranic, M., Albisser, A. M., Hanna, A. K., Minuk, H. L. and Marliss, E. B. (1978). Exercise and insulin absorption. *New Engl. J. Med. 298*, 1202.

Zull, J. E., Malbon, C. C. and Chuang, J. (1977). Binding of tritiated bovine parathyroid hormone to plasma membranes from bovine kidney cortex. *J. Biol. Chem. 252*, 1071–8.

Zumoff, B. and Hellman, L. (1977). Aggravation of diabetic hyperglycemia by chlordiazepoxide. *JAMA 237*, 1960–1.

Zusman, R. M., Keiser, H. R. and Handler, J. S. (1977a). A hypothesis for the molecular mechanism of action of chlorpropamide in the treatment of diabetes mellitus and diabetes insipidus. *Fedn. Proc. 36*, 2728–9.

Zusman, R. M., Keiser, H. R. and Handler, J. (1977b). Inhibition of vasopressin-stimulated prostaglandin E biosynthesis by chlorpropamide in the toad urinary bladder. Mechanism of enhancement of vasopressin-stimulated water flow. *J. Clin. Invest. 60*, 1348–53.

Index

oxytocin, 4; actions, 80, 293; analogues, 88, 94; antagonists, 94, 278; biosynthesis, 94; mechanism of action, 101; metabolism, 100; receptors, 103; release, 94; structure, 83; structure–activity relationships, 87; uses, 295, 303
oxytocinase, 100

pA₂, 27
Paget's disease, 393
pancreas, *see* islets of Langerhans; pancreatic acinar cells; pancreatic juices
pancreatic acinar cells, 327, 411
pancreatic juices, 398, 404, 405, 411, 414
pancreatitis, 369
pancreozymin, *see* cholecystokinin
panhypopituitarism, 49, 181
paracrine functions, 397
parafollicular cells, 378
paralytic ileus, 108
paramethasone, 159, 197
parathormone, *see* parathyroid hormone
parathyroidectomy, 133, 373, 392
parathyroid glands, 3, 372
parathyroid hormone (PTH); actions, 373, 400; antagonists, 375; biosynthesis and release, 376; mechanism of action, 377; metabolism, 377; receptors, 377; structure–activity relationships, 374; uses, 392; *see also* hypoparathyroidism; hyperparathyroidism; parathyroid glands
paraventricular nucleus, 38, 80, 97
parietal cells, *see* oxyntic cells
Parkinson's disease, 43
pars distalis (pars glandularis), 38
pars intermedia, 39, 46
pars nervosa, 38
pars tuberalis (pars infundibularis), 39
partial agonists, 27
parturition, 293; *see also* labor
parvicellular neurons, 41
PD₂, 23
penis, 186, 229
pentagastrin, 394, 401, 106
pent-4-enoic acid, 366
pepsin, 402, 408
pepstatin, 416, 425
peptic ulcer, 195, 391, 411
perchlorate, 134
perchlorate discharge test, 117, 128
periarteritis nodosa, 190
pernicious anemia, 398, 414
peroxidase, 269
perphenazine, 285, 287
phagocytosis, 170
pharmacokinetics, 7
phenformin, 359, 364
phenobarbital, 19, 118, 179, 389
phenothiazines, 68, 251, 285, 303, 358
phenoxybenzamine, 207, 213, 227, 286, 414
phentolamine, 74, 135, 207, 227, 358
phenylbutazone, 14, 118, 363, 393
phenylephrine, 207, 226
phenylethanolamine-*N*-methyltransferase (PNMT), 149, 207
phenylthiourea, 124
pheochromocytoma, 226, 400

phlorizin, 338
phosphatase, 102, 221, 340
phosphate metabolism, 373
phosphodiesterase, 102
phospholipase, 171, 289, 342
phosphorylase, 221, 329, 347
phosphorylase kinase kinase, 221, 347
phytoestrogens, 321
pigmentation, 43, 181, 200
pimozide, 206
pineal, 57
pituitary gland, 109
pituitary stalk, 38
pituitropins, 51; *see also* hypophysiotropic hormones
placenta, 147, 261, 264, 291
platelets, 205, 216
pleiotypic effects, 237
PMSG (pregnant mare's serum gonadotropin), 231
PNU, 368
polkissen, 415
polycyclic phenols, 270
polycystic ovaries, 304, 306
polydipsia, 106
polymyositis, 190
polyurea, 107
positive-feedback mechanisms, 51, 55
postpartum hemorrhage, 286, 297
potassium, 151, 159, 166, 179, 185, 193, 201, 329
potency, of drug, 24
potomania, 105
practolol, 207, 226
prednisolone, 145, 158, 197, 414
prednisone, 159, 197
pregnancy, 19, 134, 290, 362, 417
pregnane, 139
pregnanediol, 265
pregnant mare's serum gonadotropin (PMSG), 231
pregnenolone, 143
premarin, 265, 317, 320
premature (preterm) labor, 295, 299
premenstrual tension syndrome, 307
primary plexus (mantle plexus), 41
primary sex characters, 237, 262, 264
primordial follicles, 259
progesterone: actions, 263, 264, 290, 291; binding, 148; biosynthesis and release, 143, 261, 264; mechanism of action, 234, 278; metabolism, 145, 264; preparations, 274; receptors, 55, 235, 274, 278, 319; side effects, 308, 322; structure–activity relationships, 273; uses, 300, 304, 308; progestins, 264, 273, 308, 322; *see also* progesterone
proinsulin, 334
prolactin (PRL): actions, 281; biosynthesis, 47, 292; mechanism of action, 287; receptors, 289; release, 282; structure, 45, 279; *see also* hyperprolactinemia
prolactin release-inhibiting factor (PRL-RIF), 62, 282
prolactin-releasing factor, 62, 282
prolonged pregnancy, 295
proparathyroid hormone, 376
propoxyphene, 106
propranolol, 207, 213; use in diabetes mellitus, 358, 363; in hypertension, 433, in hyperthyroidism, 131, 132, in pheochromocytoma, 226, 227, in renal osteodystrophy, 390
propylthiouracil, 125, 127, 293

For EU product safety concerns, contact us at Calle de José Abascal, 56–1°, 28003 Madrid, Spain or eugpsr@cambridge.org.

www.ingramcontent.com/pod-product-compliance
Ingram Content Group UK Ltd.
Pitfield, Milton Keynes, MK11 3LW, UK
UKHW051012240426
470322UK00021B/625

Batch 470322UK00021B		
470322UKX00625B	9780521279352	Endocrine Pharmacology: Physio▶
PERFECT	8.25X11.00	514 GLOSS

www.ingramcontent.com/pod-product-compliance
Ingram Content Group UK Ltd.
Pitfield, Milton Keynes, MK11 3LW, UK
UKHW051012240426
470322UK00021B

00625B - 470322UKX00625B [1 : 1]

GLOSS

*TAT**

K

/19_LG   CONTAINS: MONO

ACKSTD

RFECT

-CONS

322UK00021B/23

625B - 470322UKX00625B [1 : 1]

UK 0 0 0 2 1 B *

APR-26 10:18 (TUE)   Promise Date: 28-APR-26 10:18 (TU

G